Thomas Stolz and Maike Vorholt

The Grammar of Maltese Prepositions

Studia Typologica

Beihefte / Supplements
STUF – Sprachtypologie und Universalienforschung
Language Typology and Universals

Editor
Thomas Stolz

Volume 35

Universität
Bremen

Thomas Stolz and Maike Vorholt

The Grammar of Maltese Prepositions

DE GRUYTER
MOUTON

The research for this monograph has been conducted as part of the project *Präpositionen und ihre Grammatik im Maltesischen / Prepositions and their grammar in Maltese* funded by the German Research Foundation (DFG, Deutsche Forschungsgemeinschaft) – Project-ID STO 186/22–1, AOBJ: 654148 - at the University of Bremen, Department of Linguistics (Allgemeine und Vergleichende Sprachwissenschaft, Fachbereich 10).

ISBN 978-3-11-133144-7
e-ISBN (PDF) 978-3-11-133159-1
e-ISBN (EPUB) 978-3-11-133169-0
ISSN 1617-2957
DOI https://doi.org/10.1515/9783111331591

Library of Congress Control Number: 2025933611

Bibliographic information published by the Deutsche Nationalbibliothek
The Deutsche Nationalbibliothek lists this publication in the Deutsche Nationalbibliografie; detailed bibliographic data are available on the internet at http://dnb.dnb.de.

Cover image: Weerayuth Kanchanacharoen / iStock / Getty Images Plus
Typesetting: Integra Software Services Pvt. Ltd.
Printing and binding: CPI books GmbH, Leck

www.degruyter.com
Questions about General Product Safety Regulation:
productsafety@degruyterbrill.com

MIX
Papier | Fördert gute Waldnutzung
FSC® C083411

Contents

Preface and acknowledgements

This book has taken us a long time to write. Prior to starting the research that was necessary to describe and evaluate the phenomena that are addressed in this study, we mistakenly assumed that it would be a matter of only a year or so to provide a full-blown account of the grammar of Maltese prepositions and related issues – a task that we felt was long overdue to be carried out. However, we soon learned that things are not that simple not the least because the COVID-19 pandemic slowed down the process of data collection as direct interaction with native speakers of Maltese was severely limited during the most crucial period of the project from 2019 to 2020. On account of the restrictions of travelling abroad and inviting people to Bremen, we decided to rely on the Korpus Malti 3.0 and additional conventionally published sources so that the results we present in this book mostly reflect the properties of prepositions in the written register. Only occasionally was it possible to complement these findings with native-speaker comments.

In retrospect, we have to admit that what we have achieved from 2019 until early 2025 covers only half of what we originally planned to achieve. The chapters on pan-Arabic comparison, cross-linguistic comparison, diachrony, grammaticalization, and language contact never materialised because the systematic account of the synchronic facts alone proved to be so time-consuming that we had to focus exclusively on the descriptive-linguistic side of our topic in order to make sure that our analyses of the facts could be made available to the general linguistic public before too long. Most of the research work and all of the writing had to be done after the financial support for the project had come to an end. This means that we could continue with the work only in irregular intervals of different lengths and on the side because the academic everyday obligations as well as other thematically unrelated projects required most of our time and energy. In 2024, we did our utmost to speed up the process of finalising the manuscript because the updated version of the electronic resource on which we relied heavily became available as Korpus Malti 4.2. For practical reasons, it would have been impossible for us to do everything all over again by way of adjusting our calculations and interpretations to the increased amount of data.

These less-than-favourable conditions notwithstanding, we are proud to present the first comprehensive monograph in English on the intricate and fascinating subject of Maltese prepositions. Investigating this topic in-depth has taught us a lot not only about Maltese and prepositions, in general, but also about how to successfully organise a large-scale research project in difficult logistic circumstances. We hope our hypotheses impel other linguists to look into Maltese prepositions to refine our findings or rectify them where we are mistaken. We happily accept comments that show us how to improve the picture we have painted of the

https://doi.org/10.1515/9783111331591-203

Maltese facts. We also look forward to this work inspiring other linguists to take up our numerous impulses for follow-up studies.

We are indebted to the Deutsche Forschungsgemeinschaft (DFG) for financially supporting the project *Präpositionen und ihre Grammatik im Maltesischen / Prepositions and their grammar in Maltese* (STO 186/22-1, AOBJ: 654148) for three years from 2019 to 2022. We would also like to show our gratitude to Nataliya Levkovych who collaborated on the project in its initial phase. Sandra Abela – our sole native-speaker consultant – deserves to be gratefully credited for discussing at length the morphosyntactic behaviour of the Maltese prepositions with us on the occasion of a short visit of hers to Bremen in 2019. Ray Fabri kindly commented on selected phenomena we discussed with him via email. We are grateful also to those colleagues who attended the talks we have given on Maltese prepositions at different places over the years. Last but not least we say thank you to Barbara Karlson of De Gruyter Mouton for helping us expertly with the publication of this book. It hardly needs to be made explicit that everything that seems to be odd with the contents is to be blamed on us alone.

List of tables

https://doi.org/10.1515/9783111331591-204

List of figures

https://doi.org/10.1515/9783111331591-205

Abbreviations

1/2/3	first/second/third person
ACT	active
ADJR	adjectiviser
ADV	adverb(ial)
AOME	absence of material exponence
ART	article
BLOMP	Bremen List of Maltese Prepositions
C	inflectional class
CAUS	causative
CMPR	comparative
COCO	coordinative construction types
COLL	collective
CONFIG	configurational
COP	copula
D	diphthongisation
DEF	definite
DEM	demonstrative
DIM	diminutive
DIR	directional
DIRECT	directionality
DIST	distal
DO	direct object
DOM	Differential Object Marking
DP	determiner phrase
DPM	Differential Place Marking
DU	dual
EXIST	existential
EXS	external sandhi
F	feminine
FUT	future
GCL	government classes
HF	high frequency
IMP	imperative
INDEF	indefinite
IO	indirect object
IPA	International Phonetic Alphabet
IPFV	imperfective
L	lengthening
LDS	Lingua Descriptive Studies
M	masculine
MONO	monosyllabic
N	noun
NEG	negation
NP	noun phrase
NUM	numeral augment
OBJ	object
O.T.	original translation
P	preposition
PASS	passive
PART	participle
PFV	perfective
PL	plural
PP	prepositional phrase
POSS	possessive
PROG	progressive
PROX	proximal
PST	past
RC	resumption clause
RFL	reflexive
SAE	Standard Average European
SEG	segments
SG	singular
VP	verb phrase
Π	preposition
Σ	sentence
Ω	(orthographic) word

https://doi.org/10.1515/9783111331591-206

PART A: **Introduction**

Towards defining and identifying Maltese prepositions

1 The long and winding road towards Maltese prepositions

The prepositions (= Πs) of Maltese, their phonological, morphological, syntactic, and semantic properties as well as sundry systematic aspects connected to them constitute the topic of this linguistic inquiry. These topics are addressed one by one in Part B from Section 2 (on phonological issues) onwards according to the order laid down in Section 1.9. The success of the descriptive sections of this study depends first and foremost on the prior establishment of the Maltese Π-inventory. Since there still exist many uncertainties in connection with the size and composition of this inventory, it is unavoidable that these problems are solved beforehand. This is the task of the introductory Part A whose length iconically reflects the intricate nature of the questions related to determining what is and what is not a Π in Maltese. Owing to the special function of the introduction, it differs stylistically from the bulk of the other sections insofar as we adopt a discursive argumentation to familiarise the reader with the topics reviewed in Part A in preparation of the subsequent parts of this study. Once these topics have been looked into, it is possible to switch to a different style starting in Part B, Section 2, which will also be applied to Part C where the conclusions are drawn.

To demonstrate why the topic of our study is worthwhile investigating in the first place, we divide this Part A into several sections each of which addresses issues which suggest that our project is a linguistic must. Section 1.1 equips us with the most basic background information about the object language. Section 1.2 marks the point of departure for the investigation of Πs by way of (impressionistically) showing that Maltese Πs do not sweepingly behave like those of selected other languages. The density of Πs in Maltese sentences is then briefly touched upon in Section 1.3. In Section 1.4, we look back on the history of research on Maltese Πs. The contributions of the Bremen research team to the debate which have arisen in the context which preceded and partly overlapped with this study are presented in Section 1.5. The list of candidates for the status of Maltese Πs, their token frequencies, and related phenomena are in the focus of Section 1.6. We characterise the database in Section 1.7 whereas theory and methodology are discussed in Section 1.8. The final part of the introduction is Section 1.9 where the general outline of the organisation of this volume is disclosed. The elaborate introduction is justified because, as alluded to above, before the properties of the Maltese Πs can be described systematically it is a necessity to determine what counts as a representative of the Maltese Π-class and why. As will transpire especially from Sections 1.4–1.5, the experts disagree on many issues related to the subject of this study so that well-argued solutions have to be found for the problems. Section 1.6, in turn,

https://doi.org/10.1515/9783111331591-001

is indicative of the Gordian knot-like nature of numerous hard to decide cases of whether a given item is or is not a Π. For some of these cases, it cannot be helped that we resort to stipulation as the issues are too controversial to be discussed at length in a primarily descriptive-linguistic study.

This study first and foremost addresses a readership composed of specialists of Maltese, general linguists, and (more indirectly also) typologists. No prior knowledge of Semitist philological traditions is presupposed not the least because we too are outsiders to this venerable discipline. There is no special focus on issues which form part of the canon of topics of Arabic linguistics. We conceive of our study as an invitation to scholars who are not necessarily versed in Maltese matters but take an interest in Π-systems to take account also of the Maltese facts when it comes to generalising over the role of Πs in human languages. In general linguistics, Maltese Πs are still a kind of terra incognita. Even in Hagège's (2010) cross-linguistic monograph on adpositions, Maltese is only mentioned four times (including an isolated sentential example) where the author mentions

- the possibility of using Italian-derived morphology on Πs such as *fuq* 'on' + *-nett* (< Italian *netto* 'clean') → *fuqnett* 'at the very top of' (Hagège 2010: 178),
- the existence of the so-called connecting Π *ta'* 'of' in Maltese which has no precedent in Classical Arabic (Hagège 2010: 178–179),
- the prepositional strategy of expressing predicative possession (Hagège 2010: 251–252), and
- the temporal function of the inessive Π *fi* 'in' (Hagège 2020: 303).

These isolated facts do not yield a coherent outline of the grammar of Maltese Πs – a task that is clearly outside the scope of Hagège's (2010) work. In the absence of a comprehensive account of the Maltese case that is accessible to linguists working outside Semitic philology,[1] we deem it necessary to document the properties of Maltese Πs with sufficient empirical detail. Thus, there will be many examples throughout this study which illustrate a given phenomenon more than just once.

Our approach is largely inductive and data-oriented. First and foremost, the data are analysed qualitatively with the occasional quantitative digression. The focus is on establishing the synchronic facts. If at all, diachronic issues are occasionally addressed too albeit only unsystematically. The diachronic perspective of Maltese Πs calls for being studied separately in more detail. More generally, in terms of methodology, we are indebted to the philosophy behind Dixon's (2010a–b, 2013) *Basic Linguistic Theory* without however subscribing sweepingly to all of the

1 Cf. Section 1.4.2.5.

author's hypotheses.[2] We situate ourselves within the framework of functional typology (Croft 1991: 1–31) and adopt a framework-free point of view (Haspelmath 2010). What all this implies for the study of Maltese Πs is elaborated upon in Section 1.8.

In what follows we presuppose a kind of rule of thumb in (1.1) which is based on the definition of Π which Stolz/Levkovych (2020: 207) have put forward in the spirit of Hagège (2010: 8). At the end of the introduction, (1.1) will be revised so that a proper definition of (Maltese) Π emerges.

(1.1) Π – rule of thumb
A Π is a function word heading a prepositional phrase (= PP) at the left edge of which the Π is situated and whose preferably (pro)nominal complement it puts in a relationship with another linguistic unit in the same sentential context by way of marking the grammatical and semantic links between them.

Elements which do not fulfil the requirements laid down in (1.1) are strictly counted out. For a Π to exist the presence of a PP is a *conditio sine qua non*. As will come to the fore in this introduction, the criteria exposed in (1.1) help us to exclude a host of pseudo-Πs from the discussion. To understand why it is necessary to apply the above rule of thumb in the first place, the reader must follow us on the long and winding path through the history of research on Maltese Πs. Section 1.1–1.2 are meant to lead us there.

1.1 Useful facts about Maltese

Before getting down to business, it is in order to provide a short overview of (selected) linguistically interesting aspects of Maltese as such. Maltese is the native language of about 0.5 million speakers the vast majority of whom live in the Republic of Malta. The earliest Maltese text dates back to the late 15th century. From the mid-18th century onwards, the use of Maltese in writing became more frequent. After an extended period of language conflict with the erstwhile dominant prestige language Italian, in the mid-1930s, Maltese gained the function of state lan-

2 This is not the place to critically assess the value of Dixon's approach. We know of course that in spite of the title *Basic Linguistic Theory* is not a proper theory. What we find attractive about it is the possibility of doing descriptive linguistics without committing ourselves to any formal model of grammar and its technical apparatus. In a way, *Basic Linguistic Theory* licenses us to work independently but with the only vaguely circumscribed domain of common linguistic practice (cf. Section 1.8).

guage of Malta where it continues to be co-official with English (Stolz 2011a: 241). A non-native variety of Maltese English has developed whose description is still on the agenda (Krug/Bonnici/Hilbert 2011). Since 2004, Maltese also has the status of a working language within the EU context. For the relatively sizable Maltese community in Australia, Bovingdon (2001) postulates the existence of a distinct variety Maltraljan which is characterised mostly by borrowings from Australian English which are unknown in Malta and the preservation of archaisms and regional traits which are no longer tolerated in the Maltese standard. The domain of Πs in Maltraljan is still unchartered territory into which we do not venture in this study. In Malta, the use of dialectal Maltese has been giving way to the Maltese standard ever since the end of World War II. Puech (1994: 17) distinguishes four dialectally distinct regions, namely

(i) the Gozitan varieties spoken on the smaller of the two major islands, Gozo,
(ii) the varieties spoken in and around the village Żurrieq on Malta,
(iii) the bulk of the other rural or village varieties on Malta, and
(iv) the (coastal) urban varieties in the north on whose basis the standard has been created.

It is certainly possible to rearrange and/or refine the above division depending on the dialectological criteria of one's choice. No dedicated study of the Π-system of any Maltese dialect has come to our notice. Pertinent information has to be retrieved from the not too numerous dialect grammars like Schabert (1976). In the absence of sufficient dialectological coverage of the phenomenon under scrutiny, we focus on the properties of Πs in standard-oriented writing[3].

In terms of genetic affiliation, the language belongs to the Semitic branch of the Afro-Asiatic language family where its closest relative seems to be Tunisian Arabic (Cohen 1988: 106). It is a Neo-Arabic variety – a peripheral dialect in Semitist parlance (Vanhove 1993: 1) – which emerged after Malta was cut off from contact with the Arabophone world in the mid-13th century (Brincat 2000). The present shape of Maltese results from agelong intensive contacts with Italo-Romance to which more recently English influence has contributed its share (Lucas/Čéplö 2020). The high degree of Italianization is especially evident in the domain of the lexicon where loanwords abound (Comrie/Spagnol 2016). However, there is also evidence of grammatical Italianisms ranging from calqued passive constructions (Stolz 1991) to borrowed derivational and inflectional morphology (Gatt/Fabri 2018). Italian Πs have also been borrowed in sizable numbers (Saari 2009). Maltese is a pro-drop language (Saade 2012) with a rich morphology that involves prefixes, infixes, suffixes,

3 Cf. Section 1.7.

and combinations thereof (Stolz 2011a: 246–247). Total reduplication is a widely common strategy to express intensification or create adverbials (Stolz/Stroh/Urdze 2011: 206–330). In the absence of infinitives, Maltese makes use of chains of finite verbs (Stolz 2009). In the domain of adnominal possession, alienable and inalienable possessees trigger different constructions (Koptjevskaja-Tamm 1996). Maltese has nominative-accusative alignment. The word order is pragmatically flexible SVO (Stolz 2011a: 251). The language is also prepositional – and this is the most important structural feature for this study.

1.2 Getting access to the topic

This study is dedicated to a very important, yet largely underexplored field of Maltese grammar. Πs constitute the backbone of numerous morphosyntactic construction types of Maltese so that it is necessary to account for their systematic aspects if one aims at describing Maltese grammar adequately. In Maltese texts of all prose genres, Πs show up with a notably high token frequency. This fact is interesting in the light of DeLancey's (2005: 199–200) hypothesis according to which adpositions do not constitute a universal category. In contrast, Givón (2021: xi) assumes that "[a]dpositions are used, universally, to mark the roles of nominal participants in the verbal clause". These different points of view suggest that it makes sense linguistically to take stock of all languages which attest to adpositions and those which do not.

To support this argument, we provide an impressionistic comparison of the frequencies with which Πs are made use of in the first chapter of the English, Latvian, and Maltese translations of Antoine de Saint-Exupéry's *Le Petit Prince* in Table 1.1.[4] The languages are ordered from left to right according to the decreasing number of Πs. In addition to Π, the symbols Ω and Σ are variables for orthographic words and sentences, respectively.

Table 1.1: Token frequencies of Πs in three languages.

	Maltese	English	Latvian
Ω	353	411	344
Σ	37	40	37
Π	59	53	22
Σ w/o Π	11	12	25

4 The translations are listed under Primary Sources at the end of the text.

In this miniature sample of three, Maltese does not only display the highest absolute number of Π-tokens but also ousts its competitors as to the shares and ratios of Π/Ω and Π/Σ which we compute in Figures 1.1–1.2 in accordance to the principles laid down by Altmann/Lehfeldt (1973). It is worth noting that the shares of Π-less Σ are almost identical for Maltese and English, namely 30%, whereas in Latvian, Π-less Σ represent 67.5% of all sentences of the sample text.

In terms of types (i.e. different Πs), Maltese and English yield identical turnouts, namely thirteen types each. In contrast, Latvian makes use of only five different Πs in the text under inspection. The token-type ratios are similar for all three languages. With 4.1 tokens per type, English has the lowest ratio. For Latvian, the ratio is 4.4 tokens per type whereas the Maltese result is slightly higher, namely 4.5 tokens per type. Maltese outranks its competitors once more albeit by a relatively narrow margin.

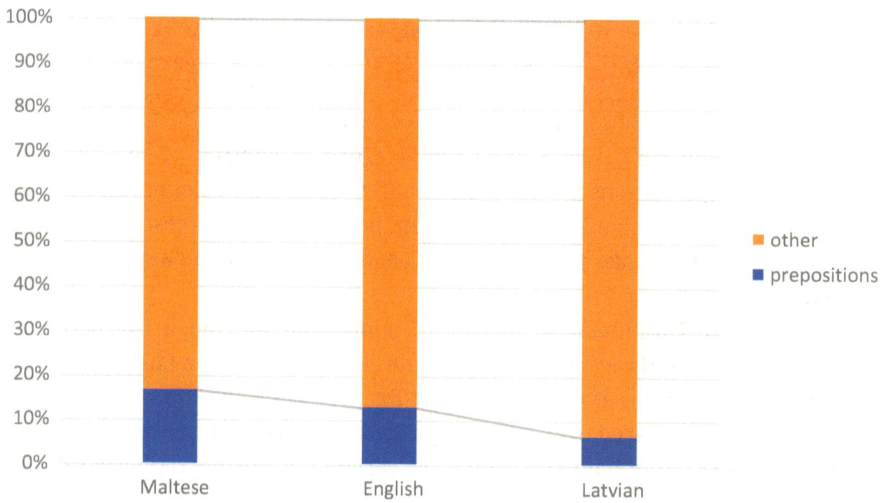

Figure 1.1: Π/Ω in three languages.

The provisional character of this comparison notwithstanding, what can be hypothesised nevertheless is that Πs are indispensable components of Maltese language structure whose proportions exceed those of at least some other individual Π-languages such as English and Latvian. It is of course no secret that some functions which are fulfilled by Πs in Maltese and English are fulfilled by morphological cases in Latvian. A larger sample of languages including Maltese has been investigated by Vorholt (2022a) with a view to determining the share of Πs

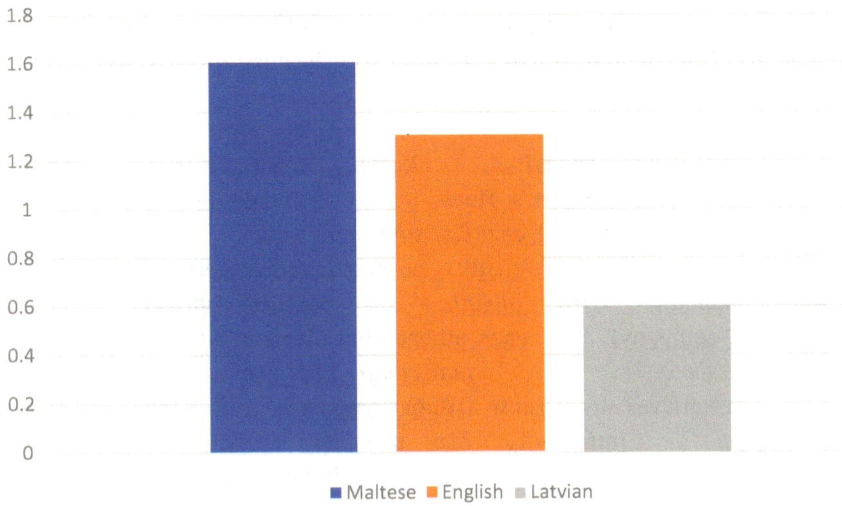

Figure 1.2: Π/Σ in three languages.

in comparable corpora[5]. We come back to cross-linguistic issues of this and other kinds in dedicated follow-up studies.[6] In the remainder of this introduction, we exclusively address selected topics of Maltese Πs. Except occasional mentions of Italian, Spanish, English or sundry languages, no mention will be made of any other language and its system of Πs. This is a strictly monolingual study.

1.3 Density

The above glance at the quantitative relations in a fragment of *Le Petit Prince* reveals that the assumed importance of Πs for the grammar of the object language does by no means imply that each and every sentence contains a Π. In contrast to

5 Cf. Section 1.5.9.

6 It is in order, at this point, to remark on the frequent use of references to future follow-up studies we will be making throughout this study. Contrary to our initial plans, in the course of the project, it turned out that the mere description of the basics of the grammar of Maltese Πs is a time-consuming endeavour whose completion does not leave enough space for the researchers to look into further issues no matter how interesting they are for Maltese linguistics and beyond. Wherever we allude to follow-up studies, linguists who are interested in Maltese and/or Πs are virtually invited to take up the issue and investigate those phenomena we were forced to abandon to make a timely publication of our results possible.

Π-less sentences, it is common in Maltese that several Πs co-occur in one and the same sentence. The phenomenon of Π-accumulation is illustrated in (1.2).[7]

(1.2) [Zammit 1976: 34]

U	*naturalment,*	***[għand***	*Kurun]*$_{PP}$	*čekček*	*it-tazza*	*aktar*
and	naturally	**[at_s.o.'s_place**	Kurun]	tinkle	DEF-glass	more
[minn *darba]*$_{PP}$	***[ma'*** *Wenzu]*$_{PP}$*;*	*tant*	*ħadu*	*pjačir*		
[from time]	**[with** Wenzu]	so_much	take:3PL.PFV	pleasure		
[bil*-kumpanija]*$_{PP}$	***[ta'*** *xulxin]*$_{PP}$	*li*	*Wenzu*	*offrielu*		
[with:DEF-company]	**[of** each_other]	that	Wenzu	offer:IO.3SG.M		
li	*jwasslu*	***[sa****č-čimiterju]*$_{PP}$	*u*	*jgħaddi*	*parti*	
that	3.IPFV:arrive:CAUS:IO.3SG.M	**[till**:DEF-cemetery]	and	3.IPFV:pass	part	
[mi*l-lejl]*$_{PP}$	***[miegħ****u]*$_{PP}$	*biex*	*iżommlu*	*kumpanija.*		
[from:DEF-night]	**[with**:3SG.M]	to	3.IPFV:hold:IO.3SG.M	company		

'And naturally [**at** Kurun's [pub]] he clinked glasses more [**than** once] [**with** Wenzu]; they liked [each other's] [company] so much that Wenzu offered to accompany him [**to** the cemetery] and pass part [**of** the night] [**with** him] to keep him company.'

In point of fact, (1.2) consists of a succession of two sentences the latter of which is a sentence made up of five clauses. There are eight PPs with six different Πs functioning as head in these PPs. If the PPs are erased from (1.2) the pragmatic and grammatical acceptability of the remaining parts is severely impaired so that the sense of the intended message is lost on the Maltese reader.

In connection to the assumed importance of Πs in Maltese, what is especially telling is the density of PPs in (1.2). There are three binary sequences of PPs. The

7 In sentential examples (including glosses and translation), PPs appear in square brackets with Πs being marked out in boldface whereas their complements are underlined. The morpheme glosses follow the *Leipzig Glossing Rules* (revised version of February 2008) and are spelled out in the list of abbreviations. Infixes do not appear in pointed brackets. In the case of the basic form of Maltese verbs (= *mamma tal-verb*) in the 3rd person singular masculine of the perfective we provide only lexical glosses. Syntactic words that consist of a personal pronoun sandwiched between the discontinuous negators *m'. . .x* are glossed as consisting of three morphological units (e.g. *mhux* = NEG:3SG.M:NEG and not MHUX = NEG). The pronominal clitics encoding the direct object or possessor are glossed only for person, number, and gender. Hyphens are usually those which are imposed by Maltese spelling rules. If additional hyphenisation is required for the purpose of analysis, we say so explicitly. We do not use the equation sign < = > to differentiate clitics from affixes. Unless otherwise stated, we respect the orthography of our sources. References to primary sources are given in square brackets. English translations are mostly ours. Those translations and glosses which are original are identified as such.

combination of two PPs by no means exhausts the possibilities of PP-chaining in Maltese. We do not exaggerate when we state that, in Maltese texts, the next PP is never too far away in either direction. Accordingly, Saari (2007: 77) makes a point, for instance, of "the extremely common usage of *ta'* ['of'] in Maltese" by way of giving examples of ternary chains of *ta'*-PPs within the same sentence. To our mind, the type and token frequency of Πs, as well as the density of PPs, are already sufficiently noteworthy to warrant an in-depth investigation of Maltese Πs and their grammar in the broadest sense of the term. As the subsequent sections will amply show many interesting new insights into Πs (both with regards to this particular language and more generally) can be gained from this project.

1.4 Previous research on Maltese Πs

We are not the first to approach Maltese Πs on a grand scale. There is no denying that up until the publication of Borg/Azzopardi-Alexander's (1997) reference grammar Πs were mostly touched upon only in passing in the literature. First of all, the descriptive grammars of Maltese from the mid-18th century onwards contain a more or less elaborate section devoted to the topic at hand. On account of the qualitative differences between the early and the more recent approaches to Maltese Πs, we dedicate a subsection (with further subdivisions) to the publications which appeared in print prior to 1990 and a separate subsection (with further subdivisions) to those published after 1990. Further major descriptive or pedagogical works on Maltese such as Borg (1978), Fenech (1978), Vanhove (1993), and Ambros (1998) are not reviewed here because the authors do not account for Maltese Πs comprehensively. However, they will be referred to wherever this is appropriate. The major dictionaries compiled by Aquilina (1987, 1991), E. Serracino-Inglott (1975a–1989), M. Serracino-Inglott (2000, 2003) and sundry lexicographic work will not be reviewed individually but repeatedly referred to as we proceed. Some of the interpretations of Maltese Πs discussed in this section are shown not to correspond to the rule of thumb in (1.1). Some violations of this kind display a remarkable persistence. Their re-emergence in more recent publications obstructs the adequate description of the Π-system. Therefore, the problematic analyses of the past call for being identified and corrected.

1.4.1 In the good old days

1.4.1.1 De Soldanis (1750)

Already De Soldanis (1750: 104–105), to whose life and work a thematic issue of the *Journal of Maltese Studies* (Vella 2010) has been dedicated, mentions Maltese

Πs whose sketchy description he splits into two separate paragraphs. The Π *ghal* (= modern Maltese *għal*) 'for' is mentioned under the rubric of adverbs together with cases like *hem* (= *hemm*) 'there', *meta* 'when', *ellùm* (= *illum*) 'today', and *issa* 'now' but also the Πs *qoddiem* (= *quddiem*) 'before, in front of' and *mal gemp* (= *maġenb*) 'at the side of, close to'. The author addresses Πs directly under the heading *Delle Proposizioni*[8] where he claims that

> [l]e Proposizioni sono certe particelle, sempre poste avanti li nomi, verbi, & articoli v.g *fuq* sopra, *qoddièm* avanti. *Geuà* dentro, cioè *fuq el beit* sulla terrazza. *Qoddièm el qnisia* avanti la Chiesa. *Geua e suq* dentro il Mercato, che noi diciam *piazza.*[9]

Note that the supposed adverb *qoddiem* (= *quddiem*) 'before, in front of' is also registered as Π. The author only mentions four Πs for which he provides PPs as examples all of which are fine from the point of view of contemporary Maltese grammar. There is no sentential example.

1.4.1.2 Vassalli (1827)

In contrast to De Soldanis, Mikiel Anton Vassalli (1827: 25–37), the acknowledged founding father of Maltese philology (Cachia 2000: 55–134), does not treat Maltese Πs as a word-class in its own right but mentions Πs in several different sections of his grammar (written in Italian) without using the Italian term *preposizioni.* Vassalli's criterion for classifying the items under inspection is morphological in the sense that he distinguishes three classes of particles, namely conjoined, independent, and compound particles. However, except *għad*[10] 'still', *il* 'ago, since', and *kif* 'how, like' (cf. below) the vast majority (= sixteen cases) of the entries in the *Sezione quinta: Delle particelle congiunte*[11] are bona fide Πs such as *bi* 'with'. For the supposed Π *l, li* (Italian meanings 'a, al, per') an interpretation as the bound or cliticised indirect object marker is preferable (Vassalli 1827: 27–28). The Π *sa* 'until' (Vassalli 1827: 32) is listed as one of the separable particles which "equivalgono alle Congiunzioni

8 Our translation: "Of the propositions."

9 All translations of quotes from languages other than English are ours. Our translation of this Italian quote is as follows: "[t]he propositions [sic!] are certain particles, always placed before the nouns, verbs and articles, e.g. *fuq* 'above', *qoddièm* [= *quddiem*] 'before'. *Geuà* [= *ġewwa*] 'inside', i.e. *fuq el beit* [= *fuq il-bejt*] 'on the terrace', *Qoddiem el qnisia* [= *quddiem il-knisja*] 'in front of the church'. *Geuà e suq* [= *ġewwa is-suq*] 'on the market', which we call *piazza.*"

10 For purely technical reasons, we cannot reproduce Vassalli's special characters. We therefore present the examples taken from his grammar of 1827 in the modern orthography of Maltese.

11 Our translation: "Fifth section: Of the conjoined particles."

delle lingue odierne emanate dalla Latina"[12] (Vassalli 1827: 29). In (1.3), we reproduce Vassali's example for *sa* in modernised spelling.

(1.3) (Vassalli 1827: 32)
Morna **[*sa*** *tarf* *il-belt]*$_{PP}$
go:1PL.PFV [**till** extremity DEF-city]
'We went [**as far as** the outskirts of the city].'
[Vassalli's Italian translation: 'Siamo andati fino all'estremità della città.']

The analysis clearly indicates that we are not dealing with a conjunction but with a genuine Π whose classification by Vassalli has to be reconsidered. The author lumps it together with adverbs like *biss* 'only', interjections like *ja* 'oh', conjunctions like *jekk* 'if', and negations like *la* 'not'. Moreover, *bil-*, the fusion of Π *bi* 'with' and the definite proclitic *il-*, is mentioned as a compound particle together with other Πs like *bejn* 'between' and *bħal* 'like' and side by side with adverbs like *flimkien* 'together', *ilbieraħ* 'yesterday', and conjunctions like *malli* 'as soon as' (Vassalli 1827: 34–35). *Barra* 'except' and *ġewwa* 'inside' are treated as "particelle separate"[13] (Vassalli 1827: 29). The use of *għand* 'at s.o.'s place' in constructions of predicative possession is explicitly stated. In contrast to De Soldanis, Vassalli's inventory excludes *quddiem* 'in front of' and *maġenb* 'close to'. For some Πs like *għajr* 'except' (in combination with the conditional conjunction *jekk* 'if'), Vassalli (1827: 27) mentions non-prepositional usages as well. The employment of several of the Πs is illustrated by sentential examples. Several of the adverbs and conjunctions which are interspersed among the Πs in Vassalli's attempt at systematisation contain erstwhile Πs in their chain of segments.

1.4.1.3 Panzavecchia (1843)

The contribution to the study of Maltese Πs of this early grammarian of our object-language deserves to be scrutinised in-depth because the many shortcomings[14] of Panzavecchia's proposal notwithstanding, this author of the mid-19th century is also a pioneer in the sense that he admits new items onboard his Π-inventory which belong to the domain of compound or complex Πs. These classes

---

12 Our translation: "are equivalent to the conjunctions of today's languages which have developed from Latin."

13 Our translation: "independent particles."

14 They are shortcomings only if judged retrospectively with our current knowledge about Πs that was not available to Panzavecchia in the mid-19th century.

re-enter the debate only a century after the publication of his grammar. A useful sketch of Panzavecchia's life and oeuvre can be found in Kontzi (1999: 6–10).

Superficially, Panzavecchia's (1843: 243–248) grammar (also published in Italian) is superior to those discussed in the foregoing paragraphs because the author not only introduces the appropriate term *preposizione* 'preposition' but he also dedicates a separate chapter to the discussion of the (general) properties of Maltese Πs. The Maltese facts are compared to those of Italian which more often than not seem to be of a similar kind to the author. More generally, Panzavecchia (1843: 243) claims that

> [s]ono nella lingua maltese, e nelle lingue orientali, le *Preposizioni* quel che sono nella lingua italiana, e nelle altre lingue europee; val a dire certe particelle indeclinabili ed invariabili, che si premettono ai nomi per esprimere le loro relazioni.[15]

In contrast to De Soldanis and Vassalli, Panzavecchia explicitly restricts the class of possible complements of Πs to nouns (which does not prevent frequent violations of this restriction by the very author). Thus, the prior problems which arose from admitting all kinds of word-classes to combine with Πs seem to disappear from the grammatical description. Panzavecchia (1843: 243–244) compares the fusion of certain Maltese Πs and the definite proclitic with the so-called *preposizioni articolate* (Rizzi 1988: 529–530) in Italian which result from a similar process of fusion (Maltese *fi* 'in' + *il-* → *fil-* ~ Italian *in* 'in' + *il* → *nel*). An argument of Panzavecchia's (1843: 244) which is hard to follow assumes that

> [h]anno anche questo di particolare le preposizioni nella lingua italiana, come nelle lingue orientali, che molto frequentemente come gli avverbj prendono i pronomi affissi.[16]

The author refers to the inflected Πs of Maltese which take pronominal suffixes if their complement is pronominal such as *fuq* 'on' + 3SG.M → ***fuqu*** 'on him'. It remains unclear to us which (most probably only imagined) parallel in Italian Panzavecchia alludes to since he does not provide Italian examples. Furthermore, Panzavecchia (1843: 244–246) makes pertinent remarks as to the semantic effects of the use of certain Πs in combination with certain verbs.[17]

Panzavecchia (1843: 247–248) also provides an ordered inventory of Πs which we reproduce in Table 1.2 because it requires comments which largely hold for

---

15 Our translation: "The *prepositions* are in the Maltese language, and in the oriental languages, what they are in the Italian language and in the other European languages; namely certain indeclinable and invariable particles which precede the nouns to express their relations."
16 Our translation: "The prepositions in the Italian language, like in the oriental languages, have also this peculiarity that they very frequently like the adverbs take the affixed pronouns."
17 Cf. Section 4.3.2.

the earlier studies in their entirety. Our critical assessment of Panzavecchia in this section only scratches the surface, in a manner of speaking. It is meant to familiarise the reader with the problems which recur not only in the history of research on Maltese Πs but which the contemporary analysts have to face when it comes to deciding whether a given element counts as Π or not. In retrospect, many of Panzavecchia's supposed Πs can be shown to belong to other word-classes. At the same time, a number of linguistically interesting cases are mentioned for the first time.

Panzavecchia's inventory bears the title *Preposizioni più ovvie*[18] which implies that there must be other less obvious cases of Πs. The Italian meanings are those given by Panzavecchia. For the English translations (until Section 1.6), we usually but not always rely on the first meaning given in Aquilina's (1987, 1991) Maltese-English dictionary.[19] Grey shading identifies those Πs which this inventory shares with those of De Soldanis and Vassalli. The asterisks * indicates that the classification of a given item as Π needs further elaboration which is provided below. Table 1.2 reflects the order in which Panzavecchia presents the different semantic categories whose tenability is not at issue in this paragraph.[20] Within Panzavecchia's categories, the Πs come in alphabetical order according to the original orthography.

Table 1.2: Panzavecchia's inventory of Πs.

category (Italian description)	Π	meaning	
		Italian	English
static ("di stato")	*barra*	'fuora'	'except'
	f, fi	'in, nel, nello, nella, nei, nelli, negli'	'in'
	fost	'tra, fra'	'between'
	fùk	'sopra'	'on'
	ghand	'presso'	'at s.o.'s place'
	hada, hdein	'vicino, appresso, presso'	'beside'
	taht	'sotto'	'under'
ablative/perlative ("di moto da luogo e per luogo")	*min*	'da, per'	'from'
	minfùk	'di sopra, per sopra'	*
	mintàht	'di sotto, per sotto'	*

18 Our translation: "More obvious prepositions."

19 This practice is motivated solely by economy. The first meanings are consistently used when individual Πs are mentioned in the main body of the text, the footnotes, or when they are glossed in numbered examples. In the English translations of sentential examples and in those paragraphs where semantic issues are discussed other meanings of the Πs are revealed.

20 The semantics of Maltese Πs will be in the focus of Section 5.

Table 1.2 (continued)

category (Italian description)	П	meaning	
		Italian	**English**
allative/directional ("di moto a luogo ossia verso luogo")	*el, il* o *l*	'verso'	'to'
	ghannahha	'verso'	*
	innahha	'verso'	*
	lbarra	'verso fuori'	*
	lein, lei	'verso'	'towards'
	lfùk	'verso sopra'	*
	lgeuua	'verso dentro'	*
	lisfel	'verso giù'	*
	sa	'fino, infino, sino, insino'	'till'
	sahansitra	'fino'	*
	sansitra	'sino'	*
causal ("di cagione")	*b*	'a motivo, a cagione'	'with'
	bil	'a motivo, a cagione'	*
	ghal	'per, in grazia, mercè, in favor'	'for'
	min	'a motivo, a cagione'	'from'
relational ("di relazione")	*ma* e *magh*	'con, in compagnia, unitamente'	'with'
	l, li e *lil*	'a, al'	*
modal ("di modo")	*bi, b, bil*	'con, col'	'with'
	la	'alla'	*
temporal ("di tempo")	*ghalhabta*	'verso, circa'	*
	hin il	'durante'	*
	kabel	'prima, avanti'	'before'
	uakt il	'durante'	*
	uara	'dopo'	'after'
exclusive ("di eccezione")	*barra*	'fuori, salvo'	'except'
	barramin	'fuori, salvo'	*
	ghair	'eccetto, fuorchè, fuori'	'except'
	hlief	'eccetto, fuori, oltre di'	'except'
prohibitive ("di proibizione")	*la*	'non'	*
explanatory ("di spiega")	*bi, b, bil*	'per, con, in'	'with'
	fi, f, fil	'per, con, in'	'in'
comparative ("di comparazione")	*bhal*	'come, al pari di ciò'	'like'
	fùk	'sopra'	'on'
	hada	'in comparazione'	*

Table 1.2 (continued)

category (Italian description)	Π	meaning	
		Italian	English
contrastive ("di opposizione")	*ghal*	'contro'	'for'
interrogative	*bièsc?*	'con che?'	*
	fein?	'dove?'	'near'
	fièsc?	'in che?'	*

Absentees from Table 1.2 are *quddiem* 'in front of', *maġenb* 'close to', *ġewwa* 'in(side)' which are registered however by Panzavecchia's predecessors. Apart from these gaps, the composition of the inventory looks like a mixed bag of genuine Πs and items which belong elsewhere as mentioned above. Comments on the doubtful cases are in order because some of the errors committed by Panzavecchia and his contemporaries show up again in later years. However, there are also cases which suggest that the author was far ahead of his times since he anticipated the discussion of types of Πs which have internal structure. What hampers the evaluation of Panzavecchia's work is the absence of phrasal and sentential examples which illustrate the use of the Πs (of Table 1.2) in 19th century Maltese. We therefore have to resort to (anachronistic) data from the 20th and 21st centuries. We start with Panzavecchia's errors and inconsistencies. Afterwards, we turn our attention to his achievements.

We first look at the Italian translations. From the point of view of modern Italian, the translation of *barra* with *fuora* in lieu of *fuori* 'outside, except' is anachronistic and typical of the poetic genre. Vassalli (1827: 29) has both *fuora* and *fuori*. More strikingly, it remains unclear to us why – except the feminine plural *nelle* – almost the entire array of forms resulting from the fusion of Italian *in* 'in' and the definite article are included in the translation of Maltese *fi* 'in'. The inclusion of the article is unmotivated because the Maltese Π is given without the definite proclitic (otherwise *fil-* would have made more sense). The equation of Maltese *bil-* (← *bi* 'with' + *il-*) with Italian *col* (← *con* 'with' + *il*) respects this requirement whereas in all other instances in which Panzavecchia adds the definite proclitic to the Π (*hin il*, *uakt il*, *fil*) there is no corresponding definite article on the Italian side. At the same time, the putative causal Π *bil* is not distinct from *bi* 'with' but involves the definite proclitic which does not form part of the Π as such. The definite proclitic *il-* has to be separated from all other combinations with Πs in Table 1.2. Thus, we cannot be sure whether the author had a clear idea of the morphological composition of the examples he used. Note that Vassalli (1827: 34) has a separate entry for *bil* translated into Italian as *col*, *collo*, *colla*, *colle*, *coi*, *co'*, *nel*, *nello*, *nei*, *negli*, and *nelle*, i.e.

with fused forms of the Πs *con* 'with' and *in* 'in' with the definite article. It cannot be ruled out that Panzavecchia's practice was influenced by Vassalli.

Table 1.2 hosts many candidates which fail to meet the criteria for being classified as Π. Amongst these cases, there are two different kinds of *la*. Interestingly, one of the two outwardly identical *la*s is originally Italian. Aquilina (1987: 718) describes the element as follows:

> **La** [. . .] (short for It[alian] *alla*) adv[erb]. Following the style of fashion of (~ *ngliża*/~ *franċiża*, following or imitating the English/French fashion/manner; ~ *branzetta*, arm in arm; ~ *moda*, fashionable; *liebsa* ~ *moda*, she is dressed in the latest fashion; ~ *ġenba*, sideways; *alla xanxè*, nonchalantly).

This means that the borrowed combination of the Italian Π *a* 'at, to' and the definite article feminine singular *la* which yields *alla* in the donor language has been subject to phonetic reduction in the replica language so that only the reflex of the Italian definite article remains. This reflex retains the erstwhile prepositional function of *alla* although the above quote only speaks of an adverb. In point of fact, (1.4)–(1.5) show that there is still variation between *la* and *a la* in modern Maltese.

(1.4) [Korpus Malti 3.0; news 136180]

ħafna	*mill-ismijiet*	*kienu*	*jitniżżlu*	*fid-dokument*
many	from:DEF-name:PL	be:3PL.PFV	3.IPFV:RFL:register:PL	in:DEF-document

tat-twelid	*miktuba*	*[**la***	*Taljana]*$_{PP}$
of:DEF-birth	PASS.PART:write:F	[**in_the_fashion**	Italian:F]

'[. . .] many of the names were registered in the birth certificate [**in** the Italian **fashion**].'

(1.5) [Korpus Malti 3.0; news 73996]

il-post	*fejn*	*normalment*	*isir*	*ħafna*	*xogħol*	*parlamentari*
DEF-place	where	normally	3.PFV:become	many	work	parliamentary

fuq	*kafè*	*[**ala***	*Taljana]*$_{PP}$
on	coffee	[**in_the_fashion**	Italian:F]

'[. . .] the place where normally a lot of parliamentary work is achieved over an [Italian] coffee.'

Whether the above analysis of *(a)la* as Maltese Π can be upheld is one of the topics of Section 1.6.6. For the time being, it suffices to emphasise that what is especially striking about this case is the fact that this is the only Italian loan-Π Panzavecchia mentions – it is even the only Italianism of this kind registered in the inventories of the earliest grammarians. In his own Maltese translation of the Bible, Panzavecchia makes use of Italian Πs such as *kontra* 'against' (Kontzi 1999: 430) which he does not include in the inventory of 1843.

The second *la* in Table 1.2 is etymologically independent of Italian since it forms part of the Semitic heritage of Maltese. This second *la* is one of the Maltese negations that occurs not only in prohibitive constructions but also in negative correlative constructions as shown in (1.6)–(1.7).

(1.6) [Galea 2001: 24]

Xbejbiet	*ħelwa*	***la***	*tibżgħux*
young_woman:DIM:PL	sweet:PL	**NEG**	2.IPFV:fear:PL:NEG

'Sweet little girls, **don't** be afraid [. . .].'

(1.7) [de Saint-Exupéry 2000: 3]

U	*billi*	*miegħi*	***la***	*kelli*	*mekkaniku*	*u*	*lanqas*
and	because	with:1SG	**NEG**	have.PFV:1SG	mechanic	and	not_even

passiġġieri
passenger:PL

'And since I had **neither** a mechanic nor passengers with me [. . .].'

The evidence speaks clearly against classifying the negation *la* as Π because none of the criteria formulated in (1.1) are met by this function word.

In contemporary Maltese, *ħada* 'beside' seems to be used only very infrequently. Where it is attested it has a spatial meaning as shown in (1.8).

(1.8) [Korpus Malti 3.0; literature 77]

il-qassis	*resaq*	*qajla*	***[ħada***	*s-sodda]*$_{PP}$
DEF-priest	approach	slowly	**[beside**	DEF-bed]

'[. . .] the priest slowly moved [closer] [**to** the bed].'

The comparative function postulated by Panzavecchia is not attested in our corpus. However, there is *ħdejn* – the formal plural of *ħada* – which is attested abundantly in the spatial and occasionally also in the comparative function. The latter is clearly derived from the former.

In Table 1.2, there are several cases of adverbs or adverbials which never take complement-NPs and therefore fall outside the category of Π. A case in point is *saħansitra* 'even' of which *sansitra* 'even' is a fully synonymous variant (Aquilina 1991: 1268). First of all, there is no reason why one is translated into Italian as *fino* 'till' and the other as *sino* 'till' since the two Italian Πs are semantically indistinguishable. Vassalli (1827: 36) claims that *sansitra* is "migliore"[21] than *saħansitra* – probably for stylistic reasons. Secondly, in contrast to the two Italian Πs, *saħansitra* 'even' is an adverbial modifier of predicates as shown in (1.9) from contemporary Maltese.

(1.9) [Galea 2001. 86]
Ġorġ u Bendu ***saħansitra*** *kienu qabbdu l-Ewnuki*
Ġorġ and Bendu **even** be:3PL.PFV order:3PL.PFV DEF-eunuch:PL
tal-Ħarem
of:DEF-harem
'Ġorġ and Bendu had **even** ordered the eunuchs of the harem [. . .].'

There is no complement-NP of *saħansitra* 'even' which however modifies the predicate to its right.

Two of the members of Panzavecchia's supposed category of interrogative Πs are easily excluded from the Π-inventory whereas the case of *fejn* 'near' calls for closer scrutiny in in the descriptive sections of this study. In modern Maltese, *biex* '(in order) to' is employed as purposive adverbial subordinator (Borg/Azzopardi-Alexander 1997: 41) as shown in (1.10).

(1.10) [Attard 2019: 221]
Ftakar dejjem ***biex*** *tkompli taġixxi daqslikieku*
remember always **to** 2SG.IPFV:continue 2SG.IPFV:act as_if
int immigrant
2SG immigrant
'Always remember **to** continue to act as if you were an immigrant [. . .].'

Similarly, *fiex* 'in what' has relativising function (Aquilina 1987: 334) as in (1.11).

(1.11) [Schembri 2002: 50]
Hekk hu, ta' min wiehed jara ***fiex*** *waslu.*
so 3SG.M of who one 3.IPFV:see **in_what** arrive:3PL.PFV
'That's it, one has to see **what** they have achieved.'

Both *biex* and *fiex* are examples of univerbation of erstwhile combinations of the Πs *bi* 'with' and *fi* 'in' with *xi(ex)* 'what'. The product of the univerbation is not a Π anymore but a semantically specified conjunction or relativiser. *Biex* and *fiex* do

21 Our translation: "better."

not pass the test for Πs because they cannot take a complement-NP. The early grammarians of Maltese frequently could not resist the temptation to consider all word forms which contained traces of acknowledged Πs to be Πs as well.

This tendency comes to the fore with several of Panzavecchia's directional Π-candidates. The initial lateral /l/ of *lbarra* 'outwards',[22] *lfuk* 'upwards', *lgeuua* 'inwards',[23] and *lisfel* 'downwards' can be identified with modern *'(i)l* which Aquilina (1987: 564) characterises as

> [a]dverbial particle indicating direction prefixed to *barra*, outside; *ġewwa*, inside; *fuq*, up; *taħt*, under; *hemm/hinn*, there; *hawn*, here; *ura* (whence *lura*), backwards, with a grammatical function similar to Eng[lish] suf[fixed] '-ward(s)' in 'upwards', 'downwards', etc. (Thus *'il barra*, *'il ġewwa*, etc.).

A case skipped in the above quote is *'l isfel* 'downwards' whose use is illustrated in (1.12). The intital /i/ of *'il* 'to' is present only if the word to its right starts with a consonant.

(1.12) [Zammit 1976: 56]

Kif	*il-ħbieb*	*tagħna*	*waslu*	*ħdejn*	*il-mitħna*	*u*
as	DEF-friend.PL	of:1PL	arrive:3PL.PFV	beside	DEF-mill	and

lwew	***'l***	***isfel***	*semgħu*	*agħjat*
bend:3PL.PFV	**to**	**down**	hear:3PL.PFV	shout:PL

'When our friends arrived at the mill and turned **downwards** they heard shouts [. . .].'

This case is representative of all instances of formation with an initial *'(i)l* 'to'. This element combines exclusively with the members of a closed set of spatial adverbs. The combination is another case of univerbation because nothing can be intercalated between *'(i)l* and the adverb so that the two elements no longer form a syntagma but a morphologically complex word. This word in turn does not belong to the class of Πs since it never takes a complement-NP. The univerbations create new spatial adverbs which modify the meaning of the predicate. The case is not closed because the exact status of *'(i)l* will require our full attention again below.[24]

Two further cases from Table 1.2 involve the noun *naħa* 'side'. This means that we are facing fully-fledged PPs for which Panzavecchia assumes Π-status. This status is doubtful. However, the rebuttal of Panzavecchia's assumption is not as

22 Cf. Vassalli (1827: 35) 'per fuori'.

23 Cf. Vassalli (1827: 35) 'per dentro'.

24 Cf. Section 1.6.5 (ii).

straightforward as in the previous cases. Under the rubrics citerior and ulterior, Borg/Azzopardi-Alexander (1997: 163–165) register several complex constructions which involve the definite NP *in-naħa* 'the side' as complement of a preceding Π (*għal* 'for' is not mentioned). If location near or motion towards the speaker is intended, the definite NP is normally accompanied by the demonstrative *din* 'this'. Otherwise, *in-naħa* is further specified by the postnominal attribute *l-oħra* 'the other' as in (1.13).

(1.13) (Borg/Azzopardi-Alexander 1997: 165)
Imxew *[sa* *n-naħa* *l-oħra* *tal-pjazza]*$_{PP}$.
walk:3PL.PFV **[till** DEF-side DEF-other:F of:DEF-square]
'The walked over [**to** the other side of the square].' [O.T.]

If we accept Panzavecchia's interpretation, we must assume a complex Π **sa n-naħa l-oħra (ta')* 'as far as the other side (of)' for (1.13). It is not primarily the complexity of this pseudo-Π which induces us to argue against this analysis. The fact that determination and attribution are possible for *in-naħa* speaks against its being decategorialised as noun (Stolz 2020: 461). *L-oħra* occupies a slot which can be filled by other attributes with which it is in a paradigmatic relation as is shown in (1.14).

(1.14) [Korpus Malti 3.0; sport 1695]
B' *xorti* *tajba,* *l-erba'* *Maltin* *li* *jinsabu*
with fortune good:F DEF-four Maltese:PL that 3.IPFV:RFL:find:PL
[fin-naħa *tat-Tramuntana* *tal-muntanja]*$_{PP}$ *ma* *ntlaqtux*
[**in**:DEF-side of:DEF-north of:DEF-mountain] NEG RFL:leave:3PL.PFV:NEG
'Fortunately, the four Maltese who are [**on** the northern side of the mountain] have not given signs of weakness.'

Example (1.14) reveals that *in-naħa* takes the attribute *tat-Tramuntana* 'of the north' to yield *in-naħa tat-Tramuntana* 'the northern side', i.e. the reference of the noun is narrowed down. The NP *in-naħa tat-Tramuntana* 'the northern side' takes a genitival attribute linked to it by *ta'* 'of'. There is a part-whole relationship between the referent of the complement of *ta'* and *in-naħa*, i.e. reference is made to the side of a given entity (in this case: the mountain). Thus, the syntactic border does not cut off *ta'* from its complement but is located between the leftmost Π and whatever follows on its righthand side. On these grounds, we discount Panzavecchia's candidates *ghannahha* and *innahha* as Πs although similar problems will haunt us again repeatedly in Part B.

Panzavecchia's candidate *ghalhabta,* too needs to be looked at more closely. According to the author, this is a temporal Π which only vaguely determines the intended point in time – a function which corresponds nicely to that of the English suffix *-ish* as in *sixish* 'at about six o'clock'. The Maltese expression is complex in the sense that it can be decomposed into three morphological units, namely the Π *għal* 'for', the polysemous noun *ħabta* one of whose meanings is 'time' (Aquilina 1987: 465), and the definite proclitic *l-* which is not represented separately in Panzavecchia's orthographic rendering of the expression. In (1.15), we illustrate how *ħabta* 'time' is used as a noun in modern Maltese.

(1.15) [Cauchi 1997: 5]

dik	*il-**ħabta***	*kienu*	*jilmħuh*	*ġej*
DEM:F:DIST	DEF-**time**	be:3PL.PFV	3.IPFV:perceive:PL:SG.M	come.ACT.PART
bla	*tbatija*	*ta'*	*xejn*	
without	hardship	of	nothing	

'[. . .] this **time** they saw him coming without any [sign of] suffering.'

It is understood that the English *on this occasion* would also be fine for the initial DP of this sample sentence. In (1.16), the same noun forms part of a PP.

(1.16) [Cooper 1985: 5]

*[**Għal**l-ħabta*	*ta'*	*l-għabex]$_{PP}$*	*f'*	*jum*	*sajfi*	*wasal*	*messaġġier*
[**for**:DEF-time	of	DEF-dusk]	in	day	summerly	arrive	messenger
Indjan	*f'*	*Forti Edwards.*					
Indian	in	Fort Edwards					

'[**At** (the time of) dusk] on a summer day an Indian messenger arrived at Fort Edwards.'

The combination of *għal*, *ħabta*, and the definite proclitic corresponds exactly to Panzavecchia's case *ghalhabta*. If Panzavecchia's analysis is correct, *(ta') l-għabex* '(of) the dusk' must be the complement of the supposed Π. The situation is similar (though not absolutely identical) to that discussed for *naħa* 'side' in the foregoing paragraph. In analogy to *il-ħabta tal-karozzi* 'the collision of the cars' (Aquilina 1987: 465), would it be possible to assume **il-ħabta ta' l-għabex* 'the time of dusk' as structurally well-formed complex NP consisting of the definite head-noun *il-ħabta* and its genitival attribute *ta' l-għabex*? This complex NP functions as complement of the Π *għal* 'for'. At the same time, there is reason to assume a complex Π **għall-ħabta (ta')* 'at the time (of)' because Aquilina (1987: 465) states that in expressions of time, *ħabta* is usually preceded by *għal* 'for'. As (1.17) suggests this rule is probably not absolutely without exception.

(1.17) [Korpus Malti 3.0; academic13]

Jmutu *fl-eqqel* *tax-xitwa,* *l-aktar* *[fil-ħabta*
3.IPFV:die:PL in:DEF-fierce.CMPR of:DEF-winter DEF-more [**in**:DEF-time
tal-Milied]$_{PP}$.
of:DEF-Christmas]
'[. . .] they die in the fiercest [period] of winter, mostly [**at** Christmas].'

In (1.17), the expected *għal* 'for' is replaced with *fi* 'in'. Replacements of this kind seem to be extremely rare. It cannot be excluded that we are dealing with an error on the part of the writer.

In Table 1.2, there are two further temporal ΠS assumed by Panazvecchia, namely *hin il* and *uakt il* both given as equivalents of Italian *durante* 'during'. Both expressions involve the definite proclitic *il-* as their second component. As stated above, the definite proclitic does not belong to the segmental chain of any of the ΠS listed in Table 1.2. What *il-* tells us is that there is a slot for a NP to be filled. The modern Maltese forms are *ħin* 'time' and *waqt* 'instance'. According to the criteria exposed in (1.1), only the latter is a Π, namely *waqt* 'at the time of' whose use is illustrated in (1.18).

(1.18) [de Saint-Exupéry 2000: 10]

Dak *in-nhar* *għamel* *wirja* *kbira* *ta'* *l-iskoperta*
DEM:M:DIST DEF-day make demonstration big:F of DEF-discovery
tiegħu *[**waqt*** *Kungress* *Internazzjonali* *ta'* *l-Astronomija]*$_{PP}$.
of:3SG.M [**during** congress international of DEF-astronomy]
'On that day he made a great demonstration of his discovery [**during** an International Congress of Astronomy].'

The absence of the definite proclitic on the initial noun of the complement has two possible explanations. NPS remain naked if indefiniteness applies. The less likely alternative reasoning assumes that the complement is a proper name and thus may lack overt definiteness marking. The details about the Π-status of *waqt* 'at the time of' will be revealed below.[25] As to the possibility of *ħin* 'time' claiming the same status, we refer to example (1.19).

(1.19) (Aquilina 1987: 545)

M' *hux* ***ħin*** *il-logħob*
NEG 3SG.M:NEG **time** DEF-play
'This is not the right **time** for playing.' [O.T.]

25 Cf. Section 1.6.5 (v).

The noun under inspection constitutes the head of the construct state illustrated in (1.19). The construct state involves two juxtaposed NPs with definiteness marking being restricted to the complement-NP (Fabri 1996). Constructions of this kind can be lexicalised and express special meanings. According to our database, there are no clear cases of *ħin* 'time' taking over prepositional functions – *pace* Aquilina (1987: 545) who speaks of prepositional usages of *ħin* 'time' but provides only examples with this noun as complement of the Π *ta'* 'of'. What is clear in connection with (1.19) is that, at least in this case, the meaning 'during' cannot be given to *ħin*.

The next three cases involve combinations of two Πs which together form a compound Π (Stolz 2020: 439) of which there are many more in Maltese than Panzavecchia's account suggests.[26] According to the modern spelling rules, the compositional structure of *minfùk*, *mintaht*, and *baramin* becomes transparent. It should not go unmentioned that *minfuq* 'in an advantageous or favourable condition' and *minn taħt* 'in a stealthy manner of an intriguer' (Aquilina 1991: 832–833) also have highly specific lexicalised meanings. Apart from these lexicalisations, the combination of *minn* 'from' and *fuq* 'on' yields *minn fuq* 'from on', that of *minn* 'from' and *taħt* 'under' yields *minn taħt* 'from under', whereas *barra* 'except' and *minn* 'from' yield *barra minn* 'except'. The brackets in (1.20)–(1.21) indicate that we are dealing with a layered combination of two Πs in lieu of a unitary (though complex) Π (Stolz 2020: 452–453)

(1.20) [Briffa 1992: 49]

Kien	*żelaq*	***[minn***	***[fuq***	*il-blata]$_{PP}$]$_{PP}$*
be	slip	**[from**	**[on**	DEF-rock]]

'He had slipped [**from** the rock] [. . .].'

(1.21) [Attard 2019: 214]

Imbagħad	*jien*	*niżgiċċa*	*b'*	*xi*	*mod*	***[minn***	***[taħt***
then	1SG	1.IPFV:scamper	with	some	manner	**[from**	**[under**
għajnejn	*l-infermier]$_{PP}$]$_{PP}$*						
eye:PL	DEF-nurse]]						

'Then I will escape somehow [**from** [**under** the nurse's eyes]] [. . .].'

For the time being, it suffices to state that Panzavecchia's inventory shrinks further because the compound Πs are no unitary Πs. As to the case of *barra minn* 'except', the situation is different because *barra minn* 'except' and *barra* 'except' are largely synonymous, meaning *barra minn* is not a compound Π. Already Vassalli (1827: 29)

---

26 We will disclose the systematic aspects of the compound Πs in some detail in Section 4.1.3.

equated *barra mil-bieb* with *barra il-bieb* both of which mean 'outside the door' (Vassalli's Italian translation being 'fuori della porta'). The presence or absence of *minn* 'from' does not alter the meaning of the PP.[27] In contrast to *barra minn*, if any of the Πs highlighted in boldface in (1.20)–(1.21) is omitted the meaning is changed to the extent that the acceptability of the utterances is jeopardised.

Finally, there is the vexing problem posed by Panzavecchia's relational Π *l, li, lil* which is translated into Italian as *a* 'to'. In modern Maltese, simplifying, the use of this Π (which is connected to the directional *'(i)l* discussed above) is restricted to the marking of direct objects with animate nouns and indirect objects with definite nouns (largely independent of animacy). More generally, the Maltese facts can be captured by way of referring to the principles of Differential Object Marking (Döhla 2016). In (1.22), the object marker *'l ~ li(l)* is featured twice.

(1.22) [Cauchi 1997: 88]

Kulħadd	*kien*	*sema'*	***['l***	*Dun Belind]*$_{PP}$	*jgħid*	***[li****x-xwejjaħ]*$_{PP}$	
everybody	be	hear	**[to**	Dun Belind]	3.IPFV:say	**[to**:DEF-old_man]	
li	*seta'*	*jagħtih*		*daqqa*	*kbira*	*kemm*	*irid*
that	can	3.IPFV:give:3SG.M		stroke	big:F	how_much	3.IPFV:want

'Everybody had heard [Dun Belind] saying [**to** the old man] that he could punch him as strongly as he wanted [. . .].'

The question of whether it is still legitimate to classify the item under inspection as genuine Π will be addressed in more detail below.[28]

Some of Panzavecchia's cases which are not in line with later ideas about Πs continue the tradition of the grammarians who worked in the 18th and early 19th centuries. The possibility of semantic change and refunctionalisation cannot be counted out completely. Still, it is unlikely that diachronic developments can explain why the negation *la* 'not' for instance, should pass as a Π. As intermediary result of the first century of Maltese grammaticographic tradition, we state that the earliest accounts of Maltese Πs are characterised by a high degree of inconsistency as to the notion of Π itself. Members of this word-class are often mixed up with those of other word-classes (first and foremost adverbs). There are sizable gaps in the Π-inventories presented by De Soldanis, Vassalli, and Panzavecchia. At the same time, some Πs are allocated to other word-classes. On this basis, it is impossible to determine how many and which Πs there were in the 18th–19th century. The authors also faced serious difficulties when it came to judging complex word

27 *Barra minn* needs to be looked into more closely in Section 4.1.2.2.2.1.3.

28 Cf. Section 1.6.5 (ii).

forms or chains of clitics correctly. Diachronic and synchronic viewpoints were not always adequately kept apart. This retrospective criticism is of course unfair to the grammarians of the past who did not have the benefit of an additional century and a half of progress in linguistics. One might even claim that, in the absence of advanced linguistic theory and methodology, De Soldanis, Vassalli, and Panzavecchia did remarkably well as philologers.

1.4.1.4 The mid-1930's: From Cremona to Sutcliffe

Cremona's (1970) multiply re-edited booklet on Maltese orthography – first published in 1935 – can also be read as a grammatical sketch of the language. On the one hand, the author still subsumed Πs under the category of particles ("il-partiċelli") together with other "kelmiet żgħar"[29] (Cremona 1970: 29) to which conjunctions, negations, and adverbs belong. On the other hand, Cremona (1970: 80) recognises Πs as a special subclass of particles and states that

> [i]l-Preposizzjoni hija kelma li turina kif in-Nom ta' warajha għandu x'jaqsam ma' kliem ieħor ta' l-istess sentenza.[30]

Only eleven Πs are mentioned, namely (in the order of appearance) *ħdejn* 'beside', *taħt* 'under', *għand* 'at s.o.'s place', *bla* 'without', *lil* 'to', *minn* 'from', *fejn* 'near', *bi* 'with', *fi* 'in', *ma'* 'with', and *ta'* 'of'. No Πs of Italian origin are mentioned. Furthermore, Cremona (1970: 81) claims that "[b]iċċa minnhom huma magħmulin minn żewġ kelmiet".[31] This observation leads him to accept *għalfejn* 'why', *sakemm* 'until', and *safejn* 'up to how long' as Πs although they are causal and temporal conjunctions (Aquilina 1991: 1245).

The situation changed moderately with Cremona's contemporary Sutcliffe (1936) whose role in the history of Maltese philology is sketched by Cachia (2000: 241–248). Sutcliffe's descriptive grammar of Maltese is more reliable than those of his predecessors. Πs are discussed extensively in a separate chapter (Sutcliffe 1936: 199–207) – and this time without inadmissible overlaps with other word-classes.

To some extent, Sutcliffe's criteria are reminiscent of those applied by Vassalli (1827) because Πs are considered to form a subclass of particles although many Πs may be "compounded with the pronominal suffixes" (Sutcliffe 1936: 199). Their morphological variability does not fit the description of (normally) indeclinable particles. Like Vassalli before him, Sutcliffe (1936: 199) divides Maltese Πs into

29 Our translation: "little words".

30 Our translation: "The preposition is a word which shows us how the noun following it is related to other words of the same sentence."

31 Our translation: "some of them are made out of two words."

separable and inseparable "or proclitic" kinds. The possibility of a given Π to fuse with the definite proclitic is a further criterion for the internal subdivision within the Π-class. In contrast to Panzavecchia, Sutcliffe refrains from classifying the Πs according to semantic principles.

Sutcliffe's compound Πs are not of the kind illustrated in (1.20)–(1.21) which involves the productive, systematic, and paradigmatically variable combination of Πs. He only mentions *minn taħt* 'stealthily' as an adverb (Sutcliffe 1936: 207). As shown in (1.23), his examples include instances of compound Πs but the author does not comment further on the cooccurrence of two Πs.

(1.23) (Sutcliffe 1936: 203)

qam ***[minn*** ***[fuq*** *is-siġġu]$_{PP}$]$_{PP}$*
rise **[from** **[on** DEF-chair]]
'He rose [**from** the chair].' [O.T.]

In Sutcliffe's interpretation compounding refers to (synchronically opaque) cases of univerbations such as *fost* 'between' which results from the fusion of *fi* 'in' "and the root which appears in the adj[ective] **wistâni** *middle*" (Sutcliffe 1936: 203). The author is the first to register *bla* 'without' as Π when he explains with reference to *bi* 'with' that

> [i]n the word **bla** *without*, lit. 'with not', it has become so fused with the second element as to have lost all separate existence, and is consequently written without the apostrophe. Similarly **bħal** *in the state of*, i.e. 'like' (Sutcliffe 1936: 200).

Most probably, the erstwhile negation *la* in *bla* 'without' is not what Panzavecchia (1843) alluded to when he declared *la* 'not' to be a prohibitive Π (cf. (1.6)–(1.7)). The most important addition to the Π-inventory is that of the highly frequent *ta'* 'of'. Sutcliffe (1936: 206) notices its multifunctionality. He also recognises its use in possessive constructions, in attributive constructions, and "in the designation of tradesmen" (Sutcliffe 1936: 206) such as *tal-ħalib* 'the milkman (lit. 'of the milk')'.

The addition of *bla* and *ta'* to the Π-list does not mean that Sutcliffe's inventory is complete. In contrast to Vassalli and Panzavecchia for instance, Sutcliffe skips mentioning *barra* 'except', *ġewwa* 'inside', *ħlief* 'except', and *sa* 'till'. Like those who worked on Maltese Πs before him, Sutcliffe does not deem non-Semitic Πs worth mentioning. At the same time, he emphasises that *lil* 'to' "has another use as a mere sign of the accusative or object of the verb" (Sutcliffe 1936: 203) (cf. (1.22)). In accordance with the spirit of his times, Sutcliffe adorns the descriptions with historically informed remarks, provides Arabic and Hebrew parallels, and adds further bits and pieces of information about the role certain Πs play in the formation of Maltese adverbial subordinators or conjunctions.

1.4.1.5 Aquilina (1959, 1976)

Joseph Aquilina was the most prominent representative of Maltese philology in the post-World War II period. He authored a number of influential studies on Maltese including the monumental English-Maltese/Maltese-English dictionary. In Aquilina (1959) – originally submitted in 1939–1940 as his PhD-thesis at the University of London – he focusses on the coexistence of Semitic and non-Semitic elements and principles in Maltese grammar. Accordingly, he divides Maltese Πs into two groups on etymological grounds. This is the first time in the history of research on our topic that the Italo-Romance component in the Maltese Π-inventory is given attention. However, Aquilina (1959: 322) remarks that

> [m]ost of these loans, when replaceable by current equivalents of Semitic Maltese, are inadmissible in the literary language.

This comment reflects the normative philosohy of *Malti Safi*[32] which was influential in intellectual circles in Malta for several decades in the mid-20th century. The purist-minded idea was to re-Semiticise Maltese by way of avoiding Italianisms (Brincat 2000: 157–158). Since this (largely failed) policy principally aimed at literary Maltese, we may assume that in the spoken register at least some of the Italian loan-Πs were firmly established. The absence of Πs of Italian origin in the work of De Soldanis, Vassalli, and Panzavecchia may partly be explicable in terms of a similar purist attitude of these writers. Note that we will learn that several Maltese Πs of Italian origin are best considered elements of the written register and certain genres and styles.[33]

If we look at Aquilina's (1959: 322) short paragraph on non-Semitic Πs which contains seven items of interest, we notice that *ċirka* (< Italian *circa*) 'about, approximately' is included although its classification as adverb (as its postnominal equivalent *inċirka* 'approximately') makes more sense. Aquilina's own examples (represented in his own phonetic transcription) ˌ***cirka*** *ɣoʃriːn ˈruːħ* = *ɣoʃriːn ˈruːħ* ***inˈcirka*** 'about twenty persons' must be analysed as follows. The syntactic words highlighted in boldface are adverbs which modify the numeral *għoxrin* 'twenty' without connecting the quantified noun to a higher external head. The case is not in line with (1.1) and thus has to be discounted as a member of the Maltese Π-inventory. Five of the remaining half dozen candidates are bona fide Πs, namely *kontra* 'against', *versu* 'towards', *skont* 'according to', *viċin* 'near', and *permezz* 'by means of'. In contrast, *sotta* 'subject to' is problematic so that it needs to be elabo-

32 Our translation: "pure Maltese".

33 Cf. Section 1.6.6.

rated upon below.[34] These loan-Πs considerably enlarge the Maltese Π-inventory. However, Aquilina's list excludes several other loan-Πs.

Amongst the Πs with a Semitic background (Aquilina 1959: 299–300), we encounter some misplaced cases such as the conjunction *bħalma* 'as'[35] which precedes predicates as shown in (1.24) as well as the (historically) complex adverb(ial)s *minn qabel* 'beforehand' and *mbagħad* 'afterwards' as shown in (1.25)–(1.26).

(1.24) [Baldacchino 1999: 2]

Niltaqgħu	*ma'*	*plejers*	*kbar*	*li*	*baqgħu*	*famużi,*
1.IPFV:meet:PL	with	player:PL	big.PL	that	remain:3PL.PFV	famous:PL

bħalma	*kienu*	*Ruġġieru Friġġieri,*	*l-aħwa*	*Samuci,*
as	be:3PL.PFV	Ruġġieru Friġġieri	DEF-brother.PL	Samuci

Salvi Troisi,	*Salvu Tabone, u*	*oħrajn.*
Salvi Troisi,	Salvu Tabone, and	other:PL.

'We will encounter great players who have remained famous **such as** Ruġġieru Friġġieri, the brothers Samuci, Salvi Troisi, Salvu Tabone and others were.'

(1.25) (Aquilina 1991: 833)

Għidli	***minn***	***qabel***	*meta*	*tkun*	*se*	*titlaq*
say:IO.1SG	**from**	**before**	when	2.IPFV:be.FUT	FUT	2.IPFV:leave

'Tell me **beforehand** when you'll be leaving.' [O.T.]

(1.26) [Zammit 1976: 81]

Imbagħad	*għandna*	*tant*	*kliem*	*li*	*jaqbel*
afterwards	at:1PL	so_much	word.COLL	that	3.IPFV:agree

mat-Taljan!
with:DEF-Italian

'**Then** we have so many words which correspond to Italian!'

As to the directional adverbials *'ilfuq* 'upwards' and *'ilquddiem* 'forwards', Aquilina (1959: 300) only accepts the initial *'il-* as Π (cf. (1.12)). Aquilina (1959: 300) makes an addition to the list of acknowledged Semitic Πs, namely *biswit* 'facing'.[36] Furthermore, he mentions the frequent combination of two Πs. Besides the already familiar *minn fuq* 'from on' and *minn taħt* 'from under', he lists *minn*

34 Cf. Section 1.6.6 (iv).
35 But: 'in the same way, for instance' according to Aquilina (1987: 113).
36 However, 'over' according to Aquilina (1959: 300).

bejnhom 'from between them', *mingħand* 'from with', and *mingħala* 'in the opinion of' (Aquilina 1959: 300). These cases represent two different classes.[37] The semantic effects of the combination of Πs with certain verbs are shortly discussed too (Aquilina 1959: 340).

In Aquilina (1976), the latter phenomenon is addressed in more detail by way of presenting an annotated list of Maltese prepositional verbs.[38] Aquilina (1976: 67) contains a list of twenty Πs for which it remains unclear whether they are meant to exhaust the list of Maltese Πs or are only those which fill a slot in the case-frame of prepositional verbs. In the Maltese-English dictionary (Aquilina 1987, 1991), Aquilina's classification, translation, and inventory of Πs look different from those of his earlier publications.

1.4.1.6 Schabert (1976)

The absence of explicit statements as to which variety of Maltese forms the object language notwithstanding, it can be taken for granted that the authors reviewed so far[39] had some kind of standard Maltese in mind. This orientation also holds for the majority of the more recent publications to be mentioned below. Schabert (1976) however is different insofar as he describes the properties of two varieties of Maltese, namely those spoken at St. Julians (as representative of urban varieties) and Marsaxlokk (as representative of the rural varieties) – both situated on the main island Malta itself. Superficially, this dialectological approach slightly impairs the compatibility of Schabert's findings with those of the other linguists working on Maltese Πs. On closer inspection however, it turns out that neither of the two dialects displays any noteworthy degree of divergence from what we already know about the system of Maltese Πs in general. If there are differences at all, they are mostly phonological.

The section on Πs (Schabert 1976: 207–215) informs about 33 candidates only two of which can be connected to Italian, namely *kontra* 'against' and the etymologically hybrid formation *flok* 'instead'. Two other Π-candidates can be shown not to belong in this word-class. There are the cases *ˀīs-* (St. Julians) ~ *ˀæys-* (Marsaxlokk) and *donn-* (both dialects) corresponding to the pseudo-verbs *qis-* 'like' and *donn-* 'as if; like; appear' in the standard (Peterson 2009: 187). Since these pseudo-verbs are mistaken for genuine Πs also in some of the posterior contributions to our topic, it makes sense to exclude them from further discussion already at this point.[40] In

37 We turn our attention to them in Section 4.1.3.

38 The integration of PPs in the valency of verbs is discussed in Section 4.3.2.

39 Cf. Sections 1.4.1.1–1.4.1.5.

40 Cf. Section 1.6.2.

(1.27) from St. Julians and (1.28) from the standard, typical uses of the pseudo-verbs are illustrated.

(1.27) (Schabert 1976: 210)

yiğri	***donnu***	*miğnūn*
3.IPFV:run	**as_if:3SG.M**	crazy

'He ran **like** crazy.'

(1.28) [Attard 2019: 162]

Żewġ	*pulizija*	***qishom***	*manigoldi*	*ħakmuh*
two	police	**like:3PL**	rogue:PL	overpower:3PL.PFV:3SG.M

'Two rogue-**like** policemen overpowered him [. . .].'

In contrast to genuine Πs, the pseudo-verbs always host a pronominal suffix which identifies their "logical subject" (Peterson 2009: 186, and with critical discussion of this notion 196–203). This means that neither *-u* in (1.27) nor *-hom* in (1.28) is the pronominal complement of a Π but rather these suffixes encode the syntactically most prominent argument of the pseudo-verb. Thus, *qis-* and *donn-* must be excluded from the Π-class.

1.4.1.7 Brother Henry (1980)

The *Grammatika Maltija* by Brother Henry (1980) – published for the first time already in 1967 – falls behind Aquilina's contribution for several reasons. First of all, all non-Semitic Πs are excluded from Brother Henry's (1980: 192–193) list of Maltese Πs. Secondly, *isfel, biex, fiex,* and *xi* are erroneously considered to be Πs. The classification of *biex* and *fiex* has already been rebutted in connection with Panzavecchia's proposal (cf. (1.10)–(1.11)). Brother Henry's fourth case calls for additional comments. *Xi* occurs frequently in Maltese discourse not the least because it is again a multifunctional morpheme which is often used as interrogative-relativiser *xi* 'what' introducing predicates. It also combines with NPs to serve as indefinite quantifier *xi* 'some' (Vanhove 2009) as shown in (1.29).

(1.29) [Attard 2019: 127]

nassistu	*lill-awtoritajiet*	*tal-Istati Uniti*	*biex*
1.IPFV:assist:PL	to:DEF:authority:PL	of:DEF-United States	to
ninterċettaw	***xi***	*terroristi*	
1.IPFV:intercept:PL	**some**	terrorist:PL	

'[. . .] we will assist the US authorities by way of intercepting **some** terrorists [. . .].'

The purpose for which *xi* is used in (1.29) is syntactically different from that of Πs. *Xi* does not link the NP over which it has scope to an external head but determines and quantifies the NP. Therefore, *xi* must remain outside the Π-category. Brother Henry's (1980: 193) own examples show exactly this: *xi* quantifies or gives an indefinite reading to a noun.

There is also the case of *isfel* 'down'. In contrast to the authors discussed in the previous sections, Brother Henry (1980: 192) treats the directional *'il* and the adverb *isfel* separately so that both figure as Πs in his inventory. Example (1.12) has already shown that *'l isfel* 'downwards' fails to stand the Π-test. This negative result also applies to *isfel* without accompanying directional as shown in (1.30).

(1.30) [Korpus Malti 3.0; academic 13]

Il-weraq	*li*	*jkun*	*hemm*	***isfel***	*ikun*
DEF-leaf.COLL	that	3.IPFV:be.FUT	there	**down**	3.IPFV:be.FUT

irid	*jitneħħa*
3.IPFV:want	3.IPFV:RFL:remove

'The leaves which are **down** there need to be removed [. . .].'

Examples of this kind strongly suggest that even after 200 years of philological work related to the topic of our choice many inconsistencies and open questions remained so that it would have been too daring to call the case of Maltese Πs closed at this point in time already. Section 1.4.2 takes us to the next phase in the development which is characterised by more rigorous efforts to describe the Π-system of Maltese in linguistically adequate ways.

1.4.2 Our contemporaries

We can only repeat. The critical remarks we direct to the work of our predecessors is largely undeserved not only because of the progress linguistics has been making in the domains of theory and methodology but also because for none of the authors featured in Section 1.4.1 have Πs been in the centre of interest. De Soldanis, Vassalli, Panzavecchia, Sutcliffe, Aquilina, Schabert, and Brother Henry aimed at producing a descriptive and in part also practical grammar of Maltese or varieties of Maltese, meaning: the Πs constitute only one among a much greater number of topics they had to address to achieve their goal. It is therefore hardly surprising that gaps remain. The situation is not much different for several of the authors whose contributions will be sketched in this section.

1.4.2.1 Procházka (1993)

Procházka's doctoral dissertation was defended in 1990 at the University of Vienna to appear in print three years later. The study is historical-comparative in the sense that the main focus is on tracing the diachronic fate of the Πs of Classical Arabic in the modern Neo-Arabic varieties. This part of the investigation is complemented by a chapter on Πs which emerged in these varieties independently of the Π-inventory of the classical language. It is clear that in a study with a Semitist orientation as strong as this Italian loan-Πs in Maltese do not receive much space. Similarly, Maltese is only one of many Neo-Arabic varieties scattered over a vast territory from Lake Chad in the south to Bukhara/Usbekistan in the north, from Mauretania in the west to Irak in the east. Procházka takes account of data from 116 localities – three of which (Marsaxlokk, La Valletta, and Gozo) are situated in the Republic of Malta. Procházka (1993: 219) mentions the case of the etymological hybrid *flok* 'instead of' which he describes as follows:

> Im Maltesischen ist eine Präposition flok (auch min flok) zur Angabe einer Substitution in Gebrauch, die [. . .] kaum mit den Personalsuffixen versehen wird, z.B.: flok 'mis „statt eines Hemdes" [. . .]. Diese Form läßt sich durch eine Verschmelzung von /fi/ mit dem sizilianischen locu „Platz" erklären.[41]

The existence of this Π had escaped the notice of the earlier grammarians. Further additions to the Maltese Π-inventory according to this comparative study are *madwar* 'around' (Procházka 1993: 218) and *kif* 'as' for which Procházka (1993: 230) assumes that it can only take pronominal complements. As to *madwar*, Procházka refers to Aquilina (1987: 206) where a (prescriptively motivated) distinction is made between spatial *mad-dwar* 'around' and thematic *madwar* 'about'. Moreover, the thematic function is given as the prerogative of *dwar* 'about'. In (1.31)–(1.32), the items under discussion are presented.

(1.31) [Attard 2019: 97]

Il-Majnas	*kien*	*jaf*	*kollox,*	***[dwar***	*<u>kulħadd</u>]$_{PP}$*
DEF-Majnas	be	3.IPFV:know	everything	**[about**	<u>everybody</u>]

'Majnas [= nickname: Minus] knew everything [**about** <u>everybody</u>].'

41 Our translation: "In Maltese, a preposition flok (also min flok) is used to express a substitution to which [. . .] the person suffixes are rarely attached, e.g.: flok 'mis 'in lieu of a shirt' [. . .]. This form can be explained as fusion of /fi/ with Sicilian locu 'place'."

(1.32) [Camilleri 2013: 99]
*Kif id-dubbien, [. . .] beda jdur dawra tond [**madwar**hom]$_{PP}$*
as DEF-fly.COLL [. . .] begin 3.IPFV:turn round round [**around**:3PL]
'When the flies [. . .] began going full-circle [**around** them] [. . .].'

In our corpus, *madwar* 'around' is overwhelmingly used in its spatial function whereas *dwar* 'about' only occurs as thematic Π[42] or with the adverbial meaning 'approximately'. The initial *kif* in (1.32) functions as temporal conjunction. Whether it makes sense to consider *kif* to be a Π at all needs to be discussed below.[43]

For several of the post-classical Πs, the author does not mention any piece of evidence from Maltese although they are attested and documented in other sources such as Maltese *tul* 'during' which goes back to Classical Arabic *ṭūl* 'length' and has been grammaticalized to a temporal Π in the Arabic varieties of Baghdad, Cairo, Bengasi, and as spatial Π in Arabic varieties in Algeria. Maltese *tul* 'during' can be used in both temporal and spatial meaning (Aquilina 1991: 1390).[44] Temporal *tul* is featured in (1.33).

(1.33) [Teuma 2015: 804]
*huwa raqad raqda waħda [**tul** il-lejl kollu]$_{PP}$*
3SG.M sleep sleep one:F [**during** DEF-night all]
'[. . .] he slept deeply [**during** the entire night] [. . .].'

Procházka's comparative study may serve as the model of a follow-up study that takes the results of our study as its starting point for situating the system of Maltese Πs in the context of the Semitic phylum – both synchronically and diachronically.

1.4.2.2 Fabri (1993)

Fabri's (1993: 186–189) doctoral dissertation on agreement in Maltese contains a section which addresses the Π-inventory of the object language. The author distinguishes three semantic classes, namely spatial Πs, temporal Πs, and a heterogeneous class of "sonstige".[45] Polysemous Πs are mentioned in each of the semantically appropriate classes as e.g. *minn* 'from' which is listed three times, namely as Π with

---

42 Cf. Section 5.1.3.13.

43 Cf. Section 1.6.7 (i).

44 The problems arising from the co-existence of *tul* and *matul* are discussed in Section 1.6.4.2 (viii).

45 Our translation: "others".

spatial, temporal, and other functions. The members of these classes are further subdivided into two groups according to the possibility to take pronominal suffixes.[46] The author accounts for 23 different Πs. A number of Πs whose existence has been acknowledged by earlier authors is absent from this inventory. These absentees would probably wind up in the other-class of Fabri's such as *bħal* 'like', *dwar* 'about', *fost* 'between', *kif* 'as', *mingħajr* 'without', etc. which are mentioned by Vassalli (1827), Panzavecchia (1847), Sutcliffe (1936), Aquilina (1987), and/or Procházka (1993). Most of the absent Πs allow for the suffixation of pronominal markers as shown for *mingħajr* 'without' in (1.34).

(1.34) [Cauchi 1997: 11]

Il-kappell	*tat-tibna*	*kien*	*ħadu*	*miegħu*	*wkoll [. . .]*
DEF-hat	of:DEF-straw	be	take:3SG.M	with:3SG.M	also

iżda	*dakinhar*	*beda*	*ta'*	*sikwit*	*ineħħih*	*minn*	*rasu*
but	that_day	begin	of	often	take_off:3SG.M	from	head:3SG.M

u	*joqgħod*	***[mingħajru]***$_{PP}$	*għal*	*xi*	*ħin*
and	3.IPFV:stay	[**without**:3SG.M]	for	some	time

'He also took the straw-hat with him [. . .] but on that day he frequently took it off and remained [**without** it] for some time [. . .].'

Note that *lil* 'to' is registered as object marker by Fabri (1993: 120–121) without Π-status. Fabri (1993: 186) adds *ġo* 'in' and *viċin* 'near' to the list of Maltese Πs. Furthermore, he observes that

> [d]ie meisten räumlichen Präpositionen können mit *minn* ‚von' kombiniert werden, um komplexe Wegpräpositionen zu bilden, die nur mit Bewegungsverben auftreten dürfen (Fabri 1993: 187).[47]

Table 1.3 is an adaptation of Fabri's own table of compound Πs[48] which contains many more examples of the phenomenon described in the above quote than were mentioned in the previous literature.

46 "Präp-Marker" in Fabri's terminology.

47 Our translation: "most of the spatial prepositions may combine with *minn* 'from' to form complex prepositions of path which can only occur together with motion verbs."

48 His "komplexe Präpositionen."

Table 1.3: Compound Πs according to Fabri (1993).

1st Π	2nd Π	translation
minn	*ġo*	'through'
	ma'	'along'
	fuq	'from on'
	taħt	'from under'
	bejn	'from between'
	ħdejn	'from beside'
	għand	'from s.o.'s place'
	quddiem	'from in front'
	wara	'from behind'

The accompanying sentential example which we reproduce in adapted form in (1.35) shows that 'through' might not always be the most adequate English translation of *minn ġo*.

(1.35) (Fabri 1993: 187)
Ħu *Brian* *ħareġ* ***[minn*** *[ġos-sodda]$_{PP}$]$_{PP}$*
Brother Brian exit **[from** [**in**:DEF-bed]
'Brian's brother got [**out of** bed].'

More importantly, Fabri is the first among the reviewed authors to recognise that the formation of compound Πs is based on systematically applicable rules. The leftmost slot in Table 1.3 is not however reserved for *minn* 'from' alone as we will see below[49] (Stolz 2020: 453).

Fabri (1993: 188) distinguishes between exclusively transitive Πs which obligatorily take a complement and Πs which may also be used intransitively, namely as adverbs as shown in (1.36).

(1.36) (Fabri 1993: 188)
Il-fatat *qiegħed* { *fuq* 'upstairs' / *quddiem* 'in front' / *wara* 'in the back' / *viċin* 'near' }
DEF-ghost stay
'The ghost is __'

49 Cf. Section 4.1.3.

Fabri (1993: 189) argues that it is necessary to assume two different but homophonous lexical entries for cases like *fuq*, namely one which reflects its status as transitive Π with the meaning 'on' and the second entry for *fuq* 'upstairs' as adverb. The pattern of the adverbial use of Πs is a recurrent theme in the literature on our subject-matter as will come to the fore in the subsequent section.

1.4.2.3 Cachia (1994)

In *Grammatika ġdida tal-Malti*,[50] Cachia (1994) mentions Πs in different sections. Apart from the short discussion of morphonological processes which affect the stems of certain Πs under suffixation (Cachia 1994: 110–111), the distinction of Πs which may take pronominal suffixes and those which cannot is made (Cachia 1994: 114–116). There is also a section dedicated to various aspects connected to Maltese Πs (Cachia 1994: 129–131). In this latter section, the author emphasises the asymmetrical relation that holds between the Π, its complement, and "partijiet oħra mis-sentenza"[51] (Cachia 1994: 129). The function of the Maltese Πs is compared to that fulfilled by morphological case in other languages such as Latin. According to Cachia (1994: 129) *lil* 'to' is a Π which "turi d-dattiv u xi drabi l-akkużattiv ukoll".[52] An important point is made by Cachia (1994: 129) when he assumes that there are genuine Πs, i.e. items which exclusively function as Πs, and "kelmiet oħra li xi drabi jintużaw bħala prepożizzjonijiet iżda huma wkoll partijiet oħra tad-diskors".[53] Accordingly, *barra* 'except / outside', *għajr* 'except / (Christian) next', *qalb* 'amongst / heart', and *waqt* 'at the time of / instance' are mentioned as cases of nouns which may also be used as Πs. In the case of *ġewwa* 'inside', *qrib* 'near', *quddiem* 'in front (of)', and *wara* 'after', Cachia (1994: 130) states that they are also nouns and adverbs, whereas *fejn* 'near / where' and *fuq* 'on / above' are said to be adverbs. In the same vein, Cachia (1994: 130) identifies functional equivalence of a non-Π and a Π in the following case:

> In-nom jew verb *qis* + suffiss pronominali jieħu t-tifsira tal-prepożizzjoni *bħal*. Għalhekk *qisu bhima* = *bħal bhima*. [. . .] L-istess użu u tifsira bħal *qisu* għandha l-frażi *donnu, donnha* u l-bqija.[54]

50 Our translation: "A new grammar of Maltese".
51 Our translation: "other parts of the sentence".
52 Our translation: "shows the dative and sometimes also the accusative".
53 Our translation: "other words which are sometimes used as prepositions but which are also other parts of speech".
54 Our translation: "The noun or verb *qis* + pronominal suffix has the meaning of the preposition *bħal*. Thus, *qisu bhima* = *bħal bhima*. [. . .] The phrase *donnu, donnha*, etc. has the same use and meaning as *qisu*."

The pseudo-verbs *qis-* 'like' and *donn-* 'as if; like; appear' which are familiar to us from the discussion above[55] are said to convey the same meaning as the Π *bħal* 'like' when they have scope over a complement NP: *qisu bhima* = *donnu bhima* = *bħal bhima* 'like an animal'. Nevertheless,[56] there are formal grounds on which these pseudo-verbs can be excluded from the list of Πs whereas the above nouns and adverbs must be included therein.[57]

On the basis of Cacchia's (1994: 130) account the Maltese Π-inventory increases by two units, namely *qalb* 'amongst' and *qrib* 'near'. The latter addition is illustrated in (1.37).

(1.37) [Teuma 2015: 365]

Kont	*nibqa'*	*hemm*	*għal*	*żmien*	*twil,*	*kieku*	*l-istorbju*
be:1SG.PFV	1.IPFV:remain	there	for	time	long	if	DEF-noise

li	*smajt*	*[qribi]*$_{PP}$	*ma*	*reġax*	*daħħalni*
that	hear:1SG.PFV	[**near**:1SG]	NEG	return:NEG	enter:CAUS:1SG

fija	*nnifsi*
in:1SG	self:1SG

'I would have stayed there for a long time had not the noise I heard [**near** me] let me come to my senses again.'

With regards to sequences of Πs, Cachia (1994: 131) argues that they are required in those cases where a given Π cannot take pronominal suffixes and therefore needs a second Π to host these suffixes. In addition, the author mentions compound Πs whose initial component is not *minn* 'from' but *għal* 'for' with the directional reading of 'towards' so that minimal pairs like *minn fuq* 'from on' ≠ *għal fuq* 'unto' arise. Moreover, Cachia (1994: 130) assumes that not only nouns, pronominals, and adjectives can fill the slot of the complement but also entire clauses.

1.4.2.4 Borg/Azzopardi-Alexander (1997)

The general knowledge about Maltese Πs has advanced considerably thanks to Borg/Azzopardi-Alexander (1997) whose reference grammar of Maltese contains a plethora of information on the subject matter at hand. The grammar is written in the format of the *Lingua Descriptive Series (LDS)* which means that spatial and other functions are prominently featured most of which are expressed by Πs in Maltese (Borg/Azzopardi-Alexander 1997: 147–173). Further aspects related to Πs

55 Cf. Section 1.4.1.6.

56 As argued in Section 1.4.1.6.

57 Cf. Section 1.6.2.

are discussed in a separate section (Borg/Azzopardi-Alexander 1997: 264–266) and passim. Since the grid employed for LDS presupposes certain semantic categories, the Maltese Πs are automatically assigned to one of these functions. Polysemous Πs therefore show up in different subsections. Moreover, the Π-status of the expressions used to encode a given function is not always clear – especially in those cases where multi-word constructions are involved (cf. (1.13) and discussion thereof).

Where the Maltese grammarians are uncertain about the correct word-class membership they resort to the relatively vague term "expression" as in the case of *daqs* 'equal to'[58] and *bħal* 'like' (Borg/Azzopardi-Alexander 1997: 111). A look at the authors' own examples for equative constructions are of interest in this context (cf. (1.38)–(1.39)).

(1.38) (Borg/Azzopardi-Alexander 1997: 111)
Żaqqu *kbira* ***[daqs*** *globu]*$_{PP}$
belly:3SG.M big:F **[like** globe]
'His belly is as big [**as** a globe].' [O.T.]

(1.39) (Borg/Azzopardi-Alexander 1997: 111)
Wiċċ *Rita* *tond* ***[bħal*** *[t'* *ommha]*$_{PP}$*]*$_{PP}$
face Rita round **[like** [**of** mother:3SG.F]]
'Rita's face is round [**like** her mother'**s**].' [O.T.]

First of all, we know from the previous discussion that most of the earlier grammarians consider *bħal* 'like' to be a Π. Given that this analysis is correct, then it makes sense to apply the same interpretation to *daqs* 'equal to' considering that both *bħal* and *daqs* may host pronominal suffixes as in *bħali* 'like me' ~ *daqsi* 'equal to me'. This reasoning receives support from the comparative construction which involves *minn* 'from' to introduce the standard of comparison. Borg/Azzopardi-Alexander (1997: 108) classify *minn* as Π as shown in (1.40).

(1.40) (Borg/Azzopardi-Alexander 1997: 109)
Bintu *iżjed* *ħelwa* ***[minn*** *bintek]*$_{PP}$
daughter:3SG.M more pretty:F **[from** daughter:POSS:2SG]
'His daughter is prettier [**than** yours].' [O.T.]

Minn also takes pronominal suffixes as in *minni* 'from me'. *Minn, bħal,* and *daqs* share the function of introducing the standard of comparison. The functional and

58 Also 'size' according to Borg/Azzopardi-Alexander (1997: 142).

formal similarities between the three function words suggest that they belong to the same word class. We follow Vanhove (1993: 339) who assumes that both *bħal* and *daqs* are Πs. At the same time, we take Borg/Azzopardi-Alexander's (1997) reluctance to do the same as an indicator of the general problems which arise when it comes to classifying items unequivocally in a context which is characterised by fuzziness. Further undecided cases for which the authors prefer the vague classification as "expression" are *minkejja* 'in spite of', *(min)barra* 'except', *ħdejn* 'beside' as well as *fejn*, *viċin*, and *qrib* all meaning 'near'.

Other Πs, which are absent from Borg/Azzopardi-Alexander (1997) but part of previous inventories, are *għajr* 'except',[59] *ħlief* 'except', *skont* 'according to', *kif* 'as',[60] and *waqt* 'at the time of'. For the latter case (cf. (1.18)), a glance at the two examples in (1.41) is in order.

(1.41) (Borg/Azzopardi-Alexander 1997: 46)

(a) "finite clause"

Ħuh	*baqa'*	*jistenna*	***waqt***	***li***	*nqara*	*l-Vanġelu*
brother:3SG.M	remain	3.IPFV:wait	**instance**	**that**	RFL:read	DEF-Gospel

'His brother waited **while** the Gospel was read.' [O.T.]

(b) "non-finite clause"

Ħuh	*baqa'*	*jistenna*	***[waqt***	*il-qari*
brother:3SG.M	remain	3.IPFV:wait	**[at_the_time**	DEF-reading
tal-Vanġelu]$_{PP}$				
of:DEF-Gospel]				

'His brother waited all [**through** the reading of the Gospel].' [O.T.]

According to Borg/Azzopardi-Alexander (1997: 46), the sentences (1.41a) and (1.41b) are related to each other derivationally because

> [it]t is possible to make [. . .] time clauses non-finite by substituting the appropriate verbal noun for the verb in the subordinate clause [. . .]. Note the following changes: the adverbial conjunctions introducing the finite form from **waqt li** [. . .] to **waqt** [. . .]. The direct object argument of the finite clause is also retained in the non-finite clause, but is now preceded by the preposition **ta'** 'of'.

We interpret the supposed reduction of the finite clause to a non-finite clause differently. In (1.41b), *waqt* 'at the time of' is not an adverbial subordinator but a

59 Although Borg/Azzopardi-Alexander (1997: 148) term *mingħajr* 'without' a "composite preposition".
60 Registered as question word only (Borg/Azzopardi-Alexander 1997: 211).

fully-fledged Π which takes the verbal noun + genitival attribute as its complement. This means that we do not subscribe to the clausal analysis of the PP.

Borg/Azzopardi-Alexander (1997) mention also several hitherto unidentified Πs such as the citerior-anterior Π *faċċata* 'opposite' (Borg/Azzopardi-Alexander 1997: 166–167) and the essive Π *bħala* 'as' whose employment is illustrated in (1.42).

(1.42) (Borg/Azzopardi-Alexander 1997: 152)

Ħadmet	***[bħala***	*infermiera]*$_{PP}$	*fi*	*sptar*	*privat.*
work:3SG.F.PFV	**[as**	nurse]	in	hospital	private

'She worked [**as** a nurse] in a private hospital.' [O.T.]

As to the status of *lil*, Borg/Azzopardi-Alexander (1997: 55) take an interesting stance when they say that

> [d]irect objects receive a case marker (preceding them) morphologically related to the directional preposition **lil** depending on their position in the animacy hierarchy. However, nominals which would not normally receive the direct object marker [. . .] can still be preceded by the preposition **lil** in conjunction with three-place verbs. [original boldface]

Accordingly, *lil* functions as marker of indirect objects, too (Borg/Azzopardi-Alexander 1997: 140). Evidence for the supposed function as directional Π is hard to find. *Lil* is scarcely attested in proper cases of concrete motion in space towards a Goal. Its primary domain is the semantically empty marking of syntactic relations. Yet, this does not mean that *lil* falls outside the category of Πs.

1.4.2.5 Saari (2003)

The hitherto most comprehensive study of Maltese Πs goes to the credit of Saari (2003). One might argue that the existence of this PhD-thesis makes our own investigation superfluous because the properties of Maltese Πs are systematically inspected and a wide range of issues connected to them is raised. We acknowledge that Saari's monograph together with three articles (Saari 2002, 2007, 2009) which tie in with some of the questions addressed in the dissertation marks a quantum leap in the research related to Maltese Πs. However, it would be wrong to assume that there is nothing left to say on Maltese Πs. To the contrary, Saari's data, analyses, and hypotheses have inspired us to modify, complement, and reinterpret his findings according to our own point of view.

Moreover, there are further aspects which strongly suggest that a follow-up to Saari (2003) is urgently needed:

(a) The PhD-thesis is written in Hebrew which renders the book almost invisible in the community of linguists outside the circle of Semitists. The English abstract counts only 24 pages and thus passes over many important points the author makes in the main body of the text.[61]

(b) Saari (2003) refers neither to Fabri (1993) nor to Borg/Azzopardi-Alexander (1997) so that he ignores the most recent progress in the investigation of Maltese ΠS. Except Vanhove (1993), the publication of the descriptive grammars of Maltese he refers to predates 1990. It is necessary to look at Saari's description of Maltese ΠS anew in the light of the ideas put forward in the more recent literature.

(c) The PhD-thesis was completed several years before the first attempts were made at creating an electronic corpus for Maltese. Accordingly, Saari's database consists only of a relatively small selection of literary prose of the 20th century. The Korpus Malti 3.0 can be used not only to test the validity of Saari's generalisations but also to discover as yet undescribed phenomena.

One aspect which lets Saari (2003: 84) stand out among the studies dedicated to Maltese ΠS is the particularly rich inventory the author presents. There are altogether 44 ΠS, i.e. two times as many ΠS as Fabri (1993) mentions. None of the previous studies yields an inventory which would be able to compete with Saari's as to its size. For three candidates, Saari (2003) is the first scholar to ever speak of ΠS, namely *għeluq* 'conclusion', *minħabba* 'on account of', and *mindu* 'since'. We agree with Saari (2003) only in the case of *minħabba* 'on account of' whose Π-status can be gathered from (1.43).

(1.43) [Teuma 2015: 394]
Meta Bakbaraħ ma rahiex aktar kellu
when Bakbaraħ NEG see:3SG.F:NEG more have.PFV:IO.3SG.M
jnaqqas mill-ġiri ***[minħabba*** *d-dlam]*$_{\text{PP}}$.
3.IPFV:reduce from:DEF-running **[on_account** DEF-darkness]
'When Bakbaraħ did not see her anymore, he had to reduce his speed [**because of** the darkness].'

The other two cases pose different problems for being admitted to the Π-category. As argued by Stolz/Levkovych (2020: 212–213), *għeluq* 'conclusion' fails to meet the criteria exposed in (1.1). The NP *għeluq is-sena* 'end of the year' in (1.44) does not involve a temporal Π with the supposed meaning 'at the end of, after'.

61 We could access the contents of the latter only because of Nataliya Levkovych's expertise in Hebrew which is herewith gratefully acknowledged.

(1.44) (Saari 2003: 43)

Għeluq	***is-sena,***	*is-sultan*	*bagħat*	*'l*	*ibnu*	*il-fustani.*
conclusion	**DEF-year**	DEF-king	send	OBJ	son:3SG.M	DEF-middle

'**At the end of the year**, the kind sent out his middle son.'

We assume that *għeluq is-sena* 'end of the year' is an adverbially employed instance of a pseudo-construct state compound (Fabri 1996: 237–238) and functions as bare adjunct in (1.44). To our minds, the leftmost slot could host the Π *fi* 'in' as in (1.45)–(1.46).

(1.45) [Korpus Malti 3.0; culture 3949]

n-numru	*ta'*	*każi*	*tal-Ebola*	*se*	*jitniżżel*
DEF-number	of	case:PL	of:DEF-Ebola	FUT	3.IPFV:RFL:descend:CAUS

għal	*żero*	*[f'*	*<u>għeluq</u>*	*<u>is-sena</u>]$_{PP}$*
for	zero	[**in**	<u>conclusion</u>	<u>DEF-year</u>]

'[. . .] the number of Ebola cases will go down to zero [**by** <u>the end of the year</u>].'

(1.46) (Saari 2003: 44)

Jagħtu	*dehra*	*tassew*	*impressjonanti*	*[f'*	*<u>għeluq</u>*	*<u>il-festa</u>]$_{PP}$.*
3.IPFV:give:PL	show	really	impressive	[**in**	<u>conclusion</u>	<u>DEF-feast</u>]

'They give a really impressive show [**at** <u>the end of the festivities</u>].'

In Maltese, there is alternation of overt marking with *fi* 'in' and zero-marking (Stolz/Lestrade/Stolz 2014: 242–244; Stolz/Levkovych/Urdze 2017: 468–471).[62] On account of this frequently attested alternation, we interpret the above facts along the lines of the formula in Figure 1.3.

$$[\left\{\begin{array}{c} f' \\ \\ Ø \end{array}\right\} \quad \underline{\textit{għeluq is-sena}}]_{PP}$$

Figure 1.3: *fi*-Ø alternation.

62 This is further discussed in Sections 1.5.2 and 4.3.3.2.

As to *mindu*, Saari (2003: 57–59) classifies it as a compound Π. However, we follow Borg/Azzopardi-Alexander (1997: 39) who speak of an "adverbial conjunction [. . .] of time which serve[s] to introduce adverbial clauses of time" (cf. also Stolz/Levkovych 2020: 211). *Mindu* never takes a complement NP but always introduces a clause as in (1.47).

(1.47) (Borg/Azzopardi-Alexander 1997: 40)

Ħuk	*kellu*	*tarbija*	*oħra*	***mindu***	*rajtu*
brother:2SG	have.PFV:IO.3SG.M	baby	other:F	**since**	see:1SG.PFV:3SG.M
l-aħħar					
DEF-last					

'Your brother had another baby **since** I last saw him.' [O.T.]

Apart from doubtful classifications like those mentioned in the foregoing paragraph, there is also the occasional gap in the sense that Saari (2003) skips Πs mentioned elsewhere in the literature relative to our subject. His inventory excludes *fejn* 'near', *viċin* 'near', *faċċata* 'opposite', *maġenb* 'close to', *versu* 'towards', and *kif* 'as'. Except *kif* 'as', these candidates are featured in Borg/Azzopardi-Alexander (1997). The missing loan-Πs of Italian origin are also absent from Saari (2009) – a study which focuses on Italian Πs in Maltese. Note that *faċċata* 'opposite' usually cooccurs with *ta'* 'of' when the spatial relation of citerior-anterior is static or allative (Borg/Azzopardi-Alexander 1997: 166–167) as shown in (1.48).

(1.48) [Camilleri 2013: 72]

Kwarta	*wara*	*deffes*	*il-karozza*	*fl-ewwel*	*kaxxa*	*bajda*	*li*
quarter	after	drive	DEF-car	in:DEF-first	box	white:F	that
lemaħ	*[**faċċata***	***tal**-blokka*	*tiegħu]*$_{PP}$				
see	[**opposite**	**of**:DEF-block	of:3SG.M]				

'A quarter of an hour later he pushed the car into the first white box he saw [**opposite** his block] [. . .].'

In contrast to cases like *naħa* 'side' and *għeluq* discussed above, *faċċata* 'opposite' displays unequivocal signs of advanced grammaticalization. There is a common noun *faċċata* 'façade' with which the Π is etymologically related. However, in contrast to the latter, the Π can never take the definite proclitic or adjectival attributes, or inflect for number. In (1.49), the common noun is shown to host the definite proclitic.

(1.49) [Korpus Malti 3.0; culture 2543]
Beda xogħol fuq ***il-faċċata*** ***ta****l-Bażilika tal-Madonna Ta' Pinu*
begin work on **DEF-façade** **of**:DEF-Basilica of:DEF-Madonna Ta' Pinu
'He had started to work on **the façade of** the Basilica of the Madonna Ta' Pinu [. . .].'

In (1.49), the noun refers to an identifiable part of a given building whereas the Π has no referential function of this kind. Its meaning is clearly detached from the object to which the common noun refers. We are facing decategorialisation and semantic reinterpretation as concomitant phenomena of the grammaticalization of an erstwhile noun to a Π.

Saari (2003) mentions another two Italian loan-Πs which his predecessors ignored, namely *favur* 'on the side of' and *rigward* 'concerning' both of which are registered in Aquilina's dictionary though (Aquilina 1987: 313, 1991: 1217–1218). Example (1.50) illustrates the use of *favur* 'on the side of'.

(1.50) [Baldacchino 1999: 59]
l-iskor kien diġà 3–0 ***[favuri****<u>hom</u>**]*** $_{PP}$
DEF-score be already 3–0 **[on_the_side**:<u>3PL</u>]
'[. . .] the score was already 3–0 **[for** <u>them</u>].'

Given that, side by side with new entries, there are gaps as well as wrongly placed items in Saari's inventory, it can be concluded that the last word about the size and composition of the Maltese Π-class has not been spoken yet.

Saari (2003) accounts for many important aspects connected to Maltese Πs. He not only devotes a chapter to the Arabic and Italian origins of Maltese Πs but he also categorises them according to their morphological structure, their syllabic structure, and the word-class from which they are derived. Moreover, the interaction of Maltese Πs with personal, relative, and indefinite pronouns as well as with the definite proclitic is discussed at length. The functional side of Maltese Πs is prominently featured too. The adverbial, adjectival, nominal, and verbal uses of PPs are granted space in the dissertation. We also find sections which address the expression of possessive, spatial, and dative relations by means of Π. The possibility of Π-drop is briefly mentioned as well as the supposed existence of a single postposition – the latter topic being elaborated upon in Saari (2007). This table of content is indicative of the impressive number of issues which are raised in the dissertation. The fact alone that so many hitherto neglected or ignored facets of the grammar of Maltese Πs are accounted for by Saari (2003) characterises his PhD-thesis as a milestone in Maltese linguistics and to some extent also in Π-research in general. We take Saari (2003) as the example for our own investigation. In the remainder of this study, we therefore have occasion time and again to refer to Saari's ideas whose tenability will be tested in the appropriate sections below.

1.5 Contributions which originated in Bremen

The above literature survey teaches us that, on the one hand, the contributions to the study of Maltese Πs are relatively numerous and on the other hand, unanimity among the descriptive linguists has not yet been achieved. The different approaches come to at least partly contradictory conclusions most of which are connected to determining the exact size and membership of the Π-class. Saari (2003) shows that there are many more and highly interesting issues beyond the clarification of this open question. Since Saari addressed some of these new areas of interest for the first time without exhausting them, there remains a lot to be done before we can claim that the final page of the grammar of Maltese Πs has been printed.

The Bremen research team has scrutinised a variety of properties of Maltese Πs in the recent past. Originally, we had no intention to provide a comprehensive account of this word-class. What raised our curiosity in this subject-matter was the lack of a generally accepted definitive description of Maltese Πs and their inventory. The absence of a description of this kind proved detrimental to the evaluation of Maltese data which were needed for a variety of typological projects. When we realised that this problem would not evaporate by itself, we decided to take over the task of filling the gap ourselves. The topics we have investigated and published about so far are sketched in Sections 1.5.1–1.5.10. Each of these topics is featured again and updated in this monograph.

1.5.1 Stolz/Stroh/Urdze (2006)

This cross-linguistic study on comitatives and related categories (such as instrumentals) contains a case-study dedicated to Maltese. Note that the project accounts for all grammaticalised expressions of the categories under review independent of the morpheme class of the given markers. Maltese happens to use Πs. Like Saari (2002) in his study of Maltese *quddiem* 'in front of', Stolz/Stroh/Urdze (2006: 282–306) focus on the properties of individual Maltese Πs by way of comparing the functional domains of *bi* 'with' and *ma'* 'with' which share the same primary English translation equivalent. In contrast to English, Maltese belongs to the cross-linguistically dominant class of languages which formally distinguish between relations of ACCOMPANIMENT (= comitative) and (EMPLOYMENT OF) TOOL(S) (= instrumental). In (1.51), *karrozza* 'car' is the tool with which the action of following somebody is carried out (Borg/Azzopardi-Alexander 1997: 148). Accordingly, the relation is expressed by the instrumental Π *bi* 'with'. In (1.52), the comitative Π *ma'* 'with' is used because the participant *Lugrezja* accompanies the participant *Mastru Gerfex* to a given destination.

(1.51) [Bartolo/Vella 2009: 75]
Reggie qabad isegwihom [bil-karrozza]$_{PP}$
Reggie begin 3.IPFV:follow:3PL [**with**:DEF-car]
'[. . .] Reggie began following them [**in** the car].'

(1.52) [Cauchi 1997: 81]
*Malli wasal Mastru Gerfex [**ma'** martu Lugrezja]$_{PP}$*
as_soon_as arrive Mastru Gerfex [**with** woman:POSS:3SG.M Lugrezja]
għajnejh marru dritt fuq Kaspru.
eye:PL:3SG.M go:3PL.PFV direct on Kaspru
'As soon as Mastru Gerfex arrived [**with** his wife Lugrezja] his eyes went directly towards Kaspru.'

The ACCOMPANIMENT-TOOL opposition does by no means exhaust the functional differences between the two Πs. In point of fact, both *bi* 'with' and *ma'* 'with' serve several purposes which transcend this opposition.[63] At this point, it suffices to say that Borg/Azzopardi-Alexander (1997: 140–173) mention eight different functions for *bi* 'with' and four for *ma'* 'with' among which prepositional objects are especially worth mentioning because the type frequency of *bi* 'with' with prepositional verbs is higher than that of *ma'* 'with'. According to Aquilina (1976), there are 25 prepositional verbs which combine with *bi* 'with' as opposed to 15 prepositional verbs which take *ma'* 'with'.[64]

1.5.2 Stolz/Lestrade/Stolz (2014) and Stolz/Levkovych/Urdze (2017)

Stolz/Lestrade/Stolz (2014) present and discuss evidence of zero-marking of spatial relations in 112 languages of all continents. Their survey includes a case-study devoted to this phenomenon in Maltese (Stolz/Lestrade/Stolz 2014: 225–273). Zero-marking is of interest for the description of the Maltese Π-system because there is the *fi-Ø* alternation in the domain of Place and Goal, i.e. the spatial Π *fi* 'in'

63 These functions will be looked at more closely in Sections 5.1.3.4–5.1.3.5.
64 We come back to this and related issues in Section 4.3.2.

competes with Ø when location at a place or motion towards a goal are expressed. The follow-up study to this chapter compares the Maltese patterns of alternation to those found in selected other languages of Europe (Stolz/Levkovych/Urdze 2017).

With reference to Krier (1975), Borg (1988: 280–295) studies expressions of static location in Maltese for which many Standard Average European (SAE) languages employ copulas and spatial relators. Borg's main focus is on the replacements Maltese employs in the absence of a fully-fledged copula in the imperfective. Yet, the author makes interesting remarks as to the *fi-Ø* alternation in connection to Place. This pattern is illustrated in (1.53)–(1.54) where two static spatial relations are described with the toponym *Malta* functioning as Ground in both examples. For the time being, we use the index X for the adverbial phrase from which the Π *fi* 'in' is absent. The examples are taken from one and the same primary source in order to show that the *fi-Ø* alternation is independent of idiolectal preferences.[65]

(1.53) [Bartolo/Vella 2009: 43]
Jason kiteb dwar il-Fiddien [f' Malta]$_{PP}$
Jason write about DEF-Silversmith [**in** Malta]
'[. . .] Jason wrote about the Silversmith [**in** Malta].'

(1.54) [Bartolo/Vella 2009: 17]
Ma rridx nemmen li qed jiġru
NEG 1.IPFV:want:NEG 1.IPFV:believe that PROG 3.IPFV:run:PL
affarijiet bħal dawn [Ø Malta]$_{X}$
affair:PL like DEM:PL:PROX [**Ø** Malta]
'I don't want to believe that things like these are happening [**in** Malta].'

Borg (1988: 282) argues that in spite of cases similar to (1.53)–(1.54) "ma nistgħux ngħidu li l-preposizzjoni 'fi' hija fakultattiva, għax iż-żewġ sentenzi jfissru ħwejjeġ differenti".[66] For Borg, the use of *fi* 'in' is indicative of the concrete physical presence of a given Figure at a given Ground whereas the Ø-strategy does not invoke these associations. In Stolz/Lestrade/Stolz (2014) and Stolz/Levkovych/Urdze (2017), it is shown that this explanation does not cover the entire domain of the *fi-Ø* alternation.

65 Cf. Figure 1.3 above.
66 Our translation: "we cannot say that the preposition *fi* is optional because the two sentences mean different things".

First of all, the phenomenon is not restricted to static relations of Place. Borg/Azzopardi-Alexander (1997: 156) argue that "[c]ertain verbs of motion, followed by a place nominal (without a preposition) express motion to a place, including arrival at a place." Saari (2003: 242–248) adduces a variety of examples of zero-marking of Place and Goal in Maltese. In (1.55)–(1.56), we provide evidence of the *fi-Ø* alternation in the expression of motion events. Like in the previous case, the examples are taken from the same source. The Ground is represented by definite NPS which involve the common noun *każin* 'club'. In both sentences, the same motion verb is used, namely *mar* 'go'. The conditions are thus almost identical for the two options.

(1.55) [Cauchi 1997: 44]
mar miegħu [fil-Każin tal-Banda Perellos]$_{PP}$
go with:3SG.M [**in**:DEF-club of:DEF-band Perellos]
'[. . .] he went with him [**to** the club of the Perellos Band].'

(1.56) [Cauchi 1997: 39]
mar [Ø il-Każin]$_{X}$ ippreparat
go [**Ø** DEF-club] prepare:PART
'[. . .] he went [**to** the club] prepared.'

The situations described in (1.55)–(1.56) are similar in the sense that the protagonist ultimately enters the clubhouse so that the criterion of concrete physical location of the Figure at the Ground does not make a difference. Saari (2003 – English abstract: 24) assumes that

> [t]he elision of Maltese prepositions testifies to some of the dynamic processes which occur in this language. Elision may occur [. . .] in dative phrases after verbs of motion. The preposition may also be omitted in locative phrases, when there is a strong linkage between the subject and the noun which expresses the location [. . .]. Unlike these cases, the preposition may not be omitted when the noun is an indefinite one or when the elision may otherwise cause ambiguity and misunderstanding.

The wording of this quoted paragraph gives rise to the idea that any kind of Π can be subject to deletion. This interpretation is certainly wrong since the only systematic pattern of alternation is that of *fi* and Ø. Saari (2003) and Borg/Azzopardi-Alexander (1997) agree insofar as they do not provide any functional, semantic, pragmatic, or syntactic explanation for the *fi-Ø* alternation in motion events whereas they make suggestions as to the reasons for the same phenomenon in the case of static situations. The label of motion events excludes spatial relations of Source, i.e. there is talk only of Goal. The explanations put forward for the *fi-Ø* alternation with Place differ between our references. As mentioned above, Borg (1988) assumes that the Ø-strategy only vaguely locates the Figure relative to the Ground whereas the *fi*-strategy presupposes physical co-presence. In contrast, Saari (2003) does not explain the use of the overt Π but assumes a higher degree of empathetic relation between Figure and Ground in the case of the absence of the Π.

We do not deny that the factors argued by Borg (1988), Borg/Azzopardi-Alexander (1997), and Saari (2003) have a say in the choice of construction. However, Stolz/Lestrade/Stolz (2014) and Stolz/Levkovych/Urdze (2017) demonstrate empirically that there is a much richer network of factors (including hitherto neglected syntactic as well as pragmatic ones) whose internal hierarchy of strength still needs to be discovered. What transpires from the contributions of the Bremen research team to the discussion is that toponyms stand out as Grounds which particularly often allow for the application of the Ø-strategy – and this especially when the predicate nucleus is a semantically rather unspecific motion verb or the copula (or one of its replacements) (Stolz/Lestrade/Stolz 2014: 228–242).[67]

1.5.3 Stolz/Ahrens (2017)

From a different angle, the omission of Πs is also addressed in Stolz/Ahrens (2017). In this study, the authors aim at providing the first systematic account of EQUI-P-deletion in Maltese. To this end, the co-ordination of two PPs with identical heads is inquired into to determine which Πs undergo deletion under which conditions and which Πs are resistant to EQUI-P-deletion for what reasons. What is especially intriguing about this issue is that there is a wide range of variation so that many Maltese Πs allow for EQUI-P-deletion but not consistently as can be gathered from (1.57)–(1.58).

(1.57) [Camilleri 2013: 63]

[B'	*pinzetta]$_{PP}$*	*u*	***[b**l-akbar*	*delikatezza]$_{PP}$,*	*Cortis [. . .]*
[with	tweezers]	and	**[with**:DEF-big.CMPR	delicacy]	Cortis [. . .]
siltet		*balla*			
pull_out:3SG.F.PFV		bullet			

'[**With** tweezers] and [**with** much delicacy] Cortis [. . .] extracted a bullet [. . .].'

67 The *fi-Ø* alternation will re-enter the scene in Section 4.3.3.2.

(1.58) [Żahra 2008: 131]

kienet *taf* *tkun* *ċajtiera* *u*
be:3SG.F.PFV 3.F.IPFV:know 3.F.IPFV:be.FUT good_humoured:F and
serja, *ferrieħa* *u* *żorra,* *dħulija* *u* *riservata*
serious:F joyous:F and rude:F agreeable:F and reserved:F
*[**b**l-istess* *qies]*$_{PP}$ *u* *[**Ø** Ø doża]*$_{X}$ *skond*
[**with**:DEF-same measure] and [**Ø** Ø dose] according_to
dak *li* *jkun* *meħtieġ.*
DEM:M:DIST that 3.IPFV:be.FUT PART:be_necessary
'[. . .] she knew how to be good-humoured and serious, joyous and rude, agreeable and reserved [**with** the same degree] and [dose] according to what would be necessary.'

In (1.57), the Π *bi* 'with' is realised on both conjuncts of the coordinative construction whereas in (1.58) the same Π occurs only once. In the latter case, we note that the second conjunct is completely naked in the sense that not only the Π is lacking but also the definite proclitic.

Borg/Azzopardi-Alexander (1997: 87) state that

> when two structurally identical prepositional phrases with adverbial function are coordinated, the second preposition can be omitted. [. . .] If the prepositional phrases are not structurally identical, neither of the prepositions is omitted.

In terms of fact, the database Stolz/Ahrens (2017) put together for their pilot-study gives exclusively evidence of forward ellipsis (= analipsis) according to Haspelmath's (2007a: 39) taxonomy of deletion under coordination. Stolz/Ahrens (2017: 137) assess the degree to which Maltese Πs are affected by EQUI-P-deletion for 21 Π-types. The quantitative results reveal that there is evidence of EQUI-P-deletion for each of these Π-types. At the same time, it turns out that the different Πs behave differently under coordination. For 20 out of 21 Π-types, EQUI-P-deletion is the majority option. Only *lil* 'to' displays a 52% preference for two overt Πs. EQUI-P-deletion applies in 75%–99% of the tokens of coordinative constructions for 17 Πs. Only four Π-types yield smaller shares from 68% down to 48%. Thus, EQUI-P-deletion is the default option for the vast majority of the 21 Πs inspected by Stolz/Ahrens (2017). In contrast, the blockage of EQUI-P-deletion constitutes the marked case – and thus calls for an explanation.

A structural factor which blocks the application of EQUI-P-deletion is the presence of a pronominal complement in the second conjunct (Stolz/Ahrens 2017: 136). In addition, there also seem to be semantic, pragmatic, and syntactic factors which regulate the applicability of EQUI-P-deletion. Since Stolz/Ahrens (2017) take account only of a subset of the Maltese Π-inventory, their results and hypotheses need to be tested on the basis of the full inventory.[68]

1.5.4 Stolz/Levkovych (2020)

When Stolz/Ahrens (2017) conducted their project on EQUI-P-deletion they became aware of the fact that no universally accepted proposal for the Maltese Π-inventory existed. This surprising experience induced the Bremen research team to establish the size and composition of this inventory in a follow-up project. The results of this follow-up project are documented in this monograph. The point of departure of the follow-up project is marked by Stolz/Levkovych (2020).

On the basis of a nuclear version of the above historiographic survey,[69] Stolz/Levkovych (2020: 201–206) show that there is no unanimity among scholars as to the size and composition of the Maltese Π-inventory. The absolute number of Πs ranges from minimally 17 to maximally 44. What is more, the smaller inventories are not fully included in the larger ones so that the Π-status of several candidates remains doubtful. The authors emphasise that to put forward hypotheses and generalisations about the system and grammar of Maltese Πs one first has to be certain about how many and which units belong to the class under scrutiny. Stolz/Levkovych (2020: 206–216) exclude several supposed Πs from the inventory they use as the basis for their project.[70]

The paper authored by Stolz/Levkovych (2020: 220–237) highlights differences between the members of their Π-inventory which come to the fore in the context of government properties. For a set of eleven selected Πs, the authors postulate six different classes according to the marking of the complement. Complements can either be lexical NPS (including proper names) or pronominal. In both complement classes, there are four patterns of government, namely accusative, nominative, genitive, and prepositional. Identity of government patterns across complement classes applies with five Πs whereas the remaining six Πs give evidence of differences between the options realised with complement NPS and complement pronouns.[71] At this point, it suffices to showcase the controversial government patterns of *kontra* 'against'.

68 We tackle this task in Section 4.1.4.

69 Cf. Sections 1.4.1.4–1.4.2.5.

70 This inventory will be presented in more detail in Section 1.6.

71 These topics will be looked into in Section 4.1.2.

In (1.59), it is shown that *kontra* takes a complement NP without additional morphological marking. In the terminology employed by Stolz/Levkovych (2020), this is an instance of nominative government.

(1.59) [Baldacchino 1999: 114]

sena	*wara*	*l-oħra*	*raw*	*it-timijiet*	*tagħna*	*jitilfu*
year	after	DEF-other:F	see:3PL.PFV	DEF-team:PL	of:1PL	3.IPFV:lose:PL
[kontra	*l-Ingliżi]*$_{PP}$					
[**against**	DEF-English:PL]					

'[. . .] year after year they saw our teams lose [**against** the English].'

In the case of pronominal complements however, there are two options, namely the use of *ta'* 'of' as host of the pronominal affix as shown in (1.60) and the direct attachment of the pronominal affix to the stem of *kontra* as shown in (1.61).

(1.60) [Teuma 2015: 256]

f'	*jum*	*il-ħaqq*	*demmhom*	*jgħajjat*	*vendetta*	***[kontra***
in	day	DEF-justice	blood:3PL	3.IPFV:shout	vendetta	[**against**
[tiegħi]$_{PP}$*]*$_{PP}$						
[**of**:1SG]]						

'[. . .] on Judgment's Day their blood will cry vendetta [**against** me].'

(1.61) [Bartolo/Vella 2009: 201]

Din	*il-potenza*	*tintuża*	***[kontrija]***$_{PP}$
DEM:F:PROX	DEF-power	3SG.F.IPFV:RFL:use	[**against**:1SG]

'This power is used [**against** me].'

Example (1.60) is representative of prepositional government whereas example (1.61) reflects the pattern of genitival government. According to Aquilina (1987: 683), the genitival pattern in (1.61) is the pragmatically neutral default strategy whereas the prepositional pattern exemplified in (1.60) is used "as a slightly more emphatic alternative". The variation is given a pragmatic explanation. As to complement NPs, Stolz/Levkovych (2020: 235) only register the prepositional strategy about which Saari (2009: 273) says that according to his research "the usage of *kontra* without *ta'* is much more common than the usage of the prepositional phrase". It is even possible that the infrequent examples of *ta'* 'of' linking *kontra* to a complement NP can be analysed differently so that the prepositional pattern disappears for good from the list of options in the domain of complement NPs of *kontra*.[72] The take-home message of this subsection is that we still do not know for certain how exactly the Maltese Πs interact with their complements.

1.5.5 Schmidt/Vorholt/Witt (2020)

In this collective study, the three authors use the same Π-inventory as Stolz/Levkovych (2020) to determine the token frequencies of the members of this inventory, classify them according to their syllabic structure, and divide them into etymologically defined subgroups. The results of the three studies constitute the input for the final cluster analysis.

The token frequencies result from a search on the Korpus Malti 3.0. There are 16 different frequency classes with enormous differences in the number of hits between the most frequent and the least frequent Πs. Two Πs are unattested in the electronic resource, namely *oltri* 'except' and *sotta* 'under'.[73] The Π with the lowest token frequency is *versu* 'towards' which is attested 228 times. At the opposite end of the hierarchy, we find *fi* 'in' which occurs 11,801,461 times. This means that *fi* is 51,760 times as frequent as *versu*. As many as 25 of 55 attested Πs fail to yield token frequencies which equal or come close to 1% of that calculated for *fi*. Only seven other Πs display token frequencies of $n \geq 1{,}000{,}000$. There are thus many relatively infrequent Πs as opposed to a small set of highly frequent Πs.

The thirty monosyllabic Πs account for 53% of all members of the Π-inventory. The second largest group is that of the bisyllabic Πs which number 21 types (~ 37%). The smallest class comprises six Πs with three or more syllables which form a minority of some 10%. The margins of the segmental chains of the Πs show a pronounced preference for hosting consonants. There are 38 Πs which end in a consonant (~ 67%) whereas 55 Πs display an initial consonant (~ 96%). Of the 30 monosyllables, 24 Π-types are covered and closed. In contrast only five monosyllabic Π-types are open and only a single case is naked.

In terms of their etymological background, the Maltese Πs are overwhelmingly Semitic. Semitic Πs cover 75% of the Π-inventory. The Romance component has a share of 22% whereas one Π each is an etymological hybrid or unclassified, respectively. Romance Πs are for the most attested with mid-range token frequencies. In contrast, the Πs with particularly high token frequencies are all of Semitic origin. The latter also stand out with regards to the number of meanings they are asso-

72 The pros and cons of this solution are discussed in Section 4.1.2.2.1.2.1 (iv).
73 But see Sections 1.6.6 (v) and 1.6.7 (iv).

ciated with in Aquilina (1987, 1991). Romance Πs are often unable to inflect for person whereas most of the Semitic Πs have this ability.

In the summary, the authors further claim that

- "[t]he frequency in the corpus does not seem to have a major effect on the ability to fuse with the definite article or personal suffixes, but tendencies for a continuous distribution can be observed where prepositions from most lower frequency classes do not fuse, while those from higher frequency classes do." (Schmidt/Vorholt/Witt 2020: 261),
- "[t]he quality of the last segment is the deciding factor whether a preposition can be fused with personal suffixes or an article. Prepositions ending with an obstruent fuse neither with the article nor with personal suffixes." (Schmidt/Vorholt/Witt 2020: 262).

The study under review inspires us to try and tie up the loose ends by way of testing the hypotheses put forward by Schmidt/Vorholt/Witt (2020) on the basis of our own data. We check the token frequencies of the Maltese Πs,[74] the phonological issues are inquired into[75] and problems related to morphology and morphonology are studied in some detail too.[76]

1.5.6 Stolz (2020)

In the edited volume on *Complex adpositions in European languages*, Stolz (2020) is responsible for the article dealing with the Maltese facts. His focus is on compound Πs, i.e. synchronically transparent and semantically compositional binary Π-Π combinations. The term compound Π is most probably a misnomer in the sense that it invokes the false impression that there is a morphologically complex but unitary Π. Alternatively, what is called compound Π may be better analysed as a syntagmatic sequence of two separate Πs which occupy different ranks in the syntactic architecture of a PP which embeds another PP as its complement. The terminological problem notwithstanding, we keep the label compound Π for reasons of practicality.

We know already from the previous literature review that compound Πs have occupied the minds of several of our predecessors. Panzavecchia (1843),[77]

74 Cf. Section 1.6.8.
75 Cf. Section 2.
76 Cf. Section 3.
77 Cf. Section 1.4.1.3, examples (1.20)–(1.21).

Aquilina (1959),[78] Fabri (1993),[79] and Cachia (1994)[80] provide examples of the compound ΠΠs and discuss their properties to some extent. Under the section headings *Complex pre-/postposition // Two prepositions*, Borg/Azzopardi-Alexander (1997: 297) claim that

> [t]he prepositions **ta'** 'of', **għal** 'for' and **minn** 'from' can be combined with (and precede) the following prepositions expressing different spatial relations: **fuq** 'on, above', **taħt** 'under, beneath', **wara** 'behind', **quddiem** 'in front of', **barra** 'out, outside', **ġo** 'in, inside', **ħdejn** 'near', **bejn** 'between'. In addition **minn** and **għal** can be combined with **għand** 'at' and **minn** can combine with **qalb** 'among'. [original boldface]

This quote enumerates the possible candidates for the formation of compound ΠΠs without making statements as to the resulting meanings except that the compound ΠΠs express spatial relations. The authors do not illustrate the use of the compound ΠΠs by way of giving sentential examples in the same section. However, there are other sections which are meant to cover all possible spatial relations. According to Borg/Azzopardi-Alexander (1997: 297), there must be 27 pairwise combinations of ΠΠs which are qualified for the status of compound ΠΠ. If we take a look at their Section 2.1.1.5 on "local semantic functions" (Borg/Azzopardi-Alexander 1997: 155), we discover additional binary combinations, namely

- *minn fejn* and *minn maġenb* for PROXIMATE MOTION FROM (Borg/Azzopardi-Alexander 1997: 157),
- *minn ma' taħt* (a combination of three ΠΠs!) for INFERIOR CONTACT MOTION PAST (Borg/Azzopardi-Alexander 1997: 162),
- *minn biswit* for LATERAL MOTION FROM/PAST (Borg/Azzopardi-Alexander 1997: 162–163),
- *minn ma'* for LATERAL CONTACT MOTION FROM (Borg/Azzopardi-Alexander 1997: 163),
- *minn madwar* for CIRCUMFERENTIAL MOTION FROM (Borg/Azzopardi-Alexander 1997: 166), and
- *minn faċċata* for CITERIOR-ANTERIOR MOTION FROM/PAST (Borg/Azzopardi-Alexander 1997: 167).

There are thus at least seven additional compound ΠΠs which are missing from the list provided by Borg/Azzopardi-Alexander 1997: 297) because the second component of the pair is not included. Since the authors argue that most of the ΠΠs they register are combinable not only with *minn* 'from' but also with *ta'* 'of' and *għal* 'for'

78 Cf. Section 1.4.1.5.
79 Cf. Section 1.4.2.2, Table 1.3, example (1.35).
80 Cf. Section 1.4.2.3).

chances are that this combinability also applies in the case of the above skipped Πs. Given that this assumption is indeed correct the number of compound Πs would increase considerably.

It is therefore hardly surprising that Saari (2003: 84–100) identifies eighty compound Πs 49 of which are of the binary type. Eight different Πs are possible fillers of the initial slot of the combination whereas 24 different Πs are admitted in the second position. What blurs the picture somewhat is that Saari (2003) does not always distinguish lexicalised univerbations from synchronic syntagmatic combinations. Stolz (2020: 449–450) emphasises that even Saari's inventory of compound Πs does not contain all attested combinations (in some cases in violation of Saari's own principles). Among the hitherto ignored compound Πs identified by Stolz (2020: 456), we find *sa waqt* 'until during' and *sa wara* 'until after' both of which are involved in (1.62).

(1.62) [Korpus Malti 3.0; news 119730]

ħafna	*nisa*	*ma*	*jkunux*	*konxji*	*minnhom*
many	women	NEG	3.IPFV:be.FUT:PL:NEG	conscious:PL	from:3PL

[sa	***[waqt***	*Ø]$_{PP}$]$_{PP}$*	*u*	***[Ø***	***[wara***	*t-tqala]$_{PP}$]$_{PP}$*
[till	**[at the time**	Ø]	and	**[Ø**	**[after**	DEF-pregnancy]]

'[. . .] many women aren't aware of them **[until [during]** and **[after** the pregnancy]].'

This example is intricate and at the same time telling. The complexity arises from the fact that under coordination we witness not only EQUI-P-deletion but also EQUI-N-deletion, i.e. the initial Π is subject to forward ellipsis whereas the complement NP is affected by backward ellipsis so that two slots remain empty. Without the two deletions, the PPs would have the following shapes: ***[sa [waqt** it-tqala]$_{PP}$]$_{PP}$* 'until during the pregnancy' and ***[sa [wara** it-tqala]$_{PP}$]$_{PP}$* 'until after the pregnancy'. The omission of *sa* 'till' on the second conjunct supports the idea that the compound Πs are not properly compounded, in the first place. We assume that there is a syntactic boundary which runs between *sa* and *waqt* as well as between *sa* and *wara*. The PPs headed by *waqt* and *wara* are incorporated into and subordinated to the PPs headed by *sa*.

Owing to the many problems which arise in connection to the analysis of compound Πs, we consider it to be advisable to start the description of the Maltese Π-system from the less complex elements thereof. Only after we know how the basics are organised do we approach the subject of compound Πs.[81]

81 Their description and evaluation forms part of Section 4.1.3.

1.5.7 Stolz/Levkovych/Vorholt (2023)

At the time of our writing this monograph, the joined paper by Stolz/Levkovych/Vorholt (2023) was still in print for an issue of the *Journal of Maltese Studies*. The topic of the paper ties in with what Stolz/Lestrade/Stolz (2014) and Stolz/Levkovych/Urdze (2017) have found out about the special morphosyntactic behaviour of toponyms in constructions expressing spatial relations. The earlier publications of the Bremen research team focus especially on zero-marking and on the *fi*-*Ø* alternation. In Stolz/Levkovych/Vorholt (2023), further patterns of variation are investigated, namely those which involve *fi* 'in', *ġewwa* 'inside', and *ġo* 'in' in combination with a toponymic complement.

The variation between the three ΠS is visible in (1.63)–(1.64). In addition to the usual boldface and underlining we also employ grey shading to mark the occurrence of *fi* and *ġewwa* as heads of PPs whose complement is not a genuine toponym. In (1.63), we assume that *Virginia* and *l-Ungerija* are appositions to the preceding toponym. Alternatively, they could be interpreted as zero-marked constructions of the type [Ø TOPONYM]$_X$ illustrated below.[82]

(1.63) [Attard 2019: 118]
Dawn l-aħħar snin kont ngħallem fl-Akkademja
DEM:PL:PROX DEF-last year.PL be:1SG.PFV 1.IPFV:teach in:DEF-academy
Nazzjonali tal-FBI [fi Quantico, Virginia]$_{PP}$ u wkoll
national of:DEF-FBI [**in** Quantico Virginia] and also
ġewwa l-International Law Enforcement Academy [f' Budapest, l-Ungerija]$_{PP}$
inside DEF-International Law Enforcement Academy [**in** Budapest DEF-Hungary]
'These last years I have taught at the National Academy of the FBI [**in** Quantico, Virginia] and also at the International Law Enforcement Academy [**in** Budapest, Hungary].'

(1.64) [Camilleri 2013: 62]
Cortis kienet għadha kemm temmet perjodu
Cortis be:3SG.F.PFV still:3SG.F how_much finish:3SG.F.PFV period
ta' sentejn u nofs ħidma fi sptar privat [ġewwa Boston]$_{PP}$
of year:DU and half work in hospital private [**inside** Boston]
'Cortis had just finished a period of two years and a half of work in a private hospital [**in** Boston].'

---

82 Cf. Section 1.6.3.

(1.65) [Attard 1999: 38]
kien il-konsul tar-Renju Unit [ġo Sao Paulo]$_{PP}$
be DEF-consul of:DEF-United Kingdom **[in** Sao Paulo]
'[. . .] he was the consul of the United Kingdom [**in** Sao Paulo] [. . .].'

The bracketed PPs in (1.63)–(1.65) express the static relation of Place and involve a toponymic Ground as complement. The toponym refers to cities outside of Malta. The largely identical structural and semantic conditions notwithstanding, three different Πs are used. According to Borg/Azzopardi-Alexander (1997: 157–158), *fi*, *ġewwa*, and *ġo* can replace each other in expressions of INTERIOR AT REST and INTERIOR MOTION TO also in combination with a common noun as complement.

Stolz/Levkovych/Vorholt (2023) try to determine whether this is a case of free variation. On the basis of the data extracted from the Korpus Malti 3.0, the authors argue that *fi* 'in' is the default Π for Place/Goal constructions with a toponymic Ground since 95% of all PPs of this kind involve *fi*. This predominance does not imply that the other two competing Πs are irrelevant. What transpires from the paper under review is that

> *ġo* and *ġewwa* display a relatively high degree of sensitivity to criteria such as the familiarity of the place name, the ontological class of the geo-object, and the complexity of the place name. This sensitivity is only relative because what results from it can only be captured with the term slight preference – a preference which is manifest more on the level of type frequencies than on that of token frequencies. None of the preferences [. . .] is strong. The choice of preposition depends on two or more factors conspiring, or so it seems (Stolz/Levkovych/Vorholt 2023: 210).

Definite conclusions cannot be drawn yet. The issue requires further scrutiny. Therefore, we will address this pattern of variation again below.[83]

1.5.8 Stolz/Vorholt (2022)

The edited volume documenting the 7th *International Conference of Maltese Linguistics* (Cracow/Poland, 2019) includes the paper by Stolz/Vorholt (2022) on WH-Πs. WH-Πs are function-words whose main function is that of an interrogative but for which prepositional usages are registered, too. This is the case for Maltese *fejn* 'near' which is connected to *fejn* 'where'. In the older literature on our subject, the

83 Cf. Sections 5.1.2.1.1 and 5.1.2.2.1.

Π-status of *fejn* 'near' is generally accepted between the lines, in a manner of speaking. Fabri (1993), Cachia (1994), and Borg/Azzopardi-Alexander (1997) explicitly acknowledge that *fejn* 'near' may function as Π whereas Aquilina (1959), Schabert (1976), Brother Henry (1980), and Saari (2003) do not assume the existence of a Π of this kind.

Stolz/Vorholt (2022) compare Maltese *fejn* to Spanish *donde* ~ *dónde* 'near ~ where' which displays similar characteristics.[84] The parallels emerge from the comparison of the Maltese sentences (1.66)–(1.67) with (1.68)–(1.69) from Spanish.

(1.66) [Bartolo/Vella 2009: 377]

Fejn	*kienet*	*marret*	*Elena?*
where	be:3SG.F.PFV	go:3SG.F.PFV	Elena

'**Where** had Elena gone?'

(1.67) (Borg/Azzopardi-Alexander 1997: 157)

It-tfal	*qegħdin*	*[**fejn**ek]$_{PP}$?*
DEF-child.PL	stay.PART:PL	[**near**:2SG]

'Are the children [**near** you]?' [O.T.]

(1.68) Spanish [De Saint-Exupéry 1995b: 72]

*¿**Dónde***	*están*	*los*	*hombres?*
where	be:3PL	DEF.M.PL	human:PL

'**Where** are the humans?'

(1.69) Spanish [Arana 2020: 60]

finalmente	*se*	*refugió*	*[**donde***	*una*	*familia*	*negra]$_{PP}$*
finally	RFL	seek_refuge:3SG.PST	[**near**	INDEF:F	family	black:F]

'[. . .] finally he found refuge [**in** a black family].'

In both languages, the spatial interrogative may also serve the purpose of a Π expressing the relation PROXIMATE REST AT. As Stolz/Vorholt (2022) demonstrate the parallels are limited in the sense that Spanish *donde* cannot serve all the purposes for which Maltese *fejn* is fit and vice versa. Nevertheless, the similarities between two genetically unrelated languages are sufficiently clear to call for an in-depth cross-linguistic study of WH-Πs.

84 Munaro/Poletto (2020: 120–124) provide evidence of the existence of WH-Πs in many Italo-Romance varieties from Sicily via Calabria and Naples to Corsica.

It is hardly surprising that the prepositional usages of Maltese *fejn* 'near' have been ignored by several of the earlier scholars.[85] First of all, it is a low frequency phenomenon. The Π-uses of *fejn* are outnumbered by far by those of interrogative *fejn*. Secondly, there is the general problem of how to handle cases of multiple word-class membership adequately. Is *fejn* exclusively or only primarily a spatial interrogative? Is the Π *fejn* a distinct function-word? Questions of this kind arise time and again when we look at the individual candidates for the status of Π in Maltese many of which are classified differently by different authors. It is not uncommon that candidates are allocated in several different word-classes by one and the same author (Stolz/Levkovych 2020: 205–206). This heterogeneity poses serious problems when it comes to determining the Maltese Π-inventory. We present our solution for these problems below.[86]

1.5.9 Vorholt (2022a)

Vorholt (2022a) compares the proportion of adpositions (i.e. not only Πs) in press corpora for a convenience sample of sixteen languages, including Maltese. The relationship between length and frequency as stated by Zipf's law of abbreviation (Zipf 1935) is examined. The comparison reveals that adposition inventories are quite divergent, both in terms of size and composition. Vorholt (2022a) finds that Maltese has the highest proportion of prepositions in the corpus across the sample languages with about 15% of all word tokens. As can be seen from Figure 1.4, adposition proportions vary significantly across corpora of different languages. Additionally, the highest frequency of occurrence is very different across the sample. Italian has the highest frequency of an individual adposition with almost 60,000 per one million words, followed by Maltese and Greek while Finnish, Lithuanian, and Hungarian show a much smaller highest frequency of 600 per million words and less.

Maltese Πs vary in length between two and eight phonemes. The most common length is four. The mean length is 4.80 segments. The results show that length can be predicted by frequency for twelve of the sixteen languages in the sample, Maltese being one of them. However, the frequency can explain only about 15% of the variation in length in Maltese. These quantitative findings will be revised for Maltese in Section 2.

85 Like those featured in Sections 1.4–1.5.

86 Cf. Section 1.6.

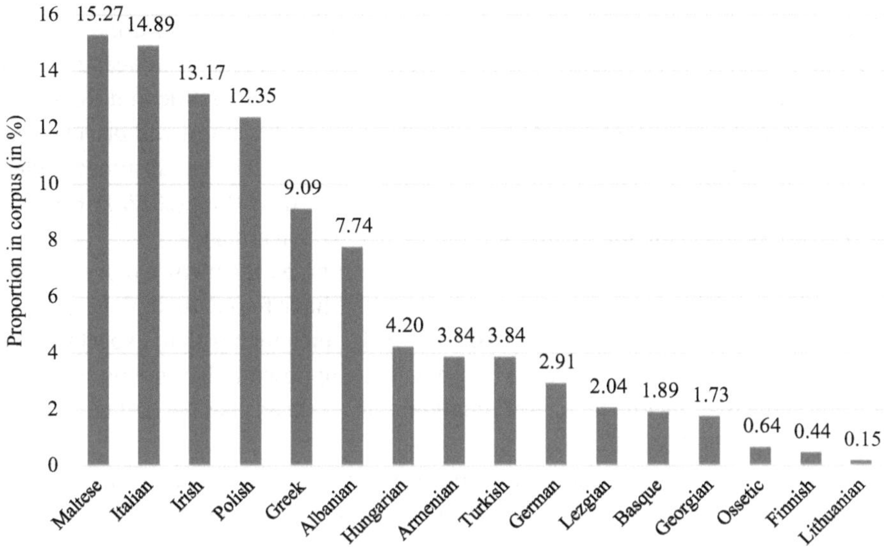

Figure 1.4: Proportion of adpositions per corpus and language.

1.5.10 Vorholt (2023a)

In her paper Vorholt (2023a) takes a closer look at the possibility of a correlation between the language of origin and token frequency of Maltese Πs. What is known about adposition borrowing in general is compared to the Maltese situation. The composition of the word class of Π is analysed and the integration of borrowed Πs into the language is described. Possible other factors influencing frequency are discussed. The study is based on a modified version of original BLOMP (presented in (1.73)) which excludes *sotta* 'under' and *oltri* 'besides', since they are not attested in the Korpus Malti 3.0.[87]

Etymological information is gathered on the basis of Aquilina's dictionary of Maltese (1987, 1991). Maltese Πs can be divided into four classes according to their language of origin. The vast majority of Maltese Πs (viz. 43 equalling almost three thirds) is Semitic. The ten Πs of Romance origin still account for a sizable proportion of 21%. The Romance-derived Πs are *apparti* 'apart from', *faċċata* 'opposite',

87 But cf. Sections 1.6.6 (v) and 1.6.7 (iv).

favur 'on the side of', *kontra* 'against', *permezz* 'by means of', *rigward* 'concerning', *sforz* 'thanks to', *skont* 'according to', *versu* 'towards', and *viċin* 'near'.[88] Additionally, there is *minflok* 'instead of, in lieu of' with mixed Semitic and Romance origin. How this case is interpreted for the purposes of this study, will be revealed below.[89] The etymologically unclear case of *daqs* 'equal to' will be subsumed under the non-Italo-Romance elements as explained below.[90] In Vorholt (2023a) the latter two Πs are responsible for slightly less than 4% of her inventory.

Vorholt (2023a) argues that, with reference to Π-types, the picture that emerges is significantly different from what previous studies have found about the composition of the Maltese lexicon. The share of loan-Πs is much higher than expected. The author concludes that Maltese does not seem to be opposed to borrowing Πs at all. We will show that the importance of Italo-Romance Πs even goes beyond the degree assumed in Vorholt (2023a).[91]

Token frequencies tell however, a different story. Πs of Semitic origin clearly dominate when it comes to overall token frequency. The token frequency of borrowed Πs does not reach that of the most frequent Πs of Semitic origin. What is more the first 17 ranks on the token frequency-based hierarchy are occupied by Semitic Πs. Semitic Πs cover almost 98% of all instances of Πs in the corpus. In terms of tokens, the highest-ranking Romance Π is *skont* 'according to' (but cf. the new count in Table 1.11).

Vorholt (2023a) looks at several aspects connected to the integration of Πs of Romance origin into the Maltese morphology and syntax. Borrowed Πs seem to be slightly longer phonologically than Semitic ones. As to their segmental length, Semitic Πs on the other hand show a wider range of variation than borrowed ones. None of the borrowed Πs can fuse with the article. However, three Romance Πs – *favur* 'in favour of', *kontra* 'against', and *skont* 'according to' – can be combined with person suffixes. An example is shown in (1.70) for *favur* parallel to the Semitic prepositions *dwar* 'about' in the same sentence (cf. (1.50)). The mixed preposition *minflok* 'instead of, in lieu of' can also be combined with suffixes.

---

88 For an updated view of the Italo-Romance component in the Maltese Π-inventory, we refer the reader to Sections 1.6.6–1.6.7.

89 Cf. Section 1.6.4.1 (v).

90 Cf. Section 2.1.

91 In the course of Section 1.6.

(1.70) [Korpus Malti 3.0; news152214]
mhux hekk il-Parlament Ewropew li dejjem kien qawwi
NEG SO DEF-parlament European that always be strong
dwar-ha u ***[favur-**<u>ha</u>**]**$_{PP}$*
about-3SG.F and **[in_favour_of-**<u>3SG.F</u>**]**
'Not so the European Parliament, which has always been strong about and **[in favour of** <u>it</u>**]**.'

Just like Maltese Πs of Semitic origin, borrowed Πs show variation in coordinative constructions (cf. Section 1.5.3) as shown in (1.71)–(1.72).

(1.71) [Korpus Malti 3.0; news208812]
ivvota ***[kontra** <u>l-ġenituri</u>**]**$_{PP}$ *u* ***[kontra** <u>t-tfal</u>**]**$_{PP}$*
vote **[against** <u>DEF-parent:PL</u>**]** and **[against** <u>DEF-child.PL</u>**]**
'They (the labor party) voted **[against** <u>parents</u> and <u>children</u>**]**.'

(1.72) [Korpus Malti 3.0; literature20]
The magic hour din ***[skont** <u>iċ-ċineasti</u>**]**$_{PP}$*
the magic hour DEM:F:PROX **[according_to** <u>DEF-filmmaker:PL</u>**]**
u [Ø <u>*l-fotografi*</u>*]*$_{PP}$
and [Ø <u>DEF-photographer:PL</u>]
'This is the magic hour, **[according to** <u>filmmakers</u> and <u>photographers</u>**]**.'

Vorholt (2023a) also determines semantic domains of Maltese Πs and finds that core meanings are not expressed by borrowed Πs. They mostly convey non-spatio-temporal functions, which fits well with the proposed hierarchies for borrowing (Matras 2011). Vorholt (2023a) finds that a regression model of the factors origin, segmental length, and semantic domain is most suitable to explain variation in frequency since this model can explain about 26% of the variation in frequency.

1.6 The *Bremen List of Maltese Prepositions (BLOMP)*

In the foregoing sections, we have put our finger on many problems in the research on Maltese Πs to conclude that none of the previous accounts is fully satisfactory in the sense that the basic question has been answered, namely how many and which Πs there are in Maltese. In contrast to the still undetermined inventory, the literature on our subject contains many very interesting and important statements in connection to other phenomena related to Maltese Πs. It remains to be seen whether these statements call for a revision if and when the vexing issue of the inventory is finally settled.

In this section, we put forward an inventory to serve as our frame of reference in the remainder of this study. This inventory goes by the name of the *Bremen List of Maltese Prepositions (BLOMP).* BLOMP emerged from our prior frustrated attempt to get a grip on the problem by way of simply lumping together each and every item which has been mentioned at least once as a Π in the relevant studies. This purely additive procedure yielded no really reliable results. We therefore had to change the approach from the hunter-gatherer strategy to one that is based on linguistic principles.

Saari (2003: 83–100) puts forward a three-layered model for the Maltese Π-inventory which heavily relies on formal criteria in the sense that monomorphic and polymorphic Πs of different degrees of complexity are distinguished. The centre of this model is occupied by the supposedly most basic Πs *bi* 'with', *lil* 'to', and *minn* 'from', the second layer hosts other morphologically unanalysable Πs such as *qabel* 'before', whereas the further subdivided third layer comprises all those Πs which can be qualified – one way or the other – as compound (such as *għal barra* 'towards the outside of'). The many interesting ideas expressed by Saari in connection with this model notwithstanding, we consider it to be unsuitable for the purpose of describing the grammar of Maltese Πs as it glosses over important functional similarities between Πs located on different layers. Therefore, we refrain from taking a purely formal criterion as the yardstick for admission to BLOMP. We prefer the functional criterion which can be captured by the notion of lexicalisation, i.e. to be a member of BLOMP a candidate must be lexicalised no matter how complex the expression is morphologically. Lexicalisation means that the Π in question does not arise from a productive derivational process which yields a semantically transparent compositional output.

In (1.73), the 57 members of the original version of BLOMP put forward by Stolz/Levkovych (2020: 217–218) are presented in alphabetical order accompanied by a single English translation equivalent and, in brackets, an ID-code involving the letter <X> which is meant to signal that the status of the candidate is still open to debate. On what principles BLOMP is based will be disclosed in the subsequent paragraphs. In the course of the ensuing discussion we will also have occasion to modify the membership of BLOMP in a number of cases. These necessary modifications are explained in the subsequent subsections. Ultimately, the definitive version of BLOMP – labelled BLOMP 2.0 – emerges from our critical review of the criteria and the candidates (cf. Section 1.6.8).

(1.73) Original BLOMP (Stolz/Levkovych 2020: 217–218)
(X1) *apparti* 'apart from'; (X2) *(min)barra* 'except'; (X3) *bejn* 'between'; (X4) *bħal* 'like'; (X5) *bħala* 'as'; (X6) *bi* 'with'; (X7) *biswit* 'facing'; (X8) *bla*

'without'; (X9) *daqs* 'equal to'; (X10) *dwar* 'about'; (X11) *faċċata* 'opposite'; (X12) *favur* 'on the side of'; (X13) *fejn* 'near'; (X14) *fi* 'in'; (X15) *(min(n))flok* 'instead of'; (X16) *fost* 'between'; (X17) *fuq* 'on'; (X18) *ġew* 'in'; (X19) *ġo* 'in'; (X20) *għajr* 'except'; (X21) *għal* 'for'; (X22) *għand* 'at s.o.'s place'; (X23) *ħdejn* 'beside'; (X24) *ħlief* 'except'; (X25) *ʕil* 'to'; (X26) *kontra* 'against'; (X27) *kif* 'as'; (X28) *lejn* 'towards'; (X29) *lil* OBJECT MARKER; (X30) *ma'* 'with'; (X31) *madwar* 'around'; (X32) *maġenb* 'close to'; (X33) *m(in)ħabba* 'on account of'; (X34) *mingħajr* 'without'; (X35) *mingħala* 'in s.o.'s opinion'; (X36) *mingħand* 'from s.o.'; (X37) *minkejja* 'in spite of'; (X38) *minn* 'from'; (X39) *mnejn* 'from where'; (X40) *oltri* 'except'; (X41) *permezz* 'by means of'; (X42) *qabel* 'before'; (X43) *qalb* 'amongst'; (X44) *qrib* 'near'; (X45) *quddiem* 'in front of'; (X46) *rigward* 'concerning'; (X47) *sa* 'till'; (X48) *sforz* 'thanks to'; (X49) *skont* 'according to'; (X50) *sotta* 'under'; (X51) *ta'* 'of'; (X52) *taħt* 'under'; (X53) *(ma)tul* 'during'; (X54) *versu* 'towards'; (X55) *viċin* 'near'; (X56) *waqt* 'at the time of'; (X57) *wara* 'after'

As already mentioned, we disclose the criteria for admission to BLOMP several of which require more or less elaborate comments. Furthermore, except the bona fide Πs like (X6) *bi* 'with', (X14) *fi* 'in', (X38) *minn* 'from', etc., we will present evidence of the Π-status of most of the items listed in (1.73) and additional candidates. On account of the modifications we are going to apply to the original BLOMP, this section closes with the revision of the frequency count applied to Maltese Πs by Schmidt/Vorholt/Witt (2020: 245–247). In the final paragraphs, our provisional definition (1.1) is modified too.

1.6.1 The canonical Π

What makes a candidate eligible for the membership in BLOMP? There are several criteria which must be fulfilled by a given item to be admitted to the inventory. The most elementary criterion is captured by the rule of thumb given in (1.1). Here is the gist of (1.1): A Π is a free morpheme which serves as subordinating link between two syntactic units. There is an external (higher) head whose complement is the PP. The Π preferably takes a complement (pro)noun thereby assuming the role of the internal head of the PP (Hagège 2010: 58). In the PP, the order is Π > COMPLEMENT. The simple dependency model in Figure 1.5 reflects this situation. We understand this model as representation of the canonical configuration. According to the philosophy of Corbetts's (2005) canonical approach in typology, the canon only serves as an abstract point of reference against which actually attested instances of linguistic phenomena can be calibrated. In this sense, Figure 1.5 allows for deviations from the canon at several points on which we will comment repeatedly in the course of the subsequent chapters and their subdivisions.

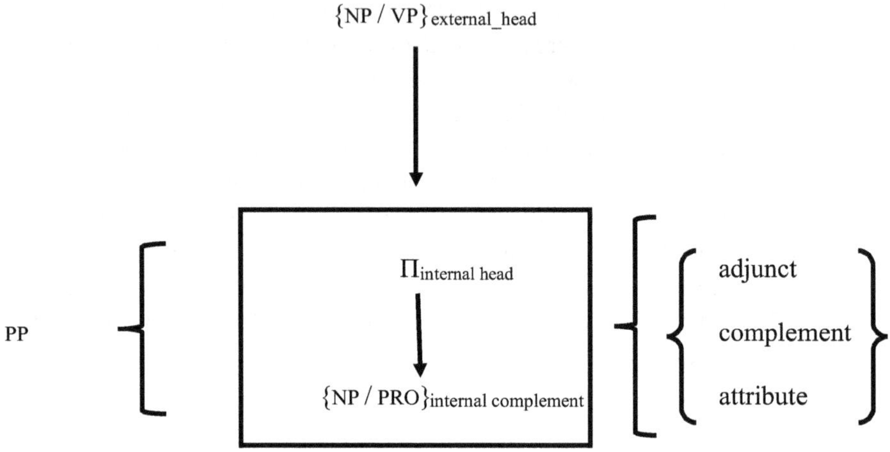

Figure 1.5: Dependency model (canonical version).

On the basis of the canon, we discuss the exclusion from and inclusion in BLOMP of candidates which have been mentioned in the literature on Maltese Πs. We start with explaining the rebuttal of Saari's idea that there is a postposition im Maltese. This rebuttal also covers pseudo-verbs.

1.6.2 No postpositions / no pseudo-verbs

The canon is not only about Maltese Πs but also about Maltese adpositions, in general. We assume that Maltese is exclusively prepositional so that there is no space for other kinds of adpositions. When we say this, we do not generally deny the possibility that predominantly prepositional languages allow for the existence of postpositions and vice versa. Cross-linguistically, the co-existence of different kinds of adpositions is amply attested. Latvian – a language mentioned already in Section 1.3 – is predominantly prepositional but makes also use of postpositions (Stolz 1984). In the case of Maltese however, there is no compelling reason to assume that the adpositional system is mixed.

Accordingly, the idea expressed by Saari (2003: 248–251, 2007) that Maltese might also attest to the existence of a temporal postposition *ilu* 'ago' must be rebutted. The criticised author is certainly right when he assumes that temporal adverbial constructions in English (e.g. *ten years **ago***) and Italian (e.g. *dieci anni*

fa 'ten years ago') have influenced the use of Maltese *ilu* (Saari 2007: 82). We interpret *ilu* as an inflected form of the pseudo-verb *il-* 'since, ago, for a time' (Fabri 1993: 198–201; Petersen 2009: 187). In (1.74)–(1.75), the paradigmatic change of the subject person of this pseudo-verb is illustrated.

(1.74) [Ellul Mercer 1986: 41]

Kien	***ilu***	*ftit*	*ma*	*jkollna*	*ħbit*	*hekk*	*qawwi*
be	**ago:3SG.M**	little	that	3.IPFV:have.FUT:1PL	attack	so	strong

bil-bombi.
with:DEF-bomb:PL
'It was a short time **ago** that we had a similarly strong bomb attack.'

(1.75) [Ellul Mercer 1986: 79]

Ma	*nafx*	*kemm*	*kont*	***ili***	*f'*
NEG	1SG.IPFV:know:NEG	how_much	be:1SG.PFV	**ago:1SG**	in

din	*il-qagħda*	*kerħa*
DEM:F:PROX	DEF-situation	ugly:F

'I don't know **for** how long I was in this ugly situation [. . .].'

Person, gender, and number are marked on the pseudo-verb (Borg/Azzopardi-Alexander 1997: 228). The use of *ilu* 'ago' in combination with expressions of time is special in the sense that "[t]he form **ilu** with the suffix of the 3 pers[on] masc[uline] sing[ular] may be used absolutely with the meaning *ago*" (Sutcliffe 1936: 197, original boldface). Aquilina (1959: 298–299) registers *il-* as an adverb of time and notes that *ilu* can be used impersonally. Saari (2007: 78) takes issue with previous attempts at etymologizing *ilu* which in his interpretation goes back to the univerbation of the erstwhile Arabic Π *'l* 'to' + the Maltese definite proclitic *il-* + the possessive suffix *-u* of the 3rd person singular masculine. Via aphaeresis of the initial lateral consonant original **lilu* fossilised to contemporary *ilu* which has become morphologically opaque. We do not comment on this diachronic scenario. What is more important for BLOMP is the clarification of the putative adpositional status of *ilu*.

Saari (2007: 80) emphasises the fact that none of his predecessors has classified this item as adposition. Most statements regard it as an adverb (Sutcliffe 1936: 196–197). Aquilina (1987: 565, original italics) assumes for instance that "*ilu* has a verbal function either when used alone [. . .] or at the beginning of a sentence [. . .]. At the end of a sentence it has an adverbial function" without assigning it to any word-class in particular. The main argument Saari (2007: 80–81) puts forward in support of his own hypothesis is the morphological disagreement in gender (and number) of *ilu* with the subject of the lexical verb as in (1.76).

(1.76) (Saari 2007: 85)

gender disagreement?

X	*iżewġ**et***	*xahrejn*	***ilu***
X	marry:**3SG.F.PFV**	month:DU	ago:**3SG.M**

'X got married two months **ago**.' [O.T.]

The subject of the lexical verb is feminine as appropriately coded on the verb by the suffix *-et* whereas *il-* hosts the usual suffix of the 3rd person singular masculine. This violates the expected pattern of agreement. According to Borg/Azzopardi-Alexander (1997: 228) "the pronominal suffix [*il-*] takes corresponds to the inflection of the lexical verb." In Saari's interpretation of this supposed case of disagreement, we have a postpositional phrase *xahrejn ilu* 'two months ago' which is a temporal adverbial adjunct to the main predicate.

We do not subscribe to this analysis because there is no need to expect subject-identity of the lexical verb and *il-*, in the first place. In constructions of this kind, *ilu* is independent of the participants encoded by NPs, pronouns, and/or verbal inflection. In temporal adverbials, *ilu* always hosts the regular suffix of the 3rd person singular *-u* which can be said to refer to the situation described in the preceding or following clause. In (1.77), the box singles out the clause which is anaphorically referred to by the person suffix of *ilu*. The arrow connects the initial clause to its anaphor. The morphology of *ilu* is not participant-oriented but serves the connection of syntactic units such as clauses or parts thereof.

(1.77) [Ellul Mercer 1986: 21]

Liġi	*bħal*	*din*	*għaddiet*	*fl-Ingilterra*
law	like	DEM:F:PROX	pass:3SG.F.PFV	in:DEF-England

xi	*tliet*	*xhur*	***ilu***
some	three	month.PL	**ago:3SG.M**

'A law like this was passed in England some three months **ago**.'

To our mind, it is still possible to analyse the final *-u* as fully functional person marker which refers to a clause in the same syntactic context. That *il-* only appears in the 3rd person masculine in these cases is unproblematic for our analysis because reference is always made to a clause and not to speech-act participants or other entities. In (1.76)–(1.77), *ilu* realises a paradigmatic form of the pseudo-verb *il-*. The notion of pseudo-verb is not referred to in Saari (2003, 2007). Pseudo-verbs have

been excluded from the Π-inventory already in connection with examples (1.27)–(1.28) and the discussion of Cachia (1994).[92] The main reason for their exclusion is the mandatory morphological agreement with their sole argument even if it is additionally represented by a lexical NP. Thus, *ilu* is counted out as Π for two reasons:
(a) it fails to meet the linearisation criterion and
(b) it retains the properties of a pseudo-verb.

More generally, no pseudo-verb is admitted to BLOMP.[93]

As an afterthought to this section, we add that if the case of *ilu* as potential postposition is reopened in the future, the discussion must necessarily take account of the syntactic behaviour of Πs like (X57) *wara* 'after' and (X42) *qabel* 'before' in temporal adverbials of the kind illustrated in Table 1.4.

Table 1.4: Selected examples of *wara* and *qabel* in final position.

Source	
Attard 1999: 74	***Tliet snin <u>wara</u>*** *mar Broken Hill fejn tkellem mal-Maltin li kienu jaħdmu fil-minjieri.* '**Three years <u>later</u>**, he went to Broken Hill where he talked to Maltese who worked in the mines.'
Attard 1999: 89	***Sebat ijiem <u>wara</u>*** *saret votazzjoni dwar it-tmexxija tal-Partit.* '**Seven days <u>later</u>**, the election for the leadership of the party took place.'
Attard 1999: 61	***Ħames xhur <u>wara</u>*** *Dun Ġorġ sar isqof ta' Puerto Rico [. . .].* '**Five months <u>later</u>**, Dun Ġorġ became bishop of Puerto Rico [. . .].'
Attard 1999: 65	***Xi tliet snin <u>qabel</u>*** *ħafna mill- irġiel kienu bla xogħol [. . .].* '**Some three years <u>back</u>**, many of the men were without work [. . .].'
Korpus Malti 3.0; news1199	***Ftit sigħat <u>qabel</u>****, l-eks Prim Ministru Silvio Berlusconi saħaq li mistenni jerġa jikkontesta l-kariga [. . .].* '**A few hours <u>earlier</u>**, ex-prime-minister Berlusconi insisted that he is expected to contest the office again [. . .].'
Korpus Malti 3.0; academic29	*[. . .]* ***ftit xhur <u>qabel</u>****, eluf ta' Maltin kienu emigraw lejn il- Kanada u l- Awstralja [. . .].* '[. . .] **a few months <u>before</u>**, thousands of Maltese had emigrated to Canada and Australia [. . .].'

In cases like these, one may assume that ellipsis applies in the sense that the slot to the right of the Π remains empty but retains an anaphorical relation to the events described in the preceding context of the sentences. Thus, *ilu* and *wara/qabel* may not be of the same kind but they are functionally related to each other. This func-

92 Cf. Section 1.4.2.3.
93 Cf. Section 1.6.5.

tional similarity calls for closer scrutiny which is a task for a separate study since it transcends the limits of this investigation of Maltese Πs.

1.6.3 No immaterial Πs

One might want to claim that a candidate is missing from (1.73), namely spatial-temporal Ø whose Π-status seems to follow from the *fi-Ø* alternation discussed already in connection with Figure 1.3.[94] Inspired by the case-study on zero-marking of spatial relations in French (Stolz/Lestrade/Stolz 2014: 188–224), Mel'čuk (2018: 274) postulates the existence of a Π Ø in this Romance language. Since the case-study dedicated to French immediately precedes that of zero-marking of spatial relations in Maltese,[95] chances are that in analogy to French, the author would propose a Π Ø also for Maltese. We do not follow Mel'čuk (2018) in his analysis of the French facts and argue against extending it to those of Maltese. The reasons for our reservations are exposed (exclusively for the Maltese case) in the next paragraph.

In a sketch of his theory of the concept of zero sign, Mel'čuk (2006: 478) argues against the unnecessary proliferation of zero signs and claims that

> [a] zero sign should be introduced only if there are no other linguistic means available to take care of the observed chunk of meaning to be expressed. Note that linguistic means include more than segmental signs (Mel'čuk 2006: 479).

Moreover, there is the concept of ellipsis (Mel'čuk 2006: 494) which covers cases of (meaningless) context-determined deletion. Mel'čuk's focus on morphology notwithstanding, his ideas are also relevant for morpho-syntactic phenomena among which we find EQUI-P-deletion.[96] To disprove the hypothesised existence of a Π Ø in Maltese, we tick off a small selection of aspects in this section.[97]

First of all, if the existence of a Π Ø is postulated on the basis of the *fi-Ø* alternation, this phonologically unrealised Π can only have the status of an allomorph of *fi*. One might then argue that there is a Π *fi* with two allomorphs *fi* and Ø whose distribution is such that the domain of Ø is fully included in that of *fi* which, however, is much larger than that of Ø. The *fi-Ø* alternation is strictly limited to the expression of spatial relations (and temporal relations derived from spatial ones). The alternation is blocked if the complement is a personal pronoun (cf. (1.78)), or *fi* is used

94 And in Sections 1.5.2 and 1.5.7.

95 As summarised in Section 1.5.2.

96 As introduced in Section 1.5.3.

97 The issue will be raised again in more detail in Section 4.3.3.2.

non-spatially (cf. (1.79)), or if a *fi*-PP is required in the case-frame of a given verb (cf. (1.80)).

(1.78) [Teuma 2015: 429]
Malli dañlet il-palazz bagñtet għand
as_soon_as enter:3SG.F.PFV DEF-palace send:3SG.F.PFV at_s.o.'s_place
Ibn Tâhir l-ilsira [. . .] li [fiha]$_{PP}$ kellha fiduċja
Ibn Tâhir DEF-slave:F [. . .] that [**in**:3SG.F] have.PFV:3SG.F faith
għamja
blind:F
'As soon as she entered the palace she sent to Ibn Tâhir the female slave [. . .] [**in** whom] she trusted blindly.'

(1.79) [Attard 1999: 67–68]
L-ittri tiegħu juru biċ-ċar kemm
DEF-letter:PL of:3SG.M 3.IPFV:show:PL with:DEF-clear how_much
kienet kbira l-ħtieġa li jkun hemm min
be:3SG.F.PFV big:F DEF-necessity that 3.IPFV:be.FUT there who
jitkellem [f' isem il-Maltin]$_{PP}$
3.IPFV:speak [**in** name DEF-Maltese:PL]
'His letters showed clearly how strong the need was to have someone speak [**in** the name of the Maltese].'

(1.80) [Attard 2019: 175]
Ftakar [fil-ħembexa]$_{PP}$ li kienet tħobb
remember [**in**:DEF-hembesha] that be:3SG.F.PFV 3SG.F.IPFV:love
tagħmel ommu naturali meta kien għadu
3SG.F.IPFV:make mother:3SG.M natural when be still:3SG.M
fl-Eritrea
in:DEF-Eritrea
'He remembered [the hembesha] his biological mother loved to make when he was still in Eritrea.'

In all three contexts, the replacement of *fi* with Ø is blocked. In (1.78), dropping *fi* would result in a construction with the free pronoun of the 3rd person singular feminine: **li Ø hija kellha fiduċja għamja.* This relative clause is acceptable only if *hija* 'she' refers to the subject and not to the participant who is trusted. Leaving out *fi* in (1.79) would produce the clause **min jitkellem Ø isem il-Maltin* 'literally: who speaks the name of the Maltese' in which the syntactic function of the NP *isem il-Maltin* changes from adjunct to direct object. The use of the Π *fi* to introduce the entity to be remembered seems to be obligatory with verbs derived from *fakar* 'remember' (Aquilina 1987: 299).

The Π *fi* can be deleted in certain contexts especially if the Ground is a toponym. If the Ground is not a genuine toponym or a comparable concept, the deletion is not licit as shown in (1.81).

(1.81) [Bartolo/Vella 2009: 215]

Xħin	*ġie*	*lura*	*kellu*	*envelowp*	*kbir*	*[f'*	*idejh]*$_{PP}$
when	come	back	have.PFV:3SG.M	envelope	big	[**in**	hand:PL:3SG.M]

'When he came back he had a big envelope [**in** his hands].'

The optional occurrence of Ø is determined by the context, i.e. ellipsis applies. This in turn means that there is no zero sign in terms of Mel'čuk's definition of the notion. If there is no zero sign it makes no sense to assume that Maltese boasts a Π Ø, in the first place.

Moreover, in contrast to a case in the domain of Maltese phonology to be discussed below,[98] the putative Π Ø does not influence the application of the laws of external sandhi. In (1.82)-(1.83), the DP in Ground function comes in two different shapes.

(1.82) [Żahra 2008: 249]

meta	*mort*	***Ø***	*id-dar*	*naqbeż*	*u*	*nogħla*
when	go:1SG.PFV	**Ø**	DEF-house	1.IPFV:jump	and	1.IPFV:rise

'[. . .] when I went home I jumped and leaped [. . .].'

(1.83) [Żahra 2008: 166]

Missieri	*beda*	*jiġi*	***Ø***	*d-dar*	*tard*
father:1SG	begin	3.IPFV:come	**Ø**	DEF-house	late

'My father began to come home late [. . .].'

In (1.82), this DP is preceded by the perfective verb-form *mort* 'I went' which ends in a binary consonant cluster. In conformity to the sandhi rules (Borg/Azzopardi-Alexander 1997: 334), the definite proclitic must take the initial vowel /i/ so that the DP comes as *id-dar* 'the house'. In contrast, in (1.83), the same DP winds up on the right of the imperfective verb-form *jiġi* 'he comes' which ends in a vowel. This final vowel requires the definite proclitic to be realised as a single consonant which means that the DP has the shape *d-dar* 'the house'. The inability of Ø to interdict

98 Cf. Section 2.2.1.2.

the application of the sandhi rules is tantamount to the absence of traces which could be interpreted as evidence of something invisible having an effect on its surroundings. Since there is no tangible proof of effects of this kind, we feel justified in claiming that the existence of Π Ø has been disproved.

1.6.4 Semantic, phonological, and morphosyntactic differences

The intricate nature of the topic of this section calls for subdividing it further into two subsections. In both subsections, we address the most important problems posed by the apparently complex Πs in (1.73). This section aims at demonstrating that in spite of their (sometimes only historical) compositionality on the expression side, these Πs can be classified as (lexicalised) units and not as multi-word syntagms of the compound Π-type. Saari (2003: 85–88) assumes ten different classes of complex Πs for some of which the distinction of syntagms and lexicalised complex units is blurred.

There are expressions which consist of several morphemes – either free or bound – whose individual semantics simply add up to yield the semantics of the entire expression. Expressions of this kind are semantically compositional. Some of them are good candidates for being included in BLOMP whereas others are not. What we exclude from BLOMP are cases of compound Πs like *minn fuq* 'from on' in (1.84).

(1.84) [Galea 2001: 46]

Dun Nurat	*rafa'*	*għajnejh*	***[minn***	*[fuq*	*il-ktieb]*$_{PP}$*]*$_{PP}$
Dun Nurat	lift	eye:PL:3SG.M	**[from**	[on	DEF-book]]

'Dun Nurat raised his eyes [**from** the book].'

The two Πs do not form an individual combination in terms of a lexicalised unit but are independent of each other. They co-occur syntagmatically in adjacent slots each of which allows for being filled by other Πs. In these cases, the semantics of the combination of Πs is always compositional.

What we accept as members of BLOMP instead are formally polymorphic Πs whose meaning does not derive directly from the combination of the meanings of the morphemes involved. There is thus a non-compositional construction meaning. The criterion of semantic compositionality can only be applied to expressions which are complex in the sense that at least two morphological units are involved in forming these expressions. In point of fact, semantic compositionality gives us a tough nut to crack. Parallel to the semantic aspects, there are also phonological and especially morphosyntactic properties to be taken account of when it comes to determining whether a formally complex Π forms a unit in its own right.

In (1.73), there are several monomorphic Πs which form also part of the segmental chain of other members of BLOMP. These Πs are *bi* 'with', *fi* 'in', *ma'* 'with', and *minn* 'from'. The question arises whether their involvement in the formation of other Πs results in semantic compositionality and whether the complex Πs obey the same morphosyntactic rules as the items which contribute to the formation. Do the component parts of the complex Π resemble their free equivalents phonologically? If all three questions are answered positively, the candidates fail the test of admission to BLOMP. Speaker awareness is of course crucial when it comes to judging whether a given expression is transparently compositional. In the absence of native-speaker statements on this issue, we have to make do with our own hypotheses. In a variety of cases, the diachronic compositionality of a Π is evident only to the eyes of the expert philologer who is knowledgeable in Semitic matters. The first cases we look at are of this kind.

1.6.4.1 Πs involving *bi* 'with' or *fi* 'in'

The following Πs belong to BLOMP although their etymology reveals that they diachronically reflect the combination of two (or, in one case, three) meaningful elements. It can be shown that synchronically their compositional nature is no longer obvious to the philologically not initiated language user also without interacting with native speakers. This is possible if at least one of the component parts of the expression has ceased to function as Maltese word outside the combination under inspection. Furthermore, the connection to an etymologically related element outside the complex expression counts as disrupted if there are morphosyntactic and/or semantic differences between the items.

(i) (X7) *biswit* 'facing'

The simplest case is that of (X7) *biswit* 'facing' < *bi* 'with' + Arabic *swit (< sawa)* 'alignment' (Aquilina 1987: 126). Both E. Serracino-Inglott (1975a: 173) and Aquilina (1987: 126) classify the item as Π. Except for the entry *swit* 'opposite, near' registered as "unused var[iant] of *biswit*" (Aquilina 1991: 1380), the Maltese-English dictionary does not provide further information on this case. Since *sawa* is not represented in the dictionary, it is unlikely that native speakers of Maltese without a background in Semitic philology are able to decompose *biswit* into its etymological components. The Korpus Malti 3.0 yields no hits for *swit* or *sawa* either.[99] There are

99 The word form *swit* occurs twice in the Korpus Malti 3.0 as 3rd person singular feminine perfective of *sewa* 'cost, be useful' whereas the three hits for *Sawa* involve the last name of a Japanese football player.

only attestations of *biswit*. On these grounds, *biswit* can be classified as synchronically unanalysable unit, i.e. as a monomorphic Π.

(ii) (X4) *bħal* 'like' and (X5) *bħala* 'as'

The line of argumentation is similar in the case of (X4) *bħal* 'like' and (X5) *bħala* 'as' < *bi* 'with' + Arabic *ḥāl* 'condition, way' (Aquilina 1987: 113; cf. also E. Serracino-Inglott 1975a: 154). According to Aquilina (1987: 482), the noun *ħal* 'condition, way' (PL *ħwiel*) occurs "as a single word [. . .] only in the prov[erb]" reproduced as (1.85).

(1.85) (Aquilina 1987: 482)

x'	***ħal**ek*	*illum,*	*mhux*	*x'*	*kont*	*ilbieraħ*
what	**condition**:2SG	today	not	what	be:2SG.PFV	yesterday

'It is your **condition** today that counts, not what you were yesterday.' [O.T.]

The noun *ħal* 'condition' is certainly an archaism in modern Maltese whose domain is limited to a very small niche. There is not a single piece of evidence for its existence in the Korpus Malti 3.0.[100] We assume that the average Maltese native speaker does not associate *bħal* 'like' and *bħala* 'as' with this isolated use of *ħal* 'condition, way'. Without knowledge of the existence and meaning of the noun *ħal* it is not possible to decompose the Π *bħal* 'like' in (1.86) into its diachronic components.

(1.86) [Camilleri 2013]

issa	*ommha*	*bdiet*	*tagħmel*	*[**bħal**ha]*$_{PP}$
now	mother:3SG.F	begin:3SG.F.PFV	3SG.F.IPFV:make	[**like**:3SG.F]

'[. . .] now her mother began to act [**like** her] [. . .].'

We therefore consider *bħal* 'like' and *bħala* 'as' to be two unalalysable (= monomorphic) Πs synchronically.

(iii) (X16) *fost* 'between'

When we look at (X16) *fost* 'between' < *fi* 'in' + Arabic *wasṭ* 'middle' (Aquilina 1987: 358; cf. also E. Serracino-Inglott 1975b: 288) we realise that this case closely resembles that of *bħal*. Aquilina (1991: 1521) mentions a noun *wast* 'middle' registered by Vassalli (1796) and Saydon (1931) which however is "[h]ardly ever used alone but

100 Note that the unrelated *Ħal* 'village' (the short form of *Raħal*) is abundantly attested in the Korpus Malti 3.0 since it forms part of many Maltese toponyms.

it occurs in *fost* 'amongst'." We have found a single hit for this noun in the Korpus Malti 3.0. which we present here as (1.87).

(1.87) [Korpus Malti 3.0; european8345]

żerriegħa	*għal*	*uċuh*	*tar-raba'*	***wast***	*is-sena*
seed	for	surface.PL	of:DEF-field	**middle**	DEF-year
tal-kontijiet					
of:DEF-count:PL					

'[. . .] seeds for the surfaces of the fields in the **middle** of the fiscal year.'

Note that the use of *wast* in (1.87) can be analysed along the lines of that of *għeluq* discussed in connection with examples (1.44)–(1.46) and Figure 1.3 in the sense that zero-marking applies in lieu of the use of *fi* 'in'.

The marginal evidence of the existence of *wast* is however unproblematic for classifying *fost* as a synchronically unanalysable Π because of the phonological distance between the initial segments /wa/ of the noun and the vowel /o/ of the Π. Synchronically, there is no regular phonological process in Maltese which would automatically transform a sequence /wa/ into /o/ if the voiceless labio-dental fricative occurs on its left (e.g. the lexical entry *fwar* 'vapour', Aquilina 1987: 371). It can safely be assumed that native speakers of Maltese do not mentally reconstruct the original segmental chain of *wast* when they utter or hear *fost* as in (1.88).

(1.88) [Attard 1999: 14]

Il-Maltin	*kienu*	***[fost***	*l-ewwel*	*Ewropej]*$_{PP}$
DEF-Maltese:PL	be:3PL.PFV	**[between**	DEF-first	European:PL]
li	*daħlu*	*fl-Alġerija.*		
that	enter:3PL.PFV	in:DEF-Algeria		

'The Maltese were [**amongst** the first Europeans] who went into Algeria.'

The bond between *fost* and *wast* has thus been disconnected on account of diachronic developments in the language's phonology.[101] Therefore, *fost* 'between' is a fully-fledged member of BLOMP.

101 The original meaning of *wast* is still discernible in the Maltese idiom *bħall-Erbgħa* ***fost*** *il-ġimgħa* 'lit. like Wednesday **in the middle of** the week' used to characterise situations in which an action is brought to an unexpected premature end to the detriment of one of the participants involved (Fenech 1984: 60).

(iv) (X8) *bla* 'without'

The situation is not as clear as that with (X8) *bla* 'without' < *bi* 'with' + Arabic *lā* 'not' (Aquilina 1987: 128–129; cf. also E. Serracino-Inglott 1975a: 177). This case is different from the foregoing because the negation *la* is still in use in Maltese. It is attested as optional negator in the prohibitive and can also be used to express negative intention or decision as well as correlative *la* . . . *la* 'neither . . . nor' (Aquilina 1987: 718). The latter use is illustrated in (1.89).

(1.89) [De Saint-Exupéry 2000: 3]

Allura	***la***	*kont*	*noqgħod*	*nitkellem*	*fuq*	*sriep*	*boa,*
SO	**NEG**	be:1SG.PFV	1.IPFV:stay	1.IPFV:speak	on	snake.PL	boa
la	*fuq*	*foresti*	*safja*	*u*	*lanqas*	*fuq*	*kwiekeb*
NEG	on	forest:PL	pure.PL	and	not_even	on	star.PL

'So, I **neither** spoke about boas **nor** about jungles and not even about stars.'

What distinguishes *bla* from the negation *la* is the distribution. Except the correlative which may be followed by NPs, the negation normally precedes a predicate whereas *bla* is combinable with (indefinite) NPs as evidenced in (1.90).

(1.90) [Ellul Mercer 1986: 58]

ta	*kull*	*għajnuna*	*lil*	*dawk*	*li*	*sfaw*
give	each	help	to	DEM.PL.DIST	that	be_reduced:3PL.PFV
*[**bla***	*dar]*$_{PP}$					
[**without**	house]					

'[. . .] he gave every kind of help to those who wound up [**without** a home].'

It is worth noting that this Π does not take pronominal complements. To allow *bla* to function as conjunction the general subordinator *ma* must be added which does not show up with the negation *la*. This situation is illustrated for *bla* in (1.91).

(1.91) [De Saint-Exupéry 2000: 1]

Is-sriep	*boa*	*jibilgħu*	*l-priża*	*tagħhom*	*sħiħa,*
DEF-snake.PL	boa	3.IPFV:swallow:PL	DEF-prey	of:3PL	whole:F
bla	***ma***	*jomogħduha*			
without	**that**	3.IPFV:chew:PL:3SG.F			

'The boas swallow their prey whole **without** chewing it.'

Chances are that these differences in the distribution over syntactic contexts preclude the possibility that native speakers associate *bla* with *la*. Admittedly, this hypothesis is not fully compelling. Its shakiness notwithstanding, we postulate the synchronically unanalysable monomorphic Π *bla* 'without'.

(v) (X15) *(min(n))flok* 'instead of' – Part I

The intricacies do not diminish when we turn our attention to (X15) *(min(n))flok* 'instead of' < *fi* 'in' + Sicilian *locu* 'place' (Aquilina 1987: 350). This has mostly to do with the fact that the long allomorph of this Π involves two other well-established Maltese Πs. E. Serracino-Inglott (1975b: 274) classifies this candidate as adverb. Like *bla*, *(min(n))flok* poses problems for an analysis which assumes non-compositionality. There is the common noun *lok* 'place' which may also be used in a very general (abstract) sense (Aquilina 1987: 757), as shown in (1.92).

(1.92) [Korpus Malti 3.0; european1962]

l-ispezzjonjiet	*fuq*	*il-**lok***	*fil-pajjiż*	*terz*	*jissodisfaw*
DEF-inspection:PL	on	DEF-**place**	in:DEF-country	third	3.IPFV:satisfy:PL

il-kondizzjonijiet	*stabbiliti*	*fl-Anness I*
DEF-condition:PL	establish:PART:PL	in:DEF-Annex I

'[. . .] the inspections on **site** in the third country meet the conditions established in Annex I.'

However, the complement of *flok* and the genitival attribute of *lok* behave differently in morphosyntactic terms. The latter requires the presence of the genitival Π *ta'* as in *lok tal-morda* 'infirmary' (literally 'place of the sick') (Aquilina 1987: 757) whereas *ta'* is not intercalated between *flok* and its complement as in (1.93).

(1.93) [Attard 1999: 25]

kellhom	*bżonn*	*ta'*	*ħaddiema*	*mill-Ewropa*
have.PFV:3PL	need	of	worker:PL	from:DEF-Europe

[flok	*dawk*	*Ċiniżi]$_{PP}$*
[instead_of	DEM:PL:DIST	Chinese:PL]

'[. . .] they needed workers from Europe [**instead of** those Chinese] [. . .].'

Semantically, *(min(n))flok* only expresses replacement and never refers to a concrete spatial object whereas the noun *lok* 'place' does not express a relation of replacement. Thus, morphosyntactic and semantic arguments support our conclusion according to which synchronically *(min(n))flok* 'instead of' forms part of BLOMP. However, this discussion is not terminated yet as will become clear from the ideas put forward below.[102]

102 Cf. Section 1.6.4.2 (i).

1.6.4.2 Πs involving *minn* 'from' or *ma'* 'with'

As can be gathered from the foregoing discussion, it is by no means always easy to prove that a given Π is non-compositional. The problems increase when we take account of those instances of complex expressions which involve (or seem to involve) the simpleton Πs *ma'* 'with' or *minn* 'from'. First of all, there are various members of BLOMP which allow for longer alloforms most of which involve optional *minn* 'from' or *ma'* 'with' as initial component. Secondly, for a long time, there was variation in the spelling of several of these cases which appeared either as one or two orthographic words (cf. Il-Kunsill Nazzjonali tal-Ilsien Malti 2008). A useful test which helps to decide whether or not these are cases of one or two Πs is coordination which however can be run only after we have learned more about the alternation of long and short allomorphs,[103] compound Πs,[104] and EQUI-P-deletion.[105] For the time being, in the subsequent paragraphs, further cases are accounted for which cannot be explained in terms of allomoprhy. We begin the presentation of the cases with two Πs for which it is doubtful whether they involve *minn* 'from' as their initial component.

(i) (X15) *(min(n))flok* 'instead of' – Part II

Example (1.93) features the short allomorph of (X15) *(min(n))flok* 'instead of'. There is however the competing long allomorph which adds *minn* 'from' to the left of the short allomorph. As mentioned in the foregoing section, E. Serracino-Inglott (1975b: 274) considers *flok* to be an adverb but he adds that it may combine with the Π *minn* to yield the Π *minflok* (E. Serracino-Inglott 1979: 151). This differentiation seems to indicate that the author puts *flok* in a different word-class from *minflok*. To our mind, both the short and the long form function as Πs. If we compare the use of *minflok* in (1.94) with that of *flok* in (1.93), we realise that there is functional equivalence or full synonymy so that it is safe to speak of allomorphy (cf. also (1.102)).

(1.94) [Korpus Malti 3.0; academic161]

Il-Gvern	*Ingliż*	*għamel*	*taxxi*	*ġodda*	***[minflok***
DEF-government	English	make	tax:PL	new.PL	**[instead_of**
<u>*id-dazju*</u>*]*$_{PP}$					
<u>DEF-excise</u>]					

'The English government made new taxes [**replacing** <u>the excise</u>] [. . .].'

103 Cf. Section 3.2.

104 Cf. Section 4.1.3.

105 Cf. Section 4.1.4.

The presence or absence of the initial *min(n)-* does not seem to influence the meaning. This leads us to assume free variation – at least for the time being. Since the allomorph *minflok* ousts its competitor *flok* in terms of token frequency in the Korpus Malti 3.0 (with 45,642 vs 13,773 matches, respectively), this Π is represented in BLOMP 2.0 by its long allomorph. This case may serve as a pattern for several of the Πs discussed in the subsequent paragraphs.

(ii) (X38) *minkejja* 'in spite of'

For E. Serracino-Inglott (1979: 154), (X38) *minkejja* 'in spite of' belongs to the class of adverbs. According to Aquilina (1991: 833), (X38) *minkejja* 'in spite of' is the mimated feminine noun of *neka* 'vex, tease'. Mimation is manifest in the prefixation of the bilabial nasal /m/ to a given CV-pattern yielding verbal nouns with a wide range of semantic functions (Aquilina 1959: 174–186), i.e. he assumes that the form *minkejja* results from the prefixation of *m-* to *nkejja* with the appropriate epenthetic /i/ to avoid a word-initial ternary consonant cluster. Since there is the verbal noun of unity *nkejja* 'vexation' which can be combined with Πs like *bi* 'with' (e.g. *bi nkejja* 'out of spite') and *għal* 'for' (e.g. *għall-inkejja ta'* 'in order to tease') (Aquilina 1991: 900), there is no reason to discard the possibility that *minkejja* 'in spite of' results from a combination of *minn* 'from' and *nkejja* 'vexation' with fusion of the denti-alveolar nasals on the morpheme boundary.

As in the cases discussed in the foregoing paragraphs, there are differences between the verbal noun *nkejja* and the Π *minkejja* on the morphosyntactic level. Example (1.95) shows that the attribute of the verbal noun is linked to it by the Π *ta'* whereas in (1.96) *minkejja* takes its complement without intervening Π.

(1.95) [Korpus Malti 3.0; news129571]
jsolvu l-problemi aktar flimkien filwaqt naqas
3.IPFV:solve:PL DEF-problem:PL more together while decrease
*l-inbix u l-**inkejja** **ta'** bejniethom*
DEF-teasing and DEF-**vexation** **of** between:PL:3PL
'[. . .] they probably solve the problems together while the teasing and **vexation** between them has grown less.'

(1.96) [Camilleri 2013: 206]
Għal waqt wieħed saħansitra kkontempla jaħrab
for moment one even contemplate 3.IPFV:escape
*mit-tieqa **[minkejja** <u>ż-żewġ</u> <u>sulari</u> <u>għoli]</u>$_{PP}$*
from:DEF-window **[in_spite_of** <u>DEF-two</u> <u>floor:PL</u> <u>height</u>]
'For a moment he even thought of escaping through the window [**in spite of** <u>the height of two floors</u>].'

In point of fact, the noun *nkejja* 'vexation' combines with an array of different Πs which serve to link it to a noun in attributive function. On account of these differences, we consider *minkejja* 'in spite of' to be a member of BLOMP.

(iii) (X34) *mħabba* 'on account of'

The next candidate is (X34) *mħabba* 'on account of' which is homophonous with the (mimated) noun *mħabba* 'love' of the verb *ħabb* 'love'. Like E. Serracino-Inglott (1979: 123–124), Aquilina (1991: 821) interprets *mħabba* 'on account of' as variant of *minħabba* 'on account of' – and vice versa (Aquilina 1991: 832). In the entry for *ħabb* 'love', only *mħabba* is mentioned as a Π (Aquilina 1987: 467). E. Serracino-Inglott (1979: 123) only speaks of a noun. We have tried to determine the quantitative relations between the two variants by way of comparing the first 100 hits we found for each of them in the Korpus Malti 3.0. Only 29% of the attestations of *mħabba* can be registered as prepositional usages whereas 81% of all attestations of *minħabba* meet the criteria of Π-hood. If we take into consideration that the overall token frequency of *mħabba* does not exceed 2,400 hits as opposed to some 160,000 hits for its competitor *minħabba* we can legitimately claim that *minħabba* is the default form of the Π and *mħabba* has the status of minor allomorph. There is however a third allomorph, *ħabba* 'on account of', which covers 20% of the first 100 attestations of the item in the Korpus Malti 3.0.[106] Aquilina (1987: 466–467) makes no mention of *ħabba* 'on account of' in Π-function.

Examples (1.97)–(1.99) are meant to show that the three candidates are indeed semantically the same. They are full synonyms of each other. We assume that they can freely replace each other unless there are stylistic reasons which favour the selection of a particular allomorph.

(1.97) [Bartolo/Vella 2009: 176]
Is-soltu l-arja f' dik is-sala enormi kienet
DEF-usual DEF-air in DEM:F:DIST DEF-hall enormous be:3SG.F.PFV
dejjem tkun kiesħa ***[minħabba*** *l-umdità li*
always 3SG.F.IPFV:be.FUT cold:F **[on_account_of** DEF-humidity that
kienet fiha]$_{PP}$
be:3SG.F.PFV in:3SG.F]
'Usually, the air in this enormous hall would always be cold [**because of** the humidity that was in there].'

106 There are also numerous instances of *ħabba* 'grain, coin; pupil (of the eye)'.

(1.98) [Korpus Malti 3.0; academic52]
F' Oxford Schrödinger kellu ħafna probemi
in Oxford Schrödinger have.PFV:3SG.M many problem:PL
***[mħabba** l-istil tal-ħajja personali]$_{PP}$*
[**on_account_of** DEF-style of:DEF-life personal]
'In Oxford, Schrödinger had many problems [**on account of** his personal style of life].'

(1.99) [Camilleri 2013: 135]
ftakar f' martu li issa ma kinitx
remember in woman:POSS:3SG.M that now NEG be:3SG.F.PFV:NEG
*ħa tibqa' ttaqtaqlu li **[ħabba***
FUT 3SG.F.IPFV:remain 3SG.F.IPFV:chatter:IO:3SG.M that [**on_account_of**
xogħlu]$_{PP}$ għad taħkmu xi trombożi jew
work:3SG.M] still 3SG.F.IPFV:affect:3SG.M some thrombosis or
puplesija
apoplexy
'[...] he remembered his wife who now would not keep chatting him up that [**on account** of his work] some thrombosis or apoplexy was to affect him.'

Given that the synonymity hypothesis holds, there is no reason to give each of the three alloforms a separate entry in (1.73). The Πs which are listed there are place-holders for the entire battery of alloforms which are context-free replacements of the Π. On account of the differences in token frequency, we modify the original BLOMP by way of promoting *minħabba* to the status of sole entry in the list from which *m(in)ħabba* is removed. In BLOMP 2.0, *minħabba* 'on account of' thus represents the set of allomorphs formed by *minħabba* ~ *mħabba* ~ *ħabba.*

One of the alloforms is segmentally identical with the noun *mħabba* 'love' which functions as subject of the third clause in (1.100).

(1.100) [Teuma 2015: 659]
*jiena tlaqtu biex niġi fejn l-**imħabba***
1SG leave:1SG.PFV:3SG.M to 1.IPFV:come where DEF-**love**
kienet issejaħli
be:3SG.F.PFV 3SG.F.IPFV:call:IO:1SG
'[...] I left him to go to where **love** was calling me [...].'

The abstract noun *mħabba* 'love' takes genitival attributes which are linked to it by the Π *ta'* 'of' whereas none of the Π alloforms requires the presence of *ta'* to connect to the complement. Either governing term and governed term are juxtaposed as in (1.97)–(1.99) or, if the latter is pronominal, the Π *fi* 'in' is intercalated as in (1.101).

(1.101) [Teuma 2015: 600]

jekk	*hu*	*jmut*	***[minħabba***	*[fija]$_{PP}$]$_{PP}$*
if	3SG.M	3.IPFV:die	**[on_account_of**	[**in**:1SG]]

'[. . .] if he dies [**on account of** me] [. . .].'

The morphosyntactic differences between the Π-alloforms and the noun *mħabba* 'love' suffice to state that the function-words are largely independent of the autosemantic lexeme. Moreover, the association with the positively connotated and emotionally loaded concept of LOVE is not invoked by the Π-alloforms whose meaning is connected to the concept of CAUSE.

What is additionally interesting about the three alloforms of this causal Π is that neither the absence of the initial *minn* nor that of the prefix *m-* have an effect on the meaning. There is thus no semantic compositionality. Similarly, *minn* can be dropped from or added to the segmental chain of two further Πs without any discernible change of meaning, namely *barra* (~ *minbarra*) 'except' and *flok* (~ *minflok*) 'instead of'. In previous subsections, examples (1.93)–(1.94) have illustrated the use of the long and the short allomorph of *minflok* 'instead of'. Example (1.102) confirms that *minflok* 'instead of' may substitute for *flok* 'instead of' without affecting the intended meaning.

(1.102) [Bartolo/Vella 2009: 392]

Jason	*għolla*	*jdejh*	*'il*	*fuq*	*bil-wirdien*
Jason	raise	hand:PL:3SG.M	to	on	with:DEF-cockroach.COLL

iferfer	***[minflok***	*difrejh]$_{PP}$*
3.IPFV:flutter	**[instead_of**	nail:PL:3SG.M]

'[. . .] Jason raised his hands with the cockroaches fluttering [**instead of** his nails].'

As mentioned above, in cases of this kind, it does not make a difference semantically whether *minn* is there or not.

(iv) (X2) *(min)barra* 'except'

Superficially, the situation seems to be the same with (X2) *(min)barra* 'except'. Aquilina (1987: 81) registers *barra* as Π with the meanings 'out (of), except, apart from'. According to this sourse, in all cases, *barra* may or may not be accompanied by *minn* 'from'. The position of *minn* is not fixed for the expression of exclusion, i.e.

both *minbarra* 'except' and *barra minn* 'except' are licit. *Minbarra* is registered as Π and conjunction in E. Serracino-Inglott (1979: 149). Is the intended meaning spatial however, *minn* can only occupy the slot to the right of *barra* yielding *barra minn* 'out (of)'. Borg/Azzopardi-Alexander (1997: 158–159) register these cases under the heading of EXTERIOR. This distribution pattern requires further scrutiny. The situation is rendered even more interesting by the existence of *mbarra* 'except, apart from' as third expression in the domain of EXCLUSION.

In (1.103)–(1.104), the spatial use of *barra* and *barra minn* is documented. As is evident from these examples, *barra* fulfils the task with or without accompanying *minn* positioned to its right.

(1.103) [Camilleri 2013: 34]

iħuf	*għal*	*kull*	*marka*	*ta'*	*swaba'*	*preżenti*
3.IPFV:gad_about	for	each	mark	of	finger	present

ġewwa	*u*	***[barra***	*l-vettura]*$_{PP}$
inside	and	**[outside**	DEF-vehicle]

'[. . .] he wanders about in search of each fingerprint inside and [**outside** the car].'

(1.104) [Żahra 2008. 93]

Ma	*rajtha*	*xejn*	***[barra***	***minn***	*postha]*$_{PP}$
NEG	see:1SG.PFV:3SG.F	nothing	**[outside**	**from**	place:3SG.F]

ħaġa	*bħal*	*din*
thing	like	DEM:F:PROX

'I did not consider a thing like this to be [**improper**] [. . .].'

In contradiction to Aquilina's above rules, example (1.105) shows that *minn* may also occur on the left side of *barra* to express the intended spatial meaning.

(1.105) [Korpus Malti 3.0; academic281]

hu	*għamel*	*xutt*	***[minn***	***barra***	*l-kaxxa]*$_{PP}$
3SG.M	make	shot	**[from**	**outside**	DEF-box]

'[. . .] he made a shot [**from outside** the box] [. . .].'

Example (1.106) shows that *barra* can express the notion of EXCLUSION on its own. Similar to *barra minn* in (1.104) and *minn barra* in (1.105), it is possible to additionally use *minn* on either side of *barra* in this function as shown in (1.107). In (1.108), we present an example of the combination of the two Πs written as one orthographic word.

(1.106) [Attard 1999: 34]
*[**Barra** n-nuqqas ta' flus]PP il-poplu kien fil-parti*
[**except** DEF-lack of money] DEF-people be in:DEF-part
l-kbira illiterat
DEF-big:F illiterate
'[**Apart from** the lack of money] the majority of the people were illiterate [. . .].'

(1.107) [Galea 2001: 27]
*[**barra** **minn** xi ftit mixtrija]PP in-nies kollha*
[**except** **from** some little PART:buy:PL] DEF-people all:3SG.F
mtlew daqs balla
fill:RFL:3PL.PFV equal_to ball
'[. . .] [**except** a few bribed ones] the people in its entirety was bottled up.'

(1.108) [Bartolo/Vella 2009: 157]
kulma kien ġarrab kien xi grif u tnittif,
whatever be endure be some scratching and plucking
*[**minbarra** l-fatt]PP li d-dinjità tiegħu*
[**except** DEF-fact] that DEF-dignity of:3SG.M
kienet qalgħet daqqa ta' ħarta iebsa
be:3SG.F.PFV receive:3SG.F.PFV blow of ploughing hard:F
'[. . .] all that he endured were some scratches and plucked feathers [**except** the fact] that his dignity had received a severe slap in the face.'

The only unambiguous alloform within the domain of *barra* is *mbarra* 'except' in (1.109) which is exclusively employed for the purpose of expressing EXCEPTION.

(1.109) [Korpus Malti 3.0; news54755]
*Żid l-ingredjenti kollha [**mbarra** d-dressing]PP*
add DEF-ingredient:PL all:3SG.F [**except** DEF-dressing]
'Add all ingredients [**except** the dressing].'

This means that singleton *barra* and the two possible combinations with *minn* are ambiguous. The exact meaning has to be inferred from the context. Orthography – one word *minbarra* vs two words *minn barra* – is helpful only in the case of *minbarra* which is never used spatially whereas *minn barra* is attested as expression of both functions under review.

Figure 1.6 gives us an idea of how the different options distribute over the two functions (spatial 'outside' and exclusive 'except/apart').

	minbarra	mbarra	barra	minn barra	barra minn
■ except/apart	70	23	23	21	29
■ outside	0	0	4	21	70

Figure 1.6: Competition between spatial and exclusive functions per expression.

The shares are calculated on the number of hits within the first 100 attestations of each expression in the Korpus Malti 3.0. The data largely confirm Aquilina in the sense that *minbarra* and *mbarra* are functionally restricted to expressing exclusion. The three remaining options are ambiguous between spatial and exclusive readings to different degrees. There is no expression which can only convey the spatial meaning although *barra minn* comes relatively close to a functional specialisation of this kind.

Figure 1.6 justifies the second modification of the original version of BLOMP. Owing to the functional specialisation of *minbarra* and *mbarra* as opposed to the ambiguities connected to *barra, minn barra*, and *barra minn*, we postulate the existence of an additional member of BLOMP, namely *minbarra* 'except' as main entry with *mbarra* 'except' as its less frequent alloform. Thus, BLOMP 2.0. does not only host a new member but the prime meaning originally attributed to *barra* has to be changed from 'except' to 'outside' to avoid confusion.

To close the discussion about *barra* and its alloforms at least for this section, we present example (1.110) which involves a case of compound ΠIs.

(1.110) [Korpus Malti 3.0; law1130]

L-oġġetti	*postali*	*kollha,*	*minbarra*	*l-ittri,*	*li*	
DEF-object:PL	postal	all:3SG.F	except	DEF-letter:PL	that	
jaslu	***[minn***	***[barra***	***minn***	*Malta]$_{PP}$]$_{PP}$*	*jistgħu*	
3.IPFV:arrive:PL	**[from**	**[outside**	**from**	Malta]		3.IPFV:can:PL

jinfetħu
3.IPFV:RFL:open:PL
'All postal objects, except letters, which arrive [**from** [**outside** Malta]] can be opened [. . .].'

The complex Π *barra minn* 'outside' forms a PP together with its toponymic complement Malta. This PP, in turn, constitutes the complement of the Π *minn* 'from' positioned on the left of *barra minn*. The involvement of *barra minn* 'outside' in a pattern of compound Πs is further indirect evidence of the unit-like character of *barra minn*. Saari (2003: 87) assumes a different analysis according to which in cases like that in (1.110) *minn barra* is the basic unit to which a second *minn* can optionally be added on the right margin.

(v) (X35) *mingħala* 'in s.o.'s opinion'
As results from the foregoing paragraphs, *minn* as part of a complex Π can sometimes be semantically empty. It is then an optional constituent of the Π whose presence is not required semantically. This is not always the case. There are Πs in (1.73) whose initial constituent is *minn* whose presence or absence is not freely negotiable. One of these Πs is (X35) *mingħala* 'in s.o.'s opinion'. E. Serracino-Inglott (1979: 152) represents this candidate without the final /a/. If the initial *minn* is omitted, the remaining segmental chain is identical to that of (X21) *għal* 'for' if the complement is pronominal as in (1.111)–(1.112).

(1.111) [Bartolo/Vella 2009: 310]

Jason	*kien*	***[mingħali**h]$_{PP}$*	*li*	*hu*	*xi*	*ħadd*
Jason	be	[**in_s.o.'s_opinion**:3SG.M]	that	3SG.M	some	someone

speċjali
special
'Jason was [**of the opinion**] that he was someone special [. . .].'

(1.112) [Bartolo/Vella 2009: 314]

Ma	*kellux*	*idea*	*għalfejn*	*kienu*	*qed*	*jistaqsu*
NEG	have.PFV:3SG.M:NEG	idea	why	be:3PL.PFV	PROG	3.IPFV:ask:PL

***[għali**h]$_{PP}$*
[**for**:3SG.M]
'He had no idea why the were asking [**for** him].'

The pronominal affixes trigger the morphonological changes on the Π-stem. The final vowel of *mingħala* is replaced with /i/ whereas the same vowel is added to the consonant-final stem of *għal*. Under omission of the initial *minn*, the two Πs would become indistinguishable with the functionally undesirable result that semantic ambiguity might arise. In this sense, *minn* in *mingħala* is no disposable junk. Note however that the authors of this study have repeatedly been told by native speaker linguists not to be overly formal in spoken Maltese by way of using *mingħala* in lieu of *għala*.[107] Comments of this kind suggest that there are stylistic differences between the spoken and the written register. Note also that *għala* 'why; cause' is registered as adverb and noun in Aquilina (1991: 953–954). *Għala* (often hosting the definite proclitic) frequently functions as indirect causal interrogative, i.e. it introduces predicates and subordinate clauses. Historically, there must be a connection between *għala* and *mingħala*. However, synchronically, they differ from each other both in terms of meaning and syntactic properties. In the same dictionary entry, Aquilina (1991: 953) postulates the existence of the binary syntagm *minn għala* 'it seems; X thought/was under the impression that'[108] which is conceptually close enough to the meaning of *mingħala* 'in s.o.'s opinion' to pass as the same. In these only orthographically distinct cases, the construction meaning cannot directly be retrieved from the combination of the meanings of the component parts. For BLOMP 2.0, we assume that *minn għala* and *mingħala* are alloforms of each other. *Mingħala* 'in s.o.'s opinion' poses a problem with regards to the criteria for the canonical Π[109] insofar as it is difficult to establish a clear dependency relation of this Π to an external head. Inflected forms of *mingħala* 'in s.o.'s opinion' seem to be freely insertable at any place in a given utterance. Some similarities with the excluded pseudo-verbs exist.[110] We nevertheless consider it a (perhaps only marginal) member of BLOMP 2.0 because it is loosely dependent on the proposition as a whole. To clarify the status of *mingħala* 'in s.o.'s opinion' once and for all however a dedicated in-depth study is required.

(vi) (X34) *mingħajr* 'without'

In the case of (X34) *mingħajr* 'without' (E. Serracino-Inglott 1979: 152; Aquilina 1991: 834), the situation is clearer insofar as stylistic considerations do not seem to

107 Ray Fabri (personal communication).

108 A typical example of this two-word combination is in the idiomatic expression *għadda minn għala widnejja* 'I heard from somebody / somebody told me' (Aquilina 1991: 953–954) as illustrated in (1.i).

(1.i) [Teuma 2015: 690]

għaddieli	***minn***	***għala***	***widnejja***	*li*	*int*	*qed*	*tixlini*
pass:IO:1SG	**from**	**reason**	**ear:PL:1SG**	that	2SG	PROG	2.IPFV:accuse:1SG

'[. . .] **somebody told me** that you are accusing me [. . .].'

Apart from this lexicalised idiom, we have not come across other usages of *minn għala*.

109 Cf. Section 1.6.1.

110 Cf. Section 1.6.2.

have a say in the choice of form. If we drop *minn* from *mingħajr* 'without', we get *għajr* 'except' which is a distinct Π with a meaning different from that of *mingħajr* as can be gathered from examples (1.113)–(1.114).

(1.113) [Bartolo/Vella 2009: 125]
*[**Mingħajr** din l-informazzjoni]$_{PP}$ il-pjan ta' Kristjan*
[**without** DEM:F:PROX DEF-information] DEF-plan of Kristjan
żgur li ma kienx jasal sa fejn wasal.
sure that NEG be:NEG 3.IPFV:arrive till where arrive
'[**Without** this information] Kristjan's plan surely would not have got as far as that.'

(1.114) [Teuma 2015: 284]
*jien ma rajt lil ħadd [**għajr** liż-żagħżugħ*
1SG NEG see:1SG.PFV to someone [**except** to:DEF-young_man
li semmejtlek]$_{PP}$
that mention:1SG.PFV:IO:2SG]
'I saw nobody [**except** the young man I mentioned to you].'

The presence/absence of *minn* is crucial for distinguishing the meanings. Neither is *mingħajr* an alloform of *għajr* nor vice versa. The meaning of *mingħajr* 'without' does not result from the combination of the meanings of its component parts (i.e. *minn* 'from' + *għajr* 'except').

(vii) (X31) *madwar* 'around' and (X10) *dwar* 'about'

There is a parallel which involves (X31) *madwar* 'around' (E. Serracino-Inglott 1979: 6; Aquilina 1991: 767[111]) and (X10) *dwar* 'about'. This time, the formal difference is based on the presence and absence of *ma'* 'with' as initial component. Like in the case of *mingħajr* 'without' and *għajr* 'except', there is a clear semantic distinction between the two formally related Πs. This difference comes nicely to the fore when we compare examples (1.115) and (1.116).

(1.115) [Bartolo/Vella 2009: 322]
*Ma jinteressanix inkun naf [**dwar***
NEG 3.IPFV:interest:1SG:NEG 1.be.FUT 1.IPF:know [**about**
il-fotografija]$_{PP}$
DEF-photography]
'I am not interested in knowing [**about** the photography].'

111 Whose translation "environs" suggests an interpretation as noun.

(1.116) [Camilleri 2013: 135]

xi	*ħadd*	*ilmenta*	*minn*	***[madwar***	*il-mejda]*$_{PP}$
some	someone	lament	from	**[around**	DEF-table]

'[. . .] someone sitting [**at** the table] complaint [. . .].'

If *ma'* is dropped from (1.116), the resulting PP *dwar il-mejda* 'about the table' would describe a completely different situation. Without *ma'* (and *minn*), the example (1.116) would refer to a situation in which someone complained about the table in lieu of indicating the whereabouts of the person who expressed the complaint. Thus, it makes a difference whether *ma'* is there or not. This means that *madwar* 'around' is not an alloform of *dwar* 'about' or vice versa. The two expressions represent two distinct Πs.

(viii) (X36) *mingħand* 'from s.o.'

There are different problems to face in connection with (X36) *mingħand* 'from s.o.' (E. Serracino-Inglott 1979: 152; Aquilina 1991: 834). From the examples (1.117)–(1.118), it results that the presence of *minn* is indispensable if the spatial relation of Source$_{HUMAN}$ is to be expressed.

(1.117) [Camilleri 2013: 142]

issa	*kien*	*wasal*	*il-mument*	*li*	*jieħu*	*r-riedni*
now	be	arrive	DEF-moment	that	3.IPFV:take	DEF-rein:PL

[mingħand	*l-Ispettur]*$_{PP}$
[from_s.o.	DEF-inspector]

'[. . .] now was the moment for him to take the reins over [**from** the inspector].'

(1.118) [Cauchi 1997: 39]

Fit-triq	*għadda*	***[għand***	*l-ispiżjar]*$_{PP}$
in:DEF-street	pass_by	**[at_s.o.'s_place**	DEF-apothecary]

'In the street, he dropped [**by** the apothecary] [. . .].'

Without *minn*, *għand* 'at s.o.'s place' alone is acceptable only if Place or Goal are to be expressed. Nevertheless, one might object to classifying *mingħand* 'from s.o.' as separate Π because, superficially, its meaning seems to be transparently compositional with *minn* 'from' contributing the ablative meaning and *għand* 'at s.o.'s place' indicating that the Ground is a human being. However, the situation is not as straightforward as it looks. In contrast to *għand* 'at s.o.'s place', the spatial reading of *mingħand* 'from s.o.' is not absolutely compelling. When the protagonist of Camilleri's novel takes over control from the inspector in (1.117) this does not mean that this happens at the inspector's home or in his office. In contrast, entering the pharmacy as implied in (1.118) involves the apothecary's work-place so that *għand* receives a spatial interpretation. For the time being, we assume that the less strict connection of *mingħand* 'from s.o.' to the spatial concept of someone's place suffices to register it as a Π in its own right because the semantics of this expression is not 100% compositional.[112]

(ix) (X53) *(ma)tul* 'during'

Equally problematic is the adequate analysis and classification of *mintul* 'throughout' (Aquilina 1991: 834). E. Serracino-Inglott (1979: 156) considers *mintul* to be an adverb. The question must be asked whether this candidate is just an alloform of (X53) *(ma)tul* 'during'. Moreover, the answer cannot be given without taking account of the allomorphic relation between *tul* and *matul* 'along', too. For a start, consider example (1.119) in which *mintul* seems to be orthographically represented as *minn tul*. In the Korpus Malti 3.0, there is no evidence of *mintul* written in one word. Since the meaning of *minn tul* in this example is determined in the ensuing discussion, the variable X is used to mark the decision as pending.

(1.119) [Korpus Malti 3.0; news 178807]

inqatlu	*ħafna*	*minnhom*	*minn*	*missili*	*sparati*
RFL:kill:3PL.PFV	many	from:3PL	from	missile:PL	fire:PART:PL

[minn tul	*il-fruntiera*	*Lebaniża]*$_{PP}$	*mill-grupp*	*militanti*
[**X**	DEF-frontier	Lebanese]	from:DEF-group	militant

Hezbollah
Hizbollah

'[. . .] many of them were killed by missiles fired [**X** the Lebanese frontier] by the militant group Hizbollah.'

There are four instances of *minn* in this example one of which is partitive (*minnhom* 'of them'), two instances mark passive agents (inanimate and human) (*minn missili* 'by missiles' and *mill-grupp* 'by the group'), and one interacts with *tul*. It is this latter case that is of interest to us here. Superficially, *minn tul* invites two interpretations, namely on the one hand 'all along' and on the other hand 'from along'. The first option would correspond relatively closely to the dictionary meaning 'throughout' (Aquilina 1991: 834) whereas the second interpretation would presuppose semantic compositionality. E. Serracino-Inglott (1979: 156) distinguishes spatial and temporal usages of *mintul*. Independent of these dimensions, his paraphrases of the meanings of *mintul* are telling. In the spatial domain, *mintul* is invested with the meaning

112 We will come back to this intricate case in Section 5.1.2.2.4.

"mit-tul jew mat-tul kollu ta' xi mkien; għal kemm hu twil il-post, l-oġġett, ecc."[113] whereas with reference to time it means "matul iż-żmien jew ħin kollu msemmi; għal kemm jieħu żmien jew ħin eċċ.".[114]

Outside the dictionaries, uncontroversial evidence of the existence of the distinct Π *mintul* 'throughout' are hard to come by not the least because, with only 24 hits, the combination of *minn* + *tul* + COMPLEMENT NOUN is generally infrequent in the Korpus Malti 3.0. Of these two dozen cases, four are indicative of the Π-status of *mintul*. Examples (1.120)–(1.123) involve *mintul* (always written as two separate words) as spatial Π.

(1.120) [Korpus Malti 3.0; law856]
l-immarkar *tas-salib* *għandu* *jkun* ***[minn tul***
DEF-marking of-cross must:3SG.M 3.IPFV:be.FUT **[throughout**
it-tarf]$_{PP}$
DEF-extremity]
'[. . .] the marking of the cross has to be [**along** the extreme end].'

(1.121) [Korpus Malti 3.0; european7407]
l-investigazzjonijiet *jiġu* *applikati* *b'* *mod* *uniformi*
DEF-investigation:PL 3.IPFV:come:PL apply:PART:PL with way uniform
u *effettiv* ***[minn tul*** *il-Komunità* *kollha]$_{PP}$*
and effective **[throughout** DEF-Community all:3SG.F]
'[. . .] the investigations are applied in a uniform and effective way [**throughout** the entire Community] [. . .].'

(1.122) [Korpus Malti 3.0.; news120389]
Iżda *huwa* *verament* *tal-mistħija* *l-kumplament* *tat-triq*
but 3SG.M truly of:DEF-shame DEF-remainder of:DEF-street
fejn ***[minn tul*** *ir- runway* *ta'* *l-ajruport]$_{PP}$* *ħadd*
where **[throughout** DEF-runway of DEF-airport] nobody
ma *jirnexxielu* *jsuq*
NEG 3.IPFV:succeed:IO.3SG.M 3.IPF:drive
'However, the remainder of the street is really a shame where [**all along** the runway of the airport] nobody is able to drive [. . .].'

113 Our translation: "through or along the entire extension of a place; for how long the place, the object, etc. is".

114 Our translation: "during the entire mentioned time; for as much time as they take". Note that Maltese has two (not fully synonymous) lexemes translating English *time*, namely *żmien* and *ħin*. We have simplified the translation by ignoring the second conjunct in the coordinations in the quote.

(1.123) [Korpus Malti 3.0; literature80]
jimxi ***[minn tul*** *ix-xelter* *kollu]*$_{PP}$
3.IPFV:walk **[throughout** DEF-shelter all:3SG.M]
'[. . .] he walks [**all along** the shelter] [. . .].'

What these four examples share is the PERLATIVE[115] function of *mintul* 'throughout' which expresses that in a (concrete or abstract) motion event the Figure moves along or through the entire extension of the Ground. What is especially important is the fact that one dimension of the Ground is affected (ideally) in its entirety. In (1.120) and (1.123), this meaning is emphasised additionally by the use of *kollha* and *kollu* 'all', respectively.

These quantifiers are also used in examples like (1.124) where we find *tul* 'during' in lieu of *mintul* 'thoughout'.

(1.124) [Teuma 2015: 804]
huwa raqad raqda waħda ***[tul*** *il-lejl* *kollu]*$_{PP}$
3SG.M sleep sleep one:F **[during** DEF-night all:3SG.M]
'[. . .] he slept deeply [all night **long**].'

Without any of the two morphemes which indicate TOTALITY, i.e. without *minn* and *koll-*, however, *tul* 'during' receives a punctual reading as in (1.125).

(1.125) [Bartolo/Vella 2009: 76]
Reggie li ***[tul*** *din* *il-konverżazzjoni]*$_{PP}$ *kien waqqaf*
Reggie that **[during** DEM:F:PROX DEF-conversation] be stop:CAUS
il-karozza
DEF-car
'Reggie who had stopped the car [**during** this conversation] [. . .].'

We thus postulate a semantic difference between *mintul* and *tul* with the former combining the functions of PERLATIVE and TOTALITY whereas the latter may also indicate punctual events. The same holds for *matul* 'along' as illustrated in (1.126).

115 In the introduction, we do not terminologically distinguish between perlative and prolative. This distinction becomes meaningful however, in Section 5.1.2.2.20.

(1.126) [Camilleri 2013: 57]

ħabbat	*wiċċu*	*magħhom*	*diversi*	*drabi*	***[matul***	*is-snin]*$_{PP}$
knock	face:3SG.M	with:3PL	several:PL	time:PL	**[along**	DEF-year.PL]

'[. . .] he encountered them several times [**over** the years].'

Several separate meetings of certain people have happened at different points in time during an extended period of time. The events take place intermittently. In the case of *mintul* 'throughout' however, the temporal function is largely underrepresented, if at all. What the examples (1.120)–(1.122) suggest is that *mintul* refers to situations in which the entire extension of the Ground – be it temporal or spatial – is affected by the event described.

There is also the occasional attestation of *mintulu*, i.e. of Π + PRONOMINAL COMPLEMENT. The examples are typically used when a situation is described in which a participant winds up stretched out full-length on the ground as in (1.127) which characterises a famous Maltese goal-keeper. Note that *għal tulu* is a possible substitute for *mintulu.*

(1.127) [Baldacchino 1999: 117]

ma	*kienx*	*jiddejjaq*	*jintefa'*	***[minntul**u]*$_{PP}$	*ma'*
NEG	be:NEG	3.IPFV:annoy	3.IPFV:RFL:throw	[**all_along**:3SG.M]	with
l-art	*iebsa*	*ta'*	*l-Istadium*		
DEF-ground	hard:F	of	DEF-Stadium		

'[. . .] he was not afraid of throwing himself [**full length**] on the hard ground of the Stadium.'

Examples like (1.127) can be classified as spatial. Owing to the infrequency of the temporal use of *mintul* in our corpus, we register this Π in BLOMP 2.0 with the spatial meaning 'all along'. This Π is distinct from both *tul* and *matul.* Thus, the number of members of BLOMP 2.0 increases further in contrast to (1.73). How *tul* and *matul* are related to each other is discussed in what follows.

Aquilina (1991: 793) translates *matul* as 'during, lengthwise', i.e. the Π has both a temporal and a spatial meaning. For the common noun *tul* 'length' however, the Maltese-English dictionary does not mention any prepositional function except in combination with *minn* 'from' and *ma'* 'with' (Aquilina 1991: 1390). E. Serracino-Inglott (1984: 510) claims that the noun *tul* can combine with the Π *ma'* to yield the adverb *matul.* This is remarkable insofar as the same scholar registers *matul* as Π in another volume of *Il-Miklem Malti* (E. Serracino-Inglott 1979: 69). Before we look at the syntactic implication of this and similar statements of Aquilina's, it is necessary to ascertain or disprove that *tul* and *matul* differ from each other semantically so that two distinct Πs can be assumed to exist. For the sake of the argument, we start from the idea that *matul* and *tul* can be distinguished on semantic and/or formal grounds. In this way, we illustrate how difficult it can be to keep allomorphic and morphemic statuses apart. At the same time, it transpires that the electronic ressources for Maltese do not automatically suffice to solve each and every problem that arises in the context of the Π-system.

To this end, we carry out a spot check involving the first 100 hits for each of the two items in the Korpus Malti 3.0. The results reveal that, in combination with a lexical complement NP, *tul* is attested as Π only 34 times whereas *matul* is exclusively attested as Π. What is more, without any exception, the first 100 hits for *matul* convey temporal meanings. In contrast, temporal meanings are associated with *tul* only in 25 (= 74%) out of 34 cases as opposed to nine instances (= 26%) of spatial usages of *tul*. If we take account of pronominal complements, the situation is not much different. In the case of the first 100 hits for the 3rd person singular masculine as complement, *matulu*, without exception, is used with temporal meaning. For the 3rd person singular feminine and the 3rd person plural, the results are almost identical, namely 99 hits with temporal meaning as opposed to a single attestation each of *matulha* and *matulhom* with spatial meaning. Written as two words *ma'* + *tul* + PRONOMINAL COMPLEMENT, the turnouts are small with *ma' tulu* yielding six temporal cases, *ma' tulha* only two, and *ma' tulhom* amounting to four cases. Spatial meanings are attested once for each of these two-word combinations. In contrast, *tul* takes pronominal complements of the 3rd persons eight times for spatial purposes as opposed to altogether 96 instances of *tulu*, *tulha*, and *tulhom* being employed to express a temporal meaning. We hypothesise that behind this distribution pattern, there might hide a peculiar kind of distinction on the basis of which it is possible to assume two distinct Πs. To prove or disprove this hypothesis, it is necessary to take a closer look at the data.

How disproportionate the relations are between *matul* and *tul*, on the one hand, and temporal and spatial meanings, on the other hand, can be gathered from Table 1.5. Table 1.5 covers all attestations of the Πs *matul* and *tul* in the primary source Baldacchino [1999] which forms part of the texts chosen for the Korpus Malti 3.0. The examples come in the shape of PPs. The examples above the horizontal dividing line are instances of temporal adverbials. The example below the dividing line is the sole case of a spatial adverbial involving one of the competing Πs. In the leftmost column, the absolute number of attestations is revealed. The page numbers for these attestations are given in the second column from the left. Grey shading identifies the isolated case of the Π *tul* in this source.

Table 1.5: Temporal and spatial uses of *matul* and *tul* in Bladacchino [1999].

n	page	example	translation
13	45, 58, 94, 96, 104, 131, 131, 159, 166, 192, 207, 210, 227	*matul il-karriera (twila) tiegħu*	'during his (long) carreer'
9	15, 61, 121, 122, 141, 171, 201, 204, 209	*matul is-snin (ta' wara)*	'over the (subsequent) years'
8	22, 26, 31, 31, 54, 58, 88, 105	*matul (dan) iż-żmien*	'during the (~this) time'
6	63, 85, 87, 88, 122, 151	*matul il-gwerra (l-Kbira)*	'during the (great) war'
4	31, 103, 219, 221	*matul dan il-perjodu*	'during this period'
3	152, 169, 219	*matul dak l-istaġun*	'during that season'
3	85, 91, 120	*matul is-sajf*	'during the summer'
2	172, 202	*matul il-lejl*	'during the night'
2	90, 142	*matul l-istorja (twila)*	'over the (long) history'
2	102, 117	*matul ħajtu (kollha)*	'during his (entire) life'
1	2	*matul dan is-seklu*	'during this century'
1	9	*matul is-seklu dsatax*	'during the 19th century'
1	5	*matul it-tieni nofs tas-seklu tmintax*	'during the second half of the 18th century'
1	2	*matul ftit anqas minn mitejn sena*	'during little less then 200 years'
1	229	*matul dawn l-aħħar għoxrin sena*	'during the last 20 years'
1	73	*matul il-ġimgħa*	'during the week'
1	151	*matul l-ewwel xhur*	'during the first months'
1	167	*matul waħda mill-isbaħ perjodi*	'during one of the best seasons'
1	54	*matul l-Ewwel Gwerra Dinjija*	'during World War I'
1	63	*matul dan l-ewwel staġun*	'during this first season'
1	152	*matul din il-kompetizzjoni*	'during this competition'
1	75	*matul it-tieni taqsima*	'during the second half'
1	230	*matul it-tieni rawnd*	'during the second round'
1	101	*matul din il-laqgħa*	'at this meeting'
1	227	*matul dawn is-suċċessi*	'during these successes'
1	54	*tul il-karriera tiegħu kollha*	'during his entire carreer'
1	18	*matul il-parade ground*	'along/across the Parade Ground'

There are 68 tokens of *matul* and only a single attestation for *tul*. Interestingly, 67 (= 99%) of the 68 instances of *matul* go to the credit of temporal adverbials. In these cases, the complements of *matul* are either calendaric terms such as *ġimgħa* 'week', generic expressions for periods of time such as *żmien* 'time', or they refer to events with a determined extension on the time-arrow such as *l-Ewwel Gwerra Dinjija* 'World War I'. There is also an unexpected case of *matul* being used to express the spatial notion of Path, namely the PP *matul il-parade ground* 'along/across the Parade Ground'. The isolated case of the Π *tul* is temporal. It is representative of 1% of all temporal adverbials in Table 1.5. This distribution speaks clearly in favour of the temporal interpretation of *matul* without however excluding spatial readings completely.

To get a better idea of the division of labour between *tul* and *matul*, we determine the uses to which the two Πs are put in a selection of sources which do not form part of the Korpus Malti 3.0. First of all, there are sources which replicate the pattern reflected in Table 1.5 in the sense that only *matul* is made use of to the (almost) total exclusion of *tul*. In such cases, *matul* assumes both temporal and spatial functions. A typical example of this monopoly of *matul* characterises the primary source Casha [2008]. In this novel, there is no trace of the Π *tul* whereas *matul* is attested 16 times. Of these 16 tokens of *matul*, 14 (= 88%) are invested with a temporal meaning and only two can be classified as spatial. The spatial tokens are clearly outnumbered by those with a temporal meaning.

However, for the purpose at hand, we are more interested in the work of authors who employ both *matul* and *tul* in the same text. The question is whether the two Πs are used interchangeably. The sources from which we take the examples are identified by capital letters as: A = [Attard 1999], B = [Bartolo/Vella 209], C = [Camilleri 2013], and D = [Abela 2016]. In these four texts, there are 17 tokens of *matul* all of which receive temporal interpretations as shown in Table 1.6.

Table 1.6: Temporal uses of *matul* in A, B, C, and D.

Source/page	example	translation
A/35	*matul il-ħakma Britannika*	'during the British domination'
A/40	*laqgħa [. . .] matulha*	'a meeting [. . .] during which'
A/183, 192	*matul (it-tieni) (il-)lejl*	'during the (second) night'
A262	*matul il-ġimgħa*	'during the week'
B/184	*matul dan kollu*	'during all this'
A/79, C/57	*matul is-snin*	'over the years'
C/188	*il-laqgħa [. . .] matulha*	'the meeting [. . .] during which'
C/190-1	*l-intervista [. . .] matulha*	'the interview [. . .] during which'

Table 1.6 (continued)

Source/page	example	translation
D 65	*matul il-jiem kollha*	'during all the days'
D72, 100, 140, 152	*matul il-gwerra*	'during the war'
D/101	*matul l-agħar bumbardamenti*	'during the worst bombardments'
D/203	*matul il-bqija tal-ġurnata*	'during the rest of the day'

What strikes the eye when we look at the data is the number of attestations of the Π *tul* which exceeds that reported above for Baldacchino [1999]. Table 1.7 hosts ten instances of *tul* as temporal Π.

Table 1.7: Temporal uses of *tul* in A, B, and C.

source/page	example	translation
A/81, B/239	*tul is-snin*	'over the years'
A335	*tul il-vjaġġ*	'during the journey'
B/273	*tul dil-konverżazzjoni kollha*	'during this conversation'
B/281	*tul dawn il-jiem*	'during these days'
B/327	*tul dan l-iterrogatorju*	'during this interrogation'
B/466	*tul ix-xhur kollha*	'during all the months'
B/467	*tul dawk ix-xhur*	'during those months'
C/136	*tul it-tmexxija tiegħu*	'during the time of his direction'
C/308	*tul dak iż-żmien kollu*	'during all that time'

Temporal *tul* is absent from source D. In contrast, there is evidence of spatial *tul* in all four of the sources as revealed in Table 1.8.

Table 1.8: Spatial uses of *tul* in A, B, C, and D.

source/page	example	transllation
A/377	*tul il-bajja*	'along the bay'
B/77	*tul il-ħajt kollu*	'all along the wall'
C/257	*tul il-kuritur*	'along the corridor'
D/118	*tul ix-xatt kollu*	'along the whole beach'
D/226	*tul il-kamra*	'through the room'

These five tokens are representative of five types provided that we equate types and complements. The absolute numbers are too small to pass as compelling evidence of anything. However, it is worth noting that if a spatial meaning has to be conveyed with one of the two competing expressions, the sources A, B, C, and D exclusively opt for *tul*. At least for three of the sources, temporal perlatives can be expressed either by *matul* or by *tul*. The Korpus Malti 3.0 glosses over individual preferences. To detect patterns of this kind, it is necessary to evaluate further texts which have not already been exploited for the Korpus Malti 3.0. On the above very small and insufficient empirical basis, we hypothesise that, in several idiolects, *matul* is semantically specialised in the sense that the default interpretation applied to it is temporal. In contrast, this preference for a temporal reading is significantly less pronounced in the case of *tul* to the extent that, again in a still undetermined number of idiolects, the latter ousts *matul* from the domain of the spatial perlative. In conclusion, the postulated distinction of $\textit{matul}_{\text{TEMPORAL}} \neq \textit{tul}_{\text{TEMPOAL/SPATIAL}}$ cannot be upheld because there is too much variation across the speech community as represented in our database. For many speakers, *matul* and *tul* are alloforms of each other. Other speakers seem to apply different default readings to the two candidates which ultimately might lead to a functional split into two distinct Πs. For the time being we turn a blind eye on these potential developments by way of postulating the existence of a Π *matul* 'during' with the alloform *tul* and a primary temporal meaning. Thus, (X53) *(ma)tul* 'along' in (1.73) changes to *matul* 'during' in BLOMP 2.0.

At this point, it is in order to address a general problem connected to the analysis of many of the complex Πs with initial *minn* or *ma'* (and sundry cases). This problem regards the syntactic status of the second constituent of the complex Π in the sense that one might challenge the assumed unity of the complex Π. We have repeatedly mentioned the outward similarity of the complex Πs with compound Πs only to discount the possibility that all *minn*-initial and *ma'*-initial cases should be classified as compound Πs.

The complex-compound dispute is not the only analytical difficulty that arises in connection with the cases under review. Aquilina's examples are often such that the final constituent of the complex Π is identical to a Maltese noun which can form a compound with another noun according to the pattern of the construct state, i.e. the two constiuents of the compound are juxtaposed with the head occupying the left slot and the attributive constitutent the right (allowing for the presence of the definite proclitic). This compound functions as the complement of the simple Π, in this case either *minn* or *ma'*. Both *minn* and *ma'* which are polysemous Πs can be used with perlative meaning without accompanying *tul* as in (1.128)–(1.129).

(1.128) [Casha 2008: 13]

trid *tgħaddi* ***[minn*** *din* *il-foresta]*$_{PP}$
2.IPFV:must 2.IPFV:pass **[from** DEM:F:PROX DEF-forest]
'[. . .] you must pass [**through** this forest].'

(1.129) [Casha 2008: 87]

s-suldati *tar-re* *kellhom* *biċċa* *xogħol* *kbira* *biex*
DEF-soldier:PL of:DEF-king have.PFV:3PL piece work big:F to
iduru ***[mat**-toroq]*$_{PP}$
3.IPFV:go_round:PL [**with**:DEF-street.PL]
'[. . .] the king's soldiers had a hard time to walk [**through** the streets] [. . .].'

On account of the possibility to employ *minn* and *ma'* for the perlative independent of the presence of *tul*, two different syntactic analyses of cases of this kind compete with each other as shown in Figure 1.7 for PPs extracted from (1.122), (1.124), and (1.126). The dotted vertical line indicates the syntactic boundary between the head of the PP and its complement.

(simple Π +) construct state				vs	(complex) Π + simple noun		
Ø	⋮	*[tul*	*il-lejl]*$_{NP}$		***[tul***$_{Π}$	⋮	*il-lejl]*$_{PP}$
[ma'$_{Π}$	⋮	*[tul*	*is-snin]*$_{NP}$*]*$_{PP}$		***[matul***$_{Π}$	⋮	*is-snin]*$_{PP}$
[minn$_{Π}$	⋮	*[tul*	*ir-runway]*$_{NP}$*]*$_{PP}$		***[mintul***$_{Π}$	⋮	*ir-runway]*$_{PP}$

Figure 1.7: Competing syntactic analyses.

If we accept the construct-state hypothesis on the left side, the number of complex Πs in (1.73) would diminish considerably because this analysis separates the component parts of the complex Π by way of assigning its final constituent to the complement NP. What arguments can be put forward in defense of the alternative interpretation of these cases as complex Πs?

The proof in support of the hypothesis according to which we are dealing with (complex) Πs is of the formal kind. There is pronominal anaphora as shown in (1.130).

(1.130) [Attard 2019: 40]

Il-President *Malti* *kellu* *laqgħa [. . .]* *ma'* *dak*
DEF-president Maltese have.PFV:3SG.M meeting with DEM:M:DIST
tal-Istati Uniti [. . .] ***[matul**ha]*$_{PP}$ *il-President* *Amerikan*
of:DEF-United States [**during**:3SG.F] DEF-president American

issuġġerixxa	*lil*	*dak*	*Malti*
suggest	to	DEM:M:DIST	Maltese

'The Maltese President had a meeting [. . .] with that of the United States [. . .] [**during** it], the American President suggested to the Maltese President [. . .].'

The pronominal anaphor referring back to *laqgħa* 'meeting' is separated from this noun by a relative clause and a sentence boundary. If *tul* was the head of a construct state compound, its appropriate place would be to the left to *laqgħa*. However, in cases like this, *matul* is never split in two. Moreover, if *tul* were part of a compound it should retain some of its categoriality as a noun. As a common noun however, *tul* requires the presence of the Π *ta'* 'of' under attribution as shown in (1.131).

(1.131) [Attard 2019: 203]

Nispera	*li*	*mhux*	*[għal*	***[tul***	***[ta'***	***żmien]***$_{PP}$*]*$_{NP}$*]*$_{PP}$
1.IPFV:hope	that	NEG	[for	**[length**	**[of**	**time**]]]

'I hope that it won't take long.'

In this sentence, *tul* 'length' is the head of the NP which is the complement of the Π *għal* 'for'. The attribute of *tul* is a PP headed by *ta'* 'of'. The presence of *ta'* would not be licit in the construct state. We assume that in cases like those covered by Figure 1.7, *tul* is stripped of its nouny properties. It is an integrated part of the complex Π or an allomorph of *matul.*

(x) (X32) *maġenb* 'close to'

The absence of the nouny properties comes more clearly to the fore in the case of (X32) *maġenb* 'close to' (E. Serracino-Inglott 1979: 7; Aquilina 1991: 767). Examples (1.132)–(1.133) stem from one and the same paragraph in the prologue to Camilleri [2013].

(1.132) [Camilleri 2013: 9]

Sakemm	*kif*	*laħaq*	*ix-xfar*	*u*	*ntebaħ*	*li*	*ma*
until	how	reach	DEF-edge.PL	and	RFL:be_aware	that	NEG

kellux	*iżjed*	*appoġġ*	*iżżerżaq*	***mal-ġenb***
have.PFV:3SGM	more	support	slide_down	**with:DEF-side**

tal-banju	*jfittex*	*ta'*	*xejn*	*minn*	*fejn*	*ħa*
of:DEF-bath_tub	3.IPFV:seek	of	nothing	from	where	FUT

jiżgiċċa
3.IPFV:scamper

'Until he reached the edges and became aware that he had no more support he slid down **along the side of the bath tub** looking in vain for a way out.'

(1.133) [Camilleri 2013: 9]
Il-magażin *bl-Ingliż* *mixħut* ***[maġenb*** *il-bedè]*$_{PP}$ *[...]*
DEF-magazine with:DEF-English PART:throw **[close_to** DEF-bidet]
sar *fellus*
become chicken
'The magazine in English thrown down [**close to** the bidet] became wet [...].'

In (1.132), the noun *ġenb* 'side' refers to an identifiable part of the bath tub, i.e. autosemanticity applies. On the formal side, *ġenb* 'side' functions as head for the genitival attribute *tal-banju* 'of the bath tub'. Together with this attributive PP and the definite proclitic, *ġenb* 'side' constitutes the complement of *ma'* 'with' with the latter being used with a perlative meaning. In contrast, example (1.133) involves *ġenb* not as an independent syntactic word but as an integrated part of the complex Π *maġenb* 'close to'. In this case, *ġenb* is not referential because no identifiable part of the bidet is identified. The Figure – the English magazine – winds up on the bathroom floor somewhere near the bidet but not necessarily at the Ground's physical rim. As part of the complex Π, *ġenb* can neither be marked for definiteness nor take a PP as attribute.

In (1.134), the complex Π occurs in a relative clause which is an attribute to *din il-kappella* 'this chapel' in the main clause. The morphosyntactic properties of the noun *ġenb* 'side' do not play a role for the behaviour of *maġenb* 'close to'.

(1.134) [Korpus Malti 3.0; news122332]
Ma' *din* *il-kappella* *hemm* *ukoll* *marbut* *fatt* *storiku*
with DEM:F:PROX DEF-chapel there also PART:tie fact historical
li ***[maġenbha]***$_{PP}$ *kienet* *saret* *ġlieda* *qalila*
that **[close_to:**3SG.F] be:3SG.F.PFV become:3SG.F.PFV fight fierce:F
bejn *il-Maltin* *u* *l-Franċiżi*
between DEF-Maltese:PL and DEF-French:PL
'To this chapel the historical fact is connected that [**close to** it] a fierce fight between the Maltese and the French had taken place [...].'

Otherwise we would expect the PP **mal-ġenb tagħha* 'at its side' which refers to a particular part of the building. The meanings of the noun *ġenb* 'side' and of the Π *maġenb* 'close to' are metonymically interconnected but dissociated from each other. That *maġenb* 'close to' is a Π like any other Π mentioned in (1.73) is evident from (1.135).

(1.135) [Bartolo/Vella 2009: 406]
Is-sħaba *kkulurita* *ta'* *friefet* *kienet* ***[taħt***$_{Π}$ *Ø]*$_{PP}$,
DEF-cloud colour:PART:F of butterfly.PL be:3SG.F.PFV **[under** Ø]

[maġenb$_{\Pi}$ ***Ø]***$_{PP}$***,*** ***[wara***$_{\Pi}$ ***Ø]***$_{PP}$ *u* ***[fuq***$_{\Pi}$ *Estella]*$_{PP}$ *kullimkien*
[close_to Ø] **[behind** Ø] and **[on** Estella] everywhere
'The coloured cloud of butterflies was [**under**, **at the side**, **behind** and **above** Estella] – everywhere.'

There are four Πs which share the proper name *Estella* as complement. The Πs form an enumerative coordination whose second conjunct is *maġenb* 'close to', i.e. this expression is preceded and followed by other members of BLOMP whose status as Π is beyond doubt. Under coordination, *maġenb* 'close to' is treated like an expression with the same status as the other conjuncts. Thus, it too is a legitimate member of BLOMP and not an instance of a compound Π.

1.6.5 Word-classes / morpheme-classes

As will become clear from Part B, Maltese Πs do not behave absolutely homogeneously on all parameters. The morphological properties differ widely across BLOMP. Similarly, there are also differences between Πs in the morphosyntactic domain. What *apparti* 'apart from' and *fi* 'in', for instance, have in common nevertheless is the fact that they meet the criteria of (1.1). The very same criteria are also met by expressions which in other contexts do not behave like genuine Πs. Many members of BLOMP belong to several word-classes at the same time. Their use as Π might turn out to cover only a small segment of their functional domain. In (1.73), we find items which (sometimes predominantly) also function as noun, adverb, or conjunction. Multiple word-class membership is however no criterion for excluding a candidate from the Π-inventory. It suffices that the candidate is systematically made use of in contexts which correspond to the definition given in (1.1) and the canonical Π sketched in Figure 1.5.

(i) (X22) *għand* 'at s.o.'s place'
Pseudo-verbs have categorically been excluded from BLOMP.[116] This does not imply that the Π *għand* 'at s.o.'s place' (E. Serracino-Inglott 1976: 175; Aquilina 1991: 967–968) has to be removed from BLOMP on account of the diachronic connection to the predicative possessive pseudo-verb *għand-* 'have' in the imperfective. The contrast of pseudo-verb and Π is a favourite topic in Maltese linguistics (Fabri 1993: 198–202, Peterson 2009: 199–200). Comrie (1982) and Borg/Azzopardi-Alexander (1997: 112) argue convincingly for a split into two distinct homophonous morphemes whose morphosyntactic behaviour differs markedly (Stolz et al. 2008: 96–97). The differ-

116 Cf. Section 1.6.2.

ences between pseudo-verb and homophonous Π are clearly visible in the three examples in (1.136). The original hyphenisation and the glosses have been modified according to the principles applied in this study. All English translations are original. The examples (1.136a–c) are also given in Peterson (2009: 190).

(1.136) *għand*: pseudo-verb vs Π

(a) (Comrie 1982: 13)

Il-bajda	*[**għand***	*<u>Pawlu</u>]*$_{PP}$
DEF-egg	[**at_s.o.'s_place**	<u>Paul</u>]

'The egg is [**at** <u>Paul's</u> **place**].'

(b) (Comrie 1982: 14)

Pawlu	***għandu***	*bajda*
Paul	**have.IPFV:3SG.M**	egg

'Paul **has** an egg.'

(c) (Peterson 2009: 190)

Għandu	*Pawlu*	*bajda*
have.IPFV:3SG.M	Paul	egg

'Paul **has** an egg.'

Only in (1.136a) does *għand* 'at s.o.'s place' function as a Π. It is morphologically invariable in combination with the proper name *Pawlu* as complement noun. In contrast, in (1.136b–c) *għand-* hosts the pronominal suffix of the 3rd person singular masculine which agrees with the proper name *Pawlu* which represents the possessor. In predicative possessive constructions in the imperfective, the suffix *-u* is indispensable if the possessor is masculine singular no matter whether the lexical NP is copresent or not. This holds for the pragmatically neutral SVO order in (1.136b) as well as for the pragmatically marked (Peterson 2009: 190) VSO order in (1.136c). As argued above,[117] the mandatory pronominal inflection is a typical property of pseudo-verbs that distinguishes them from Πs which do not permit pronominal inflection in the presence of a lexical complement NP. Superficially, (1.137) seems to contradict this rule.

(1.137) [Żahra 2008: 54]

Dik	*il-ħabta*	*[**fost**<u>na</u>*	*<u>it-tfal</u>]*$_{PP}$	*kien*	*hemm*	*żewġ*
DEM:F:DIST	DEF-blow	[**between**:<u>1PL</u>	<u>DEF-child.PL</u>]	be	there	two

partiti
party:PL

'At that time, there were two parties [**amongst** <u>us</u> <u>children</u>] [. . .].'

117 Cf. Section 1.6.2.

The presence of the lexical NP within the PP is no problem to our above analysis because *it-tfal* 'the children' is not (directly) governed by the Π *fost* 'between' but serves as apposition to the pronominal complement in the 1st person plural. The appositive function of *it-tfal* is marked by double underlining in the example. Were this NP the complement of *fost*, the suffix *-na* would not be allowed to attach to the Π. Admissible PPs are *fost it-tfal* 'amongst the children' and *fosthom* 'amongst them'. In these cases, however, the reference to the 1st person plural is made impossible. On these grounds and in reference to *għand*, we assume two functionally distinct expressions, namely Π *għand* 'at s.o.'s place' ≠ pseudo-verb *għand-* 'have$_{\text{IPFV}}$' with the former being a full-blown member of BLOMP.

(ii) (X29) *lil* OBJECT MARKER

Another bone of contention in Maltese linguistics regards (X29) *lil* OBJECT MARKER. The two major opinions found in the literature disagree as to the classification of *lil*. For Fabri (1993: 114–122) *lil* is an object marker sui generis which does not count among the Maltese Πs[118] whereas Borg/Azzopardi-Alexander (1997: 137) classify *lil* as a basically directional Π which is also employed for marking indirect objects and definite animate direct objects.[119] In their interpretation *lil* can be translated with English *to*. This interpretation is in line with that of E. Serracino-Inglott (1978: 296) and Aquilina (1987: 750). With reference to the functional homology of adpositions and morphological case, Hagège (2010: 17–18) discusses the idea that what can be done with morphological case can also be done with adpositions. This means that major syntactic functions such as direct and indirect object can be marked either by bound morphological means or free morphemes, i.e. adpositions. Givón (2021: 1) goes as far as claiming that while adpositions may occasionally mark object/patient they "most commonly mark the case-roles [. . .] of various indirect objects (so-called 'obliques')". In a way, this means that *lil* can be a Π and an object marker at the same time. The function for which the expression of choice is employed is irrelevant to the classification of the formal means. Accordingly, it does not matter whether *lil* is ascribed the function of object-marking or that of the directional, it is always a Π albeit one displaying traits of its own. Discounting these details, *lil* does not behave sufficiently different from other Πs mentioned in (1.73) to justify being excluded from BLOMP. Granting full Π-status to *lil* does not require of us to blindly accept the above directional meaning assumed by Borg/Azzopardi-Alexander (1997).

Directionality re-emerges as a possibility if we take account of (X25) *'il* 'to'. First of all, *lil* has two allomorphs *'il* ~ *'l* (Borg/Azzopardi-Alexander 1997: 137). The examples (1.138)–(1.140) which are taken from one and the same primary source provide proof of this allomorphy.

118 Cf. Section 1.4.2.2.
119 Cf. Section 1.4.2.4.

(1.138) [Zammit 1976: 102]
ħadet *[**lil*** *Zolli]*$_{PP}$ *paqq* *fuq* *għonqha*
take:3SG.F.PFV [**to** Zolli] bang on neck:3SG.F
'[. . .] she took [Zolli] – bang – by her neck [. . .].'

(1.139) [Zammit 1976: 103]
ħataf *[**'il*** *Zolli]*$_{PP}$ *mill-pexxul* *u* *qalibha*
grab [**to** Zolli] from:DEF-calf and turn_over:3SG.F
'[. . .] he grabbed [Zolli] by the calf and turned her over.'

(1.140) [Zammit 1976: 100]
ma *nċediha* *[**'l*** *ħadd]*$_{PP}$
NEG 2.IPFV:surrender:3SG.F [**to** someone]
'[. . .] I won't surrender [**to** anybody].'

The phonologically-conditioned reduced allomorphs of *lil* are optional. What renders them interesting in the context of this section is that *'(i)l* is frequently attested in binary adverbials. As mentioned already,[120] Aquilina (1987: 564) speaks of an "[a]dverbial particle indicating direction" which is "prefixed" to certain hosts to yield word forms "with a grammatical function similar to Eng[lish] suf[fixed] '-ward(s)' in 'upwards', 'downwards' etc." The directional adverbials highlighted in boldface in (1.141)–(1.142) are typical examples of this kind of formation.

(1.141) [Cauchi 1997: 24]
poġġa *idejh* *ma'* *zaqqu* *u* *pprova* *jiġbidha*
place hand:PL:3SG.M with belly:3SG.M and try 3.IPFV:pull:3SG.F
'l ġewwa
to inside
'[. . .] he put his hands on his belly and tried to push it **inwards**.'

(1.142) [Cauchi 1997: 18]
rah *wiċċu* *ma'* *l-art* *bil-warrani*
see:3SG.M face:3SG.M with DEF-ground with:DEF-posterior
ppuppat ***'il fuq***
stick_out:PART **to on**
'[. . .] he saw him with his face on the ground and his posterior stuck out **upwards**.'

120 Cf. Section 1.4.1.5. The quote is partially repeated from the earlier mention.

The adverbials involving *'il* form a sizable paradigm as shown in Figure 1.8. Except one, all of these combinations of *'(i)l* + X are attested in the primary source Ellul Mercer [1986]. Those highlighted in bold belong to the set of licit combinations acknowledged by Aquilina (1987: 564) as does the underlined case which is absent from the primary source however.

'(i)l +	***barra***	'outwards'
	bogħod	'at a distance'
	fuq	'upwards'
	ġewwa	'inwards'
	hawn	'hither'
	hemm	'thither'
	hinn	'thither'
	isfel	'downwards'
	quddiem	'onwards'
	taħt	'beneath'
	ura (< *wara* 'behind')	'backwards'

Figure 1.8: Directional adverbials with initial *'(i)l* 'to'.

Sutcliffe (1936: 199) analyses these cases as combinations of the Π *l* and certain adverbs, i.e. to his mind *'(i)l* is a genuine Π which is identical to the initial segment of *lil* (Sutcliffe 1936: 202). The Π-status of *'(i)l* is also assumed by Aquilina (1959: 300) and Brother Henry (1980: 192). Fabri (1993: 186) classifies the same item as particle which synchronically has a function distinct from that of the object marker *lil*. Saari (2003: 178) registers cases like those mentioned in Figure 1.8 as instances of adverbs whose structure is that of an erstwhile PP with a synchronically obsolete Π as (diachronic) head.

There are several reasons to doubt that we are dealing with a full-blown Π. If Aquilina (1987) is right when he claims that *'(i)l* is prefixed to a host, then the free-morpheme condition is not met. According to (1.1), a Π is a function-word. The function-word status presupposes that the item under review is a free morpheme. As prefix however, *'(i)l* would be a bound morpheme – most probably one with derivational function. Moreover, the adverbials in Figure 1.8 do not fulfil the prerequisites for a canonical PP because the majority of the units which combine with *'(i)l* are not complement NPs. There are deictic adverbs such as *hawn* 'here', *hemm* 'there', and *hinn* 'there'. According to Aquilina (1991: 1249), *isfel* 'down' is exclusively used as adverb whereas *bogħod* 'distance; far away' can be both a noun and an adverb (Aquilina (1987: 119). Many fillers of the right slot of the binary adverbial are elements which are also featured in (1.70), namely *barra* 'outside', *fuq* 'on', *ġewwa* 'inside', *quddiem* 'in front of', *taħt* 'under', and *wara* 'behind'. What adds to the doubts as to the existence of a Π *'(i)l* is the closed set of potential "complements" of this supposed Π. The right slot of the pattern involving initial *'(i)l* can be occupied only by those elements which are listed in Figure 1.8. The bulk of the members of BLOMP however is not subject of restrictions of this kind. In Hagège's (2010: 58) hierarchy of the classes of terms which can be governed by adpositions, adverbs do not show up. Isolated Πs are also absent from this hierarchy whereas PPs are featured therein. Since the above adverbials take no complements but modify the predicate, *'(i)l* is disqualified for the status of Π.

One might want to object against this conclusion on account of cases like (1.143) which seem to involve *'(i)l*-initial adverbials which are linked to a potential complement by the intercalated Π *minn* 'from'.

(1.143) [Ellul Mercer 1986: 78]

Kelli	*surġent-maġġur*	*rieqed*	*ftit*	*'il*	*bogħod*	
have.PFV:1SG	sergeant-major	sleep.PART	little	DIR	distance	

[*minni*]$_{PP}$
[from:1SG]
'I had a sergeant-major sleeping at a little distance [**from** me].'

We assume that the adverbial *ftit 'il bogħod* 'at a little distance' in the box primarily modifies the predicate nucleus *rieqed* 'sleeping' and that the PP *minni* 'from me' is a secondary adjunct which forms no constituent with the adverbial to its left. The vertical line indicates that there is a syntactic boundary. Note that within the adverbial there is another modifying relation, namely that between *ftit* 'little' and the remainder of the construction. According to this interpretation, there is no Π **'(i)l bogħod*. In analogy, similar cases like *'il barra* can be ruled out as candidates for the Π-status. Whether there are compound Πs which involve *'(i)l*-initial adverbs and *minn* is a different matter.

On account of the exclusion of *'(i)l* from BLOMP 2.0, *lil* has lost a potential allomorph which could be responsible for the assumed spatial meaning of the Π. Thus, nothing prevents us from glossing *lil* as OBJECT MARKER in lieu of assuming an allative meaning for it. At the same time, there is no other Π which would lay claim to the same allative meaning. Therefore, we continue to gloss *lil* as before (i.e. as 'to') without denying however that its function is that of introducing direct or indirect objects.

The adverbial use of certain Maltese Πs has been acknowledged by several authors addressed in the above literature review. As mentioned there[121] in connection to (1.36), Fabri (1993: 188) distinguishes transitive and intransitive Πs with the latter allowing for being employed adverbially, i.e. without complement. The possibility of certain Πs to also function as adverbs in the appropriate syntactic context forms no obstacle for these Πs to be admitted to BLOMP. For the purpose of this study, we do not adopt Fabri's transitivity-based distinction of different classes of Πs. We speak of Πs only if the criteria laid down in (1.1) are met. This means that an intransitively used Π of Fabri's is an adverb in our terminology.

(iii) (X46) *qalb* 'amongst'

The cases of (X11) *faċċata* 'opposite' (< noun *faċċata* 'façade'); (X13) *fejn* 'near' (< interrogative *fejn* 'where') have been addressed already.[122] There are another three candidates in (1.70) which display a connection between a common noun and a Π. (X12) *favur* 'in favour of' will be addressed below.[123] The body-part term *qalb* 'heart' and (X46) *qalb* 'amongst' (E. Serracino-Inglott 1981: 174; Aquilina 1991: 1111) are connected to each other etymologically via grammaticalization. In (1.144), the possessed noun *qalbu* 'his heart' is used to describe the state of mind of the protagonist.

(1.144) [de Saint-Exupéry 2000: 79]

Iżda	*xi*	*ħaġa*	*qawwietlu*	***qalb**u*
but	some	thing	strengthen:3SG.F.PFV:IO:3SG.M	**heart**:3SG.M

'However, something strengthened his **heart**.'

At least metonymically, reference is made to the body-part of the Little Prince. The situation is different for both (1.145) and (1.146). In neither case does *qalb* refer to a concrete body-part. In (1.145), the origin of the sprouting grapes is identified. There is no heart of the petals, in the first place, but a region which is delimited by the surrounding petals.

---

121 Cf. Section 1.4.2.2.

122 In Section 1.4.2.5 (examples (1.48)–(1.49)) and Section 1.5.8 (examples (1.66)–(1.67)), respectively.

123 Cf. Section 1.6.6.

(1.145) [Bartolo/Vella 2009: 85]

il-fjura *safranija* *li* *kellha* *ħafna* *xniexel* *ħerġin*
DEF-flower yellowish:F that have.PFV:3SG.F many grape.PL exit.PART:PL
[minn ***[qalb*** *il-petali]$_{PP}$]$_{PP}$*
[from **[amongst** DEF-petal:PL]]
'[. . .] the yellowish flower which had many grapes sprouting [**from** [**amongst** the petals]].'

In (1.146), there are three syntactically complex PPs all of which involve the Π *qalb* 'amongst' as head. In none of the three PPs do we recognise any reference to a body-part. What is expressed by the Π instead is the location of the character Gallo and several socially defined groups of people at the same time at the same place.

(1.146) [Camilleri 2013: 123]

[qalb *tfajliet* *jippruvaw* *jimxu* *dritt* *bi* *tkaken*
[amongst girl:PL 3.IPFV:try:PL 3.IPFV:walk:PL straight with heel.PL
għoljin]$_{PP}$ ***[qalb*** *ġuvintur* *jippruvaw* *ta' xejn*
high:PL] **[amongst** young_man.PL 3.IPFV:try:PL of nothing
jagħmluha *tal-ħelwin]$_{PP}$* ***[qalb*** *tfajliet* *li*
3.IPFV:make:PL of:DEF-sweet:PL] **[amongst** girl:PL that
saru *ommijiet* *qabel* *iż-żmien]$_{PP}$* *Gallo* *lagħab*
become:3PL.PFV mother:PL before DEF-time] Gallo play
il-parti *tiegħu* *wkoll*
DEF-part of:3SG.M also
'[**Amongst** girls who were trying to walk straight on high heels], [**amongst** young men who were trying in vain to appear sweet], [**amongst** girls who had become mothers prematurely], Gallo also played his part [. . .].'

It can therefore be taken for granted that the noun *qalb* 'heart' and the Π *qalb* 'amongst' are semantically dissociated from each other to justify postulating two distinct homophonous lexicon entries of which only the Π is of interest for this study.

(iv) (X47) *qrib* 'near'

As to (X47) *qrib* 'near', the line of argumentation repeats what has been said already time and again in the foregoing paragraphs. Aquilina (1991: 1162) mentions *qrib* 'nearness, near' as verbal noun related to *qorob* 'approach'. In (1.147), *qrib* is used as noun.

(1.147) [Ellul Mercer 1986: 35]

Ix-xelters	*li*	*qed*	*jinqatgħu*	*fil-blat*	*fil-**qrib***
DEF-shelter:PL	that	PROG	3.IPFV:RFL:cut:PL	in:DEF-rock	in:DEF-**nearness**
tad-dar	*tiegħi*				
of:DEF-house	of:1SG				

'[. . .] the shelters which are being cut from the rock in the **vicinity** of my house [. . .].'

In this case, *qrib* displays those properties which are usally associated with Maltese common nouns. First of all, it is made definite overtly. Secondly, it takes a genitival attribute. The attribute is linked to its head by the Π *ta'*. Thirdly, it functions as complement noun in a PP. In (1.148), *qrib* can no longer be analysed as noun.

(1.148) [Teuma 2015: 612]

Għalhekk	*il-kalif*	*niżżlu*	*ma'*	*dawk*
therefore	DEF-calif	descend:CAUS:3SG.M	with	DEM:PL:DIST
in-nies	*tal-qorti*	*li*	*kienu*	[*l-aktar* ***qrib*** *tiegħu* DEF-more **near** of:3SG.M]
DEF-people	of:DEF-court	that	be:3PL.PFV	

'Thus, the calif let him go down with those people of the court that were **close**st to him.'

The box contains what is predicated in the relative clause over *dawk in-nies tal-qorti* 'those people of the court'. The presence of the copula *kienu* 'they were' does not tell us whether we are dealing with an essive, locative, or qualitative predicate. There is the analytical comparative marker *aktar* 'more' (also: *iktar* ~ *iżjed*) which is preceded by the definite proclitic. This combination is the most common way of expressing the superlative degree of adjectives in Maltese (Borg/Azzopardi-Alexander 1997: 75). Normally (but not exclusively) the analytical definite comparative marker precedes the adjective. In the case of (1.148), this could mean that *qrib* is adjectival which is also the classification of this candidate given by E. Serracino-Inglott (1981: 247). Predicative adjectives agree in gender and number with the subject-NP of a copula clause (Borg/Azzopardi-Alexander 1997: 260). Since the copula *kienu* 'they were' is in the plural, we expect the supposed adjective to formally reflect the same number value. However, the at least theoretically possible plural *qraba* ~ *qorob* 'relatives' is probably blocked because it is used for the adjective *qarib* 'near'. The existence of this adjective in turn, renders the putative adjectival status of *qrib* doubtful. Furthermore, *qrib* is followed by the possessive PP *tiegħu* 'of him = his'. Borg/Azzopardi-Alexander (1997: 145) state that "[a]djectives also govern objects indirectly through a number of different prepositions" without however mentioning *ta'* 'of'. The *ta'*-PP is typical of genitival attributes within a NP.

Thus, in (1.148), *qrib* seems to display properties of a noun as well. It is difficult to classify it as full-blown Π in this sentence because it is in the scope of the comparative marker. Gradability is certainly not a prototypical property of Πs. In contrast, many Maltese adverbs can have a comparative and superlative (Borg/Azzopardi-Alexander 1997: 45). To cut a potentially long story short, we conclude that in cases like (1.148) the exact status of *qrib* is hard to determine.

This is not always the case. In (1.149) we see for instance that *qrib* can function as spatial Π in accordance with the prerequistes formulated in (1.1).

(1.149) [Camilleri 2013: 145]

It-titjira	*charter*	*ma*	*kinitx*	*mill-iktar*	*sempliċi*
DEF-flight	charter	NEG	be:3SG.F.PFV:NEG	from:DEF-more	simple

b'	*turbolenzi*	*strambi*	***[qrib***	*il-mitjar]*$_{PP}$
with	turbulence:PL	strange	**[near**	DEF-airport]

'The charter flight was not the simplest with strange turbulences [**near** the airport] [. . .].'

The behaviour of *qrib* is reminiscent of that reported for several other Maltese Πs such as those discussed in Section 1.6.4.2. As a noun, *qrib* requires the presence of *ta'* under attribution. If *qrib* is used as Π, it combines directly with its complement provided this is a lexical NP. It is clear that the Π-function bars the possibility of definiteness marking on *qrib*. The usual behaviour of nouns is blocked for the Π *qrib*.[124]

(v) (X58) *waqt* 'at the time of'

Very much the same can be said in relation to (X58) *waqt* 'at the time of'. This Π has been alluded to already in the context of the examples (1.18) and (1.62). *Il-Miklem Malti* registers *waqt* as noun and adverb (E. Serracino-Inglott 1989: 25). In the Maltese-English dictionary, Aquilina (1991: 1517) gives the noun *waqt* 'instance, point in time'. This common noun is featured in example (1.150).

(1.150) [Ellul Mercer 1986: 41]

Mur	*ara*	*x'*	*kien*	*jiġri*	*li*	*kieku*	*ġemgħat*	*ta'*	*nies*
go	see	what	be	3.IPFV:happen	that	if	crowd:PL	of	people

sabu	*ruħhom*	*hemm*	*f'*	*dan*	*il-**waqt**!*
find:3PL.PFV	soul:3PL	there	in	DEM:M:PROX	DEF-**moment**

'Imagine what would have happened if crowds of people had been there at this **moment**.'

124 As to the possibility to inflect *qrib* for person, the reader is referred to Section 3.1.2.

The PP *fil-waqt* 'punctually, on time' can be used adverbially. In the orthographic shape *filwaqt* 'while' it functions as conjunction (Aquilina 1987: 339) the function of which is not always properly temporal but may also have a contrastive tinge as in (1.151). Borg/Azzopardi-Alexander (1997: 39) have *waqt li* 'while' in the same function.

(1.151) [Korpus Malti 3.0; academic203]

il-kliem	*ġie*	*miktub*	*minn*	*Goffredo Mameli,*	***filwaqt***
DEF-word.COLL	come	PART:write	from	Goffredo Mameli	**while**

li	*l-kompożitur*	*kien*	*Michele Novaro*
that	DEF-composer	be	Michele Novaro

'[. . .] the lyrics were written by Goffredo Mameli **whereas** Michele Novaro was the composer.'

In this function, *filwaqt* is normally accompanied by the general complementiser *li* 'that' which also serves as relativiser. It is conceivable that the conjunction *filwaqt* originates from an erstwhile relativisation of *waqt*. The nouny properties of *waqt* still shine through in this construction. There is definiteness marking, embedding in a PP, and a relative clause as attribute. Example (1.152) shows that none of these units surfaces when *waqt* is used prepositionally.

(1.152) [Bartolo/Vella 2009: 290]

Dik	*sibtha*	*mwaħħla*	*ma'*	*wiċċ*	*Kristjan*
DEM:F:DIST	find:1SG.PFV:3SG.F	PART:fix:F	with	face	Kristjan

[waqt	*il-ġlieda]$_{PP}$*
[at_the_time_of	DEF-fight]

'I found it tied around Kristjan's face [**at the time of** the fight].'

In (1.152) and comparable cases, *waqt* can be neither made definite overtly nor relativised. It can take no attributes nor undergo morphological manipulation. Not all of these restrictions correspond to those imposed on the heads of the construct state. Thus, like in the case of noun *qrib* ≠ Π *qrib*, we assume the coexistence of the noun *waqt* and the homophonous Π *waqt*.

1.6.6 Borrowings

In the literature, it comes to the fore that the existence of Maltese Πs with an Italian etymology was acknowledged only relatively recently. Apart from individual mentions of Italian-derived Πs by Panzavecchia (1843) and Sutcliffe (1936), Aquilina

(1959) is the first to accept several Πs of this kind in what he calls non-Semitic or mixed Maltese. In point of fact, it is impossible to adequately account for the systematic aspects of the Maltese Π-system without taking the borrowings from Italian (or Sicilian) into account. Schmidt/Vorholt/Witt (2020: 253) conclude that

> [the] analysis of Maltese prepositions in relation to etymology shows that there are relevant differences between prepositions of Romance and Semitic etymology.

These differences are manifest inter alia in the different type and token frequencies of the members of the two etymologically defined sub-groups of Maltese Πs. Those Πs which emerged in the course of language contact also display a lower functional load than their Semitic counterparts. The massive impact of Italian on the lexicon of Maltese is an undisputed fact among linguists. Given that probably more than half of the modern lexicon of Maltese has an Italian background, it can hardly surprise to find that Italian has also contributed its share to the Π-inventory.

When we refer to borrowings we mean fully integrated loanwords. In journalistic Maltese, Fenech (1978: 59) identifies

> [a]nother sub-class of prepositions [which] is made up of those prepositions borrowed from Italian, Latin or English in direct loans involving foreign expressions. Such prepositions have a very limited function in Maltese because their use is restricted to the expressions in which they occur. [. . .] In literary Maltese, such prepositions are replaced by Semitic Maltese prepositions as far as possible.

The quoted author mentions Italian *di* 'of', *a(d)* 'to', Latin *sub* 'under', English *for* 'for', and the etymologically ambiguous *in* 'in' which can have an Italian/Latin or English history. These items form part of larger chunks of foreign language material within a Maltese context. The three English words marked in boldface in (1.153) illustrate the phenomenon.

(1.153) [Camilleri 2013: 76]

U	*sab*	*iktar*	***in bad taste***	*l-ħafna*	*kummenti*
and	find	more	**in bad taste**	DEF-many	comment:PL

'And he also found the many coumments to be **in bad taste** [. . .].'

The English Π *in* is part of a larger expression from which it cannot be extracted to serve other structural purposes elsewhere in the Maltese utterance. For the purpose of this study, we exclude cases of this type from the further discussion because the items under scrutiny cannot be productively used in combination with non-English elements. They are thus no members of BLOMP.

There are however borrowings from Italian which synchronically are full-blown Maltese Πs (Stolz/Levkovych 2020: 226–230). Saari (2009: 271) assumes that there are four Πs in modern Maltese which have an Italian origin, namely (X12) *favur* 'in favour of' (cf. (1.50)), (X26) *kontra* 'against', (X49) *rigward* 'concerning', and (X52) *skont* 'according to'. Note that only two of this group of four are Πs also in the donor language. These are *kontra* < Italian/Sicilian *contra* 'against' (standard Italian *contro*) and *skont* < Italian *secondo* 'according to'. For *favur*, Saari (2009: 272) assumes that the "Italian prepositional phrase *a favore di*" has served as pattern. Beside (X12) *favur* 'in favour of', there are also the synonymous alloforms *a favur* and *affavur* which Aquilina (1987: 313) marks as "Italianate usage". E. Serracino-Inglott (1975: 215) only mentions *affavur ta'*. There is a further variant *għaffavur* which seems to be an orthographic alternative to *affavur*. These so-called variants are stylistically marked and much less frequent than *favur* itself so that it makes sense to subsume all alloforms under (Π13) *favur* 'in favour of' in BLOMP 2.0.[125] Note also that the reflexes of the Italian Π *a* 'to' are devoid of functions of their own. They are only meaningless phonological segments. Similarly, Maltese *rigward* is connected diachronically to Italian *riguardo a* ~ *a riguardo di* (Saari 2009: 274). For the latter Π, Saari (2009: 274) observes an uneven distribution over styles and genres. *Riġward* seems to be especially frequent in legal texts.[126] What is relevant for the subsequent paragraphs here however is the fact that we have identified more than twice as many Maltese Πs with Italian etymology as Saari (2009) does.

We start with a negative case. Panzavecchia (1843) assumed *la* to be an Italian-derived Π in Maltese.[127] E. Serracino-Inglott (1978: 238) speaks of an invariable adverbial particle with modal functions. In our own opinion, the Π-status of *la* is as doubtful as that of the directional *'(i)l* addressed in the foregoing section. To the best of our knowledge, *la* cannot freely combine with all kinds of NPs so that its productivity is limited. As far as we can see, *la* is incompatible with complex NPs/DPs as well as with pronouns. These restrictions still need to be verified empirically. For the time being, we assume that if *la* can be given any function at all the most convincing would be that of a derivational prefix which turns simple NPs (mostly singleton nouns or proper names) into modal adverbials with the meaning 'in the fashion of X'.

We already suggested[128] that *faċċata* 'opposite' is a Maltese Π whose origin can be traced back to the Italian noun *facciata* 'façade' which also exists with the same meaning in Maltese (E. Serracino-Inglott 1975b: 184; Aquilina 1987: 294). Thus, *faċċata* 'opposite' rightfully claims a place in BLOMP. The same claim can be made

125 Cf. Section 3.2.6.

126 We dedicate a separate subsection to this Π where the problems it poses are highlighted in some detail (= Section 4.1.2.2.1.2.1 (ix)) so that *rigward* is not discussed further in this section.

127 As discussed in Section 1.4.1.3. in connection to examples (1.4)–(1.5).

128 Cf. Section 1.4.2.5, examples (1.48)–(1.49).

for half a dozen further candidates which have been borrowed from Italian. Most of the candidates we are going to discuss briefly have been mentioned by one or two of our predecessors in their accounts of the Maltese Π-inventory but none of these authors covered all of the cases.

(i) (X51) *sforz* 'thanks to'

For a start, we introduce (X51) *sforz* 'thanks to' – a borrowing connected to the Italian common noun *sforzo* 'effort' (Aquilina 1991: 1305). E. Serracino-Inglott (1984: 133) characterises *sforz* as noun and adverb. Stolz/Levkovych (2020: 218) are the first to acknowledge the Π-status of this expression by way of admitting it as member of the original version of BLOMP. In (1.154), we encounter *sforz* 'effort' as Italian loan-noun in Maltese.

(1.154) [Baldacchino 1999: 19]

Jiena għamilt ***sforzi*** *kbar biex nipprova nsib*
1SG make:1SG.PFV **effort**:PL big.PL to 1.IPFV:try 1.IPFV:find
kopja ta' l-Athletic News.
copy of DEF-Athletic News
'I made great **effort**s to find a copy of the Athletic News.'

In (1.155), the homophonous Π *sforz* 'thanks to' is the head of a PP whose complement is a complex NP with an apposition (containing the product name) and PP-attribute.

(1.155) [Korpus Malti 3.0; literature20]

Kellu tbissima brikkuna fuq wiċċu ***sforz***
have:3SG.M smile scoundrel:F on face:3SG.M **[thanks_to**
vodka red bull f' idu]$_{PP}$
DEF-vodka Red Bull in hand:3SG.M]
'He had a mean smile on his face [**thanks to** the vodka Red Bull in his hand].'

In this context, *sforz* can neither undergo pluralisation nor take an attribute so that it behaves differently from the common noun *sforz* in (1.154) in terms of morphosyntax. Semantically, the Π *sforz* does not seem to automatically invoke the notion of (physical) effort. What the Π *sforz* expresses is a (perhaps qualified) causal relation the stimulus of which is represented by the complement NP whereas the entity or event which is the undergoer of the effect is mentioned in the predicate. In Italian, no equivalent use of *sforzo* 'effort' or the corresponding transitive verb *sforzare* 'force' could be identified.

(ii) (X44) *permezz* 'by means of'

The next candidate on our list of Italianisms is (X44) *permezz* 'by means of'. To the mind of E. Serracino-Inglott (1981: 55), we are dealing with an adverb. Aquilina (1991: 1048) has the following lexicon entry for *per* which also mentions this expression:

> It[alian] prep[osition] 'per' occurring in certain M[altese] phrases. Ex[amples]: ~ *eżempju*, for example; [. . .] ~ *każu*, for instance; ~ *eċċellenza*, par excellence; [. . .] ~ *mezz*, by means of; ~ *sald*/ [. . .] ~ *saldu*, adv[erb] full settlement (payment of a bill or money due); and in other occasional phrases.

The examples given in this quote exclusively consist of borrowings from Italian. They thus resemble the chunks of foreign language material discussed at the beginning of this section. Πs which are solely involved in chunks of this kind are excluded from BLOMP. However, in the case of *permezz* 'by means of' the situation is different. There is the Maltese noun *mezz* 'means' borrowed from Italian (*mezzo* 'means') (Aquilina 1991: 816). In (1.156), the common noun *mezz* takes a prenominal (quantifying) attribute and the NP is overtly marked for definiteness. These are typically nouny properties which never apply to the Π *permezz*.

(1.156) [Attard 1999: 71]

l-uniku	***mezz***	*biex*	*jasal*	*l-Awstralja*	*kien*	*bil-baħar*
DEF-only	**means**	to	3.IPFV:arrive	DEF-Australia	be	with:DEF-sea

'The only **way** to reach Australia was by sea [. . .].'

Maltese *permezz* corresponds closely to the Italian multi-word construction *per mezzo di* 'by means of' with Maltese *ta'* 'of' taking over the role of Italian *di* 'of'. As will be discussed below,[129] we do not consider *ta'* to form part of the Π itself. According to our analysis, *permezz* belongs to those Maltese Πs whose complement normally comes in the shape of a PP (Stolz/Levkovych 2020: 226–227). Note that this is a matter of government and not a case of a compound Π. It is also worth noting that the distribution of *permezz* is not restricted to combinations with other lexical Italianisms. In (1.157)–(1.158), we illustrate the prepositional use of *permezz*. Example (1.158) is especially telling since the PP headed by *permezz* takes a complement NP whose nucleus is the definite common noun *mezz* in the plural which is further modified by a postnominal adjective.

129 Cf. Section 4.1.2.2.1.1 (ii).

(1.157) [Camilleri 2013: 285]

*[**Permezz** [**ta'** walkie-talkie]$_{PP}$]$_{PP}$ ordna lill-erba'*
[**by_means_of** [**of** walkie-talkie]] order to:DEF-four
stakkamenti [. . .] idawru l-inħawi
detachment:PL 3.IPFV:surround:PL DEF-side.PL
'[**Via** a walkie-talkie] he ordered the four detachments to surround the neighbourhood.'

(1.158) [Korpus Malti 3.0; academic1][130]

*[**Permezz** tal-mezzi elettroniċi]$_{PP}$ l-istudenti*
[**by_means_of** of:DEF-means:PL electronic:PL] DEF-student:PL
{j}ingħataw l-opportunità li {j}iktbu b' mod
{3.IPFV}RFL:give:PL DEF-opportunity that {3.IPFV}write:PL with manner
ġurnalistiku
journalistic
'[**By means of** the electronic tools] the students are given the opportunity to write journalistically [. . .].'

The common nouns and the Π behave pronouncedly different in the morphosyntactic domain. Since *per* on its own does not have the status of a Π in Maltese, *permezz* cannot be interpreted as a semantically transparent construction. We therefore register *permezz* as instrumental Π in BLOMP – old and new version. Only Aquilina (1959: 322) classifies this expression similarly.

(iii) (X56) *versu* 'towards'

Prior to the publication of the original version of BLOMP, Aquilina (1959: 322) was also the only grammarian of Maltese to consider (X56) *versu* 'towards' to be a Π. Aquilina assumes a second meaning 'for' for this Π. E. Serracino-Inglott (1989: 9–10) also grants Π-status to *versu*. The corresponding Italian Π *verso* 'in the direction of' also displays concrete spatial meanings which seem to be absent from the semantics of Maltese *versu*.[131] Aquilina (1991: 1505) characterises *versu* stylistically as belonging to the rhetorical genre. A typical example of the Π *versu* in contemporary Maltese is given in (1.159).

130 In this example, the prefixes of the 3rd person imperfective had to be restored (in curly brackets) since the corresponding *j-* was erroneously absent from the original source.
131 Cf. Sections 5.1.2.2.6 and 5.1.3.11.

(1.159) [Korpus Malti 3.0; news100659]

dan	*mhux*	*atteġġjament*	*li*	*x-xufier*	*kellu*
DEM:M:PROX	NEG	attitude	that	DEF-driver	have

jġib	*ruħu*	*bih*	***[versu***	*l-pubbliku]$_{PP}$*
3.IPFV:behave	soul:3SG.M	with:3SG.M	**[towards**	DEF-public]

'[. . .] this is not the way the driver should have behaved [**towards** the public].'

There is nothing that impedes the inclusion of *versu* in the Maltese Π-inventory. Accordingly, it is a full-blown member of all versions of BLOMP.

(iv) (X1) *apparti* 'apart from'

The final four cases to be discussed in this section are interconnected in not immediately obvious ways. The links between the candidates will become visible in the course of the subsequent discussion. As to its phonological shape, *apparti* can be connected to Italian *a parte* 'apart' but its function and syntax are better explained with reference to English *apart from* (Aquilina 1987: 35–36). The initial /a/ of the segmental chain of (X1) *apparti* 'apart from' is inseparable from the expression and has no morphological status of its own. For E. Serracino-Inglott (1975a: 41), this is an adverb. Example (1.160) illustrates the use of (X1) *apparti* 'apart from'. In the original, this sentence contains an extended enumeration of objects in Reggie's room. To avoid rendering the reading too difficult, we refrain from representing the coordination of PPs as instances of EQUI-P-deletion.

(1.160) [Bartolo/Vella 2009: 21]

[Apparti	*pjanti*	*u*	*xitel*	*u*	*qsari*	*u*	
[apart_from	plant:PL	and	plant.PL	and	flower_pot:PL	and	

ħxejjex]$_{PP}$	*fil-kamra*	*Reggie*	*kellu*	*karti*	*u*
vegetable.COLL]	in:DEF-room	Reggie	have.PFV:3SG.M	paper:PL	and

kotba	*u*	*gazzetti*
book.PL	and	magazine:PL

'[**Apart from** plants and plants and flower-pots and vegetables] Reggie had papers and books and magazines [. . .] in the room.'

The Π-status of *apparti* 'except' was assumed first by Stolz/Levkovych (2020: 217–218). The reason for granting *apparti* membership in BLOMP is analogy. Since *minbarra* 'except' (sometimes as variant of *barra*) is accepted by several linguists (Aquilina 1991; Procházka 1993; Borg/Azzopardi-Alexander 1997; Saari 2003) as fully-fledged Π, the same classification must be applied in the case of *apparti* because there are no principled differences between the two candidates in the morphosyntactic or semantic domain. Both have similar functions and both take complement NPs. Their semantic closeness makes it possible – in certain contexts – to replace the one with the other. Thus, if *minbarra* is a Π, then *apparti* is a Π, too.

(v) (X43) *oltri* 'except'

Superficially, this line of argumentation should also hold for (X43) *oltri* 'except'. However, the supposed English translation leads us astray. The Italian *oltre* 'beyond' is involved in *oltre a* 'except' which is the most probable source for the Maltese Π. E. Serracino-Inglott (1979: 339) mentions *oltri* as adverb. Aquilina (1991: 1014) characterises *oltri* 'besides' as Π "of limited use amongst the educated classes". Apart from Stolz/Levkovych (2020), no other author has included *oltri* in the Maltese Π-inventory. E. Serracino-Inglott (1979: 339) classifies *oltri* as invariable adverb. This exclusion seems to be justified since if we search for *oltri* in the Korpus Malti 3.0, the turnout is nil. However, if we replace *oltri* with the Italian form *oltre* we get 2,984 matches. Amongst these matches we find (1.161). Note that Sandra Abela[132] assumes that *oltri* and *oltre* are dialectal varaiants of each other.

(1.161) [Korpus Malti 3.0; european5897]

għandha	*spejjeż*	*li*	*jmorru*	*[**oltre***	*l-bżonnijiet*
have.IPFV:3SG.F	expense.PL	that	3.IPFV:go:PL	[**except**	DEF-need:PL
tal-fabbrika]$_{PP}$					
of:DEF-factory]					

'[. . .] she has expenses which go [**beyond** the needs of the factory].'

The vast majority of the hits found for *oltre* invite a translation into English as either 'besides' or 'beyond'; in contrast, the meaning 'except' in the sense of excluding the participant represented by the complement NP cannot be pinpointed in the Korpus Malti 3.0.[133] On account of the absence of the word form *oltri* and the supposed meaning 'except' from the electronic resource, it cedes its former place in the original version of BLOMP to *oltre* 'beyond' in BLOMP 2.0. Since *oltre* abounds in texts produced in the context of the E.U. administration, it is conceivable that an erstwhile Π *oltri* 'except' has been superseded by a secondary loan from Italian, namely *oltre* 'beyond'. The recency of the latter is reflected by the failure of the mid-high front vowel /e/ to be raised to /i/ in final position – a process to which older borrowings were regularly subjected.

Even if *oltri* has to be cancelled from BLOMP, there still remains a sizable group of Πs which belong together semantically in the sense that, one way or another,

132 Personal communication, 11 February, 2018.
133 Cf. Section 5.1.3.2.

they serve the purpose of expressing exclusion, viz. (X1) *apparti* 'apart from', (X8) *bla* 'without', (X20) *għajr* 'except', (X24) *ħlief* 'except', (X34) *minbarra* 'except', (X35) *mingħajr* 'without', and marginally also (X43) *oltre* 'beyond' if we take account of the reading 'besides'. This long list of Πs which are employed for excluding a participant from a situation raises the question why the opposite function, i.e. that of including a participant in a situation, is not represented by a Π in BLOMP. We take up this issue by way of discussing the possibility of promoting *inkluż* 'including' to the rank of Π.

(vi) *inkluż* 'including'

In previous accounts, *inkluż* 'including' is never mentioned as a Π-candidate. It is easy to understand why it has never occurred to anyone that we might have another Π right before our eyes. Borg/Azzopardi-Alexander (1997: 154) classify *inkluż* as adjective. E. Serracino-Inglott (1977: 318) and M. Serracino-Inglott (2000: 244) register *inkluż* as adjective and passive participle of the verb *inkluda* 'include'. It is sensitive to gender and number of the head noun it modifies like its Italian equivalent *incluso* – the passive participle of *includere* 'include'. In Maltese, there is a paradigm of three forms, namely M.SG *inkluż* ≠ F.SG *inkluża* ≠ PL *inklużi* (with Italian offering the four forms M.SG *incluso* ≠ F.SG *inclusa* ≠ M.PL *inclusi* ≠ F.PL *incluse*). This morphological property is illustrated in (1.162)–(1.164).

(1.162) [Camilleri 2013: 75–76]

iżjed	*minn*	*nofs*	*it-team*	*kien*	*prodott*	*tan-nursery*	***inkluż***
more	from	half	DEF-team	be	product	of:DEF-nursery	**including**
Mizzi	*stess*						
Mizzi	self						

'[. . .] more than half of the team were the product of the nursery **including** Mizzi himself.'

(1.163) [Attard 1999: 35]

Bħalma	*soltu*	*jiġri*	*fl-ambjent*	*Malti*	*kollox*
as	usual	3.IPFV:run	in:DEF-environment	Maltese	everything

iċċappas	*mal-kwistjoni*	*tal-lingwa*	***inkluża***
be_stained	with-DEF:question	of:DEF-language	**including:F**
l-emigrazzjoni			
DEF-emigration			

'As it usually happens in the Maltese context everything was tinged with the language question, **including** the emigration.'

(1.164) [Camilleri 2013: 189]
filmati bijografiċi dwar il-vittma ***(inklużi***
film_insert:PL biographic:PL about DEF-victim **(including:PL**
intervisti ma' xi tnejn minn sħabu)
interview:PL with some two from pal.PL:3SG.M
'[. . .] biographic film inserts about the victim (**including** interviews with some of his pals) [. . .].'

In (1.162), *inkluż* is followed by the proper name *Mizzi* which refers to a male human being. Thus, the form of the M.SG is used. Similarly, the form *inkluża* is required by the presence of the feminine common noun *emigrazzjoni* 'emigration' in (1.163). The pluralised common noun *intervisti* 'interviews' triggers the PL *inklużi* in (1.164). Like the equivalent Italian adjectives/participles, the gender-number-sensitive Maltese word forms may also occur in postnominal position as shown in (1.165)–(1.167). This is the preferred position of attributive adjectives in Maltese (Borg 1996: 15).

(1.165) [Korpus Malti 3.0; news147329]
F' dik il-ġurnata jiltaqgħu tliet vacant posts
in DEM:F:DIST DEF-day 3.IPFV:RFL:meet:PL three vacant post:PL
b' post ta' day nurse ***inkluż***
with post of day_nurse **including**
'On that day, they come across three vacant posts, with a post for a day nurse **included.**'

(1.166) [Korpus Malti 3.0; parl8264]
Ġie diskuss jekk fil-pajjiżi ż-żgħar Malta
come discuss.PART if in:DEF-country:PL DEF-small.PL Malta
inkluża *għandux ikun hemm xi forma*
including:F have.IPFV:3SG.M:NEG 3.IPFV:be.FUT there some form
ta' strateġija differenti mill-pajjiżi kbar
of strategy different from:DEF-country:PL big.PL
'It was discussed whether in small countries – Malta **included** – there should be a kind of strategy different from the big countries.'

(1.167) [Korpus Malti 3.0; news210237]
appella sabiex jintemmu l-vjolenzi minn kull naħa,
appeal so_that 3.IPFV:RFL:finish:PL DEF-violence:PL from each side
bil-forzi tal-Gvern u anke tal-Oppożizzjoni
with-force:PL of:DEF-government and also of:DEF-opposition

inklużi
including:PL
'[. . .] he requested that the violent acts from all sides be finished, with the forces of the government and those of the opposition **included**.'

In (1.165), the postnominal *inkluż* agrees with the common noun *post* 'job, position' which is masculine singular. Since the place name *Malta* is feminine the modifying adjective following it can only take the shape of *inkluża* in (1.166). Similarly, the noun *forzi* 'forces' is pluralised in (1.167) and thus the adjective on the right edge of the phrase has to take the form of *inklużi*.

These agreement patterns speak against considering *inkluż* to be a Π. On the very same grounds, even a postpositional analysis like that Saari (2007) suggested for *ilu* is ruled out. We subscribe to this argument for properly adjectival *inkluż*. There is however ample evidence of *inkluż* defying the usual morphosyntactic rules to which well-behaved adjectives obey in Maltese. We are alluding to prenominal invariable *inkluż*.

In (1.162)–(1.167), *inkluż* behaves like an adjective and not like a Π. In (1.168)–(1.169), it is exactly the other way around, i.e. *inkluż* fails to display typical adjectival behaviour but fulfils the expectations we have of a Π. EQUI-P-deletion under coordination is not taken account of.

(1.168) [Camilleri 2013: 83]
bl-email baghtuha lill-kuntatti kollha taghhom
with:DEF-email send:3PL.PFV:3SG.F to:DEF:contact:PL all:3SG.F of:3PL
*[**inkluż** ġurnali u stazzjonijiet]$_{PP}$*
[**including** journal:PL and station:PL]
'[. . .] with the email they sent to all of their contacts [**including** journals and stations].'

(1.169) [Abela 2016: 270]
Il-bniet qalu lill-mamà u lill-papà kulma kien
DEF-girl.PL say:3PL.PFV to:DEF-mum and to:DEF-dad all_that be
*ġara [**inkluż** id-darba meta marsupjal Awstraljan tela'*
happen [**including** DEF-time when marsupial Australian climb_up
mit-tieqa taghhom billejl]$_{PP}$
from:DEF-window of:3PL at_night]
'The girls told mum and dad everything that had happened [**including** the episode when an Australian marsupial climbed up their window at night].'

In these sentences, *inkluż* remains morphologically invariable although it precedes pluralised nouns in (1.168) and a noun in the feminine singular in (1.169). According to the agreement patterns of Maltese adjectives, we would have expected to find *inklużi* in (1.168) and *inkluża* in (1.169). Except invariable adjectives in *-li* like *oriġinali* 'original' which are insensitive to gender and number distinctions, adjectival gender-number agreement is never suspended in Maltese. This means that the above behaviour of *inkluż* violates the morphosyntactic rules which are valid in the domain of adjectives.

It is important to understand that we are not dealing with an occasional phenomenon. Disagreement of prenominal *inkluż* with the NP(s) it has scope over is frequently attested in Maltese. To prove this empirically, we check the first 100 attestations of *inkluż* in the Korpus Malti 3.0. Of these first 100 cases, seven involve *inkluż* as predicative adjective. There are also 17 instances of *inkluż* combining with syntactic units other than NPs (such as PPs, subordinate clauses, adverbs, adjectives). Seven cases are unclear to us and thus have to be omitted from the discussion. There remain 68 cases in which *inkluż* combines with NPs four of which involve postnominal *inkluż*. The postnominal *inkluż* always agrees correctly in gender and number with the preceding NP which is masculine singular. If, however, *inkluż* occupies the slot to the left of the NP(s), agreement is by no means the rule. Of the 64 instances of prenominal *inkluż*, only 22 (= 34%) involve a noun which is masculine singular. With 42 tokens, the vast majority or two thirds (= 66%) of all cases of *inkluż* fail to agree in gender and/or number. On the basis of the first 100 matches found in the Korpus Malti 3.0, we hypothesise that disagreement is normal (though not mandatory) for *inkluż*. This preference casts serious doubts on the sweeping classification of all cases of *inkluż* as adjectival. The implicit assumption that not every *inkluż* is an adjective is tantamount to assuming a split into *inkluż*$_1$ ≠ *inkluż*$_2$ which differ from each other in terms of their word-class membership. We propose that *inkluż*$_2$ is a Π. This hypothesis calls for being scrutinised more closely.

Table 1.9 covers the 34 cases found in the Korpus Malti 3.0 which give evidence of disagreement in number between *inkluż* and the following NP(s). The examples are ordered top-down according to the increasing morphosyntactic complexity of the NPs. Grey shading singles out an example that will be discussed separately below.

Table 1.9: Instances of number disagreement between *inkluż* and NP.

Source	example + translation
[culture2399]	***inkluż*** *familji* '**including** families'
[culture2771]	***inkluż*** *ritratti* '**including** photos'
[culture3089, 3090 + 3091]	***inkluż*** *tfal* '**including** children'
[culture3159]	***inkluż*** *studenti* '**including** students'
[culture3791]	***inkluż*** *l-infermieri* '**including** the nurses'
[academic508]	***inkluż*** *il-partiċelli* '**including** the particles'
[culture1104]	***inkluż*** *il-klieb* '**including** the dogs'
[culture979]	***inkluż*** *tfal żgħar* '**including** little children'
[culture1012]	***inkluż*** *ħames baned* '**including** five bands'
[culture303]	***inkluż*** *20 Għawdxi u tminn nisa* '**including** 20 Gozitans and eight women'
[culture792]	***inkluż*** *siti storiċi, persuni u xeni naturali* '**including** historical sites, persons, and natural scenes'
[culture3823]	***inkluż*** *operazzjonijiet, testijiet, radjoloġija, servizz ta' out-patient u servizzi oħra* '**including** operations, tests, radiology, out-patient service, and other services'
[culture519]	***inkluż*** *dawk spritiwali* '**including** those spiritual ones'
[culture585]	***inkluż*** *dawk mhux eletturi* '**including** those non-electors'
[culture 1092]	***inkluż*** *dawk morda u anzjani* '**including** the sick and elderly'
[culture584]	***inkluż*** *dawk akkademiċi u multi-dixxiplinari* '**including** those academic and multi-disciplinary ones'
[culture4000]	***inkluż*** *dawk morda b' mard serju* '**including** those suffering from a serious illness'
[culture2157]	***inkluż*** *dawk li ma jafuhiex personalment* '**including** those they do not know personally'
[culture24]	***inkluż*** *dawk li kienu parti mill-ġurija ta' Malta* '**including** those who were part of the jury of Malta'
[culture3859]	***inkluż*** *dawk li jaħdmu fis-Sala Sant' Anna* '**including** those who worked in Sant' Anna's Ward'
[culture3753]	***inkluż*** *dawk li jaħdmu fl-Isptar Karen Grech* '**including** those who worked in the Karen Grech Hospital'
[culture3960]	***inkluż*** *dawk li jbatu b' kundizzjonijiet respiratorji* '**including** those who suffer from respiratory conditions'
[culture3254 + 3430]	***inkluż*** *żjarat fil- Forti* '**including** visits in the Fort'
[culture3238]	***inkluż*** *masterclasses u kunċerti* '**including** masterclasses and concerts'
[academic242]	***inkluż*** *ittri anonimi u allegazzjonijiet ta' frodi* '**including** anonymous letters and allegations of fraud'
[culture2515]	***inkluż*** *l-abbiltà, ir-razza u l-rientazzjoni sesswali* '**including** the ability, the race, and the sexual orientation'

Table 1.9 (continued)

Source	example + translation
[european10005]	***inkluż*** *logo, tikketta, stiker, opuskuli, istruzzjonijiet għall-użu jew dokument ta' garanzija li ikun iġorr dak is-simbolu* '**including** a logo, a label, a sticker, pamphlets, instructions of use or a document of guarantee which could bear that symbol'
[culture3738]	***inkluż*** *l-għalliema tal-primarja u s-sekondarja* '**including** the teachers of the primary and secondary school'
[european100]	***inkluż*** *il-proċeduri għall-ġbir ta' dawn id- dażji* '**including** the procedure for collecting these customs'
[culture1679]	***inkluż*** *il-pizez b' għaġina magħmula minn dqiq bla glutina* '**including** the pizzas with dough made from gluten-free flour'
[culture415]	***inkluż*** *is-seba' żjarat li għamel Gaddafi f' Malta* '**including** Ghaddafi's seven visits in Malta'

Table 1.9 reveals that *inkluż* remains morphologically invariable no matter whether the NP is a singleton indefinite noun in the plural (like *tfal* 'children') or an enumeration of several definite and pluralised NPs (like *ittri anonimi u allegazzjonijiet ta' frodi* 'anonymous letters and allegations of fraud'). Pluralisation should be triggered either by NPs overtly marked for plural as well as by two or more coordinated NPs independent of the number values of the individual conjuncts. Under these circumstances, proper Maltese adjectives in prenominal position would inflect for plural although Borg (1996: 14) characterises the prenominal position of adjectives as "generally marked". A rare example of a pluralised prenominal adjective is given in (1.170).

(1.170) [Korpus Malti 3.0; news158078]

jirnexxielu	*joħroġ*	*il-potenzjal*	*maqful*
3.IPFV:succeed:IO:3SG.M	3.IPFV:bring_out	DEF-potential	PART:shut_in

f'	*dawn*	*iċ-**ċkejknin***	*gżejjer*
in	DEM:PL:PROX	DEF-**small:PL**	island.PL

'[. . .] he succeeds in bringing out the potential that is shut in on these **small** islands [. . .].'

Table 1.9 contains a particularly intriguing case – marked out by grey shading. This example is interesting because it attests to a potential morpho-syntactic conflict which might have a bearing on the (dis)agreement of *inkluż*. In (1.171), the immediately preceding syntactic context of *inkluż* is revealed.

(1.171) [Korpus Malti 3.0; culture303]
kien hemm 91 rekluti ġodda **[*inkluż*** *20 Għawdxi u*
be there 91 recruit:PL new.PL **[including** 20 Gozitan and
tmien nisa]PP
eight woman.PL]
'[. . .] there were 91 new recruits **[including** 20 Gozitans and eight women].'

This example is special in the sense that it violates the rules of Maltese number marking in the part of the clause preceding the square brackets. With a complex cardinal numeral above ten we excpect the quantified NP to be formally singular (Borg 1996: 13), so that the pluralised NP *rekluti ġodda* 'new recruits' should be replaced by *rekluta ġdid* 'new recruit' after 91 (= *wieħed u disgħin* 'ninety-one'). Within the bracketed part of (1.171) however, the rules of number marking are strictly obeyed to. The cardinal numeral 20 (= *għoxrin* 'twenty') requires that the quantified noun *Għawdxi* 'Gozitan' remains in the singular whereas *tmien* (misspelled as *tminn* in the source) – the prenominal form of *tmienja* 'eight' – correctly takes the quantified noun in the plural, namely *nisa* 'women' (the suppletive plural of *mara* 'woman'). Since two coordinated NPs are in the scope of *inkluż* the conditions are favourable for plural agreement on *inkluż*. However, *inkluż* does not behave accordingly. One might want to argue that *inkluż* is sensitive to number only locally in the sense that it agrees preferably with the NP on its immediate right which is formally singular. However, the cardinal numeral shows that the NP is semantically plural. Moreover, example (1.172) proves that the formal singular of the NP does not automatically trigger the singular of *inkluż*.

(1.172) [Korpus Malti 3.0; academic363]
il-Premier League kien l-aktar kampjonat rappreżentat
DEF-Premier_League be DEF-more championship represent:PART
b' aktar minn tmenin plejer fil-kompetizzjoni ***inklużi***
with more from eighty player in:DEF-competition **including:PL**
wkoll 21 tat-23 plejer fl-iskwadra Ingliża
also 21 of:DEF-23 player in:DEF-team English:F
'[. . .] the Premier League was the biggest championship represented by more than eighty players in the competition, **including** also 21 of the 23 players of the Engish team.'

Here we find semantic number agreement because the formal singular *plejer* 'player' is quantified by the partitive construction *21 tat-23* '21 of the 23' and thus clearly refers to a group of people and not to a single individual. The singular of the quantified noun is required by the cardinal numerals *21* (= *wieħed u għoxrin* 'twenty-one') and *23* (= *tlieta u għoxrin* 'twenty-three') in conformity to the above rule. In this case at least *inkluż* displays adjectival behaviour. Whatever the explanation is for the differential behaviour of *inkluż* in (1.171) and (1.172) it is clear that the item under review does not always conform to the morpho-syntactic patterns associated with adjectives in Maltese.

The violation of the prerequisites of number agreement is not the only phenomenon that supports our idea that *inkluż*$_2$ is not an adjective. Table 1.10 lists the eight cases of failed gender agreement we have identified amongst the first 100 attestations of *inkluż* in the Korpus Malti 3.0.

Table 1.10: Instances of gender disagreement between *inkluż* and NP.

Source	example + translation
[culture432]	***inkluż*** *kampanja ta' informazzjoni* '**including** an information campaign'
[academic396]	***inkluż*** *il-pjaneta Art* '**including** the planet Earth'
[culture3034]	***inkluż*** *il-promozzjoni tiegħu* '**including** his promotion'
[culture3231 + 4048]	***inkluż*** *l- istazzjon tal-Ħamrun* '**including** the station of Ħamrun'
[culture3826]	***inkluż*** *sospensjoni mill-post tax-xogħol* '**including** the suspension from the job'
[academic328]	***inkluż*** *l-istamperija f' tal-Pietà* '**including** the printing press in Pietà'
[culture3276]	***inkluż*** *ir-riċerka li jkun għamel* '**including** the research he will conduct'

In all these cases, the formally masculine singular *inkluż* has scope over a single NP which has the number value singular and the gender value feminine. Meaning: if *inkluż* was an adjective, it should be marked for feminine, too, and thus take the shape *inkluża*.

The examples presented in Tables 1.9–1.10 gain in significance if we complement them with further quantitative data. In Figure 1.9, we account for the frequency of the patterns which arise for the first 100 attestations in the Korpus Malti 3.0 of each of the three word forms, namely *inkluż*, *inkluża*, and *inklużi*. The seven unclear cases have been excluded from the calculations.

The disproportions are striking. Disagreement and the combination with syntactic units other than NPs are monopolies of *inkluż*. This means that wherever *inkluża* and *inklużi* are used the agreement is correct whereas this is untrue for *inkluż*. Postnominal attestations are infrequent for all three elements. These postnominal cases can be classified as instances of correct agreement without changing much of the general picture. What comes additionally to the fore is the relative scarcity of the predicative use of *inkluż* (7/92 = 8%) as compared to that of *inklużi* (29/96 = 30%) and *inkluża* (43/100 = 43%). In the copular sentences (1.173)–(1.175), the predicative adjective agrees with the relative NP in gender in number.

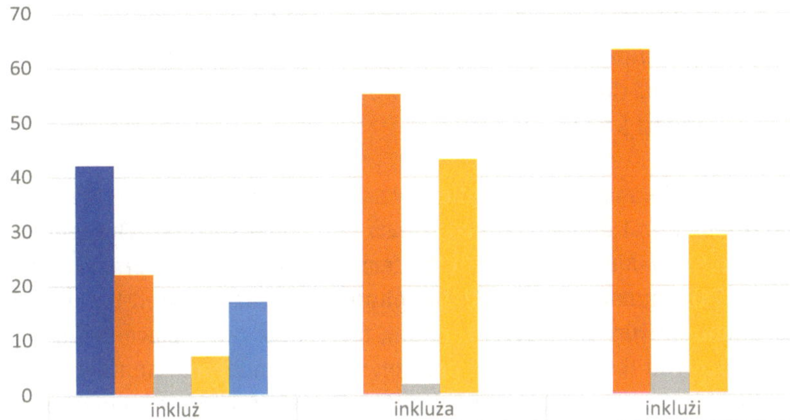

	inkluż	inkluża	inklużi
disagreement	42	0	0
agreement	22	55	63
postnominal	4	2	4
predicative	7	43	29
other	17	0	0

Figure 1.9: Shares of agreement with *inkluż/a/i.*

(1.173) [Korpus Malti 3.0; culture3355]
Fid-diskors *tal-baġit* *kien* ***inkluż*** *wkoll* *il-qasam*
in:DEF-discourse of:DEF-budget be **including** also DEF-field
kulturali
cultural
'In the speech on the budget, the cultural domain was also **included**.'

(1.174) [Korpus Malti 3.0; news112164]
l-integrazzjoni *tal-immigranti* *kienet* ***inkluża*** *f'*
DEF-integration of:DEF-immigrant:PL be:3SG.F.PFV **including:F** in
portfolio *ministerjali*
portfolio ministerial
'[. . .] the integration of the immigrants was **included** in a ministerial portfolio [. . .].'

(1.175) [Korpus Malti 3.0; news159730]
Dawn *kollha* *kienu* *miżuri* *li* *kienu*
DEM:PL:PROX all:3SG.F be:3PL.PFV measure:PL that be:3PL.PFV

inkluži *fir-rapport* *tal-Prof. Scicluna.*
including:PL DEF-report of:DEF-professor Scicluna
'All of these were measures that were **included** in professor Scicluna's report [. . .].'

In (1.173), the masculine singular form *inkluż* agrees with the NP *il-qasam kulturali* 'the cultural domain' which is also masculine singular. The feminine gender marking on *inkluża* fits in with the feminine gender of *integrazzjoni* 'integration' in (1.174). Similarly, in (1.175), the plural *inkluži* is required because the relativised noun *miżuri* 'measures' is plural, too. *Inkluż* can replace *inkluża* and *inkluži* exclusively in the prenominal position where gender and number distinctions on the item in question are thus neutralised. Outside this context however, no replacement of *inkluża* and *inkluži* with *inkluż* is grammatical. These facts further add to the special character of prenominal *inkluż*$_2$, meaning *inkluż*, *inkluża*, and *inkluži* cannot simply be lumped together. The latter two display more similarities among each other than the two have in common with *inkluż*.

On these grounds, we postulate the continuum in Figure 1.10. At the left edge, the criteria for Π-hood are met to a relatively high degree whereas at the opposite extreme those typical of adjectives prevail.

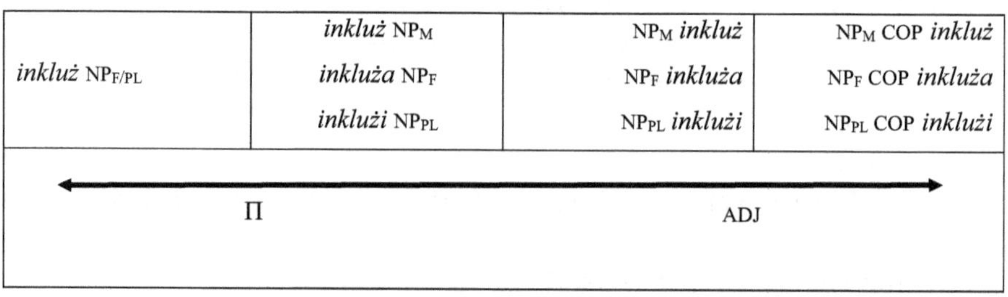

Figure 1.10: Continuum – *inkluż* between Π and adjective.

Inkluż comes closest to a genuine Π if it fills the slot to the left of a NP which is either feminine or plural. The Π-like properties are not as strong in the case of prenominal *inkluż*, *inkluża*, and *inkluži* if there is gender and number agreement with the NP. At the same time, from the point of view of linearisation, the prenominal position is atypical for adjectives. Moreover, *inkluż*, *inkluża*, and *inkluži* become more like adjectives if they follow their head-noun. The highest degree of adjectivehood is reached if the elements are used predicatively.

We are aware of the daring nature of what we conclude on the basis of the above facts. We assume the existence of the Π *inkluż*$_2$ 'including' which is a member of BLOMP 2.0. This Π *inkluż*$_2$ 'including' is morpho-syntactically distinct from the homophonous and synonymous *inkluż*$_1$ 'including' which belongs to a different word-class, namely that of adjectives where however it stands out because of its preference for being placed to the left of the modified noun.[134]

(vii) (X57) *viċin* 'near'

The last candidate we discuss in this section is (X57) *viċin* 'near' (< Italian *vicino* 'near'). It is a synonym of (X47) *qrib* 'near'.[135] Unlike its synonym, *viċin* 'near' can function as fully-fledged adjective with a paradigm of three word forms, viz. M.SG *viċin* ≠ F.SG *viċina* ≠ PL *viċini*. The adjectival use is illustrated in (1.176)–(1.178). Underlining is used to identify the controller in relation to which the different forms of *viċin* function as target (Corbett 2006: 35 and 40).

(1.176) [Attard 1999: 10]

Il-Gżejjer	*Jonji*	*huma*	*arċipelagu*	*Grieg*	***viċin***	*ħafna*
DEF-island.PL	Ionian:PL	3PL	archipelago	Greek	**near**	very
ta'	*l-Albanija*					
of	DEF-Albania					

'The Ioanian Islands are a Greek archipelago very **close** to Albania [. . .].'

(1.177) [Korpus Malti; academic140]

Veru	*li*	*l-Italja*	*hija*	***viċina***
true	that	DEF-Italy	3SG.F	**near:F**

'It is true that Italy is **close by**.'

(1.178) [Korpus Malti 3.0; european2041]

magħżul	*minn*	*fost*	*it-tliet*	*ċentri*	*l-aktar*	***viċini***
PART:choose	from	amongst	DEF-three	centre:PL	DEF-more	**near:PL**
tal-post						
of:DEF-place						

'[. . .] chosen from amongst the three centres **close**st to the place [. . .].'

In (1.176), the controller *arċipelagu* 'archipelago' is masculine singular and therefore the attributive adjective has the corresponding shape *viċin*. In (1.177), we find the predicative adjective in the feminine singular *viċina* because it agrees with the subject *l-Italja* 'Italy' which is feminine singular, too. The plural *viċini* is required because the adjective modifies the pluralised noun *ċentri* 'centres' in (1.178).

---

134 For reasons of econonmy, we drop the index when we mention Π *inkluż*$_2$ 'including' in the remainder of this study because the adjectival *inkluż*$_1$ 'including' will not concern us further.

135 Cf. Section 1.6.5 and the discussion in connection to examples (1.146)–(1.148).

Like in the previous case of *inkluż* 'including', we argue that there are two morphossyntactically distinct homophonous and synonymous expressions *viċin*$_1$ ≠ *viċin*$_2$ one of which is a bona fide adjective whereas the other a Π. The Π-status of *viċin* has been acknowledged by Aquilina (1959: 300), Fabri (1993: 186), and Borg/ Azzopardi-Alexander (1997: 157). In contrast, E. Serracino-Inglott (1989: 10) accepts only the classification as adjective. Gender-number disagreement is again a good indicator of this Π-status. In (1.179), *viċin* is formally masculine singular although – if it were an adjective – it should be inflected for feminine since the noun *belt* 'city' belongs to this gender. The participant represented by this noun is described as being located near the sea.

(1.179) [Attard 1999: 9]
Ġeneralment il-Maltin kienu jippreferu belt li
generally DEF-Maltese:PL be:3PL.PFV 3.IPFV:prefer:PL city that
*kienet ukoll port jew għallinqas [**viċin** il-baħar]*$_{PP}$
be:3SG.F.PFV also port or at_least [**near** DEF-sea]
'The Maltese generally preferred a city that also was a port or at least [**near** the sea].'

The failure of *viċin* to agree with *belt* 'city' in gender is paralleled by the absence of number agreement in (1.180).

(1.180) [Korpus Malti 3.0; academic376]
L-Assemblea Nazzjonali tal-Irlanda ta' Fuq għandha
DEF-assembly national of:DEF-Ireland of above have.IPFV:3SG.F
*poteri [**viċin** dawk li huma stabiliti*
power:PL [**near** DEM:PL:DIST that 3PL establish:PART:PL
fl-Iskozzja]$_{PP}$
in:DEF-Scotland]
'The National Assembly of Northern Ireland has powers [**similar to** those which are established in Scotland].'

All syntactic units which are anaphorically[136] related to the pluralised noun *poteri* 'powers' are overtly marked for plural number, namely the demonstrative *dawk* 'those', the pronominal *ersatz*-copula *huma* 'they', and the participle *stabiliti* 'established'. If *viċin* was an adjective, it would have joined in by way of taking plural

136 The relation to *poteri* 'powers' is not genuinely anaphoric but involves an elliptical DP **dawk (il-poteri)* 'those (powers)' whose bracketed NP is omitted.

inflection to yield *viċini*. In the absence of the plural marker, *viċin* does not meet the criteria for being classified as adjective. In analogy to our above line of argumentation in connection to *inkluż*, we consider those instances of *viċin* which seem to violate the requirements of agreement to realise the Π *viċin*$_2$ which is morpho-syntactically distinct from the homophonous and synonymous adjective *viċin*$_1$.[137]

(viii) *a skapitu* 'at the expense'

Another Maltese Π with an Italian background is *a skapitu* 'at the expense'. This candidate is special insofar as it has never before been mentioned by anyone in the literature dealing with Maltese Πs. Aquilina (1991: 1327) vaguely characterises *skapitu* as a "word" which (only) occurs in the construction *a skapitu ta'* 'at the expense of, to the detriment of'. It originates from Italian *a scapito di* 'at the expense of'. Its syntactic behaviour is illustrated in (1.181).

(1.181) [Korpus Malti 3.0; news117641]

jimtlew	*il-vakanzi*	*eżistenti*	*biex*	*ebda*	*servizz*
3.IPFV:RFL:fill:PL	DEF-vacancy:PL	existing	so_that	no	service

ma	*jsofri*	**[*a skapitu***	***ta'***	*servizz*	*ieħor*]$_{PP}$
NEG	3SG.IPFV:suffer	**[at_the_expense**	**of**	service	other]

'[...] the existing vacancies are being filled so that no service suffers [**at the expense of** another service].'

This candidate poses a number of problems. First and foremost, Aquilina (1991: 1327) assumes that it can be inflected for number. However, no traces of the supposed form *skapiti* could be found in the Korpus Malti 3.0. That is why we consider the Π to be indeclinable. Moreover, the initial vowel /a/ (which is often represented as <à> in writing) seems to be unstable. Apart from merely allographic versions like *għas-skapitu*, there are also instances of *bi skapitu, għal skapitu*, and bare *skapitu* without preceding segments. As far as we can judge, this formal variability does not trigger any semantic differences. Occasionally, *skapitu* seems to retain nominal properties insofar as it can take attributes. These aspects notwithstanding, we admit this candidate to BLOMP 2.0 where it is represented by its most frequently attested alloform as (Π2) *a skapitu* 'at the expense of'. The blank which orthographically gives the impression that we are dealing with a two-word syntagm can be ignored. As in the above case of *a favur* 'in favour of', we assume that *a skapitu* constitutes a unit which disallows for intercalations of any kind between its component parts and thus counts as monomorphemic. Throughout this study, we will

137 In analogy to our practice in the case of Π *inkluż*$_2$ 'including', the index on adjectival Π *viċin*$_2$ is not used in the descriptive parts of this study because there is no further mention of *viċin*$_1$.

have occasion repeatedly to talk about this Π because it is a kind of notorious troublemaker, in a manner of speaking, very often disobeying the rules to which the bulk of the Maltese Πs are subjected. Its Π-status might be put in question but in contrast to other cases to be introduced below (Π2) *a skapitu* 'at the expense of' is not a phantom since it is actually attested in the Korpus Malti 3.0.

The foregoing paragraphs have shown that it is by no means always a straightforward business to prove the Π-status of a given candidate. Some of the cases are largely unproblematic whereas other interpretations leave a margin of doubt. The problems increase when we look at the last quartet of candidates in 1.6.7.

1.6.7 Rarities

To guarantee that we are not talking about too many phantom-Πs, we have stipulated the following criterion. A candidate is accepted as member of BLOMP if it fulfils the requirements of (1.1) and is attested at least once in a sentential context in one of our sources of the 20th or 21st century – be they primary or secondary. Except *oltri*, the candidates we have discussed so far are attested frequently enough to pass the test successfully. There are however, four cases which pose problems since convincing pieces of proof of their Π-status are very hard to come by. These problems can be solved for each case individually.

(i) (X27) *kif* 'as'

The first case to be discussed in this section is (X27) *kif* 'as'. It was touched upon already in Section 1.4.2.1 with reference to Procházka (1993). This candidate is classified as adverb, noun, and Π by Aquilina (1987: 651) whereas E. Serracino-Inglott (1978: 108) only uses the label adverb. Example (1.182) suggests that *kif* can also have the function of a temporal conjunction as mentioned also by Aquilina (1987: 652).

(1.182) [Casha 2008: 157]

Kif	*daħlu*	*poġġew*	*madwar*	*mejda*
as	enter:3PL.PFV	sit_down:3PL.PFV	around	table

'**As soon as** they entered, they sat down at a table [. . .].'

We have been unable to find any evidence of the following construction in the Korpus Malti 3.0 or any other specimen of Maltese prose. According to Aquilina (1987: 652), *kif* may take person inflection when the meaning 'the equal of X' (= 'like X') is intended. Similarly, E. Serracino-Inglott (1978: 108) mentions inflected forms like *kifek* 'the equal of you'. Fenech (1984: 154) has the example *m'hawnx kifu* 'there is no one like him'. We reproduce one of Aquilina's own examples as (1.183) which are identical to those given in *Il-Miklem Malti*.

(1.183) (Aquilina 1987: 652)
Min *hu* ***[kifna]***$_{PP}$ *fid-dinja?*
who 3SG [**as**:1PL] in:DEF-world
'Who is our equal [= **like** us] in the world?' [O.T.]

Semantically, *kif* resembles (X4) *bħal* 'like' in this example. According to a philologically versed native speaker, this Π-usage of *kif* is absolutely unknown to our informant.[138] To her mind, *kif* can only be used adverbially or as conjunction. It is very likely that the examples given in Aquilina (1987) are anachronistic. They have become obsolete by now and are no longer familiar even to educated speakers of Maltese. Moreover, there is no evidence of a PP involving *kif* and a lexical complement NP. Our above self-imposed rule however, is not violated because (1.182) provides the required sentential example. We therefore keep *kif* 'as' as Π also for BLOMP 2.0 knowing that synchronically the Π-status is doubtful. As the discussion throughout Part B will show, *kif* 'as' is a good choice as a potential victim of a future revision of BLOMP 2.0.

(ii) (X42) *mnejn* 'from where'
Similarly, the case of (X42) *mnejn* 'from where' hinges on the evidence given in Aquilina (1991: 847) who explicitly terms this candidate a Π. There is no evidence of the spatial interrogative of Source being used as a Π anywhere else in our database. Like with *kif*, Aquilina only mentions instances of *mnejn* in combination with pronominal complements as in (1.184).

(1.184) (Aquilina 1991: 847)
titħarrekx ***[mnejni]***$_{PP}$
2.IPFV:move.CAUS:NEG [**from_where**:1SG]
'Do not move [**from near** me].' [O.T.]

According to Sandra Abela,[139] the use of *mnejn* in cases like (1.183) is acceptable but hardly known to everybody in the Maltese speech community. As a native speaker, she personally prefers to exchange *mnejn* with the synonymous *minn fejn* 'from where'. This preference for the binary combination over *mnejn* is also discussed in Stolz/Vorholt (2022). There is no evidence of *mnejn* taking a lexical complement

138 Sandra Abela (personal communication, 11 February, 2018).
139 Personal communication, 11 February, 2018.

NP. Sandra Abela's comments correspond closely to what E. Serracino-Inglott (1979: 183) remarks on *mnejn* (which he classifies as Π and adverb):

> Falz[on] u ta' wrajh jagħtuh ukoll bis-suff[issi] pron[ominali] pers[onali], spec[jalment] tal-pl[ural], eż[empju] Mnejna, flok 'mnejn aħna', iżda ma nafx li qatt smajt il-part[iċella] tingħad hekk! F'dan il-każ, in-nies tgħid: 'minn fejnna' jew, aħjar, 'minn ħdejna' eċċ.[140]

This quote suggests that the pronominal suffixation is artificial to some extent in the sense that a grammarian of the late 19th century seems to have propagated this construction contrary to the preferences of the speech community.

This and other problems notwithstanding, we treat *mnejn* analogously to *kif*, i.e. it remains a member of BLOMP also in the revised version BLOMP 2.0 but only on probation for the same reasons as *kif*. On account of the connections to (X13) *fejn* 'near' however, the meaning assumed for *mnejn* is changed from 'from where' to 'from near'.

(iii) (X18) *ġew* 'in'

In the original version of BLOMP, (X18) *ġew* 'in' was included solely because, when the provisional list of Maltese Πs was compiled, the ideal was to primarily register the shortest alloform of a candidate. Aquilina (1987: 393) mentions *ġew* as shortened form of *ġewwa*. In *Il-Miklem Malti* (E. Serracino-Inglott 1976: 33), the same relation between the two expressions is assumed with the addition that *ġew* is sometimes further shortened to *ġo* 'in'. This is certainly counterfactual since *ġo* is much more common than *ġew*. In the meantime, we have abolished the criterion of segmental extension and replaced it with that of the most frequent alloform. As argued in Stolz/Levkovych/Vorholt (2023), the distribution of *ġew* 'in' is severely restricted to deictic adverbials such as *hawn ġew* 'in here' in (1.185), whereas only the segmentally more complex alloform *ġewwa* 'inside' can claim Π-status as in (1.186).[141] Note that *ġew* follows the syntactic unit it has scope over and thus takes a position which is excluded for Πs.

(1.185) [Korpus Malti 3.0; literature10]

Tiftakar	*meta*	*konna*	*ninqaflu*	***hawn***	***ġew***
2.IPFV:RFL:remember	when	be:1PL.PFV	1.IPFV:RFL:shut:PL	**here**	**in**

140 Our translation: "Falzon and those after him use it also with the pronominal suffixes, especially in the plural, example: Mnejna ['from where we are / from near us'], instead of 'mnejn aħna', but I do not know to have ever heard the particle being uttered like that! In this case, the people say: 'minn fejnna' or better 'minn ħdejna', etc." The name Falzon refers to G.B. Falzon, the author of the *Breve esposizione grammaticale della lingua maltese* included in his *Dizionario Italiano-Inglese-Maltese* dating back to 1882.

141 Cf. Section 1.5.7.

biex *nistudjaw* *għall-O Levels*
to 1.IPFV:study:PL for:DEF-O_Levels
'Do you remember when we locked us up **in here** to study for the O-Levels?'

(1.186) [Attard 2019: 138]
Bud kien ġie inkarigat jagħtiha l-lezzjonijiet f' parti
Bud be come entrust:PART 3.IPFV:give:3SG.F DEF-lesson:PL in part
mill-kantina tal-kwartieri ta' SIKTA ***[ġewwa*** *l-Palazz*
from:DEF-cellar of:DEF-quarter:PL of SIKTA **[inside** DEF-Palace
ta' Verdala]$_{PP}$
of Verdala]
'Bud had been entrusted with giving her lessons in a section of the cellar of the headquarters of SIKAT [**inside** Verdala Palace] [. . .].'

Accordingly, in BLOMP 2.0, *ġewwa* 'inside' takes in the place formerly held by *ġew*.

(iv) (X50) *sotta* 'under'

To close this section, we address the case of (X50) *sotta* 'under'. According to Aquilina (1959: 322), there is the Π *sotta* (< Italian *sotto* 'under') which takes a pronominal complement with the help of the intercalated Π *ta'* 'of' to yield constructions like *sotta tiegħek* 'subject to you'. In the Maltese-English dictionary, Aquilina (1991: 1346) reviews the purposes to which *sotta* and its many alloforms (*sutta* ~ *sotto* ~ *sottu* ~ *sott*) are put in Maltese. It seems that *sotta* is mainly employed as derivational morpheme as in *sottakap* 'assistant manager'. The only phrase which seems to support the Π-analysis is *sotta ta' xi ħadd* 'under s.o.'s thumb'. Neither this supposed idiom in particular nor *sotta* as a Π in general is known to our native-speaker consultant.[142] The only sentential example we have found is (1.187) from E. Serracino-Inglott's *Il-Miklem Malti*.

(1.187) (E. Serracino-Inglott 1984: 226)
Jien m' jiniex ***[sotta*** *tiegħek]*$_{PP}$
1SG NEG 1SG:NEG **[under** of:2SG]
'I am not [your underling]!'

This is the sole sentence which can be taken to prove the Π-status of this candidate. The uncertainty of the reliability of this piece of evidence notwithstanding, it seems to meet the criteria for being admitted to the Maltese Π-inventory. Therefore, we accept it as member of BLOMP 2.0. This does not preclude the alternative analy-

142 Sandra Abela (personal communication, 11 February, 2018).

sis according to which (1.187) is an adnominal possessive construction with *sotta* being the head noun with the meaning 'subaltern'. As will become evident very soon in the empirical part of this study, (X50) *sotta* 'under' most probably would not survive a future revision of BLOMP 2.0.

1.6.8 BLOMP 2.0

It has taken us long to ultimately determine how many and which Πs constitute BLOMP 2.0. The discussion in the foregoing sections yields an inventory of 60 Πs as shown in (1.188). However, BLOMP 2.0 differs from its predecessor not only because it contains three members in excess. Several of the Πs are represented by different alloforms. The meanings of a number of Πs has been adjusted. Unless special reasons are given, only these new meanings will be employed throughout the remainder of this study. Furthermore, a small number of erstwhile candidates has vanished from BLOMP 2.0. whereas an equally small number of new entries has been admitted to the list. To mark the Π-status of the members of BLOMP 2.0. as definitively confirmed, the letter <X> of the previous ID-code is replaced with the Greek letter <Π>. Grey shading identifies those Πs which, on closer scrutiny, might be removed from BLOMP 2.0 in a dedicated follow-up study.

(1.188) BLOMP 2.0

(Π1) *apparti* 'apart from'; (Π2) *a skapitu* 'at the expense of'; (Π3) *barra* 'outside'; (Π4) *bejn* 'between'; (Π5) *bħal* 'like'; (Π6) *bħala* 'as'; (Π7) *bi* 'with'; (Π8) *biswit* 'facing'; (Π9) *bla* 'without'; (Π10) *daqs* 'equal to'; (Π11) *dwar* 'about'; (Π12) *faċċata* 'opposite'; (Π13) *favur* 'in favour of'; (Π14) *fejn* 'near'; (Π15) *fi* 'in'; (Π16) *fost* 'amongst'; (Π17) *fuq* 'on'; (Π18) *ġewwa* 'inside'; (Π19) *ġo* 'in'; (Π20) *għajr* 'except'; (Π21) *għal* 'for'; (Π22) *għand* 'at s.o.'s place'; (Π23) *ħdejn* 'beside'; (Π24) *ħlief* 'except'; (Π25) *inkluż* 'including'; (Π26) *kif* 'as'; (Π27) *kontra* 'against'; (Π28) *lejn* 'towards'; (Π29) *lil* 'to'; (Π30) *ma'* 'with'; (Π31) *madwar* 'around'; (Π32) *maġenb* 'close to'; (Π33) *matul* 'during'; (Π34) *minbarra* 'except'; (Π35) *minflok* 'instead of'; (Π36) *mingħajr* 'without'; (Π37) *mingħala* 'in s.o.'s opinion'; (Π38) *mingħand* 'from s.o.'; (Π39) *minħabba* 'on account of'; (Π40) *minkejja* 'in spite of'; (Π41) *minn* 'from'; (Π42) *mintul* 'all along'; (Π43) *mnejn* 'from near'; (Π44) *oltre* 'beyond'; (Π45) *permezz* 'by means of'; (Π46) *qabel* 'before'; (Π47) *qalb* 'amidst'; (Π48) *qrib* 'near'; (Π49) *quddiem* 'in front of'; (Π50) *rigward* 'concerning'; (Π51) *sa* 'till'; (Π52) *sforz* 'thanks to'; (Π53) *skont* 'according to'; (Π54) *sotta* 'under'; (Π55) *ta'* 'of'; (Π56) *taħt* 'under'; (Π57) *versu* 'towards'; (Π58) *viċin* 'near'; (Π59) *waqt* 'at the time of'; (Π60) *wara* 'after'

We have reached the point where it makes sense to refine our rule of thumb (1.1) in order to provide a definition of the Maltese Π on the basis of which all subsequent investigations included in this study can be carried out. This definition is given in (1.189).

(1.189) The Maltese Π – definition
In Maltese, a Π is a simple or semantically non-compositional lexicalized syntactic word which functions as head of a PP. Whether this syntactic word is also a member of other word-classes is irrelevant for its classification as Π in the appropriate syntactic context. It must however give evidence of decategorialisation and/or differ from its non-Π uses as to the semantic and/or morpho-syntactic properties. The Π is situated at the left edge of the PP. The Π takes a complement which is either nominal or pronominal. The complement is put in a relationship with another linguistic unit in the same sentential context by way of marking the grammatical and semantic links between them.

For the remainder of this study, this definition is presupposed when we talk about a given element as being a member of BLOMP 2.0.

The alphabetical order in which the Πs are presented in (1.188) does not tell us much about the interrelations among the members of BLOMP 2.0. The different Πs are not absolutely equal. This can be shown impressionistically by way of comparing the token frequencies which result from a check of the Korpus Malti 3.0. The results are disclosed in Table 1.11. For reasons to be exposed in Section 1.7, we approach the data from an angle which differs from that of Schmidt/Vorholt/Witt (2020).

To rank the Πs we have determined the token frequency of Π-tokens for the first 100 attestations of the segmental chain in the Korpus Malti 3.0 (= column A). In column B, we present the total number of tokens of the segmental chain in the Korpus Malti 3.0. The turnout in column A yields the share of Π-tokens. Column C contains the estimated number of Π-tokens in the Korpus Malti 3.0 in accordance with the share determined for column A.[143] The Πs are ordered top-down according to the decreasing number of tokens estimated for column C. Grey shading marks out the eleven Πs which exceed the average number of tokens as of column C per Π which amounts to 214,568.63

143 This method of ours is a purposefully crude adaptation of the much more sophisticated approach defended by Saade (2020) in the spirit of (Gaeta/Ricca 2006). For purely practical reasons, we have simplified the corpus-linguistically ambitious model by way of stripping off the theory-oriented aspects of the calculus.

Table 1.11: Members of BLOMP 2.0. ranked according to estimated token frequency.

rank	Π	A	B	C
		1st 100	total	estimate
1	(Π55) *ta'* 'of'	97	5,096,341	4,943,450
2	(Π41) *minn* 'from'	90	1,256,789	1,131,110
3	(Π17) *fuq* 'on'	93	1,151,992	1,071,352
4	(Π21) *għal* 'for'	96	1,036,925	995,448
5	(Π11) *dwar* 'about'	89	651,499	579,834
6	(Π29) *lil* 'to'	96	404,833	388,639
7	(Π30) *ma'* 'with'	95	355,717	337,931
8	(Π6) *bħala* 'as'	100	328,932	328,932
9	(Π15) *fi* 'in'	100	319,840	319,840
10	(Π60) *wara* 'after'	56	444,289	248,801
11	(Π4) *bejn* 'between'	100	236,899	236,899
12	(Π56) *taħt* 'under'	92	203,834	187,527
13	(Π28) *lejn* 'towards'	100	171,669	171,669
14	(Π27) *kontra* 'against'	100	167,406	167,406
15	(Π33) *matul* 'during'	100	149,416	149,416
16	(Π7) *bi* 'with'	100	141,102	141,102
17	(Π5) *bħal* 'like'	83	157,848	131,013
18	(Π39) *minħabba* 'on account of'	80	159,460	127,568
19	(Π45) *permezz* 'by means of'	100	115,212	115,212
20	(Π51) *sa* 'till'	78	130,402	101,713
21	(Π16) *fost* 'amongst'	100	92,813	92,813
22	(Π50) *rigward* 'concerning'	74	98,661	73,009
23	(Π3) *barra* 'outside'	53	136,814	72,511
24	(Π9) *bla* 'without'	90	80,061	72,054
25	(Π36) *mingħajr* 'without'	66	107,066	70,663
26	(Π46) *qabel* 'before'	29	229,852	66,657
27	(Π53) *skont* 'according to'	93	70,443	65,511
28	(Π13) *favur* 'in favour of'	88	60,542	53,276
29	(Π49) *quddiem* 'in front of'	31	154,607	47,928
30	(Π40) *minkejja* 'in spite of'	52	82,667	42,986
31	(Π22) *għand* 'at s.o.'s place'	94	35,070	32,965
32	(Π59) *waqt* 'at the time of'	19	170,317	32,360
33	(Π38) *mingħand* 'from s.o.'	100	32,167	32,167
34	(Π35) *minflok* 'instead of'	65	45,642	29,667
35	(Π31) *madwar* 'around'	26	95,815	24,911
36	(Π34) *minbarra* 'except'	70	30,050	21,035
37	(Π18) *ġewwa* 'inside'	46	45,315	20,844
38	(Π25) *inkluż* 'including'	64	32,143	20,571
39	(Π48) *qrib* 'near'	67	29,384	19,687
40	(Π19) *ġo* 'in'	100	17,759	17,759
41	(Π23) *ħdejn* 'beside'	100	16,489	16,489

Table 1.11 (continued)

rank	Π	A	B	C
		1st 100	total	estimate
42	(Π1) *apparti* 'apart from'	56	28,630	16,032
43	(Π24) *ħlief* 'except'	44	25,346	11,152
44	(Π10) *daqs* 'equal to'	38	28,592	10,864
45	(Π58) *viċin* 'near'	53	16,751	8,878
46	(Π47) *qalb* 'amidst'	32	18,935	6,059
47	(Π20) *għajr* 'except'	62	8,489	5,263
48	(Π14) *fejn* 'near'	1	489,321	4,893
49	(Π8) *biswit* 'facing'	88	2,958	2,603
50	(Π44) *oltre* 'beyond'	75	2,984	2,238
51	(Π12) *faċċata* 'opposite'	26	7,415	1,927
52	(Π32) *maġenb* 'close to'	99	1,720	1,702
53	(Π2) *a skapitu* 'at the expense of'	95	1,367	1,327
54	(Π52) *sforz* 'thanks to'	3	7,730	231
55	(Π57) *versu* 'towards'	98	228	223
56	(Π37) *mingħala* 'in s.o.'s opinion'	1	1	1
57	(Π26) *kif* 'as'	0	787,519	0
58	(Π43) *mnejn* 'from near'	0	1,949	0
59	(Π42) *mintul* 'all along'	0	0	0
60	(Π54) *sotta* 'under	0	0	0

We need to explain how we have come by the quantities in Table 1.11. The Korpus Malti 3.0 was searched for full syntactic words only, i.e. no allomorphs, cliticised, or inflected word forms have been taken account of. Consequently, certain types of Πs are overrepresented in the above count because they are neither subject to allomorphic alternations nor do they undergo cliticisation or host pronominal suffixes. Still, except #8 (Π6) ***bħala*** 'as', the first seventeen ranks in Table 1.11 are monopolised by Πs whose frequency would further increase if allomorphy, cliticisation, and pronominal suffixation had been included in the probe. Thus, Table 1.11 tells only part of the story without distorting it beyond recognition. The missing chapters will be added in Part B of this study, in a manner of speaking.[144] The above frequency count is of limited use. However, on its basis, it is possible to form an idea of the hierarchy of importance among the members of the Maltese Π-inventory.

144 The refined and final version of the frequency count is presented in Section 3.4.

The individual estimates in column C of Table 1.11 add up to a total of 12,874,118 Π-tokens in the shape of full syntactic words. Figure 1.11 reveals that only nine Πs have shares of n ≥ 2%. The majority of 51 Πs covers less than a quarter of all tokens, namely 22%. In contrast, the nine top-ranking Πs together have a share of 78% of all tokens. The most outstanding case is (Π55) *ta'* 'of' which alone is responsible for 38% of all Π-tokens whereas none of its fellow-Πs reaches the 10%-mark.

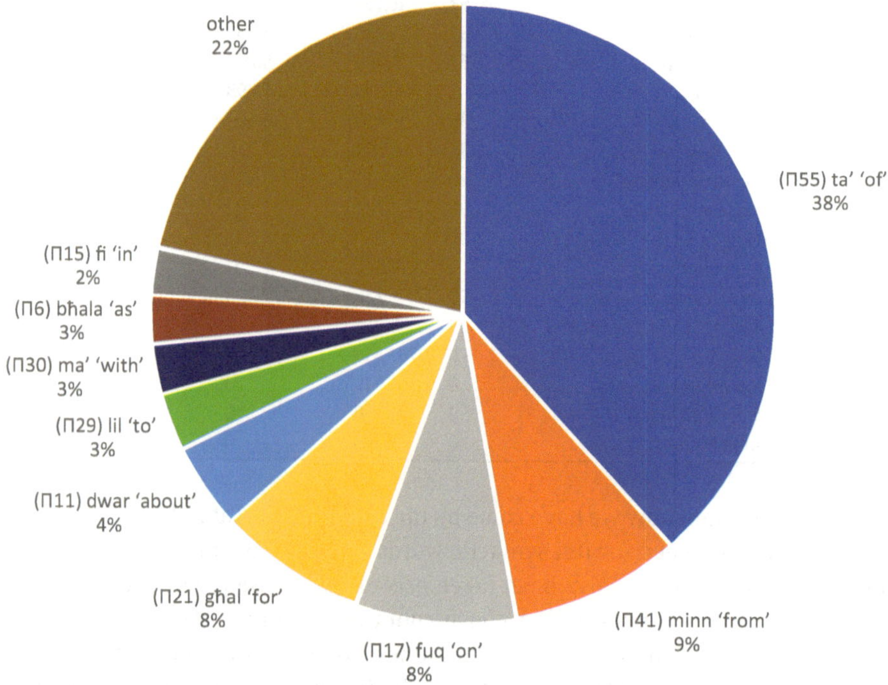

Figure 1.11: Shares of Πs of all tokens in the Korpus Malti 3.0.

Part B will not only update the picture painted in Figure 1.11 by way of integrating more quantitative information into it but also by investigating potential correlations between the token frequencies and shares of the Πs on the one hand and their phonological, morphological, syntactic, and semantic properties on the other hand. To better understand these correlations and to identify them in the first place, it is necessary for us to conduct the empirical work on the basis of a reliable database.

1.7 The database

The examples we have presented in the foregoing sections stem from a variety of different sources, namely

- the major electronic resource of Maltese – the Korpus Malti 3.0,[145]
- conventionally published literary and academic prose,
- the extant descriptive linguistic literature on Maltese.

In addition, native-speaker judgments have been referred to occasionally in order to decide especially intricate cases of doubt. This practice of extracting data from as many sources as possible will be continued throughout this study. Therefore, our principal source types are sketched in this section.

We have argued that one of the advantages of our investigation over that of Saari (2003) for instance, results from the extensive use we can make of the Korpus Malti 3.0 which saw the light of day only in 2016, i.e. much later than the most recent contributions of Saari (2007, 2009) to the debate. At the end of the first decade of the new millennium, the predecessors of Korpus Malti 3.0 were still in preparation (Rosner 2009). With a size of 249,256,855 words and a coverage of 253,068 texts, the Korpus Malti 3.0 offers many opportunities for linguistic research. It is a must for a project like ours to make the most of these opportunities. At the same time however, one has to be cautious not to rely exclusively on the Korpus Malti 3.0.[146]

In the course of previous studies dedicated to Maltese, members of the research team at the university of Bremen have experienced a number of problems which arise when working with the Korpus Malti 3.0 (e.g. Saade 2018, 2019, 2020) and this is by no means different in our case. The difficulties encountered range from the variation in spelling of given items via homographs and the multiple occurrence of one and the same sentence and general disagreement as to the tagged categories to the erroneous identification of items with elements from languages other than Maltese. By way of example, we mention the investigations on the competition between long and short forms of personal pronouns (such as e.g. long *jiena* vs

145 We have not made use of the Bulbulistan corpus (http://www.bulbul.sk/bonito2/) because it has at first been designed specifically for the purposes of the compiler's PhD-project but steps have been taken to integrate the two corpora so that the Korpus Malti 3.0 remains "the focal point of Maltese corpus linguistics" whereas the Bulbulistan corpus continues to serve as "data source for a number of special projects" (Čéplö 2018: 57).

146 When the manuscript version of this book reached its final stages, Korpus Malti 4.0 entered the scene in Spirng 2024. Owing to the long duration of our project and the exhaustions caused during it, we purposefully ignored the existence of Korpus Malti 4.0. Future work on Maltese IIs must determine whether the general picture of the grammar of Maltese IIs looks different if the more recent version of the Korpus Malti had been taken account of, too.

short *jien* in 1st person singular) carried out by Stolz/Saade (2016) and Saade (2018). These studies bring to the fore a very strong levelling effect of the Korpus Malti 3.0 as opposed to the differential results of the analysis of selected literary texts. One might want to argue that the size of the Korpus Malti 3.0 guarantees that its users get a much more realistic picture of the Maltese language than it is possible on the basis of a smaller and manually analysed text corpus. This objection would be valid, had the representation of text genres in the Korpus Malti 3.0 been different. Schmidt/Vorholt/Witt (2020: 242) correctly characterise the Korpus Malti as "an opportunistic collection and not a balanced representation of the Maltese language". In point of fact, the internal composition of the Korpus Malti 3.0 is skewed in the sense that certain genres are vastly overrepresented to the detriment of others whose share is relatively small.

This imbalance is inherited from earlier versions of the Korpus Malti. In the Korpus Malti 2.0 Beta as of 2013 (Gatt/Fabri 2018: 157), there is an overwhelming 90%-majority of texts representing either journalistic Maltese or Maltese as used in parliamentary debates. In Figure 1.12, religious texts (0.32%), belles lettres (0.3%), academic texts (0.14%), miscellaneous (0.1%), and speeches (0.0 1%) together have a share below 1%.

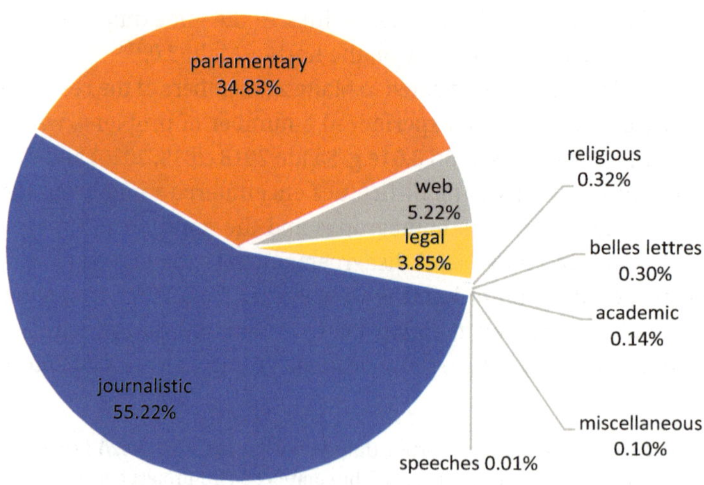

Figure 1.12: Representation of genres in the Korpus Malti 2.0 Beta.

Since the Bulbulistan corpus compiled by Čéplö (2018: 62) reflects a very similar distribution of text genres (journalistic = 52%, parliamentary = 45%, other = 3%), there is the danger that working exclusively with an unbalanced corpus of this kind might seriously distort the picture in the sense that the peculiarities of certain privileged genres are inadvertently generalised over the entire language so that the system of Maltese ΠIs is described as it works in the printed media and in parliament although it might follow different rules in other domains. With reference to the Korpus Malti 2.0 Beta, Gatt/Čéplö (2013) emphasise that there is an urgent need to increase the shares of hitherto underrepresented text genres.

In Table 1.12, the number of words per genre in the Korpus Malti 2.0 Beta and in the Korpus Malti 3.0 are compared. The total size of the Korpus Malti 3.0 is twice as big as that of its predecessor.

Table 1.12: Genres in the Korpus Malti 2.0 Beta and Korpus Malti 3.0.

Korpus Malti 2.0		**Korpus Malti 3.0**		**Difference**
genre	**words**	**genre**	**words**	
parliamentary	43.400.000	parliamentary	119.733.592	76.333.592
journalistic	68.800.000	journalistic	89.235.823	20.435.823
		sports news	1.079.198	1.079.198
		culture news	761.317	761.317
legal	4.800.000	European law	22.821.688	24.171.538
		Maltese law	6.149.850	
web	6.500.000	web	6.468.544	-31.456
belles lettres	375.000	belles lettres	1.882.985	1.507.985
academic	170.000	academic	683.919	513.919
religious	403.700	religious	439.939	36.239
miscellaneous	123.000	miscellaneous	0	-123.000
speeches	18.000	speeches	0	-18.000
total	124,589,700	total	249,256,855	124,667,155

Apart from the disappearance of the categories miscellaneous and speeches, the differentiation of genres has increased. Of those genres which were grossly underrepresented in the Korpus Malti 2.0 Beta, academic texts and belles lettres have considerably grown in size in terms of the number of words by which they are represented in the Korpus Malti 3.0. There are now four times as many words as before under the rubric academic. In the case of belles lettres, the increase is even more substantial since there are now five times as many words in this category. However, as Figure 1.13 reveals, neither belles lettres nor academic texts cover more than a negligible percentage of the entire database.

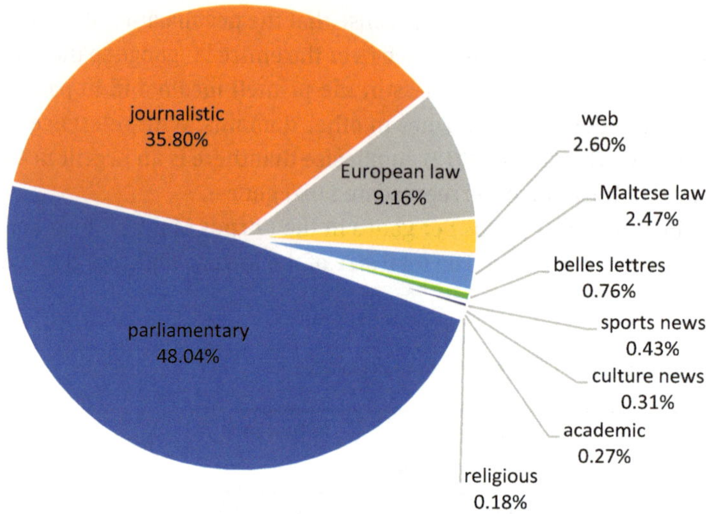

Figure 1.13: Representation of genres in the Korpus Malti 3.0.

The dominance of the three top-ranking genres – parliamentary texts, journalistic texts, and European law – is overwhelming. This group of three covers 93% of all words in the Korpus Malti 3.0. The dominance becomes even more pronounced as it rises to 96.21% if we lump together all texts with legal content – European Law and Maltese Law – on the one hand, and journalistic texts, sports news, and culture news, on the other hand. This means that academic texts and belles lettres are still massively underrepresented in the Korpus Malti 3.0. No language – Maltese included – can be claimed to be adequately documented if the genres are represented as unevenly as in the case under scrutiny. We therefore take the above statement of Gatt/Čéplö (2013) to define a mission which is still far from being accomplished. Accordingly, we make ample use of the Korpus Malti 3.0 butcomplement the findings with and check them against the data collected outside the electronic resource with special emphasis on belles lettres and academic texts. This complementation is applied in the phonological and morphological chapters.[147] It is however, used much less in connection to morpho-syntax and ceases when we turn our attention to semantics.[148] This means that wherever we needed countable data and comparable contexts, we could not help making use of the Korpus Malti 3.0 not the least because other sources were not automatically searchable.

147 Cf. Section 2–3.
148 Cf. Sections 4–5.

The reasons for the lack of balance in the Korpus Malti 3.0 (and the Bulbulistan corpus) are mostly of a practical nature (Čéplö 2018: 58). The accessibility of the documents was always the most crucial factor so that difficult to access texts made it less often into the database. Accessibility also plays an important role for collecting data from outside the Korpus Malti 3.0. The Malta Centre at the University of Bremen hosts the Maltese Library (https://maltacentre.eu/maltese-in-bremen/maltese-library/) with more than 1,100 volumes[149] which, apart from a small number of specimens of poetry and drama, covers the genres of belles lettres (prose), religious, and academic texts – both original and translated – which are still underrepresented digitally. Since the Maltese Library was founded seven years before the Korpus Malti 3.0 began to operate, there is some overlap in the coverage of texts. We draw extensively on the volumes on the shelves of the Maltese Library thereby taking care to use also texts whose date of publication is posterior to 2016. The texts selected from the Maltese Library are identified in the Maltese section of the primary sources below. The total of words covered by these examples of belles lettres and academic texts is approximately 490,000. The books taken from the Maltese Library are representative of fictional and nonfictional prose published in the course of the 20th century and the first two decades of the 21st century. This means that texts of artificially re-Semiticised Malti Safi occur side by side with those which have not been written in the spirit of purism. We have imposed no restrictive criteria as to the admittance of certain sub-genres such as traditional legends, books for children, autobiographies, etc. The texts taken from the Maltese Library serve mostly the purpose of qualitative research whereas the Korpus Malti 3.0 is better suited for quantitative analyses. Thus, the two source types complement each other in terms of methodology (cf. Section 1.8.2). In contrast to Čéplö (2018: 58), we do not exclude translations from our database. Excluding translations sweepingly would have meant that the sizable section of documents in the Korpus Malti 3.0 resulting from the publication of E.U. regulations had to be ignored. Some uses of Maltese Πs are however tightly connected to exactly this kind of texts as has come to the fore for instance, in the course of the discussion on (Π44) *oltre* 'beyond'.[150] The exclusion of the E.U.-related texts would have resulted in the cancelation from the Maltese Π-inventory of this Π and all those which for whatever reason are absent from the texts in the Korpus Malti 3.0. Since the period in which the literary and academic texts were produced spans some 120 years, it is only to be expected that certain Πs

149 As of October, 2023.
150 Cf. Section 1.6.6.

attested early on in this timeframe have become obsolete by the end of the period. Several candidates of this kind have been mentioned.[151] We take account of them nevertheless since the above 120 years are coextentive with a sequence of three generations of speakers which equals our idea of synchrony.

Our database may be partly different from that represented by the Korpus Malti 3.0 or Čéplö's (2018) Bulbulistan corpus. What all these data collections have in common nevertheless is their focus on the written register. What we describe in this study is nothing else and nothing more than the grammar of the Πs in written Maltese. No genuine spoken-language data is taken account of, since there is as yet no corpus of the oral Maltese register. We exploit the extant descriptive (and normative) linguistic literature on Maltese including the contributions of our predecessors and contemporaries[152] as well as the major dictionaries (Aquilina 1987, 1991; E. Serracino-Inglott 1975a,b–1989; M. Serracino-Inglott 2000, 2003) and sundry secondary sources. In these secondary sources, the colloquial usage of certain elements is at times mentioned. However, these hints are given far too unsystematically to allow for generalisations over Πs in spoken Maltese. We therefore have to accept the limitations of our database in the sense that when we speak about Maltese Πs we always and exclusively make statements about Πs in written Maltese.

Owing to the problems caused by the COVID-19 pandemic, the originally planned fieldwork sessions on Malta had to be postponed repeatedly and ultimately cancelled for good because the funding organisation required of us to finish the project on time, i.e. within two years and an extension of nine months with the final deadline being scheduled for June 2022. This meant that the questionnaire we designed early in 2018 could be used only for a single (four days) test-run with an individual native speaker of Maltese who visited Bremen in February 2018. Sandra Abela, a philologically trained and experienced teacher (including Maltese as a foreign language), kindly answered all questions, commented on some of the issues raised in the questionnaire, and corrected the original version wherever the need came up. As already practiced in Part A, Sandra Abela's statements will occasionally be used in Part B to support or modify hypotheses of ours. Only in case a given phenomenon is attested neither in the Korpus Malti 3.0 nor in the Maltese Library nor in the grammars and dictionaries of Maltese, examples from the questionnaire will be integrated to illustrate what is at issue.

151 Cf. Section 1.6.7.
152 As reviewed in Sections 1.4–1.5.

1.8 Theory and methodology

This section is logically divided in two. In 1.8.1, we elaborate on the deliberately sketchy information given in the opening paragraphs of Part A as to the backbone of this study, i.e. the theories whose axioms and hypotheses guide us through the empiry. Section 1.8.2 adds further information about the methodology which we employ to describe and evaluate the Maltese data. Those aspects which are relevant for theory and/or methodology but are skipped in Sections 1.8.1–1.8.2 will be touched upon at the appropriate places in Parts B–C.

1.8.1 Theory

In contrast to the four readings of the term theory distinguished by Haspelmath (2010: 289), we make do with two different but interrelated kinds of theory. On the one hand, there is grammatical theory as overarching framework of any approach. On some 700 pages, Müller (2016) covers the most influential modern grammatical theories which have emerged since the times of Transformational Grammar. We refrain from picking out any of the theories which are on offer there because we doubt that we could do justice to the system of Maltese Πs by way of describing it according to a preconceived model. To our mind, the literature review[153] has shown that before this system can be approached from the point of view of a given grammatical theory it first needs to be presented in a format that is independent of exclusively theory-borne assumptions (Haspelmath 2010: 288–289). Only after we know what exactly we are talking about when we refer to the Maltese Π-system does it make sense to subject it to a (re-)analysis along the lines of a particular grammatical theory. We are confident that this will be possible after the completion of this study.

This is why we lean towards Haspelmath's (2010: 303) concept of framework-free grammatical theory within the heterogeneous school-of-thought of linguistic functionalism. Haspelmath (2010: 304–307) is critical of *Basic Linguistic Theory*. As explained above,[154] we agree that the assumed universal applicability of some of the concepts put forward by Dixon (2010a–b, 2013) is doubtful. However, the philosophy behind *Basic Linguistic Theory* appeals to us in the sense that it allows us to employ traditional (i.e. mostly structuralist) terminology and concepts to describe the phenomena which are important for this study without having to fit

153 Cf. Sections 1.4–1.5.
154 Cf. Footnote 2.

them into a formalistic framework. We do not accept Dixon's analyses, hypotheses, and ideas sweepingly. What we accept instead is the conceptual liberation from any kind of overly rigid theory. This liberation is absolutely justified because we do not intend to prove any grammatical theory to be right or wrong. This study aims at giving linguists of different convictions access to a hitherto understudied component of the grammar of a language which has only recently begun to be taken notice of by scholars who are not interested primarily in Neo-Arabic varieties or language-contact phenomena.

The second kind of theory to be addressed in this section is more narrowly dedicated to adpositions in general or Πs in particular. There is a plethora of studies which deal with the adpositional systems of individual languages or genetically defined groups of languages (Guimier 1981). Some of these studies have the format of longish linguistically annotated lists such as Stoebke (1968) on adpositions in the Balto-Finnic branch of the Uralic language family. An example of a by far more ambitious project is Hagège's (1975) monograph on the Π-category in Chinese which contains a typologically-minded chapter – the predecessor of Hagège (2010). Even for individual languages, different approaches to Πs compete with each other as shown for instance, by the different contributions to the edited volumes dedicated to French Πs (Schwarze 1981) and German Πs (Rauh 1991). In other publications, subtypes of Πs are in the focus of attention such as French spatial Πs (Vandeloise 1991) or complex Πs in Romance languages addressed in a special issue of the *Revue Romane* (Fagard/Lima/Stosic 2019). A comprehensive account of English Πs, their meanings and uses is provided by Dixon (2021). Recent comparative attempts are documented inter alia in Garzonio/Rossi (2020). Giving a full account of the rich literature on adpositions is not only beyond the scope of our project but also hardly feasible on account of the numbers and the diversity in the domains of methodology and theory which characterise this field of linguistic research. What this short and impressionistic paragraph is meant to show is that the description of the Maltese Π-system will fill a gap that needs to be filled in order for the research dedicated to adpositions to advance further.

Saari (2003) approaches his subject matter from the implicit point of view of traditional Semitic philology without taking any stance as to a general Π-related theory. Our gateway to the Maltese facts is different since we do not adopt a Semitist perspective – we even go as far as avoiding this perspective to circumvent the pitfalls of tradition while being aware that many advantages of the philological approaches to Πs might be lost as well. We address the Maltese Π-system with the equipment of functional typology. In this sense, Hagège (2010) provides guidance to us insofar as the author generalises over the structural and functional properties of adpositions in human languages on a very robust cross-linguistic basis. His findings, implications, and hierarchies are helpful when it comes to interpreting the characteristics of Maltese Πs.[155]

Since exhaustively describing the grammar of Maltese Πs of necessity requires that we give our full attention also to PPs, their structure, and their functions, it is clear that syntactic and semantic issues are prominent in Part B.[156] The syntactic issues will also be tackled in accordance to the usual functional-typological practice (Givón 2001) and the occasional recourse to ideas developed within the (heterogeneous) framework of *(Radical) Construction Grammar* (Croft 2013).

For *Basic Linguistic Theory*, Dixon (2010a: 224–225) emphasises that all kinds of adpositions belong to the larger class of markers of grammatical functions and that "[t]he terms 'preposition' and 'postposition' are not strictly necessary [but] useful if employed in a careful manner, with the realisation that they code grammatical elements." The functional equivalence of adpositions and case morphology is stressed time and again by Dixon (2010a: 230–231). At the same time, Dixon (2010a: 231–232) takes issue with the double syntactic relationship of Πs which are "saddled not only with being a governor, but also a head" although the complement NP may be lexical (Dixon 2010a: 231). Similarly, on the basis of his model of dependency relations, Čéplö (2018: 7 and 92) argues against the head-status of Πs (not only) in Maltese exactly because they are not autosemantemes. This interpretation follows the ideas of dependency grammar à la Tenière who subsumed Πs under the label of empty words which are barred from the function of syntactic nucleus (Weber 1992: 23). In contrast to these opinions, we follow Nichols (1993: 165) who assumes that PPs are always headed by Πs but they may differ as to the parameter of endocentrictiy / exocentricity, i.e. in (predominantly) head-marking languages PPs are endocentric because they have the same distribution as their head whereas in (predominantly) dependent-marking languages PPs may be exocentric so that head and phrase differ as to their distribution. Maltese combines head-marking and dependent-marking properties so that both options would be fine. Since in Fabri's (1993) terms there are transitive and intransitive Πs in Maltese, we have exocentric PPs (= those whose head can never stand alone) and endocentric PPs (= those which can be used adverbially on their own) side by side in the same language.

In accordance with the canonical Π and its structural relations sketched in Figure 1.5, we assume that a given Π generally forms a syntactic bridge between a higher external head and an internal complement. Lehmann/Stolz (1992: 12–13)

155 Moreover, Hagège's work allows us to situate Maltese typologically in relation to other Π-languages. However, the originally planned cross-linguistic perspective had to be abandoned in the course of the project when it became clear that a full-blown typological comparison would constitute a separate project in its own right.

156 Especially in Sections 4–5.

distinguish PPs which are properly governed from those which are merely (adverbially modifying) adjuncts. In both these cases, the bridging function of the Π seems to be the same – independent of the exact spellout of the syntactic relations. Thus, our analyses always presuppose head-function for any Π. Whether we are dealing with adjuncts or complements will be clarified below.[157]

1.8.2 Methodology

The mixed nature of the database precludes the possibility of conducting a fully-fledged corpus linguistic evaluation of the texts we employ in this study. Our study is neither corpus-based nor corpus-driven (Biber 2010). As a matter of fact, the quantitative methods required for a proper corpus linguistic project would be applicable only to the Korpus Malti 3.0 which, as argued above,[158] does not suffice when it comes to comprehensively describing the grammar of the Maltese Π-system. This means that in spite of working with a relatively sizable digitised resource, we are not doing proper corpus linguistics. Our research is quality-oriented so that on the quantitative side, we content ourselves with relatively unsophisticated frequency counts (Johnson 2008: 5–12) to support the qualitative analyses.

What we do instead is describing the grammar of Maltese Πs in a typologically-informed way. To this end, we make use of the analytical methods and categories which are grounded in our structuralist heritage as circumscribed by Heine/Narrog (2010: 7) in their introduction to the *Oxford Handbook of Linguistic Analysis*. This handbook itself proves that there is an incredible wealth of descriptive-linguistic approaches which often appear to be largely incompatible with each other. However, mostly without explicit discussion, there is an abundance of shared terminology and procedures. We rely on this common core. This means that when we talk about phonological, morphological, syntactic, and/or semantic aspects in connection to Maltese Πs we always base our analysis on the established mainstream in the domains involved.

1.9 Internal organisation of this study

To close the introduction, we shortly reveal in this section how the remainder of this monograph is organised. The study is divided into four parts, namely

157 Cf. Section 4.
158 Cf. Section 1.7.

- Part A *Introduction* which terminates at the end of this very section,
- Part B *Grammar* which contains Section 2 *Phonology*, Section 3 *Morphology*, Section 4 *Syntax*, and Section 5 *Semantics*,
- Part C *Terminus* which presents the conclusions in Section 6, where generalisations, hypotheses, the impact of the Maltese findings on Π-research, and topics for future research are featured,
- Part D *Lexicon* in which each member of BLOMP 2.0 is presented according to a format which will be explained in situ (in the Appendix).

The list of abbreviations, the references for primary and secondary sources, and the index complement this study. The individual sections are further subdivided. Their organisation is revealed at the beginning of each section. All examples, tables, and figures are identified by numbers to which the section number is prefixed and followed by a full stop to facilitate retrieval.

PART B: **Grammar**

From phonology to semantics

2 Phonology

The synchronic description of the grammar of Maltese Π follows the traditional patterns in the sense that we proceed from phonology via morphology and syntax to semantics. For all of these sections, the division of the members of BLOMP 2.0 in (1.188) into two major classes based on the language of origin of the Πs – Semitic vs Italo-Romance – serves as a common reference point. Saari (2003: 15–16 and 44–45) distinguishes 29 Semitic from four Italo-Romance Πs, i.e. in his account, 88% of the Maltese Πs have a Semitic background as opposed to only 12% which are borrowings from Italo-Romance. Our own investigations yield strikingly different percentages.

In Figure 2.1, a third and a fourth class labelled MIXED and UNCERTAIN are featured, respectively. The sole representative of MIXED is (Π35) *minflok* 'instead of' whose hybrid character historically results from the combination of Semitic *minn* 'from', Semitic *fi* 'in' and the Italo-Romance noun *lok* 'place'.[1] UNCERTAIN too contains only a single element, namely (Π10) *daqs* 'equal to' which is assumed to have either a dialectal Arabic or a Greek origin (Aquilina 1987: 249).

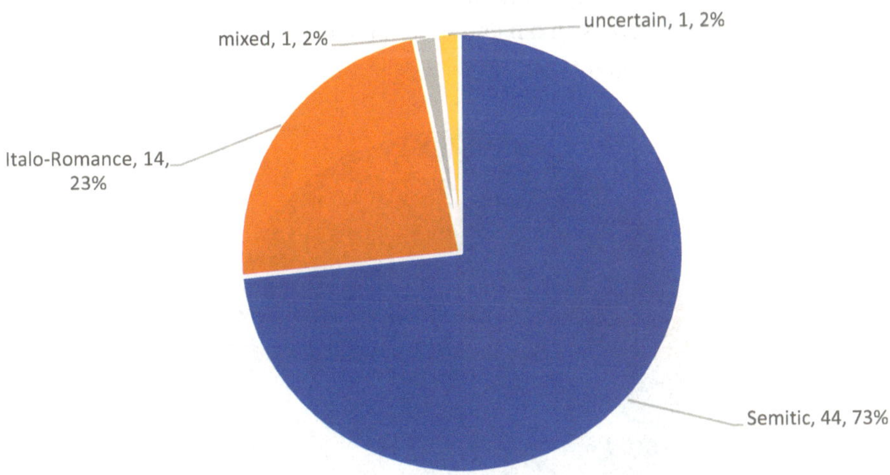

Figure 2.1: Share of Maltese Πs according to language of origin I.

The picture which emerges from Figure 2.1 closely resembles that presented in Schmidt/Vorholt/Witt (2020: 249) and Vorholt (2023a). For the purpose of the

1 As argued in Section 1.6.4.1 (v).

https://doi.org/10.1515/9783111331591-002

ensuing discussion, it is, however, unnecessary to keep the minor classes MIXED and UNCERTAIN. Since in the latter case, the possibility of a dialectal Arabic source cannot be disproved, we classify (Π10) *daqs* 'equal to' as a Semitic Π. Serracino-Inglott (1975b: 19) claims that the proposed Arabic and Greek origins of *daqs* are too far off the mark semantically. His own proposal assumes the univerbation of the PP *ta' qies* 'of (a) measure' to *daqs*. This diachronic development requires several intermediate steps, namely first the proper univerbation of Π and complement which is then followed by the shift of the stress site from *qies* to *ta'*. This shift is necessary to motivate the reduction and deletion of the long high front vowel in the second syllable. Independent of the disappearance of this vowel, the initial voiceless denti-alveolar plosive /t/ must have undergone sonorization to yield /d/. This means that Serracino-Inglott's hypothesis rests on the chronology of several subprocesses none of which can be proved by philological evidence – and thus remains hypothetical. Even if the Greek hypothesis is correct, it is still possible to argue that this Π belongs to the non-Italo-Romance component of the inventory. This gives us the opportunity to subsume (Π35) *minflok* 'instead of' under the rubric of the Italo-Romance Πs of Maltese on the understanding that a Maltese Π is considered Italo-Romance in origin if at least an identifiable part of its form can be traced back etymologically to an Italo-Romance variety. In this way, UNCERTAIN is added to Semitic whereas MIXED and Italo-Romance are united to yield the distribution presented in Figure 2.2.

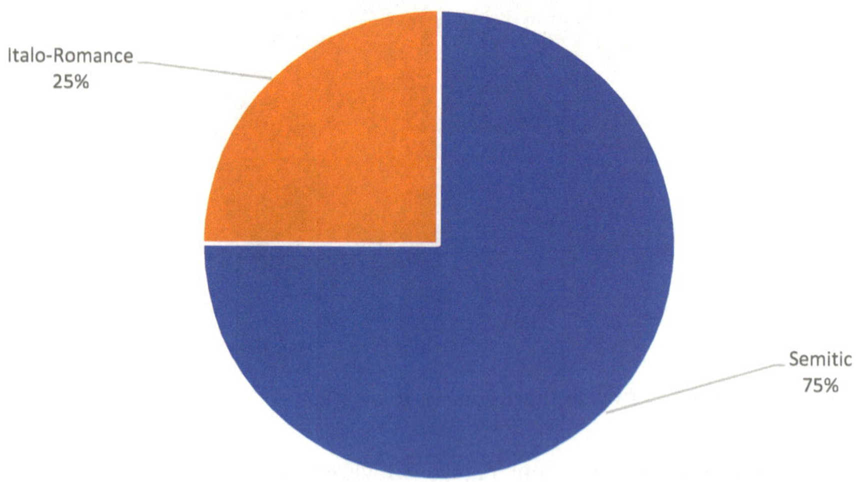

Figure 2.2: Share of Maltese Πs according to language of origin II.

There is a now a majority of 45 Semitic Πs alongside a minority of 15 Italo-Romance Πs – these absolute numbers correspond to a three-to-one ratio. For the interpretation of the phenomena presented and discussed subsequently, this quantitative difference provides a kind of yardstick insofar as we check for each phenomenon whether the above shares are replicated. The Italo-Romance share of the Maltese Π-inventory is for instance smaller than the Italo-Romance share in the Maltese lexicon in general but much bigger than the Italo-Romance share in the domain of function words (Schmidt/Vorholt/Witt 2020: 250–251). According to Comrie/Spagnol (2016: 325 and 328), some 30% of the Maltese lexicon can be attributed to Italo-omance (as opposed to 56% classified as Arabic = Semitic) whereas with 85% the Semitic component predominates in the domain of function words where Italo-Romance covers only 6% of the inventory. Thus, Πs occupy a middle position between the sizable Italo-Romance share in the general lexicon and the almost negligible Italo-Romance impact on function words. This means that Πs differ from other subclasses of function words in terms of the processes of contact-induced transfer. In this section, among other things, we test whether Semitic and Italo-Romance Πs behave differently as to their phonological properties.

To address this and other issues, we start from the suprasegmental level in Section 2.1 which is dedicated to syllables. Section 2.2 looks into the segmental side of the phonological form of Maltese Πs. The facts are analysed both qualitatively and quantitatively. In the latter case, we try to correlate the token frequency of the Maltese Πs to their complexity in terms of the number of syllables and segments which constitute their lexical form.[2] To facilitate understanding of what we are about to argue in the subsequent sections, we provide an overview over the Πs and some of their relevant properties in Table 2.1. The Πs are ordered according to alphabetical principles. For each Π, the phonetic realisation is disclosed in broad IPA-based transcription. In those cases where final devoicing neutralises the phonation contrast to the detriment of [+voice], the cell is divided in two with the right sub-cell hosting the word-final phoneme. For polysyllables, we add the syllable boundaries separately using standard orthography. The number of syllables is indicated as well as that of segments. Note that long vowels count as two segments in Table 2.1. In the rightmost column, the language of origin of the Π is identified.

2 Morphonological processes (including external sandhi) which result in allophony and segmental reduction of the Πs are discussed in Section 3.

Table 2.1: Overview over Πs and their phonologically relevant properties.

#	Π	IPA		syllabified	syllables	segments	origin
Π01	*apparti*	[ɐppɐrtɪ]		ap.par.ti	3	7	Romance
Π02	*a skapitu*	[ɐskɐpɪtʊ]		as.ka.pi.tu	4	8	Romance
Π03	*barra*	[bɐrrɐ]		bar.ra	2	5	Semitic
Π04	*bejn*	[bɛɪn]			1	4	Semitic
Π05	*bħal*	[phɐːl]			1	5	Semitic
Π06	*bħala*	[phɐːlɐ]		bħa.la	2	6	Semitic
Π07	*bi*	[bɪ]			1	2	Semitic
Π08	*biswit*	[bɪswiːt]		bis.wit	2	7	Semitic
Π09	*bla*	[blɐ]			1	3	Semitic
Π10	*daqs*	[dɐʔs]			1	4	Semitic
Π11	*dwar*	[dwɐr]			1	4	Semitic
Π12	*faċċata*	[fɐʧʧɐtɐ]		faċ.ċa.ta	3	7	Romance
Π13	*favur*	[fɐvʊr]		fa.vur	2	5	Romance
Π14	*fejn*	[fɛɪn]			1	4	Semitic
Π15	*fi*	[fɪ]			1	2	Semitic
Π16	*fost*	[fɔst]			1	4	Semitic
Π17	*fuq*	[fʊʔ]			1	3	Semitic
Π18	*ġewwa*	[ʤɛʊwɐ]		ġew.wa	2	5	Semitic
Π19	*ġo*	[ʤɔː]			1	3	Semitic
Π20	*għajr*	[ɐːɪr]			1	5	Semitic
Π21	*għal*	[ɐːl]			1	4	Semitic
Π22	*għand*	[ɐːnt]	/d/		1	5	Semitic
Π23	*ħdejn*	[hdɛɪn]			1	5	Semitic
Π24	*ħlief*	[hlɪːf]			1	5	Semitic
Π25	*inkluż*	[ɪnklʊːs]	/z/	in.kluż	2	7	Romance
Π26	*kif*	[kɪf]			1	3	Semitic
Π27	*kontra*	[kɔntrɐ]		kon.tra	2	6	Romance
Π28	*lejn*	[lɛɪn]			1	4	Semitic
Π29	*lil*	[lɪl]			1	3	Semitic
Π30	*ma'*	[mɐ]			1	2	Semitic
Π31	*madwar*	[mɐdwɐr]		ma.dwar	2	6	Semitic
Π32	*maġenb*	[mɐʤɛmp]	/b/	ma.ġenb	2	6	Semitic
Π33	*matul*	[mɐtʊːl]		ma.tul	2	6	Semitic
Π34	*minbarra*	[mɪnbɐrrɐ]		min.bar.ra	3	8	Semitic
Π35	*minflok*	[mɪnflɔk]		min.flok	2	7	Romance
Π36	*mingħajr*	[mɪnɐːɪr]		min.għajr	2	8	Semitic
Π37	*mingħala*	[mɪnɐːlɐ]		min.għa.la	3	8	Semitic
Π38	*mingħand*	[mɪnɐːnt]	/d/	min.għand	2	8	Semitic
Π39	*minħabba*	[mɪnhɐbbɐ]		min.ħab.ba	3	8	Semitic
Π40	*minkejja*	[mɪnkɛjjɐ]		min.kej.ja	3	8	Semitic
Π41	*minn*	[mɪnn]			1	4	Semitic
Π42	*mintul*	[mɪntʊːl]		min.tul	2	7	Semitic
Π43	*mnejn*	[mnɛɪn]			1	5	Semitic

Table 2.1 (continued)

#	Π	IPA		syllabified	syllables	segments	origin
Π44	*oltre*	[oltrɛ]		ol.tre	2	5	Romance
Π45	*permezz*	[pɛrmɛʦʦ]		per.mezz	2	7	Romance
Π46	*qabel*	[ʔɐbɛl]		qa.bel	2	5	Semitic
Π47	*qalb*	[ʔɐlp]	/b/		1	4	Semitic
Π48	*qrib*	[ʔrɪp]	/b/		1	4	Semitic
Π49	*quddiem*	[ʔʊddɪ:m]		qud.diem	2	7	Semitic
Π50	*rigward*	[rɪgwɐrt]	/d/	ri.gward	2	7	Romance
Π51	*sa*	[sɐ]			1	2	Semitic
Π52	*sforz*	[sfɔrʦ]			1	5	Romance
Π53	*skont*	[skɔnt]			1	5	Romance
Π54	*sotta*	[sɔttɐ]		sot.ta	2	5	Romance
Π55	*ta'*	[tɐ]			1	2	Semitic
Π56	*taħt*	[tɐħt]			1	4	Semitic
Π57	*versu*	[vɛrsu]		ver.su	2	5	Romance
Π58	*viċin*	[vɪʧɪn]		vi.ċin	2	5	Romance
Π59	*waqt*	[wɐʔt]			1	4	Semitic
Π60	*wara*	[wɐrɐ]		wa.ra	2	4	Semitic

The grey-shaded rows host Πs whose segmental chain is phonetically by one unit shorter than its phonological equivalent because of the hidden presence of the abstract phoneme *għajn*[3] How we calculate the number of segments is explained below.[4]

2.1 Syllables

We look at the syllable in connection to the Maltese Π from two different angles. For a start, we determine the patterns which arise in the domain of syllabicity, i.e. the number of syllables which are contained in the lexicon form of a given Π. This is the task of Section 2.1.1. In Section 2.1.2 the focus is on the syllable structure, especially of those Πs which only count one syllable. The latter section forms the bridge to the discussion of the segmental structure of the Πs to be addressed in Section 2.2.

---

3 This abstract phoneme is defined in Section 2.2.1.

4 Cf. in Section 2.2.2.

2.1.1 Syllabicity

Saari (2003) does not treat syllabicity specifically. Schmidt/Vorholt/Witt (2020: 248–249) provide a ternary taxonomy of Maltese ΠIs which comprises the categories monosyllabic, disyllabic (= "bisyllabic"), and polysyllabic (for ΠIs with more than two syllables). According to their count, monosyllables form the largest group with thirty members followed by disyllabic ΠIs with 21 cases. The smallest turnout goes to polysyllables which account for only six cases. In contrast to Schmidt/Vorholt/Witt (2020), we revert to a binary taxonomy with monosyllables and polysyllables. Figure 2.3 suggests that the shares are absolutely identical. Both monosyllables and polysyllables cover half of BLOMP 2.0. The average syllabicity ratio is 1.63 per ΠI.

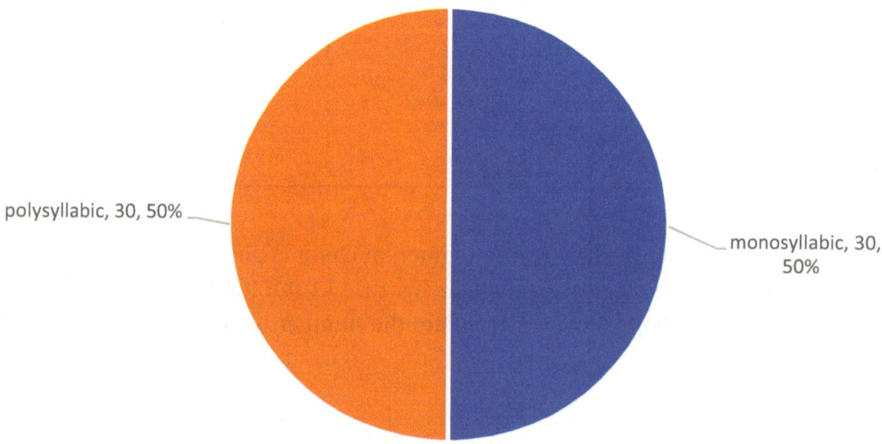

Figure 2.3: Shares of monosyllabic and polysyllabic ΠIs.

Superficially, one could mistake these identical shares to mean that nothing much can be gained from inquiring further into syllabicity. Figure 2.4 however, proves this assumption wrong.

It transpires from Figure 2.4 that, in terms of syllabicity, it makes a difference whether a ΠI stems from Semitic or from Italo-Romance. About 62% of all Semitic ΠIs are monosyllables. As to those ΠIs which have been borrowed from Italo-Romance, the share of monosyllables is down to slightly more than 13%. The importance of this imbalance is further strengthened by the shares the two languages of origin have of the different categories of syllabicity as shown in Figure 2.5.

The Italo-Romance contribution to the subclass of monosyllabic ΠIs is small since it hardly reaches the 7%-mark. In contrast, Italo-Romance is responsible for 43% of all polysyllabic ΠIs. This share is remarkably bigger than that of Italo-Romance

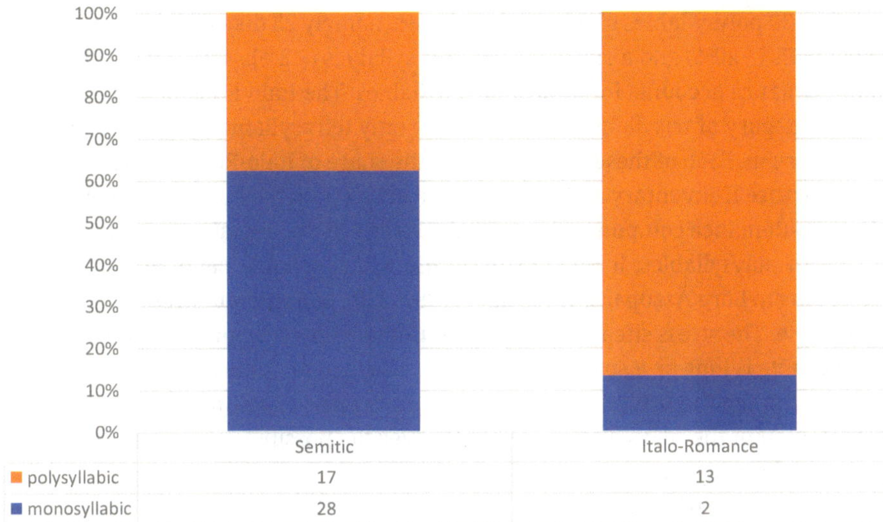

Figure 2.4: Shares of monosyllabic and polysyllabic Πs per language of origin.

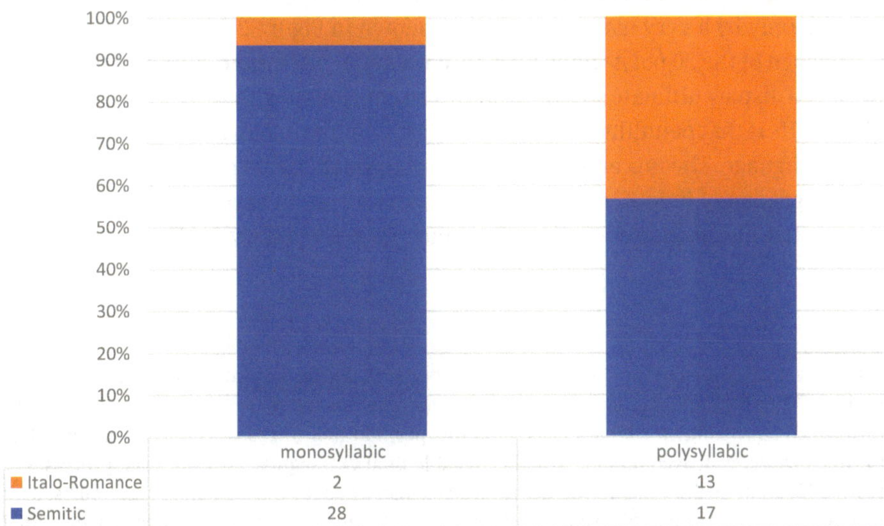

Figure 2.5: Shares of language of origin per syllabicity class.

in the general inventory of Πs as shown in Figure 2.1. This means that Italo-Romance origin is strongly associated with polysyllabicity whereas monosyllabicity and Semitic origin go together very well. Note that we are talking about preferences and not about categorical differences.

The 30 polysyllables divide into two types, namely 23 disyllabic (~ 77%), six trisyllabic Πs (~ 20%) and a single Π with four syllables (~ 3%). With ten disyllabic Πs, Italo-Romance accounts for 43% of all disyllables. The Italo-Romance share is 33% in the category of trisyllables. Moreover, the only tetrasyllabic Π is also of Italo-Romance origin. Each of these shares exceeds the share of Italo-Romance as computed for the entire Π-inventory and thus, they are in line with the above observation that the Italo-Romance component is especially strong in the domain of polysyllables.

With polysyllables, it remains to decide which syllable bears primary stress. According to Borg/Azzopardi-Alexander (1997: 320), penultimate stress is the default in Maltese. The stress site is located in the ultimate however, provided the vowel in this syllable is long or a stress-shifting suffix is added or there is a heavy coda with heavy meaning the presence of a geminate consonant or a consonant cluster.[5] The distribution of the polysyllabic Maltese Πs over the possible two stress sites can be gathered from Figure 2.6.

All six of the trisyllabic Πs and the tetrasyllabic Π give evidence of penultimate stress. The remaining ten cases of Πs with penultimate stress divide into six Semitic and four Italo-Romance Πs. There are altogether thirteen Πs with stress on the final syllable. In this subclass of the disyllables, the Italo-Romance share is even higher (though only by a very narrow margin) as shown in Figure 2.7.

Eleven of the 30 polysyllables are Πs with initial *ma'*- or *minn-*.[6] Six of this group of eleven display ultimate stress. What can be concluded with regard to polysyllabic Maltese Πs is that penultimate stress is far less dominant than in the general lexicon of the language. The increased importance of ultimate stress results primarily from the integration of Italo-Romance Πs and the lexicalisation of complex Πs. It is worth noting that ultimate stress in Italo-Romance Πs is secondary in the sense that the

5 Borg/Azzopardi-Alexander (1997: 320) assume tautosyllabic ternary consonant clusters in CODA position. This assumption contradicts the same authors' model of the canonical syllable structure which allows only for binary clusters in this position (Borg/Azzopardi-Alexander 1997: 313) to which we refer in Section 2.1.2. Whether syllable-final CCC-sequences result from processes of resyllabification at the syllable contact in polysyllables cannot be ascertained since our source does not provide examples of ternary clusters in polysyllables. A possible pattern from which final CCC-sequences arise is discussed however in connection with the negative clitic *-x* /ʃ/ which can be attached to words ending in CC. Since this situation arises typically with perfective verbs in the 1st/2nd singular marked by the suffix *-t*, Borg/Azzopardi-Alexander (1997: 307) argue that the phonological sequence /t/ + /ʃ/ yields the affricate [ʧ] on the phonetic level so that it is possible to reanalyse the ternary cluster as a binary one with the affricate constituting the offset. In any case, ternary clusters in the coda are possible only if a combination of inflection and cliticization occurs. From the lexicon however, extra-heavy CODAS seem to be excluded.

6 As discussed in Section 1.6.4.2.

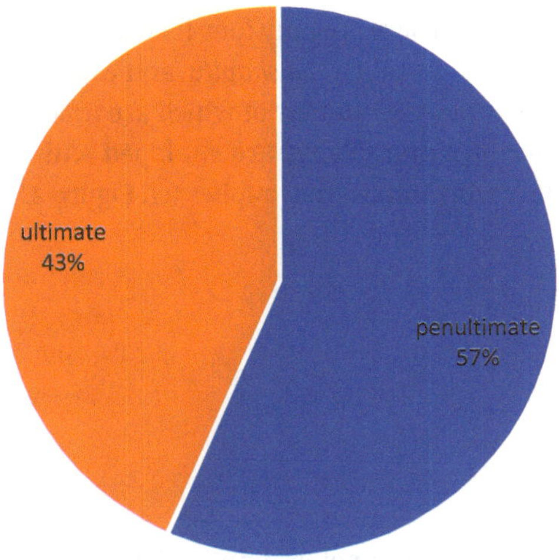

Figure 2.6: Distribution of polysyllabic Πs over stress sites.

	penultimate	ultimate
Italo-Romance	5	6
Semitic	6	7

Figure 2.7: Shares of language of origin per stress site in disyllables.

original penultimate stress remained in place also after the unstressed final vowel underwent apocope.[7]

7 Cf. Section 1.6.6 on this and related processes of adaptation.

What else can be said about polysyllables? There are only four Πs whose initial segment is a vowel, namely (Π1) *apparti* 'apart from', (Π2) *a skapitu* 'at the expense of', (Π25) *inkluż* 'including', and (Π44) *oltre* 'beyond' all four of which are Italo-Romance. All other Πs, be they polysyllabic or monosyllabic, are equipped with an initial consonant. This means that the following implication applies (cf. Figure 2.8).

initial vowel ⊃ { Italo-Romance origin / polysyllabicity }

Figure 2.8: Implicational pattern of vowel-initial Πs.

Furthermore, it is possible to complement this with another implication which we disclose in Figure 2.9.

Semitic origin ⊃ initial consonant

Figure 2.9: Implicational pattern of Semitic Πs.

Why there are no genuine cases of vowel-initial Semitic Πs is discussed below.[8]

When we extend the discussion to the right margin of the Πs, we notice that no tri- or tetrasyllabic Π ends in a consonant. Thus, Figure 2.10 features a third implication.

σσσ(σ) ⊃ final vowel

Figure 2.10: Implicational pattern of tri- and tetrasyllabic Πs.

BLOMP 2.0 contains seven monosyllables which end in a vowel. All of these Πs are Semitic, viz. (Π7) *bi* 'with', (Π9) *bla* 'without', (Π15) *fi* 'in', (Π19) *ġo* 'in', (Π30) *ma'* 'with', (Π51) *sa* 'till', and (Π55) *ta'* 'of'. Accordingly, monosyllabicity, final vowel, and Semitic origin are correlated to each other implicationally as shown in Figure 2.11.

monosyllabic & final vowel ⊃ Semitic origin

Figure 2.11: Implicational pattern of monosyllabic Πs with final vowel.

This implication guides us directly to Section 2.1.2 where we scrutinise among other things the structure of monosyllables.

8 Cf. Section 2.2.

2.1.2 Syllable structure

Saari (2003: 101), as well as Schmidt/Vorholt/Witt (2020: 248), do not differentiate between the different syllabicity classes. The structures of the ΠIs are represented as sequences of C-symbols and V-symbols resulting in a plethora of different patterns glossing over syllable boundaries. For the purpose of this study, we prefer to separate monosyllables from polysyllables in order to make meaningful statements about the syllable structure patterns of the former. In what follows in this section, we only make statements about ΠIs whose segmental extension amounts to exactly a syllable.

We take Vennemann (1988: 5–6) as our frame of reference for the analysis of syllable structures. This means that we assume the hierarchical model of the syllable presented in Figure 2.12. The model is meant to hold universally. Thus, it should also be applicable to Maltese. The immediate constituents of the syllable are HEAD and RHYME. The latter can be complex provided there is a CODA. The red circles and the green triangle single out pairs of constituents of the syllable which are believed to interact with each other systematically but which do not form proper constituents according to the logic of the architecture of Vennemann's model. The syllable BODY comprises HEAD and NUCLEUS. HEAD and CODA form the syllable shell. The only obligatory constituent is the NUCLEUS whose slot is normally filled by a vowel or a syllabic consonant whereas the positions to the right and the left of the NUCLEUS are optionally occupied by consonants which may give rise to clusters. The leftmost consonant is labelled ONSET, the consonant at the opposing extreme is called OFFSET. Consonants positioned between ONSET/OFFSET and the NUCLEUS constitute the SLOPE. In polysyllables, the syllable CONTACT involves those segments between which the boundary separating two consecutive syllables in the same segmental chain has to be drawn.

Borg/Azzopardi-Alexander (1997: 313) postulate the schema in Figure 2.13 as the canonical syllable type of Maltese. We have added the square brackets and indexes.

This canon is largely in line with the model proposed by Galea/Ussishkin (2018: 56) whose additions to the list of possible syllable structures do not affect our subsequent interpretation of the facts attested within the domain of ΠIs. A Maltese syllable thus consists minimally of a vocalically filled syllable NUCLEUS whereas both HEAD and CODA can optionally be filled by three or two consonants, respectively. According to Maddieson's (2005: 54) ternary typology, Maltese – though being absent from Maddieson's sample – belongs to the (second most frequently attested) class of languages with complex syllable structure. This further means that there are uncovered-open, covered-open, open-closed, and uncovered-closed syllables in the general lexicon of Maltese. Word-finally, only binary consonant clusters are

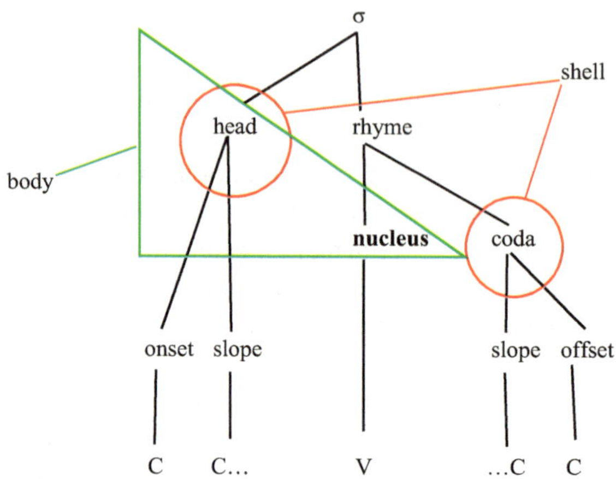

Figure 2.12: Internal structure of syllables according to Vennemann (1988).

$[[(C_{ONSET})\ (C_{SLOPE})\ (C_{SLOPE})]_{HEAD}\ [V_{NUCLEUS}\ [(C_{SLOPE})\ (C_{OFFSET})]_{CODA}]_{RHYME}]_{\sigma}$

Figure 2.13: Maltese canonical syllable structure.

legit whereas the maximum size of initial clusters is three with severe restrictions as to the onset consonant in ternary clusters (Borg/Azzopardi-Alexander 1997: 308–310). Purely vocalic monosyllables like the conjunction *u* 'and' are rare. Some of the other patterns occur more frequently. Examples of monosyllables realising the above patterns in the general lexicon are: VC *id* 'hand', VCC *art* 'ground', CV *la* NEGATION, CCV *fra* 'lay brother', CCCV *stma* 'esteem', CVC *dar* 'house', CVCC *borġ* 'heap', CCVC *flus* 'money', CCCVC *xprun* 'spur', CCVCC *spint* 'too uninhibited', and CCCVCC *żgwinċ* 'oblique' (cf. also Borg/Azzopardi-Alexander 1997: 306). Not all of these patterns are reflected in the Π-inventory.

From Figure 2.5, we know already that there are thirty monosyllables amongst the Maltese Πs. Figure 2.8 tells us that no monosyllabic Π starts with a vowel. This means that uncovered monosyllables are counted out generally so that V, VC, and VCC are impossible patterns for Maltese Πs. The tetrachoric table in Table 2.2 reveals the facts on the highest possible level of syllable structure. We will zoom in on the details below.

The overwhelmingly preferred pattern is that of the covered-closed syllable which accounts for 77% of all monosyllabic Πs. These monosyllables represent the set of Πs which are equipped with a syllable SHELL. According to Figure 2.11, the seven instances of monosyllables with empty CODA (= Πs which consist only of a syllable BODY) must refer to Semitic items. As to the complexity of the syllable margins, Figure 2.14 shows that HEAD and CODA have the same leanings towards the presence

Table 2.2: Head-coda combinations in monosyllables.

		coda		sum
		empty	filled	
head	empty	0	0	0
	filled	7	23	30
total		7	23	30

of a single consonant. Clusters have a slightly stronger position in the domain of CODAS than they have in the domain of the syllable HEAD.

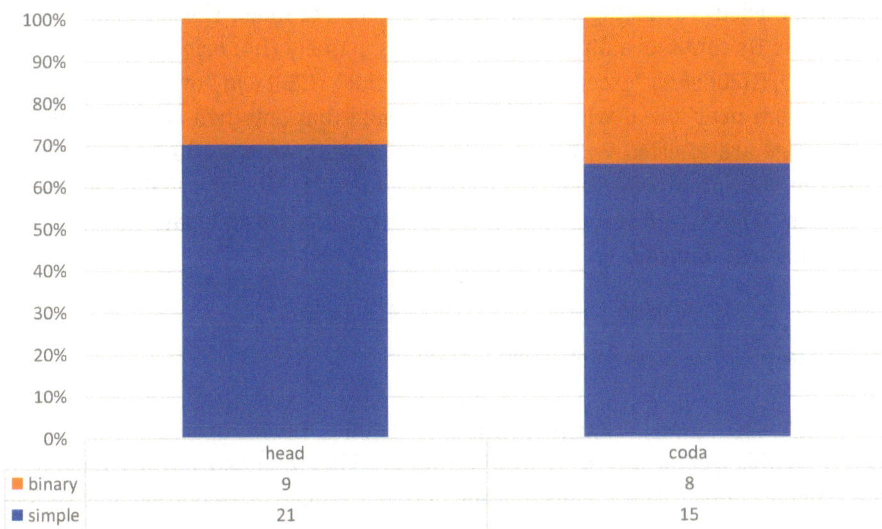

Figure 2.14: Simple vs binary heads and codas of monosyllables.

Of the thirty monosyllables which attest to a filled HEAD nine (= 30%) give evidence of an initial binary cluster, i.e. there is a SLOPE. As to the coda, we find eight (= 35%) instances of binary clusters of the SLOPE-OFFSET kind. Only two Πs allow for clusters at both margins, namely the Italo-Romance borrowings (Π52) *sforz* 'thanks to' and (Π53) *skont* 'according to'. No Semitic Π tolerates two clusters in the syllable SHELL. In Figure 2.15 SHELLS of this kind are classified as heavy.

[CC. . .CC]$_{SHELL}$ ⊃ Italo-Romance origin

Figure 2.15: Implicational pattern of monosyllabic Πs heavy SHELLS.

We emphasise again the fact that the final cluster in the two Italo-Romance Πs is secondary because the original final vowel /o/ (raised to /u/) was lost in the borrowing process. The most common SHELL contains one consonant per margin. The combination C_{HEAD}-C_{CODA} covers nine different Πs. The SHELLS CC_{HEAD}-C_{CODA} and C_{HEAD}-CC_{CODA} with asymmetrically filled margins are attested six times each. No Π with a ternary cluster on either side of the NUCLEUS has been identified. Monosyllables with a syllable SHELL outnumber monosyllables which consist only of the syllable BODY by a ratio of three-to-one.

The NUCLEUS does not pose many problems. Two dozen monosyllabic Πs involve a simple nucleus consisting of a vowel (either short or long). The segmental chain of only six Πs contains a diphthong as nucleus, namely (Π4) *bejn* 'between', (Π14) *fejn* 'near', (Π20) *għajr* 'except', (Π23) *ħdejn* 'beside', (Π28) *lejn* 'towards', and (Π43) *mnejn* 'from near' all of which are Semitic. Note that only two of seven Maltese diphthongs are attested with monosyllabic Πs (Borg/Azzopardi-Alexander 1997: 299). The same diphthongs /ɛɪ/ and /ɐɪ/ are also attested with polysyllabic Semitic Πs in Maltese. Figure 2.16 reveals that there is another interesting pattern connected to the language of origin.

diphthong ⊃ Semitic origin

Figure 2.16: Implicational pattern of Πs with diphthong.

Turning our attention back to polysyllables we notice that there are sixteen Πs with heterosyllabic medial clusters (excluding ambisyllabic geminates) which emerge at the syllable contact. This means that the majority of 55% of all polysyllabic Πs displays word-internal consonant clusters. Seven (= 44%) of these sixteen cases have an Italo-Romance origin. The share of Italo-Romance in this class of polysyllables is higher than the share Italo-Romance Πs have of all polysyllables (cf. Figure 2.5). The Italo-Romance impact on the shape of Maltese Πs is even stronger if we differentiate the medial clusters according to the number of segments they involve. Only CC and CCC are attested. There are twelve binary clusters which cover 75% of all medial clusters. Of these dozen Πs, nine are Semitic and three Italo-Romance. As to the four ternary clusters, the Semitic share is nil since all four Πs are classified as Italo-Romance according to our criteria. These Πs are (Π25) *inkluż* 'including', (Π27) *kontra* 'against', (Π35) *minflok* 'instead of', and (Π44) *oltre* 'beyond' (cf. Figure 2.17).

CCC_{medial} ⊃ Italo-Romance origin

Figure 2.17: Implicational pattern of polysyllabic Πs with ternary medial clusters.

In all four cases, the syllable boundary runs between the single CODA consonant of the initial syllable and the binary HEAD of the following (= final) syllable (lower case $ marks the internal syllable boundary): *in$_\$$kluż, kon$_\$$tra, min$_\$$flok, ol$_\$$tre.* The CODA position of the initial syllable is filled by sonorants (nasal, lateral approximant) whereas the HEAD of the final syllables consists of an obstruent (plosive, fricative) as ONSET and a liquid (lateral approximant, rhotic) as SLOPE. These patterns raise the question of whether the Maltese Πs display any recurrent properties on the segmental level too.

2.2 Segments

In comparison to the suprasegmental level, the segments which are involved in the phonology of Maltese require a number of additional explanations before the analysis can be executed. To this end, we divide this section into two parts. We start with the individual phonemes in Section 2.2.1 and round off the description with Section 2.2.2 which takes account of the length of the segmental chains.

2.2.1 Individual phonemes

For a start, we look at the Maltese consonants in Section 2.2.1.1. Maltese vowels are discussed further below in Section 2.2.1.2. Since the approaches to Maltese phonology reflect different theories and models they tend to disagree upon the exact spell-out of the phonemics of Maltese. To circumvent the theory-related pitfalls of the debate, we adopt again the neutral position of phonological typology which is indebted largely to the ideas expressed by Gordon (2016: 43–122).

2.2.1.1 Consonants

Borg/Azzopardi-Alexander (1997: 299) postulate an inventory of phonemic consonants which comprises either 22 or 23 units. This oscillation between two options has to do with the uncertain status of the Italian loan phoneme /ʣ/. In Table 2.3 we present our version of the inventory of Maltese consonantal phonemes.

Table 2.3: Maltese consonantal phonemes.

	bilabial	labio-dental	denti-alveolar	post-alveolar	palatal	velar	glottal
plosive	p b		t d			k g	ʔ
nasal	m		n				
trill			r				
fricative		f v	s z	ʃ		*	h
affricate			ts	tʃ dʒ			
approximant	w				j		
lateral approximant			l				

This inventory counts 23 units although we have excluded all recent loan phonemes of which there are more than the above descriptive grammar assumes (Stolz/Levkovych 2021: 583).[9] Apart from the fact that no loan phoneme is relevant for the phonology of Maltese IIs, Table 2.3 raises another and more important issue. There is the grey shaded cell for the velar fricative (of unspecified phonation). In lieu of the expected IPA-symbols /x/ or /ɣ/ the asterisk * is used. Why is that?

Alexander Borg (1997: 249) claims that the Maltese "sound system has an 'abstract phoneme' represented by the orthographical symbol *għ* (called [a:yn]) in the M[altese] alphabet) and representing the reflex of the [Old Arabic]fricatives */ʕ/ and */ɣ/". This abstract phoneme has the effect of lengthening adjacent low (= /ɐ/ → /ɐ:/) or mid-high vowels (= /ɛ/ > /ɛ:/, /ɔ/ →/ɔ:/) whereas it diphthongises the high vowels /ɪ/ and /ʊ/ to /ɛɪ/ ~ /ɐɪ/ and /ɐʊ/ ~ /ɔʊ/, respectively. Hume et al. (2009: 43) conclude that the lengthening effect of the *għajn* is restricted to certain environments under certain prosodic conditions. Cassola (2014: 16) assumes that at least the voiced velar fricatives was still "alive and kicking" in the 16th century. A full account of the diachrony and diatopic variation of the *għajn* and sundry lost phonemes is given in Puech (2018). According to Alexander Borg (1997: 261), the voiced velar fricative realisation [ɣ] has survived in "[a] few conservative rural communities" on Gozo. What is more, the phoneme under scrutiny is realised as voiceless velar fricative [x] or voiceless pharyngeal fricative [ħ] under certain morphonological conditions (Alexander Borg 1997: 276) some of which are also relevant for Maltese IIs.[10] For the time being, we follow Cohen (1970: 129–131) and Vanhove

9 Alexander Borg (1997: 249) admits not only /ʣ/ but also the equally marginal loan phoneme /ʒ/ (borrowed from English) on the list of phonemes of Maltese.

10 These cases will be discussed in Section 3.

(1993: 9) who treat the *għajn* as fully-blown consonant even in those cases where it has no segmental status of its own. We consider this case to be different from the supposed zero-Π rebutted above[11] because the *għajn* produces identifiable effects on neighbouring segments whereas the absence of the Π has no such morpho-syntactic repercussions.

Not only does our treatment of the *għajn* explain why there are no vowel-initial Πs of Semitic origin in BLOMP 2.0 but this abstract consonant is also responsible for the relative homogeneity of the syllable structures. Moreover, it has to be counted on a par with other segments when we try to determine to what extent (if at all) the differences in segmental length of the Maltese Πs reflect systematic patterns. The role played by *għajn* in this context will come to the fore subsequently.[12]

Every consonant featured in Table 2.3 is attested in word-initial position in the general lexicon of Maltese (Borg/Azzopardi-Alexander 1997: 308). The situation is different in the domain of Πs. In Table 2.4, we mark those consonants which occupy the leftmost slot of the segmental chains of Maltese Πs. The colour code is as follows: pink identifies consonants which are admissible in this position exclusively with Semitic Πs whereas yellow highlights the opposite case, i.e. those consonants which function as word-initial segments only in Italo-Romance Πs. Green is reserved for consonants which are attested word-initially with Semitic as well as Italo-Romance Πs. Unmarked cells host consonants which are never attested as onset or head of any Π.

Table 2.4: Consonants in word-initial position.

	bilabial	labio-dental	denti-alveolar	post-alveolar	palatal	velar	glottal
plosive	p b		t d			k g	ʔ
nasal	m		n				
trill			r				
fricative		f v	s z	ʃ		*	h
affricate			ʦ	ʧ ʤ			
approximant	w				j		
lateral approximant			l				

With sixteen consonants being attested on the left margin, Maltese Πs make use of only 70% of the options offered by the inventory of phonemic consonants. If

11 Cf. Section 1.6.3.
12 In Sections 2.2.2 and 2.3.

we subtract the three consonants which can occupy the leftmost slot only in Italo-Romance ΠIs, the share goes down to 57%. We interpret this as a clear sign of the special behaviour of ΠIs in comparison to the general lexicon of the object language.

The individual consonants which are eligible for the word-initial position do not distribute evenly over the 56 consonant-initial Maltese ΠIs. Figure 2.18 reveals that the three most frequently attested initial consonants share the place feature [labial]. Word-initial consonants with the same frequency share a bar.

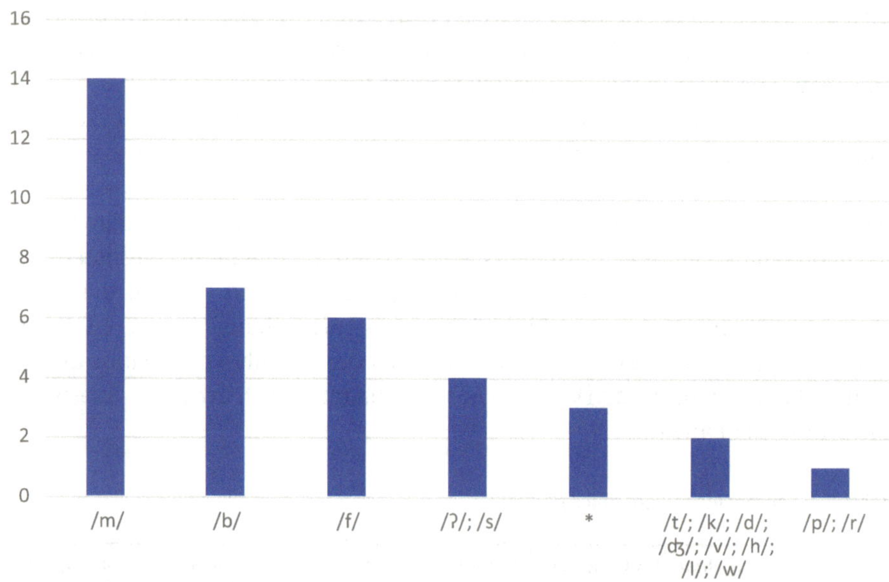

Figure 2.18: Frequency of word-initial consonants.

The bilabial nasal /m/, the voiced bilabial plosive /b/, and the voiceless labiodental fricative /f/ account for 27 of the 56 cases under scrutiny. If we add the results for the other consonants which display the same place feature – the voiced labio-dental fricative /v/, the bilabial approximant /w/,[13] and the voiceless bilabial plosive /p/ – the

13 The exact classification of Maltese /w/ is difficult. The phoneme chart provided by Alexander Borg (1997: 249) does not identify places and manners of articulation terminologically. Its organisation is however suggestive of the symbol /w/ being used to represent the bilabial approximant. The author characterises each of the sound classes except that of the approximants. With reference to /w/, Borg/Azzopardi-Alexander (1997: 303) speak of a voiced labio-velar approximant (for which the IPA has the symbol /ʋ/, however). It is likely that /w/ represents the voiced labial-velar approximant. Whatever the correct phonological analysis might be our argument remains unaffected because the feature [labial] is present either way.

share of labial consonants in word-initial position rises to 57% (with 32 out of 56 ΠIs). No other place feature can compete with [labial] because [denti-alveolar] with eleven cases, [glottal] with six cases, [velar] with five cases, and [postalveolar] with only two cases can lay claim only to considerably smaller shares. These differences are visible in Figure 2.19. The absence of the palatal place of articulation is worth noting.

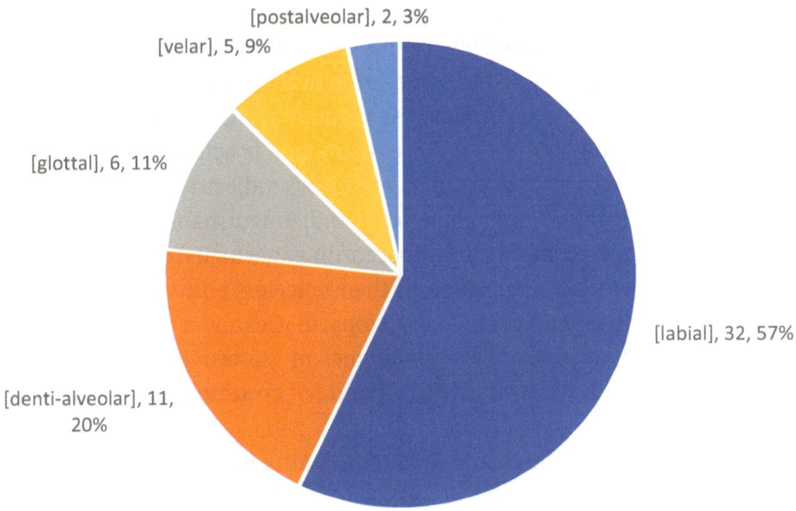

Figure 2.19: Shares of place features with initial consonants.

It is true that in the general lexicon of Maltese too, the distribution of initial consonants over the places of articulation is not egalitarian. However, if we impressionistically judge the relations by way of counting the pages in Aquilina's (1991) dictionary which feature /m/-initial entries (pp. 765–879 = 114 pages) and those containing /s/-initial entries (pp. 1244–1380 = 136 pages) it becomes doubtful that the phonology of the Maltese ΠIs simply replicates the phonology of the general lexicon of the same language.

Twelve of the fourteen /m/-initial ΠIs are lexicalised complex ΠIs with *ma-* or *minn-* as left constituent. This means that the synchronically biggest group of ΠIs is secondary in the sense that most of its members have arisen in the course of the univerbation of erstwhile sequences of different ΠIs or other syntagms. As a by-product of this development, /m/-initial ΠIs are preferably polysyllables and, except (Π35) *minflok* 'instead of', also of Semitic origin. Similarly, many /b/-initial ΠIs (with four out of seven cases) and to a much lesser extent also those with initial /f/ (one isolated case) result from univerbation processes. The preference for labial onsets/heads therefore is not so much phonologically motivated but the effect of diachronic processes in the morphosyntactic domain. Italo-Romance contributes five Πs with an initial labial to the inventory. Borrowing is also responsible for the participation of the voiceless bilabial plosive /p/ and the voiced labiodental fricative /v/ in the dominant pattern.

In the general lexicon of Maltese, the voiced obstruents /b/, /d/, /g/, /z/, /v/, and /ʤ/ are barred from occurring in word-final position – this is the opinion of Borg/ Azzopardi-Alexander (1997: 307). One might take issue with this interpretation because the above ban on the voiced obstruents can also be described in terms of obligatory final devoicing (mentioned by the same authors on the same page). If vowel-initial affixes are added to words whose segmental chain ends phonetically in a voiceless obstruent, the neutralised voice distinction is activated again as e.g. in minimal pairs like (noun) SG *salt* 'jump → PL *salti* vs (adjective) SG *sald* 'sound' → PL *saldi* with both singulars being phonetically indistinguishable, namely [sɐlt]. We assume a classical neutralisation analysis according to which the voiced obstruents are represented in word-final position by their voiceless positional allophones (Hall 2011: 97–99). Accordingly, pace Borg/Azzopardi-Alexander (1997: 307), we claim that, except *għajn*, all consonantal phonemes of Maltese are admissible to the word-final position with the proviso that /j/ and /w/ occur word-finally only as closing part of diphthongs.[14]

For the representation of the word-final consonants in Table 2.5 we employ the same colour code as before. With thirteen out of the 22 word-final consonants of the general lexicon, Maltese Πs select only 59% for this position in their inventory. Half a dozen of these phonemes are attested only in Semitic Πs. Four consonants occur both in Semitic and Italo-Romance Πs whereas three consonants are restricted to Italo-Romance Πs. There is no evidence of a Π hosting an approximant in word-final position. Likewise, the postalveolar and palatal places of articulation are unattested in this position. Were it not for the isolated case of (Π35) *minflok* 'instead of', the entire region between the denti-alveolar place of articulation and the glottal place of articulation would not be involved in word-final consonants of Maltese Πs. Italo-Romance influence is responsible for disallowing this generalisation.

There are 37 Πs which end in a consonant. These Πs cover 63% of the Π-inventory. Like in the case of the word-initial consonants, those which qualify as offset or coda do not distribute equally over the Πs as shown in Figure 2.20. Consonants with identical frequencies share a bar.

14 Cf. Section 2.2.1.2.

Table 2.5: Consonants in word-final position.

	bilabial	labio-dental	denti-alveolar	post-alveolar	palatal	velar	glottal
plosive	p b		t d			k g	ʔ
nasal	m		n				
trill			r				
fricative		f v	s z	ʃ			h
affricate			ʦ	ʧ ʤ			
approximant	w				j		
lateral approximant			l				

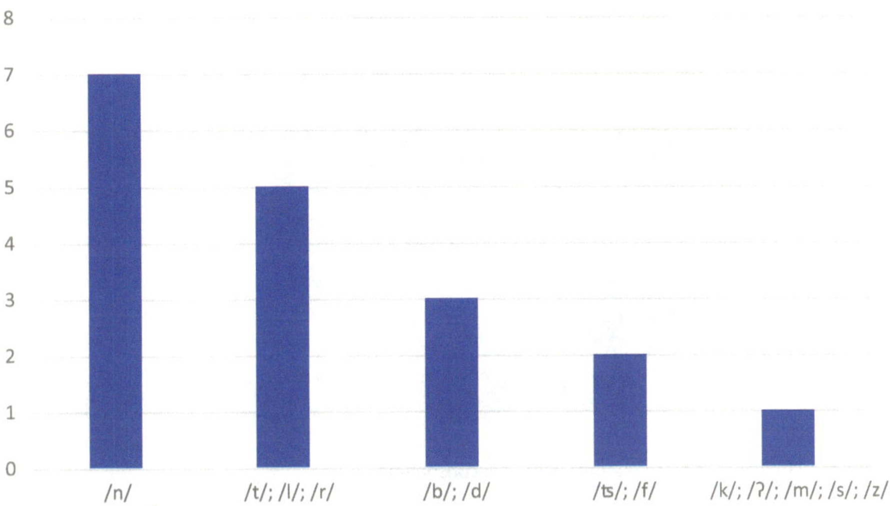

Figure 2.20: Frequency of word-final consonants.

The highest rank goes to the denti-alveolar nasal /n/ which is attested on the right margin of 19% of all IIs which end in a consonant. This is remarkable insofar as this phoneme is not among those consonants which are accepted in the word-initial position. The distribution of the denti-alveolar nasal is thus restricted. It is almost needless to say that, in the general lexicon of Maltese, /n/ is relatively common as onset or head of lexemes. In Aquilina's (1991: 879–929) dictionary, the representation of /n/-initial words requires fifty printed pages. Moreover, the four most frequently encountered word-final phonemes are all produced at the denti-alveolar place of articulation, namely the nasal /n/, the voiceless plosive /t/, the lateral approximant /l/, and the trill /r/. In word-initial position, none of these four consonants is attested with more than two ΠIs. In contrast, their type frequency is 5–7 word-finally. Together they account for 22 ΠIs which is equivalent of 59% of all ΠIs with a final consonant. If we add up the frequencies of all word-final consonants which share the place feature [denti-alveolar] the number of affected ΠIs increases by seven. With 29 cases, the place feature [denti-alveolar] is present in 78% of all ΠIs ending in a consonant. Figure 2.21 visualises the outstanding position of this privileged place feature.

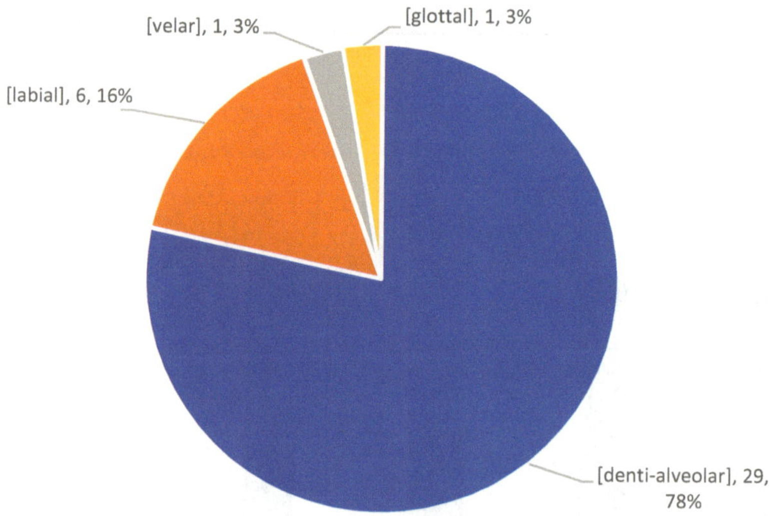

Figure 2.21: Shares of place features with final consonants.

The comparison of Figures 2.19 and 2.21 clearly shows that the opposing edges of the segmental chains of words differ as to the preferences they have in the domain of the places of articulation. It strikes the eye that the nasal manner of articulation ousts its competitors both on the left margin and on the right margin. ΠIs with a final labial are Semitic. Both Semitic and Italo-Romance contribute to the strong position of the denti-alveolar place of articulation.

Besides the word-initial and word-final consonants discussed in the previous paragraphs, there are also those consonants which occur in the SLOPE either to the right or the left of the NUCLEUS. If there is a SLOPE, there is also a consonant cluster. Borg/Azzopardi-Alexander (1997: 308–310) report on the restrictions to which clusters in the syllable HEAD and syllable CODA are subject. If we understand the authors correctly they assume that the voice harmony rules do not permit that phonemes with different phonation can be direct neighbours of each other in the segmental chain. There cannot be any doubt as to the existence of the voice harmony rule in Maltese. However, in contrast to our source, we conceive of voice harmony in terms of regressive assimilation which affects phonemes in such a way that they are represented by different allophones according to the phonological context. As in the above case of final devoicing, the otherwise phonemic phonation values are neutralised under assimilation. Assimilation is often triggered morphonologically as e.g. *kiteb* [kɪtɛp] 'he wrote' → *jiktbu* [jɪgdbʊ] 'they write'. For the purposes of this study, we distinguish between the abstract phonological and the concrete phonetical level. Only the former is of interest to us in the present context.

In the word-initial HEAD of Maltese ΠS, the following nine types of binary clusters are attested (only the first is attested twice): /b/ + /h/, /b/ + /l/, /h/ + /d/, /h/ + /l/, **/s/** + **/f/**, **/s/** + **/k/**, /ʔ/ + /r/, /d/ + /w/, and /m/ + /n/ all of which are also attested more or less frequently in the general lexicon. The sibilant-initial clusters are Italo-Romance (marked out in boldface). The cluster /s/ + /k/ is also the only example which involves a voiceless plosive in the SLOPE. In the other cases, the SLOPE contains segments with a higher degree of sonority (Vennemann 1988: 9), namely the voiced plosive /d/, the voiceless fricatives /f/ and /h/, the liquids /l/ and /r/, or the approximant /w/. These SLOPE consonants constitute only a small part of the consonants which are allowed in this position in the general lexicon where voiceless plosives for instance are frequent in SLOPES (Borg/Azzopardi-Alexander 1997: 309). In the word-final CODA, Maltese ΠS attest to ten different binary clusters (only the first is attested twice): /n/ + /d/, /n/ + /b/, **/n/** + **/t/**, **/r/** + **/d/**, **/r/** + **/ts/**, /ʔ/ + /t/, /ʔ/ + /s/, /l/ + /b/, /s/ + /t/, and /h/ + /t/. Boldface identifies those clusters which are exclusive to the Italo-Romance component of the inventory. The remaining seven patterns are Semitic. These clusters fit the description of word-final consonant clusters given in Borg/Azzopardi-Alexander (1997: 310) except that the Maltese grammarians consider final voiced obstruents to be representatives of the corresponding voiceless phonemes. What we learn from the descriptive grammar is that the number of different phonotactic combinations on the right margin of Maltese words is much higher than the small number of clusters attested in the Π-inventory.

Borg/Azzopardi-Alexander (1997: 311) state that "[a]ll the consonant clusters permissible as word-final clusters are permissible word-medially." We assume that this quote refers to binary clusters. The word-medial clusters mentioned in our source are heterosyllabic sequences. The ΠS give evidence of eleven different types of word-medial clusters (only the first is attested three times), namely /n/ + *, /n/ + /t/, /n/ + /k/, /n/ + /h/, /n/ + /w/, /d/ + /w/, /s/ + /w/, **/s/** + **/k/**, **/g/** + **/w/**, **/r/** + **/m/**, and **/r/** + **/s/**. Boldface is used to identify the Italo-Romance cases. The denti-alveolar nasal /n/ occurs in combination with five different consonants, the approximant /w/ is attested in four different clusters, the trill /r/ is the first consonant in two different clusters. Except /n/ + /t/ none of these clusters is also attested in the word-final position of Πs. Figure 2.17 has revealed already that in polysyllables ternary clusters occur only in Italo-Romance borrowings. There are four different CCC-clusters, namely /n/ + /f/ + /l/, /n/ + /k/ + /l/, /n/ + /t/ + /r/, and /l/ + /t/ + /r/. The syllable boundary separates the first consonant from the following two consonants so that the former functions as CODA of the first syllable and the other two consonants form the HEAD of the second syllable. None of these HEADs is attested as word-initial cluster in the Π-inventory.

These findings are hardly surprising since the limited size of BLOMP 2.0 does not provide many opportunities for clusters to emerge. Nevertheless, it is interesting to see that the denti-alveolar nasal /n/ is generally strong in post-nuclear positions whereas it is only marginally attested as SLOPE consonant in pre-nuclear position. The bilabial nasal /m/ is almost the mirror image of its fellow nasal in the sense that the stronghold of /m/ is on the pre-nuclear side as opposed to its negligible status on the right of the nucleus. At the extremes of the segmental chain, only the place features [labial] and [denti-alveolar] display a certain prominence. The denti-alveolar nasal also stands out in word-medial clusters. If the trill forms part of a cluster word-medially or word-finally, the Π is Italo-Romance (cf. Figure 2.22).

(C)/r/(C)$_{\text{medial/final}}$ ⊃ Italo-Romance origin

Figure 2.22: Implicational pattern of Πs with clusters involving /r/.

Both Maltese and Italian have geminate consonants. It is therefore unsurprising to find geminates also in the Maltese Π-inventory. The Maltese Πs of this kind are polysyllabic so that another unilateral implication can be put forward (cf. Figure 2.23).

geminate ⊃ polysyllabic

Figure 2.23: Implicational pattern of Πs with geminates.

Except for (Π45) *permezz* 'by means of' the geminate is always ambisyllabic across syllable boundaries. Of the ten Πs whose segmental chain involves geminates four are of Italo-Romance, namely (Π1) *apparti* 'apart from', (Π12) *faċċata* 'opposite', (Π45) *permezz* 'by means of', and (Π54) *sotta* 'under'. The six Semitic Πs are (Π3) *barra* 'outside', (Π18) *ġewwa* 'inside', (Π34) *minbarra* 'except', (Π39) *minħabba* 'on account of', (Π40) *minkejja* 'in spite of', and (Π49) *quddiem* 'in front of'. Πs with geminates account for 34% of all polysyllables in BLOMP 2.0.

2.2.1.2 Vowels

In Table 2.6, we present the vowel phonemes (monophthongs) of Maltese according to Borg/Azzopardi-Alexander (1997: 299).

Table 2.6: Maltese vowel phonemes (monophthongs).

	front	near-front	central	near-back	back
close	iː				uː
near-close		ɪ ɪː		ʊ	
open-mid	ɛ ɛː				ɔ ɔː
near-open			ɐ ɐː		

Except for the close high vowels and the near-close near-back vowel, all vowel qualities come in short and long versions. Quantity distinctions are phonemic. In addition to the above monophthongs, Maltese has a system of seven diphthongs which Borg/Azzopardi-Alexander (1997: 299) classify as phonetic: [ɐʊ], [ɐɪ], [ɛʊ], [ɛɪ], [ɪʊ], [ɔɪ], and [ɔʊ]. On the phonological level, they assume that the closing component of the diphthong allophonically represents the palatal or labial-velar approximant. Thus, all diphthongs with final [ɪ] are understood as sequences of V + /j/ while those ending in [ʊ] count as sequences of V + /w/. This decomposition of the phonetic diphthongs into a vocalic and a consonantal component produces a higher number of complex CODAS because in the Π-inventory a diphthong is always followed by a tautosyllabic sonorant. In contrast to Borg/Azzopardi-Alexander (1997: 299) but in accordance with Alexander Borg (1997: 270–271), we assume complex nuclei for these cases, i.e. the diphthongs are not only a phonetic phenomenon but also constitute a phonological unit.

There are seven Πs whose segmental chain contains a diphthong. This is equivalent to only 12% of BLOMP 2.0. Moreover, only the diphthongs /ɐɪ/ and /ɛɪ/ are attested. Diphthongs which involve (near-)back vowels are alien to the Π-inventory. The seven Πs which give evidence of diphthongs are Semitic without exception. On this basis, it is possible to formulate the complex implication in Figure 2.24.

$$\text{Diphthong} \supset \left\{ \begin{matrix} /\text{ɐɪ}/ \\ \\ /\text{ɛɪ}/ \end{matrix} \right\} \supset \_\_ \mathrm{C}_{[\text{SONORANT}]} \supset \text{Semitic}$$

Figure 2.24: Implicational pattern of Πs with diphthongs.

As to the monophthongs, it turns out that nine of the eleven phonemes featured in Table 2.6 also occur in the Π-inventory. As Table 2.7 reveals only the long quantities of the open-mid vowels (marked by grey shading) do not turn up in BLOMP 2.0.

Table 2.7: Attested vs unattested vowels (monophthongs).

	front	near-front	central	near-back	back
close	iː				uː
near-close		ɪ ɪː		ʊ	
open-mid	ɛ ɛː				ɔ ɔː
near-open			ɐ ɐː		

The segmental chains of Maltese ΠIs yield a total of ninety monophthongs 55 of which bear stress whereas 33 occur in atonic positions. We simplify matters by way of lumping together several categories of Table 2.7 to determine the impact of the language of origin on the frequency of the attested vocalic monophthongs in Figure 2.25.

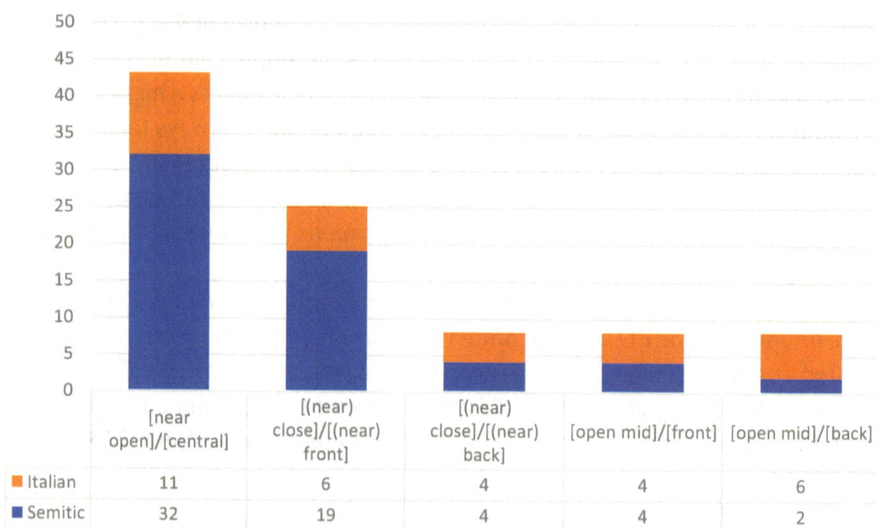

Figure 2.25: Vowel qualities (monophthongs) and language of origin.

The Semitic ΠIs overwhelmingly dictate the distribution of /ɐ(ː)/, /ɪ(ː)/, and /iː/. On the understanding that polysyllables account for several vowels at one go, there are 51 cases which involve Semitic ΠIs whereas only seventeen involve Italo-Romance ΠIs. Semitic outnumbers Italo-Romance by a ratio of 3-to-1. In contrast, the distribution of the less frequently attested vowels /uː/, /ʊ/, /ɔ/, and /ɛ/ is marked by an increase of Italo-Romance ΠIs. The latter account for thirteen cases as opposed to only ten cases which are found in Semitic ΠIs. Thus, the ratio of 1.3-to-1 is to the advantage of the Italo-Romance ΠIs.

In the foregoing sections, we have learned that the segmental phonemes of Maltese do not behave as one when it comes to forming ΠIs. Some phonemes are more often involved than others so that preferences emerge which are not necessarily the same as in the general Maltese lexicon. The differences which characterise the individual phonemes qualitatively tell only part of the story of the phonology of Maltese ΠIs. An important episode in the story remains to be told, namely the size of the segmental chains.

2.2.2 Segmental chains

With a view to determining the size of the segmental chains, we have to stipulate certain basics whose in-depth discussion would lead us too far away from the main topic of this study. We measure the size of a segmental chain by way of adding up the slots which are – virtually or phonetically – filled by a phonological unit. To keep the procedure as simple as possible, we adopt a very crude arithmetic according to which each slot-filler pair has the value '1'. Geminates, diphthongs, and long vowels spread over two consecutive slots so that they are counted as 2 x 1 = 2. The abstract phoneme *għajn* is assumed to fill a slot too.[15] Thus, the Italo-Romance ΠI (ΠI45) *permezz* 'by mean of' for instance, is registered as comprising seven segments.

The size of the segmental chains ranges from minimally two segments to a maximum of eight. If we correlate the different sizes and their type frequency with the languages of origin an interesting picture emerges as shown in Figure 2.26.

The 23 ΠIs with less than five segments are the monopoly of Semitic. Beyond this limit, the shares of Italo-Romance (15 ΠIs) and Semitic (22 ΠIs) are relatively close to each other with 48% and 52%, respectively. Accordingly, the implicational pattern in Figure 2.27 can be postulated.

15 Cf. Section 2.2.1.1.

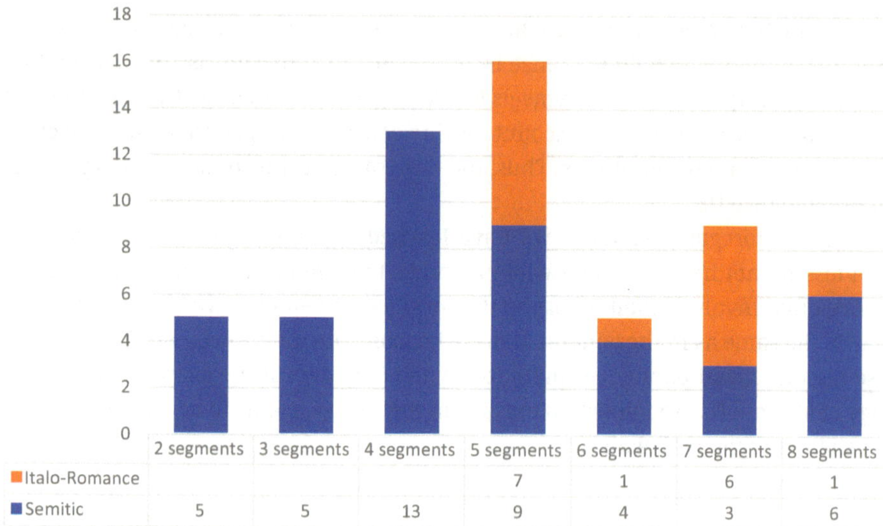

Figure 2.26: Length of segmental chains and language of origin.

Italo-Romance ⊃ n > 4 segments

Figure 2.27: Implicational pattern of Italo-Romance Πs with n = 5–8 segments.

The longest segmental chains – i.e. those which comprise more than five segments – yield polysyllables. Since some polysyllables count less than six segments, it is possible again to put forward a unilateral implication shown in Figure 2.28.

n > 5 segments ⊃ polysyllabic

Figure 2.28: Implicational pattern of Πs with n = 6–8 segments and syllabicity.

Furthermore, Πs whose phonological form involves geminates are never shorter than five segments (cf. Figure 2.29).

geminate ⊃ n > 4 segments

Figure 2.29: Implicational pattern of Πs with geminates and length.

The average number of segments per Π is 5.1. With 21 Πs about 35% of the members of BLOMP 2.0 reach or exceed this mark whereas the majority of 39 Πs (= 65%) remains below the average. In terms of types, it is therefore more normal for a

Maltese ΠΙ to involve less than six segments. In the subsequent section, we investigate whether this preference for shorter segmental chains is also reflected in the domain of token frequency.

2.3 Frequency and length

Table 1.11 in Part A informs about the token frequency of the members of BLOMP 2.0. in the Korpus Malti 3.0. We repeat that it was necessary to apply double book keeping to the items under scrutiny because many of them are commonly used also in functions other than those of proper ΠΙs. In this section, however, it is possible to return to the undifferentiated general token frequency. This is possible because we want to determine to what extent (if at all) token frequency and length are correlated to each other. Given that many expressions are members of several word classes, we have to abstract away from the focus on ΠΙ-uses. The correlatability of token frequency and length is independent of the word-class membership of the elements. Thus, in what follows, the frequencies given refer to the overall occurrences of the expressions and not to their occurrences as ΠΙs alone.

Schmidt/Vorholt/Witt (2020: 245–247) distinguish ten frequency classes in their study of the token frequency of the members of the original version of BLOMP in the Korpus Malti 3.0. Vorholt (2022) looks into the relationship between token frequency and length of ΠΙs for a sample of languages to which Maltese belongs.[16] We take up the quantitative issues raised in these previous publications by way of comparing the token frequency of the members of BLOMP 2.0 with the syllabicity and number of segments of the same ΠΙs. In Table 2.8, the ΠΙs are ranked top-down according to their decreasing frequency. The two columns on the right disclose how many syllables and how many segments form part of the word form which represents a given ΠΙ in the lexicon. Like in the case of Table 1.11, only full syntactic words are taken account of. Reduced and cliticised allomorphs are excluded from the count. Grey shading highlights the first dozen ranks of the hierarchy which host those ΠΙs which exceed the average of 262,900 tokens as calculated for column B in Table 1.11. The rows marked out in yellow feature Italo-Romance ΠΙs.

16 Cf. Section 1.5.9.

Table 2.8: Token frequencies of Πs in the Korpus Malti 3.0, syllabicity, and number of segments.

rank	Π	tokens	syllables	segments
1	(Π55) *ta'* 'of'	5,096,341	1	2
2	(Π41) *minn* 'from'	1,256,789	1	4
3	(Π17) *fuq* 'on'	1,151,992	1	3
4	(Π21) *għal* 'for'	1,036,925	1	4
5	(Π26) *kif* 'as'	787,519	1	3
6	(Π11) *dwar* 'about'	651,499	1	4
7	(Π14) *fejn* 'near'	489,321	1	4
8	(Π60) *wara* 'after'	444,289	2	4
9	(Π29) *lil* 'to'	404,833	1	3
10	(Π30) *ma'* 'with'	355,717	1	2
11	(Π6) *bħala* 'as'	328,932	2	6
12	(Π15) *fi* 'in'	319,840	1	2
13	(Π4) *bejn* 'between'	236,899	1	4
14	(Π46) *qabel* 'before'	229,852	2	5
15	(Π56) *taħt* 'under'	203,834	1	4
16	(Π28) *lejn* 'towards'	171,669	1	4
17	(Π59) *waqt* 'at the time of'	170,317	1	4
18	(Π27) *kontra* 'against'	167,406	2	6
19	(Π39) *minħabba* 'on account of'	159,460	3	8
20	(Π5) *bħal* 'like'	157,848	1	5
21	(Π49) *quddiem* 'in front of'	154,607	2	6
22	(Π33) *matul* 'during'	149,416	2	6
23	(Π7) *bi* 'with'	141,102	1	2
24	(Π3) *barra* 'outside'	136,814	2	5
25	(Π51) *sa* 'till'	130,402	1	2
26	(Π45) *permezz* 'by means of'	115,212	2	7
27	(Π36) *mingħajr* 'without'	107,066	2	8
28	(Π50) *rigward* 'concerning'	98,661	2	7
29	(Π31) *madwar* 'around'	95,815	2	6
30	(Π16) *fost* 'amongst'	92,813	1	4
31	(Π40) *minkejja* 'in spite of'	82,667	3	8
32	(Π9) *bla* 'without'	80,061	1	3
33	(Π53) *skont* 'according to'	70,443	1	5
34	(Π13) *favur* 'in favour of'	60,542	2	5
35	(Π35) *minflok* 'instead of'	45,642	2	8
36	(Π18) *ġewwa* 'inside'	45,315	2	5
37	(Π22) *għand* 'at s.o.'s place'	35,070	1	5
38	(Π38) *mingħand* 'from s.o.'	32,167	2	8
39	(Π25) *inkluż* 'including'	32,143	2	7
40	(Π34) *minbarra* 'except'	30,050	3	8
41	(Π48) *qrib* 'near'	29,384	1	4
42	(Π1) *apparti* 'apart from'	28,630	3	7
43	(Π10) *daqs* 'equal to'	28,592	1	4

Table 2.8 (continued)

rank	Π	tokens	syllables	segments
44	(Π24) *ħlief* 'except'	25,346	1	5
45	(Π47) *qalb* 'amidst'	18,935	1	4
46	(Π19) *ġo* 'in'	17,759	1	3
47	(Π58) *viċin* 'near'	16,751	2	5
48	(Π23) *ħdejn* 'beside'	16,489	1	5
49	(Π20) *għajr* 'except'	8,489	1	5
50	(Π52) *sforz* 'thanks to'	7,730	1	5
51	(Π12) *faċċata* 'opposite'	7,415	3	7
52	(Π44) *oltre* 'beyond'	2,984	2	5
53	(Π8) *biswit* 'facing'	2,958	2	7
54	(Π43) *mnejn* 'from near'	1,949	1	5
55	(Π32) *maġenb* 'close to'	1,720	2	6
56	(Π2) *a skapitu* 'at the expense of'	1,367	4	8
57	(Π57) *versu* 'towards'	228	2	5
58	(Π37) *mingħala* 'in s.o.'s opinion'	1	3	8
59	(Π42) *mintul* 'all along'	0	2	7
60	(Π54) *sotta* 'under	0	2	5

Superficially, the quantitative data in Table 2.8 fail to yield a clear picture. Monosyllabic Πs for instance, can be found in the upper half as well as in the lower half of the table. Phonological chains which comprise five segments share the same fate, in a manner of speaking. At the same time, we notice that of the twelve Πs which are attested more frequently than the average only one – (Π6) *bħala* 'as' on rank #11 – exceeds the average of 5.1 computed for segments. The same Π and (Π60) *wara* 'after' on rank #8 are the only examples of disyllables in the top section of the hierarchy where monosyllables are the rule.

The bird's eyes view shows that by and large the Maltese Πs confirm Zipf's Law in the sense that high token frequency goes hand in hand with relative shortness of an expression (Zipf 1935: 38–39) whereas particularly long segmental chains are found more often with low-frequency Πs. Figure 2.30 reveals that we are dealing with (strong) tendencies rather than with strict rules. The x-axis measures the token frequency whereas the y-axis determines the length of the expressions in terms of syllables (blue dots) and segments (orange dots). As to syllabicity, the blue dots remain close to the value 1 from the highest frequency down to token frequencies of half a million or less. Below this threshold, syllabicity starts to oscillate first between 1 and 2 and at the bottom of the hierarchy between 1 and 3. This means that polysyllables are by far more common with low-frequency Πs than with high-frequency Πs. Monosyllables are typical of the high-frequency section. Similarly, the orange dots first rise from the value 2 to that of 4 with the top-ranking Πs. With less frequent Πs, the number of segments is bigger than 2 in most of the cases. At the bottom of the hierarchy, particularly long segmental chains dominate the mean values.

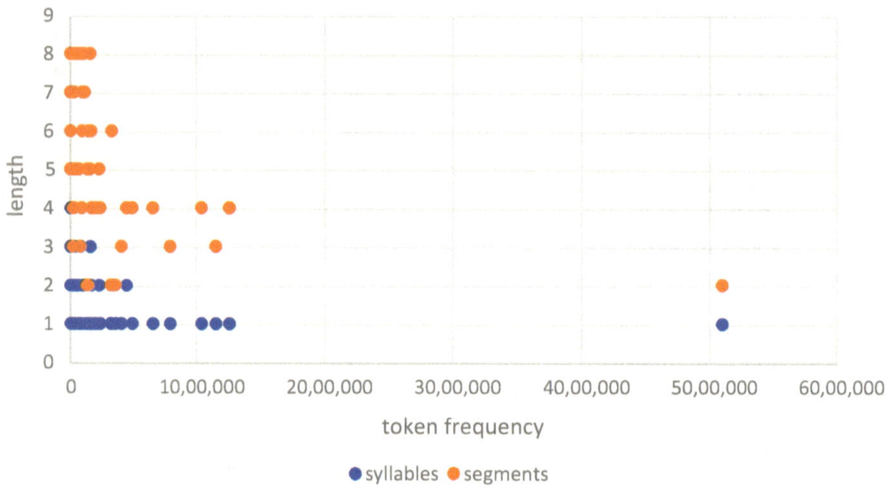

Figure 2.30: Frequency correlated to length (syllabicity and segments).

The Maltese Πs as a class are thus well-behaved because they meet the expectations of Zipf's Law.

Schmidt/Vorholt/Witt (2020: 251) state that "the four highest frequency classes [. . .] are only represented by Semitic prepositions". Our study replicates this finding since the highest-ranking Italo-Romance Π is (Π27) *kontra* 'against' on rank #18. Except for this Π and (Π45) *permezz* 'by means of' on rank #26, none of the Italo-Romance Πs has a token frequency of more than 100,000. The vast majority of the Italo-Romance Πs (twelve out of fifteen) have token frequencies below 80,000. There are only fifteen Semitic Πs in this low-frequency section of the hierarchy, i.e. it is typical for Italo-Romance Πs to be used relatively infrequently. This observation also holds for the two Italo-Romance monosyllables (Π53) *skont* 'according to' on rank #33 and (Π52) *sforz* 'thanks to' on rank #50. This means that monosyllabicity alone is not enough to promote a Π to the status of high-frequency Π. It is the other way around: high frequency increases the probability that a given Π is monosyllabic.

2.4 The canonical Π – phonology

In this section, we use the insights we have gained in the previous sections on phonological issues to complement the characterization we have given of the canonical Maltese Π in Section 1.6.1. The suprasegmental and segmental properties of Maltese Πs vary. Nevertheless, it is possible to formulate a number of generalisations which capture the most typical aspects of the members of BLOMP 2.0. The following preferences circumscribe some of the properties which are associated with the canonical Π:

1. Semitic > Italo-Romance
2. monosyllabic > polysyllabic
3. covered > uncovered
4. closed > open
5. [n < 5 segments] > [n > 5 segments]
6. [n < 262,900 tokens] > [n > 262,900 tokens]

According to preferences 2–4, the canonical Π should be a monosyllable equipped with both a HEAD and a CODA. Figure 2.14 suggests that SLOPES are generally not preferred so that the margins should host only single consonants. According to Figure 2.25, the best candidate for the NUCLEUS is a monophthong whose backness is zero. The most frequent vowel employed in segmental chains of Πs has the feature [low]. As to the quantity of the nuclear segment, short and long vowels obey a "complimentary quantity restriction" (Galea/Ussishkin 2018: 55) according to which a doubly filled CODA requires the vowel to be short whereas long vowels fit in with single CODA consonants. Since the canon has the structure CVC, the nucleus should bear the feature [long]. Note that this automatic rule adds a slot to the segmental chain so that CVC has to be reinterpreted as CV:C with the length symbol occupying a slot of its own. For the head, the feature [labial] is crucial whereas in the CODA [denti-alveolar] is the preferred choice. Since at both margins the most frequently attested manner of articulation is [nasal] the canonical Π looks like the hypothetical form in Figure 2.31.

C	V	C
m	*a:*	*n*

Figure 2.31: Segmental chain of the canonical Π.

The fact that there is no Maltese Π **ma:n* is perfectly in line with the philosophy of the canonical approach propagated by Corbett (2005). The canon is only an abstract yardstick which need not be attested at all. What we can do with this yardstick is measure the distances which separate the realised word forms from the canon.

To this end, we separate monosyllables from polysyllables since only the former display the same syllabicity as the canon. We stipulate eleven criteria which can be either fulfilled or not. Fulfilment is signalled by '1' in Table 2.9. If a given criterion is not met the corresponding cell hosts a '0'. Since all monosyllables are consonant-initial, the presence of a HEAD is not distinctive and can thus be ignored as such. What makes a difference for HEADS in relation to the canon however is whether there is a singleton consonant or a cluster. If there is a consonant, the question arises whether it realises the place feature [labial] and the manner feature [nasal]. For the NUCLEUS it is important to know whether it is a monophthong, lacks backness, can be characterised as [low] and [long]. Not all monosyllables are closed. Thus, it must be determined whether the CODA is filled or not. If there is a CODA, we need to know whether it is simple or complex. The consonant under inspection must have the place feature [denti-alveolar] and the manner feature [nasal]. If there is a SLOPE, the criteria are tested only for the onset or offset. The rightmost column in Table 2.9 reveals how many features a given Π shares with the canon. The bottom row gives the total of how many Πs share a given feature with the canon. Grey shading identifies those Πs which count among the most frequently attested ones in Table 2.8. The colour yellow marks the two Italo-Romance Πs. The Πs are ordered top-down according to the decreasing number of features shared with the canon.

Table 2.9: Canonicity of monosyllables.

	head			nucleus				coda				sum
	simple	labial	nasal	monophthong	-backness	low	long	filled	simple	denti-alveolar	nasal	
(Π41) *minn* 'from'	1	1	1	1	1	0	0	1	0	1	1	8
(Π21) *għal* 'for'	1	0	0	1	1	1	1	1	1	1	0	8
(Π5) *bħal* 'like'	0	1	0	1	1	1	1	1	1	1	0	8
(Π59) *waqt* 'at the time of'	1	1	0	1	1	1	0	1	1	1	0	8
(Π22) *għand* 'at s.o.'s place'	1	0	0	1	1	1	1	1	0	1	0	7
(Π43) *mnejn* 'from near'	0	1	1	0	1	0	0	1	1	1	1	7
(Π4) *bejn* 'between'	1	1	0	0	1	0	0	1	1	1	1	7
(Π14) *fejn* 'near'	1	1	0	0	1	0	0	1	1	1	1	7

Table 2.9 (continued)

	head			nucleus				coda				sum
	simple	**labial**	**nasal**	**monophthong**	**-backness**	**low**	**long**	**filled**	**simple**	**denti-alveolar**	**nasal**	
(Π29) *lil* 'to'	1	0	0	1	1	0	1	1	1	1	0	7
(Π11) *dwar* 'about'	0	0	0	1	1	1	1	1	1	1	0	7
(Π17) *fuq* 'on'	1	1	0	1	0	0	1	1	1	0	0	6
(Π26) *kif* 'as'	1	0	0	1	1	0	1	1	1	0	0	6
(Π30) *ma'* 'with'	1	1	1	1	1	1	0	0	0	0	0	6
(Π28) *lejn* 'towards'	1	0	0	0	1	0	0	1	1	1	1	6
(Π20) *għajr* 'except'	1	0	0	0	1	1	0	1	1	1	0	6
(Π10) *daqs* 'equal to'	1	0	0	1	1	1	0	1	0	0	1	6
(Π23) *ħdejn* 'beside'	0	0	0	0	1	0	0	1	1	1	1	5
(Π24) *ħlief* 'except'	0	0	0	1	1	0	1	1	1	0	0	5
(Π48) *qrib* 'near'	0	0	0	1	1	0	1	1	1	0	0	5
(Π56) *taħt* 'under'	1	0	0	1	1	1	0	1	0	0	0	5
(Π16) *fost* 'amongst'	1	1	0	1	0	0	0	1	0	1	0	5
(Π47) *qalb* 'amidst'	1	0	0	1	1	1	0	1	0	0	0	5
(Π15) *fi* 'in'	1	1	0	1	1	0	0	0	0	0	0	4
(Π55) *ta'* 'of'	1	0	0	1	1	1	0	0	0	0	0	4
(Π7) *bi* 'with'	1	1	0	1	1	0	0	0	0	0	0	4
(Π51) *sa* 'till'	1	0	0	1	1	1	0	0	0	0	0	4
(Π9) *bla* 'without'	0	1	0	1	1	1	0	0	0	0	0	4
(Π19) *ġo* 'in'	1	0	0	1	0	0	1	0	0	0	0	3
(Π53) *skont* 'according to'	0	0	0	1	0	0	0	1	0	1	0	3
(Π52) *sforz* 'thanks to'	0	0	0	1	0	0	0	1	0	0	1	3
total	21	12	3	24	25	13	10	23	15	15	8	

No Π reaches the potential maximum of eleven shared features. Four Πs realise eight of the eleven features of the canon. The Italo-Romance Πs yield low turnouts in terms of canonical features so that both candidates occupy ranks near the bottom of the hierarchy. At the top end of the hierarchy, there are two representatives of the highly frequent Πs. Six further members of this class share six to seven features with the canon. None of the features is shared by all thirty Πs. The absence of backness in the NUCLEUS is typical of 25 Πs. The nasal manner of articulation of the CODA consonant is the least common shared feature since it is realised only with eight Πs.

A different choice of criteria might have brought about slightly different results. We insist however on the fact that our own choice is not completely arbitrary. The outcome is interesting insofar as it shows that no Π comes very close to the canon because the highest number of shared features leaves a gap of three between shared and canonical features. Low frequency seems to correspond relatively often with a reduced number of shared features whereas high frequency results more often in degrees of canonicity which oscillate between 36% and 73%. It is clear that the grammar of Maltese IIs is by no means exhausted because apart from phonological properties, there are also morphological, syntactic, and semantic aspects to be taken account of before the canonical Maltese II can ultimately be determined. This means that Figure 2.31 is but a small ingredient which needs to be combined with others to yield the desired result. The next ingredient will be presented in Section 3 on morphological issues.

3 Morphology

This section looks at the morphological variability of Maltese Πs from two different angles. For a start, the focus is on the inflectional properties of the members of BLOMP 2.0. Section 3.1 and its further subdivisions survey the inflected Πs, the inflectional classes they constitute, the allomorphic processes they are subject to, and sundry issues which belong to word-based morphology. Section 3.2 is dedicated to the competition between long and short forms of certain Πs which is neither explicable with reference to word-based morphology nor in terms of sandhi phenomena. We look beyond the word-boundary in Section 3.3 where external sandhi and further aspects of a sentence's phonetic nature are highlighted. In Section 3.4, the results are made use of to provide a new summary account of allomorphy in the domain of Maltese Πs. Saari (2003: 111–152) addresses these issues for his own inventory of Maltese Πs in some detail but does not strictly differentiate between word-morphology and morphosyntax. The latter topic forms part of the contents of Section 4. Several of the phenomena to be discussed in the subsequent sections are also attended to in Schmidt/Vorholt/Witt (2020) for the previous version of BLOMP.

3.1 Inflectional morphology

3.1.1 Inflected vs invariant Πs

With 37 out of 60 Πs, a majority of almost two-thirds of the members of BLOMP 2.0 can be inflected whereas the minority of 23 Πs remains morphologically invariant. For the time being, (Π26) *kif* 'as', (Π42) *mintul* 'all along', and (Π43) *mnejn* 'from where' for which it is uncertain that they are subject to inflection are included in the latter category. In what follows we focus on type frequency.[1] In (3.1), we provide a list of all inflected Πs whereas the invariant Πs are presented in (3.2). Grey shading highlights doubtful cases.

(3.1) Inflected Πs
(Π4) *bejn* 'between'; (Π5) *bħal* 'like'; (Π7) *bi* 'with'; (Π8) *biswit* 'facing'; (Π10) *daqs* 'equal to'; (Π11) *dwar* 'about'; (Π13) *favur* 'in favour of'; (Π14) *fejn* 'near'; (Π15) *fi* 'in'; (Π16) *fost* 'amongst'; (Π17) *fuq* 'on'; (Π20) *għajr* 'except'; (Π21) *għal* 'for'; (Π22) *għand* 'at s.o.'s place'; (Π23) *ħdejn* 'beside'; (Π24) *ħlief* 'except'; (Π27) *kontra* 'against'; (Π28) *lejn* 'towards'; (Π29) *lil* 'to'; (Π30) *ma'*

1 General information about the token frequency of inflected Πs is provided in Section 3.4.

https://doi.org/10.1515/9783111331591-003

'with'; (Π31) *madwar* 'around'; (Π32) *maġenb* 'close to'; (Π33) *matul* 'during'; (Π35) *minflok* 'instead of'; (Π36) *mingħajr* 'without'; (Π37) *mingħala* 'in s.o.'s opinion'; (Π38) *mingħand* 'from s.o.'; (Π41) *minn* 'from'; (Π46) *qabel* 'before'; (Π47) *qalb* 'amidst'; (Π48) *qrib* 'near'; (Π49) *quddiem* 'in front of'; (Π53) *skont* 'according to'; (Π55) *ta'* 'of'; (Π56) *taħt* 'under'; (Π59) *waqt* 'at the time of'; (Π60) *wara* 'after'

(3.2) Invariant Πs (including doubtful cases)
(Π1) *apparti* 'apart from'; (Π2) *a skapitu* 'at the expense of'; (Π3) *barra* 'outside'; (Π6) *bħala* 'as'; (Π9) *bla* 'without'; (Π12) *faċċata* 'opposite'; (Π18) *ġewwa* 'inside'; (Π19) *ġo* 'in'; (Π25) *inkluż* 'including'; (Π26) *kif* 'as'; (Π34) *minbarra* 'except'; (Π39) *minħabba* 'on account of'; (Π40) *minkejja* 'in spite of'; (Π42) *mintul* 'all along'; (Π43) *mnejn* 'from near'; (Π44) *oltre* 'beyond';[2] (Π45) *permezz* 'by means of'; (Π50) *rigward* 'concerning'; (Π51) *sa* 'till'; (Π52) *sforz* 'thanks to'; (Π54) *sotta* 'under'; (Π57) *versu* 'towards'; (Π58) *viċin* 'near'

It strikes the eye that the three uncertain cases in (3.2) are properly included in the class of doubtful Πs of whose membership they cover three quarters with the unlikely (Π54) *sotta* 'under' contributing the fourth member.[3]

The shares of the morphologically variable and morphologically invariable Πs are disclosed in Figure 3.1. Morphological invariability can be associated with Doleschal's (1997) concepts of uninflectedness and uninflectability. Invariable Maltese Πs always instantiate the latter option.

This distribution corresponds closely to that presented by Schmidt/Vorholt/Witt (2020: 254) on the basis of the original version of BLOMP.

Before we address the issue of what it means to be an inflected Π in Maltese, we first investigate possible factors which correlate with morphological (in)variability. As in the previous section on phonological issues, we start from the idea that the language of origin might make a difference when it comes to inflecting Πs. If we go by the shares of Semitic and Romance Πs as featured in Figure 2.2, we expect very different results from those computed for the languages of origin in relation to the types of inflected and invariant Πs. Figure 3.2 reveals that Romance contributes only a relatively small number of Π to the class of inflected Πs whereas the share of Πs with a historical background in Romance is strikingly high in the class of invar-

2 Sandra Abela (personal communication) argues that (Π44) *oltre* 'beyond' inflects for person in analogy to the Πs in (3.1). We nevertheless keep it as one of the invariant Πs. This case will be looked into again in Section 4.1.2.2.2.1.

3 Their cases will be looked at specifically in Section 3.1.2.3. Until then they are always tacitly included when we refer to invariant Πs.

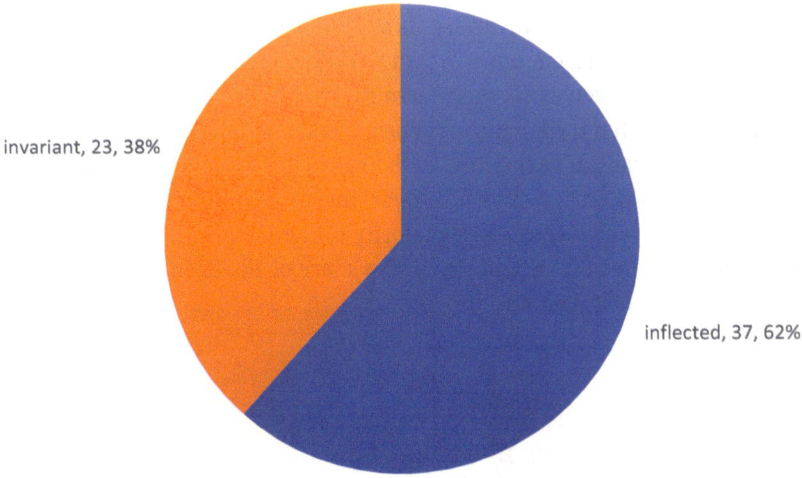

Figure 3.1: Shares of inflected vs invariant Πs.

iant Πs. It goes without saying that the reverse holds for the Semitic component of the Π-inventory: Semitic is strong in the domain of inflected Πs but relatively weak in that of invariant Πs.

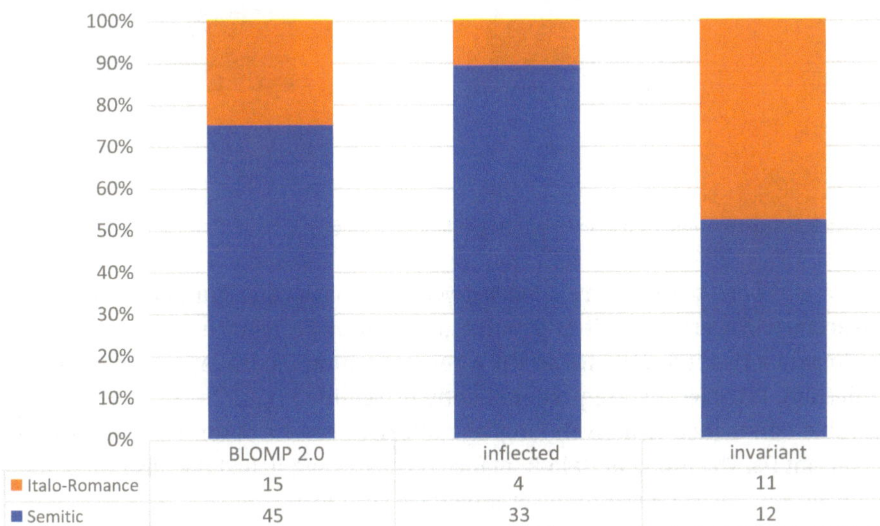

Figure 3.2: Shares of Semitic/Italo-Romance per inflected vs invariant Πs.

As is clearly visible, the Italo-Romance share shrinks considerably from a quarter of the entire Π-inventory to just 11% of the number of inflected Πs. At the same time, the Italo-Romance share of invariant Πs amounts to 47.8% and thus reflects a remarkable increase as opposed to the 25% in BLOMP 2.0.

The above picture receives further corroboration by the values featured in Figure 3.3. Here we see that the languages of origin display opposing preferences in the domain of morphological variability. In the case of Semitic, 73% of the Πs can be inflected as opposed to 27% which are invariant. The relation is the other way around for Italo-Romance since a 73.3%-majority of the Πs derived from Italo-Romance belong to the class of invariant Πs. With a share of 26.7%, inflected Italo-Romance Πs form the minority option.

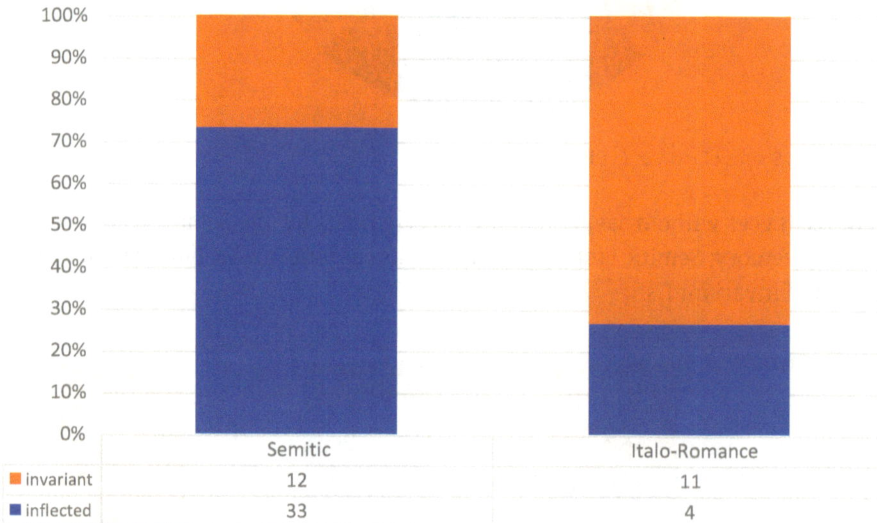

Figure 3.3: Shares of inflected vs invariant Πs per language of origin.

There is no one-to-one correlation between language of origin and inflection or invariance. Nevertheless, it is possible to speak of pronounced preferences. Italo-Romance Πs tend to be invariant whereas Semitic Πs display a preference for inflection. Inflected Πs are predominantly of Semitic origin. In contrast, no comparable dominance of a given language of origin can be observed in the case of invariant Πs. The influx of Italo-Romance Πs has strengthened the position of morphologically invariable Πs in the grammatical system of Maltese.

We know[4] that there is a similar connection between syllabicity and language of origin. As shown in Figure 2.3, BLOMP 2.0 contains balanced shares of monosyl-

4 From Section 2.1.1.

labic ПIs (= 50%) and polysyllabic ПIs (= 50%). This identity of the shares is replaced by more asymmetric distributions when we take account of the morphological (in) variability of the ПIs. From Figure 3.4, it transpires that, as registered in the lexicon, inflected ПIs tend to be monosyllabic rather than polysyllabic. Monosyllables account for 62% of all inflected ПIs leaving a share of 38% for polysyllables. The distribution resembles that of inflected vs invariant ПIs disclosed in Figure 3.1 above. Only one inflected ПI is trisyllabic, namely (ПI37) *mingħala* 'in s.o.'s opinion'. The minor role polysyllabic ПIs play in the domain of inflected ПIs fits in with the prominence Italo-Romance ПIs display in the realm of polysyllabic ПIs. According to the pattern reflected in Figure 3.5, monosyllabic ПIs overwhelmingly inflect whereas polysyllabic ПIs show no clear preference for morphological variability or invariability although there is a slight preference for the latter.

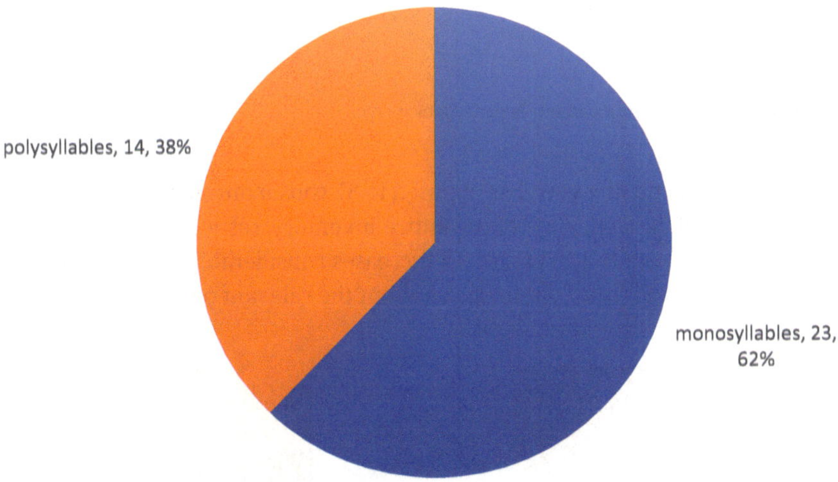

Figure 3.4: Shares of monosyllables/polysyllables with inflected ПIs.

As in the case of the language of origin, syllabicity is not a hard criterion which determines whether a given ПI inflects or not. Yet, again in parallel to the origin of the ПIs, we can state that there are pronounced preferences. Monosyllabicity and inflection go together well. Polysyllabicity is clearly disfavoured in the domain of inflection. Polysyllables cover 69.5% of all invariant ПIs so that monosyllables account for less than a third of the same syllabicity class.

In terms of syllable structure of the form in which the ПI is registered in the lexicon, final closed syllables are typical of ПIs which allow for inflection although final open syllables are by no means counted out. In BLOMP 2.0, there are alto-

	monosyllables	polysyllables
invariant	7	16
inflected	23	14

Figure 3.5: Shares of inflected/invariant Пs per syllabicity type.

gether 23 Πs which end in a vowel as opposed to 37 consonant-final Πs. This corresponds to shares of 38% and 62% of the entire inventory, respectively. The prominent position of Πs with the structure -C] increases considerably if we restrict the scope to inflected Πs. Figure 3.6 informs us about the relevant percentages.

	BLOMP 2.0	inflected
C]	37	30
V]	23	7

Figure 3.6: Shares of V-final vs C-final Пs.

With inflected ΠIs, the share of vowel-final ΠIs decreases to 19%. Final consonants can thus be considered to be particularly favourable to inflection whereas final vowels and inflection are less easy to accommodate. Note also that the majority of ΠIs (n = 10) with final consonant clusters or geminates allow for inflection whereas only three ΠIs with these properties are invariant. The assumption that C] facilitates inflection is further supported by the fact that far less than a third of all vowel-final ΠIs inflect as opposed to 81% of all consonant-final ΠIs as shown in Figure 3.7.

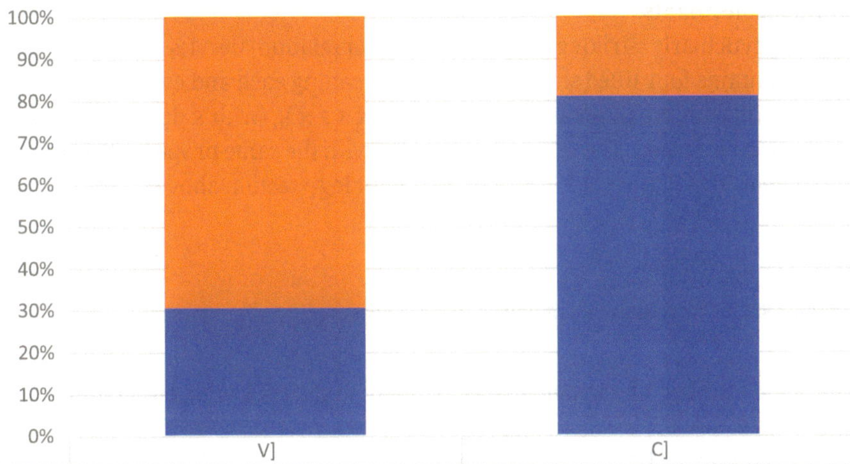

	V]	C]
invariant	16	7
inflected	7	30

Figure 3.7: Shares of invariant/inflected ΠIs per V-final vs C-final ΠIs.

Schmidt/Vorholt/Witt (2020: 256) assume that the possibility of a ΠI to inflect "seems to depend neither on the quality of the last segment nor on the number of syllables". Strictly speaking, there is indeed no 100% dependency. However, as we have noticed in connection with syllabicity, there are undisputable preferences which connect monosyllables with morphological variability more strongly than polysyllables are associated with inflection. Similarly, the presence of a final consonant in the lexicon form of the ΠI yields a remarkably high probability that the ΠI will also be inflected. In contrast, the probability of inflection is severely limited with vowel-final ΠIs as Schmidt/Vorholt/Witt (2020: 257) tell us when they conclude that "most prepositions which cannot be combined with personal suffixes end with a vowel."

On account of these preferences, it is interesting to determine whether the quality of the final consonant is in any way significant for the choice of invariance or inflection. According to Figure 2.20, there are only thirteen different consonantal phonemes which occupy the final slot in the segmental chain of the 37 consonant-final Πs. The seven consonant-final Πs which are invariant share the place feature [denti-alveolar] which is also the majority option for final consonants throughout the Π-inventory as shown in Figure 2.21. This means that phonologically there is nothing special about the quality of the final segments of invariant Πs because the same consonants are also permitted in the word-final slot of inflected Πs. Thus, Schmidt/Vorholt/Witt (2020) are right when they discard the possibility of a correlation between morphological (in)variability and the quality of final consonants.

A particularly striking unidirectional correlation ties Πs whose segmental chain contains four filled slots to inflection, meaning each and every of the thirteen Maltese Πs with the phonological structure [x x x x]$_{\Pi}$ (with x denoting a filler of a slot) can be inflected. There is no invariant Π with the same property. Figure 3.8 can be compared to Figures 2.27–2.29 in the phonology section above.

[x x x x] ⊃ inflected

Figure 3.8: Implicational pattern of Πs with four segments.

No other number of segments is as strongly associated with either inflection or invariance.

It remains to be seen whether frequency and morphological variability are interconnected in any meaningful way. Table 1.11 accounts for the token frequency of the lexicon form of the Πs in the Korpus Malti 3.0. As explained above,[5] inflected, cliticised, or reduced forms of Πs are excluded on purpose from the frequency count since the quantities associated with these forms will be looked at separately in dedicated sub-sections below. To determine whether morphological (in)variability and high or low token frequency have anything to do with each other generally, it suffices to take the information contained in Table 1.11. What one recognises immediately is that ten of the eleven Πs which exceed the average of 214,568.63 matches[6] in the Korpus Malti 3.0 belong to the class of inflected Πs. Only morphologically invariant (Π6) *bħala* ‘as’ is an exception to this strong tendency. This means that 91% of all Πs displaying high token frequency are inflected Πs. Below the average, we find 49 Πs 27 (or 55%) of which can be inflected but 22 (or 45%) cannot. Figure 3.9 helps to visualise these remarkable differences.

5 Cf. Section 1.6.8.

6 Based on the estimates in column C of Table 1.11.

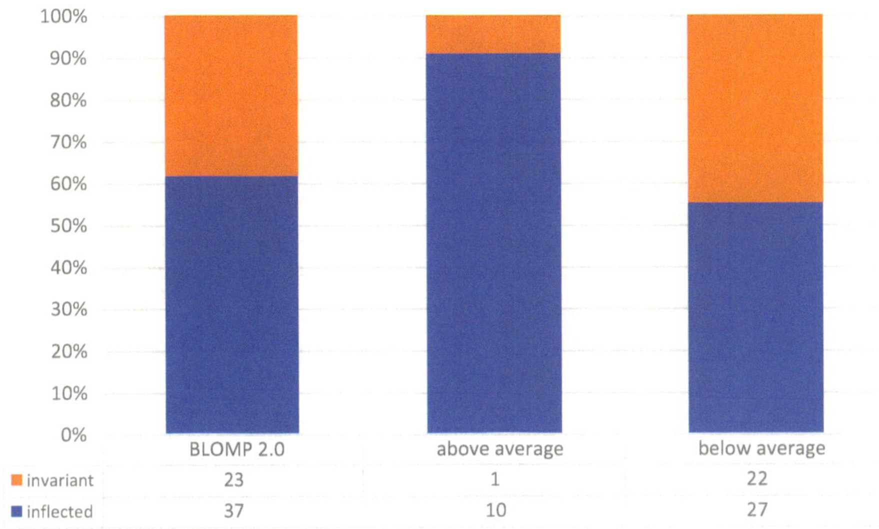

Figure 3.9: Shares of invariant/inflected Πs per token frequency classes.

High token frequency and morphological variability associate with each other relatively strongly whereas there is no discernible preference for either inflection or invariance below the average although the share of invariant Πs is bigger in the domain of low frequency than it is in the entire inventory. This result is further supported by the distribution captured in Figure 3.10.

What we see here is that 22 out of 23 invariant Πs fail to reach the average so that 95.6% of the invariant Πs yield relatively low frequency rates. In the case of inflected Πs, the share of low frequency Πs equals 73%. Invariance is thus rather strongly associated with low token frequency; high token frequency of an invariant Π being exceptional. High token frequency is also a minority phenomenon with inflected Πs but cannot be termed exceptional since more than a quarter of all inflected Πs count as highly frequent. This fits well into the trend observed by Schmidt/Vorholt/Witt (2020). They report that the invariant Πs are not present in the three highest frequency classes while inflected Πs are present in all frequency classes but the last one (Schmidt/Vorholt/Witt 2020: 256).

On the basis of the above findings, we are now in a position to zoom in on the central component of the canonical pattern shown in Figure 1.5.[7] The central component is of course the canonical Maltese Π. In accordance with the practice

7 As discussed in Section 1.6.1.

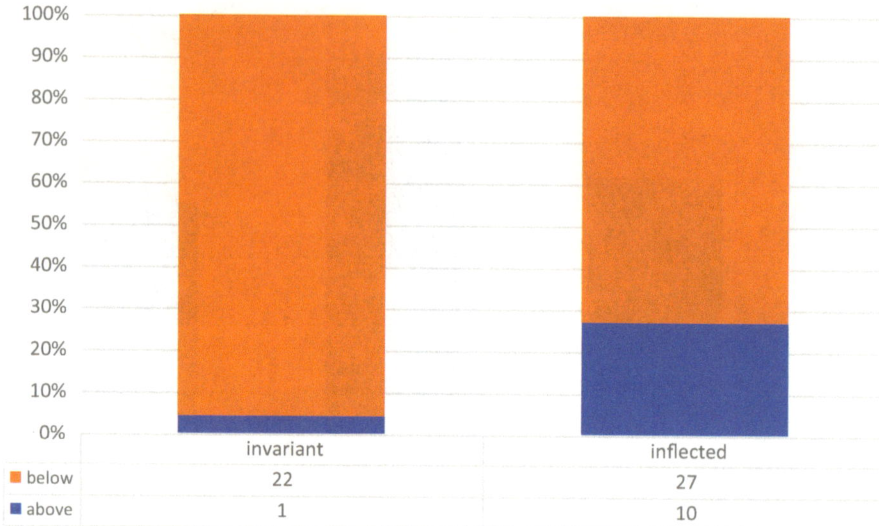

Figure 3.10: Shares of token frequency classes of invariant/inflected Πs.

employed above,[8] Figure 3.11 identifies the two combinations of properties which favour either inflection or invariance of a Maltese Π.

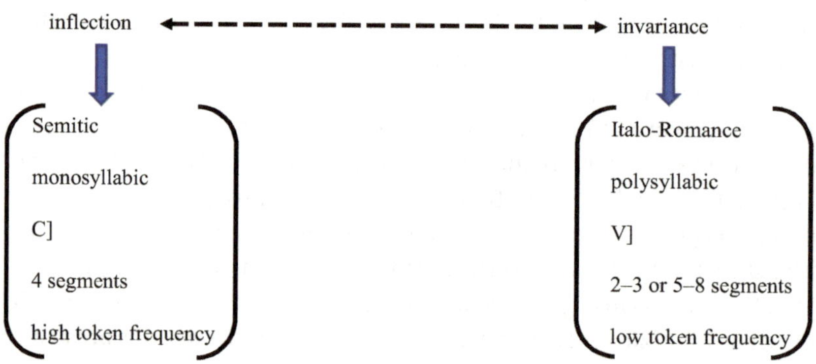

Figure 3.11: Continuum of preferred characteristics between inflection and invariance.

Since the majority of 61% of the Maltese Πs (cf. Figure 3.1) are inflected, we postulate that the left combination of properties is most typical for the members of BLOMP 2.0. The combination of properties on the right of the continuum, however,

8 Cf. Section 2.4 where the phonology of the canonical Maltese Π is described.

defines the most atypical kind of Maltese Π. In Table 3.1, we check which of the 37 inflected Πs fulfils how many of the criteria assumed to be prototypical for an inflected Π. MONO is short for monosyllabic, SEG abbreviates segments, and HF spells out as high frequency. Only if a criterion is fulfilled do we mark this in the appropriate cell. Yellow shading is indicative of an Italo-Romance Π. The Πs are ordered according to the decreasing number of fulfilled criteria which is revealed in the rightmost column.

Table 3.1: Number of prototypical criteria met by inflected Πs.

	Semitic	C]	MONO	4 SEG	HF	sum
(Π4) *bejn* 'between'	yes	yes	yes	yes	yes	5
(Π11) *dwar* 'about'	yes	yes	yes	yes	yes	5
(Π21) *għal* 'for'	yes	yes	yes	yes	yes	5
(Π41) *minn* 'from'	yes	yes	yes	yes	yes	5
(Π17) *fuq* 'on'	yes	yes	yes		yes	4
(Π29) *lil* 'to'	yes	yes	yes		yes	4
(Π5) *bħal* 'like'	yes	yes	yes	yes		4
(Π10) *daqs* 'equal to'	yes	yes	yes	yes		4
(Π14) *fejn* 'near'	yes	yes	yes	yes		4
(Π16) *fost* 'amongst'	yes	yes	yes	yes		4
(Π20) *għajr* 'except'	yes	yes	yes	yes		4
(Π22) *għand* 'at s.o.'s place'	yes	yes	yes	yes		4
(Π28) *lejn* 'towards'	yes	yes	yes	yes		4
(Π47) *qalb* 'amidst'	yes	yes	yes	yes		4
(Π48) *qrib* 'near'	yes	yes	yes	yes		4
(Π56) *taħt* 'under'	yes	yes	yes	yes		4
(Π59) *waqt* 'at the time of'	yes	yes	yes	yes		4
(Π15) *fi* 'in'	yes		yes		yes	3
(Π30) *ma'* 'with'	yes		yes		yes	3
(Π55) *ta'* 'of'	yes		yes		yes	3
(Π60) *wara* 'after'	yes			yes	yes	3
(Π23) *ħdejn* 'beside'	yes	yes	yes			3
(Π24) *ħlief* 'except'	yes	yes	yes			3
(Π7) *bi* 'with'	yes		yes			2
(Π8) *biswit* 'facing'	yes	yes				2
(Π31) *madwar* 'around'	yes	yes				2
(Π32) *maġenb* 'close to'	yes	yes				2
(Π33) *matul* 'during'	yes	yes				2

Table 3.1 (continued)

	Semitic	C]	MONO	4 SEG	HF	sum
(Π36) *mingħajr* 'without'	yes	yes				2
(Π38) *mingħand* 'from s.o.'	yes	yes				2
(Π46) *qabel* 'before'	yes	yes				2
(Π49) *quddiem* 'in front of'	yes	yes				2
(Π53) *skont* 'according to'		yes	yes			2
(Π13) *favur* 'in favour of'		yes				1
(Π35) *minflok* 'instead of'		yes				1
(Π37) *mingħala* 'in s.o.'s opinion'	yes					1
(Π27) *kontra* 'against'						0
total	33	30	24	16	10	113

There is a range from the maximum of five fulfilled criteria to the absence of any correspondence to the combination of properties which associate with inflected Πs. It is worth noting that the four Italo-Romance Πs which inflect are situated at or near the bottom of the ranking order in Table 3.1. (Π27) *kontra* 'against' is the only case of an inflected Π to violate all five of the criteria. The four Πs at the top of the hierarchy which stand out from the bulk of their competitors because they fulfil all five criteria are not too distant from the segmental chain of the canonical Π in Figure 2.31. The resemblance is especially strong since (Π4) *bejn* 'between', (Π11) *dwar* 'about', (Π21) *għal* 'for', and (Π41) *minn* 'from' constitute covered-closed syllables (provided, in the case of (Π21) *għal* 'for', that the abstract phoneme *għajn* is accepted as a consonantal element). Like the phonological canon, these Πs are represented by segmental chains with four slots. Apart from the largely identical syllable shells /m/ + /n/ of the canon in Figure 2.31 and (Π41) *minn* 'from' as well as the low vowel in (Π11) *dwar* 'about' and (Π21) *għal* 'for', the qualities of the segments involved in the phonological shape of the above top-ranking Πs do not directly reflect those of **maːn*.

To assess the validity of the above patterns, we scrutinise the invariant Πs according to their fulfilment of the same criteria as those used in Table 3.1. The results can be seen in Table 3.2. As usual, yellow shading is indicative of Italo-Romance Πs. In addition, we shade those rows grey which host the three Πs whose ability to inflect is doubtful. The green colour is used for the sole invariant Π whose token frequency surpasses the average.

Table 3.2: Number of prototypical criteria met by invariant Πs.

	Semitic	C]	MONO	HF	4 SEG	sum
(Π26) *kif* 'as'	yes	yes	yes			3
(Π43) *mnejn* 'from near'	yes	yes	yes			3
(Π6) *bħala* 'as'	yes			yes		2
(Π9) *bla* 'without'	yes		yes			2
(Π51) *sa* 'till'	yes		yes			2
(Π42) *mintul* 'all along'	yes	yes				2
(Π52) *sforz* 'thanks to'		yes	yes			2
(Π3) *barra* 'outside'	yes					1
(Π18) *ġewwa* 'inside'	yes					1
(Π19) *ġo* 'in'	yes					1
(Π34) *minbarra* 'except'	yes					1
(Π39) *minħabba* 'on account of'	yes					1
(Π40) *minkejja* 'in spite of'	yes					1
(Π25) *inkluż* 'including'		yes				1
(Π45) *permezz* 'by means of'		yes				1
(Π50) *rigward* 'concerning'		yes				1
(Π58) *viċin* 'near'		yes				1
(Π1) *apparti* 'apart from'						0
(Π2) *a skapitu* 'at the expense of'						0
(Π12) *faċċata* 'opposite'						0
(Π44) *oltre* 'beyond'						0
(Π54) *sotta* 'under'						0
(Π57) *versu* 'towards'						0
total	12	8	5	1	0	26

No invariant Π meets all five criteria. There are six invariant Πs which do not comply to a single criterion. These six Πs stem from Italo-Romance. The remaining four Italo-Romance Πs fulfil just one criterion, namely that of involving a word-final consonant. The highest number of fulfilled criteria is three – reached only by (Π26) *kif* 'as' and (Π43) *mnejn* 'from near' which are doubtful candidates for the status of inflected Πs. As expected, none of the invariant Πs is represented by a segmental chain with four slots. We interpret the differences which come to the fore when we compare Tables 3.1 and 3.2 as proof of the general validity of the set of criteria we have proposed for the characterisation of the canonical inflected Π in our object language.

Inflected and invariant Πs do not only differ on the parameter of morphological (in)variability but also display markedly different preferences for other properties which are not genuinely morphological. This means that there is a network of factors – from language of origin via syllabicity, syllable structure, and segmental length to token frequency – which interact such that a given Π is made fit for inflection (or, the other way around, unfit for it). Now that we have familiarised ourselves with these aspects of the inflection-invariance contrast in BLOMP 2.0, we can focus more specifically on the class of inflected Πs.

3.1.2 What it means to be an inflected Π in Maltese

For a start, we take a detour via language typology – though only for the first paragraphs of this section. In his chapter on person marking on adpositions, Bakker (2005: 198) reviews a worldwide sample of 378 languages 63 of which do not attest to the category of adposition. With 209 languages, the majority of the sample languages do not mark person on adpositions at all. In contrast, 106 languages show person marking either for pronouns alone (83 languages) or for pronouns and nouns (23 languages). The shares are revealed in Figure 3.12.

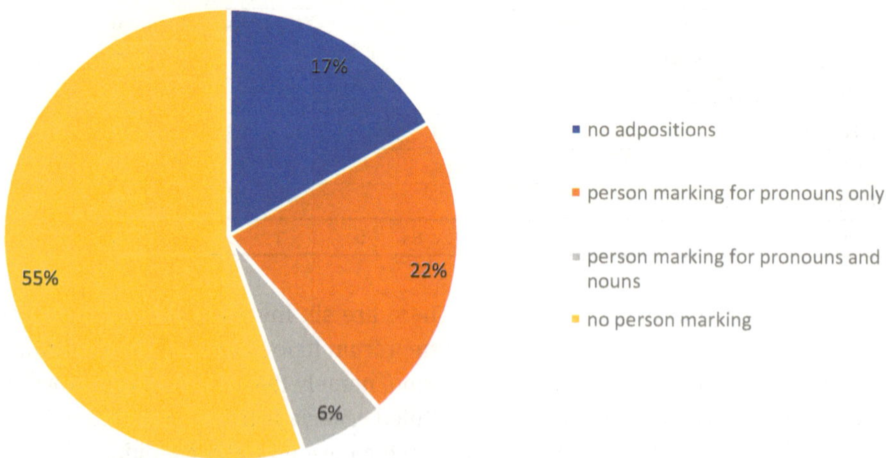

Figure 3.12: Share of person marking on adpositions in Bakker's (2005) sample.

Person marking thus occurs in about 28% of the sample languages. Maltese (which is absent from Bakker's sample) belongs to this sizable minority group – more precisely to those 22% of the sample for which person marking on adpositions is restricted to pronominal complements. A more fine-grained typology would probably have differentiated further according to prepositional and postpositional languages as well as according to the use of prefixes or suffixes to encode person. In the case of Maltese, we are facing a language which uses the suffixal strategy to mark person on its Πs. In what follows we review the patterns that arise in connection with this strategy. Throughout Section 3, we approach the subject matter exclusively from the synchronic point of view of descriptive linguistics.

According to the model of canonical inflection (Corbett 2007a: 23–24), it would be possible to produce a grid that represents the canonical paradigm of an inflected Maltese Π. To this end, the categories and their respective values that could ideally be expressed overtly on the word forms of the members of this paradigm must be multiplied with each other. For Maltese inflected Πs the relevant categories are person (1st, 2nd, 3rd), number (singular and plural), and gender (masculine and feminine). In combination, the three categories with their seven values yield a grid with twelve cells of which only eight host a word form in realised Maltese.[9] To avoid lengthy (though in principle interesting) discussions about the four phantom cells, we relegate the confrontation of the canonical paradigm with its attested counterpart to a separate follow-up study. In what follows we study the paradigms of Maltese Πs from the point of view of person marking. Number and gender will be discussed wherever this is called for.

3.1.2.1 Encoding person

Siewierska (2004: 145) states that "[i]n contrast to person agreement on predicates and possessed nouns, person agreement on adpositions has not been extensively studied." As we will see shortly, it is not only agreement that is understudied but

9 The discrepancy between the canonical paradigm and the realised paradigm is visible from the table in this footnote.

Table 3.I: Canonical vs realised paradigm.

person	**number**		**gender**	
	singular	**plural**	**masculine**	**feminine**
1st	X	X		
2nd	X	X		
3rd	X	X	X	X

The cells that host an X identify that part of the canonical paradigm that is identical with the realised paradigm whereas yellow shading indicates the excess cells that are only there because the formula according to which the canon is calculated requires their presence.

also person marking on adpositions, in general. Hagège (2010: 172–174) touches upon this phenomenon only in the context of supposedly verbal properties shared by adpositions. In cross-linguistic perspective, we are thus still largely underinformed as to person-inflection on adpositions.

In contrast, inflected Πs are usually given ample space in the descriptive and normative grammars of Maltese. In the course of the historical retrospection,[10] we repeatedly had the opportunity to refer to the attention various authors have paid to inflected Πs. Discounting minor details, there seems to be a general consensus as to what facts need to be mentioned in connection with the inflection of Maltese Πs (Borg/Azzopardi-Alexander 1997: 265–266, 274–277). An aspect we want to highlight especially is that of the allomorphy which characterises person marking.

Person inflection on Maltese Πs is triggered for the members of the class of inflected Πs if and only iff the complement of the Π is pronominal. In the presence of a lexical complement that does not function as apposition of a pronoun person inflection on the Π is barred. Person marking on the Π is thus not a case of agreement. This is an important trait that distinguishes Πs from pseudo-verbs since the latter host person-sensitive agreement markers.[11] Accordingly, person marking on Maltese Πs follows the principles of head-marking in as far as it instantiates person indexing in lieu of realizing person agreement (Haspelmath 2013: 197).[12] The basic structure of a Maltese Π which is inflected for person is schematised in Figure 3.13.

$\Pi_{\text{STEM-}\_\_\_\text{PERSON}}$

Figure 3.13: Basic structure of a Maltese Π inflected for person.

The structure is basically binary in the default. The left slot hosts the stem which is often but not always identical with the citation form of the Π. The right slot is reserved for the affix which encodes person. It is this affix that we are about to investigate in this sub-section.

In Maltese, the category of person is grammatically relevant for finite verbs, inalienably possessed nouns, and inflected Πs – typical targets of person marking according to Siewierska's (2004: 127–128) typological study. The sets of person exponents overlap in the sense that the affixes which attach to Πs are not unique but identical with those that mark the pronominal possessor in inalienable posses-

10 Cf. Section 1.4

11 Cf. Section 1.6.2.

12 How invariant Πs behave in combination with pronominal complements is discussed in some detail in Section 4.1.2.2. This is also the section in which the token frequencies of the different person inflections on Πs will be presented.

sion and, for a subset of the Πs, with the direct object clitics on transitive verbs. This is interesting insofar as each person value has its distinct morphological exponent under possession and for object marking whereas subject person markers on verbs display syncretism (Baerman/Brown/Corbett 2005: 4–7) both in the imperfective (*t-mur* 2SG.IPFV = 3SG.F.IPFV 'you$_{SG}$ go/she goes') and in the perfective aspect (*mor-t* 1SG.PFV = 2SG.PFV 'I/you$_{SG}$ went'). Since gender is formally distinguished only in the 3rd person singular where masculine and feminine constitute an opposition there is a paradigm of seven pronominal forms as shown in Table 3.3 for free-standing pronouns, inalienable possessors, and direct object clitics.

Table 3.3: Free pronouns and bound (non-subject) person markers.

person	free pronoun	possessor	object clitic
1SG	*jien(a)*	*-i ~ -ja*	*-ni*
2SG	*int(i)*	*-ek ~ -ok ~ -k*	
3SG.M	*hu(wa)*	*-h ~ -u*	
3SG.F	*hi(ja)*	*-ha*	
1PL	*aħna*	*-na*	
2PL	*intom*	*-kom*	
3PL	*huma*	*-hom*	

All the bound person markers mentioned in Table 3.3 are also employed for person inflection on Πs. There are no dedicated pronominal affixes which are exclusive to Πs. In example (3.3) for instance, we find the suffix *-ha* of the 3rd person singular feminine in three different functions. Grey shading is meant to help identify the relevant cases.

(3.3) [Buhagiar 2001: 7]

Is-soru *marret* *tiġri* ***[lejha]***$_{PP}$ *tgħinha,*
DEF-sister go:3SG.F.PFV 3SG.F.IPFV:run **[towards**:3SG.F] 3SG.F.IPFV:help:3SG.F
idha *sħuna* *fuq* *ġbin* *ix-xebba*
hand:3SG.F hot:F on brow DEF-young_woman
'The Sister went quickly [**towards** her] to help her, (putting) her hot hand on the young woman's brow [. . .].'

The inflected Π *lejha* 'towards her' consists of the stem *lej-* (of (Π28) *lejn* 'towards') and the suffix *-ha* for the pronominal complement. The finite verb *tgħinha* 'she helps her' involves *-ha* as object clitic whereas in the case of *idha* 'her hand' *-ha* indicates that the possessor of the body part is a 3rd person feminine.

Moreover, the obligatory phonologically conditioned allomorphy of person markers in the singular reported for inalienable possession and object clitics follows exactly the same rules in the domain of inflected Πs. The set of person markers which may fill the right slot in the schema presented in Figure 3.13 are identified once again in Figure 3.14.

		-i ~ *-ja* ~ *-ni*	/ 1SG
		-ek ~ *-ok* ~ *-k*	/ 2SG
		-u ~ *-h*	/ 3SG.M
PERSON	→	*-ha*	/ 3SG.F
		-na	/ 1PL
		-kom	/ 2PL
		-hom	/ 3PL

Figure 3.14: Person markers on Πs.

The choice of the possessor-like inflection vs the object clitic version in the 1st person singular does not depend on the phonological context.[13] What we need to know right here are the phonological conditions which trigger the selection of the allomorphs except for *-ni*.[14] Allomorphy is restricted to the speech-act participants in the singular and the 3rd person singular masculine. This distribution should be kept in mind because this pattern recurs also in other contexts to be discussed below.

There are two allomorphs each for the 1st person singular and the 3rd person singular masculine whereas there are three allomorphs for the 2nd person singular.[15] We take these allomorphs as basis for the distinction of two sets A and B of person markers.

– Set A → {*-i; -ek/-ok; -u; -ha; -na; -kom; -hom*}
– Set B → {*-ja/-ni; -k; -h; -ha; -na; -kom; -hom*}

13 We will address this issue separately in Section 4.1.2.2.2.2.3.

14 Whose status will keep us busy repeatedly in Sections 3.1.2.3.2 and 4.1.2.2.2.2.3.

15 Arguably this distribution of cells that are subject to allomorphy and those that are immune against it in Π-paradigms could be connected to Round's (2015: 42) notion of meromorphome. The issue is however, too complex to be discussed at any length in the context of a descriptive-linguistic study like ours.

These sets overlap as to the 3rd person singular feminine and all plural persons. In the 1st and 2nd person singular as well as in the 3rd person singular masculine the choice of the correct allomorph depends on the kind of Π-stem. Set A combines with consonantal Π-stems (except those with final /j/) whereas Set B is the choice for vocalic Π-stems and /j/-final stems as shown in Section 3.1.2.2. In this section, we only address the distribution rules for the allomorphs of the 1st person singular and the 3rd person singular and those of the 2nd person singular. The former two are rather straightforward examples of inward sensitivity of the person marker to the stem-final segments. We start with the 3rd person singular masculine. Any consonant except /j/ in this position triggers the vocalic allomorph of the person marker. In contrast, if the stem ends in a vowel or in /j/, only the person marker which involves a consonant is licit. This general rule is directly reflected in the allomorphy for the person markers of the 3rd person singular masculine as shown in Figure 3.15.

$$3\text{SG.M} \rightarrow \begin{cases} \text{-}u \quad / \quad C_{[\text{-palatal\&-approximant}]}\,\_\_ \\ \text{-}h \quad / \quad \begin{Bmatrix} V \\ j \end{Bmatrix}\,\_\_ \end{cases}$$

Figure 3.15: Allomorphy of 3SG.M person markers.

We illustrate the application of the rule with three sentences taken from one and the same primary source. In (3.4), (Π41) *minn* 'from' takes a pronominal complement in the 3rd person singular masculine. Since the Π-stem is identical with the citation form of (Π41) *minn* 'from' the phonological context offers a stem-final consonant other than /j/ so that the allomorph *-u* is chosen.

(3.4) [Bartolo/Vella 2009: 54]
Reggie ma kellux idea x' kien qed
Reggie NEG have.PFV:3SG.M:NEG idea what be PROG
jitlob ***[minnu]***$_{PP}$
3SG.M.IPFV:request **[from:3SG.M]**
'Reggie had no idea what he was asking **[of** him].'

In example (3.5), the stem-final segment of (Π15) *fi* 'in' is the vowel /i/ which requires the person marker to come in the shape of the allomorph /h/.

(3.5) [Bartolo/Vella 2009: 54]

id-dlam *ċappa* *li* *kien* *jinsab* ***[fih]***$_{PP}$ *kien*
DEF-darkness lump that be 3SG.M.IPFV:RFL:find [**in**:3SG.M] be
tant *iswed*
so black
'[. . .] the total darkness [**in** which] he found himself was so black [. . .].'

The same allomorph must be used for the 3rd person singular masculine in combination with (Π28) *lejn* 'towards' whose Π-stem is *lej-* ending in the palatal approximant /j/. This is the other phonological condition that calls for the use of the person marker allomorph *-h* as shown in (3.6). In what ways the Π-stem may differ from the segmental chain the Π is equipped with in the lexicon will be discussed below.[16]

(3.6) [Bartolo/Vella 2009: 27]

Iż-żewġt *it-tfal* *ħarsu* ***[lejh]***$_{PP}$ *bla* *ma*
DEF-two DEF-child.PL look:3PL.PFV [**towards**:3SG.M] without that
lissnu *kelma*
utter:3PL.PFV word
'The two children looked [**towards** him] without uttering a word.'

These rules apply strictly. The domains of the allomorphs are disjunct. Neither can replace the other in a given context.

The same holds for allomorphy for the encoding of a pronominal complement of a Π in the 1st person singular. The rule for the distribution of the two allomorphs of the 1st person singular is given in Figure 3.16. It is a replica of the previous rule shown in Figure 3.15.

$$\text{1SG} \rightarrow \left\{ \begin{array}{lll} \textit{-i} & / & C_{[\text{-palatal\&-approximant}]} \;\_\_ \\ \textit{-ja} & / & \left\{ \begin{array}{c} V \\ j \end{array} \right\} \;\_\_ \end{array} \right.$$

Figure 3.16: Allomorphy of 1SG person markers.

16 Cf. Section 3.1.2.2.

Sentence (3.7) features (П41) *minn* 'from' again. It is clear from the above that the only permissible allomorph encoding the 1st person singular is *-i*.

(3.7) [Abela 2016: 39]
Isma' ***[minni]***$_{PP}$
listen [**from**:1SG]
'Listen [**to** me]!'

Similarly, (П28) *lejn* 'towards' is combinable only with the allomorph *-ja* since the П-stem ends in the palatal approximant /j/. This is shown in (3.8).

(3.8) [Abela 2016: 72]
Għaliex qed iħares ***[lejja]***$_{PP}$
why PROG 3SG.M.IPFV:look [**towards**:1SG]
'Why is he looking [**at** me]?'

In (3.9), we provide an example of a vowel-final П-stem which takes the same allomorph as shown in (3.8) in connection with /j/ in stem-final position. If a pronominal complement in the 1st person singular is to be marked on (П7) *bi* 'with', the only possible choice is the allomorph *-ja*.[17]

(3.9) [Korpus Malti 3.0; literature14]
Forsi se tidħak ***[bija]***$_{PP}$ *Amy*
perhaps FUT 2SG.IPFV:laugh [**with**:1SG] Amy
'Perhaps you will laugh [**at** me], Amy.'

The rule regulating allomorphy with the 2nd person singular resembles those schematised in Figures 3.15–3.16. However, this time, the rule is responsible for three different outputs in lieu of just two. The formula is presented in Figure 3.17.

17 But cf. Sections 3.1.2.3.2 and 4.1.2.

$$2\text{SG} \rightarrow \begin{cases} \textit{-ek} & / \quad V_{[\text{-mid\&-back}]}(C)C_{[\text{-palatal\&-approximant}]}\ \_\_ \\ \textit{-ok} & / \quad o(C)C_{[\text{-palatal\&-approximant}]}\ \_\_ \\ \textit{-k} & / \quad \begin{Bmatrix} V \\ j \end{Bmatrix}\ \_\_ \end{cases}$$

Figure 3.17: Allomorphy of 2SG person markers.

In contrast to the allomorphy in the 1st person singular and the 3rd person singular masculine, there is an additional condition which makes it necessary to differentiate between two kinds of consonant-final stems. The vast majority of Π-stems trigger the suffixation of *-ek* as illustrated in (3.10) where (Π38) *mingħand* 'from s.o.' ends in a denti-alveolar plosive.

(3.10) [Buhagiar 2001: 92]

Għandi	*bżonn*	*pjaċir*	***[mingħandek]**$_{PP}$*	*Kurun*
have.IPFV:1SG	need	favour	**[from_s.o.**:2SG]	Kurun

'I need a favour [**from** you], Kurun.'

However, the employment of *-ek* is only possible because the last vowel in the Π-stem is not /o/. The mid high back vowel /o/ is unique because it is the sole vowel to trigger vowel harmony on suffixes which have the structure *-eC* (with C = either /t/ or /k/).[18] Vowel harmony is possible only if /o/ and the suffix are separated by at least a consonant. This condition is met by (Π53) *skont* 'according to' as shown in (3.11).

18 Maltese seems to instantiate a kind of backness harmony in the sense that the mid-high front vowel of the suffix gives way to the mid-high back vowel of th stem. Backness harmony is common cross-linguistically (Gordon 2016: 134). Its restriction to exactly one pair of vowels is interesting. As far as we know, there is as yet no study that focusses specifically on vowel-harmonic systems of this kind.

(3.11) [Korpus Malti 3.0, parl11592]

imbagħad	*għidilna*	*għalfejn*	***[skontok]**$_{PP}$*	*Karl Camilleri*
then	tell:IO:1PL	why	**[according_to**:2SG]	Karl Camilleri
mhux	*korrett*			
NEG:3SG.M:NEG	correct			

'[. . .] then tell us why [**in** your opinion] Karl Camilleri is not correct.'

This vowel-harmonic pattern is by no means restricted to Maltese Πs (Cachia 1994: 111). Borg (1997: 276) speaks of /o/ being "subject to a rule of progressive roundness assimilation both intra- and intermorphemically" which affects the 2nd person singular possessor suffix (e.g. *ommok* 'your mother' vs. *darek* 'your house'), the direct and indirect object clitics (e.g. *nitolbok* 'I beg you' vs *nibgħatek* 'I send you' or *nagħtihomlok* 'I give them to you' with vowel harmony caused by the /o/ vowel in the direct object clitic *-hom*[19]), and the person marker on Πs. For verbs like *romol* 'become a widower' the 3rd person singular feminine in the perfective is *romlot* 'she became a widow' with *-ot* in lieu of *-et* as portmanteau suffix co-encoding the subject. Thus, vowel harmony is not limited to contexts where the 2nd person singular is involved.

In (3.12), an example of the use of the monoconsonantal allomorph *-k* on a Π-stem ending in /j/ is provided. The citation form of (Π60) *wara* 'after' contains no palatal approximant.

(3.12) [Abela 2016: 49]

jien	*inkun*	***[warajk]**$_{PP}$*
1SG	1SG.IPFV:be.FUT	[**after**:2SG]

'[. . .] I will be [**behind** you].'

The same allomorph *-k* is triggered by a vowel as final segment of the Π-stem. This is illustrated in (3.13) with (Π21) *għal* 'for'. As in the previous case, the segment /i/ which is crucial for the choice of the allomorph of the person marker is absent from the citation form of the Π.

(3.13) [Buhagiar 2001: 32]

U	*nibqa'*	*nitlob*	***[għalik]**$_{PP}$*	*sa*	*ma*	*mmut*
and	1SG.IPFV:stay	1SG.IPFV:request	[**for**:2SG]	until	that	1SG.IPFV:die

'And I keep asking [**for** you] until I die.'

19 Borg/Azzopardi-Alexander (1997: 315) argue that vowel harmony is relatively productive in Maltese exactly because of the frequent combination of the direct object clitic *-hom* and the indirect object clitic of the 2nd person singular. The vowel /o/ in the former enforces vowel harmony on the suffix to its right which can only take the shape *-lok*.

No other person value is subject to allomorphic variation. We therefore have reached the point where we can address the existence of different morphological classes among the inflected Πs. In the foregoing paragraphs, we have repeatedly hinted at the non-identity of the lexicon form of certain Πs and their Π-stem to which the person affixes are attached. Since only a subset of the inflected Πs attests to special Π-stems, the predictability of the shape of the Π-stem is limited. Furthermore, not all special Π-stems reflect the same phonological properties. On the basis of these stem-related uncertainties, it suggests that there are different classes of inflected Πs. Whether these classes can properly be called inflectional classes will be discussed below.[20] Until then we make do with employing the uncommitted term class.

3.1.2.2 Π-stems

In this section, we briefly look at processes which affect the Π-stem so that for certain Πs, there is non-identity of citation form and Π-stem. The full paradigms of the individual Πs will be presented below.[21] The majority of the Maltese inflected Πs attests to the identity of citation form and Π-stem. The 27 Πs are enumerated in (3.14). The ten cases of Πs which have a special Π-stem are presented in (3.15).

(3.14) Π-stem = citation form
(Π4) *bejn* 'between'; (Π5) *bħal* 'like'; (Π7) *bi* 'with'; (Π8) *biswit* 'facing'; (Π10) *daqs* 'equal to'; (Π11) *dwar* 'about'; (Π14) *fejn* 'near'; (Π15) *fi* 'in'; (Π16) *fost* 'amongst'; (Π17) *fuq* 'on'; (Π20) *għajr* 'except'; (Π22) *għand* 'at s.o.'s place'; (Π24) *ħlief* 'except'; (Π29) *lil* 'to'; (Π31) *madwar* 'around'; (Π32) *maġenb* 'close to'; (Π33) *matul* 'during'; (Π35) *minflok* 'instead of'; (Π36) *mingħajr* 'without'; (Π38) *mingħand* 'from s.o.'; (Π41) *minn* 'from'; (Π47) *qalb* 'amidst'; (Π48) *qrib* 'near'; (Π49) *quddiem* 'in front of'; (Π53) *skont* 'according to'; (Π56) *taħt* 'under'; (Π59) *waqt* 'at the time of'

(3.15) Π-stem ≠ citation form
(Π13) *favur* 'in favour of'; (Π21) *għal* 'for'; (Π23) *ħdejn* 'beside'; (Π27) *kontra* 'against'; (Π28) *lejn* 'towards'; (Π30) *ma'* 'with'; (Π37) *mingħala* 'in s.o.'s opinion'; (Π46) *qabel* 'before'; (Π55) *ta'* 'of'; (Π60) *wara* 'after'

20 Cf. Section 3.1.2.3.2.
21 Cf. Section 3.1.2.3.

Figure 3.18 can be taken as an indicator for the identity of Π-stem and citation form to be another preferred property which can be added to those which have been identified as favourable to inflection in Figure 3.11.

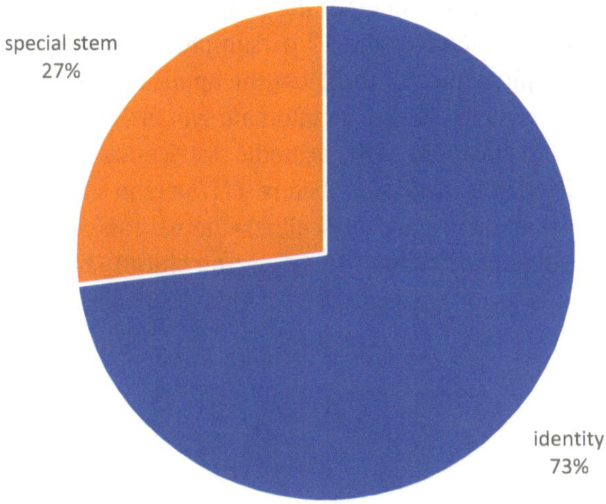

Figure 3.18: Share of special Π-stems vs identity with citation form.

Slightly more than a quarter of the inflected Πs give evidence of special stems.[22] Five of the ten Πs in (3.15) end in a consonant whereas the other five Πs display a final vowel. Five Πs are monosyllabic and five are polysyllabic. The length of the segmental chains ranges from two to eight segments. Going by the quantities exposed in Table 1.11, (Π21) *għal* 'for', (Π30) *ma'* 'with', (Π55) *ta'* 'of', and (Π60) *wara* 'after' boast high token frequencies whereas (Π13) *favur* 'in favour of', (Π23) *ħdejn* 'beside', (Π27) *kontra* 'against', (Π28) *lejn* 'towards', (Π37) *mingħala* 'in s.o.'s opinion', and (Π46) *qabel* 'before' do not. There are two Italo-Romance Πs as opposed to eight Semitic Πs. If we discount the dominance of Semitic Πs, it is hardly possible to pinpoint any criterion which is decisive for the existence of a special Π-stem. Since two Italo-Romance inflected Πs are characterised by stem-citation form identity, the language of origin does not seem to be sufficiently strong a factor to determine whether a Π has a special Π-stem or not.

22 One might argue that (Π4) *bejn* 'between' also belongs here because of its distinct plural forms. However, we argue that these plural forms do not constitute a special Π-stem. On account of the many other questions this Π raises, we postpone its detailed discussion until Section 3.1.2.3.

The ten Πs in (3.15) are not of the same kind. First of all, we have to single out (Π46) *qabel* 'before' because we are not dealing with a proper Π-stem but with an automatic morphonological process which deletes post-tonic mid high /e/ before a stem-final consonant if a vowel-initial suffix is attached.[23] If the suffix has an initial consonant, deletion is not possible but /e/ is raised to /i/. Note that the orthographic <h> of *-ha* in the 3rd person singular feminine and *-hom* in the 3rd person plural is silent but represents an abstract consonantal phoneme which blocks the application of deletion.[24] Besides migration of the stress site to the new penultimate syllable, deletion is a relatively common strategy to avoid violations of the prosodic restrictions imposed on polysyllables in Maltese. The deletion is mandatory before *-i* (1st person singular), *-ek* (2nd person singular), and *-u* (3rd person singular masculine) whereas raising must be realised before *-ha* (3rd person singular feminine), *-na* (1st person plural), *-kom* (2nd person plural), and *-hom* (3rd person plural) as schematised in Figure 3.19.

$$\textit{-e-} \;\rightarrow\; \begin{cases} \emptyset & /\_\_\ \mathrm{C}\ \text{-}\{\mathrm{V(C)}\} \\ \\ \textit{-i-} & /\_\_\ \mathrm{C}\ \text{-}\{\mathrm{CV(C)}\} \end{cases}$$

Figure 3.19: Deletion/raising of *-e-*.

The only Maltese inflected Π which fits this pattern is indeed (Π46) *qabel* 'before'. Deletion of /e/ to yield the stem *qabl-* is illustrated in (3.16) whereas raising of /e/ to yield the stem *qabil-* is presented in (3.17).

(3.16) [Korpus Malti 3.0; literature59]

Taħseb	*li*	***[qablek]***$_{PP}$	*ma*	*kellix*	*nisa*
2SG.IPFV:think	that	[**before**:2SG]	NEG	have.PFV:1SG:NEG	woman.PL
oħra	*jew*				
other:PL	or				

'You think that I did not have other women [**before** you], do you?'

23 The rules described in this paragraph are also valid beyond the domain of Πs. They are especially prominent in verb morphology (Ambros 1998: 204–205). For more details, the reader is referred to Aquilina (1959: 290–294).

24 We assume that this silent phoneme cannot be lumped together with the abstract *għajn*-phoneme since it has not the same effect as the latter on its phonological context (for instance, it does not trigger diphthongization of high vowels). However, this issue belongs to the domain of Maltese phonology whose internal catalogue of problems is too remote from our own topic to justify a dedicated digression on our part.

(3.17) [Korpus Malti 3.0; culture3071]

Jieħdu	*sehem*	***[qabil**<u>hom</u>]_{PP}*	*Freddie Portelli*	*u*	*l-Isvediżi*
3.IPFV:take:PL	part	[**before**:<u>3PL</u>]	Freddie Portelli	and	DEF-Swede:PL

The Rock Orchestra
The Rock Orchestra

'[**Before** <u>them</u>] Freddie Portelli and the Swedes *The Rock Orchestra* will participate.'

For the Π-stems of the other nine Πs mentioned in (3.15), no comparable general rules can be invoked. To illustrate these cases, we have picked one example for each Π (except (Π13) *favur* 'in favour of') from the ten pages 81–90 of our primary source [Bartolo/Vella 2009]. The opener is (Π28) *lejn* 'towards'. Examples (3.6) and (3.8) have prepared us already for the deletion of stem-final /n/ under suffixation. The loss of /n/ before suffixes is pervasive paradigmatically. Example (3.18) shows that it also applies in the 3rd person plural.

(3.18) [Bartolo/Vella 2009: 85]

Tommy	*resaq*	*[**lej**<u>hom</u>]_{PP}*
Tommy	approach	[**towards**:<u>3PL</u>]

'Tommy came [**towards** <u>them</u>].'

More importantly, /n/-deletion is not exclusive to (Π28) *lejn* 'towards' but is attested also with (Π23) *ħdejn* 'beside' as comes to the fore in (3.19).

(3.19) [Bartolo/Vella 2009: 86]

Estella	*waslet*	*[**ħdej**<u>hom</u>]_{PP}*
Estella	arrive:3SG.F.PFV	[**beside**:<u>3PL</u>]

'Estella arrived [**at** <u>their</u> <u>side</u>].'

According to Borg/Azzopardi-Alexander (1997: 333, original boldface), "the word-final **-n** is often deleted in words like prepositions" for which the authors mention (Π23) *ħdejn* 'beside', (Π28) *lejn* 'towards', (Π14) *fejn* 'near', and (Π43) *mnejn* 'from near' explicitly. The latter can be discarded from further discussion since this Π is considered invariant for the time being. Interestingly, (Π14) *fejn* 'near' never displays /n/-deletion under suffixation in our database.[25] Moreover, (Π4) *bejn* 'between' which is in a rhyming relation with the aforementioned Πs is also immune against /n/-deletion (Cachia 1994: 110). This clearly means that identical phonological con-

25 For a differently motivated pattern of *n*-deletion cf. Section 3.3.2.

ditions notwithstanding, the Πs behave differently from each other. It is therefore not possible to put forward a rule which holds for more than two Πs. However, (Π23) *ħdejn* 'beside' and (Π28) *lejn* 'towards' constitute the sole members of a small class.[26]

Another miniature class emerges in connection with the special Π-stems of (Π30) *ma'* 'with' and (Π55) *ta'* 'of'. In the paradigms of these Πs, the abstract phoneme *għajn* resurfaces not only orthographically but also phonetically because it is pronounced as voiceless velar fricative [x] in the 3rd person singular feminine and in the 3rd person plural, i.e. in those word forms where suffixes with initial orthographic <h> are attached.[27] This is the case for instance, in examples (3.20)–(3.21).

(3.20) [Bartolo/Vella 2009: 89]

Ma	*kellhiex*	*aptit*	*tiddiskuti*	***[magħha]**$_{PP}$*
NEG	have.PFV:3SG.F:NEG	appetite	3SG.F.IPFV:discuss	[**with**:3SG.F]
l-pjanijiet				
DEF-plan:PL				

'She did not long for discussing the plans [**with** her] [. . .].'

(3.21) [Bartolo/Vella 2009: 90]

telqet	*tiġri*	*żżur*	*il-ġardina*	*żgħira*
leave:3SG.F.PFV	3SG.F.IPFV:run	3SG.F.IPFV:visit	DEF-garden	small:F
***[tagħha]**$_{PP}$*				
[**of**:3SG.F]				

'[. . .] she left quickly to visit [**her**] little garden [. . .].'

The stems alternate between citation forms with a CV-structure and a special Π-stem with CVC-structure. The *għajn* has its usual effects – diphthongisation and vowel lengthening – so that the parallel paradigms of the two Πs appear relatively heterogeneous phonologically. They too will be presented in the subsequent section.

The previous cases discussed in this section involve pairs of Πs whose morphological behaviour is largely identical. We now enter a less homogeneous territory. The four cases of (Π13) *favur* 'in favour of', (Π21) *għal* 'for', (Π27) *kontra* 'against', and (Π37) *mingħala* 'in s.o.'s opinion' have in common that their Π-stem ends in

26 Cf. Section 3.1.2.3.

27 This case is intriguing insofar as – at least orthographically – *għajn* and the silent <h> form a cluster so that it is not at all clear to which of the two members of the cluster the phonetic realization can be ascribed. Note that both <għ> and <h> are realised as [x] word-finally. In this study we stipulate that what is phonetically realised is the *għajn*.

the vowel /i/ which does not form part of the segmental chain of the citation form. In connection to (Π27) *kontra* 'against', Saari (2009: 273) mentions that "Maltese also uses the Arabic possessive suffixation with this concrete preposition" to add that this also holds for (Π13) *favur* 'in favour of' and (Π53) *skont* 'according to'. However, the two former Italo-Romance Πs boast special Π-stems whereas (Π53) *skont* 'according to' does not. As to (Π13) *favur* 'in favour of', the citation form is consonant-final, i.e. the Π-stem counts a syllable more because of the additional stem vowel.[28] Since the Π-stem ends in /i/ the person marker allomorph for the 1st person singular is *-ja* as shown in (3.22).

(3.22) [Korpus Malti 3.0; news73391]
Naħseb *li* *din* *taħdem* ***[favurija]***$_{PP}$
1SG.IPFV:think that DEM:F:PROX 3SG.F.IPFV:work **[in_favour_of:1SG]**
għax *jien* *politiku* *tal-poplu* *mhux* *ta'* *partit* *biss*
because 1SG politician of:DEF-people NEG:3SG.M:NEG of party only
'I think that this works [**in** my **favour**] because I am a politician of the people and not of a party alone.'

The same pattern is attested for Semitic (Π21) *għal* 'for'. As in the previous case, the consonant-final citation form changes to a Π-stem which ends in the high front vowel /i/ as shown in (3.23).

(3.23) [Bartolo/Vella 2009: 81]
Naħseb *li* *r-riħa* *kienet* *wisq* *qawwija*
1SG.IPFV:think that DEF-wind be:3SG.F.PFV too_much strong:F
[għalija]$_{PP}$
[**for**:1SG]
'I think that the wind was too strong [**for** me].'

The behaviour of these two Πs runs counter to the usual behaviour of consonant-final inflected Πs such as e.g. (Π31) *madwar* 'around' and (Π5) *bħal* 'like' which are phonologically comparable to (Π13) *favur* 'in favour of' and (Π21) *għal* 'for' because of identical syllabicity and identical single word-final liquids. The former two cases suggest that it is perfectly feasible for a Π of this kind to host the person affixes without creating a special Π-stem: *madwari* 'around me' and *bħali* 'like me' are the only legitimate forms in the 1st person singular whereas **għali* 'for me' is unac-

28 Deviations from this pattern are discussed in Section 3.1.2.3.3.2 where the *i*-stems are looked into.

ceptable. As to the supposedly equally unacceptable **favuri* 'in my favour' further discussion is needed.[29] Thus, on the basis of the citation form, it is not possible to predict which stem is chosen under inflection.

The other two Πs with Π-stems ending in /i/ display citation forms which are equipped with a final /a/. This vowel is replaced with /i/ under suffixation, i.e. citation form and Π-stem show identical syllabicity. What makes the case of (Π37) *mingħala* 'in s.o.'s opinion' especially remarkable is the etymological closeness to (Π21) *għal* 'for'. In (3.24), (Π37) *mingħala* 'in s.o.'s opinion' is inflected for the 2nd person singular whereas in (3.25), (Π27) *kontra* 'against' combines with the person marker of the 1st person plural.

(3.24) [Bartolo/Vella 2009: 81]

U	*int*	***[mingħalik]***$_{PP}$	*li*	*jien*	*nerġa'*
and	2SG	**[in_s.o.'s_opinion**:2SG]	that	1SG	1SG.IPFV:return
nidħol	*hemm*	*taħt?*			
1SG.IPFV:enter	there	under			

'And [**according to** you] I shall re-enter down there?'

(3.25) [Bartolo/Vella 2009: 81]

wegħditni	*li*	*mhux*	*se*	*tmur*
promise:3SG.F.PFV:1SG	that	NEG:3SG.M:NEG	FUT	3SG.F.IPFV:go
[kontrina]$_{PP}$				
[**against**:1PL]				

'[. . .] she has promised me that she will not turn [**against** us].'

That there is no automatism which requires the replacement of a final /a/ with /i/ transpires from the next example which shows that (Π60) *wara* 'after' preserves the final vowel of the citation form by way of creating a Π-stem in /j/ as shown in (3.26).

(3.26) [Bartolo/Vella 2009: 82]

[Warajhom]$_{PP}$	*daħal*	*jiġri*	*Erkole.*
[**after**:1PL]	enter	3SG.IPFV:run	Erkole

'[**Behind** them] Erkole entered quickly.'

The unpredictable behaviour of the nine Πs reviewed in the foregoing paragraphs is hard to explain synchronically. Whether the diachronic perspective promises better results in connection to these cases will be assessed in a follow-up study. For

29 Cf. Section 3.1.2.3.3.2.

the time being, however, it suffices to state that Maltese inflected Πs can be distributed over several classes. These classes are in the focus of the subsequent section.

3.1.2.3 Paradigms

In this section, we review the paradigms of all inflected Πs. The presentation and discussion extend to the doubtful cases mentioned in (3.2) above. We proceed as follows. For each class, a separate subsection is reserved. The classes receive distinctive labels. Their properties and the problems they raise are discussed in some detail. The inflectional paradigm of each member of a given inflectional class is featured in the appropriate subsection. We start with the biggest class in Section 3.1.2.3.1. This class comprises all those Πs whose Π-stem is identical to the citation form.

3.1.2.3.1 Class I (= C I)

The criterion for a Π to be allocated in C I is the identity of a consonant-final Π-stem and the corresponding citation form. Superficially, this criterion is valid for most of the Πs listed in (3.14). Since not all of the Πs included in (3.14) behave identically under inflection, the picture is less homogeneous than one might think. First of all, it is necessary to separate bona fide instances of membership in C I from problematic cases. Accordingly, Section 3.1.2.3.1.1 features the unproblematic majority of the members of this class. Section 3.1.2.3.1.2 is dedicated to the discussion of the problematic minority.

3.1.2.3.1.1 Straightforward members of C I

C I hosts twenty Πs which is equivalent to 54% of all inflected Πs. It is therefore relatively normal for an inflected Π to behave according to the following principles. The basic rule for person inflection of Πs which belong to C I is given in Figure 3.20.

$$\Pi_{\text{INFLECTED}} \quad \rightarrow \quad C]_{\text{CITATION FORM}} + \text{person marker}_{\text{SET A}}$$

Figure 3.20: Rule for inflection of Πs in C I.

The schema is reminiscent of the principles of agglutination. However, we know already that a distinction must be made between citation forms with the mid-high back vowel /o/ in the last syllable and those without /o/ because the vowel determines which allomorph is licit in the 2nd person singular. There is thus inward sensitivity of the person markers. The automatic character of this allomorphic choice notwithstanding, we divide C I into two subclasses, namely neutral stems (= C Ia) and vowel-harmonic stems (= C Ib).

The two subclasses differ in size. The subclass C Ia hosts 18 Πs whereas there are only two members in subclass C Ib. Neutral stems are thus nine times as frequent as vowel-harmonic stems. In Table 3.4, we provide the complete survey of the paradigms of all uncontroversial members of C Ia. The vowel-harmonic stems of C Ib are given in Table 3.5. Grey shading highlights the cells of the 2nd person singular which is affected by vowel harmony.

Table 3.4: Paradigms of Πs in C Ia.

Π	singular				plural		
	1	2	3.M	3.F	1	2	3
(Π5) *bħal* 'like'	*bħali*	*bħalek*	*bħalu*	*bħalha*	*bħalna*	*bħalkom*	*bħalhom*
(Π8) *biswit* 'facing'	*biswiti*	*biswitek*	*biswitu*	*biswitha*	*biswitna*	*biswitkom*	*biswithom*
(Π10) *daqs* 'equal to'	*daqsi*	*daqsek*	*daqsu*	*daqsha*	*daqsna*	*daqskom*	*daqshom*
(Π11) *dwar* 'about'	*dwari*	*dwarek*	*dwaru*	*dwarha*	*dwarna*	*dwarkom*	*dwarhom*
(Π14) *fejn* 'near'	*fejni*	*fejnek*	*fejnu*	*fejnha*	*fejnna*	*fejnkom*	*fejnhom*
(Π17) *fuq* 'on'	*fuqi*	*fuqek*	*fuqu*	*fuqha*	*fuqna*	*fuqkom*	*fuqhom*
(Π20) *għajr* 'except'	*għajri*	*għajrek*	*għajru*	*għajrha*	*għajrna*	*għajrkom*	*għajrhom*
(Π22) *għand* 'at s.o.'s place'	*għandi*	*għandek*	*għandu*	*għandha*	*għandna*	*għandkom*	*għandhom*
(Π24) *ħlief* 'except'	*ħliefi*	*ħliefek*	*ħliefu*	*ħliefha*	*ħliefna*	*ħliefkom*	*ħliefhom*
(Π29) *lil* 'to'	*lili*	*lilek*	*lilu*	*lilha*	*lilna*	*lilkom*	*lilhom*
(Π31) madwar 'around'	*madwari*	*madwarek*	*madwaru*	*madwarha*	*madwarna*	*madwarkom*	*madwarhom*
(Π32) *maġenb* 'close to'	*maġenbi*	*maġenbek*	*maġenbu*	*maġenbha*	*maġenbna*	*maġenbkom*	*maġenbhom*
(Π36) *mingħajr* 'without'	*mingħajri*	*mingħajrek*	*mingħajru*	*mingħajrha*	*mingħajrna*	*mingħajrkom*	*mingħajrhom*

Table 3.4 (continued)

П	singular				plural		
	1	2	3.M	3.F	1	2	3
(П38) *mingħand* 'from s.o.'	*mingħandi*	*mingħandek*	*mingħandu*	*mingħandha*	*mingħandna*	*mingħandkom*	*mingħandhom*
(П41) *minn* 'from'	*minni*	*minnek*	*minnu*	*minnha*	*minna*	*minnkom*	*minnhom*
(П48) *qrib* 'near'	*qribi*	*qribek*	*qribu*	*qribha*	*qribna*	*qribkom*	*qribhom*
(П49) *quddiem* 'in front of'	*quddiemi*	*quddiemek*	*quddiemu*	*quddiemha*	*quddiemna*	*quddiemkom*	*quddiemhom*
(П56) *taħt* 'under'	*taħti*	*taħtek*	*taħtu*	*taħtha*	*taħtna*	*taħtkom*	*taħthom*

Table 3.5: Paradigms of Пs in C Ib.

П	singular				plural		
	1	2	3.M	3.F	1	2	3
(П35) *minflok* 'instead of'	*minfloki*	*minflokok*	*minfloku*	*minflokha*	*minflokna*	*minflokkom*	*minflokhom*
(П53) *skont* 'according to'	*skonti*	*skontok*	*skontu*	*skontha*	*skontna*	*skontkom*	*skonthom*

What all these twenty Πs have in common on the expression side is that their citation forms end in a consonant. Superficially, this seems to be a trivial fact. However, as will become evident below,[30] the fact is much more relevant than it might appear.

To better understand why several of the Πs are problematic,[31] it is helpful to show that the Πs in Tables 3.4–3.5 are equipped with complete paradigms. The crucial test for completeness (as will be explained below) involves the speech-act participants in the singular. A Π has a complete paradigm if there is proof of inflectional forms for the 1st and 2nd person singular. Figure 3.21 presents the implication whose tenability will be addressed in the subsequent section.

30 Cf. Section 3.1.2.3.2.
31 To be discussed in Section 3.1.2.3.1.2.

$$1/2\text{SG} \supset \left\{ \begin{array}{l} 3\text{SG.M} \\ 3\text{SG.F} \\ 1\text{PL} \\ 2\text{PL} \\ 3\text{PL} \end{array} \right\}$$

Figure 3.21: Implicational relation between person inflections.

Evidence of inflection in the persons corresponding to the speech-act participants has been provided already for the following five members of C I: (Π14) *fejn* 'near' in (1.67), (Π38) *mingħand* 'from s.o.', (Π41) *minn* 'from' in (1.143), (Π48) *qrib* 'near' in (1.37), and (Π53) *skont* 'according to' in (3.11). There are thus fifteen Πs whose compliance with the implicational pattern in Figure 3.21 must still be substantiated. This is what we attempt to do in the subsequent paragraphs.

To prove our point, dialogues and direct speech are the most promising sources for the use of Πs in combination with a speech-act participant as their complement. By no means do all of our primary sources contain direct speech. Thus, the choice of sources is more restricted than in the case of other phenomena which are also attested in narrative, etc. Examples (3.27)–(3.41) illustrate the inflection of the remaining fifteen Πs for either the 1st person singular or the 2nd person singular.

(3.27) [Camilleri 2013: 222]

Mela	*taħdem*	*biss*	***[bħalek]**$_{PP}$*
well_then	3SG.F.IPFV:work	only	[**like**:2SG]

'Does she only work then [**like** you]?'

(3.28) [Korpus Malti 3.0; literature23]

Qatt	*ma*	*rajt*	*lil*	*ħadd*	*jiġri*	***[daqsek]**$_{PP}$*
ever	NEG	see:1SG.PFV	to	someone	3SG.M.IPFV:run	[**equal_to**:2SG]

'I never saw anybody run [**like** you].'

(3.29) [Bartolo/Vella 2009: 62–63]

Estella	*kuljum*	*tistaqsi*	***[dwarek]**$_{PP}$*
Estella	everyday	3SG.F.IPFV:ask	[**about**:2SG]

'Everyday Estella asks [**about** you].'

(3.30) [Teuma 2015: 799]

Kif	*tasal*	*hemm*	***[biswitek]***$_{PP}$	*tara*	*niċċa*	*b'*
how	2SG.PFV:arrive	there	[**facing**:2SG]	2SG.IPFV:see	niche	with
lampa	*tixgħel*	*fiha*				
lamp	3SG.F.IPFV:kindle	in:3SG.F				

'When you arrive there, you see a niche with a lighted lamp [**facing** you].'

(3.31) [Camilleri 2013: 101]

Rajtu	*x'*	*kiteb*	***[fuqi]***$_{PP}$	*dak*	*il-purċinell*
see:2SG.PFV:3SG.M	what	write	[**on**:1SG]	DEM:M:DIST	DEF-punch

'Have you seen what this punch has written [**about** me]?'

(3.32) [Korpus Malti 3.0; literature77]

u	*hawn*	*ma*	*jkollhom*	*'il*	*ħadd*	***[għajrek]***$_{PP}$
and	here	NEG	3SG.M.IPFV:have.FUT:IO.3PL	to	someone	[**except**:2SG]

'[. . .] and here they will have no-one [**except** you].'

(3.33) [Camilleri 2013: 131]

Leħnek	*mis-sema*	*sejjaħli*	*Mulejja,*	*biex*	*niġi*
voice:2SG	from:DEF-heaven	call:IO:1SG	Lord:1SG	to	1SG.IPFV:come
[għandek]$_{PP}$	*ħa*	*ngħixu*	*flimkien*		
[**at_s.o.'s_place**:2SG]	FUT	1.IPFV:live:PL	together		

'Your voice from heaven has called me, oh my Lord, to come [**to** you] so that we will live together.'

(3.34) [Bartolo/Vella 2009: 63]

il-verità	*ngħidhielha*	*jien*	*u*	*ħadd*	***[ħliefi]***$_{PP}$
DEF-truth	1SG.IPFV:say:IO:3SG.F	1SG	and	someone	[**except**:1SG]

'[. . .] I will tell her the truth and nobody else [**but** me].'

(3.35) [Camilleri 2013: 166]

Tgħajjarnix	*giddieb*	***[lili]***$_{PP}$	*għax*	*inxoqqok*
2SG.IPFV:call_names:1SG:NEG	liar	[**to**:1SG]	because	1SG.IPFV:cleave:2SG

'Don't call [**me**] a liar, because I will cut you to pieces!'

(3.36) [Bartolo/Vella 2009: 55]

dawk	*li*	*jħobbuk*	*u*	*jduru*	***[madwarek]***$_{PP}$
DEM:PL:DIST	that	3.IPFV:love:PL:2SG	and	3.IPFV:turn:PL	[**around**:2SG]

'[. . .] those who love you and care [**for** you].'

(3.37) [Korpus Malti 3.0; literature3]
Jien se mmur ***[minflokok]***$_{PP}$ *l-iskola*
1SG FUT 1SG.IPFV:go **[instead_of**:2SG] DEF-school
'[**Instead of** you] I will go to the school [. . .].'

(3.38) [Bartolo/Vella 2009: 51]
kif tista' tibqa' kalm b' persuna tant
how 2SG.IPFV:can 2SG.IPFV:remain calm with person so
irritant ***[maġenbek]***$_{PP}$
irritating **[close_to**:2SG]
'[. . .] how can you remain calm with such an irritating person [**next to** you]?'

(3.39) [Bartolo/Vella 2009: 195]
Tgħid mhux se jkomplu ***[mingħajri]***$_{PP}$
2SG.IPFV:say NEG:3SG.M:NEG FUT 3.IPFV:continue:PL [**without**:1SG]
'They must go on [**without** me]!'

(3.40) [Bartolo/Vella 2009: 26]
Int kont bilqiegħda bil-ktieb magħluq ***[quddiemek]***$_{PP}$
2SG be:2SG.PFV sitting with:DEF-book PART:close [**in_front_of**:2SG]
'You were sitting with the closed book [**in front of** you].'

(3.41) [Korpus Malti 3.0; literature58]
Mit-tieqa tista' tara ***[taħtek]***$_{PP}$ *il-bjut*
from:DEF-window 2SG.IPFV:can 2SG.IPFV:see [**under**:2SG] DEF-roof.PL
tar-raħal
of:DEF-village
From the window you can see the roofs of the village [**below** you].'

We now know that all of the above members of C I can be inflected for speech-act participants and thus (according to the implication in Figure 3.21) are furnished with complete paradigms. This is important to keep in mind because there are further potential members of C I whose paradigms are defective.

3.1.2.3.1.2 Problem children in C I

The ΠS which show certain affinities with C I but do not seem to fulfil all of the necessary criteria are problematic for different reasons. We look at each of them one by one starting with a morphonologically motivated case of deviation from the required pattern in subsection (i). Subsection (ii) addresses the particularly intriguing case of (Π4) *bejn* 'between'. Plural number is not only relevant for this Π but also for the Πs discussed in subsection (iii). Subsection (iv) is dedicated to Πs whose semantics block their use with speech-act participants as complements. The Πs in (iii)–(iv) instantiate defectiveness (Baerman/Corbett 2010). Whether defectiveness also applies in the case of the three doubtful inflected Πs as of (3.2) will be determined in subsection (v).

(i) (Π46) *qabel* 'before'

This Π has already been discussed in connection with the rule in Figure 3.19 which affects unstressed post-tonic /e/. Examples of deletion and raising are given in (3.16) and (3.17), respectively. The former example additionally shows that (Π46) *qabel* 'before' can take a pronominal complement of a speech-act participant in the shape of affixal morphology. The reason why this Π is not included among the straightforward members of C I is the outward sensitivity of the Π-stem to the person markers. One might object by way of claiming that we are facing regular synchronic morphonological processes which affect all inflected words whose segmental chain ends in the unstressed mid-high front vowel /e/ + single consonant. Given this automatism, the case of (Π46) *qabel* 'before' is not much different from those of the vowel-harmonic Πs in C Ib. In point of fact, the allomorph *-ok* of the 2nd person singular of (Π35) *minflok* 'instead of' and (Π53) *skont* 'according to' is also chosen on the basis of a morphonological automatism. Thus, we accept (Π46) *qabel* 'before' as a further (slightly special and therefore unique) member of C I. Since it neither fits into C Ia nor C Ib, we register it as the sole member of subclass C Ic. Its full paradigm is given in Table 3.6. Grey shading highlights those cases where deletion applies. All other cells of the paradigm give evidence of raising.

Table 3.6: Paradigm of (Π46) *qabel* 'before'.

person	form	process
1SG	*qabli*	DELETION
2SG	*qablek*	DELETION
3SG.M	*qablu*	DELETION
3SG.F	*qabilha*	RAISING
1PL	*qabilna*	RAISING
2PL	*qabilkom*	RAISING
3PL	*qabilhom*	RAISING

The pattern according to which 1SG, 2SG, and 3SG.M behave similarly to each other whereas 3SG.F, 1PL, 2PL, and 3PL form a group of word forms which reflect other properties is by no means an isolated trait of (Π46) *qabel* 'before'.

(ii) (Π4) *bejn* 'between'

This Π is remarkable for several reasons. What strikes the eye most is the use of a plural marker on the Π when the pronominal complement is one of the three persons in the plural. This is illustrated in (3.42) for the 3rd person plural. The additional number morpheme is highlighted in grey.

(3.42) [Camilleri 2013: 90]

Dik *il-btala* *kellha* *timmarka* *t-tieni*
DEM:F:DIST DEF-vacation have.PFV:IO:3SG.F 3SG.F.IPFV:mark DEF-second
fażi *tar-relazzjoni* *ta'* ***[beijnie**t**hom**]*$_{PP}$
phase of:DEF-relation of [**between**:PL:3PL]
'This vacation must have marked the second phase of the relation [**between** them].'

The inflected form *bejniethom* 'between them' consists of three morphemes, namely the Π-stem *bejn* which is identical with the citation form and thus in line with the requirements for allocation in class C I, the plural marker *-iet*, and the person marker *-hom*. No other Maltese Π boasts plural markers. The absence of plural markers from all other inflected Πs isolates (Π4) *bejn* 'between' from its fellow-Πs.

However, as shown in (3.43), this Π can combine the Π-stem directly with the person markers in the plural also in the absence of *-iet*.

(3.43) [Korpus Malti 3.0; academic460]

il-Montenegro *ivvota* *biex* *l-għaqda* *politika* ***[bejn**hom**]***$_{PP}$
DEF-Montenegro vote to DEF-union political:F [**between**:3PL]
u *bejn* *is-Serbja* *tispiċċa*
and between DEF-Serbia 3SG.F.IPFV:end
'Montenegro voted for terminating the political union [**between** them] and Serbia.'

The competition between two options – plurals with vs plurals without extra plural marker – is another unique trait of (Π4) *bejn* 'between' because there is no other Maltese Π with this kind of overabundance in the plural.[32] Thornton

32 But cf. Section 3.1.2.3.2.

(2019) defines overabundance as the morphological mismatch which is obtained if one and the same cell in a paradigm is occupied by more than just one word form which are largely synonymous functionally. Sutcliffe (1936: 202) claims that the forms which involve the overt plural marker "are used to denote distinction between several members of a group". According to this author, the plural marker is absent when a contrast between groups is expressed. Examples (3.42)–(3.43) confirm this hypothesis.[33]

What Sutfcliffe's assumption entails further is that forms which involve *-iet* are excluded from coordinative constructions like that exemplified in (3.43). This is corroborated by a search on the Korpus Malti 3.0 which yielded no matches at all for constructions of this kind involving *bejnietna* 'between us', *bejnietkom* 'between you', or *bejniethom* 'between them' (Saari 2003: 117). This means that (Π4) *bejn* 'between' displays another unique property within the Maltese Π-inventory, namely overdifferentiation in the sense that in a subset of the cells of the paradigm formal distinctions are consistently made in order to mark out categories which are not registered in other paradigms of the same word-class (Corbett 2007a: 28). According to the principles of Canonical Morphology, overdifferentiation constitutes a morphological mismatch. Like overabundance (from which it is not always easily distinguished), overdifferentiation is a noncanonical phenomenon.

Before we take a look at the paradigm itself, it is worth noting that under coordination (Π4) *bejn* 'between' may occur on both conjuncts though not as often as example (3.43) might suggest (Stolz/Ahrens 2017: 138) as most of the Maltese Πs do.[34] At this point, it is sufficient to state that this Π is frequently inflected for speech-act participants as in (3.44).

(3.44) [Korpus Malti 3.0; news7055][35]

llum	*inżammet*	*laqgħa*	***[bejn**<u>i</u>]$_{PP}$*	*u*	***[bejn**<u>ek</u>]$_{PP}$*
today	RFL:hold:3SG.F.PFV	meeting	[**between**:<u>1SG</u>]	and	[**between**:<u>2SG</u>]
fil-Kwartieri	*tal-Partit*	*Laburista*			
in:DEF-quarters	of:DEF-party	Labour			

'[. . .] today a meeting was held [**between** <u>me</u>] and [<u>you</u>] in the quarters of the Labour Party [. . .].'

33 The semantic side of this case will be looked into in Section 5.1.2.1.13.

34 We inquire into this issue more thoroughly in Section 4.1.4.

35 Missing special Maltese characters have been added by us.

Not only does this sentence illustrate the doubling of the Π under coordination but it also involves two inflected forms of (Π4) *bejn* 'between' which refer to speech-act participants. This is remarkable insofar as conceptually the BETWEEN-relation always involves two participants so that an inflected form like *bejni* 'between me' does not make sense – superficially. However, it makes perfect sense because *bejni* 'between me' and *bejnek* 'between you' can never be used on their own, i.e. they are always part of a coordinative construction. The two or more conjunctions of this construction guarantee that there are more participants involved than just the 1st or 2nd person singular. That this is again something that makes (Π4) *bejn* 'between' special will transpire from the discussion of the next two Πs.

The paradigm of this special case is given in Table 3.7. We distinguish plain inflection from overt pluralisation.

Table 3.7: Paradigm of (Π4) *bejn* 'between'.

person	word form	
	plain	**pluralised**
1SG	*bejni*	
2SG	*bejnek*	
3SG.M	*bejnu*	
3SG.F	*bejnha*	
1PL	*bejnhom*	*bejnietna*
2PL	*bejnkom*	*bejnietkom*
3PL	*bejnhom*	*bejniethom*

As to the word forms in column PLAIN, it is obvious that they meet the criteria for the Π under scrutiny being admitted to class C Ia. The Π-stem and the citation form are identical. There is a word-final consonant to which the regular person affixes are attached. Were the paradigm as simple as that, (Π4) *bejn* 'between' would pass the test for class-membership in C Ia. The co-presence of two forms for each person in the plural mark out this Π as a special case. This special status notwithstanding, (Π4) *bejn* 'between' can be considered to be a marked member of C I. In analogy to the previous case, (Π4) *bejn* 'between' forms a subclass of its own, namely C Id. The reasons for the markedness assumed for (Π4) *bejn* 'between' are different (morphological and functional) from those assumed for (Π46) *qabel* 'before' (phonological). The plural marker *-iet* is alien to the bulk of the inflected Πs of Maltese. Outside the domain of Πs, *-iet* is relatively common in the number system of feminine nouns ending in *-a* such as SG *gremxula* 'lizard' → PL *gremxuliet* (Borg/Azzopardi-Alexander 1997: 176). Synchronically, the presence of *-iet* in the paradigm of a Maltese Π is inexplicable. For the time being, we conclude that those forms of (Π4) *bejn* 'between' which host the overt plural marker mix elements of different word-classes and are thus instances of heteroclisis (Baerman 2007: 16).

(iii) (Π16) *fost* 'amongst' and (Π47) *qalb* 'amidst'

These two Πs are similar to (Π4) *bejn* 'between' conceptually in the sense that their complement must refer to a multitude of participants. In contrast to the foregoing case however, neither (Π16) *fost* 'amongst' nor (Π47) *qalb* 'amidst' makes a distinction along the lines of that postulated by Sutcliffe for the two plurals of (Π4) *bejn* 'between'. Moreover, neither of the two Πs under inspection is attested bearing the person markers of any speech-act participant in the singular. Cachia (1994: 115, original italics) claims that "*fost* tieħu biss suffissi pronominali plurali"[36] which is also Sutcliffe's (1936: 203) as well as Saari's (2003: 137) point of view. Accordingly, examples like (3.45a–b) meet the expectations.

(3.45) (a) [Camilleri 2013: 136]

Min	*[**fost**<u>kom</u>]$_{PP}$*	*huwa*	*ħerqan*	*daqstant*	*illi*
who	[**amongst**:<u>2PL</u>]	3SG.M	keen	so_much	that

jeħodli	*post*
3SG.M.IMPFV:take:IO:1SG	post

'Who [**amongst** <u>you</u>] is so keen as to take away my job?'

(b) (Aquilina 1991: 1111)

rajtu	*[**qalb**<u>hom</u>]$_{PP}$*
see:1SG.PFV:3SG.M	[**amongst**:<u>3PL</u>]

'I saw him [**amongst** <u>them</u>].' [O.T.]

Given the semantics of the two Πs, it is easy to understand that forms like **fostok* 'amongst you$_{SG}$' and **qalbek* 'amongst you$_{SG}$' are absent from the database.[37] Once more in contrast to (Π4) *bejn* 'between', coordination does not come into play here either. If the Πs are repeated on both (or all) conjuncts, there is never a form of the 1st person singular or 2nd person singular involved.

However, the assumed restriction to the morphological plural is not correct. As shown in (3.46), (Π16) *fost* 'amongst' may be inflected in the 3rd person singular if the person marker is coreferential with a noun invested with collective meaning. In this example, the noun fulfilling this criterion is *familja* 'family' which we highlight in grey in (3.46).

36 Our translation: "*fost* takes only plural pronominal suffixes".

37 *Qalbek* 'your heart' is of course attested frequently in the Korpus Malti 3.0 – but only as inalienably possessed body part (often used metaphorically).

(3.46) [Camilleri 2013: 146]

Wara kollox seta' jistenna mod ieħor minn wild familja
after all can 3SG.M.IPFV:wait manner other from child family
ultranazzjonalista mill-fortizza ta' Zemun tħaddan
ultranationalist from:DEF-fortress of Zemun 3SG.F.IPFV:embrace
*[**fostha**]$_{PP}$ bi kburija veterani tal-ġenoċidji*
[**amongst**:3SG.F] with pride veteran:PL of:DEF-genocide:PL
fil-Bosnja
in:DEF-Bosnia
'After all what else could he expect from a child of an ultranationalist family from the fortress of Zemun which proudly counts [**amongst** them] veterans of the genocides in Bosnia?'

Familja 'family' is a feminine and formally (i.e. morphologically) singular. It is therefore anaphorically represented by the suffix *-ha* of the 3rd person singular feminine on the Π. In (3.47), we provide a parallel case in the masculine singular. The person suffix *-u* of the 3rd person singular attached to (Π16) *fost* 'amongst' refers back to the definite noun *il-Malta Labour Party* whose Maltese equivalent *il-Partit Laburista* has masculine gender and collective meaning.

(3.47) [Korpus Malti 3.0; parl1560]

Naħseb li l-Malta Labour Party jagħmel sew
1SG.IPFV:think that DEF-Malta Labour Party 3SG.M.IPFV:make right
*jekk jisma' lin-nies [**fostu**]$_{PP}$ li wkoll*
if 3SG.M.IPFV:listen to:DEF-people [**amongst**:3SG.M] that also
huma favur li dan isir
3PL in_favour_of that DEM:M:PROX 3SG.M.IPFV:become
'I think that the Malta Labour Party does the right thing if they listen to the people [**amongst** them] who are also in favour of this going to happen.'

We have not been able to find evidence of formally equivalent cases for the 3rd person singular masculine/feminine for (Π47) *qalb* 'amidst'. On account of its synonymy with (Π16) *fost* 'amongst' and parallel morphosyntactic behaviour of the two Πs, we stipulate that (Π47) *qalb* 'amidst' can host the suffixes *-u* and *-ha*, too provided there is an anaphorical relation to a noun with collective meaning. On account of their special behaviour, the two Πs under consideration are assigned to a separate subclass C Ie.

In Table 3.8, it is shown that we are facing two cases of semantically induced defectiveness of paradigms. Grey shading marks those cells which remain empty for (Π16) *fost* 'amongst' and (Π47) *qalb* 'amidst'.

Table 3.8: Defective paradigms of (Π16) *fost* 'amongst' and (Π47) *qalb* 'amidst'.

person	(Π16) ***fost*** 'amongst'	(Π47) ***qalb*** 'amidst'
1SG		
2SG		
3SG.M	*fostu*	*qalbu*
3SG.F	*fostha*	*qalbha*
1PL	*fostna*	*qalbna*
2PL	*fostkom*	*qalbkom*
3PL	*fosthom*	*qalbhom*

Corbett (2007a: 27) registers defectiveness among the morphological mismatches. It is clear that neither of the two Πs is defective for phonological or morphological reasons. The inflections of the speech-act participants are blocked only for semantic reasons. A similar motivation for defectiveness applies to the cases to be discussed in the subsequent paragraphs.

(iv) (Π33) *matul* 'during' and (Π59) *waqt* 'at the time of'

From the discussion above,[38] we are already familiar with some of the problems (Π33) *matul* 'during' poses. (Π59) *waqt* 'at the time of', too was scrutinised above.[39] Is there anything else that could be added to the information we already have? The answer to this rhetorical question is positive.

In the previous discussion of (Π33) *matul* 'during', we have tried to establish whether there are significant differences as to its spatial or temporal readings. In this context, we have analysed examples in which this Π is inflected for the 3rd person singular masculine and the 3rd person singular feminine. Sentence (3.48) proves that (Π33) *matul* 'during' can also take the person marker of the 3rd person plural.

38 Cf. Section 1.6.4.2 (ix).

39 Cf. Section 1.6.5 (v).

(3.48) [Korpus Malti 3.0; academic13]
Il-pjanta jrid ikollha
DEF-plant 3SG.M.IPFV:require 3SG.M.IPFV:have.FUT:IO.3SG.F
mill-inqas sentejn li ***[matulhom]***$_{PP}$ *ma*
from:DEF-little.CMPR year:DU that **[during**:3PL] NEG
nkunux bdilna l-qasrija
1.IPFV:be.FUT:PL:NEG change:1PL.PFV DEF-flower_pot
'The plant needs to have at least two years [**during** which] we will not have changed the flower-pot.'

Semantically surprising is the possibility of (Π33) *matul* 'during' inflecting for the 2nd person singular as in (3.49).

(3.49) [Korpus Malti 3.0; news85507]
Illum lanqas norqod aktar minn erba' sigħat
today not_even 1SG.IPFV:sleep more from four hour:PL
[matulek]$_{PP}$ *ja lejl*
[**during**:2SG] oh night
'Nowadays I do not even sleep four hours [**during**] the entire night [. . .].'

E. Serracino-Inglott (1979: 69) characterises this usage as poetic. The idiomatic construction involves the inflected Π and the interjection *ja* 'oh' as specified elements. To the right of the interjection there is a slot that is open to fillers which designate periods of time. The meaning of the construction is non-compositional insofar as it refers to the entire period of time represented by the paradigmatically variable noun. The Korpus Malti 3.0 gives evidence of twenty tokens of the construction with five different types as shown in Figure 3.22.

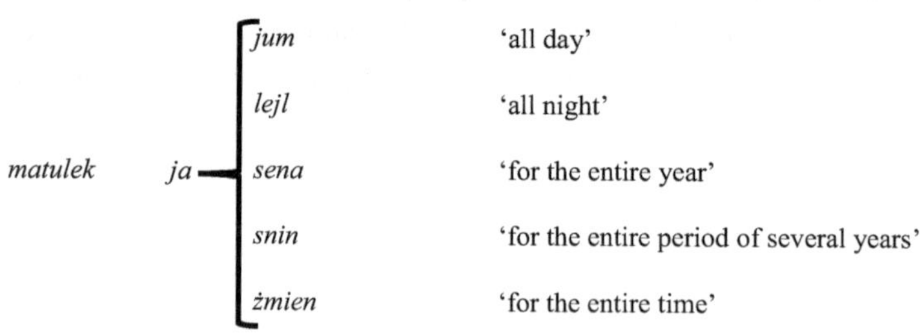

Figure 3.22: Poetic constructions with *matulek*.

We assume that originally the *-ek* suffix was possible on (Π33) *matul* 'during' because the time expression underwent personification in the poetic style. The possibility of using the plural *snin* 'years' in combination with *matulek* suggests that this construction is no longer fully transparent for at least some of those who employ it. *Matulek* is exclusively attested in this idiomatic construction. There are no matches for **matuli*, **matulna*, and **matulkom* in the Korpus Malti 3.0. The paradigm of (Π33) *matul* 'during' is thus more defective than the paradigms of (Π16) *fost* 'amongst' and (Π47) *qalb* 'amidst' because it displays three empty cells as opposed to only two empty cells in the latter two cases. Moreover, the cell of the 2nd person singular is filled only under certain highly specific stylistic conditions.

The temporal semantics of (Π33) *matul* 'during' make it plausible that the Π does not freely inflect for the speech-act participants of both numbers. Temporal semantics are also at issue with (Π59) *waqt* 'at the time of'. Examples (3.50)–(3.51) document the use of this Π with person markers of the 3rd person singular feminine and the 3rd person plural, respectively.

(3.50) [Korpus Malti 3.0; european5320]
fil-laqgħa *tal-grupp* *ta'* *ħidma* *leġislazzjoni* *fl-20* *ta'*
in:DEF-meeting of:DEF-group of work legislation in:DEF-20 of
Marzu *1996* *li* ***[waqtha]***$_{PP}$ *l-Istati* *Membri*
March 1966 that **[at_the_time_of**:3SG.F**]** DEF:state:PL member:PL
ikkonfermaw *li* *rċevew* *id-dokument*
confirm:3PL.PFV that receive:3PL.PFV DEF-document
'[. . .] at the meeting of the legislation working group on 20 March, 1996 [**during** which] the member states confirmed that they had received the document [. . .].'

(3.51) [Korpus Malti 3.0; european4017]
Wara *konsultazzjonijiet* *li* ***[waqthom]***$_{PP}$ *wieħed*
after consultation:PL that **[at_the_time_of**:3PL**]** one
mill-Istati *Membri* *qajjem* *oġġezzjonijiet* *għal*
from:DEF-state:PL member:PL rise:CAUS objection:PL for
din *is-soluzzjoni*
DEM:F:PROX DEF-solution
'After consultations [**during** which] one of the member states raised objections against this solution [. . .].'

As expected, inflected forms of (Π59) *waqt* 'at the time of' in the 1st/2nd persons of both numbers are not attested. This means that there are four gaps in the paradigm of this Π – a fact which renders (Π59) *waqt* 'at the time of' the most defective Π of all bona fide inflected Πs. The paradigms of the two Πs discussed in this subsection are given in Table 3.9.

Table 3.9: Defective paradigms of (Π33) *matul* 'during' and (Π59) *waqt* 'at the time of'.

person	(Π33) *matul* 'during'	(Π59) *waqt* 'at the time of'
1SG		
2SG	*matulek*	
3SG.M	*matulu*	*waqtu*
3SG.F	*matulha*	*waqtha*
1PL		
2PL		
3PL	*matulhom*	*waqthom*

On account of the special status of *matulek* as (perhaps) frozen unanalysable part of an idiomatic construction, we consider these Πs to constitute the subclass C If.

(v) (Π26) *kif* 'as', (Π42) *mintul* 'all along', and (Π43) *mnejn* 'from where'

The final words of this section are dedicated to three of the notorious cases of doubt. Owing to their phonological properties, (Π26) *kif* 'as', (Π42) *mintul* 'all along', and (Π43) *mnejn* 'from where' are candidates for the membership in C I. Whether they are members of the class of consonantal stems hinges on the question whether they can be inflected at all. We have put forward arguments for their inclusion in BLOMP 2.0.[40] We defended their inclusion inter alia by way of referring to sources which assume that these would-be Πs can undergo inflection. However, the morphological variability of (Π26) *kif* 'as', (Π42) *mintul* 'all along', and (Π43) *mnejn* 'from where' remains questionable since neither the Korpus Malti 3.0 nor our corpus of contemporary written prose provides compelling evidence of inflected word forms of these doubtful cases. Moreover, the sources which suggest that we are dealing with inflected Πs are themselves not entirely trustworthy. Accordingly, we have registered the three candidates amongst the invariant Πs in (3.2). In this way, the paradoxical situation has arisen in which candidates which were originally counted as Πs because of their ability to inflect are at the same time classified as invariant Πs. It is therefore high time to determine once and for all the exact status of (Π26) *kif* 'as', (Π42) *mintul* 'all along', and (Π43) *mnejn* 'from where'.

An example of inflection is given for (Π42) *mintul* 'all along' in (1.127), for (Π26) *kif* 'as' in (1.182), and for (Π43) *mnejn* 'from where' in (1.183). The latter two stem from Aquilina's (1987, 1991) dictionary whereas inflected (Π42) *mintul* 'all along' has been found in the Korpus Malti 3.0. The main problems connected to these examples are of two kinds. First of all, we have to ask whether the dictionary

40 Cf. Sections 1.6.4.2 (ix) and 1.6.7 (i)–(ii).

entries are reliable. Secondly, it remains to be seen whether we are dealing with isolated lexicalised word forms or genuine paradigms.

This is what Aquilina (1987: 652) has to say about (Π26) *kif* 'as':

> **5.** ~ + pers[onal] suffixes, the equal of; [Vass[alli]] *m'hemmx ~u/~ha*, there is no one like him/her; *m'hawnx ~hom*, there is nobody like them (as good) [K[otba tal-]M[ogħdija taż-]Ż[mien] 14, p. 33] *intom il-qbejjel għaż-żwiemel ma hawnx ~kom*, there is nobody as expert as you are in the knowledge of equine species; *min hu ~na fid-dinja?*, who is our equal (like us) in the world?

Aquilina's sources for the examples of inflected (Π26) *kif* 'as' are Vassalli's dictionary *Ktyb yl Klym Mâlti 'mfysser byl-Latin u byt-Talyân* published in 1796 and the 14th volume of the folkloristic series *Kotba tal-Mogħdija taż-Żmien* which existed between 1899 and 1915. This means that there is a considerable time-gap of some 75 to 180 years between the publication of the first volume of Aquilina's dictionary in 1987 and that of the two sources. To prove that (Π26) *kif* 'as' hosts person affixes, E. Serracino-Inglott (1978: 108) employs the sample sentence *Min hu kifna fid-dinja?* which is identical to the last one mentioned in the above quote so that we assume that both dictionary compilers have exploited the same source. An additional idiomatic use is mentioned as well, namely *Għala kifu minn ħaddieħor* 'act according to one's own liking (not caring about others)'. Like Aquilina, E. Serracino-Inglott does not list the complete set of possible forms of inflected (Π26) *kif* 'as'. He mentions *kifek* 'like you', *kifu* 'like him', and *kifna* 'like us'. Aquilina provides examples of *kifkom* 'like you$_{PL}$' and *kifhom* 'like them'. We have to go back in time as far as Falzon (1882: 183–184) to find a list of word forms which resemble a (defective) paradigm:

> *Si costruisce cogli affissi nella maniera seguente;*[41] Kìfu, kifha, kifek, kifna, kifkom, kifhom, kifna [sic!]. Fid-dinja kolha, ma hemx kifna. *In tutto il mondo, non v'è persona al par di noi,* in the whole world there is nobody equal to us.

From the second edition of Falzon's trilingual Maltese-Italian-English dictionary, a next-to-complete paradigm emerges from which only the hypothetical **kifi* 'like me' is missing. Both E. Serracino-Inglott and Aquilina have made use of Falzon (1882). It is therefore not surprising to see that one of the sample sentences in the three dictionaries is almost identical. Aquilina (1987: xxv) is critical of the quality of Falzon's work especially in relation to the second edition from which the above quote has been drawn. According to Aquilina (1987: xxv), "Falzon's trilingual dictionary [. . .] reproduces Vassalli's dictionary with a good number of additions." We assume that Falzon has taken over inflected (Π26) *kif* 'as' from Vassalli. Since there

41 Our translation: "It combines with the affixes in the following manner".

is no trace of inflected (Π26) *kif* 'as' in the Korpus Malti 3.0 and evidence of morphologically variable (Π26) *kif* 'as' is also absent from our corpus of contemporary literary Maltese, it is legitimate to ask whether the inflection of this Π was only an ephemeral possibility in the past, if at all. Given these uncertainties, we adopt the following compromise. Diachronically, (Π26) *kif* 'as' could be inflected for person at least in some idiolects of Maltese. We assume that, for an extended period of time in the past, it competed with quasi-synonymous (Π5) *bħal* 'like' which occurred also in fixed expressions similar to those reported above for (Π26) *kif* 'as'.[42] The possibility to inflect for person must have gradually become obsolete for (Π26) *kif* 'as' in the early 20th century so that, synchronically, this Π is now invariant.

The story of (Π42) *mintul* 'all along' is not exactly the same as that of the previous Π. Aquilina (1991: 834) considers this Π to result from the univerbation of a former PP *minn* 'from' + *tul* 'length, duration'. In the lexicon entry for *tal* 'become long', Aquilina (1991: 1390) provides a sentential example of this PP which we reproduce as (3.52).

(3.52) (Aquilina 1991: 1390)

waqa'	*[minn*	*tulu]*$_{PP}$	*fl-art*
fall	**[from**	length:3SG.M]	in:DEF-ground

'He fell [**lengthwise**] on the ground.' [O.T.]

The orthographic separation of the Π and its possessed complement seems to speak against considering (Π42) *mintul* 'all along' to be univerbated at all. In the same lexicon entry, Aquilina (1991: 1390) mentions that Vassalli used to write *mintul* + PERSON SUFFIXES in one word. The example he probably copies from Vassalli is *ma' mintulek ja sena* 'all the year round/throughout the year'. This example corresponds to the poetic construction with (Π33) *matul* 'during' illustrated in Figure 3.22. E. Serracino-Inglott (1979: 156) takes issue with Vassali's and Falzon's practice of restricting the use of person suffixes with (Π42) *mintul* 'all along' to the 2nd person singular. He claims that the Π can combine "bis-suff[issi] pron[ominali] pers[onali] kollha skond il-każ".[43] He provides a single example for the use of the suffix of the 3rd person singular masculine which we reproduce in (3.53).

(3.53) (E. Serracino-Inglott 1979: 156)

Somtu	*[mintulu*	*kollu*	*r-Randan]??*
fast:2SPL.PFV	**[all_along:3SG.M**	all:3SG.M	DEF-Lent]

'You have fasted [**throughout** the entire Lent]!'

42 Such as *min bħalu* 'who is like him?' (E. Serracino-Inglott 1976a: 154).

43 Our translation: "with all of the person suffixes according to what is required".

This sentence is problematic because the person inflection in the presence of a lexical complement NP violates the criteria we have laid down for Π-hood.[44] The other example given in E. Serracino-Inglott's (1979: 156) dictionary fits the description of the poetic construction in Figure 3.22. Falzon (1882: 251) offers a lexicon entry only for the inflected form *mintulek* which he classifies as adverb. The examples he provides are again of the type *mintulek ja sena* translated into Italian as *lungo l'anno* and into English as *during the whole week* without mention of (Π33) *matul* 'during'. Since there is no match for uninflected (Π42) *mintul* 'all along' in the Korpus Malti 3.0, it comes as no surprise that none of the theoretically possible inflected forms of this Π is attested either. What is attested in the electronic resource is the PP *minn tul-* as shown (with added token frequencies) in Figure 3.23.

minn tul- +	*-i*	6
	-ek	1
	-u	135
	-ha	19
	-na	0
	-kom	0
	-hom	12

Figure 3.23: Token frequency of person inflection with *minn tul-*.

More often than not these forms are attested in combination with verbs like *(n)tefa'* 'throw (oneself)', *x(t)eħet* 'cast (oneself)', *firex* 'spread', *waqa'* 'fall', etc. A typical example is given in (3.54).

(3.54) [Korpus Malti 3.0; literature75]
L-imsejkna kienet mixħuta **[minn** tulha]$_{PP}$ *fuq is-sodda*
DEF-pitiful:F be:3SG.F.PFV PART:cast:F **[from** length:3SG.F] on DEF-bed
'The poor [girl] was thrown **[lengthwise]** on the bed [. . .].'

There is nothing that prevents us from analysing cases of this kind as PPs which consist of two syntactic words, i.e. there is no synchronic evidence proving that (Π42) *mintul* 'all along' is an inflected Π. As in the foregoing case, it is possible that (Π42) *mintul* 'all along' enjoyed this status in an earlier phase of the diachronic

44 Cf. Sections 1.6.1–1.6.2.

development of Maltese. To our minds, this status must have been lost at some point in the early 20th century. As to contemporary Maltese, we assume that (Π42) *mintul* 'all along' is invariant.

The final doubtful candidate is (Π43) *mnejn* 'from where'. Aquilina (1991: 847) provides examples of person inflection of this Π in the 1st person singular and plural the latter being drawn from Caruana (1903) reproduced as (3.55) here.

(3.55) (Aquilina 1991: 847)

naf	*dak*	*ir-raġel*	*għax*	***[mnejnna]***$_{PP}$
1SG.IPFV:know	DEM:M:DIST	DEF-man	because	**[from_where**:1PL]

'I know that man because he lives [**near** us].' [O.T.]

Sutcliffe (1936: 195) who uses exactly the same example as Aquilina speaks of (Π43) *mnejn* 'from where' as an adverb that "is used with the plural pronominal suffixes: *mnejna, mnejkom, mnejhom*", i.e. with loss of the final nasal of the citation form under inflection. As mentioned above,[45] E. Serracino-Inglott (1979: 183) casts doubt upon the naturalness of the inflected forms of this Π which he denounces as an invention of Falzon's. Falzon (1882: 264) mentions only the 1st person plural *mnejnna* for which he provides the Italian translation 'dalle parti o contorni nostri' and the English translation 'from our place or neighbourhood'. E. Serracino-Inglott (1979: 183) claims that it would be better to replace *mnejnna* with *mnejn aħna*, i.e. with the free pronoun of the 1st person plural as complement of the Π. However, even this alternative is judged unnatural because native speakers usually opt for *minn fejnna* or *minn ħdejna* which are combinations of two Πs, namely (Π41) *minn* 'from' + (Π14) *fejn* 'near' / (Π23) *ħdejn* 'beside' with the rightmost Π hosting the person inflection. The search in the Korpus Malti 3.0 has not yielded any match for any of the theoretically possible inflected forms of (Π43) *mnejn* 'from where'. Moreover, our native-speaker consultant Sandra Abela never uses inflected forms of this Π but concedes the possibility that there might be Maltese idiolects for which person inflection on (Π43) *mnejn* 'from where' is acceptable. Without hard proof of this possibility, we assume that like in the other two doubtful cases, (Π43) *mnejn* 'from where' is synchronically invariant whereas it might have been an inflected Π at least for some speakers of Maltese in the past.

In sum, no matter what their morphological behaviour was in the 19th century, (Π26) *kif* 'as', (Π42) *mintul* 'all along', and (Π43) *mnejn* 'from where' must be classified as invariant in contemporary Maltese because their morphological variability cannot be proved empirically.

---

45 Cf. Section 1.6.7. (ii).

3.1.2.3.2 Class II (= C II) – a provisional arrangement

C II comprises those Πs in (3.14) which are not covered by the rule exposed in Figure 3.20. This means that we are now confronted with vocalic stems which fulfil the criterion of identical Π-stem and citation form. The appropriate rule is schematised in Figure 3.24.

$$\Pi_{\text{INFLECTED}} \rightarrow V]_{\text{CITATION FORM}} + \text{person marker}_{\text{SET B}}$$

Figure 3.24: Rule for inflection of Πs in C II.

In point of fact, there are only two Πs which behave according to this rule, namely (Π7) *bi* 'with' and (Π15) *fi* 'in'. Phonologically, the two resemble each other closely as their citation forms are in a rhyming relation. With a covered-open syllable of two segments, they are rather short. Other Πs with a citation form in the shape of a CV-syllable inflect differently. This is the case with (Π30) *ma'* 'with' and (Π55) *ta'* 'of' to be discussed below.[46] Others remain invariant as e.g. (Π19) *ġo* 'in' and (Π51) *sa* 'till'.

The paradigms of the two Πs are structured in identical manner. This morphological parallelism includes a property which renders (Π7) *bi* 'with' and (Π15) *fi* 'in' unique within BLOMP 2.0. Borg/Azzopardi-Alexander (1997: 330, original boldface) state that "[t]he preposition takes the nominal suffixes but **bi** and **fi** can take both the nominal and the verbal suffixes" so that in the 1st person singular *fija* and *fini* 'in me' as well as *bija* and *bini* 'with me' co-exist. This co-existence if illustrated for (Π15) *fi* 'in' in (3.56)–(3.57).

(3.56) (Borg/Azzopardi-Alexander 1997: 265)

Il-karozza	*għodda*	*daħlet*	*[fija]*$_{PP}$
DEF-car	almost:3SG.F	enter:3SG.F.PFV	[**in**:1SG]

'The car almost bumped [**into** me].' [O.T.]

(3.57) (Borg/Azzopardi-Alexander 1997: 266)

[Fini]$_{PP}$	*x'*	*tara*	*mxarrab*	*kif*	*jien*
[**in**:1SG]	what	2SG.IPFV:see	PART:wet	how	1SG

'I'm a pretty sight to see, wet as I am!' [O.T.]

46 Cf. Section 3.1.2.3.5.

The important point about this allomorphy is that *fini* cannot replace *fija* nor vice versa (Borg/Azzopardi-Alexander 1997: 266). Sutcliffe (1936: 201, original boldface and italics) assumes that "**[f]ini** does not mean simply *in me*, but *there is in me, I have*, having thus a verbal meaning attached to it." In analogy, the same author claims that "[f]or the [1st person singular] the form used is **bini** when the meaning is verbal and equivalent to *I have*." (Sutcliffe 1936: 201, original boldface and italics). However, in contrast to *fija* and *fini*, *bija* and *bini* seem to be alternative options in one and the same context as transpires from (3.58).

(3.58) (Sutcliffe 1936: 200)

jisgħob	{ [***bija***]PP / [***bini***]PP }	*għax*	*rgħextek*
3SG.M.IPFV:repent	[**with**:1SG]	because	offend:1SG.PFV:2SG

'I am sorry because I have offended thee.' [O.T.]

Aquilina (1991: 1342) uses the same example identifying it as a religious formula to be used during the Act of Contrition. Fabri (1993: 200) classifies *fini* as pseudo-verb whereas *bini* is not mentioned in this context. The latter is also not included in Peterson's (2009: 187) list of pseudo-verbs. In contrast to Fabri, Peterson speaks of (Π15) *fi* 'in' in general as a pseudo-verb, meaning all forms of its paradigm are involved and not just the 1st person singular.[47] Since in this section we are only interested in properly morphological phenomena, we feel entitled to skip the semantico-syntactic implications of the above allomorphy.

We therefore present the paradigms of the Πs under consideration in Table 3.10. This reveals how the paradigms look like if we take the statement of Borg/Azzopardi-Alexander (1997: 330) as our point of departure. According to this statement, the competition between nominal and verbal person affixes is limited to the 1st person singular.

47 We will have to come back to this issue in Section 4.1.2.2.2.2.3.

Table 3.10: Paradigms of (П7) *bi* 'with' and (П15) *fi* 'in' – version I.

person	(П7) *bi* 'with'		(П15) *fi* 'in'	
1SG	*bija*	*bini*	*fija*	*fini*
2SG	*bik*		*fik*	
3SG.M	*bih*		*fih*	
3SG.F	*biha*		*fiha*	
1PL	*bina*		*fina*	
2PL	*bikom*		*fikom*	
3PL	*bihom*		*fihom*	

For all other persons, there is only a single person inflection each. In this situation, the question arises whether from the 2nd person singular down to the 3rd person plural the person affixes are indeed those which occur also as possessor affixes on nouns as the quoted authors seem to suggest. The problem is that this question cannot be answered as clearly as one might think.

Table 3.3 reveals that, except in the 1st person singular, it is simply impossible to decide – on the basis of the morphological form of the affixes alone – whether *-k, -h, -ha, -na, -kom*, or *-hom* marks the possessor or the direct object. Playing the devil's advocate for a while, given the indistinction of possessor marking and object marking for the majority of the affixes, version II of the paradigms as given in Table 3.11 is a relatively likely alternative interpretation.

While Table 3.10 depicts the situation as one in which overabundance applies to only one out of seven values, Table 3.11 gives evidence of syncretism. Accordingly, the functional distinction assumed for the pair *fija* ≠ *fini* is neutralised in all other persons. Another possibility could be to analyse the two cellmates in the 1st person singular to represent a case of overdifferentiation with clear-cut functional differences between the cellmates. Does this entail that there is large-scale defectiveness in the paradigm because none of the other persons has a dedicated marker for one of the functions? On top of that, we have explicitly excluded pseudo-verbs from BLOMP 2.0.[48] On the assumption that *fini* is a pseudo-verb, it thus can be eliminated from the paradigm of (П15) *fi* 'in' which then takes an entirely regular shape. This solution however is possible only for (П15) *fi* 'in' but not for (П7) *bi* 'with' since

48 Cf. Section 1.6.2.

Table 3.11: Paradigms of (Π7) *bi* 'with' and (Π15) *fi* 'in' – version II.

person	(Π7) ***bi*** 'with'		(Π15) ***fi*** 'in'	
1SG	*bija*	*bini*	*fija*	*fini*
2SG	*bik*		*fik*	
3SG.M	*bih*		*fih*	
3SG.F	*biha*		*fiha*	
1PL	*bina*		*fina*	
2PL	*bikom*		*fikom*	
3PL	*bihom*		*fihom*	

pseudo-verb status has not yet been claimed for this Π as decidedly as in the case of (Π15) *fi* 'in'. A third version of the paradigms could therefore look like Table 3.12.

Table 3.12: Paradigms of (Π7) *bi* 'with' and (Π15) *fi* 'in' – version III.

person	(Π7) ***bi*** 'with'		(Π15) ***fi*** 'in'
1SG	*bija*	*bini*	*fija*
2SG	*bik*		*fik*
3SG.M	*bih*		*fih*
3SG.F	*biha*		*fiha*
1PL	*bina*		*fina*
2PL	*bikom*		*fikom*
3PL	*bihom*		*fihom*

These little *Gedankenspiele* are only meant to suggest that whatever interpretation is applied to the members of C II, morphological mismatches and noncanonical behaviour cannot be avoided. To create two parallel well-behaved paradigms, ***bini*** must be argued not to belong there in the first place. To this end, the morphosyntax and semantics of the cellmates have to be looked into.[49]

49 This is a task for Section 4.1.2.2.2.2.3 where a fourth version of the paradigms (= Table 4.13) will be presented to solve the riddle these Πs pose.

3.1.2.3.3 Class III (= C III)

We now enter the domain of those inflected Π-stems that do not equal the citation form. Different processes may operate on the citation form to create the Π-stem if we assume that the latter is indeed derived from the former. We distinguish three subclasses of C III which we label with reference to the final segment of the Π-stem. We discuss *j*-stems in Section 3.1.2.3.3.1, *i*-stems in Section 3.1.2.3.3.2, and conclude this overview with the presentation of *għajn*-stems in Section 3.1.2.3.3.3. We emphasise that there are alternative ways of class-formation based on different principles such as the final segment of the citation form or the process which generates the Π-stem. Our choice is solely motivated by practicality.

3.1.2.3.3.1 *j*-stems (= C IIIa)

There are three inflected Πs whose Π-stem ends in the palatal approximant /j/ although this is not the final segment of the citation form. To start, this is the case for (Π23) *ħdejn* 'beside'. Sutcliffe (1936: 203) and Aquilina (1987: 471) postulate the existence of a Π *ħada* 'beside' which cannot host the person suffixes. Cachia (1994: 115, original italics) specifies that "*[ħ]ada* li tintuża flok *ħdejn* ma tiħux suffissi".[50] The exact synchronic relation between *ħada* and (Π23) *ħdejn* 'beside' is unclear. Aquilina (1959: 300) sketches the situation as follows: *ħada* is the free form of the Π but *ħdejn* is used in combination with person markers. E. Serracino-Inglott (1977: 107) only registers (Π23) *ħdejn* 'beside' for which he assumes a historical connection to *ħeda* (= *ħada*). In the Korpus Malti 3.0, one comes across nineteen occurrences of *ħada* eighteen of which result from a recurrent spelling error: the intended verb-form in the 3rd person singular masculine perfective + direct object clitic *-ha* of the 3rd person singular feminine is *ħadha* 'he took her'. The only case of a prepositional use of *ħada* in the Korpus Malti 3.0 is example (1.8).[51] According to our estimate in Table 1.11, (Π23) *ħdejn* 'beside' as a free form has a token frequency of 16,489. Clearly, (Π23) *ħdejn* 'beside' has ousted *ħada* in contemporary Maltese. The latter has probably been more widely in use in the 19th century when Falzon (1882: 106 and 119) described it as the citation form of the Π under inspection. The absence of tangible evidence of the current use of *ħada* induces us to exclude this historical form from further discussion in the morphology section.

As to the formation of the *j*-stem, in two cases – (Π23) *ħdejn* 'beside' and (Π28) *lejn* 'towards' – /j/ belongs to the segmental chain of the citation form whereas (Π60) *wara* 'after' lacks this segment in the citation form (Borg/Azzopardi-Alexander 1997: 265). This means that two very different processes are required to derive the Π-stem. The final nasal /n/ of the citation form of (Π23) *ħdejn* 'beside' and (Π28) *lejn* 'towards'

50 Our translation: "*ħada* which is used instead of *ħdejn* does not take suffixes".

51 Presented and discussed in Section 1.4.1.3.

is deleted yielding the Π-stems *ħdej-* and *lej-* whose final segment determines that the person suffixes are taken from the B-set (cf. Section 3.1.2.1). The stem-formation process is thus subtractive. In contrast, (Π60) *wara* 'after' leaves the segmental chain of the citation form intact but adds /j/ to the final vowel to create the *j*-stem. This is an additive process. Since other Πs which rhyme with (Π23) *ħdejn* 'beside' and (Π28) *lejn* 'towards' keep their final /n/ and other Πs with citation forms ending in /a/ behave differently from (Π60) *wara* 'after' no generalisations can be put forward. The members of C IIIa are specified for their behaviour in the lexicon.

Examples for the inflection of these Πs in the 3rd person plural have been given already in (3.18)–(3.19) and (3.26). We still have to prove that they can also be inflected for speech-act participants. To this end, we present examples (3.59)–(3.61).

(3.59) [Bartolo/Vella 2009: 50]

tista'	*tpoġġi*	*[**ħdejja**]$_{PP}$*
2SG.IPFV:can	2SG.IPFV:sit	[**beside**:1SG]

'[. . .] you can sit [**beside** me].'

(3.60) [Bartolo/Vella 2009: 319]

Ħares	*[**lejja**]$_{PP}$*	*meta*	*nkellmek*
look	[**towards**:1SG]	when	1SG.IPFV:speak:2SG

'Look [**at** me] when I speak to you!'

(3.61) [Bartolo/Vella 2009: 53]

Se	*toqgħod*	*toffrilhom*	*it-tazzi*	*tat-te*
FUT	2SG.IPFV:stay	2SG.IPFV:offer:IO:3PL	DEF-glass:PL	of:DEF-tea

afrodiżijaku	*lin-nies*	*biex*	*iġġenninhom*
aphrodisiac	to:DEF-people	in_order_to	2SG.IPFV:instigate:3PL

*[**warajk**]$_{PP}$*
[**after**:2SG]

'Will keep offering the aphrodisiac tea to the people to make them become fond [**of** you]?'

Table 3.13 features the paradigms of the three Πs in C IIIa. Note that (Π60) *wara* 'after' may also appear as *ura* under inflection (Sutcliffe 1936: 207). E. Serracino-Inglott (1989: 26) assumes that *ura* is possible only in combination with *'l* (as in *'l ura* > *lura* 'back'). Except for the single occurrence of *urajha*, there is no evidence of this alternative Π-stem in the Korpus Malti 3.0. It is likely that the alternative Π-stem *ura* has become obsolete while it enjoyed a much wider use in the past. For the sake of comprehensiveness, we mention the word forms which involve *ura* under the proviso that they are no longer in common use.

Table 3.13: Paradigms of (Π23) *ħdejn* 'beside', (Π28) *lejn* 'towards', and (Π60) *wara* 'after'.

person	(Π23) *ħdejn* 'beside'	(Π28) *lejn* 'towards'	(Π60) *wara* 'after'	
1SG	*ħdejja*	*lejja*	*warajja*	*urajja*
2SG	*ħdlejk*	*lejk*	*warajk*	*urajk*
3SG.M	*ħdejh*	*lejh*	*warajh*	*urajh*
3SG.F	*ħdejha*	*lejha*	*warajha*	*urajha*
1PL	*ħdejna*	*lejna*	*warajna*	*urajna*
2PL	*ħdejkom*	*lejkom*	*warajkom*	*urajkom*
3PL	*ħdejhom*	*lejhom*	*warajhom*	*urajhom*

3.1.2.3.3.2 *i*-stems (= C IIIb)

In C IIIb, we find four Πs which reflect two different patterns for the formation of the Π-stem. To the Π-stem, the person suffixes of Set B attach. There are two Πs whose citation form qualifies them for membership in C I because the final segment is a consonant. These are (Π13) *favur* 'in favour of' and (Π21) *għal* 'for'. From potential consonantal stems they change to vocalic stems by way of adding the vowel /i/ to the citation form as shown in (3.22)–(3.23). These examples also prove that both Πs are compatible with the person suffixes of speech-act participants. The Π-stem of (Π13) *favur* 'in favour of' and (Π21) *għal* 'for' results from an additive process.

This is different with (Π27) *kontra* 'against' and (Π37) *mingħala* 'in s.o.'s opinion'.[52] Their citation form is polysyllabic and ends in the low vowel /a/. Thus, they resemble (Π60) *wara* 'after' discussed in the foregoing subsection. However, this phonological resemblance does not entail identical morphological behaviour. In the case of (Π27) *kontra* 'against' and (Π37) *mingħala* 'in s.o.'s opinion' the final /a/ of the citation form is replaced with /i/ under suffixation as shown in (3.24)–(3.25). The compatibility of (Π37) *mingħala* 'in s.o.'s opinion' with person suffixes of speech-act participants has been documented in (3.24) already. This possibility can also be proved for (Π27) *kontra* 'against' as shown in (3.62).

(3.62) [Bartolo/Vella 2009: 201]

Din	*il-potenza*	*tintuża*	***[kontrija]**$_{PP}$*
DEM:F:PROX	DEF-power	3SG.F.IPFV:RFL:use	**[against**:1SG]

'This power is used [**against** me].'

52 The unlikely possibility of (Π44) *oltre* 'beyond' being a further member of this subclass is rebutted in Section 4.1.2.2.2.2.2 (ii).

The paradigms of the four members of C IIIb are presented in Table 3.14.

Table 3.14: Paradigms of (Π13) *favur* 'in favour of', (Π21) *għal* 'for', (Π27) *kontra* 'against', and (Π37) *mingħala* 'in s.o.'s opinion'.

person	(Π13) *favur* 'in favour of'	(Π21) *għal* 'for'	(Π27) *kontra* 'against'	(Π37) *mingħala* 'in s.o.'s opinion'
1SG	*favurija*	*għalija*	*kontrija*	*mingħalija*
2SG	*favurik*	*għalik*	*kontrik*	*mingħalik*
3SG.M	*favurih*	*għalih*	*kontrih*	*mingħalih*
3SG.F	*favuriha*	*għaliha*	*kontriha*	*mingħaliha*
1PL	*favurina*	*għalina*	*kontrina*	*mingħalina*
2PL	*favurikom*	*għalikom*	*kontrikom*	*mingħalikom*
3PL	*favurihom*	*għalihom*	*kontrihom*	*mingħalihom*

The story of (Π13) *favur* 'in favour of' does not end here. The Korpus Malti 3.0 gives ample evidence of violations of the rules for the formation of the Π-stem of this Π. In point of fact, beside the expected *i*-stem hosting the person inflections, there is a not negligible number of cases where *favur* is treated as a member of C Ia, i.e. the stem formative -*i* does not show up so that the person suffixes attach directly to the final consonant of the citation form. This alternative pattern is attested for all persons except the 1st and 2nd person singular. The resulting person inflected forms are identical to those of the pronominally possessed noun *favur* 'favour'.

Table 3.15 compares the turnouts of the competing forms of person-inflected (Π13) *favur* 'in favour of' in the Korpus Malti 3.0.

Table 3.15: Competing forms of person-inlected (Π13) *favur* 'in favour of'.

person	*i*-stem		consonant stem	
	form	tokens	form	tokens
3SG.M	*favurih*	291	*favuru*	59
3SG.F	*favuriha*	198	*favurha*	44
1PL	*favurina*	41	*favurna*	3
2PL	*favurikom*	1	*favurkom*	2
3PL	*favurihom*	158	*favurhom*	130

Except in the 2nd and 3rd person plural, the regular *i*-stems outnumber the alternative Π-stem by a wide margin. The ratios are 4.5-to-one, 4.9-to-one, and 13-to-one for the 3rd person singular feminine, the 3rd person singular masculine, and the 1st person plural, respectively. In the case of the 2nd person plural, the token frequencies of the two contenders are too low to justify generalizing on their basis. The *i*-stem of the 3rd person plural is only 1.2 times more frequent than the consonant stem. What strikes the eye immediately is the distribution of the consonant stem over genres. There is not a single instance in belles-lettres, academic, or religious texts. In contrast, the attestations cluster in the domain of parliamentary speeches, news, E.U.-related documents, and occasionally show up in legal texts, too. The frequency is too high to sweepingly classify the cases as typographical errors. Table 3.16 provides examples of the use of the consonant stem for each of the person categories mentioned above.

Table 3.16: Examples of usage with consonant stem of (Π13) *favur* 'in favour of'.

person	example /translation	source
3SG.M	*għamlu sforzi kbar biex jimmobilizzaw il-vot* ***favuru*** *u daħħlu wkoll lil-Liberali fl-alleanza tagħhom* '[. . .] they made big efforts to mobilise a **favourable** vote **for him** and brought the Liberals too into their alliance [. . .]'	news113113
3SG.F	*ma naqbilx ma' din il-mozzjoni u mhux se nivvota* ***favurha*** '[. . .] I don't agree with this motion and will not vote **for it** [. . .].'	parl9797
1PL	*Dawn m'humiex se jivvotaw* ***favurna*** *għax iridu jbiegħu l-prodott tagħhom* 'They will not vote **for us** because they want to sell their product.'	news108657
2PL	*se jmexxi xi kruċjata* ***favurkom*** '[. . .] he will lead a crusade **to your benefit** [. . .].'	news79026
3PL	*Hafna paroli kontrihom jew* ***favurhom*** *minghajr l-edukazzjoni u informazzjoni bażika u tajba* 'Many empty words against them and **in their favour** without basic and good education and information [. . .].'	news60666

Especially the co-occurrence of the *i*-stem *kontrihom* 'against them' and the consonant stem *favurhom* 'in their favour' suggests that the *i*-stem is not generally avoided in the genres under inspection. On account of the consonant stem's relatively strong position in certain genres, it makes sense to study the competition of the stems in connection to the possibility of the emergence of a Maltese Eurolect according to the ideas expressed in Portelli/Caruana (2018). We relegate this investigation to a follow-up study in the not too distant future. The tokens resulting from the occurrences of the consonant stem of (Π13) *favur* 'in favour of' will be integrated in the overall frequency count in Section 3.4.

3.1.2.3.3.3 *għajn*-stems (= C IIIc)

C IIIc counts only two members which however are very interesting morphologically. We are speaking of (Π30) *ma'* 'with' and (Π55) *ta'* 'of'. Were it not for the orthographic apostrophe, the citation forms would call to mind vowel stems. However, the apostrophe conventionally indicates that a historically lost *għajn* re-emerges elsewhere in the paradigm of a given word. This is also the case with (Π30) *ma'* 'with' and (Π55) *ta'* 'of'.

The resurrection of the *għajn* is not simply an orthographic phenomenon. First of all, the addition of <għ> to the (superficially) vowel-final citation forms of (Π30) *ma'* 'with' and (Π55) *ta'* 'of' creates a consonantal stem. Accordingly, the person suffixes of Set A have to be employed. Under suffixation, the abstract phoneme *għajn* interacts with the segments in its neighbourhood so that considerable phonological changes affect the segmental chain of the individual word forms. Diphthongisation occurs in the 1st person singular and the 3rd person singular masculine, lengthening of the preceding /a/ occurs in the word forms of the 1st and 2nd persons plural, *għajn* is realised as voiceless velar or uvular fricative [x] or [χ] in 3rd person singular feminine and the 3rd person plural where it is adjacent orthographically to the silent <h> of the suffixes *-ha* and *-hom*, the suffix vowel in the 2nd person singular undergoes lengthening. These processes are identified in the paradigms in Table 3.17. Note that the diphthongs have two different phonetic realisations which depend on the individual speaker's preferences. The abbreviations in Table 3.17 are D = diphthongisation, L = lengthening, X = velar fricative.

Table 3.17: Paradigms of (Π30) *ma'* 'with' and (Π55) *ta'* 'of'.

person	(Π30) *ma'* 'with'		(Π55) *ta'* 'of'		process
	orthography	IPA	orthography	IPA	
1SG	*miegħi*	[mɪ.ɛɪ] ~ [mɪ.eɪ]	*tiegħi*	[tɪ.ɛɪ] ~ [tɪ.eɪ]	D
2SG	*miegħek*	[mɪ.ɛːk] ~ [mɪ.eːk]	*tiegħek*	[tɪ.ɛːk] ~ [tɪ.eːk]	L
3SG.M	*miegħu*	[mɪ.ɔʊ] ~ [mɪ.eʊ]	*tiegħu*	[tɪ.ɔʊ] ~ [tɪ.eʊ]	D
3SG.F	*magħha*	[mɐxxɐ]	*tagħha*	[tɐxxɐ]	X
1PL	*magħna*	[meːna]	*tagħna*	[teːna]	L
2PL	*magħkom*	[meːkɔm]	*tagħkom*	[teːkɔm]	L
3PL	*magħhom*	[mɐxxɔm]	*tagħhom*	[tɐxxɔm]	X

While the cellmates in the IPA columns constitute cases of free variation, the paradigm gives ample evidence of phonological (or weak) suppletion (Corbett 2007b). The Π-stem is affected by several processes which can nevertheless be considered regular insofar as they are triggered by *għajn* according to patterns which hold throughout the morphonological system of Maltese. It would be possible therefore to integrate C IIIc into C I as subclass C Ig if one assumes an underlying final *għajn* for the citation forms of the two Πs. However, we have reason to object this integration because both (Π30) *ma'* 'with' and (Π55) *ta'* 'of' are involved in external sandhi patterns which make it highly improbable that the citation forms involve a silent final *għajn*.[53]

Sample sentences for the use of inflected forms of (Π30) *ma'* 'with' and (Π55) *ta'* 'of' were given in (3.20)–(3.21) for persons other than the speech-act participants. To fill the gap, we provide examples of the latter in (3.63)–(3.64).

(3.63) [Attard 2019: 139]
tridux *tiġu* ***[miegħi]***$_{PP}$ *mela*
2.IPFV:require:PL:NEG 2.IPFV:come:PL **[with**:1SG] then
'[. . .] don't you want to come [**with** me] then [. . .]?'

(3.64) [Attard 2019: 84]
dik *hi* *l-intenzjoni* ***[tiegħi]***$_{PP}$
DEM:F:DIST 3SG.F DEF-intention **[of**:1SG]
'[. . .] that is [**my**] intention.'

3.1.2.3.4 Intermediate summary

The density and richness of the information given in the previous subsections calls for a short recapitulation of some of the major insights we have gained so far. Table 3.18 provides a survey of the morphological classes, their label, and their members.

Table 3.18: Survey of morphological classes of inflected Πs.

class	subclass	label	members	n
C I	C Ia	plain consonantal	(Π5) *bħal* 'like'; (Π8) *biswit* 'facing'; (Π10) *daqs* 'equal to'; (Π11) *dwar* 'about'; (Π14) *fejn* 'near'; (Π17) *fuq* 'on'; (Π20) *għajr* 'except'; (Π22) *għand* 'at s.o.'s place'; (Π24) *ħlief* 'except'; (Π29) *lil* 'to'; (Π31) *madwar* 'around'; (Π32) *maġenb* 'close to'; (Π36) *mingħajr* 'without'; (Π38) *mingħand* 'from s.o.'; (Π41) *minn* 'from'; (Π48) *qrib* 'near'; (Π49) *quddiem* 'in front of'; (Π56) *taħt* 'under'	18
	C Ib	vowel-harmonic	(Π35) *minflok* 'instead of'; (Π53) *skont* 'according to'	2
	C Ic		(Π46) *qabel* 'before'	1
	C Id		(Π4) *bejn* 'between'	1
	C Ie	defective I	(Π16) *fost* 'amongst'; (Π47) *qalb* 'amidst'	2
	C If	defective II	(Π33) *matul* 'during'; (Π59) *waqt* 'at the time of'	2

53 The data which support our interpretation will be discussed in Section 3.3.

Table 3.18 (continued)

class	subclass	label	members	n
C II		plain vocalic	(Π7) *bi* 'with'; (Π15) *fi* 'in'	2
C III	C IIIa	*j*-stems	(Π23) *ħdejn* 'beside', (Π28) *lejn* 'towards'; (Π60) *wara* 'after'	3
	C IIIb	*i*-stems	(Π13) *favur* 'in favour of'; (Π21) *għal* 'for'; (Π27) *kontra* 'against';(Π37) *mingħala* 'in s.o.'s opinion'	4
	C IIIc	*għajn*-stems	(Π30) *ma'* 'with'; (Π55) *ta'* 'of'	2

Except C Ia none of the classes can be called big. At a considerable distance, C IIIb with just four Πs is second best. Given the number of small classes with two to three members together with C Ic and C Id which host only one Π each, the catalogue of morphological classes might look relatively uneconomical. However, it was not our intention to organise the system of inflected Πs in terms of descriptive economy. Our guiding principle has been qualitative in nature. Whether our classifications always represent the best choice is a completely different story.

What needs to be kept in mind is the importance of morphological mismatches in the system. We had occasion to discuss phenomena such as overabundance, overdifferentiation, defectiveness, and suppletion. If the cellmates *bija* and *bini* of the 1st person singular of (Π7) *bi* 'with' in C II can be shown to be functionally (more or less) the same insofar as *bini* can also be used in contexts where an interpretation as pseudo-verb fails, one might even consider the possibility that some kind of deponency-like mismatch is also attested (Baerman 2007). This is one of the many open questions that remain after we have reviewed the above morphological classes. Some of these questions will receive an answer in due course in the remainder of the morphology section and/or subsequently. In spite of the fact that most of the differences between the above classes are neither of the affixal nor the prosodic kind (Corbett/Baerman 2006), we assume that the use of deviant inflectional material in C Id and C II justifies our speaking of inflectional classes. If (Π4) *bejn* 'between' and (Π7) *bi* 'with' and (Π15) *fi* 'in' display inflections that are not shared by the bulk of the Πs, there is reason to assume four inflectional classes, namely C I (except C Id), C Id, C II, and C III. Given the morphological intricacies caused by the interaction of several mismatches, it might turn out to make sense to increase the number of inflectional classes further. This however is a subject matter for a separate study.

Other problems have dissolved because it could be shown that certain inflected forms, Π-stems, or citation forms are archaisms, especially in extant Maltese dictionaries. The dictionary compilers have relied substantially on primary and secondary sources which date back to the period which ranges from the late 18th century through the 19th century to the beginning of the 20th century. The Korpus Malti 3.0 and our separate corpus of contemporary Maltese prose give no evidence of the current existence of these elements of the past. Therefore, it was possible to exclude these items from further discussion in the synchronic part of this study.

3.2 Long vs short forms: morphological reduction

This section takes up an issue which repeatedly came up above,[54] namely the possibility of certain dimorphemic Πs to dropping their initial constituent which is usually either *ma-* or *minn-*. The allomorphy of long forms with *ma-/minn-* and short forms without *ma-/minn-* can be subsumed neither under the rubric of stem allomorphy addressed in the foregoing section nor under that of sandhi phenomena to be discussed in Section 3.3. It is therefore in order to dedicate a separate section to the long-short alternation attested for a small number of Maltese Πs.[55]

In this section, the Πs of interest and their corresponding short allomorphs are (Π33) *matul* 'during' vs *tul* (cf. examples (1.124) and (1.126)); (Π34) *minbarra* 'except' vs *mbarra* (cf. examples (1.108)–(1.109)); (Π35) *minflok* 'instead of' vs *flok* (cf. examples (1.93)–(1.94)); and (Π39) *minħabba* 'on account of' vs *mħabba* vs *ħabba* (1.97)–(1.99). In addition, we also look at the competition of alloforms for (Π2) *a skapitu* 'at the expense of' and (Π13) *favur* 'in favour of'.[56] On account of the many intricacies connected to a potential sixth case, the discussion of (Π50) *rigward* 'concerning' is relegated to the chapter on morphosyntax.[57] What we want to show in what follows is that the choice of alloform is not contextually determined. This means that there is no phonological or morphosyntactic trigger which requires the employment of any particular allomorph. The empirical illustration of the phenomenon stems exclusively from the Korpus Malti 3.0 to guarantee compatibility of the data. Owing to its exemplary character, we start with the competition between the two alloforms of (Π34) *minbarra* 'except' to which Section 3.2.1 is dedicated. In the section on (Π34) *minbarra* 'except' and those following it, we will conduct very simple tests in order to determine whether there are any structural reasons for the choice of a given alloform. The format of these tests is decribed in Section 3.2.1. Section 3.2.7 evaluates the findings.

54 Cf. Section 1.6.4.2.

55 A particularly intricate and at the same time doubtful additional case of short-long alternation is discussed at length in Section 4.1.2.2.1.2.1 (x).

56 Cf. Sections 3.2.5 and 3.2.6, respectively.

57 Cf. Section 4.1.2.2.1.2.1 (ix).

3.2.1 Alloforms of (П34) *minbarra* 'except'

As Figure 1.6 has shown for the two allomorphs of (П34) *minbarra* 'except' (= *minbarra* and *mbarra*), they are exclusively used to encode EXCEPTION whereas *barra*, *minn barra* and *barra minn* are far less dedicated to this function. We therefore restrict the comparison of the alternative expressions of EXCEPTION to the two first mentioned one-word alloforms. In Figure 3.25, the turnouts resulting from the quantitative evaluation of the first 100 matches in the Korpus Malti 3.0 are repeated below in a different format from Figure 1.6.

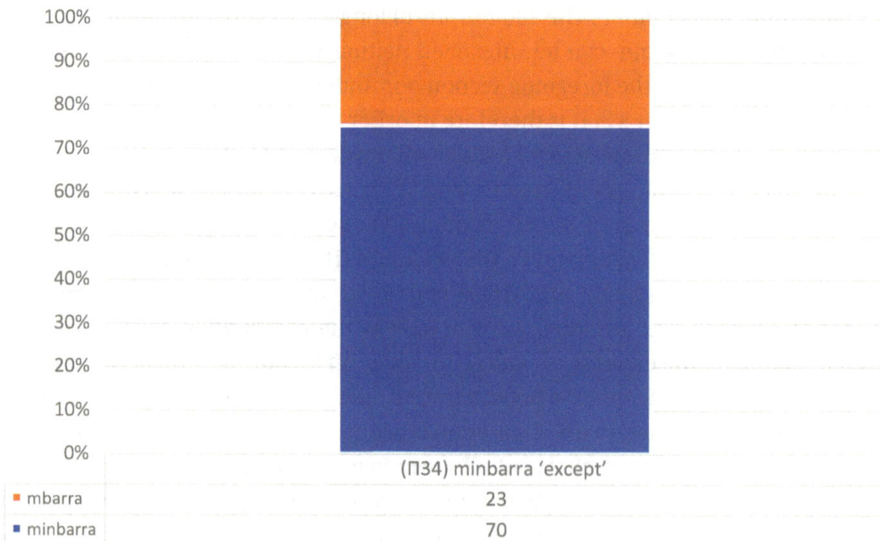

Figure 3.25: Token frequency (1st 100) of competing allomorphs of (П34) *minbarra* 'except'.

The tokens in Figure 3.25 add up to the total 93 matches. With 75%, *minbarra*'s share of this total is 3.5 times the size of the share of its competitor. However, the picture is vastly distorted because the token frequency of *mbarra* in the entire Korpus Malti 3.0 is the same as that given in Figure 3.25. The estimated token frequency of *minbarra* is much higher, namely 21,035. Thus, going by the values in Figure 3.25, the allomorph *mbarra* is grossly overrepresented. If we take account of the estimated figures, *minbarra* with 99.9% leaves only a very narrow margin for *mbarra* (n < 0.1%). At least in terms of token frequency, *mbarra* can thus be considered to be exceptional. The marked status of *mbarra* offers us a perfect starting point for the identification of possible factors in the choice of allomorphs. The small number of attestations of this allomorph in the Korpus Malti 3.0 is helpful for recognising potential patterns.

Examples of *mbarra* can be found in only four genres. E.U.-related texts are responsible for twelve matches and thus constitute a majority of 52% of all cases of *mbarra*. Parliamentary speeches are the source for seven cases, news reports are responsible for three tokens and legal texts contribute one token. This allomorph seems to be alien to belles lettres, academic, and religious genres. This fact alone is already interesting because it is indicative of the possibility of stylistic preferences for or against *mbarra*. In eight out of twelve matches in E.U.-related texts, *mbarra* heads a PP which is surrounded by brackets in writing. We interpret the use of brackets as a sign of the PP's status as addition to the clause or constituents thereof. Table 3.19 contains all bracketed *mbarra*-PPS.

Table 3.19: Bracketed *mbarra*-PPS.

source	example / translation
european3436	*Il-pitrava zokkrija (**mbarra** ż-żrieragħ)* 'the sugar-beet (**except** the seeds)'
european3436	*Egħruq tal-għalf u l-brassicas (**mbarra** ż-żrieragħ)* 'feed roots and brassicas (**except** the seeds)'
european3436	*Fjuri u pjanti ornamentali (**mbarra** tal-mixtla)* 'ornamental flowers and plants (**except** the nursery bed)'
european3436	*Żrieragħ u nebbiet f' raba' li jinħarat (**mbarra** ċ-ċereali, il-ħxejjex niexfa, il-patata u l-pjanti ta' żrieragħ li minnhom jingħasar iż-żejt)* 'seeds and seedlings on farmland under the plough (**except** cereals, dry grass, potatoes and plants of seeds from which oil is pressed)'
european3436	*Uċuħ tar-raba' sekondarji suċċessivi (**mbarra** uċuħ tar-raba' f'ġonna tas- suq u wċuħ tar-raba' taħt il-ħġieġ)* 'successive secondary arable land (**except** arable land in market gardens and arable land under glass protection)'
european651	*dawk il-prodotti speċifikament esklużi taħt id-Direttiva ta' Vultaġġ Baxx (**mbarra** apparat mediku)* 'those products which are specifically excluded under the Low Voltage Directive (**except** medical apparatus)'
european8298	*Ħass ikkaboċċjat (**mbarra** ħass tat-tip "Iceberg")* 'lettuce in the form of cabbage (**except** lettuce of the type "Iceberg")'
european8298	*ħass Cos jew Romaine (**mbarra** ħass b' weraq oħxon)* 'Cos or Romaine lettuce (**except** lettuce with thick leaves)'

The first five examples stem from one and the same text which is also the source for two further instances of *mbarra*-PPS. These additional cases and another one from a different E.U.-related text resemble the bracketed examples in Table 3.19 insofar as the *mbarra*-PP is clearly marked as an insertion – this time not by brackets but by commas as in (3.65) where we keep the interpunction.

(3.65) [Korpus Malti 3.0; european3436]

Mergħa u art miżrugħa ħafur, ***[mbarra*** *mergħat*
pasture and land PART:SOW:F wild_oat **[except** pasture:PL
ħorox]$_{PP}$
rugged.PL]
'[. . .] pasture and land sown with wild oat, **[except** rugged pastures] [. . .]'

Almost all of the examples with inserted *mbarra*-PPs stem from longish lists which enumerate those items to which certain E.U. regulations must be applied and which exceptions must be made. The entries in these lists do not display a full-blown sentence structure because they lack a proper predicate. Thus, we are confronted with enumerations of more or less complex NPs. Outside syntactically fragmentary contexts of this kind, *mbarra*-PPs have been identified, too.

Since the presence/absence or shape of the initial constituent of the above ΠS is crucial for the distinction of long and short allomophs, it makes sense to take stock of the final segment of the syntactic word to the immediate left of the Π in the same sentence. There are three sentence-initial instances of *mbarra* which have to be excluded from the count. Of the remaining twenty attestations of *mbarra*, eight are preceded by a vowel-final word whereas twelve follow a consonant-final word. The gap between 40% and 60% is not big enough to prove beyond doubt that the choice of *mbarra* depends on the surrounding phonological context. *Mbarra* starts with a binary cluster which would normally trigger the use of the prosthetic /i/ if the preceding word ends in a consonant (Borg/Azzopardi-Alexander 1997: 334). This is the case in (3.66) where the word to the left of *mbarra* ends in the voiceless postalveolar fricative /ʃ/.

(3.66) [Korpus Malti 3.0; news56537]

Għax ***[imbarra*** *leħinha]*$_{PP}$ *dawk ix-xufftejn tagħha*
because **[except** voice:3SG.F] DEM:PL:DIST DEF-lip:PL of:3SG.F
mimlija u ħomor f' wiċċ qamħi mfaqqa' kienu
full:PL and red.PL in face corn_coloured PART:burst be:3PL.PFV
jħarrku d-demm tal-irġiel kollha li
3.IPFV:produce_ardour:PL DEF-blood of:DEF:man.PL all that
kienu jikluha bil-ħars
be:3PL.PFV 3.IPFV:eat:PL:3SG.F with:DEF-look
'Since **[apart from** her voice], these full and red lips of hers in a shining corn-coloured face made the blood of all men boil who devoured her with their eyes.'

The expected *i*-prosthesis does not always take place as shown in (3.67) where the final /n/ of *snin* 'years' should require the presence of /i/ on *mbarra* but the prosthetic vowel fails to appear.

(3.67) [Korpus Malti 3.0; parl566]
F' dawn l-aħħar tliet snin ***[mbarra*** *għajnuna]*$_{PP}$
in DEM:PL:PROX DEF-last three year.PL **[except** help]
dan il-Gvern ivvota Lm 85,000 għall-iżvillupp
DEM:M:PROX DEF-government vote £m 85,000 for:DEF-development
ta' elite sport
of elite sport
'In the last three years [**apart from** help], this government voted £m 85,000 for the development of elite sports.'

We assume that the violation of the sandhi rules of Maltese which is a recurrent phenomenon with *mbarra* has a sentence prosodic explanation. As emphasised above, *mbarra*-PPs more often than not come as insertions which are probably prosodically marked as such. The orthographic means – interpunction and brackets – signal potential intonation breaks. The minimal pauses separate the insertion from the preceding context such that the phonological properties of the final element of the word immediately preceding the insertion fails to activate the usual sandhi processes. This means that the *mbarra*-PPs behave as if uttered in isolation which, in turn, is the same as occupying the clause-initial or sentence-initial position. However, of the three sentence-initial attestations of *mbarra* in this position, only two give evidence of *i*-prosthesis. This is the case in (3.68). Example (3.69) illustrates the absence of the prosthetic /i/.

(3.68) [Korpus Malti 3.0; parl9830]
[Imbarra *l-mentalità li l-ministru jista'*
[except DEF-mentality that DEF-minister 3SG.M.IPFV:can
jagħmel li jrid]$_{PP}$ *hemm il-mentalità*
3SG.M.IPFV:do that 3SG.M.IPFV:want] there DEF-mentality
l-oħra wkoll
DEF-other:F too
'[**Apart from** the mentality according to which the minister can do what he wants] there is also the other mentality [. . .].'

(3.69) [Korpus Malti 3.0; parl4231]

*[**Mbarra** dan]$_{PP}$ isir ukoll xogħol*
[**except** DEM:M:PROX] 3SG.M.IPFV:become too work

mal-familji ta' dawn iż-żgħażagħ biex
with:DEF-family:PL of DEM:PL:PROX DEF-adolescent.PL in_order_to

jiġu ntegrati lura fil-familji
3.IPFV:come:PL integrate:PART:PL back in:DEF-family:PL

'[**Apart from** this] there will also be work with the families of these youngsters to reintegrate them into the families.'

Interestingly, in both cases the final word of the preceding sentence ends in a vowel, namely *jagħmlu* 'they do' and *pariri* 'advice$_{PL}$', respectively. The vacillation as to the appropriate alloform in a given context should not be overvalued because the small number of cases and the possibility of typos preclude generalisations. It should not go unmentioned nevertheless that *mbarra* occurs three times in sentence-initial position. These cases account for 13% of all tokens registered for *mbarra*. As to the seventy tokens counted for *minbarra* functioning as Π (of the 1st 100 matches in the Korpus Malti 3.0), this long alloform occupies the leftmost slot in 37 sentences which equals a share of 53%. We can only speculate that this relative propensity of the long alloform to occur sentence-initially might be connected causally to the phonotactic "disadvantage" of the initial cluster in the short alloform. *Minbarra* needs no prosthetic /i/ and may combine with vowel-final and consonant-final words to the left of it. What is additionally noteworthy about the long alloform is that amongst the above seventy matches for *minbarra*, there is only a single instance of an inserted *minbarra*-PP surrounded by brackets. Three further cases are marked as insertions by means of commas. Thus, only 6% of the *minbarra*-PPs function as (orthographically marked) insertions. On top of this, *minbarra* is attested in all of the genres which are represented in the Korpus Malti 3.0.

Moreover, we have also looked at the complements of *mbarra* with focus on definiteness. There are eight indefinite NPs which function as complement. If we discount cases of coordination and enumeration of several NPs overtly marked for definiteness, we have detected thirteen definite NPs as complements of *mbarra*. Nine of these involve nouns which host the definiteness pro-clitic. Three complements are pronominal, one an inalienably possessed noun, and one is a PP with a definite complement. To simplify, we disregard the latter case for the time being as it requires a dedicated discussion which will be started only later.[58] After removing the problematic case, we have twelve cases of definite complements (= 60%) as

58 In Section 4.1.

opposed to eight cases of indefinite complements (= 40%) combining with *mbarra*. This means that (in)definiteness is neither an obstacle nor an instigator for the employment of *mbarra*. In contrast, the first seventy matches of *minbarra* involve sixty definite (= 86%) and ten indefinite NPs (= 14%) as complements. In spite of the considerable differences in the shares of (in)definite complements with the two competing alloforms, we refrain from drawing definitive conclusions since the turnout of tokens is far too small in the case of *mbarra*. The impossibility of pinpointing any structural factors in our quest for the logic behind the choice of *mbarra* from the list of allomorphs of (Π34) *minbarra* 'except' supports the hypothesis that we are dealing with a matter of style and not with a grammatical rule. This conclusion is not too far-fetched in the case of an only marginally attested allomorph. Is the situation similar for the other patterns of allomorphy?

3.2.2 Alloforms of (Π35) *minflok* 'instead of'

In Section 1.6.4.2. (i) we have argued that the long alloform *minflok* is the best representative of (Π35) *minflok* 'instead of' because it is much more frequent than *flok* in the Korpus Malti 3.0. The number of matches we referred to on that occasion did not differentiate between uses as Π and other functions of the alloforms. In Figure 3.26, we only take account of the Π-uses of *minflok* and *flok*.

For the 1st 100 matches of the competitors, *minflok* yields 65 matches as opposed to the 61 matches counted for *flok*. The small gap between the two turnouts widens considerably if we look at the estimated token frequencies. The estimates are given in Figure 3.26. According to these hypothetical values, the alloform *minnflok* is used 3.4 times as often as *flok*. If we add up the token frequencies, we get the total of 38,068 tokens 22% of which go to the credit of *flok* whereas *minflok* claims a share of 78%. The quantitative differences are thus significant although they are by no means comparable to those computed for the previous pair *minbarra-barra*.

Before we tick off the same parameters as in the previous section we have to take account of a property which the alloforms of (Π34) *minbarra* 'except' lack, namely the ability to inflect for person. There are altogether 5,563 matches of inflected forms of (Π35) *minflok* 'instead of' in the Korpus Malti 3.0. Figure 3.27 shows that the two alloforms are unevenly involved in person inflection.

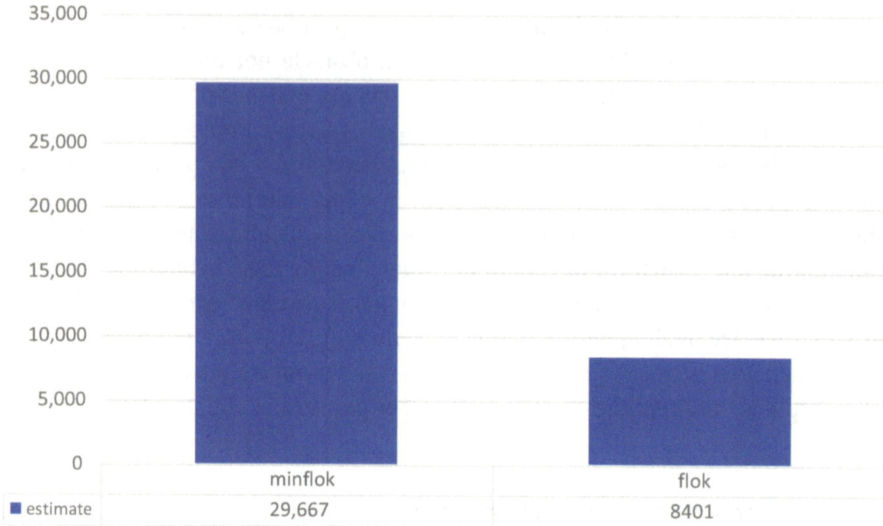

Figure 3.26: Token frequency of competing allomorphs of (П35) *minflok* 'instead of'.

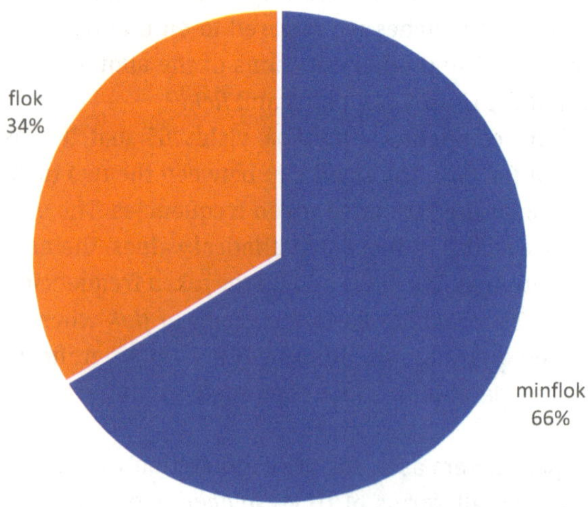

Figure 3.27: Shares of alloforms of (П35) *minflok* 'instead of' under inflection.

Under inflection, the share of *flok* increases visibly. From less than a quarter of the uninflected tokens of (П35) *minflok* 'instead of' the alloform *flok* claims slightly more than a third of all inflected tokens of (П35) *minflok* 'instead of'. This increase notwithstanding, the long alloform still outnumbers the short alloform also where person inflection comes into play. The question arises whether person inflection generally is a favourable condition for the choice of the alloform *flok*. In Figure 3.28, we determine the frequencies of inflected *minflok* and inflected *flok* for each of the persons. It immediately comes to the fore that independent of the person *minflok* always remains the most frequently chosen option. At the same time, Figure 3.28 reveals that the shares of *minflok* and *flok* differ from person to person.

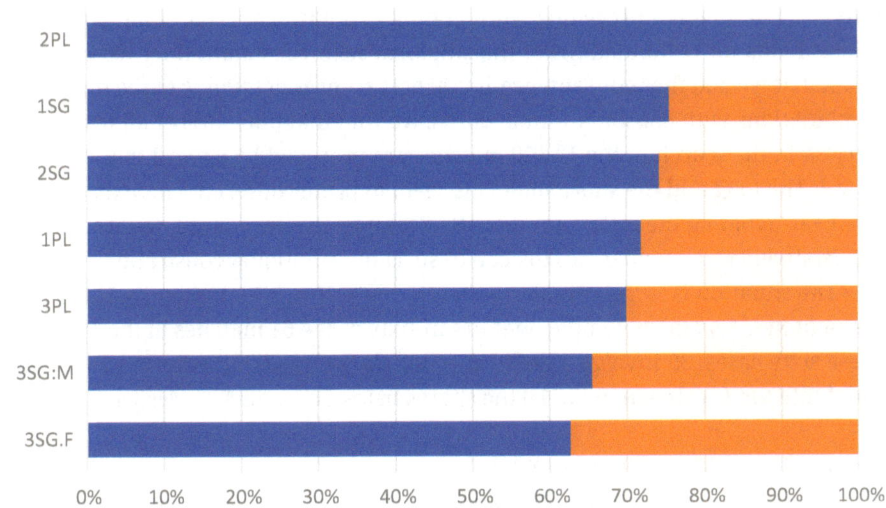

	3SG.F	3SG.M	3PL	1PL	2SG	1SG	2PL
■ minflok	803	1750	972	23	49	96	1
■ flok	476	920	416	9	17	31	0

Figure 3.28: Inflected alloforms of (Π35) *minflok* 'instead of' per person.

We discount the 2nd person plural because it is a hapax for this Π in the Korpus Malti 3.0. For the remaining six person values, the shares of *flok* range from 28% for the 1st person singular to 37% for the 3rd person singular feminine. More generally, *flok* is relatively strong in the 3rd persons of both numbers with shares of n ≥ 30% whereas it is relatively weak with speech-act participants in the singular. However, these gradual differences do not alter the fact that in each person (except the 2nd plural) the share of *flok* is bigger than that given in Figure 3.26 for the uninflected word forms. The differential behaviour of *flok* with and without person inflection is certainly interesting but no hard-and-fast rules emerge from these differences. The most we learn from the above quantitative data is that both alloforms of (Π35) *minflok* 'instead of' go together well with person inflection.

At this point, we turn out attention to checking the data which result from the search for the 1st 100 matches in the Korpus Malti 3.0. Of the 61 hits for *flok*, 36 feature this alloform in combination with a preceding vowel-final word in the same sentence. There are 21 tokens of *flok* which combine with a consonant-final word on the left. In addition, we have identified four cases of sentence-initial *flok*. This means that 59% of all instances of *flok* in the search of the 1st 100 matches are preceded by a word that ends in a vowel. As to the long alloform, the phonological quality of the final segment of the preceding word is irrelevant. *Minflok* combines 22 times with a preceding word ending in a vowel and the same number of tokens holds for combinations of this alloform with consonant-final words to the left of it. If we add the four sentence-initial occurrences of *minflok* to the number of combinations with consonant-final words, we find that only 46% of all instance of *minflok* in the search of the 1st 100 matches are preceded by a word that ends in a vowel. The long and the short alloforms seem to prefer different contexts without, however, avoiding the disfavoured context completely. Once more, we are witnessing relatively weak tendencies instead of strict distributional constraints.

The situation is largely similar with regards to the (in)definiteness of the complement NPs. *Flok* takes complement NPs in sixty of the 61 matches in the 1st 100. Of these sixty cases, 35 involve a definite NP (as shown in (3.70)) whereas 25 involve indefinite NPs (as shown in (3.71)) the shares being 58% and 42%, respectively.

(3.70) [Korpus Malti 3.0; culture1115]

William	*u*	*Kate*	*iltaqgħu*	*mal-President*	*Tan*	*u*
William	and	Kate	meet:3PL.PFV	with-president	Tan	and
martu	***[flok***	*ir-reġina]*$_{PP}$	*f'*	*lukanda*	*lussuża*	
woman:POSS:3SG.M	**[instead_of**	DEF-queen]	in	hotel	luxurious:F	

'[**Instead of** the Queen] William and Kate met President Tan and his wife in a luxurious hotel [. . .].'

(3.71) [Korpus Malti 3.0; culture2054]

Din	*kienet*	*tkun*	*storja*	*differenti*	*kieku*
DEM:F:PROX	be:3SG.F.PFV	3SG.F:be.FUT	story	different	if
[flok	*tifla]*$_{PP}$	*kellhom*	*tifel*		
[instead_of	child:F]	have.PFV:IO:3PL	child		

'This would have been a different story if [**instead of** a girl] they had a boy.'

Minflok is attested 45 times with a complement NP. In 21 cases the NP is definite as in (3.72) and in 24 cases it is indefinite as in (3.73). The shares are 47% for definite NPs and 53% for indefinite NPs.

(3.72) [Korpus Malti 3.0; academic13]

aħjar *li* *nużaw* *il-kompost* ***[minflok*** *il-ħamrija* *komuni]_{PP}*
good.CMPR that 1.IPFV:use:PL DEF-compost **[instead_of** DEF-red_soil ordinary]

'[. . .] it is better that we use compost [**instead of** ordinary red soil].'

(3.73) [Korpus Malti 3.0; academic46]

ippublika *edizzjoni* *privata* *għall-abbonati* *biss* ***[minflok*** *ktieb* *għall-publiku]*$_{PP}$
publish edition private:F for:DEF-subscriber:PL only **[instead_of** book for:DEF-public]

'[. . .] he published a private edition exclusively for subscribers [**instead of** a book for the general public].'

We do not intend to read too much into these relatively subtle differences. It is likely that a more thorough search of the Korpus Malti 3.0 will show that the shares level out for both alloforms. One should not jump to the conclusion that *minflok* has a leaning towards the indefiniteness of its complement whereas *flok* prefers definite complements over indefinite ones.

On the basis of the parameters we have defined for the present purpose it is not possible to decide what exactly determines the choice of the alloforms of (Π35) *minflok* 'instead of'. This unsatisfactory result might change if more and different criteria are applied to the cases under scrutiny. For the time being, however, we must assume that the alloforms constitute a case of free variation. Most probably, the preference for the one or the other is a matter of style or idiolectal choices.

3.2.3 Alloforms of (Π33) *matul* 'during'

The possibility of a semantically motivated split of the alloforms of (Π33) *matul* 'during' resulting in two distinct candidates for the status of Π has been discussed at length and finally rebutted.[59] Given that semantics can largely be ruled out as a determining factor, the distribution of the long alloform *matul* and the short alloform *tul* needs to be looked at afresh.

59 Cf. Section 1.6.4.2. (ix).

We approach the problems first from the perspective of quantities. In Figure 3.29 we compare the estimated token frequencies of *matul* and *tul* in the Korpus Malti 3.0. Figure 3.29 takes account only of uninflected free forms.

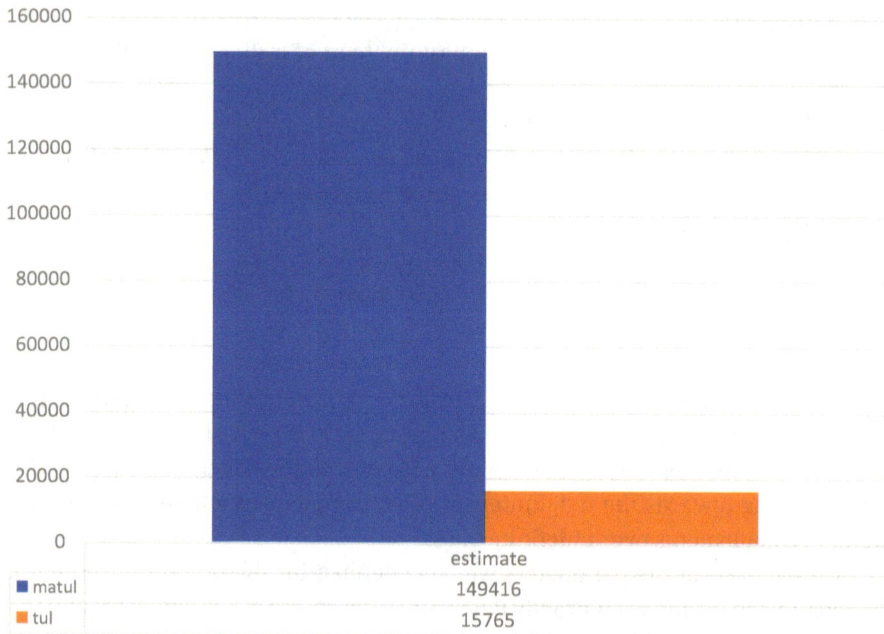

Figure 3.29: Token frequency of competing allomorphs of (П33) *matul* 'during'.

Matul yields almost ten times as many tokens as *tul*. The latter is thus a marginal option in comparison to the predominant long allomorph. The discrepancy between the frequencies of the two alloforms is even more pronounced under inflection. Figure 3.30 reveals that the long alloform clearly outranks the short alloform. Of the 6,220 inflected forms of (П33) *matul* 'during', 6,069 go to the credit of *matul* whereas only 151 matches have been counted for *tul*.

We know that (П33) *matul* 'during' is subject to paradigmatic defectiveness in the sense that inflected forms are attested almost exclusively for the 3rd persons of both genders and both numbers.[60] The 2nd person singular *matulek* occurs only infrequently and in a particular idiomatic construction type. The same holds for the short alloform *tul* as shown in Figure 3.31.

60 Cf. Section 3.1.2.3.1.2 (iv).

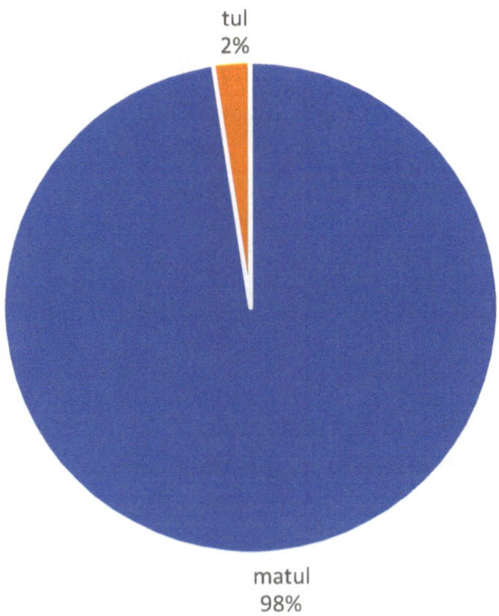

Figure 3.30: Shares of alloforms of (П33) *matul* 'during' under inflection.

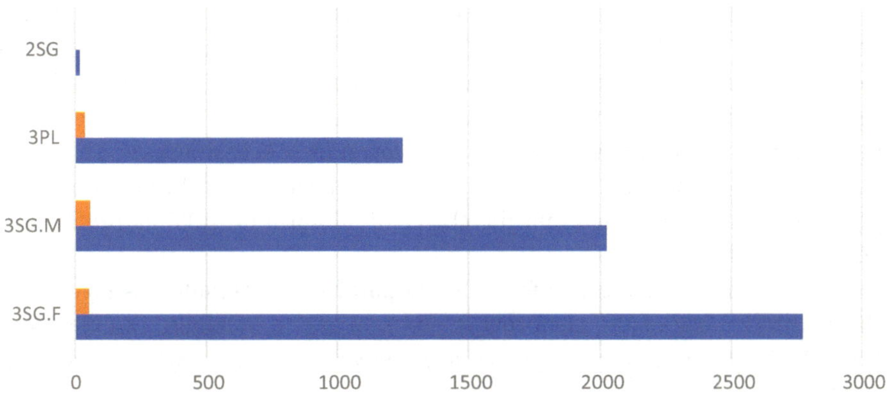

	3SG.F	3SG.M	3PL	2SG
tul	53	59	38	1
matul	2772	2025	1252	20

Figure 3.31: Inflected alloforms of (П33) *matul* 'during' per person.

The short alloform *tul* displays shares which range from 1.8% in the 3rd person singular feminine to 4.7% in the 2nd person singular. The latter relatively high percentage results from the sparing use of the idiomatic construction type presented in Figure 3.22. The isolated case of the short alloform *tul* being used in this construction frame is given in (3.74) (cf. (3.49) above).

(3.74) [Korpus Malti 3.0; news123284]

fil-ħarifa *nikkonsolidaw* *dak* *kollu* *li*
in:DEF-autumn 1.IPFV:consolidate:PL DEM:M:DIST all:M that
nkunu *ottjenejna* *bis-sagrifiċċji* *kollha*
1.IPFV:be.FUT:PL obtain:1PL.PFV with:DEF-sacrifice:PL all:PL
*[**tulek**]$_{PP}$* *ja* *sena*
[**during**:2SG] oh year
'[. . .] in the autumn we consolidate all that we will have obtained by means of all the sacrifices [all year long].'

In contrast to the competition between *minflok* and *flok* addressed in the foregoing section, person inflection does not increase the shares of the short allomorph of (Π33) *matul* 'during'. Its frequency remains marginal independent of the presence of person inflection.

For the remaining tests we limit the scope to the hits within the 1st 100 returns in the Korpus Malti 3.0. We had to remove four tokens for each of the competing alloforms on account of unclear structural properties. Thus, in what follows, we assume thirty tokens for *tul* and 96 tokens for *matul*. According to Figure 3.32, the distribution of the alloforms over the three contexts – after a vowel-final word, after a consonant-final word, and in the leftmost slot of the sentence – is almost identical. If we lump together C] __ and [S, we immediately recognise that the chance factor makes itself felt. *Matul* and *tul* are attested to the right of a preceding word with a final vowel in 50% of the cases. The other half of their attestations is covered by the non-vocalic conditions. On these grounds, we rebut the idea that the choice of alloform of (Π33) *matul* 'during' depends on phonology.

The situation is not much different for the parameter of definiteness. Figure 3.33 shows that neither of the alloforms is prone to taking indefinite complements. For technical reasons, only the percentages between 90% and 100% are visible in Figure 3.33.

Both *matul* and *tul* prefer definite complements over indefinite ones in nine out of ten cases. There is thus no discernible difference between the two alloforms on this parameter either. This leads us to the conclusion that in analogy to the previous cases of competing long and short alloforms, the alloforms of (Π33) *matul* 'during' are in free variation. In the absence of structural reasons, their choice must depend wholly on idiolectal and stylistic preferences.

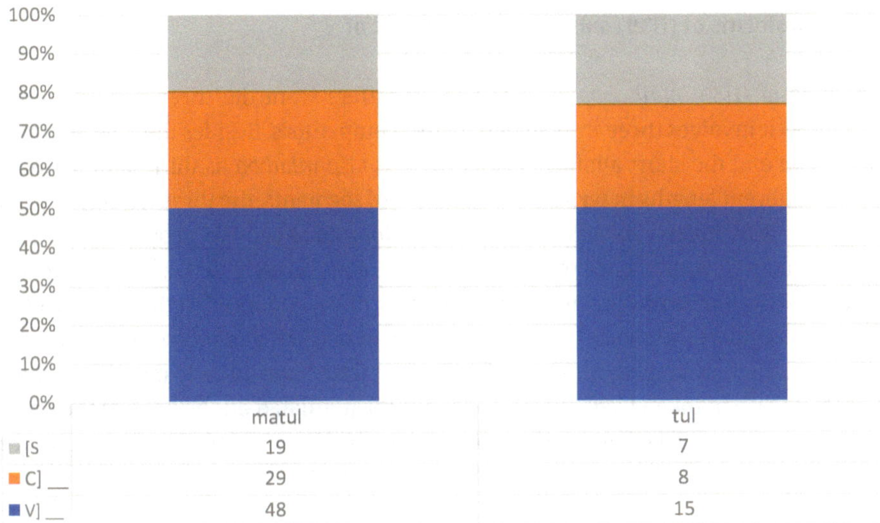

Figure 3.32: Distribution of alloforms of (П33) *matul* 'during' over phonological contexts.

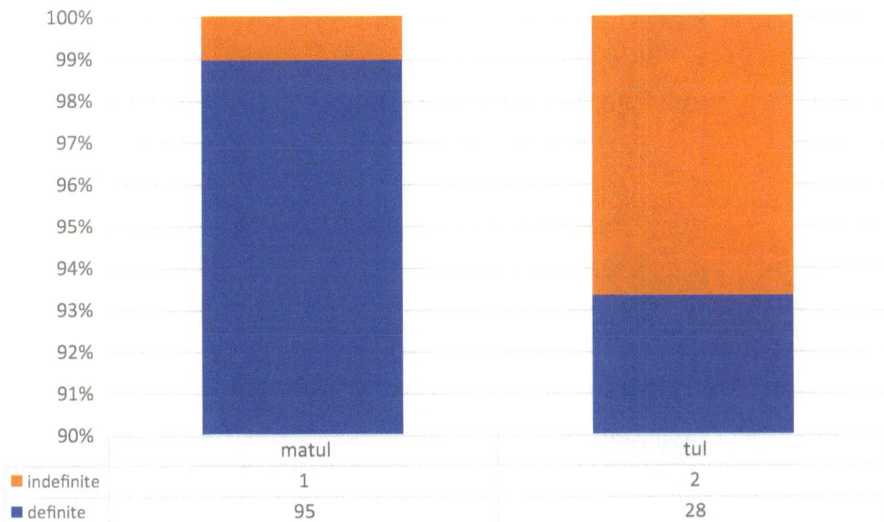

Figure 3.33: Alloforms of (П33) *matul* 'during' and (in)definite complements.

3.2.4 Alloforms of (Π39) *minħabba* 'on account of'

The case of (Π39) *minħabba* 'on account of' differs from the three previous ones insofar as it involves three instead of only two competitors. Besides the long alloform *minħabba* and the short alloform *ħabba* there is also *mħabba* as third option. Going by the different lengths in terms of the number of segments, the three alloforms yield the order *minħabba* > *mħabba* > *ħabba* from long to short. We provided evidence of the semantic equivalence of the three expressions already above.[61] Since we are dealing with allomorphy and not with distinct Πs, one wonders what the factors might be which regulate the choice of the three alloforms of (Π39) *minħabba* 'on account of'. When we addressed these alloforms for the first time,[62] we already mentioned their widely divergent token frequencies. These are recapitulated on the basis of the usual estimate in Figure 3.34. For the 1st 100 returns in the Korpus Malti 3.0. the results are eighty tokens for *minħabba*, 35 tokens for *mħabba*, and 25 tokens for *ħabba*.

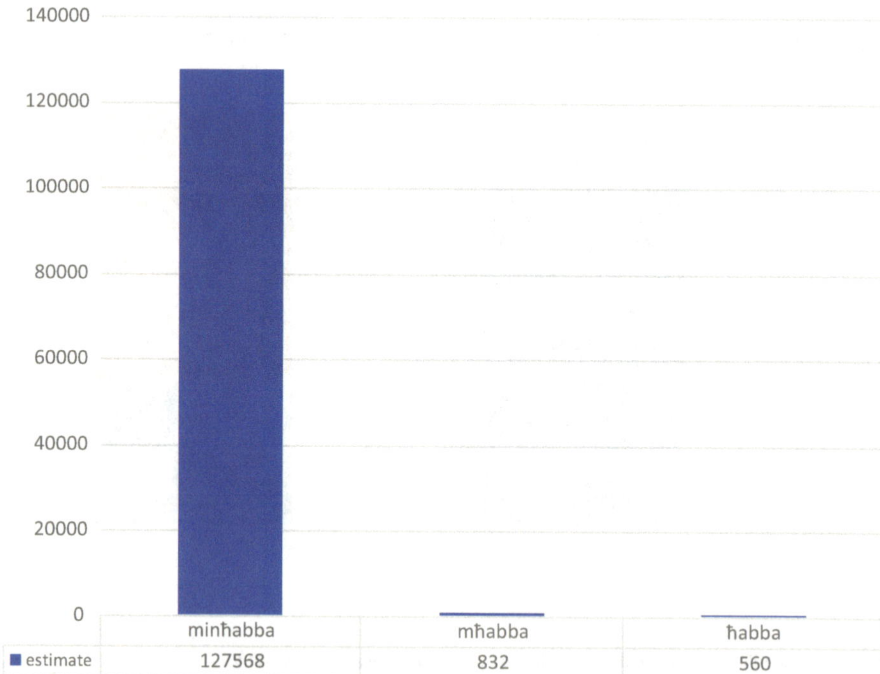

Figure 3.34: Estimated token frequency of alloforms of (Π39) *minħabba* 'on account of'.

61 Cf. Section 1.6.4.2 (iii).

62 Cf. Section 1.6.4.2 (iii).

The long alloform is 153 to 182 times more frequent than the two shorter alloforms. The short alloforms are thus only marginal alternatives for the realisation of (Π39) *minħabba* 'on account of'. The marginal status of the short alloform as opposed to the dominant long alloform is a leitmotif of this kind of allomorphy in Maltese.

We turn our attention to testing the possibility of whether phonology has a say in the choice of alloforms of (Π39) *minħabba* 'on account of'. To this end, we compare the results for the combinations of the three alloforms with preceding vowel-final and consonant-final words as well as the occurrences in absolute sentence-initial position. The data which serve as input for Figure 3.35 stem from the search of the 1st 100 matches.

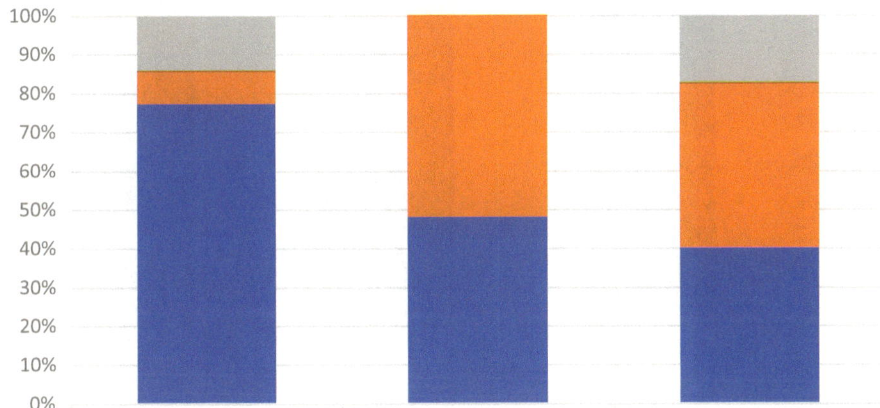

	mħabba	ħabba	minħabba
[S	5	0	14
C]__	3	13	34
V]__	27	12	32

Figure 3.35: Distribution of alloforms of (Π39) *minħabba* 'on account of' over phonological contexts.

The alloforms are ordered from left to right according to the increase of the number of combinations with words to their left which ends in a consonant. The alloforms do not behave homogeneously. *Mħabba* occurs nine times as frequently with vowel-final words as with consonant-final-words. If we reduce the three contexts to the binary opposition of vocalic vs consonantal context we notice that 77% of the tokens of *mħabba* can be found in vocalic contexts whereas this context is relevant only for 40% of the tokens of the long allomorph. Interestingly, the shortest allomorph *ħabba* attests to almost equal shares for vocalic and consonantal contexts, namely 48% and 52%, respectively. This means that, for the first time in our review of this kind of allomorphy, there emerges a relatively strong divergence in the distribution of alloforms over phonological contexts. The long allomorph clearly favours consonantal contexts whereas the preference of *mħabba* for vocalic contexts is even more pronounced. In contrast, the shortest alloform appears to be uncommitted, in a manner of speaking.

In Figure 3.36, we try to establish whether a comparable situation arises when the alloforms take (in)definite complements. The data are again those which result from the search of the 1st 100. For structural reasons, the absolute numbers given for *minħabba* and *ħabba* are smaller than those reported in connection to Figure 3.35.

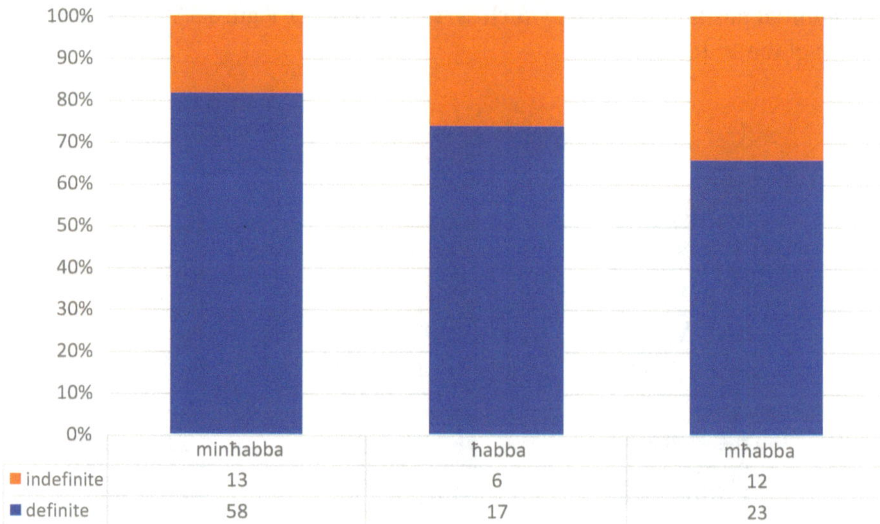

Figure 3.36: Alloforms of (П39) *minħabba* 'on account of' and (in)definite complements.

The alloforms are ordered from left to right according to the decreasing share claimed by definite complements. Like in the previous test, *minħabba* and *mħabba* wind up at the opposite extremes of the scale with *ħabba* occupying the intermediate position. Definite complements are in the majority for all three allomorphs. In the case of *minħabba*, definite complements account for 82% of all registered instances. The share drops to 66% for *mħabba*, whereas *ħabba* has a percentage of 74% for definite complements. These differeces are certainly not as striking as those which come to the fore in the analysis of the distribution over phonological contexts.

In point of fact, the following shares in Table 3.20 are indicative of a given alloform being overrepresented or underrepresented on a given parameter. The yellow colour is indicative of significantly low percentages whereas a cell highlighted in grey hosts an unexpectedly high percentage.

Table 3.20: Overrepresentation/underrepresentation per parameter.

alloform	1st 100		phonology				(in)definiteness			
			C]/[S		V]		definite		indefinite	
minħabba	80	57%	48	69%	32	45%	58	59%	13	42%
mħabba	35	25%	8	12%	27	38%	23	23%	12	39%
ħabba	25	18%	13	19%	12	17%	17	17%	6	19%
total	140		69		71		98		31	

From this table it is obvious that *ħabba* is largely well-behaved in the sense that its shares remain close to the share it has within the 1st 100. This is different for *minħabba* and *mħabba*. In combination with definite complements, the three allomorphs meet our expectations. In contrast, *minħabba* is overrepresented in non-vocalic contexts where *mħabba* is widely underrepresented. The opposite holds for vocalic contexts where the size of *minħabba*'s share is considerably smaller than that of this alloform in the 1st 100. *Mħabba* however has a share that exceeds that of the same alloform in the 1st 100. The same pattern applies to combinations with indefinite complements. Here, *minħabba* is again widely underrepresented whereas *mħabba* is overrepresented.

The observed differences are interesting but fail to characterise the alloforms as being categorically different from each other. It is better by far to assume preferences which are shaped by stylistic and/or idiolectal choices. There is no hard evidence of genuinely grammatical rules which dictate the choice of a given alloform. Rare examples like (3.75) even preclude the possibility that the idiolectal choices are always strict.

(3.75) [Korpus Malti 3.0; literature59]

Ġenninhom u dan mhux biss ***[minħabba***
instigate:3PL and DEM:M:PROX NEG:3SG.M:NEG only **[on_account_of**

l-ħwejjeġ żmattati u x-xagħar twil]$_{PP}$
DEF-thing.PL become_untidy:PART:PL and DEF-hair long]

iżda anke ***[ħabba*** *l-ħarsa t' għajnejh]$_{PP}$*
but also **[on_account_of** DEF-look of eye:PL:3SG.M]

'He drove them crazy and this not only [**because of** the untidy clothes and the long hair] but also [**because of** the look of his eyes].'

In this sentence, the two alloforms *minħabba* and *ħabba* are used in two consecutive adverbial phrases. Both take definite complements. *Minħabba* is preceded by a consonant-final word whereas *ħabba* follows a vowel-final word. Since

Figure 3.35 characterises *ħabba* as being equally compatible with consonant-final and vowel-final words to its left, the sequence *anke* + *ħabba* is not surprising. That *ħabba* and not *minħabba* is chosen for this combination is perhaps more telling since the long allomorph prefers consonantal contexts over vocalic contexts (though only gradually). The fact remains that in one and the same sentence several alloforms of (Π39) *minħabba* 'on account of' may peacefully co-exist without creating confusion. It cannot be ruled out that the author from whose work example (3.75) has found its way into the database of the Korpus Malti 3.0 was guided by the principle of *variatio delectat* according to which it is stylistically preferable not to use identical word forms invested with the same meaning in the same context.

3.2.5 Alloforms of (Π2) *a skapitu* 'at the expense of'

Like in the case of (Π39) *minħabba* 'on account of', (Π2) *a skapitu* 'at the expense of' comes with an array of alloforms. The shortest alloform *skapitu* is statistically insignificant. All other alloforms are longer by one (= *a skapitu*) to three segments if *għajn* is counted as full consonant (= *bi skapitu, għas-skapitu, għal skapitu*). There are no morphological or phonological criteria according to which one of the all-forms is chosen instead of others in a given context. Figure 3.37 takes account of post-vocalic and post-consonantal occurrences of the alloforms under scrutiny for the 1st 100 hits of each alloform in the Korpus Malti 3.0. The sentence-initial position is occupied only once by *a skapitu*, all other attestations of no matter which alloform are sentence-medial. The overall frequency of bare *skapitu* is too small to justify the representation of the respective values in Figure 3.37. The alloforms are ordered from left to right according to the decreasing shares of the C]__ context. As can be seen immediately, the two alloforms to the left have a slight preference for following a consonant-final word whereas the two alloforms on the right display a similar preference for occurring after a word which ends in a vowel. The percentages are still relatively close to the 50%-mark so that the chance factor cannot be ruled out.

As to the combinability with definite and indefinite complements, Figure 3.38 informs us about the behaviour of the same four alloforms of (Π2) *a skapitu* 'at the expense of'.

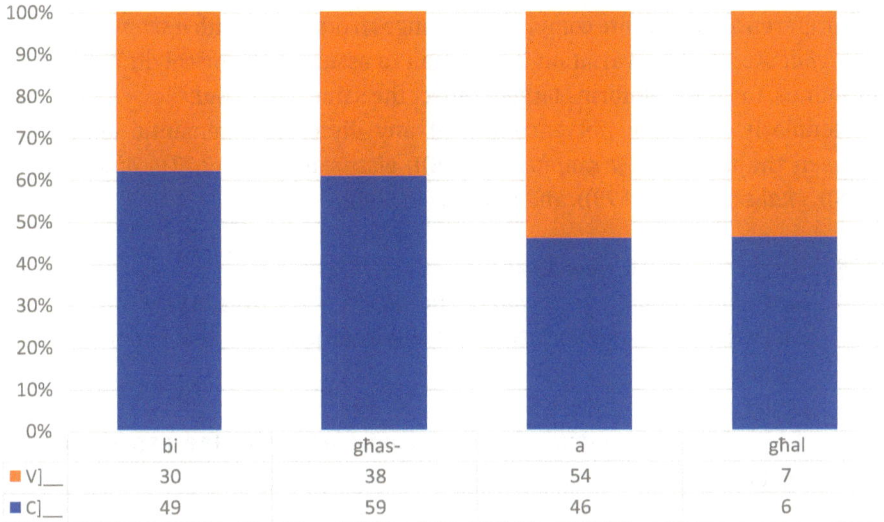

Figure 3.37: Distribution of alloforms of (П2) *a skapitu* 'at the expense of' over phonological contexts.

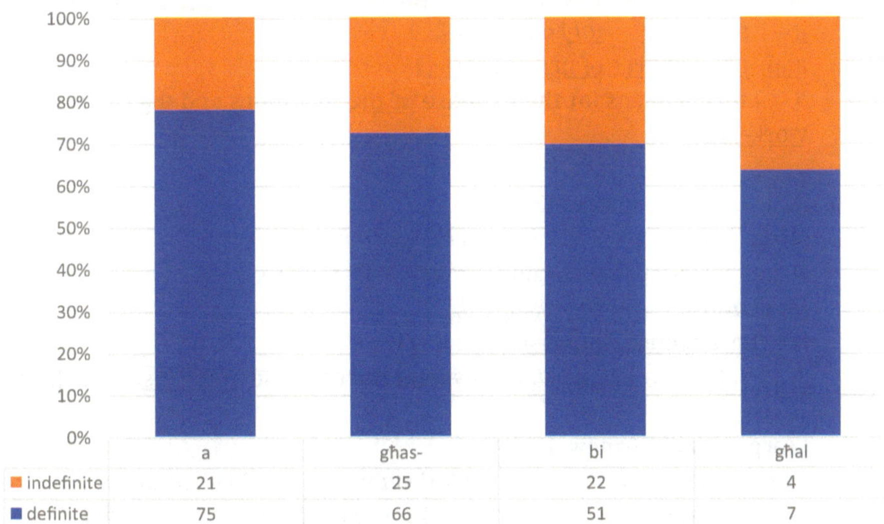

Figure 3.38: Alloforms of (П2) *a skapitu* 'at the expense of'and (in)definite complements.

The preference for definite complements ranges from 75.1% with *a skapitu* to 63.6% with *għal skapitu*. The variation is too small to assume categorical differences. To our minds, the four alloforms largely reflect the same behaviour.

Similarly, we have not encountered any discernible semantic differences between the alloforms *a skapitu* (= (3.76)), *għas-skapitu* (= (3.77)), *bi skapitu* (= (3.78)), *għal skapitu* (= (3.79)), and *skapitu* (= (3.80)).[63]

(3.76) [Korpus Malti 3.0; news136050]
jagħmlu *l-flus* *mill- ambjent* ***[à skapitu***
3.IPFV:do:PL DEF-money from:DEF-environment **[at_the_expense**
tal-poplu]$_{PP}$
of:DEF-people]
'[. . .] they make money from the environment [**at the expense** **of** the people].'

(3.77) [Korpus Malti 3.0; news124496]
dan *isir* ***[għas-skapitu*** *tal-kundizzjonijiet*
DEM:M:PROX 3SG.M.IPFV:become **[at_the_expense** **of**:DEF-condition:PL
u *d-drittijiet* *tal-ħaddiema]*$_{PP}$
and DEF-right:PL of:DEF-worker:PL]
'[. . .] this happens [**at the expense** **of** the conditions and the rights of the workers].'

(3.78) [Korpus Malti 3.0; news126217]
sejrin *igawdu* *biss* *erbgħa* *min-nies*
FUT:PL 3.IPFV:enjoy:PL only four from:DEF-people
[bi skapitu *tar-residenti]*$_{PP}$
[at_the_expense **of**:DEF-respdent:PL]
'[. . .] only a few people will enjoy it [**at the expense** **of** the residents] [. . .].'

(3.79) [Korpus Malti 3.0; news115614]
jagħmel *aktar* *qliegħ* ***[għal skapitu*** *tal-ħaddiema]*$_{PP}$
3SG.M.IPFV:do more profit **[at_the_expense** **of**:DEF-worker:PL]
'[. . .] he makes more profit [**at the expense** **of** the workers].'

63 For similarities with (II50) *rigward* 'concerning' the reader is referred to Section 4.1.2.2.1.2.1 (ix).

(3.80) [Korpus Malti 3.0; news105051]

Progress	*mhux*	***[skapitu***	*tal-passat]$_{PP}$*
progress	NEG:3SG.M:NEG	**[at_the_expense**	**of**:DEF-past]

'Progress is not [**to the detriment of** the past].'

What the alloforms have in common is that (Π55) *ta'* 'of' always forms part of the PPs – a topic to which we come back below.[64] To conclude the discussion of the Π under inspection, we look at the overall token frequencies of the four alloforms featured in the previous figures in the Korpus Malti 3.0 in Figure 3.39.

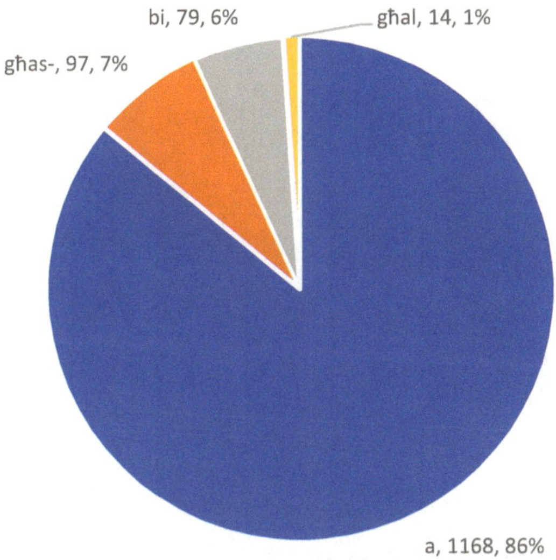

Figure 3.39: Token frequency of alloforms of (Π2) *a skapitu* 'at the expense of'.

The major option, *a skapitu*, outnumbers all other alloforms by far. Not even added up do the lesser options come anywhere near the dominant competitor. Whether and to what extent stylistic or idiolectal preferences are responsible for the choice of some of the minority options cannot be determined on the basis of our data.

64 Cf. Section 4.1.2.2.1.1.

3.2.6 Alloforms of (Π13) *favur* 'in favour of'

For this Π, three alloforms have to be taken account of, namely the shortest *favur* and two which are one (*a favur*) and three segments including *għajn* (*affavur* with the probably only orthographic variant *għaffavur*).[65] Figure 3.40 makes it clear that the longer alloforms are only marginally attested in the Korpus Malti 3.0 whereas the short alloform *favur* is represented by an abundance of tokens.

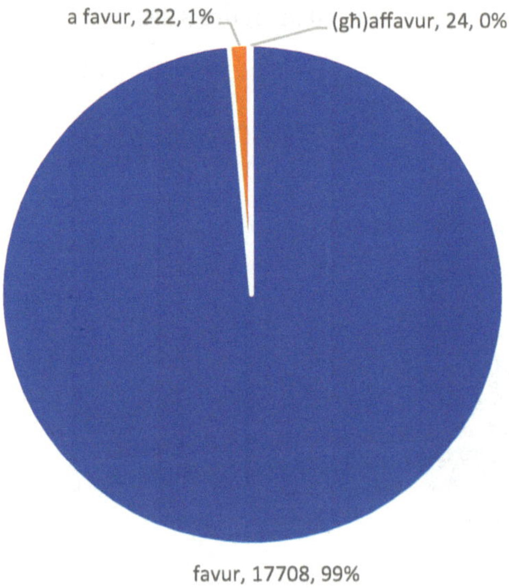

Figure 3.40: Token frequency of alloforms of (Π13) *favur* 'in favour of'.

65 Saari (2009: 272) additionally speaks of a "modern variant" *għal-favur ta'* 'in favour of'. It cannot be ruled out that this is a reanalysis of *għaffavur* which in written form appears often as *għaf-favur* suggesting the presence of a cliticised combination of (Π21) *għal* 'for' and the definiteness marker although /f/ is never the product of assimilation in Maltese. The search in the Korpus Malti 3.0 yielded a mixed bag of fifteen instances four of which are bona fide cases of nominative government of a nominal complement (such as [news147040] *Forsi vvintawha biex jiġbru aktar simpatija **[għal favur** id-divorzju]$_{PP}$?* 'Have they perhaps invented it to receive more sympathy [in favour of divorce]?'). There is also one instance of genitive government ([news117772] *jiżdied **[għal favur** ta' kull min għandu x' jaqsam mas- suq tax-xogħol]$_{PP}$* '[. . .] it adds something [to the benefit of everyone who has to do with the work market].') and a doubtful case involving (Π29) *lil* 'to' as linker ([news106593] *Dik ommissjoni inumana **[għal favur** lil tal- klikka]$_{PP}$*. 'This is an omission **[in favour of** those of the clique].'). On account of many uncertainties, we refrain from discussing these cases because this so-called modern variant does not seem to be fully established in the language.

Since none of the alloforms is attested sentence-initially in the 1st 100-test in the Korpus Malti 3.0, it suffices to look at the final segment of the preceding word to determine whether there are any favourable phonological conditions for the choice of the alloforms. The results of the test are exposed in Figure 3.41.

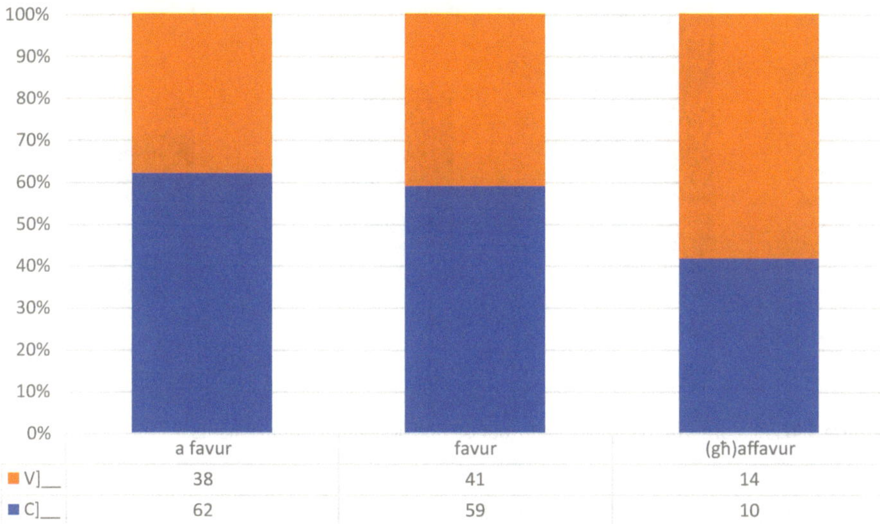

Figure 3.41: Distribution of alloforms of (П13) *favur* 'in favour of' over phonological contexts.

With 62% and 59%, *a favur* and *favur*, respectively display a moderate preference for the context C]__. The leanings of *(għ)affavur* towards combinations with vowel-final words amount to 59.4%. However, in none of the cases, the relation between majority and minority are especially skewed. It is therefore recommendable not to postulate any phonologically motivated distribution of the above alloforms.

The situation is much the same if we take the (in)definiteness of the complement into account. As Figure 3.42 suggests the alloforms display similar predilections for definite complements. The range of variation is from the maximum of 91.9% with definite complements to 86.8% for the same complement type. The gap between these two values is not huge enough to declare (in)definiteness of complements a factor which determines which alloform is given preference over the others.

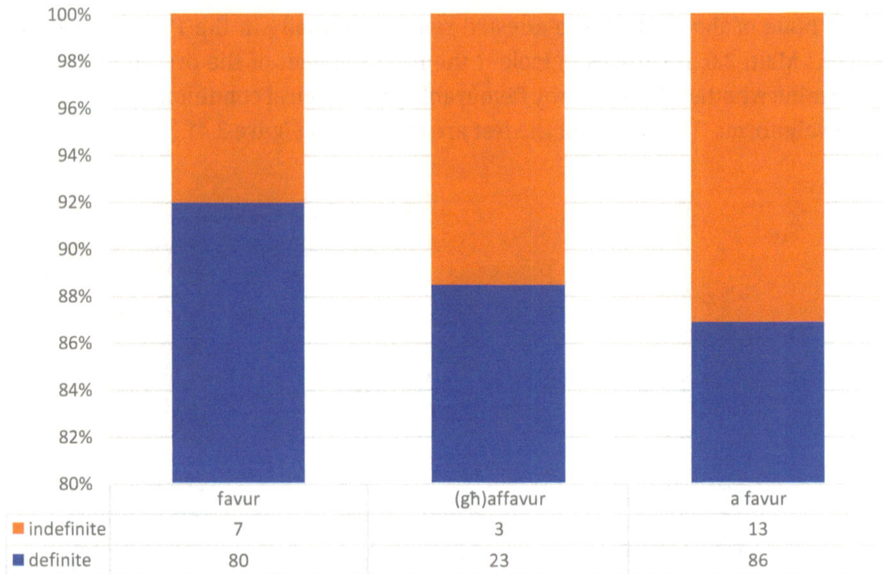

Figure 3.42: Alloforms of (П13) *favur* 'in favour of' and (in)definite complements.

The examples (3.81)–(3.84) additionally prove that the alloforms are indeed semantically equivalent. They may replace each other without affecting the meaning of the utterance. What is expressed by any of the alloforms is a BENEFACTIVE relation.

(3.81) [Korpus Malti 3.0; academic12]
Id- direttur il-ġdid beda b' enerġija kbira jaħdem
DEF-director DEF-new begin with energy:F big:F 3SG.M.IPFV:work
[favur l-emigranti u l-familji tagħhom]$_{PP}$
[**in_favour_of** DEF-emigrant:PL and DEF-family:PL of:3PL]
'The new director began very energetically to work [**for the benefit of** the emigrants and their families].'

(3.82) [Korpus Malti 3.0; news172198]
jintużaw il- fondi pubbliċi [a favur persuni bi
3.IPFV:RFL:use:PL DEF-fund:PL public:PL [**in_favour_of** person:PL with
bżonn ta' appoġġ u ħarsien partikolari]$_{PP}$
need of support and protection particular:PL]
'[. . .] the public funds are used [**to the benfit of** persons in need of special support and protection].'

(3.83) [Korpus Malti 3.0; news153235]

bdew	*jippjanaw*	*strateġiji*	*kif*	*ixaqilbu*
begin:3PL.PFV	3.IPFV:plan:PL	strategy:PL	how	3.IPFV:turn:PL

l-miżien	*elettorali*	***[affavur***	*sidhom]*$_{PP}$
DEF-balance	electoral:PL	**[in_favour_of**	master:3PL]

'[. . .] they began to plan strategies how to tip the electoral balance [**in favour of** their master].'

(3.84) [Korpus Malti 3.0; law1337]

jiġu	*kkonfiskati*	***[għaf-favur***	*tal-Gvern]*$_{PP}$
3.IPFV:come:PL	confiscate:PART:PL	**[in_favour_of**	of:DEF-government]

'[. . .] they are confiscated [**to the benefit** of the government].'

3.2.7 Free variation

The six Πs whose allomorphic patterns involving long and short forms have been reviewed above yield similar results. One of the alloforms predominates quantitatively so much so that the competitor(s) is/are severely marginalised. The alloform with the highest token frequency is most of the time the long(est) form. This fact is interesting in itself because it contradicts Zipf's Law which postulates an inverse relationship of segmental length and frequency.[66] Short forms and high frequency should thus go together well. The alloforms of (Π33) *matul* 'during', (Π34) *minbarra* 'except', (Π35) *minflok* 'instead of', and (Π39) *minħabba* 'on account of' do not confirm this correlation because they attest to long forms which are particularly frequent as opposed to short forms which are much less frequently attested. The distribution of the alloforms of (Π13) *favur* 'in favour of' is however in line with Zipf's Law. In the case of (Π2) *a skapitu* 'at the expense of', the privileged alloform is neither the shortest (which turns out to be only marginally attested) nor the longest. What one can say to accommodate this situation to the expectations according to Zipf's Law is that of the three alloforms attested with more than marginal frequency, *a skapitu* is definitely the shortest alloform.

Furthermore, in none of the cases of allomorphic competition is there any tangible proof of structural reasons for the choice of alloform. There is no denying that in some of the cases certain preferences are discernible. However, these preferences are never absolute, meaning, by and large, the domains of the alloforms overlap to a considerable extent so that each can be used in every context albeit

66 As discussed in Section 2.3.

with different frequency. Phonological factors seem to be relatively strong for the distribution of the alloforms of (Π39) *minħabba* 'on account of' whereas person inflection has different effects in the cases of (Π33) *matul* 'during', (Π35) *minflok* 'instead of', and (Π13) *favur* 'in favour of'. Especially in the case of (Π34) *minbarra* 'except', the text genre also seems to influence the choice of alloforms.

These interesting different tendencies on the micro-level notwithstanding, we hypothesise that the alternation between long and short alloforms of the above Πs can best be captured by way of referring to style. Grammatically, they are instances of free variation. This conclusion rests however on shaky ground because the parameters we have controlled and the quantitative method we have applied are certainly far too simplistic to call the case closed. We are convinced that a more finegrained and methodologically more sophisticated in-depth study of this type of allomorphy will reveal linguistically meaningful patterns which have escaped our notice.

3.3 From segmental reduction to external sandhi

In contrast to the opposition of short and long forms discussed above, the phenomena highlighted in this section do not reduce originally dimorphemic Πs to monomorphemic short forms. In point of fact, by no means all of the processes are connected to the loss of segmental material on the part of the Πs. In the interaction of certain Πs with the definite proclitic it is often the latter which suffers reduction. The target of the reductive process is not a morph but a phonological segment. Moreover, the Πs which are featured in this section are monosyllabic (= σ) and (synchronically) monomorphic (= μ). As will come to the fore in what follows, the loss of single segments more often than not is caused by the phonological context created by the words with which the Π interacts syntagmatically. We are confronted with external sandhi (Andersen 1986). No Italo-Romance Π is affected by external sandhi (= EXS). In what follows we use external sandhi as a terminological cover for the slightly heterogeneous class of phenomena addressed below. To sum up, the three following unilateral implicational patterns hold for all cases discussed in this section (cf. Figures 3.43–3.45).

EXS $\supset$ σ

Figure 3.43: External sandhi implies monosyllabicity.

EXS $\supset$ μ

Figure 3.44: External sandhi implies monomorphism.

EXS ⊃ Semitic

Figure 3.45: External sandhi implies Semitic origin.

In Section 3.3.1, (Π29) *lil* 'to' represents the only case of a Maltese Π which is subject to phonological reduction at the left margin. The vast majority of the Πs which are sensitive to external sandhi are affected by reduction at the right margin of their segmental chains. The first of these cases is *n*-deletion presented in Section 3.3.2. The phenomena investigated in Sections 3.3.1–3.3.2 are largely optional and thus only infrequently attested in the Korpus Malti 3.0. We assume that the segmental reduction of (Π29) *lil* 'to' and *n*-deletion occur more frequently in everyday oral conversation. The spoken register is underrepresented in our database.[67] The conclusions we draw must therefore be taken with a grain of salt. This is different for the topics of Section 3.3.3–3.3.4 because in these cases we address phenomena which for the most are mandatory. Section 3.3.3 is dedicated to the protocliticisation of Πs in syntagms from which the definite proclitic is absent. The last scenario, i.e. the fusion of Πs and the definite proclitic, is explored in Section 3.3.4.

3.3.1 Segmental reduction of (Π29) *lil* 'to'

Diachronically, (Π29) *lil* 'to' can be analysed as resulting from the univerbation of two morphemes (Saari 2003: 33).[68] Sutcliffe (1936: 171, original boldface) argues that this Π

> is itself composed of the fragmentary preposition **l** and the article **il**, and meant originally *to the*. The force of the article became obscured [. . .] and the word became what it still remains, a preposition meaning *to*.

While E. Serracino-Inglott (1978: 296) adheres to this hypothesis too, Cachia (1994: 131) takes issue with this interpretation. According to this author (Π29) *lil* 'to' goes back to the combination of two Arabic Πs *li* + *ila*. This possibility is also mentioned by Procházka (1993: 148). We are uncommitted as to the correct diachronic analysis. What counts for us in the present context is the fact that the erstwhile morphological compositionality of (Π29) *lil* 'to' is no longer transparent to the average native speaker of Maltese for whom the Π can only be monomorphemic. This synchronic opacity distinguishes (Π29) *lil* 'to' from the morphologically transparent cases discussed in the previous section.

67 As we know from Section 1.7.

68 For the (im)possible connection of (Π29) *lil* 'to' to the putative postposition *ilu* 'ago', we refer the reader back to Section 1.6.2.

The three alloforms of (Π29) *lil* 'to' have been presented already above where their competition is illustrated with examples (1.138)–(1.140).[69] What especially distinguishes the short forms from the long form is the disability of the former to inflect. If (Π29) *lil* 'to' is inflected for person, only the long alloform is admissible as host of the person suffixes. The paradigm of this Π in Table 3.4 is only possible if the long alloform serves as Π-stem. Thus, a pronominal complement in the 3rd person singular masculine is always represented as *lilu* and never as *'ilu* or *'lu*. Accordingly, it is entirely sufficient for the purpose of this section to look into the behaviour of the three alloforms outside the domain of pronominal complements. In Figure 3.46 we provide the overall token frequencies of the competing uninflected candidates computed in accordance to the practice applied to the cases discussed in Section 3.2.

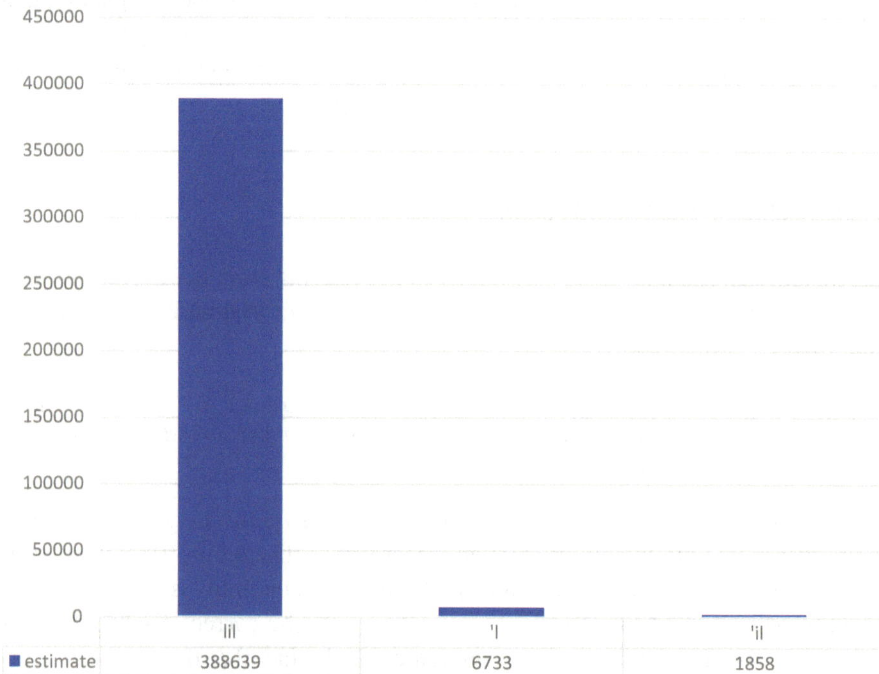

Figure 3.46: Token frequency of competing allomorphs of (Π29) *lil* 'to'.

The estimates speak a very clear language. The long alloform is responsible for the bulk of attestations of (Π29) *lil* 'to' in the Korpus Malti 3.0. The long form is 58 times as frequent as *'l*. In the case of *'il*, the discrepancy is even much bigger since *lil*

69 Cf. Section 1.6.5 (ii).

occurs 209 times more often than *'il*. The two short alloforms are clearly marginal in comparison to the dominant long alloform. As mentioned repeatedly already in connection with the competition of short and long forms in the previous section, the overwhelming quantitative predominance of *lil* and the general infrequency of *'l* and *'il* cast doubt on the sweeping applicability of Zipf's law. The validity of this oft-invoked law is disproved for the contrast between long and short forms. However, it seems to work much better if we only take account of the two short forms. The shortest alloform *'l* is 3.6 times more frequent than the bisegmental *'il*. As will become obvious in the course of the further discussion in this section, the two short forms relate to each other differently from their relation to the common long form *lil*. At this point a clarification is called for. The candidates *'il* and *'l* boast many more matches in the Korpus Malti 3.0. These additional matches, however, do not involve these alloforms as genuine Πs but result from their frequent use as prefixes of spatial adverbs. This use has been excluded from the grammar of Maltese Πs and thus can be discounted in the ensuing discussion.[70]

Figure 3.47 gives us a clearer idea about the compatibility of the alloforms with final vowels and consonants of words positioned to their immediate left.

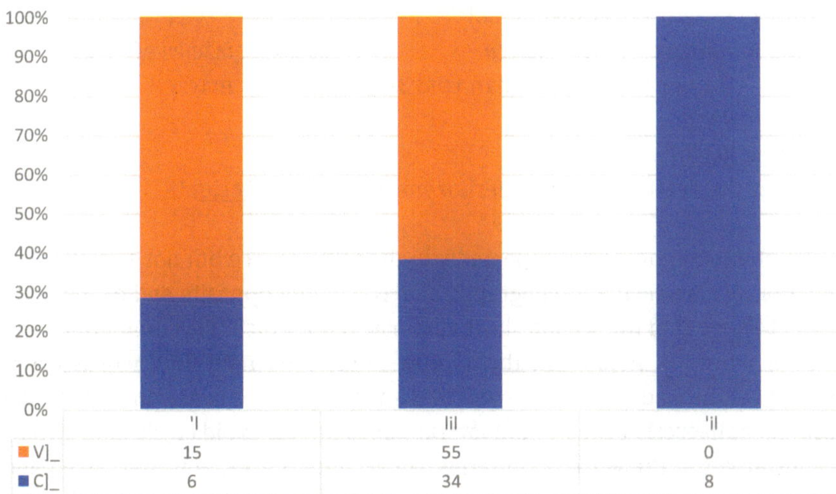

Figure 3.47: Distribution of alloforms of (Π29) *lil* 'to' over phonological contexts.

The three alloforms are characterised by different preferences. The shortest alloform *'l* clearly has a leaning towards combinations with preceding vowel-final

70 Cf. Section 1.6.5 (ii).

words which account for 71% of the token of *'l* in the 1st 100. From the point of view of sentence phonetics, the tendency of *'l* to follow a vowel is only logical in the sense that *'il* would have created a hiatus in this context. Only 29% of the *'l*-tokens combine with consonant-final words. For these cases, we would have expected to find that the complement of the Π starts with a vowel as in (3.85).

(3.85) [Korpus Malti 3.0; academic14]
warrab minn quddiemna għax i<m>morru nugħawk
move from before:1PL because 1.IPFV:go:PL 1.IPFV:accuse:PL:2SG
['l ommok]$_{PP}$ u 'l missierek
[to mother:2SG] and to father:2SG
'[. . .] leave us because we are going to accuse you [**before** your mother] and your father.'

Note that the second *'l* in (3.85) is triggered by the preceding monovocalic conjunction *u* 'and'. Surprisingly the shortest – monoconsonantal – alloform shows up also in a position where it is sandwiched between consonants as in (3.86)

(3.86) [Korpus Malti 3.0; academic138]
Wara l-mewt ta' martu Berlioz iżżewweġ ['l
after DEF-death of woman:POSS:3SG.M Berlioz marry [to
Marie Recio]$_{PP}$
Marie Recio]
'After the death of his wife, Berlioz married [Marie Recio].'

The same phenomenon occurs repeatedly in the same source but not elsewhere. It is likely that we are facing a typographical error. At the opposite end of the scale in Figure 3.47, we find the second short alloform *'il* for which no combination with vowel-final words is attested in the 1st 100. Thus, consonant-final words cover 100% of all instances of *'il*. This unequivocal situation makes sense phonologically because the aphaeresis of the initial lateral approximant avoids a clash of two or more consonants at the word boundaries. On account of this, *'il* would have been the better choice for the representation of (Π29) *lil* 'to' in (3.86). Between the two extremes in Figure 3.47, there is *lil* which also attests to a majority for V] __ (= 62%) and a minority for C] __ (= 38%). The preponderance of the vocalic contexts might result from our selfmade quantitative method of checking only the 1st 100 for the estimates. Interestingly, the long allomorph is the only possible choice for a *lil*-PP in sentence-initial position. There are four cases of this kind amongst the 1st 100 of *lil*. Neither *'il* nor *'l* is attested at the left margin of sentences in the 1st 100. In contrast to most of the pairs of short and long forms discussed in the foregoing section, the alloforms of (Π29) *lil* 'to' display a distribution which seems to be sensitive to the phonological context. It is important to note that consonant-final words do not automatically require the alloform *'il* whereas the use of *'il* presupposes the co-presence of a consonant-final word to its left.

(In)definiteness of the complement too plays a more important role in the distribution of the alloforms of (Π29) *lil* 'to' than in the previously discussed cases. Figure 3.48 shows that indefinite complements are of minor importance for the choice of the short alloforms. With only 12% and 19%, *'il* and *'l* display relatively small shares for indefinite complements in comparison to the 43%-share indefinite complements claim in the case of the long alloform. It should not go unmentioned that ten of the seventeen definite complements of *'l* and three of the seven definite complements of *'il* are proper names as in (3.87)–(3.88).

(3.87) [Korpus Malti 3.0; academic138][71]

jidher	*ċar*	*li*	*kien*	*influwenza*	*lil*	*Liszt*	*u*	***['l***
3SG.M.IPFV:appear	clear	that	be	influence	to	Liszt	and	**[to**

Wagner]$_{PP}$
Wagner]
'[. . .] it seems clear that he had influenced Liszt and [Wagner].'

(3.88) [Korpus Malti 3.0; academic135]

tħobb	*il-letteratura*	*klassika*	*l-iżjed*	***['il***	*Shakespeare]*$_{PP}$
3SG.F.IPFV:love	DEF-literature	classic:F	DEF-more	**[to**	Shakespeare]

'[. . .] she loves the classical literature, especially [Shakespeare].'

As the PP *lil Liszt* 'to Liszt' in example (3.87) proves, the long alloform also combines with proper names. However, in contrast to the two short alloforms proper names are not as prominent a class of complements for the long alloform.

Our findings corroborate Borg/Alexander-Azzopardi's (1997: 137) claim according to which the short alloforms of (Π29) *lil* 'to' are optional in the sense that they can always be replaced with the long form (Akkademja tal-Malti 1998: 54). On the other hand, certain conditions transpire from the search of the Korpus Malti 3.0 which seem to be particularly favourable for the selection of the one or the other of the short alloforms. Both *'il* and *'l* give preference to definite complements (especially proper names). Furthermore, *'il* exclusively occurs after consonant-final words whereas *'l* displays a clear preference for vowel-final words. Both short alloforms are banned from the sentence-initial position. They are also morphologically

71 In this example, only the *lil*-PP with the short allomorph appears in brackets.

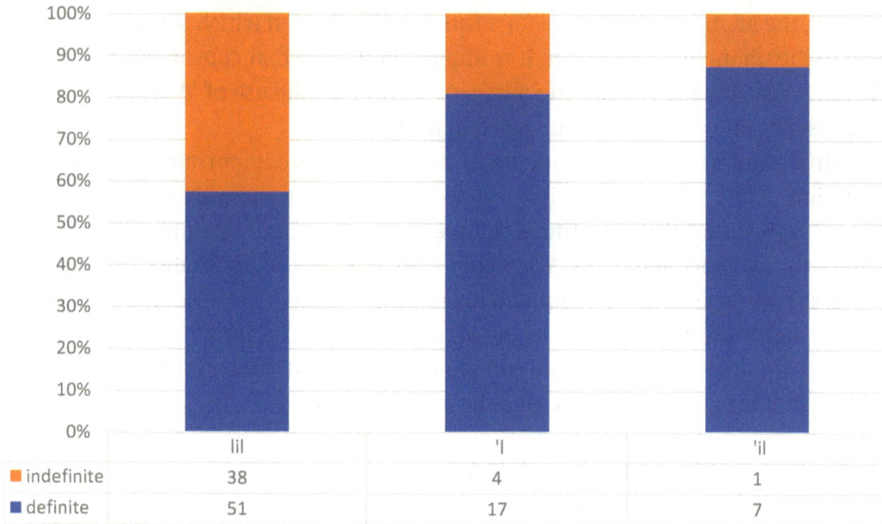

Figure 3.48: Alloforms of (П29) *lil* 'to' and (in)definiteness.

invariant. Their optional status thus does not preclude the existence of more or less pronounced preferences. Since *'il* cannot replace *'l* in vocalic contexts, the allomorphic patterns of (П29) *lil* 'to' cannot simply be regarded as straightforward instances of free variation. Short and long forms are in free variation with the latter constituting the default case. The two short forms cannot freely be exchanged the one for the other. In Figure 3.49, we recapitulate the preferences of the short forms discounting minor exceptions (Cassar 2000: 25).

$$(\Pi 29)\ \textit{lil}\ \text{`to'}_{\text{SHORT FORM}} \rightarrow \begin{cases} \textit{'il} & / \text{C}\ \_\_\ \text{NP}_{\text{DEF}} \\ \textit{'l} & / \text{V}\ \_\_\ \text{NP}_{\text{DEF}} \end{cases}$$

Figure 3.49: Preferred contexts of short forms of (П29) *lil* 'to'.

While *'il* and *'l* are in free variation with *lil*, the distribution of the two short forms over contexts is such that one might want to speak of phonologically conditioned allomorphy in the making.

3.3.2 *n*-deletion

Beginning with this section, we turn our attention exclusively to the right margin of the segmental chains of the Πs. This means that the final segments of words which precede the Πs syntagmatically are no longer of any relevance for the identification of allomorphic patterns. This criterion is therefore replaced with the properties of the initial segment(s) of the word which is placed to the right of the Π. The phenomenon addressed in this section is far too sparingly attested to show that this change of perspective may yield interesting results.

We have learned above about *n*-deletion in connection to the formation of Π-stems in general and the *j*-stems in particular.[72] Less than a handful of Πs display the necessary phonological shape for *n*-deletion to apply. Those two Πs which create their Π-stem by way of subtracting the final nasal of their citation form account for half of all potentially *n*-deleting Πs. The number of Πs which delete /n/ under inflection is small. This number increases only minimally when we count in cases of *n*-deletion with uninflected Πs. Table 3.21 tells us that the phenomenon is generally exceptional in the Korpus Malti 3.0. Only those cases of *n*-deletion are counted which involve the items under inspection in Π-function.

Table 3.21: Instances of *n*-deletion in the Korpus Malti 3.0.

	ħdej	*bej*	*lej*	sum
_[V	1	1	3	5
_[C	0	1	1	2
definite	1	2	3	6
indefinite	0	0	1	1
total	1	2	4	7

There are seven Πs whose segmental chain ends in a denti-alveolar nasal, namely (Π4) *bejn* 'between', (Π14) *fejn* 'near', (Π23) *ħdejn* 'beside', (Π28) *lejn* 'towards', (Π41) *minn* 'from', (Π43) *mnejn* 'from near', and (Π58) *viċin* 'near'. As transpires from Table 3.21, only three of those are subject to *n*-deletion. Besides the two representatives of C IIIa (Π23) *ħdejn* 'beside' and (Π28) *lejn* 'towards', there is also (Π4) *bejn* 'between' – a member of C Ia – which never undergoes *n*-deletion under inflection. Sentence (3.89) shows that the final /n/ is occasionally dropped also in the case of this Π when used with a lexical complement NP.

72 Cf. Sections 3.1.2.2 and 3.1.2.3.3.1.

(3.89) [Korpus Malti 3.0; european333]
jista' *jeffetwa* *l-kummerċ* ***[bej*** *il-Komunità*
3SG.M.IPFV:can 3SG.M.IPFV:affect DEF-trade **[between** DEF-E.U.
u *ir-Russja]*$_{PP}$
and DEF-Russia]
'[. . .] it can affect the trade [**between** the E.U. and Russia].'

More than half of the few *n*-deleted tokens go to the credit of (Π28) *lejn* 'towards' whose use is illustrated in (3.90) followed suit by example (3.91) for *n*-deletion with (Π23) ***ħdejn*** 'beside'.

(3.90) [Korpus Malti 3.0; news169997]
It-titjira *imbagħad* *tkompli* *fi* *triqitha* ***[lej***
DEF-flight then 3SG.F.IPFV:continue in street:POSS:3SG.F **[towards**
Dubai]$_{PP}$
Dubai]
'Then the flight continues on its course [**towards** Dubai].'

(3.91) [Korpus Malti 3.0; news119927]
jmur *jiltaqa'* *miegħu* ***[ħdej*** *l-għalqa*
3SG.M.IPFV:go 3SG.M.IPFV:meet with:3SG.M **[beside** DEF-field
fil-Qala]$_{PP}$
in:DEF-Qala]
'[. . .] he goes to meet him [**near** the field in Qala].'

Five out of seven times, the final /n/ is dropped in the presence of a vowel-initial word on the right. Twice *n*-deletion is attested in the context __ [C. The matches are too few to allow for far-reaching generalisations. The phonological logic behind the dropping of /n/ in front of a vowel remains unclear to us. Similarly, the preponderance of *n*-deletion in combination with a definite complement is hard to explain and may be nothing but a negative frequency effect.

More generally, in all of the above contexts the *n*-deleted forms are outranked by the full citation forms of the Πs. The token frequency of the *n*-deleted allomorphs is absolutely negligible in comparison to the long forms. However, the small turnouts of reductive processes reported below should not be misunderstood as proof of the general scarcity of external sandhi affecting Maltese Πs.[73]

---

73 Cf. Sections 3.3.3.1–3.3.3.2.

3.3.3 Procliticisation

Under this heading we subsume all cases of segmental reduction of Πs in those contexts where the syntagmatic neighbour on the right side of the Π is not the definite proclitic.[74] Procliticisation of the kind discussed in this section affects only monosyllabic Πs with CV-structure. However, not all Maltese Πs which meet this criterion are subject to segmental reduction. Five of the six candidates procliticise, namely the inflected Πs (Π7) *bi* 'with' and (Π15) *fi* 'in' representing C II, the *għajn*-stems (Π30) *ma'* 'with' and (Π55) *ta'* 'of' from C IIIs, and the invariant Π (Π51) *sa* 'till'. In contrast, the invariant (Π19) *ġo* 'in' does not participate in procliticisation. Figure 3.50 supports the impression that segmental reduction is generally infrequent in terms of type frequency.

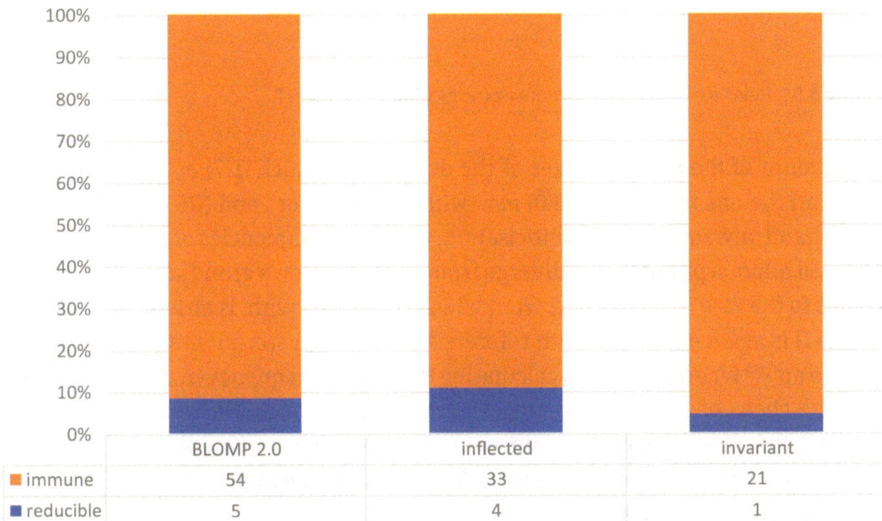

Figure 3.50: Type frequency of segmentally reducible Πs.

Only in the class of the inflected Πs does the number of types which are reducible exceed the 10%-mark albeit only minimally (n = 11%). As we will see shortly, the low type frequency is counterbalanced by relatively high token frequencies of individual short forms.

Under procliticisation the Πs drop their vowel to form a phonological word with the host to their right. Consequently, the Πs loose their syllabic status. The

74 The interaction of Π and definite proclitic is the topic of Section 3.3.4.

susceptibility of the Πs to procliticisation is not the same for the above group of five. They all undergo procliticisation if they precede a vowel-initial word. However, only the two Πs whose citation form ends in the high front vowel /i/ also procliticise if the word to their right starts with a single consonant. We postulate a hierarchy of vowels /i/ > /a/ > /o/ reflecting the decreasing number of contexts in which they are deleted. Figure 3.51 captures the conditions which trigger vowel deletion under procliticisation. To what extent these rules are optional will be discussed in due course below.

$$\mathrm{V} \quad \rightarrow \quad \emptyset \left\{ \begin{array}{l} /\ [\mathrm{i}] \ - \left\{ \begin{array}{l} \mathrm{CV} \\ \\ \mathrm{V} \end{array} \right. \\ \\ /\ [\mathrm{a}] \quad \mathrm{V} \end{array} \right.$$

Figure 3.51: Rules for vowel deletion under procliticisation – version I.

On account of the different sizes of the domains in which (Π7) ***bi*** 'with' and (Π15) *fi* 'in', on the one hand, and (Π30) *ma'* 'with', (Π55) *ta'* 'of', and (Π51) *sa* 'till', on the other hand, are subject to procliticisation, the token frequencies of the segmentally reduced allomorphs of the Πs diverge from each other on a grand scale. This clearly comes to the fore in Figure 3.52 from which (Π30) *ma'* 'with' is absent. The absence of this Π is again explicable with reference to the simplistic quantitative methodology we apply when it comes to calculating the overall frequency in the Korpus Malti 3.0. The check of the 1st 100 matches of *m'* yields a turnout of zero instances for the reduced allomorph of the Π since they exclusively represent the homophonous shortened form *m'* of the negation *ma*. This does not mean however that (Π30) *ma'* 'with' never procliticises.

In the cases of (Π7) ***bi*** 'with' and (Π15) *fi* 'in', the token frequencies given in Figure 3.52 considerably exceed those estimated for the corresponding long forms as of Table 1.11. The reduced allomorph *f'* is five times as frequent as the long allomorph *fi*. Even more strikingly, the tokens of *b'* outnumber those of *bi* by a ratio of 6.7-to-1. Accordingly, the members of inflection class C II climb up several steps in the quantitative ranking order of the Maltese Πs. In contrast, *m'*, *s'*, and *t'* are attested by far more infrequently than the respective long forms *ma*, *sa*, and *ta*. The additional tokens of the procliticised allomorphs do not alter the rank position of these Πs.

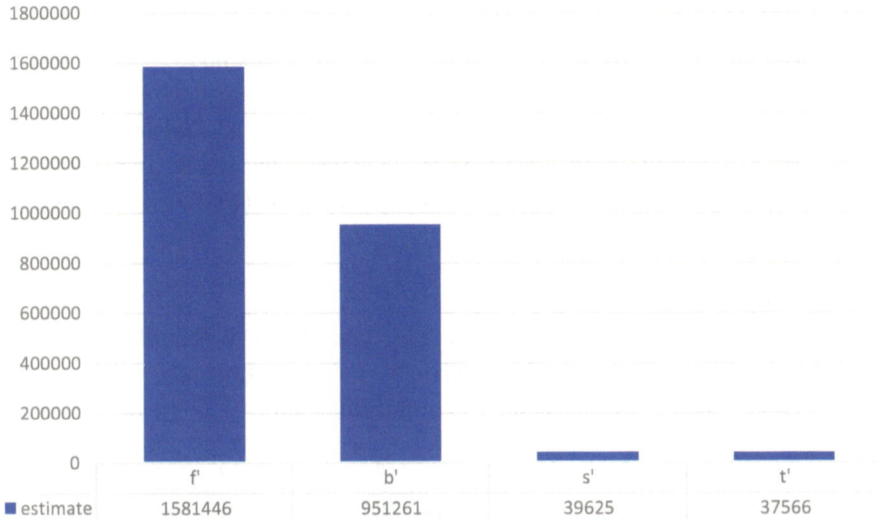

Figure 3.52: Token frequency of procliticised ПIs in the Korpus Malti 3.0.

We illustrate the process of procliticisation for each of the ПIs with examples mostly taken from the literary sources and only occasionally from the Korpus Malti 3.0. We begin with prevocalic procliticisation. Sentences (3.92)–(3.96) feature the reduced allomorphs of the five ПIs in combination with different vowel-initial words.

(3.92) [Cauchi 1997: 38]
x-xwejjaħ Gerfex tfaċċa fil-Każin ***[b'*** *umbrella*
DEF-old:DIM Gerfex emerge in:DEF-band_club **[with** umbrella
sewda tassew pulita]$_{PP}$
black:F really neat:F]
'[. . .] the old man Gerfex appeared suddenly in the band club [**with** a really neat black umbrella].'

(3.93) [Cauchi 1997: 20]
seta' jieħu l-għasfur ***[f'*** *idu]$_{PP}$*
can 3SG.M.IPFV:take DEF-bird **[in** hand:3SG.M]
'[. . .] he could take the bird [**in** his hand].'

(3.94) [Cauchi 1997: 6]

Ommu *kienet* *mietet* *u* *għalhekk* *ġie*
mother:3SG.M be:3SG.F.PFV die:3SG.F.PFV and thus come
joqgħod ***[m'*** *oħtu* *armla]$_{PP}$* *f'* *Ħal Busbies.*
3SG.M.IPFV:stay **[with** sister:3SG.M widow] in Ħal Busbies
'His mother had died and therefore he came to live [**with** his widowed sister] in Ħal Busbies.'

(3.95) [Korpus Malti 3.0; academic69]

kienet *mifruxa* *sew* *matul* *in-Nofs in-nhar*
be:3SG.F.PFV PART:spread:F precisely during DEF-South
tal-katina *tal-muntanji* *tal-Ħimalaja* *f'* *arja* *vasta*
of:DEF-chain of:DEF-mountain:PL of:DEF-Himalaya in area vast:F
minn *Uttar Pradesh* *fin-Nepal* *Sikkim* *Bengal* *u* *Butan*
from Uttar Pradesh in:DEF-Nepal Sikkim Bengal and Bhutan
[s' *Assam]$_{PP}$*
[**till** Assam]
'[. . .] it was quite common along the southern part of the Himalaya Mountain range in a vast area from Uttar Pradesh, Nepal, Sikkim, Bengal, and Bhutan [**as far as** Assam].'

(3.96) [Cauchi 1997: 47]

Niġi *nagħti* *daqqa* ***[t'*** *id]$_{PP}$* *jien*
1SG.IPFV:come 1SG.IPFV:give stroke **[of** hand] 1SG
'I will come to lend you [a hand] [. . .].'

The vowel deletion is mandatory with those ΠS which end in /i/. For those which have a final low vowel, the picture is not as clear. There are examples like (3.97)–(3.99) where the final vowel resists being deleted.

(3.97) [Bartolo/Vella 2009: 116]

kienet *għadha* *irrabjata* ***[ma'*** *oħtha]$_{PP}$*
be:3SG.F.PFV still:3SG.F anger:PART:F **[with** sister:3SG.F]
'[. . .] she still was angry [**with** her sister].'

(3.98) [Korpus Malti 3.0; academic2]

*Meta jasal il-monsoon minn Mejju **[sa** Awwissu]$_{PP}$*
when 3SG.M.IPFV:arrive def-monsoon from May **[till** August]
u minn Novembru sa Jannar id-dar kollha
and from November till January DEF-house all:3SG.F
tiddedika ħinha għall-kannela
3SG.F.IPFV:dedicate time:3SG.F for:DEF-cinnamon
'When the monsoon arrives – from May [**until** August] and from November until January – the entire house dedicates its time to the cinnamon.'

(3.99) [Korpus Malti 3.0; culture1084]

*m' għandux l-użu **[ta'** idu x-xellugija]$_{PP}$*
NEG have:3SG.M:NEG DEF-use **[of** hand:3SG.M DEF-left:F]
'[. . .] he could not make use [**of** his left hand].'

The realisation of the rules set by Akkademja tal-Malti (1998: 53–54) as to the procliticisation of Πs in front of vowel-initial words is by no means homogeneous across the texts of our literary corpus. In [Bartolo/Vella 2009] for instance, procliticisation of *a*-final Πs is mostly but not entirely avoided whereas *i*-final Πs regularly procliticise to words which fulfil the criteria laid down in Figure 3.51. A rare example of the employment of *t'* instead of *ta'* in this primary source is given in (3.100). There are two other cases, namely [Bartolo/Vella 2009: 379] ***t'**ommi* '**of** my mother' and [Bartolo/Vella 2009: 467] ***t'**ommu* '**of** his mother' with, on the same page, [Bartolo/Vella 2009: 379] ***ta'** ommu* '**of** his mother' showing that procliticisation is not obligatory.

(3.100) [Bartolo/Vella 2009: 248]

ma' għonqu kellu mdendel il-basket tad-drapp
with neck:3SG.M have.PFV:IO:3SG.M PART:suspend DEF-basket of:DEF-cloth
***[t'** oħtu]$_{PP}$*
[of sister:3SG.M]
'[. . .] around his neck he had suspended the cloth basket [**of** his sister].'

In Table 3.22, we register all instances of person names (= first names) starting with a vowel as complement of the four of the five Πs under inspection. No example for the use of (Π51) *sa* 'till' with a following person name could be identified in the novel. Table 3.22 exclusively features PPS without further sentential context. Instances of blocking are presented in the upper part of the table. At the bottom, the four instances of procliticisation can be found. The cases are ordered according to the decreasing number of tokens.

Table 3.22: Blocked and permitted procliticisation with proper names in [Bartolo/Vella 2009].

page	example	translation
13, 33, 46, 49, 50, 53, 62, 80, 225, 232, 253, 259, 264, 339, 431, 478	*ta' Estella*	'Estella's'
97, 99 (3x), 102, 137, 338, 339, 436	*ta' Efrem*	'Efrem's'
28, 74, 240, 259, 283, 308, 411, 441	*ma' Estella*	'with Estella'
224, 229, 231 (2x), 377	*ta' Erkole*	'Erkole's'
16, 70, 210, 445	*ta' Antida*	'Antida's'
37, 131, 212, 234	*ma' Antida*	'with Antida'
99, 389, 429	*ma' Efrem*	'with Efrem'
30, 369	*ta' Anita*	'Anita's'
39, 44	*ta' Elena*	'Elena's'
29	*ta' Annie*	'Annie's'
459	*ma' Elaine*	'with Elaine'
471	*ma' Elena*	'with Elena'
155	*b'Efrem*	'with Efrem'
52	*b'Estella*	'with Estella'
211	*f'Antida*	'in Antida'
61	*f'Estella*	'in Estella'

The evidence is unequivocal. Πs with final /a/ never procliticise to vowel-initial person names whereas Πs with final /i/ always do. If we look beyond this particular primary source, the picture becomes less straightforward. An informal and unsystematic check of the Korpus Malti 3.0 yielded an admittedly small number of matches for the procliticisation of Πs which never procliticise with names in [Bartolo/Vella 2009]. However, we are convinced that a more systematic search of the electronic resource will yield a much more sizable harvest than the few cases listed in Table 3.23.

Table 3.23: Procliticised *m'* and *t'* with first names in the Korpus Malti 3.0.

source	example	translation
parl8675	*m'Angela Abela*	'with Angela Abela'
literature19	*m'Albert*	'with Albert'
literature11	*m'Ernest*	'with Ernest'
news218043	*t'Angela Bodin*	'of Angela Bodin'
literature19	*t'Albert*	'of Albert'
literature11	*t'Ernest*	'of Ernest'

The scarcity of the matches notwithstanding, it is evident that there are no insurmountable obstacles to the procliticisation of the two members of C IIIc. In contrast, no evidence of the short allomorph of (Π51) *sa* 'till' being procliticised to person names could be identified in our informal search in the Korpus Malti 3.0. If we abstract away from human names by way of extending the search to the name of God *Alla*, the results confirm what we have seen above. Figure 3.53 reflects the usual partition of the five Πs: (Π51) *sa* 'till' never combines with *Alla* whereas (a) the *i*-final Πs always procliticise and (b) the *a*-final Πs are divided between procliticisation and its blockage. What is especially interesting is the different preferences of (Π30) *ma'* 'with' and (Π55) *ta'* 'of'. Where the latter gives preference to the short form, (Π30) *ma'* 'with' opts more frequently for the long form. These preferences do not translate into patterns of strong dominance because with shares of 36% for *m'* and 43% for *ta'*, the two minority solutions are by no means marginal phenomena.

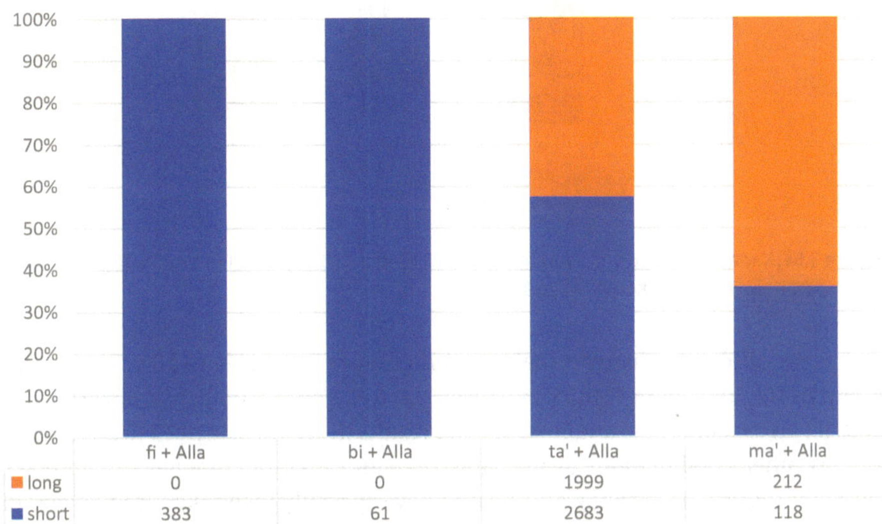

Figure 3.53: Procliticisation vs blockage with *Alla* 'God' in the Korpus Malti 3.0.

The above vacillation of the Πs from C IIIc is not restricted to the domain of combinations with proper names. We do not deny that it might turn out to be the case that blockage of procliticisation is especially strong in the realm of names. However, the competition between blockage and procliticisation pervades the entire inventory of nouns of Maltese.

Since the above primary source gives evidence of variation with kinship terms as to the possibility of *a*-final Πs dropping their vowel before a vowel-initial host, it makes sense to look at the patterns which emerge from the comparison of the combination of the five Πs with selected vowel-initial complements. Figure 3.54 takes stock of the cooccurrences of the allomorphs *bi* and *b'* with *umbrella* 'umbrella', *fi* and *f'* with *idu* 'his hand', *ma'* and *m'* with *oħtu* 'his sister', *ta'* and *t'* with *id* 'hand', and *sa* and *s'* with *issa* 'now' in the Korpus Malti 3.0.

	(Π29) ma' 'with'	(Π54) ta' 'of'	(Π50) sa 'till'	(Π6) bi 'with'	(Π14) fi 'in'
■ deleted V	23	515	39094	9	1297
■ long form	54	190	3939	0	0

Figure 3.54: Variation of long and short forms of (Π7) *bi* 'with', (Π15) *fi* 'in', (Π30) *ma'* 'with', (Π55) *ta'* 'of', and (Π51) *sa* 'till' in the Korpus Malti 3.0.

The share of long forms with the *a*-final Πs ranges from 10% to 70%, i.e. for this group of Πs the rule in Figure 3.51 turns out to be only a tendency of certain Πs and not an exceptionless automatism. The results in Figure 3.54 need to be checked against other combinations of the same Πs with different complements to make sure that there is no skewing effect caused by especially high token frequencies of collocations such as *s'issa* 'until now'. For the kinship term *omm* 'mother' with a 3rd person masculine possessor (*ommu* 'his mother') the results are as follows: (Π7) *bi* 'with' and (Π15) *fi* 'in' are exclusively represented by their short allomorphs *b'* (14 times) and *f'* (7 times) whereas (Π30) *ma'* 'with' and (Π55) *ta'* 'of' give evidence of a majority of long forms, namely 37 times *ma'* vs a single match for *m'* and 59 times *ta'* vs three times *t'*. (Π51) *sa* 'till' is not attested in combinations with this kinship term. Going by these ratios, we assume that procliticisation is not a must for (Π30) *ma'* 'with', (Π55) *ta'* 'of', and (Π51) *sa* 'till' while it is definitely compulsory in the case of (Π7) *bi* 'with' and (Π15) *fi* 'in'.

Examples of preconsonantal procliticisation are given in (3.101)–(3.102) for (Π7) *bi* 'with' and (Π15) *fi* 'in'.

(3.101) [Cauchi 1997: 59]
Ma' dan il-kliem ix-xwejjaħ fetaħ il-miżwet
with DEM.M.PROX DEF-word.COLL DEF-old:DIM open DEF-shell
*[**b'** difrejh]*$_{PP}$
[**with** fingernail:PL:3SG.M]
'This said the old man opened the bean shell [**with** his fingernails] [. . .].'

(3.102) [Cauchi 1997. 26]
Lil martu Lugrezja ma qallhiex x'
to woman:POSS:3SG.M Lugrezja NEG say:IO:3SG.F:NEG what
*kellu [**f'** rasu]*$_{PP}$
have.PFV:IO:3SG.M [**in** head:3SG.M]
'He did not tell his wife Lugrezja what was [**on** his mind] [. . .].'

It makes no difference what vowel or what consonant occupies the initial slot of the host, procliticisation cannot be escaped. Similarly, the word-class to which the host belongs is irrelevant for the rule exposed in Figure 3.51. In the case of the representatives of C II this means that they procliticise also to hosts whose initial segment is identical with the consonant of the Π as shown in (3.103)–(3.104).

(3.103) [Cauchi 1997: 18]
*Mar għand Peppi tal-Ħaxix [**b'** borża*
go at_s.o.'s_place Peppi of:DEF-vegetable [**with** bag
tal-plastik f' idu]$_{PP}$
of:DEF-platsic in hand:3SG.M]
'He went to Peppi the greengrocer [**with** a plastic bag in his hand] [. . .].'

(3.104) [Camilleri 2013: 65]
*qallu iżjed [**f'** forma ta' kmand]*$_{PP}$ *milli b'*
say:IO:3SG.M more [**in** form of command] from:that with
korteżija
courtesy
'[. . .] he said to him more [**in** the form of a command] than with politeness.'

In this way, geminates arise in the initial position of the phonological word. Galea/ Ussishkin (2018: 70) state that "[w]ord-initial geminates occur due to morphonological processes" which they assume to be disallowed phonologically. According to their analysis word-initial geminates are underlyingly ambisyllabic and thus require the presence of an epenthetic vowel /i/. The orthographic representation of the phenomenon in cases like (3.103)–(3.104) does not confirm this hypothesis although the reduced allomorphs of the Πs are surrounded by consonants on both sides: *ʃ* + *b* + *b* in (3.103) and *d* + *f* + *f* in (3.104). Galea and Ussishkin's rule holds for word-internal morphonological rules. Its applicability to geminates resulting from external sandhi remains doubtful. We concede however that this rule was never intended to also cover word-external processes.

Independent of the solution to this perhaps only imagined problem, there is a more serious difficulty to confront. We have defined the abstract *għajn* as a consonantal phoneme of Maltese.[75] This status is challenged by the rule in Figure 3.51 because as a full-blown consonant *għajn* should block procliticisation of those Πs which end in the low vowel /ɐ/. This however is not the case as transpires from examples (3.105)–(3.109) which prove that all of the five Πs under inspection drop their final vowel before *għajn*.

(3.105) [Camilleri 2013: 200]
poġġiet *maġenbu* *u* *ħalliet* *idha*
sit:3SG.F.PFV close_to:3SG.M and let:3SG.F.PFV hand:3SG.F
tibda *vjaġġ* *ta'* *sejba* *minn* *irkopptu*
3SG.F.IPFV:start journey of treasure_trove from knee:POSS:3SG.M
'l fuq ***[b'*** *għajnejha]*$_{PP}$ *jħeġġu*
upwards [**with** eye:PL:3SG.F] 3.IPFV:shine:PL
'[. . .] she sat close to him and let her hand start a discovery trip from his knees upwards [**with** her eyes] shining [. . .].'

(3.106) [Cauchi 1997: 44]
qasqas *minnha* *tliet* *biċċiet* ***[f'*** *għamla* *ta'*
cut_with_scissors from:3SG.F three piece:PL [**in** shape of
tbajja *tal-linka]*$_{PP}$
stain.PL of:DEF-ink]
'[. . .] with the scissors he cut three pieces from it [**in** the shape of ink stains].'

Examples (3.105)–(3.106) do not pose any problems since (Π7) *bi* 'with' and (Π15) *fi* 'in' procliticise no matter whether the next word starts with a vowel or a single consonant. In contrast, (Π30) *ma'* 'with', (Π55) *ta'* 'of', and (Π51) *sa* 'till' are excluded from procliticisation in front of a consonant. Nevertheless, their short forms are attested too if the host displays an initial *għajn*.

75 Cf. Section 2.2.1.

(3.107) [Korpus Malti 3.0; literature21]
[M' għonqu]$_{PP}$ *kellu* *ġiżirana* *tal-fidda*
[**with** neck:3SG.M] have.PFV:IO:3SG.M necklace of:DEF-silver
'[**Around** his neck] he wore a silver necklace [. . .].'

(3.108) [Camilleri 2013: 65]
Jekk *tgħaddihomli* *[s' għada]*$_{PP}$ *aħjar*
if 2SG.IPFV:pass:3PL:IO:1SG [**till** tomorrow] good.CMPR
'It would be better if you pass them on to me [**until** tomorrow].'

(3.109) [Camilleri 2013: 253]
jneħħi *r-riħa* *[t' għaraq]*$_{PP}$
3SG.M.IPFV:take_off DEF-smell [**of** sweat]
'[. . .] it takes off the smell [**of** sweat].'

However, there is variation in the behaviour of (Π30) *ma'* 'with', (Π55) *ta'* 'of', and (Π51) *sa* 'till' before *għajn*-initial words as becomes clear from the comparison of examples (3.107)–(3.109) above with examples (3.110)–(3.112) below.

(3.110) [Korpus Malti 3.0; literature1]
Ġina *qamet* *u* *dawret* *idejha* *[ma'*
Gina rise:3SG.F.PFV and surround:3SG.F.PFV hand:PL:3SG.F [**with**
għonqu]$_{PP}$
neck:3SG.M]
'Gina rose and put her hands [**around** his neck] [. . .].'

(3.111) [Korpus Malti 3.0; literature29]
[Sa għada]$_{PP}$ *għad* *baqa'* *erbgħa u għoxrin* *siegħa*
[**till** tomorrow] still remain twenty_four hour
'There are still twenty-for hous [**until** tomorrow] [. . .].'

(3.112) [Korpus Malti 3.0; news18317]
hemm *ukoll* *riħa* *qawwija* *[ta' għaraq]*$_{PP}$
there also smell strong:F [**of** sweat]
'[. . .] there is also a strong smell [**of** sweat].'

The Πs under inspection combine with the same complements in the two ternary sets of examples. However, in (3.107)–(3.109) the initial *għajn* of *għonqu* 'his neck', *għada* 'tomorrow', and *għaraq* 'sweat' causes segmental reduction whereas the same words fail to do so in (3.110)–(3.112). This variation is relatively common for

(Π30) *ma'* 'with', (Π55) *ta'* 'of', and (Π51) *sa* 'till' but unattested for (Π7) *bi* 'with' and (Π15) *fi* 'in'. Consider the token frequencies given in Figure 3.55.

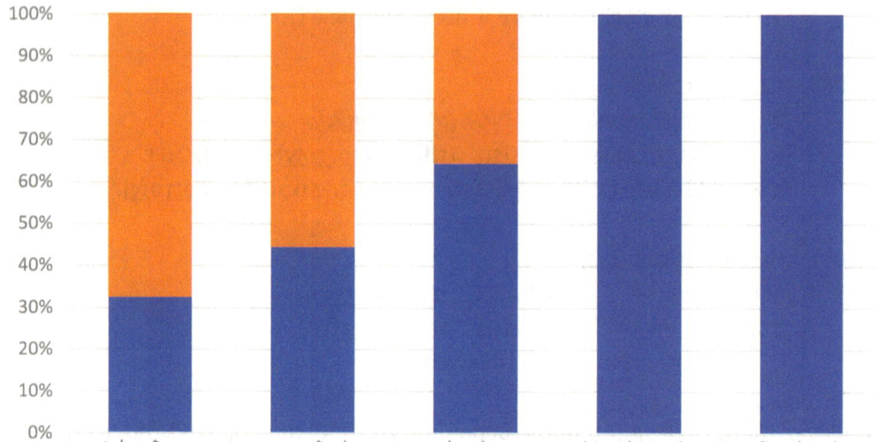

	ta' + għaraq	sa + għada	ma' + għonqu	bi + għajnejha	fi + għamla
■ long	23	259	81	0	0
■ short	11	206	146	214	538

Figure 3.55: Range of variation with *għajn*-initial complements.

We have searched the Korpus Malti 3.0 for evidence of variation involving the five Πs featured in this section as heads of PPs with the complements we know from examples (3.103)–(3.112). The members of C II never show up in their long form when followed by the selected *għajn*-initial complement. As to (Π30) *ma'* 'with', (Π55) *ta'* 'of', and (Π51) *sa* 'till', the turnouts speak a different language insofar as short and long allomorphs compete with each other to different degrees. (Π55) *ta'* 'of' and (Π51) *sa* 'till' prefer long forms over short forms with 68% to 32% and 56% to 44%, respectively. (Π30) *ma'* 'with' is represented by the short allomorph in 64% of the cases and by the long allomorph in 36% of the cases. We assume that the patterns of variation would not be much different had we chosen different *għajn*-initial complements. We conclude on the above basis that the phonological status of *għajn* is undecided. Sometimes it is treated like a consonantal segment triggering the long allomorphs of (Π30) *ma'* 'with', (Π55) *ta'* 'of', and (Π51) *sa* 'till' and at other times it does not have segmental status so that the above Πs are represented by their short allomorphs.

We cannot solve this problem satisfactorily for Maltese phonology. To circumvent the intricacies of the status of *għajn*, we take a shortcut by way of modifying the rule in Figure 3.51. Figure 3.56 features an additional branch for the subrule defining the domain of procliticisation of *a*-final Πs. The brackets around the lower branch indicate that this part of the subrule is optional. We know from the above exemplification that *a*-final Πs may or may not undergo procliticisation. There is no denying that this solution is ad hoc and might not stand the test in a proper phonological study. Nevertheless, it suffices for the purpose of this section. The five Πs whose procliticisation has been reviewed in the foregoing paragraphs are going to occupy our minds again since they are also involved in processes of fusion with the definite proclitic.[76]

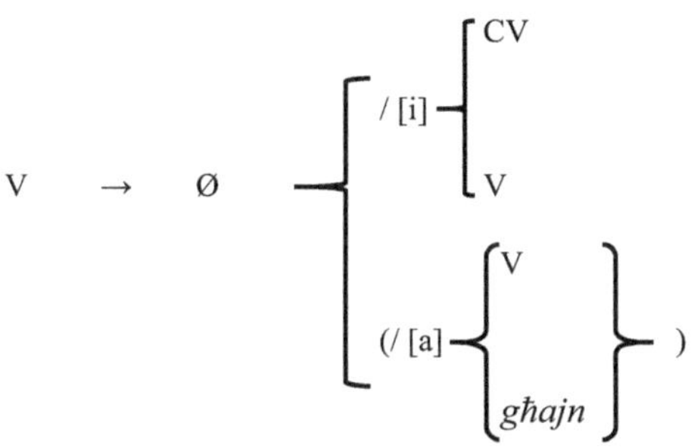

Figure 3.56: Rules for vowel deletion under procliticisation – version II.

3.3.4 Fusion (Π + definite proclitic)

The following ten Πs fuse with the definite proclitic: (Π5) *bħal* 'like', (Π7) *bi* 'with', (Π15) *fi* 'in', (Π19) *ġo* 'in', (Π21) *għal* 'for', (Π29) *lil* 'to', (Π30) *ma'* 'with', (Π41) *minn* 'from', (Π51) *sa* 'till', and (Π55) *ta'* 'of'. The two Πs highlighted in grey are invariant whereas four times as many Πs affected by fusion inflect for person. Figure 3.57 exposes the type frequency of Πs which are sensitive to combinations with the definite proclitic.

Compared to the picture painted in Figure 3.50 for the type frequency of segmentally reducible Πs the shares for Πs involved in fusion are visibly higher but that of the invariant Πs fails to come close to the 20%-mark. We are again facing a minority phenomenon in terms of the number of Π-types which participate in fusion. Like in the case of the segmentally reducible Πs the low type frequency goes along with an extremely high token frequency of certain Πs as will transpire from the subsequent paragraphs.

76 Cf. Section 3.3.4.

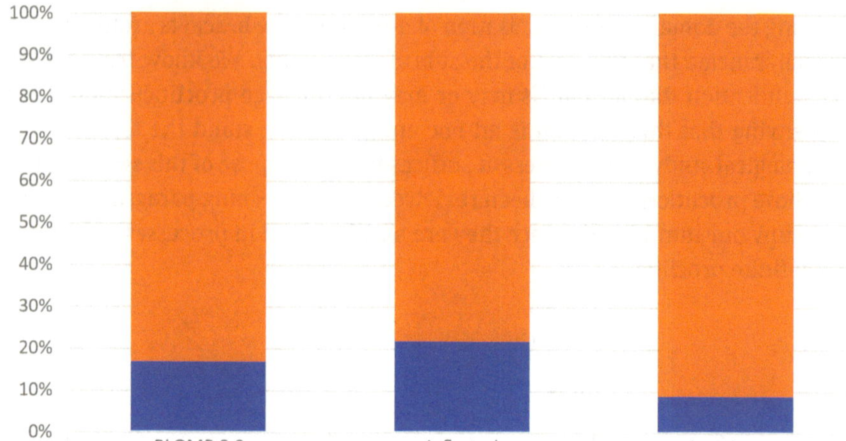

	BLOMP 2.0	inflected	invariant
immune	50	29	21
fusion	10	8	2

Figure 3.57: Type frequency of Πs undergoing fusion.

The five Πs which procliticise also participate in fusion. Moreover, all Maltese monosyllabic Πs with CV-structure fuse with the definite proclitic. The implicational patterns are again unilateral as shown in Figures 3.58–3.59.

procliticisation ⊃ fusion

Figure 3.58: Procliticisation implies fusion.

CV ⊃ fusion

Figure 3.59: CV-structure implies fusion.

Πs which have the shape of a closed syllable end in a sonorant which is preferably the lateral approximant /l/.

The fusion of Π and definite proclitic is not homogeneous in the sense that there exist several patterns. The Πs behave differently from each other. This differential behaviour extends to the definite proclitic so that it is not possible to generalise sweepingly over all Πs which are subject to fusion. As we will see below, the Πs can be divided into several groups according to their behaviour in interaction with the definite proclitic (henceforth: ART(icle)). Before we look at what it means to fuse in terms of the qualities involved, it makes sense to show how important fusion is for the token frequency of the Πs under scrutiny. The Korpus Malti 3.0 hosts almost 15,000,000 instances of Π-ART fusion. These matches outnumber the slightly less than 13,000,000 instances of free forms of Πs found in the same source.[77] The ten Πs which participate in fusion are those privileged in the sense that their overall frequency increases considerably under fusion whereas the frequency count for the other 50 Maltese Πs remains unaltered. More importantly, the ten privileged Πs do not have even shares of the 15,000,000 matches as is shown in Figure 3.60.

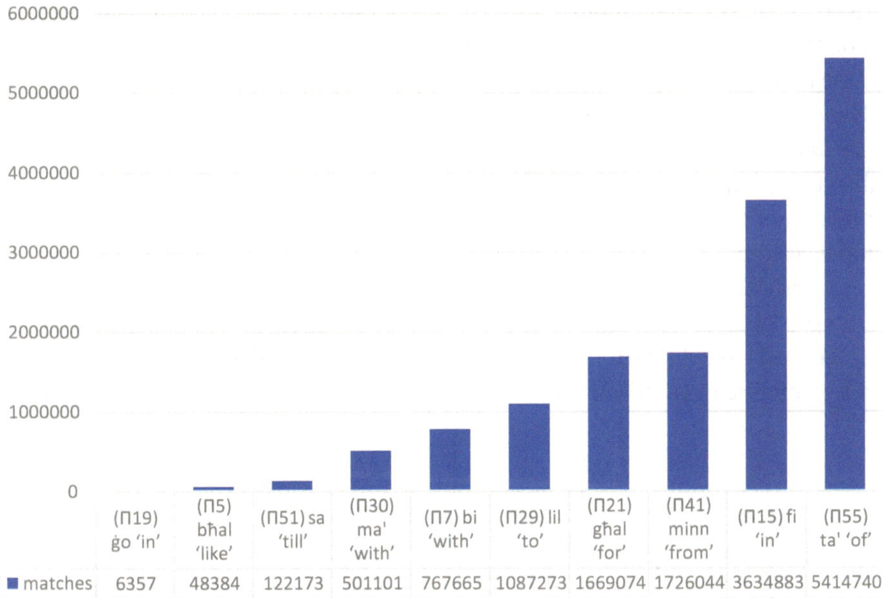

Figure 3.60: Token frequency of fused forms in the Korpus Malti 3.0.

More than a third of the tokens of fused forms are registered for (Π55) *ta'* 'of'. Fused forms of this Π are 852 times as frequent as those of (Π19) *ġo* 'in' at the bottom of the scale. (Π55) *ta'* 'of', (Π15) *fi* 'in', (Π41) *minn* 'from', and (Π21) *għal* 'for' exceed the average of 1.5 mio matches. It can be stated that Πs under the top four in Figure 3.60 are also among the top eleven in Table 1.11, i.e. their already established high token frequency is further increased under fusion. The three Πs at the bottom end of the scale in Figure 3.60 occupy also only lower ranks in Table 1.11.

We now know that the members of a selected set of Maltese Πs literally thrive in the context of fusion. Fusion applies if a Π procliticises to a host whose left-most morpheme is ART. The latter is a phonologically changeful proclitic with (V) C-structure whose obligatory but versatile segment is the lateral approximant /l/

77 As argued in Section 1.6.8.

(Aquilina 1976: 13). The optional vocalic segment is the high front vowel /i/ which is present only if ART is surrounded by consonants or occupies the sentence-initial position while procliticised to a consonant-initial host. The default form of ART is *(i)l-*. However, the lateral approximant regularly assimilates to a following coronal consonant (except /ʤ/) (Borg/Azzopardi-Alexander 1997: 136 and 328). Furthermore, if the host displays an initial consonant cluster starting with the voiceless fricatives /s/ (so-called impure /s/), /ʃ/ or a sonorant, an epenthetic /i/ separates the /l/ of ART from the first consonant of the host (Cassar 2000: 23). Interestingly, none of the descriptive-linguistic sources mentions the fact that the cluster-initial voiced alveolar fricative /z/ may also trigger /i/-epenthesis as in *żvilupp* 'development' → *l-iżvilupp* (Aquilina 1991: 1631). Cachia (1994: 61) mentions a number of (unpredictable) exceptions from these rules whereas Sutcliffe (1936: 18–19) assumes that the insertion of the epenthetic /i/ is generally possible if the host is equipped with an initial cluster.

Since the phonologically conditioned allomorphy of ART is crucial in the context of fusion, we recapitulate the above observations in the format of rules in Figures 3.61–3.63. In the formulas, the symbol L is employed as representative of the consonantal segment of ART which is subject to assimilation to coronal consonants.

$$\text{ART} \rightarrow \left[\begin{array}{lll} iL\text{-} & / \left\{\begin{array}{c} \text{C]} \\ \# \end{array}\right\} & \_\_\,[\text{C} \\ l\text{-} & / & \_\_\,[\text{V} \end{array}\right.$$

Figure 3.61: Default rule for the choice of allomorph of ART.

$$L \rightarrow \left[\begin{array}{lll} \text{X} = [\text{C}_{\text{HOST}} & /\_\_\,[\text{C}_{\text{CORONAL\&-/ʤ/}} & \\ /\text{l}/ & /\_\_ \left\{\begin{array}{l} [\text{C}_{\text{-CORONAL}} \\ [\text{ʤ} \\ [\text{V} \end{array}\right\} & \end{array}\right.$$

Figure 3.62: Assimilation rule for ART.

$$\text{CC} \rightarrow i\text{CC} \;/\; \text{ART} + \text{C1}_{\text{SONORANT/S(H)IBILANT}}$$

Figure 3.63: /i/-epenthesis.

The open questions related to the morphonological behaviour of ART cannot be answered in this study. For the purpose at hand, it suffices to work on the basis of the rules presented in Figures 3.61–3.63. These rules guide us through Sections 3.3.4.1–3.3.4.3 which review the behaviour of the ten Πs in three different contexts which are defined by the initial segment of the host of the sequence of proclitics.

3.3.4.1 Π + ART + [V

The output of the above rules is relevant for the behaviour of the ten Πs under fusion. These Πs can be subdivided in three subclasses according to the quality of the final segments (Akkademja tal-Malti 1998: 53–54). The two Πs ending in /i/ behave differently from those whose segmental chain ends in /a/. Both *i*-final and *a*-final Πs differ from consonant-final Πs when interacting with ART. To outline the major differences between these subclasses, we start from the default situation as circumscribed in Figure 3.61. If the host of ART has an initial vowel, ART is represented by the lateral approximant alone. In this case, fusion is blocked for the Πs ending in a vowel other than /i/ as shown in Figure 3.64.

$V_{a/o}$] → ~~FUSION~~ / _ ART-[V

Figure 3.64: Blocking of fusion.

Whether this is only a purely orthographic convention is difficult to determine in the absence of spoken language data. We therefore have to take the evidence for the blocking of fusion at face value. The Korpus Malti 3.0 features 980,203 cases of blocked fusion of which 887,219 (= 91%) involve (Π55) *ta'* 'of'. There are 73,623 matches of (Π30) *ma'* 'with' combining with ART without fusion, 18,011 matches of the same constellation for (Π51) *sa* 'till', and only 1,350 for (Π19) *ġo* 'in'. These turnouts for blocked fusion are accounted for in Table 1.11.

The above means that *i*-final and consonant-final Πs behave similarly to each other since they fuse with ART if the latter precedes a vowel. However, *i*-final and consonant-final Πs fuse with ART in different ways. In the former case, the rule for procliticisation exposed in Figure 3.56 applies, i.e. the lateral approximant of ART is treated like a single word-initial consonant which means that the vowel of the Π is dropped and no ancillary vowel is required for ART. The rule is specified in Figure 3.65.

C_i] → C] / _ ART-[V

Figure 3.65: *i*-deletion with C II Πs under fusion.

The sequence of proclitics resulting from this rule is either *bl-* for (Π7) *bi* 'with' or *fl-* for (Π15) *fi* 'in' as illustrated in (3.113)–(3.114).

(3.113) [Żahra 2008: 33]

u	*l-kompożizzjoni*	*kienet*	*issir*	*kollha*
and	DEF-composition	be:3SG.F.PFV	3SG.F.IPFV:become	all:F

[***b**l-idejn*]$_{PP}$
[**with**:DEF-hand:PL]

'[. . .] the type-setting was executed entirely [**by** hand].'

(3.114) [Agius 2002: 30]

Apollo	*kien*	*mixħut*	[*fl-art*]$_{PP}$	*ma'*	*ġenbu*
Apollo	be	PART:throw	[**in**:DEF-ground]	with	side:3SG.M

'Apollo was stretched out [**on** the floor] at his side.'

The nouns *idejn* 'hands' and *art* 'ground' host vowels in the initial slot of their segmental chains. This vowel triggers the allmoprh {/l/} of ART. This allomorph, in turn, selects the short i.e. monoconsonantal allomorph of the C II Πs. This holds also for cases of /i/-epenthesis according to Figure 3.63. Examples are the combination with the noun *skop* 'purpose' in (3.115) and the place name *Svezja* 'Sweden' in (3.116).

(3.115) [Korpus Malti 3.0; academic7]

jiffaċilita	*l-investimenti*	*tal-kapital*	[***b**l-iskop*]$_{PP}$
3SG.M.IPFV:facilitate	DEF-investment:PL	of:DEF-capital	[**with**:DEF-purpose]

li	*jtejjeb*	*il-produzzjoni*
that	3SG.M.IPFV:CAUS:good	DEF-production

'[. . .] it facilitates capital investment [**with** the aim] of ameliorating the production.'

(3.116) [Gauci 1994: 43]

Is-Sitt	*Edizzjoni*	*tat-Tazza*	*tad-Dinja*	*kien*	*deċiż*
DEF-sixth	edition	of:DEF-cup	of:DEF-world	be	decide.PART

li	*tintlagħab*	[*fl-Isvezja*]$_{PP}$
that	3SG.F.IPFV:RFL:play	[**in**:DEF-Sweden]

'It was decided that the sixth edition of the World Cup should take place [**in** Sweden].'

No other Π's behaviour is covered by the above rule.

Turning our attention to the consonant-final Πs, we notice that all four of them fuse with ART such that a geminate lateral approximate arises. In the case of (Π5) *bħal* 'like', (Π21) *għal* 'for', and (Π29) *lil* 'to', this is by no means remarkable since these Πs are already equipped with a final /l/. The resulting fused forms are *bħall-*, *għall-*, and *lill-*. In contrast, (Π41) *minn* 'from' whose final segment is the geminate denti-alveolar nasal /n/ gives evidence of total assimilation /n/ > /l/ in combination with ART yielding the fused form *mill-*.[78] Figure 3.66 shows that different inputs have identical outputs.

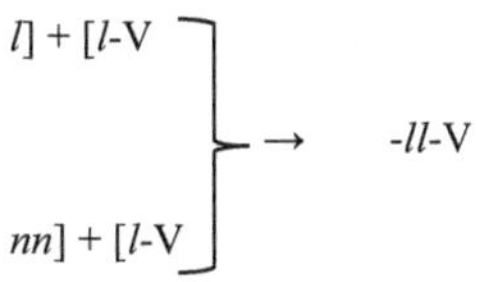

Figure 3.66: Gemination of /l/ under fusion.

We provide examples for the behaviour of each of the CVC-Πs before ART + [V in (3.117)–(3.120).

(3.117) [Camilleri 2013: 144]

Beża' joħroġ ta' bigotta jew ta' razzista
fear 3SG.M.IPFV:come_out of hypocrite or of racist
*[**bħal**l-oħrajn]$_{PP}$*
[**like**:DEF-other.PL]
'He was afraid to prove to be a hypocrite or a racist [**like** the others].'

(3.118) [Bartolo/Vella 2009: 21–22]

Iżda min jidħol f' din il-kamra-ġungla
but who 3SG.M.IPFV:enter in DEM:F:PROX DEF-room-jungle
*[**għal**l-ewwel darba]$_{PP}$ kien ikun xi ftit*
[**for**:DEF-first time] be 3SG.M.IPFV:be.FUT what a_little
maħsud
PART:surprise
'However, whoever enters this jungle kind of room [**for** the first time] would be a bit surprised [. . .].'

78 The assimilation of the final segment of (Π41) *minn* 'from' will be discussed further in Section 3.3.4.3.

(3.119) [Cauchi 1997: 30]
beda *jgħid* *[lill-irġiel* *l-oħra]*$_{PP}$
begin 3SG.M.IPFV:say [**to**:DEF-man.PL DEF-other.PL]
'[. . .] he started to say [**to** the other men].'

(3.120) [Abela 2016: 94]
Xogħli *li* *nara* *li* *jaslu* *qawwijin* *u*
work:1SG that 1SG.IPFV:see that 3.IPFV:arrive:PL strong:PL and
sħaħ *[**mill**-Ingilterra]*$_{PP}$ *sal-Awstralja.*
whole.PL [**from**:DEF-England] till:DEF-Australia
'It is my job to see that they arrive safe and dry [**from** England] in Australia.'

3.3.4.2 Π + ART + [C$_{-CORONAL}$

The rule given in Figure 3.66 is specified for hosts with an initial vowel. In point of fact, gemination occurs wherever a CVC-Π encounters the *l*-allomorph of ART. This is also possible with consonant-initial hosts which, according to Figure 3.61, require the allomorph *il-* of ART. However, under fusion, this prosthetic vowel does not surface. In the absence of /i/ the final consonant of the Π and the lateral approximant of ART become direct neighbours. In this context, gemination applies and, in the case of (Π41) *minn* 'from', the final nasal assimilates to the lateral. Superficially, Figures 3.66–3.67 resemble each other closely (with X = either V or C).

$$\left\{\begin{array}{l} l] + [il\text{-C} \\ \\ nn] + [il\text{-C} \end{array}\right\} \rightarrow \quad \text{-}ll\text{-C(X)}$$

Figure 3.67: Gemination of /l/ and blockage of *i*-prosthesis under fusion.

However, the rule in Figure 3.67 is not as simple as that in Figure 3.66 because the former involves the additional requirement of deleting or blocking the prosthetic /i/. Once more we cannot properly judge whether we are confronted with phonologically irrelevant orthographic conventions. The realisation of the rule presented in Figure 3.67 is illustrated in (3.121)–(3.124).

(3.121) [Bartolo/Vella 2009: 181]

Kollox	*kien*	*magħmul*	*mill-istess*	*kulur*	*skur*
everything	be	PART:make	from:DEF-same	colour	dark

***[bħall-blat]**$_{PP}$*
[**like**:DEF-rock]
'Everything was made of the same dark colour [**as** the rock].'

(3.122) [Gauci 1994: 63]

diġà	*kienu*	*bbukkjaw*	*l-akkomodazzjoni*	*[għall-kwarti*
already	be:3PL.PFV	book:3PL.PFV	DEF-accomodation	[**for**:DEF-quarter

tal-finali]$_{PP}$
of:DEF-final:PL]
'[. . .] they had already booked the accommodation [**for** the quarter finals].'

(3.123) [Korpus Malti 3.0; academic10]

tiġi	*proposta*	*liġi*	*populari*	*[lill-gvern*
3SG.F.IPFV:come	propose:PART:F	law	popular	[**to**:DEF-government

nazzjonali]$_{PP}$
national]
'[. . .] a popular law is proposed [**to** the national government].'

(3.124) [Cauchi 1997: 28]

ħarġet	*flixkun*	*żgħir*	*[mill-basket]*$_{PP}$
come_out:3SG.F.PFV	bottle	small	[**from**:DEF-basket]

'[. . .] she pulled a small bottle [**from** the basket].'

In Figure 3.67 it is shown that the ancillary /i/ of ART does not surface under fusion with consonant-final Πs. In combinations of (Π7) *bi* 'with' and (Π15) *fi* 'in' with ART two identical vowels meet at the morpheme boundary one of which is suppressed. /i/ + /i/ does not result in a long vowel. It is not absolutely clear which of the two high front vowels appears in the surface form. We assume that we are dealing with the ancillary /i/ of ART because this vowel is automatically required if ART attaches to a consonant-initial host. Only in a second step does the Π enter the scene, i.e. chronologically after the ancillary vowel. According to this logic, the members of C II drop their vowel in front of the vowel of ART. This in turn means that in this context, the short allomorphs *b'* and *f'* are used. This situation is captured schematically in Figure 3.68.

C*i*] → C] / __ ART-C(X)

Figure 3.68: Loss of /i/ before ancillary /i/.

Examples (3.125)–(3.126) present two cases in which the rule exposed in Figure 3.68 is realised.

(3.125) [Camilleri 2013: 90]

Biswit	*l-istazzjon*	*sabu*	*ħanut*	*ibigħ*	*umbrelel*
facing	DEF-station	find:3PL	shop	3SG.M.IPFV:sell	umbrella.PL
*[**b**il-gzuz]*$_{PP}$					
[**with**:DEF-stack.PL]					

'Opposite the station they found a store which sold umbrellas [**by** the score].'

(3.126) [Bartolo/Vella 2009: 105]

*[**f**il-fatt]*$_{PP}$	*ħija*	*Tommy*	*dejjem*	*jgħid*	*li*	
[**in**:DEF-fact]	brother:1SG	Tommy	always	3SG.M.IPFV:say	that	
Melchior	*huwa*	*l-aktar*	*tifel*	*bravu*	*tal-iskola*	
Melchior	3SG.M	DEF-more	boy	capable	of:DEF-school	

'[. . .] [**in** fact] my brother Tommy always says that Melchior is the cleverest boy in the school.'

In contrast to the cases featured in (3.125)–(3.126), the CV-ΠS ending in either /a/ or /o/ keep their original vowel under fusion so that there is no slot which could be filled by the ancillary /i/ of ART. The formula for the corresponding sandhi process is disclosed in Figure 3.69.

il- → *l-* / $V_{a/o}$] __ C(X)

Figure 3.69: Blockage of ancillary /i/ under fusion.

We provide examples of this process in (3.127)–(3.130).

(3.127) [Abela 2016: 86]

Il-vapur	*kien*	*sorġut*	*[**ma**l-moll]*$_{PP}$	*bħal*	*bestja*
DEF-ship	be	lie_at_anchor:PART	[**with**:DEF-jetty]	like	beast
ġganteska	*rieqda*				
gigantic:F	sleep.PART:F				

'The ship lay at anchor [**at** the mole] like a gigantic sleeping beast.'

(3.128) [Camilleri 2013: 17]
*U minn [**ġo**l-high chair]*$_{PP}$ *Eleanor [. . .] innutat*
and from [**in**:DEF-high chair] Eleanor note:3SG.F.PFV
il-bidla f' wiċċ ommha
DEF-change in face mother:3SG.F
'And from [the high chair] Eleanor [. . .] took notice of the change in her mother's face.'

(3.129) [Bartolo/Vella 2009: 18]
imma ħi dak diġà sar idum
but brother DEM:M:DIST already become 3SG.M.IPFV:remain
*barra [**sa**l-ħdax ta' billejl]*$_{PP}$
outside [**till**:DEF-eleven of night]
'[. . .] however, the brother of that one was already allowed to stay out [**until** eleven at night].'

(3.130) [Cauchi 1997: 27]
*ħa miegħu xi ħaġa [**ta**l-flus]*$_{PP}$
take with:3SG.M what thing [**of**:DEF-money]
'[. . .] he took some [**of** the money] with him [. . .].'

The differential morphonological behaviour of the final vowels of CV-Πs is in line with the pattern we have observed already in connection with procliticisation.[79] /i/ is prone to being deleted whereas /a/ and /o/ are much more resistant to reductive processes.

There is another problem we already mentioned in the previous sections, namely the role of *għajn*. With regards to the interpretation of this abstract phoneme either as an equivalent of a consonantal segment or as a phonological non-entity, there is again variation. Akkademja tal- Malti (1998: 53–54) twice uses the wording "jekk trid"[80] in connection to the supposedly silent graphemes <għ> and <h>. In this way, the authors of the booklet treating of the orthographic rules of Maltese leave it to the individual writers to decide whether they classify *għajn* as consonant or not. This flexibility explains why there is heterogeneity in the first place because the authors of the texts of our database are divided – and it is by no means exceptional that one and the same writer alternates between the two available options. The expected behaviour of (Π7) *bi* 'with' and (Π15) *fi* 'in' in (3.131)–(3.132) where *għajn*

79 Cf. Section 3.3.3.
80 Our translation: "if you want".

is treated like a consonant compete with cases where the *għajn*-initial host is taken to start with a vowel as shown in (3.133)–(3.134).

(3.131) [Bartolo/Vella 2009: 68]
*kien maħnuq u mejjet [**b**il-għatx]*$_{PP}$
be PART:make_hoarse and dead [**with**:DEF-thirst]
'[. . .] he was hoarse and dying [**of** thirst].'

(3.132) [Bartolo/Vella 2009: 159]
*bdew iduru [**f**il-għoli]*$_{PP}$
begin:3PL.PFV 3.IPFV:turn:PL [**in**:DEF-high]
'[. . .] they began to circle [**in** the sky].'

(3.133) [Bartolo/Vella 2009: 342]
*[**b**l-għajnuna ta' Tommy u sħabu]*$_{PP}$ *dewwiet*
[**with**:DEF-help of Tommy and companion.PL] heal:3SG.F.PFV
il-feriti kollha ta' Steve
DEF-wound:PL all of Steve
'[. . .] [**with** the help of Tommy and his companions] she healed all Steve's wounds.'

(3.134) [Bartolo/Vella 2009: 476]
*Il-gwida żamm l-umbrella ħamra [**f**l-għoli]*$_{PP}$ *fuq rasu*
DEF-guide hold DEF-umbrella red:F [**in**:DEF-high] on head:3SG.M
'The guide held the red umbrella [high] above his head [. . .].'

It is striking that this pattern of variation reflects the variable practice of the authors of the same primary source. Figure 3.70 suggests that no overall statement can be made as to the general preferences in the choice of allomorphs of (Π7) *bi* 'with' and (Π15) *fi* 'in'. We have searched the Korpus Malti 3.0 for combinations of *bi/b'* with *għajjat* 'shouting', *għajnuna* 'help', *għaqal* 'prudence', *għatx* 'thirst' and *fi/f'* with *għajjat* 'shouting', *għajnejn* 'eyes', *għerien* 'caves', *għoli* 'high'.

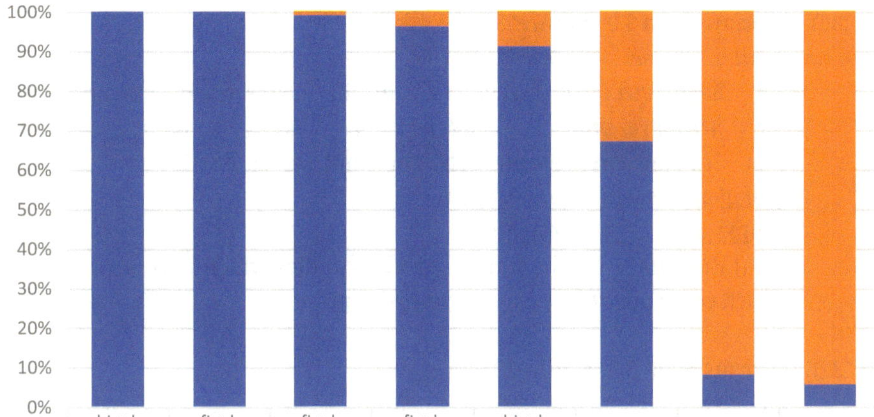

	bi + l-għajjat	fi + l-għajjat	fi + l-għajnejn	fi + l-għerien	bi + l-għajnuna	fi + l-għoli	bi + għaqal	bi + l-għatx
long	0	0	4	3	489	523	2619	285
short	107	11	477	76	5044	1072	232	17

Figure 3.70: Variation of allomorphs of C II-ΠS in combination with *għajn*-initial nouns.

For several combinations, there is a clear preference for the short alloform – the long one being excluded completely in the two cases on the left of the scale. However, at the opposite end of the scale, we find two combinations which involve majorities of 90% and more for the long allomorph. This problem needs further study which is beyond the scope of our present project.

Given that the status of *għajn* is controversial among native speakers (or: writers), we would expect to find evidence for it being treated as phonologically non-existent such that the host is interpreted as vowel-initial in the case of CV-ΠS which end in a vowel other than /i/. However, this expectation is not corroborated empirically. In (3.135)–(3.138), the four CV-ΠS which end in a vowel other than /i/ fuse with ART so that the latter has no need for the ancillary vowel. This means that the rule exposed in Figure 3.69 applies also in these cases.

(3.135) [Bartolo/Vella 2009: 212]
[Mal-għajbien tal-missier]$_{PP}$ kulħadd kien qisu sab
[**with**:DEF-disappearance of:DEF-father] everybody be as_if:3SG.M find
il-ħin tiegħu differenti minn tal-oħrajn biex
DEF-time of:3SG.M different from of:DEF-other:PL to
jinżel jiekol
3SG.M.IPFV:descend 3SG.M.IPFV:eat
'[**With** the father gone] everybody seemed to have found their own time different from that of the others to come down to eat.'

(3.136) [Korpus Malti 3.0; culture239]
ir-raħal se jiddawwal permezz tax-xemgħat
DEF-village FUT 3SG.M.IPFV:CAUS:light by_means_of of:DEF-candle:PL
li jirrappreżenta d-dawl li ħareġ minn
that 3SG.M.IPFV:represent DEF-light that come_out from
*[ġo**l**-għar ta' Betlehem]$_{PP}$*
[**in**:DEF-cave of Bethlehem]
'[. . .] the village will be illuminated by candles which represent the light which came out of [the cave of Bethlehem].'

(3.137) [Korpus Malti 3.0; culture205]
Il-Ġimgħa filgħaxija l-fiera tiftaħ sa nofsillejl
DEF-Friday evening DEF-fair 3SG.F.IPFV:open till midnight
*is-Sibt [**sa**l-għaxra ta' filgħaxija]$_{PP}$ filwaqt li*
DEF-Saturday [**till**:DEF-ten of evening] while that
l-Ħadd il-fiera se tkun miftuħa matul
DEF-Sunday DEF-fair FUT 3SG.F.IPFV:be.FUT PART:open:F during
il-jum kollu
DEF-day all:3SG.M
'Friday evening the fair will be open until midnight, on Saturday [**until** ten in the evening] whereas on Sunday, the fair will be open during the whole day.'

(3.138) [Bartolo/Vella 2009: 18]
*l-istorja [**ta**l-għajbien tas-subien mid-dehra]$_{PP}$*
DEF-story [**of**:DEF-disappearance of:DEF-boy.PL from:DEF-sight]
kienet tafha bl-amment
be:3SG.F.PFV 3SG.F.IPFV:know:3SG.F with:DEF-memory
'[. . .] she knew by heart the story [**of** the disappearance out of sight of the boys].'

Examples (3.139)–(3.142) demonstrate that consonant-final IIs behave according to the prerequisites of the rule given in Figure 3.67 when combining with ART and a host which starts with *għajn*.

(3.139) [Korpus Malti 3.0; academic176]
prodotti li ġejjin mill-annimali bħall-ġobon
product:PL that come.PART:PL from:DEF-animal:PL like:DEF-cheese
*u l-mudullun jew sustanzi ħelwin [**bħal**l-għasel]*$_{PP}$
and DEF-marrow or substance:PL sweet:PL [**like**:DEF-honey]
'[. . .] products which stem from animals like cheese and marrow or sweets substances [**like** honey] [. . .].'

(3.140) [Bartolo/Vella 2009: 181]
Liċ-ċkejken maskot tagħna għad għandna bżonnu
to:DEF-little mascot of:1PL still have.IPFV:1PL need:3SG.M
*[**għal**l-għeluq tat-turnament]*$_{PP}$
[**for**:DEF-conclusion of:DEF-tournament]
'We still need our little mascot [**for** the closure of the tournament].'

(3.141) [Korpus Malti 3.0; academic4]
taw tagħrif importanti kemm lill-istudenti
give:3PL.PFV information important how_much to:DEF-student:PL
*kif ukoll [**lil**l-għalliema]*$_{PP}$
how also [**to**:DEF-teacher:PL]
'[. . .] they gave information which was important not only for the students but also [**for** the teachers].'

(3.142) [Camilleri 2013: 158]
lemaħ lil Eleanor iċċapċap fuq spallejn missierha
perceive to Eleanor 3SG.F.IPFV:slap on shoulder:PL father:3SG.F
*[**mil**l-għoli ta' karru trijonfali]*$_{PP}$
[**from**:DEF-high of wagon triumphal]
'[. . .] he caught sight of Eleanor slapping her father's shoulders [**from** a wagon of the triumphal procession].'

To sum up the above observations, we state that *i*-final, *a/o*-final, and consonant-final ΠS display different behavioural properties when it comes to fusing with ART in front of a host whose initial segment is a non-coronal consonant. The differences are also visible in the case of *għajn* which is treated as consonantal segment in combinations with *a/o*-final and consonant-final ΠS but has uncertain status in combinations with the two members of C II. Further differences come to the fore in connection to the fate of the ancillary vowel of ART. In the case of the *i*-final ΠS, the ancillary vowel is present and triggers the use of the short allomorphs of the Πs. From all other combinations the prosthetic /i/ is excluded. The long allomorphs/citation forms of (Π5) *bħal* 'like', (Π19) *ġo* 'in', (Π21) *għal* 'for', (Π29) *lil* 'to', (Π30) *ma'* 'with', (Π41) *minn* 'from', and (Π51) *sa* 'till' are used in these contexts. In the subsequent section, the feature [+CORONAL] will diversify the picture further.

3.3.4.3 Π + ART + [$C_{+CORONAL}$

The sandhi processes to be discussed in this section are relatively unspectacular if we take account only of the vowel final Πs. Figure 3.62 has familiarised us with the assimilation that affects the lateral approximant of ART if it is in contact with an initial coronal of the host. The assimilation is full in the sense that the consonant of ART copies the entire set of phonological properties of the segment to its right. This means that, discounting the presence/absence of the ancillary /i/, ART can come in ten different realisation forms as shown in Figure 3.71.

		/t/	/ __ [t
		/d/	/ __ [d
		/ʦ/	/ __ [ʦ
		/ʧ/	/ __ [ʧ
/l/	→	/s/	/ __ [s
		/z/	/ __ [z
		/ʃ/	/ __ [ʃ
		/n/	/ __ [n
		/r/	/ __ [r
		/l/	/ __ [l

Figure 3.71: Realisation forms of ART.

In the case of the vowel-final Πs, the assimilation to which the lateral approxminant is subject does not affect the rules exposed in Figures 3.68–3.69. There is thus no need to fully document the interaction of the assimilated ART with either (Π7) *bi* 'with' and (Π15) *fi* 'in' or (Π19) *ġo* 'in', (Π30) *ma'* 'with', (Π51) *sa* 'till', and (Π55) *ta'* 'of'. For the present purpose it is sufficient to provide one example for each of these six Πs in combination with an assimilated form of ART in (3.143)–(3.148).

(3.143) [Cauchi 1997: 8]
sab ruħu bilqiegħda fil-kannestru
find soul:3SG.M sitting in:DEF-wicker_basket
*[**b**it-tadam imgħaffeġ taħtu]*PP
[**with**:DEF-tomato.COLL PART:crush under:3SG.M]
'[. . .] he found himself sitting in the wicker basket [**with** the tomatoes crushed under him].'

(3.144) [Gauci 1994: 64]
*[**F**is-semi finali l-oħra]*PP *il-Portugiżi offrew*
[**in**:DEF-semi_final DEF-other:F] DEF-Portuguese:PL offer:3PL.PFV
oppożizzjoni kbira lill-Ingliżi
opposition big:F to:DEF-English:PL
'[**In** the other semi-final], the Portuguese opposed the Englishmen strongly.'

(3.145) [Korpus Malti 3.0; literature21]
*flimkien imxew minn [**ġ**ox-xagħri]*PP *lejn*
together walk:3PL.PFV from [**in**:DEF-rocky_fields] towards
l-irdumijiet
DEF-cliff:PL
'[. . .] they walked together from [the rocky fields] towards the cliffs.'

(3.146) [Bartolo/Vella 2009: 168]
*Il-wirdien [**ma**s-swaba' tiegħu]*PP
DEF-cockroach.COLL [**with**:DEF-finger.PL of:3SG.M]
bdew iferfru ġwenħajhom
begin:3PL.PFV 3.IPFV:flap:PL wing:PL:3PL
'The cockroaches [**with** their fingers] began to flap their wings [. . .].'

(3.147) [Korpus Malti 3.0; news61637]
x' fiha ħażin li l-istatwa tibqa'
what in:3SG.F bad that DEF-statue 3SG.F.IPFV:remain
*tinħareġ [**sa**z-zuntier tal- knisja]*PP
3SG.F.IPFV:RFL:come_out [**till**:DEF-churchyard of:DEF-church]
'[. . .] what bad is there when the statue continues to be brought out [**as far as** the churchyard]?

(3.148) [Abela 2016: 96]

kien	*fi*	*triqtu*	*lejn*	*in-naħa*	*l-oħra*
be	in	street:POSS:3SG.M	towards	DEF-side	DEF-other:F

*[**ta**d-dinja]*$_{PP}$
[**of**:DEF-world]
'[. . .] it was on its way to the other side [**of** the world].'

In combination with the assimilated ART, the vowel-final Πs behave exactly the same as they would in combination with an unassimilated ART. This is different for the consonant-final Πs because their final segment is affected by the assimilation process, too.

To get a clear picture of the behaviour of (Π5) *bħal* 'like', (Π21) *għal* 'for', (Π29) *lil* 'to', and (Π41) *minn* 'from', it is necessary to remember that the lateral approximant of ART bears the feature [+CORONAL], too. According to Akkademja tal-Malti (1998: 54) the application of gemination along the lines of the rule in Figure 3.67 is prohibited if the host boasts an initial /l/ because "[h]ekk nevitaw tliet konsonanti L wara xulxin".[81] In point of fact, this convention holds also for all cases of assimilation of ART since the orthographic representation of three identical consonants in a row is generally avoided. We assume that the triplication of the consonants would have no direct phonological correlate since extralong consonants beyond geminates are not reported for Maltese.

More importantly, the regressive or anticipatory assimilation to which the lateral approximant of ART is exposed is inherited by the final consonant of the Πs. More precisely, the Π's CODA consonant assimilates to the consonant of ART. This explains why the final nasal /n/ of (Π41) *minn* 'from' regularly becomes /l/ when in direct contact with the lateral approximant of ART. Since all other consonant-final Πs investigated in this section end in /l/ already in their citation form, the assimilatory force of the ART-consonant is not immediately visible. If, however, ART is represented by a different coronal consonant the influence on the segmental chain of the Π is more transparent. It must be kept in mind though that gemination is blocked so that the final segment of the Π and the assimilated ART-consonant become one. Table 3.24 surveys the different shapes that the fusion of consonant-final Πs and ART can take.

81 Our translation: "in this way we avoid three L consonants in a sequence".

Table 3.24: Survey of fused forms of assimilated ART and consonant-final Π.

ART		**Π**			
		bħal	***għal***	***lil***	***minn***
/t/	<t>	*bħat-*	*għat-*	*lit-*	*mit-*
/d/	<d>	*bħad-*	*għad-*	*lid-*	*mid-*
/ʦ/	<z>	*bħaz-*	*għaz-*	*liz-*	*miz-*
/ʧ/	<ċ>	*bħaċ-*	*għaċ-*	*liċ-*	*miċ-*
/s/	<s>	*bħas-*	*għas-*	*lis-*	*mis-*
/z/	<ż>	*bħaż-*	*għaż-*	*liż-*	*miż-*
/ʃ/	<x>	*bħax-*	*għax-*	*lix-*	*mix-*
/n/	<n>	*bħan-*	*għan-*	*lin-*	*min-*
/r/	<r>	*bħar-*	*għar-*	*lir-*	*mir-*
/l/	<l>	*bħal-*	*għal-*	*lil-*	*mil-*

The different combinations are exploited quantitatively to widely different extents as becomes clear from a comparison of the absolute token frequencies in the Korpus Malti 3.0 given in Table 3.25. The presentation of the data is ordered as follows. The ART-consonants are ordered top-down according to the decreasing token frequency. The Πs are ordered from left to right also according to the decreasing number of matches. Cells are shaded grey if they host the highest number of tokens in a given column. Yellow is indicative of the highest quantity within a row. Where the number of tokens is the highest for both column and row the colour green is used. Green and yellow show up only in the domain of (Π21) *għal* 'for' and (Π41) *minn* 'from' which boast the two highest token frequency values. Together they account for 83% of all cases of fusion identified in the Korpus Malti 3.0.

Table 3.25: Token frequencies for fused consonant-final Πs and ART in the Korpus Malti 3.0.

ART	**Π**				**sum**
	għal	***minn***	***lil***	***bħal***	
<s>	128,680	88,019	39,185	5732	261,616
<t>	142,998	79,127	31,419	2,930	256,474
<d>	68,622	92,328	25,281	1,957	188,188
<n>	26,699	110,898	25,315	797	163,709
<r>	8,4748	47,611	14,868	1,279	148,506
<l>	32,822	43,737	5,991	31,106	113,656

Table 3.25 (continued)

ART	П				sum
	għal	*minn*	*lil*	*bħal*	
<x>	31,550	18,536	3,249	583	53,918
<ż>	21,618	15,840	9,484	497	47,439
<ċ>	11,303	14,002	8,570	774	34,649
<z>	613	577	471	135	1,796
total	549,653	510,675	163,833	45,790	1,269,951

The four top-ranking alveolar obstruents resulting from assimilation cover 69% of the instances of fusion. The liquids /r/ and /l/ together are responsible for 21% whereas the remaining s(h)ibilants and especially the affricates make do with much smaller shares. These differences are to some extent explicable with reference to the different numbers of nouns in the Maltese lexicon which take their initial segment from this array of consonants.

The assimilation to the ART-consonant requires a thorough modification of the rule presented in Figure 3.67. This modification results in the formula given in Figure 3.72.

C] + [X → X / X = [+coronal]

Figure 3.72: Assimilation to coronal ART-consonant.

The process can be interpreted in different ways. One possibility is to assume that the final consonant of the Π is deleted so that no assimilation takes place at all. An alternative assumes that after assimilating to the coronal ART-consonant the final consonant of the Π coalesces with the ART-consonant and the resulting geminate is degeminated in turn. We are not committed to any of the options although the former appeals to us because of its simplicity. If we spell out this interpretation those Πs whose input form ends in a consonant have a special allomorph for fusion with coronal resulting from assimilation, namely *bħa-*, *għa-*, *li-*, and *mi-* – all of them deprived of the final segment of the citation form. A further possibility which we will not argue for in this study is that Π30 *ma'* 'with' and Π55 *ta'* 'of' loose their final *għajn* under fusion and thus behave similarly to the above class of Πs.

The ten examples (3.149)–(3.158) are meant to illustrate the process circumscribed in Figure 3.72 for each of the consonant-final Πs and each of the coronal consonants so that some Πs are represented more than once in this set of examples.

(3.149) [Gauci 1994. 53]
Iżda mbagħad ġew miżmuma draw
but then come:3PL.PFV PART:hold:F draw
*[**mi**t-team modest tal-Kolumbja]*$_{PP}$
[**from**:DEF-team modest of:DEF-Columbia]
'Then, however, a draw was defended [**by** the mediocre team of Columbia].'

(3.150) [Attard 2019: 15]
*ma deherx li kien wieħed sinjifikanti [**għa**d-dinja]*$_{PP}$
NEG appear:NEG that be one significant [**for**:DEF-world]
'[. . .] it did not appear to be a significant one [**for** the world].'

(3.151) [Korpus Malti 3.0; news117.883]
*Waħda minnhom baqgħet ħelwa [**bħa**z-zokkor]*$_{PP}$
one:F from:3PL remain:3SG.F.PFV sweet:F [**like**:DEF-sugar]
'One of them remained sweet [**like** sugar].'

(3.152) [Korpus Malti 3.0; culture2223]
Iċ-ċikkulata fiha biss 38 kalorija u għaldaqstant
DEF-chocolate in:3SG.F only 38 calory and for_this_reason
*ma tħaxxinx [**bħa**ċ-ċikkulata normali]*$_{PP}$
NEG 3SG.F.IPFV:CAUS.become_thick [**like**:DEF-chocolate normal]
'This chocolate only has 38 calories and thus does not make you fat [**like** normal chocolate].'

(3.153) [Camilleri 2013: 209]
*tat xeħta iktar surreali [**li**s-sitwazzjoni]*$_{PP}$
give:3SG.F.PFV throw more surreal [**to**:DEF-situation]
'[. . .] it gave a more surreal touch [**to** the situation].'

(3.154) [Bartolo/Vella 2009: 102]
*se ssibuh [**li**ż-żgħir tagħna]*$_{PP}$
FUT 2.IPFV:find:PL:3SG.M [**to**:DEF-little of:1PL]
'[. . .] are you going to find [our little one]?'

(3.155) [Attard 2019: 211]
*Iddeċieda li jidħol [**għa**x-xogħol]*$_{PP}$
decide that 3SG.M.IPFV:enter [**for**:DEF-work]
'He decided to got [**to** work].'

(3.156) [Zammit 1977. 122]

għalhekk kellu miegħu lill-ispiżjar u
therefore have.PFV:IO:3SG.M with:3SG.M to:DEF-pharmacist and
*[**li**n-Nutar]*$_{PP}$
[**to**:DEF-notary]

'[. . .] he therefore had the pharmacist and [the notary] with him [. . .].'

(3.157) [Korpus Malti 3.0; academic107]

*l-għana kien jiġi kantat [**mi**r-raħħala*
DEF-folk_song be 3SG.M.IPFV:come sing:PART [**from**:DEF-villager:PL
u ħaddiema komuni]$_{PP}$
and worker:PL common:PL]

'[. . .] the folk song was being sung [**by** common villagers and workers].'

(3.158) [Korpus Malti 3.0; academic105]

r-regoli kienu jvarjaw kemm mill-perjodi
DEF-rule:PL be:3PL.PFV 3.IPFV:vary:PL how_much from:DEF-period:PL
*kif ukoll [**mi**l-lokazzjoni]*$_{PP}$
how also [**from**:DEF-location]

'[. . .] the rules varied not only between periods of time but also [**between** locations].'

We have reached the terminus of the empirical illustration of those phenomena which are connected to the morphology and morphonology of the Maltese ΠS. Section 3.4 summarises the major findings of the foregoing sections and contains an update of the frequency count of the members of BLOMP 2.0.

3.4 Allomorphy revisited

Looking back on the morphology, morphonology, and the external sandhi phenomena which Maltese ΠS are subject to, we notice that the relatively densely populated C Ia constitutes the only case of a sizable class. There is no example of the emergence of further classes whose membership is as numerous as that of the major inflectional class. This statement does not only hold for inflectional classes as such but also extends to the competition between long and short forms, segmental reduction, procliticisation, and fusion. In Table 3.26, we survey those alloforms which deviate from the citation forms of the ΠS. The ΠS are ordered alphabetically whereas the alloforms are presented chronologically, i.e. they follow the order of the sections in which the phenomena have been discussed in this part of the study.

Table 3.26: Survey of alloforms.

Π	inflection	short/long	reduced	procliticised	fused
(Π4) *bejn* 'between'			*bej*		
(Π5) *bħal* 'like'					*bħa-*
(Π7) *bi* 'with'				*b'*	*b*
(Π13) *favur* 'in favour of'	*favuri-*	*a favur* ~ *(għ) affavur*			
(Π15) *fi* 'in'				*f'*	*f-*
(Π21) *għal* 'for'	*għali-*				*għa-*
(Π23) *ħdejn* 'beside'	*ħdej-*		*ħdej-*		
(Π27) *kontra* 'against'	*kontri-*				
(Π28) *lejn* 'towards'	*lej-*		*lej-*		
(Π29) *lil* 'to'			*'il* ~ *'l*		*li-*
(Π30) *ma'* 'with'	*magħ-* ~ *miegħ-*			*m'*	
(Π33) *matul* 'during'		*tul*			
(Π34) *minbarra* 'except'		*mbarra*			
(Π35) *minflok* 'instead of'		*flok*			
(Π37) *mingħala* 'in s.o.'s opinion'	*mingħali-*				
(Π39) *minħabba* 'on account of'		*mħabba* ~ *ħabba*			
(Π2) *a skapitu* 'at the expense of'		*għas-skapitu* ~ *bi skapitu* ~ *għal skapitu*			
(Π41) *minn* 'from'					*mil-*
(Π51) *sa* 'till'				*s'*	
(Π55) *ta'* 'of'	*tagħ-* ~ *tiegħ-*			*t'*	*t-*
(Π60) *wara* 'after'	*waraj-*				

There are twenty-one Πs which give evidence of allomorphy. The highest number of allomorphs – including the citation form – is four. This range of allomorphy applies in the cases of (Π2) *a skapitu* 'at the expense of', (Π29) *lil* 'to', (Π30) *ma'* 'with', and (Π55) *ta'* 'of'. Several allomorphs outside the domain of genuine inflection are used only optionally. The favourable conditions for their employment are mostly phonological. An especially strong factor, which contributes to the optional use of certain allomorphs, is the uncertain status of *għajn*. We are not absolutely sure whether the patterns we have discovered in connection to this abstract phoneme can be mapped onto utterances in the spoken register.

It has turned out that the members of the small inflectional classes C II and CIIIc behave in identical ways also outside the realm of inflection in the sense that they follow the same rules under external sandhi. Their parallel behaviour in different contexts lends support to our original classification. Further parallels can be found among a subset of the inflected consonant-final Πs. More generally, morphonology affects only 34% of the members of BLOMP 2.0. The picture is far from being homogeneous. The rules we have put forward in the previous sections more often than not only cover a small number of Πs.

The different susceptibility of Maltese Πs to inflection and other morpho(no) logical forces has an impact on the token frequency. In Table 1.11, the estimated token frequency of the free forms of the Πs was calculated for the entire Korpus Malti 3.0. We are now in a position to revise the count by way of accounting also for the occurrences of inflected forms, short allomorphs, reduced allomorphs, procliticised, and fused forms of Πs (i.e. the values are not based on the 1st 100-test). The quantitative data are revealed in Table 3.27. Column A gives the new rank position whereas B reminds us of the old rank position in Table 1.11. Grey shading marks out those Πs whose rank position has remained the same. For Πs which have climbed upwards in the hierarchy the colour yellow is used. The unmarked rows host Πs which have lost their former higher position. The new total amounts to 31,583,068 which means that the average token frequency per Π is 526,380. The first ten positions are filled by Πs which exceed the average (marked by an extra-fat line).

Table 3.27: Revised ranking order according to token frequencies in the Korpus Malti 3.0.

A	B	Π	free	inflection	short/ long	reduced	procliticised	fused	sum
1	1	(Π55) *ta'* 'of'	4,943,450	1,252,814			37566	3634883	9868713
2	9	(Π15) *fi* 'in'	319,840	173,887			1581446	3634883	5710056
3	2	(Π41) *minn* 'from'	1,131,110	192,515				1726044	3049669
4	4	(Π21) *għal* 'for'	995,448	164,576				1669074	2829098
5	16	(Π7) *bi* 'with'	141,102	63,282			951261	767665	1923310
6	6	(Π29) *lil* 'to'	388,639	69,022		8591		1087273	1553525
7	3	(Π17) *fuq* 'on'	1,071,352	85,647					1156999
8	7	(Π30) *ma'* 'with'	337,931	119,088				501101	958120
9	17	(Π5) *bħal* 'like'	131,013	14,200				501101	646314
10	5	(Π11) *dwar* 'about'	579,834	50,899					630733
11	8	(Π6) *bħala* 'as'	328,932						328932

Table 3.27 (continued)

A	B	Π	free	inflection	short/ long	reduced	procliticised	fused	sum
12	20	(Π51) *sa* 'till'	101,713				39625	122173	263511
13	10	(Π60) *wara* 'after'	248,801	13,918					262719
14	11	(Π4) *bejn* 'between'	236,899	20,123		2			257024
15	12	(Π56) *taħt* 'under'	187,527	14,422					201949
16	13	(Π28) *lejn* 'towards'	171,669	16,778		4			188451
17	14	(Π27) *kontra* 'against'	167,406	8,405					175811
18	15	(Π33) *matul* 'during'	149,416	6,487	15765				171668
19	21	(Π16) *fost* 'amongst'	92,813	57,149					149962
20	18	(Π39) *minħabba* 'on account of'	127,568		1392				128960
21	19	(Π45) *permezz* 'by means of'	115,212						115212
22	26	(Π46) *qabel* 'before'	66,657	12,334					78991
23	25	(Π36) *mingħajr* 'without'	70,663	2,854					73517
24	29	(Π49) *quddiem* 'in front of'	47,928	29,817					77745
25	22	(Π50) *rigward* 'concerning'	73,009						73009
26	23	(Π3) *barra* 'outside'	72,511						72511
27	24	(Π9) *bla* 'without'	72,054						72054
28	27	(Π53) *skont* 'according to'	65,511	306					65817
29	28	(Π13) *favur* 'in favour of'	53,276	954	246				54476
30	34	(Π35) *minflok* 'instead of'	29,667	5,640	8401				43708

Table 3.27 (continued)

A	B	П	free	inflection	short/ long	reduced	procliticised	fused	sum
31	30	(П40) *minkejja* 'in spite of'	42,986						42986
32	33	(П38) *mingħand* 'from s.o.'	32,167	4,395					36562
33	35	(П31) *madwar* 'around'	24,911	10,833					35744
34	31	(П22) *għand* 'at s.o.'s place'	32,965	1,746					34711
35	32	(П59) *waqt* 'at the time of'	32,360	2,220					34580
36	40	(П19) *ġo* 'in'	17,759					6357	24116
37	36	(П34) *minbarra* 'except'	21,035		23				21058
38	37	(П18) *ġewwa* 'inside'	20,844						20844
39	38	(П25) *inkluż* 'including'	20,571						20571
40	39	(П48) *qrib* 'near'	19,687	413					20100
41	41	(П23) *ħdejn* 'beside'	16,489	3,583		1			20073
42	42	(П1) *apparti* 'apart from'	16,032						16032
43	46	(П47) *qalb* 'amidst'	6,059	6,776					12835
44	44	(П10) *daqs* 'equal to'	10,864	1,836					12700
45	43	(П24) *ħlief* 'except'	11,152	194					11346
46	45	(П58) *viċin* 'near'	8,878						8878
47	47	(П20) *għajr* 'except'	5,263	434					5697
48	48	(П14) *fejn* 'near'	4,893	412					5305
49	56	(П37) *mingħala* 'in s.o.'s opinion'	1	4,965					4966

Table 3.27 (continued)

A	B	Π	free	inflection	short/ long	reduced	procliticised	fused	sum
50	49	(Π8) *biswit* 'facing'	2,603	188					2791
51	52	(Π32) *maġenb* 'close to'	1,702	930					2632
52	50	(Π44) *oltre* 'beyond'	2,238						2238
53	51	(Π12) *faċċata* 'opposite'	1,927						1927
54	53	(Π2) *a skapitu* 'at the expense of'	1,168	190					1358
55	54	(Π52) *sforz* 'thanks to'	231						231
56	55	(Π57) *versu* 'towards'	223						223
57	57	(Π26) *kif* 'as'	0						0
58	58	(Π43) *mnejn* 'from near'	0						0
59	59	(Π42) *mintul* 'all along'	0						0
60	60	(Π54) *sotta* 'under	0						0
	total		12,873,959	24,14,232	25,827	8,598	2,609,898	13,650,554	31,583,068

The most remarkable cases of promotion in the ranking order are (Π7) *bi* 'with' with eleven ranks, (Π5) *bħal* 'like' and (Π51) *sa* 'till' with eight ranks each, (Π15) *fi* 'in' with seven ranks, and (Π37) *mingħala* 'in s.o.'s opinion' equally with seven ranks. The latter Π stands out additionally because it not only represents one of two cases where the inflected forms outnumber the free forms in terms of tokens but for (Π37) *mingħala* 'in s.o.'s opinion', the inflected forms account for 99% of all attestations in the Korpus Malti 3.0. The isolated case of (Π37) *mingħala* 'in s.o.'s opinion' taking a complement other than a pronominal one can be ignored for the discussion of the syntactic properties of this Π.[82]

82 Cf. Section 4.1.2.1.

The different categories referred to in Table 3.27 contribute differently to the total of Π-tokens in the Korpus Malti 3.0. It results from Figure 3.73 that the bulk of the tokens consists of free forms and fused forms which together account for 84% of all Π-tokens.

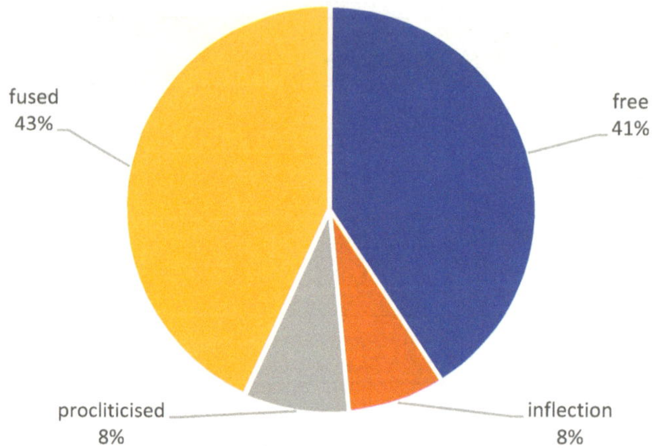

Figure 3.73: Shares of morphological categories.

The 8%-share of prociliticised forms is based almost in its entirety on the occurrences of the short allomorphs of the two members of C II. Inflected forms of Πs have the same share as the procliticised forms, namely 8%. This humble share notwithstanding, the inflected Πs will necessitate our full attention twice again when we address the issue the patterns of government of certain Πs, respectively.[83] Not before we take up these issues in the subsequent sections do we disclose the token frequencies of the different person and number values of the inflected Πs. Short/long and reduced forms are by far too infrequent to reach the 1%-mark and are thus not displayed in Figure 3.73.

We close the review of the morphological properties of Maltese Πs by way of repeating that one cannot treat the morphologically variable members of BLOMP 2.0 like a structural monolith. What we have seen in the previous paragraphs is a heavy dose of heterogeneity with only relatively small subsets of the Π-inventory displaying similar properties. It is of course not true that each Maltese Π behaves absolutely individually in the morphological domain. Yet, there is a plethora of parameters on which several Πs deviate in different ways from other Πs. It is one of the tasks of Section 4 to determine whether this tendency towards (limited) individuality also holds in the domain of morphosyntax.

83 Cf. Section 4.1.2.2.2.

4 Morphosyntax

This section is dedicated to the description of the morphosyntactic behaviour of Maltese Πs. Properly morphological and morphonological phenomena have been the topic of the previous Section 3. As to the semantic factors which are relevant for explaining the distribution of the Πs, Section 5 will provide ample information. In what follows below, we focus on the syntagmatic interaction of Πs and other units on the phrasal, clausal, and sentential levels without recapitulating the already discussed effects of external sandhi on the segmental chains of some of the members of BLOMP 2.0 or anticipating the particulars of their semantics. To this end, we take the PP itself as our starting point in Section 4.1 where the relation of Πs and their different classes of complements is the central issue. The behaviour of attributive Πs within the NP (including relative clauses) is scrutinised in Section 4.2. The topic of Section 4.3 is the behaviour of Πs in VPs whereas Section 4.4 wraps up the insights gained in the course of the discussion in Sections 4.1–4.3.

4.1 Inside the PP

All those Maltese Πs whose existence is confirmed in the Korpus Malti 3.0 can take syntactically light complements in the sense of single word units. Except for the notoriously doubtful Πs, we have not discovered any differences across BLOMP 2.0 in terms of the complexity of the complement the Πs can combine with. More generally, there seems to be no upper limit as to the size of complements. Multi-word complements are widely common including those involving multiple embedding. In (4.1), we provide an example of a complex complement NP. Further examples of complement complexity can be found throughout this study.

(4.1) [Zammit 1977: 73]

ftakar	*[fil-landa*	*mimlija*	*liri*	*ta'*	*Grigor]$_{PP}$*
<RFL>remember	[**in**:DEF-tin:F	full:F	pound:PL	of	Grigor]

'[. . .] he remembered [Grigor's tin full of pounds].'

The complement noun *landa* 'tin' does not only host the definiteness pro-clitic but is modified by a complex attribute which consists of the adjective *mimlija* 'full' which is itself modified by the common noun *liri* 'pounds' which in turn has a GENITIVAL attribute *ta' Grigor* 'of Grigor'. Despite the number of syntactic dependencies which relate the constituents of the complement to each other, the case in (4.1) is not a particularly extreme one as can be gathered from other examples such as (4.7).

https://doi.org/10.1515/9783111331591-004

Looking further into the issue of complement complexity makes sense when we address compound Πs and EQUI-P-deletion below.[1] For the time being, however, it suffices to state that no Maltese Π is principally excluded from syntactic contexts which involve complex complements. This compatibility of Maltese Πs with complex complements presupposes their general combinability with simple complements – a property which no longer needs to be proved since there is ample evidence of it in the previous sections.

There are however other properties of complements which have repercussions on the choice of Π and/or the shape of PP. With a view of preparing the ground for the discussion of the government properties of Maltese Πs[2] we first have to eliminate a potential problem arising in connection with (Π9) *bla* 'without' because this Π behaves differently from the bulk of the members of BLOMP 2.0 or so it seems. The clarification of this issue and its entailments will be attempted in the next section.

4.1.1 Definiteness

The definiteness of complements has not been studied in-depth so far in the literature dealing with Maltese Πs. The issue is usually raised indirectly in connection with the fusion of certain Πs with the definite pro-clitic and the accompanying morphonological processes.[3] There is however more to the topic than the usual focus on morphonology makes us believe. In what follows, we try to determine whether there are constraints which affect the combinability of Πs with (in)definite complements (independent of the presence/absence of the definite pro-clitic). This endeavour is justified because, in the extant literature, the illustration of the properties of Maltese PPs in syntax is skewed in the sense that in the function of complement, definite common nouns outnumber by far indefinite complements and proper names in the same function. An informal cursory look at the exemplification of PPs in two of our secondary sources suffices to prove our point.[4] In

1 In Sections 4.1.3–4.1.4.

2 Cf. Section 4.1.2.

3 As discussed in Sections 3.3.3–3.3.4.

4 For this informal count, we have accepted all those PPs which are featured in Borg/Azzopardi-Alexander (1997) in sections which are dedicated to PPs as a constituent or address functions which are predominantly fulfilled by Πs in Maltese. In the case of Saari (2003), only those sentential examples have entered the check which contain a PP marked by underlining – which is Saari's way of highlighting Πs whose properties are discussed in the paragraphs immediately preceding or following the example. Only those Πs have been accounted for which also belong to BLOMP 2.0.

the relevant paragraphs of Borg/Azzopardi-Alexander (1997), we find 125 tokens of PPs. This total comprises 102 PPs taking definite common nouns as complements (= 81.6%) whereas indefinite common nouns are attested only 19 times in this function (= 15.2%) and proper names even less frequently with three attestations (= 3.2%). As to Saari (2003), the turnout is similar. Of the total of 139 PPs, 101 (= 72.6%) involve definite nouns as complements, 29 (= 20.8%) PPs take an indefinite common noun as complement and proper names as complements are attested only in nine (= 6.4%) cases. If those studies which pay Πs considerable attention prefer to illustrate Π-related syntactic phenomena in combination with definite complements one might ask whether this overrepresentation of definite complements is purely incidental.[5] To approach this issue, we formulate the (impressionistic) hierarchy in Figure 4.1 based on the above informal check of Borg/Azzopardi-Alexander (1997) and Saari (2003).

definite common nouns > indefinite common nouns > proper names

Figure 4.1: Hypothetical frequency hierarchy of complement types.

The formula postulates that the complement type to the left of the arrow is attested more frequently than the complement type to the right of the same arrow in the Korpus Malti 3.0. We use this section to put the hierarchy to the test. As a rule of thumb, Πs which display preferences for complements which deviate from the hierarchy in Figure 4.1 call for an explanation whereas those Πs which conform to the hierarchy can be waved through without further discussion. To this end, it is necessary to ideally account for the behaviour of each Π. The examples to be presented below additionally serve as preparation for the detailed discussion of the government classes of Maltese Πs below.[6]

4.1.1.1 (Π9) *bla* 'without' and (Π36) *mingħajr* 'without'

There are two Πs with an identical lexical meaning corresponding to English *without*, namely (Π9) *bla* 'without' and (Π36) *mingħajr* 'without'.[7] In spite of supposedly being synonymous the two Πs display strikingly different morphosyntactic behaviour. The major difference between (Π9) *bla* 'without' and (Π36) *mingħajr*

5 The domain of ART in Maltese is comparatively big since "[i]n addition to prototypical count nouns the nouns that require a definite article in definite noun phrases include those denoting substances [. . .], colours [. . .], and abstract notions" (Plank/Moravcsik 1996: 190). ART is also used for generic reference so that the likelihood of a Π meeting a noun that hosts the ART is relatively high.

6 Cf. Section 4.1.2.

7 The semantic side of this case is taken up again in Section 5.1.3.2.

'without' is the questionable possibility of the former to take a definite complement no matter whether definiteness is expressed overtly or remains implicit (Saari 2003: 139–140). Aquilina (1987: 128–129) lists numerous uses of (Π9) *bla* 'without' none of which features the Π as head of a PP whose complement is definite. The same picture results from a look at the entry for (Π9) *bla* 'without' in E. Serracino-Inglott (1975a: 177). Similarly, Borg/Azzopardi-Alexander (1997: 149–151) and Saari (2003: passim) only give examples of (Π9) *bla* 'without' combining with indefinite nouns. In none of the extant descriptions of Maltese is the morphosyntactic behaviour of (Π9) *bla* 'without' discussed at any noticeable length. Whether or not it can take definite complements at all remains an open question although the examples provided in the secondary sources do not contain a single instance of ***bla*** + NP_{DEF} whereas there is a plethora of instantiations of ***bla*** + NP_{INDEF}. This turnout is unsurprising since (Π9) *bla* 'without' likewise cannot be inflected pronominally (see (3.2) above) nor can it combine with free pronouns be they personal or demonstratives. It is therefore justified to hypothesise that the use of definite complements is generally blocked for (Π9) *bla* 'without'.

In contrast, (Π36) *mingħajr* 'without' is not subject to any comparable constraints as transpires from examples (4.2)–(4.5).

(4.2) [Bartolo/Vella 2009: 125]
*[**Mingħajr** din l-informazzjoni]$_{PP}$ il-pjan ta' Kristjan*
[**without** DEM:F:PROX DEF-information] DEF-plan of Kristjan
żgur li ma kienx jasal sa fejn wasal.
sure that NEG be:NEG 3SG.M.IPFV:arrive till where arrive
'[**Without** this information] Kristjan's plan would surely not have worked out as it did.'

(4.3) [Bartolo/Vella 2009: 144]
*issa [**mingħajr**hom]$_{PP}$ kien iħossu biss raġel mitluf*
now [**without**:3PL] be 3SG.M.IPFV:feel:3SG.M only man PART:lose
'[. . .] now [**without** them] he felt himself to be but a lost man [. . .].'

(4.4) [Korpus Malti 3.0; academic425]
*kienu qed jilgħabu [**mingħajr** Roy Keane]$_{PP}$*
be:3PL.PFV PROG 3.IPFV:play:PL [**without** Roy Keane]
'[. . .] they were playing [**without** Roy Keane] [. . .].'

(4.5) [Camilleri 2013: 123]

Meta ghallimha tghum ***[minghajr*** *flippers]*$_{PP}$ *fil-bahar*
when teach:3SG.F 3SG.F.IPFV:swim **[without** flipper:PL] in:DEF-sea
jobrom ta' Hondoq ir-Rummien
3SG.M.IPFV:turn of Hondoq ir-Rummien
'When he taught her to swim [**without** flippers] in the agitated waters of Hondoq ir-Rummien.'

The first three examples illustrate the combination of (Π36) *minghajr* 'without' with definite complements. The complement NP in (4.2) is doubly marked for definiteness since there is not only the definiteness pro-clitic but also the proximal demonstrative. In (4.3), the Π takes a suffixed pronominal complement which is inherently definite. Inherent definiteness also applies in the case of the person name as complement of (Π36) *minghajr* 'without' in (4.4). In contrast to these examples, the complement of the same Π in (4.5) is a bare common noun which, in the absence of indefinite articles in Maltese, is interpreted as indefinite (Fabri 1993: 42). Thus, (Π36) *minghajr* 'without' is not sensitive to definiteness distinctions.

According to our above hypothesis, however, (Π9) *bla* 'without' could replace (Π36) *minghajr* 'without' only in (4.5) since it is assumed to be excluded from combinations with definite complements. We repeat that there are no pronominal forms of (Π9) *bla* 'without'. Typically, (Π9) *bla* 'without' is attested in contexts such as that shown in (4.6).

(4.6) [Camilleri 2013: 157]

Jekk trid li naqbdu l-qattiela ta' żewġek
if 2SG.IPFV:want that 1.IPFV:catch:PL DEF-murderer of husband:2SG
mhux ahjar tghidilna ***[bla*** *telf ta' żmien]*$_{PP}$
NEG:3SG.M:NEG better 2SG.IPFV:say:IO:1PL **[without** loss of time]
fejn kien jahbihom il-karti?
where be 3SG.M:hide:3PL DEF-paper:PL
'If you want us to catch the murderer of your husband would it not be better to tell us [**without** delay] where he hides the papers?'

What contradicts the assumed incompatibility of definite complements and (Π9) *bla* 'without' is (4.7) which involves the sole case of this Π combining with a definite NP in our additional texts from literary Maltese.

(4.7) [Camilleri 2013: 290]

J.J. Theuma qatta' l-bqija ta' jiemu f' siġġu
J.J. Theuma spend_time DEF-rest of day.PL:3SG.M in chair
*tar-roti [**bla** l-ġieħ għall-karriera li tant*
of:DEF-wheel:PL [**without** DEF-honour for:DEF-career that so_much
ħolom dwarha]$_{PP}$
dream about:3SG.F]
'J.J. Theuma spent the rest of his days in a wheel-chair [**without** the honour of the career he dreamed so much about].'

What violates the incompatibility hypothesis is the definite noun *(i)l-ġieħ* 'the honour' adjacent to the Π under inspection. Is the hypothesis absolutely invalid? We argue that the hypothesis is not entirely wrong.

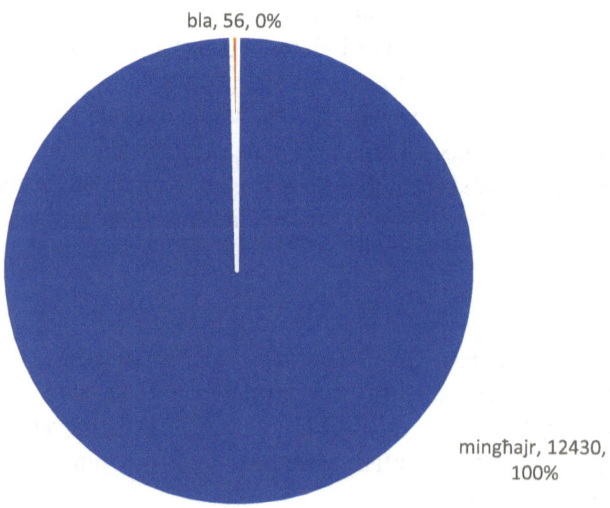

Figure 4.2: Shares of (Π9) *bla* 'without' and (Π36) *mingħajr* 'without' of privative/abessive PPs with definite complements.

Our refusal to accept defeat in this case rests on the following facts. According to Table 3.27, (Π9) *bla* 'without' and (Π36) *mingħajr* 'without' have relatively similar token frequencies in the Korpus Malti 3.0 with 72,054 hits and 73,517 hits, respectively. This quantitative similarity does not hold for the combinations of these Πs with definite complements. We have searched the Korpus Malti 3.0 for combinations of (Π9) *bla* 'without' and (Π36) *mingħajr* 'without' with complements whose initial element is the definite pro-clitic. With only 56 hits for (Π9) *bla* 'without' in this context as opposed to 12,430 hits for its synonymous competitor, it is absolutely clear that (Π9) *bla* 'without' is not the preferred option when it comes to expressing privative/abessive relations with definite complements. According to Figure 4.2, (Π36) *mingħajr* 'without' ousts (Π9) *bla* 'without' by a ratio of 223-to-one in the context under inspection.

As to (Π9) *bla* 'without', the share is as low as 0.4% and thus almost negligible. Definite complements in turn, are attested only marginally (0.005%) among all *bla*-tokens in the Korpus Malti 3.0 whereas their share reaches 16.9% in the case of (Π36) *mingħajr* 'without' as results from Figure 4.3.

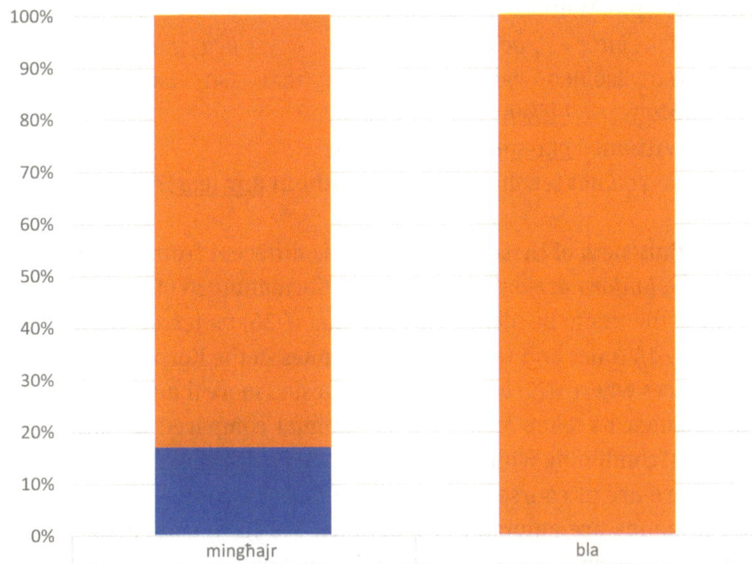

	mingħajr	bla
other	61087	71998
definite complement	12430	56

Figure 4.3: Definite complements in PPs per privative/abessive Π.

In both Figure 4.2 and Figure 4.3, the existence of *bla* + NP_{DEF} is almost invisible. In this case, *bla* + NP_{INDEF} is the uncontested majority option. In contrast, definite complements constitute a noticeable minority option for PPs headed by (Π36) *mingħajr* 'without'. Moreover, (Π36) *mingħajr* 'without' is attested in combination with all alloforms of the definite pro-clitic whereas evidence of *bla* + NP_{DEF} is limited to combinations with the lateral approximant /l/ (= <l>) (54 tokens out of 56) and the voiceless postalveolar fricative /ʃ/ (= <x>). This restriction speaks against the free availability of (Π9) *bla* 'without' as filler of the head slot in PPs with definite complements. The bulk of the cases of *bla* + NP_{DEF} (34 instances) found in the Korpus

Malti 3.0 involves the prenominal negation *ebda* 'no' (21 tokens as in (4.8)) and the comparative *iċken* of the adjective *ċkejken* 'small' which frequently occupies the position to the left of the noun it modifies (thirteen tokens as in (4.9)).

(4.8) [Korpus Malti 3.0; literature56]

Soltu	*kien*	*jirnexxilu*	***[bla***	*l-ebda*	*sforz]$_{PP}$*
usually	be	3SG.M.IPFV:succeed:IO:3SG.M	**[without**	DEF-no	effort]

'Usually he succeeded [**without** any effort].'

(4.9) [Korpus Malti 3.0; news65246]

M'	*intix*	*qed*	*tara*	*li*	*jien*	*nirkeb*	*fuqu*
NEG	2SG:NEG	PROG	2SG.IPFV:see	that	1SG	1SG.IPFV:mount	on:3SG.M

[bla	*l-iċken*	*biża']$_{PP}$?*
[without	DEF-small.CMPR	fear]

'Do you not see that I board it [**without** any fear]?'

The definiteness of these complements is different from for instance, that of *min-għajr din l-informazzjoni* 'without this information' in (4.2) because there is no referent of the NPs to be identified. Note that (Π36) *mingħajr* 'without' co-occurs with *l-ebda* 1,619 times and with *l-iċken* 205 times in the Korpus Malti 3.0. Thus, even in those cases where (Π9) *bla* 'without' is attested more than once with a given definite complement its token frequency is minimal compared to that of (Π36) *mingħajr* 'without' combining with the same definite complement.

There are of course other kinds of definite complements. Pronominally possessed nouns are automatically definite. We have compared the occurrences of the kinship-terms *missier* 'father' and *omm* 'mother' hosting the possessor suffixes and functioning as complement of the two privative/abessive Πs. As results from Table 4.1, the turnout is small since the search in the Korpus Malti 3.0 yielded only thirteen hits. It is symptomatic however, that twelve tokens (= 92.3%) involve (Π36) *mingħajr* 'without' and only one (= 7.7%) (Π9) *bla* 'without'.

Table 4.1: Partitive/abessive Πs + possessed kinship terms in the Korpus Malti 3.0.

text	example	POSSESSOR
news91430	*din ma riditx tmur* ***mingħajr missierha*** 'she did not want to go **without her father**'	3SG.F
news74871	*Iżda tweġġagħni li naf li ħuti għandhom il- maġġoranza ta' ħajjithom* ***mingħajr missierna*** 'It hurt me to know that my brothers lived most of their life **without our father**.'	1PL

Table 4.1 (continued)

text	example	POSSESSOR
opinion608	*It-tfal tiegħu ma jifilħux jistennew minuta oħra* ***mingħajr missierhom***. 'His children could not wait another wait another minute **without their father**.'	3PL
news59080	*imma jien kif noħroġ minn hawn irrid inkompli nħabbat wiċċi ma' realtà kiefra li ngħix* ***mingħajr ommi*** *l-għażiża*. '[. . .] but on coming out of here I must continue to face the cruel fact that I live **without my** beloved **mother**.'	1SG
news34807 + news92151	*jgħidli li ma kienx se jgħix* ***mingħajr ommu*** '[. . .] he told me that he would not live **without his mother** [. . .].'	3SG.M
parl8313	*min jiċħad 'l ommu m' għandux dritt jgħix, għax* ***mingħajr ommu*** *ma kienx ikun hawn hu* '[. . .] who denies his mother has no right to live because **without his mother** he would not be here [. . .].'	3SG.M
news112612	*It-tifla qalet li l-ħajja fid-dar* ***mingħajr ommha*** *hija ħafna aħjar minn meta kienet tkun id-dar.* 'The girl said that the life in the house **without her mother** is much better than when she would be at home.'	3SG.F
news124223	***mingħajr ommhom*** *żgur ma kinux jagħmluha* '[. . .] **without their mother** they would certainly be unable to manage [. . .].'	3PL
news43518	*Hija qalet li l-ħajja d-dar* ***mingħajr ommhom*** *hija ferm aħjar.* 'She said that life at home **without their mother** is much better.'	3PL
news97722 + news97723	***mingħajr ommhom*** *dawn jispiċċaw imutu bil-ġuħ fil-bejta wkoll.* '[. . .] **without their mother** they too will starve to death in the nest.'	3PL
news69810	*Ma tistħix li ibnu ta' taħt l-għaxar snin kellu jgħaddi* ***bla missieru*** *għal ħamsa minnhom?* 'Isn't she ashamed that his son who is not yet ten years old had to pass five thereof **without his father**?'	3SG.M

Only the grey shaded example at the bottom of Table 4.1 involves (Π9) *bla* 'without' in combination with a possessed kinship term. (Π36) *mingħajr* 'without' does not only combine more frequently with possessed kinship terms but it allows for combinations with different possessors, too.

In addition, we have scrutinised the first 750 hits in the Korpus Malti 3.0. for each of the two Πs. The result is telling insofar as (Π36) *mingħajr* 'without' is attested in combination with five different proper names whereas the combination *bla* + PROPER NAME is unattested. The five cases involving (Π36) *mingħajr* 'without' are presented in Table 4.2.

Table 4.2: (Π36) *mingħajr* 'without' in combination with proper names (1st 750 hits in the Korpus Malti 3.0).

text	example
academic425	*Franza lagħbu* ***mingħajr Zinedine Zidane*** '[. . .] France played **without Zinedine Zidane**.'
academic425	*kienu qed jilgħabu* ***mingħajr Roy Keane*** '[. . .] they were playing **without Roy Keane** [. . .].' (see (4.4))
academic77	*marret lura Parigi fi stat miżeru u* ***mingħajr Moustaki****, li kien telaqha* '[. . .] she went back to Paris in a miserable state and **without Moustaki** who had left her.'
culture2886	*Kważi biljun u nofs persuna madwar id-dinja spiċċaw* ***mingħajr Facebook*** 'Almost a billion and a half people around the world were **without** (access to) **Facebook** [. . .].'
culture732	***Mingħajr Kristu*** *m'hemmx imħabba.* '**Without Christ** there is no love.'

In Table 4.3, we take up the example at the bottom of Table 4.2 by way of presenting all hits of the religious proper names *Ġesù* 'Jesus' and *Kristu* 'Christ' as complements of a privative/abessive Π in the Korpus Malti 3.0. This Π is always (Π36) *mingħajr* 'without' and never (Π9) *bla* 'without'.

Table 4.3: (Π36) *mingħajr* 'without' in combination with *Ġesù* 'Jesus' and *Kristu* 'Christ' in the Korpus Malti 3.0.

text	example
news51121	***mingħajr Ġesù*** *l-Knisja tkun biss NGO* '[. . .] **without Jesus** the Church will be only an NGO.'
news56880	*U mitlufa tassew għax maqtugħa għal rasha u* ***mingħajr Ġesù****.* 'And (she is) lost indeed because (she is) isolated on her own and **without Jesus**.'
news59423	*il-Milied qed jiġi ridott għal festa* ***mingħajr Ġesù*** '[. . .] Christmas is being reduced to a festivity **without Jesus** [. . .].'
news62480	*Fuq id-dgħajsa kien waħdu,* ***mingħajr Ġesù.*** 'On the boat, he was alone, **without Jesus**.'
culture732	***Mingħajr Kristu*** *m'hemmx imħabba.* '**Without Christ** there is no love.' (see Table 4.2)
news22502	*Għaldaqstant il- Knisja ma tistax teżisti* ***mingħajr Kristu.*** 'For this reason, the Church cannot exist **without Christ**.'
news135017	*Min hu Nisrani jiftaħar kemm bis-salib* ***mingħajr Kristu*** *fuqu u kemm bil- kurċifiss li hu s-salib bil-korp ta' Kristu fuqu.* 'Whoever is a Christian praises not only the cross **without Christ** on it but also the crucifix which is the cross with Christ's corpus on it.'

Table 4.3 (continued)

text	example
news158236	*jikkonċentraw biss fuq ir-realtajiet ta' din id-dinja* ***mingħajr Kristu!*** '[. . .] they concentrate only on the realities of this world without Christ!'
news65048	*Għax x'inhi l-Knisja* ***mingħajr Kristu*** *ħaj u rebbieħ!* 'Since what is the Church **without Jesus** alive and victorious.'
parl113	*anke* ***mingħajr Kristu*** *għandek il-Kristjaneżimu.* '[. . .] also **without Christ** you have Christianity.'
religion178	*iżda xejn minn dan ma kien jagħmel sens mingħajr iċ-ċelebrazzjoni ta' l- Ewkaristija ta' kuljum;* ***mingħajr Kristu*** *nnifsu.* '[. . .] but nothing of this makes sense without the daily celebration of the Holy Eucharist **without Christ** himself.'

On this basis, it can be hypothesised that (Π9) *bla* 'without' is excluded from combinations with proper names. It might be objected however, that the Korpus Malti 3.0 provides 24 tokens of (Π9) *bla* 'without' taking *Alla* 'God' as its complement as opposed to 55 tokens for the same complement in the case of (Π36) *mingħajr* 'without'. Two examples from one and the same text prove that the competing Πs occur in almost identical contexts, namely that of a common noun being modified by a prepositional attribute as shown in (4.10)–(4.11).

(4.10) [Korpus Malti 3.0; news58778]
ma naqbilx man-Nażi li kellhom ideoloġija ***[bla***
NEG agree:NEG with:DEF-Nazi that have.PFV:3PL ideology **[without**
Alla]$_{PP}$
God]
'[. . .] he did not agree with the Nazis who had a [godless] ideology.'

(4.11) [Korpus Malti 3.0; news58778]
Hija biss ideologija ***[mingħajr*** *Alla]*$_{PP}$ *li tista'*
3SG.F only ideology **[without** God] that 3SG.F.IPFV:can
tippjana u tesegwixxi l-qerda ta' poplu
3SG.F.IPFV:plan and 3SG.F.IPFV:execute DEF-annihilation of people
sħiħ
whole
'It was only a [godless] ideology that could plan and realise the annihilation of an entire people.'

If *Ġesù* 'Jesus' and *Kristu* 'Christ' are classified as proper names should not the same classification apply to *Alla* 'God' as well? The problem is tricky insofar as the proper-name status of *Alla* 'God' remains uncertain in a number of cases where an interpretation as generically employed common noun seems to make sense, too. This is probably the case in (4.10)–(4.11) because in these examples *Alla* 'God' does not necessarily refer to any particular supreme being such as the Christian God but metonymically invokes certain ethical and moral principles of human behaviour.

For the purpose of comparability with other Πs and their interaction with (in) definite complements, we make use of the already familiar 1st 100-test which consists in checking the occurrences of a given feature across the first one hundred hits of a Π in the Korpus Malti 3.0. In Figure 4.4, we show how often the two synonymous Πs focused upon in this section take definite common nouns, indefinite common nouns, and proper names as complements. The two Πs featured in Figure 4.4 already pose a challenge to the hypothetical hierarchy in Figure 4.1 since both (Π9) *bla* 'without' and (Π36) *mingħajr* 'without' favour indefinite common nouns over definite common nouns.

	bla	mingħajr
proper name	0	0
definite NP	0	25
indefinite NP	90	40

Figure 4.4: Shares of types of (in)definite complements with (Π9) *bla* 'without' and (Π36) *mingħajr* 'without' in the 1st 100 in the Korpus Malti 3.0.

Figure 4.4 will be referred back in the subsequent sections when we determine the combinability of other Maltese Πs with the above three complement types. Proper names as complements are absent from Figure 4.4 (but cf. Table 4.2). Indefinite common nouns are most important quantitatively but (Π36) *mingħajr* 'without' allows for a sizable minority share of 38.4% (= 25 out of 65) for definite common nouns.

We conclude that the complements of (Π9) ***bla*** 'without' display a very strong tendency to being indefinite or generic. Figure 4.5 captures the indefiniteness preference which holds for (Π9) *bla* 'without'. The arrow "→" in inverted commas indicates that there is a strong preference which is almost but not entirely without exception.

Π	"→"	~*bla*	/ ___ complement$_{DEF}$
complement	"→"	indefinite	/ *bla* ___

Figure 4.5: Indefiniteness preference.

In contrast, (Π36) ***mingħajr*** 'without' is not as picky as to the (in)definiteness of its complements.

4.1.1.2 Compatibility of Maltese Πs with (in)definite complements

If we discount those Maltese Π whose frequency is especially low and which are therefore difficult to interpret, we can state that there is no other member of BLOMP 2.0 whose distribution is restricted in the same way as that of (Π9) *bla* 'without'. The bulk of the Maltese Πs side with (Π36) *mingħajr* 'without' in the sense that they are roughly equally compatible with definite and indefinite complements. We illustrate the general compatibility of the vast majority of the Maltese Πs with indefinite and definite complements.[8] For different reasons to be exposed below,[9] (Π2) *a skapitu* 'at the expense of', (Π12) *faċċata* 'opposite', (Π14) *fejn* 'near', (Π26) *kif* 'as'; (Π37) *mingħala* 'in s.o.'s opinion', (Π42) *mintul* 'all along'; (Π43) *mnejn* 'from near', (Π45) *permezz* 'by means of', (Π52) *sforz* 'thanks to', and (Π54) *sotta* 'under' are excluded from the discussion below.

For most of the remaining Πs, we provide an example for taking an indefinite complement (= the (a)-sentence) and another example featuring the same Π in combination with a definite complement (= the (b)-sentence). Only those combinations are accounted for in which Π and complement are adjacent.[10] We will inspect other cases where complement and head are (obligatorily or optionally) linked to each

8 In Section 4.1.1.2.1 on Πs which are subject to fusion with the definite pro-clitic and Section 4.1.1.2.2 on Πs which are exempt from fusion.

9 Cf. Section 4.1.2.

10 For indefinite complements, this means that naked nouns as well as those which are preceded by an indefinite or general quantifier, a cardinal numeral or a prenominal adjective are counted in. As to definite complements other than proper names, the range of constructions involves all kinds of NPs (simple or complex) which are overtly marked for definiteness including possessor inflection and stand-alone demonstrative pronouns.

other by an intervening element below.[11] Definite complements usually involve a common noun hosting either the definite pro-clitic or a possessor suffix. Definite multi-word NPs and demonstrative pronouns are also used as representatives of definite complements. Pronominal complements are looked at separately.[12] Proper names[13] as complements are given only in those cases where the pattern Π + PROPER NAME could be identified at least once in the 1st 750 hits of the Π in the Korpus Malti 3.0 (= the (c)-sentence). For a handful of Πs, the documentation is fragmentary. Comments accompany only those examples which call for elaborate explanations. Issues which arise in connection with polysemy or multifunctionality of Πs and their combinability with complements are discussed in the semantics section. For (Π9) *bla* 'without' and (Π36) *mingħajr* 'without', the reader is referred to the previous section. Inside the following two sections, the Πs are ticked off in the alphabetical running order established for BLOMP 2.0.

4.1.1.2.1 Πs subject to fusion

In this section, we treat of those Πs which undergo fusion with the definite pro-clitic.[14] They have to be separated from the remainder of the Πs because it is technically impossible to apply the 1st 100-test to the fusable Πs. To conduct the 1st 100-test a single common output file must be created which potentially accounts for all realization forms of a given Π. Owing inter alia to the assimilation processes which affect the definite pro-clitic and Π under fusion, no single file can be produced. This is only possible with those Πs which escape fusion.[15] For the fusable Πs we have to content ourselves with estimates based on the figures given in Table 3.27. Direct comparability with our provisional yardstick in Figure 4.4 is possible only for the Πs to be discussed in the subsequent section. This methodological incongruity is no

---

11 Accordingly, the issue of (in)definiteness of complements needs to be reconsidered for a subset of the Πs discussed in this section where the focus is on asyndetic syntagmatic relations between Π and complement. In Section 4.1.2, syndesis plays an important role whose adequate description requires the revision of some of the ideas expressed previously.

12 Cf. Section 4.1.2.

13 There is a long list of sub-categories forming part of the class of proper names. Beside personal names, there are place names, animal names, event names, calendaric names, names of institutions, works of art, commercial products, natural phenomena, etc. with further sub-divisions (Nübling/Heuser/Fahlbusch 2015: 98–336). For practical reasons, we cannot study the behaviour of each and every category in this rich phenomenology not the least because it is not always easy to draw the line between common-noun status and proper-name status. In case of doubt, we give preference to classifying a given item as common noun.

14 As discussed in-depth in Section 3.3.4.

15 Cf. Section 4.1.1.2.2.

serious flaw to our line of argumentation because the Πs featured in Figure 4.4 are also excluded from fusion with the definite pro-clitic.

(i) (Π5) *bħal* 'like': Like (Π36) *mingħajr* 'without', (Π5) *bħal* 'like' meets our expectations in the sense that there is ample evidence of this Π taking indefinite complements and definite complements including proper names. The examples in (4.12) reflect this oligo-compatibility of (Π5) *bħal* 'like'.

(4.12) (Π5) *bħal* 'like'

(a) [Korpus Malti 3.0; academic109]

L-għoti *tal-penalty* *isir* *l-istess* ***[bħal***
DEF-donation of:DEF-penalty 3SG.M.IPFV:become DEF-same [**like**
penalty *normali*]$_{PP}$
penalty normal]
'The penalty shoot-out happens to be the same [**like** a normal penalty].'

(b) [Korpus Malti 3.0; academic 104]

Ħafna *operazzjonijiet* *binarji* *ta'* *l-aritmetika* ***[bħal***
much operation:PL binary:PL of DEF-arithmetic [**like**
l-addizzjoni *u* *l-multiplikazzjoni*]$_{PP}$ *huma* *funzjonijiet*
DEF-addition and DEF-multiplication] 3PL function:PL
mill-prodott *karteżjan*
from:DEF-product Cartesian
'Many binary arithmetic operations [**like** addition and multiplication] are function of the Cartesian product.'

(c) [Korpus Malti 3.0; academic10]

F' *għajnejn* *ekonomisti* ***[bħal*** *Adam Smith*]$_{PP}$ *[. . .]* *l-idea* *li*
in eye:PL economist:PL [**like** Adam Smith] DEF-idea that
d-dinja *hija* *enormi* *b'* *riżorsi* *mingħajr* *limitu*
DEF-world 3SG.F enormous with resource:PL without limit
iffurmat *is-sies* *għat-teorija* *ekonomika.*
form:3SG.F.PFV DEF-foundation for:DEF-theory:F economic:F
'In the eyes of economists [**like** Adam Smith], the idea that the world is enormous with limitless resources formed the foundation of economic theory.'

The qualitative evidence of the insensitivity of (Π5) *bħal* 'like' to different degrees of (in)definiteness does not answer the question of whether there are quantitative preferences in the choice of complement of (Π5) *bħal* 'like'. If we go by the token frequency of *bħal*-ART fusion in Table 3.27, we can determine the lower limit for the share of definite complements for this Π at 77.5%.

(ii) (Π7) *bi* 'with': We know from the frequency count in Table 3.27 that (Π7) *bi* 'with' is one of the most frequently employed Πs of Maltese which also attests to numerous instances of Π-ART fusion. There can thus be no doubt that this Π can take all kinds of complements in terms of (in)definiteness as the examples in (4.13) suggest.

(4.13) (Π7) *bi* 'with'

(a) [Cauchi 1997: 10]

int lanqas [b' qargħa ħamra]$_{PP}$ taħt imnieħrek ma
2SG not_even [**with** pumpkin:F red:F] under nose:2SG NEG
tinduna
2SG.IPFV:notice:F
'[. . .] you do not even notice [a red pumpkin] right under your nose.'

(b) [Gauci 1994: 63]

Kreitlin ma kienx jaf jitkellem
Kreitlin NEG be:NEG 3SG.M.IPFV:know 3SG.M.IPFV:RFL:speak
[bl-Ispanjol]$_{PP}$
[**with**:DEF-Spanish]
'Kreitlin did not know [Spanish] [. . .].'

(c) [Korpus Malti 3.0; news121387]

ħassewhom skomdi [bi Kristu]$_{PP}$
feel:3PL.PFV:3PL uncomfortable:PL [**with** Christ]
'[. . .] they felt uncomfortable [**with** Christ].'

Based on the figures given in Table 3.27, we calculate that at least 41.2% of all attestations of (Π7) *bi* 'with' involve a definite common noun as complement. It is highly probable that the share of definite complements is considerably bigger than this but we cannot determine the exact size of the share. Similar problems arise with all other Πs which fuse with the definite pro-clitic.

(iii) (Π15) *fi* 'in': What we have said as to (Π7) *bi* 'with' and the examples in (4.13) is largely true also for (Π15) *fi* 'in'. This Π is the second most frequent in the Korpus Malti 3.0 with an abundance of tokens for both definite and indefinite complementation. The examples in (4.14) represent the patterns *fi* + NP_{INDEF}, *fi* + NP_{DEF}, and *fi* + PROPER NAME.

(4.14) (Π15) *fi* 'in'

(a) [Camilleri 2014: 145]

Ma setgħux iħallu din is-sitwazzjoni
NEG can:3PL.PFV:NEG 3.IPFV:leave:PL DEM:F:PROX DEF-situation

[f' kolonja Ingliża]ₚₚ
[**in** colony:F English:F]
'They could not allow this situation [**in** an English colony] [. . .].'

(b) [Gauci 1994: 73]
[Fil-kwarti-finali]ₚₚ il-Messikani kellhom imorru
[**in**:DEF-quarter_final:PL] DEF-Mexican:PL have.PFV:3PL 3.IPFV:go:PL
Toluca
Toluca
'[**In** the quarter-finals], the Mexicans had to go to Toluca [. . .].'

(c) [Korpus Malti 3.0; literature39]
[fi Kristu]ₚₚ hemm żewġ naturi
[**in** Christ] there two nature:PL
'[. . .] [**in** Christ], there are are two natures [. . .].'

Going by the token frequencies in Table 3.27, it can be assumed that at least 63.6% of all *fi*-PPs involve a definite complement which hosts the definite pro-clitic because (Π15) *fi* 'in' usually fuses with the latter.

(iv) (Π19) *ġo* 'in': Since (Π19) *ġo* 'in' and (Π18) *ġewwa* 'inside' are semantically close to each other and probably also diachronically related (E. Serracino-Inglott 1976: 33; Aquilina 1987: 409), we should not be surprised to find similarities between the two Πs also in the domain under inspection. Of the two Πs, only (Π19) *ġo* 'in' is affected by fusion.[16] (Π19) *ġo* 'in' is attested in all three of the patterns as shown in (4.15).

(4.15) (Π19) *ġo* 'in'
(a) [Attard 2019: 213]
kif wasalna Malta ħaduni [ġo post ikrah]ₚₚ
how arrive:1PL.PFV Malta take:1SG.PFV [**in** place ugly]
'[. . .] when we arrived in Malta they took me [**to** an ugly place].'

(b) [Camilleri 2013: 97]
Xeħithom [ġol-kafè]ₚₚ
throw:3PL [**in**:DEF-coffee]
'He threw them [**into** the coffee].'

16 For information on (Π18) *ġewwa* 'inside, the reader is referred to Section 4.1.1.2.2 (xi). The finer semantic points related to the competition between these two Πs (and further contenders) will be looked into in Section 5.1.2.1.1.

(c) [Korpus Malti 3.0; academic12]
Il-President Kennedy ġie maqtul ***[ġo** Texas]$_{PP}$*
DEF-president Kennedy come PART:kill **[in** Texas]
'President Kennedy was killed [**in** Texas] [. . .].'

The underrepresentation of definite common nouns as complements of (П19) *ġo* 'in' results from an estimate based on the token frequencies given in Table 3.27. The calculated minimum share for definite complements is as low as 26.3%. Note that (П19) *ġo* 'in' is exceptional – together with (П51) *sa* 'till' – that it undergoes fusion without being one of the high frequent Пs.

(v) (П21) *għal* 'for': We know from Table 3.27 that (П21) *għal* 'for' belongs to the top-ranking Пs as to the token frequency in the Korpus Malti 3.0. Therefore, what has been said in connection to (П7) *bi* 'with' and (П15) *fi* 'in' is also valid for (П21) *għal* 'for'. Expectedly, (П21) *għal* 'for' is attested in combination with each of the complement types reviewed in this section. The examples in (4.16) give evidence of this.

(4.16) (П21) *għal* 'for'
(a) [Gauci 1994: 64]
Wara 10 minuti wasal il-goal l-aktar diskuttibbli
after 10 minute:PL arrive DEF-goal DEF-more debatable
***[għal** snin twal]$_{PP}$*
[for year.PL long.PL]
'After ten minutes the [**for** many years] most questionable goal happened [. . .].'
(b) [Attard 2019: 118]
Ma ġejniex hawn ***[għall**-flus]$_{PP}$*
NEG come:1PL.PFV:NEG here **[for**:DEF-money]
'We have not come here [**for** the money].'
(c) [Korpus Malti 3.0; academic117]
Piqué għamel id-debutt fil-Premier League f' Ottubru bħala
Piqué do DEF-debut in:DEF-Premier League in Octobre as
sostitut ***[għal** John O'Shea]$_{PP}$*
substitute **[for** John O'Shea]
'Piqué made his debut in the Premier League in October as a substitute [**for** John O'Shea].'

For the same reasons as those mentioned previously, it is not feasible for us to apply the 1st 100-test to (П21) *għal* 'for. We have to make do with stating that at least 58.9%

of all *għal*-PPs involve definite complements (most probably common nouns in their majority).

(vi) (Π29) *lil* 'to': Since (Π29) *lil* 'to' mostly fulfils the functions of object marker – both of direct and indirect objects according to a system of differential object marking based on animacy and definiteness[17]– it comes as no surprise that this Π does not only combine with all complement types (as shown in (4.17)) but also counts among the most frequent Πs of Maltese.

(4.17) (Π29) *lil* 'to'

(a) [Camilleri 2014: 25]

nagħtu	*kollox*	***[lil***	*ftit*	*nies]$_{PP}$*	*biex*	*nagħmluhom*
1.IPFV:give:PL	all	**[to**	few	people]	in_order_to	1.IPFV:do:3PL

sinjuri
rich_man:PL
'[. . .] we give everything [**to** a few people] to make them rich.'

(b) [Galea 2001: 50]

Daħal	*fil-kastell*	*ifittex*	***[lil**l-Kontessa]$_{PP}$*
enter	in:DEF-castle	3SG.M.IPFV:search_for	[**to**:DEF-countess]

'He entered the castle to look [**for** the countess].'

(c) [Bartolo/Vella 2009: 12]

Sib	***[lil***	*Tommy]$_{PP}$*	*u*	*ejjew*	*f'*	*kamarti*
find.IMP	**[to**	Tommy]	and	come:2PL.IMP	in	room:POSS:1SG

'Find [Tommy] and come to my room!'

If we divide the value given for (Π29) *lil* 'to' in the column fusion in Table 3.27 and divide this value by the total frequency of this Π in the Korpus Malti 3.0 the minimum share of definite complements is 69.9%.

(vii) (Π30) *ma'* 'with': Their identical English translation notwithstanding, (Π7) *bi* 'with' and (Π30) *ma'* 'with' have only partially overlapping functional domains which will be scrutinised in some detail below.[18] For the time being, it suffices to state that (Π30) *ma'* 'with' like (Π7) *bi* 'with' gives evidence of combinations with all three complement types as comes to the fore in (4.18).

---

17 Cf. Section 1.6.5 (ii).
18 In Sections 5.1.3.4–5.1.3.5.

(4.18) (П30) *ma'* 'with'

(a) [Bartolo/Vella 2009: 75]

*Fi triqithom il<t>aqgħu **[ma' ragħaj xiħ]**$_{PP}$*
in street:POSS:3PL <RFL>receive:3PL.PFV [**with** shepherd old]
'On their way, the met [**with** an old shepherd] [. . .].'

(b) [Bartolo/Vella 2009: 48]

*Serraħ rasu **[mal-ħġieġa]**$_{PP}$*
give_rest head:3SG.M [**with**:DEF-window_pane]
'He rested his head [**against** the window-pane].'

(c) [Bartolo/Vella 2009: 37]

*[**Ma'** Tommy]$_{PP}$ ma riditx titkellem*
[**with** Tommy] NEG want:3SG.F.PFV:NEG 3SG.F.IPFV:RFL:speak
*dwar ir-relazzjoni tagħha [**ma'** Antida]$_{PP}$ u [**ma'** Ġingo]$_{PP}$.*
about DEF-relation:PL of:3SG.F [**with** Antida] and [**with** Ġingo]
'She did not want to talk [**to** Tommy] about her relationships [**with** Antida] and [Ġingo].'

The share claimed by definite complements is at least as big as 52.3% for *ma'*-PPS. Even if this estimate remains uncertain for the known reasons it is clear that the importance of definite complements can only increase but never diminish if we take other kinds of definite complements into account. What is further interesting is the discrepancy between the relatively low percentage of 41.2% computed for definite complements of (П7) *bi* 'with' and the much bigger share definite complements boast in the case of (П30) *ma'* 'with'.

(viii) (П41) *minn* 'from': From the examples in (4.19) it is immediately clear that (П41) *minn* 'from' is blind to the (in)definiteness of its complements. Definite common nouns, indefinite common nouns, and proper names are eligible for the complement slot in *minn*-PPS.

(4.19) (П41) *minn* 'from'

(a) [Camilleri 2013: 200]

*daqslikieku stenbħu [**minn** ngħsa twila]$_{PP}$*
as_if wake_up:3PL.PFV [**from** sleep:F long:F]
'[. . .] as if they woke up [**from** a long sleep].'

(b) [Abela 2016: 170]

*rat lil Mrs Swan [**mi**t-tieqa tal-librerija]$_{PP}$*
see:3SG.F.PFV to Mrs Swan [**from**:DEF-window of:DEF-library]
'[. . .] she saw Mrs Swan [**from** the library window].'

(c) [Bartolo/Vella 2009: 43]

Imma	*jien*	*nixtieq*	*inkun*	*naf*	*x'*
but	1SG	1SG.IPFV:want	1SG.IPFV:be.FUT	1SG.IPFV:know	what

jiġri	***[minn***	*Estella]$_{PP}$*
3SG.M.IPFV:run	**[from**	Estella]

'But I want to know what is happening [**to** Estella].'

The provisional calculation of the lower limit of the potentially much bigger share of definite complements of (Π41) *minn* 'from' yields 56.5%.

(ix) (Π51) *sa* 'till': Like mentioned for (Π19) *ġo* 'in' in Paragraph (iv) of this section, (Π51) *sa* 'till' fails to produce a token frequency in the Korpus Malti 3.0 which could compete with those of the top-ranking Πs. The lower token frequency notwithstanding, there is evidence of (Π51) *sa* 'till' taking complements of all three types as shown in (4.20).

(4.20) (Π51) *sa* 'till'

(a) [Korpus Malti 3.0; academic107]

It-temi	*ta'*	*dan*	*l-għana*	*jvarjaw*	*minn*
DEF-topic:PL	of	DEM:M:PROX	DEF-folk_song	3.IPFV:vary:PL	from

stejjer	*u*	*leġġendi*	*tradizzjonali [. . .]*	***[sa***	*avvenimenti*
story.PL	and	legend:PL	traditional	**[till**	event:PL

storiċi]$_{PP}$
historical:PL]

'The topics of this folk-singing range from traditional stories and legends [**to** historical events].'

(b) [Bartolo/Vella 2009: 47]

mexa	*l-aħħar*	*ftit*	*passi*	***[sa**l-bieb*	*ta'*	*barra]$_{PP}$*
walk	DEF-last	few	step:PL	**[till:**DEF-door	of	outside]

'[. . .] he walked the last few steps [**to** the front-door].'

(c) [Korpus Malti 3.0; academic253]

F'	*din*	*l-intrapriża*	*introduċa*	*t-teatru*	*Russu*	*minn*
in	DEM:F:PROX	DEF-enterprise	introduce	DEF-theatre	Russian	from

Cekov	***[sa***	*Vampilov]$_{PP}$*
Chekov	**[to**	Vampilov]

'In this enterprise, he introduced the Russian theatre from Chekov [**to** Vampilov].'

We determine the smallest possible size of the share of definite complements in *sa*-PPs in the same way as in the eight previous cases in this section. The result computed in this way is relatively small, namely 46.3%.

(x) (Π55) *ta'* 'of': The final case is that of the most frequent Π in the Korpus Malti 3.0, namely (Π55) *ta'* 'of'. Owing to the primary function of linking GENITIVAL (and similar) attributes to head nouns within NPs, this Π faces no obstacles when it comes to taking complements which are either definite or indefinite. All combinations are legitimate as shown in (4.21).

(4.21) (Π55) *ta'* 'of'

(a) [Bartolo/Vella 2009: 45]

*beda jikber aktar fih is-sens [**ta'** avventura]$_{PP}$*

begin 3SG.M.IPFV:grow more in:3SG.M DEF-sense [**of** adventure]

'[. . .] the sense [**of** adventure] began to grow more in him.'

(b) [Bartolo/Vella 2009: 95]

*konna qed niżvujtaw il-kamra [**ta**t-tifel]$_{PP}$*

be:1PL.PFV PROG 1.IPFV:empty:PL DEF-room [**of**:DEF-boy]

'[. . .] we were clearing out the [boy'**s**] room.'

(c) [Bartolo/Vella 2009: 29]

*Estella ukoll qagħdet kwieta mal-kliem [**ta'** Reggie]$_{PP}$*

Estella also stay:3SG.F.PFV quiet:F with:DEF-word [**of** Reggie]

'Estella too remained silent at [Reggie'**s**] words.'

Interestingly, the high token frequency of (Π55) *ta'* 'of' goes along with a particularly low (potential) share of definite complements. According to our simple arithmetic, the share is as low as 36.8%.

Under the proviso that the quantities cannot be taken at face value, we summarise the above estimates in Figure 4.6. The Πs are ordered top-down according to the increasing potential share of definite complements. What emerges from this ordering is a continuum. At the one extreme of the continuum, there is (Π19) *ġo* 'in' with the smallest share of 26.3% whereas at the opposite end of the scale we find (Π5) *bħal* 'like' with a potential share of 77.5% for definite complements. We learn from the huge gap between the two extremes that Maltese Πs do not behave homogeneously as to their preference for (in)definite complements. None of the Πs reviewed in this section exclusively accepts either definite complements or indefinite ones. For four Πs, it cannot be ruled out beforehand that indefinite complements are more popular than definite complements with them. On account of this theoretical possibility, we will specifically take note of Πs which display a clear preference for only one type of complements.[19]

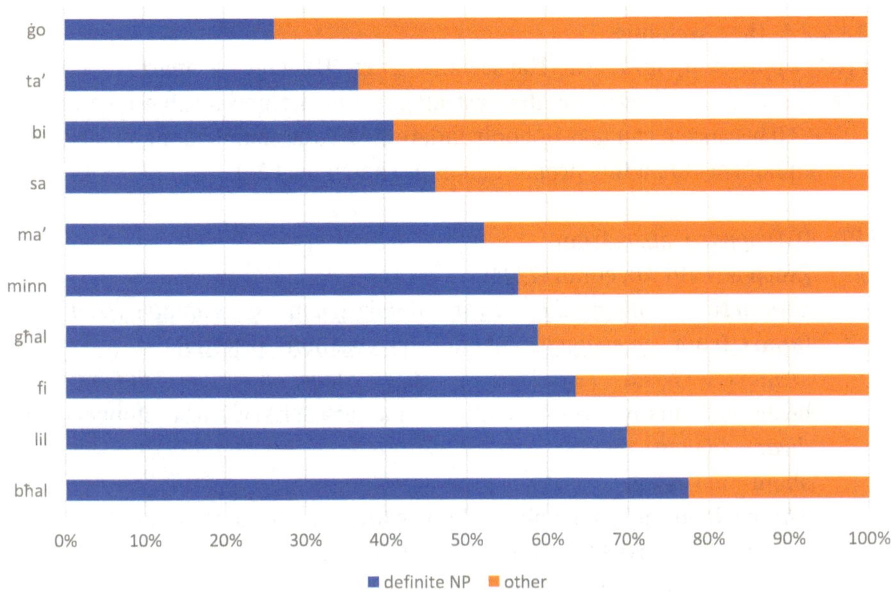

Figure 4.6: Potential shares of definite complements with fusable Πs.

4.1.1.2.2 Πs barred from fusion

This section is organised along the same lines as the previous one. Each non-fusable Maltese Πs is featured in a numbered paragraph of its own.[20] Where possible, examples of the syntactic employment of the patterns Π + NP_{INDEF}, Π + NP_{DEF}, and Π + PROPER NAME are provided. The Πs are addressed in alphabetical order. For the four Πs which have long and short alloforms,[21] we compare the token frequencies of the competing alloforms but illustrate only the use of the most frequent (long) alloform with sentential examples. Most of the paragraphs contain a short quantitative account of the occurrences of the above patterns in the Korpus Malti 3.0. These individual frequency counts are integrated into a general survey.[22] In the

19 In Section 4.1.1.2.2.
20 The catalogue of Πs is presented in Section 4.1.1.2.2.1.
21 Cf. Section 3.2.
22 Cf. Section 4.1.1.2.2.2.

general survey, only the long alloforms of those Πs with long and short forms are taken account of.

4.1.1.2.2.1 The catalogue

(i) (Π1) *apparti* 'apart from': The scene is set by (Π1) *apparti* 'apart from' in the sense that this Π is typical for the vast majority of the non-fusable members of BLOMP 2.0 because it can take not only indefinite NPs as complement but also definite complements including proper names as shown in (4.22).

(4.22) (Π1) *apparti* 'apart from'

(a) [Korpus Malti 3.0; culture2204]

*[**Apparti** mistoqsijiet dwar il-vokazzjoni saċerdotali]$_{PP}$ [. . .]*
[**apart_from** question:PL about DEF-vocation priestly]
kienu bosta l-istudenti li staqsewh fil-fond
be:3PL.PFV many DEF-student:PL that ask:3PL.PFV:3SG.M in:DEF-depth
dwar l-Iżlam
about DEF-Islam
'[**Apart from** questions about the vocation of priesthood] many were the students who asked him thoroughly about Islam.'

(b) [Korpus Malti 3.0; culture43]

*[**Apparti** l-kotba]$_{PP}$ għandek ukoll passjoni kbira lejn*
[**apart_from** DEF-book.PL] have:2SG also passion big:F towards
il-films
DEF-film:PL
'[**Apart from** books] you also have a great passion for films [. . .].'

(c) [Korpus Malti 3.0; culture755]

*Sadanittant [**apparti** Jessie J]]$_{PP}$ [. . .] se jkun hemm*
meanwhile [**apart_from** Jessie J.] FUT 3SG.IPFV:be.FUT there
ukoll il-kantanta Ingliża ta' 22 sena Rita Ora
also DEF-singer:F English:F of 22 year Rita Ora
'In the meantime [**apart from** Jessie J.] there will also be 22 years old English singer Rita Ora.'

The ternary set of examples in (4.22) tells us that (Π1) *apparti* 'apart from' is compatible with complements independent of their (in)definiteness. What we do not learn from the above illustration is how common the patterns (*apparti* + NP_{INDEF}, *apparti* + NP_{DEF}, and *apparti* + PROPER NAME) are in the corpora. In point of fact, the distribution is uneven as results from Figure 4.7. On the basis of the 1st 100 hits of *apparti* + NP_{DEF} in the Korpus Malti 3.0, the pattern *apparti* + NP_{DEF} can be considered the default option since it accounts for 90% of all countable attestations. The preference for definite complements increases to 94% if we lump together the patterns *apparti* + NP_{DEF} and *apparti* + PROPER NAME. This preference for definite complements is in stark contrast to the leaning to indefinite complements observed with (Π9) *bla* 'without'. It remains to be seen whether definite complements are also the favourites of other non-fusable Maltese Πs.

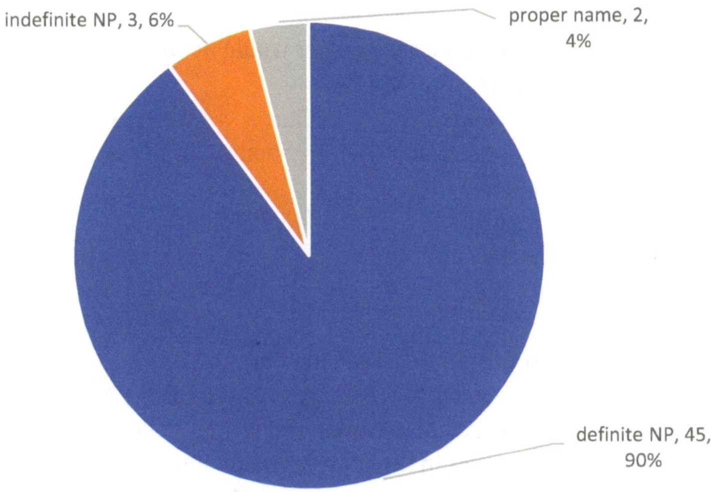

Figure 4.7: Token frequency of (in)definite complements of (Π1) *apparti* 'apart from' in 1st 100 (Korpus Malti 3.0).

(ii) (Π3) *barra* 'outside': As comes to the fore from the examples in (4.23), (Π3) *barra* 'outside' too is attested with definite and indefinite complements alike.

(4.23) (Π3) *barra* 'outside'

(a) [Korpus Malti 3.0; culture766]

L-idea ġiet mhux biss mill-imħabba tiegħu
DEF-idea come:3SG.F.PFV NEG:3SG.M:NEG only from:DEF-love of:3SG.M
għall-klieb iżda wkoll minn memorja ta' meta
for:DEF-dog.PL but also from memory of when
kellu madwar erba' snin u tħalla waħdu
have.PFV:IO:3SG.M around four year.PL and RFL:leave one:3SG.M
*fil-karozza [**barra** supermarket]$_{PP}$*
in:DEF-car [**outside** supermarket]

'The idea did not only arise from his love for dogs but also from the memory of being about four years old and left alone in the car [**outside** a supermarket].'

(b) [Korpus Malti 3.0; academic122]

l-avversarju jiskorja meta l-gowler ikun
DEF-adversary 3SG.M.IPFV:score when DEF-goalie 3SG.M.IPFV:be.FUT
*[**barra** l-pożizzjoni]*$_{PP}$
[**outside** DEF-position]
'[. . .] the adversary scores when the goal-keeper would not be [**in** position].'

(c) [Korpus Malti 3.0; academic97]

kienu jżejnu l-ħitan tad-Dar tat-Trux li
be:3PL.PFV 3.IPFV:adorn:PL DEF-wall:PL of:DEF-house of:DEF-deaf that
*kienet [**barra** Madrid]*$_{PP}$
be:3SG.F.PFV [**outside** Madrid]
'[. . .] they decorated the walls of the Quinta del Sordo which was [**outside** Madrid].'

If we take account only of the 1st 100 hits for *barra* in the Korpus Malti 3.0, we notice immediately that the majority of the tokens (24 hits) goes to the credit of the short form *barra* of (П34) *minbarra* 'except'.[23] In contrast, the spatial meaning of (П3) *barra* 'outside' is attested only four times in the same segment of the corpus. These four cases exclusively instantiate the pattern *barra* + NP_{DEF}. Thus, the above preference for definite common nouns as complements receives further support also in the case of (П3) *barra* 'outside'.

(iii) (П4) *bejn* 'between': The situation is more diverse in the next case. First of all, the story seems to be repeating itself since (П4) *bejn* 'between' can be shown to be insensitive to the (in)definiteness of its complements as transpires from the comparison of the examples in (4.24).

(4.24) (П4) *bejn* 'between'

(a) [Korpus Malti 3.0; academic7]

*jiffaċilitaw ir-reżoluzzjoni ta' kontroversji [**bejn***
3.IPFV:facilitate:PL DEF-resolution of controversy:PL [**between**
stati differenti]$_{PP}$
state:PL different]
'[. . .] they facilitate the resolution of controversies [**between** different states].'

23 As discussed in Section 3.2.1.

(b) [Korpus Malti 3.0; academic10]
reġgħet *ġiet* *imtennija* *r-rabta* ***[bejn***
repeat:3SG.F.PFV come:3SG.F.PFV PART:repeat:F DEF-tie **[between**
l-iżvilupp *tal-bniedem* *u* *l-ħarsien* *ta'* *l-ambjent]$_{PP}$*
DEF-development of:DEF-man and DEF-protection of DEF-ambient]
'[. . .] the connection [**between** the development of man and the protection of nature] was repeated again [. . .].'

(c) [Korpus Malti 3.0; academic12]
Trasport ***[bejn*** *Malta* *u* *l-Awstralja]$_{PP}$* *kien* *nieqes* *ħafna*
Transport **[between** Malta and DEF-Australia] be wanting much
'Transport [**between** Malta and Australia] was very defective.'

Like in the previous cases, (П4) *bejn* 'between' prefers definite common nouns as complements over proper names and indefinite NPs. However, the latter two types of complements are not as marginalised as in the case of (П3) *barra* 'outside'. Figure 4.8 reveals that the pattern *bejn* + NP_{DEF} ousts the two contenders but not as strongly as is the case with (П1) *apparti* 'apart from'.

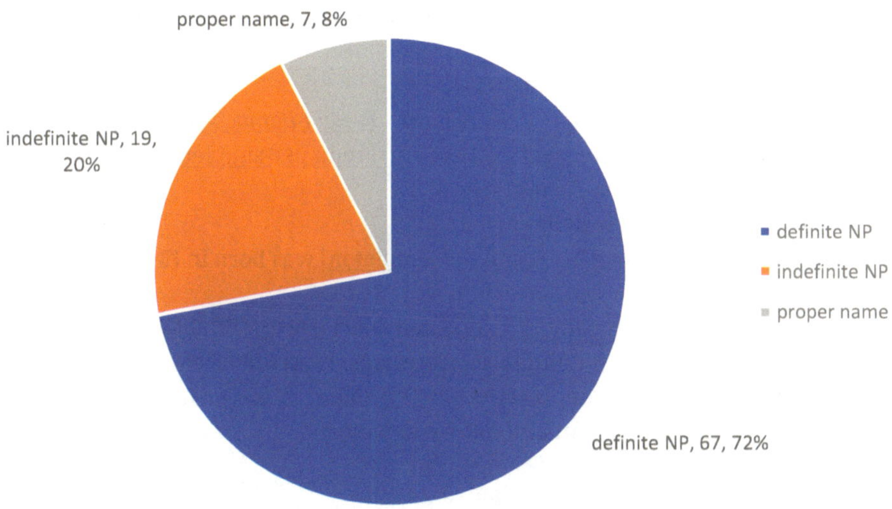

Figure 4.8: Token frequency of (in)definite complements of (П4) *bejn* 'between' in 1st 100 (Korpus Malti 3.0).

Proper names and definite common nouns taken together yield a robust majority of 80%. Yet, with a share of 20% of the hits, indefinite complements form a sizable minority option. The next paragraph is going to prove that, depending on the individual П, the share of this minor option may easily increase.

(iv) (Π6) *bħala* 'as': It is interesting to see whether our next candidate (Π6) *bħala* 'as' which is semantically and diachronically related to (Π5) *bħal* 'like' (Aquilina 1987: 112–113) displays similar characteristics as those described for the latter.[24] As a matter of fact, the role of indefinite complements is very strong with (Π6) *bħala* 'as' than can be hypothesised for (Π5) *bħal* 'like'. The examples in (4.25) illustrate that the three patterns *bħala* + NP$_{\text{INDEF}}$, *bħala* + NP$_{\text{DEF}}$, and *bħala* + PROPER NAME are empirically attested in our database.

(4.25) (Π6) *bħala* 'as'

(a) [Vella 2014: 183]

Kellhom	*wieħed*	*impjegat*	***[bħala***	*reporter*	*tal-qorti]*$_{PP}$	*biss*
have.PFV:3PL	one	employee	**[as**	reporter	of:DEF-court]	only

'They had only one employee [**as** court reporter].'

(b) [Korpus Malti 3.0; academic107]

L-għana	*minn*	*dejjem*	*kien*	*ikkunsidrat*	***[bħala***	*l-mużika*
DEF-folk_song	from	always	be	consider:PART	**[as**	DEF-music
tan-nies	*komuni]*$_{PP}$					
of:DEF-people	common]					

'The folk song was always considered to be [the music of the common people].'

(c) [Korpus Malti 3.0; academic103]

Fredu Abela	*magħruf*	***[bħala***	*Fredu l-Bamboċċu]*$_{PP}$ *[. . .]*
Fredu Abela	PART:recognise	**[as**	Fredu DEF-simpleton]
twieled	*Ħaż-Żabbar*		
RFL:give_birth	Ħaż-Żabbar		

'Fredu Abela known [**as** Fredu the Simpleton] was born in Ħaż-Żabbar.'

In both Borg/Azzopardi-Alexander (1997) and Saari (2003), (Π6) *bħala* 'as' appears only in combination with indefinite complements. Figure 4.9 clearly suggests that in stark contrast to many other Maltese Πs (Π6) *bħala* 'as' has an undeniable predilection for cooccurring with an indefinite complement.

24 Cf. Section 4.1.1.2.1 (i).

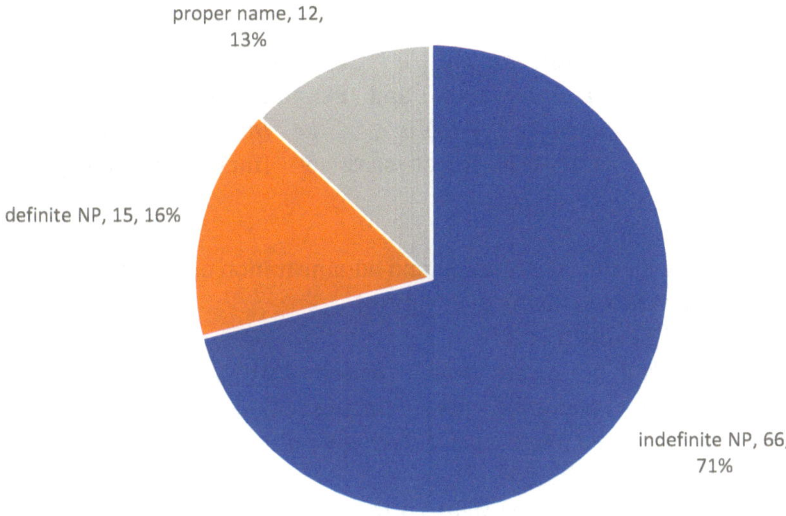

Figure 4.9: Token frequency of (in)definite complements of (Π6) *bħala* 'as' in 1st 100 (Korpus Malti 3.0).

The preference for indefiniteness of complements might not be as strong as in the case of (Π9) *bla* 'without' (see Figure 4.2). However, indefinite complements of (Π6) *bħala* 'as' outnumber definite ones by a ratio of 2.4-to-one. A noticeable difference between (Π9) *bla* 'without' and (Π6) *bħala* 'as' is the questionable status of definite complements taken by the former. The pattern *bħala* + NP_{DEF}, however, gives only rise to acceptable constructions.

(v) (Π8) *biswit* 'facing': No comparable difficulties arise in connection with (Π8) *biswit* 'facing'. We have evidence of this Π combining with definite common nouns, indefinite common nouns, and proper names. Three examples of this range of possibilities are given in (4.26).

(4.26) (Π8) *biswit* 'facing'

(a) [Korpus Malti 3.0; european8679]

Hi *għandha* *dan* *ir-rwol* ***[biswit*** *ħidmiet* *li*
3SG.F have:3SG.F DEM:M:PROX DEF-role **[facing** activity:PL that
jsiru *fil-kuntest* *tal-Prioritajiet* *tematiċi]$_{PP}$*
3.IPFV:become:PL in:DEF-context of:DEF-priority:PL thematic:PL]
'She has this role [**regarding** activities which happen in the context of the thematic priorities].'

(b) [Korpus Malti 3.0; european46]
jassiguraw *il-konservazzjoni* *u* *l-amministrazzjoni* *razzjonali*
3.IPFV:secure:PL DEF-conservation and DEF-administration rational
tal-ħażniet *tal-ħut* *fl-ibħra* *ta'* ***[biswit*** *ix-xtut*
of:DEF-provision of:DEF-fish in:DEF-sea.PL of **[facing** DEF-shore.PL
tagħhom $]_{PP}$
of:3PL]
'[. . .] they secure the conservation and administration of the provisions of the fish in the waters [**facing** their shores].'

(c) [Korpus Malti 3.0; literature21]
Xon *qam* *ukoll* *u* *sab* *post* ***[biswit*** *Riku* $]_{PP}$
Xon rise too and find place **[facing** Riku]
'Xon stood up too and found a place [**facing** Riku].'

Among the 1st 100 hits in the Korpus Malti 3.0, there is not a single case of a proper name functioning as complement of (П8) *biswit* 'facing'. Figure 4.10 shows that the pattern *biswit* + NP_{DEF} dominates the scene.

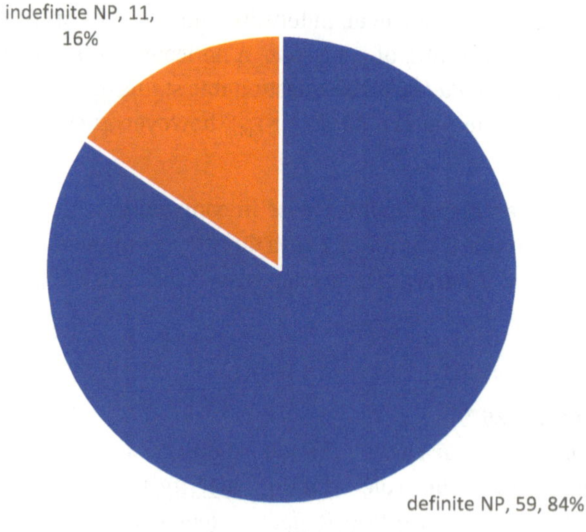

Figure 4.10: Token frequency of (in)definite complements of (П8) *biswit* 'facing' in 1st 100 (Korpus Malti 3.0).

The absence of countable occurrences of proper names as complements notwithstanding, the picture reflected by Figure 4.10 looks familiar because it features definite common nouns as the majority option for the status of complement of (П8) *biswit* ‘facing’. Thus, this Π largely conforms to the patterns established for (Π1) *apparti* ‘apart from’, (Π3) *barra* ‘outside’, (Π4) and *bejn* ‘between’.

(vi) (Π10) *daqs* ‘equal to’: Another Maltese Π which can be shown to combine freely with complements irrespective of their (in)definiteness is (Π10) *daqs* ‘equal to’. The general ability of (Π10) *daqs* ‘equal to’ to fill the slot of the head in PPs involving definite common nouns, indefinite common nouns, or proper names can be gathered from the examples in (4.27).

(4.27) (Π10) *daqs* ‘equal to’

(a) [Korpus Malti 3.0; academic13]

Ma nistgħux ngħidu li l-euonimus ma
NEG 1.IPFV:can:PL:NEG 1.IPFV:say:PL that DEF-eunymus NEG
tħobbx l-umdità anzi din
3SG.F.IPFV:love:NEG DEF-humidity on_the_contrary DEM:F:PROX
*hi meħtieġa għalkemm forsi mhux **[daqs***
3SG.F PART:need:F although perhaps NEG:3SG.M:NEG **[equal_to**
pjanti oħra]$_{PP}$
plant:PL other.PL]
‘We cannot say that the eunymus does not love humidity; on the contrary, it is needed although perhaps not [**as much as** with other plants].’

(b) [Korpus Malti 3.0; academic12]

*l-ebda poplu ma hu milqugħ **[daqs** il-poplu*
DEF-no people NEG 3SG.M PART:receive **[equal_to** DEF-people
Malti]$_{PP}$
Maltese]
‘[. . .] no people are welcomed [**as much as** the Maltese people].’

(c) [Korpus Malti 3.0; academic169]

Hu skorja l-istess numru ta' gowls fil-Premiership
3SG.M score DEF-same number of goal:PL in:DEF-premiership
***[daqs** Andry Shevchenko]$_{PP}$*
[equal_to Andry Shevchenko]
‘He scored the same number of goals in the premiership [**as** Andry Shevchenko].’

Quantitatively, (Π10) *daqs* ‘equal to’ lends only weak support to the testable assumption that definite complements are generally preferred over indefinite ones. Figure 4.11 shows that the gap between the shares claimed by definite complements and indefinite complements is not extraordinarily huge. In this figure, the majority option exceeds the 50%-mark only by a relatively narrow margin whereas indefinite complements cover slightly more than a third of all countable PPs among the 1st 100 hits in the Korpus Malti 3.0. The relatively strong position of the indefinite complements adds further weight to the hypothesis that there is a continuum ranging from the dominance of definite complements to that of indefinite complements.

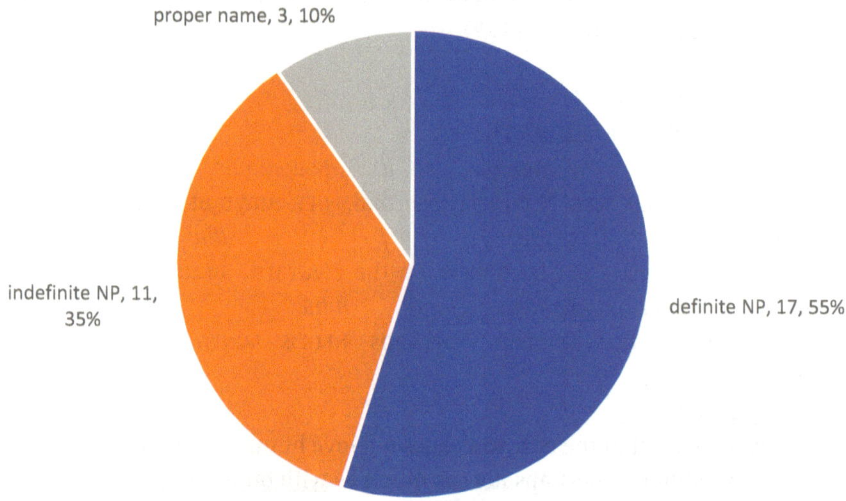

Figure 4.11: Token frequency of (in)definite complements of (Π10) *daqs* 'equal to' in 1st 100 (Korpus Malti 3.0).

(vii) (Π11) *dwar* 'about': We are back to the uncontested dominance of definite common nouns as complements when we look at (Π11) *dwar* 'about'. It is true that this Π too gives evidence of combinations with all three types of complements under inspection as shown in (4.28). However, indefinite common nouns and proper names function only infrequently as complements of (Π11) *dwar* 'about'.

(4.28) (Π11) *dwar* 'about'

(a) [Bartolo/Vella 2009: 12]

għandna	*bżonn*	*nitkellmu*	***[dwar***	*affarijiet*	*oħra]*$_{PP}$
have:1PL	need	1.IPFV:RFL:speak:PL	**[about**	matter:PL	other.PL]

'[. . .] we need to talk [**about** other matters].'

(b) [Bartolo/Vella 2009: 10]

setgħet	*turi*	*kemm*	*kienet*	*studjat*
can:3SG.F.PFV	3SG.F.IPFV:show	how_much	be:3SG.F.PFV	study:3SG.F.PFV
*[**dwar***	*<u>dawn</u>*	*<u>il-friefet</u>]$_{PP}$*		
[**about**	<u>DEM:PL:PROX</u>	<u>DEF-butterfly.PL</u>]		

'[. . .] she could show how much she had learned [**about** <u>these butterflies</u>].'

(c) [Korpus Malti 3.0; academic139]

kien	*kiteb*	*artiklu*	*[**dwar***	*<u>Herbert Ganado</u>]$_{PP}$*
be	write	article	[**about**	<u>Herbert Ganado</u>]

'[. . .] he had written an article [**about** <u>Herbert Ganado</u>].'

The quantitative disproportion is visible in Figure 4.12. The pattern *dwar* + NP$_{DEF}$ is seven times as frequent as the pattern *dwar* + NP$_{INDEF}$ and fourteen times as numerous as that of *dwar* + PROPER NAME. Added up the definite complements account for 88% of all attestations of countable *dwar*-PPs among the 1st 100 hits in the Korpus Malti 3.0.

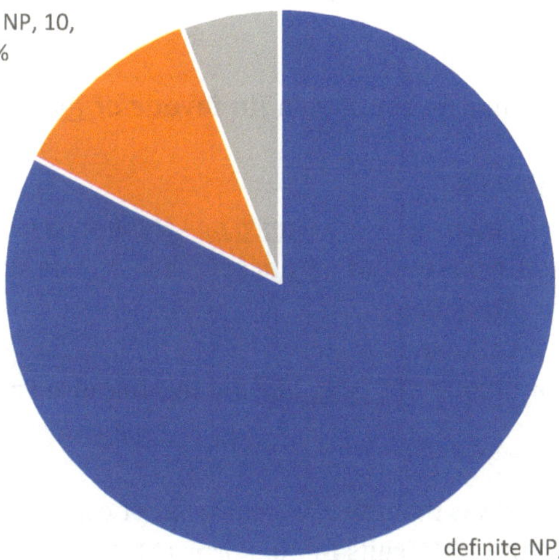

Figure 4.12: Token frequency of (in)definite complements of (Π11) *dwar* 'about' in 1st 100 (Korpus Malti 3.0).

(Π11) *dwar* 'about' provides another example of those Maltese Πs whose preferred type of complement is definite.

(viii) (Π13) *favur* 'in favour of': The dominance of definite common complements over the other two types of complements is similarly strong in the case of (Π13) *favur* 'in favour of'. In (4.29), we provide one example each for the patterns *favur* + NP$_{INDEF}$, *favur* + NP$_{DEF}$, and *favur* + PROPER NAME.

(4.29) (Π13) *favur* 'in favour of'

(a) [Korpus Malti 3.0; academic7]

Il-mod kif jinqasam il-ġid kummerċjali
DEF-manner how 3SG.M.IPFV:RFL:divide DEF-wealth commercial
*huwa żbilanċjat għall-aħħar **[favur** il-pajjiżi*
3SG.M unbalanced for:DEF-last **[in_favour_of** DEF-country:PL
l-iżjed industrijalizzati]$_{PP}$
DEF-more industrialise:PART:PL]
'The way the economic wealth is distributed is unbalanced to the utmost [**in favour of** the most industrialised countries].'

(b) [Korpus Malti 3.0; academic161]

*kien qed jagħmel wisq propaganda **[favur***
be PROG 3SG.M.IPFV:do too_much propaganda **[in_favour_of**
il-lingwa Taljana]$_{PP}$
DEF-language:F Italian:F]
'[. . .] he was making too much propaganda [**in favour of** the Italian language].'

(c) [Korpus Malti 3.0; academic140]

*Meta Ganado kiteb **[favur** Franco]$_{PP}$ għamel dan*
when Ganado write **[in_favour_of** Franco] do DEM:M:PROX
għax Franco kien Kattoliku
because Franco be Catholic
'When Ganado wrote [**in favour of** Franco] he did this because Franco was a Catholic.'

A look at Figure 4.13 strongly suggests that the different types of complements are not of the same calibre. Among the 1st 100 hits in the Korpus Malti 3.0, the pattern *favur* + NP$_{DEF}$ occurs ten times more frequently than *favur* + PROPER NAME and even twelve times as often as the pattern *favur* + NP$_{INDEF}$.

We do not want to read too much into the fact that (Π13) *favur* 'in favour of' is the first of the members of BLOMP 2.0 for which proper names are attested more often as complements than indefinite common nouns are. What is more important is that the shares of the two sub-types of definite complements cover 93% of all *favur*-PPs in the segment of the Korpus Malti 3.0 we have looked at.

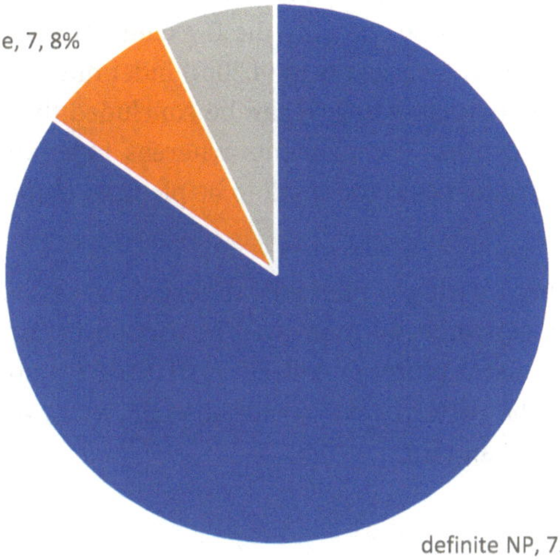

Figure 4.13: Token frequency of (in)definite complements of (Π13) *favur* 'in favour of' in 1st 100 (Korpus Malti 3.0).

(ix) (Π14) *fejn* 'near': This Π poses a number of problems which render it difficult to determine to what extent definiteness plays a role for the choice of complements. In (4.30), we provide the usual three examples also for (Π14) *fejn* 'near' – under the proviso that indefinite complements are absent from the printed books which form part of the primary sources employed for this study.

(4.30) (Π14) *fejn* 'near'

(a) [Korpus Malti 3.0; sport1417]
*għadna lura [**fejn** timijiet oħra]*$_{PP}$
still:1PL back [**near** team:PL other:PL]
'[. . .] we are still lagging behind [**in comparison to** other teams].'

(b) [Bartolo/Vella 2009: 453]
il-qattusa ħabiba tal-qalb ta' Anita kienet
DEF-cat:F friend:F of:DEF-heart of Anita be:3SG.F.PFV
*tinsab midfuna [**fejn** seħbitha]*$_{PP}$
3SG.F.IPFV:RFL:find PART:bury:F [**near** companion:POSS.F:3SG.F]
'[. . .] the cat, Anita's dearest friend, was buried [**near** its companion].'

(c) [Vella 2014: 162]
*jitlajja [**fejn** Bieb il-Belt]*$_{PP}$
3SG.M.IPFV:linger [**near** City Gate]
'[. . .] he lingers [**near** the City Gate].'

Stolz/Vorholt (2022: 178–179) argue that (Π14) *fejn* 'near' is subject to a semantically motivated split in the sense that if used spatially the Π exclusively takes definite complements whereas in non-spatial usage (as in (4.30b)) indefinite complements are licit too. This possibility notwithstanding, it can be concluded that (Π14) *fejn* 'near' overwhelmingly takes definite complements whereas indefinite complements are exceptional. Their frequency is too small to be pictured in the format of a diagram.

(x) (Π16) *fost* 'amongst': The situation is markedly different in the case of (Π16) *fost* 'amongst'. Owing to the semantically motivated restriction to combinations with complements which are either plural or collective, (Π16) *fost* 'amongst' lacks uncontroversial evidence of the pattern **fost* + PROPER NAME. Accordingly, (4.31) comprises only the sentences (a) and (b).

(4.31) (Π16) *fost* 'amongst'

(a) [Korpus Malti 3.0; academic10]

l-impenn	*sar*	*ukoll*	*fuq*	*livell*	*lokali*	***[fost***	*ċittadini*
DEF-engagement	become	also	on	level	local	**[amongst**	citizen:PL
familji	*u*	*assoċjazzjonijiet]$_{PP}$*					
family:PL	and	association:PL]					

'[. . .] the engagement happened also on the local level [**amongst** citizens, families, and associations].'

(b) [Korpus Malti 3.0; academic12]

[Fost	*l-Eġizzjani]$_{PP}$*	*qam*	*ferment*	*kontra*	*l-preżenza*	
[amongst	DEF-Egyptian:PL]	rise	agitation	against	DEF-presence	
ta'	*truppi*	*Ingliżi*	*f'*	*pajjiżhom.*		
of	troop:PL	English:PL	in	country:3PL		

[**Amongst** the Egyptians] agitation arose against the presence of English troops in their country.'

The pluralised ethnonym *l-Eġizzjani* 'the Egyptians' in (4.31b) is a borderline case since it could be understood as a proper name whereas we treat it as a common noun here (cf. below).

It might be objected that examples like (4.32) prove that the Π under scrutiny is by no means excluded from combinations with proper names as complements. However, combinations of this kind are licit only if certain conditions apply, namely the loss of monoreferentiality and pluralization.

(4.32) [Korpus Malti 3.0; culture1319]

Ir-rampa	*tal-Lufthansa*	*se*	*tkun*	*żona*
DEF-ramp	of:DEF-Lufthansa	FUT	3SG.F.IPFV:be.FUT	zone

tas-supercars	*bl-attrazzjonijiet*	*internazzjonali*	***[fost***
of:DEF-supercar:PL	with:DEF-attraction:PL	international	**[amongst**

il-Ferraris	*u*	*l-Porsches*	*lokali]$_{PP}$*
DEF-Ferrari:PL	and	DEF-Porsche:PL	local]

'The Lufthansa ramp will be a zone of supercars with international attractions [**amongst** the local Ferraris and Porsches].'

In cases of this kind, we are facing generic reference, i.e. the proper name acquires properties of common nouns (overt marking of definiteness and pluralization) which are normally absent from inherently definite and individualizing proper names which are prototypically in the singular (Nübling/Fahlbusch/Heuser 2015: 17–27, 61–63). Given these special conditions, it is not surprising that the category of proper names is absent from Figure 4.14. However, even in the absence of this class of complements, the dominance of definite complements is still strong. The 1st 100 hits in the Korpus Malti 3.0 feature (Π16) *fost* 'amongst' as head of a PP with a definite common noun as complement in 93 out of 100 cases. Only 7% of the turnout involve indefinite complements. In other words, (Π16) *fost* 'amongst' comes close to monopolizing definite complements.

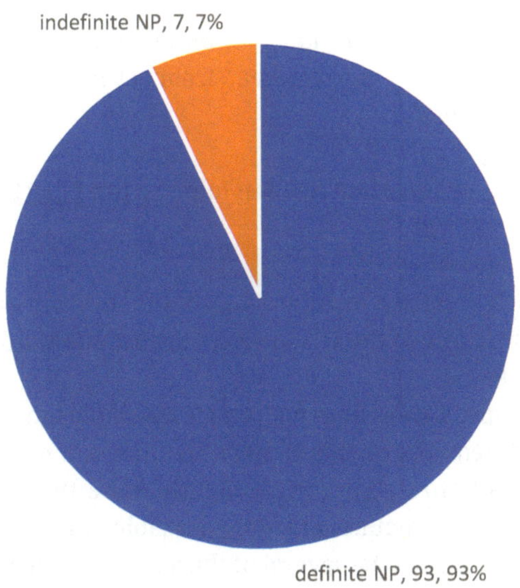

Figure 4.14: Token frequency of (in)definite complements of (Π16) *fost* 'amongst' in 1st 100 (Korpus Malti 3.0).

What transpires from the sketch of (Π16) *fost* 'amongst' should be compared to paragraph (xxix) where the partial synonym (Π47) *qalb* 'amidst' is discussed.

(xi) (Π17) *fuq* 'on': The three sample sentences in (4.33) make it clear to us that (Π17) *fuq* 'on' can take any kind of nouny complement in the sense that (in)definiteness of the latter is neither an obstacle nor a precondition for the combination with this Π.

(4.33) (Π17) *fuq* 'on'

(a) [Attard 2019: 175–176]

Irġiel	*t'*	*etá*	*avvanzata*	*kienu*	*miġburin*	*barra*	*kull*
man.PL	of	age	advanced:F	be:3PL.PFV	PART:gather	outside	each

ħanut	*tat-te*	*bilqiegħda*	***[fuq***	*banketti*	*żgħar]*$_{PP}$
shop	of:DEF-tea	sitting	**[on**	stool:PL	small.PL]

'Outside each tea parlour, men of an advanced age had gathered sitting [**on** little stools].'

(b) [Korpus Malti 3.0; academic10]

Il-kriżi	*ta'*	*l-ilma*	*tinsab*	*f'*	*ħafna*	*oqsma*
DEF-crisis	of	DEF-water	3SG.F.IPFV:RFL:find	in	many	sector.PL

tal-ħajja	***[fuq***	*il-pjaneta]*$_{PP}$
of:DEF-life	**[on**	DEF-planet]

'The water crisis can be found in many sectors of life [**on** the planet].'

(c) [Korpus Malti 3.0; academic12]

Ir-regolamenti	*ġodda*	*ġew*	*iffirmati*	*mill-President*
DEF-regulation:PL	new.PL	come:3PL.PFV	sign:PART:PL	from:DEF-president

Lyndon Johnson	***[fuq***	*Liberty Island]*$_{PP}$
Lyndon Johnson	**[on**	Liberty Island]

'The new regulations were signed by president Lyndon Johnson [**on** Liberty Island].'

Proper names as complements are back on board in Figure 4.15 but their share is minimal. According to the results of our search of the 1st 100 hits in the Korpus Malti 3.0, they account for only 1% of all *fuq*-PPs.

Yet, the marginal role played by proper names does not reduce the importance of definiteness in the domain of complements of this Π since definite common nouns cover almost three-quarters of all attestations. With slightly more than a quarter of the same count, indefinite common nouns constitute a sizable minority. (Π17) *fuq* 'on' conforms to the pattern established by several of the above Πs which in the majority of cases opt for definite complements.

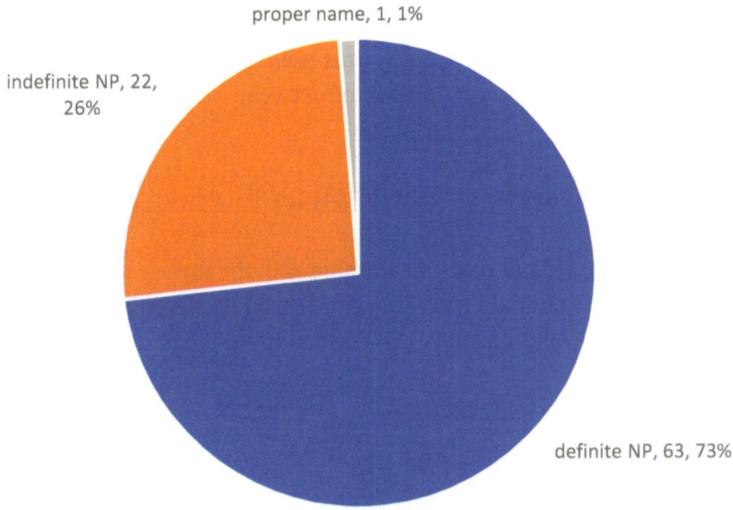

Figure 4.15: Token frequency of (in)definite complements of (П17) *fuq* 'on' in 1st 100 (Korpus Malti 3.0).

(xii) (П18) *ġewwa* 'inside': Our database provides examples of (П18) *ġewwa* 'inside' taking definite common nouns, indefinite common nouns, and proper names as complements.[25] Superficially, the examples in (4.34) seem to recount an already familiar story.

(4.34) (П18) *ġewwa* 'inside'

(a) [Korpus Malti 3.0; academic10]

il-ġid *u* *s-servizzi* *li* *jiġu* *prodotti*
DEF-wealth and DEF-service:PL that 3.IPFV:come:PL produce:PART:PL
[ġewwa *pajjiż]*$_{PP}$ *matul* *is-sena*
[inside country] during DEF-year
'[. . .] the wealth and the supply that are produced [**in** a country] during the year.'

(b) [Galea 2001: 143]

[ġewwa *dak* *il-blat]*$_{PP}$ *kien* *hemm* *għar* *mudlam* *u*
[inside DEM:M:DIST DEF-rock] be there cave PART:darken and
ikrah *iwaħħxek*
ugly 3SG.M.IPFV:scare:2SG
'[. . .] [**inside** this rock] there was a dark and scaringly ugly cave [. . .].'

25 The synonym (П19) *ġo* 'in' is featured in Section 4.1.1.2.1 (iv).

(c) [Korpus Malti 3.0; academic102]
Twieled *nhar* *il-5* *ta'* *Mejju* *tas-sena* *1944* ***[ġewwa***
RFL:give_birth day DEF-5 of May of:DEF-year 1944 **[inside**
San Pawl-il-Baħar]$_{PP}$
San Pawl-il-Baħar]
'He was born on the 5th of May of the year 1944 [**in** San Pawl-il-Baħar].'

The 1st-100 test reveals however that this time the shares of the three complement types display a different distribution. Figure 4.16 gives evidence of the preponderance of the pattern *ġewwa* + PROPER NAME.

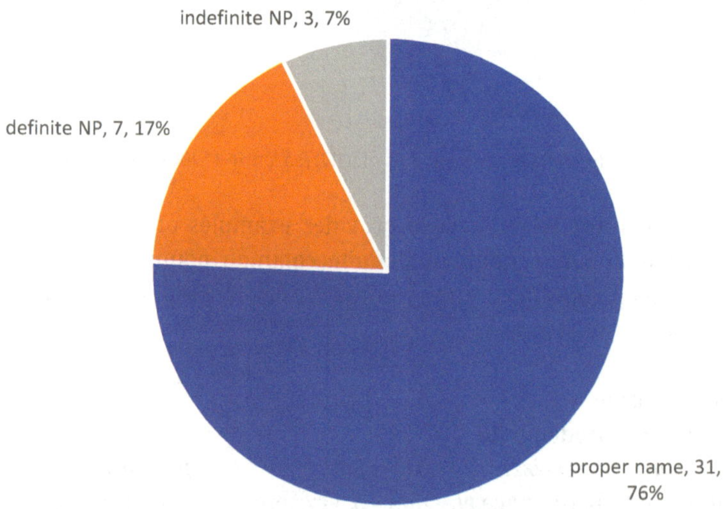

Figure 4.16: Token frequency of (in)definite complements of (П18) *ġewwa* 'inside' in 1st 100 (Korpus Malti 3.0).

It is especially remarkable that definite common nouns boast only a relatively small share whereas more than three-quarters of all *ġewwa*-PPs in the 1st-100 test involve a proper name, more precisely: a place name.[26] The prominence of proper-name complements fits the definiteness preference observed with many other Maltese Пs since 93% of all *ġewwa*-PPs registered for the above test are equipped with a definite complement.

26 We will have occasion to focus on the interaction of this П (and several competitors) with place names in Section 5.1.2.1.1.

(xiii) (Π20) *għajr* 'except': With (Π20) *għajr* 'except' we are almost back to normal, in a manner of speaking. This Π displays the expected majority of complements which have the shape of definite common nouns. This majority does not exclude the other complement categories from combining with (Π20) *għajr* 'except' as results from (4.35).

(4.35) (Π20) *għajr* 'except'

(a) [Korpus Malti 3.0; academic139]
Hekk twieled il-Partit Demokratiku Nazzjonalista li
so RFL:give_birth DEF-party democratic nationalist that
ma kien xejn ***[għajr*** *satellita tal-Kurja]$_{PP}$*
NEG be nothing **[except** satellite of:DEF-Curia]
'In this way the Natioanl-Democratic Party was born which was [**nothing but** a satellite of the Curia].'

(b) [Korpus Malti 3.0; European10876]
m' hemmx kawża dovuta jew ġustifikazzjoni biżżejjed
NEG there:NEG cause:F due:F or justification enough
għal din il-bidla ***[għajr*** *l-imposizzjoni tad-dazju]$_{PP}$*
for DEM:F:PROX DEF-change **[except** DEF-imposition of:DEF-excise]
'[. . .] there is no due cause nor sufficient justification for this change [**except** the imposition of the excise].'

(c) [Korpus Malti 3.0; european1065]
It-territorju tar-Renju tan-Norveġja ***[għajr*** *Svalbard]$_{PP}$*
DEF-territory of:DEF-Kingdom of:DEF-Norway **[except** Svalbard]
'The territory of the Kingdom of Norway [**except** Svalbard] [. . .].'

Compared to several previously discussed Πs, the majority of the pattern *għajr* + NP_{DEF} is not particularly strong. Based on the 1st 100-test in the Korpus Malti 3.0, Figure 4.17 shows that indefinite complements are relatively frequent. Definite complements – proper names + definite common nouns – account for 62% of the *għajr*-PPs.

The shares featured in Figure 4.17 resemble those established for (Π10) *daqs* 'equal to' in Figure 4.11.

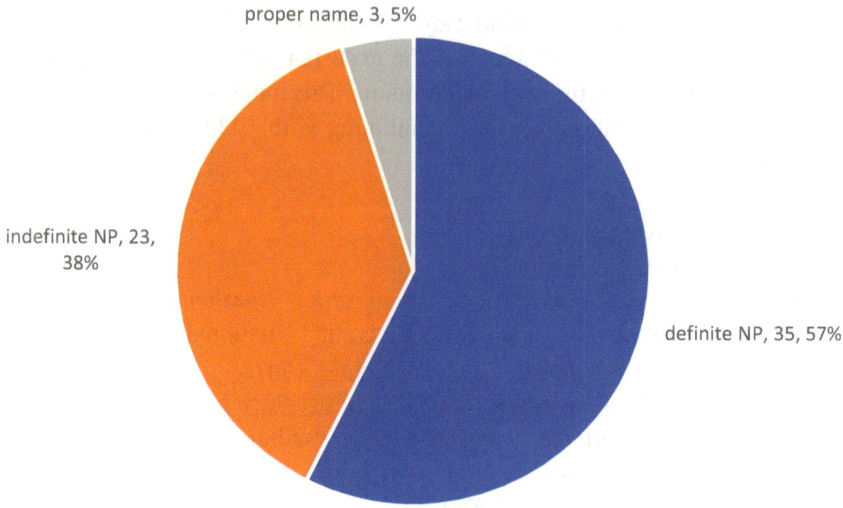

Figure 4.17: Token frequency of (in)definite complements of (Π20) *għajr* 'except' in 1st 100 (Korpus Malti 3.0).

(xiv) (Π22) *għand* 'at s.o.'s place': Next in line is (Π22) *għand* 'at s.o.'s place'. The sentences (a), (b), and (c) in (4.36) reflect the usual distribution in the sense that this Π too is insensitive to the (in)definiteness of the complement.

(4.36) (Π22) *għand* 'at s.o.'s place'

(a) [Camilleri 2013: 185]

żjarat	***[għand***	*persuni*	*midħla*	*tal-vittma]$_{PP}$*
visit:PL	**[at_s.o.'s_place**	person:PL	PART:enter.PL	of:DEF-victim]

'[. . .] visits [**at the places** of persons familiar with the victim] [. . .].'

(b) [Bartolo/Vella 2009: 112]

kienu	*diġà*	*marru*	***[għand***	*il-ġenituri*
be:3PL.PFV	already	go:3PL.PFV	**[at_s.o.'s_place**	DEF-parent:PL
tas-subien	*l-oħra]$_{PP}$*			
of:DEF-boy.PL	DEF-other.PL]			

'[. . .] they had already gone [**to the homes** of the other boys' parents].'

(c) [Bartolo/Vella 2009: 98]

int	*qatt*	*ma*	*kont*	*mort*	***[għand***	*Steve]$_{PP}$*
2SG	ever	NEG	be:2SG.PFV	go:2SG.PFV	**[at_s.o.'s_place**	Steve]

'[. . .] you have never gone [**to** Steve's].'

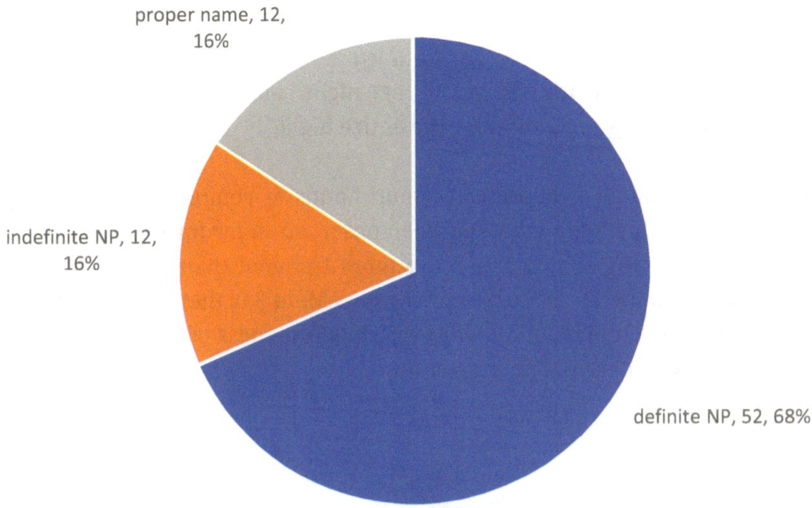

Figure 4.18: Token frequency of (in)definite complements of (Π22) *għand* 'at s.o.'s place' in 1st 100 (Korpus Malti 3.0)..

The 1st 100-test yields the following results for (Π22) *għand* 'at s.o.'s place'. Definite common nouns occur 2.1 times as often as the other two complement classes taken together. In 84% of all *għand*-PPs, the complement is definite. Figure 4.18 also shows that proper names and indefinite common nouns display identical shares.

(xv) (Π23) *ħdejn* 'beside': The distribution of (Π23) *ħdejn* 'beside' over complement types largely follows the same path as those Πs which give preference to definite complements. In (4.37), examples for the patterns *ħdejn* + NP_{INDEF}, *ħdejn* + NP_{DEF}, and *ħdejn* + PROPER NAME are provided.

(4.37) (Π23) *ħdejn* 'beside'

(a) [Bartolo/Vella 2009: 476]
Waqaf xħin waslu ***[ħdejn*** *roundabout żgħira* $]_{PP}$
stop when arrive:3PL.PFV **[beside** roundabout small:F]
'He stopped when they arrived [**at** a small roundabout].'

(b) [Bartolo/Vella 2009: 190]
kien ħallieh fuq il-komodina ***[ħdejn*** *is-sodda fil-kamra*
be leave:3SG.M on DEF-cupboard **[beside** DEF-bed in:DEF-room
tiegħu $]_{PP}$
of:3SG.M]
'[. . .] he had left it on the cupboard [**beside** the bed in his room].'

(c) [Bartolo/Vella 2009: 342]
*Efrem baqa' [**ħdejn** Steve]PP il-lejl kollu.*
Efrem remain [**beside** Steve] DEF-night all:3SG.M
'Efrem stayed [**at** Steve's **side**] for the entire night.'

The numerical dominance of definite common nouns as complements resulting from Figure 4.19 is in line with what has been observed so far for the majority of the Maltese Πs. Since proper names too are more frequent than indefinite complements according to the 1st 100-test in the Korpus Malti 3.0, the share of definite complements increases to 92% of all *ħdejn*-PPs. Indefiniteness of the complement thus becomes almost a marginal phenomenon.

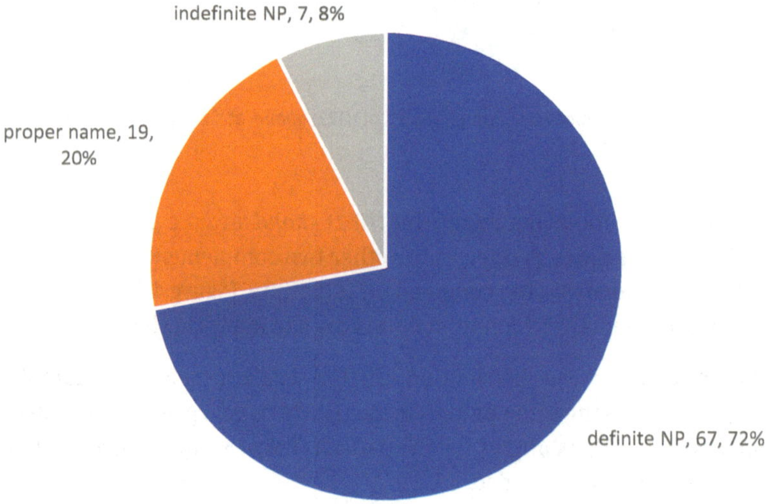

Figure 4.19: Token frequency of (in)definite complements of (Π23) *ħdejn* 'beside' in 1st 100 (Korpus Malti 3.0).

(xvi) (Π24) *ħlief* 'except': The next case is interesting because what we say about (Π24) *ħlief* 'except' can be compared to our observations as to (Π20) *għajr* 'except' above because of the apparent synonymity between the two Πs. Before we have a look at the quantitative side of the issue, we provide examples of the oligo-compatibility of (Π24) *ħlief* 'except' with (in)definite complements in (4.38).

(4.38) (Π24) *ħlief* 'except'
(a) [Bartolo/Vella 2009: 46]
Tommy ma kellu għalfejn jieħu xejn
Tommy NEG have.PFV:IO.3SG.M why 3SG.M.IPFV:take nothing

miegħu *[ħlief* *flixkun* *ilma]*$_{PP}$
with:3SG.M **[except** bottle water]
'Tommy had no reason to take anything with him [**except** a bottle of water].'

(b) [Bartolo/Vella 2009: 42]
Ma kellha lil ħadd ma' min titkellem
NEG have.PFV:IO.3SG.F to anybody with who 3SG.F.IPFV:RFL:speak
[ħlief *is-siġar* *u* *l-għasafar]*$_{PP}$
[except DEF-tree.PL and DEF-bird.PL]
'She had no one to talk to [**except** the trees and the birds].'

(c) [Bartolo/Vella 2009: 37]
Kienet fehmet li ħadd *[ħlief* *Antida]*$_{PP}$
be:3SG.F.PFV understand:3SG.F.PFV that anybody **[except** Antida]
ma seta' jgħidilha l-verità
NEG can 3SG.M.IPFV:say:IO:3SG.F DEF-truth
'She had understood that nobody [**except** Antida] could tell her the truth.'

In Figur 4.19, we see that indefinite complements dominate the scene. Even added up the shares of definite complements remain inferior to that of the pattern *ħlief* + NP_{INDEF}.

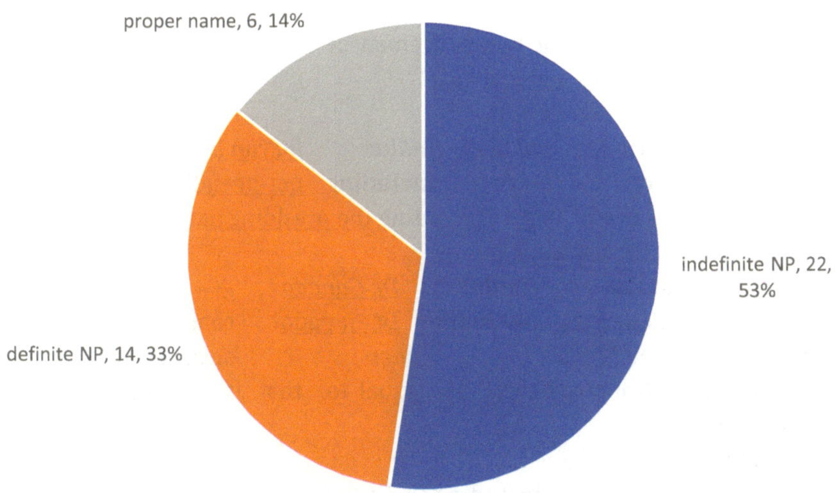

Figure 4.20: Token frequency of (in)definite complements of (Π24) *ħlief* 'except' in 1st 100 (Korpus Malti 3.0).

In a way, Figure 4.20 is the mirror-image of Figure 4.17 where the results of the 1st 100-test are disclosed for (Π20) *għajr* 'except'. In the latter case, definite complements claim 57% of all tokens whereas indefinite complements cover 38% of the tokens. It cannot be denied that the two synonymous Πs are insensitive to the (in) definiteness of their complements. Nevertheless, there seems to be a (perhaps only very weakly developed) division of labour according to which (Π20) *għajr* 'except' is the better choice if the complement is definite while (Π24) *ħlief* 'except' is at its best in combination with indefinite complements. The differences are not as pronounced as between (Π9) *bla* 'without' and (Π36) *mingħajr* 'without'.[27] This scenario will occupy us again when we address the third synonymous Π of exclusion – (Π34) *minbarra* 'except' – in paragraph (xxiii).

(xvii) (Π25) *inkluż* 'including': We continue the survey much in the same vein with (Π25) *inkluż* 'including'. As is evident from the examples in (4.39), this Π is compatible with all three types of complements.

(4.39) (Π25) *inkluż* 'including'

(a) [Camilleri 2013: 285]

kull wieħed magħmul minn żewġ pulizjotti tal-SAG
each one PART:do from two policeman:PL of:DEF-SAG
armati kif suppost ***[inkluż*** *taser guns u gass*
arm:PART:PL as suppose:PART **[including** taser_gun:PL and gas
tad-dmugħ]$_{PP}$
of:DEF-tear.PL]
'[. . .] each composed of two policemen of the SAG armed as supposed [**including** taser guns and tear gas].'

(b) [Korpus Malti 3.0; culture203]

F' Malta hawn 22 Imħallef ***[inkluż*** *il-Prim Imħallef]*$_{PP}$
in Malta here 22 judge **[including** DEF-presiding judge]
'In Malta, there are 22 judges [**including** the presiding judge].'

(c) [Korpus Malti 3.0; academic139]

grupp ta' Maltin ***[inkluż*** *Dr. Ganado]*$_{PP}$ *ġew*
group of Maltese:PL **[including** Dr. Ganado] come:3PL.PFV
ittrattati mill-Ingliżi daqs li kieku kienu
treat:PART:PL from:DEF-English:PL equal_to that if be:3PL.PFV
kriminali.
criminal:PL
'[. . .] a group of Maltese [**including** Dr. Ganado] were treated by the English as if they were criminals.'

27 As discussed in Section 4.1.1.1.

The usual check of the 1st 100 hits in the Korpus Malti 3.0 yields the results disclosed in Figure 4.21.

Two thirds of the turnout go to the credit of definite common nouns. In comparison to several other Maltese Πs the share of indefinite common nouns is remarkably big. Proper names occupy a niche at the margins only, in a manner of speaking.

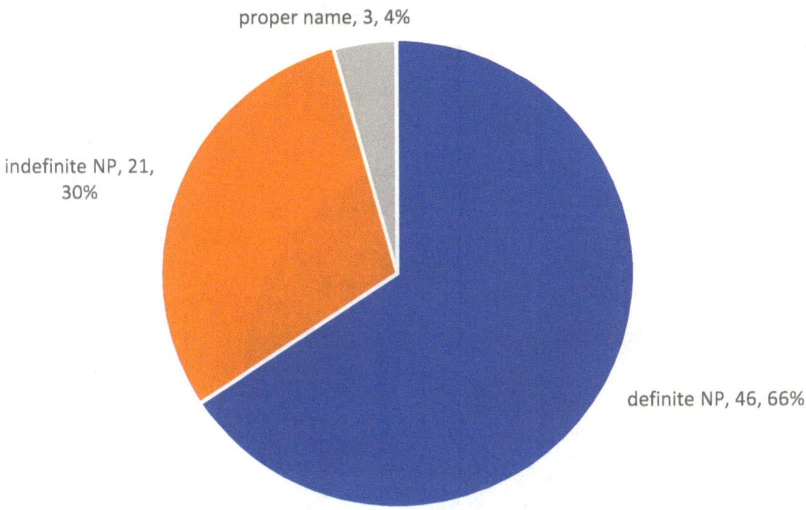

Figure 4.21: Token frequency of (in)definite complements of (Π25) *inkluż* 'including' in 1st 100 (Korpus Malti 3.0).

(xviii) (Π27) *kontra* 'against': The omni-compatibility of (Π27) *kontra* 'against' with the three complement types is illustrated in (4.40).

(4.40) (Π27) *kontra* 'against'

(a) [Bartolo/Vella 2009: 161]

X' qatt seta' biċċa qanfud jagħmel ***[kontra*** *barri*
what ever can piece hedgehog 3SG.M.IPFV:do **[against** bull
ors brimba kbira ajkla u lupu]$_{PP}$?
bear spider:F big:F eagle and wolf]
'What can a little hedgehog ever do [**against** a bull, a bear, a huge spider, an eagle, and a wolf]?'

(b) [Bartolo/Vella 2009: 295]

jirribella ***[kontra*** *dik l-imġiba mqanżħa*]$_{PP}$
3SG.M.IPFV:rebel **[against** DEM:F:DIST DEF-behaviour:F PART:bungle:F]
'[. . .] he rises up [**against** that incompetent behaviour].'

(c) [Bartolo/Vella 2009: 130]
Reggie qatt ma kien qal xejn **[kontra** Antida]$_{PP}$
Reggie ever NEG be say nothing **[against** Antida]
'Reggie had never said anything [**against** Antida].'

What strikes us as worth noting is the relatively strong position of proper names in Figure 4.22. This complement type accounts for slightly less than a quarter of all occurrences of *kontra*-PPs in the 1st 100-test.

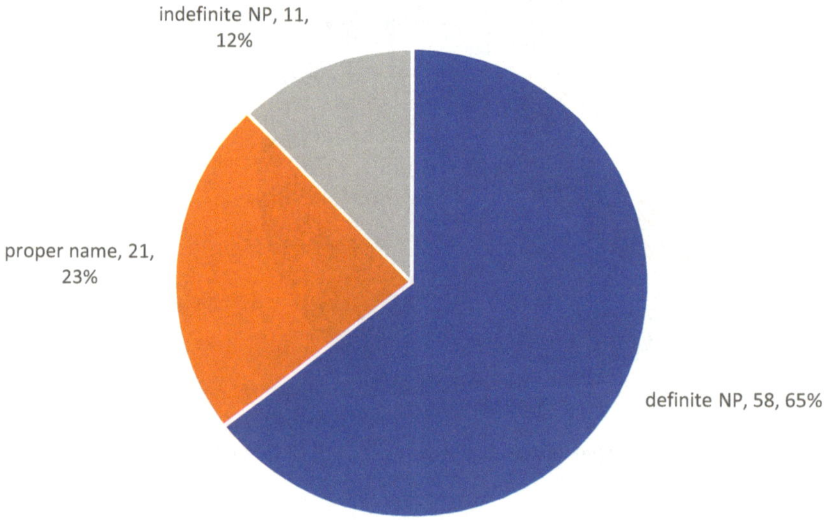

Figure 4.22: Token frequency of (in)definite complements of (Π27) *kontra* 'against' in 1st 100 (Korpus Malti 3.0).

The exceptional size of the share of proper names is more remarkable if we compare it to the marginal role played by proper names in the domain of (Π13) *favur* 'in favour of' – the antonym of (Π27) *kontra* 'against' – discussed in paragraph (viii), illustrated in (4.29), and counted in Figure 4.13 above.[28]

(xix) (Π28) *lejn* 'towards': The importance of proper names increases further in the case of (Π28) *lejn* 'towards'. First of all, (4.41) gives evidence of the possibility of this Π to take complements from each of the three types under review.

28 We will address this issue again in Sections 4.1.1.2.2.2.

(4.41) (Π28) *lejn* 'towards'

(a) [Bartolo/Vella 2009: 22]

Kienet	*qed*	*tiċċassa*	***[lejn***	*inkwatru*	*li*
be:3SG.F.PFV	PROG	3SG.F.IPFV:stare	**[towards**	framed_picture	that
kien	*hemm*	*imdendel*	*mal-ħajt]PP*		
be	there	PART:hang	with:DEF-wall]		

'She was staring [**at** the picture that was hanging on the wall].'

(b) [Vella 2014: 91]

Jiftaħ	*għajnejh*	*u*	*jerġa'*	*jdawwar*
3SG.M.IPFV:open	eye:PL:3SG.M	and	3SG.M.IPVF:do_again	3SG.M.IPFV:turn
rasu	***[lejn***	*it-taraġ]PP*		
head:3SG.M	**[towards**	DEF-stair.PL]		

'He opened hos eyes and turned his head again [**towards** the stairs].'

(c) [Bartolo/Vella 2009: 17]

U	*t-tnejn*	*li*	*huma*	*qagħdu*	*jiċċassaw*	***[lejn***
and	DEF-two	that	3PL	stay:3PL.PFV	3.IPFV:stare:PL	**[towards**
Antida]PP	*għal*	*tletin*	*sekonda*	*sħaħ.*		
Antida]	for	thirty	second	continually		

'And both of them were staring [**at** Antida] for thirty seconds without interruption.'

Moreover, proper names are attested so often as complements of (Π28) *lejn* 'towards' in the 1st 100-test that their share comes very close to that of the definite common nouns as can be seen in Figure 4.23.

What is the cause of this unexpected strength of proper names in *lejn*-PPS? What are the differences between (Π28) *lejn* 'towards' and its near synonym (Π57) *versu* 'towards' in paragraph (xxxvi)? We try to answer this question along with similar questions in the evaluative part of this section and the semantics section.[29]

(xx) (Π31) *madwar* 'around': Another surprise is waiting for us in this paragraph on (Π31) *madwar* 'around'. As the examples in (4.42) suggest anything goes with this Π insofar as it is compatible with all three complement types.

29 Cf. Sections 4.1.1.2.2.2 and 5.1.2.2.6.

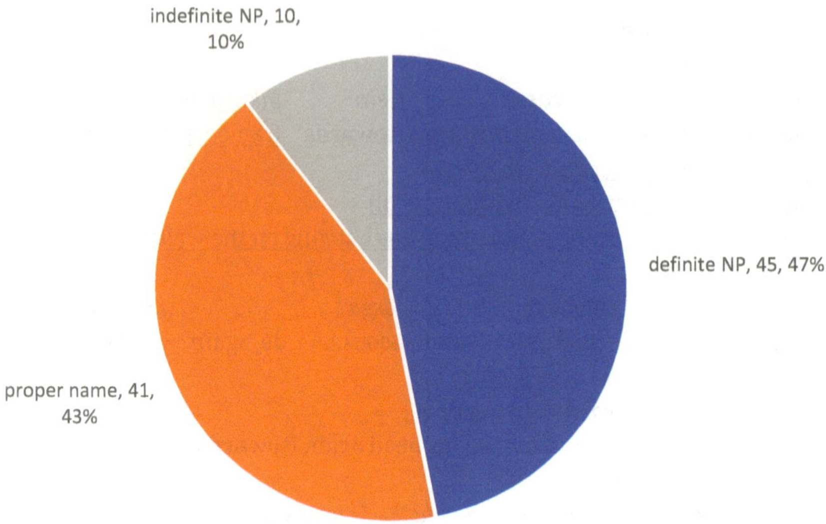

Figure 4.23: Token frequency of (in)definite complements of (П28) *lejn* 'towards' in 1st 100 (Korpus Malti 3.0).

(4.42) (П31) *madwar* 'around'

(a) [Camilleri 2013: 101]

żewġ irġiel jostru ħsibijiethom wara nuċċali
two man.PL 3.IPFV:hide:PL thought:PL:3PL after spectacle:PL
*tax-xemx [**madwar** mejda tonda]$_{PP}$*
of:DEF-sun [**around** table:F round:F]
'[. . .] two men were hiding their thoughts behind sunglasses [**at** a round table].'

(b) [Bartolo/Vella 2009: 119]

kienet tiftakar f' binha jiġri fuq
be:3SG.F.PFV 3SG.F.IPFV:remember in son:3SG.F 3SG.M.IPFV:run on
*ir-rota [**madwar** it-toroq tal-Mosta]$_{PP}$*
DEF-wheel [**around** DEF-street.PL of:DEF-Mosta]
'[. . .] she remembered her son riding the bicycle [**all over** the streets of Mosta].'

(c) [Bartolo/Vella 2009: 209]

Waqaf jidħak minnufih u l-morsa ta' driegħu
stop 3SG.M.IPFV:laugh immediately and DEF-vice of arm:3SG.M
*[**madwar** Steve]$_{PP}$ inħallet xi ftit.*
[**around** Steve] RFL:leave:3SG.F.PFV some little
'He stopped laughing immediately and the pressure of his arm [**around** Steve] loosened a fraction.'

The quantities are interesting because the 1st 100-test yields a shattering majority of 76% for the pattern *madwar* + NP_{INDEF}. The usual top-ranking definite common nouns are outnumbered by 3.5-to-one. Figure 4.24 is suggestive of a clear preference for indefinite complements that exceeds the same preference as observed for (Π36) *mingħajr* 'without' above without reaching the maximum marked by (Π9) *bla* 'without' in Figure 4.4. This extraordinary behaviour of (Π31) *madwar* 'around' finds its explanation in semantics.[30]

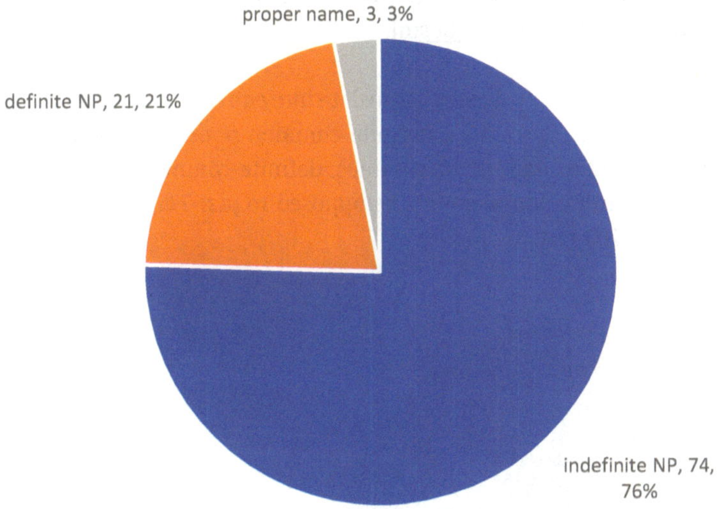

Figure 4.24: Token frequency of (in)definite complements of (Π31) *madwar* 'around' in 1st 100 (Korpus Malti 3.0).

(xxi) (Π32) *maġenb* 'close to': The taps turn once again when we look at (Π32) *maġenb* 'close to'. In terms of its general compatibility with the different complement types, the situation is identical to that reported for the majority of the Maltese Π, i.e. it does not matter much what type the complement represents since (Π32) *maġenb* 'close to' accepts all kinds of complements as shown in (4.43).

(4.43) (Π32) *maġenb* 'close to'

(a) [Camilleri 2013: 94]

fl-aħħar	*iljieli*	*hemm*	*f'*	*soddithom*	*kienet*	*rieqda*
in:DEF-last	night.PL	there	in	bed:POSS:3PL	be:3SG.F.PFV	sleep.PART:F
[maġenb	*qattiel]*$_{PP}$*?*					
[close_to	murderer]					

'[...] il the last nights, had she been sleeping in their bed [**close to** a murderer]?'

30 As will be shown in Section 5.1.2.1.12.

(b) [Bartolo/Vella 2009: 48]
Ntefa' *bilqiegħda* ***[maġenb*** *it-tieqa]$_{PP}$*
RFL:throw sitting **[close_to** DEF-window]
'He sat down [**next to** the window].'

(c) [Bartolo/Vella 2009: 124]
kienu *jikkonfoffaw* ***[maġenb*** *Roberto]$_{PP}$*
be:3PL.PFV 3.IPFV:conspire:PL **[close_to** Roberto]
'[. . .] they conspired [**close to** Roberto].'

However, the general compatibility does not translate into equal shares for the complement types. According to Figure 4.25, definite common nouns are the default type of complement. Together with proper names, definite common nouns yield a majority of 93% for definite complements as opposed to just 7% calculated for indefinite ones in the 1st 100-test.

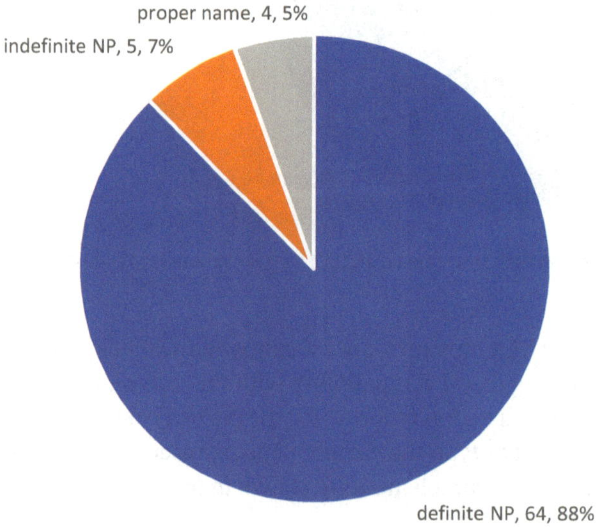

Figure 4.25: Token frequency of (in)definite complements of (Π32) *maġenb* 'close to' in 1st 100 (Korpus Malti 3.0).

(xxii) (Π33) *matul* 'during': The predominance of definite common nouns in the domain of complements holds also for (Π33) *matul* 'during'. As (4.44) proves, other complement types are not excluded from *matul*-PPS.

(4.44) (Π33) *matul* 'during'

(a) [Korpus Malti 3.0; academic109]
il-Major League Soccer fil-bidu kellha wkoll
DEF-Major League Soccer in:DEF-beginning have.PFV:IO:3SG.F also
l-għoti tal-penalties wara t-tmiem tal-partita anki
DEF-donation of:DEF-penalty:PL after DEF-end of:DEF-match also
***[matul** partiti tal-kampjonat nazzjonali]*$_{PP}$
[**during** match:PL of:DEF-championship national]
'[. . .] in the beginning, the Major League Soccer too had the penalty shoot-out after the ende of the match also [**during** matches of the national championship].'

(b) [Camilleri 2013: 213]
***[Matul** l-ewwel ġimgħat tagħha taħt prova]*$_{PP}$ *Gallo ma*
[**during** DEF-first week:PL of:3SG.F under probation] Gallo NEG
sabhiex diffiċli jintebaħ kif dak kien
find:3SG.F:NEG difficult 3SG.M.IPFV:be_aware how DEM:M:DIST be
il-mod tagħha
DEF-manner of:3SG.F
'[**During** her first weeks on probation] Gallo did not find it difficult to realise that this was her way.'

(c) [Korpus Malti 3.0; academic 13]
dan għandu jsir ***[matul** Jannar jew*
DEM:M:PROX have:3SG.M 3SG.M.IPFV:become [**during** January or
Frar]$_{PP}$
February]
'[. . .] this must take place [**in** January or February].'

In point of fact, definite common nouns are complements of 94% of all *matul*-PPs in the 1st 100-test. There is hardly any space left for indefinite common nouns and proper names. To determine whether this result is robust, we have conducted a second 1st 100-test for the much less frequent short alloform *tul*. In Figure 4.26 the results of the two tests are contrasted with each other.

The preponderance of definite common nouns is not as strong with the short alloform as it is in the case of the long alloform. Nevertheless, with a share of 82%, this complement type is still the unchallenged default option. Proper names are marginal in both tests whereas indefinite common nouns are slightly more popular with *tul* than with *matul*. We keep these facts in mind to refer back to them when we discuss the three other Πs with long and short alloforms below.

	matul	tul
proper name	3	1
indefinite NP	3	5
definite NP	94	28

Figure 4.26: Token frequency of (in)definite complements of (П33) *matul* 'during' and its short alloform *tul* in 1st 100 (Korpus Malti 3.0).

(xxiii) (П34) ***minbarra*** **'except'**: Owing to the existence of the synonymous Пs (П20) *għajr* 'except' and (П24) *ħlief* 'except' which we have looked at in paragraphs (xiii) and (xvi), respectively, (П34) *minbarra* 'except' deserves special attention. Like its synonyms, (П34) *minbarra* 'except' takes complements from all three types as shown in (4.45).

(4.45) (П34) *minbarra* 'except'

(a) [Bartolo/Vella 2009: 394–395]

kulħadd spiċċa mal-art fuq xulxin kulħadd
everybody finish with:DEF-floor on each_other everybody
[minbarra *figura waħda biss]$_{PP}$*
[except figure:F one:F only]
'[. . .] everybody finished on the floor one over the other – everybody **[except** one little figure].'

(b) [Bartolo/Vella 2009: 157]

kulma kien ġarrab kien xi grif u tnittif
whatever be experience be some scratching and plucking
[minbarra *l-fatt li d-dinjità tiegħu kienet qalgħet*
[except DEF-fact that DEF-dignity of:3SG.M be:3SG.F.PFV earn

daqqa ta' ħarta iebsa]$_{PP}$
hit:F of slap hard:F]
'[. . .] all he went through was some scratching and plucking [**except** the fact that his dignity received a hard slap in the face].'

(c) [Bartolo/Vella 2009: 353]
U kulħadd infexx iċapċap u jgħajjat
and everybody give_vent 3.IPFV.3SG.M:applaud and 3.IPFV:shout
*bil-ferħ kulħadd **[minbarra** Verdra]*$_{PP}$
with:DEF-joy everybody **[except** Verdra]
'And everybody broke out in applause and shouts of joy – everybody [**except** Verdra].'

We know from Figure 4.20 that (Π24) *ħlief* 'except' combines most often with indefinite complements whereas (Π20) *għajr* 'except' favours combinations with definite common nouns (see Figure 4.17). Since (Π34) *minbarra* 'except' also has a short alloform *barra*, we have again conducted two tests on the basis of the 1st 100 hits in the Korpus Malti 3.0 the results of which are presented in Figure 4.27.

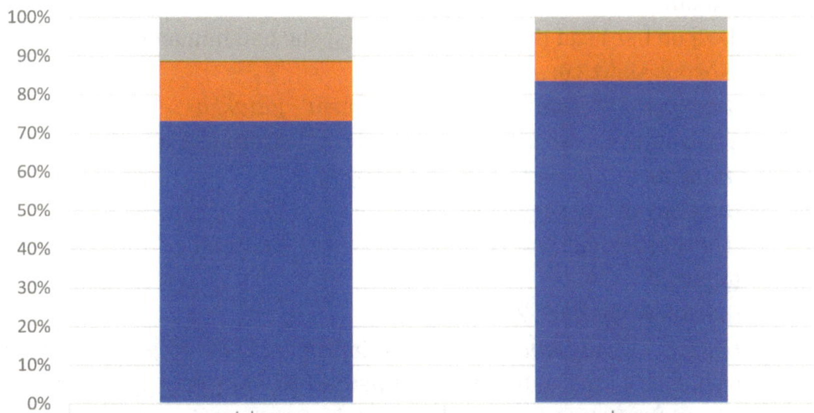

	minbarra	barra
proper name	8	1
indefinite NP	11	3
definite NP	51	20

Figure 4.27: Token frequency of (in)definite complements of (Π34) *minbarra* 'except' and its short alloform *barra* in 1st 100 (Korpus Malti 3.0).

In both tests, definite common nouns constitute the clear majority. In the case of the long alloform, 72.8% of all relevant PPs involve a definite common noun whereas the short alloform occurs in 83.3% of its attestations with a definite noun as complement. Of the three Πs of EXCLUSION, (Π34) *minbarra* 'except' is thus the one which has the strongest leaning towards definiteness. (Π34) *minbarra* 'except' however prefers indefiniteness whereas (Π20) *għajr* 'except' occupies a position between the two extremes. What motivates this differential behaviour of semantically identical Πs needs to be investigated further.[31]

(xxiv) (Π35) *minflok* 'instead of': The next Π behaves special because of the different preferences of the long and the short alloform as to the choice of complement type. The examples in (4.46) indicate that there is in principle no obstacle for (Π35) *minflok* 'instead of' to take an indefinite complement or a definite complement be they common nouns or proper names.

(4.46) (Π35) *minflok* 'instead of'

(a) [Bartolo/Vella 2009: 299]

fuq rasha ***[minflok*** *rix]*$_{PP}$ *kellha xagħar*
on head:3SG.F **[instead_of** feather.PL] have.PFV:IO:3SG.F hair
uman
human
'[. . .] on her head [**instead of** feathers] she had human hair.'

(b) [Korpus Malti 3.0; academic13]

Il-ħanfus jista' faċilment jattakka l-għeruq
DEF-beetle.PL 3SG.M.IPFV:can easily 3SG.M.IPFV:attack DEF-root.PL
[minflok *iz-zkuk u l-weraq]*$_{PP}$
[instead_of DEF-trunk.PL and DEF-leaf.PL]
'The beetles can easily attack the roots [**instead of** the trunks and the foliage].'

(c) [Bartolo/Vella 2009: 213]

Użajt id-Datura ferox ***[minflok*** *il-Brugmansia sanguinea]*$_{PP}$
use:1SG.PFV DEF-datura ferox **[instead_of** DEF-Brugmansia sanguinea]
'I used the Datura Ferox [**instead of** the Brugmansia Sanguinea].'

The comparison of the turnouts for short and long alloforms of (Π33) *matul* 'during' and (Π34) *minbarra* 'except' in the two previous paragraphs has shown that the outstanding position of definite common nouns is not affected by the choice of alloform. This is strikingly different in the case of (Π35) *minflok* 'instead of' as shown in Figure 4.28.

---

31 Cf. Section 4.1.1.2.2.2.

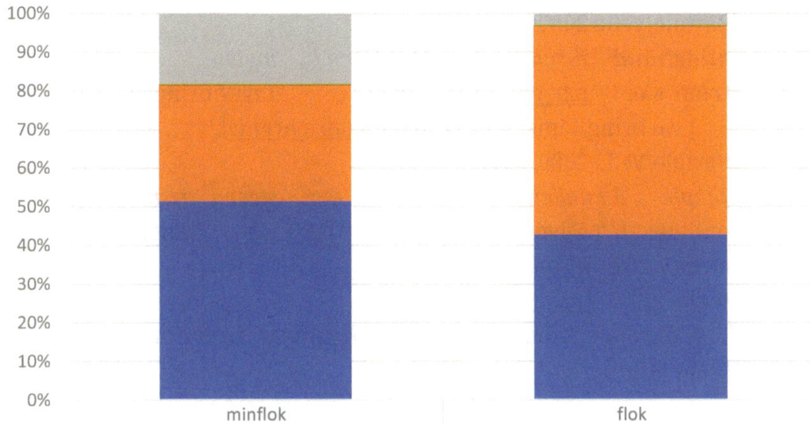

	minflok	flok
proper name	8	2
definite NP	13	33
indefinite NP	22	26

Figure 4.28: Token frequency of (in)definite complements of (Π35) *minflok* 'instead of' and its short alloform *flok* in 1st 100 (Korpus Malti 3.0).

With the long alloform, definite common nouns form only a 30%-minority whereas indefinite common nouns cover 51.1% of all countable PPs in the 1st 100-test in the Korpus Malti 3.0. The two complement types change places if we exclusively take account of the short alloform. In this case, definite common nouns are in the majority with 54% leaving indefinite common nouns behind in second position with 42.6%. Across the two tests, indefinite complements retain a strong position. In the 1st 100-test, the short alloform *flok* yields more countable PPs than is the case with the long alloform. Whether frequency effects are responsible for the differential behaviour cannot be determined in this study because a predominantly quantitative approach is needed to tackle this problem.

(xxv) (Π38) *mingħand* 'from s.o.': Owing to their shared restriction to animate (preferably human) complements, (Π38) *mingħand* 'from s.o.' should be compared to (Π22) *għand* 'at s.o.'s place' which is the topic of paragraph (xiv). This preference can be recognised in (4.47) below and (4.35) above. For both Πs, (in)definiteness of complements is irrelevant.

(4.47) (Π38) *mingħand* 'from s.o.'

(a) [Bartolo/Vella 2009: 469]

għandna rigal għalik [***mingħand*** *ħabib kbir tiegħek*]$_{PP}$

have:1PL present for:2SG [**from_s.o.** friend big of:2SG]

'[. . .] we have a present for you [**from** a good friend of yours].'

(b) [Bartolo/Vella 2009: 289]

[mingħand	*il-ġenituri*	*tagħkom]_{PP}*	*inġibu*	*xi*	*oġġetti*
[from_s.o.	DEF-parent:PL	of:2PL]	1.IPFV:bring:PL	some	object:PL

'[. . .] we bring some objects [**from** your parents].'

(c) [Bartolo/Vella 2009: 194]

kompla	*il-Fiddien*	*bla*	*ma*	*stenna*	*risposta*	*jew*	*reazzjoni*
continue	DEF-Silversmith	without	that	wait	answer	or	reaction
[mingħand	*Jason]$_{PP}$*						
[from	Jason]						

'[. . .] the Silversmith continued without waiting for an answer or reaction [**from** Jason].'

The shared conditions notwithstanding, the two Πs display widely different behaviours as to their preferences in the domain of complements. From Figure 4.29 we learn that definite common nouns fail to exceed the 50%-mark although this complement type has the biggest share. Furthermore, proper names double their share compared to that reported for (Π22) *għand* 'at s.o.'s place' in Figure 4.18.

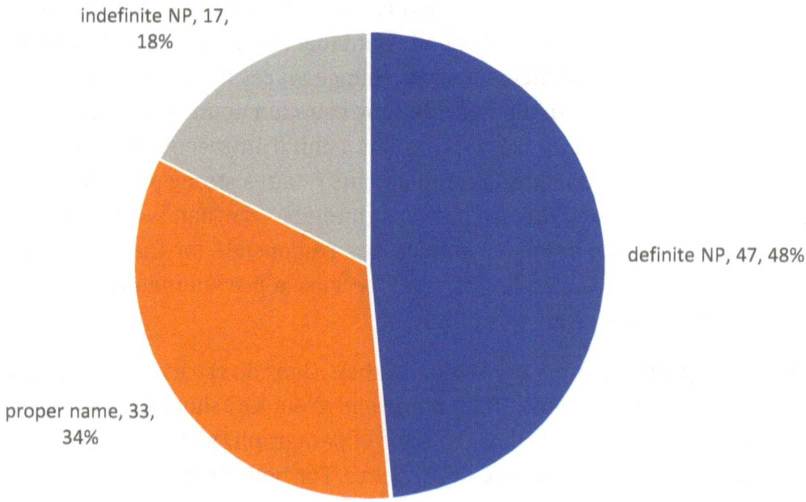

Figure 4.29: Token frequency of (in)definite complements of (Π38) *mingħand* 'from s.o.' in 1st 100 (Korpus Malti 3.0).

The motivation for this dissimilarity is not self-evident and thus needs to be explored further.[32]

32 Cf. Section 4.1.1.2.2.2.

(xxvi) (П39) *minħabba* 'on account of': The causal (П39) *minħabba* 'on account of' follows the general tendency of being open to combinations with all types of complements as shown in (4.48).

(4.48) (П39) *minħabba* 'on account of'

(a) [Camilleri 2013: 133]

fil-konġura *ta* *sehmu* *wkoll* *it-traffiku* ***[minħabba***
in:DEF-conspiracy take part:3SG.M also DEF-traffic **[on_account_of**
ħabta *ħdejn* *l-Addolorata]*$_{PP}$
collision beside DEF-Addolorata]
'[. . .] in the conspiracy, the traffic too played its part [**on account of** a collision near the Addolorata].'

(b) [Bartolo/Vella 2009: 123]

Efrem *kien* *dak* *li* *jibża'* *ħafna* *mill-oħrajn*
Efrem be DEM:M:DIST that 3SG.M.IPFV:fear much from:DEF-other:PL
[minħabba *ċ-ċokon* *tiegħu]*$_{PP}$
[on_account_of DEF-smallness of:3SG.M]
'Efrem was the one who was more afraid than the others [**on account of** his small size].'

(c) [Bartolo/Vella 2009: 220]

Kien *inkwitat* *ukoll* ***[minħabba*** *Erkole]*$_{PP}$
be trouble:PART also **[on_account_of** Erkole]
'[. . .] he was worried also [**on account of** Erkole].'

We have introduced the three alloforms of (П39) *minħabba* 'on account of' above.[33] Figure 4.30 provides the opportunity to compare the three alloforms *minħabba, mħabba,* and *ħabba* in terms of the token frequency the three complement types have in combinations with these alloforms. As one might expect, definite common nouns are in the majority in all three of the 1st 100-tests. Proper names occur only in combination with *ħabba.* Indefinite complements have their stronghold in PPS which involve the alloform *mħabba* as head. However, in none of the cases do indefinite complements claim more than a third of the countable PPS. It can therefore be concluded that the coexistence of several alloforms is irrelevant to the behaviour of (П39) *minħabba* 'on account of'.

33 In Section 3.2.4.

	minħabba	(i)mħabba	ħabba
proper name	0	0	2
indefinite NP	16	11	6
definite NP	64	22	15

Figure 4.30: Token frequency of (in)definite complements of (Π39) *minħabba* 'on account of' and its short alloforms *mħabba* and *ħabba* in 1st 100 (Korpus Malti 3.0).

(xxvii) (Π40) *minkejja* 'in spite of': The most remarkable fact about (Π40) *minkejja* 'in spite of' is the absence of examples of this Π taking a proper name as complement. Accordingly, (4.49) only gives evidence of the patterns *minkejja* + NP_{DEF} and *minkejja* + NP_{INDEF}.

(4.49) (Π40) *minkejja* 'in spite of'

(a) [Camilleri 2013: 144]

l-aħħar tfakkira t' imgħoddi li ***[minkejja*** *sforzi*
DEF-last reminder:F of past that **[in_spite_of** effort:PL
kosmetiċi]$_{PP}$ ħadd ma rnexxielu jiknes
cosmetic:PL] anybody NEG succeed:IO:3SG.M 3SG.M.IPFV:sweep
għalkollox taħt it-tapit
completely under DEF-carpet
'[. . .] the last reminder of a past which [**in spite of** cosmetic efforts] nobody succeeded in sweeping completely under the carpet.'

(b) [Camilleri 2013: 144]

stennieha ikreh imma skoprielha ċertu seħer [. . .]
wait:3SG.F ugly.CMPR but discover:IO:3SG.F certain magic

[minkejja | *t-tiġrib* | *u* | *t-taqlib* | *maż-żminijiet]$_{PP}$*
[in_spite_of DEF-experiencing and DEF-changing with:DEF-time:PL]
'[. . .] he expected it to be uglier but he discovered in it a certain magic **[in spite of** the experiences and the sudden changes over time].'

In the 1st 100-test, the overwhelming majority (= 96%) of the *minkejja*-PPs involve a definite common noun as complement as shown in Figure 4.31.

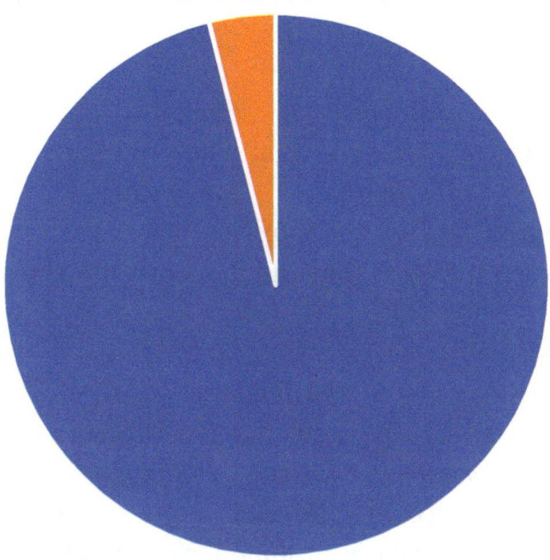

Figure 4.31: Token frequency of (in)definite complements of (Π40) *minkejja* 'in spite of' in 1st 100 (Korpus Malti 3.0).

The reason for the absence of proper names has to be sought in semantics.

(xxviii) (Π44) *oltre* 'beyond': For (Π44) *oltre* 'beyond' it is difficult to identify uncontroversial instances of the pattern *oltre* + PROPER NAME. The close connection of the use of this Π and the Maltese eurolect has been mentioned already.[34] We interpret *Komunità* 'community' in (4.50c) as autonym that is employed in E.U. legal texts to refer by name to the European Union. In this sense then we are facing a proper name as complement of (Π44) *oltre* 'beyond'.

34 In Section 1.6.6 (v).

(4.50) (Π44) *oltre* 'beyond'

(a) [Korpus Malti 3.0; news36898]

l-maltatoday	*kienet*	*ikkundannata*	*mill-Qorti*
DEF-Maltatoday	be:3SG.F.PFV	condemn:PART:F	from:DEF-court

tħallasni	*s-somma*	*ta'*	*€ 18,000*	*[**oltre***	*imgħax]$_{PP}$*	*bħala*
3SG.F.IPFV:pay:1SG	DEF-sum	of	€ 18,000	[**beyond**	interest]	as

danni
damage:PL

'[. . .] Maltatoday was sentenced to pay me the sum of € 18,000 [**beyond** interests] as compensation.'

(b) [Korpus Malti 3.0; european629]

ma	*għandux*	*jagħti*	*deċiżjoni*	*li*	*tmur*
NEG	have:3SG.M:NEG	3SG.M.IPFV:give	decision	that	3SG.F.IPFV:go

*[**oltre***	*l-iskop*	*stipulat*	*f'*	*dan*	*l-Artikolu]$_{PP}$*
[**beyond**	DEF-scope	stipulate:PART	in	DEM:M:PROX	DEF-article]

'[. . .] he should not take a decision that goes [**beyond** the scope stipulated in this article].'

(c) [Korpus Malti 3.0; european9744]

jiddeċiedi	*sabiex*	*jaċċetta*	*kwalunkwe*	*rigal*
3SG.M.IPFV:decide	to	3SG.M.IPFV:accept	any	present

legati	*jew*	*għotjiet*	*minn*	*sorsi*	*[**oltre***	*l-Komunità]$_{PP}$*
legacy:PL	or	donation:PL	from	source:PL	[**beyond**	DEF-Community]

'[. . .] he decides to accept any present, legacies or donations from sources [**outside** the Community].'

Figure 4.32 reflects not only the scarcity of proper names as complements of (Π44) *oltre* 'beyond' in the 1st 100-test in the Korpus Malti 3.0 but, at the same time, it also shows the uncontested dominance of definite common nouns as complements of the same Π.

(xxix) (Π46) *qabel* 'before': As can be gathered from (4.51), (Π46) *qabel* 'before' is well-behaved insofar as it can take complements of any type.

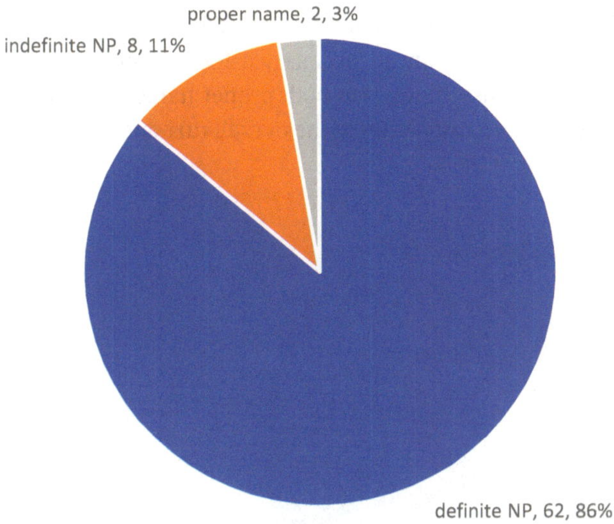

Figure 4.32: Token frequency of (in)definite complements of (П44) *oltre* 'beyond' in 1st 100 (Korpus Malti 3.0).

(4.51) (П46) *qabel* 'before'

(a) [Korpus Malti 3.0; academic13]

Żgur li ma jkunx hemm ħtieġa li
sure that NEG 3SG.M.IPFV:be.FUT:NEG there necessity that
nibdlu l-qasrija ***[qabel*** *tlett snin]*$_{PP}$
1.IPFV:change:PL DEF-flower_pot **[before** three year.PL]
'There surely is no need for us to change the flower-pot [**for** three years].'

(b) [Camilleri 2013: 70]

Għamlet tabirruħha li dieħla sat-toilet minsub
do:3SG.F.PFV as_if that enter.PART:F till:DEF-toilet PART:RFL:find
eżatt ***[qabel*** *it-turġien]*$_{PP}$
exactly **[before** DEF-stair:PL]
'She pretended to go to the toilet which was located exactly [**in front of** the stairs].'

(c) [Camilleri 2013: 241]

Bħal Napuljun ***[qabel*** *Warerloo]*$_{PP}$ *issa kellu*
like Napoleon **[before** Waterloo] now have.PFV:IO:3SG.M
jgħaddi mill-fatti għall-attakk.
3SG.M.IPFV:pass from:DEF-fact:PL for:DEF-attack
'Like Napoleon [**before** Waterloo] he now had to move on from the facts to the attack.'

The 1st 100-test additionally supports the idea that (Π46) *qabel* 'before' agrees with many other Maltese Πs as to the pronounced preference for definite common nouns as complements. Figure 4.33 also shows that both proper names and indefinite complements contribute only insignificantly to the overall turnout.

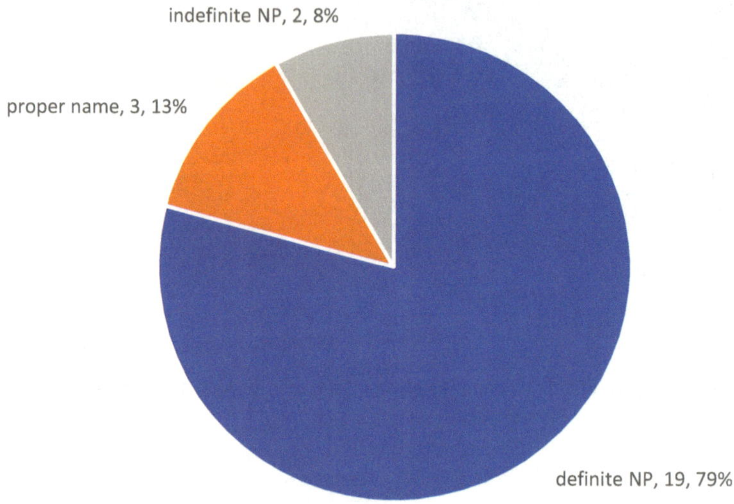

Figure 4.33: Token frequency of (in)definite complements of (Π46) *qabel* 'before' in 1st 100 (Korpus Malti 3.0).

(xxx) (Π47) *qalb* 'amidst': The behaviour of (Π47) *qalb* 'amidst' is largely the same as that of its near synonym (Π16) *fost* 'amongst' discussed in paragraph (x). First of all, both Πs share the disability of taking genuine proper names as complements. As in (4.31), the examples in (4.52) only involve (in)definite common nouns. The (c)-sentence is missing.

(4.52) (Π47) *qalb* 'amidst'

(a) [Bartolo/Vella 2009: 75]

kien miexi ***[qalb*** *merħla żgħira ta' nagħaġ]$_{PP}$*

be walk.PART **[amidst** flock:F small:F of sheep]

'[. . .] he was walking **[in the middle of** a small flock of sheep].'

(b) [Bartolo/Vella 2009: 81–82]

huma lestew ruħhom biex jidħlu ***[qalb***

3PL prepare:3PL.PFV soul:3PL to 3.IPFV:enter:PL **[amidst**

il-friegħi tal-ħarruba]$_{PP}$

DEF-branch:PL of:DEF-carob]

'[. . .] they prepared themselves for entering **[into** the branches of the carob-tree].'

The parallel behaviour of (Π47) *qalb* 'amidst' and (Π16) *fost* 'amongst' transpires also from the comparison of the results of the 1st 100-test in Figures 4.14 and 4.34.

The shares of definite common nouns and indefinite common nouns are divided in absolutely identical ways. Both (Π47) *qalb* 'amidst' and (Π16) *fost* 'amongst' display a 93%-majority of definite common nouns as opposed to a 7%-minority of indefinite common nouns. We assume that this parallelism is not incidental but cannot offer a ready explanation at this point.

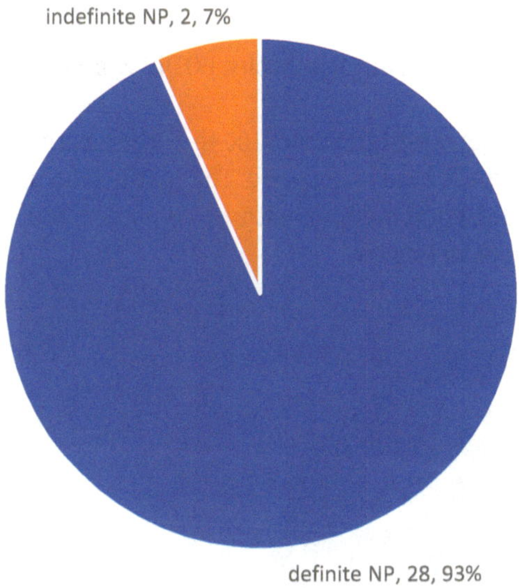

Figure 4.34: Token frequency of (in)definite complements of (Π47) *qalb* 'amidst' in 1st 100 (Korpus Malti 3.0).

(xxxi) (Π48) *qrib* 'near': Our next item on the list has no problems combining with proper names. (Π48) *qrib* 'near' is compatible with each of the three complement types as can be seen in (4.53).

(4.53) (Π48) *qrib* 'near'

(a) [Korpus Malti 3.0; culture1270]

dan *tal-aħħar* *segwihom* *sakemm* *ipparkjaw* ***[qrib***
DEM:M:PROX of:DEF-last follow:3PL until park:3PL.PFV **[near**

supermarket $]_{PP}$
supermarket]

'[. . .] this last one followed them until they parked [**near** a supermarket].'

(b) [Camilleri 2013: 171]
sab ruħu fi Triq il-Pont ***[qrib*** *l-iskola tal-bniet* $]_{PP}$
find soul:3SG.M in Triq il-Pont [**near** DEF-school of:DEF-girl.PL]
'[. . .] he found himself in Bridge Street [**near** the girlschool].'

(c) [Korpus Malti 3.0; academic2]
Fil-punent tas-Sri Lanka ***[qrib*** *Colombo* $]_{PP}$ *il-belt kapitali*
in:DEF-west of:DEF-Sri Lanka [**near** Colombo] DEF-city capital
taħdem il-PODIE
3SG.F.IPFV:work DEF-PODIE
'In the west of Sri Lanka [**near** Colombo], the capital city, PODIE works [. . .].'

Figure 4.35 features proper names as second best in the 1st 100-test. The relatively big share claimed by this complement type is easily explained. The spatial meaning of (Π48) *qrib* 'near' makes it eligible for combinations with place names which serve as Ground for allocating the PP-external figure.

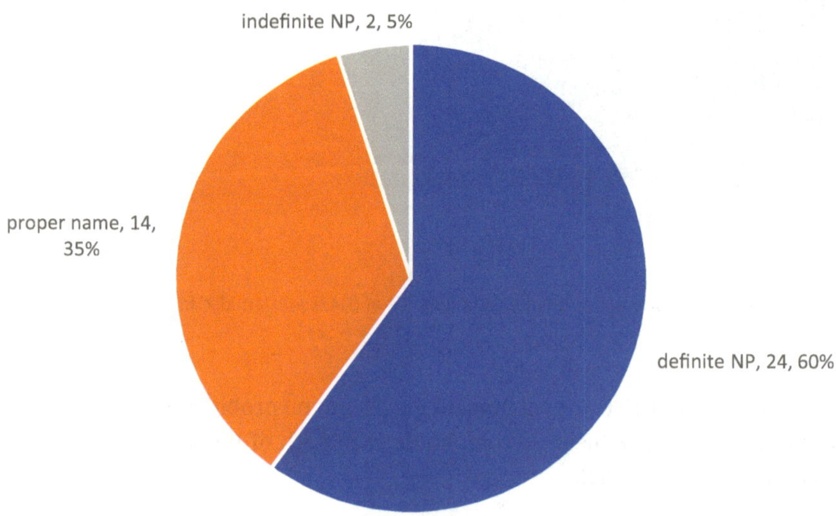

Figure 4.35: Token frequency of (in)definite complements of (Π48) *qrib* 'near' in 1st 100 (Korpus Malti 3.0).

As will come to the fore in paragraph (xxxvii), the synonymous (Π58) *viċin* 'near' yields strikingly similar results (see Figure 4.41) so that the idea suggests itself that the high incidence of proper-name complements is semantically motivated.

(xxxii) (Π49) *quddiem* 'in front of': In the case of (Π49) *quddiem* 'in front of', too, we cannot help repeating ourselves since this Π conforms to the general pattern according to which a Maltese Π prototypically allows for combinations with representatives of each complement type. This ability is reflected in (4.54).

(4.54) (Π49) *quddiem* 'in front of'

(a) [Bartolo/Vella 2009: 77]
waqfet ***[quddiem*** *ħajt* *għoli]*$_{PP}$
stop:3SG.F.PFV **[in_front_of** wall high]
'[. . .] she stopped [**in front of** a high wall].'

(b) [Bartolo/Vella 2009: 78]
Seta' jxomm biss riħa qawwija li kienet
can 3SG.M.IPFV:smell only odour:F strong:F that be:3SG.F.PFV
ġejja mis-siġar li kien hemm ***[quddiem***
come:PART:F from:DEF-tree.PL that be there **[in_front_of**
il-ħajt tas-sejjieħ]$_{PP}$
DEF-wall of:DEF-rubble]
'He could only smell a strong odour coming from the trees which were there [**in front of** the rubble wall].'

(c) [Bartolo/Vella 2009: 24]
indunat aħjar ma tgħidhiex ***[quddiem***
note:3SG.F.PFV better NEG 3SG.F.IPFV:say:3SG.F:NEG **[in_front_of**
Reggie]$_{PP}$
Reggie]
'[. . .] she noticed that she should better not say it [**in front of** Reggie].'

The spatial meaning of (Π49) *quddiem* 'in front of' notwithstanding, Figure 4.36 is no replica of Figure 4.35. Proper names do not stand out in the 1st 100-test. Does this mean that different spatial functions have different repercussions on the choice of complement type? The answer to this question must be sought in a follow-up study that grants more space to native-speaker judgements.

(xxxiii) (Π50) *rigward* 'concerning': As to the combinability with different complement types, (Π50) *rigward* 'concerning' takes side with the majority of the Maltese Πs. In (4.55), the usual triple of sentences (a), (b), and (c) illustrates the omni-compatibility of this Π.

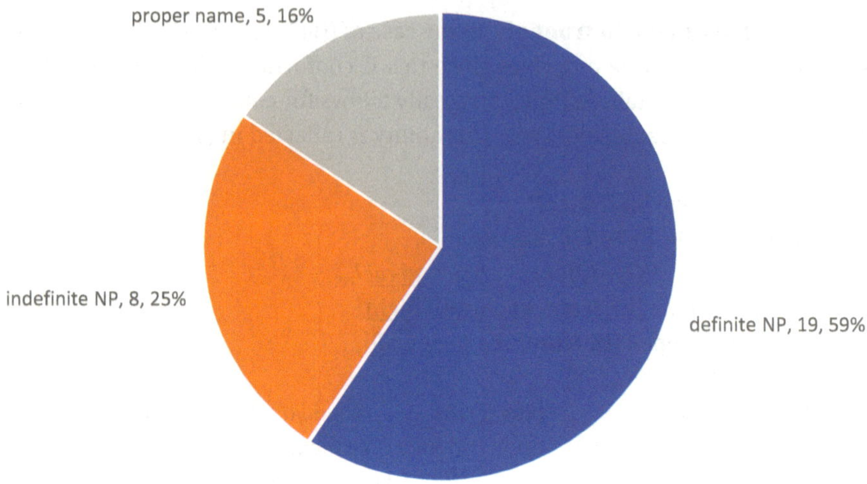

Figure 4.36: Token frequency of (in)definite complements of (Π49) *quddiem* 'in front of' in 1st 100 (Korpus Malti 3.0).

(4.55) (Π50) *rigward* 'concerning'

(a) [Korpus Malti 3.0; academic165]

Ser nipprova ndaħħal artikli ***[rigward*** *kompjuters]PP*

FUT 1SG.IPFV:try 1SG.IPFV:CAUS:enter article:PL **[concerning** computer:PL]

'I will try to introduce articles [**concerning** computers].'

(b) [Korpus Malti 3.0; academic273]

nikteb artiklu tajjeb ***[rigward*** *id-drittijiet ta' l-awtur]PP*

1SG.IPFV:write article good **[concerning** DEF-right:PL of DEF-author]

'[. . .] I write a good article [**concerning** the rights of the author].'

(c) [Korpus Malti 3.0; academic30]

Punt pożittiv ***[rigward*** *Mintoff]PP kien il-fatt tar-Repubbllika*

point positive **[concerning** Mintoff] be DEF-fact of:DEF-republic

'A positive point [**concerning** Mintoff] was the matter of the republic [. . .].'

Likewise, the 1st 100-test in Figure 4.37 helps to classify (Π50) *rigward* 'concerning' as one of those Πs which confirm the hierarchy postulated in Figure 4.1. Definite common nouns are attested five times as frequently as indefinite common nouns whereas proper names only come in third place.

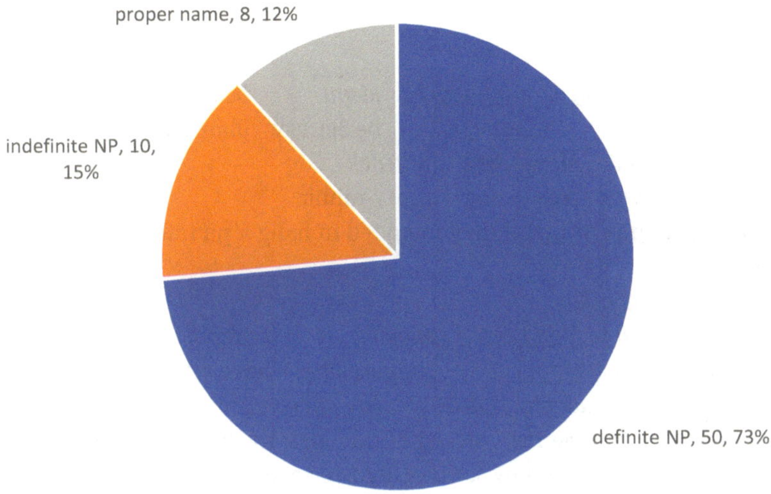

Figure 4.37: Token frequency of (in)definite complements of (Π50) *rigward* 'concerning' in 1st 100 (Korpus Malti 3.0).

(xxxiv) (Π52) *sforz* 'thanks to': This Π stands out from the bulk of the members of BLOMP 2.0 insofar as we have found no evidence of (Π52) *sforz* 'thanks to' taking indefinite complements or proper names as complements. Neither the Korpus Malti 3.0 nor our literary corpus are helpful with regards to the filling of the gaps. Thus, (4.56) illustrates only the combination with a definite common noun.

(4.56) [Chetcuti 1990: 2]

ommi *dak* *il-mument* *qaltli* *hekk*
mother:1SG DEM:M:DIST DEF-moment say:3SG.F.PFV:IO:1SG so
[sforz *id-dwejjaq* *li* *kellha* *fuqha* $]_{PP}$
[thanks_to DEF-worries that have.PFV:3SG.F on:3SG.F]
'[. . .] in that moment my mother told me so [**because of** the worries she had].'

In the absence of two of the three complement classes, it is unnecessary to carry out any calculations as to the frequency of the constructions.

(xxxv) (Π53) *skont* 'according to': The fact that (Π53) *skont* 'according to' allows for the three patterns *skont* + NP_{INDEF}, *skont* + NP_{DEF}, and *skont* + PROPER NAME is illustrated in (4.57).

(4.57) (П53) *skont* 'according to'

(a) [Camilleri 2013: 37]

*[**Skont** wieħed minnhom]$_{PP}$ kienu beħsiebhom*
[**according_to** one from:3PL] be:3PL.PFV plan:3PL
idendlu ma' siġra kelb tal-fenek
3.IPFV:hang:PL with tree dog of:DEF-rabbit
'[**According** to one of them] they intended to hang a pharao's hound from a tree.'

(b) [Bartolo/Vella 2009: 72]

*[**Skont** il-mappa]$_{PP}$ [. . .] suppost kellhom*
[**according_to** DEF-map] suppose:PART have.PFV:IO:3PL
jibqgħu għaddejjin dritt fuq din l-istess triq
3.IPFV:remain:PL pass:PART:PL straight on DEM:F:PROX DEF-same street
'[**According** to the map] they supposedly had to keep going straight on the same road.'

(c) [Bartolo/Vella 2009: 87]

*[**skont** Estella]$_{PP}$ il-ħarruba kienet qaltilhom*
[**according_to** Estella] DEF-carob:F be:3SG.F.PFV say:3SG.F.PFV:IO:3PL
biex ma jdumux fil-ġnien.
to NEG 3.IPFV:take_time:PL:NEG in:DEF-garden
'[. . .] [**according to** Estella] the carob-tree had told them not to remain too long in the garden.'

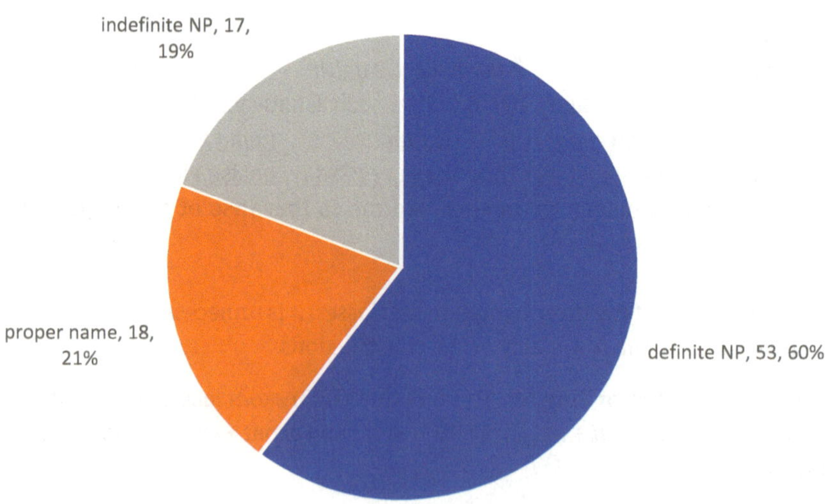

Figure 4.38: Token frequency of (in)definite complements of (П53) *skont* 'according to' in 1st 100 (Korpus Malti 3.0).

The 1st 100-test yields results which show a surprisingly strong position of proper names in comparison to indefinite complements. Since (Π53) *skont* 'according to' functions as relator whose task it is to introduce the source of a given opinion, it is only logical that this source is not only predominantly human but also a participant who can be identified by name. Accordingly, the noteworthy size of the share of proper names in Figure 4.38 is semantically motivated.

(xxxvi) (Π56) *taħt* 'under': The three complement types are also attested in combination with (Π56) *taħt* 'under' as shown in (4.58).

(4.58) (Π56) *taħt* 'under'

(a) [Camilleri 2013: 156]

meta jkun f' sitwazzjoni kkumplikata jew
when 3SG.M.IPFV:be.FUT in situation complicate:PART:F or
[taħt pressjoni]$_{PP}$
[**under** pressure]
'[. . .] when he will be in a complicated situation or [**under** pressure] [. . .].'

(b) [Abela 2016: 211]

Teresa daħħlet rasha [taħt il-mejda]$_{PP}$
Teresa CAUS:enter:3SG.F.PFV head:3SG.F [**under** DEF-table]
'Teresa put her head [**under** the table] [. . .].'

(c) [Korpus malti 3.0; academic12]

[Taħt Meħmet Ali]$_{PP}$ *kien inbeda perijodu ta' żvilupp*
[**under** Mehmet Ali] be RFL:begin period of development
'[**Under** Mehmet Ali] a period of devolpment had begun.'

Having the possibility of combining with each of the complement types once more does not imply that the three patterns *taħt* + NP_{INDEF}, *taħt* + NP_{DEF}, and *taħt* + PROPER NAME occur equally frequently in the 1st 100-test. Figure 4.39 gives evidence of the overwhelming 83%-majority formed by *taħt*-PPs involving definite common nouns. Proper names are lagging behind whereas the small size of the share of indefinite complements suffices to be placed in second position.

(xxxvii) (Π57) *versu* 'towards': The general compatibility of (Π57) *versu* 'towards' with the three complement types under inspection is no surprise to us. In (4.59), we find the relevant examples.

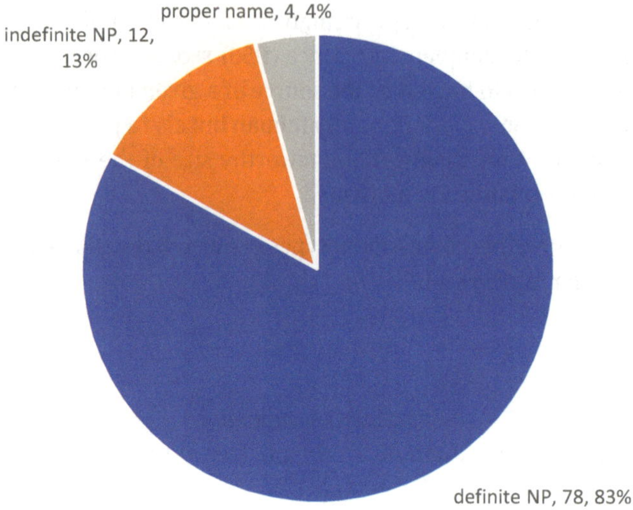

Figure 4.39: Token frequency of (in)definite complements of (П56) *taħt* 'under' in 1st 100 (Korpus Malti 3.0).

(4.59) (П57) *versu* 'towards'

(a) [Korpus Malti 3.0; news80225]

naraw *l-imġieba [. . .]* *tad-diversi* *stati* *tal-Ewropa*
1.IPFV:see:PL DEF-behaviour of:DEF-diverse state:PL of:DEF-Europe
[versu *gvernijiet* *li* *ma* *jafux* *fejn*
[**towards** government:PL that NEG 3.IPFV:know:PL:NEG where
toqgħod *id-demokrazija]*$_{PP}$
3SG.F.IPFV:stay DEF-democracy]
'[. . .] we see the behaviour of several states of Europe [**towards** governments which do not know where to apply democracy].'

(b) [Korpus Malti 3.0; european9744]

il-Kummissjoni *trid* *tkun* *responsabbli*
DEF-commission 3SG.F.IPFV:require 3SG.F.IPFV:be.FUT responsible
[versu *ċ-ċittadini]*$_{PP}$
[**towards** DEF-citizen:PL]
'[. . .] the Commission needs to be responsible [**towards** the citizens] [. . .].'

(c) [Korpus Malti 3.0; news81933]

Dan *ferm* *differenti* *mill-imġiba* *ta'* *Dr Gonzi*
DEM:M:PROX very different from:DEF-behaviour of Dr Gonzi
[versu *Austin Gatt]*$_{PP}$
[**towards** Austin Gatt]
'This is very different from the behaviour of Dr Gatt [**towards** Austin Gatt].'

The conformity of this Π with the dominant pattern among the members of BLOMP 2.0 should not be taken to render the case of (Π57) *versu* 'towards' uninteresting. To the contrary, this case becomes intriguing if the results of the 1st 100-test disclosed in Figure 4.40 are compared to those of (Π28) *lejn* 'towards' presented in Figure 4.23.

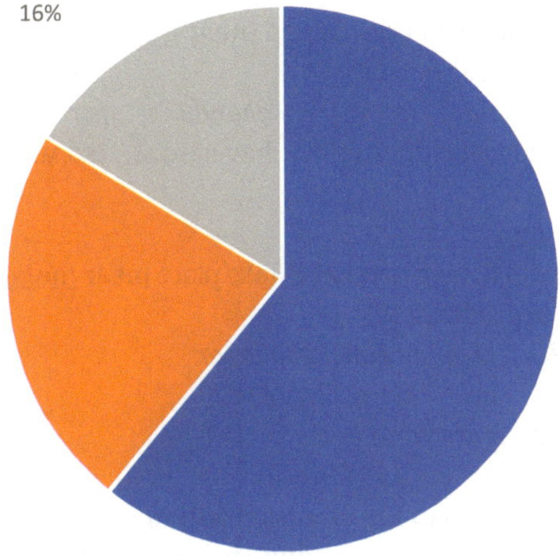

Figure 4.40: Token frequency of (in)definite complements of (Π57) *versu* 'towards' in 1st 100 (Korpus Malti 3.0).

Going by the English translation, the two Πs are synonymous. This supposed synonymy does not preclude differential behaviour of the Πs. Where (Π57) *versu* 'towards' closely follows the hierarchy proposed in Figure 4.1, (Π28) *lejn* 'towards' is a paradigm example of a Maltese Π for which proper names are extraordinarily important as complements. We hypothesise that the behavioural differences between (Π57) *versu* 'towards' and (Π28) *lejn* 'towards' might originate in (perhaps not very subtle) semantic differences.[35]

(xxxviii) (Π58) *viċin* 'near': Definite common nouns, indefinite commons, and proper names are attested as complements of (Π58) *viċin* 'near'. This is the expected situation which is reflected by the examples in (4.60).

35 To which we return in Section 5.1.2.2.6.

(4.60) (П58) *viċin* 'near'

(a) [Korpus Malti 3.0; culture512]

marru *joqgħodu* *f'* *Queensland* ***[viċin*** *fabbrika*
go:3PL.PFV 3.IPFV:stay:PL in Queensland **[near** factory
taz-zokkor]$_{PP}$
of:DEF-sugar]
'[. . .] they went to settle in Queensland [**near** a sugar-factory].'

(b) [Korpus Malti 3.0; european10763]

is-sajd *bix-xbieki* *għandu* *jsir*
DEF-fishing with:DEF-fishing_net.PL have:3SG.M 3SG.M.IPFV:happen
[viċin *il-qiegħ]*$_{PP}$
[near DEF-bottom]
'[. . .] the fishing with fishing-nets has to take place [**near** the bottom of the sea].'

(c) [Korpus Malti 3.0; academic12]

[viċin *Margo]*$_{PP}$ *kien* *hemm* *għadajjar*
[near Margo] be there pool.PL
'[. . .] [**near** Margo] there were pools [. . .].'

What is far more interesting than the omni-compatibility with complement types are the results of the 1st 100-test as shown in Figure 4.41. It is not only the unexpectedly big share of proper names which catches our attention but also the almost full identity of Figures 4.41 and 4.35. The latter accounts for the results of the 1st 100-test for (П48) *qrib* 'near'.

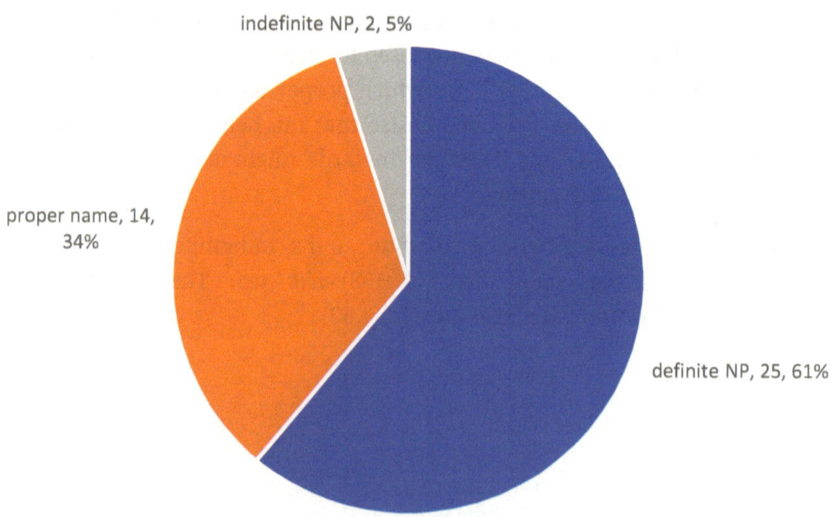

Figure 4.41: Token frequency of (in)definite complements of (П58) *viċin* 'near' in 1st 100 (Korpus Malti 3.0).

The differences are minimal (1%). The two Πs are synonyms of each other and display overrepresentation of proper names. We assume that the parallel behaviour of (Π58) *viċin* 'near' and (Π48) *qrib* 'near' is causally connected to their semantic equivalence.[36]

(xxxix) (Π59) *waqt* 'at the time of': Nothing special can be reported as to (Π59) *waqt* 'at the time of'. No complement type is excluded from combinations with this Π as is shown in (4.61).

(4.61) (Π59) *waqt* 'at the time of'

(a) [Camilleri 2013: 214]

*iddikjara Drasko Pavlovic **[waqt** filmat editjat ta'*
declare Drasko Pavlovic **[at_the_time_of** film_insert edit:PART of
ħames minuti]_{PP}
five minute:PL]
'[. . .] declared Drasko Pavlovic [**during** an edited film insert of five minutes].'

(b) [Bartolo/Vella 2009: 128]

Bħas-soltu kienet ħalliet il-mowbajl
like:DEF-custom be:3SG.F.PFV leave:3SG.F.PFV DEF-mobile_phone
*mixgħul **[waqt** il-lezzonijiet]$_{PP}$*
PART:switch_on **[at_the_time_of** DEF-lesson:PL]
'As usual she had left the mobile-phone switched on [**during** the lessons].'

(c) [Korpus Malti 3.0; academic3]

***[Waqt** l-Intifada [. . .] tal-1990]$_{PP}$ ħafna nisa*
[at_the_time_of DEF-Intifada of:DEF-1990] many woman.PL
Palestinjani saru l-kapijiet de facto tal-familji
Palestinian:PL become:3PL.PFV DEF-boss:PL de facto of:DEF-family:PL
tagħhom
of:3PL
'[**During** the Intifada of 1990] many Palestinian women became the de facto heads of their families.'

The results of the 1st 100-test presented in Figure 4.42 are remarkable only for the relatively big share of indefinite complements which amounts to almost a third of all countable *waqt*-PPS.

36 This will be paid further attention in Section 5.1.2.1.4.

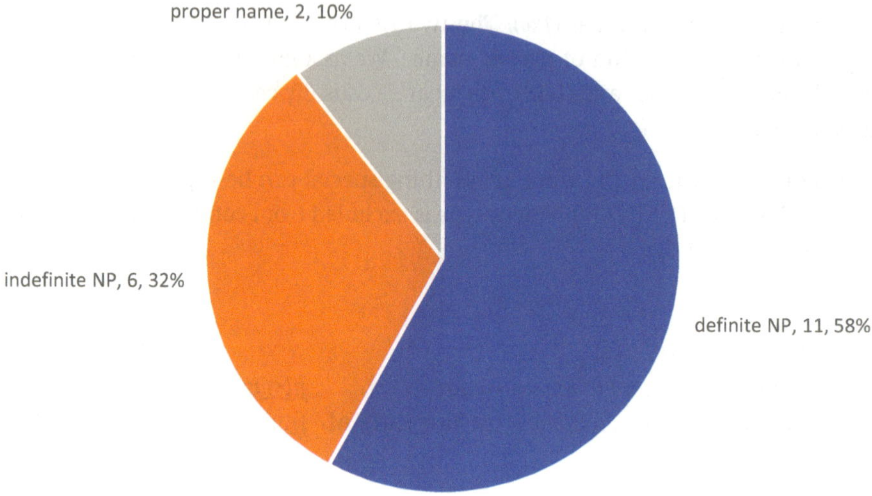

Figure 4.42: Token frequency of (in)definite complements of (Π59) *waqt* 'at the time of' in 1st 100 (Korpus Malti 3.0).

(xl) (Π60) *wara* 'after': The last Π to comment upon is (Π60) *wara* 'after'. The examples in (4.62) are indicative of the three patterns *wara* + NP_{INDEF}, *wara* + NP_{DEF}, and *wara* + PROPER NAME being licit.

(4.62) (Π60) *wara* 'after'

(a) [Bartolo/Vella 2009: 28]

Ħadd mit-tnejn l-oħra ma seta' jgħid xejn
anybody from:DEF-two DEF-other.PL NEG can 3SG.M.IPFV:say nothing
***[wara** monologu bħal dak]$_{PP}$*
[after monologue like DEM:M:DIST]
'None of the other two could say anything [**after** a monologue like that].'

(b) [Abela 2016: 65]

***[Wara** l-ikel]$_{PP}$ kien jaqbad id il-mamà*
[after DEF-meal] be 3SG.M.IPFV:seize hand DEF-mother
'[**After** the meal] he took mother's hand [. . .].'

(c) [Bartolo/Vella 2009: 77]

*qabad miexi **[wara** Estella]$_{PP}$*
seize walk.PART **[after** Estella]
'[. . .] he started to walk [**behind** Estella].'

Interestingly, the results of the 1st 100-test for (Π60) *wara* 'after' presented in Figure 4.43 resemble those given for (Π59) *waqt* 'at the time of' in Figure 4.42.

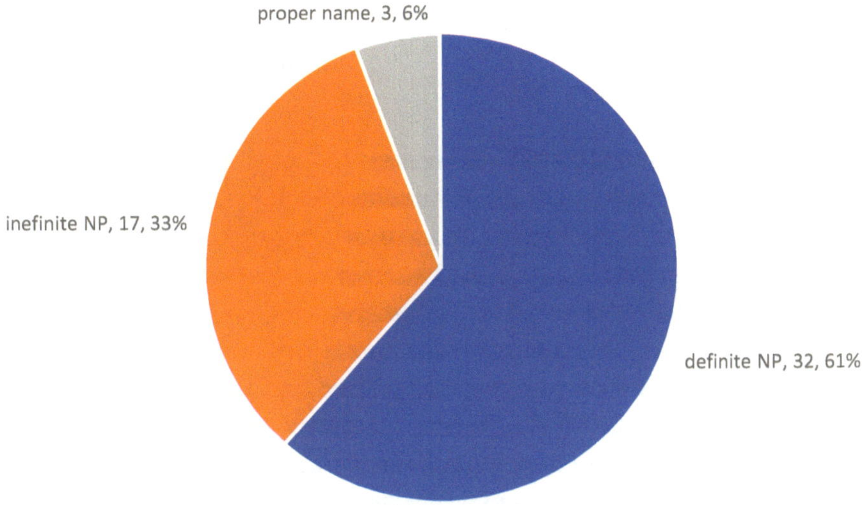

Figure 4.43: Token frequency of (in)definite complements of (П60) *wara* 'after' in 1st 100 (Korpus Malti 3.0).

The resemblance is particularly obvious as to the percentage of indefinite NPS. Whether the similarity between the two ΠIs can be attributed to the fact that both – predominantly or exclusively – express temporal relations has to be investigated below.[37]

The catalogue of non-fusable ΠIs has reached its terminus. We now have the task to systematise the above individualised facts to determine whether the differential behaviour of the ΠIs is random or conditioned by certain (probably semantic) factors.

4.1.1.2.2.2 General survey and evaluation

In the foregoing section, a range of variation has emerged which suggests that the hierarchy in Figure 4.1 covers only a subset of the non-fusable ΠIs of Maltese. As a matter of fact, of the forty non-fusable ΠIs only 24 conform to the hierarchy in the sense that definite common nouns outnumber indefinite common nouns which in turn are more frequent than proper names. With a share of 60%, this pattern dominates but it remains doubtful whether we are witnessing something else than chance frequency. However, the limited importance of the hypothetical hierarchy does not mean that the situation is absolutely chaotic. In 35 (= 87.5%) out of 40 cases, the most frequently attested complement type is that of definite common

37 Cf. Section 5.1.2.1.11.2.

nouns. Similarly, definite complements taken together are numerically stronger than indefinite complements in 36 cases (= 90%). Thus, there is a very strong affinity of complementation to definiteness.

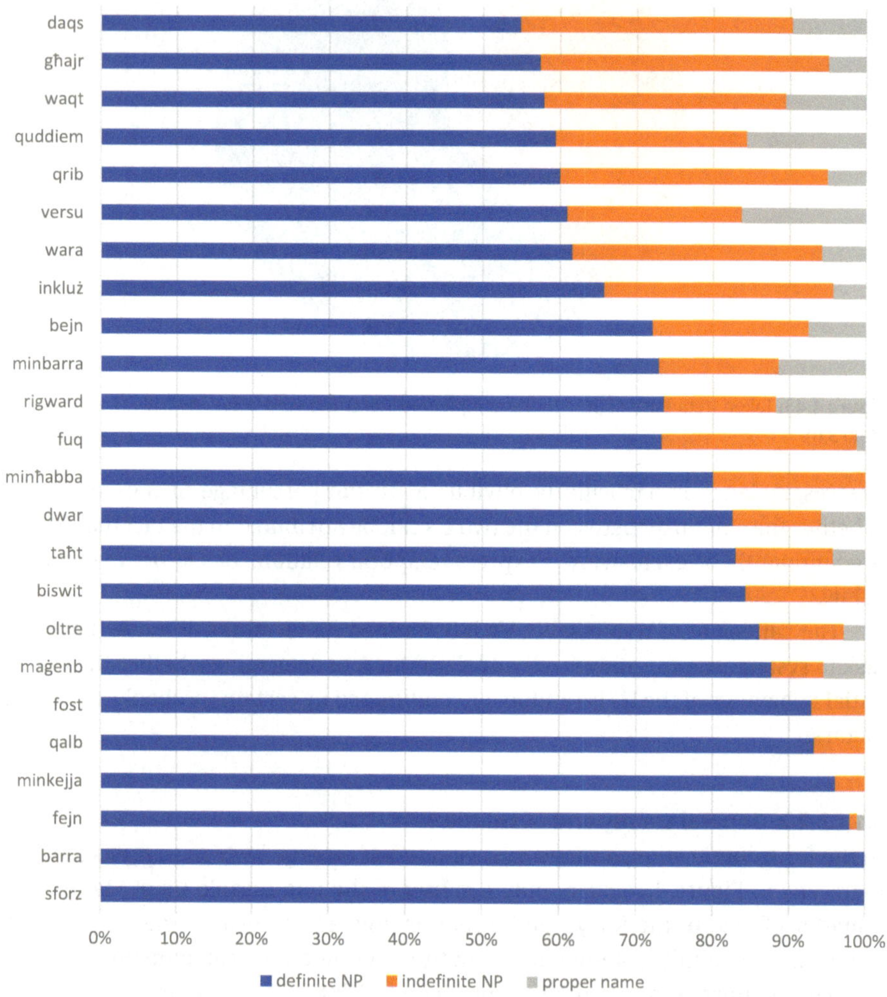

Figure 4.44: Non-fusable Пs corroborating the hypothetical hierarchy of complement types.

Against this backdrop, we put forward the trivial assumption that there is no general preference for definite complements shared by all members of BLOMP 2.0 although definite complements are dominant. What we will probably find instead is a continuum or scale ranging from a clear preference for definite complements to an equally clear preference for indefinite complements. The idea that there is a continuum finds corroboration already with those Πs which by and large confirm the hierarchy of complements. Consider Figure 4.44 for that matter. The non-fusable Πs which are in harmony with Figure 4.1 are ordered top-down according to the increasing share of the definite common nouns. At the bottom of Figure 4.44, (Π9) *bla* 'without' is the sole case of a Π which allows only for one complement type in the 1st 100-test. A further five Πs lack evidence of proper-name complements. The share of definite common nouns is always bigger than half of the countable PPs. The list hosts Semitic Πs alongside Italo-Romance Πs. There are monosyllables as well as polysyllables, Πs which end in a vowel and others which display a final consonant. It is unlikely that any formal factors are decisive for the appearance of a given Π in Figure 4.44. Since the Πs featured in Figure 4.44 also cover a wide range of functions some of which are spatial and others of different kinds it would be too daring to blindly assume that semantics can be held responsible for the compliance of these Πs to the hierarchical pattern. If it is impossible to find an explanation for the observed facts with this group of Πs, it may still be possible to detect motivating factors in those cases which deviate from the supposed hierarchy.

Six Πs violate the hypothetical hierarchy by putting indefinite complements on the topmost rank. In these cases, definite common nouns are always second best whereas proper names are not affected by the change of ranking order as can be gathered from Figure 4.45. The Πs are ordered from left to right according to the decreasing share of indefinite complements. In all cases covered by Figure 4.45, indefinite complements have shares of 51% and more. The composition of this group of Πs is not much different from that of the previous one. There are five Semitic and one Italo-Romance Π. Four Πs fall under the rubric of polysyllabicity as opposed to two monosyllables. Two Πs are equipped with a final vowel, four Πs end in a consonant. In terms of their semantics, the five of the Πs featured in Figure 4.45 can be characterised ex negativo in the sense that they have no spatial function.

Superficially, the semantic argument seems to be rather weak because there is (Π31) *madwar* 'around' on the list which expresses the circumferential relation (Borg/Azzopardi-Alexander 1997: 166). This is the intended meaning of (Π31) *madwar* 'around' when it combines with a definite complement in the 1st 100-test. However, most of the indefinite complements of the same Π are involved in estimates in which the indefinite common noun is quantified with a margin of imprecision. This use is clearly outside the proper spatial domain. In (4.63), the quantified common noun *spetattur* 'spectator' has to be inferred from the context.

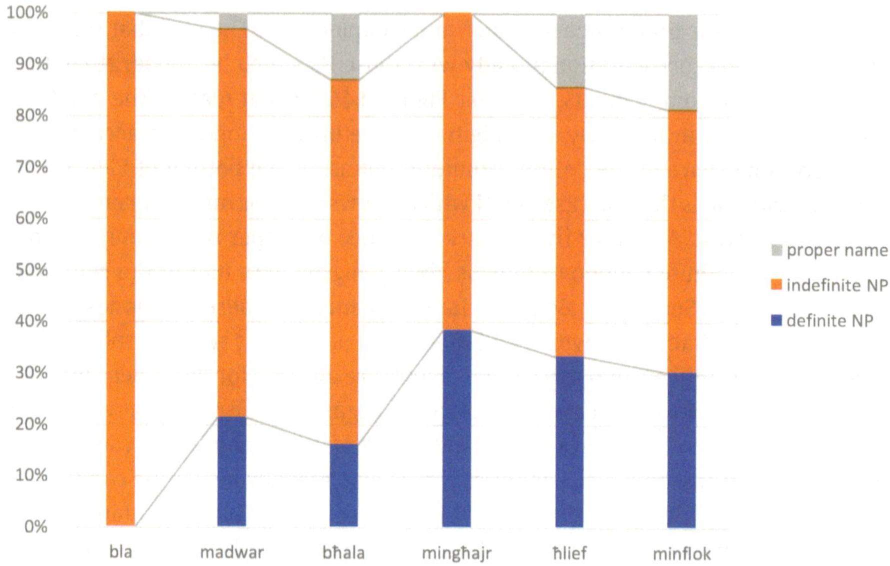

Figure 4.45: Non-fusable Πs with preference for indefinite complements.

(4.63) [Gauci 1994: 26]

Bnew	*grounds*	*ġodda*	*f'*	*Rio*	*wieħed*	*minnhom*
build:3PL.PFV	ground:PL	new.PL	in	Rio	one	from:3PL

jesa'	***madwar***	*200,000*
3SG.M.IPFV:contain	**around**	200,000

'They built new grounds in Rio, one of them sitting **around** 200,000.'

As argued in connection with Aquilina's candidate *(in)ċirka* 'approximately'[38] this usage of *madwar* can be analysed as adverbial and thus falls outside the domain of (Π31) *madwar* 'around'. Adverbial *madwar* does not connect the numeral to a higher head. This fact motivates the absence of square brackets in (4.63). The brackets are not needed because there is no PP, in the first place. The underlined numeral is not the complement of any Π but the head of a quantifier phrase that is adverbially modified by *madwar*.

In the case of (Π6) *bħala* 'as', the preference for indefinite complements cannot be explained with reference to polysemy. However, this Π is frequently used in contexts in which the complement is generic as in (4.25a). The noun in the complement often refers to a profession or function for which a given participant in the situ-

38 Cf. Section 1.4.1.5.

ation is qualified. The already discussed privative/abessive Πs (Π9) *bla* 'without' and (Π36) *mingħajr* 'without' more often than not introduce complements which are irrelevant to the discourse since they do not participate in what happens next. These absentees are frequently represented by generic terms. It is not the absence of a particular known referent which is stated but the absence of any member of a class of referents.

Figure 4.46 surveys the twelve Πs which display an extra-high incidence of proper names as complements. This means that in these cases the share of proper names is either as big as or bigger than that of indefinite complements.

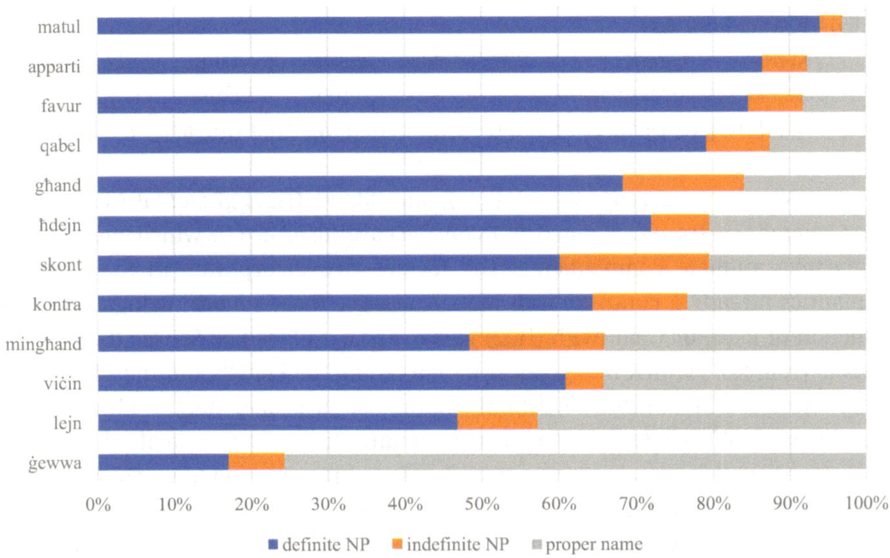

Figure 4.46: Non-fusable Πs with higher ranking proper name complements.

Only with (Π18) *ġewwa* 'inside' do proper names also oust definite common nouns (which are always stronger quantitatively than indefinite complements). In Paragraph (xi), we have already alluded to the possibility that this overrepresentation of (Π18) *ġewwa* 'inside' is based on the widely common use of this Π with place names (see (4.33c)). To some extent this explanation also holds for (Π28) *lejn* 'towards' and (Π58) *viċin* 'near' though in these cases personal names also have a say in generating the extraordinarily big shares of proper names. We assume that (Π13) *favur* 'in favour of' and (Π27) *kontra* 'against' are also logical candidates for combining with proper names because they are involved in descriptions of situations in which something happens to the benefit or to the detriment of a competitor. The participant representing the competitor is prototypically human as shown in (4.29c) and

(4.39c) or associated with institutions run by humans. The latter can be political parties, football clubs, religious denominations, countries, or the like which are identified by name. Typical examples from the history of the football World Championship are given in (4.64)–(4.65).

(4.64) [Gauci 1994: 84]

r-referee	*Taylor*	*ordna*	*penalty*	***[favur***	*Olanda]$_{PP}$*
DEF-referee	Taylor	order	penalty	[**in_favour_of**	Holland]

'[. . .] the referee Taylor gave a penalty [**for** the Netherlands].'

(4.65) [Gauci 1994: 62]

kienu	*nieqsa*	*għall-ewwel*	*partita*	***[kontra***	*l-Bulgarja]$_{PP}$*
be:3PL.PFV	miss.PART.PL	for:DEF-first	match	[**against**	DEF-Bulgaria]

'[. . .] they were absent from the first match [**against** Bulgaria] [. . .].'

As mentioned in Paragraphs (xiv), (xxv), and (xxxv), (Π22) *għand* 'at s.o.'s place', (Π38) *mingħand* 'from s.o.', and (Π53) *skont* 'according to' are Πs which overwhelmingly take animate complements. The former two incorporate in their lexical semantics already a connection to human beings. These human beings have the function of Place or Goal (with (Π22) *għand* 'at s.o.'s place') or Source ((Π38) *mingħand* 'from s.o.') which in a given situation tend to be known and therefore definite. These are favourable conditions for the use of proper names. Much in the same vein, (Π53) *skont* 'according to' usually introduces the source of information in the sense that a participant shares his/her opinion on a given topic. Opinions can only be expressed by human beings or institutions metonymically associated with [+human]. If the origin of an opinion is traced the source becomes definite because it can be identified and individualised. A name can be put to them.

The many open questions that remain notwithstanding, we conclude that at least for certain non-fusable Πs the deviation from the hypothetical hierarchy finds an explanation in semantics. Since there are also several (seemingly) synonymous Πs which do not behave in identical ways, we have to take a second look at the qualitative and quantitative data throughout Section 5. Before we get there, it is necessary however to provide an integrated view of the government properties of Maltese Πs. As will come to the fore shortly, the phenomena reviewed in this section on definiteness do by no means exhaust the range of variation in the domain of Π-complement interaction.

4.1.2 Government

Section 4.1.1 has shown us that most of the members of BLOMP 2.0 can directly take a complement NP without any further linker. However, this pattern is not the only possible choice in Π-complement interaction. Not only is there variation to some extent as to the shape of the PPs which involve lexical or proprial complements but pronominal complements also display an array of patterns. This section is dedicated to systematizing the structural facts comprehensively by way of identifying the government classes which result from the behaviour of the Maltese Πs in combination with two different complement types, namely nominal and pronominal complements (in the restricted sense of personal pronouns and their bound forms). This section elaborates on the hypotheses originally put forward by Stolz/ Levkovych (2020: 220–236).

We notionally distinguish the following types of government for Maltese Πs (with partially adjusted terminology; examples will be provided and discussed in the subsections to follow):

- **NOMINATIVE**: NOMINATIVE government applies if Π and complement combine asyndetically with the latter being identical with the form it would have in subject function. NOMINATIVE government is possible with nominal and pronominal Πs though to different extents.
- **ACCUSATIVE**: ACCUSATIVE government refers to PPs which involve a pronominal complement that is expressed in the form of the direct object suffix of verbs. There is no equivalent pattern in the domain of nominal complements. As will become clear from the discussion, ACCUSATIVE government is a strong candidate for being cancelled from the list of government patterns.[39]
- **POSSESSIVE**: POSSESSIVE government is also restricted to pronominal complements. The latter comes in the shape of the possessor affix attached to the Π. POSSESSIVE and ACCUSATIVE government can be told apart only for the 1st person singular.
- **prepositional**: If the head and the complement of a given PP are linked to each other by an intercalated Π, we speak of prepositional government. We stipulate that the intercalated Π (termed linker) forms a binary constituent with the nominal complement according to the multi-level construction pattern $[\Pi_{\text{HEAD}}\ [\Pi_{\text{LINKER}}\ [\text{NP}_{\text{COMPLEMENT}}]]_{PP}]_{PP}$. The linker marks the relation between head and complement according to the principles of dependent marking. The pattern is

39 Cf. Section 4.1.2.2.2.2.3.

attested with nominal and pronominal complements.[40] There are six classes of prepositional government, namely

- GENITIVE: GENITIVE government requires the presence of (Π55) *ta'* 'of' as linker.
- LOCATIVE: LOCATIVE government requires the presence of (Π15) *fi* 'in' as linker.
- ABLATIVE: ABLATIVE government requires the presence of (Π41) *minn* 'from' as linker.
- ALLATIVE: ALLATIVE government requires the presence of (Π28) *lejn* 'towards' as linker (only attested with personal pronominal complements).
- BENEFACTIVE: BENEFACTIVE government requires the presence of (Π21) *għal* 'for' (doubtful and probably only attested with nominal complements).
- PROXIMATE: PROXIMATE government requires the presence of either (Π14) *fejn* 'near' or (Π23) *ħdejn* 'beside' as linker. This pattern is unique to pronominal complements.

The government classes to which individual Πs are assigned result from the combination of choices they make from the above list in the interaction with nominal and pronominal complements. The patterns emerging from the combinations of choices are multifold not the least because some Πs give evidence of the competition of several types of government. The intricacies of the system are disclosed in Section 4.1.2.2 and its further subdivisions. In Section 4.1.2.1 however, we meet again with (Π9) *bla* 'without' and a small number of other cases to reconfirm their special status amidst the bulk of the members of BLOMP 2.0.

4.1.2.1 On missing complement types

As argued above,[41] (Π9) *bla* 'without' is in principle incompatible with definite nominal complements although there is spurious and most probably also controversial evidence to the contrary. The situation is similar if we look at pronominal complements. To make our point, we exceptionally look at demonstrative pronouns, too. In the Korpus Malti 3.0, there are nine *bla*-PPs which contain a demonstrative pronoun as complement. In Table 4.4, we compare the token frequencies of the PPs *bla* + DEMONSTRATIVE and *mingħajr* + DEMONSTRATIVE to additionally support the idea that the former pattern is absolutely exceptional.

40 Its internal logic will be exposed in Section 4.1.2.3.
41 Cf. Section 4.1.1.1.

Table 4.4: Token frequencies of PARTITIVE/ABESSIVE Πs with demonstrative pronouns as complements in the Korpus Malti 3.0.

categories	form	*bla*	*mingħajr*	sum
M.SG.PROX	*dan*	0	17	17
F.SG.PROX	*din*	1	7	8
PL.PROX	*dawn*	4	23	27
M.SG.DIST	*dak*	3	14	17
F.SG.DIST	*dik*	0	2	2
PL.DIST	*dawk*	1	7	8
total		9	70	79

The picture which emerges from Table 4.4 is in line with what we have argued already.[42] Of the two partitive/abessive Πs, (Π36) *mingħajr* 'without' is the first choice when it comes to taking definite complements. In contrast, the turnout for (Π9) *bla* 'without' is meagre and outnumbered by seven-to-one by that of (Π36) *mingħajr* 'without'. Moreover, PPs of the shape **bla* + *dan* (proximal masculine singular) or **bla* + *dik* (distal feminine singular) have not been found. The attested cases stem from the genres news reports and parliamentary debates. This is also the context from which (4.66) is taken.

(4.66) [Korpus Malti 3.0; parl1087]

Jista'	*l-Ministru*	*jikkonsidra*	*li*	*jitwaħħlu*			
3SG.M.IPFV:can	DEF-minister	3SG.M.IPFV:consider	that	3SG.M.IPFV:fix			
bozoz	*fi*	*Triq il-Għasel*	*f'*	*parti*	*minn*	*Triq San Silvestru*	*u*
light_bulb.PL	in	Triq il-Għasel	in	part	from	Triq San Silvestru	and
fi	*Triq Camillo Sceberras*	*kollha*	*fil- Mosta*	*għaliex*	*dawn*		
in	Triq Camillo Sceberras	all:3SG.F	in.DEF-Mosta	because	DEM:PL:PROX		
it- toroq	*għadhom*	***[bla***	*<u>dawn</u>]$_{PP}$*				
DEF-street.PL	still:3PL	**[without**	<u>DEM:PL:PROX</u>]				

'Would the Minister consider fixing light-bulbs in Triq il-Għasel, in parts of Triq San Silvestru and in Triq Camillo Sceberras, all in Mosta, because these streets are still [**without** <u>those</u>] [. . .].'

We assume that the occasional employment of (Π9) *bla* 'without' with definite complements in general is an idiolectal characteristic of individual speakers but no commonly followed practice across the Maltese speech-community. Accord-

42 Cf. Section 4.1.1.

ingly, only (Π36) *mingħajr* 'without' allows for personal pronominal complements whereas this complement type is completely excluded from *bla*-PPs.

The general incompatibility of (Π9) *bla* 'without' singles this Π out from the bulk of its fellow-Πs because no other Maltese Π is excluded completely from combinations with definite complements. This holds also for (Π52) *sforz* 'thanks to'.[43] In contrast to (Π9) *bla* 'without', (Π52) *sforz* 'thanks to' frequently heads a PP whose complement is a definite common noun as in (4.67).

(4.67) [Korpus Malti 3.0; parl995]

[sforz	*l-istipendju]$_{PP}$*	*llum*	*għandna*	*8,000*	*student*
[**thanks_to**	DEF-stipend]	today	have:1PL	8,000	student
fl-Università					
in:DEF-university					

'[. . .] [**thanks to** the stipend] we have today 8,000 students at the university.'

This pattern of NOMINATIVE government by (Π52) *sforz* 'thanks to' recurs also with indefinite complements such as the one in (4.68).

(4.68) [Korpus Malti 3.0; parl11658]

jakkuża	*lill-Oppożizzjoni*	*li*	*qed*	*nitkellmu*	*u*
3SG.M.IPFV:accuse	to:DEF-opposition	that	PROG	1.IPFV:RFL:speak:PL	and
nwasslu	*dawn*	*il- messaġġi*	***[sforz***	*xi*	
1.IPFV:CAUS:arrive:PL	DEM:PL:PROX	DEF-message:PL	[**thanks_to**	some	
tip	*ta'*	*paniku]$_{PP}$*			
type	of	panic]			

'[. . .] he accuses the opposition that we are talking about and report these messages [**thanks to** some kind of panic] [. . .].'

In this latter aspect of syntactic behaviour, (Π52) *sforz* 'thanks to' and (Π9) *bla* 'without' resemble each other. However, they immediately go separate ways because variation comes into play affecting solely (Π52) *sforz* 'thanks to'. As the examples in (4.69) show, NOMINATIVE government has a rival in prepositional government of the GENITIVE brand with the latter Π. The linker is highlighted doubly in bold combined with underlining.

43 Cf. Section 1.6.6 (i) and example (1.155).

(4.69) (Π52) *sforz* 'thanks to'

(a) [Korpus Malti 3.0; news111598]

jidher *li* *ċ-ċansijiet* *li* *l-aċċident* *ġara*
3SG.M.IPFV:appear that DEF-chance:PL that DEF-accident happen
*[**sforz*** ***ta'*** *bomba]$_{PP}$* *huma* *għoljin*
[**thanks_to** **of** bomb] 3PL high:PL
'[. . .] it appears that the chances that the accident happened [**because** of a bomb] are high [. . .].'

(b) [Korpus Malti 3.0; news119166]

Beijing *anke* *[**sforz*** ***ta'*** *l-Olimpjadi]$_{PP}$* *qed*
Beijing also [**thanks_to** **of** DEF-Olympics] PROG
tesperjenza *twelid* *mill- ġdid* *kulturali* *u*
3SG.F.IPFV:experience birth from:DEF-new cultural and
arkitettoniku
architectonical
'Also [**thanks to** the Olympic Games], Beijing is experiencing a cultural and architectonical rebirth.'

What happens in the case of pronominal complementation? As in the case of (Π9) *bla* 'without', we first have a look at demonstrative pronouns which function as complements of (Π52) *sforz* 'thanks to'. In the Korpus Malti 3.0, there are only five instances of *sforz* + DEMONSTRATIVE PRONOUN four of which involve the distal masculine singular *dak* 'that' and one is the proximal masculine singular *dan* 'this'. Examples like (4.70) are thus definitely marginal.

(4.70) [Korpus Malti 3.0; news64393]

*[**sforz*** *dak* *li* *seħħ* *fid-dinja* *din*
[**thanks_to** DEM:M:DIST that happen in:DEF-world DEM:F:PROX
is-sena]$_{PP}$ *dan* *il-baġit* *għandu* *importanza* *kruċjali*
DEF-year] DEM:M:PROX DEF-budget have:3SG.M importance crucial
'[. . .] [**owing to** what happened in the world this year] this budget has a crucial importance [. . .].'

It is worth noting that all instances of *sforz* + DEMONSTRATIVE PRONOUN feature demonstrative pronouns which are modified by an attributive relative clause, i.e. the demonstrative on its own is never a complement of the Π under inspection. Furthermore, there is no evidence of prepositional government in these constructions so that **sforz* + *ta'* + DEMONSTRATIVE PRONOUN can be discounted as an alternative to NOMINATIVE government. Going by the Korpus Malti 3.0, (Π52) *sforz* 'thanks to' never cooccurs with personal pronominal complements no matter whether we look for NOMINATIVE government or prepositional government. Were it not for the example in (4.71) from our questionnaire, we could be absolutely certain that the combination of (Π52) *sforz* 'thanks to' and personal pronominal complements is blocked completely.

(4.71) [Q – Sandra Abela, personal communication]

il-bravura	*li*	***[sforz***	***tagħħa*** *]$_{PP}$*	*it-tim*	*Malti*	*tilef*	*biss*
DEF-feat	that	**[thanks_to**	**of**:3SG.F]	DEF-team	Maltese	lose	only

2–0
2–0
'the feat [**thanks to** which] the Maltese team lost only 2–0'

This example stems from a Cloze-test part of the questionnaire. Sandra Abela who expressed serious doubts as to the Π-status of (Π52) *sforz* 'thanks to' prior to the test provided herself *tagħħa* 'of her' as suitable filler for the gap left in the stimulus. The use of the filler was artificially enforced by the set-up of the Cloze-test. It is therefore daring to postulate on the basis of this kind of doubtful evidence that (Π52) *sforz* 'thanks to' displays prepositional government of the GENITIVAL brand with personal pronominal complements. At the same time, the possibility that exactly this is correct cannot be discarded sweepingly without further data from native speakers of Maltese.

There are further cases of Maltese Πs which lack evidence of being compatible with one of the complement types. The most straightforward case is that of (Π37) *mingħala* 'in s.o.'s opinion' since this Π can never take nominal complements (Sandra Abela, personal communication).[44] (Π37) *mingħala* 'in s.o.'s opinion' is exclusively attested with personal pronominal complements which are represented by the possessor suffixes on the Π so that POSSESSIVE government applies. Semantically, the strict exclusion of (Π37) *mingħala* 'in s.o.'s opinion' from combinations with nominal complements seems to be unmotivated. This issue needs to be explored further in due course. The problems posed by (Π42) *mintul* 'all along' have been discussed already.[45] It was concluded there that this Π is morphologically invariant. On closer inspection, it turns out that (Π42) *mintul* 'all along' generally fails to show up in combination with pronominal complements. *Mintul*-PPS can only take nominal complements with which they combine asyndetically as shown in (1.119)–(1.123). In contrast to (Π9) *bla* 'without', (Π42) *mintul* 'all along' is not restricted to combinations with indefinite complements.

44 Cf. Section 1.6.4.2 (v).
45 Cf. Sections 1.6.4.2 (ix) and 3.1.2.3.1.2 (v).

Three additional cases involve Πs whose membership in BLOMP 2.0 rests on shaky grounds, namely (Π26) *kif* 'as', (Π43) *mnejn* 'from near', and (Π54) *sotta* 'under'. We know that the evidence of their Π-status is scarce and doubtful.[46] Moreover, their distribution over complement types cannot be determined satisfactorily. If we go by the little information we have found in the major dictionaries of Maltese, it is possible – with many reservations – to hypothesise that these three Πs are like (Π37) *mingħala* 'in s.o.'s opinion' since they never cooccur with nominal complements. (Π26) *kif* 'as' and (Π43) *mnejn* 'from near' resemble (Π37) *mingħala* 'in s.o.'s opinion' further because they also attest to POSSESSIVE government whereas the only example of the use of (Π54) *sotta* 'under' in syntax points to GENITIVAL government.

We conclude this section by way of stating that BLOMP 2.0. has three members which are barred from taking personal pronominal complements. For one of these Πs – (Π9) *bla* 'without' – the inability to take complements of this kind has been confirmed by empirical proof. As to (Π52) *sforz* 'thanks to', the situation is not as clear. If this Π is not categorically excluded from combinations with personal pronominal complements, *sforz*-PPs with complements of this kind are at least extremely rare. There are also four Πs for which no evidence for combinations with nominal complements could be found. Three of these Πs are candidates for being dropped from BLOMP 2.0 under a future revision. Table 4.5 summarises our above results.

Table 4.5: Maltese Πs attested with only one complement type.

Π	government		
	nominal		pronominal
bla	NOMINATIVE		Ø
kif	Ø		POSSESSIVE
mingħala	Ø		POSSESSIVE
mintul	NOMINATIVE		Ø
mnejn	Ø		POSSESSIVE
sforz	NOMINATIVE	GENITIVE	(GENITIVE?)/Ø
sotta	Ø		GENITIVE

In the subsequent paragraphs, a number of further problematic cases will be mentioned and discussed.

46 Cf. Sections 1.6.7 (i), (ii), and (iv) and Section 3.1.2.3.1.2 (v).

4.1.2.2 Πs with nominal and pronominal complements

Since important aspects of the interaction between Πs and nominal complements have been introduced already,[47] we continue the presentation of the facts of Maltese Π-grammar by way of looking into the variation of government types in this domain in Section 4.1.2.2.1. The range of variation in PPs which involve personal pronominal complements is the topic of Section 4.1.2.2.2. Section 4.1.2.2.3 provides an integrated view of the government types attested with the complement types under scrutiny.

4.1.2.2.1 Government types with nominal complements

In this section, the focus is on variation. To better understand the range of variation, we start with a short overview of those bona fide cases of Πs which are exempt from variation. As will become clear shortly, not every supposedly straightforward case is as uncontroversial as it might seem on the surface.

Based on Table 4.5, we assume that 55 Πs can take nominal complements. This is a share of 91.7% of BLOMP 2.0. Of these 55 Πs, 35 (= 63.6%) are not subject to variation as to the government of their nominal complements. The members of this group give evidence of NOMINATIVE government. These Πs are listed in (4.72).

(4.72) Maltese Π with mandatory NOMINATIVE government of nominal complements (Π4) *bejn* 'between'; (Π5) *bħal* 'like'; (Π6) *bħala* 'as'; (Π7) *bi* 'with'; (Π9) *bla* 'without'; (Π10) *daqs* 'equal to'; (Π11) *dwar* 'about'; (Π14) *fejn* 'near'; (Π15) *fi* 'in'; (Π16) *fost* 'amongst'; (Π17) *fuq* 'on'; (Π19) *ġo* 'in'; (Π20) *għajr* 'except'; (Π21) *għal* 'for'; (Π22) *għand* 'at s.o.'s place'; (Π23) *ħdejn* 'beside'; (Π24) *ħlief* 'except', (Π25) *inkluż* 'including'; (Π28) *lejn* 'towards'; (Π29) *lil* 'to'; (Π30) *ma'* 'with'; (Π32) *maġenb* 'close to'; (Π38) *mingħand* 'from s.o.'; (Π40) *minkejja* 'in spite of'; (Π41) *minn* 'from'; (Π46) *qabel* 'before'; (Π47) *qalb* 'amidst'; (Π49) *quddiem* 'in front of'; (Π51) *sa* 'till'; (Π53) *skont* 'according to'; (Π55) *ta'* 'of'; (Π56) *taħt* 'under'; (Π57) *versu* 'towards'; (Π59) *waqt* 'at the time of'; (Π60) *wara* 'after'

4.1.2.2.1.1 From consistency to variation

The role of the outsider is played by two Πs, viz. (Π2) *a skapitu* 'at the expense of' and (Π45) *permezz* 'by means of'. Since (Π2) *a skapitu* 'at the expense of' has never before caught the attention of linguists who are interested in Πs in general and those of Maltese in particular, it makes perfect sense to inspect this Π first. We already know

47 Cf. Section 4.1.1.

what happens at the left edge of this Π[48] but we are still in the dark about the interaction with nominal complements to the right of (Π2) *a skapitu* 'at the expense of'.

(i) (Π2) *a skapitu* 'at the expense of': This Π, as will be shown in this section, is the only member of BLOMP 2.0 which gives evidence of consistent GENITIVE government with nominal complements. According to the model of presentation employed in the foregoing sections, we illustrate the government properties of (Π2) *a skapitu* 'at the expense of' with a set of three examples in (4.73).

(4.73) (Π2) *a skapitu* 'at the expense of'

(a) [Korpus Malti 3.0; news121866]

Ma nistgħux niffokaw fuq settur ***[à skapitu***
NEG 1.IPFV:can:PL:NEG 1.IPFV:focus:PL on sector **[at_the_expense**
ta' *settur ieħor]*$_{PP}$
of sector other]
'We cannot focus on a sector [**at the expense of** another sector].'

(b) [Korpus Malti 3.0; news104263]

kienet saret kampanja aggressiva sabiex
be:3SG.F.PFV become:3SG.F.PFV campaign:F aggressive:F so_that
kulħadd jistudja l-informatika ***[a skapitu***
everybody 3SG.IPFV:study DEF-information_science **[at_the_expense**
***tal**-lingwi]*$_{PP}$
of:DEF:language:PL]
'[. . .] an aggressive campaign had taken place for everybody to study information science [**to the detriment of** the languages] [. . .].'

(c) [Korpus Malti 3.0; news136899]

Ried jeleva lil Chris Said fil-kariga ta' ministru
want 3SG.M.IPFV:elevate to Chris Said in:DEF-office of minister
[a skapitu ta' *Debono]*$_{PP}$
[at_the_expense of Debono]
'He wanted to promote Chris Said to the office of minister [**at the expense of** Debono].'

It is not just a possibility or preference of (Π2) *a skapitu* 'at the expense of' to take (Π55) *ta'* 'of' as linker in combination with nominal complements (and, as we will see below; pronominal ones too). The presence of the linker is mandatory. It is attested in 100% of all relevant *a skapitu*-PPS (n = 1,232) in the Korpus Malti 3.0.

48 Cf. Section 3.2.5.

No other type of government can replace GENITIVE government with (Π2) *a skapitu* ‘at the expense of’. The small number of problematic cases can be explained more often than not with reference to the retention of at least certain properties which define nominal categoriality. Their combined presence might occasionally cast doubt on the Π-status of (Π2) *a skapitu* ‘at the expense of’. Of all doubtful cases, there are only two sentences which superficially invoke NOMINATIVE government both being of the kind illustrated in (4.74).

(4.74) [Korpus Malti 3.0; parl10787]
l- Oppożizzjoni tipprova tirbaħ vantaġġ u vot
DEF-opposition 3SG.F.IPFV:try 3SG.F.IPFV:win advantage and vote
kontra l-Gvern ***[a skapitu*** ***Ø*** *u minn fuq*
against DEF-government **[at_the_expense** **Ø** and from on
dahar membru parlamentari tal-Gvern]$_{PP}$
back member parliamentary of:DEF-government]
‘[…] the opposition tries to get advantage and win a vote against the government [**at the expense of** and owing to a parliamentary member of the government].’

Two PPs are joined in coordination. The leftmost PP contains (Π2) *a skapitu* ‘at the expense of’ which is coordinated with the second PP by the conjunction *u* ‘and’. The second PP is complex in the sense that it involves a sequence of two Πs (Π41) *minn* ‘from’ and (Π17) *fuq* ‘on’.[49] For the time being, it suffices to notice that the complement in this second NP is *dahar membru parlarmentari tal-Gvern* ‘the back of a parliamentary member of the government’ with *minn fuq dahar* + NP constituting an idiomatic expression roughly corresponding to English *owing to the help of.* In contrast, the complement of (Π2) *a skapitu* ‘at the expense of’ would be only *membru parlarmentari tal-Gvern* ‘a parliamentary member of the government’ which should have followed the linker whose absence from (4.74) is marked by Ø. Dropping the first linker under coordination is licit in Maltese if the second conjunct also requires the presence of the same linker which is then placed only once, namely after the second conjunct as shown in (4.75).

(4.75) [Korpus Malti 3.0; parl417]
se jkun hemm żieda fl-importazzjoni
FUT 3SG.M.IPFV:be.FUT there increase in:DEF-importation
[a skapitu ***Ø*** *u għad-dannu* ***tal****-prodotti]*$_{PP}$
[at_the_expense **Ø** and for:DEF-damage **of**:DEF-product:PL]
‘[. . .] there will be an increase of importation [**at the expense** and to the detriment **of** the products].’

49 This instantiates a morphosyntactic strategy to be focused upon in Section 4.1.3.

In (4.76), we see that the linker (Π55) *ta'* 'of' gives way to (Π21) *għal* 'for' as required by the second coordinand.

(4.76) [Korpus Malti 3.0; parl210]

Mela	*l-attività*	*ta'*	*żvilupp*	*bilfors*	*trid*
well	DEF-activity	of	development	perforce	3SG.F.IPFV:need

tkun	*hemm*	*imma*	*dan*	*m'*	*għandux*
3SG.F.IPFV:be.FUT	there	but	DEM:M:PROX	NEG	have:3SG.M:NEG

ikun	***[a skapitu***	*u*	*ta'*	*ħsara*
3SG.M.IPFV:be.FUT	**[at_the_expense**	and	of	damage

għall-ambjent	*Għawdxi]$_{PP}$*
for:DEF-environment	Gozitan]

'Well then, the activities of development must perforce be there but this must not happen [**at the expense** and damage of the Gozitan environment].'

Examples of these kinds question whether the membership of (Π2) *a skapitu* 'at the expense of' in BLOMP 2.0 is fully justified. The absence of the linker in the slot to the immediate right of (Π2) *a skapitu* 'at the expense of' in (4.75) is unproblematic since we can classify this as a case of catalipsis or backward ellipsis under coordination (Haspelmath 2007a: 39). In (4.76), the Π required by the second coordinand wins out against that required by the first coordinand.

The story does not end here because one might expect that the two coordinands should be of the same kind, meaning: they should belong to one and the same word-class. This requirement is difficult to meet since *dannu* 'damage' and *ħsara* 'damage' are common nouns whereas the first coordinand represents the class of Π. Furthermore, *dannu* 'damage' and *ħsara* 'damage' are themselves the nominal complements of (Π21) *għal* 'for' and take (Π55) *ta'* 'of', respectively, which means that we have the coordination of a Π ((Π2) *a skapitu* 'at the expense of') and a PP (*għad-dannu* 'to the detriment', *ta' ħsara* 'to the detriment'). Haspelmath (2007a: 19) does not consider the syntactic differences between the coordinands insurmountable problems when he says that

> [i]t is sometimes said that the coordinands must belong to the same phrasal category [. . .]. However, coordination of different phrasal categories is often possible when both have the same semantic role.

(Π2) *a skapitu* 'at the expense of' and both *għad-dannu* 'to the detriment' and *ta' ħsara* 'to the detriment' are indeed largely synonymous expressions with different word-class membership. Alternatively, one could assume that (Π2) *a skapitu* 'at the expense of' has an internal structure in analogy to *għad-dannu* 'to the detriment' and *ta' ħsara* 'to the detriment', namely that of a PP. In this case {*a*} would be the Π and *skapitu* its nominal complement. We have already argued against Italian *a* 'to' being a fully-blown Π in Maltese.[50] It normally forms part of the morphologically unanalysable segmental chain of segments of Maltese expressions some of which happen to be Πs.

Under coordination, (Π2) *a skapitu* 'at the expense of' most often occupies the leftmost slot as in the examples above and those presented in Table 4.6.

Table 4.6: (Π2) *a skapitu* 'at the expense of' as first coordinand.

source	example / translation
parl12127	*a* ***skapitu*** *u prezz tad-dritt tal-privatezza* '**at the expense** and at the price of privacy'
parl2701	*a* ***skapitu*** *u a spejjeż tal-kwalità tal-ħajja tan-nies* '**to the detriment** and at the expense of the quality of the people's life'
parl328	*a* ***skapitu*** *jew b' kapriċċ ta' xi ħadd* '**at the expense** and for someone's whim'
parl586	*a* ***skapitu*** *u spejjeż żejda ta' kull min ikun irid jagħmel kuntratti* '**at the expense** and with higher costs for everyone who will want to make contracts'

Except (4.77), an *a skapitu*-PP as second coordinand is possible only if the first coordinand is an *a skapitu*-PP, too as shown in Table 4.7 which covers all pertinent cases found in the Korpus Malti 3.0. All instances of (Π2) *a skapitu* 'at the expense of' are highlighted in bold. The behaviour of (Π2) *a skapitu* 'at the expense of' under coordination is marked.[51]

Table 4.7: (Π2) *a skapitu* 'at the expense of' as second coordinand.

source	example / translation
parl10310	***a skapitu ta****l-istess votanti Laburisti u* ***a skapitu ta****l-istess ċittadini* '**at the expense of** those who vote Labour and **at the expense of** the citizens'
parl11718	***a skapitu ta****l-familji Maltin u Għawdxin u* ***a skapitu ta****l-anzjani* '**at the expense of** the Maltese and Gozitan families and **at the expense of** the elderly people'
parl2051	***a skapitu ta****l-ħaddiem u* ***a skapitu ta'*** *min jiflaħ l-inqas* '**at the expense of** the workers and **at the expense of** the weaker'
parl562	***a skapitu ta'*** *kull trasparenza fit-tmexxija u* ***a skapitu ta'*** *kull ħaġa li suppost nemmnu fiha* '**at the expense of** every transparency of direction and **at the expense of** everything we are supposed to believe in'
parl5877	***a skapitu ta'*** *kollox u* ***a skapitu ta'*** *kull pajjiż u organizzazzjoni* '**at the expense of** everything and **at the expense of** each country and organisation'

50 In Sections 1.6.6 (iv) and (viii),

51 Cf. Section 4.1.4.

Table 4.7 (continued)

source	example / translation
parl6418	***a skapitu ta'*** *ħidma aktar sfiqa parlamentari [. . .] u* ***a skapitu ta****l-verità sagrosanta* '**at the expense of** a more focused parlarmentary activity and **at the expense of** the sacred truth'
parl6619	***a skapitu ta****l-lokal,* ***a skapitu ta****r- residenti u* ***a skapitu ta'*** *dawk il- Maltin li jmorru fir-raħal tal-Qala* '**at the expense of** the location, **at the expense of** the residents and **at the expense of** those Maltese who go to the village of Qala'
parl6749	***a skapitu ta****l-pensjoni tas-servizz u* ***a skapitu ta****l-pensjoni ta' l-irtirar* '**at the expense of** the pensions of the service and **at the expense of** the pensions of retirement'
parl7716	***a skapitu ta'*** *l-effiċjenza u* ***a skapitu ta****s-serjetà* '**at the expense of** effectiveness and **at the expense of** seriousness'
parl797	***a skapitu ta****l-funzjoni li tikko<o>rdina u* ***a skapitu ta****l-funzjoni li tgħin lill- Gvern* '**at the expense of** the coordinating function and **at the expense of** the function which helps the government'
parl8206	***a skapitu ta****l-għajnuna u* ***a skapitu ta'*** *dawn is-sussidji li qegħdin jirċievu* '**at the expense of** the help and **at the expense of** those subsidies which they receive'
parl866	***a skapitu ta'*** *l-obbligi soċjali li għandha l-kumpannija u* ***a skapitu ta'*** *l-iżvilupp soċjo-ekonomiku tal-gżira Għawdxija* '**at the expense of** the social obligations the company has and **at the expense of** the socio-economical development of the Gozitan island'

In the Korpus Malti 3.0, the only case of (П2) *a skapitu* 'at the expense of' as second coordinand without a preceding *a skapitu*-PP is (4.77).

(4.77) [Korpus Malti 3.0; parl1784]

qed	*nassistu*	*għal*	*favoritiżmu*	*sfrenat*	
PROG	1.IPFV:assist:PL	for	favouritism	throw_off_all_restraint:PART	
[favur	*il-lobby*	*ta'*	*l-insurance*	*agencies]$_{PP}$*	*u*
[in_favour_of	DEF-lobby	of	DEF-insurance	agency:PL]	and
[a skapitu	*tal-persuni*	*bi*	*bżonnijiet*	*speċjali]$_{PP}$*	
[at_the expense	**of**:DEF-person:PL	with	need:PL	special]	

'[. . .] we are assisting unrestrained favouritism [**in favour of** the lobby of the insurance agencies] and [**at the expense of** persons with special needs].'

This sentence is of special interest for the topic under examination because the coordination involves coordinands which are semantically antonymous. Since the П-status of (П13) *favur* 'in favour of' is uncontroversial, it may be argued that this status also holds for (П2) *a skapitu* 'at the expense of' because it forms a syntactic parallel to the former.

However, there are other properties of (Π2) *a skapitu* 'at the expense of' which connect this Π to common nouns. It is possible to modify (Π2) *a skapitu* 'at the expense of' as shown in Table 4.8. The examples are drawn from the Korpus Malti 3.0. The modifier is highlighted in grey.

Table 4.8: (Π2) *a skapitu* 'at the expense of' and its alloforms taking modifiers.

source	example / translation
parl851	***a skapitu*** *żejjed tad- difiża* '**at the** excessive **expense** of the defense'
news101690	***għas-skapitu*** *personali tagħhom* '**at** their personal **expense**'
news7440	*tfal sfruttati* ***għal skapitu*** *sesswali* 'children exploited **for** sexual **damage**'
news129013	***għal skapitu*** *politiku* ***'at the cost*** *of political* ***damage'***
news144265	***bi skapitu*** *inuman u utilitarju tal-ħajja ta' ħaddieħor* '**at the** inhuman and utilitarian **expense** of the life of others'
news82164	***bi skapitu*** *kbir tal- ekonomija Maltija* '**to the** great **damage** of the Maltese economy'
news154151	***għal skapitu*** *aħħari tiegħu* '**to** his final **damage**'

On account of these examples, we assume *pace* Aquilina (1991: 1327) that there is a common noun *skapitu* 'expense' which is used infrequently but almost only with a preceding Π which replaces the expected initial /a/. In these rare cases, it is also possible that the linker fails to show up. In (4.78)–(4.79), *ta' skapitu* 'detrimental' requires a different Π to identify the participants to whose detriment something is done. In these examples, we refrain from using square brackets since we do not assume that *skapitu* has Π-status in either (4.78) or (4.79).

(4.78) [Korpus Malti 3.0; news178234]

dan	*m'*	*għandux*	*ikun*	***ta'***	***skapitu***
DEM:M:PROX	NEG	have:3SG.M:NEG	3SG.M.IPFV:be.FUT	**of**	**expense**
għas-saħħa	*tal-ħaddiem*				
for:DEF-health	of:DEF-worker				

'[. . .] this must not be **detrimental** for the health of the worker.'

(4.79) [Korpus Malti 3.0; parl10127]

d-drittijiet	*tal-individwu*		*jistgħu*	*jiġu*	
DEF-right:PL	of:DEF-individual		3.IPFV:can:PL	3.IPFV:come:PL	
mħarsa	*mingħajr*	*ma*	*jkunu*	***ta'***	***skapitu***
PART:protect.PL	without	that	3.IPFV:be.FUT:PL	**of**	**expense**

għad-drittijiet *ta'* *ħaddieħor*
for:DEF:right:PL of someone_else
'[. . .] the individual rights can be protected without being **detrimental** to the rights of others.'

Furthermore, the examples in Table 4.9 suggest that Π and linker do not form an intimate syntagmatic unit since other words might be inserted between the two constituents.

Table 4.9: Insertion of material between (Π2) *a skapitu* 'at the expense of' and the linker.

source	example / translation
news91827	***għas-skapitu*** *mhux biss* ***ta'*** *bwiet il- Maltin* '**at the expense** not only **of** the Maltese pockets'
parl10487	***a skapitu*** *mhux* ***ta'*** *partit jew ieħor* '**at the expense** not **of** some party or other'
parl8942	***a skapitu*** *mhux biss* ***ta****l-liġijiet ta' pajjiżna* '**at the expense** not only **of** the laws of our country'
parl134	***a skapitu*** *anke* ***ta'*** *l-operat tal-Gvern innifsu* '**at the expense** too **of** the deeds of the government itself'
parl165	*U* ***a skapitu*** *– biex nieħdu każ estrem –* ***ta****d-dritt tal-leġittima tiegħi* 'And **at the expense** – to take an extreme case – **of** my legitimate share of the bequest
parl206	***a skapitu*** *– din hija l-agħar ħaġa –* ***ta'*** *nies oħra* '**at the expense** – this is the worst thing – **of** other people'

Figure 4.47 summarises the above findings by way of identifying the position between Π and linker into which attributes as well as other syntactic words can be inserted.

a skapitu ___ *ta'* NP

Figure 4.47: Free slot between Π and linker.

To conclude our review of the properties of (Π2) *a skapitu* 'at the expense of', we confirm that this Π is the only member of BLOMP 2.0 which obligatorily displays GENITIVE government. At the same time, there are several signs of this Π being a very special one in the sense that, in a number of contexts, it could better be interpreted as a common noun. In the majority of the attestations of (Π2) *a skapitu* 'at the expense of' in the Korpus Malti 3.0 however, no such nominal properties come to the fore.

(ii) (П45) *permezz* 'by means of': According to the literature, (П45) *permezz* 'by means of' is the only Maltese П which consistently and without exception applies GENITIVE government.[52] Saari (2003: 62–63) mentions no variation for (П45) *permezz* 'by means of'. The same author even skips the discussion of this П in his paper on Italo-Romance Пs in Maltese although other Пs with similar properties are sketched appropriately (Saari 2009). With reference to (П45) *permezz* 'by means of', E. Serracino-Inglott (1981: 55) notes that "[l]-avv[erbju] jitlob dejjem wara l-part[iċella] prep[osizzjonali] 'ta'!".[53] This requirement is obeyed by the 58,777 hits of the pattern *permezz* + *ta'* + NP in the Korpus Malti 3.0. The case could be claimed closed were it not for the counter-evidence attested in the same electronic resource. In these cases, NOMINATIVE government is exerted by (П45) *permezz* 'by means of'. A particularly interesting case in shown in (4.80).

(4.80) [Korpus Malti 3.0; news108389]

(a) NOMINATIVE government

bejn	*ix- xhur*	*ta'*	*April*	*u*	*Mejju*	*2012*
between	DEF-month.PL	of	April	and	May	2012

ETM General Imports Limited	*kienet*	*ittrasferiet*	*€ 83,779*
ETM General Imports Limited	be:3SG.F.PFV	transfer:3SG.F.PFV	€ 83,779

minn	*Malta*	***[permezz***	*il-Western Union Business Malta]$_{PP}$*
from	Malta	**[by_means_of**	DEF- Western Union Business Malta]

'[. . .] between April and May 2012, ETM General Imports Limited had transferred € 83,779 from Malta [**by means of** Western Union Business Malta].'

(b) GENITIVE government

Din	*is-somma*	*għaddiet*	***[permezz***	***ta'***	*disa*
DEM:F:PROX	DEF-sum	pass:3SG.F.PFV	**[by_means_of**	**of**	nine

tranżazzjonijiet]$_{PP}$
transaction:PL]

'This sum went through [**by means of** nine transactions].'

The two sentences (4.80a) and (4.80b) follow each other directly within one and the same paragraph of a news report. The author of the report opts for NOMINATIVE government with (П45) *permezz* 'by means of' in (4.80a) but switches to GENITIVE government in the immediately following (4.80b). The difference between the two complements is that *disa tranżazzjonijiet* 'nine transactions'[54] is indefinite whereas

52 Cf. (1.157)–(1.158) in Section 1.6.6 (ii).

53 Our translation: "the adverb requires always the prepositional particle *ta'* behind it!"

54 Recte: *disa' transazzjonijiet*.

il-Western Union Business Malta is a proper name overtly marked for definiteness. However, neither (in)definiteness nor proper-name status are decisive for the choice of government type. The search of the Korpus Malti 3.0 yielded 24,453 hits for *permezz* + *tal*-NP as opposed to five hits for *permezz* + *il*-NP. As (4.81) proves GENITIVE government also happens with proper-name complements.

(4.81) [Korpus Malti 3.0; news108664]
marru fil-vantaġġ fis-16-il minuta logħob
go:3PL.PFV in:DEF-advantage in:DEF-16-NUM minute play
[permezz ta' *Peter Paul Cassar]$_{PP}$*
[by_means_of of Peter Paul Cassar]
'[. . .] they went in front in the 16th minute of the match [**thanks to** Peter Paul Cassar] [. . .].'

As to indefinite complements, we have found a single case of *permezz* + *xi* NP, but 399 instances of *permezz* + *ta' xi* NP. Examples (4.82)–(4.83) contain three PPs headed by (П45) *permezz* 'by means of' which takes an indefinite complement. The variation is telling.

(4.82) [Korpus Malti 3.0; european2829]
dan il-laħam huwa importat ***[permezz ta'*** *xi*
DEM:M:PROX DEF-meat 3SG.M import:PART **[by_means_of of** some
sengħa partikolari]$_{PP}$
trade particular]
'[. . .] this meat is imported [**by means of** a particular trade] [. . .].'

(4.83) [Korpus Malti 3.0; law1517]
kull wieħed mid-Direttorati jista'
each one from:DEF-directorate:PL 3SG.M.IPFV:can
jeżerċita xi wieħed jew iktar mill-funzjonijiet
3SG.M.IPFV:exercise some one or more from:DEF-function:PL
tagħhom sew direttament sew ***[permezz*** *xi wieħed jew*
of:3PL either directly or **[by_means_of** some one or
iktar mill-uffiċjali jew impjegati tagħhom]$_{PP}$ [. . .] jew
more from:DEF-official:PL or employee:PL of:3PL] or
[permezz ta' *kuntrattur jew persuna oħra]$_{PP}$*
[by_means_of of contractor or person:F other:F]
'[. . .] each of the directorates can exercise any or several of their functions either directly or [**by means of** one or several of their officials or employees] or [**by means of** a contractor or other person] [. . .].'

Not only do we see that both NOMINATIVE and GENITIVE government is possible with indefinite complements but it becomes additionally clear that the variation must be firmly established because it occurs in one and the same complex sentence in (4.82) which is extracted from a legal document. The fact that variation of this kind is tolerated in a text which must meet the standards of jurisprudence suggests that the alternation between *permezz ta'* + COMPLEMENT and *permezz* + COMPLEMENT is nothing out of the ordinary. Both constructions are equally acceptable to the authors of the legal document. We therefore assume that it is not true that (Π45) *permezz* 'by means of' only allows for GENITIVE government. There certainly is variation between GENITIVE government and NOMINATIVE government. Quantitatively the two options are on different levels. GENITIVE government is overwhelmingly more common than NOMINATIVE government as is shown in Figure 4.48. There are 101,474 *permezz*-PPS with nominal complements 82% of which instantiate GENITIVE government and only 18% give evidence of NOMINATIVE government. The minority share is too small to challenge the dominance of GENITIVE government. However, the minority is big enough to require us to register it dutifully. On account of this case, it makes sense to distinguish major options and minor options for those Πs which are subject to variation in the domain of government.

The Korpus Malti 3.0 surprises us further with a putative third option of government. Superficially, example (4.84) seems to illustrate the existence of ABLATIVE government of (Π45) *permezz* 'by means of'.

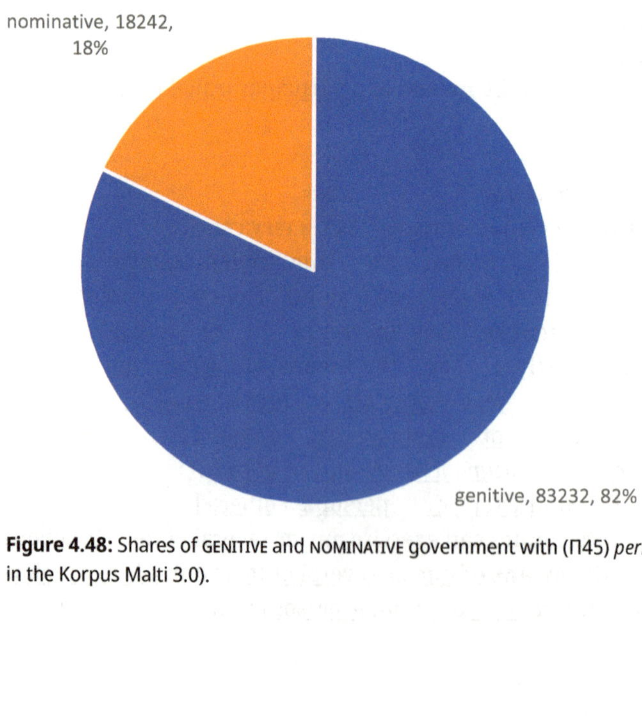

Figure 4.48: Shares of GENITIVE and NOMINATIVE government with (Π45) *permezz* 'by means of' (tokens in the Korpus Malti 3.0).

(4.84) [Korpus Malti 3.0; european10234]

l-Istat	*Membru*	*għandu*	*jibgħat*	*lill-Kummissjoni*
DEF-state	member	have:3SG.M	3SG.M.IPFV:send	to:DEF:commission
l-informazzjoni	*kollha*	*meħtieġa*	*biex*	*isir*
DEF-information:PL	all:3SG.F	PART:need.PL	so_that	3SG.M.IPFV:become
valutazzjoni	*ta'*	*l-iskema*	*introdotta*	***[permezz***
evaluation	of	DEF-scheme:F	introduce:PART:F	**[by_means_of**
minn	*dan*	*il-Kapitolu]PP*		
from	DEM:M:PROX	DEF-chapter]		

'[. . .] the member state has to send all the necessary information to the commission so that an evaluation of the scheme introduced [**by means of** this chapter] can take place [. . .].'

This employment of (Π41) *minn* 'from' as linker in *permezz*-PPs is absolutely unique. Since the example stems from the large body of E.U.-related texts, we assume that we are facing a simple translation mishap. We ignore the language from which the text has been translated into Maltese. If it was Italian the preposition *di* 'of, from' might have induced the translator to go for (Π41) *minn* 'from' instead of (Π55) *ta'* 'of'. We cannot unequivocally prove our point. However, we feel entitled nevertheless to exclude this isolated case from the inventory of government types of (Π45) *permezz* 'by means of'. We will apply this practice also to similar cases provided the evidence speaks against classifying a given case as genuine.

In this section, we have learned that the absence of competition in the domain of government types with nominal complements is only possible if the sole realised pattern is that of NOMINATIVE government. Every other government type must co-exist with at least one competitor. We will see in the course of the further discussion that NOMINATIVE government is possible with each of the Πs either as major option or as minor alternative.

4.1.2.2.1.2 Patterns of variation

Having clarified the issue posed by (Π45) *permezz* 'by means of', we can now present the full list of Maltese Πs with variable government of nominal complements in (4.85).

(4.85) Maltese Π with variable government of nominal complements
(Π1) *apparti* 'apart from'; (Π3) *barra* 'outside'; (Π8) *biswit* 'facing'; (Π12) *faċċata* 'opposite'; (Π13) *favur* 'in favour of'; (Π18) *ġewwa* 'inside'; (Π27) *kontra* 'against'; (Π31) *madwar* 'around'; (Π33) *matul* 'during'; (Π34) *minbarra* 'except'; (Π35) *minflok* 'instead of'; (Π36) *mingħajr* 'without'; (Π39)

minħabba 'on account of'; (Π44) *oltre* 'beyond'; (Π45) *permezz* 'by means of'; (Π48) *qrib* 'near'; (Π50) *rigward* 'concerning'; (Π52) *sforz* 'thanks to'; (Π58) *viċin* 'near'

We review the different variation patterns separately starting with that of NOMINATIVE and GENITIVE since we are familiar with them from the discussion in the foregoing section. Variation patterns which involve more than two options come last. For evidence of NOMINATIVE government, we regularly refer back to the relevant examples mentioned above[55] unless particular phenomena come up which need to be illustrated separately.

4.1.2.2.1.2.1 NOMINATIVE VS GENITIVE

The cases of (Π45) *permezz* 'by means of' and (Π52) *sforz* 'thanks to' need not be illustrated again because they have been discussed already previously. Further Πs which attest to NOMINATIVE-GENITIVE variation with nominal complements are presented one after the other in alphabetical order below.

(i) (Π8) *biswit* 'facing': NOMINATIVE government exerted by (Π8) *biswit* 'facing' is documented in (4.26). E. Serracino-Inglott (1976: 173), Aquilina (1987: 126), and Borg/Azzopardi-Alexander (1997: 162–163) exclusively provide examples of this Π and NOMINATIVE government. In the Korpus Malti 3.0 however, cases like (4.86) can be found where (Π55) *ta'* 'of' serves as linker in a construction instantiating GENITIVE government.

(4.86) [Korpus Malti 3.0; news81478]

Dak	*in-nhar*	*kien*	*qiegħed*	*isir*	*xogħol*	*fuq*
DEM:M:DIST	DEF-day	be	PROG	3SG.M.IPFV:become	work	on

id-dar	*[**biswit***	***ta'***	*dik*	*ta'*	*Buhagiar]*$_{PP}$
DEF-house	[**facing**	**of**	DEM:F:DIST	of	Buhagiar]

'On that day, construction work was going on on the house [**facing** that of Buhagiar] [. . .].'

Figure 4.49 reveals that GENITIVE government is quantitatively hardly of significance. The tokens for NOMINATIVE government are 103 times as frequent as those of GENITIVE government. In a way, for purely technical reasons, Figure 4.49 slightly overestimates the share of GENITIVE government. There can be no doubt that NOMINATIVE government is the major option for (Π8) *biswit* 'facing' – even considerably

55 In Section 4.1.2.2.

more so than in the case of (Π45) *permezz* 'by means of' in Figure 4.48. On account of the insignificant turnout for GENITIVE government triggered by (Π8) *biswit* 'facing', the question arises whether it makes sense to speak of variation in the first place. Is it not more likely that the occasional instance of *biswit* + *ta'* NP can be filed away as a slip of the tongue of a given speaker or as a slip of the pen of a given author/translator? To settle this issue, we first have to scrutinise the remaining cases of NOMINATIVE-GENITIVE competition.

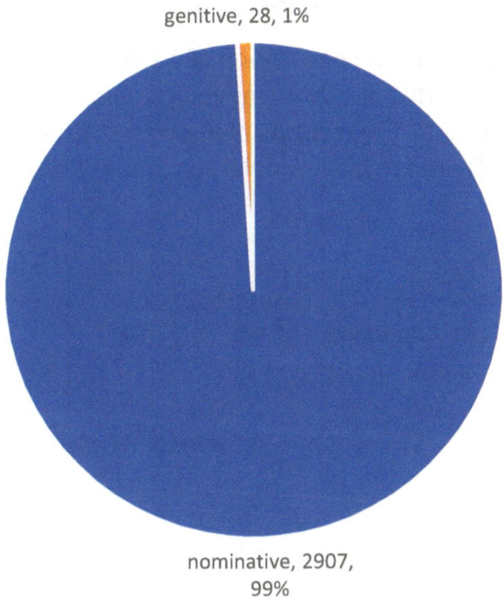

Figure 4.49: Shares of GENITIVE and NOMINATIVE government with (Π8) *biswit* 'facing' (tokens in the Korpus Malti 3.0).

(ii) (Π12) *faċċata* 'opposite': This Π is absent from the above catalogue[56] because it should usually be followed by the linker (Π55) *ta'* 'of' (Aquilina 1987: 294). However, usually does not necessarily mean always. As mentioned early on already,[57] Borg/Azzopardi-Alexander (1997: 167) assume that the linker is omitted in ABLATIVE and PERLATIVE constructions where (Π12) *faċċata* 'opposite' is preceded by (Π41) *minn* 'from' so that contrasts of the kind illustrated in (4.87)–(4.88) arise. Note that (Π12) *faċċata* 'opposite' in (4.88) is not an empty linker but a semantically loaded Π.

56 Cf. Section 4.1.1.2.2.1.

57 Cf. Section 1.4.2.5 in the context of examples (1.48)–(1.49).

(4.87) with linker (Borg/Azzopardi-Alexander 1997: 166)
Il-bank *qiegħed* ***[faċċata*** *tal-knisja]*$_{PP}$
DEF-bank be_located **[opposite** **of**:DEF-church]
'The bank is situated [**opposite** the church].' [O.T.]

(4.88) without linker (Borg/Azzopardi-Alexander 1997: 167)
L-għajjat *kien* *ġej* ***[minn*** ***faċċata*** *il-każin]*$_{PP}$
DEF-shout:PL be come.PART **[from** **opposite** DEF-band_club]
'The shouting was coming [**from opposite** the club].' [O.T.]

It is again the Korpus Malti 3.0 which impedes the postulation of a strict rule. We have found 65 instances of ABLATIVE/PERLATIVE PPS involving the combination of (Π41) *minn* 'from' and (Π12) *faċċata* 'opposite'. The majority of these PPS (54 = 83%) follow the pattern illustrated in (4.88). There are however eleven cases (= 17%) in which the linker turns up as shown in (4.89).

(4.89) [Korpus Malti 3.0; news152940]
jaqbdu *tal-linja* ***[minn*** ***faċċata*** *tal-Università]*$_{PP}$
3.IPFV:catch:PL of:DEF-line **[from** **opposite** **of**:DEF-university]
'[. . .] they catch the bus [**from opposite** the university].'

The absence of the linker is not restricted to PPS with ABLATIVE/PERLATIVE function as transpires from example (4.90).

(4.90) [Korpus Malti 3.0; european6276]
l-isem *u* *l-indirizz* *tar-rappresentant* *stabbilit*
DEF-name and DEF-address of:DEF-representative establish:PART
ġewwa *l-Komunità* *jidher* ***[faċċata*** *l-isem* *u*
inside DEF-comunity 3SG.M.IPFV:appear **[opposite** DEF-name and
l-indirizz *tal-manifattur]*$_{PP}$ *fuq* *ir-reġistru*
DEF-address of:DEF-manufacturer] on DEF-register
'[. . .] the name and the address of the representative established in the E.U. appears [**opposite** the name and the address of the manufacturer] in the register [. . .].'

Owing to the frequency of the homophonous noun *faċċata* 'façade' in the Korpus Malti 3.0, it would be necessary to determine the quantitative relation between GENITIVE government and NOMINATIVE government manually. As a short-cut, we use the procedure of extrapolating from the result gained in the course of the 1st 100-test. In this small segment, the Korpus Malti 3.0 yields 27 *faċċata*-PPS with nominal complements 24 of which (= 88.8%) involve the linker (Π55) *ta'* 'of' whereas the remaining minority of three (= 11.2%) gives evidence of NOMINATIVE government. The three hits for NOMINATIVE government are registered in documents which fall under the rubric of E.U.-related texts. In conclusion, (Π12) *faċċata* 'opposite' definitely prefers GENITIVE government over NOMINATIVE government with the latter being relatively marginal in our database.

(iii) (Π13) *favur* 'in favour of': NOMINATIVE government of (Π13) *favur* 'in favour of' is featured in (4.29). Saari (2009: 272) who also mentions the variants[58] *affavur* and *a favur* (which would regularly require the presence of the linker) emphasises that

> Aquilina argues that *favur*, too, is often accompanied by *ta'*, but the corpus of my research indicates that the form *favur* without *ta'* is much more common.

Accordingly, NOMINATIVE government should be numerically stronger than GENITIVE government while the latter has the monopoly with the two aforementioned variants. On closer inspection, the situation looks different since the shares of the two complement types are almost even for *a favur* Figure 4.50. The dominance of GENITIVE government with *(għ)affavur* notwithstanding, it is clear that Saari's assumption cannot be upheld as is.

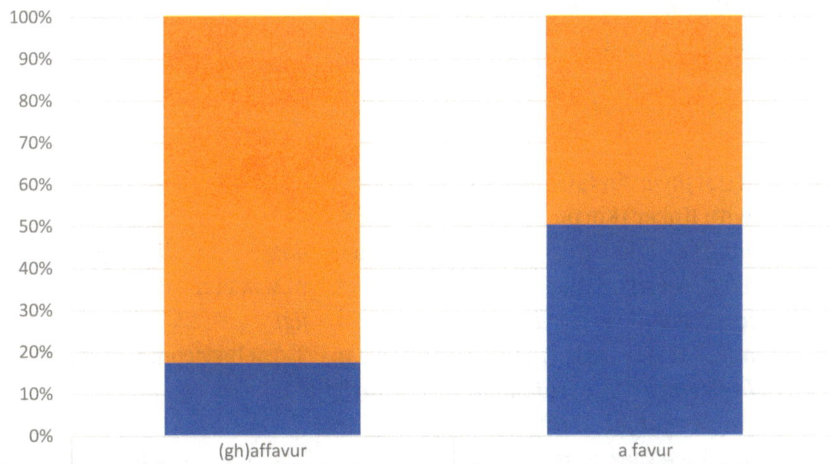

	(gh)affavur	a favur
genitive	19	100
nominative	4	101

Figure 4.50: Shares of NOMINATIVE and GENITIVE government with variants of (Π13) *favur* 'in favour of'.

58 All 201 instances of the two-word variant *a favur* are integrated into the frequency counts relating to (Π13) *favur* 'in favour of'. In contrast, the seven tokens going to the credit of *affavur* have not been counted in.

The coexistence of both government types with the two variants is illustrated in (4.91) for *affavur* and in (4.92) for *a favur*.

(4.91) (П13) *favur* 'in favour of': variant *affavur*

(a) with linker [Korpus Malti 3.0; parl12824]

din il-proċedura qiegħdha tevita t-twelid ta'
DEM:F:PROX DEF-procedure PROG:F 3SG.F.IPFV:avoid DEF-birth of
trabi b' ċertu kundizzjonijiet ***[affavur** ta' trabi f'*
baby.PL with certain condition:PL **[in_favour_of** of baby.PL in
saħħithom]$_{PP}$
health:POSS:3PL]
'[. . .] this procedure avoids the birth of bybies with certain conditions [**in favour of** babies in good health].'

(b) without linker [Korpus Malti 3.0; parl3786]

Jista' il-Ministru jgħid jekk jeżistix xi
3SG.M.IPFV:can DEF-minister 3SG.M.IPFV:say if 3SG.M.IPFV:exist some
tip ta' pjan biex isiru skemi ta' ipparkjar
type of plan so_that 3SG.IPFV:become:PL scheme:PL of parking
***[affavur** ir-residenti ta' Paceville]$_{PP}$*
[in_favour_of DEF-resident:PL of Paceville]
'Can the minister say whether there exists any kind of plan for developing parking space [**to the benefit of** the residents of Paceville] [. . .].'

(4.92) (П13) *favur* 'in favour of': variant *a favur*

(a) with linker [Korpus Malti 3.0; parl2658]

infittxu li nimmassimizzaw dak li jeżisti
1.IPFV:seek:PL that 1.IPFV:maximise:PL DEM:M:DIST that 3SG.M.IPFV:exist
u dak li jista' jiġi żviluppat
and DEM:M:DIST that 3SG.M.IPFV:can 3SG.M.IPFV:come develop:PART
***[a favur** tal-poplu tagħna]$_{PP}$*
[in_favour_of of:DEF-people of:1PL]
'[. . .] we seek to maximise that which exists and that which can be developed [**to the benefit of** our people] [. . .].'

(b) without linker [Korpus Malti 3.0; parl1610]

Sfortunatament hemm min jaħseb li jien nikkumbatti
unfortunately there who 3SG.M.IPFV:think that 1SG 1SG.IPFV:fight
***[a favur** il-lingwa Maltija]$_{PP}$ a skapitu ta'*
[in_favour_of DEF-language:F Maltese:F] at_the_expense of

lingwi *oħra*
language:PL other.PL
'Unfortunately, there are people who think that I fight [**for** the Maltese language] at the expense of other languages [. . .].'

The illustration of the use of GENITIVE government with (Π13) *favur* 'in favour of' is provided in (4.93) which contains a sentence taken from legal regulations in the domain of the right of inheritance.

(4.93) [Korpus Malti 3.0; law1240]
tista' *ssir* *disposizzjoni* *b'* *titolu* *partikolari*
3SG.F.IPFV:can 3SG.F.IPFV:become disposition with title particular
*[**favur*** ***ta'*** *persuna* *li* *għandha* *tiġi*
[**in_favour_of** **of** person that have:3SG.F 3SG.F.IPFV:come
magħżula *mill-werriet]*$_{PP}$
PART:choose:F from:DEF:heir:PL]
'[. . .] a disposition under a special title can be made [**in favour of** a person that has to be chosen by the heirs] [. . .].'

The competition between GENITIVE government and NOMINATIVE government holds for all alloforms of (Π13) *favur* 'in favour of'. Saari (2009) is proved right as to the dominance of NOMINATIVE government with (Π13) *favur* 'in favour of' outside the domain of the above variants as is visible from Figure 4.51.

The picture is a replica of Figure 4.49 for (Π8) *biswit* 'facing'. NOMINATIVE government is pervasive leaving only a tiny niche for GENITIVE government. In contrast to the former, GENITIVE government is a negligible quantity. NOMINATIVE government is once more the uncontested major option.

(iv) (Π27) *kontra* 'against': Three cases of NOMINATIVE government of this Π have been given already in (4.40). Saari (2009: 273) argues that it is "possible to combine *kontra* with its nominal complement straight ahead or with the preposition *ta'* in front of the complement." This is tantamount to assuming competition between NOMINATIVE and GENITIVE government.[59] The author goes on to specify that "the

59 We analyse Saari's (2009: 273) example (5) differently from the quoted author. According to his interpretation, *kontra tan-naħa l-oħra* is an instance of GENITIVE government of (Π27) *kontra* 'against'. In our analysis, the supposed linker (Π55) *ta'* 'of' in this phrase does not connect Π and complement but is not a syntactic constituent of the PP in the first place. We are facing a case of the elliptical construction ~~*dawk*~~ *tan-naħa l-oħra* 'those of the other side' where the head can be left out. Thus, Saari's example illustrates NOMINATIVE government instead of GENITIVE government.

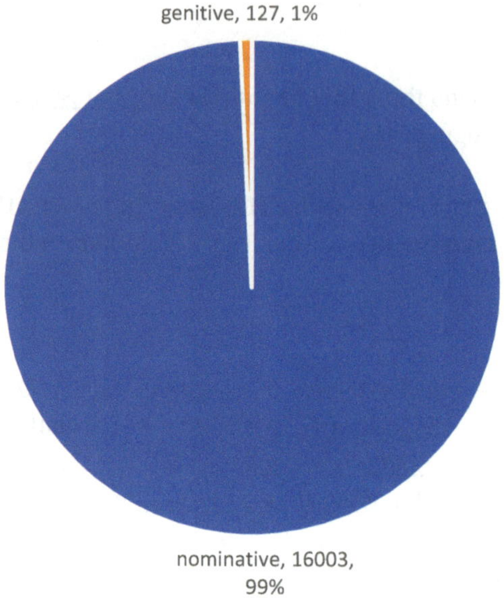

Figure 4.51: Shares of GENITIVE and NOMINATIVE government with (Π13) *favur* 'in favour of' (tokens in the Korpus Malti 3.0).

usage of *kontra* without *ta'* is much more common than the usage of the prepositional phrase." We therefore expect to find that NOMINATIVE government represents the main option. This assumption receives empirical support if we look at the quantitative discrepancy between the two patterns of government shown in (4.94).

(4.94) (Π27) *kontra* 'against' – variation

(a) [Korpus Malti 3.0; news141676]

żamm lil Josef Aquilina ta' 30 sena mill-Marsa ***[kontra***
hold to Josef Aquilina of 30 year from:DEF-Marsa **[against**
ta' *l-volontà tiegħu]$_{PP}$*
of DEF-will of:3SG.M]

'[. . .] he held Josef Aquilina, 30 years old, from Marsa, [**against** his will].'

(b) [Korpus Malti 3.0; news10294]

Saleban Shide ġie akkużat ukoll li ħebb għat-tfajla li
Saleban Shide come accuse:PART too that attack for:DEF-girl:F that
heddidha u żammha arrestata ***[kontra*** *l- volontà tagħha]$_{PP}$*
threaten:3SG.F and hold:3SG.F arrest:PART:F **[against** DEF-will of:3SG.F]

'Saleban Shide was also accused of attacking the girl, threatening her and holding her prisoner [**against** her will].'

The PP *kontra ta' l-volontà* 'against the will' (cf. (4.94a)) is a hapax in the Korpus Malti 3.0 whereas its *ta'*-less equivalent *kontra l-volontà* 'against the will' (cf. (4.94b)) is attested 493 times in the same electronic resource. If we look at the 1st 100 hits for *kontra* and *kontra ta'* in the Korpus Malti 3.0, the differences come clearly to the fore since 99 out of 100 attestations of *kontra* are clearly prepositional whereas only nine out of 100 attestations of *kontra ta'* fulfil the criteria of Π-hood. In point of fact, there is only a single piece of evidence for *kontra ta'* being used as Π among the 1st 100 hits which involve (Π27) *kontra* 'against'. Accordingly, Figure 4.52 reflects the overwhelming majority of NOMINATIVE government as opposed to the marginally attested GENITIVE government.

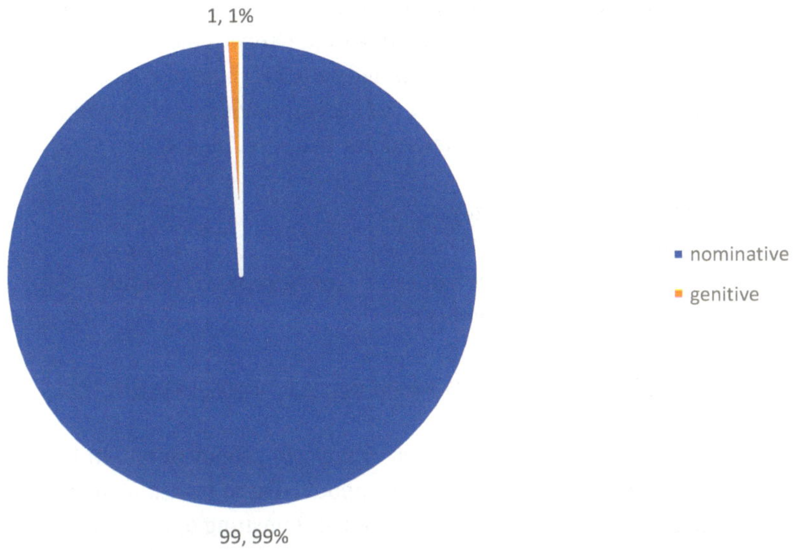

Figure 4.52: Shares of GENITIVE and NOMINATIVE government with (Π27) *kontra* 'against' (tokens in the Korpus Malti 3.0; 1st 100).

It is worth noting that the combination of *kontra* and *ta'* is registered 691 times in the Korpus Malti 3.0. However, in the vast majority of these cases, *kontra* displays the properties of a common noun, i.e. more often than not it hosts the definite proclitic *il-* and functions as complement in a PP as in (4.95).

(4.95) [Korpus Malti 3.0; literature59]

Tania	*kienet*	*fieqet*	*għalkollox*	*u*
Tania	be:3SG.F.PFV	recover:3SG.F.PFV	completely	and

ħassitha	*kompletament*	*imqajma*	*bil-**kontra***
feel:3SG.F.PFV:3SG.F	completely	PART:wake_up:F	with:DEF-**against**

ta' *Chris* *li* *kien* *qed* *jongħos* *u*
of Chris that be PROG 3SG.M.IPFV:doze_off and
jistenbaħ *kull* *ftit* *minuti*
3SG.M.IPFV:wake_up every little minute:PL
'Tania had completely recovered and felt herself awake **in contrast to** Chris who was dozing off and waking up every other minute.'

Aquilina (1987: 683) writes of a Π "used nominally" to be translated into English as 'the opposite'. In this function, *kontra* combines most often with (Π7) *bi* 'with' and (Π21) *għal* 'for' (Saari 2003: 47).

(v) (Π31) *madwar* 'around' / (Π11) *dwar* 'about': We have to integrate two Πs in this section because the evidence that is available in the Korpus Malti 3.0 is skewed in the sense that GENITIVAL government is attested for (Π31) *madwar* 'around' only indirectly, namely via the use as synonym of (Π11) *dwar* 'about' as shown in (4.96). NOMINATIVE government is amply documented, for instance, in (4.42).

(4.96) [Korpus Malti 3.0; european4116]
Huwa *rakkomandat* *li* *wieħed* *ikun* ***[madwar***
3SG.M recommend:PART that one 3SG.M:be.FUT **[about**
ta' *ċiklu* *ta'* *ċellola* *waħda* $]_{PP}$
of cycle of cell:F one:F]
'It is recommended that one (test) is [**about** the cycle of one particular cell] [. . .].'

Of the 1st 100 hits for *madwar ta'*, only two are properly prepositional. In both of these cases, the meaning is that of (Π11) *dwar* 'about'. There is also a single case of adverbial *madwar ta'* 'approximately' (cf. (4.62) and ensuing discussion). In the remaining 97 cases, *madwar* 'environs' (Aquilina 1991: 767) is either employed nominally as definite complement in a PP headed by Πs such as (Π15) *fi* 'in', (Π30) *ma'* 'with', or (Π55) *ta'* 'of', or the linker *ta'* forms part of a place name such as *Ta' Qali*. The list of the 1st 100 hits for (Π31) *madwar* 'around' does not contain any example of GENITIVAL government so that we conclude that NOMINATIVE government is the default for (Π31) *madwar* 'around' and the little proof of GENITIVAL government with this Π is spurious. For cases of this kind, it does not make sense to account for the occurrences with diagrams and tables.

(vi) (Π33) *matul* 'during': The situation is similar with (Π33) *matul* 'during'. There is overwhelming evidence of NOMINATIVE government as illustrated in (4.44). At the same time, GENITIVAL government is attested only five times in the Korpus Malti 3.0 with three hits remaining controversial. Example (4.97) represents one of two uncontroversial cases.

(4.97) [Korpus Malti 3.0; parl10001]

fid-dawl ta' dak li kien qed jiġi
in:DEF-light of DEM.M.DIST that be PROG 3SG.M.IPFV:come
diskuss fil-pajjiż ***[matul ta'*** *dawn l-aħħar*
discuss.PART in:DEF-country **[during of** DEM:PL:PROX DEF-last
xhur]$_{PP}$
month.PL]
'[. . .] in light of what was being discussed in the country [**during** these last months] [. . .].'

GENITIVAL government is absent from the 1st 100 hits for (Π33) *matul* 'during'. In analogy to the previous case of (Π31) *madwar* 'around', we assume that the evidence of GENITIVAL government involving the linker *ta'* in combination with (Π33) *matul* 'during' is not only exceptional but also questionable. Similar cases of doubtful GENITIVAL government are awaiting us below.

(vii) (Π35) *minflok* 'instead of': The ability of (Π35) *minflok* 'instead of' to trigger NOMINATIVE government has been shown in (4.46) already. Besides NOMINATIVE government, this Π also allows for GENITIVAL government as shown for the long and short alloforms in (4.98).

(4.98) (Π35) *minflok* 'instead of' – GENITIVAL government

(a) [Korpus Malti 3.0; parl142]

In-numru ta' sentenzi mogħtija tant kien żgħir li
DEF-number of sentence:PL PART:give:PL so_much be small that
l-proċedura kriminali ***[minflok ta'*** *deterrent]*$_{PP}$ *kienet qed*
DEF-procedure criminal **[instead_of of** deterrent] be:3SG.F.PFV PROG
issir sabiex is-sidien jinqdew biha
3SG.F.IPFV:become in_order_to DEF-owner:PL 3.IPFV:RFL:serve:PL with:3SG.F
'The number of convictions was so small that the criminal procedure [**instead of** a deterrent] was becoming something which the owners could utilise to their advantage.'

(b) [Korpus Malti 3.0; news105515]

Francis Zammit Dimech irsista kemm felaħ
Francis Zammit Dimech be_persistent how_much have_sufficient_energy
biex il-kuntratt jieħdu hu ***[flok ta'*** *Henley]*$_{PP}$
so_that DEF-contract 3SG.M.IPFV:take 3SG.M **[instead_of of** Henley]
'Francis Zammit Dimech tried as hard as he could so that [**instead of** Henley] he could take the contract.'

In the Korpus Malti 3.0, the co-occurrence of the two alloforms of (Π35) *minflok* 'instead of' and *ta'* more often than not is of a different kind than that illustrated in (4.98a). Most of the hits involve a *ta'*-PP whose head is separated from *minflok*. A typical example is given in (4.99).

(4.99) [Korpus Malti 3.0; news138039]

Din *tal-aħħar* *tista'* *tieħu* *forma* *ta'*
DEM.F.PROX of:DEF-last 3SG.F.IPFV:can 3SG.F.IPFV:take form of
ħakma ***[flok*** *[Ø* *ta'* *gvern]*$_{PP}$*]*$_{PP}$
domination **[instead_of** [Ø of government]]
'This last one can take the form of domination [**instead of** [(that) of government]].'

The adjacency of *flok* and *ta'* results from the EQUI-deletion of the noun *forma* 'form' which is the head of the attributive PP *ta' gvern*. Cases of this kind abound in the Korpus Malti 3.0 and thus skew the picture. If we subtract these false friends from the list, the importance of GENITIVAL government diminishes considerably. There is not a single example of GENITIVAL government in the 1st 100 hits for either of the alloforms of (Π35) *minflok* 'instead of'. In the entire Korpus Malti 3.0, there are only eight instances of GENITIVAL government triggered by (Π35) *minflok* 'instead of' (two with the short alloform *flok* and six with the long alloform *minflok*). Thus, NOMINATIVE government is the rule with the occasional and exceptional instance of GENITIVAL government.

(viii) (Π36) *mingħajr* 'without': A striking parallel to (Π33) *matul* 'during' is the putative GENITIVAL government with (Π36) *mingħajr* 'without'. As shown in (4.2), this Π normally takes its complements without intervening linker so that NOMINATIVE government applies. In the Korpus Malti 3.0, we find however twelve instances of *mingħajr ta'*. In (4.100), we provide an example of a pattern that is responsible for exactly half of all these attestations. Three other cases of *mingħajr ta'* have to be excluded because they involve the subordination of dependent clauses. In two further cases, *ta'* forms part of a headless GENITIVE construction to be addressed below[60] and thus is not required by (Π36) *mingħajr* 'without'.

60 Cf. Section 4.2.

(4.100) [Korpus Malti 3.0; parl10555]

Jista' *l-Ministru* *jara* *li* *tiġi*
3SG.M.IPFV:can DEF-minister 3SG.M.IPFV:see that 3SG.F.IPFV:come
rimedjata ***[mingħajr ta'*** *aktar telf ta' żmien]*$_{PP}$
remedy:PART:F **[without of** more loss of time]
il-kwestjoni li teżisti bejn il-gvern u
DEF-question that 3SG.F.IPFV:exist between DEF-government and
l-Kunsill Lokali tar-Rabat
DEF-local_council of:DEF-Rabat
'Can the Minister see that the controversy that exists between the government and the local council of Rabat will be remedied [**without** further delay]?'

In the Korpus Malti 3.0, there are six hits for *mingħajr ta' aktar telf ta' żmien* 'without further loss of time' but 57 hits for the synonymous expression *mingħajr aktar telf ta' żmien / ħin*. The seventh case involving *mingħajr ta'* is structurally comparable to (4.99). The PP *mingħajr ta' iktar bżonn (ta' għajnuna)* 'without further need (of help)' involves *iktar* 'more' which is an alloform of *aktar* 'more'. The combination *mingħajr iktar NP* 'without further NP' is attested 97 times in the electronic resource. The PP *mingħajr iktar / aktar bżonn* 'without further need' is attested twice. The picture is relatively clear. NOMINATIVE government prevails to such an extent that there is hardly any space left for GENITIVAL government. Moreover, it seems that the latter is restricted to complements which are introduced by *aktar / iktar* 'more' – and even in these combinations the presence of the linker *ta'* is a quantitatively negligible phenomenon. We doubt that it is firmly established throughout the Maltese speech community.

(ix) (Π48) ***qrib*** **'near'**: The situation is markedly different in the case of (Π48) *qrib* 'near'. We know from (4.53) that this Π can take complements without the linker *ta'*. NOMINATIVE government is common. However, there are also numerous instances of GENITIVAL government as in (4.101).

(4.101) [Korpus Malti 3.0; european8793]

għandu *jitpoġġa* *wieħed* ***[qrib ta'*** *kull post*
have:3SG.M 3SG.M.IPFV:place one **[near of** each post
tal-ħruġ]$_{PP}$
of:DEF-exit]
'[. . .] one [= facility of manual help] has to be placed [**near** every exit].'

NOMINATIVE government ousts GENITIVAL government by a ratio of 11.25-to-one according to the 1st 100 hits of (Π48) *qrib* 'near'. Figure 4.53 reveals that only 49% of all attestations of *qrib* can be considered to be prepositional.

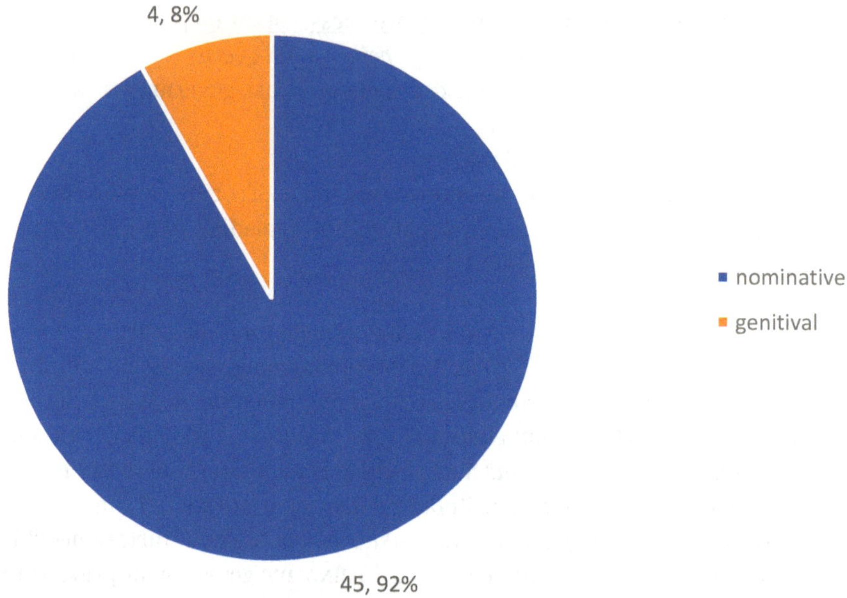

Figure 4.53: Shares of GENITIVE and NOMINATIVE government with (Π48) *qrib* 'near' (tokens in the Korpus Malti 3.0; 1st 100).

In contrast to (Π31) *madwar* 'around', (Π33) *matul* 'during', and (Π36) *mingħajr* 'without', the competition between the two options of government is relatively substantial because the examples of GENITIVAL government are not restricted to specific collocations although NOMINATIVE government is clearly the majority solution.

(x) (Π50) *rigward* 'concerning': The case of (Π50) *rigward* 'concerning' bears resemblance to that of (Π2) *a skapitu* 'at the expense of'.[61] It makes sense therefore to apply similar criteria when we investigate the case at hand, i.e., that of (Π50) *rigward* 'concerning'. The examples in (4.55) have proved that NOMINATIVE government is possible with this Italo-Romance Π. In (4.102), we present a piece of evidence for GENITIVAL government. The ensuing discussion will show that we are dealing with a particularly difficult case.

61 As described in Section 3.2.6.

(4.102) [Korpus Malti 3.0; european10680]

jadottaw	*approċċi*	*komuni*	***[rigward***	***ta'***	*metodi*
3.IPFV:adopt:PL	approach:PL	common:PL	**[concerning**	**of**	method:PL

ta'	*ġbir*	*u*	*rappurtar*	*ta'*	*informazzjoni*	*strateġikament*
of	collecting	and	reporting	of	information	strategically

rilevanti]$_{PP}$
important:PL]

'[. . .] they adopt common approaches [**concerning** methods of colleting and reporting strategically relevant information].'

Saari (2003: 48) mentions the alternation of *rigward*-PPs with and without the linker *ta'*. His example for *rigward ta'* however, features *rigward* as definite noun complement of (Π15) *fi* 'in'. The examples are repeated in Saari (2009: 274–275) where the author emphasises that "the occurrences of *rigward* in contemporary Maltese are not rare at all." The presence of the linker *ta'* in Maltese might be explicable in terms of a PAT-copy of Italian *riguardo a / a riguardo di* 'concerning; with regards to' which involve the Πs *a* 'to' and *di* 'of', respectively.[62]

Of the 18,407 hits for *rigward ta'* in the Korpus Malti 3.0, 13,787 (= 75%) involve *rigward* as definite noun complement of (Π15) *fi* 'in' yielding the PP *fir-rigward ta' NP* 'with regards to NP'. The remaining quarter of cases which are unaccounted for raises several questions. In point of fact, (Π50) *rigward* 'concerning' poses a number of problems because of the range of variation it associates with. First of all, there are potential long alloforms to be taken account of which have not been addressed above for reasons which will become clear below.[63]

In Figure 4.54, the different (pseudo-)Πs with which *rigward ta'* combines in the Korpus Malti 3.0 are ordered top-down according to their decreasing token frequency. The differences are drastic. The three least frequent options account only for a fraction of the occurrences of *rigward*. We have already excluded *a* and *in* from the inventory of Maltese Πs so that they can best be considered optional submorphemic,[64] i.e. meaningless components of the segmental chain of the (Π50) *rigward* 'concerning'. None of these potential long alloforms is mentioned in the

62 Cf. Section 1.6.6.

63 In Section 3.2.

64 This is why isolated examples like (4.i) are not taken account of in the discussion.

(4.i) [Korpus Malti3.0; european4229]

L-ebda	*argument*	*ġdid*	*ma*	*tressaq*	***in***	***dan***	***ir-rigward***
DEF-no	argument	new	NEG	RFL:CAUS:approach	**in**	**DEM.M.PROX**	**DEF-regard**

'No new argument was put forward **with regard to this** [issue].'

three reference dictionaries used in this study. The problem remains to be solved whether we are dealing with fully-blown but only marginal alloforms of the Π under inspection.

f(ir-)		13,801
Ø		4,585
in		19
	+ *rigward ta'* NP	
għal		13
a		9
b(ir-)		8

Figure 4.54: Potential allomorphs of (Π50) *rigward* 'concerning'.

It is possible to exclude all those instances from further consideration in which *rigward* displays nouny properties such as definiteness marking and taking a GENITIVAL attribute.[65] In this way, all instances of *f' / b' (dan) ir-rigward ta'* NP 'with regards to this NP' can be removed from the list. The number of problematic cases considerably diminishes. However, there remain enough riddles to solve because even in the absence of overt definiteness marking and/or the absence of the linker *ta'* do we still find variation and serious difficulties when it comes to evaluating the turnout of our query in the Korpus Malti 3.0.

It is clear nevertheless, that the combination with demonstrative and definiteness marker characterizes *rigward* as a common noun in this example. Cf. also in Section 1.6.6.

65 In the Korpus Malti 3.0, there are also 21 tokens of *rigwardi* – the regular plural of *rigward*. Twenty of these cases function as complements in PPs headed by (Π15) *fi* 'in' nineteen of which stem from E.U. documents including the sole example of bare *rigwardi* which is neither quantified or modified. We assume that (4.ii) below is a typo and not a genuine example of a plural form being used prepositionally.

(4.ii) [Korpus Malti 3.0; european8839]

jeħtieġu	*kjarifika*	***[rigwardi***	*l-qasam*	*u*	*ċerti*
3.IPFV:need:3PL	clarification	**[concerning:PL**	DEF-distribution	and	certain:PL
definizzjonijiet	*tad-Direttiva]*$_{PP}$				
definition:PL	of:DEF-directive]				

'[. . .] they need a clarification [**concerning** the distribution and certain definitions of the Directive] [. . .].'

For a start, we look at the involvement of *rigward* in PPs headed by (Π7) *bi* 'with'. There are 171 *bi*-PPs which take *rigward* as a complement only thirteen of which also involve the definite pro-clitic on the complement. Table 4.10 surveys these cases which are ordered to the principles which we disclose in the next paragraph.

Table 4.10: *bi*-PPs involving definite *rigward* in the Korpus Malti 3.0.

source	example / translation
europe-an10149	*l-investigazzjoni kopriet biss is-sitwazzjoni ekonomika ta' l-IK* ***bir-rigward ta****s-suq Komunitarju* '[. . .] the investigation covered only the economic situation of the IK **with regards to** the Common Market.'
europe-an10149	*Din l-analiżi kopriet ukoll l-importazzjoni mit-Tajlandja* ***bir-rigward ta'*** *volumi u prezzijiet*. 'This analysis covered also the imports from Thailand **with regards to** volume and prices.'
european3690	***bir-rigward ta'*** *razza partikolari* '[. . .] **with regards to** a particular [animal] race [. . .]'
european8761	*Il- Kunsill adotta l- Pożizzjonijiet Komuni 97 [. . .]* ***bir-rigward ta'*** *l- Angola* 'The council adopts the Common Position 97 [. . .] **with regards to** Angola [. . .].'
european2996	***bir-rigward*** *dovut, fejn xieraq,* ***għa****d-disposizzjonijiet tal-punt 9.5* '[. . .] **with** due **regards**, where desirable, **to** the dispositions of point 9.5.'
european3898	***bir-rigward*** *dovut, fejn xieraq,* ***għa****ll-effetti ta' fatturi atmosferiċi* '[. . .] **with** due **regards**, where desirable, **to** the effects of atmospheric factors.'
european3898	*jkun iddisinjat* ***bir-rigward*** *dovut* ***għa****ll-prinċipji ergonomiċi* '[. . .] it will be designed **with** due **regards to** the ergonomic principles.'
european9385	***bir-rigward*** *dovut* ***għa****ll-prijoritajiet stipulati fir-rapport* '[. . .] **with** due **regards to** the priorities stipulated in the report.'
european5678	*din id- Deċiżjoni għandha tapplika* ***bir-rigward*** *dovut* ***għa****d-dispożizzjonijiet* '[. . .] this decision must be applied **with** due **regards to** the dispositions [. . .].'
european5678	*Din id-Deċiżjoni għandha tapplika* ***bir-rigward*** *dovut* ***lejn*** *id-dispożizzjonijiet* 'This decision must be applied **with** due **regards to** the dispositions [. . .].'
european8744	***bir-rigward*** *dovut* ***lejn*** *ir-responsabbiltajiet* '[. . .] **with** due **regards to** the responsibilities [. . .].'
european9915	***bir-rigward lejn*** *dak li l-industrija setgħet kienet qegħda tistenna* '[. . .] **with regards to** what the industry could have been expecting [. . .].'
european8761	*iżżomm taħt għajnejha s-sanzjonijiet imponuti fuq l-UNITA* ***bir- rigward*** *l-implimentazzjoni tal-Memorandum ta' Ftehim* '[. . .] it keeps an eye on the sanctions imposed on UNITA **with regards to** the implementations of the Memorandum of Understanding [. . .].'

What needs to be mentioned first in connection with Table 4.10 is that each and every example stems from the genre of E.U. documents, i.e., we are facing translations. No other genre gives evidence of the *bir-rigward (Π) NP*. Second, the grey-shaded part of the table marks out those cases in which *rigward* is modified by the postnominal attribute *dovut* 'due' and thus treated like a common noun. Third, *bir-rigward* combines with three different Πs, namely (Π21) *għal* 'for', (Π28) *lejn* 'towards', and (Π55) *ta'* 'of'. The resulting constructions seem to be synonymous. Note especially that [european5678] gives evidence of the parallel use of (Π21) *għal* 'for' and (Π28) *lejn* 'towards'. Fourth, there is only one example of asyndesis in the sense that *bir-rigward* directly combines with the demonstrative *dak* 'that'. In the majority of cases, a Π serves as linker.

We do not want to jump to conclusions yet because there are several other phenomena to be taken account of first. There are, for instance, six *bi*-PPs which combine with *rigward* in the absence of the definiteness marker. Five of the cases are given in Table 4.11.

Table 4.11: *bi*-PPs involving bare *rigward* in the Korpus Malti 3.0.

source	example / translation
european2279	*jikkonformaw ma' l-ispeċifikazzjonijiet ta' Stat Membru għall-użu* ***b' rigward ta'*** *ċerti kategoriji ta' vetturi* '[. . .] they are in accordance with a Member State's specifications for the use **with regards to** certain categories of vehicles [. . .].'
european7502	*l-introjtu kollu miġbur* ***b' rigward ta'*** *dik is-sena finanzjarja* '[. . .] the entire revenue collected **with regards to** this financial year [. . .]'
european8022	***b' rigward ta'*** *riproduzzjoni privata analoga* '[. . .] **with regards to** a private analogical reproduction [. . .]'
european3912	***b' rigward tan-****natura ta' l- attivitajiet li huma jwettqu* '[. . .] **with regards to** the nature of the activities which they realise.'
news135815	***B' rigward tas-****sullokazzjonijiet* '**With regards to** the sublettings [. . .]'

The absent definite article is suggestive of a reduced degree of categoricity of *rigward* in these constructions. In connection to this observation, example (4.103) is especially noteworthy because it seems to reflect an even higher degree of grammaticalization of *b' rigward ta'*.

(4.103) [Korpus Malti3.0; european8121]

B'	***rigward***	***ta'***	*din*	*id-disparità*	*u*	*[**mħabba***
with	**regard**	**of**	DEM.F.PROX	DEF-disparity	and	[**on_account_of**

n-nuqqas	*totali*	*ta'*	*koperazzjoni*	*mill-esportaturi*	*Ċiniżi*
DEF-lack	total	of	co-operation	from:DEF-exporter:PL	Chinese:PL

jew *mill-produtturi* *esportaturi]$_{PP}$* *hemm* *possibiltà*
or from:DEF-producer:PL exporter:PL] there possibility
qawwija *li* *jiġi* *evitat* *l-ħlas*
strong:F that 3SG.M.IPFV:come avoid:PART DEF-payment
tad-dazji *eżistenti*
of:DEF-excise:PL existing:PL
'**With regards to** this disparity and [**on account of** the total lack of co-operation of the Chinese exporters or producers-exporters], there is the strong possibility that payment of the existing excises will be avoided.'

The complex sentence starts with a coordinating construction whose second conjunct is a bona fide PP headed by the causal (Π39) *minħabba* 'on account of' which is represented by one of its short alloforms, viz. *mħabba*. If the status of the *mħabba*-PP is uncontroversial, we can assume that under co-ordination two conjuncts of similar syntactic status are joined to each other, meaning: if there is a *mħabba*-PP nothing speaks against granting PP-status also to ***b' rigward ta'*** *din id-disparità* '**with regards to** this disparity'. The simple formula in Figure 4.55 summarises this line of argumentation.

b' rigward ta' = Π / *mħabba* = Π

Figure 4.55: Π-status of *b' rigward ta'*.

This is probably only an illicit short-cut because there are further issues to be looked into. Most notably, in the Korpus Malti 3.0, there are 158 hits of *b' rigward* occurring without the linker *ta'*. The absence of both the definiteness marker and the linker *ta'* seems to indicate that an even higher degree of grammaticalization is reached because *rigward* seems to lack all those properties which are usually expected of a common noun. An example like (4.104) supports this opinion.

(4.104) [Korpus Malti3.0; news140983]
B' *referenza* *għall-artiklu* *li* *deher* *fil-ħarġa* *ta'*
with reference for:DEF-article that appear in:DEF-edition of
L-Orizzont *tal-Ħamis* *15 ta' Lulju 2010* ***[b'*** ***rigward***
L-Orizzont of:DEF-Thursday 15 July 2010 **[with** **regard**
l-aċċident *li* *seħħ* *nhar* *it-Tnejn* *li* *għadda]$_{PP}$*
DEF-accident that happen day DEF-Monday that pass]
'With reference to the article which appeared in the edition of L-Orizzont on Thursday, 15 July 2010, [**with regards to** the accident which happened last Monday] [. . .].'

In contrast to the previously discussed constructions in which *rigward* is involved, the bracketed part of (4.104) is a good candidate for a PP and thus *b' rigward* would automatically be entitled to the label Π. Things are not as straightforward as it seems since (4.104) is only one of altogether thirteen cases in the Korpus Malti 3.0 in which *b' rigward* behaves like a Π. This turnout equals some 8% of all hits of *b' rigward* without *ta'* in the electronic resource. In the bulk of the cases, however, *rigward* fails to display similarly uncontroversial Π-properties. In 131 cases, the role of the linker *ta'* is taken over by (Π21) *għal* 'for' whereas combinations with (Π28) *lejn* 'towards' (three instances), (Π15) *fi* 'in' (once), and (Π29) *lil* 'to' (once) are only attested exceptionally. In addition, there are three cases which cannot be classified. Moreover, there is ample evidence of *rigward* being modified by an attribute. The forty instances of attribution involve the following adjectives and participles (cf. Figure 4.56 with token frequencies).

b' rigward	+	*partikolari*	'particular'	10
		xieraq	'suitable'	10
		(kif) mistħoqq	'(as) deserved'	9
		dovut	'due'	6
		debitu	'due'	2
		speċjali	'special'	2
		sħiħ	'full'	1

Figure 4.56: Attributes modifying *(b') rigward.*

In each of these 40 cases, the modifier is followed by a Π. We exclude all instances of attribution from the domain of (Π50) *rigward* 'concerning'.

With 151 out of 158 tokens, E.U. documents account for 95.5% of all attestations of *b' rigward* without *ta'* in the Korpus Malti 3.0. News and parliamentary speeches contribute three and four cases, respectively. There is thus a very strong association with exactly that genre which holds the monopoly also for other constructions involving *rigward* as shown above. The limited distribution over genres is also visible in the case of

(a) *in rigward* 'concerning' whose 63 tokens result from 49 attestations in E.U. documents, 13 hits in parliamentary speeches, and a single hit in a legal text, and
(b) *a rigward* 'concerning' whose 52 occurrences comprise forty hits in parliamentary speeches, nine hits in E.U. documents, and three hits in the news genre,

(c) *għal rigward* 'concerning' 36 occurrences twenty-two hits go to the credit of parliamentary speeches, nine are from E.U. documents, and five originate form news texts.[66]

According to Figure 4.54, *in rigward* combines with *ta'* nineteen times whereas this combination is attested nine times for *a rigward.* The only other Π attested with *in rigward* is (Π21) *għal* 'for' which is employed eight times. In contrast, *a rigward* does not tolerate the use of a Π other than (Π55) *ta'* 'of'. Neither *in rigward* nor *a rigward* allows for *rigward* being modified by an attribute. In (4.105)–(4.106), we present examples of the employment of both expressions being asyndetically combined with a following noun.

(4.105) [Korpus Malti3.0; parl929]
Jista' *l-Ministru* *jinforma* *lill-Kamra*
3SG.M.IPFV:can DEF-minister 3SG.M.IPFV:inform to:DEF-chamber
*[**in rigward** taxpayer reġistrat bin-numru [. . .] li*
[**concerning** taxpayer register:PART with:DEF-number that
hija non-resident]$_{PP}$
3SG.F non-resident]
'Can the Minister inform the Chamber [**concerning** a taxpayer registered with the number [. . .] who is a non-resident]?'

(4.106) [Korpus Malti 3.0; parl9860]
Tista' *l-Ministru* *tgħid* *u* *tagħti*
3SG.F.IPFV:can DEF-minister 3SG.F.IPFV:say and 3SG.F.IPFV:give
*rendikont [**a rigward** iċ-childcare subsidy scheme]*$_{PP}$
statement [**concerning** DEF-childcare subsidy scheme]
'Can the Minister speak and give a statement [**concerning** the childcare subsidy scheme]?'

What speaks further in favour of analysing the above usages as prepositional is example (4.107) where two PPS come in a row.

(4.107) [Korpus Malti 3.0; parl9664]
Jista' *l-Ministru* *jgħid* *x'* *arranġamenti*
3SG.M.IPFV:can DEF-minister 3SG.M.IPFV:say what arrangement:PL

66 Cf. discussion in Section 4.1.3.2.

kien *hemm* *bejn* *il-Gozo Channel* *u* *l-Malta Drydocks*
be there between DEF-Gozo_Channel and DEF-Malta_Drydocks
[rigward *il-manutenzjoni* *tal-vapuri* *tal-Gozo Channel*
[concerning DEF-maintenance of:DEF-ship:PL of:DEF-Gozo_Channel
[anke ***a rigward*** *ħlasijiet* *għal* *tali* *xogħol* $]_{PP}]_{PP}$
[also **concerning** payment:PL for such work]]
'Can the Minister say what arrangements there are between Gozo Channel and Malta Drydocks [**concerning** the maintenance of the ships of Gozo Channel [also **with regards to** payments for such work]]?'

In (4.107), *rigward* and *a rigward* co-occur as heads of two separate PPS. There is no reason to analyse the two elements differently from each other. There is also no reason to assume any semantic differences. Thus, one might want to conclude that *a rigward* and, in analogy, *in rigward* too are in an allomorphic relation to bare *rigward* the three of them being realization forms of (Π50) *rigward* 'concerning'.

However, a word of caution is in order. If we accept *a rigward* and *in rigward* as long alloforms of (Π50) *rigward* 'concerning' we also accept competition of NOMINATIVE and GENITIVAL government in the case of *a rigward* and NOMINATIVE, GENITIVAL, and BENEFACTIVE government in the case of *in rigward*. In the Maltese Π-inventory, BENEFACTIVE government is confined to combinations with *in rigward*. It is true that there are four hits for bare *rigward* combining with (Π21) *għal* 'for' superficially yielding a Π-like combination. However, not only are three of the four cases identical but all four combinations are followed by clauses. There is no evidence of *rigward* + *għal* + NP. In point of fact, there is a syntactic boundary separating *rigward* from *għal*.[67] Thus, no BENEFACTIVE government applies in the case of bare *rigward*.

67 The repeated example is given in (4.iii)

(4.iii) [Korpus Malti 3.0; parl11574]
Jista' *jagħti* *din* *l-informazzjoni* ***rigward***
3SG.M.IPFV:can 3SG.M.IPFV:give DEM.F.PROX DEF-information **concerning**
[għal *liema* *organi* *hemm* *stennija]* *u* *kemm* *hemm*
[for which organ:PL there expectation] and how_many there
pazjenti *jistennew* *kull* *organu*
patient:PL 3.IPFV:wait:PL each organ
'Can he give this information **concerning** [**for** which organs there are expectations] and how many patients are there waiting for each organ?'

As the square bracket is meant to show, *rigward* and *għal* belong to two different syntactic units. The latter Π depends on the noun *stennija* 'expectation' whereas *rigward* takes the entire sequence of two clauses as its complement.

At the same time, the restriction of *a rigward* and *in rigward* to only a small subset of the text genres represented in the Korpus Malti 3.0 strongly suggests that we are dealing with marked options. The bare form *rigward* is attested in many more genres from academic, via culture, E.U. documents, news, opinion, parliamentary speeches, and religious texts to sports. It ousts the long alloforms quantitatively to such an extent that it is impossible to picture the discrepancy in a diagram. We calculate 70,049 tokens for bare *rigward*. This means that this alloform is between 1,347 and 1,111 times more frequent than *a rigward* and *in rigward*, respectively. The bare form accounts for 99.8% of all instances of (Π50) *rigward* 'concerning'. We therefore treat the competition of the three forms under inspection as special.

Moreover, *a rigward, in rigward*, and *rigward* do not seem to follow a context-dependent distribution since all three are attested with preceding vowel-final and consonant-final words as well as with definite and indefinite complements to their right. As comes to the fore in Figure 4.57, *a rigward* has a stronger leaning towards combinations with consonant-final words whereas *in rigward* favours combinations with vowel-final words. The sentence-initial position is subsumed under C]_.

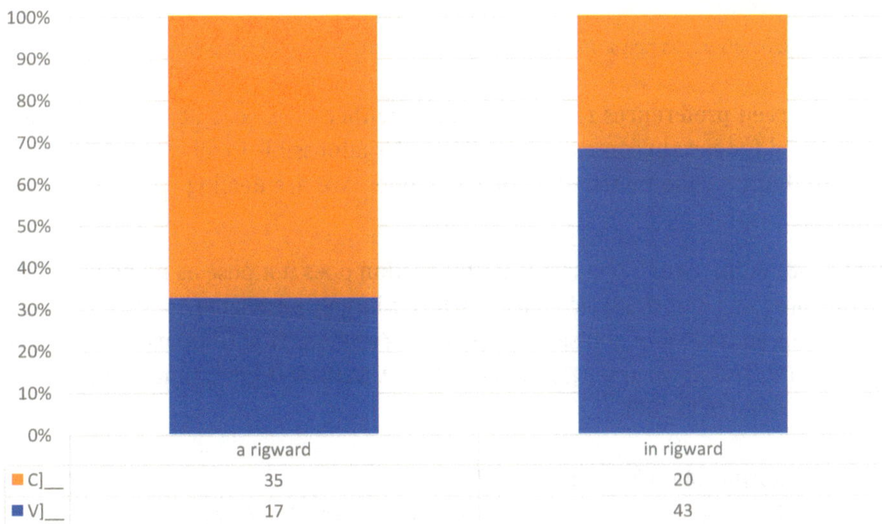

Figure 4.57: *a rigward / in rigward* over phonological contexts in the Korpus Malti 3.0.

The combinability with definite and indefinite complements must be further differentiated because the linkers *ta'* and *għal* come into play. Figure 4.58 reveals that *a rigward* is especially strong with definite complements. Similarly, *għal* is used predominantly with definite complements. The absolute numbers differ from those given above because unclear cases have been removed from the count.

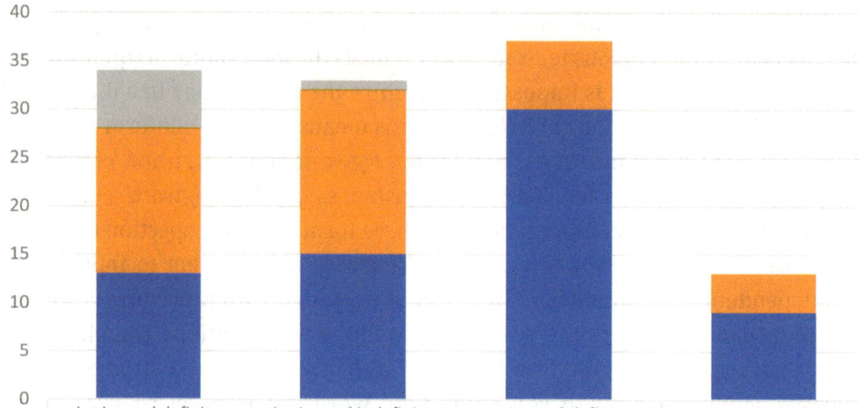

	in rigward definite	in rigward indefinite	a rigward definite	a rigward indefinite
għal	6	1		
ta'	15	17	7	4
Ø	13	15	30	9

Figure 4.58: *a rigward / in rigward* with (in)definite complements in the Korpus Malti 3.0.

The observed preferences notwithstanding, neither *a rigward* nor *in rigward* are specialised for certain contexts. Bare *rigward* is admissible in all of the contexts as well so that no conditioned allomorphy applies. We are dealing with stylistically motivated variation.

(xi) (П58) *viċin* 'near': The last П in this section does not pose as many problems as the foregoing one. (П58) *viċin* 'near' behaves largely parallel to its synonym (П48) *qrib* 'near' addressed in sub-section (ix) in this section. That (П58) *viċin* 'near' triggers NOMINATIVE government is shown in (4.60). GENITIVAL government of the same П is illustrated in (4.108).

(4.108) [Korpus Malti 3.0; academic32]
Hija *tgħix* *fi* *Bray* ***[viċin*** ***ta'*** *Dublin*$]_{PP}$
3SG.F 3SG.F.IPFV:live in Bray [**near** **of** Dublin]
'She lives in Bray [**near** Dublin].'

The competition between the two kinds of government does not yield as extreme results as in many of the previously discussed cases. Figure 4.59 shows that for (П58) *viċin* 'near' NOMINATIVE government is the majority option but GENITIVAL government retains a numerically strong position too.

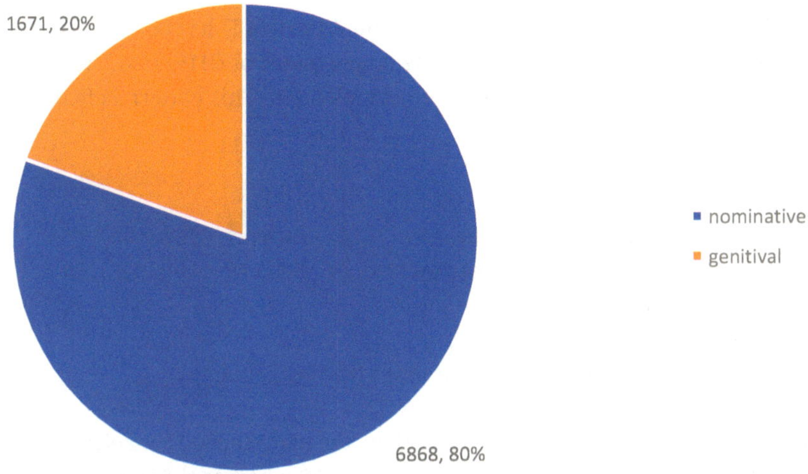

Figure 4.59: Shares of GENITIVE and NOMINATIVE government with (Π58) *viċin* 'near' (tokens in the Korpus Malti 3.0).

On the basis of a check of the 1st 100 hits for each of the competitors, it is impossible to determine whether there are structural conditions which determine the choice of government. Both competitors are attested with definite and indefinite complements as well as with words to their right which start with a vowel or a consonant.

4.1.2.2.1.2.2 Further patterns of government

The bulk of the patterns of variation in the domain of government with nominal complements goes to the credit of NOMINATIVE and GENITIVE as shown in the previous section. In addition to this widely dominant competition of NOMINATIVE and GENITIVE, there are three further contestants, namely LOCATIVE government, ABLATIVE government, and ALLATIVE government which are also subject to variation as shown below.[68]

4.1.2.2.1.2.2.1 Patterns involving LOCATIVE government

Two Πs allow for LOCATIVE government as an alternative to NOMINATIVE government (and doubtful cases of GENITIVAL government). These Πs are (Π18) *ġewwa* 'inside' and (Π39) *minħabba* 'on account of' whose behaviour is sketched in (i) and (ii), respectively.

68 Cf. Sections 4.1.2.2.1.2.2.1–4.1.2.2.1.2.2.3.

(i) (Π18) *ġewwa* 'inside': NOMINATIVE government for this Π has been illustrated in (4.34). LOCATIVE government involves the employment of (Π15) *fi* 'in' as linker between (Π18) *ġewwa* 'inside' and the noun complement. This pattern is featured in (4.109).

(4.109) [Korpus Malti 3.0; parl9636]

ħames	*serqiet*	*li*	*seħħew*	*fi*	*ħwienet*	***[ġewwa***	***f'***
five	theft:PL	that	happen:3PL.PFV	in	shop:PL	**[inside**	**in**
Ħaż–Żebbuġ]$_{PP}$							
Ħaż-Żebbuġ]							

'[. . .] five thefts that took place in shops [**in** Ħaż-Żebbuġ] [. . .].'

To express the spatial relations of Place and Goal, the place name *Ħaż-Żebbuġ* combines with *ġewwa fi* only once in the Korpus Malti 3.0 whereas there are two hits for *ġewwa Ħaż-Żebbuġ* and 138 matches for *fi Ħaż-Żebbuġ*.[69] For the time being, it is sufficient to state that LOCATIVE government of (Π18) *ġewwa* 'inside' is an infrequent phenomenon yielding a turnout of 45 hits in the Korpus Malti 3.0. With only four hits in the same electronic corpus, GENITIVAL government is even rarer than LOCATIVE government. The combination of *ġewwa* and *ta'* is illustrated in (4.110).

(4.110) [Korpus Malti 3.0; european190]

jagħti	*lill-istudenti*	*universitarji*	*l-opportunita*			
3SG.M.IPFV:give	to:DEF-student:PL	universitarian:PL	DEF-opportunity			
sabiex	*jagħmlu*	*perjodu*	*sħiħ*	*ta'*	*studju*	*rikonoxxut*
in_order_to	3.IPFV:make:PL	period	whole	of	study	recognise:PART
[ġewwa	***ta'***	*l-anqas*	*Stat Membru*	*wieħed]$_{PP}$*	*bħala*	*parti*
[inside	**of**	DEF-less	member state	one]	as	part
integrali	*mid-diploma*	*tagħhom*				
integral	from:DEF-diploma	of:3PL				

'[. . .] to give the university students the opportunity to do a full period of studies which is recognised [**in** at least one member state] as integral part in their diploma [. . .].'

Among the 1st 100 hits, (Π18) *ġewwa* 'inside' is attested 42 times with NOMINATIVE government as opposed to four hits for LOCATIVE government. No evidence of GENITIVAL government could be found in the test-run. According to Table 3.27, the overall token frequency of (Π18) *ġewwa* 'inside' amounts to 20,844 matches, meaning LOC-

69 We come back to the issue of semantic equivalence of these constructions in Section 5.1.2.1.1.1.

ATIVE and GENITIVAL government account only for a fraction of the possible cases in which (Π18) *ġewwa* 'inside' takes a noun as complement. NOMINATIVE government must be taken to represent the default pattern.

(ii) (Π39) *minħabba* 'on account of': In (4.48), we have provided examples for NOMINATIVE government of (Π39) *minħabba* 'on account of'. The possibility of (Π39) *minħabba* 'on account of' taking a complement in the GENITIVE is demonstrated in (4.111).

(4.111) [Korpus Malti 3.0; european15]

dawn	*id-diskrepanzi*	*ma*	*kienux*
DEM:PL:PROX	DEF-discrepancy:PL	NEG	be:3PL.PFV:NEG

[minħabba	***fi***	*żbalji*	*kommessi*	*apposta]$_{PP}$*
[on_account_of	**in**	mistake:PL	commit:PART:PL	on_purpose]

'[. . .] these discrepancies did not happen [**on account of** purposefully committed errors] [. . .].'

The absolute tokens for LOCATIVE government with (Π39) *minħabba* 'on account of' sum up to 1,760 hits whereas those for GENITIVAL government – exemplified in (4.112) – yield a meagre total of 31 hits.

(4.112) [Korpus Malti 3.0; parl186]

sirt	*naf*	*aktar*	*bil-problema*	*ta'*	*nies*
become:1SG.PFV	1SG.IPFV:know	more	with:DEF-problem	of	people

li	*għandna*	*hawn*	*Malta*	***[minħabba***	***ta'***	*l-inkwiet*	*li*
that	have:1PL	here	Malta	**[on_account_of**	**of**	DEF-trouble	that

għandhom	*f'*	*pajjiżhom]$_{PP}$*
have:3PL	in	country:3PL]

'I have come to know more about the problem of people we have here in Malta [**because of** the trouble they have in their country].'

The turnout for LOCATIVE government notwithstanding, there can be no doubt that NOMINATIVE government is the default solution for nominal complements in PPS headed by (Π39) *minħabba* 'on account of'. Figure 4.60 speaks a clear language with regard to the quantitative relation between the three options. The share of GENITIVAL government is so small that it is invisible in the diagram. LOCATIVE government, however, claims about 1% and thus constitutes only a very marginal alternative to the overwhelmingly dominant NOMINATIVE government.

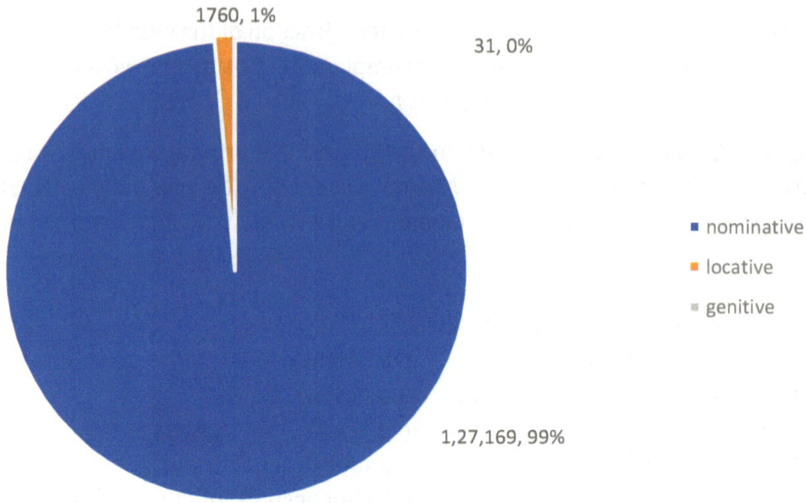

Figure 4.60: Shares of LOCATIVE, GENITIVE, and NOMINATIVE government with (Π39) *minħabba* 'on account of' (tokens in the Korpus Malti 3.0).

4.1.2.2.1.2.2.2 Patterns involving ABLATIVE government

ABLATIVE government in the domain of nominal complements is attested with three Πs, namely (Π1) *apparti* 'apart from', (Π34) *minbarra* 'except', and (Π44) *oltre* 'beyond' as discussed in sections (i)–(iii) below. Interestingly, these three Πs are semantically related to each other.[70]

(i) (Π1) *apparti* 'apart from': The proof that this Π too allows for NOMINATIVE government was given in (4.22). The fact that ABLATIVE government is another option becomes evident from (4.113) where (Π41) *minn* 'from' functions as linker.

(4.113) [Korpus Malti 3.0; culture2529]

[Apparti	***mill-**kompetizzjoni*	*tad-dehra]$_{PP}$*	*Katrina*
[apart_from	**from:DEF-**competition	of:DEF-appearance]	Katrina

trid	*taħdem*	*ukoll*	*fuq*	*kif*	*titkellem*
3SG.F.IPFV:require	3SG.F.IPFV:work	also	on	how	3SG.F.IPFV:speak

u	*fuq*	*kemm*	*tkun*	*sportiva*
and	on	how_much	3SG.F.IPFV:be.FUT	sportive:F

'[**Apart from** the beauty competition] Katrina also needs to work on how she speaks and her sportivity.'

70 This fact will be focussed upon in Sections 5.1.3.3. and 5.1.3.15.

ABLATIVE government accounts for 3,195 *apparti*-PPs in the Korpus Malti 3.0. ABLATIVE government is sixty times more common than GENITIVAL government for which there are only 52 hits among which we find (4.114).

(4.114) [Korpus Malti 3.0; parl2748]

waħda *mill-miri* *tiegħu* ***[apparti*** ***ta'*** *l-armata*
one:F from:DEF-goal:PL of:3SG.M **[apart_from** **of** DEF-army
Ewropea *eċċ]*$_{PP}$ *hija* *li* *jarmonizza* *wkoll* *is-sistema*
European:F etc.] 3SG.F that 3SG.M.IPFV:harmonise also DEF-system
tat-taxxi
of:DEF-tax:PL
'[. . .] one of his goals [**apart from** the European army, etc.] is to harmonise the tax system [. . .].'

NOMINATIVE government is still what applies in the vast majority of the cases as can be gathered from Figure 4.61.

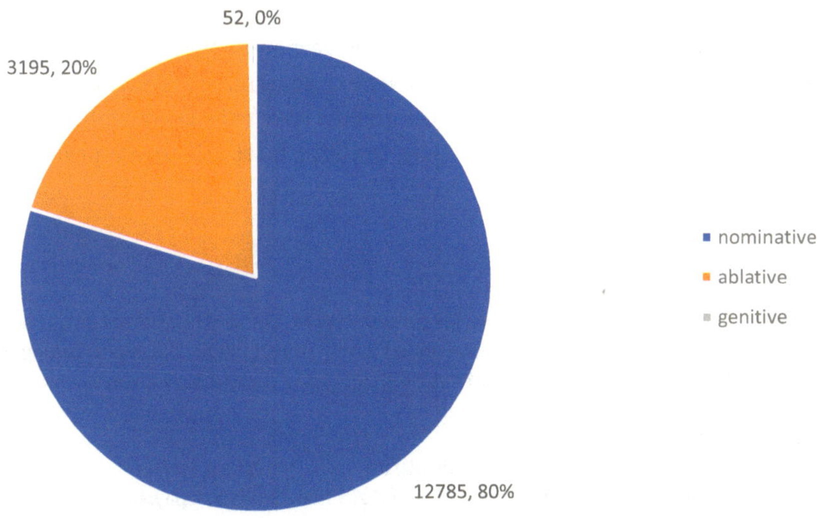

Figure 4.61: Shares of ABLATIVE, GENITIVE, and NOMINATIVE government with (П1) *apparti* 'apart from' (tokens in the Korpus Malti 3.0).

GENITIVAL government covers only the smallest segment of the total of the *apparti*-PPS. In contrast, ABLATIVE government forms a sizable minority option which cannot simply be filed away.[71]

(ii) (Π34) *minbarra* 'except': This Π has been the focus of attention already above.[72] Its NOMINATIVE government is documented in (4.45). ABLATIVE government can be found in (4.115).

(4.115) [Korpus Malti 3.0; european2522]
servizzi li kellhom ikunu pprovduti
service:PL that have.PFV:3PL 3.IPFV:be.FUT:PL provide:PART:PL
fit-territorju tal-pajjiż ***[minbarra*** ***minn*** *servizzi li*
in:DEF-territory of:DEF-country **[except** **from** service:PL that
hemm referenza għalihom fl-Artikolu 1]$_{PP}$
there reference for:3PL in:DEF Article_1]
'[. . .] services which should be provided in the territory of the country [**except** services which are referred to in Article 1].'

There are not many cases which reflect ABLATIVE government. We have counted 74 hits in the Korpus Malti 3.0. The total for GENITIVAL government is slightly lower since there are only 62 hits.[73] These absolute numbers have to be taken with a grain of salt though because they comprise many doubtful cases which are probably

71 We return to this issue in Section 4.1.2.2.2.
72 Cf. Sections 1.6.4.2 (iv) and 3.2.1.
73 The figures are based on a count of the long alloform *minbarra*. The short alloform *barra* is not taken account of since its inclusion would have added more interpretative challenges. Example (4.iv) is typical of the problems which come up if the distribution of *barra* is investigated.

(4.iv) [Korpus Malti 3.0; law1663]
profitti li jirriżultaw minn dividendi li jitħallsu
profit:PL that 3.IPFV:result:PL from dividend:PL that 3.IPFV:RFL:pay:PL
mill-kont **ta'** *income* ***[barra ta'*** *kumpanniji oħra residenti f'*
from:DEF-account **of** income **[except of** company:PL other:PL resident:PL in
Malta]$_{PP}$
Malta]
'[. . .] profits resulting from dividends paid from the income account [**except** (that of) other companies resident in Malta].'

Grey shading highlights a possible pitfall. Is *ta'* required by *barra* or is it part of the GENITIVAL attribute of the EQUI-deleted noun *kont* 'account' or a pronominal substitute thereof? This is a recurrent issue which needs to be looked into in Section 4.2.

better analysed differently. Interpretative problems arise inter alia in connection with the role of (П55) *ta'* 'of'.[74] For the time being, (4.116) may serve as illustration of GENITIVAL government.

(4.116) [Korpus Malti 3.0; news114027]
kulħadd *bil-vaganzi* ***[minbarra*** *tal-Partit Nazzjonalista]$_{PP}$*
everybody with:DEF-holiday:PL **[except** **of**:DEF-National Party]
'[. . .] everybody is on holiday [**except** the National Party] [. . .].'

As alluded to above, *tal-Partit Nazzjonalista* can also be understood as meaning '(those) of the National Party' so that no GENITIVAL government applies, in the first place. Turning a blind eye to this and related problems, we can summarise the results in Figure 4.62.

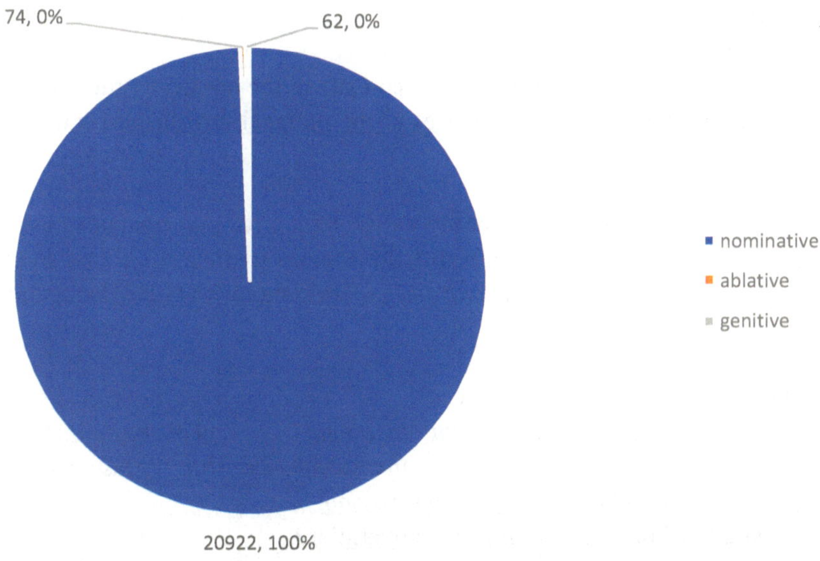

Figure 4.62: Shares of ABLATIVE, GENITIVE, and NOMINATIVE government with (П34) *minbarra* 'except' (tokens in the Korpus Malti 3.0).

The shares of ABLATIVE and GENITIVAL government are far too small to matter, in a manner of speaking. The picture would not change much if on revising the quantitative data the token frequencies of the two minor options diminish further. They

74 We address this problem in some detail in Section 4.2.

are already negligible. The uncontested default option for (Π34) *minbarra* 'except' is NOMINATIVE government.

(iii) (Π44) *oltre* 'beyond': According to (4.50), NOMINATIVE government is licit with (Π44) *oltre* 'beyond'. However, there is also ample evidence of ABLATIVE government as example (4.117) suggests.

(4.117) [Korpus Malti 3.0; parl101]

r- riżorsi	*minerali*	*kbar*	*li*	*hemm*	*fl-oċeani*
DEF-resource:PL	mineral	big.PL	that	there	in:DEF-ocean:PL
fil-qiegħ	*tal-baħar*	***[oltre***	***mil**-limiti*		
in:DEF-bottom	of:DEF-sea	**[beyond**	**from**:DEF-limit:PL		
tal- ġurisdizzjoni	*nazzjonali]$_{PP}$*	*għandhom*	*ikunu*		
of:DEF-jurisdiction	national]	have:3PL	3.IPFV:be.FUT:PL		
sfruttati					
exploit:PART:PL					

'[. . .] the big mineral resources which exist in the oceans on the bottom of the sea [**beyond** the boundaries of the national jurisdiction] must be exploited [. . .].'

The Korpus Malti 3.0 gives evidence of 505 cases of ABLATIVE government exerted by (Π44) *oltre* 'beyond'. This pattern is 101 times more frequent than GENITIVAL government for which there are only five pieces of evidence with (4.118) being one of them.

(4.118) [Korpus Malti 3.0; european252]

ma	*kienx*	*ġie*	*proprjament*	*investigat*	*għal*	*raġunijiet*
NEG	be:NEG	come	properly	investigate:PART	for	reason:PL
[oltre	***tar**-rifjut*	*għall-koperazzjoni]$_{PP}$*				
[beyond	**of**:DEF-refusal	for:DEF-cooperation]				

'[. . .] it was not properly investigated for reasons [**beyond** the refusal to co-operate] [. . .].'

The share of ABLATIVE government is almost as big as a quarter of all *oltre*-PPS in the Korpus Malti 3.0. NOMINATIVE government nevertheless outnumbers ABLATIVE government by a ratio of 3.4-to-one. GENITIVAL government is almost non-existent as shown in Figure 4.63.

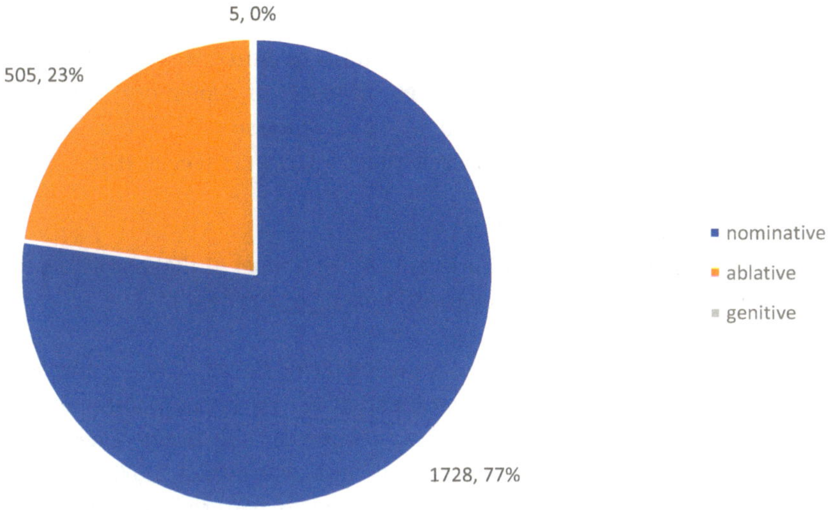

Figure 4.63: Shares of ABLATIVE, GENITIVE, and NOMINATIVE government with (Π44) *oltre* 'beyond' (tokens in the Korpus Malti 3.0).

4.1.2.2.1.2.2.3 A pattern possibly involving ALLATIVE government

The fact that (Π51) *sa* 'till' may govern lexical NPs according to the NOMINATIVE pattern has been shown in (4.20). In addition, there is the infrequently exploited possibility of (Π28) *lejn* 'towards' functioning as linker between (Π51) *sa* 'till' and the complement noun. It is not immediately obvious whether the two PPs in (4.119) differ semantically.[75]

(4.119) (Π51) *sa* 'till'

(a) [Korpus Malti 3.0; academic12]

*[**Sa** <u>tmiem</u> <u>is-seklu</u> <u>dsatax</u>]$_{PP}$ il-Maltin fl-Awstralja*
[**till** <u>end</u> <u>DEF-century</u> <u>nineteen</u>] DEF-Maltese:PL in:DEF-Australia
kienu ftit
be:3PL.PFV few
'[**Until** <u>the end of the 19th century</u>] the Maltese in Australia were not many.'

75 We address this possibility in more detail in Section 5.2.1.4.3.1.

(b) [Korpus Malti 3.0; parl3600]

l-impatt	*negattiv*	*tal-ġrajjiet*	*tal-11 ta' Settembru*	*li*
DEF-impact	negative	of:DEF-event:PL	of:DEF-11_of_September	that

skond	*id-WTO*	*jibqgħu*	*jinħassu*	***[sa***
according_to	DEF-WTO	3.IPFV:remain:PL	3.IPFV:RFL:feel:PL	**[till**

lejn	*tmiem*	*l-2003]$_{PP}$*
towards	end	DEF-2003]

'[. . .] the negative impact of the events of September 11 that according to the WTO will continue to be felt [**until** the end of 2003] [. . .].'

If we assume for the time being that we are indeed facing the variation of fully synonymous patterns we can state that the combination *sa* + *lejn* is vastly outnumbered by the NOMINATIVE pattern. The ALLATIVE pattern is attested only twelve times in the Korpus Malti 3.0 whereas NOMINATIVE government boasts almost 22,000 times as many hits as its competitor. This means that the minority pattern is too infrequent to be adequately made visible in a diagram.[76]

4.1.2.2.1.3 Synopsis – nominal complements

NOMINATIVE government has been proved to dominate in the domain of PPS with nouns as complement. The default character of NOMINATIVE government holds also for all the cases of variation we have reviewed in the foregoing sections. In many cases, alternative kinds of government are insignificant because they only cover a small fraction of the tokens which reflect NOMINATIVE government. There are some exceptions – most notably (Π12) *faċċata* 'opposite', the only Maltese Π for which GENITIVAL government ousts NOMINATIVE government. However, in the remaining cases, the share of the strongest competitor of NOMINATIVE government fails to reach the 25%-mark. If there are more than two options, the third kind of government yields minimal results. Note that each Π that gives evidence of variation of government patterns allows for GENITIVAL government besides NOMINATIVE government.

In Figure 4.64, the Πs are ordered top-down according to the decreasing share of NOMINATIVE government. Only those Πs are represented in the diagram which attest to variation.

Two facts strike the eye, namely that

(a) eight of seventeen Πs which are subject to variation in the domain of government with nominal complements are of Italo-Romance origin, i.e., Italo-Romance Πs are considerably overrepresented since they cover 47% of the Πs

76 As to the co-occurrence of *sa* + *fejn* and *sa* + *ħdejn*, we postpone the discussion until Section 4.1.2.2.2.1.5 below.

with variable government although they are entitled to only 25% of BLOMP 2.0; at the same time, the eight Italo-Romance Πs are responsible for 53% of all Italo-Romance Πs registered in BLOMP 2.0;

(b) at the bottom of the scale, there are four Italo-Romance Πs, meaning: Italo-Romance Πs are especially strong where the dominance of NOMINATIVE government is the least pronounced or even absent.

However, Italo-Romance Πs are not banned from higher ranks. The second position, for instance, goes to (Π35) *minflok* 'instead of'. Another interesting fact is that only two of the Πs in Figure 4.64 are monosyllabic, viz. (Π48) *qrib* 'near' and (Π51) *sa* 'till'. We remind the reader of the fact that especially in those cases where the second or third option of the government patterns is vastly outnumbered by the cases which reflect the majority solution there is a strong specialization in terms of style and genre. It remains to be seen whether prepositional government of pronominal complements yields comparable results and how these can be integrated into the definition of the canonical Maltese Π.

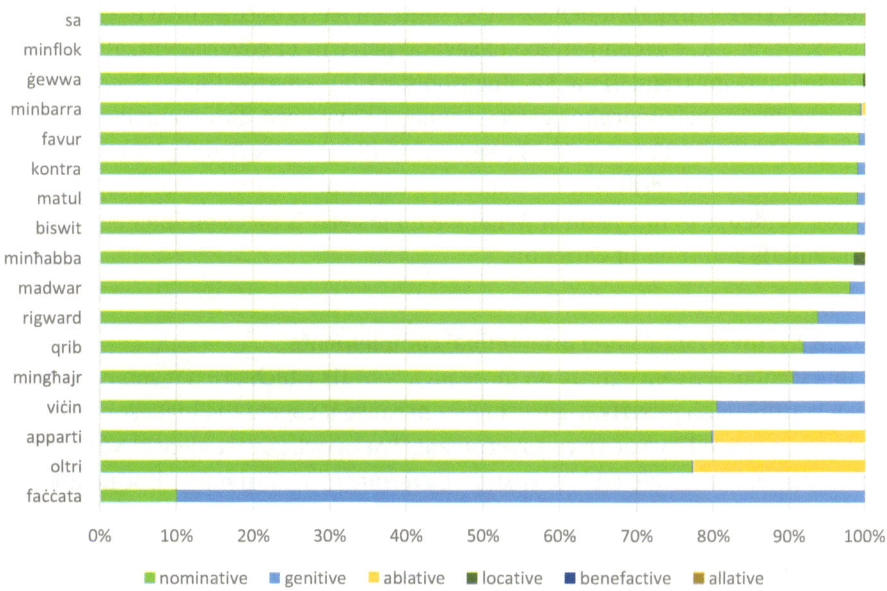

Figure 4.64: Overview of shares of government patterns in competition.

4.1.2.2.2 Government types with personal pronominal complements

The basic bipartition of the Maltese Пs as to their interaction with pronominal complements is reflected in the opposition of inflected Пs vs invariant Пs presented in (3.1)–(3.2) at the beginning of the morphology section.[77] For the 37 inflected Пs we assume POSSESSIVE government[78] which implies that the П is inflected for person encoded by the appropriate suffix also used in inalienable possessive constructions with pronominal possessors. POSSESSIVE government thus covers 62% of the members of BLOMP 2.0. This is the majority option. It is clear that the remaining 23 morphologically invariant Пs (= 28%) must either opt for different kinds of government or fail to take pronominal complements in the first place. From Table 4.5 we know already that (П9) *bla* 'without', (П42) *mintul* 'all along', and perhaps also (П52) *sforz* 'thanks to' do not take pronominal complements or display special behaviour with regards to this class of complements. GENITIVAL government of the latter П is illustrated in (4.68) but absent from the Korpus Malti 3.0. For POSSESSIVE government with (П37) *mingħala* 'in s.o.'s opinion', the reader is referred to example (1.111). The same government pattern for the controversial cases (П26) *kif* 'as', П43) *mnejn* 'from near', and (П42) *mintul* 'all along' is illustrated above,[79] respectively, whereas GENITIVAL government of (П54) *sotta* 'under' is documented in (1.187).

Variation, i.e., the competition of several patterns of government, affects both inflected and invariant Пs alike albeit only within narrow limits. Pronominally inflected Пs in their morpho-syntactic context have been documented already in examples (3.3)–(3.13), (3.16)–(3.53), and (3.55)–(3.64). These examples amply illustrate POSSESSIVE government so that there is no need to systematically provide further evidence of this pattern. What we do instead is focus on the behaviour of invariant Пs and cases of variation.[80]

4.1.2.2.2.1 Invariant Пs and pronominal complements

In this section, we concentrate on those invariant Пs which are not subject to variation starting with GENITIVAL government because it covers the highest number of П-types[81] followed by LOCATIVE government,[82] ABLATIVE government,[83] NOMINATIVE government,[84] ALLATIVE and PROXIMATE government with the two last mentioned

77 Cf. Section 3.1.1.
78 Cf. Section 4.1.2.
79 Cf. Sections 1.6.7 (i)–(ii) and 1.6.4.2 (ix), respectively.
80 Cf. Section 4.1.2.2.2.1 and Section 4.1.2.2.2.2, respectively.
81 Cf. Section 4.1.2.2.2.1.1.
82 Cf. Section 4.1.2.2.2.1.2.
83 Cf. Section 4.1.2.2.2.1.3.
84 Cf. Section 4.1.2.2.2.1.4.

patterns being summarily treated.[85] The PPs which involve invariant Πs and their pronominal complements come in two shapes as shown in Figure 4.65.

NOMINATIVE government:	$[\Pi_{\text{INVARIANT}}\ \text{PRO}]_{PP}$
Prepositional government:	$[\Pi_{\text{INVARIANT}}\ \Pi\text{-PRO}]_{PP}$

Figure 4.65: PPs involving invariant Πs and pronominal complements.

NOMINATIVE government means juxtaposition of the invariant Π and the personal pronoun in the form it has when it functions as (emphatic) subject pronoun or predicatively in the absence of a copula. The second option – prepositional government – disallows the juxtaposition of Π and pronoun. The construction is syndetic in the sense that a second Π is intercalated between head and complement. This second Π belongs to the class of inflected Πs and thus serves as host of the pronominal suffixes which represent the complement. It is important to understand that the second Π does not have an impact on the meaning of the construction. It is only a structurally necessary element whose sole task is to provide a stem to which the pronominal affixes can be attached.[86] Another important aspect of prepositional government is that those Πs which function as host of the pronominal suffixes are always Semitic. There is no Italo-Romance Π that fulfils the same function in Maltese.

4.1.2.2.2.1.1 GENITIVAL government

There are five Πs which require the presence of (Π55) *ta'* 'of', namely (Π2) *a skapitu* 'at the expense of', (Π12) *faċċata* 'opposite', (Π45) *permezz* 'by means of', (Π50) *rigward* 'concerning', and (Π58) *viċin* 'near'. Examples of their behaviour in combination with pronominal complements is illustrated in (4.120).

(4.120) GENITIVAL government

(a) (Π2) *a skapitu* 'at the expense of' [Korpus Malti 3.0; news75046]

irid	*jagħmel*	*ftit*	*propaganda*	*rħisa*
3SG.M.IPFV:want	3SG.M.IPFV:make	few	propaganda	cheap:F
[a skapitu	***tiegħi]***$_{PP}$			
[at_the_expense_of	**of:1SG]**			

'[. . .] he wants to make a bit of cheap propaganda [**at** my **expense**].'

85 Cf. Section 4.1.2.2.2.1.5.

86 The lack of a semantic effect distinguishes these combinations of Πs from those which are addressed in Section 4.1.4.

(b) (Π12) *faċċata* 'opposite' [Korpus Malti 3.0; literature15]
Kont qiegħed nimla l-karozza bil-petrol meta
be:1SG.PFV PROG 1SG.IPFV:fill DEF-car with:DEF-petrol when
nara żewġ ħaddiema ***[faċċata tiegħi]**$_{PP}$*
1SG.IPFV:see two worker:PL **[opposite of:**1SG]
'I was filling the car with petrol when I saw two workers [**opposite** me] [. . .].'

(c) (Π45) *permezz* 'by means of' [Korpus Malti 3.0; news156093]
Il-kuntatti bdew ***[permezz tiegħi** u ta' ħabib*
DEF-contact:PL begin:3PL.PFV **[by_means_of of:**1SG and of friend
Taljan tiegħi]$_{PP}$
Italian of:1SG]
'The contacts began [**via** me and an Italian friend of mine] [. . .].'

(d) (Π50) *rigward* 'concerning' [Korpus Malti 3.0; law1388]
fil-kelmiet parti tad-dominji Britanniċi 'l barra
in:DEF-word:PL part of:DEF-dominion:PL British:PL to outside
mir-Renju Unit jitqies li jidħol kull
from:DEF-UK 3SG.M.IPFV:RFL:measure that 3SG.M.IPFV:enter each
territorju li qiegħed taħt il-protezzjoni Britannika
territory that sit.PART under DEF-protection British:F
jew li ***[rigward tiegħu]**$_{PP}$ qiegħed jiġi*
or that **[concerning of:**3SG.M] PROG 3SG.M.IPFV:come
eżerċitat mandat minn Gvern ta' xi parti
exercise:PART mandate from government of some part
tad-dominji Britanniċi
of:DEF-dominion:PL British:PL
'[. . .] each territory which is under British protection or [**with regards to** which] a mandate of a part of the British dominions is exercised is considered to be covered by the words "part of the British dominions outside of the UK".'

(e) (Π58) *viċin* 'near' [Korpus Malti 3.0; news452]
f' mument minnhom tant resaq ***[viċin tiegħi]**$_{PP}$ li*
in moment from:3PL so_much approach **[near of:**1SG] that
kelli nimbuttah
have.PFV:1SG 1SG.IPFV:push:3SG.M
'[. . .] at one moment he got so [**close to** me] that I had to push him away.'

These Πs are known from the previous discussion[87] to allow for GENITIVAL government with nominal complements. Figure 4.64 informs us that in the cases of (Π50)

87 Cf. Sections 4.1.2.2.1.1 and 4.1.2.2.1.2.1.

rigward 'concerning' and (Π58) *viċin* 'near' GENITIVAL government is only a minor or even only marginal option whereas it predominates with (Π2) *a skapitu* 'at the expense of', (Π12) *faċċata* 'opposite', and (Π45) *permezz* 'by means of' – in the latter case it is even mandatory. The match between the behaviour of these five Πs in combination with pronominal complements and that in combination with nominal complements is not perfect but at least there is an overlap.

What is further remarkable about this group of five is that they are all polysyllabic Italo-Romance Πs. Since there are neither monosyllabic nor Semitic Πs with obligatory GENITIVAL government in PPs with pronominal complements, it is possible to postulate the implicational pattern in Figure 4.66.

GENITIVAL government ⊃ Italo-Romance$_{polysyllabic}$ /__ + PRO

Figure 4.66: Implicational pattern for GENITIVAL government.

4.1.2.2.2.1.2 LOCATIVE government

Compulsory LOCATIVE government with pronominal complements is attested for four Πs, namely (Π18) *ġewwa* 'inside', (Π19) *ġo* 'in', (Π39) *minħabba* 'on account of', and (Π40) *minkejja* 'in spite of'. The examples in (4.121) reveal that these Πs need the presence of (Π15) *fi* 'in' to be able to take a pronominal complement.

(4.121) LOCATIVE government

(a) (Π18) *ġewwa* 'inside' [Korpus Malti 3.0; news200350]
kienet inbidlet xi ħaġa ***[ġewwa*** *fija]*$_{PP}$
be:3SG.F.PFV RFL:change:3SG.F.PFV some thing **[inside** in:1SG]
'[. . .] something **[inside of** me] had changed [. . .].'

(b) (Π19) *ġo* 'in' [Korpus Malti 3.0; literature26]
Ħassejt ***[ġo*** *fija]*$_{PP}$ *sodisfazzjon kbir*
feel:1SG.PFV **[in** in:1SG] satisfaction big
'I felt much satisfaction **[inside of** me].'

(c) (Π39) *minħabba* 'on account of' [Korpus Malti 3.0; literature73]
Imsejkna oħti imsejken int ukoll kemm
poor:F sister:POSS.F:1SG poor 2SG too how_much
kellkom tbatu ***[minħabba*** *fija]*$_{PP}$
have.PFV:2PL 2.IPFV:suffer:PL **[on_account_of** in:1SG]
'My poor sister and poor you too, how much you had to suffer **[on account of** me].'

(d) (Π40) *minkejja* 'in spite of' [Teuma 2015: 375]
[minkejja *fija nnifsi]*$_{PP}$ *ma stajtx*
[in_spite_of in:1SG self:1SG] NEG can.1SG.PFV:NEG

ma *nidħakx* *b'* *dak* *il-miġnun*
NEG 1SG.IPFV:laugh:NEG with DEM.M.DIST DEF-crazy
'[**In spite of** myself] I could not help laughing about that madman.'

In (4.122), examples are provided which show that the shorter alloforms of (Π39) *minħabba* 'on account of' may also combine with (Π15) *fi* 'in' in PPs with pronominal complements.

(4.122) Alloforms of (Π39) *minħabba* 'on account of'

(a) [Korpus Malti 3.0; literature77]
dagħa *lit-Torok* *li* ***[mħabba*** *fihom]*$_{PP}$
curse to:DEF-Turk.PL that [**on_account_of** **in**:3PL]
Lippu *qagħad* *żmien* *twil* *fir-Rabat*
Lippu stay time long in:DEF-Rabat
'[. . .] he cursed the Turks [**on account of** whom] Lippu stayed for a long time in Rabat [. . .].'

(b) [Camilleri 2022: 187]
Dustin *beda* *jogħxa* *jara* *din* *id-drama*
Dustin begin 3SG.M.IPFV:faint 3SG.M.IPFV:see DEM:F:PROX DEF-drama
kollha *li* *kienet* *inqalgħet* ***[ħabba***
all:3SG.F that be:3SG.F.PFV RFL:pull_up:3SG.F.PFV [**on_account_of**
fih]$_{PP}$
in:3SG.M]
'Dustin began to faint seeing all this drama that had happened [**on account of** him].'

Allomorphy is no obstacle for LOCATIVE government to apply. It is known already that (Π18) *ġewwa* 'inside' and (Π39) *minħabba* 'on account of' allow for LOCATIVE government as a minor option also in combination with nominal complements whereas (Π19) *ġo* 'in' and (Π40) *minkejja* 'in spite of' are cases of NOMINATIVE government with nominal complements.[88] There is thus heterogeneity within this group of Πs.

What distinguishes the Πs with LOCATIVE government from those with GENITIVAL government is the language of origin. We have noticed above that GENITIVAL government of pronominal complements implies that the Π is of the Italo-Romance brand. In the case of LOCATIVE government with pronominal complements, however, Italo-Romance Πs are counted out because all four of the Πs in (4.121)

88 Cf. Section 4.1.2.2.1.2.2.1.

are Semitic. On this basis, it is possible to formulate the implicational pattern in Figure 4.67.

LOCATIVE government ⊃ Semitic / __ + PRO

Figure 4.67: Implicational pattern for LOCATIVE government.

In contrast to Figure 4.66, syllabicity is irrelevant for Figure 4.67. Semantically, the employment of (Π15) *fi* 'in' together with the two spatial Πs (Π18) *ġewwa* 'inside' and (Π19) *ġo* 'in' makes sense. As to (Π39) *minħabba* 'on account of' and (Π40) *minkejja* 'in spite of' however, the choice of the linking Π is not immediately clear.

4.1.2.2.2.1.3 ABLATIVE government

With (Π1) *apparti* 'apart from' and (Π3) *barra* 'outside', only two Πs attest to mandatory ABLATIVE government with pronominal complements – a pattern which involves the inflected (Π41) *minn* 'from' as host of the pronominal suffixes as shown in (4.123).

(4.123) ABLATIVE government

(a) (Π1) *apparti* 'apart from' [Korpus Malti 3.0; parl4942]

Jien kont qiegħed nirrappreżenta l-Kunsill ta' l-Ewropa
1SG be:1SG.PFV PROG 1SG.IPFV:represent DEF-European_Council
u ***[apparti*** ***<u>minni</u>]_{PP}*** *ma kien hemm ħadd Malti*
and **[apart_from** **<u>from</u>:1SG]** NEG be there nobody Maltese
'I represented the European Council and **[apart from** <u>me</u>] there was no Maltese.'

(b) (Π3) *barra* 'outside' [Korpus Malti 3.0; literature58]

Ħadt l-ewwel belgħa mill-kikkra safra u
take:1SG.PFV DEF-first swallowing from-cup yellow:F and
s-sħana tax-xarba nxterdet ma' ġismi ħjiel
DEF-heat of:DEF-drink RFL:disperse:3SG.F.PFV with body:1SG hint
ta' dinja oħra ***[barra*** ***<u>minni</u>]_{PP}*** *u barra*
of world other:F **[outside** **<u>from</u>:1SG]** and outside
mill-bjuda tal-kamra
from:DEF-whiteness of:DEF-room
'I took the first sip from the yellow cup and the heat of the drink spread through my body, a sign of another world [**outside of** <u>me</u>] and outside of the whiteness of the room.'

The possibility of (Π1) *apparti* 'apart from' triggering ABLATIVE government with nominal complements has been discussed already.[89] For (Π3) *barra* 'outside', the situation is by far less straightforward because of the partial overlap of alloforms with (Π34) *minbarra* 'except'.[90] We will therefore relegate the discussion of this issue until later.[91]

4.1.2.2.2.1.4 NOMINATIVE government

NOMINATIVE government with pronominal complements seems to be obligatory only for two Πs,[92] namely (Π6) *bħala* 'as' and (Π25) *inkluż* 'including' as shown in (4.124).

(4.124) NOMINATIVE government

(a) (Π6) *bħala* 'as' [Korpus Malti 3.0; parl926]

qed nitkellem ***[bħala*** *jien]$_{PP}$ u mhux qed*
PROG 1SG.IPFV:speak **[as** 1SG] and NEG:3SG.M:NEG PROG
nitkellem għall-Gvern
1SG.IPFV:speak for:DEF-government
'[. . .] I am speaking [**for** myself] and not for the government.'

(b) (Π25) *inkluż* 'including' [Korpus Malti 3.0; news136754]

Nammetti li kull min hu fil-politika ***[inkluż***
1SG.IPFV:admit that each who 3SG.M in:DEF-politics **[including**
jien]$_{PP}$ jieħu gost li jingħata pubbliċità
1SG] 3SG.M.IPFV:take pleasure that 3SG.M.IPFV:RFL:give publicity
'I admit that everybody who is in politics [**including** me] enjoys that he is given publicity [. . .].'

---

89 Cf. Section 4.1.2.2.1.2.2.2 (i).

90 Cf. Section 3.2.1.

91 Cf. Section 4.1.2.2.2.2.4 whose topic is variation of government patterns involving ABLATIVE government.

92 There are many false friends which outwardly look like instances of NOMINATIVE government but, on closer inspection, turn out to involve a clause so that the putative Π can be shown to be a conjunction or an adverb as in (4.v) where the pronoun of the 1st person singular *jien* is used predicatively so that *kif jien* like 'I am' forms a clause:

(4.v) [Korpus Malti 3.0; literature71]

Li nibqa' ***kif jien*** *niswa xi ħaġa għall-ġid*
that 1SG.IPFV:remain **like 1SG** 1SG.IPFV:be_worth some thing for:DEF-benefit
ta' Malta
of Malta
'If I remain **like I am** I am worth something to the benefit of Malta.'

As in the case of ABLATIVE government addressed in the foregoing section, NOMINATIVE government involves a Semitic and an Italo-Romance Π. It is also mandatory for both (Π6) *bħala* 'as' and (Π25) *inkluż* 'including' to take nominal complements in the NOMINATIVE.

4.1.2.2.2.1.5 PROXIMATE and ALLATIVE government

The two remaining Πs are isolated cases in the sense that no other Maltese Π shares their government patterns. PROXIMATE government is exclusively attested for (Π51) *sa* 'till' which needs the presence of (Π14) *fejn* 'near' or (Π23) *ħdejn* 'beside' as host of the pronominal suffixes representing the complement as shown in (4.125).

(4.125) PROXIMATE government

(a) [Korpus Malti 3.0; literature77]

Pietru	*għandu*	*jasal*	*[sa <u>fejni</u>]$_{PP}$*
Pietru	have.IPFV:3SG.M	3SG.M.IPFV:arrive	[**till** **<u>near</u>**:<u>1SG</u>]
jgħarrafni	*x'*	*għandi*	*nagħmel*
3SG.M.IPFV:inform:1SG	what	have.IPFV:1SG	1SG.IPFV:do

'Pietru has to come [**to** <u>me</u>] to tell me what I have to do.'

(b) [Korpus Malti 3.0; literature20]

Qam	*bilwieqfa*	*mar*	*sat-tieqa*	*kbira*	*daqs*	*ħajt*	*u*
stand_up	standing	go	till:DEF-window	big:F	equal_to	wall	and
stiednek	*tmur*	*[sa*	*<u>ħdejh</u>]$_{PP}$*				
invite:2SG	2SG.IPFV:go	[**till**	**<u>beside</u>**:<u>3SG.M</u>]				

'He got to his feet, went to the wall-sised window and invited you to come [**to** <u>his</u> **<u>side</u>**].'

Sandra Abela (personal communication) considers the patterns *sa fejn* and *sa ħdejn* to be fully synonymous. Outside the domain of pronominal complementation, (Π51) *sa* 'till' triggers NOMINATIVE government. The case of (Π14) *fejn* 'near' vs (Π23) *ħdejn* 'beside' as linkers for *sa*-PPs will keep us busy again.[93] For the time being, Figure 4.68 gives us an approximate idea of how the two competitors fare in the Korpus Malti 3.0.

---

93 Cf. Section 4.1.3.

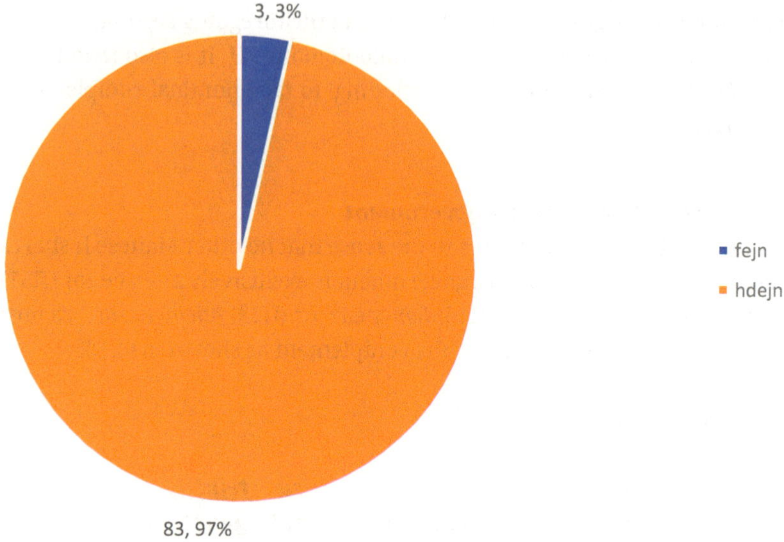

Figure 4.68: Competition of (Π14) *fejn* 'near' vs (Π23) *ħdejn* 'beside' as linkers in *sa*-PPS with pronominal complements in the Korpus Malti 3.0.

There are only 86 *sa*-PPS with pronominal complements in the Korpus Malti 3.0. The employment of (Π14) *fejn* 'near' as linker is very infrequent. This Π is attested once each in the 1st person singular, 2nd person singular, and the 3rd person singular feminine. In contrast, (Π23) *ħdejn* 'beside' functions as linker in combination with all person categories. It can therefore be considered the typical linker within *sa*-PPS with pronominal complements.

ALLATIVE government involves (Π28) *lejn* 'towards' as host of the pronominal suffixes. The pattern applies only in PPS headed by (Π57) *versu* 'towards' as in (4.126).

(4.126) ALLATIVE government [Korpus Malti 3.0; parl3134]

U ngħidu bl-ikbar sinċerità ***[versu*** ***lejk*** *]*$_{PP}$

and 1.IPFV:say:PL with:DEF-big.CMPR sincerity **[towards** **towards**:2SG]

għaliex taf li kulħadd jgħid it-tajjeb fik

because 2SG.IPFV:know that everybody 3SG.M:say DEF-good in:2SG

'And we speak [**to** you] with the greatest sincerity because you know that everybody says that you are efficient.'

It is worth noting that the lexical meaning of (Π57) *versu* 'towards' and (Π28) *lejn* 'towards' is identical so that synonymy applies, or so it seems. The parallel to the cases of (Π18) *ġewwa* 'inside' and (Π19) *ġo* 'in' combining with (Π15) *fi* 'in' cannot be denied.[94] Note also that both cases presented in this section are associated with (abstract) motion towards a Goal.[95] Since (Π57) *versu* 'towards' is an Italo-Romance Π whereas (Π51) *sa* 'till' stems from Semitic, there is again no direct connection to the language of origin that would explain the morpho-syntactic behaviour of the Πs.

4.1.2.2.2.2 Variation with pronominal complements

There is variation galore when it comes to governing pronominal complements. Several of the cases that we address in the subsections below are controversial whereas others seem to be bona fide instances of the competition of licit patterns. Not in every case is it possible to register the competition as free variation because the choice of government pattern may – more or less strictly – correlate with the disambiguation of otherwise polysemous Πs. The classes that can be defined on the basis of the patterns of variation are rather small. In several cases, only one Π-type realises a given pattern. We start the presentation with those inflected Πs which do not only attest to POSSESSIVE government but also allow for GENITIVAL government with pronominal complements.[96] Another subsection is dedicated to two cases of competition between POSSESSIVE and NOMINATIVE government one of which is a bona fide example of this alternation whereas the second case poses a number of problems not the least because GENITIVAL government is involved as a third pattern.[97] The competition of POSSESSIVE and ACCUSATIVE government is in the focus subsequently[98] whereas that of ABLATIVE VS NOMINATIVE government is looked at afterwards.[99] The synopsis of variation in the domain of pronominal complementation is given after the presentation of the individual patterns.[100]

4.1.2.2.2.2.1 POSSESSIVE VS GENITIVAL government

Four Πs – (Π8) *biswit* 'facing', (Π13) *favur* 'in favour of', (Π27) *kontra* 'against', and (Π48) *qrib* 'near' – attest to the competition of POSSESSIVE government and GENITIVAL government, i.e., they can be inflected either directly or only indirectly via the intercalation of (Π55) *ta'* 'of'. Two Πs which reflect this competition are Italo-Romance. Their combinability with (Π55) *ta'* 'of' connects them to the cases of oblig-

94 Cf. Section 4.1.2.2.2.1.2.
95 We will come back to this issue in Section 5.2.1.4.3.1.
96 Cf. Section 4.1.2.2.2.2.1.
97 Cf. Section 4.1.2.2.2.2.2.
98 Cf. Section 4.1.2.2.2.2.3.
99 Cf. Section 4.1.2.2.2.2.4.
100 Cf. Section 4.1.2.2.2.2.5.

atory GENITIVAL government discussed above.[101] Since (Π13) *favur* 'in favour of' and (Π27) *kontra* 'against' are polysyllabic Italo-Romance Πs, they are in line with the implicational pattern in Figure 4.66. However, this time there are two Semitic Πs one of which is a monosyllabic Π. Obligatory GENITIVAL government is tightly and exclusively connected to Italo-Romance and polysyllabicity. Furthermore, both (Π13) *favur* 'in favour of' and (Π27) *kontra* 'against' belong to CIIIb, i.e., they are inflected Πs with *i*-stems. In addition, Figure 4.64 shows that the two Πs favour NOMINATIVE government with nominal complements leaving only a narrow margin for GENITIVAL government with the same type of complement. The further particulars of the competing government patterns of (Π13) *favur* 'in favour of' and (Π27) *kontra* 'against' are discussed separately in subsections (i)–(ii). The two Semitic Πs with POSSESSIVE and GENITIVAL government are discussed in (iii) and (iv).

(i) (Π13) *favur* 'in favour of': Saari (2009: 272) qualifies the situation with regards to this Π as follows:

> The variants *affavur* and *a favur* are always used together with the preposition *ta'* 'of'. [. . .] Aquilina argues that *favur*, too, is often accompanied by *ta'*, but the corpus of my research indicates that the form *favur* without *ta'* is much more common.

Interestingly, the quoted author does not mention explicitly how (Π13) *favur* 'in favour of' interacts with pronominal complements. His examples only cover PPs which involve nominal complements. It is known that (Π13) *favur* 'in favour of' belongs to CIIIb, i.e., it is an inflected Π with an *i*-stem.[102] Accordingly, example (4.127a) meets our expectations whereas (4.127b) strikingly differs from the expected pattern.

(4.127) (Π13) *favur* 'in favour of'

(a) [Baldacchino 1999: 163]

L-Ispartans skurjaw l-ewwel iżda l-Ajax b' ħeffa
DEF-Spartans score:3PL.PFV DEF-first but DEF-Ajax with quickness
u b' futbol tal-għaġeb qalbu r-riżultat għal 2-1
and with football of:DEF-wonder turn:3PL.PFV DEF-result for 2-1
[favurihom]$_{PP}$
[in_favour_of:3PL]
'The Spartans scored first but Ajax quickly and with miraculous football turned the result to 2-1 [**in** their **favour**].'

101 Cf. Section 4.1.2.2.2.1.1.
102 Cf. Section 3.1.2.3.3.2.

(b) [Baldacchino 1999: 158]

l-Ispartans	*attakkaw*	*b'*	*kollox*	*biex*	*qalbu*
DEF-Spartans	attack:3PL.PFV	with	everything	in_order_to	turn:3PL.PFV

r-riżultat	*għal*	*wieħed*	*ta'*	*4-2*	***[favur***	***taghhom**]$_{PP}$*
DEF-result	for	one	of	4-2	**[in_favour_of**	**of:3PL]**

'[. . .] the Spartans attacked with all they had to turn the result to one of 4-2 [**in** their **favour**].'

The two examples stem from the same primary source where they are printed only a few pages apart from each other. Moreover, they form part of declarative sentences which describe very similar situations from famous matches in the history of Maltese football. The choice of POSSESSIVE government or GENITIVAL government does not seem to make a difference semantically. It remains doubtful whether there is a difference on the pragmatic level. One might assume that the separation of the head from its complement by way of using (Π55) *ta'* 'of' as host of the latter makes it possible to put the focus accent on *tagħhom* 'of them' and thus facilitates emphasizing the complement independently of (Π13) *favur* 'in favour of'. Whether this functional explanation is valid needs to be tested in a future study dedicated to pragmatic issues connected with the grammar of Maltese Πs.

In Tables 3.15–3.16, we have provided examples of the infrequent inflection of (Π13) *favur* 'in favour of' as a consonant stem to which the pronominal suffixes are attached directly although in the vast majority of the cases the Π inflects as a vowel stem (*i*-stem). This occasional variation as to the choice of stem is of no avail in the context of this section. However, the quote from Saari (2009) at the beginning of this section mentions two of the long alloforms of (Π13) *favur* 'in favour of' claiming that they obligatorily trigger GENITIVAL government. Examples (3.82)–(3.83) disprove this hypothesis since both *a favur* and *affavur* may take nominal complements without intervening (Π55) *ta'* 'of'. What happens if the complement of these alloforms of (Π13) *favur* 'in favour of' is pronominal? In point of fact, there is no evidence of the long alloforms themselves hosting the pronominal suffixes. Moreover, combinations with pronominal complements seem to be extremely rare generally. The only examples we could find in the Korpus Malti 3.0 involve *a favur* as shown in (4.128).

(4.128) [Korpus Malti 3.0; parl16]

irrid	*nara*	*jekk*	*il-Knisja*	*tistax*
1SG.IPFV:want	1SG.IPFV:see	if	DEF-Church	3SG.F.IPFV:can:NEG

tiddisponi	*mill-proprjetà*	*tagħna*	***[a favur***	***tiegħi**]$_{PP}$*
3SG.F.IPFV:dispose	from:DEF-property	of:1PL	**[in_favour_of**	**of:1SG]**

'[. . .] I want to see whether the Church can dispose of our property [**in favour of** me].'

The pattern *a favur* + *ta'*-PRO is attested twenty-one times in the Korpus Malti 3.0. The 2nd person of both numbers does not show up among the hits. Neither *affavur* nor *għaf-favur* yield any hits for combinations with pronominal complements. The quantitative basis is relatively small but we dare putting forward the following rule in Figure 4.69.

government → { POSSESSIVE / *favur* + PRO
GENITIVAL / *(a) favur* + PRO }

Figure 4.69: POSSESSIVE VS GENITIVAL government with alloforms of (Π13) *favur* 'in favour of'.

Speaking of quantities, we may now look at the turnout of hits that result from the competition of POSSESSIVE and GENITIVAL government with (Π13) *favur* 'in favour of' in combination with pronominal complements. In the Korpus Malti 3.0, GENITIVAL government is clearly the dominant option since with 1,589 matches it accounts for 69% of all cases of pronominal government of this Π (n = 2,305). Only 716 hits (= 31%) involve POSSESSIVE government. In Figure 4.70, it is shown that GENITIVAL government ousts POSSESSIVE government in each cell of the paradigm of person categories with shares (from left to right) ranging from 89% for the 2nd person plural down to 64% for the 3rd person singular masculine. At least for what concerns the representation of the Maltese language in the Korpus Malti 3.0, GENITIVAL government is by far more characteristic of (Π13) *favur* 'in favour of' as compared to POSSESSIVE government. The higher frequency of the GENITIVAL pattern does not lend support to the hypothesis that this kind of government is connected to the pragmatically marked focus function.

(ii) (Π27) *kontra* 'against': In contrast to the Π discussed in (i), Saari (2009: 273) observes that "*kontra* may take *ta'* before the POSSESSIVE suffix, though Maltese also uses the Arabic POSSESSIVE suffixation with this concrete preposition." This formulation suggests that GENITIVAL government is more common than POSSESSIVE government with (Π27) *kontra* 'against' in combination with pronominal complements. Before we determine whether this interpretation is corroborated by the quantitative facts, we present examples of the two options of government in (4.129).

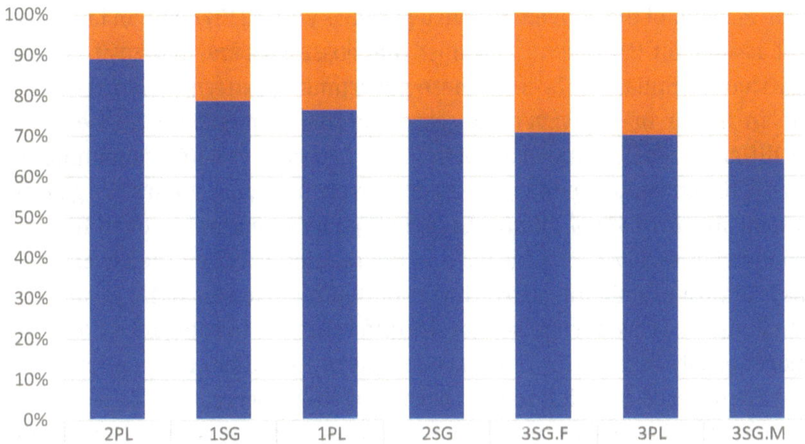

	2PL	1SG	1PL	2SG	3SG.F	3PL	3SG.M
possessive	1	9	41	18	198	158	291
genitival	8	33	132	51	477	369	519

Figure 4.70: POSSESSIVE VS GENITIVAL government across person categories with (Π13) *favur* 'in favour of'.

(4.129) (Π27) *kontra* 'against'

(a) [Attard 1999: 86]

Ipprotestaw ukoll minħabba d-diskriminazzjoni ***[kontrihom]$_{PP}$***
protest:3PL.PFV also on_account_of DEF-discrimination **[against:3PL]**
'They also protested because of the discrimination [**against** them] [. . .].'

(b) [Attard 1999: 73]

kien hemm xi ġurnali lokali li bdew kampanja
be there some journal:PL local:PL that begin:3PL.PFV campaign
[kontra tagħhom]$_{PP}$
[against of:3PL]
'[. . .] there were some local journals which strated a campaign [**against** them].'

The two *kontra*-PPs stem from the same primary source. In this historiographic study, they are employed in similar morphosyntactic contexts because they are both prepositional attributes of nouns (*d-diskriminazzjoni kontrihom* 'the discrimination against them' and *kampanja kontra tagħhom* 'a campaign against them'). Semantically too the contexts resemble each other closely. We can therefore assume that there are no differences in meaning between POSSESSIVE and GENITIVAL government of (Π27) *kontra* 'against'.

On account of the abovementioned parallels with (Π13) *favur* 'in favour of', one might assume for (Π27) *kontra* 'against' that POSSESSIVE and GENITIVAL government claim shares similar to those of these government patterns calculated for (Π13) *favur* 'in favour of'. However, the search on the Korpus Malti 3.0 yields surprisingly different results. First of all, of the 13,101 *kontra*-PPs with pronominal complements, 8,405 go to the credit of POSSESSIVE government, i.e., POSSESSIVE government is responsible for 64% of all hits. GENITIVAL government has a token frequency of 4,696 matches (= 36%). This means that the relation between the government patterns is almost exactly the mirror image of that reported for (Π13) *favur* 'in favour of'. In Figure 4.71 we recognise that the different person categories do not behave homogeneously as to their preference for one or the other of the options. The categories are ordered from left to right according to the decreasing share of GENITIVAL government. What we see is that GENITIVAL government is relatively strong with speech-act participants three of which even attest to a majority of hits for GENITIVAL government. In all other cases (= 3rd persons), POSSESSIVE government is more common than GENITIVAL government. Accordingly, (Π27) *kontra* 'against' and (Π13) *favur* 'in favour of' prefer different options when it comes to governing pronominal complements.

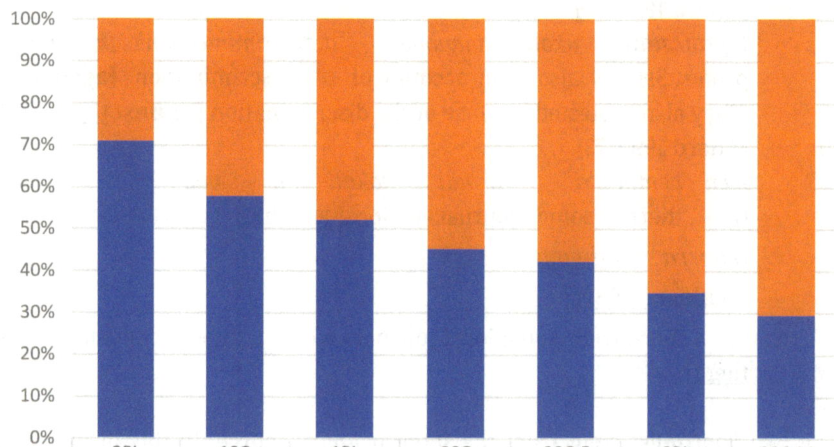

	2PL	1SG	1PL	2SG	3SG.F	3PL	3SG.M
possessive	14	223	360	203	1261	2069	4275
genitival	34	304	391	167	920	1105	1775

Figure 4.71: POSSESSIVE VS GENITIVAL government across person categories with (Π27) *kontra* 'against'.

(iii) (Π8) *biswit* 'facing': This Π is one of those which attest to NOMINATIVE-GENITIVE variation with nominal complements as shown above.[103] In combination with pronominal complements, there is variation, too, namely between POSSESSIVE government and GENITIVAL government as shown in (4.130).

(4.130) (Π8) *biswit* 'facing'
(a) [Korpus Malti 3.0; news110045]
Kristu jinsab ***[biswitek]***$_{PP}$ *fil-flat ta' maġenbek*
Christ 3SG.M.IPFV:RFL:find **[facing**:2SG] in:DEF-flat of close_to:2SG
'Christ is to be found [**facing** you] in the flat close to yours.'
(b) [Korpus Malti 3.0; parl10673]
m' għandekx bini daqshekk ***[biswit tiegħek]***$_{PP}$
NEG have.IPFV:2SG:NEG building of_this_size [**facing of**:2SG]
'[. . .] you don't have a building of this size [**facing** you] [. . .].'

The two options divide the 211 *biswit*-PPs in the Korpus Malti 3.0 among each other as follows. The majority of 188 hits which equal 89% of all cases is the responsibility of POSSESSIVE government. Only 23 cases (= 11%) attest to GENITIVAL government. As Figure 4.72 shows, the preponderance of POSSESSIVE government holds for all attested person categories. Note that there is no *biswit*-PP involving a 2nd person plural as complement. The share of GENITIVAL government is so small that the diagram only covers the stretch from 75% upwards to 100%. The person categories are ordered from left to right according to the decreasing share of GENITIVAL government.

(iv) (Π48) *qrib* 'near': Like the previous Π (Π48) *qrib* 'near' has already been mentioned in connection with NOMINATIVE-GENITIVE competition.[104] That two patterns compete with each other also in the pronominal domain can be seen in (4.131).

(4.131) (Π48) *qrib* 'near'
(a) [Korpus Malti 3.0; news2333]
Bosta familjari tiegħu nġabru fl-isptar
many family_member:PL of:3SG.M RFL:gather:3PL.PFV in:DEF-hospital
biex ikunu ***[qribu]***$_{PP}$
in_order_to 3.IPFV:be.FUT:PL [**near**:3SG.M]
'Many members of his family gathered in the hospital to be [**near** him].'

103 Cf. Section 4.1.2.2.1.2.1 (i).
104 Cf. Section 4.1.2.2.1.2.1 (ix).

(b) [Korpus Malti 3.0; literature32]

Ħares	*lejn*	*nanntu*	*bilqiegħda*	***[qrib***	***<u>tiegħu</u>*** *]*$_{PP}$
watch	towards	grandmother:POSS.F:3SG.M	sitting:F	**[near**	**of**:<u>3SG.M</u>]

'He looked at his grandmother sitting [**near** <u>him</u>] [. . .].'

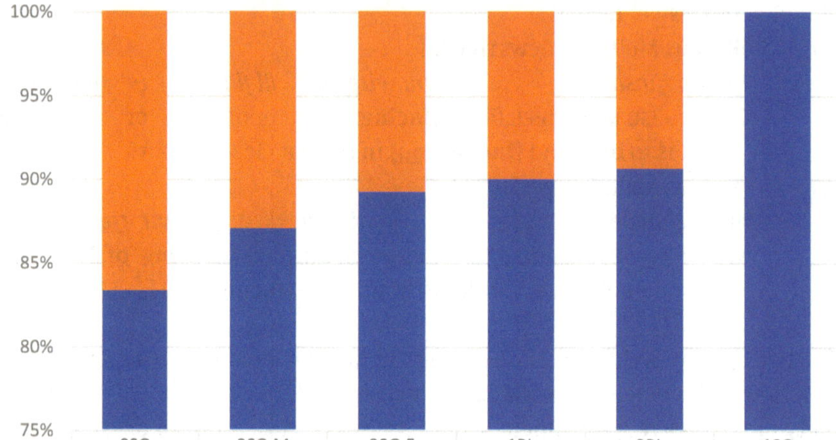

	2SG	3SG.M	3SG.F	1PL	3PL	1SG
genitival	1	11	7	1	3	0
possessive	5	74	58	9	29	13

Figure 4.72: POSSESSIVE VS GENITIVAL government across person categories with (Π8) *biswit* 'facing'.

In contrast to the case of (Π8) ***biswit*** 'facing', it is GENITIVAL government that represents the majority of the *qrib*-PPs in the Korpus Malti 3.0. With 1,134 hits, GENITIVAL government accounts for 73% of all cases of pronominal complementation of this Π (n = 1,547). POSSESSIVE government occurs 413 times (= 27%). As transpires from Figure 4.73, GENITIVAL government is the majority option in each person category. The person categories are ordered from left to right according to the increasing share of GENITIVAL government.

The four Πs do not behave homogeneously as to the preference for one of the competing patterns. In the case of (Π48) *qrib* 'near', the preference for the GENITIVAL pattern fits in with the obligatory GENITIVAL government reported for the synonymous (Π58) *viċin* 'near'.[105]

105 Cf. Section 4.1.2.2.2.1.1.

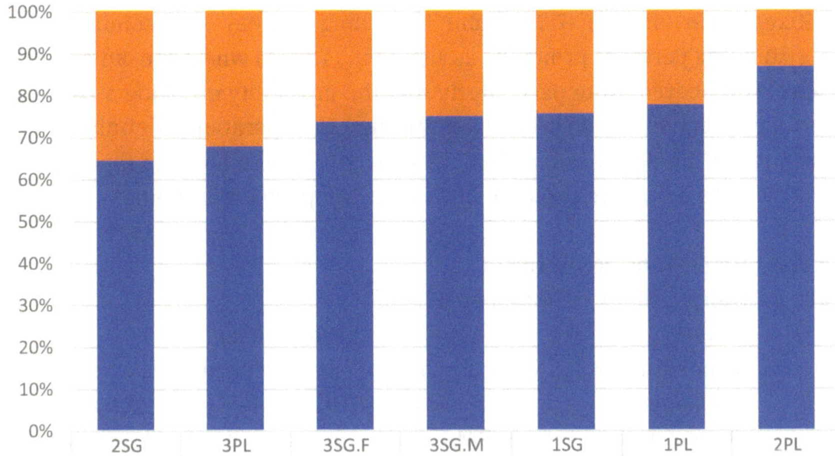

	2SG	3PL	3SG.F	3SG.M	1SG	1PL	2PL
possessive	31	111	43	130	23	73	2
genitival	56	233	120	388	71	253	13

Figure 4.73: POSSESSIVE VS GENITIVAL government across person categories with (Π48) *qrib* 'near'.

4.1.2.2.2.2.2 POSSESSIVE VS NOMINATIVE (and GENITIVAL government)

This section ties in with the previous one since it features two Italo-Romance Πs one of which gives evidence of GENITIVAL government as one of supposedly three options this Π has to take with pronominal complements. In (i), we first address the uncontroversial case of (Π53) *skont* 'according to' reserving section (ii) for the discussion of the by far more complicated case of (Π44) *oltre* 'beyond'. The latter is registered as invariant Π in (3.2) whereas (Π53) *skont* 'according to' counts among the inflected Πs. (Π44) *oltre* 'beyond' is a polysyllabic Π as opposed to monosyllabic (Π53) *skont* 'according to'.

(i) (Π53) *skont* 'according to': This Π has an interesting history as it used to be written in two different ways, namely *skond* when no inflections were added and *skont* when it served as host of the pronominal suffixes – a purely orthographic practice which ceased to be applied recently (Stolz/Levkovych 2020: 222–225).[106] (Π53) *skont* 'according to' inflects for person as shown in Table 3.5. It belongs to CIIb, i.e., to the set of two Πs which trigger vowel-harmony in the 2nd person singular. However, the Π attests to variation because besides hosting the pronominal

106 Since final devoicing is the rule in Maltese phonology (Borg/Azzopardi-Alexander 1997: 307), the graphematic difference of <d> ≠ <t> is phonetically and phonologically meaningless. Cf. also Aquilina (1991: 1331–1332).

suffixes as in (4.132a) (П53) *skont* 'according to' may also combine with the free-standing personal pronouns as shown in (4.132b) where the old orthographic rule is still obeyed. Note particularly that the possibility of taking a free-standing personal pronoun as complement contradicts the operational definition of the PP put forward by Borg/Azzopardi-Alexander (1997: 69) who state that the complement "can be a suffixed pronoun but not an independent pronoun."

(4.132) (П53) *skont* 'according to'

(a) [Korpus Malti 3.0; *news144064]*

***[skontu]**$_{PP}$*	*l-iżbalji*	*ma*	*nħolqux*	
[**according_to**:3SG.M]	DEF-error:PL	NEG	RFL:creat:3PL.PFV:NEG	
minħabba	*l-ARMS*	*imma*	*minħabba*	*s-sistema*
on_account_of	DEF-ARMS	but	on_account_of	DEF-system
l-ġdida	*tal-kontijiet*			
DEF-new:F	of:DEF-account:PL			

'[**According to** him] the errors were not created because of ARMS but because of the new accounting system.'

(b) [Attard 1999: 113]

imma	***[skond***	*huwa]*$_{PP}$	*din*	*kienet*	*proċedura*	
but	[**according_to**	3SG.M]	DEM:F:PROX	be:3SG.F.PFV	procedure	
normali	*ma'*	*ħaddiema*	*barranin*	*li*	*jieħdu*	*l-ewwel*
normal	with	worker:PL	foreign:PL	that	3.IPFV:take:PL	DEF-first
impjieg	*fil-Kanada*					
job	in:DEF-Canada					

'[. . .] but [**according to** him] this was a normal procedure with foreign workers who took their first job in Canada.'

To our mind, there is no semantic difference between the two *skont*-PPs in (4.132). The use of two separated syntactic words in (4.132b) might again be taken as an advantage for pragmatic purposes since the free-standing personal pronoun can serve as landing site of emphatic stress in order to realise contrastive focus – a possibility that is counted for the inflected form *skontu* because penultimate accent applies. In the absence of reliable spoken language data this remains mere speculation.

We face a serious difficulty when we want to determine the frequency of the two options of government. There is the common noun *skont* 'discount' whose regular plural is either *skonti* or *skontijiet* (Aquilina 1991: 1331) which means that the form *skonti* is ambiguous as to the two readings 'discounts' and 'according to me'. The Korpus Malti 3.0 yields 118 returns for this word form only one of which represents the inflected П as shown in the short dialogue in (4.133) which, by the way, forms part of a text of metalinguistic reflexion on the use of correct Maltese. The turns of the partners in dialog are seprated by a slash.

(4.133) [Korpus Malti 3.0; news74341]

Jiġifieri *skontok* *il-ħajja* *kompliet*
3SG.M.IPFV:mean according_to:2SG DEF-life continue:3SG.F.PFV
togħla / ***[skonti]***$_{PP}$ *iva*
3SG.F.IPFV:become_dearer / **[according_to:**1SG] yes
'This means that, according to you, life continued to become more expensive? / [**To** my **mind**] yes.'

The scarcity of evidence for the inflected form of the 1st person singular is already telling. We do not expect many returns for any of the inflected forms of (Π53) *skont* 'according to'. The Korpus Malti 3.0 yields a total of only 124 hits which covers but 2% of the 5,362 *skont*-PPs with pronominal complement in the electronic resource. In contrast, there are 5,238 tokens of NOMINATIVE government (= 98%). POSSESSIVE government is thus a marginal phenomenon. The peripheral status of POSSESSIVE government also holds for each of the person categories as comes to the fore in Figure 4.74. The shares of POSSESSIVE government are so small that the diagram only displays the range between 90%–100%. The person categories are ordered from left to right according to the decrease in the share of NOMINATIVE government.

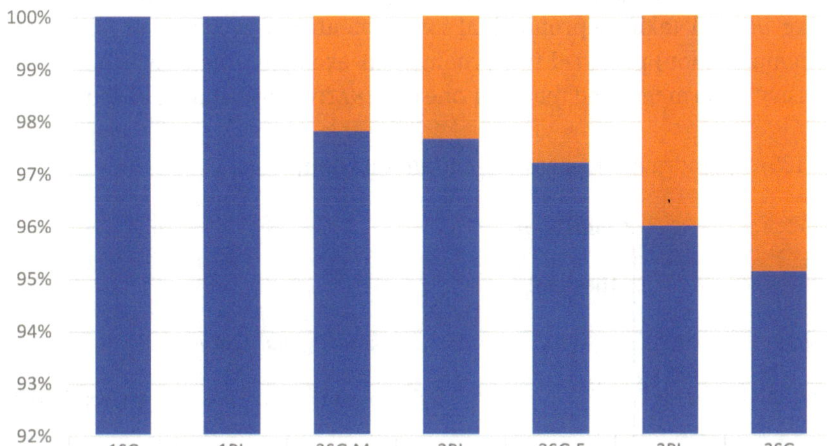

	1SG	1PL	3SG.M	3PL	3SG.F	2PL	2SG
possessive	0	0	76	22	19	1	6
nominative	113	25	3382	917	660	24	117

Figure 4.74: POSSESSIVE VS NOMINATIVE government across person categories with (Π53) *skont* 'according to'.

Interestingly, there is no evidence of POSSESSIVE government for the 1st person in both numbers. Shares of 4%–5% are reached by POSSESSIVE government only for the 2nd person of both numbers. As to the 3rd persons, they boast shares of POSSESSIVE government which range between slightly over 2% and slightly under 3%. There is no denying: (Π53) *skont* 'according to' overwhelmingly opts for NOMINATIVE government and marginalises POSSESSIVE government.

(ii) (Π44) *oltre* 'beyond': This Π is particularly problematic in terms of its interaction with pronominal complements. The problems start with the almost total absence of suitable *oltre*-PPs in the Korpus Malti 3.0. The only example we have come by is reproduced in (4.134).

(4.134) [Korpus Malti 3.0; parl12769]

niddiskutu	*iktar*	*x'*	*inhuma*	*r-red lines*	*li*	***[oltre***	
1.IPFV:discuss:PL	more	what	3PL	DEF-red_lines	that	**[beyond**	
***tagħhom**]_{PP}*	*wieħed*	*m'*	*għandux*	*jintervjeni*			
of:3PL]	one	NEG	have.IPFV:3SG.M:NEG	3SG.M.IPFV:intervene			

'[. . .] we will discuss further what the red lines are [**beyond** which] one must not intervene [. . .].'

This isolated example illustrates the GENITIVAL pattern of which there is no further proof in the Korpus Malti 3.0. Similarly, the literary corpus on which we rely to complement the findings in the electronic corpus does not contain a single case of (Π44) *oltre* 'beyond' taking a pronominal complement. However, Sandra Abela (personal communication) provided two synonymous examples one of which involves POSSESSIVE government and the other one NOMINATIVE government as shown in (4.135)

(4.135) [Sandra Abela, personal communication]

it-tfal *li* { *[**oltri**hom]PP* [**beyond**:PL] / *[**oltre** huma]PP* [**beyond** 3PL] } *ħadd* *ma* *jorqod* *fil-kamra*

DEF-child.PL that { [**beyond**:PL] / [**beyond** 3PL] } nobody NEG 3SG.M.IPFV:sleep in:DEF-room

'the children [**except** whom] nobody sleeps in the room'

According to our informant, the two options are absolutely synonymous. She did not mention the possibility of GENITIVAL government. We have been unable to ascertain the existence of either of the government patterns (NOMINATIVE and POSSESSIVE) in independent sources. It is therefore doubtful that the paradigms in Table 4.12 reflects general usage across the Maltese speech-community. The two columns on the right (highlighted in grey) are perhaps only idiolectal phenomena.

Table 4.12: (Putative) paradigms of (Π44) *oltre* 'beyond'.

person	GENITIVAL	NOMINATIVE	POSSESSIVE
1SG	*oltre tiegħi*	*oltre jien(a)*	*oltrija*
2SG	*oltre tiegħek*	*oltre int(i)*	*oltrik*
3SG.M	*oltre tiegħu*	*oltre hu(wa)*	*oltrih*
3SG.F	*oltre tagħha*	*oltre hi(ja)*	*oltriha*
1PL	*oltre tagħna*	*oltre aħna*	*oltrina*
2PL	*oltre tagħkom*	*oltre intom*	*oltrikom*
3PL	*oltre tagħhom*	*oltre huma*	*oltrihom*

On account of the remaining doubts we refrain from classifying (Π44) *oltre* 'beyond' as inflected Π. We keep treating it as invariant Π in lieu of counting it in as a new member of CIIIb (*i*-stems). It cannot be ruled out that we are facing a general uncertainty of Maltese native speakers as to the behaviour of this Π in combination with pronominal complements. Example (4.134) is of course much too small a basis to generalise from. However, we do not want to pass over the fact that the only corpus-based piece of evidence for (Π44) *oltre* 'beyond' taking a pronominal complement involves GENITIVAL government. We stress this point because it is a further (albeit only shaky) corroboration of the implicational pattern in Figure 4.66 according to which the vast majority of the Πs allowing for GENITIVAL government with pronominal complements are Italo-Romance polysyllables.

4.1.2.2.2.2.3 POSSESSIVE vs ACCUSATIVE government

In Section 3.1.2.3.2, we have introduced the two members of C II – (Π7) *bi* 'with' and (Π15) *fi* 'in' – which stand out from the bulk of the inflected Πs because of their ability to host not only the possessor suffixes but, alternatively, also the pronominal markers of the direct object (= ACCUSATIVE government), i.e., typically nominal inflections compete with typically verbal inflections (Borg/Azzopardi-Alexander 1997: 133). However, the two options are formally distinct only in the 1st person singular. This is the cell of the paradigm on which we focus in the remainder of this section. The question is whether the pairs of forms *fini* / *fija* 'in me' and *bini* / *bija* 'with me' consist of synonyms or is there a functional difference between POSSESSIVE government and ACCUSATIVE government.

We remind the reader that Peterson (2009: 186–187) lists (Π15) *fi* 'in' under the rubric of pseudo-verbs from which (Π7) *bi* 'with' is absent. As pseudo-verb, (Π15) *fi* 'in' is translated as 'contain' cf. also Vanhove (1993: 393–396). Most probably, it is this pseudo-verbal function that licenses the use of *-ni* as pronominal marker in (3.57)[107] whereas *-ja* in (3.56) signals that we are dealing with a genuine Π (Stolz/Levkovych 2020: 220–221). If we let things stand as illustrated above,[108] we are forced to cancel *fini* from the list of Π-forms because according to the principles laid down in the introduction[109] pseudo-verbs are generally excluded from BLOMP 2.0 since they are no Πs according to our own criteria. As mentioned earlier, the cancellation of *fini* would entail that (Π15) *fi* 'in' that the foundations of class C II have to be reformulated. Before we jump to a radical conclusion of this kind, we need to see whether the change of class also applies to *bini*. Moreover, one might want to ask whether any change of class is absolutely necessary, in the first place.

Borg/Azzopardi-Alexander (1997: 266) emphasise that it is not possible to replace *fini* with *fija* in (3.57) and *fija* with *fini* in (3.56). As to the relation between *bini* and *bija* however, Saari (2003: 122) assumes full equivalence of the two inflected forms as shown in (4.136) which differs only slightly from (3.58) taken from Sutcliffe (1936: 200).

---

107 If *fija* is used in a comparable function, it usually co-occurs with *hemm* 'there' in a kind of existential construction as in (4.vi):

(4.vi) [Korpus Malti 3.0; news58482]

Illum spiċċajt niċċassa lejja nnifsi fil-mera u
today finish:1SG.PFV 1SG.IPFV:stare towards:1SG self:1SG in:DEF-mirror and

bla ma nkun stennejtha niskopri kemm
without that 1SG.IPFV:be.FUT expect:1SG.PFV:3SG.F 1SG.IPFV:discover how_much

[*fija*]$_{PP}$ *hemm minn ommi*
[**in**:1SG] there from mother:1SG

'Today I finally looked at myself in the mirror and without expecting it discovered how much of my mother there is [**in** me].'

108 Cf. Section 3.1.2.3.2.

109 Cf. Section 1.6.2.

(4.136) Adapted from Saari (2003: 122)

jisgħob	{ ***bija*** [**with**:1SG]$_{PP}$ / ***bini*** [**with**:1SG]$_{PP}$ }	*minn*	*dnubieti*
3SG.M.IPFV:repent		from	sin:PL:1SG

'I am sorry for my sins.'

Two remarks have to be made with regards to this example. As mentioned above,[110] *jisgħob bija* ~ *bini* represents formulaic language (Aquilina 1991: 1342–1343). Furthermore, Saari (2003: 122) identifies the novel *Ineż Farruġ* written by Anton Manwel Caruana as the primary source for (4.136). The book appeared in print for the first time in 1887. Saari quotes from the 1978 edition to which we had no access. We have not been able to find the kind of variation in our edition dating to 2001 where only *jisgħob bija* occurs. We therefore assume that the alternative *bini* is an addition of Saari's based on what is said in Sutcliffe (1936) and Aquilina (1991).

Since the form *bini* is ambiguous as to the prepositional reading 'with me' and the highly frequent homophonous common noun *bini* 'building' (Aquilina 1987: 123), the search for appropriate examples of prepositional *bini* in the Korpus Malti 3.0 is too demanding. We make do with Aquilina's (1987: 114) lexicon entry for (Π7) *bi* 'with' where *bija* is given as the first option for combinations with pronominal suffixes whereas *bini* is mentioned only as second best. It is used for expressing physico-mental states as in (4.137).

(4.137) Physico-mental states

(a) hunger – [Korpus Malti 3.0; opinion1485]

Fis-sitta	*ta'*	*filgħaxija*	*reġa'*	***bini***	*l-ġuħ*
in:DEF-six	of	evening	return	**with:DO.1SG**	DEF-hunger

'At six p.m. I was hungry again.'

(b) thirst – [Korpus Malti 3.0; *news61776]*

bini	*l-għatx*
with:DO.1SG	DEF-thirst

'I am thirsty.'

110 Cf. Section 3.1.2.3.2.

In neither of the two cases is it possible to replace *bini* with *bija*. At least, this is what the zero turnouts in the Korpus Malti 3.0 suggest. Similarly, there is no evidence in the Korpus Malti 3.0 of *bini* replacing *bija* with Maltese prepositional verbs whose complement is a *bi*-PP. Aquilina (1976) lists 24 verbs which conform to this pattern. We have limited the search in the Korpus Malti 3.0 to cooccurrences of the *mamma tal-verb* of these verbs with *bija* and *bini*. The results are telling. Only four of the 24 verbs are attested with a *bi*-PP whose pronominal complement is the 1st person singular. All of the 26 tokens go to the credit of *bija*. There is no evidence of *bini* taking over the function of *bija*.

If we extend the search to the pair *fija* vs *fini*, similar problems come up since *fini* can also be the plural of the adjective *fin* 'fine, sharp' or the singular of the common noun *fini* 'aim' (Aquilina 1987: 339–340). As a shortcut, we have decided to look only at those Πs which attest to LOCATIVE government with pronominal complements.[111] In the Korpus Malti 3.0, only three of the four candidates are attested with the 1st person singular as pronominal complement. There are altogether 374 *fi*-PPs none of which involves *fini*. Only *fija* is attested in the returns. The results are suggestive of the spilt captured in Figure 4.75.

$$\text{government} \rightarrow \begin{cases} \text{ACCUSATIVE} & /\ \text{pseudo-verb} \\ \text{POSSESSIVE} & /\ \Pi \end{cases}$$

Figure 4.75: Functional split with (Π7) *bi* 'with' and (Π15) *fi* 'in'.

Bini and *fini* are thus pseudo-verbs which cannot be lumped together with *bija* and *fija* which are genuine Πs. Only morphologically do *bini* and *fini* look like Πs. The morphologically motivated C II does not necessarily disappear from the scene only because we extend the scope of our research to syntax. Class C II differs from C I because of the use of the B-set of pronominal suffixes. Even if *-ni* has to be cancelled from the B-set because it is attested only with pseudo-verbs, there remains *-ja* in the 1st person singular which distinguishes the B-set from the A-set. (Π7) *bi* 'with' and (Π15) *fi* 'in' remain primary vowel stems and thus fit in neither with C I (= consonant stems) nor with C IIIb (= secondary *i*-stems). Their original paradigms given in Tables 3.10–3.12 now look differently as shown in Table 4.13.

111 Cf. Section 4.1.2.2.2.1.2.

Table 4.13: Paradigms of (Π7) *bi* 'with' and (Π15) *fi* 'in' – version IV.

person	(Π7) *bi* 'with'	(Π15) *fi* 'in'
1SG	*bija*	*fija*
2SG	*bik*	*fik*
3SG.M	*bih*	*fih*
3SG.F	*biha*	*fiha*
1PL	*bina*	*fina*
2PL	*bikom*	*fikom*
3PL	*bihom*	*fihom*

On this basis, we can conclude that there is no ACCUSATIVE government of Maltese Πs, in the first place, and that (Π7) *bi* 'with' and (Π15) *fi* 'in' consistently realise POSSESSIVE government only like the bulk of the inflected Πs.

4.1.2.2.2.2.4 ABLATIVE VS NOMINATIVE government

The alternation between ABLATIVE and NOMINATIVE government with pronominal complements is associated with two Πs that display overlapping alloforms, namely (Π3) *barra* 'outside' and (Π34) *minbarra* 'except'. The intricacies of the relation between these two Πs have been addressed already twice.[112] Moreover, the previous subsection on ABLATIVE government contains example (4.123b) which features (Π3) *barra* 'outside' in its spatial function. We assume that with this Π ABLATIVE government is compulsory if the spatial meaning is expressed and the complement is pronominal. However, it is by no means always easy semantically to tell the spatial meaning and that of exclusion apart. In the latter case, *barra* is a short alloform of (Π34) *minbarra* 'except'. We assume that the cases in (4.138) involve this alloform although (4.138a) is ambiguous in the sense that the translation of *barra* as 'outside of' could be a viable alternative to 'except'.

(4.138) *barra* as short alloform of (Π34) *minbarra* 'except'

(a) [Korpus Malti 3.0; news162163]

Jien l-omnipotenza assoluta u ***[barra*** ***minni****]$_{PP}$* *ma*
1SG DEF-omnipotence absolute:F and **[except** **from**:1SG] NEG
jeżisti ħadd u xejn aktar
3SG.M.IPFV:exist nobody and nothing more
'I am the absolute omnipotence and [**except** me] nobody and nothing else exists.'

112 Cf. Sections 1.6.4.2 (iv) and 3.2.1.

(b) [Korpus Malti 3.0; news49395]

Biżżejjed ngħidlek li meta ħarġet tqila
enough 1SG.IPFV:say:IO:2SG that when exit:3SG.F.PFV heavy:F
bit-tielet tarbija kulħadd kien jaf ***[barra***
with:DEF-third child eveyrbody be 3SG.M.IPFV:know **[except**
jiena]$_{PP}$
1SG]
'It suffices that I tell you that when she was pregnant with the third child everybody knew [**except** me].'

The second short alloform of (Π34) *minbarra* 'except' is *mbarra* which is generally infrequent. In the Korpus Malti 3.0, we have found only a single case which involves this alloform and a pronominal complement, viz. (4.139). On account of its marginal frequency, *mbarra* will not be taken account of further in this section. For similar reasons we skip discussing the *minn barra (minn)* construction.

(4.139) *mbarra* as short alloform of (Π34) *minbarra* 'except' [Korpus Malti 3.0; law343]

password sigrieta uffiċjali maħruġa għall-użu ta' persuna
password secret:F official PART:exit:F for:DEF-use of person
oħra ***[mbarra*** *huwa nnifsu]*$_{PP}$
other:F **[except** 3SG.M self:3SG.M]
'[. . .] an official secret password issued for the use of another person [**except** himself] [. . .]'

As transpires from (4.140), the long alloform of (Π34) *minbarra* 'except' triggers either ABLATIVE or NOMINATIVE government.

(4.140) long alloform of (Π34) *minbarra* 'except'

(a) [Korpus Malti 3.0; parl54]

[minbarra minni]$_{PP}$ *kien hemm disa' Ministri li*
[except from:1SG] be there nine minister:PL that
siefru għall-btala tagħhom mindu waqaf
travel:3PL.PFV for:DEF-holidays of:3PL since stop
il-Parlament
DEF-parliament
'[. . .] [**besides** me] there were nine ministers who travelled abroad for their holidays since the parliamentary sessions stopped.'

(b) [Camilleri 2022: 29]

għandi bżonn website ħi għax kulħadd
have.IPFV:1SG need website bro' because everybody

għandu *[**minbarra*** *jien]*$_{PP}$
have.IPFV:3SG.M **[except** 1SG]
'[. . .] I need a website, bro', because everybody has one [**except** me].'

How does the above co-existence of different alloforms and government patterns translate into quantities? In the Korpus Malti 3.0, there are 155 PPs which involve alloforms of (Π34) *minbarra* 'except' in combination with pronominal complements. As results from Figure 4.76, there is a slight majority for the short alloform *barra* which counts nine tokens in excess of the long alloform *minbarra*.

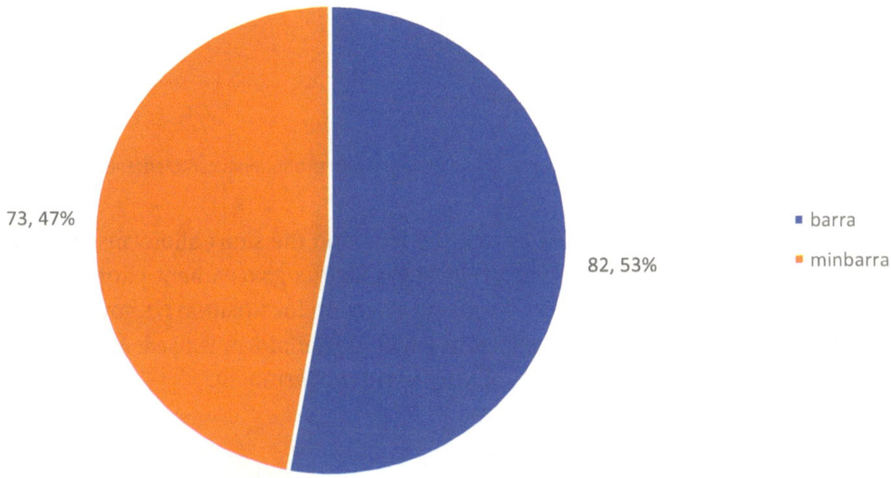

Figure 4.76: Alloforms of (Π34) *minbarra* 'except' with pronominal complements in the Korpus Malti 3.0.

Superficially, neither the absolute numbers nor the percentages suggest that the distribution of the two options is determined by specific factors. Before we prove this impression wrong, it makes sense to mention that there are also 406 tokens of (Π3) *barra* 'outside', i.e., the spatial Π occurs almost five times as often with pronominal complements than the homophonous short alloform *barra* of (Π34) *minbarra* 'except' does.

In Figure 4.77, we recognise that NOMINATIVE government ousts ABLATIVE government since the NOMINATIVE pattern is attested twice as frequently as the ABLATIVE pattern. The question arises whether this discrepancy correlates in any meaningful way with the competition of the two alloforms of (Π34) *minbarra* 'except'.

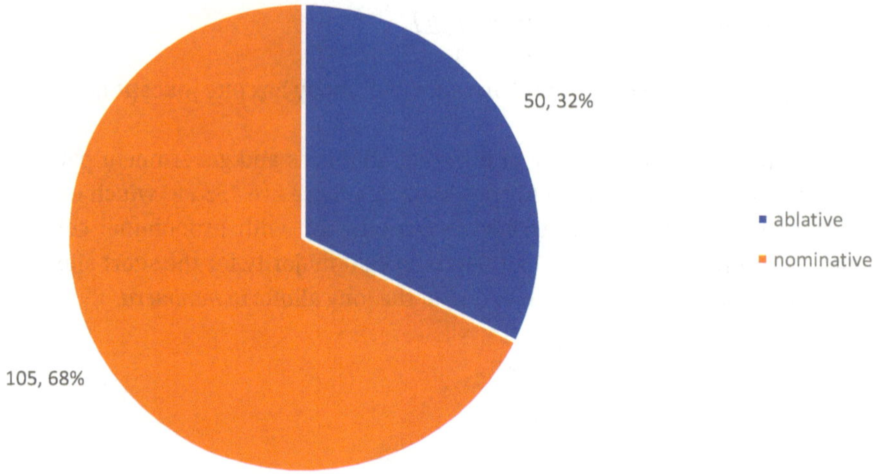

Figure 4.77: Shares of government patterns in *minbarra*-PPs with pronominal complements.

Figures 4.78 and 4.79 clearly tell us that the long and the short alloforms of (П34) *minbarra* 'except' differ as to their preferences for the government patterns. The long alloform *minbarra* has a very strong predilection for NOMINATIVE government since this pattern covers 97% of all PPs in which this alloform is used. *Minbarra* is responsible for 68% of all instances of NOMINATIVE government.

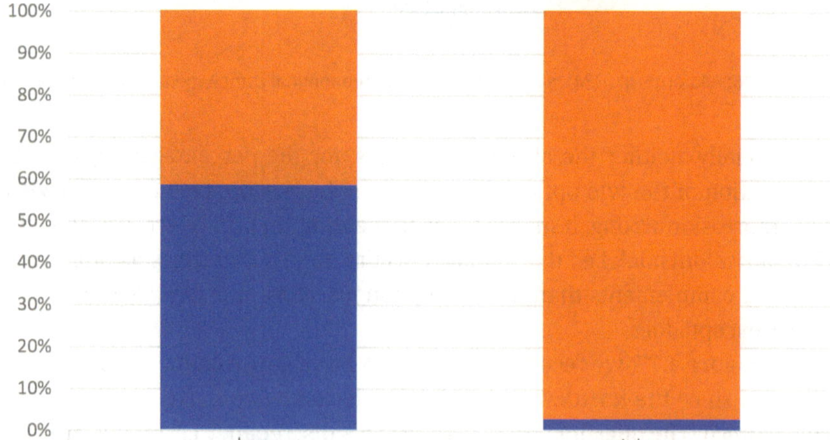

	barra	minbarra
nominative	34	71
ablative	48	2

Figure 4.78: Government patterns over alloforms of (П34) *minbarra* 'except'.

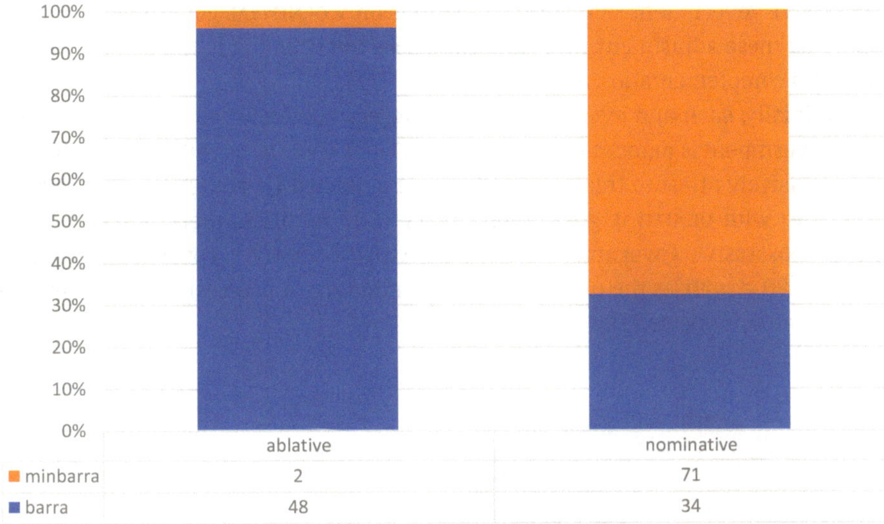

Figure 4.79: Alloforms of (Π34) *minbarra* 'except' over government patterns.

In contrast, the short alloform *barra* accounts for 96% of all cases of ABLATIVE government. This government pattern accounts for 58.5% of all combinations of *barra* with pronominal complements. Note that the abovementioned 406 hits for (Π3) *barra* 'outside' exclusively involve ABLATIVE government; there is no proof of this Π triggering NOMINATIVE government with pronominal complements. A conceivable explanation for this difference could be avoidance in the sense that the long alloform already involves the string *min(n)-* so that a second (Π41) *minn* 'from' as linker tends to be blocked whereas in the case of the short allomorph there is no such blocking effect since the string *min(n)-* is absent.

4.1.2.2.2.2.5 Synopsis of variation with pronominal complements

We have reached the point where it is possible to present a final synoptic account as to the range of government patterns in the domain of pronominal complementation. The previously assumed existence of ACCUSATIVE government has been disproved so that the picture resulting from the foregoing discussion is less rich than the catalogue of government patterns presented at the beginning of Section 4.1.2. The synopsis excludes (Π9) *bla* 'without', (Π26) *kif* 'as', (Π42) *mintul* 'all along', (Π43) *mnejn* 'from near', (Π52) *sforz* 'thanks to', and (Π54) *sotta* 'under' because they either do not take pronominal complements or are uncertain Π-candidates anyway. Furthermore, we doubt that (Π44) *oltre* 'beyond' displays NOMINATIVE and

POSSESSIVE government so that this Π exhibits only GENITIVAL government.[113] After applying these adjustments, we can summarise the situation in the domain of pronominal complementation as follows.

Trivially, each and every inflected Π triggers POSSESSIVE government. POSSESSIVE government is mandatory for 32 out of 37 inflected Πs, i.e., 86% of the inflected Πs exclusively attest to this pattern. For four inflected Πs POSSESSIVE government competes with GENITIVAL government (= 11%). In addition, there is also a Π for which POSSESSIVE government alternates with NOMINATIVE government (= 3%). Figure 4.80 visualises the uncontested predominance of POSSESSIVE government in the domain of inflected Πs.

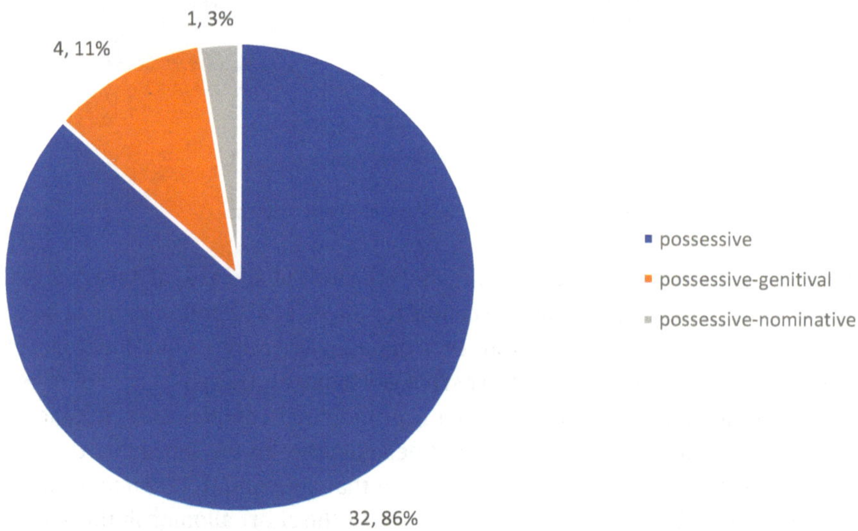

Figure 4.80: Shares of government patterns with inflected Πs (pronominal complementation).

POSSESSIVE government is counted out when we take account of the invariant Πs. There are seventeen invariant Πs for which we have reliable data as to their behaviour in the interaction with pronominal complements. These seventeen Πs distribute over seven different government patterns including one of variation. Figure 4.81 shows that GENITIVAL government and LOCATIVE government add up to 58% of all cases whereas the remaining five patterns have only relatively small shares. In three cases, a given pattern is valid only for a single Π. It is possible therefore to speak of a high degree of individual behaviour of the invariant Πs. Across the two classes of inflected and invariant Πs, GENITIVAL government is only second best

113 As explained in Section 4.1.2.2.2.2.2 (ii).

after the overwhelmingly dominant POSSESSIVE government. Figure 4.82 discloses the distribution by way of dissolving the variation patterns so that each member of a pattern of this kind is counted separately. The next subsection is meant to integrate the findings in the domain of pronominal complementation with those insights we gained in the domain of nominal complementation.

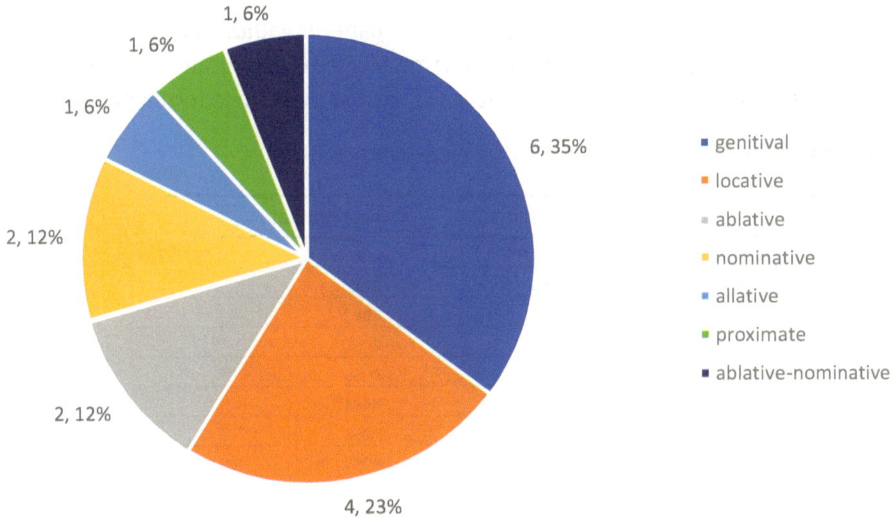

Figure 4.81: Shares of government patterns with invariant Пs (pronominal complementation).

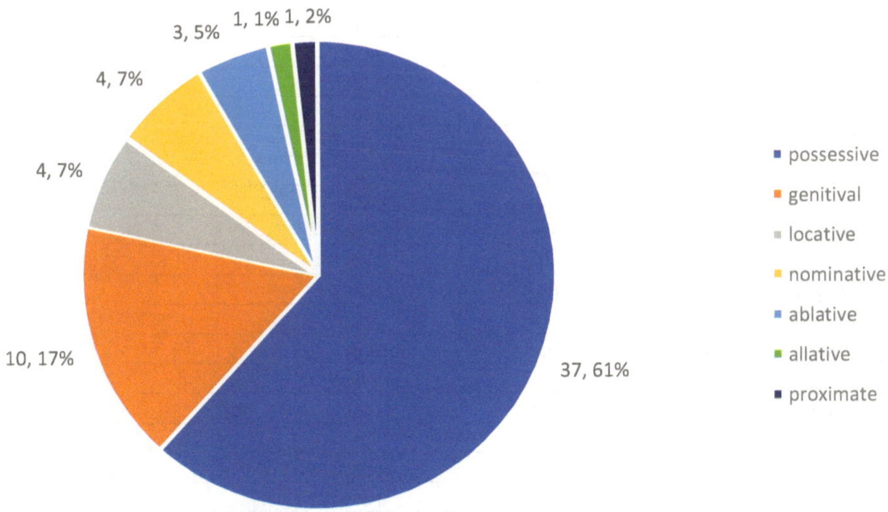

Figure 4.82: Shares of government patterns in BLOMP 2.0 (pronominal complementation).

4.1.2.2.3 Integration of government types

This section closes the discussion on government properties of Maltese Πs by way of correlating the patterns identified in the domain of nominal complementation with those found in the domain of pronominal complementation. To this end, we present the full list of members of BLOMP 2.0 together with their government patterns in Table 4.14. The Πs come in the established order. The uncertain candidates are kept on the list for comprehensiveness. We also mention all government patterns without differentiating them according to their type and token frequency. Grey shaded cells indicate that a given complementation type is unattested. Brackets and question marks identify doubtful cases.

Table 4.14: The members of BLOMP 2.0 and their government patterns.

Π	nominal		pronominal	
(Π1) *apparti* 'apart from'	NOM	ABL	ABL	
(Π2) *a skapitu* 'at the expense of'	GEN		GEN	
(Π3) *barra* 'outside'	NOM		ABL	
(Π4) *bejn* 'between'	NOM		POSS	
(Π5) *bħal* 'like'	NOM		POSS	
(Π6) *bħala* 'as'	NOM		NOM	
(Π7) *bi* 'with'	NOM		POSS	
(Π8) *biswit* 'facing'	NOM	GEN	POSS	GEN
(Π9) *bla* 'without'	NOM			
(Π10) *daqs* 'equal to'	NOM		POSS	
(Π11) *dwar* 'about'	NOM	GEN	POSS	
(Π12) *faċċata* 'opposite'	NOM	GEN	GEN	
(Π13) *favur* 'in favour of'	NOM	GEN	POSS	GEN
(Π14) *fejn* 'near'	NOM		POSS	
(Π15) *fi* 'in'	NOM		POSS	
(Π16) *fost* 'amongst'	NOM		POSS	
(Π17) *fuq* 'on'	NOM		POSS	
(Π18) *ġewwa* 'inside'	NOM	LOC	LOC	
(Π19) *ġo* 'in'	NOM		LOC	
(Π20) *għajr* 'except'	NOM		POSS	
(Π21) *għal* 'for'	NOM		POSS	
(Π22) *għand* 'at s.o.'s place'	NOM		POSS	
(Π23) *ħdejn* 'beside'	NOM		POSS	
(Π24) *ħlief* 'except'	NOM		POSS	
(Π25) *inkluż* 'including'	NOM		NOM	
(Π26) *kif* 'as'			(POSS?)	
(Π27) *kontra* 'against'	NOM	GEN	POSS	GEN
(Π28) *lejn* 'towards'	NOM		POSS	
(Π29) *lil* 'to'	NOM		POSS	
(Π30) *ma'* 'with'	NOM		POSS	
(Π31) *madwar* 'around'	NOM	GEN	POSS	

Table 4.14 (continued)

Π	nominal		pronominal		
(Π32) *maġenb* 'close to'	NOM		POSS		
(Π33) *matul* 'during'	NOM	GEN	POSS		
(Π34) *minbarra* 'except'	NOM	ABL	NOM	ABL	
(Π35) *minflok* 'instead of'	NOM	GEN	POSS		
(Π36) *mingħajr* 'without'	NOM	GEN	POSS		
(Π37) *mingħala* 'in s.o.'s opinion'			POSS		
(Π38) *mingħand* 'from s.o.'	NOM		POSS		
(Π39) *minħabba* 'on account of'	NOM	LOC	LOC		
(Π40) *minkejja* 'in spite of'	NOM		LOC		
(Π41) *minn* 'from'	NOM		POSS		
(Π42) *mintul* 'all along'	NOM		(POSS?)		
(Π43) *mnejn* 'from near'	(NOM?)		(POSS?)		
(Π44) *oltre* 'beyond'	NOM	ABL	GEN	(POSS?)	(NOM?)
(Π45) *permezz* 'by means of'	GEN		GEN		
(Π46) *qabel* 'before'	NOM		POSS		
(Π47) *qalb* 'amidst'	NOM		POSS		
(Π48) *qrib* 'near'	NOM	GEN	POSS	GEN	
(Π49) *quddiem* 'in front of'	NOM		POSS		
(Π50) *rigward* 'concerning'	NOM	GEN	GEN		
(Π51) *sa* 'till'	NOM		PROX		
(Π52) *sforz* 'thanks to'	NOM				
(Π53) *skont* 'according to'	NOM		POSS	NOM	
(Π54) *sotta* 'under'			(GEN?)		
(Π55) *ta'* 'of'	NOM		POSS		
(Π56) *taħt* 'under'	NOM		POSS		
(Π57) *versu* 'towards'	NOM		ALL		
(Π58) *viċin* 'near'	NOM	GEN	GEN		
(Π59) *waqt* 'at the time of'	NOM		POSS		
(Π60) *wara* 'after'	NOM		POSS		

We compare the government patterns of the individual Πs in order to form classes of those Πs which attest to the same combination of patterns across complementation types. The classes that emerge in this way are termed government classes (short: GCL). In (4.141), fifteen GCLs are distinguished. Only those Πs are included in the catalogue which are attested both in the nominal and the pronominal domain.[114] GCL 9–15 are isolated cases in the sense that only one Π realises a given pattern.

114 The following patterns have fallen victim to the exclusion: NOMINATIVE-Ø [n = 2]: (Π9) *bla* 'without'; (Π52) *sforz* 'thanks to'; Ø-(POSSESSIVE) [n = 2]: (Π26) *kif* 'as'; (Π37) *mingħala* 'in s.o.'s opinion'; Ø-GENITIVE [n = 1]: (Π54) *sotta* 'under'

(4.141) GCLs

(a) GCL 1: NOMINATIVE-POSSESSIVE [n = 28]: (Π4) *bejn* 'between'; (Π5) *bħal* 'like'; (Π7) *bi* 'with'; (Π10) *daqs* 'equal to'; (Π14) *fejn* 'near'; (Π15) *fi* 'in'; (Π16) *fost* 'amongst'; (Π17) *fuq* 'on'; (Π20) *għajr* 'except'; (Π21) *għal* 'for'; (Π22) *għand* 'at s.o.'s place'; (Π23) *ħdejn* 'beside'; (Π24) *ħlief* 'except'; (Π28) *lejn* 'towards'; (Π29) *lil* 'to'; (Π30) *ma'* 'with'; (Π32) *maġenb* 'close to'; (Π38) *mingħand* 'from s.o.'; (Π41) *minn* 'from'; (Π46) *qabel* 'before'; (Π47) *qalb* 'amidst'; (Π49) *quddiem* 'in front of'; (Π55) *ta'* 'of'; (Π56) *taħt* 'under'; (Π59) *waqt* 'at the time of'; (Π60) *wara* 'after' + **doubtful** (Π42) *mintul* 'all along'; (Π43) *mnejn* 'from near'

(b) GCL 2: NOMINATIVE/GENITIVE-POSSESSIVE [n = 5]: (Π11) *dwar* 'about'; (Π31) *madwar* 'around'; (Π33) *matul* 'during'; (Π35) *minflok* 'instead of'; (Π36) *mingħajr* 'without'

(c) GCL 3: NOMINATIVE/GENITIVE-POSSESSIVE/GENITIVE [n = 4]: (Π8) *biswit* 'facing'; (Π13) *favur* 'in favour of'; (Π27) *kontra* 'against'; (Π48) *qrib* 'near'

(d) GCL 4: NOMINATIVE/GENITIVE-GENITIVE [n = 3]: (Π12) *faċċata* 'opposite'; (Π50) *rigward* 'concerning'; (Π58) *viċin* 'near'

(e) GCL 5: NOMINATIVE-NOMINATIVE [n = 2]: (Π6) *bħala* 'as'; (Π25) *inkluż* 'including'

(f) GCL 6: NOMINATIVE-LOCATIVE [n = 2]: (Π19) *ġo* 'in'; (Π40) *minkejja* 'in spite of'

(g) GCL 7: NOMINATIVE/LOCATIVE-LOCATIVE [n = 2]: (Π18) *ġewwa* 'inside'; (Π39) *minħabba* 'on account of'

(h) GCL 8: GENITIVE-GENITIVE [n = 2]: (Π2) *a skapitu* 'at the expense of'; (Π45) *permezz* 'by means of'

(i) GCL 9: NOMINATIVE-ABLATIVE [n = 1]: (Π3) *barra* 'outside'

(j) GCL 10: NOMINATIVE-ALLATIVE [n = 1]: (Π57) *versu* 'towards'

(k) GCL 11: NOMINATIVE-PROXIMATE [n = 1]: (Π51) *sa* 'till'

(l) GCL 12: NOMINATIVE-NOMINATIVE/POSSESSIVE [n = 1] (Π53) *skont* 'according to'

(m) GCL 13: NOMINATIVE/ABLATIVE-ABLATIVE [n = 1]: (Π1) *apparti* 'apart from'

(n) GCL 14: NOMINATIVE/ABLATIVE-NOMINATIVE/ABLATIVE [n = 1]: (Π34) *minbarra* 'except'

(o) GCL 15: NOMINATIVE/ABLATIVE-GENITIVE(/NOMINATIVE/POSSESSIVE) [n = 1]: (Π44) *oltre* 'beyond'

The size of the different GCLs varies considerably. With 28 members GCL 1 outranks all other GCLs. The combination of NOMINATIVE government with nominal complements and POSSESSIVE government with pronominal government is more frequent than any other combination. It is worth noting that GCL 1 consists only of Semitic Πs whereas GCL 4 and GCL 8 exclusively host Italo-Romance Πs. In Figure 4.83, the shares of the combinations of patterns valid for pronominal complementation with NOMINATIVE government in the nominal domain are revealed.

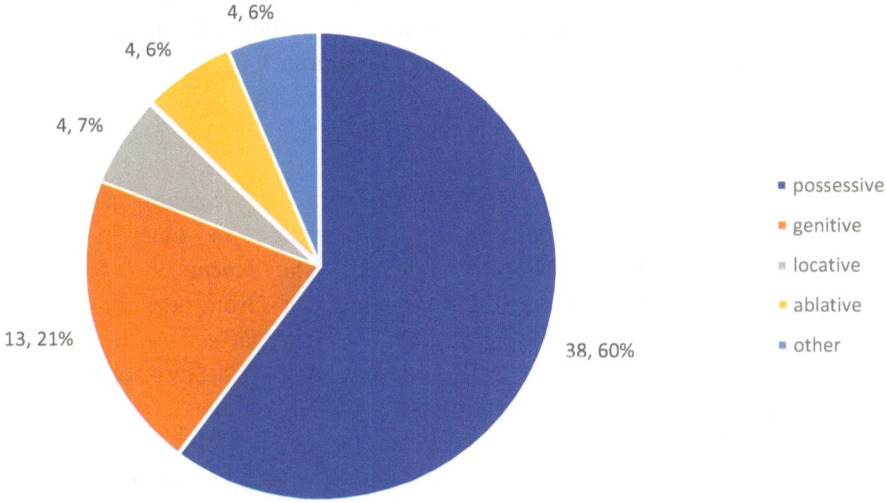

Figure 4.83: Shares of pronominal government patterns combining with NOMINATIVE government in the nominal domain.

The lion's share goes to the combination NOMINATIVE-POSSESSIVE which is therefore the combination we adopt for the canonical Maltese Π in Figure 4.84.

government → { NOMINATIVE / nominal complement
POSSESSIVE / pronominal complement }

Figure 4.84: Government patterns of the canonical Maltese Π.

The system of nominal and pronominal government of Maltese Πs has been described in some detail in the foregoing paragraphs. The many pieces of information we have provided above do not exhaust the topic completely. In the subsequent sections, issues relating to government will come up again every once in a while without however changing the general picture.

4.1.3 Compound Πs

Maltese gives evidence of sequences of Πs the analysis of which raises a number of problems. What is the syntactic relation between the different Πs and how is their combination to be interpreted semantically? To approach this subject, we have a look at example (4.142).

(4.142) [Teuma 2015: 1019]

Is-sultan bla ma qagħad jitlef iż-żmien fi
DEF-king without that sit 3SG.M.IPFV:lose DEF-time in
kliem fieragħ tellgħu fi kmajra fuq mnejn
word empty raise:3SG.M in room.DIM on from_near
***[minn [wara** l-persjani]$_{PP}$]$_{PP}$ bla ma jidher*
[from [after DEF-blind:PL]] without that 3SG.M.IPFV:appear
seta' jara lill-prinċipessa
can 3SG.M.IPFV:see to:DEF-princess
'Without wasting time with empty words, the king brought him upstairs into a small room from where he could see the princess [**from [behind** the blinds]] without being visible.'

The combination of (Π41) *minn* 'from' and (Π60) *wara* 'after' in (4.142) is different from the cases of prepositional government discussed in several subsections above.[115] Neither of the two Πs in (4.142) requires the presence of a linker to take a complement. Both (Π41) *minn* 'from' and (Π60) *wara* 'after' attest to NOMINATIVE government with nominal complements, i.e., the Π and its complement are asyndetically juxtaposed to each other.[116] Likewise, both Πs are inflected and thus realise POSSESSIVE government with pronominal complements.[117] Thus, *minn wara NP* belongs in a different category from *qrib ta' NP*. As argued above,[118] the former represents the class of compound Πs which should be confounded neither with cases of prepositional government nor with those of morphologically complex but lexicalised Πs like (Π33) *matul* 'during'. The necessity of making these distinctions is motivated by semantics. Prepositional government involves a second Π as linker – an element which is semantically empty in the sense that it does not contribute anything to the meaning of the construction. Lexicalised complex Πs are similar insofar as the semantic autonomy of the univerbated morphemes is reduced. This

115 Cf. Section 4.1.2.2.1.
116 Cf. Sections 4.1.1.2.1 (viii) and 4.1.1.2.2.1 (xl), respectively.
117 Cf. Sections 3.1.2.3.1.1 and 3.1.2.3.3.1.
118 Cf. Section 1.6.4.

is different for compound Πs because the Πs which enter the combination remain semantically autonomous and contribute substantially to the constructional meaning. Compound Πs do not constitute fossilised units but paradigmatically flexible but syntagmatically strictly organised sequences of Πs which reflect a certain division of labour to be explained in this section.

Compound Πs differ also from phrasal Πs because compound Πs are combinations of several members of BLOMP 2.0 whereas phrasal Πs are prepositionally employed phrases which might or might not involve a member of BLOMP 2.0. Stolz (2020: 444–451) reviews previous accounts of compound Πs, namely Aquilina (1959: 300), Fabri (1993: 187), Cachia (1994: 131), Borg/Azzopardi-Alexander 1997: 147–172 and 297), and Saari (2003: 84–100), several of which can be shown not always to distinguish adequately between lexicalised complex Πs, prepositional government, and proper compound Πs. In his cross-linguistic study of adpositions, Hagège (2010: 130–131) adopts an all-embracing definition of compound Πs which covers all of these and further cases. We seriously doubt that refraining from discriminating further is helpful when it comes to describing what is the case synchronically in Maltese. It is impossible to account for the phenomena in one go because this would mean mixing up diachrony and synchrony as well as different kinds of functional motivation. Therefore, we restrict our reading of the term compound Π to "the association of two or more [Πs] to render time and space specifications" (Hagège 2010: 130). Note however that the quoted author assumes that the Πs in constructions of this kind form a unit, i.e., he conceives of a compound Π as a single syntactic element. To our mind however, it is probably more adequate to speak of productive Π-Π combinations to avoid treating proper compound Πs terminologically on a par with unproductive univerbations of erstwhile syntagms and prepositional government (Stolz 2020: 452–457).

4.1.3.1 Π-Π combinations

The bracketed PP in (4.142) is a good representative of the category under scrutiny. The two Πs do not arbitrarily interact but realise a constructional pattern. This pattern has the following architecture (cf. Figure 4.85).

$[Π_1 \quad [Π_2 \quad NP]_{PP}]_{PP}$

Figure 4.85: Architecture of Π-Π combinations.

According to Figure 4.85, a PP is embedded in another PP. The two Πs heading the PPs do not form a syntactic unit but share the same complement. Why can we say this? There are two major reasons for separating the Πs from each other. On the one hand, the slots they occupy can be filled by various other Πs giving rise to a plethora of combinations. On the other hand, each of the Πs contributes a distinct meaning component to the constructional meaning. The Πs are thus largely independent of each other. Their autonomy transpires from the examples we are going to discuss in the subsequent paragraphs.

Lestrade (2010: 65–134) assumes a template for the basic dimensions which are involved when it comes to expressing a spatial situation, namely DIRECTIONALITY, CONFIGURATION, and GROUND. These dimensions can be applied to the analysis of Π-Π combinations as follows. The ternary distinction of Path, Place, Goal, and Source is covered by the notion of DIRECTIONALITY, i.e., on this dimension dynamic vs static spatial relations are distinguished; in the case of dynamic relations, movement from is distinguished from movement towards the GROUND. The label CONFIGURATION specifies the region of or relative to the GROUND which is of relevance in the spatial situation. GROUND is the landmark relative to which the FIGURE is positioned or moved. The GROUND is represented by the complement NP whereas DIRECTIONALITY and CONFIGURATION are separately expressed by Π_1 and Π_2 in the formula given in Figure 4.85. The order is always DIRECTIONALITY > CONFIGURATION > GROUND, meaning: the leftmost Π_1 has the task of specifying whether we are facing the spatial relation Place, Goal, or Source, Π_2 is responsible for the identification of the region of the GROUND which is involved in the spatial situation, and the complement NP represents the GROUND. For *minn wara l-persjani* 'from behind the blinds' this means that (Π41) *minn* 'from' tells us that the spatial relation of Source applies, (Π60) *wara* 'after' indicates that the region behind the GROUND is involved, and *l-persjani* 'the blinds' identifies the GROUND. Only Π_2 can be inflected in case the complement is pronominal. Insertions between Π_1 and Π_2 are prohibited. It is equally important to note that the slot of Π_1 generally remains empty if the static spatial relation of Place is to be expressed. This in turn means that only motion towards or originating from the GROUND is encoded overtly.

For a start, we focus on binary combinations of Πs. In (4.143), we contrast Π-Π combinations with a lexicalised univerbation (= (4.143a)) and a case of prepositional government (= (4.143b)).

(4.143) Π-Π combinations in contrast

(a) [Korpus Malti 3.0; literature77]

medd	*riġlu*	***[għal***	***[maġenb***	*il-mejda*	*fejn*	*kien*	*qiegħed*	*Wenzu]$_{PP}$]$_{PP}$*
stretch	leg:3SG.M	**[for**	**[close_to**	DEF-table	where	be	sit.PART	Wenzu]]

'[. . .] he stretched his leg [**close to** the table where Wenzu was sitting] [. . .].'

(b) [Bartolo/Vella 2009: 254]

Il-ħoss *kien* *ħiereġ* ***[minn*** ***[ġo*** ***fih**]$_{PP}$]$_{PP}$*

DEF-sound be exit.PART **[from** **[in** in:3SG.M]]

'The sound was coming [**from** [**inside of** him]] [. . .].'

There are two cases of embedding. In (4.143a), the morphologically complex (Π32) ***maġenb*** 'close to' functions as Π_2 of a Π-Π combination. In accordance with Figure 4.85, it specifies the region relative to the GROUND – in this case the immediate vicinity of the table. The historical compositionality of (Π32) ***maġenb*** 'close to' notwithstanding, it nowadays forms an unseparable unit with a unitary meaning. This meaning is independent of that contributed by the Π_1 which is (Π21) ***għal*** 'for' which specifies the DIRECTIONALITY as Goal-oriented. The examples in Table 4.15 prove that as Π_1 (Π21) ***għal*** 'for' can combine with a number of different fillers of the slot of Π_2. The examples in Table 4.15 represent types and are taken from the primary source [Bartolo/Vella 2009]. The token frequency of ***għal***-Π_2 combinations is much higher.

Table 4.15: (Π21) *għal* 'for' + Π_2 in [Bartolo/Vella 2009].

page	example	Π_2
59	*Id-dawl tax-xemx dieħel mit-tieqa **[għal [fuq** is-sodda]$_{PP}$]$_{PP}$ kien qawwi* 'The sunlight entering through the window [**unto** the bed] was strong.'	(Π17) *fuq* 'on'
66	*imbagħad qabżet dritt **[għal [ġo**l-basket]$_{PP}$]$_{PP}$* '[. . .] then she jumped [**into** the basket] [. . .].'	(Π19) *ġo* 'in'
85	*Estella marret tiġri **[għal [ħdej**h]$_{PP}$]$_{PP}$* 'Estella ran [**to** his **side**] [. . .].'	(Π23) *ħdejn* 'beside'
232	*U baqa' jidwi f'widnejhom l-għajjat u l-inbiħ ta' Erkole minuti qabel ma ġera **[għal [taħt** ir-roti tal-karrozza]$_{PP}$]$_{PP}$* 'And Erkole's cries and the barking minutes before he ran [**under** the wheels of the car] kept echoing in their ears.'	(Π56) *taħt* 'under'
74	*telaq bil-ġirja **[għal [waraj**ha]$_{PP}$]$_{PP}$* '[. . .] he quickly ran [**after** her] [. . .].'	(Π60) *wara* 'after'

As Π_2, (Π32) ***maġenb*** 'close to' is in a paradigmatic relation with the other Π_2-candidates so that it can be replaced by any of them (and vice versa). The replacement affects (Π32) ***maġenb*** 'close to' in its entirety. No part of this Π escapes being replaced. Syntagmatically, (Π32) ***maġenb*** 'close to' may combine with a different Π_1 under the condition that the spatial relation is not that of Goal. In (4.144), this possibility is illustrated for the spatial relation of Path.

(4.144) [Bartolo/Vella 2009: 226]

Estella	*ħasset*	*xi*	*ħaġa*	*għaddejja*	*bis-saħħa*
Estella	feel:3SG.F.PFV	some	thing	pass:PART:F	with:DEF-strength
[minn	***[maġenbha**]$_{PP}$]$_{PP}$*				
[from	**[close_to**:3SG.F]				

'Estella felt something passing with force [**close to** her] [. . .].'

The Π1 (Π41) *minn* 'from' is thus in a paradigmatic relation with (Π21) *għal* 'for'. Accordingly, (Π41) *minn* 'from' occurs in combinations with Π_2s many of which are also attested with (Π21) *għal* 'for' as Π_1. Table 4.16 contains the types of *minn*-Π_2 combinations found in the same primary source as used for the previous table.

Table 4.16: (Π41) *minn* 'from' + Π_2 in [Bartolo/Vella 2009].

page	example	Π_2
100	*għaddiet **[minn [bejn** il-purtieri]$_{PP}$]$_{PP}$* '[. . .] she passed [**through** the curtains] [. . .]'	(Π4) *bejn* 'between'
429	***[minn [fost** il-folla]$_{PP}$]$_{PP}$ tela' gidi żgħir* '[. . .] [**from [amongst** the crowd]] a small kid emerged [. . .].'	(Π16) *fost* 'amongst'
67	*kienet taqa' **[minn [fuq** għonqu]$_{PP}$]$_{PP}$* '[his head] would have fallen [**from** his neck].'	(Π17) *fuq* 'on'
60	*Ħarġet b'qabża ħafifa **[minn [ġo**l-basket]$_{PP}$]$_{PP}$* 'She jumped lightly [**out of** the basket].'	(Π19) *ġo* 'in'
54	*telaq **[minn [ħdejn** Estella]$_{PP}$]$_{PP}$* '[. . .] he went away [**from** Estella].'	(Π23) *ħdejn* 'beside'
27	*Tommy qam **[minn [ma**l-mejda]$_{PP}$]$_{PP}$* 'Tommy stood up [**from** the table].'	(Π30) *ma'* 'with'
220	*U reġgħet inkisret biċċa silġ oħra **[minn [madwar** qalb Estella]$_{PP}$]$_{PP}$* 'And again a piece of ice broke [**from [around** Estella's heart]].'	(Π31) *madwar* 'around'
85	*kellha ħafna xniexel ħerġin **[minn [qalb** il-petali]$_{PP}$]$_{PP}$* '[. . .] it had many grapes emerging [**from [amidst** the petals]].'	(Π47) *qalb* 'amidst'
232	*La Tommy u lanqas Estella ma setgħu jneħħu **[minn [quddiem** għajnejhom]$_{PP}$]$_{PP}$ ix-xena tal-biża'* 'Neither Tommy nor Estella could take the freightening scene [**from** their eyes] [. . .].'	(Π49) *quddiem* 'in front of'
25	*Ħarġet dik il-profezija **[minn [taħt** il-ġilda tal-qoxra]$_{PP}$]$_{PP}$* 'She pulled out this prophecy [**from [under** the leather cover]].'	(Π56) *taħt* 'under'
90	***[minn [warajha**]$_{PP}$]$_{PP}$ semgħet lil Lucia tieħu nifsijiet kbar.* '[. . .] [**from [behind** her]] she heard Lucia breathing deeply.'	(Π60) *wara* 'after'

The Π-Π combinations differ from prepositional government insofar as the latter is sensitive to the type of complement – a criterion which is irrelevant for the Π-Π combinations. In (4.143b), the pronominal complement of (Π19) *ġo* 'in' requires the use of (Π15) *fi* 'in' as linker since (Π19) *ġo* 'in' cannot be inflected. In contrast, the Π_1

needs neither (П19) *ġo* 'in' nor (П15) *fi* 'in' to take a pronominal complement since *minnu* 'from him' would be perfectly well-formed. The П-П combination is not motivated by the choice of complement but by the necessity to express DIRECTIONALITY and CONFIGURATION overtly and separately.

In Stolz (2020: 453), (П51) *sa* 'till' is mentioned as the third candidate for the role of $П_1$ in П-П combinations. As (4.145) shows, this П meets the expectations insofar as it is responsible for DIRECTIONALITY whereas its partner in the function of $П_2$ focusses the REGION of the GROUND.

(4.145) [Bartolo/Vella 2009: 103]

mara ta' xi 35 sena b' xagħarha pulit u libsa
woman of some 35 year with hair:3SG.F neat and dress
griża **[*sa*** **[*taħt*** *irkopptejha* $]_{PP}]_{PP}$
grey:F **[till** **[under** knee:PL:3SG:F]
'[. . .] a woman of about 35 years with neat hair and a grey dress reaching [**beneath** her knees] [. . .].'

In the reference text, in terms of types and tokens, (П51) *sa* 'till' is not as frequently employed as $П_1$ as (П41) *minn* 'from' and (П21) *għal* 'for' are. Yet, this П forms a paradigm of three together with (П41) *minn* 'from' and (П21) *għal* 'for'. Table 4.17 reveals which combinations of these $П_1$s with which $П_2$s are attested in the Korpus Malti 3.0. The П-П combinations are ordered according to the decreasing number of $П_1$s combining with a given $П_2$. Empty cells are additionally marked out in grey. The entire column of (П51) *sa* 'till' needs to be treated with caution since the combinations registered there might in their vast majority be interpreted in such a way that the П-status of the putative $П_1$ becomes doubtful. Further explanations are required and will be given below (e.g. with regard to the use of colours other than grey).

Table 4.17: П-П combinations with (П41) *minn* 'from', (П21) *għal* 'for', or (П51) *sa* 'till' as $П_1$.

$П_2$	$П_1$		
	(П41) *minn* 'from'	**(П21) *għal* 'for'**	**(П51) *sa* 'till'**
(П3) *barra* 'outside'	YES	YES	YES
(П4) *bejn* 'between'	YES	YES	YES
(П5) *bħal* 'like'	YES	YES	YES
(П8) *biswit* 'facing'	YES	YES	YES
(П12) *faċċata* 'opposite'	YES	YES	YES
(П14) *fejn* 'near'	YES	YES	YES
(П15) *fi* 'in'	YES	YES	YES
(П17) *fuq* 'on'	YES	YES	YES

Table 4.17 (continued)

Π₂	Π₁		
	(Π41) *minn* 'from'	**(Π21) *għal* 'for'**	**(Π51) *sa* 'till'**
(Π18) *ġewwa* 'inside'	YES	YES	YES
(Π19) *ġo* 'in'	YES	YES	YES
(Π23) *ħdejn* 'beside'	YES	YES	YES
(Π32) *maġenb* 'close to'	YES	YES	YES
(Π33) *matul* 'during'	YES	YES	YES
(Π46) *qabel* 'before'	YES	YES	YES
(Π48) *qrib* 'near'	YES	YES	YES
(Π49) *quddiem* 'in front of'	YES	YES	YES
(Π56) *taħt* 'under'	YES	YES	YES
(Π58) *viċin* 'near'	YES	YES	YES
(Π59) *waqt* 'at the time of'	YES	YES	YES
(Π60) *wara* 'after'	YES	YES	YES
(Π13) *favur* 'in favour of'	YES	YES	
(Π16) *fost* 'amongst'	YES	YES	
(Π27) *kontra* 'against'	YES	YES	
(Π30) *ma'* 'with'	YES	YES	
(Π31) *madwar* 'around'	YES	YES	
(Π47) *qalb* 'amidst'	YES	YES	
(Π51) *sa* 'till'	YES	YES	
(Π28) *lejn* 'towards'	YES		YES
(Π22) *għand* 'at s.o.'s place'		YES	YES
(Π41) *minn* 'from'		YES	YES
(Π6) *bħala* 'as'	YES		
(Π50) *rigward* 'concerning'		YES	
total: 32	29	30	23

The totals for the three Π_1s differ markedly from the count provided by Saari (2003: 87–88). We have identified many more cases in excess of the combinations registered in Saari's dissertation. The most striking difference comes to the fore in connection to (Π21) *għal* 'for'. This Π is Π_1 only in eight Π-Π combinations in Saari's study whereas we have identified twenty-two additional combinations. Similarly, our turnouts for (Π41) *minn* 'from' and (Π51) *sa* 'till' are bigger by twelve cases each than those provided by Saari (Stolz/Levkovych 2021: 448–449). From Table 4.17, 28

Maltese Πs are absent. We list the absentees in (4.146). The use of the different colours is explained below.

(4.146) Πs excluded from combinations with (Π41) *minn* 'from', (Π21) *għal* 'for', or (Π51) *sa* 'till' as Π_1
(Π1) *apparti* 'apart from'; (Π2) *a skapitu* 'at the expense of'; (Π7) *bi* 'with'; (Π9) *bla* 'without'; (Π10) *daqs* 'equal to'; (Π11) *dwar* 'about'; (Π20) *għajr* 'except'; (Π21) *għal* 'for'; (Π24) *ħlief* 'except'; (Π25) *inkluż* 'including'; (Π26) *kif* 'as'; (Π29) *lil* 'to'; (Π34) *minbarra* 'except'; (Π35) *minflok* 'instead of'; (Π36) *mingħajr* 'without'; (Π37) *mingħala* 'in s.o.'s opinion'; (Π38) *mingħand* 'from s.o.'; (Π39) *minħabba* 'on account of'; (Π40) *minkejja* 'in spite of'; (Π42) *mintul* 'all along'; (Π43) *mnejn* 'from near'; (Π44) *oltre* 'beyond'; (Π45) *permezz* 'by means of'; (Π52) *sforz* 'thanks to'; (Π53) *skont* 'according to'; (Π54) *sotta* 'under'; (Π55) *ta'* 'of'; (Π57) *versu* 'towards';

The list in (4.146) has to be taken with a grain of salt because there are isolated examples of some of the Πs being employed as Π_2 in combination with one of the Π_1-candidates but the examples are too scarce and mostly doubtful to justify the inclusion of the case among the Π-Π combinations in Table 4.17. Some of the doubtful cases will nevertheless be presented and discussed in the course of this section. The discussion will additionally reveal that even among those combinations which have been admitted to Table 4.17 on account of their relatively high token frequency or their combinability with at least two different Π_1s, there are doubtful cases which on revision might vanish from the inventory.

Given that our division of the Maltese Πs into two groups is acceptable for the time being, we can state that with 53% slightly more than half of the members of BLOMP 2.0 are attested as Π_2. Five of these attested Π_2s are Italo-Romance which is equivalent to a share of 16% in this group. Italo-Romance Πs are thus clearly underrepresented as compared to the 25% share they have in BLOMP 2.0. The ten Italo-Romance Πs in (4.146) however yield a share of some 36% which exceeds the expected 25% by far. This is a case of overrepresentation. It is also worth noting that only two of the five Italo-Romance Πs in Table 4.17 allow for combinations with each of the three Π_1 candidates. (Π50) *rigward* 'concerning' as Π_2 is exceptional insofar as it can combine only with (Π21) *għal* 'for'. The three Π_1s in Table 4.17 are of Semitic origin. The differences between the languages of origin of the Πs are captured in Figures 4.86–4.87. The red line marks the 25% threshold for Italo-Romance Πs in BLOMP 2.0.

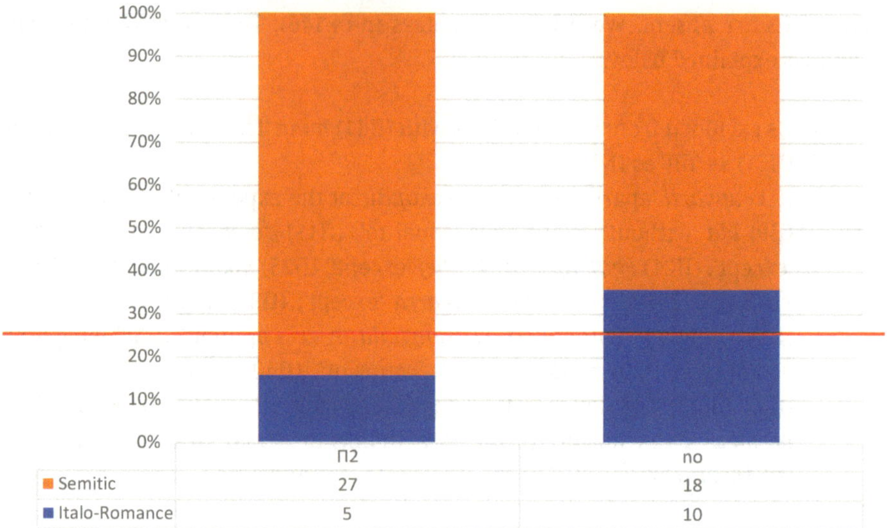

Figure 4.86: Language of origin over (im)possible Π_2s.

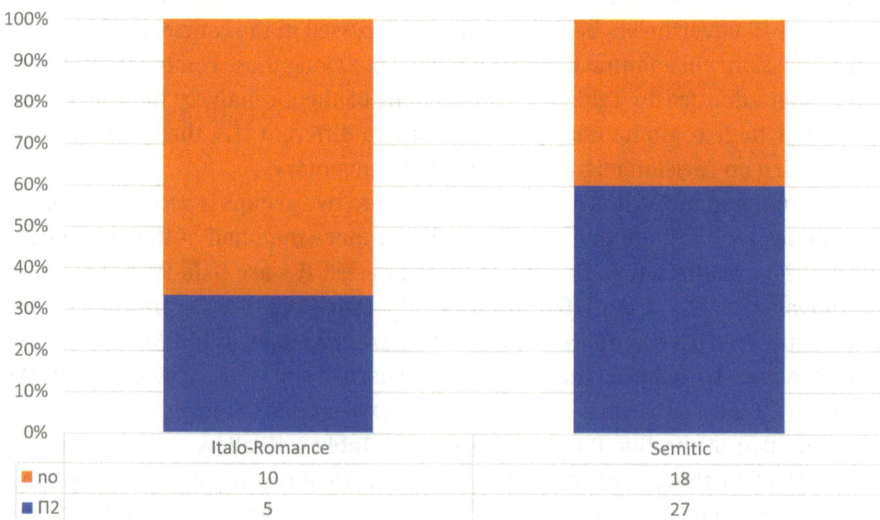

Figure 4.87: (Im)possible Π_2s over language of origin.

It is clear that the language of origin matters although there is only a strong tendency but no 100% predictability of the (im)possibility of a given Π to function as Π_2. The story of compound Πs and Π-Π combinations does not stop at this point.

Several comments are necessary to get a better idea of the system of Π-Π combinations in Maltese. In the subsequent section, a number of problematic issues will be addressed.

4.1.3.2 Problems and how to solve them

The thrity-two Π-Π combinations in Table 4.17 comprise twenty Π_2s which are possible partners for all three of the Π_1 candidates. For two Π_2s there is only one possible Π_1. If a Π_2 combines with (Π51) *sa* 'till' it always combines with at least one other Π_1. The number of Π_2s which are attested in combinations not only with (Π41) *minn* 'from' but also with (Π21) *għal* 'for' amounts to 27. The *minn-għal* pairing forms the basic functional opposition in the system of Π-Π combinations. However, the polysemy of Maltese Πs, in general and of (Π41) *minn* 'from', (Π21) *għal* 'for', and (Π51) *sa* 'till', in particular precludes a straightforward explanation of the distribution of Π_1s over Π_2s and vice versa. Since semantic issues will be scrutinised in some detail in Section 5, we do not want to anticipate too much already at this point. Suffice it to say that the three Π_1s are registered with lexical meanings which are bona fide examples of spatial concepts. The same holds for twenty-two of the Π_2s (marked by yellow in Table 4.17) to which two further Π_2s have to be added whose secondary spatial meanings are relevant for the Π-Π combinations (marked by orange in Table 4.17). Three quarters of all Π_2s in Table 4.17 can thus be classified as spatial. There are also six spatial Πs in (4.146) – equally highlighted in yellow – that account for 21% of those Πs which are unattested as Π_2s in combination with (Π41) *minn* 'from', (Π21) *għal* 'for', or (Π51) *sa* 'till'. Three of this group of six are notorious problem children within BLOMP 2.0. The number of Italo-Romance Πs with spatial meaning in (4.134) is also three. Superficially, the quantitative differences might not appear to be conclusive. Yet, they clearly indicate that the division of labour according to the formula in Figure 4.85 is the main task for the majority of the Π-Π combinations under review.

The next issue to raise is also situated in the spatial domain. (Π51) *sa* 'till' boasts the highest number of empty cells in Table 4.17. This might seem questionable since there is empirical evidence of this Π combining with some of the putatively excluded Π_2s as shown in (4.147).

(4.147) (Aquilina 1991: 1244)

sa	*ma'*	*ħutu*	*kellu*	*xi*	*jgħid*
till	with	brother:PL:3SG.M	have.PFV:3SG.M	what	3SG.M.IPFV:say

'He **even** quarrelled with his brothers.' [O.T.]

In syntactic contexts of this kind, (Π51) *sa* 'till' is not used prepositionally but adverbially. Its best translation equivalent in English is *even*. As adverb *sa* 'even' does not interact with the Π to its right but with the complement NP. In (4.147) for instance, *sa* is not responsible for the expression of DIRECTIONALITY. If *sa* was dropped from the sentence it would still be grammatical. Only in terms of pragmatics would something be missing from the utterance. Combinations of adverbial *sa* with other Πs have not been admitted to Table 4.17.

As we know from the discussion above,[119] (Π51) *sa* 'till' is the only Maltese Π which realises PROXIMATE government in the presence of a pronominal complement. In contrast, NOMINATIVE government applies in *sa*-PPs with nominal complements. The Π is the sole member of GCL 11 as shown in (4.141k). In the case of pronominal complementation, (Π51) *sa* 'till' has the choice between (Π14) *fejn* 'near' and (Π23) *ħdejn* 'beside' as linker. Unexpectedly the latter Π is involved also in (4.145).

(4.148) [Bartolo/Vella 2009: 84]

Mexa	***[sa***	***[ħdejn***	*ħuh]$_{PP}$]$_{PP}$*
walk	**[till**	**[beside**	brother:3SG.M]

'He walked [**to** his brother's **side**] [. . .].'

Does the presence of (Π23) *ħdejn* 'beside' in (4.145) call for a revision of our account of government in the domain of nominal complementation? Does PROXIMATE government alternate with NOMINATIVE government if the *sa*-PP contains a complement noun? We can only give negative answers to these rhetorical questions. In point of fact, (4.148) is perfectly in line with the Π-Π hypothesis because the presence of (Π23) *ħdejn* 'beside' is not motivated by any need to have a linker which connects (Π51) *sa* 'till' to its complement. The two examples in (4.149) suggest that the presence / absence of (Π23) *ħdejn* 'beside' makes a difference in terms of semantics.

(4.149) With and without (Π23) *ħdejn* 'beside'

(a) [Korpus Malti 3.0; news11100]

il-familja	*Cassar*	*tixtieq*	*li*	*tieħu*	*lil*	*Julian*
DEF-family	Cassar	3SG.F.IPFV:want	that	2SG.F:take	to	Julian

***[sa**l-garaxx*	*tal-pulizija*	*tat-traffiku]$_{PP}$*
[till:DEF-garage	of:DEF-police	of:DEF-traffic]

'[. . .] the Cassar family wants you to take Julian [**to** the garage of the traffic police] [. . .].'

119 Cf. Section 4.1.2.2.2.1.5.

(b) [Korpus Malti 3.0; *literature20]*
ġera [sa [ħdejn il-garaxx tad-dar ta' ħdejna]$_{PP}$]$_{PP}$
run [**till** [**beside** DEF-garage of:DEF-house of beside:1PL]]
'[. . .] he ran [**to the side of** the garage of the house next to ours] [. . .].'

In (4.149a), the absence of (Π23) *ħdejn* 'beside' invites the interpretation of the spatial situation as involving the Figure (*Julian*) moving into the GROUND (= the garage of the traffic police). In contrast, in (4.149b), the presence of (Π23) *ħdejn* 'beside' signals that the GROUND (= the neighbours' garage) is not entered since the Figure remains on its outside. In (4.149b), (Π51) *sa* 'till' encodes DIRECTIONALITY (= Goal) while (Π23) *ħdejn* 'beside' identifies the REGION of the GROUND that is relevant for determining the CONFIGURATION. Thus, (Π23) *ħdejn* 'beside' is not a linker whose presence is required for the purpose of government with nominal complements.

It is no surprise to see that Table 4.17 does not feature autocombinations of the $Π_1$s with themselves as $Π_2$, i.e., there is no evidence of combinations like **minn-minn*, **għal-għal*, or **sa-sa*. Moreover, (Π21) *għal* 'for' is completely ruled out as $Π_2$ so that there is also no evidence of **minn-għal* and **sa-għal*. As to (Π41) *minn* 'from' and (Π51) *sa* 'till' however, they can combine with each other and with (Π21) *għal* 'for'. In (4.150), we provide examples for the combinations *sa-minn* and *minn-sa*. For the time being, we follow the practice of bracketing employed hitherto for regular Π-Π combinations. Arguments against this analogy are presented below.

(4.150) *sa-minn* vs *minn-sa*
(a) [Korpus Malti 3.0; academic378]
[Sa [minn żmien l-Eġizzjani u l-Griegi]$_{PP}$]$_{PP}$ l-mużika
[**till** [**from** time DEF-Egyptian:PL and DEF-Greek:PL]] DEF-music
kienet tiġi kkunsidrata bħala terapija
be:3SG.F.PFV 3SG.F.IPFV:come consider:PART:F as therapy
għall-fejqan ta' mard
for:DEF-healing of illness
'[**Since** the times of the Egyptians and Greeks] music has been considered to be a therapy for the healing of an illness.'
(b) [Korpus Malti 3.0; news148251]
Kemm hawn persuni li [minn [sa ċkunithom]$_{PP}$]$_{PP}$
how_much here person:PL that [**from** [**till** childhood:POSS.F:3PL]

kienu *jew* *għadhom* *vittmi* *tal-bullying*
be:3PL.PFV or still:3PL victim:PL of:DEF-bullying
'How many people are here who [**since** their childhood] were or still are victims of bullying?'

Obviously, the different linearization of the two Πs in (4.150a) and (4.150b) is semantically irrelevant at least for this pair of sentences. However, the order *sa minn* is attested 4,342 times in the Korpus Malti 3.0 whereas the inverse order occurs only 202 times. Moreover, *minn sa* – in stark contrast to *sa minn* – does not always correspond to English *since* as shown in (4.151).

(4.151) [Korpus Malti 3.0; european9387]
Il-pajjiż *jista'* *jibbenenfika* ***[minn*** ***[sa*** *ħames*
DEF-country 3SG.M.IPFV:can 3SG.M.IPFV:benefit **[from** **[till** five
mandati *u* *faċilitajiet* *differenti* $]_{PP}]_{PP}$
mandate:PL and facility:PL different:PL]
'The country can benefit [**from** [**up to** five different mandates and facilities]] [. . .].'

The majority of the hits for the order *minn sa* invites the reading 'up to' for (Π51) *sa* 'till'. The meaning 'since' for the same Π in the combination *minn sa* illustrated in (4.150b) seems to be exceptional. For Aquilina (1991: 1244), the order *sa minn* seems to be emphatic as suggested by his own example ***sa*** *minn tfulitu* 'from his **very** childhood'. Accordingly, *sa minn* is a kind of emphatic (Π41) *minn* 'from'. (4.152) shows that (Π41) *minn* 'from' can fulfil the task of expressing the temporal relation without (Π51) *sa* 'till' being present too.

(4.152) [Korpus Malti 3.0; *academic57]*
bidla *li* *kontinent* *kien* *ilu* *ma* *jara* *bħalha*
change that continent be ago that 3SG.M.IPFV:see like:3SG.F
[minn *żmien* *l-Imperu* *Ruman* $]_{PP}$
[from time DEF-empire Roman]
'[. . .] a change that a continent has seen nothing like it [**since** the time of the Roman Empire].'

It is very likely that many of the supposed Π-Π combinations with initial *sa* whose existence is assumed in Table 4.17 turn out to involve adverbial *sa* 'even' and thus have to be counted out as Π-Π combinations. The clarification of this question requires not only closer scrutiny but also the co-operation of native speakers. We will take up this issue again in a follow-up study.[120]

A Π-Π combination for which the Π-status of (Π51) *sa* 'till' is out of the question involves (Π22) *għand* 'at s.o.'s place' as Π_2. The same Π_2 combines with (Π21) *għal* 'for' but not with (Π41) *minn* 'from'. The existence of lexicalised (Π38) *mingħand* 'from s.o.' blocks the possibility of a Π-Π combination of (Π41) *minn* 'from' and (Π22) *għand* 'at s.o.'s place'.[121] On the basis of the three examples of (Π22) *għand* 'at s.o.'s place' in (4.153) we will discuss the combinations with (Π51) *sa* 'till' and (Π21) *għal* 'for'.

(4.153) (Π22) *għand* 'at s.o.'s place' as Π_2

(a) [Korpus Malti 3.0; literature71]
Ix-xwejjaħ Bażilju mar ***[sa [għand*** *Pietru* $]_{PP}]_{PP}$
DEF-old_man Bażilju go **[till [at_s.o.'s_place** Pietru]]
'The old man Bażilju went [**to** Pietru's place] [. . .].'

(b) [Korpus Malti 3.0; literature4]
naqbdu l-karozza u nitilqu ***[għal [għand***
1.IPFV:seize:PL DEF-car and 1.IPFV:leave:PL **[for [at.s.o.'s_place**
il-girlfriends tagħna $]_{PP}]_{PP}$
DEF-girlfriend:PL of:1PL]]
'[. . .] we take the car and leave [**for** our girlfriends]].'

(c) [Korpus Malti 3.0; literature22]
Ħa t-tfal u mar ***[għand*** *ommu* $]_{PP}$ *waħdu*
take DEF-child.PL and go **[at_s.o.'s_place** mother.3SG.M] one:3SG.M
'He took the children and went [**to** his mother] alone.'

To express the spatial relation of Goal, it is unnecessary to do this overtly by way of adding a Π_1 to (Π22) *għand* 'at s.o.'s place'. Bare (Π22) *għand* 'at s.o.'s place' suffices as (4.153c) proves. This is also the usual construction we encounter in the Korpus Malti 3.0 where the (3rd person singular masculine perfective of the motion verb *mar* 'go' combines 707 times directly with (Π22) *għand* 'at s.o.'s place'. The combinations *mar għal għand* and *mar sa għand* are attested only twice and once, respectively. Emphasis does not seem to play a role so that the reinterpretation of *sa* in

120 There are several lexicalised univerbations of function words with initial *sa* in Maltese such as the conjunction *sakemm* 'until the time; as long as', the adverbs *saħansitra* 'even', *safrattant* 'in the meantime', etc. explained as PAT-borrowings from Italian (Aquilina 1991: 1245).

121 Orthographically, (Π38) *mingħand* 'from s.o.' is frequently written as two words <*minn*> and <*għand*> separated by a blank. We count these cases as instances of (Π38) *mingħand* 'from s.o.' and not as examples of a Π-Π combination.

(4.150a) as adverb is ruled out. At the same time, it is also impossible to correlate the presence or absence of a Π_1 with semantic distinctions except that the employment of either (Π21) *għal* or (Π51) *sa* 'till' precludes the (unlikely) interpretation of the spatial situation as being static.

The frequent co-occurence of (Π21) *għal* 'for' as Π_1 and (Π41) *minn* 'from' as Π_2 in the Korpus Malti 3.0 results from a number of factors. First of all, a multitude of cases involves the interrogative pronoun *min* 'who' misspelled as <*minn*>.[122] What is usually intended is the PP *għal min VP* 'for who VP'. Almost equally frequent is the combination of (Π21) *għal* 'for' with the phrasal adverbial *minn tal-anqas* 'at least' used to quantify the minimum of a given measurement.[123] In this case, no Π-Π combination applies, in the first place. Even (4.154) fails to fulfil the requirements of the formula give in Figure 4.85.

(4.154) [Korpus Malti 3.0; european2603]
f' din it-temperatura din is-soluzzjoni
in DEM:F:PROX DEF-temperature DEM:F:PROX DEF-solution
tibqa' stabbli ***[għal [minn*** *jum]$_{PP}$* ***[sa*** *tlett ijiem]$_{PP}$]$_{PP}$*
3SG.F.IPFV:remain stable **[for [from** day] **[till** three day.PL]]
'[. . .] at this temperature this solution stays stable **[for** a day **[up to** three days]] [. . .].'

The binary construction [*minn* X *sa* Y] 'from X (up) to Y' and (Π21) *għal* 'for' do not form a close unit. All three of the Πs express DIRECTIONALITY so that the expected division of labour does not apply. Therefore, it is possible to cancel the combination *għal-minn* from the inventory of Π-Π combinations in Table 4.17.

122 A case in point is given in (4.vii):

(4.vii) [Korpus Malti 3.0; news203141]
jiġbru il-fondi ***għal min<n>*** *hu fil-bżonn*
3.IPFV:collect:PL DEF-fund:PL **for who** 3SG.M in:DEF-need
'[. . .] they collect the funds **for those who** are in need.'

123 We illustrate this frequent and incidental co-occurrence with (4.viii):

(4.viii) [Korpus Malti 3.0; news218778]
l-prezz tal-gass ma jinbidilx ***għal minn tal-anqas***
DEF-price of:DEF-gas NEG 3SG.M.IPFV:RFL:change:NEG **for from of:DEF-least**
għaxar snin
ten year.PL
'[. . .] the price for gas does not change **for at least** ten years.'

The same can also be said as to the combination *għal-rigward*. Many problems arising in connection with (Π50) *rigward* 'concerning' have been discussed already.[124] The Korpus Malti 3.0 produces thirty-six hits for *għal-rigward* three of which fall outside the definition of PP. In (4.155), it is shown that the combination either can be followed directly by the complement NP or the linker (Π55) *ta'* 'of'. These options are in line with the competition of NOMINATIVE and GENITIVAL government of (Π50) *rigward* 'concerning'.

(4.155) *għal-rigward* combinations

(a) [Korpus Malti 3.0; parl10171]

Jidher	*li*	*hemm*	*proposta*	***[għal***	***rigward***
3SG.M.IPFV:appear	that	there	proposal	**[for**	**concerning**
ir- referendum]$_{PP}$					
DEF-referendum]					

'It seems that there is a proposal [**concerning** the referendum] [. . .].'

(b) [Korpus Malti 3.0; parl2425]

[Għal	***rigward***	***t'***	*Għawdex]*$_{PP}$	*l-Onor. Interpellant*		*huwa*
[for	**concerning**	**of**	Gozo]	DEF-honorable interpellant		3SG.M
mitlub	*jirriferi*		*din*	*il-mistoqsija*	*lill-Ministru*	
PART:ask	3SG.M.IPFV:refer		DEM:F:PROX	DEF-question	to:DEF-minister	
kkonċernat						
concern:PART						

'[**As to** Gozo], the honorable interpellant is asked to refer this question to the respective minister.'

The presence of (Π21) *għal* 'for' does not appear to have any effect on the semantics of the construction. From (4.153), it can be deduced that either (Π21) *għal* 'for' or (Π50) *rigward* 'concerning' alone could fulfil the same function as their combination.

(4.156) No second Π

(a) [Korpus Malti 3.0; news9864]

[Għal	*Għawdex]*$_{PP}$	*Joseph Muscat*	*qal*	*li*	*l-Partit*	*għandu*
[for	Gozo]	Joseph Muscat	say	that	DEF-party	have.IPFV:3SG.M

124 Cf. Section 4.1.2.2.1.2.1 (x) where the combinability of (Π21) *għal* 'for' and (Π50) *rigward* 'concerning' is mentioned as one of the potential alloforms of the latter.

pjan biex jinħolqu impjiegi ġodda
plan in_order_to 3.IPFV:RFL:create:PL job:PL new.PL
'[**As to** Gozo], Joseph Muscat said that the party has a plan in order to creat new jobs.'

(b) [Korpus Malti 3.0; news163074]
***[Rigward** Għawdex]$_{PP}$ Joe Vella Bonnici qal li spiss*
[**concerning** Gozo] Joe Vella Bonnici say that often
nitkellmu fuq l-Għawdxin bħala żvantaġġjati
1.IPFV:RFL:speak:PL on DEF-Gozitan:PL as disadvantaged:PL
'[**Concerning** Gozo], Joe Vella Bonnici said that we often talk about Gozitans as being disadvantaged.'

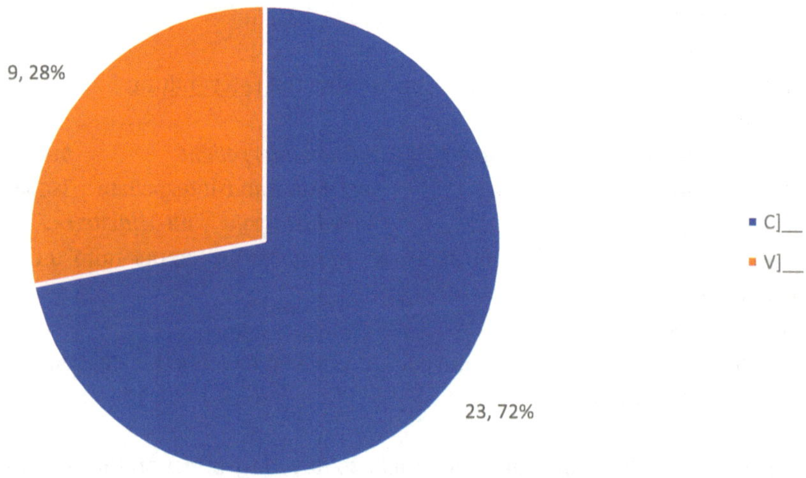

Figure 4.88: Phonological contexts of *għal-rigward* (tokens in the Korpus Malti 3.0).

There is thus no combination of meaning components to add up to a distinct constructional meaning. We learn from Figure 4.88 that the majority of *għal-rigward* co-occurrences is attested after a preceding consonant-final word (including fifteen sentence-initial cases).

The combination behaves similarly to *a rigward* in Figure 4.57. The similarity to *a rigward* is also discernible from the fact that there is no evidence of *għal-rigward* being followed by (Π21) *għal* 'for' (cf. Figure 4.58) but this possibility is unlikely in any event since it would involve the double occurrence of (Π21) *għal* 'for' in one and the same construction. On account of the above arguments, we cancel *għal-rigward* from Table 4.17 since it does not instantiate a genuine Π-Π combination.

To conclude this section, we cursorily look at three further cases which are attested only marginally in the Korpus Malti 3.0. In (4.157), (Π21) *għal* 'for' precedes (Π29) *lil* 'to' which functions as object marker.

(4.157) [Korpus Malti 3.0; parl4841]

Qed	*nirriferi*	***[għal***	***lil***	*Miss Hella Jean Bartolo]$_{PP}$*
PROG	1SG.IPFV:refer	**[for**	**to**	Miss Hella Jean Bartolo]

'I am referring [**to** Miss Hella Jean Bartolo] [. . .].'

The co-occurrence of these two Πs is attested only once in the Korpus Malti 3.0. Since no other Π_1 candidate combines with (Π29) *lil* 'to' we have refrained from registering this unique example with the other Π-Π combinations in Table 4.17. We assume that the combination of (Π21) *għal* 'for' and (Π29) *lil* 'to' in (4.157) results from a contamination of two constructions. These constructions are featured separately in (4.158).

(4.158) [Korpus Malti 3.0; news113321]

Irrid	*nirreferi*	***[lil***	*Matthew Mamo]$_{PP}$*	***[għal***	*minuti*
1SG.IPFV:want	1SG.IPFV:refer	**[to**	Matthew Mamo]$_{DO}$	**[for**	minute:PL
tas-seduta	*numru*	*G 19]$_{PP}$*			
of:DEF-session	number	G 19]$_{IO}$			

'I want to refer [Matthew Mamo] [**to** the minutes of session #G 19].'

There are two PPS, viz. one headed by (Π29) *lil* 'to' which introduces the direct object of the verb *irrefera* 'refer' and one headed by (Π21) *għal* 'for' that introduces what we for convenience term the indirect object. The direct object represents the participant who is referred to something. This something is represented by the indirect object. In (4.157), the speaker intends to refer the audience to *Miss Hella Jean Bartolo* so that (Π21) *għal* 'for' alone would have been sufficient. As a matter of fact, in terms of grammatical acceptability, there is no space for (Π29) *lil* 'to' in this context at all. A possible explanation for the presence of (Π29) *lil* 'to' is that it is usually used for direct as well as indirect objects with human referents. In a way, animacy is in conflict with the case frame of *irrefera* 'refer' and triggers the overuse of (Π29) *lil* 'to'. The superfluous use of (Π21) *għal* 'for' alongside other

Πs listed in (4.146) is occasionally attested in the Korpus Malti 3.0. We skip discussing these cases because we do not assume that they reflect any rule-governed behaviour.

In (4.159), two further cases which involve (Π41) *minn* 'from' as Π_1 are genuine cases of Π-Π combinations in the sense of Figure 4.85 although they might not appear to be promising candidates for this status on the surface.

(4.159) (Π41) *minn* 'from' as Π_1

(a) [Korpus Malti 3.0; news166640]

forsi	*dan*	*il-benefiċċju*	*jibda*	*jitgawda*
perhaps	DEM:M:PROX	DEF-benefit	3SG.M.IPFV:start	3SG.M.IPFV:RFL:enjoy
[minn	***[lejn***	*l-aħħar*	*tas-sena* $]_{PP}]_{PP}$	
[from	**[towards**	DEF-last	of:DEF-year]]	

'Perhaps this benefit begins to be enjoyed [**from** [**around** the end of the year]] [. . .].'

(b) [Korpus Malti 3.0; news170268]

il-Prim Ministru	*Ariel Sharon*	*ħabbar*	*r-riżenja*	*tiegħu*	
DEF-primeminister	Ariel Sharon	announce	def-resignation	of:3SG.M	
[minn	***[bħala***	*l-Kap*	*tal-Partit Likud* $]_{PP}]_{PP}$		
[from	**[as**	DEF-head	of:DEF-Likud party]]		

'The primeminister Ariel Sharon announced his resignation [**as** leader of the Likud party].'

In (4.156a), the semantic incompatibility of Source-oriented (Π41) *minn* 'from' and Goal-oriented (Π28) *lejn* 'towards' is no obstacle to combining the two Πs since the use of (Π28) *lejn* 'towards' with time expressions reflects the English usage of *towards* as *in towards the end of the year*. Metonymically, (Π28) *lejn* 'towards' indicates the REGION of the GROUND *l-aħħar tas-sena* 'end of the year' whereas (Π41) *minn* 'from' does what it is supposed to do, namely expressing DIRECTIONALITY (i.e. Source). A similar (albeit not absolutely identical) analysis can be applied to (4.159b). We interpret *l-Kap tal-Partit Likud* 'the leader of the Likud party' as GROUND with (Π6) *bħala* 'as' serving as expression of CONFIGURATION (without however specifying a particular REGION) – and once more (Π41) *minn* 'from' does what it is supposed to do, namely expressing DIRECTIONALITY.

4.1.3.3 Further combinations

The Π-Π combinations discussed in the foregoing sections do not exhaust all instances of two Maltese Πs occurring side by sid e in a given utterance. Those cases which have not already been focussed upon[125] are however not of the same kind as the Π-Π combinations with initial (Π41) *minn* 'from', (Π21) *għal* 'for', or (Π51) *sa* 'till'.[126] In what follows here we summarily look at other relatively uncommon prepositional sequences which do not reflect the formula presented in Figure 4.85.

We must distinguish incidental co-occurrences from systematic ones. To this end, we look at binary Π-sequences whose second Π is (Π29) *lil* 'to'. This Π is representative of the bulk of the members of BLOMP 2.0 the behaviour of which thus needs not be documented in all details. We start with the incidental cases.[127] (4.160) features the sequence of (Π11) *dwar* 'about' and (Π29) *lil* 'to'. No bracketing is applied in this example.

(4.160) [Korpus Malti 3.0; news103070]

Wara	*l-membri*	*tal-kumitat*	*kienu*	*jieħdu*
after	DEF-member:PL	of:DEF-committee	be:3PL.PFV	3.IPFV:take:PL

d-deċiżjoni	***dwar***	***lil***	*min*	*jagħtu*	*l-kuntratt*
DEF-decision	**about**	**to**	who	3.IPFV:give:PL	DEF-contract

'Afterwards, the members of the committee took the decision **about** whom they would give the contract.'

The two Πs depend on different heads. There is a clausal boundary between them. (Π11) *dwar* 'about' belongs to *deċiżjoni* (*deċiżjoni dwar* 'a decision about') and forms part of the predicate *kienu jieħdu d-deċiżjoni* 'they took the decision'. In contrast, (Π29) *lil* 'to' marks the indirect object (recipient) of the ditransitive verb

125 Cf. Sections 4.1.3.1–4.1.3.2.

126 Sequences of Πs which involve (Π55) *ta'* 'of' either as first or second Π are discussed separately in Section 4.2.2.

127 Since several Πs can be used as conjunctions, there is a plethora of cases in which a supposed sequence of Πs turns out to consist of an initial conjunction which has scope over the entire clause and a Π which only has scope over its complement. In (4.ix), (Π40) *minkejja* 'in spite of' is the conjunction from which (Π29) *lil* 'to' is independent (the *lil*-PP depends on the verb *trattah* 'he treated him' whose object clitic *-h* is coreferential with the complement of the *lil*-PP):

(4.ix) [Korpus Malti 3.0; news217358]

Lawrence Gonzi	*semma*	*kif*	***minkejja***	***lil***	*Anġlu Farrugia*	*Joseph Muscat*
Lawrence Gonzi	mention	how	**in_spite_of**	**to**	Anġlu Farrugia	Joseph Muscat

trattah	*bi*	*qtil*	*politiku*	*Muscat*	*ma*	*mexiex*	*bl-istess*
treat:3SG.M	with	murder	political	Muscat	NEG	walk:NEG	with:DEF-same

mod	*ma'*	*Toni Abela*
manner	with	Toni Abela

'Lawrence Gonzi mentioned how Joseph Muscat, **although** he treated Anġlu Farrugia with political murder, did not act in the same manner with Toni Abela.'

For obvious reasons, there is no need to discuss these cases in any detail here or elsewhere in this study.

ta 'give' which is the predicate nucleus (*jagħtu* 'they give') of the second clause. The two Πs cooperate neither syntactically nor semantically. This scenario recurs across a variety of other incidental Π-sequences a selection of which is given in Table 4.18. The three examples are taken from the Korpus Malti 3.0 and are mostly hapaxes.

Table 4.18: Fake Π-Π combinations.

source	example / translation	initial Π
parl348	*Kemm-il fabbrika ħadet lura l-MDC* ***mingħand lil*** *min kienet kritha?* 'How many factories has the MDC taken back **from** those it rented them **to**?'	(Π38) *mingħand* 'from s.o.'
news33115	*f' diversi entitajiet qegħdin jitpoġġew nies li mhumiex kompetenti u li qegħdin hemm iktar* ***minħabba lil*** *min jafu milli għal kemm jafu* '[. . .] in several corporations people are being installed who are not competent and who are there **because of** whom they know rather than because of how much they know.'	(Π39) *minħabba* 'on account of'
news91019	*konna nafu li f' pajjiżna tirnexxi* ***skont lil*** *min taf u mhux skont kemm taf* '[. . .] we knew that in our country you succeed **according to** whom you know and not according to how much you know.'	(Π53) *skont* 'according to'

In (4.161), we present an example of a combination which can be classified as systematic in the sense that both Πs are in a relation to the same head but independently of each other. We again refrain from using brackets in the example.

(4.161) [Korpus Malti 3.0; news160604]

Azzopardi	*kien*	*ċempel*	*lil*	*diversi*	*persuni*	***apparti***	***lil***	*Camilleri*
Azzopardi	be	phone	to	diverse:PL	person:PL	**apart_from**	**to**	Camilleri

'Azzopardi had phoned several persons **except** Camilleri.'

(Π29) *lil* 'to' serves as the direct-object marker for the verb *ċempel* 'phone'. The marker's presence is required with each NP in object function. In contrast, (Π1) *apparti* 'apart from' is in no way connected to object-marking as such. Its sole function is to signal that a participant is excluded from the situation. (Π1) *apparti* 'apart from' and (Π29) *lil* 'to' are thus syntactically and semantically unrelated to each other. As the selection of examples in Table 4.19 shows, this unrelatedness is frequently encountered with Πs that express exclusion, inclusion, or replacement. As before, the examples are taken from the Korpus Malti 3.0.

Table 4.19: Sequences of independent Πs involving (Π29) *lil* 'to'.

source	example / translation	initial Π
news36454	*L-avukat qal li dan juri li din ma rat xejn* ***għajr lil*** *Farrugia mal-art* 'The lawyer said that this shows that she saw nothing **but** Farrugia on the floor.'	(Π20) *għajr* 'except'
academic145	*il-poeta ma jsemmi l-ebda kavallier* ***ħlief lil*** *La Valette* '[. . .] the poet mentions no knight **except** La Valette.'	(Π24) *ħlief* 'except'
literature3	*Iddeċieda li jibgħat lil kulħadd id-dar* ***minbarra lil*** *Mandy u Wendy* 'He decided to send everybody home except Mandy and Wendy [. . .].'	(Π34) *minbarra* 'except'
news136781	*d-delegati Laburisti għażlu lil Dr Muscat* ***minflok lil*** *Dr George Abela* '[. . .] the Labour delegates elected Dr Muscat **instead of** Dr George Abela [. . .].'	(Π35) *minflok* 'instead of'
news155006	*Il-pulizija akkużat lil xi Laburisti* ***inkluż lil*** *Rokku Abdilla* 'The police accused several Labourites **including** Rokku Abdilla [. . .].'	(Π25) *inkluż* 'including'

In none of the examples do (Π29) *lil* 'to' and the preceding Π relate to each other semantically or syntactically. These and similar sequences of Πs thus cannot be lumped together with the previously discussed Π-Π combinations where the expression of DIRECTIONALITY and CONFIGURATION is distributed over the two Πs of the construction. In sum, functionally motivated Π-Π combinations are categorically different from incidental or systematic Π-sequences. At the same time, the functional motivation of the genuine Π-Π combinations does not render these combinations unitary (complex) Πs. We emphasise that Π-Π combinations are combinations of distinct Πs and not compound Πs like the lexical ised univerbations nor are they cases of prepositional government.

4.1.4 EQUI-P-deletion

The phenomenon to be studied in this section is no structural monopoly of Maltese since it is attested in many languages worldwide where it goes under the name of EQUI-P-deletion. So far, EQUI-P-deletion has not received much attention among typologists. That is why the Maltese data might serve as a starting-point for future comparative studies. EQUI-P-deletion is a kind of ellipsis which occurs under coordination and in the Maltese case, it affects the second of two identical Πs thus realising what Haspelmath (2007a: 39) terms analipsis or forward ellipsis. Analipsis applies if the second Π is not real ised phonologically whereas the first one is in place. Previous research focussing on the Maltese situation largely corroborates Borg/Azzopardi-Alexander (1997: 87) who claim that

> when two structurally identical prepositional phrases with adverbial function are coordinated, the second preposition can be omitted. [. . .] If the prepositional phrases are not structurally identical, neither of the prepositions is omitted.

It remains to be seen what exactly is meant by structural identity. For the purpose of this section, we focus on the two coordinative construction types (= COCO) given in Figure 4.89.

Conjunction: $[[\Pi_i \text{ NP}]_{PP}$ *u* $[(\Pi_i) \text{ NP}]_{PP}]_{COCO}$

Disjunction: $[[\Pi_i \text{ NP}]_{PP}$ *jew* $[(\Pi_i) \text{ NP}]_{PP}]_{COCO}$

Figure 4.89: Conjunctive and disjunctive constructions.

With exclusive reference to COCOs of the conjunctive kind, Stolz/Arens (2017: 121) observe that "Maltese allows for the coexistence of constructions with two overt identical PREPs and constructions which make do with only one PREP [original small caps]". In continuation of their research which she extends to disjunctive COCOs as well, Vorholt (2022b, 2023b–d) shows that the possibility of a Π being affected or not by EQUI-P-deletion is not entirely random but is influenced by a variety of semantic, pragmatic, syntactic, and stylistic factors whose hierarchy and interaction have not yet been explored satisfactorily. What particularly caught the linguist's attention is the co-existence of cases of coordination with and without EQUI-P-deletion although all structural conditions seem to be the same. This intriguing situation is captured by the two sentences in (4.162). The coordinating conjunction *u* 'and' is highlighted in yellow.

(4.162) EQUI-P-deletion applies vs is blocked

(a) [Korpus Malti 3.0; literature4]

*292 **[quddiem** grupp minn sħabna]$_{PP}$ u **[quddiem***
292 **[in_front_of** group from friend.PL:1PL] and **[in_front_of**

il-Brother jew lecturer li kien jgħallimna dak
DEF-Brother or lecturer that be 3SG.M.IPFV:teach:1PL DEM:M:DIST

is-suġġett]$_{PP}$
DEF-subject]

'292 [pupils] [**in front of** a group of our friends] and [(**in front of**) the Brother or lecturer who taught this subject to us].'

(b) [Korpus Malti 3.0; news61500]

Din	*id-darba*	*ħassejtni*	*tqil*	*biex*	*nitkellem*
DEM:F:PROX	DEF-time	feel:1SG.PFV:1SG	heavy	in_order_to	1SG.IPFV:speak

[quddiem	*ommi*	*u*	***(Ø)***	*missieri]*$_{PP}$	*fuq*	*il-veru*	*jien*
[in_front_of	mother:1SG	and	**(Ø)**	father:1SG]	on	DEF-true	1SG

'This time I felt it difficult to speak [**in front of** my mother and (**Ø**) my father] about the real me.'

In (4.162a), each of the two conjuncts hosts (Π49) *quddiem* 'in front of' so that we can speak of two distinct but coordinated *quddiem*-PPs. This analysis is not felicitous in the case of (4.162b) because (Π49) *quddiem* 'in front of' occurs only once and the second slot in which the same Π could show up remains empty. Arguably, (4.162b) could alternatively be interpreted as a single *quddiem*-PP with two coordinated nouns as complement in lieu of assuming the coordination of two *quddiem*-PPs the second of which undergoes EQUI-P-deletion, i.e., it loses its separate head. In this case, we are facing what is called group-formation by semanticists (Landmann 1989, Lasersohn 1990) with the coordinated nouns *ommi u missieri* 'my mother and my father' being replaceable with *il-ġenituri tiegħi* 'my parents' since mother and father constitute a kind of collective unit whose members happen to be present at the same place at the same time. Interestingly, spatial and temporal co-presence also applies to the referents of the complements of the two *quddiem*-PPs in (4.162a) because the pupils and the instructor are participants in one and the same situation. However, these participants do not seem to form as tight a unit as mother and father which are typical and cross-linguistically well-established instances of co-compounds and natural coordination (Wälchli 2005: 44). Thus, the difference between (4.162a) and (4.162b) in terms of the presence / absence of the second (Π49) *quddiem* 'in front of' might be explicable with reference to subtle semantic (or conceptual) distinctions between natural and occasional groups.

The second type of coordination to account for is *or*-co-ordination (aka disjunction (Haspelmath 2007a: 1–2)) which in Maltese involves the conjunction *jew* 'or'. The chunk of text reproduced in (4.163) contains a sequence of three *dwar*-PPs none of which is subject to EQUI-P-deletion since in all of the three cases (Π11) *dwar* 'about' is phonologically realised. Another interesting aspect of this example is that the second *dwar*-PP involves a complement with two coordinated GENITIVAL attributes headed by (Π55) *ta'* 'of'. The *or*-coordination notwithstanding, EQUI-P-deletion fails to affect the second *ta'*-PP.

(4.163) [Korpus Malti 3.0; european1951]

dak	*li*	*għandu*	*x'*	*jaqsam*	*ma'*
DEM:M:DIST	that	have.IPFV:3SG.M	what	3SG.M.IPFV:concern	with

kwalunkwe deċiżjoni ***[dwar*** *id-dħul]PP* *jew* ***[dwar*** *ir-rifjut*
whatever decision **[about** DEF-entry] or **[about** DEF-refusal
tal-ħruġ jew tat-tiġġdid tal-permess tar-residenza]PP
of:DEF-exit or of:DEF:renewal of:DEF-permit of:DEF-residence]
jew ***[dwar*** *l-ordni għat-tkeċċija mit-territorju]PP*
or **[about** DEF-order for:DEF-expulsion from:DEF-territory]
'[...] that what has to do with whatever decision [**about** the entry] or [(**about**) the refusal of departure or of the renewal of the permit of residence] or [**about** the order of expulsion from the territory] [...].'

Disjunction seems to be less favourable to EQUI-P-deletion than *and*-conjunction. As is shown in (4.164b), it is nevertheless possible to drop an identical Π from a series of PPS. If the disjunctive coordination is strictly binary, however, the blockage of EQUI-P-deletion seems to be relatively common as suggested by the presence of the second (Π9) *bla* 'without' in (4.164a).

(4.164) Disjunction
(a) [Korpus Malti 3.0; news36866]
uħud mit-tfal imorru l- iskola ***[bla*** *ikel]PP*
some from:DEF-child.PL 3.IPFV:go:PL DEF-school **[without** food]
jew ***[bla*** *uniformi]PP*
or **[without** uniform:PL]
'[. . .] some of the children go to school [**without** food] or [(**without**) uniforms] [. . .].'
(b) [Korpus Malti 3.0; law1240]
jmut ***[bla*** *tfal]PP* *jew* ***[bla*** *mara* jew ***(Ø)***
3SG.M.IPFV:die **[without** child.PL] or **[without** woman or **(Ø)**
raġel]PP
man]
'[. . .] he dies [**without** children] or [**without** wife or (Ø) husband].'

Disjunction does not strike us as particularly well suited for group-formation – at least not as straightforwardly as *and*-conjunction – since it is a strategy to separate concepts from each other which otherwise might form a complex unit. However, (4.164b) proves that disjunction too gives evidence of the effects of group-formation. The second *bla*-PP involves a coordinated complement consisting of the common nouns *mara* 'woman' and *raġel* 'man'. The *and*-coordination *mara u raġel* 'woman and man = wife and husband' is a translation equivalent of English *spouses*. A single lexeme translating English *spouse(s)* into Maltese does not exist. The case of *mara u raġel* 'spouses' is another example of a cross-linguistically common co-compound type (Wälchli 2005: 138). This co-compound allows for the replacement of the conjunction *u* 'and' with the disjunction *jew* 'or' so that in (4.164b), *bla mara jew raġel* could also be translated as 'without a spouse'. Thus, group-formation is a strong factor to be accounted for when we want to understand why and when EQUI-P-deletion applies in Maltese (Stolz/Ahrens 2017: 145).

Whatever the adequate semantic interpretation of the above differences might be, there seem to be other reasons which additionally condition the application of EQUI-P-deletion or call for the presence of the second Π. On the basis of a subset of twenty-one members of what later became BLOMP 2.0, Stolz/Ahrens (2017: 136–137) identify the following factors which play a role when it comes to dropping or keeping a second Π.

(i) **semantics**: Do the complements in the coco form a collective unit or two separate individual entities? Collectives favour EQUI-P-deletion, cf. examples in (4.162)–(4.164).
(ii) **morpheme class**: Are both complements free NPs or is at least one of them a bound pronoun? Πs which are inflected for person tend to block EQUI-P-deletion as shown in (4.165).
(iii) **noun class**: Is the complement noun a place name or a common noun? Place names are prone to EQUI-P-deletion under certain conditions to be spelt out below.[128]
(iv) **syntactic weight**: Is the complement syntactically heavy, i.e., does it involve several syntactic words, or is it syntactically light, i.e., (preferably) a single syntactic word? Heavy weight seems to be detrimental to the application of EQUI-P-deletion as can be gathered from (4.166).
(v) **number of PPs**: Is the coco binary or do we face a chain of three or more coordinated PPs? The longer the chain gets the more likely it becomes that none of the non-initial Πs is omitted as can be seen in (4.167).

In (4.165a), the first conjunct comes in the shape of an inflected Π. In coordination with PPs which take a lexical NP as complement, inflected Πs generally tend to occupy the leftmost slot of the coco. If both conjuncts are PPs with lexical NPs as complements – as in (4.165b) – EQUI-P-deletion is free to apply.

(4.165) Morpheme class
(a) two Πs [Friggieri 2000: 119]

Il-ħarsa	*tan-nies*	*kien*	*jaraha*	*verdett*	*qalil*
DEF-look	of:DEF-people	be	3SG.M.IPFV:see:3SG.F	verdict	harsh

128 Cf. Section 4.3.3.2.

[fuqu]$_{PP}$ u [fuq familtu kollha]$_{PP}$
[**on**:3SG.M] and [**on** family:POSS.F:3SG.M all:3SG.F]
'He considered the look people gave to be harsh judgment [**about** him] and [(**about**) his entire family].'

(b) one Π [Friggieri 2000: 197]
L-imwiet tat-trabi kienu komuni u kien hemm
DEF-death.PL of-infant.PL be:3PL.PFV common:PL and be there
min iqishom għajb [fuq l-omm u (Ø)
who 3SG.M.IPFV:measure:3PL disgrace [**on** DEF-mother and (**Ø**)
l-familja]$_{PP}$
DEF-family]
'The deaths of infants were common and there were those who considered them to be a disgrace [**for** the mother and (**Ø**) the family].'

The differential behaviour of the second Π in contexts similar to those documented in (4.165) is not without (infrequent) exception but it reflects a very strong trend. We therefore feel entitled to postulate the split rule in Figure 4.90 according to which EQUI-P-deletion does not normally apply if Π_1 hosts a pronominal suffix.

$$\Pi_2 \rightarrow \begin{cases} \text{optional} & / \quad \Pi_1 = \text{uninflected} \\ \text{realised} & / \quad \Pi_1 = \text{inflected} \end{cases}$$

Figure 4.90: Split (preference) rule (morpheme class-induced behaviour of Π_2).

What is illustrated in (4.166) equally reflects a rather strong tendency. EQUI-P-deletion affects coordinated PPs with simple complements more easily than coordinated PPs whose complements are syntactically heavy because the nouns are accompanied by adjectival attributes so that the NP consists of several syntactic words.

(4.166) Syntactic weight
(a) two Πs [Friggieri 2000: 43]
[b' nuċċali żgħir żgħir u tond]$_{PP}$ u [b' għajnejn
[**with** spectacles small small and round] and [**with** eye:PL
ċassi imħassbin]$_{PP}$
staring:PL thoughtful:PL]
'[. . .] [**with** really small and round spectacles] and [(**with**) thoughtful, staring eyes] [. . .].'

(b) one Π [Friggieri 2000: 42]

It-tielet	*siġġu*	*[bil-kikkra*	*u*	*(Ø)*	*l-plattina]$_{PP}$*
DEF-third	chair	[**with**:DEF-cup	and	(Ø)	DEF-saucer]

quddiemu	*kien*	*għadu*	*vojt*
in_front_of:3SG.M	be	still:3SG.M	empty

'The third chiar [**with** the cup and (Ø) the saucer] in front of it was still empty.'

Deviations from the pattern given in Figure 4.91 are relatively numerous in the case of EQUI-P-deletion with syntactically light complements. In contrast, the heavier a complement NP gets the more likely it is that the second Π is phonologically realised. Syntactic heaviness of the first conjunct causes the two Πs to be separated from each other by a sizable number of syntactic words so that the repetition of the Π_1 as Π_2 makes sense functionally to facilitate the parsing of the utterance by the hearer.

Π_2 → realised / Π_1 $[N\ X\ (Y\ (Z))]_{COMPLEMENT}$ Π_2

Figure 4.91: Heaviness triggering realisation of Π_2.

The application of EQUI-P-deletion in (4.166b) is not only licensed by the syntactic lightness of the complements but also – and perhaps much more so – by the co-compound status of *kikkra u platina* 'cup and saucer' which might be understood to as translation of English *cover* or *place* with reference to a set table (similar to the spoon-and-plate cases cited in Wälchli (2005: 141–142)).

In (4.167a) a sequence of four *bi*-PPs constitutes a lengthy enumeration whose final component is added to the list by way of using the copulative conjunction *u* 'and'. In (4.167b) however, there are only two *bi*-PPs the second of which is subject to EQUI-P-deletion.

(4.167) Number of PPs

(a) no EQUI-P [Friggieri 2000: 220]

Issa	*laħqu*	*kkuntentaw*	*ruħhom*	*[bir-razzett]$_{PP}$*
now	reach:3PL.PFV	be_content:3PL.PFV	self:3PL	[**with**:DEF-farmhouse]

[bil-misraħ	*ta'*	*quddiemu]$_{PP}$*	*[bil-kappella]$_{PP}$*	*u*
[**with**:DEF-square	of	in_front_of:3SG.M]	[**with**:DEF-chapel]	and

[bil-bihem]$_{PP}$
[**with**:DEF-cattle]

'Now they had come so far to content themselves [**with** the farmhouse], [(**with**) the square in front of it], [(**with**) the chapel] and [(**with**) the cattle] [. . .].'

(b) one Π [Friggieri 2000: 144]

Ħassitha *qisha* *Ali Baba* *fl-għar* *imdawwar*
feel:3SG.F.PFV like:3SG.F Ali Baba in:DEF-cave PART:surround
*[**b**id-dehbijiet* *u* *(Ø)* *t-teżori]*$_{PP}$
[**with**:DEF-gold:PL and (Ø) DEF-treasure:PL]
'She felt like Ali Baba surrounded [**by** the gold and (Ø) the treasures].'

Very extended sequences of PPs headed by identical Πs can be compared to the case of syntactic heaviness because with each added PP the distance between the initial Π and the last PP increases. The preceding PP is of course adjacent but the size of the COCO seems to impose limits on the applicability of EQUI-P-deletion. In Figure 4.92 the third strong tendency is captured although there are again exceptions to the rule.

Π_{1+n} → realised / conjuncts = n ≥ 3

Figure 4.92: Number of conjuncts favouring realisation of non-initial Πs.

Not only do we find exceptions but it is often the case that several factors conspire to block EQUI-P-deletion or to trigger it. The strongest force among the above factors is probably the semantic distinction between collectives and individuals. Even in the case of (4.167b) one might want to argue that Ali Baba's riches form a kind of conceptual unit – his gold and treasures belong together like say, cup and saucer, at least in the context of the fairy story in which Ali Baba is the protagonist.

For their small selection of Maltese Πs identified in (4.168), Stolz/Ahrens (2017) observed that all Πs which are attested under coordination with themselves allow both for EQUI-P-deletion and the presence of the second Π albeit to different extent and except (Π29) *lil* 'to', it holds that EQUI-P-deletion is (often by far) more common than the preservation of the second Π. Can these observations be generalised over BLOMP 2.0 in its entirety?

(4.168) Πs studied by Stolz/Ahrens (2017)
(Π4) *bejn* 'between'; (Π5) *bħal* 'like'; (Π7) *bi* 'with'; (Π9) *bla* 'without'; (Π11) *dwar* 'about'; (Π15) *fi* 'in'; (Π17) *fuq* 'on'; (Π19) *ġo* 'in'; (Π21) *għal* 'for'; (Π22) *għand* 'at s.o.'s place'; (Π28) *lejn* 'towards'; (Π29) *lil* 'to'; (Π30) *ma'* 'with'; (Π36) *mingħajr* 'without'; (Π41) *minn* 'from'; (Π46) *qabel* 'before'; (Π49) *quddiem* 'in front of'; (Π51) *sa* 'till'; (Π55) *ta'* 'of'; (Π56) *taħt* 'under'; (Π60) *wara* 'after'

As of now, BLOMP 2.0 counts thirty-nine Πs more than are mentioned in (4.168). If we omit the four problematic candidates from the inventory, there are the thirty-five Πs in (4.169) whose behaviour under coordination urgently calls for being described.

(4.169) Πs yet to be studied as to their behaviour under coordination
(Π1) *apparti* 'apart from'; (Π2) *a skapitu* 'at the expense of'; (Π3) *barra* 'outside'; (Π6) *bħala* 'as'; (Π8) *biswit* 'facing'; (Π10) *daqs* 'equal to'; (Π12) *faċċata* 'opposite'; (Π13) *favur* 'in favour of'; (Π14) *fejn* 'near'; (Π16) *fost* 'amongst'; (Π18) *ġewwa* 'inside'; (Π20) *għajr* 'except'; (Π23) *ħdejn* 'beside'; (Π24) *ħlief* 'except'; (Π25) *inkluż* 'including'; (Π27) *kontra* 'against'; (Π31) *madwar* 'around'; (Π32) *maġenb* 'close to'; (Π33) *matul* 'during'; (Π34) *minbarra* 'except'; (Π35) *minflok* 'instead of'; (Π37) *mingħala* 'in s.o.'s opinion'; (Π38) *mingħand* 'from s.o.'; (Π39) *minħabba* 'on account of'; (Π40) *minkejja* 'in spite of'; (Π44) *oltre* 'beyond'; (Π45) *permezz* 'by means of'; (Π47) *qalb* 'amidst'; (Π48) *qrib* 'near'; (Π50) *rigward* 'concerning'; (Π52) *sforz* 'thanks to'; (Π53) *skont* 'according to'; (Π57) *versu* 'towards'; (Π58) *viċin* 'near'; (Π59) *waqt* 'at the time of'

A further limitation of Stolz/Ahrens's (2017) pioneering exploration is that disjunction has not been taken account of. Nevertheless, the authors observe that there are striking differences between conventionally published literary sources and the Korpus Malti 3.0 as to the numerical dominance of EQUI-P-deletion over the retention of Π_2 with the latter being more strongly represented in contemporary Maltese prose. Accordingly, there is ample opportunity for follow-up studies to fill the remaining gaps. Vorholt (2022b) seizes this opportunity to complete the picture of the systematicity of EQUI-P-deletion in Maltese not only for conjunction but also for disjunction. In contrast to her predecessors, however, she exclusively looks at data found in the Korpus Malti 3.0 but differentiates her findings in correlation to the genres which are distinguished in the Korpus Malti 3.0.[129]

Vorholt (2022b) takes account of fifty-four members of BLOMP 2.0 including those listed in (4.169). An initial search in the electronic resource revealed that fifty-one Πs occurred in copulative coordination with themselves whereas only forty-eight Πs were attested in disjunctive constructions involving two (or more) identically headed PPs. The average share of EQUI-P-deletion Πs differs between *and*-conjunction and *or*-disjunction although EQUI-P-deletion is by far the preferred

129 This constitutes part of her PhD-thesis which aims to identify, describe and evaluate the factors influencing EQUI-P-deletion in Maltese.

strategy with both kinds of coordination. The boxplot in Figure 4.93 shows that the average of EQUI-P-deletion per Π amounts to 86% for disjunction but rises to 91% for conjunction.

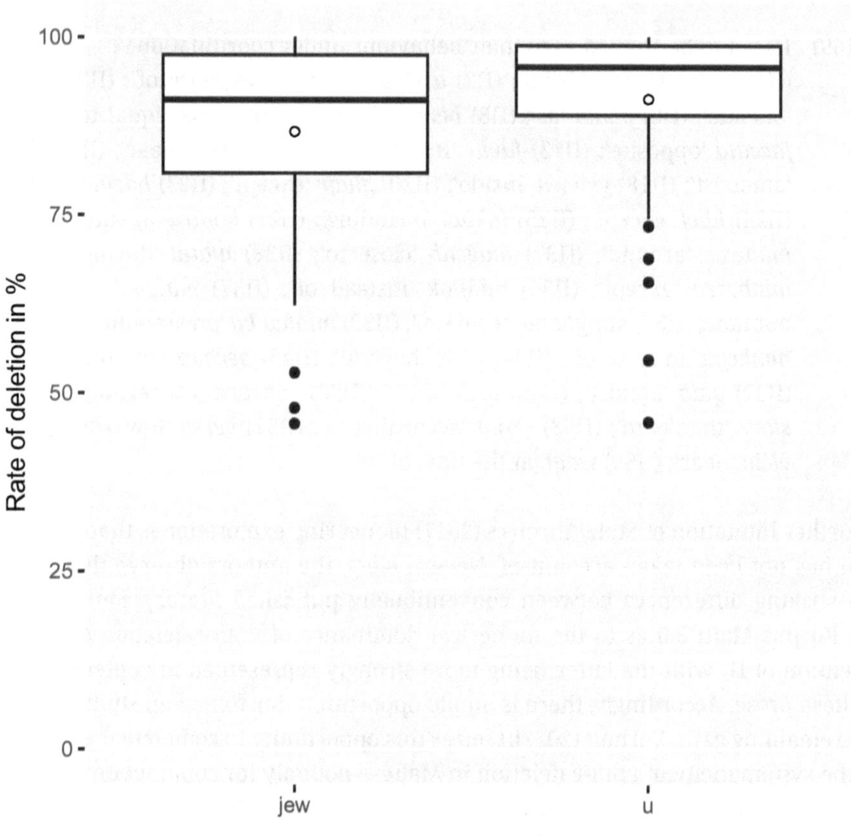

Figure 4.93: Shares of EQUI-P-deletion per Π in conjunctive/disjunctive constructions.

As we will see shortly, the individual Πs do not behave homogeneously. Figure 4.94 surveys the results for the competition of applied and blocked EQUI-P-deletion in the Korpus Malti 3.0. The figure accounts for *and*-coordination only. The continuum that results from the quantitative evaluation largely resembles that gained on the basis of a much smaller subset of Πs in Stolz/Ahrens (2017: 139–140 and 143). The five Πs at the bottom end of the scale, i.e., those which display relatively low percentages of EQUI-P-deletion are the same in both studies, namely (starting from the bottom-line) (Π29) *lil* 'to', (Π9) *bla* 'without', (Π15) *fi* 'in', (Π7) *bi* 'with', and (Π36) *mingħajr* 'without' with (Π29) *lil* 'to' being the only Π for which EQUI-P-deletion is

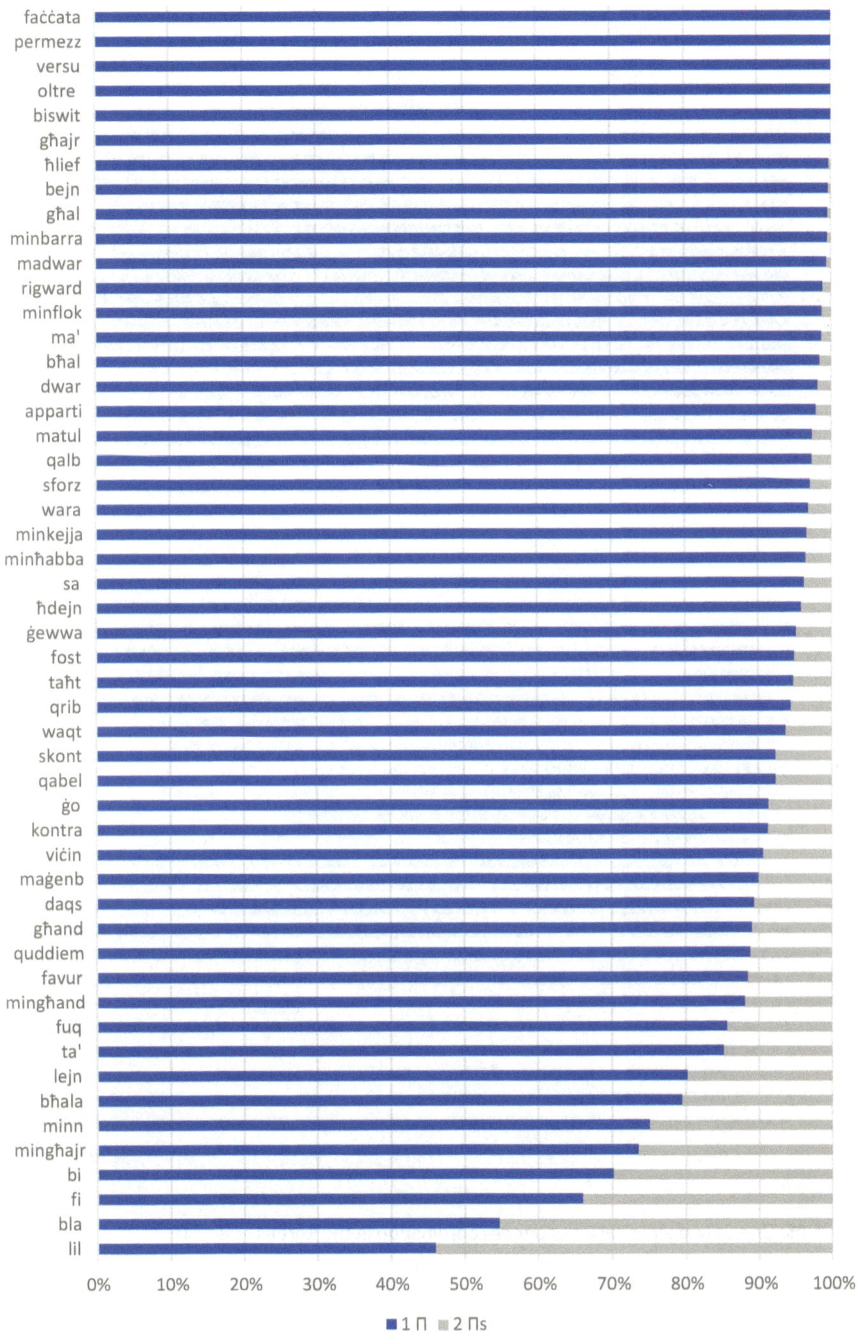

Figure 4.94: One vs two identical Пs under *and*-coordination in the Korpus Malti 3.0.

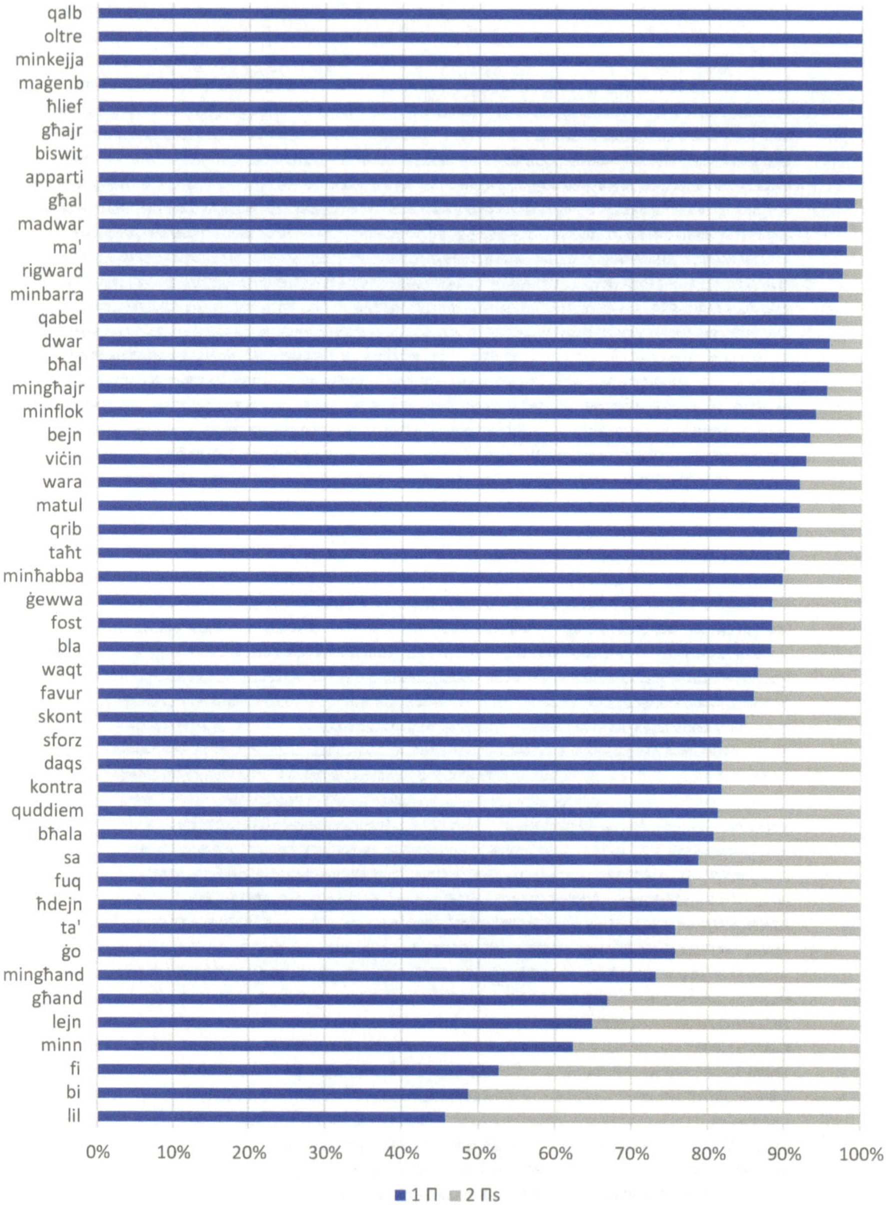

Figure 4.95: One vs two identical Πs under *or*-coordination in the Korpus Malti 3.0.

not the majority option. The most striking differences between Stolz/Ahrens (2017) and Vorholt (2022b) involve (Π21) *għal* 'for' and (Π30) *ma'* 'with' whose shares of Π_2-retention diminish considerably to approximate zero in the more recent study. Both studies agree again as to the almost complete absence of Π_2-retention with (Π4) *bejn* 'between' in the Korpus Malti 3.0 which strikingly contrasts with the frequent presence of the Π_2 in coordinated *bejn*-PPs in Maltese literature.

Interesting differences come to the fore when we compare the above results for *and*-coordination for those which are connected to *or*-coordination. Figure 4.95 shows among other things that there are two Πs which favour Π_2-retention over EQUI-P-deletion, namely the expected (Π29) *lil* 'to' and (rather unexpectedly also) (Π7) *bi* 'with'.

From afar the diagrams in Figures 4.94 and 4.95 might look too diverse to lend themselves to easy comparison. However, if we zoom in on the five positions at the top and the five positions at the bottom of the scales we notice that there are remarkable similarities. To this end Table 4.20 is helpful. Note that we do not go by ranks but only by listing, meaning: some Πs which have identical turnouts with those mentioned in Table 4.20 are not included in the comparison.

Table 4.20: Top and bottom five compared for conjunction and disjunction.

position	conjunction	disjunction
TOP	(Π20) *għajr* 'except'	(Π1) *apparti* 'apart from'
	(Π8) *biswit* 'facing'	(Π8) *biswit* 'facing'
	(Π44) *oltre* 'beyond'	(Π20) *għajr* 'except'
	(Π57) *versu* 'towards'	(Π24) *ħlief* 'except'
	(Π45) *permezz* 'by means of'	(Π32) *maġenb* 'close to'
BOTTOM	(Π36) *mingħajr* 'without'	(Π28) *lejn* 'towards'
	(Π7) *bi* 'with'	(Π41) *minn* 'from'
	(Π15) *fi* 'in'	(Π15) *fi* 'in'
	(Π9) *bla* 'without'	(Π7) *bi* 'with'
	(Π29) *lil* 'to'	(Π29) *lil* 'to'

Πs which show up both with conjunction and disjunction are marked out in colour. What needs to be stated right away is that there are correspondences between conjunction and disjunction whereas there are none between top and bottom, i.e., no Π with a top position for conjunction is registered in the bottom section of disjunction and vice versa. The outstanding role of (Π29) *lil* 'to' is reflected by its homogeneous behaviour with both kinds of coordination insofar as it always prefers Π_2-retention over EQUI-P-deletion. In the bottom section of both *and*-coordination and *or*-coordination, we also find (Π7) *bi* 'with' and (Π15) *fi* 'in' whose parallel behaviour fits in very well with their constituting a morphological class of their own.[130] It further intrigues us that the synonymous Πs (Π36) *mingħajr* 'without' and (Π9) *bla* 'without' (which are antonyms of (Π7) *bi* 'with') belong to the group of Πs which occupy the five positions at the bottom for conjunction. Looking towards the top-end of the scales, we find (Π20) *għajr* 'except' and (Π8) *biswit* 'facing' on both sides of the divide between the one and the other kind of coordination. Interestingly, seven of the ten top positions (= five for conjunction and three for disjunction) host polysyllabic Πs. At the bottom, however, nine of the ten positions feature monosyllabic Πs. Two further aspects are of interest. On the one hand, four of the ten top positions are occupied by Πs whose meaning is associated with exception, namely (Π1) *apparti* 'apart from', (Π24) *ħlief* 'except', and twice (Π20) *għajr* 'except'. To this group of four (Π44) *oltre* 'beyond' could be added as the fifth member on account of its secondary meaning. On the other hand, four positions at the top go to Italo-Romance Πs whereas there are only Semitic Πs at the bottom. The high-ranking Italo-Romance Πs are (Π1) *apparti* 'apart from', (Π44) *oltre* 'beyond', (Π57) *versu* 'towards', and (Π45) *permezz* 'by means of'. The distribution of the Πs over positions and types of coordination is thus not entirely random.

There are also pronounced differences with regard to the application of EQUI-P-deletion for one and the same Π across text genres. Figure 4.96 illustrates this with the shares of EQUI-P-deletion of (Π9) *bla* 'without' in seven distinct genres included in the Korpus Malti 3.0. The quantities exclusively refer to tokens of *and*-coordination.

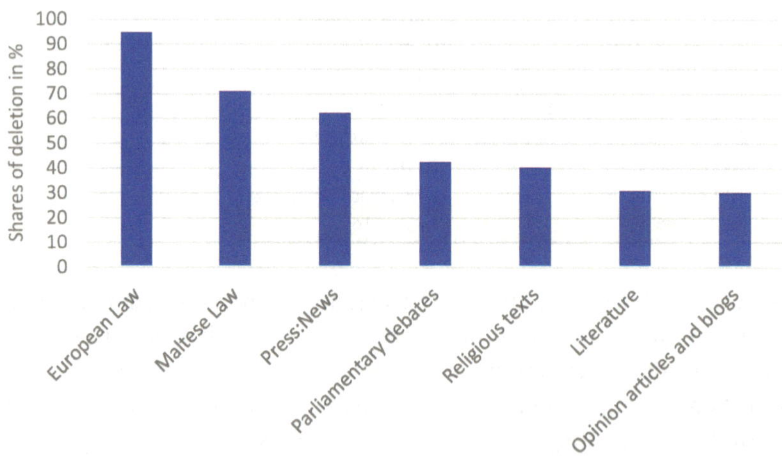

Figure 4.96: Shares of EQUI-P-deletion of (Π9) *bla* 'without' across genres in the Korpus Malti 3.0.

130 Cf. Sections 3.1.2.3.2 on Class II and 4.1.2.2.2.2.3 on ACCUSATIVE government.

The deletion of the second (Π9) *bla* 'without' is the preferred option only in three genres, namely European Law, Maltese Law, and Press News. Most remarkably, EQUI-P-deletion is almost mandatory in texts relating to European Law whereas especially in Literature and Opinion Articles / Blogs Π_2-retention ousts EQUI-P-deletion by a ratio of 2.3-to-1. Even between those genres which favour EQUI-P-deletion, the differences can be very pronounced as comes to the fore in Figure 4.97.

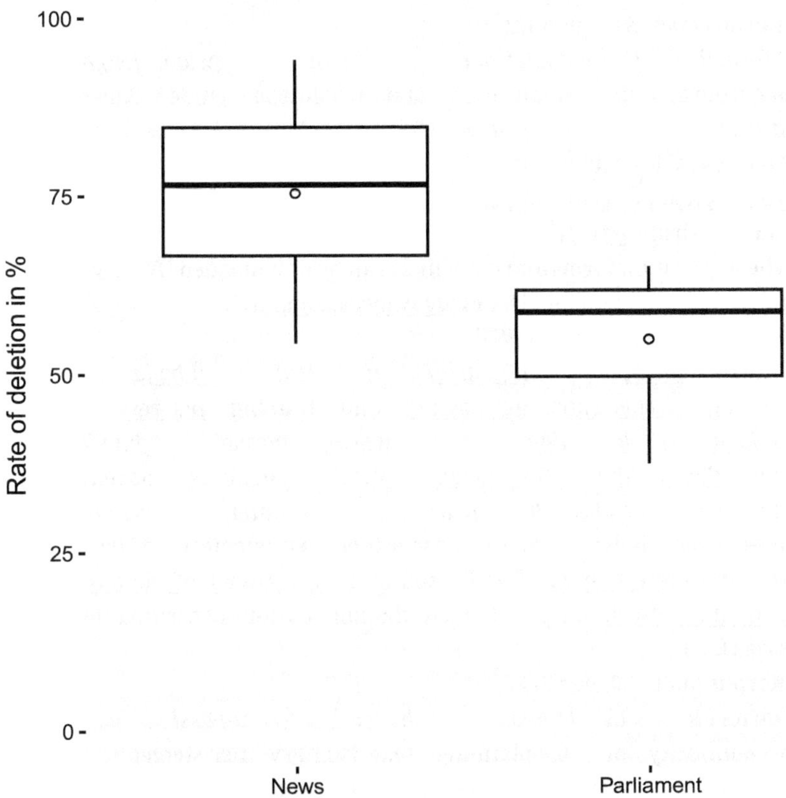

Figure 4.97: Shares of EQUI-P-deletion in the genres News and Parliamentary Speeches in the Korpus Malti 3.0.

In news texts, the average share of EQUI-P-deletion across Πs is 75% whereas it is only 55% in speeches in the Maltese parliament. We expect that similar differences affect most of the Πs in BLOMP 2.0.

To terminate this section, it is necessary to look at the behaviour of so-called compound Πs under coordination.[131] The phenomenon has been studied by Vorholt (2023c). She observes that Π-Π sequences are affected by EQUI-P-deletion too but in ways which differ noticeably from other kinds of synchronically or diachronically complex Πs. In (4.170), we provide a selection of pertinent cases which illustrate similarities and dissimilarities of Πs in this domain.

(4.170) Different kinds of complex Πs under coordination

(a) [Korpus Malti 3.0; news57227]
il-formoli tal-applikazzjoni u l-linji gwida jistgħu
DEF-from:PL of:DEF-application and DEF-line:PL guide 3.IPFV:can:PL
jitniżżlu ***[minn*** ***[fuq*** *www.skyparksbusiness.com]$_{PP}$ u*
3.IPFV:CAUS:descend:PL **[from** **[on** website_name] and
[fuq *www.mcca.com.mt]$_{PP}$]$_{PP}$*
[on website_name]]
'The application forms and guidelines can be downloaded [**from** www.skyparksbusiness.com] and [**from** www.mcca.com.mt].'

(b) [Korpus Malti 3.0; news67353]
[Matul *dawn il-jumejn]$_{PP}$ u* ***[tul*** *il-bqija*
[**during** DEM:PL:PROX DEF-day:DU] and [**during** DEF-rest
tal-jiem li jiġu wara]$_{PP}$ niesna għandhom
of:DEF-day.PL that 3.IPFV:come:PL after] people:1PL have.IPFV:3PL
id-drawwa qaddisa li jżuru ċ-ċimiterji tagħna
DEF-custom holy:F that 3.IPFV:visit:PL DEF-cemetery of:1PL
'[**During** these two days] and [**during** the remainder of the days which come thereafter], our people have the holy custom of visiting our cemeteries [. . .].'

(c) [Korpus Malti 3.0; *parl2727]*
l-awtorità tà l-ippjanar ħadet il-passi
DEF-authority of DEF-planning take:3SG.F.PFV DEF-step:PL
neċessarji skond il-liġi fuq xogħol mhux
necessary:PL according_to DEF-law on work NEG:3SG.M:NEG
awtorizzat fil-madwar u ***[maġenb*** *razzett*
authorise:PART in:DEF-surrounding and **[close_to** farm_house

131 Cf. Section 4.1.3.

*eżistenti fil-wied tal-Wardija]*PP
existing in:DEF-valley of:DEF-Wardija]
'[. . .] the planning authority took the necessary steps according to the law as to unauthor ised work in the surroundings and [**close to** a farm-house in the Wardija valley].'

(d) [Korpus Malti 3.0; news215531]
l-osservazzjonijiet li kienu reġistrati
DEF-observation:PL that be:3PL.PFV register:PART:PL
fl-inħawi ***[maġenb** Ø] u* ***[madwar** il-lokalita*
in:DEF-neighbourhood.PL **[close_to** Ø] and **[around** DEF-locality
*ta' El Ażiżja]*PP
of El Ażiżja]
'[. . .] the observations which were registered in the neighbourhood [**close to** and **around** the locality of El Ażiżja].'

(e) [Korpus Malti 3.0; news107212]
dawn il-prodotti qegħdin jinġiebu
DEM:PL:PROX DEF-product:PL PROG:PL 3.IPFV:REFL:bring:PL
***[mill**-istess post]*PP *u* ***[mingħand** l-istess ditta]*PP
[**from**:DEF-same place] and [**from_s.o.** DEF-same firm]
'[. . .] these products are being fetched [**from** the same place] and [**from** the same firm].'

In (4.170a), we see that the binary combination *minn fuq* 'from on' is not fully repeated on the second conjunct where the presence of (Π17) *fuq* 'on' alone suffices because (Π41) *minn* 'from' has scope over both PPs. Along the lines of our argumentation in the foregoing section, this disproves the hypothesis which assumes that Π-Π sequences of this kind are unitary Πs. Superficially, the same could be claimed for (4.170b) where the first conjunct involves the long allomorph of (Π33) *matul* 'during' whereas the second conjunct involves the short allomorph of the same Π. This looks like a parallel to (4.170a). However, we do not consider this conclusion to be correct since the presence or absence of the initial *ma-* of (Π33) *matul* 'during' is semantically irrelevant. In the case of (Π41) *minn* 'from' in (4.170a) however, there is the possibility that one of the conjuncts involves a competing leftmost Π such as (Π21) *għal* 'for' so that the scope of (Π41) *minn* 'from' would be limited to the first PP. Moreover, while the short allomorph *tul* may replace the long allomorph *matul* as first conjunct in (4.170b) without semantic consequences, the replacement of *minn fuq* with simple *fuq* in (4.170a) is impossible because (Π17) *fuq* 'on' means something else than *minn fuq* 'from on'. The short-long alternation is thus of a different kind from the Π-Π sequences. Furthermore, (4.170c–d) show that the copresence of an identical initial *ma-* in a coordinative construction does not automatically trigger the omission of the initial *ma-* on the second conjunct. In (4.170c), we are not witnessing the coordination of two Πs because the first conjunct contains nominalised (Π31) *madwar* 'around'. Finally, (4.170e) shows that the cooccurrence of (Π41) *minn* 'from' on the first conjunct and of fossilised *minn-* as initial component of (Π38) *mingħand* 'from s.o.' on the second conjunct does not have any effect on the form of the latter. The two elements under inspection are not identified with each other so that neither EQUI-P nor the suppression of the initial component of the second Π is licit.

Not all questions which arise from these facts can be answered in this study. There is thus an urgent need for an in-depth study of the genre-dependent behaviour of Πs under coordination – conjunction as well as disjunction. Whether any of the subcategories which are assumed for coordination (such as additive coordination, non-exhaustive list coordination, inclusive disjunctive coordination, and exclusive list coordination mentioned by Croft (2022: 470–473)) have to be factored in to explain the competition of patterns can only be ascertained on the basis of a considerably larger database.[132] For this study, we provisionally conclude that the competition between EQUI-P-deletion and Π_2-retention is subject to a multitude of factors which are responsible for a highly variegated picture. This colourful picture notwithstanding, the quantities speak in favour of EQUI-P-deletion as functioning as the default option whereas Π_2-retention is possible or even required if certain conditions apply.

4.2 Inside the NP

This section is divided into three parts. In Section 4.2.1, We start with an account of PPs functioning as attributes of nouns in NPs. Section 4.2.2 is exclusively dedicated to *ta'*-PPs because of their special behaviour within NPs. Resumptive constructions are discussed in Section 4.2.3. Further interesting aspects of Πs inside the NP are touched upon only in passing in the subsequent subsections. The problems which are raised additionally in this way constitute subjects for future research in the domain of Maltese Πs.

132 The creation and evaluation of this future database are tasks that will be tackled in Vorholt's PhD-thesis titled "Quantitative aspects of Maltese prepositions".

4.2.1 Prepositional attribution

The task of this section is to determine which Maltese Π can head a PP which is an attribute of a head noun within a NP. This means that we are looking at constructions whose internal structure corresponds to the schema given in Figure 4.98. Here and in the examples that follow in this section, the head noun appears in bold whereas the prepositional attribute is marked out in yellow.

[NP$_{\text{head}}$ [PP]$_{\text{attribute}}$]$_{\text{NP}}$

Figure 4.98: Construction pattern of prepositional attribution.

A typical example of a prepositional attribute is involved in the complex NP at the very beginning of the sentence in (4.171).

(4.171) [Korpus Malti 3.0; news1828]

[Kull	***raġel***	*[bejn*	*it-18*	*u*	*l-34*	*sena]$_{PP}$]$_{NP}$*
[each	**man**	[between	DEF-18	and	DEF-34	year]]

jingħata	*senter*	*jew*	*pistola*	*wara*	*li*	*jservi*
3SG.M.IPFV:REFL:give	rifle	or	pistol	after	that	3SG.M.IPFV:serve

fil-militar
in:DEF-military
'[Each **man** [between 18 and 34 years of age]] is given a rifle or pistol after serving in the military.'

The attributive PP occupies the slot on the right of the head noun. The attributive PP modifies the head noun semantically by way of specifying it qualitatively which in turn means that the reference of the head noun is narrowed down to a subset of the possible referents of the bare noun itself. The attributive PP restricts the reference of the entire NP. In the case of (4.168), the noun *raġel* 'man' (and more so in combination with the total quantifier *kull* 'each') is referentially unrestricted. Only the added *bejn*-PP makes it possible to identify the subset of men for whom the predication holds. The question is whether all members of BLOMP 2.0 can function like (Π4) *bejn* 'between'. As we will learn in the subsequent paragraphs this is not the case.

To avoid analytical difficulties, we restrict the data collection to clear cases, namely those in which the NP which involves a prepositional attribute functions as argument of a verb (preferably as subject) or as apposition to a NP. This restriction is necessary because adjacency of a NP and a PP does not automatically mean that the two phrases belong to the same constituent. The PPs often adverbially modify the verb in lieu of qualifying the NP. To be on the safe side, we focus on examples which feature the constructions as (subconstituents of) subject-NPs, as (direct/indirect) object-NPs, as appositions, or as (mostly GENITIVAL) attributes. An object-NP which embeds an attributive PP is involved in (4.172).

(4.172) [Korpus Malti 3.0; news210621]

*Il-Pulizija li marru fuq il-post sabu [r-**raġel***
DEF-police that go:3PL.PFV on DEF-place find:3PL.PFV [DEF-**man**
[bla ħajja]$_{PP}$]$_{NP}$
[without life]]
'The police who went to the place found [the lifeless **man**].'

An alternative analysis is possible insofar as the *bla*-PP could be a depictive secondary predication which then would not be part of the NP. The appositive use of the construction of interest is illustrated in (4.173).

(4.173) [Korpus Malti 3.0; news178760]

*L-avukat Rob Moodie [**raġel** [mingħajr xagħar]$_{PP}$ u*
DEF-lawyer Rob Moodie [**man** [without hair] and
[bil-mustaċċi]$_{PP}$]$_{NP}$ apparti li jilbes ta' mara
[with:DEF-moustache]] apart_from that 3SG.M.IPFV:wear of woman
iżomm ukoll handbag miegħu meta jmur
3SG.M.IPFV:hold also handbag with:3SG.M when 3SG.M.IPFV:go
il-qorti
DEF-court:PL
'The lawyer Rob Moodie, [**a man** [without hair] and [with a moustache]], apart from dressing like a woman, keeps also a handbag with him when he goes to court.'

In this example, the NP *raġel mingħajr xagħar u bil-mustaċċi* 'a man without hair [or: a bald man] and with a moustache' does not impose any restrictions on the reference of the personal name *Rob Moodie* but adds further information about the referent which is supposed to contrast with what is predicated about the subject. The NP is an apposition which is only loosely integrated into the syntax of the sentence in which it behaves like an optional insertion. The NP is additionally interesting because it embeds two coordinated prepositional attributes.

Example (4.174) features a NP which even takes three prepositional attributes in a row together with which it forms the GENITIVAL attribute of the initial NP *l-idea* 'the idea'.

(4.174) [Korpus Malti 3.0; literature59]

*L-idea ta' [dar-**raġel** [qalb il-hippies]$_{PP}$ [ġo tinda [fit-tajn]$_{PP}$]$_{PP}$ u [f' dik l-iskumdità kollha]$_{PP}$]$_{NP}$ għal xi raġuni ma rahiex implawsibbli*

DEF-idea of [DEM:M:DEF-**man** [amongst DEF-hippie:pl] [in tent [in:DEF-mud]] and [in DEF:F:DIST DEF-inconvenience all:3SG.F]] for some reason NEG see:3SG.F:NEG implausible

'For some reason, he did not consider the idea of [this **man** [amongst the hippies], [in a tent [in the mud]] and [in all this trouble]] to be implausible [. . .].'

The examples (4.172)–(4.174) do not only demonstrate in which syntactic functions NPs with prepositional attributes are licit but they also prove that the six Πs (Π7) *bi* 'with', (Π9) *bla* 'without', (Π15) *fi* 'in', (Π19) *ġo* 'in', (Π36) *mingħajr* 'without', and (Π47) *qalb* 'amidst' resemble (Π4) *bejn* 'between' in the sense that they too are admissible as head of an attributive PP embedded in a NP. There are another fifty-three Πs in BLOMP 2.0 not all of which turn up in attributive PPs. Apart from the problematic candidates (Π26) *kif* 'as', (Π42) *mintul* 'all along' (Π43) *mnejn* 'from near' and (Π54) *sotta* 'under' whose absence from attributive PPs is largely unsurprising, there are further Πs which do not seem to be fit for the function of head of prepositional attributes.

To explain this inability, we first have to see which Πs are attested in constructions of the type featured in Figure 4.98. The semantics of the head noun restricts the choice of admissible prepositional attributes. In Table 4.21 we provide examples for those Πs which are involved in attributive PPs modifying the head noun *raġel* 'man' as in the previous examples (4.172)–(4.173). It transpires from Table 4.21 that by no means all members of BLOMP 2.0 are attested in combination with this head noun in the Korpus Malti 3.0.

Table 4.21: Attributive PPs modifying *raġel* 'man' in the Korpus Malti 3.0.

source	example / translation	Π
news175427	*s-surfing innifshu huwa attraenti qalet Paris Hilton iżda [**raġel** [bħal Johnson]$_{PP}$]$_{NP}$ huwa sempliċiment mhux ta' min jitilfu* '[. . .] surfing in itslef is attractive, said Paris Hilton, but [**a man** [like Johnson]] simply is someone you should not miss [. . .].'	(Π5) *bħal* 'like'
news100657	*Maġenbu nzerta bilqiegħda [**raġel** [daqs ġgant]$_{PP}$]$_{NP}$* 'Next to him [**a man** [like a giant]] happened to sit.'	(Π10) *daqs* 'equal to'
news174500	*Valentinu, li kien [**raġel** [favur l-imħabba u ż-żwieġ]$_{PP}$]$_{NP}$ ma tax kas ta' din l-ordni* 'Valentinu who was [**a man** [in favour of love and marriage]] did not care about this order.'	(Π13) *favur* 'in favour of'

Table 4.21 (continued)

source	example / translation	Π
news137825	*Jum l-umiltà u l-ubbidjenza ta[l-akbar* ***raġel*** *[fost il-bnedmin tal-imgħoddi]*$_{PP}$*]*$_{NP}$ 'a day of humility and obedience for [the greatest **man** [among the people of the past]] [. . .].'	(Π16) *fost* 'amongst'
literature45	*[**raġel** [fuq l-għatba tal-mewt]*$_{PP}$*]*$_{NP}$ *ma tikkonfrontahx b'dat-tip ta' mistoqsijiet urġenti* '[. . .] you do not confront [**a man** [on the brink of death]] with this kind of urgent questions [. . .].'	(Π17) *fuq* 'on'
news181205	*[**Raġel** [ġewwa l-Florida]*$_{PP}$*]*$_{NP}$ *qatel lil martu* '[**a man** [in Florida]] killed his wife [. . .].'	(Π18) *ġewwa* 'inside'
academic177	*[**raġel** [minn Ħal-Tarxien]*$_{PP}$*]*$_{NP}$ *ġie akkużat li pprova joqtol lil martu u lil ġenituri tagħha* '[. . .] [**a man** [from Ħal-Tarxien]] was accused of trying to kill his wife and her parents [. . .].'	(Π41) *minn* 'from'
news123688	*Nibda biex ngħid li jien [**raġel** [qrib it-80 sena]*$_{PP}$*]*$_{NP}$ 'I start by saying that I am [**a man** [close to 80 years of age]] [. . .].'	(Π48) *qrib* 'near'
news182803	*[Ir-**raġel** [ta' nazzjonalita' Ingliża]*$_{PP}$*]*$_{NP}$ *li jgħix f'Malta instab mitluf minn sensih* / '[The **man** [of English nationality]] who lives in Malta was found unconcious [. . .].'	(Π55) *ta'* 'of'
news986	*[**raġel** [taħt l-effett tax-xorb]*$_{PP}$*]*$_{NP}$ *spiċċa ferut wara li prova jirkeb kukkudrill twil ħames metri* '[. . .] [**a man** [under the influence of alcohol]] was injured after he tried to ride on a five metres long crocodile.'	(Π56) *taħt* 'under'

The lack of evidence for combinations of *raġel* 'man' with other Πs is probably only an effect of the composition of the Korpus Malti 3.0. As can be learned from Table 4.22, the vast majority of the Maltese Πs can head attributive PPs within NPs. The examples are again exclusively taken from the Korpus Malti 3.0.

Table 4.22: Attributive PPs modifying further nouns in the Korpus Malti 3.0.

source	example / translation	Π
news95592	*[**Il-ħajja** [barra mill-UE]*$_{PP}$*]*$_{NP}$ *tkun perfettament tollerabbli* '[**The life** [outside of the EU]] will be perfectly tolerable [. . .].'	(Π3) *barra* 'outside'
news47155	*[**Il-ħajja** [bħala President]*$_{PP}$*]*$_{NP}$ *kienet relattivament kalma u trankwilla.* '[**The life** [as president]] was relatively calm and tranquil.'	(Π6) *bħala* 'as'
news173737	*[**Il-pjazza** [biswit il-Knisja]*$_{PP}$*]*$_{NP}$ *ukoll se tingħata dehra ġdida.* '[**The square** [facing the church]] too will be give a new look.'	(Π8) *biswit* 'facing'
news80516	*[**Il-kotba** [dwar Dimech]*$_{PP}$*]*$_{NP}$ *li ħarġu f'dawn l-aħħar għaxar snin kienu kollu dwar x'qal u x'ma qalx.* '[**The books** [about Dimech]] which came out these last ten years had all something to talk about and something to keep silent about.'	(Π11) *dwar* 'about'

Table 4.22 (continued)

source	example / translation	Π
academic420	*L-ewwel kwadru kien juri [**ix-Xatt** [faċċata ta' Manoel Island]$_{PP}$]$_{NP}$.* 'The first painting showed [**the beach** [opposite Manoel island]].'	(Π12) *faċċata* 'opposite'
news111631	*Ma ddispjaċini xejn li tlift [**il-vapur** [għal Għawdex]$_{PP}$]$_{NP}$.* 'I do not feel sorry at all that you missed [**the ferry** [to Gozo]].'	(Π21) *għal* 'for'
news33623	*[**ż-żjara** [għand groomer]$_{PP}$]$_{NP}$ spiċċat il-kawża tal-mewt tal-kelb u l-qattus tagħhom* '[. . .] [**the visit** [at the groomer's]] closed the case of the death of their dog and their cat.'	(Π22) *għand* 'at s.o.'s place'
culture3294	*L-attività [. . .] se tiġi organizzata nhar il-Ġimgħa 25 ta' Lulju mit-8.00 p.m. 'il quddiem f[**il-pjazza** [ħdejn il-Knisja ta' Ħad–Dingli]$_{PP}$]$_{NP}$* 'The activity [. . .] will be organ ised on Friday, 25 July from 8 o'clock p.m. onwards in [**the square** [beside the church of Ħad–Dingli]].'	(Π23) *ħdejn* 'beside'
news14327	*Il-palk prinċipali se jilqa'* [**l-aqwa artisti Maltin** [inkluż Red Electrick]$_{PP}$]$_{NP}$ 'The main stage will welcome [**the strongest Maltese artists** [including Red Electrick]] [. . .].'	(Π25) *inkluż* 'including'
news106826	*[**Il-gwerra** [kontra l-ġuħ]$_{PP}$]$_{NP}$ se tkun il-prijorità tan-nofs is-seklu.* '[**The war** [on hunger]] will be the priority of the mid-century.'	(Π27) *kontra* 'against'
news115587	*[**il-vapur** [lejn il-Libja]$_{PP}$]$_{NP}$ kien jitlaq minn Malta* '[. . .] [**the ship** [to Libia]] left from Malta [. . .].'	(Π28) *lejn* 'towards'
news107049	*F' [**ittra** [lil missieru]$_{PP}$]$_{NP}$ kien iddeskriva l-logħob tiegħu mal-lebbrużi fil-Perù.* 'In [**a letter** [to his father]] he described his playing with the leprous in Peru.'	(Π29) *lil* 'to'
news10821	*Maria tgħid li [**din il-laqgħa** [ma' Papa Franġisku]$_{PP}$]$_{NP}$ hi l-isbaħ rigal li setgħet taqla' f'ħajjitha kollha* 'Maria says that [**this meeting** [with Pope Franciscus]] is the most beautiful present she could get in her entire life.'	(Π30) *ma'* 'with'
culture992	*Graham beda [l-**vjaġġ** [madwar id-dinja]$_{PP}$]$_{NP}$ fl-1 ta' Jannar, 2009* 'Graham began [**the journey** [around the world]] on 1st January, 2009 [. . .].'	(Π31) *madwar* 'around'
literature21	*Baqa' tiela' 'l quddiem u poġġa fuq [**bank** [maġenb it-tebut tan-nannu]$_{PP}$]$_{NP}$.* 'He kept climbing and sat down on [**a bank** [close to the coffin of the grandfather]].'	(Π32) *maġenb* 'close to'
news122900	*[Id-diskussjoni [matul din il-laqgħa]$_{PP}$]$_{NP}$ ċċentrat fuq is-sostenibbiltà tal-finanzi pubbliċi* '[**The discussion** [during this meeting]] focussed on the sustenability of the public finances [. . .].'	(Π33) *matul* 'during'
culture1771	*[**Ittra** [mingħand tifla żgħira titlob għal ftit ħin liberu ma' missierha]$_{PP}$]$_{NP}$ ġabet is-simpatija tal-kumpanija Google li magħha jaħdem missier din l-istess tifla.* '[**A letter** [from a little girl asking for some free time with her father]] was received with sympathy by Google where the father of this girl works.'	(Π38) *mingħand* 'from s.o.'

Table 4.22 (continued)

source	example / translation	Π
news13674	*Ċifri maħruġa mill-Eurostat juru stampa ftit li xejn sabiħa fejn jidħol [**il-mewt** [minħabba l-kanċer fl-UE]$_{PP}$]$_{NP}$* 'Numbers public ised by Eurostat show a fairly ugly picture where [**the deaths** [on account of cancer in the EU]] comes in [. . .].'	(Π39) *minħabba* 'on account of'
european222	*ma jikkawżawx danni lill-ambjent ta' Stat oħrajn jew ta' [**żoni** [oltre l-limiti tal- ġurisdizzjoni nazzjonali]$_{PP}$]$_{NP}$* '[. . .] they do not cause damage of the ambient of another state or of [**zones** [outside the national jurisdiction]].'	(Π44) *oltre* 'beyond'
parl6057	*il-qassis jitkellem fuq is-serq mill-benefiċċji soċjali, [**is-serq** [permezz tal-korruzzjoni]$_{PP}$]$_{NP}$ u s-serq mill-fondi pubbliċi* '[. . .] the priest speaks about the theft of social benefits, [**the theft** [by means of corruption]] and the theft from the public funds.'	(Π45) *permezz* 'by means of'
news85632	*[**Il-mewt** [qabel żmienha tat-tabib Karl Chircop]$_{PP}$]$_{NP}$ nisslet niket lil martu* '[**The** untimely **death** of the physician Karl Chircop] caused grief for his wife [. . .].'	(Π46) *qabel* 'before'
news125384	*[**il-pjazza** [quddiem il-Kon-Katidral ta' San Ġwann]$_{PP}$]$_{NP}$ issir sit ta' kostruzzjoni għas-snin li ġejjin.* '[. . .] [**the square** [in front of the Co-Cathedral of San Ġwann]] will become a construction site for the years to come.'	(Π49) *quddiem* 'in front of'
parl7790	*[**Il-mistoqsija** [rigward id-dgħajjes kollha tas-sajd u l-klassifikazzjoni tagħhom (full-time, part-time etc)]$_{PP}$]$_{NP}$ għandha tiġi riferuta lill-Ministeru konċernat.* '[**The question** [concerning all the boats of the fishermen and their classification (full-time, part-time, etc.)]] must be referred to the Minister concerned.'	(Π50) *rigward* 'concerning'
literature20	*[**Il-vjaġġ** [sa Ta' Giorni mit-toroq battala]$_{PP}$]$_{NP}$ kien qasir wisq* '[**The journey** [till Ta' Giorni through the empty streets]] was too short [. . .].'	(Π51) *sa* 'till'
parl12031	*Jekk verament irridu nagħmlu analiżi serja ta' din il-policy, ma nistgħux naqbdu ngħidu li jkun hawn [**id-diskussjoni** [skont il-policies li jidher li għandna oġġezzjoni fuqhom]$_{PP}$]$_{NP}$* 'If we really want to conduct a serious analysis of this policy, we cannot start by saying that there will be '[**the discussion** [according to the policies which we seem to have objections about]].'	(Π53) *skont* 'according to'
parl547	*meta qamet il-kwestjoni ta' [**l-embargo tal-Ġnus Magħquda** [versu l-Libja]$_{PP}$]$_{NP}$* '[. . .] when the question of [**the embargo of the United Nations** [against Libia]] was raised.'	(Π57) *versu* 'towards'
news63151	*[**Persuni** [viċin il-Knisja]$_{PP}$]$_{NP}$ ħeġġewha biex [. . .] tqis mill-ġdid id-deċiżjoni tagħha* '[**Persons** [close to the Church]] urged her to reconsider her decision.'	(Π58) *viċin* 'near'

Table 4.22 (continued)

source	example / translation	Π
parl7764	*[**inċident** [waqt l-inżul tal-helicopter]$_{PP}$]$_{NP}$ kien ikollu konsegwenzi serji* '[. . .] [**an incident** [during the landing of the helicopter]] could have serious consequences [. . .].'	(Π59) *waqt* 'at the time of'
academic376	*[**L-ewwel snin** [wara l-Gwerra]$_{PP}$]$_{NP}$ raw it-twaqqif ta' l-iStat ta' Għajnuna Soċjali Brittanniku* '[**The first years** [after the war]] saw the creation of the British Social Aid State.'	(Π60) *wara* 'after'

On the one hand, Tables 4.21–4.22 show that in principle it is possible for the bulk of the members of BLOMP 2.0 to function as head of a NP-internal attributive PP. There is thus a good reason to declare this function a trait of the canonical Π. We hesitate to do so because, on the other hand, the pieces of evidence more often than not stem from the news genre. According to Table 1.12 and Figure 1.13, journalistic prose is vastly overrepresented in the Korpus Malti 3.0 where it accounts for more than a third of all texts. For many (if not all) of the cases of attribution in Tables 4.21–4.22 there are no examples from other genres. Some of the examples from the news section read like slightly more elaborated headlines of newspapers. Chances are therefore that we are facing a genre-based stylistic preference along the lines of similar observations made in the foregoing section in connection with EQUI-P-deletion. Even if there is a genre bias, this does not alter the picture completely because the fact remains that prepositional attribution is a structural potential of Maltese which, however, is not activated to similar extents across all registers, styles, and genres.

In addition to the cases covered by Tables 4.21–4.22 and the previous sentential examples, there are four Πs (Π1) *apparti* 'apart from', (Π20) *għajr* 'except', (Π24) *ħlief* 'except', and (Π34) *minbarra* 'except' which form a semantically coherent class in the sense that they all belong to the functional domain of EXCLUSION. Note that (Π44) *oltre* 'beyond' is only attested in prepositional attribution in its spatial meaning and never with the meaning 'except' in the construction frame under discussion. The four first mentioned Πs are admissible as heads of attributive PPs provided the NP they modify is globally quantified as in (4.175) where *kollha* 'all' is additionally underlined.

(4.175) [Korpus Malti 3.0; parl5931]

[Il-vetturi	***kollha***	*[apparti*	*dik*	*tar-Rettur]$_{PP}$]$_{NP}$*
[DEF-vehicle:PL	all:3SG.F	[apart_from	DEM:F:DIST	of:DEF-rector]]
jiġu	*pparkjati*	*fuq*	*il-Campus*	
3.IPFV:come:PL	park:PART:PL	on	DEF-campus	

'[**All the cars** [apart from the rector's]] are parked on the campus.'

The three (partly) synonymous Πs which follow the same pattern are illustrated in Table 4.23.

Table 4.23: Attributive PPs modifying quantified nouns in the Korpus Malti 3.0.

source	example / translation	Π
news182915	*jirriżulta li [**l-barranin** __kollha__ [għajr Stuart Roberts]$_{PP}$]$_{NP}$ ilkoll sofrew ġrieħi ħfief* '[. . .] it results that [__all__ **the foreigners** [except Stuart Roberts]] suffered light injuries too.'	(Π20) *għajr* 'except'
academic240	*kaxkru [**l-Għawdxin** __kollha__ [ħlief ix-xjuħ u l-morda]$_{PP}$]$_{NP}$* '[. . .] they dragged away [__all__ **the Gozitans** [except the elderly and the sick]].'	(Π24) *ħlief* 'except'
culture2466	*[**il-ħwienet** __kollha__ [minbarra r-ristoranti]$_{PP}$]$_{NP}$ jagħlqu* '[. . .] [__all__ **the shops** [except the restaurants]] close.'	(Π34) *minbarra* 'except'

Seven other Πs are not attested in the construction frame [**NP**$_{head}$ [PP]$_{attribute}$]$_{NP}$ introduced in Figure 4.98. They cover 12% of BLOMP 2.0 and are listed alphabetically in (4.176).

(4.176) Πs not attested as heads of attributive PPs
(Π2) *a skapitu* 'at the expense of'; (Π14) *fejn* 'near'; (Π29) *lil* 'to'; (Π35) *minflok* 'instead of'; (Π37) *mingħala* 'in s.o.'s opinion'; (Π40) *minkejja* 'in spite of'; (Π52) *sforz* 'thanks to'

We cannot completely count out the possibility that the absence of evidence for these Πs being used in prepositional-attributive constructions is causally connected to the general properties of the Korpus Malti 3.0. We can very briefly comment on those cases which do not partake in the formation of semantic groups. There is for instance (Π14) *fejn* 'near' whose token frequency as Π is negligible in comparison to its use as an interrogative adverb or relativiser. We have already addressed some of the problems posed by this candidate.[133] It appears that when (Π14) *fejn* 'near' modifies a head noun it normally introduces a relative clause. (Π37) *mingħala* 'in s.o.'s opinion' cannot occur in the construction frame presented in Figure 4.98 because it takes an obligatory pronominal complement together with which it forms a PP which is not fully integrated into the syntax, meaning: a *mingħala*-PP is not governed by any other constituent of the clause.[134] The case of (Π29) *lil* 'to' is different because this Π – as object marker – always needs a verb to depend on. The occur-

133 Cf. Section 4.1.1.2.2.1 (ix).
134 Cf. Section 1.6.4.2 (v).

rence of (Π29) *lil* 'to' in the construction frame $[\mathbf{NP}_{\text{head}}\ [\text{PP}]_{\text{attribute}}]_{\text{NP}}$ is barred. (Π35) *minflok* 'instead of' following a noun is attested mostly as conjunction introducing a clause but never as head of an attributive PP. As to (Π2) *a skapitu* 'at the expense of', (Π40) *minkejja* 'in spite of', and (Π52) *sforz* 'thanks to', we are surprised not to find any uncontroversial evidence of their being used as heads of attributive PPS.

As the above paragraphs have shown that prepositional attribution is possible with 82% of the members of BLOMP 2.0 although the frequency with which a given Π functions as head of an attributive PP is largely dependent on the text genre. It is possible that journalistic Maltese is responsible for the diffusion of the patterns. Moreover, even within this genre, the occurrences of attributive PPS are not numerous. Further research into this phenomenon is required to determine unequivocally whether prepositional attribution is a characteristic trait of the news genre or a common feature across all styles of Maltese.

4.2.2 *ta'*-PPS

At this point, we enter new territory insofar as (Π55) *ta'* 'of' displays a number of properties which are alien to the other Maltese Πs. A remarkable trait which is almost unique to (Π55) *ta'* 'of' is its supposed derivational function. The same function is claimed for (Π7) *bi* 'with' by Aquilina (1987: 114). A variety of expressions which put labels to professions involve initial (Π55) *ta'* 'of' as shown for *tal-posta* 'postman' (literally 'of the post') in (4.177). This and similar kinds of formations involving an initial reflex of (Π55) *ta'* 'of' are discussed in more detail as so-called lexical *ta'*-constructions by Fabri (2022).

(4.177) Professions

(a) [Korpus Malti 3.0; news79819]

Tal-posta	*baqa'*	*jħares*	*lejha*	*qisha*	*qiegħda*
postman	remain	3SG.M.IPFV:look	towards:3SG.F	like:3SG.F	PROG:F
tkellmu	*bil-Grieg*				
3SG.F.IPFV:speak	with:DEF-Greek				

'**The postman** kept looking at her as if she was speaking Greek.'

(b) [Korpus Malti 3.0; opinion1649]

Dalgħodu	*domt*	*xi*	*siegħa*	*u*	*nofs*	*nistenna*
this_morning	take_time:1SG.PFV	some	hour	and	half	1SG.IPFV:wait
lil	***tal-posta***	*imma*	*ma*	*ġiex*		
to	**postman**	but	NEG	come:NEG		

'This morning I spent an hour and a half waiting **for the postman** but he did not come.'

(c) [Korpus Malti 3.0; news174192]

l-individwu *jista'* *jċempel* *l-uffiċċju* *tal-posta*
DEF-individual 3SG.M.IPFV:can 3SG.M.IPFV:phone DEF-office of:DEF-post
li *jkun* *indikat* *fin-nota* *mħollija* ***minn***
that 3SG.M.IPFV:be.FUT indicate:PART in:DEF-note PART:leave:F **from**
tal-posta
postman
'The individual may phone the post-office that is indicated in the note left **by the postman.**'

The same pattern is reflected by a number of other formations all of which refer to professions[135] important for the daily provisions in a village society. Our analysis assumes that, strictly speaking, (Π55) *ta'* 'of' is not a derivational morpheme but the head of a PP which functioned as the GENITIVAL attribute of an unrealised head-noun. This is shown in Figure 4.99.

Ø	*tal-*	*ħobż*	'baker'	('of the bread')
		ħanut	'publican'	('of the shop')
		ħaxix	'green-grocer'	('of the vegetable')
		laħam	'butcher'	('of the meat')
		merċa	'grocer'	('of the merchandise')

Figure 4.99: *ta'*-formations as names of professions.

Whether the Ø-slot was ever occupied by a phonologically real ised element – perhaps a demonstrative like *dan* 'this' and its paradigmatic forms[136] – cannot be ascertained in this study. What is important however is that

---

135 The pattern is not restricted to professions as shown in (4.x) where *tal-familja* 'relative(s)' is employed as indirect object:

(4.x) [Korpus Malti 3.0; news151585]
L-emigranti *kitbu* ***lil*** ***tal-familja*** *f'* *Għawdex*
DEF-emigrant:PL write:3PL.PFV **to** **relative** in Gozo
'The emigrants wrote **to their relatives** in Gozo.'

136 This is different in the case of *tal-linja* '(public) bus (service)' which is probably short for *karrozza tal-linja* 'bus' as in (4.xi) where the feminine agreement morphology on the first finite verb might reflect the feminine gender of the omitted *karrozza* 'car':

(a) the *ta'*-formation cannot be made definite,
(b) the *ta'*-formation is transnumeral in the sense that it can refer to an individual or several individuals (but cf. (d)),
(c) the *ta'*-formation may further refer generically to an institution or the locality in with which a given profession is normally associated,
(d) the nouns involved in the *ta'*-formation may be plural ised but cannot be modified by attributes (but cf. (b)),
(e) under coordination, the initial *ta'* is immune against EQUI-P-deletion,
(f) *ta'*-formations are eligible as complements for all kinds of PPS,
(g) in GENITIVAL PPS only one *ta'* is licit.

On the basis of these properties, we conclude that in *ta'*-formations, *ta'* is no longer a Π nor is it a fully-blown syntactic word of its own.[137] Cases like *tal-posta* 'postman' are lexical ised. The lexicalisation does not invalidate our hypothesis according to which *ta'* is not a proper derivational morpheme.

Having clarified the status of *ta'*-formations, we now know that some putative cases of PP-sequences seemingly involving *ta'* as second Π turn out to be no Π-sequences at all. This negative judgment applies also to other constructions which involve (Π55) *ta'* 'of'. In (4.178), we present examples of GENITIVAL *ta'*-phrases which illustrate the second kind of look-alikes of Π-sequences.

(4.xi) [Korpus Malti 3.0; literature7]

Tal-linja	*waslet*	*ftit*	*minuti*	*wara*	*li*	*wasal*	*fuq*	*l-istejġ*
bus	arrive:3SG.F.PFV	little	minute:PL	after	that	arrive	on	DEF-bus_stop

'**The bus** arrived a few minutes after he arrived at the bus-stop.'

137 Several place names start with (Π55) *ta'* 'of' which is an inseparable part thereof. Place names like *Tas-Sliema* reflect erstwhile headless GENITIVAL PPS. In (4.xii) it is shown that the integrated *ta'* does not take over functions (such as that of introducing the possessor) required by the context in which the place name occurs.

(4.xii) [Korpus Malti 3.0; academic420]

Il-popolazjoni	***[ta'***	*tas-Sliema]$_{PP}$*	*minn*	*25000*	*fl-1950*	*niżlet*
DEF-population	**[of**	Tas-Sliema]	from	25000	in:DEF-1950	descend:3SG.F.PFV
għall	*13000*					
for	13000					

'In 1950, the population [**of** Tas-Sliema] went down from 25000 to 13000.'

(4.178) GENITIVAL *ta'*-phrases

(a) [Korpus 3.0; culture72]

L-istil mużikali u l-istil tal-kant ta' Federica huma
DEF-style musical and DEF-style of:DEF-singing of Federica 3PL
totalment differenti ***minn ta'*** *Gaia*
totally different **from of** Gaia
'Federica's styles of music and singing are totally different **from** Gaia**'s**.'

(b) [Korpus Malti 3.0; news9081]

l-pagi ta' dawn il-ħaddiema jkunu
DEF-wage:PL of DEM:PL:PROX DEF-worker.PL 3.IPFV:be.FUT:PL
ekwivalenti ***għal tal-****ħaddiema tal- Gvern*
equivalent:PL **for of**:DEF-worker.PL of:DEF-government
'[. . .] the wages of these workers will be equivalent **to** those **of** the workers of the government.'

In cases of this kind, there is coordination of NPs both of which involve a GENITIVAL attribute. The NP which represents the possessee is the same for both conjuncts so that it is unnecessary to repeat it with the second conjunct. This means that *minn* and *ta'* in (4.178a) and *għal* and *ta'* in (4.178b) are only incidentally adjacent. Virtually the slot between the two Πs is occupied by *l-istil mużikali u l-istil tal-kant* 'the styles of music and singing' and *l-pagi* 'the wages', respectively. The two Πs do not belong together either syntactically or semantically. This semantic and syntactic separation is also visible in the case of pronominal possessors as shown in (4.179).

(4.179) Pronominal possessor

(a) [Korpus Malti 3.0; culture3050]

Fil-każ ta' din l-iskola s-sid ta' l-art
in:DEF-case of DEM:F:PROX DEF-school DEF-owner of DEF-land
ordna lill-amministrazzjoni biex tevakwa l-art sa
order to:DEF-administration to 3SG.F.IPFV:evacuate DEF-land till
l-aħħar ta' Frar minħabba li hi ta' tribù differenti
DEF-last of February on_account_of that 3SG.F of tribe different
minn tiegħu
from of:3SG.M
'In the case of this school, the land-owner ordered the administration to evacuate the land until the end of February because it belongs to a different tribe **from his**.'

(b) [Korpus Malti 3.0; parl485]
Aħna l-partit alternattiv ***għal tagħkom*** *fil-gvern*
1PL DEF-party alternative **for of:2PL** in:DEF-government
'We are the alternative party **for yours** in the government.'

The NPs which represent the possessees in (4.179a–b) are not repeated in the PPs governed by (Π41) *minn* 'from' and (Π21) *għal* 'for' in front of the inflected forms of (Π55) *ta'* 'of'. On the surface, there is a binary Π-sequence which however incidentally results from the absence of the NPs *tribù* 'tribe' and *l-partit* 'the party' from the slot to the left of *tiegħu* 'his' and *tagħkom* 'yours', respectively. This phenomenon – the absence of the second NP – is possible with practically every member of BLOMP 2.0 (except the usual problematic candidates) so that the erroneous impression might come up that (Π55) *ta'* 'of' is a very frequent Π_2 in a Π-sequence. This however is not the case. Examples like (4.180) lend support to this interpretation.

(4.180) [Korpus Malti 3.0; academic356]
F' Plovdiv fadal ukoll żewġ moskei ***minn*** ___ ***ta'*** *żmien*
in Plovdiv remain also two mosque:PL **from** ___ **of** time
l-Imperu Ottoman
DEF-Empire Ottoman
'In Plovdiv too, there remain two mosques **from** those **of** the times of the Ottoman Empire.'

The empty slot marked by underlining in the example shows us where an anaphor of the noun *moskei* 'mosques' would be placed. This could be either the repetition of the lexical noun itself or a pronominal substitute (such as *dawk* 'those'). If we omit (Π55) *ta'* 'of' from (4.180) we get *miż-żmien l-Imperu Ottoman* 'from the time of the Ottoman Empire' which qualifies the mosques in terms of their age. If (Π55) *ta'* 'of' is kept in place the relation expressed by (Π41) *minn* 'from' is partitive, i.e., the two remaining mosques are characterised as belonging to an erstwhile much larger group of mosques. In neither case is there a Π-sequence of the kind described above.[138]

The position of (Π55) *ta'* 'of' can be reversed relative to another Π to yield the order NP *ta'* Π NP. A typical example of this order is given in (4.181).

(4.181) [Korpus Malti 3.0; academic12]
Renju Unit ***ta' wara*** *l-gwerra kellu bżonn huwa*
United_Kingdom **of after** DEF-war have.PFV:3SG.M need 3SG.M

138 Cf. Section 4.1.3.

wkoll ta' ħafna ħaddiema għar-rikostruzzjoni tal-pajjiż
too of many worker.PL for:DEF-reconstruction of:DEF-country
'The postwar United Kingdom too needed many workhands for the reconstruction of the country.'

In cases like this (П55) *ta'* 'of' functions as linker which connects the postnominal attribute to the head noun. Owing to the presence of (П55) *ta'* 'of' in (4.181), it is clear that *wara l-gwerra* 'after the war' qualifies *Renju Unit* 'United Kingdom'. If we drop (П55) *ta'* 'of' from the construction the PP *wara l-gwerra* 'after the war' may be interpreted as a temporal adverbial modifying the predicate – and for that reason would have to be placed elsewhere in the sentence, preferably at the left or right margin. From Table 4.22 we know however that attributive PPs almost unrestrictedly combine with their head nouns without the attribute marker (П55) *ta'* 'of'. There are indeed two constructions – one with and the other without (П55) *ta'* 'of' – in the domain of prepositional attribution. The two examples in (4.182) seem to illustrate free variation but the brackets tell a different story.

(4.182) With and without (П55) *ta'* 'of'
(a) [Korpus Malti 3.0; european6492]
***[Il-kejl ta' l-iżbalji** [waqt l-installazzjoni]$_{PP}$]$_{NP}$*
[DEF-measuring of DEF-mistake:PL [at_the_time_of DEF-installation]]
u matul l-użu għandu jsir
and during DEF-use have.IPFV:3SG.M 3SG.M.IPFV:become
skond dawn il-kondizzjonijiet
according_to DEM:PL:PROX DEF-condition:PL
'[**The measuring of the mistakes** [at the time of the installation]] and during the use has to be carried out according to these conditions [. . .].'
(b) [Korpus Malti 3.0; european10576]
biex inaqqsu kemm jista' jkun
in_order_to 3.IPFV:reduce:PL how_much 3SG.M.IPFV:can 3SG.M.IPFV:be.FUT
***[l-iżbalji** [ta' [waqt it-tħaddim]$_{PP}$]$_{PP}$]$_{NP}$*
[DEF-mistake:PL [of [at_the_time_of DEF-operation]]]
'[. . .] in order to reduce as much as possible [**the mistakes** [[at the time of the operation]]] [. . .].'

In (4.182a), the *waqt*-PP modifies the complex NP to its left the head of which is *il-kejl* 'the measurement'. Had (П55) *ta'* 'of' been inserted after *l-iżbalji* 'the mistakes' the mistakes would have happened at the time of the installation. However, it is the procedure of measurement that is carried out during the installation. It is not implied that any mistake will actually happen at that point in time. This means that the absence / presence of (Π55) *ta'* 'of' makes a difference semantically. The Π is there in (4.182b) because the entire NP has a kind of generic reference to mistakes that might come up eventually during the operation. We assume that the absence / presence of (Π55) *ta'* 'of' correlates with the distinction between qualitative attributive PPs and restrictive attributive PPs. If (Π55) *ta'* 'of' is missing from the construction as is the case throughout the examples in Tables 4.21–4.23 we are facing qualitative attribution whereas in the presence of (Π55) *ta'* 'of' the construction receives a restrictive interpretation as suggested by the rule in Figure 4.100.

$$PP_{attributive} \rightarrow \begin{cases} ta' + \Pi & / \text{ restrictive} \\ \Pi & / \text{ qualitative} \end{cases}$$

Figure 4.100: Rule for presence / absence of (Π55) *ta'* 'of'.

An attributive PPs without additional (Π55) *ta'* 'of' shares information about properties of the referent of the head noun without defining the referent in this way. In contrast, the presence of (Π55) *ta'* 'of' identifies the referent by way of setting it apart from possible other referents. Thus, the constructions in Figure 4.101 may also differ pragmatically since the restrictive construction can be used for contrastive purposes as well.

$[NP_{head}\ [PP]_{attribute}]_{NP}$ = qualitative / neutral / predicative

$[NP_{head}\ [ta'\ [PP]]_{attribute}]_{NP}$ = restrictive / contrastive / attributive

Figure 4.101: Two construction types compared.

The rule is not strict. Restrictive (Π55) *ta'* 'of' may be absent from a NP which hosts a prepositional attribute as is the case with the first NP *żwieġ bla tfal* 'a marriage without children' in (4.183a). In the same example, childless marriages are compared to marriages with children. The contrast between the two types of marriages facilitates the use of (Π55) *ta'* 'of' in the final NP. The presence of (Π55) *ta'* 'of' is not only triggered by the contrastive function but also by the distal demonstrative *dak* 'that' which replaces the noun *żwieġ*. In (4.183b), (Π55) *ta'* 'of' must be present to link attribute and head noun since in its absence the sentence would become contradictory: the main clause *Arsenal spiċċaw iż-żmien bla telfa* 'Arsenal finished the time without defeat' would no longer fit in with the adverbial clause because the latter mentions a defeat of the team.

(4.183) Mandatory vs optional *ta'* I

(a) [Korpus Malti 3.0; parl8185]

*[**żwieġ** [bla tfal]$_{PP}$]$_{NP}$ huwa żwieġ tal-istess*
[**marriage** [without child.PL]] 3SG.M marriage of:DEF-same
*kwalità bħal [**dak** [<u>ta'</u> [bit-tfal]$_{PP}$]$_{PP}$]$_{NP}$*
quality like [**DEM:M:DIST** [<u>of</u> [with:DEF-child.PL]]]
'[. . .] [**a marriage** [without children]] is a marriage of the same quality as [**that** [with children]].'

(b) [Korpus Malti 3.0; academic443]

*Arsenal spiċċaw [**iż-żmien** [<u>ta'</u> [bla telfa]$_{PP}$]$_{PP}$]$_{NP}$*
Arsenal terminate:3PL.PFV [**DEF-time** [<u>of</u> [without defeat]]]
meta tilfu 2-0 kontra Manchester United
when lose:3PL.PFV 2-0 against Manchester United
'Arsenal finished [**the period** [without defeat]] when they lost 2-0 against Manchester United [. . .].'

In (4.184), we present the final three examples of this section. The sentences are again taken from the Korpus Malti 3.0.

(4.184) Mandatory vs optional *ta'* I

(a) [Korpus Malti 3.0; academic254]

*[**it-temperatura** [matul il-jum]$_{PP}$]$_{NP}$ rarament titla'*
[**DEF-temperature** [during DEF-day]] rarely 3SG.F.IPFV:rise
iżjed minn 273 K
more than 273 K
'[. . .] [**the temperature** [during the day]] rarely rises more than 273 K [. . .].'

(b) [Korpus Malti 3.0; academic356]

*[**It-temperatura medja** [<u>ta'</u> [matul il-jum]$_{PP}$]$_{PP}$]$_{NP}$ f' Lulju*
[**DEF-temperature average:F** [<u>of</u> [during DEF-day]]] in July
tkun 36 grad
3SG.F.IPFV:be.FUT 36 degree
'In July, [**the average temperature** [during the day]] will be 36 degrees [. . .].'

(c) [Korpus Malti 3.0; news107026]

*[**It-temperatura medja** [matul dan ix-xahar]$_{PP}$]$_{NP}$*
[**DEF-temperature average:F** [during DEM:M:PROX DEF-month]]
kienet ta' 14.1 grad Celsius
be:3SG.F.PFV of 14.1 degree Celsius
'[**The average temperature** [during this month]] was 14.1 degrees Celsius.'

These sentences indicate that there is considerable fuzziness involved if one tries to properly analyse the alternation of constructions with and without (Π55) *ta'* 'of'. The examples describe the temperature during specified periods generalizing about the meteorological facts. The adjacency of the PP *matul il-jum* 'during the day' and the NP *it-temperatura* in (4.184a) notwithstanding, uncertainty arises with regards to the syntactic relation between the two phrases because semantically the PP *matul il-jum* 'during the day' could also be understood as adverbial modification of the predicate with the finite verb *titla'* 'she rises' from which it is separated however by the adverb *rarament* 'rarely'. In (4.184b), the employment of (Π55) *ta'* 'of' makes it clear that this is an attributive prepositional construction. However, the presence of (Π55) *ta'* 'of' does not seem to be compulsory since in (4.184c) there is no trace of this Π. Does the absence of (Π55) *ta'* 'of' mean that the PP *matul dan ix-xahar* 'during this month' modifies the predicate in lieu of modifying the NP to its left? Semantico-syntactic borderline cases of this kind are abundantly represented in our primary sources.

In the Korpus Malti 3.0, there are numerous tokens of combinations of (Π55) *ta'* 'of' with a PP headed by a different Π. Almost all Πs shown to be involved in prepositional attribution in Tables 4.21–4.23 and examples (4.172)–(4.174) may also combine with preceding (Π55) *ta'* 'of'. It is even difficult to find impossible combinations when we tick off the members of BLOMP 2.0. Since pragmatics are crucial for the interpretation of many of these cases, the thorough investigation of the phenomenon has to be relegated until the not-too-distant future when it will be possible to conduct a follow-up project with native-speaker informants.

4.2.3 Resumptive PPs in the relative clause

Relative clauses are attributes of head nouns with which they form a complex NP. Maltese makes use of morphologically invariable relativisers like *li* 'that' which may better be termed relativising conjunctions. To encode the semantic relation which applies to the relativised NP within the relative clause Maltese employs the strategy of pronominal resumption. The pronoun which anaphorically refers back to the external head is the complement of a Π heading a PP which occurs inside the relative clause. A brief characterisation of the Maltese relative clause is provided in Borg/Azzopardi-Alexander (1997: 37–38). A functionally specified kind of relative clause is the possessive relative clause described in some detail in Stolz (2011) which however is outside the scope of this section. Camilleri (2014) is a detailed study of the restrictive relative clause in Maltese which takes issue with accounts of the same phenomenon in previous studies.

The issue of pronominal resumption has been touched upon already in relation to (П52) *sforz* 'thanks to' in (4.71) and in example (4.134) for (П44) *oltre* 'beyond' above.[139] In this section, we elaborate on this topic and start by illustrating the phenomenon with three examples in (4.185). The NP in the main clause which provides the head of the NP in which the relative clause is embedded is printed in bold, the relative clause is marked in yellow and the resumptive PP is underlined.

(4.185) pronominal resumption I

(a) [Korpus Malti 3.0; academic12]

*Malta ġiet ikklassifikata ma' [**erba' pajjiżi***
Malta come:3SG.F.PFV classify:PART:F with [**four country:PL**
***żgħar oħra** [li [bejniethom]$_{PP}$ setgħu jibagħtu*
small.PL other.PL [that [between:PL:3PL] can:3PL.PFV 3.IPFV:send:PL
86 emigrant fis-sena]$_{RC}$]$_{NP}$
86 emigrant in:DEF-year]
'Malta was classified with [**four other small countries** [that [between them] could send 86 emigrants per year].'

(b) [Korpus Malti 3.0; academic12]

*wieħed mill-uffiċjali qara [**digriet** [li [bih]$_{PP}$ wera*
one from:DEF-officer:PL read [**decree** [that [with:3SG.M] show
li Lampedusa kienet parti mis-saltna tar-re
that Lampedusa be:3SG.F.PFV part from:DEF-kingdom of:DEF-king
ta' Napli]$_{RC}$]$_{NP}$
of Naples]]
'[. . .] one of the officers read a [**decree** [[with] which he showed that Lampedusa was a part of the kingdom of the king of Naples]][. . .].'

(c) [Korpus Malti 3.0; european2098]

il-pretendent jibgħat il-pretensjoni tiegħu lill-istituzzjoni
DEF-claimant 3SG.M.IPFV:send DEF-claim of:3SG.M to:DEF-institution
*ta' [**l-Istat Membru** [li [tiegħu]$_{PP}$ huwa ċittadin*
of [**DEF-state member** [that [of:3SG.M] 3SG.M citizen
nazzjonali]$_{RC}$]$_{NP}$
national]]
'[. . .] the claimant sends his claim to the institution of [**the member state** [whose national citizen he is]].'

139 Cf. Section 4.1.2.2.2.2.2 (ii).

The position of the resumptive PP on the immediate right of the relativiser *li* 'that' is very common although other options are also open for the placement of the resumptive PP as shown in the three examples in (4.186) where the resumptive PP follows the finite verb or a noun within the relative clause.

(4.186) pronominal resumption II

(a) [Korpus Malti 3.0; parl9957]
*Jiena se nġib eżempju ta' [**każ** [li għext*
1SG FUT 1SG.IPFV:bring example of [**case** [that live:1SG.PFV
[fih]$_{PP}$ *professjonalment bħala avukat]*$_{RC}$*]*$_{NP}$
[in:3SG.M] professionally as lawyer]
'I will give the example of a [**case** [that I was involved [in] professionally as lawyer]] [. . .].'

(b) [Korpus Malti 3.0; news106914]
*kien [**il-proġett politiku** [li ħadem [għalih]*$_{PP}$ *Mintoff]*$_{RC}$*]*$_{NP}$
be [DEF-project political [that work [for:3SG.M] Mintoff]]
'[. . .] it was [the political project [that Mintoff worked [for]]].'

(c) [Korpus Malti 3.0; *news64396]*
qalli li kliemi hu Malti antikwat ta'
say:IO:1SG that word.COLL:1SG 3SG.M Maltese antiquated of
*żmien [**il-Professur Saydon** [li l- kitba [tiegħu]*$_{PP}$
time [**DEF-Professor Saydon** [that DEF-writing [of:3SG.M]
trid dizzjunarju mill-Malti għall-Malti
3SG.F.IPFV:want dictionary from:DEF-Maltese for:DEF-Maltese
biex tifhimha]$_{RC}$*]*$_{NP}$
in_order_to 2SG.IPFV:understand:3SG.F]]
'[. . .] he told me that my wording was antiquated Maltese of the time of [**Professor Saydon** [whose writing requires a Maltese-Maltese dictionary to understand it]].'

In (4.185a), (Π4) *bejn* 'between' is inflected for plural and hosts the suffixal exponent *-hom* of the 3rd person plural.[140] This pronominal suffix serves as anaphor of the NP *erba' pajjiżi żgħar oħra* 'four other small countries' in the main clause. Except for the number inflection, the behaviour of (Π4) *bejn* 'between' is representative of the behaviour of other inflected Πs in resumptive PPs (such as (Π7) *bi* 'with' in (4.184b), (Π55) *ta'* 'of' in (4.185c) and (4.186c), (Π15) *fi* 'in' in (4.185a), and (Π21) *għal* 'for' in (4.185b)).

140 Cf. Section 3.1.2.3.1.2 (ii).

Since resumption involves the use of pronominal complements it is clear that (Π9) *bla* 'without' is banned from use in relative clauses to allow for anaphorically referring back to the head noun in the main clause.[141] Discounting most but not all of the usual suspects whose Π-status is difficult to determine empirically, this leaves us numerous candidates for pronominal resumption. As we will see shortly 52 Maltese Πs occur in the construction in question in the Korpus Malti 3.0, i.e., 87% of all Maltese Πs can be shown to partake in resumption. In Table 4.24, we list all those cases of resumption which our native-speaker informant Sandra Abela has confirmed (except (Π4) *bejn* 'between', (Π7) *bi* 'with', (Π15) *fi* 'in', (Π21) *għal* 'for', (Π44) *oltre* 'beyond', (Π52) *sforz* 'thanks to'and (Π55) *ta'* 'of' for which examples have been provided above already). The structural pattern involves a definite NP as external head followed by a relative clause introduced by *li* to the immediate right of which the resumptive PP is placed. The PP contains a pronominal complement which is referentially identical with the head noun. The construction schematically looks like the pattern in Figure 4.102.

$[NP_{DEF}\ [\textit{li}\ PP\ (X)]_{RC}]_{NP}$

Figure 4.102: Pattern of pronominal resumption in relative clauses (questionnaire).

Some of the examples in Table 4.24 are based on (modified) sentences found in our primary sources. In The table, only the resumptive PPs and their English translations are bracketed and highlighted in bold. Grey shading marks those cases in which prepositional government applies. Where two alternatives are given both were suggested by Sandra Abela.

Table 4.24: Relative clauses containing resumptive PPs.

Π	example	translation
(Π1) *apparti* 'apart from'	*it-tfal, li* ***[apparti minnhom]***$_{PP}$ *ħadd ma jorqod fil-kamra*	'the children, **[apart from whom]** nobody sleeps in the room'
(Π5) *bħal* 'like'	*it-tfal, li* ***[bħalhom]***$_{PP}$ *ħadd ma rajt qabel*	'the children **[the likes of which]** I have never seen before'
(Π8) *biswit* 'facing'	*it-tfal, li* ***[biswithom]***$_{PP}$ *ħadd ma kien joqgħod bil-qiegħda'*	'the children **[facing whom]** nobody was sitting'
(Π10) *daqs* 'equal to'	*l-istorbju li* ***[daqsu]***$_{PP}$ *qatt ma kont sema'*	'the noise I never heard anything like it'
(Π11) *dwar* 'about'	*it-tfal, li* ***[dwarhom]***$_{PP}$ *ma naf xejn*	'the children **[that]** I know nothing **[about]**'

141 Cf. Sections 4.1.1.1 and 4.1.2.2.2.

Table 4.24 (continued)

Π	example	translation
(Π12) *faċċata* 'opposite'	*il-lukanda li* ***[faċċata tagħha]**$_{PP}$ *kien hemm ir-roqgħa*	'the hotel **[opposite which]** there was the plot'
(Π13) *favur* 'in favour of'	*it-tim, li* ***[favurih]**$_{PP}$ *ir-referì ħa deċiżjoni kontroversjali*	'the team **[in favour of which]** the referee took a controversial decision'
(Π14) *fejn* 'near'	*it-tifel li* ***[fejnu]**$_{PP}$ *ma riditx toqgħod*	'the boy **[near whom]** she did not want to be'
(Π16) *fost* 'amongst'	*it-tfal, li* ***[fosthom]**$_{PP}$ *kien hemm ukoll ħija*	'the children **[amongst whom]** there was also my brother'
(Π17) *fuq* 'on'	*il-wiċċ li* ***[fuqu]**$_{PP}$ *kien hemm tbissima*	‚the face **[on which]** there was a smile'
(Π18) *ġewwa* 'inside'	*l-isptar li* ***[ġewwa fih]**$_{PP}$ *miet in-nannu*	'the hospital **[in which]** the grandfather died'
(Π19) *ġo* 'in'	*l-ittri li* ***[ġo fihom]**$_{PP}$ *tkellimt dwar ħajtek*	'the letters **[in which]** you spoke about your life'
(Π20) *għajr* 'except'	*in-nisa li* ***[għajrhom]**$_{PP}$ *ħadd ma seta' jagħmel ix-xogħol*	'the women **[except whom]** nobody could do the work'
(Π22) *għand* 'at s.o.'s place'	*oħti li* ***[għandha]**$_{PP}$ *mort kuljum*	'my sister **[to whom]** I went daily'
(Π23) *ħdejn* 'beside'	*ir-roundabout li* ***[ħdejnha]**$_{PP}$ *waqaf*	'the roundabout **[beside which]** he stopped'
(Π24) *ħlief* 'except'	*it-tfal, li* ***[ħliefhom]**$_{PP}$ *ħadd ma jorqod fil-kamra*	'the children, **[except whom]** nobody sleeps in the room'
(Π27) *kontra* 'against'	*il-barri li* ***[kontrih]**$_{PP}$ *Steve se jkollu jiġġieled*	'the bull Steve would have to fight **[against]**'
(Π28) *lejn* 'towards'	*l-arloġġ li* ***[lejnha]**$_{PP}$ *kien iħares*	'the clock **[towards which]** he looked'
(Π29) *lil* 'to'	*it-tfal li* ***[lilhom]**$_{PP}$ *oħti tħallihom jorqodu fil-kamra*	'the children that my sister lets sleep in the room'
(Π30) *ma'* 'with'	*it-tfal li* ***[magħhom]**$_{PP}$ *tkellimt il-bieraħ*	'the children **[with whom]** I spoke yesterday'
(Π31) *madwar* 'around'	*il-kċina li* ***[madwarha]**$_{PP}$ *Erkole biss kien qed idur*	'the kitchen **[in which]** Erkole was only turning around'
(Π32) *maġenb* 'close to'	*ix-xiħa li* ***[maġenbha]**$_{PP}$ *Estella tpoġġi bilqiegħda*	'the elderly lady **[close to whom]** Estella sat down'
(Π33) *matul* 'during'	*il-vjaġġ li* ***[matulu]**$_{PP}$ *ħadd ma qal kelma*	'the journey **[during which]** nobody said a word'
(Π34) *minbarra* 'except'	*it-tfal, li* ***[minnbarra minnhom]**$_{PP}$ *ħadd ma jorqod fil-kamra*	'the children, **[except whom]** nobody sleeps in the room'
(Π35) *minflok* 'instead of'	*il-karrozza li* ***[flokha]**$_{PP}$ *użajt il-biċikletta*	'the car **[instead of which]** I used the bicycle'

Table 4.24 (continued)

Π	example	translation
(Π36) *mingħajr* 'without'	*it-tfal li* ***[mingħajrhom]****$_{PP}$ l-iskola kienet vojta*	'the children **[without whom]** the school was empty'
(Π37) *mingħala* 'in s.o.'s opinion'	*it-tfal li* ***[mingħalihom]****$_{PP}$ kienu maħbubin*	'the children **[according to whose opinion]** they were loved'
(Π38) *mingħand* 'from s.o.'	*ommok li* ***[mingħandha]****$_{PP}$ ġibtlek messaġġ*	'your mother **[from whom]** I brought you a message'
(Π39) *minħabba* 'on account of'	*it-tfal, li* ***[minħabba fihom]****$_{PP}$ ħadd ma jista' jorqod fil-kamra*	'the children, **[on account of whom]** nobody can sleep in the room'
(Π40) *minkejja* 'in spite of'	*il-bieb magħluq, li* ***[minkejja fih]****$_{PP}$ huwa seta' jagħraf ilħna maskili*	'the closed door **[in spite of which]** he could recognise male voices'
(Π41) *minn* 'from'	*il-klassi, li* ***[minnha]****$_{PP}$ it-tlieta kienu*	'the class **[to which]** the three belonged'
(Π43) *mnejn* 'from near'	*il-persuna li* ***[mnejnha]****$_{PP}$ hu qatt ma ħarrek*	'the person **[from whom]** he never moved away'
(Π45) *permezz* 'by means of'	*il-metal detectors li* ***[permezz tagħhom]****$_{PP}$ Martin Mizzi seta' sab il-balal l-oħra*	'the metal detectors **[by means of which]** Martin Mizzi could find the other bullets'
(Π46) *qabel* 'before'	*it-turġien li* ***[qabilhom]****$_{PP}$ jinsab it-toilet*	'the stairs **[before which]** there is the toilet'
(Π47) *qalb* 'amidst'	*it-tfal li* ***[qalbhom]****$_{PP}$ l-għalliem qagħad bil-wieqfa*	'the children **[amidst whom]** the teacher stood'
(Π48) *qrib* 'near'	*il-mara li* ***[qribha]****$_{PP}$ kien lagħab it-tifel*	'the woman **[near whom]** the boy had played'
(Π49) *quddiem* 'in front of'	*it-tfal li* ***[quddiemhom]****$_{PP}$ ħadd ma ried jitkellem*	'the children **[in front of whom]** nobody wanted to speak'
(Π50) *rigward* 'concerning'	*il-kwistjoni li* ***[rigward tagħha]****$_{PP}$ kellna naraw xi tgħid il-liġi*	'the problem **[concerning which]** we had to see what the law says'
(Π51) *sa* 'till'	*il-palazz li* ***[sa ħdejh / sa fejnu]****$_{PP}$ it-triq twassalna*	'the palace **[to which]** the street leads us'
(Π53) *skont* 'according to'	*il-mappa, li* ***[skontha / skont hija]****$_{PP}$ ikollna nimxu fuq din it-triq*	'the map **[according to which]** we will have to walk on this street'
(Π56) *taħt* 'under'	*l-art, li* ***[taħtha]****$_{PP}$ kien hemm id-deheb*	'the ground **[beneath which]** there was the gold'
(Π58) *viċin* 'near'	*il-mara li* ***[viċin tagħha]****$_{PP}$ kien dejjem joqgħod*	'the woman **[near whom]** he always stayed'
(Π59) *waqt* 'at the time of'	*l-interrogatorji li* ***[waqthom]****$_{PP}$ il-ħati ma qal xejn*	'the interrogations **[during which]** the culprit said nothing'
(Π60) *wara* 'after'	*it-tfal li* ***[warajhom]****$_{PP}$ ħadd ma kien jimxi*	'the children **[behind whom]** nobody was walking'

We have been unable to find evidence for pronominal resumption in relative clauses for (Π3) *barra* 'outside', (Π6) *bħala* 'as', (Π9) *bla* 'without', (Π25) *inkluż* 'including', (Π26) *kif* 'as', (Π42) *mintul* 'all along', (Π54) *sotta* 'under', and (Π57) *versu* 'towards'. It is possible that at least some of these eight Πs are absent from the domain of pronominal resumption only because of the limitations of the Korpus Malti 3.0. At the same time, many of the attested cases of pronominal resumption found in the same corpus exclusively occur in legal texts, E.U. documents, news reports, or parliamentary speeches. It is debatable whether the constructions sound natural to the average native speaker of Maltese. For several of the relative clauses in Table 4.24, Sandra Abela remarked that the constructions are very unusual.

What can additionally be said as to the Πs whose participation in resumptive constructions is uncontroversial is that under resumption they behave exactly the same as they would in combination with pronominal complements. The government patterns inside and outside the relative clause are the same. This means that their morphosyntactic behaviour is consistent over sentence types.

Pronominal resumption terminates the section dedicated to the morphosyntax of Πs within the NP. The gist of the insights gained in the foregoing subsections will be presented together with those from Sections 4.1 and 4.3 in the final wrap-up of the syntax chapter in Section 4.4. Before we get there it is necessary, however, to scrutinise Πs and their behaviour inside the VP.

4.3 Inside the VP

In this section, we focus on three topics, namely argument marking (i.e. object marking) via Πs,[142] so-called prepositional verbs (i.e. verbs whose valency or case-frame requires or allows for the presence of a PP as complement),[143] and finally PPs as adjuncts (or adverbials).[144] This is of course only a small selection of issues which could be raised in connection with the VP-internal behaviour of Πs. Those phenomena which are not covered in the subsequent sections need to be addressed in future studies dedicated to Maltese Πs. Further subject matters which seem to be missing from this section are treated separately throughout Section 5 where semantics come more forcefully into play.

142 Cf. Section 4.3.1.
143 Cf. Section 4.3.2.
144 Cf. Section 4.3.3.

4.3.1 (Differential) object marking

Borg/Azzopardi-Alexander (1997: 140–143) describe object marking in Maltese distinguishing between direct, indirect, and "[o]ther objects governed by verbs" (Borg/Azzopardi-Alexander 1997: 142) in accordance with the internal logic of the LDS-format for descriptive grammars. Especially in the latter subsection, a plethora of Πs is mentioned whose function it is to introduce the object of a given verb. Since the majority of these putative object markers are semantically transparent it is questionable whether their classification as object markers is felicitous in the first place. Following Fabri (1993: 114–121), we assume that the most grammatical ised object marker is (Π29) *lil* 'to'. The other Πs Borg/Azzopardi-Alexander (1997) mention in addition to this bona fide object marker will be discussed in the section on prepositional verbs.[145] As to the possibility of replacing (Π29) *lil* 'to' as indirect object marker with (Π21) *għal* 'for' (Borg/Azzopardi-Alexander 1997: 141), we will take up this issue below.[146]

Saari (2003: 32–40) dedicates an extraordinarily long section to (Π29) *lil* 'to' in which many of the points are raised to which we refer in what follows. The fact that (Π29) *lil* 'to' displays special behaviour which sets it apart from the bulk of the other Πs in BLOMP 2.0 has been mentioned time and again in the foregoing parts of this study.[147] Most strikingly, (Π29) *lil* 'to' is the Π which is least prone to EQUI-P-deletion as argued above.[148] The retention rate of this Π on the second conjunct exceeds that of other Πs by far. (Π29) *lil* 'to' is the only Maltese Π for which Π_2-retention is more frequent than EQUI-P-deletion. This divergence from the usually preferred pattern of EQUI-P-deletion is causally connected to the object-marking function of the Π under inspection because each NP which fulfils the syntactic function of object tends to be marked as such. However, there is variation.

First of all, the employment of (Π29) *lil* 'to' depends on a conspiracy of factors according to which especially animacy and definiteness are crucial for triggering overt object marking so that the object marker should never occur with indefinite inanimate direct objects (Borg/Mifsud 2002: 35). Furthermore, for definite animate (preferably human) objects the Π neutralises the distinction of direct and indirect objects – a classical study to be referred to with regards to object diffuseness in Maltese is Borg/Comrie (1984). Borg/Mifsud (2002: 34–36) assume that the object marker is obligatory with personal names and stressed personal pronouns as well as with possessed human nouns (i.e., kinship terms hosting the possessor suffix) whereas the use of (Π29) *lil* 'to' is optional with definite animate common nouns and indefinite-specific human nouns

145 Cf. Section 4.3.2.

146 In Section 5.1.3.17 where the semantic role of BENEFACTIVE is discussed.

147 E.g. in Sections 1.6.5 (ii), 3.3.1, and 4.1.1.2.1 (vi).

148 Cf. Section 4.1.4.

as well as with topicalised animate objects provided the verb hosts the object clitic (Borg/Azzopardi-Alexander 2009: 74). Just/Čéplö (2022: 118–119) and Čéplö/Just (2023) provide a long list of criteria which might come into play in the domain of object-marking by way of indexing on the verb. Döhla (2016: 154) observes that there is a hierarchy of object marking which we adapt for Figure 4.103.

personal pronoun > personal name > human noun [definite > specific] > other

Figure 4.103: Hierarchy of object marking according to Döhla (2016).

On the basis of his fieldwork data, the quoted author further assumes that personal names as topics are always overtly marked as objects. Definite human nouns are marked as objects in more than half of the cases whereas specific human nouns display overt object marking in less than 50% of the cases.

Since we do not intend to explore the question of objecthood any further to which we cannot add anything particularly new because the topic is beyond the scope of this study on Maltese ΠS, we make do with illustrating some of the above phenomena with examples from the Korpus Malti 3.0. The bulk of the cases presented below are connected to direct-object marking. Indirect object-marking will be discussed only briefly at the end of the illustration. In (4.187), the predicted behaviour of direct pronominal objects is shown alongside that of a pronominally possessed kinship term. The transitive verb *ra* 'see' is involved in both examples.

(4.187) Pronominal objects and possessed kinship terms

(a) [Korpus Malti 3.0; opinion1929][149]

Rajt	*[lilu]$_{PP}$*	*u*	*[lil*	*missieru]$_{PP}$*	*waqt*
see:1SG.PFV	[**to**:3SG.M]	and	[**to**	father:3SG.M]	at_the_time_of
l-iffirmar	*tal-kuntratt*				
DEF-signing	of:DEF-contract				

'I saw [him] and [his father] at the time of the signing of the contract.'

(b) [Korpus Malti 3.0; news72775]

rajt	*l-aħbarijiet*	*u*	*[lilu]$_{PP}$*	*rajtu*
see:1SG.PFV	DEF-news:PL	and	[**to**:3SG.M]	see:1SG.PFV:3SG.M
jiċċelebra	*r-rebħa*	*elettorali*		
3SG.M.IPFV:celebrate	DEF-victory	electoral		

'[. . .] I watch the news and [it was him] celebrating the electoral victory I saw.'

149 We had to correct the example because in the original the second *lil*-PP appeared as <*lill-missieru*> which is ungrammatical since possessed nouns which host the possessor suffix do not combine with the definiteness pro-clitic.

In (4.187a), we have a copulative coordination whose first conjunct is a pronominally inflected Π. This is one of the factors which tend to block EQUI-P-deletion on the second PP as we already know.[150] The Korpus Malti 3.0 does not provide any hits for **lilu u missieru* or similar combinations of pronominally inflected (Π29) *lil* 'to' as first conjunct and a possessed kinship term as second conjunct (cf. also Saari 2003: 36–37). In (4.187b), the *lil*-PP is positioned to the left of the predicate nucleus yielding a pragmatically marked linear order whose effect is that of contrastive focus or topicalisation – an issue we will come back to below. Note also that the definite inanimate object *l-aħbarijiet* 'the news' in (4.187b) does not require (Π29) *lil* 'to' as object marker.

The three sentences in (4.188) involve personal names as direct objects (cf. also (4.17c)).

(4.188) Personal names

(a) [Korpus Malti 3.0; literature1]

rajt	***[lil***	*Marika]*$_{PP}$	*ddur*	*il-kantuniera*
see:1SG.PFV	**[to**	Marika]	3SG.F.IPFV:turn	DEF-corner

'[. . .] I saw [Marika] turning around the corner [. . .].'

(b) [Korpus Malti 3.0; culture70]

[Lil	*Paula]*$_{PP}$	*niddeskriviha*	*bħala*	*persuna*
[to	Paula]	1SG.IPFV:describe:3SG.F	as	person

ferriħija	*u*	*pożittiva*
good_humoured:F	and	positive:F

'[(As to) Paula] I describe her as a good-humoured and positive person.'

(c) [Korpus Malti 3.0; literature10][151]

[Mandy]$_{NP}$	*x'*	*nambiha*
[Mandy]	what	1SG.IPFV:have_use_for:3SG.F

'[As to Mandy], what use do I have for her?'

The pragmatically neutral situation is illustrated in (4.188a). In contrast, both (4.188b) and (4.188c) can be classified as pragmatically marked since the sentence-initial personal name is coreferential with the object marker *-ha* on the finite verb, meaning: the personal name occurs in a slot that is not normally occupied by the lexical object-NP. As mentioned above, Borg/Azzopardi-Alexander (2009: 74) assume that under topicalisation the object marker (Π29) *lil* 'to' can be dropped

---

150 Cf. Section 4.1.4.

151 The verb *amba* 'have some use for' is defective and occurs only in the imperfective (Aquilina 1987: 24).

because as topic it is extrasentential and thus exempt from the usual dependency and government relations that hold within a sentence.

Two things come to our mind in connection to the optional use of the object marker in pragmatically marked contexts. The remarks we are about to make have to be taken with a grain of salt since we refer to written Maltese whereas Borg/Azzopardi-Alexander (1997, 2009) have analysed spoken data. First, the absence of (П29) *lil* 'to' is not restricted to pragmatically highlighted lexical NPs. The personal pronoun can be topicalised in its lexical citation form independent of the syntactic function its coreferential expression has in the sentence as shown in (4.189a).

(4.189) *Jien* vs *lili*

(a) [Farrugia 2022: 127]

Mikiel [jien]$_{NP}$ *tawni transfer vendikattiv*
Mikiel [1SG] give:3PL.PFV:1SG transfer vendicative
'Mikiel, [as for me], they gave me a vendicative transfer.'

(b) [Korpus Malti 3.0; *news86074]*

[Lili]$_{PP}$ *tawni ċ-ċans li nistabbilixxi*
[**to**:1SG] give:3PL.PFV:1SG DEF-chance that 1SG.IPFV:establish
ruħi bħala kowċ
self:1SG as coach
'They gave [me] the chance of establishing myself as a coach.'

In (4.189a), the topic *jien* 'I' is coreferential with the object marker *-ni* on the finite verb. The fact that the grammatically possible form *lili* 'to me' is not chosen is evidence of the extrasentential status of the topic. This argument seems to be circular since topics are supposed to be extrasentential by definition. The absence of the object marker is thus what we expect to find. However, what about cases like (4.188b) and (4.189b) where the left-dislocated personal name is preceded by (П29) *lil* 'to'? Does it make a difference pragmatically whether or not (П29) *lil* 'to' is present?

If the absence of (П29) *lil* 'to' is a symptom of the extrasentential status of the topic, then it must be the case that the presence of (П29) *lil* 'to' indicates that the PP is not extrasentential because the object marker is required by the transitive verb, meaning the supposed topic obeys to the rules which hold within the clause. What

(4.187b), (4.188b–c) and (4.189a–b) have in common is the cooccurrence with object marking on the verb. Borg/Azzopardi-Alexander (1997: 126) state that

> an enclitic object and/or indirect object pronoun coreferential with the direct and indirect object respectively, is obligatorily suffixed to the verb where these elements are topicalised by movement to the initial position.

If no indexation of the object on the verb applies although the object-NP occurs before the finite verb Fabri/Borg (2002: 359–360) assume that the fronted object-NP is focussed but not topicalised. Just/Čéplö (2022: 106–109) review the extant hypotheses with regards to the distinction of topic and focus on the one hand, and the possibility of the existence of very fine-grained hierarchies of topic-worthiness, on the other hand. The authors specifically address the indexation of the object on the verb under particular pragmatic conditions. They conclude that "it is impossible to account for [optional object indexing] on the basis of rules" (Just/Čéplö 2022: 127). The presence vs absence of (Π29) *lil* 'to' is mentioned as one of the parameters taken account of in the study (Just/Čéplö 2022: 119) but in the presentation of the results, this parameter is not mentioned again. We hypothesise that the presence vs absence of (Π29) *lil* 'to' in cooccurrence with object indexation on the verb correlates with different degrees of topicality of the pragmatically highlighted NP. In Figure 4.104, we propose a simple scale of four degrees.

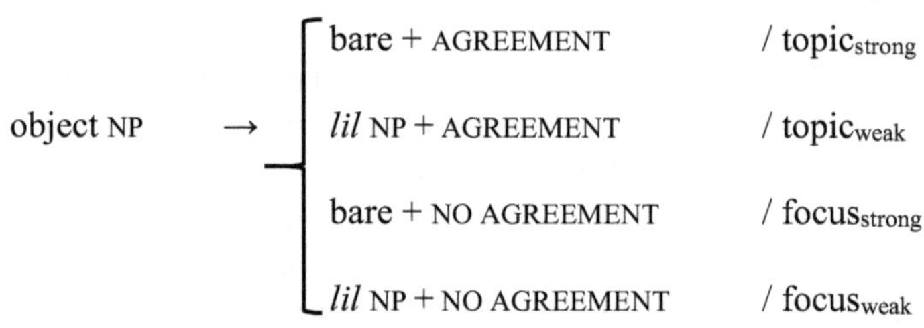

Figure 4.104: Binary scale of topicality.

This tentative scale calls for corroboration on the basis of a more reliable data set reflecting the language practice in spoken Maltese. Our own exclusively written material is insufficient to clarify pragmatic problems once and for all. Information structure is not a prominent theme in this study since prosodic phenomena cannot be accounted for. What is important for the study of Maltese Πs is the variable affectedness of (Π29) *lil* 'to' to pragmatic categories. To what extent other Πs resemble (Π29) *lil* 'to' with regards to their sensitivity to pragmatic requirements is a topic for future research.

In (4.190), kinship terms in object function are featured.

(4.190) Kinship terms

(a) [Korpus Malti 3.0; literature15]

rajt	*missier*	*ħiereġ*	*mill-crying room*	*bil-qmis*
see:1SG.PFV	father	exit.PART	from:DEF-crying_room	with:DEF-shirt
imqattgħa				
PART:torn				

'[. . .] I saw a father coming out of the crying room with his shirt torn [. . .].'

(b) [Korpus Malti 3.0; news75800]

Rajt	***[lil**l-missier]*$_{PP}$	*jiżvojta*	*ż-żibel*
see:1SG.PFV	[**to**:DEF-father]	3SG.M.IPFV:empty_out	DEF-rubbish

'I saw [the father] emptying out the waste.'

(c) [Korpus Malti 3.0; literature12]

[Lil	*ommi]*$_{PP}$	*ma*	*ridtx*	*ninkwetaha*
[**to**	mother:1SG]	NEG	want:1SG.PFV:NEG	1SG.IPFV:trouble:3SG.F

'[As for my mother] I did not want to trouble her.'

The examples in (4.190) conform to our expectations. (Π29) *lil* 'to' is present in (4.190b) where it introduces a definite human object and in (4.190c) where it combines with a pragmatically prominent possessed kinship term. (4.190a) is in accordance with the literature on the subject too because an indefinite object does not require the presence of the Π. However, we have learnt from (4.17) that (Π29) *lil* 'to' is not absolutely barred from indefinite contexts. It is possible that the PP *lil ftit nies* 'to a few people' is specific in the sense that the speaker knows the recipients. Moreover, (Π29) *lil* 'to' functions as marker of an indirect object in this earlier example. It may also be used as marker of the direct object (albeit infrequently) if it is indefinite but human as shown in (4.191).

(4.191) [Korpus Malti 3.0; news82798]

Din	*hi*	*sistema*	*li*	*hija*	*maħduma*	*biex*
DEM:F:PROX	3SG.F	system	that	3SG.F	PART:make:F	in_ortder_to
iżomm	***[lil***	*ftit*	*nies]*$_{PP}$	*fis-setgħa*		
3SG.M.IPFV:hold	[**to**	little	people]	in:DEF-power		

'This is a system that is made to keep [a few people] in power.'

Like in the case of (4.17), it can be assumed that the speaker knows who the few people are that the system keeps in power. The object NP is therefore specific. Since it also contains a human noun the employment of (Π29) *lil* 'to' is facilitated although it is by no means mandatory. As comes to the fore in (4.192c), the Π can be absent even if the object NP is both definite and human. Note however that definite NPs frequently have generic reference in Maltese.

(4.192) Human nouns[152]

(a) [Korpus Malti 3.0; european3576]
jipproteġi ***[lil**l-ħaddiema]$_{PP}$* *kontra* *r-riskju*
3SG.M.IPFV:protect **[to**:DEF-worker:PL] against DEF-risk
'[. . .] it protects [the workers] against the risk [. . .].'

(b) [Korpus Malti 3.0; parl23]
***[Lil**l-ħaddiema]$_{PP}$* *għandna* *nħalluhom* *jaħdmu*
[to:DEF-worker:PL] have.IPFV:1PL 1.IPFV:let:PL:3PL 3.IPFV:work:PL
bis-serjetà
with:DEF-seriousness
'[As to the workers] we have to let them work seriously.'

(c) [Korpus Malti 3.0; academic8]
fabbriki *li* *jimpjegaw* *it-tfal* *jew* *jisfruttaw*
factory:PL that 3.IPFV:employ:PL DEF-child.PL or 3.IPFV:exploit:PL

152 As to supposed cases of indefinite human nouns combining with the object marker, we emphasize that, in the Korpus Malti 3.0, there are numerous false friends caused by typos. Example (4.xiii) is representative of the frequent omission of the definite pro-clitic in writing which occurs mostly in contexts in which no assimilation of the definiteness marker applies. We have added the missing elements between pointed brackets.

(4.xiii) [Korpus Malti 3.0; news110695]
il-Ministru *Cardona* *rringrazzja* *lilhom* *u* ***[lil**<l-> ħaddiema* *kollha*
DEF-minister Cardona thank to:3PL and **[to**<DEF->worker:PL all:3SG.F
Maltin]$_{PP}$ *tal-impenn* *u* *l-kontribut* *li* *jgħatu*
Maltese:PL] of:DEF-engagement and DEF-contribution that 3.IPFV:give:PL
'The minister Cardona thanked them and [all the Maltese workers] for the engagement and the contribution they give.'

The postnominal universal quantifier *kollha* 'all' supports the interpretation that the object NP must be definite. The use of (Π29) *lil* 'to' with the human noun is additionally motivated by the pronominally inflected form of the same Π as first conjunct in the coordination.

[il-ħaddiema]_{NP} *fil-proċess* *produttiv*
[DEF-worker:PL] in:DEF-process productive
'[. . .] factories which employ children or exploit [the workers] in the production process [. . .].'

The examples (4.192a) and (4.192b) are nothing out of the ordinary since definite human nouns are accompanied by the object marker. In the pragmatically marked sentence-initial position the fronted object-NP is in an anaphoric relation to the object marker *-hom* on the finite verb in (4.190b). In the light of (4.192a) however, we would have expected to find the object marker also in (4.189c) because the same definite human noun functions as object. There is a parallel in the same sentence, namely the absence of (Π29) *lil* 'to' in *jimpiegaw it-tfal* 'they employ (the) children'. As the examples in (4.193) show the same verb may take an indefinite human object with or without object marker.

(4.193) With or without object marker
(a) [Korpus Malti 3.0; news125215]
l-kumpaniji *b'* *mod* *ċar* *jippreferu* *jimpjegaw*
DEF-company:PL with manner clear 3.IPFV:prefer:PL 3.IPFV:employ:PL
[lil *student]_{PP}*
[**to** student]
'[. . .] the companies clearly prefer to employ [a student].'
(b) [Korpus Malti 3.0; *news182643]*
qed *jistieden* *aktar* *kumpaniji* *ta'* *l-ICT* *biex*
PROG 3SG.M.IPFV:invite more company:PL of DEF-ICT in_order_to
jimpjegaw *[studenti]_{NP}* *għax-xhur* *tas-Sajf*
3.IPFV:employ:PL [student:PL] for:DEF-month.PL of:DEF-summer
'[. . .] he is inviting more companies to employ [students] for the summer months.'

The variation is obviously strong enough to preclude the predictability of the behaviour of (Π29) *lil* 'to' in combination with human nouns. Genericity, number, and sundry factors might be involved in determining the behaviour. Yet, we do not know how to hierarchise these factors on a scale of strength.

In (4.194), we can see that even NPS which represent inanimate but highly individualised referents attest to variation as to object marking.

(4.194) Inanimate nouns
(a) [Korpus Malti 3.0; academic13]
iżda *ma* *ninsewx* *biex* *jekk* *jista'*
but NEG 1.IPFV:forget:PL:NEG in_order_to if 3SG.M.IPFV:can

jkun *nevitaw* *[ix-xemx* *ta'* *bejn* *l-10 a.m.*
3SG.M.IPFV:be.FUT 1.IPFV:avoid:PL [DEF-sun of between DEF-10 a.m.
u *t-3 p.m.]*$_{NP}$
and DEF-3 p.m.]
'[. . .] however, we should not forget to avoid if possible [the sun between 10 a.m. and 3 p.m.]

(b) [Korpus Malti 3.0; news88481]
Il-qamar *għatta* *kważi* *għal* *kollox* *[**li**x-xemx]*$_{PP}$
DEF-moon cover almost for all [**to**:DEF-sun]
'[. . .] the moon almost completely covers [the sun].'

(c) [Korpus Malti 3.0; academic13]
[Ix-xemx *diretta]*$_{NP}$ *ma* *jriduhiex* *iżda* *jħobbu*
[DEF-sun direct:F] NEG 3.IPFV:want:PL:3SG.F:NEG but 3.IPFV:love:PL
jkunu *fid-dawl*
3.IPFV:be.FUT:PL in:DEF-light
'[As to the direct sun] they do not like it but they love to be in the light [. . .].'

The remarkable example is (4.194b) because the definite inanimate noun *ix-xemx* 'the sun' is preceded by the object marker. In contrast, the object marker is absent from (4.194c) where the same definite inanimate noun is topicalised.

The two examples in (4.195) provide us with further pieces of evidence of variation in the domain of object marking of inanimate nouns. Note that the examples are taken from the same source text and describe very similar situations.

(4.195) Further inanimate nouns

(a) [Korpus Malti 3.0; european9696]
l-pendlu *jolqot* *[il-mera]*$_{NP}$ *fil-mument* *meta*
DEF-pendulum 3SG.M.IPFV:hit [DEF-mirror] in:DEF-moment when
l-pendlu *jilħaq* *il-pożizzjoni* *vertikali*
DEF-pendulum 3SG.M.IPFV:reach DEF-position vertical
'[. . .] the pendulum hits [the mirror] in the moment when the pendulum reaches the vertical position [. . .].'

(b) [Korpus Malti 3.0; european9696]
l-martell *jolqot* *[**lil**l-mera]*$_{PP}$ *fuq* *in-naħa* *tal-wiċċ*
DEF-hammer 3SG.M.IPFV:hit [**to**:DEF-mirror] on DEF-side of:DEF-face
li *jirrifletti*
that 3SG.M.IPFV:reflect
'[. . .] the hammer hits [the mirror] on the side of the reflecting surface [. . .].'

We have no ready explanation for this and similar patterns of variation. We would have to guess why *il-mera* 'the mirror' is the direct object of *jolqot* 'it hits' (< *laqat* 'hit') without object marker in (4.195a) but takes the object marker in (4.195b) although the verb remains the same. We agree with Just/Čéplö (2022) that it is necessary to conduct a thorough bottom-up analysis of the distributional facts before putting forward an explanatory model for (not only differential) object marking in Maltese.

To close this section, we briefly look at the blurred distinction between direct and indirect objects. In (4.196), the verb *bagħat* 'send' takes a coordinated direct object in the first example but an indirect and a direct object in the second example.

(4.196) Direct vs indirect object
(a) [Korpus Malti 3.0; literature80]
bagħat ***[lil*** *martu]*$_{PP}$ *u* ***[lil*** *uliedu]*$_{PP}$ *Għawdex*
send **[to** wife:POSS.F:3SG.M] and **[to** child.PL:3SG.M] Gozo
'[. . .] he sent [his wife] and [his children] to Gozo.'
(b) [Korpus Malti 3.0; news83955]
kien bagħat ittra ***[lil*** *ommu]*$_{PP}$
be send letter **[to** mother:3SG.M]
'[. . .] he had sent a letter [**to** his mother].'

(Π29) *lil* 'to' as such is neutral as to the functional differentiation between direct and indirect object. This means that given the appropriate conditions, both object NPs of a verb can be introduced by the same Π as in (4.197). The *lil*-PPs are additionally put into curly brackets which are indexed for their object functions.

(4.197) [Korpus Malti 3.0; news208382]
kellu dritt bħala ċittadin Ewropew jirrapporta
have.PFV:3SG.M right as citizen European 3SG.M.IPFV:report
*{[**lil**ha]*$_{PP}$*}*$_{DO}$ *{[**lil** OLAF]*$_{PP}$*}*$_{IO}$ *iżda ma kellu ebda dritt*
{[**to**:3SG.F]} {[**to** OLAF]} but NEG have.PFV:3SG.M no right
li jheddidha
that 3SG.M.IPFV:threaten:3SG.F
'[. . .] as European citizen he had the right to report {[her]} {[**to** OLAF]} but he had no right to threaten her.'

The use of (Π29) *lil* 'to' as marker of the indirect object is not subject to special conditions as when it is used for direct objects. Borg/Azzopardi-Alexander (1997: 140–141) mention systematic exceptions to the otherwise compulsory employment of (Π29) *lil* 'to' for indirect objects.[153] Table 4.25 contains examples of animate and inanimate indirect objects with varying definiteness cooccurring with the object marker. All examples stem from the Korpus Malti 3.0. Different forms of the ditransitive verb *ta* 'give' are involved throughout the list of examples.

Table 4.25: (Π29) *lil* 'to' as marker of the indirect object.

source	example / translation	values
news11971	*F'kumment li ta {[**lil** ġurnal Taljan]$_{PP}$}$_{IO}$ il-kaptan qal li kien sfurzat biex jammetti li hu ħati ta' dan.* 'In a comment he made {[**to** an Italian journal]}, the captain said that he was forced to admit that he was guilty of that.'	inanimate, indefinite
literature51	*kien mingħalih ta {[**li**d-dinja]$_{PP}$}$_{IO}$ l-invenzjoni tas-seklu* '[. . .] to his mind he had given {[the world]} the invention of the century.'	inanimate, definite
european10045	*s-sitwazzjoni deboli finanzjarja tagħha li ma kienetx tippermetti li aktar kreditu jkun mogħti {[**lil**ha]$_{PP}$}$_{IO}$ minn xi bank ieħor* '[. . .] its weak financial situation that did not permit that more credit would be given {[**to** it]} from any other bank.'	inanimate (bank), pronominal
literature34	*{[**Lil** oħtu]$_{PP}$}$_{IO}$ ma qalilhiex li hu kien ivvota għall-bidla* '{[As to his sister]} he did not tell her that he voted for the change [. . .].'	human, possessed, topic
parl4824	*Kont jien li tajt {[**lil** Dr Alfred Sant]$_{PP}$}$_{IO}$ l-ittri ta' korrispondenza* 'It was me who gave {[Dr Alfred Sant]} the letters of correspondence [. . .].'	human, personal name

To sum up the above observations some of which remain inconclusive or fragmentary we propose the following preference rule which captures the main factors which regulate the use of (Π29) *lil* 'to' as object marker. It almost goes without saying that Figure 4.105 simplifies matters considerably and that the preference rule will certainly undergo substantial modification in the future. For the time being it suffices to see that the presence and absence of the object marker is determined by an intricate network of different factors. It remains to be seen whether similarly complex systems of rules and subrules emerge in the context of other VP-internal phenomena.

153 These exceptions will be discussed in Sections 4.3.2–4.3.3.

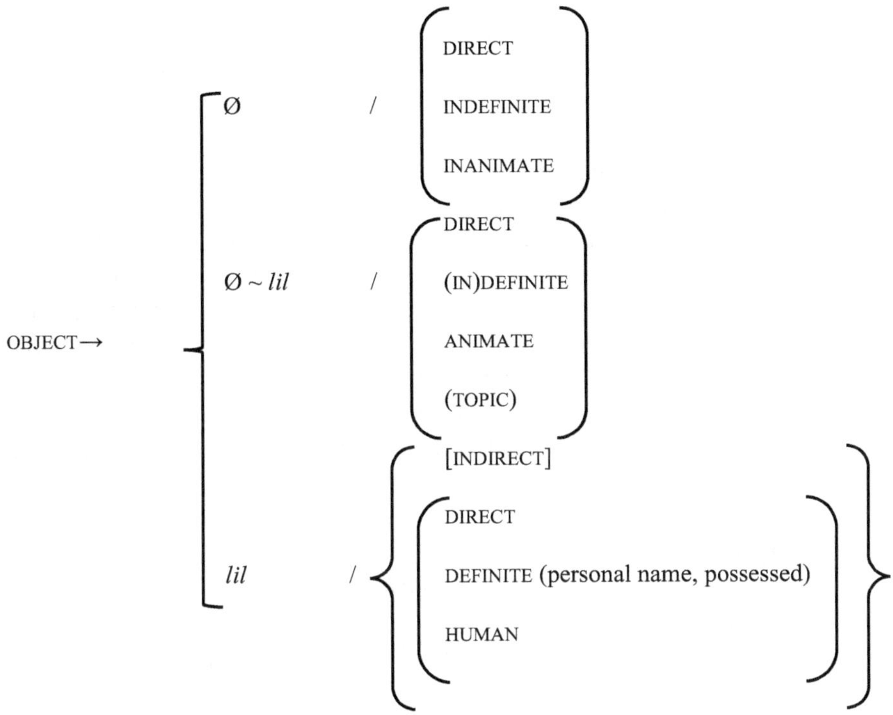

Figure 4.105: Preference rule for the use of the object marker.

4.3.2 Prepositional verbs: from valency enhancement to object marking

In a paper which was originally published as early as 1967, Aquilina (1976: 67–68) assumes an inventory of Maltese ΠS which contains only twenty members about which the author says that

> [e]ach of the [. . .] twenty prepositions has its own idiomatic collocation in the sentence structure of the spoken language covering a range of situations from literalness (physical location) to figurativeness (figurative illustration). To these one must add an additional semantic function which occurs when they enter into the verbal structure becoming, so to say, an integral part of it affecting its meaning, either by adding something to the original meaning, (that is, the original meaning is retained but in a modified form) or by introducing a completely new idea. This happens when the addition of the preposition brings with it a complete change of meaning.

In cases of complete meaning change, Aquilina (1976: 67) speaks of prepositional verbs of which there supposedly exist many albeit not as many as in English (Aquilina 1976: 67). The above quote clearly shows that Aquilina's notion of prepositional verbs is exclusively semantic. One might argue therefore that the discussion of his proposal would be better placed in Section 5. However, in the light of Borg/Azzopardi-Alexander's (1997: 142–143) assumption that there are Πs other than (Π29) *lil* 'to' which serve as object-markers in Maltese it makes sense to look at the candidate verbs and Πs from the point of view of syntax not the least because Aquilina's (1976) catalogue and the case-book of Borg/Azzopardi-Alexander (1997) largely overlap. As the reader will notice soon the issues raised below more often than not make it necessary to make a reference to semantics, too.

We approach the subject by way of assessing Aquilina's (1976: 71) example for the combination of *ħabb* 'love' and (Π15) *fi* 'in' which we reproduce as (4.198) with added glosses.

(4.198) (Aquilina 1976: 71)

M'	*għandekx*	*xi*	*tħobb*	*[fiħ]*$_{PP}$
NEG	have.IPFV:2SG.NEG	what	2SG.IPFV:love	[**in**:3SG.M]

'There is nothing about him you can love him for.' [O.T.]

Aquilina seems to assume the existence of a prepositional verb **ħabb fi* 'love s.o.' (which would nicely explain why (Π39) *minħabba* 'on account of' is a Π with locative government[154]). The object pronoun that is marked in yellow in the English translation would correspond to this analysis. We take however, issue with this assumption because there is no direct syntactic relation between *tħobb* 'you love' and *fih* 'in him' although the two syntactic words are neighbours in (4.198). We argue that the PP *about him* highlighted in grey in the English translation is what counts. The object of *tħobb* 'you love' is *xi (fih)* 'something (in him)'. In a way, this object is a shared argument of the two clauses (*you do not have* ***anything*** + *you love* ***something***). The PP *fih* 'in him' only serves to locate the referent of *xi* in (figurative) space. It is a spatial adverbial with the status of an adjunct.[155] Its presence is optional in the sense that the verb *ħabb* 'love' does not require the presence of a *fi*-PP to take an object. Thus, we rebut Aquilina's analysis since there is no prepositional verb **ħabb fi* 'love s.o.' either semantically or syntactically, in the first place. In terms of syntax, several of Aquilina's examples form parallels to this case insofar as they involve PPs whose function is that of adverbial modification. Their combination with a given verb often

154 As discussed in Section 4.1.2.2.2.1.2.
155 Cf. Section 4.3.3.

affects the semantics in such a way that a new constructional meaning arises. In these cases, we are confronted with idioms which we cannot satisfactorily account for in this study. What we do instead is focus on recurrent patterns of valency-changing uses of ΠIs.

Intransitive verbs may take an additional NP which comes as the complement of a PP. This is the case for instance, with *daħak* 'laugh' in (4.199). In the (a)-sentence, the expected use of the verb as intransitive is documented. In the (b)-sentence, there is a second participant besides the agent *Teodoro*, namely the 1st person singular which is the undergoer in the laughing event. This undergoer is introduced into the situation by way of adding a *bi*-PP to the clause.

(4.199) Valency change

(a) [Korpus Malti 3.0; literature20]

Marco	*daħak*	*u*	*s-Sur Tabone*	*wkoll*
Marco	laugh	and	DEF-Mr Tabone	too

'Marco laughed and Mr Tabone also.'

(b) [Korpus Malti 3.0; literature18]

Kemm	*daħak*	[***bija***]$_{PP}$	*Teodoro*	*dakinhar*
how_much	laugh	[**with**:1SG]	Teodoro	that_day

'How much Teodoro laughed [**at** me] on that day!'

Aquilina (1976: 69) also lists this verb in his catalogue. For him however, the most important aspect is that there are contexts in which *daħak bi* must be translated into English as 'deceive', i.e., the construction does no longer have a compositional meaning, or so it seems. In a way, Aquilina's interest in cases of this kind is motivated similarly to the classical approach of early Construction Grammar (Fillmore/Kay/O'Connor 1988) where idioms were in the focus of attention. Since *daħak bi* can always be translated as 'laugh at' (with a metonymical interpretation where necessary) however, the argument for a fully-blown meaning change is not compelling. However, we leave semantic issues aside for the moment. In (4.199a), *daħak* 'laugh' is monovalent since it only allows for the presence of the subject argument. As mentioned above, in (4.199b), a second participant is involved in the situation namely the undergoer in the 1st person singular. The *bi*-PP is the only way to integrate this additional participant into the situation. In contrast to transitive verbs which would require the employment of the object marker (Π29) *lil* 'to' or the object indexation on the finite verb with pronominal objects as shown in the foregoing section, many intransitive verbs enhance their valency by way of admitting a PP headed by (Π7) *bi* 'with' or other ΠIs to the construction. This strategy of valency enhancement is not always easy to distinguish from adverbial uses of identically headed PPs. Multifunctional (Π7) *bi* 'with' is a paradigm case with regards to problems of this kind. We anticipate minimally what is the major theme of Section 4.3.3 when we illustrate the interpretative problems with the examples in (4.200).

(4.200) Different kinds of *bi*-PPs

(a) [Bartolo/Vella 2009: 38]

imissni	*għedtilkom*	*[**bi**ha]$_{PP}$*	*qabel*
touch:1SG	say:1SG.PFV:IO:2PL	[**with**:3SG.F]	before

'[. . .] I should have told you [**about** it] before [. . .].'

(b) [Bartolo/Vella 2009: 55]

Xi	*trid*	*tgħid*	*[**bi**ha]$_{PP}$*	*din*
what	3SG.F.IPFV:want	3SG.F.IPFV:say	[**with**:3SG.F]	DEM:F:PROX

'What does she want to say [**with** it]?'

(c) [Bartolo/Vella 2009: 60]

Le	*Ġingo*	*qaltilha*	*[**b'***	*vuċi*	*telepatika*	*dgħajfa*
no	Ġingo	say:3SG.F.PFV:IO:3SG.F	[**with**	voice	telepathic	weak
ħafna]$_{PP}$	*Antida*					
very]	Antida					

'"No, Ġingo", said Antida to her [**with** a very weak telepathic voice].'

The verb *qal* 'say' (Aquilina 1976: 74) combines with three different kinds of *bi*-PPs in the above examples. The PPs cannot be told apart on formal grounds in the sense that there is no morphological means that indicates for exactly what function the PP is used. In point of fact, in (4.200a), (Π7) *bi* 'with' introduces the participant which should have been mentioned earlier by the speaker. In (4.200b) however, the *bi*-PP refers to an earlier speech-act of the female antagonist. The content of this speech-act is depicted as the instrument for achieving a still unknown goal. Thus, (4.200b) illustrates the instrumental function of (Π7) *bi* 'with'. In contrast, (4.200c) features a *bi*-PP which adverbially modifies the predicate by way of characterising the loudness and style of the way Antida utters her negative reaction. The *bi*-PP can be interpreted either as instrumental or modal (or as a combination of both functions). This means that the cooccurrence of the verb *qal* 'say' (and others like it) with (Π7) *bi* 'with' in a clause cannot sweepingly be interpreted in one and only one way.

In the remainder of this section, we will try to work out in what way (if at all) Πs other than (Π29) *lil* 'to' can be classified as object markers. Since the section Borg/Azzopardi-Alexander (1997: 142–143) treats of so-called other objects it is not possible to apply tests like passivisation to determine the objecthood of a given PP because this would only be feasible with direct objects. In the subsequent paragraphs, we will repeatedly emphasise that the putative object markers reflect only a very low degree of grammaticalization (Lehmann 1995), if at all. Their paradigmaticity is limited. Very often there is competition between several Πs for fulfilling the same function. Moreover, the Πs display hardly any signs of desemanticisation. In yet other cases the employment of a given Π depends on a specific verb and cannot be generalised over other verbs. The supposedly object-marking Πs are responsible for participants which are normally not subject to a change-of-state in the situation and thus fail to meet the criteria of the prototypical transitivity (Croft 2022: 183–186). We are therefore reluctant to declare these Πs to be object markers on a par with (Π29) *lil* 'to'. The best status they can be given is that of oblique objects whose objecthood is grounded more in semantics than in syntax and whose behavioural properties cannot be captured by applying syntactic rules. A possible exception is the already mentioned (Π7) *bi* 'with' which is involved regularly in a variety of patterns which come close to object marking. This case will be discussed at the end of this section.

For a start, we take a look at Aquilina's (1976: 68) very first verb, namely *bata* 'suffer'. Three Πs are listed as combining with *bata* 'suffer' to form prepositional verbs one of these Πs being (Π7) *bi* 'with' the other two being (Π21) *għal* 'for' and (Π41) *minn* 'from'. In (4.201), we provide examples of four different syntactic contexts in which this verb is attested in the Korpus Malti 3.0. The verb is represented by the form *ibati* 'he suffers', i.e. by the 3rd person singular masculine in the imperfective. The verb, its gloss, and the translation are marked in yellow colour.

(4.201) *bata* 'suffer'

(a) intransitive – [Korpus Malti 3.0; news108177]

Ma nistgħux inġgħelu lil xi ħadd
NEG 1.IPFV:can:PL:NEG 1.IPFV:comple:PL to some somebody
ibati
3SG.M.IPFV:suffer
'We cannot compel anybody to suffer.'

(b) transitive – [Korpus Malti 3.0; *culture3912]*

l-ivvizzjat ibati {l-effett dirett tal-abbuż
DEF-depraved 3SG.M.IPFV:suffer {DEF-effect direct of:DEF-abuse
tad-droga}$_{DO}$
of:DEF-drug}
'[. . .] the depraved suffers {the direct effect of the drug abuse}.'

(c) *bi*-PP – [Korpus Malti 3.0; news77994]

Hu allegat li l-istess raġel qed ibati
3SG.M allege:part that DEF-same man PROG 3SG.M.IPFV:suffer
***[bi** problemi mentali]*$_{PP}$
[**with** problem:PL mental]
'Allegedly the same man is suffering [**from** mental problems].'

(d) *minn*-PP – [Korpus Malti 3.0; culture2014]
*Is-seminarista kien qed ibati [**minn** kanċer fi stat avvanzat]PP*
DEF-seminarist be PROG 3SG.M.IPFV:suffer [**from** cancer in state advance:PART]
'The seminarist was suffering [**from** advanced stage cancer].'

Aquilina (1987: 85–86) registers *bata* 'suffer' as an ambitransitive verb, i.e., it can be used intransitively as in (4.201a) and transitively as in (4.201b). Furthermore, there are also so-called prepositional usages as those shown in (4.201c–d). The above sentences describe situations as experiential events (Croft 2022: 226–234). The subject in all four examples is always the experiencer. The experience is negative. In (4.201b–d), the cause for the suffering is mentioned overtly, namely in the guise of a direct object in (4.201b) and in the shape of prepositional adjuncts in (4.201c–d). The omission of the cause in (4.201b–d) would not put the grammaticality of the sentences in jeopardy, meaning: the overt mention of the cause is in no way mandatory. The transitivity of (4.201b) largely corresponds to the transitivity of English *suffer sth* whereas the *bi*-PP and *minn*-PP in (4.201c) and (4.201d), respectively, are in line with English *suffer from sth.*

In Figure 4.106, it is shown how often *ibati* 'he suffers' occurs with or without cause-NP in the Korpus Malti 3.0.

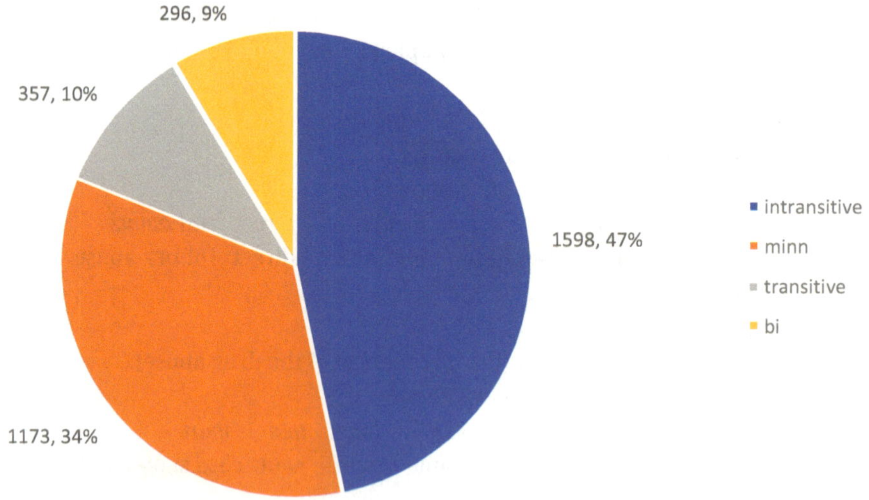

Figure 4.106: *ibati* 'he suffers' with and without cause-NP in the Korpus Malti 3.0.

The intransitive usage without cause-NP covers slightly less than half of the 3,424 tokens. The fully transitive option with the cause functioning as direct object is responsible for only 10% of all attestations of *ibati* 'he suffers'. Slightly more than a third of the hits involve a *minn*-PP which contains the cause whereas *bi*-PPs fulfilling the same function account only for 9% of the returns. The quantitative differences between the different options notwithstanding, we see that no possibility is too infrequent to be ignored. The variation as to the marking of the cause poses difficulties when we want to decide whether we are confronted with different ways of object-marking or something else. Functionally, the fully transitive option and the two PPs are largely identical insofar as they serve the purpose of expressing the cause. In terms of semantics, the meaning differences between the three cause-introducing constructions seem to be relatively subtle. It is doubtful whether any clear semantic distinction can be pinpointed for the two PPs. Both Πs occur with the complement *(il-)kanċer* '(the) cancer' in the Korpus Malti 3.0, namely seven and seventeen times for (Π7) *bi* 'with' and (Π41) *minn* 'from', respectively. However, in (4.201b), the NP *l-effett dirett tal-abbuż tad-droga* 'the direct effect of the drug abuse' is a core argument of $\textbf{\textit{bata}}_{\text{transitive}}$ 'suffer sth' whereas the PPs in (4.201c–d) introduce peripheral participants to the situation built around $\textit{bata}_{\text{intransitive}}$ 'suffer'. Neither *problemi mentali* 'mental problems' in (4.201c) nor *kanċer fi stat avvanzat* 'advanced stage cancer' in (4.201d) is a core argument. Note that Croft (2022: 178) states that the semantic role of cause belongs to the common peripheral semantic roles. We conclude that in sentences like (4.201c–d), (Π7) *bi* 'with' and (Π41) *minn* 'from' enhance the valency of basically monovalent verbs like $\textit{bata}_{\text{intransitive}}$ 'suffer' by way of introducing a peripheral participant to the situation without promoting it to the rank of a core argument.

This valency-enhancing function of PPs applies to several of the cases mentioned by Borg/Azzopardi-Alexander (1997: 142) as "prepositions mark[ing] a number of other objects". This quote refers to oblique objects as mentioned above. The class of these other objects is not further explained or terminologically differentiated. It is very likely that many of the candidates turn out to be instances of valency enhancement epiphenomenally caused by the integration of peripheral participants in a given situation. This interpretation receives support from the fact that in stark contrast to object marking of direct and indirect objects (either asyndetically or via (Π29) *lil* 'to'[156]) different Πs compete with each other in the domain of peripheral participants with identical or similar semantic roles. This is the case

156 According to the rules laid down in Section 4.3.1.

for instance, with (Π17) *fuq* 'on' which is supposed to mark an object in (4.202) according to Borg/Azzopardi-Alexander (1997: 143).

(4.202) (Borg/Azzopardi-Alexander 1997: 143)

Resqet	***[fuqu]***	*u*	*saqsietu*	*l-ħin*
approach:3SG.F.PFV	**[on**:3SG.M]	and	ask:3SG.F.PFV:3SG.M	DEF-time

'She came up [**to** him] and asked him the time.' [O.T.]

The verb *resaq* 'approach' is an inherently Goal-oriented motion verb. The Ground which the Figure is approaching can be introduced by different Πs. The choice of (Π17) *fuq* 'on' in (4.202) is only one among a number of options as results from the selection of examples from the Korpus Malti 3.0 in Table 4.26. Yellow marks the forms of *resaq* 'approach' which are always a 3rd person singular in the perfective and their English translations. Grey shading highlights those cells which showcase *resaq* 'approach' as a transitive verb. The three examples at the bottom of the table illustrate the use of the same verb in the absence of any PP.

Table 4.26: *Resaq* 'approach' with and without PPs in the Korpus Malti 3.0.

#	source	example / translation	Π / comment
1	literature80	*Xħin resaq [**biswit** Marsamxett]PP, kont sejjer bil-lanċa mill-Belt għal Tas-Sliema.* 'When he approached [(the shore) **facing** Marsamxett], I was going with the ferry from Valetta to Tas-Sliema.'	(Π8) *biswit* 'facing'
2	literature70	*Resaq [**fejn** il-marid]PP, flieh tajjeb, kitiblu xi duwa u qallu li jerġa' jarah filgħodu.* 'He approached [the sick], examined him well, prescribed some medicine and told him that he would see him again the next morning.'	(Π14) *fejn* 'near'
3	literature45	*Il-waitress reġgħet resqet [**fi**l-viċinanzi tal-mejda tiegħu]PP* 'The waitress came close again [**in** the vicinity of his table] [. . .].'	(Π15) *fi* 'in'
4	news161761	*ħadd minnhom ma resaq [**għal** imqar eżami wieħed tal-O Level]PP* '[. . .] none of them sat [**for** even a single exam of the O-level].'	(Π21) *għal* 'for'
5	news114405	*resaq [**għand** ħabib tal-familja]PP* '[. . .] he approached [a friend of the family (**at his place**)] [. . .]'	(Π22) *għand* 'at s.o.'s place'
6	literature20	*kienet resqet [**ħdej**kom]PP l-omm Victoria* '[. . .] the mother Victoria had approached [you] [. . .].'	(Π23) *ħdejn* 'beside'
7	literature11	*Reġgħet resqet [**lejn** il-kuker]PP biex tkompli ssajjar* 'She went [**towards** the cooker] again to continue cooking [. . .].'	(Π28) *lejn* 'towards'

Table 4.26 (continued)

#	source	example / translation	Π / comment
8	literature21	*Mikiel resaq [**ma**l-mejda]$_{PP}$* 'Mikiel approached [the table] [. . .].'	(Π30) *ma'* 'with'
9	literature77	*u fis resqet [**maġenb** ommha]$_{PP}$ għall-biża' li tħossha ħażin* / '[. . .] and quickly she went [**close to** her mother] for fear that she was feeling bad [. . .].'	(Π32) *maġenb* 'close to'
10	culture3653	*Il-mara resqet [**qrib** it-tifla]$_{PP}$* 'The woman went [**near** the girl] [. . .].'	(Π48) *qrib* 'near'
11	european9992	*Il-Kummissjoni għalhekk resqet [**quddiem** il-Kunsill]$_{PP}$ proposta dwar il-miżuri kkonċernati* 'The commission therefore brought a proposal [**before** the council] about the measures concerned [. . .].'	(Π49) *quddiem* 'in front of'
12	news80786	*Mark resaq [**taħt** it-tined]$_{PP}$ fejn klen hemm in-nies* 'Mark went [**under** the tents] where there were the people [. . .].'	(Π56) *taħt* 'under'
13	culture920	*Wara dan il-miraklu, jidher li l-familja tal-mara reġgħet resqet [**viċin** il-Fidi]$_{PP}$* 'After this miracle it seems that the woman's family approached again [the Faith].'	(Π58) *viċin* 'near'
14	literature78	*dar il-kantuniera bil-ħeffa u resaq [**wara** l-bieb ta' Pawlu]$_{PP}$* / '[. . .] he turned around the corner quickly and went [**in front of** Pawlu's door] [. . .].'	(Π60) *wara* 'after'
15	news9807	*Illum il-Pulzija resqet [**Ø** il- Qorti] lil Omar Hassan Ahmed* 'Today the police brought Omar Hassan Ahemd [**to** the court] [. . .].'	zero-marked
16	news119241	*L-Oppożizzjoni Laburista resqet vot ta' sfiduċja fil-Ministru Jesmond Mugliett* 'The Labour opposition brought forward a vote of non-confidence in Minister Jesmond Mugliett [. . .].'	no overt Ground
17	literature29	*Minnufih resaq pastur xwejjaħ jiġbed warajh ħmara magħluba* 'Suddenly a shepherd approached pulling behind him an emaciated she-ass.'	no overt Ground

If we include (Π17) *fuq* 'on' in the count, there are fifteen different Πs which are combinable with *resaq* 'approach'. Except the combination with (Π21) *għal* 'for' which occurs only in the context of examination events (Aquilina 1976: 76), all usages of Πs together with the verb under inspection form part of the description of situations in which the Figure physically moves closer to the Ground. The choice of Π seems to be partly dependent upon the CONFIGURATION of the Ground that is of interest in the event (such as (Π8) *biswit* 'facing' and (Π56) *taħt* 'under' among others). This means that these Πs are put to service in the event construction because of their meaning. Some of the Πs cannot easily replace each other whereas others such as the near synonyms (Π23) *ħdejn* 'beside' and (Π32) *maġenb* 'close to', (Π48) *qrib* 'near' and (Π58) *viċin* 'near' do not pose any problems to being replaced the one with the other. What is more, (Π17) *fuq* 'on' is not a suitable replacement for most of the other Πs unless the complement has a human referent. The specialisation on human referents is not necessarily an obstacle to the interpretation of (Π17) *fuq* 'on' as object marker since we know that in the domain of direct-object marking, (Π29) *lil* 'to' is activated if the referent of the object NP ranks high(est) on the animacy hierarchy.[157] Yet, (Π29) *lil* 'to' does not have to compete with a plethora of other Πs that are equipped with a largely transparent meaning. Therefore, we conclude that the chances are minimal that (Π17) *fuq* 'on' is not even a standardised marker of an oblique but one of many markers of this kind.

One Π at least stands out from the bulk of the members of BLOMP 2.0, namely (Π7) *bi* 'with'. The special role of this Π results from the fact that it is used obligatory with certain verbs when an undergoer is part of the scene. In (4.203), there is a minimal pair of two sentences which involve the verb *jaf* 'know'.[158] Yellow marks out the verb and underlining is used for the pronominal affix.[159]

(4.203) Knowing
(a) 'know s.o./s.th.' I [Korpus Malti 3.0; culture1740]

Tmur	*waħdek*	*f'*	*pajjiż*	*fejn*	*ma*	*jafek*
2SG.IPFV:go	one:2SG	in	country	where	NEG	3SG.M.IPFV:know:2SG
ħadd						
nobody						

'You go alone to a country where nobody knows you [. . .].'

157 Cf. Section 4.3.1.

158 Aquilina (1976: 72) registers *jaf* 'know' as prepositional verb but mentions only possible idiomatic combinations with (Π21) *għal* 'for'.

159 The verb *jaf* 'know' can be used also without an overt complement (but with implicit undergoer) as in (4.xiv).

(4.xiv) [Korpus Malti 3.0; news109202]

Busuttil	*spiċċa*	*kiser*	*ir-rekord*	*negattiv*	*tiegħu*	*bla*	*ma*
Busuttil	end	break	DEF-record	negative	of:3SG.M	without	that
jaf							
3SG.M.IPFV:know							

'Busuttil ended up breaking his negative record without knowing it.'

(b) 'know s.o./s.th.' II [Korpus Malti 3.0; literature82]
[Lilek]$_{PP}$ ili nafek ħafna qabel ma int
[**to**:2SG] ago:1SG 1SG.IPFV:know:2SG much before that 2SG
sirt taf [lili]$_{PP}$
become:2SG.PFV 2SG.IPFV:know [**to**:1SG]
'[As to you] I knew you already long before you came to know [me].'

(c) 'know about s.o./s.th.' [Korpus Malti 3.0; *news64413]*
irbaħ u kulħadd iħobbok itlef u ħadd ma
win and everybody 3SG.M.IPFV:love:2SG lose and nobody NEG
jrid jaf [bik]$_{PP}$
3SG.M.IPFV:want 3SG.M.IPFV:know [**with**:2SG]
'[. . .] win and everybody loves you, lose and nobody wants to know [**about** you].'

The direct-object construction (either by object-marking on the verb or via a *lil*-PP) is chosen if the undergoer is affected comparatively strongly by the personal acquaintance with other people who then know the undergoer. This is the case in (4.203a–b). In (4.203c), the undergoer is introduced in the shape of the complement of the *bi*-PP although the verb is the same as in (4.203a–b). This prepositional strategy is employed because the undergoer is known to others not via direct personal contact but by word of mouth, i.e. indirectly as someone who is talked about but not a personal acquaintance. Thus, (Π7) *bi* 'with' depicts the undergoer as only weakly affected by the knowledge of others. What can be said generally is that wherever verbs display the same contrast of treating the undergoer as direct object or as oblique object, the *bi*-PP is indicative of a relatively low degree of affectedness of the undergoer. In cases like that shown in (4.203c), the *bi*-PP does not simply serve to enhance the valency of the verb but its use contrasts with that of the direct-object construction and this contrast correlates with differences on the semantic side. It is possible therefore that in this case (Π7) *bi* 'with' is an object-marker for less affected undergoers.

The next verb that calls for closer scrutiny is *induna* 'realise'. There are 17,148 hits for this verb (including all aspectual forms as well as the frequent cases of aphaeresis of the initial /i/) in the Korpus Malti 3.0. A realising-event normally involves something that is realised, meaning: we expect the presence of an undergoer. This expectation can be checked for 13,326 hits which equal 78% of all tokens. For various reasons, 3,822 cases have to remain outside the following evaluation. What we find in terms of quantities is revealed in Figure 4.107.

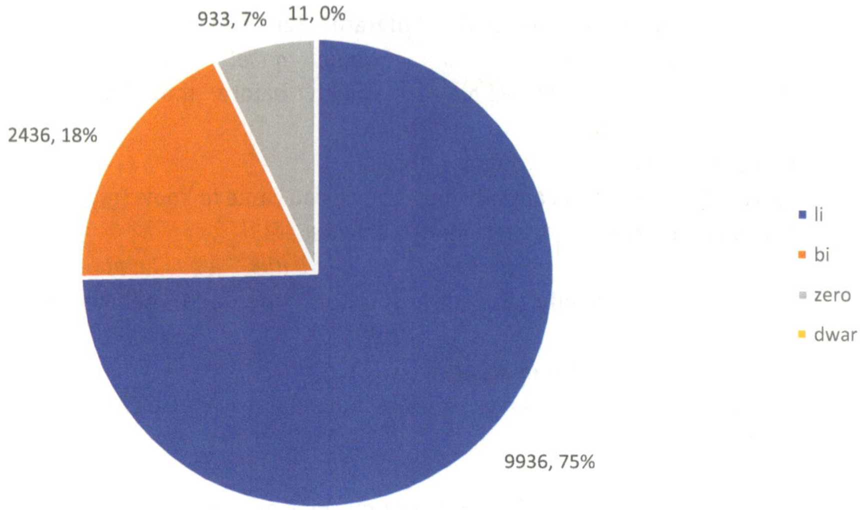

Figure 4.107: Complements of *induna* 'realise' in the Korpus Malti 3.0.

Three-quarters of all interpretable instances of *induna* 'realise' are followed by a *li*-clause be it a relative clause or an object clause. Second best is the employment of a *bi*-PP to link the undergoer to the verb. This strategy accounts for 18% of all cases. What we simplifyingly call zero strategy covers 7%. The smallest share goes to a pattern that involves (Π11) *dwar* 'about' which holds for less than 1% of the tokens. As to the zero strategy, examples like (4.204) which leave the undergoer unexpressed are unexpected. The absence of the overt realisation of the undergoer is marked by Ø.

(4.204) [Korpus Malti 3.0; news39387]

waqgħet	*minn*	*fuq*	*trakk*	*mingħajr*	*ma*	*x-xufier*	*induna*	*Ø*
fall:3SG.F.PFV	from	on	truck	without	that	DEF-driver	realise	Ø

'[. . .] it [= a piece of wood] fell from the truck without the driver noticing (it).'

The verb seems to behave like an intransitive verb. At the same time, it is clear that the driver should have noticed the fall from the truck of the piece of wood. A pronominal affix on the verb or a PP could have facilitated the anaphorical reference.

A very small number of other zero cases looks like the same verb is transitive and takes its direct object without any additional marker. This exceptional constellation is illustrated in (4.205).

(4.205) Asyndetic vs syndetic

(a) [Korpus Malti 3.0; news269]

iżda	*qallu*	*li*	*l-poplu*	*Malti*	*induna*	*Ø*	*dan*
but	say:IO:3SG.M	that	DEF-people	Maltese	realise	Ø	DEM:M:PROX

'[. . .] but he said to him that the Maltese people realised this.'

(b) [Korpus Malti 3.0; academic29]

Iż-żminijiet	*kienu*	*qed*	*jinbidlu*	*u*	*l-knisja*
DEF-time:PL	be:3PL.PFV	PROG	3.IPFV:REFL:change.PL	and	DEF-church
kattolika	*indunat*	*[b'*	*dan]*$_{PP}$		
catholic:F	realise:3SG.F.PFV	**[with**	DEM:M:PROX]		

'The times were changing and the catholic Church realised [this].'

In the Korpus Malti 3.0, the evidence for this zero-pattern is scarce. There are exactly four instances which involve a demonstrative pronoun as object. In contrast, the demonstrative combines 68 times with (Π7) *bi* 'with' according to the pattern illustrated in (4.205b). *Bi*-PPs seem to be the normal option when it comes to expressing the undergoer participant as shown in (4.206) with a lexical NP and a pronominal complement.

(4.206) Lexical and pronominal undergoers

(a) [Korpus Malti 3.0; news210094]

Meta	*Camilleri*	*induna*	*[bil-preżenza*	*tal-Pulizija]*$_{PP}$	*ħarab*
when	Camilleri	realise	**[with**:DEF-presence	of:DEF-police]	flee
min-naħa	*ta'*	*wara*			
from:DEF-side	of	behind			

'When Camilleri noticed [the presence of the Police] he fled through the back [. . .].'

(b) [Korpus Malti 3.0; news150506]

Id-differenza	*malajr*	*induna*	*[biha]*$_{PP}$
DEF-difference	quickly	realise	**[with**:3SG.F]

'As to the difference, he realised [it] quickly [. . .].'

The eleven tokens of *dwar*-PPs employed as an alternative to the expected *bi*-PPs display an interesting distribution. First, no forms of *induna* 'realise' with a subject in the 1st person singular or 2nd person plural are attested with (Π11) *dwar* 'about'. Similarly, the only subject person that is attested with this Π in the perfective aspect is the 3rd person plural. What is more, there is no example of (Π11) *dwar* 'about' taking a pronominal complement. In (4.207), we provide examples of (Π11) *dwar* 'about' combining with a demonstrative and a lexical NP.

(4.207) *dwar*-PPS

(a) [Korpus Malti 3.0; news127976]

Biżżejjed wieħed iħares lejn ir-rapport
enough one 3SG.M.IPFV:look towards DEF-report
tal-Awditur Ġenerali biex jinduna ***[dwar***
of:DEF-Auditor_General in_order_to 3SG.M.IPFV:realise **[about**
dan]$_{PP}$
DEM:M:PROX]
'It suffices that one looks at the report of the Auditor General to realise [this].'

(b) [Korpus Malti 3.0; culture1203]

mhux kulħadd jinduna ***[dwar***
NEG:3SG.M:NEG everybody 3SG.M.IPFV:realise **[about**
l-implikazzjonijiet]$_{PP}$
DEF-implication:PL]
'[. . .] not everybody realises [the implications] [. . .].'

In connection with the demonstrative functioning as complement of the Π in the (a)-sentence we refer back to the examples in (4.207). The NP *l-implikazzjonijiet* 'the implications' in the (b)-sentence finds a parallel in (4.208) but this time as complement of (Π7) *bi* 'with'.

(4.208) [Korpus Malti 3.0; news115856]

ħadd s' issa għadu ma nduna ***[b****l-implikazzjonijiet*
nobody until now still:3SG.M NEG realise **[with**:DEF-implication:PL
serji tagħha]$_{PP}$
serious:PL of:3SG.F]
'[. . .] hitherto nobody has realised [its (= the law's) serious implications], yet.'

On account of the marginal frequency of the *dwar*-PPS in combination with *induna* 'realise', we postulate that the established strategy of linking this verb to the expression used for representing the undergoer is the employment of a *bi*-PP. Thus, we are facing a case of (Π7) *bi* 'with' behaving morphosyntactically very much like a marker of an object which is not an oblique.

There are other verbs which behave similarly to *induna* 'realise' such as its near synonym *ntebaħ* 'become aware'. The usual way of encoding the stimulus in an experiential event expressed by this verb is by way of employing a *bi*-PP as shown in (4.209).

(4.209) [Camilleri 2013: 133]

intebaħ	*[b'*	*parkeġġ*	*vojt]*$_{PP}$
become_aware	**[with**	parking_lot	empty]

'[. . .] he noticed [an empty parking lot] [. . .].'

The empty parking lot stimulates the perception of the story's main character who is searching for a place to leave his car. As in the previous case of *induna* 'realise' (cf. (4.204)), *ntebaħ* 'become aware' may stand alone with any overtly realised stimulus. This however is a minor option since of the 1,162 classifiable hits (of altogether 1,401 attestations) for the 3rd person singular and plural forms of *ntebaħ* 'become aware' in the perfective found in the Korpus Malti 3.0 only sixty-four attest to the absence of a stimulus (= 5.5%). With 74% (= 857 hits), object clauses introduced by *li* form the majority – a picture that resembles the situation represented in Figure 4.107 albeit with a less extreme predominance of the *li*-clauses. *Bi*-PPs are attested 241 times and thus claim a share of 21% of all cases. The similarities with *induna* 'realise' also extend to demonstrative pronouns as stimulus (cf. (4.205)) because there are thirty-two tokens of *bi*-PPs with a demonstrative as complement as opposed to just a single instance of *dan* 'this' being the direct object of *ntebaħ* 'become aware' without any linking element.

Formally, *ntebaħ* (occasionally also: *ntnebaħ*) 'become aware' is a reflexive-passive stem VII of **nebaħ* of which only the causative stem II *nebbaħ* 'make s.o. aware of' exists (Aquilina 1991: 897, 925–926). We have not glossed the verb for reflexivity because of the obsolete stem I. Reflexivity means that there is actor-undergoer identity and that the verb already has two participants (Croft 2022: 207). The *bi*-PP opens a slot for an additional, non-identical participant. The pattern is recurrent since it is commonly attested with other reflexive verbs such as *nqeda* 'be served' the reflexive-passive of *qeda* 'serve'. In (4.210), the verb is followed by a *bi*-PP which has a noticeable semantic effect.

(4.210) [Camilleri 2013: 132]

Il-messaġier	*ma*	*jinqediex*				*[bil-pustiera]*$_{PP}$
DEF-messenger	NEG	3SG.M.IPFV:REFL:serve:NEG				**[with**:DEF-postman:F]
biex	*wassal*	*messaġġ*	*dirett*	*u*	*daqstant*	*importanti*
in_order_to	arrive.CAUS	message	direct	and	so_much	important

'The messenger does not make use **[of** the postwoman] to deliver a direct and so important message.'

This pattern is almost exceptionless. In the Korpus Malti 3.0, there is only an isolated case of (П41) *minn* 'from' being used with the 3rd person masculine in the perfective *nqeda* 'he made use of' whereas there are 120 hits for this verb form being followed by a *bi*-PP. It is therefore safe to assume that (Π7) *bi* 'with' is the Π which normally introduces the additional participant. Its use is even mandatory if the intended meaning is reflexive and not passive.

The systematicity of the interrelation of the use of *bi*-PPs and reflexive verbs needs to be studied in-depth separately. In (4.211), we provide further examples with three verbs which host the morphological marker of the passive-reflexive (here the prefix *t*-) and combine with (Π7) *bi* 'with', namely *tmejjel* 'be inclined' (stem V of *miel* 'be inclined), *tnejjek* 'be contemptuous' (stem V of *niek* 'fornicate'), *tnellaħ* 'be frivolous' (stem V of **nelaħ* 'destroy'), and *tpaxxa* 'be delighted' (stem V of *paxxa* 'make happy') (Aquilina 1991: 825, 901, 909, and 1043).

(4.211) Reflexives

(a) [Bartolo/Vella 2009: 83]

għajnejhom *tpaxxew* *[biż-żfin* *grazzjuż*
eye:PL:3PL REFL:delight:3PL.PFV [**with**:DEF-dance gracious
tal-eluf *ta'* *fjuri* *kkuluriti]*$_{PP}$
of:DEF-thousand.PL of flower:PL colour:PART:PL]
'[. . .] their eyes were delighted [**with** the gracious dance of the thousands of colourful flowers] [. . .].'

(b) [Bartolo/Vella 2009: 230]

Mhux *biżżejjed* *li* *ilu* *jitmejjel*
NEG:3SG.M:NEG enough that ago:3SG.M 3SG.M.IPFV:REFL:incline
[bina]$_{PP}$ *das-sekli* *kollha?*
[**with**:1PL] DEM:M:DEF-centrury:PL all:3SG.F
'Is it not enough that he has been making a fool [**of** us] all these centuries?'

(c) [Camilleri 2013: 12]

tnejjek *[bih]*$_{PP}$ *is-samrani*
REFL:damage.CAUS [**with**:3SG.M] DEF-rather_dark_skinned
'[. . .] the rather dark-skinned [boy] made a fool [**of** him] [. . .].'

(d) [Korpus Malti 3.0; news85407]

Strickland *biex* *jitnellaħ* *[bihom]*$_{PP}$
Strickland in_order_to 3SG.M.IPFV:REFL:damage [**with**:3PL]
qalhielhom *bit-Taljan*
say:3SG.F:IO:3PL with:DEF-Italian
'Strickland, in order to mock [them], said it to them in Italian [. . .].'

The situations presented in (4.211) largely belong to the sphere of emotional and moral experience. In the (a)-sentence, the *bi*-PP introduces the stimulus or cause for the delight whereas in (4.211b–d) the *bi*-PP introduces the experiencer who is made a fool of or mocked about.

Those verbs which combine with a *bi*-PP to govern a NP need not be reflexives. In (4.212), *qarraq* 'deceive' illustrates this possibility.

(4.212) [Bartolo/Vella 2009: 359]

Naf	*li*	*ommok*	*qarrqet*	*[b'*	*missierek]*$_{PP}$
1SG.IPFV:know	that	mother:2SG	deceive:3SG.F.PFV	**[with**	father:2SG]
biex	*jiżżewweġ*	*lilha*			
in_order_to	3SG.M.IPFV:marry	to:3SG.F			

'I know that your mother deceived [your father] in order for him to marry her [. . .].'

According to Aquilina (1991: 1133), *qarraq* 'deceive' is the stem II of an unattested verb **qaraq*. The attested stem is formally a causative. In the same lexicon entry, Aquilina registers two competing options as to object-marking with this verb, namely on the one hand the employment of a *bi*-PP and on the other hand the use of (Π29) *lil* 'to'. If the latter is made use of Aquilina assumes that the meaning is 'be unfaithful to'. Examples like (4.213) suggest, however, that the two Πs yield constructions which are largely synonymous.

(4.213) [Korpus Malti 3.0; news128400]

qarraq	*[lid-dinja]*$_{PP}$	*bis-somma*	*ta'*	*50 elf miljun*	*dollaru*
deceive	**[to**:DEF-world]	with:DEF-sum	of	5 billion	dollar

'[. . .] he cheated [the world] out of five billion dollars.'

The discrepancy between the token frequencies of the two strategies with *qarraq* 'deceive' is telling. If we limit the search in the Korpus Malti 3.0 to the combinations of *bi*-PPs and *lil*-PPs with the 3rd person singular masculine in the perfective, we find 275 instances of (Π7) *bi* 'with' as opposed to only seven hits for (Π29) *lil* 'to'. *Bi*-PPs occur 39 times as often as *lil*-PPs with *qarraq* 'deceive'.

It makes sense to compare these quantities with those connected to the competition of the same two Πs with a partial synonym of *qarraq* 'deceive', namely *inganna* 'swindle'. The examples in (4.214) show that *inganna* 'swindle' allows for more ways to mark the undergoer.

(4.214) undergoers of *inganna* 'swindle'

(a) zero-marking [Korpus Malti 3.0; news164417]

Inganna Ø *għadd* *ta'* *persuni* *b'* *wegħdi* *ta'* *xogħol*
swindle Ø number of person:PL with promise of work
'He cheated Ø a number of persons with a promise of work.'

(b) direct object marking on verb [Korpus Malti 3.0; parl8233]

Issa *l-poplu* *aktar* *minn* *qatt* *qabel* *jaf* *li*
now DEF-people more from ever before 3SG.M.IPFV:know that
l-partit *fil-gvern* *ingannah* *daħak* *bih* *u*
DEF-party in:DEF-government swindle:3SG.M laugh with:3SG.M and
ġab *lill-pajjiż* *fi* *kriżi* *serja*
bring to_DEF-country in crisis serious:F
'Now the people know more than ever before that the ruling party cheated it, deceived it and brought the country into a serious crisis.'

(c) *lil*-PP [Korpus Malti 3.0; news57358]

fil-Bibbja *naraw* *kif* *Ġuda* *inganna* ***[lil*** *Kristu]*$_{PP}$
in:DEF-Bible 1.IPFV:see:PL how Judas swindle **[to** Christ]
'[. . .] in the Bible, we see how Judas deceived [Christ].'

(d) *bi*-PP [Korpus Malti 3.0; opinion346]

naslu *għall-konklużjoni* *li* *Mugliett* *inganna*
1.IPFV:arrive:PL for:DEF-conclusion that Muglietti swindle
***[bil**-Parlament]*$_{PP}$
[**with**:DEF-parlament]
'[. . .] we arrive at the conclusion that Mugliett deceived [the parlament].'

The four options for *inganna* 'swindle' display an uneven distribution over the Korpus Malti 3.0. If we go solely by the expression of the undergoer with this verb in the 3rd person singular masculine perfective, an interesting pattern emerges which can be compared to the smaller range of options for *qarraq* 'deceive'. Figure 4.108 strongly suggests that in the case of *qarraq* 'deceive', (Π7) *bi* 'with' is the default option whereas (Π29) *lil* 'to' is only marginally attested. Even more striking is the complete absence of (Π29) *lil* 'to' in combination with pronominal complements. Since the verb *qarraq* 'deceive' does not take pronominal suffixes for direct objects either, (Π7) *bi* 'with' is the only way to link a pronominal object to the verb. As to *inganna* 'swindle', *lil*-PPs are responsible for 55% of all undergoers with this verb whereas *bi*-PPs account only for 3%. The absence of object markers has a share of 5%. With forty-eight hits (= 30%), pronominal object marking on the verb is second best.

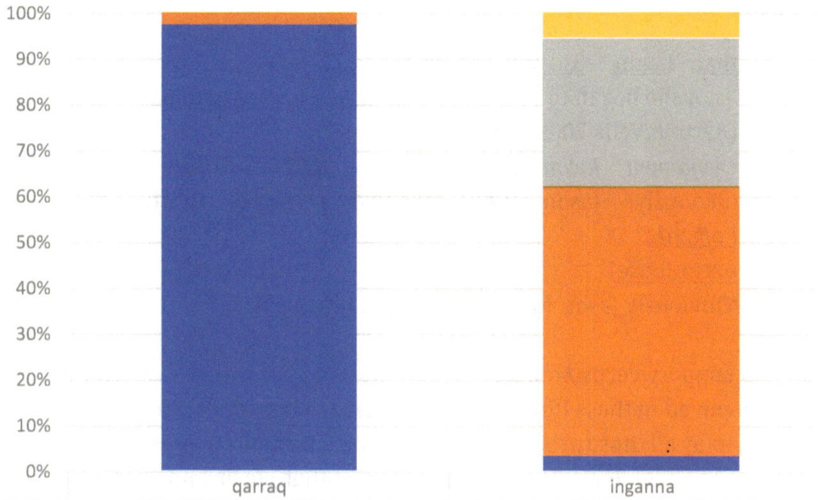

	qarraq	inganna
zero	0	8
pronominal	0	48
lil	7	87
bi	275	5

Figure 4.108: Frequency of *bi*-PPS vs *lil*-PPS with *qarraq* 'deceive' and *inganna* 'swindle' in the Korpus Malti 3.0.

The preferences of the two verbs are diametrically opposed although their semantics are almost the same. In the case of *qarraq* 'deceive', *lil*-PPS hardly play a role whereas with *inganna* 'swindle' they outnumber *bi*-PPS by a ratio of 17-to-one. On this basis, we assume that Maltese verbs may vary considerably as to undergoer marking. There seem to be certain preferences with verbs of similar semantics and diathesis but these preferences allow for a wide margin of variation.

The final paragraph devoted to the role of (Π7) ***bi*** 'with' in the domain of undergoer marking concerns so-called support verb constructions (Croft 2022: 419–424). In these constructions, a relatively desemanticised verb combines with a noun to express a given meaning. The combination of VERB + NOUN can be transitive and thus opens a slot for an undergoer. Typical examples from Maltese are given in (4.215). The support verb construction is surrounded by curly brackets.

(4.215) Support verbs
(a) [Camilleri 2013: 296]

bdiet	*{tieħu*	*pjaċir}*	*[**b**l-attenzjoni*
begin:3SG.F.PFV	{3SG.F.IPFV:take	pleasure}	[**with**:DEF-attention

li *beda* *jitiha]*$_{PP}$
that begin 3SG.M.IPFV:give:3SG.F]
'[. . .] she began to {enjoy} [the attention he began to pay to her] [. . .].'

(b) [Bartolo/Vella 2009: 171]
Ovvjament *Jason* *{ma* *ħax* *gost}* ***[b'*** *din*
obviously Jason {NEG take:NEG pleasure} **[with** DEM:F:PROX
l-offiża]$_{PP}$
DEF:offense]
'Obviously Jason {did not like} [this offense] [. . .].'

The two support verbs *ħa pjaċir* 'enjoy' and *ħa gost* 'enjoy' are synonymous. The nouns involved in these binary constructions are largely decategorialised, i.e., they do not display all those properties which nouns normally possess. Thus, the support verbs are complex lexical units. It is doubtful that *pjaċir* and *gost* 'pleasure' have object status in support-verb constructions. The only permissible marking for the undergoer with the above support verb constructions is a *bi*-PP.[160]

The foregoing paragraphs provide only the fragments of a full account of object marking and related issues in Maltese. We admit that we have unsystematically scratched the surface. Many issues have been skipped and others need to be looked into more deeply. These are tasks for the future when a dedicated valency dictionary of Maltese will be compiled. We are content with drawing the attention of others to the richness of the network of phenomena we have alluded to in this section. What has come to the fore is the particular importance of (П7) *bi* 'with' in the domain under scrutiny. To what extent other Maltese Пs are also involved in undergoer marking and oblique object marking remains to be investigated in times to come.

4.3.3 Adverbials

Maltese is a language with a relatively small inventory of primary adverbs, i.e., dedicated underived Semitic function words which exclusively serve to modify the

---

160 Both *ħa pjaċir* 'enjoy' and *ħa gost* 'enjoy' may be used without undergoer. There is also the occasional combination with a *fi*-PP. However, in these cases, the *fi*-PP serves as spatial adverbial to indicate where the actor enjoys themselves as in (4.xv).

(4.xv) [Korpus Malti 3.0; news138649]
l-Papa *{ħa* *gost}* *[f'* *pajjiżna]*$_{PP}$
DEF-Pope {take pleasure} **[in** country:1PL]
'[. . .] the Pope {enjoyed} himself **[in** our country] [. . .].'

predicate (Borg/Azzopardi-Alexander 1997: 65). In Aquilina (1959: 298–299), we find a mixed bag of supposed adverbs which comprises deictics (*hemm* 'there'), conjunctions (*mindu* 'since'), interrogatives ((Π14) *fejn* 'near' = 'where'), quantifiers (*ħafna* 'many'), pseudo-verbs (*għad* 'still'), negations (*ma* 'not'), and also Πs ((Π56) *taħt* 'under'). The boundaries between these word-classes are permeable. Basic adverbs like *biss* 'only', *dari* 'fromerly', *issa* 'now', *malajr* 'quickly' are scarce. The scarcity is particularly pronounced in the domain of manner adverbials. Aquilina (1959: 299) states that "[m]ost adverbs of manner in Maltese are obtained phrasally by means of **bi** ('with') + (i) Nominal, or (ii) Descriptive" [original boldface]. Italian is a major source of loan-adverbs (Saade 2019, 2020) because

> Semitic Maltese does not easily express adverbs of manner. In this case, in the colloquial, it copies Italian by adding **ment** to the last consonant, or to **a** after a semivowel **y** of the foreign adjectives. This usage, however, is still very much restricted, rare in the villages, though frequently employed by the educated classes [original boldface] (Aquilina 1959: 321).

Whether the Italian pattern has made it into spoken Maltese also outside the urban agglomerations during the 64 years after the publication of the above statement is outside the scope of this study. What is important is that PPs constitute an important resource for the formation of adverbials in Maltese (Saari 2003: 155–193). The PPs compete with a handful of original Semitic adverbs, loans from Italian, imperfective verbs (Borg/Azzopardi-Alexander 1997: 67–69), bare NPs, and total reduplication of members of different word-classes (Stolz/Stroh/Urdze 2011: 303–310). The competing strategies are illustrated and indexed appropriately in (4.216).

(4.216) Other adverbial strategies

(a) [Farrugia 2022: 134]

is-soltu$_{ADV}$	***qajla***$_{ADV}$	*kien*	*joħroġ*	*għerf*	*minn*
DEF-usual	**hardly**	be	3SG.M.IPFV:exit	wisdom	from
ħalqu					
mouth:3SG.M					

'[. . .] **usually** there came **hardly** any wisdom out of his mouth.'

(b) [Camilleri 2022: 53]

Triq Hompesch	*kienet*	*tieqaf*	*għal*	***litteralment***$_{ADV}$		
Hompesh_Street	be:3SG.F.PFV	3SG.F.IPFV:stop	for	**literally**		
ftit	*sigħat*	*u*	*taqa'*	*fis-silenzju*	*miksur*	
little	hour:PL	and	3SG.F.IPFV:fall	in:DEF-silence	PART:break	
okkażjonalment$_{ADV}$	*minn*	*karozza*	*għadejja*	*gas down*		
occasionally	from	car	pass:PART:F	gas_down		

'[. . .] Hompesh Street stopped for **literally** a few hours and fell silent interrupted **occasionally** by a car passing gas down [. . .].'

(c) [Agius 2002: 54]
il-kelb *niżel* ***{dritt*** ***dritt}***$_{ADV}$ *f'* *nofshom*
DEF-dog descend **{straight straight}** in half:3PL
'[. . .] the dog descended **{directly}** in their midst.'

(d) [Chetcuti 1990: 60]
ġie ***dritt***$_{ADV}$ ***jiġri***$_{ADV}$ *fuq* *missieri*
come **straight** **3SG.M.IPFV:run** on father:1SG
'[. . .] he hurried **straight** towards my father.'

In (4.216a), we have two adverbials, namely the bare NP *is-soltu* 'the usual' and the basic adverb *qajl(a)* 'hardly'. Two Italian loan adverbs in *-ment* are featured in (4.216b) whereas (4.216c) shows how the adjective *dritt* 'straight' is made into an adverbial by way of totally reduplicating it. *Dritt* may also be used adverbially without being reduplicated as can be gathered from (4.216d) where we also find the imperfective verb *jiġri* 'he runs' which modifies the more general deictic motion verb *ġie* 'he came' as to the speed with which the motion event was realised.

We do not intend to account for all phenomena connected to the grammar of adverbials in Maltese. We are exclusively interested in the role Πs play in this domain. Since the semantic side of the phenomenon is touched upon in Section 5, we concentrate on two aspects in the remainder of this section. As mentioned at the beginning of this section, (Π7) *bi* 'with' is of special importance for the formation of adverbial PPs – a special status which reminds us of the Π's outstanding position in the realm of obliques discussed in the previous section. Adverbial *bi*-PPs are inspected in Section 4.3.3.1. In Section 4.3.3.2, the alternation between bare spatial adverbials and *fi*-PPs is discussed.

4.3.3.1 *Bi*-adverbials

Adverbial modification can be either participant-oriented or event-oriented with the former strategy distinguishing between depictive and resultative secondary predication; event orientation goes along with manner secondary predication (Croft 2022: 432–433). The task of the subsequent paragraphs is to determine whether all three kinds of secondary predications are compatible with *bi*-PPs as adverbial modifiers. Further aspects of secondary predication in Maltese are discussed by Fabri (2023).

In accordance with the extant literature, the manner function is unproblematic because examples thereof abound in literary Maltese (Borg/Azzopardi-Alexander 1997: 151). To keep the empirical documentation within reasonable bounds, we limit the illustration of *bi*-PPs as manner adverbials to the cases presented in

(4.217). In the English translations, the adverbs which correspond to the *bi*-PPs in the original are highlighted in yellow.

(4.217) Manner adverbials

(a) [Friggieri 2000: 60]

Susanna tbissmet u ħarset lejh
Susanna smile:3SG.F.PFV and look:3SG.F.PFV towards:3SG.M
*[**b**il-ħlewwa]$_{PP}$*
[**with**:DEF-sweetness]
'Susanna [sweetly] smiled and looked at him.'

(b) [Camilleri 2021: 12]

Mintoff u Saydon iqararhom
Mintoff and Saydon 3SG.M.IPFV:hear_confession:3PL
*[**b**il-moħbi]$_{PP}$ fl-istess kappella minħabba l-interdett*
[**with**:DEF-PART:hide] in:DEF-same chapel on_account_of DEF-interdict
'[. . .] as to Mintoff and Saydon, he hears their confession [secretly] in the same chapel on account of the interdict.'

(c) [Agius 2002: 42]

*Samwel beda jisgħol [**b**il-qawwi]$_{PP}$*
Samwel begin 3SG.M.IPFV:cough [**with**:DEF-strong]
'Samwel began to cough [heavily].'

In the above cases, the *bi*-PPs are examples of event-oriented modification since in the (a)-sentence it is the smiling and looking that is qualified as happening with sweetness. Similarly, the hearing of the confessions takes place secretly in (4.217b) just as in (4.217c) the coughing is strong. In none of these cases do we find participant-oriented modification.

This is different with the examples in (4.218) all of which characterise a participant involved in the event.

(4.218) Depictives

(a) [Bartolo/Vella 2009: 95]

*Le le ħi qalet omm Steve [**bi** tbissima miksura]$_{PP}$*
no no brother:1SG say:3SG.F.PFV mother Steve [**with** smile PART:break:F]
'"No, no, my brother!", said Steve's mother [smiling obliquely].'

(b) [Farrugia 2022: 43]

*Daħal fis-sodda [**b**il-buri]$_{PP}$ u [**b**il-buri]$_{PP}$ m'*
enter in:DEF-bed [**with**:DEF-mood] and [**with**:DEF-mood] NEG

għalaqx għajn m' għajn
close:NEG eye with eye
'He went to bed [**in** a bad mood] and [**in** a bad mood] he could not fall asleep.'

(c) [Camilleri 2013: 89]
Cheese għajtet il-gwida [b' wiċċha miġbud]$_{PP}$
Cheese shout:3SG.F.PFV DEF-guide [**with** face:3SG.F PART:pull]
'"Cheese!", shouted the guide [**with** her distorted face].'

The *bi*-PP in (4.218a) describes the facial expression of Steve's mother while she utters her negative statement. In (4.218b), *bil-buri* 'in a (bad) mood' occurs twice. This PP describes the psycho-mental state the protagonist is in. The (c)-sentence can be compared to the (a)-sentence because what the *bi*-PP describes is again the facial expression of the protagonist. Depictives as well as event-oriented modification by *bi*-PPs are highly frequent in Maltese. They constitute productive patterns which are freely made use of in all genres and styles of the language.

As to resultatives, the situation is strikingly different. In (4.219), we present examples of the two common resultative constructions that involve *bi*-PPs.

(4.219) Resultatives
(a) [Korpus Malti 3.0; literature50]
mbagħad reġa' poġġa [bil-qiegħda]$_{PP}$ fuq il-bank
then return sit [**with**:DEF-stay.PART:F] on DEF-bank
'[. . .] then he sat [down] again on the bank.'

(b) [Korpus Malti 3.0; european5034]
Il-ħnieżer kollha jridu għalinqas ikunu
DEF-pig.PL all:3SG.F 3.IPFV:want:PL at_least 3.IPFV:be.FUT:PL
jistgħu jimteddu u jqumu
3.IPFV:can:PL 3.IPFV:REFL:stretch:PL and 3.IPFV:rise:PL
[bil-wieqfa]$_{PP}$ fil-posizzjoni naturali tagħhom
[**with**:DEF-stand.PART:F] in:DEF-position natural of:3PL
'All pigs at least want to be able to stretch out and stand [up] in their natural position.'

These constructions are special. In contrast to event-oriented and depictive *bi*-PPs, the resultatives form a closed set of two. The complement of the Π is a definite active participle hosting the suffix of the feminine singular. The form *qiegħda* 'staying (F)' is the regular active participle of *qagħad* 'stay' which inflects for gender and number yielding the ternary paradigm M.SG *qiegħed* ≠ F.SG *qiegħda* ≠ PL *qegħdin*. However, the usual agreement with the NP representing the subject is blocked so that, in (4.219a), a masculine subject can cooccur with the feminine form of the participle. In point of fact, the form of the participle in the resultative construction is fossilised. The fossilisation also holds for the participle in (4.219b) because *wieqfa* 'standing (F)' is the regular active participle of *waqaf* 'stand upright' and belongs to the set of three forms M.SG *wieqaf* ≠ F.SG *wieqfa* ≠ PL *wiqfin*. The gender-number agreement is suspended as in the previous case. In (4.219b), the subject is in the 3rd person plural but the participle remains feminine singular. The two participles are insensitive to the categories which are relevant for the subject. They are morphologically invariant. It may be claimed therefore that both *bil-qiegħda* and *bil-wieqfa* are lexicalised units meaning something like 'in a sitting position' and 'in a standing position'. The pattern is not synchronically productive so we conclude that of the three kinds of adverbial modification addressed in this section, resultative participant-oriented modification, in terms of types, is only marginally attested with *bi*-PPs in Maltese whereas the two other kinds of adverbial modification are very common. In the Korpus Malti 3.0, the token frequency of *bil-qiegħda* and *bil-wieqfa* is moderately high as shown in Figure 4.109 where we compare those of the above *bi*-PPs with each other that involve the definite article. The frequencies range from a minimum of ten tokens for *bil-buri* to a maximum of 1,652 tokens for *bil-qiegħda*. The quantitative discrepancy between the two extremes is noteworthy. If one wants to draw conclusions as to the degree of lexicalisation of the constructions on the basis of their token quantities, the little information given in Figure 4.109 is certainly not sufficient. It is necessary to compare the frequencies of many more adverbial *bi*-PPs. This is a task we relegate to future research.

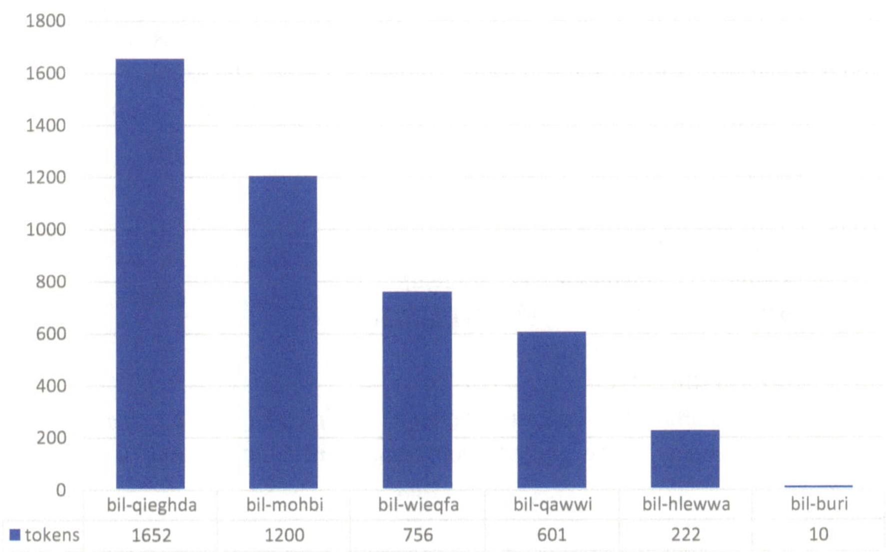

Figure 4.109: Token frequencies of adverbial *bi*-PPs compared (Korpus Malti 3.0).

4.3.3.2 The Ø-*fi* alternation

The phenomenon to be discussed in this section has been treated by Saari (2003: 242–248) and more systematically by Stolz/Lestrade/Stolz (2014: 225–273), Stolz/Levkovych/Urdze (2017),[161] and more recently and in a different context by Azzopardi (2023). Borg/Azzopardi-Alexander (1997: 155–156) claim that

> [f]or general location in a place which is fairly predictable from the context no preposition is used [. . .]. Certain verbs of motion, followed by the place nominal (without a preposition) express motion to a place, including arrival at a place.

The grammarians of Maltese restrict the possibility of spatial adverbials occurring without prepositional head to the domain of general location to the exclusion of the relation of Source. In point of fact, except for semantically specialised verbs such as *telaq* 'leave (a place)', the explicit marking of Source is compulsory. The mandatory expression of Source contributes heavily to the high token frequency of (Π41) *minn* 'from'. What we learn further from the above quote is that context-dependent predictability is crucial for choosing the zero-marking strategy in the case of the static spatial relation of Place. As to the dynamic spatial relation of Goal, Borg/Azzopardi-Alexander (1997) assume that the motion verb semantically incorporates the Goal-orientation and thus makes the overt expression thereof superfluous. The use of Ø in spatial situations alternates with the employment of (Π15) *fi* 'in' which is insensitive to the distinction Place vs Goal.[162] The Ø-*fi* alternation is illustrated in (4.220) for Place and in (4.221) for Goal. The zero-marked adverbials are bracketed and indexed with X to distinguish them from genuine PPs. The examples are taken from one and the same primary source with those illustrating the Goal-relation being printed on the same page of the original.

(4.220) Place

(a) [Attard 2019: 197]

kienu	*għadhom*	*[Ø*	*il-Libja]$_X$*
be:3PL.PFV	still:3PL	[Ø	DEF-Libya]

'[. . .] they were still [**in** Libya].'

(b) [Attard 2019: 319]

Hawn	*ħafna*	*iktar*	*pjanti*	*komuni*	*[f' Malta]$_{PP}$*	*li*	*jekk*
here	many	more	plant:PL	common:PL	[**in** Malta]	that	if

161 Cf. Section 1.5.2.

162 Whether other Πs that compete with (Π15) *fi* 'in' as to the expression of Place and Goal should be understood as further contestants in the domain under inspection will be discussed in Section 5.1.2.1.1.1.

jittieklu *jistgħu* *joqtluk*
3.IPFV:REFL:eat:PL 3.IPFV:can:PL 3.IPFV:kill:PL:2SG
'There are many more plants here [**in** Malta] that, if they are eaten, can kill you.'

(4.221) Goal
(a) [Attard 2019: 56]
Tawhieli *mill-Kwartieri Ġenerali* *f'* *Washington*
give:3PL.PFV:3SG.F:IO:1SG from:DEF-Headquarters in Washington
qabel *ma* *ġejt* *[Ø* *Malta]$_X$*
before that come:1SG.PFV [**Ø** Malta]
'They gave it [= the letter] to me from the headquarters in Washington before I came [**to** Malta].'
(b) [Attard 2019: 56]
ġieli *jiġu* *aġenti* ***[f'*** *Malta]$_{PP}$* *u* *lanqas*
sometimes 3.IPFV:come:PL agent:PL [**in** Malta] and not_even
biss *ikollhom* *country clearance* *mill-ambaxxata*
only 3SG.M.IPFV:have.FUT:3PL country_clearance from:DEF-embassy
'[. . .] sometimes agents come [**to** Malta] without even having the country clearance from the embassy.'

In all four examples, the Ground is represented by a place name so that the spatial nature of the situation is clear. At the same time, the predicates are sufficiently generic. In the case of Place, (4.220a) past reference is achieved by way of using the perfective copula *kienu* 'they were'. In (4.220b), we face an existential construction. The uncontroversial spatial character of both scenarios notwithstanding, the adverbial in the (a)-sentence is complete without Π whereas there is (Π15) *fi* 'in' in the (b)-sentence. The same puzzling situation arises in connection with Goal. The same motion verb *ġie* 'come' is involved in both situations that also share the Ground *Malta*. These similarities do not preclude that the zero-strategy applies only once, namely in (4.221a), whereas the overt Π is featured in (4.221b). Semantic factors do not seem to determine the choice of construction. We strongly assume that all four of the adverbials are grammatical with and without (Π15) *fi* 'in'. The absence of a semantic motivation for the choices made in these and similar cases justifies the discussion of the Ø-*fi* alternation in this section and not in the semantics section. Haspelmath (2019) compares alternations of this kind – labelled Differential Place Marking (DPM) – with Differential Object Marking (DOM).[163] In the same

163 Cf. Section 4.3.1.

publication, the author introduces the term toponouns for those common nouns which share morphosyntactic properties with genuine place names. The class of toponouns is also relevant in Maltese.

Stolz/Lestrade/Stolz (2014: 272–273) conclude their detailed study of the Ø-*fi* alternation in Maltese by way of stating that no hard and fast rules for the presence or absence of (Π15) *fi* 'in' can be postulated. The variation is strong. Both options are frequently made use of. The genericity of the static or dynamic verb does not seem to be overly decisive. What appears to be more important is the Ground-NP. Place names are more prone to zero-marking than common nouns although certain common nouns referring to institutions or similar concepts also allow for zero-marking of spatial relations. The latter is possible if the NP is generic and syntactically light. These conclusions were arrived at when the Korpus Malti 3.0 was still in the making. We seize the opportunity to test selected results of the earlier studies now with the help of the electronic resource.

In what follows we carry out a spot check for spatial situations of Place and Goal with involve the place names *Malta* or *Għawdex* as Ground. Only the 3rd person singular masculine perfective of the verbs *kien* 'be', *baqa'* 'remain', *qagħad* 'stay, and *twieled* 'be born' for Place and *ġie* 'come', *mar* 'go', *daħal* 'enter', and *wasal* 'arrive' for Goal is taken account of. Moreover, we restrict the search to those cases in which the verb and Ground-NP are immediate neighbours with the verb preceding the Ground. In Figures 4.110–4.111, the results for Place are revealed separately for *Malta* and *Għawdex* whereas those for Goal are presented in Figures 4.112–4.113.

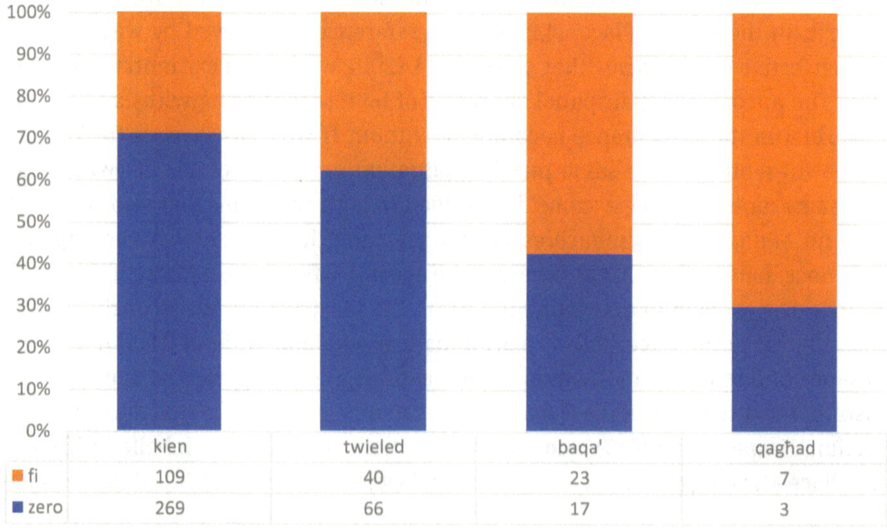

Figure 4.110: Ø-*fi* alternation with *Malta* as Ground (Place).

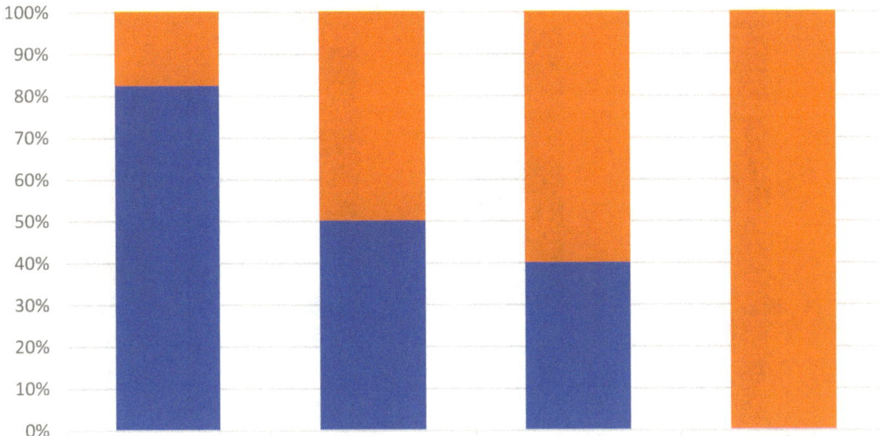

	kien	twieled	baqa'	qagħad
fi	9	3	3	9
zero	42	3	2	

Figure 4.111: Ø-*fi* alternation with *Għawdex* as Ground (Place).

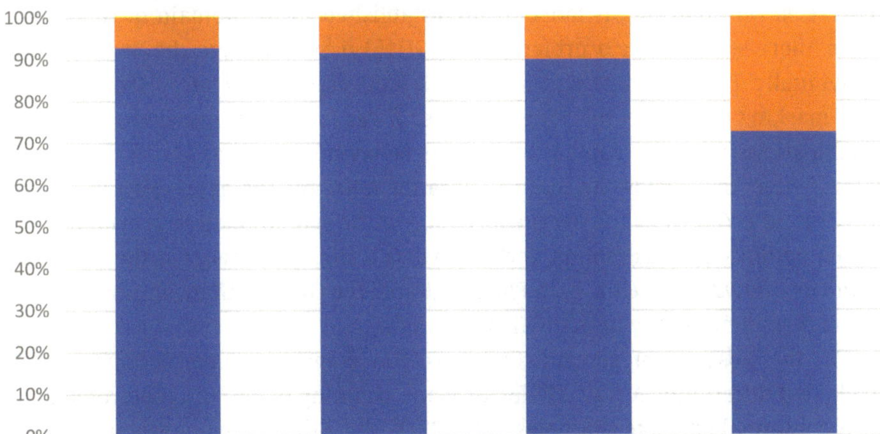

	ġie	wasal	mar	daħal
fi	70	77	1	69
zero	888	824	9	182

Figure 4.112: Ø-*fi* alternation with *Malta* as Ground (Goal).

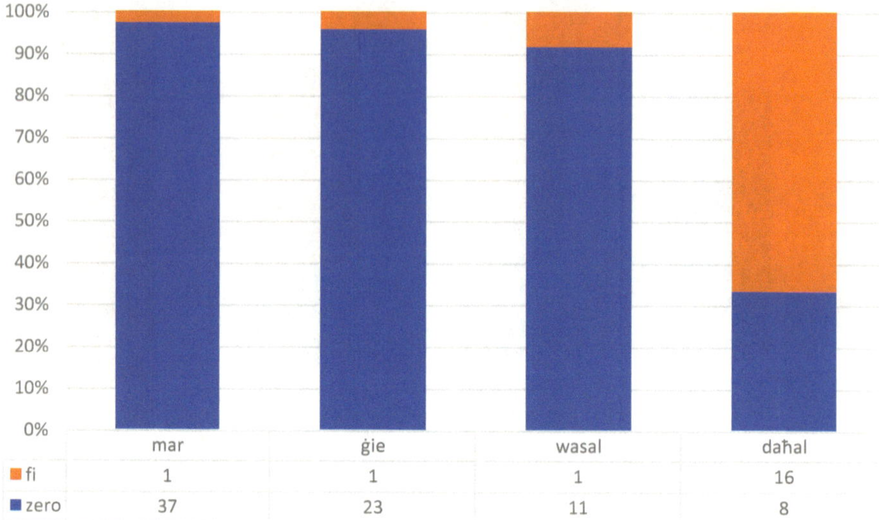

Figure 4.113: Ø-*fi* alternation with *Għawdex* as Ground (Goal).

The picture that emerges from Figures 4.110–4.113 confirms earlier findings insofar as zero-marking is more firmly established in the domain of Goal. As to Place, there seems to be a preference for (Π15) *fi* 'in' although the verbs behave individually. The most desemanticised verb – the copula *kien* 'be' – clearly prefers zero-marking over overt marking. However, *fi*-PPS are not absolutely banned from combinations with the copula. Individual behaviour also applies in the case of *daħal* 'enter' that yields the biggest share for *fi*-PPS for the expression of Goal. If the Ground is *Għawdex* the overt expression of the spatial relation is even the preferred option with this motion verb. It makes a slight difference if the Ground is *Malta* or *Għawdex* since for both Place and Goal the Ground *Għawdex* tolerates the use of *fi*-PPS more easily than the Ground *Malta*.

In the absence of semantic reasons, the choice of construction must be explained differently. A possible factor is the chaining of PPS and their avoidance. As the minimal pair-like examples in (4.222) suggest there are no principles which would obstacle the use of either option if the Ground of the motion event *wasal* 'arrive' is followed by a *minn*-PP which indicates the place of origin of the Figure. The head of the Source-PP is doubly underlined.

(4.222) PP-chaining

(a) [Korpus Malti 3.0; news12151]

kien	*għadu*	*kif*	*wasal*	*[Ø*	*Malta]$_X$*	*minn*	*Brussell*
be	still:3SG.M	how	arrive	[Ø	Malta]	from	Brussels

'[. . .] he had only just arrived [**in** Malta] from Brussels [. . .].'

(b) [Korpus Malti 3.0; news201376]

kien għadu kif wasal [f' Malta]$_{PP}$ *minn Zurich*

be still:3SG.M how arrive [**in** Malta] from Zurich

'[. . .] he had only just arrived [**in** Malta] from Zurich [. . .].'

In line with what we have argued above there is no reason to assume that coming from Brusseles automatically triggers the choice of the zero-strategy as opposed to Zurich which would require the presence of (Π15) *fi* 'in'. This explanation is nonsensical. What we see when we look at the quantities is that the use of (Π15) *fi* 'in' is exceptional under PP-chaining. In Figure 4.114, this is shown for the motion verb *wasal* 'arrive' and the Ground *Malta* followed by a *minn*-PP. The presence of (Π15) *fi* 'in' is only marginally attested. Its absence on the other hand is the default option. That the avoidance of PP-chains does not represent the final word with regards to the Ø-*fi* alternation results from the striking fact that in combination with the stative verb *baqa'* 'remain' the Ground *Malta* only allows for a following *fi*-PP (no other PPs being attested) if the Ground itself is preceded by (Π15) *fi* 'in'. Things are complicated further since many of these cases seem to stem from identical quotes from one and the same original text in different files in the Korpus Malti 3.0.

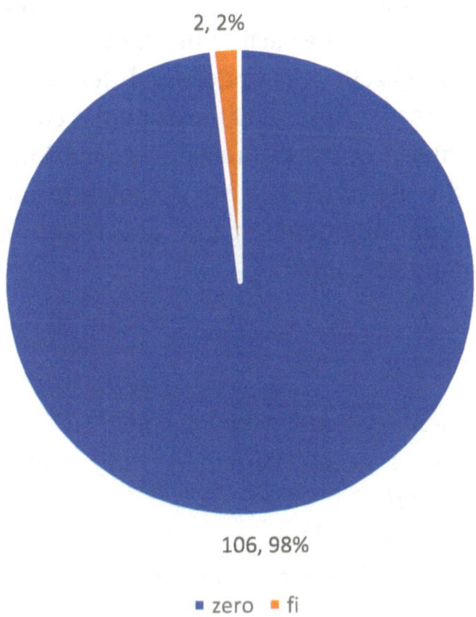

Figure 4.114: The Ø-*fi* alternation under PP-chaining.

Certain preferences emerge nevertheless as to the choice between the two strategies. Glossing over many still open questions it can be stated that zero-marking is preferred over *fi*-PPs (without ousting the latter completely) under the following conditions:

(a) the spatial situation is dynamic = Goal > Place, cf. (4.223),
(b) the Ground is a place name = place name > common noun, cf. (4.224),
(c) the Ground is a single word = light > heavy, cf. (4.225),
(d) the Ground has a single referent = singular > dual/plural, cf. (4.226),
(e) the Ground is (inherently) definite = definite > indefinite, cf. (4.227),
(f) the Ground is generic = generic > specific, cf. (4.228),
(g) the Ground receives a metonymical/metaphorical reading = abstract > concrete, cf. (4.229),
(h) the verb is semantically unspecific = general > specific, cf. (4.230).

Some of the conditions contradict each other such as (b) and (e) vs (f). Place names are definite per definition since they are mono-referential. Together with overtly definite Grounds (common nouns), they fail to fulfil the criterion of genericity in (f). It is worth mentioning at this point that in Maltese genericity more often than not goes along with definiteness marking of the generic noun. Criterion (g) also seems to be difficult to accommodate especially with criterion (b). A network of very different factors arises which conspire to block the possibility of formulating one rule to cover all phenomena in one go. We illustrate each of the above binary hierarchies with pairs of sentences in (4.223)–(4.230). It is understood that there are also many counterexamples for each of the cases. Several of the examples could be classified differently according to the criteria in (a)–(h) above. Zero-marking is illustrated in the (a)-sentences whereas the use of (Π15) *fi* 'in' is shown in the (b)-sentences. Several of the sentences contain additional spatial adverbials of interest. Each pair of sentences is briefly commented upon.

(4.223) Goal > Place

(a) [Korpus Malti 3.0; news135725]

Studja fl-iskola kattolika St Mary ta' Mumbai u wara
study in:DEF-school Catholic St Mary of Mumbai and after

mar [Ø l-Università ta' Stanford]$_x$
go [Ø DEF-University of Stanford]

'He studied at the Catholic school St Mary in Mumbai and afterwards, he went [**to** the University of Stanford].'

(b) [Korpus Malti 3.0; academic265]

fl-1985 beda jistudja l-Liġi [fl-Università ta'
in:DEF-1985 begin 3SG.M.IPFV:study DEF-law [**in**:DEF-University of

Malta]$_{PP}$
Malta]
'[. . .] in 1985 he began to study law [**at** the University of Malta].'

In (4.223a–b), the Ground is a given university identified by a GENITIVAL postnominal *ta'*-PP. In spite of these structural similarities, the zero-strategy is applied in (4.223a) because the predicate kernel is an ALLATIVE motion verb whereas in (4.223b) the spatial relation is static. The opposition dynamic vs static gives preference to zero-marking for Goal as opposed to overt marking of Place. Note that the first *fi*-PP in (4.223a) is a perfect match for *fl-Università ta' Malta* 'at the University of Malta' in (4.223b) since the static character of the spatial relation justifies the use of (Π15) *fi* 'in'. The use of (Π15) *fi* 'in' might also be motivated by the syntactic heaviness of the complement because the noun is accompanied by an attribute.

(4.224) Place name > common noun
(a) [Korpus Malti 3.0; culture48]
Jiena mir-Raħal Ġdid imma noqgħod [Ø Ħaż–Żebbuġ]$_X$
1SG from:DEF-Raħal Ġdid but 1SG.IPFV:stay [Ø Ħaż–Żebbuġ]
'I am from Raħal Ġdid, but I live [**in** Ħaż–Żebbuġ].'
(b) [Korpus Malti 3.0; news130522]
Għal snin jien għixt l-Ingilterra u kont
for year.PL 1SG live:1SG.PFV DEF-England and be:1SG.PFV
noqgħod [fil-kampanja]$_{PP}$
1SG.IPFV:stay [**in**:DEF-countryside]
'For years I lived in England and I lived [**in** the countryside].'

The imperfective form of the verb *qagħad* 'stay' occurs in both (4.224a) and (4.224b). Yet, in the (a)-sentence the Ground is a place name and thus needs not be explicitly marked for the spatial relation of Place. This is different for *il-kampanja* 'the countryside' in the (b)-sentence because it is a common noun which lacks the inherent spatial character of a place name and thus its function as Ground has to be formally established.

(4.225) light > heavy
(a) [Korpus Malti 3.0; parl9543]
Ħdimt fl-isptar Monte Carmeli u fl-isptar
work:1SG.PFV in:DEF-hospital Monte Carmeli and in:DEF-hospital
San Luqa ħdimt [Ø l-Università]$_X$
San Luqa work:1SG.PFV [Ø DEF-university]

'I worked in the Monte Carmeli Hospital and in the San Luqa Hospital, I worked [**at** the university] [. . .].'

(b) [Korpus Malti 3.0; news100368]

noqgħod *[f'* *dan* *ir-raħal* *żgħir]$_{PP}$*
1SG.IPFV:stay [**in** DEM:M:PROX DEF-village small]
'[. . .] I live [**in** this small village].'

The first sentence of this pair is particularly revealing because it hosts three comparable clauses which contain identical verbs, namely the 1st person singular perfective of *ħadem* 'work'. The speaker tells us where she worked at different times of her career. In the case of the two hospitals, she uses a *fi*-PP to introduce the Ground. In the third clause, no Π is used. The first two NPs are heavy because the NPs consist of several words whereas *l-università* 'the university' is a mono-word complement. (4.225b) repeats the pattern of (4.225a) insofar as the heavy NP (with prenominal determiner and postnominal adjective) which functions as Ground triggers the employment of (Π15) *fi* 'in'.

(4.226) singular > dual/plural

(a) [Korpus Malti 3.0; academic12]

Lil *uliedhom* *kienu* *jibagħtuhom* *[Ø* *l-iskola]$_{X}$*
to child.PL:3PL be:3PL.PFV 3.IPFV:send:PL:3PL [Ø DEF-school]
'As to their children, they send them [**to** school].'

(b) [Korpus Malti 3.0; culture1791]

tijm *ta'* *persuni* *b'* *diżabilità* *jmorru* *[fl-iskejjel]$_{PP}$*
team of person:PL with handicap 3.IPFV:go:PL [**in**:DEF-school.PL]
'[. . .] a team of persons with a handicap go [**to** the schools] [. . .].'

In the above examples, two motion events are described. The Ground is the noun *skola* 'school' once in the singular and once in the plural (*skejjel* 'schools'). The singular allows for the absence of (Π15) *fi* 'in' whereas the plural requires its presence. The difference between the behaviour of the singular and the plural form of one and the same noun might receive a different interpretation because in the (a)-sentence, *l-iskola* does not refer to an identifiable school(-building) but to primary/secondary education in general. In the (b)-sentence, however, *l-iskejjel* can be understood as referring to identifiable schools which take part in a special program of the Ministry of Education. Thus, chances are that we are facing the opposition of generic vs specific or abstract vs concrete.

(4.227) definite > indefinite

(a) [Korpus Malti 3.0; news7171]

Hu rrappurtat li 68,000 turist ġew [Ø
3SG.M report:PART that 68,000 tourist come:3PL.PFV [Ø
pajjiżna]ₓ għal vaganza filwaqt li ftit inqas minn 12,000
country:1PL] for holiday while that little less from 12,000
persuna ġew fil-gżejjer Maltin fuq negozju
person come:3PL.PFV in:DEF-island.PL Maltese:PL on business
'It is reported that 68,000 tourists came [**to** our country] for holidays while little less than 12,000 persons came to the Maltese islands on business.'

(b) [Korpus Malti 3.0; culture53]

Qed ngħixu [f' pajjiż żgħir b' ħafna limitazzjonijiet]$_{PP}$
PROG 1.IPFV:live:PL [**in** country small with many limitation:PL]
'[. . .] we are living [**in** a small country with many limitations].'

In (4.227a), the Ground-noun is made definite by way of possessor suffixation. In (4.227b), the same noun *pajjiż* 'country' remains bare and is thus formally indefinite. It is clear however that the person who uttered the (b)-sentence was referring to Malta, i.e., the Ground noun is specific. The possessed noun *pajjiżna* 'our country' is contextually a synonym of the place name *Malta*. Interestingly, *il-gżejjer Maltin* 'the Maltese islands' in (4.227a) require the presence of the Π although this expression is also often used as a synonym of *pajjiżna* and *Malta* (if the latter refers to the state and not the main island of the republic). The reason for choosing the *fi*-PP is the multi-word structure of the expression.

(4.228) generic > specific

(a) [Korpus Malti 3.0; parl31]

U dan jidher ferħan qisu xi ħadd
and DEM:M:PROX 3SG.M:appear happy like:3SG.M some somebody
li mar [Ø il- knisja]ₓ biex jiżżewweġ
that go [Ø church] in_order_to 3SG.M.IPFV:marry
'And this one seems to be happy as if he was someone who went [**to** church] for getting married [. . .].'

(b) [Korpus Malti 3.0; news150096]

dakinhar il-kaptan Anton Kurashvili kien mar [fil-Knisja
on_that_day DEF-captain Anton Kurashvili be go [**in**:DEF-church
magħrufa bħala Ta' Liesse fil-Belt Valletta]$_{PP}$
PART:know:F as Ta' Liesse in:DEF-city Valletta]
'[. . .] on that day, Captain Anton Kurashvili had gone [**to** the church known as Ta' Liesse in the city of Valletta].'

In these two sentences both the motion verb *mar* 'go' and the Ground *il-knisja* 'the church' are the same. (Π15) *fi* 'in' is absent from (4.228a) because no particular church is referred to. The entire situation is depicted as a generic scenario that could happen in any church. The (b)-sentence involves the Π because the Figure went to a particular church identified by name and further coordinates. Since mentioning the church's name and its location make the complement heavy, the explanation for the use of the *fi*-PP could also be linked to the syntactic weight of the complement.

(4.229) abstract > concrete

(a) [Korpus Malti 3.0; academic12]

Il-missier kien jaħdem u t-tfal kollha
DEF-father be 3SG.M.IPFV:work and DEF-child.PL all:3SG.F
marru [Ø l- iskola]$_X$
go:3PL.PFV [Ø DEF-school]
'The father worked and all the children went [**to** school].'

(b) [Korpus Malti 3.0; culture1371]

Meta l-Pulizija waslu [fl-iskola]$_{PP}$ Brandon Hill ma
when DEF-police arrived [**in**:DEF-school] Brandon Hill NEG
rreżistiex l-arrest iżda kkopera magħhom
resist:NEG DEF-arrest but cooperate with:3PL
'When the police arrived [**at** the school], Brandon Hill did not resist the arrest but cooperated with them.'

In (4.229a), the situation is again generic. The sentence pictures the social conditions of the old times in which fathers usually had a job and the children attended school. In this case, going to school does not mean that referentially identifiable children went to a referentially identifiable school. The intended meaning is that children generally received primary education. In (4.229b) however, the police arrive on the premises of a particular school on occasion of a historical event. The school is not only known as such but the spatial situation described in the sample sentence refers to the school as a concrete physical object. This is why (Π15) *fi* 'in' is employed.

(4.230) generic verb > specific verb

(a) [Korpus Malti 3.0; literature14]

"So happy for you!" qaltlu Abigail li kienet
"So happy for you!" say:3SG.F.PFV:IO:3SG.M Abigail that be:3SG.F.PFV
[Ø l-università]$_X$ miegħu
[Ø DEF-university] with:3SG.M
'"So happy for you!" said Abigail to him, who was [**at** university] with him.'

(b) [Korpus Malti 3.0; news143479]
kienet *tgħallem* *[fl-Università]*$_{PP}$
be:3SG.F.PFV REFL:teach [**in**:DEF-university]
'[. . .] she had learned [**at** university] [. . .].'

In the (a)-sentence of this pair of examples the semantically almost devoid copula *kien* 'be' is used. We term this a generic verb. The genericity of the verb licences the application of the zero-strategy. In contrast, the verb *tgħallem* 'learn' in (4.230b) is semantically much more loaded than the copula. Therefore, the *fi*-PP seems to be the better choice over the zero-strategy. Note that both cases of *l-università* 'the university' as Ground in these sentences are generic or abstract in the sense that no particular university is identified but a certain period of the participant's tertiary education is referred to.

As the brief discussion of the above examples shows, it is often difficult to determine the motivating factor for the choice of construction. Frequently, several factors seem to be active jointly. Intricacies of this and other kinds strongly suggest that this phenomenon – the Ø-*fi* alternation – deserves to be studied separately and in-depth. Pragmatics, word order, syntactic weight, and further factors such as genericity and definiteness must properly be accounted for on a sufficiently large empirical basis. The many other problems that call for being addressed in this study do not allow us to investigate the intriguing world of the Ø-*fi* alternation here and now. Answers to the questions it raises must be found in future research.

4.4 Πs in Maltese morphosyntax – wrap-up

We have reached the endpoint of the syntax chapter of this study. In several of the sections devoted to particular subject matters, we have been able only to show that inspecting the phenomenon at hand in more detail is a very promising project both for the descriptive linguistics of Maltese and linguistics in general. On account of the many properties of Maltese Πs that we have described throughout Section 4, it makes sense to recapitulate the major findings. This will be done summarily in order to avoid lengthy repetitions of special traits which only apply in a minority of cases or only with individual Πs.

The summary can be connected directly to the concept of the Canonical Maltese Π.[164] We proceed chronologically according to the order of the sections and subsections on syntactic issues. What is typical of a Maltese Π in syntax? We start with those properties that can be considered positive features. Those which involve the lack of certain traits are presented separately below. The Canonical Π

164 Cf. Section 1.6.1 and passim.

(i) takes complements irrespective of their definiteness including names,
(ii) applies the pattern of NOMINATIVE government with lexical NPs as complement,
(iii) applies the pattern of POSSESSIVE government with pronominal complements,
(iv) may fill the second slot in Π-Π sequences provided it has spatial functions,
(v) is subject to EQUI-P-deletion deletion,
(vi) can be used NP-internally to form prepositional attributes of head nouns,
(vii) can be used in resumptive constructions within relative clauses,

The thirty-six Πs which fulfil all of the above criteria of canonicity are listed in (4.231). They represent 60% of BLOMP 2.0. There is thus a sizable majority for the canon.

(4.231) Canonical Πs in the domain of syntax
(Π4) *bejn* 'between'; (Π5) *bħal* 'like'; (Π7) *bi* 'with'; (Π8) *biswit* 'facing'; (Π10) *daqs* 'equal to'; (Π11) *dwar* 'about'; (Π13) *favur* 'in favour of'; (Π14) *fejn* 'near'; (Π15) *fi* 'in'; (Π16) *fost* 'amongst'; (Π17) *fuq* 'on'; (Π20) *għajr* 'except'; (Π22) *għand* 'at s.o.'s place'; (Π23) *ħdejn* 'beside'; (Π24) *ħlief* 'except'; (Π27) *kontra* 'against'; (Π28) *lejn* 'towards'; (Π29) *lil* 'to'; (Π30) *ma'* 'with'; (Π31) *madwar* 'around'; (Π32) *maġenb* 'close to'; (Π33) *matul* 'during'; (Π35) *minflok* 'instead of'; (Π36) *mingħajr* 'without'; (Π38) *mingħand* 'from s.o.' (Π41) *minn* 'from'; (Π46) *qabel* 'before'; (Π47) *qalb* 'amidst'; (Π48) *qrib* 'near'; (Π49) *quddiem* 'in front of'; (Π53) *skont* 'according to'; (Π55) *ta'* 'of'; (Π56) *taħt* 'under'; (Π58) *viċin* 'near'; (Π59) *waqt* 'at the time of'; (Π60) *wara* 'after'

	Semitic	Italo-Romance
noncanonical	14	10
canonical	31	5

Figure 4.115: Morphosyntactic (non)canonicity over language of origin.

Five of the Πs featured in (4.231) are of Italo-Romance origin. This means that two-thirds of all borrowed Πs show noncanonical behaviour in the domain of morphosyntax. The relation is the other way around in the case of Semitic Πs 69% of which are morphosyntactically in conformity with the canon. The Italo-Romance share rises to some 42% under the rubric of noncanonicity. This means that loan-Πs are overrepresented in this category whereas with only 14% they are clearly underrepresented in the canonical class. This can be gathered from Figures 4.115–4.116.

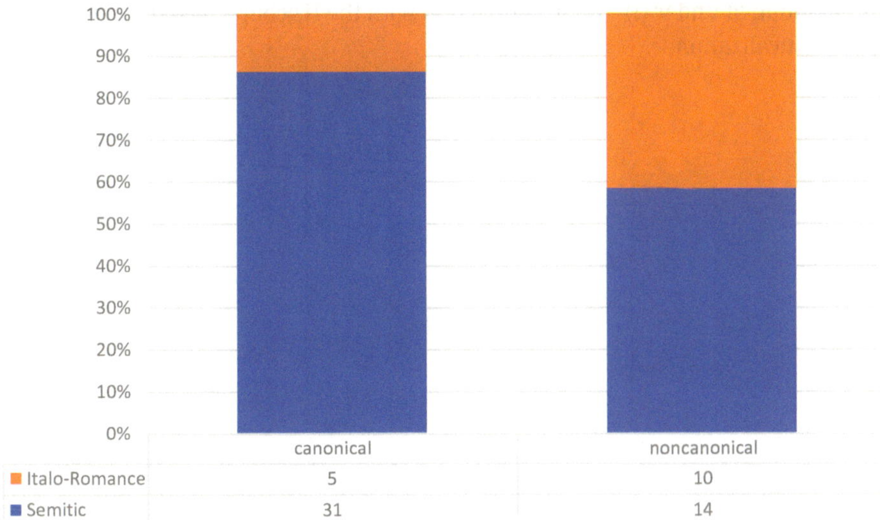

Figure 4.116: Language of origin over morphosyntactic (non)canonicity.

In addition to the seven positive traits mentioned in (i)–(vii) above, there are further properties which are unique to a given Π, namely

(i) derivational functions,
(ii) marking of core arguments,
(iii) mandatory oblique object marking,
(iv) alternation with zero as head of adverbial PPS.

These are criteria that fail to apply to the bulk of the Πs in (4.231). The criteria single out four of the canonical Πs. (Π7) *bi* 'with' is the only Π which comes close to the status of an obligatory oblique object marker. Except for (Π29) *lil* 'to', (Π15) *fi* 'in' is the only Π which can be replaced with zero outside coordinative constructions. (Π29) *lil* 'to' is subject to DOM and thus also attests to alternations with zero. Moreover, it has the uncontested privilege of marking direct and indirect objects. And (Π55) *ta'* 'of' alone can be used for derivational purposes. These four Πs are Semitic monosyllables counting among the most frequently employed Πs of Maltese. They deviate from the bulk of the canonical Πs because of their individual properties. These properties may tentatively be subsumed under the label of advanced grammaticalization. The advanced grammaticalization of (Π7) *bi* 'with', (Π15) *fi* 'in', (Π29) *lil* 'to', and (Π55) *ta'* 'of' renders them less canonical than the majority of the Maltese Π. Put another way, this means that the eighth property of morphosyntactic canonicity of Maltese Πs is their relatively weak degree of grammaticalization. In contrast, one looks in vain for Italo-Romance Πs that realise any of the properties mentioned under (i)–(iv) in connection with the Πs characterised by advanced grammaticalization.

5 Semantics

The previous sections on syntactic issues are replete with anticipatory references to this section which is dedicated to the semantic side of Maltese Πs. We repeatedly emphasised that the differential behaviour of many members of BLOMP 2.0 is explicable most often with reference to their polysemy. We will test and modify this hypothesis below.[1] As function words, Maltese Πs are synsemantic elements in the sense that the appropriate reading of their possible meanings depends largely on the constructional context, i.e. the words with which a given Π is syntagmatically related determine how the Π is interpreted semantically. The synsemantic nature of the Πs does not require the Πs to be completely desemanticised. In point of fact, some Πs display a relatively strong basic meaning component of their own whereas others seem to be almost devoid of proper semantic features. We know from the historical review[2] that the semantic domains of Maltese Πs are widely understudied. Panzavecchia (1843) is an early exception. Saari (2003: 238–248) highlights only a small selection of semantic issues whereas Borg/Azzopardi-Alexander (1997: 139–173) provide a rich catalogue of functions fulfilled by Maltese Πs. It is our task now to complement the received knowledge and systematise the findings. This is achieved by way of looking at the data twice from two different points of view, namely first from the onomasiological perspective which takes the functions as its point of reference[3] and then from the semasiological perspective which departs from the concrete Πs (Lehmann 2004: 149–154).[4] Section 5.1 largely follows the lines laid out in Hagège (2010) and Borg/Azzopardi-Alexander (1997) whereas in Section 5.2 the results of the previous section will be systematised according to a modified taxonomy. Section 5.3 will integrate the findings into the model of the canonical Π.

5.1 Domains

Hagège (2010: 261–262) proposes a semantic system for adpositions in general. This system is divided into two major domains. There is the class of core meanings which is differentiated from the large class of non-core meanings with the latter being subdivided into spatio-temporal meanings (static vs non-static) and

1 In Section 5.2.1.4.4 for spatio-temporal meanings and in Section 5.2.2.1 for all other meaning categories.

2 Cf. Sections 1.4.1–1.4.2.

3 Cf. Section 5.1.

4 Cf. Section 5.2.

https://doi.org/10.1515/9783111331591-005

non-spatio-temporal meanings. We adapt Hagège's system in Table 5.1 since we use it for orientation in our subsequent account of the situation in Maltese. Wherever we diverge from this system we will mention and explain our decision to do so. A number of issues arise especially in connection with Hagège's terminology which is reproduced in Table 5.1 but is not consistently adhered to in the subsequent sections. The cells marked out in grey in Table 5.1 contain subdivisions of the COMPARATIVE of equality and the COMPARATIVE of inequality.

Table 5.1: Semantic system of adpositions according to Hagège (2010: 261–262).

core meanings	**non-core meanings**				
	spatio-temporal				**non-spatio-temporal**
	spatial	**temporal**	**spatial**	**temporal**	
AGENTIVE	INESSIVE	INESSIVE	ILLATIVE		PROPRIETIVE
PATIENTIVE	APUDESSIVE		ALLATIVE		EXCLUSIVE
ATTRIBUTIVE	ADESSIVE		TERMINATIVE	TERMINATIVE	EXCEPTIVE
POSSESSIVE	ABESSIVE	ABESSIVE	ABLATIVE		COMITATIVE
	OBESSIVE		OBVERSIVE		INSTRUMENTIVE
	SURESSIVE		SURLATIVE		MEDIATIVE
	SUPERESSIVE		SUPERVERSIVE		MOTIVATIVE
	SUBESSIVE		SUBLATIVE		CONCESSIVE
	PREESSIVE	PREESSIVE	PRELATIVE		EQUATIVE
	POSTESSIVE	POSTESSIVE	POSTLATIVE		ASSIMILATIVE
	CIRCUMESSIVE		CIRCUMLATIVE		INEQUALITY
	MEDIOESSIVE		MEDIOLATIVE		ESSIVE/ TRANSLATIVE/ MUTATIVE
	INTERESSIVE		INTERLATIVE		PURPOSIVE
			PERLATIVE	PERLATIVE	ADVERSATIVE
			PROLATIVE		PERTENTIVE
			SECUTIVE		ROBORATIVE
					ADNUMERATIVE
					ADDITIVE
					SUBSTITUTIVE
					HYPOTHETICAL

According to Hagège's own count, there are fifty-one distinct functions. This number increases to fifty-nine if we take account of the distinction of spatial and temporal subcategories as well as those of the COMPARATIVE function. With fifty-nine functions the distinctions seem to map rather closely upon the sixty Πs in BLOMP 2.0. However, we will see that one-to-one relations of function and form are not necessarily the rule. On the contrary, one-to-many as well as many-to-one relations abound in the system under scrutiny. The functions will be discussed as far as possible in the order and with the labels introduced by Hagège (2010), i.e., we start with the so-called core meanings,[5] continue with spatio-temporal relations,[6] and conclude the survey with a review of non-spatio-temporal relations.[7]

A word of caution is in order at this point. Since our main reference (i.e. Hagège 2010) does not reveal explicitly which model of semantic roles is adhered to, we cannot always be sure whether we interpret the category labels correctly. Things are made worse by the frequent absence of empirical data supporting some of Hagège's distinctions. What is more, it is occasionally also necessary that we adjust some of the categories stipulated by Hagège either because their scope is too wide or too narrow semantically. As far as it is possible, we follow Hagège's sometimes rather doubtful practice of using Latinate labels for the meaning categories which implies that, because of problems with the solutions proposed in our reference, we are forced to coin labels of our own either because those introduced by Hagège are inappropriate or because Hagège does not account for a given category, in the first place. In numerous cases, we make use of a terminological mix by way of employing the LDS-based terms of the reference grammar (or new descriptive labels built on the same pattern) side by side with those proposed by Hagège (2010). These terminological solutions are not meant to be permanent. Its ad-hoc character calls for a thorough revision in a dedicated follow-up study. In Section 5.1, we limit the invention of category labels to those instances in which a member of BLOMP 2.0 needs no other Π to express a meaning. For Π-Π combinations with Π_1 expressing directionality and Π_2 expressing configuration, we refrain from creating distinctive category labels (apart from compounds involving already established terms). The expression of directionality is most often the task of (Π21) *għal* 'for', (Π41) *minn* 'from', and (Π51) *sa* 'till'. These Πs are presented separately in the sections on the directional, the TERMINATIVE, the ELATIVE, and the ABLATIVE.[8] Wherever these Πs combine with other Πs to form a directionality-configuration pair the resulting combination is discussed in the appropriate section on the dynamic spatio-temporal category of a given configuration. In most of these cases, no Latinate label is used. Special treatment is also reserved for the SECUTIVE, PERLATIVE, PROLATIVE and

5 Cf. Section 5.1.1.

6 Cf. Section 5.1.2.

7 Cf. Section 5.1.3.

8 Cf. Sections 5.1.2.2.2, 5.1.2.2.6, 5.1.2.2.7, and 5.1.2.2.8, respectively.

related categories.[9] In what follows wherever we talk about categories the term refers to meanings and/or functions of Πs without determining in any way how far a given construction has advanced in the process of grammaticalization.

5.1.1 The core

The core meanings cover those relations which display a particularly high degree of abstraction. Put differently, they belong to the proper grammatical domain and are strongly desemanticised (Hagège 2010: 283). This includes syntactic functions such as subject and object. Hagège (2010: 260) assumes that with core participants adpositions "are only rarely used as markers" cross-linguistically. As will transpire from our argumentation,[10] Maltese offers examples of Πs being used for exactly these functions. It is clear that in these cases, a particularly high degree of grammaticalization can be assumed.

5.1.1.1 Agentive

For the AGENTIVE function, Hagège (2010: 261) claims that there is no English Π which could illustrate this function. We assume that the AGENTIVE function applies if a participant acts as agent in a given situation. This agent need not correspond to the subject. If we accept that passive agents are participants with AGENTIVE function, then Maltese is a case in point. Possibly, Hagège (2010) classifies passive agents as instances of the INSTRUMENTIVE category.[11]

Maltese has two formally distinct passive constructions. There is the construction which involves a verb which inflects for the reflexive-passive by way of prefixing and/or infixing a dedicated marker as in ACT *qal* 'say' → PASS *n(t)qal* 'be said'. The second option requires the presence of a passive participle and, if used predicatively, that of the passive auxiliary *ġie* 'come' (for the dynamic passive) or *kien* 'be' (for the stative passive) as in *(il-vittma ġie/kien) marbut (mas-sodda)* '(the victim was (being)) tied (to the bed)' with ACT *rabat* 'tie' → PART *marbut*. In both cases, it is possible to express the agent overtly as complement in a *minn*-PP (Borg/Azzopardi-Alexander 1997: 212–216). Borg/Azzopardi-Alexander (1997: 146) make a distinction between animate passive agents which require (Π41) *minn* 'from' and inanimate passive agents for which (Π7) *bi* 'with' is the appropriate choice. We disagree with this rule on two accounts. First, inanimate passive agents marked by

---

9 Cf. Section 5.1.2.2.20 on Path.

10 Cf. below Sections 5.1.1.1–5.1.1.2.

11 Cf. Section 5.1.3.5.

(П7) *bi* 'with' can better be classified as cases of the INSTRUMENTIVE[12] and, second, examples like (5.1a) and (5.1c) show that inanimate nouns are not excluded from *minn*-PPs which express passive agents. In (5.1b), the employment of the participle in an attributive passive construction is illustrated.

(5.1) Passive agent

(a) [Camilleri 2021: 55]

Quddiem dik il-purċissjoni mnikkta ta' vetturi
in_front_of DEM:F:DIST DEF-procession PART:sadden:F of vehicle:PL
suwed Gallo qabbad sigarett u nħakem [minn malinkonija
black.PL Gallo light cigarette and REFL:rule [from melancholy
liema bħalha]$_{PP}$.
which like:3SG.F]
'In front of this sorrowful procession of black cars, Gallo lit a cigarette and was dominated [**by** an unequalled melancholy].'

(b) [DeBono 2001: 81]

nies u vetturi miġbuda [minn żwiemel]$_{PP}$ kienu
people and vehicle:PL PART:pull:PL [from horse.PL] be:3PL.PFV
jidhru għaddejjin
3.IPFV:appear:PL pass:PART:PL
'[. . .] people and vehicles pulled [**by** horses] were seen passing by [. . .].'

(c) [Korpus Malti 3.0; news5687]

In-Nepal reġa' ntlaqat [minn terremot ieħor qawwi]$_{PP}$
DEF-Nepal return REFL:hit [from earthquake other strong]
'Nepal was hit again [**by** another strong earthquake] [. . .].'

Natural forces as in (5.1c) and physico-mental states as in (5.1a) can take on the role of passive agent. AGENTIVE is thus a category which embraces different semantic roles which can be subsumed under the meta-role actor. Maltese fits into the general picture. Many languages across the globe employ markers which otherwise function as expressions of ABLATIVE relations to encode the passive agent – and Пs meaning 'from' are especially prominent among the known examples (Kuteva et al. 2019: 34–36).

5.1.1.2 PATIENTIVE

Like in the case of the AGENTIVE, Hagège (2010: 261) claims that no English П serves as marker of the PATIENTIVE function. Our interpretation of the term PATIENTIVE links it to the semantic role of patient (which perhaps needs to be extended such

12 Cf. Section 5.1.3.5.

that it covers further semantic roles in the domain of the Undergoer). In Maltese, the object-marking (Π29) *lil* 'to' is responsible for the expression of the PATIENTIVE function. However, as we know already from the previous discussion,[13] the situation is not absolutely straightforward since the use of (Π29) *lil* 'to' is dependent on animacy and definiteness. DOM applies in the sense that only animate (and preferably definite) direct objects require the presence of this Π. Furthermore, (Π29) *lil* 'to' is also the marker of the DATIVE relation as shown below.[14] In (5.2), we give two examples of the use of (Π29) *lil* 'to' with an action verb in the (a)-sentence and a perception verb in the (b)-sentence.

(5.2) PATIENTIVE I

(a) [Attard 2019: 186]

libbsu	***[lill-mara]***$_{PP}$
dress.CAUS:3PL.PFV	**[to:**DEF-woman]

'[. . .] they dressed [the woman].'

(b) [Attard 2019: 148]

Fil-bogħod	*stajt*	*nara*	***[lil***	*missieri]*$_{PP}$
in:DEF-distance	can:1SG.PFV	1SG.IPFV:see	**[to**	father:1SG]

'In the distance, I could see [my father] [. . .].'

Above we have argued that (Π7) *bi* 'with' is the other Π which comes closest to object-marking without however fully meeting our necessary criteria for being a bona fide object marker (Stolz/Stroh/Urdze 2006: 292–293).[15] (Π7) *bi* 'with' introduces the patient of a huge (but still undetermined) number of originally intransitive verbs which in this way undergo valency enhancement. Very frequently it is difficult, even next to impossible to tell these cases apart from adverbial usages of the instrumental or other brands. If we apply as wide as possible an interpretation of the term PATIENTIVE, it is possible to take it to embrace many kinds of undergoer-like relations. Under this condition, what we find is that (Π7) *bi* 'with' occurs frequently as a PATIENTIVE marker with verbs of perception, speaking, or expression of physico-mental reactions. In (5.3), three different verbs of perception – *induna* 'perceive', *ntebaħ* 'be aware', and *ittenda* 'notice' – combine with a *bi*-PP to allow for the perceived entity to be mentioned.

---

13 Cf. Section 4.3.1.

14 To be discussed in Section 5.1.1.3.

15 Cf. Sections 4.3.2 and 4.3.3.1.

(5.3) PATIENTIVE II

(a) [Cauchi 1997: 10]

Kif jista' jkun ma nindunax
how 3SG.M.IPFV:can 3SG.M.IPFV:be.FUT NEG 1SG.IPFV:perceive:NEG
***[b'** qargħa ħamra li tkun qrib tiegħi]$_{PP}$*
[**with** pumpkin red that 3SG.F.IPFV:be.FUT near of:1SG]
'How could it be that I won't notice [a red pumpkin in my vicinity]?'

(b) [Bartolo/Vella 2009: 9]

*lanqas biss kienet intebħet **[bih]**$_{PP}$*
not_even only be:3SG.F.PFV be_aware:3SG.F.PFV [**with**:3SG.M]
'[. . .] and she even had not been aware [**of** him].'

(c) [Korpus Malti 3.0; literature78]

U billi dik il-laqgħa kienet għoġbitu,
and since DEM:F:DIST DEF-meeting be:3SG.F.PFV please:3SG.F.PFV:3SG.M
kompla jgħidlu biex jurih li kien
continue 3SG.IPFV:say:IO:3SG.M to 3SG.IPFV:show:3SG.M that be
*ittenda [**b**il-biċċa]$_{PP}$*
notice [**with**:DEF-piece]
'"And since that meeting has pleased him," he continued to tell him to show him that he had noticed [the fact].'

This parallel behaviour of different verbs with similar semantics is striking. Since all of them are compatible with (Π7) *bi* 'with' when it comes to mentioning the perceived entity, one may assume that there is a conceptual link which facilitates associating perception events with the Π under inspection. That Πs with a basic meaning equivalent to English *with* come to be employed as PATIENTIVE markers seems to be infrequent cross-linguistically since the second edition of the *World Lexicon of Grammaticalization* (Kuteva et al. 2019) does not mention any cases of this kind. That (Π29) *lil* 'to' serves as PATIENTIVE marker is however plausible since in cross-linguistic perspective, the grammaticalization of ALLATIVE markers to patient markers is attested frequently (Kuteva et al. 2019: 50).

5.1.1.3 DATIVE ("Attributive")

Hagège (2010: 283–284) goes to some length to explain his terminological choice for the label of this core meaning. He discards DATIVE as a term because of its association with certain grammatical (i.e. morphological) case categories. However, to our mind, attributive is not the best solution either because the term evokes something completely different, namely a marker of attribution in the sense of a linker between head and modifier like in the case of attributive adjectives accompanying a noun in the NP. Therefore, we discard Hagège's term by way of reverting to DATIVE. Another more substantial point of disagreement is connected to Hagège's (2010: 283–284) decision to subsume the BENEFACTIVE under attributive. In our opinion, the notions have to be separated. We understand the DATIVE as the semantic role which applies to recipients in situations of transfer be it of concrete objects or reified other concepts. BENEFACTIVE/MALEFACTIVE however refers to situations in which something is done to the benefit or to the detriment of a participant. This is why we add a section on the BENEFACTIVE and another one for the MALEFACTIVE to complement Hagège's system.[16] The English Π illustrating the attributive is *to* (Hagège 2010: 261).

There is again only one candidate in BLOMP 2.0 that can be accepted as fulfilling the DATIVE function, viz. (Π29) *lil* 'to' – the same Π which is employed for the PATIENTIVE function discussed in the previous section. The syntactic properties of this Π have been investigated above.[17] We know therefore that by no means does every instance of (Π29) *lil* 'to' count as a DATIVE. In (5.4), typical examples of the DATIVE usage of this Π are provided.

(5.4) DATIVE

(a) [Attard 2019: 127–128]

Imbagħad	*għidt*	***[lil***	*Whitfield]$_{PP}$*	*li*	*fil-korp*
then	say:1SG.PFV	**[to**	Whitfield]	that	in:DEF-corps
għandna	*pulizija*	*Eritrean*			
have:1PL	police	Eritrean			

'Then I told [Whitfield] that we have an Eritrean policeman in the force [. . .].'

(b) [Camilleri 2021: 202]

fit-tieni	*ittra*	*ta'*	*San Pawl*	*mibgħuta*	***[lil**l-Korintin]$_{PP}$*
in:DEF-second	letter	of	Saint Paul	part:send:F	[**to**:DEF-Corinthian:PL]
għandna	*niżżu*		*ħajr*	***[lil***	*Alla]$_{PP}$*
have:1PL	1.IPFV:render:PL		gratitude	**[to**	God]

'[. . .] in Saint Paul's second letter, sent [**to** the Corinthians], we have to show gratitude [**towards** God] [. . .].'

The *World Lexicon of Grammaticalization* tells us that ALLATIVES and BENEFACTIVES are the most prominent source concepts for the grammaticalization of DATIVES ("recipient" in the terminology of the *World Lexicon*). Πs with the meaning 'to'

16 Cf. Sections 5.1.3.17–5.1.3.18.
17 Cf. Section 4.3.1.

are abundantly represented in the database Kuteva et al. (2019: 52–53) draw on. Maltese thus behaves like many other languages in different parts of the world. What is remarkable about the Maltese case is that the grammaticalization of (Π29) *lil* 'to' has progressed so far that the original ALLATIVE meaning of the Π is obsolete now and has to be fulfilled by other Πs.[18]

5.1.1.4 POSSESSIVE

The POSSESSIVE relation is again one for which Hagège (2010: 284) refuses to use established terms like genitive, etc. His decision to label the relation under inspection POSSESSIVE is unproblematic so that we can keep it for the purpose of this investigation. The POSSESSIVE in this sense covers adnominal genitival attributes. Hagège (2010: 261) claims the English *s*-genitive and its near synonym *of* are good representatives of the POSSESSIVE relation although the English *s*-clitic does not fulfil our criteria of Π-hood. Going by Hagège's (2010: 261) own words, POSSESSIVE is an umbrella term for a variety of relations not all of which are properly POSSESSIVE. Other POSSESSIVE relations not covered by POSSESSIVE are discussed under the heading PROPRIETIVE further below.[19]

If we think back to Section 4.2.2., it is clear that the established POSSESSIVE Π can only be (Π55) *ta'* 'of' (Borg/Azzopardi-Alexander 1997: 111–113). Since the use of this Π has been abundantly illustrated throughout the foregoing sections, we give only one additional pair of examples of NP-internal (Π55) *ta'* 'of' in (5.5). The (a)-sentence is representative of adnominal possession of the prototypical kind (i.e. with an animate possessor) whereas the (b)-sentence reflects a part-whole relationship of two inanimate concepts (Borg/Azzopardi-Alexander 1997: 152).

(5.5) POSSESSIVE

(a) [Cauchi 1997: 21]

Għajnejh	*waqgħu*	*fuq*	*il-kappell*	***[ta'***	*Mastru Gerfex]*$_{PP}$
eye:PL:3SG.M	fall:3PL.PFV	on	DEF-hat	**[of**	Mastru Gerfex]

'His gaze caught the hat [**of** Mastru Gerfex] [. . .].'

(b) [DeBono 2001: 63]

dan	*ġie*	*jħabbat*	*fuq*	*il-bieb*	***[ta'***	*kamritha]*$_{PP}$
DEM:M:PROX	come	3SG.M.IPFV:knock	on	DEF-door	**[of**	room:POSS:3SG.F]

'[. . .] he came to knock on the door [**of** her room].'

18 As discussed in Section 5.1.2.2.6.

19 Cf. Section 5.1.3.1.

In the above, we have encountered cases of two Πs serving one core meaning and one Π being used for two different core meanings according to Hagège's system. Maltese employs Πs also for those relations which supposedly have no prepositional expression in English again according to Hagège. We have also seen that animacy and definiteness are crucial factors determining the use of some of the Πs. Do these patterns also hold outside the domain of core meanings? The following two sections are meant to provide an answer to this question.

5.1.2 Spatio-temporal relations

In the Maltese reference grammar (Borg/Azzopardi-Alexander 1997: 155–173) spatio-temporal relations are illustrated in much detail. Hagège (2010: 285) argues that

> [t]he spatial domain is the richest in [adpositions], in practically all languages having more than one, two, or three [adpositions]. The inessive, illative, and the ablative are often considered as the three fundamental spatial meanings, and this triad is sometimes enriched with a fourth member, the perlative [. . .].

The quoted author provides a by far more fine-grained catalogue of spatial relations accompanied by a considerably smaller set of temporal distinctions. His INESSIVE, ILLATIVE, ABLATIVE, and PERLATIVE largely correspond to Place, Goal, Source, and Path in the LDS grammar format (and beyond). As mentioned above, except for situations in which misunderstanding might arise, we use both terminologies side by side in the subsequent sections. Note that Hagège (2010: 291) also claims that

> in languages with both case affixes and [adpositions], spatial *directions*, such as "into", "out of", "across", etc. will often be expressed by bound morphemes, while spatial *dimensions*, such as "inside", "above", "in front of", "beside", will tend to be expressed by independent morphemes like [adpositions] [original italics].

If we replace "directions" with "directionality" and "dimension" with "configuration", we have the equivalent of the situation found in Maltese with the sole differences that no case morphology is made use of. Both directionality and configuration need Πs to be made explicit. This necessity will come to the fore below where dynamic spatio-temporal relations are at issue.[20]

20 Cf. Section 5.1.2.2 and its subdivisions.

5.1.2.1 Static relations

The distinction of static vs dynamic spatial relations is neutralised for Place and Goal for most of the ΠIs in this domain. This is the reason why some ΠIs are featured not only in this section but also in Section 5.1.2.2 dedicated to dynamic relations. Hagège's (2010: 286) terminologically infelicitous category of "abessive" is redefined and given the new label EXTRAESSIVE.[21] What is more, there are further (potential) candidates for the status of static spatial relation which are missing from Hagège's system, but which need to be discussed in the wider context of Path.[22]

5.1.2.1.1 INESSIVE

Hagège (2020: 261) distinguishes between a properly spatial INESSIVE (associated with English *in* and *within*) and a temporal INESSIVE (for which English *in, on, at,* and *as of* are given as examples). We first look at the spatial INESSIVE.

5.1.2.1.1.1 Spatial INESSIVE

It seems that Hagège presupposes that the spatial INESSIVE applies in situations where the Figure is located inside a container. As transpires from the examples in (5.6), there is competition between several Maltese ΠIs in the domain of the spatial INESSIVE.

(5.6) Spatial INESSIVE I

(a) [Abela 2016: 101]

Qalli li jien ħalliela baxxa u qafilni [fl-istalla]$_{PP}$
say:IO:1SG that 1SG thief:F low:F and lock:1SG [**in**:DEF-stable]
għal ġimgħa
for week
'He said that I was a lowly thief and he locked me up [**in** the stable] for a week.'

(b) [Camilleri 2013: 168]

Kien diġà stħajjel il-katavru tiegħu stess ħiereġ
be already imagine DEF-corpse of:3SG.M self exit.PART
imgeżwer [ġo borża taż-żibel]$_{PP}$
PART:wrap_up [**in** bag of:DEF-rubbish]
'He imagined already his own body appearing wrapped up [**in** a rubbish sack].'

---

21 Cf. Section 5.1.2.1.17.

22 Their description is thus relegated to Section 5.1.2.2.20, i.e., the paragraphs addressing them are integrated into the section dedicated to dynamic relations.

(c) [Korpus Malti 3.0; academic379]
hemm drawwa li kunċert mużikali imqar jekk ***[ġewwa***
EXIST custom that concert musical even if **[inside**
nitfa' ta' pabb]$_{\text{PP}}$ *jispiċċa bl-innu nazzjonali*
a_little of pub] 3SG.M.IPFV:end with:DEF-hymn national
'[. . .] there is the custom that a concert even [**in** a small pub] terminates with the national anthem.'

In all of these cases, the Figure is located inside a container. In (5.6a), the ego is locked up in a stable, in (5.6b) the body of the speaker is imagined to be wrapped up in a rubbish sack, and in (5.6c) the concert takes place inside a small pub. We can thus assume that in these contexts the three Πs are largely synonymous (Borg/Azzopardi-Alexander 1997: 157–158). As Stolz/Levkovych/Vorholt (2023) show, this synonymy also holds for constructions with a place name as Ground. We already have addressed the Ø-*fi* alternation.[23] Owing to the equivalence of (Π15) *fi* 'in', (Π18) *ġewwa* 'inside', and (Π19) *ġo* 'in' as markers of the spatial INESSIVE, the binary alternation can be extended to cover all replacements of (Π15) *fi* 'in'. In (5.7), this possibility is illustrated with constructions that involve a toponymic Ground (in this case a settlement name).

(5.7) Spatial INESSIVE II
(a) [Debono 2001: 106]
Bħal ***[f'*** *Bombay]*$_{\text{PP}}$ *daqsxejn ta' karrozella wasslithom*
like **[in** Bombay] a_little of small_coach arrive.CAUS:3SG.F.PFV:3PL
fuq il-moll tal-port
on DEF-pier of:DEF-port
'Like [**in** Bombay] a small carriage took them to the pier of the harbour.'
(b) [DeBono 2001: 98]
L-S.S. Osiris għamel waqfa qasira ta' ftit sigħat ***[ġo***
DEF-S.S. Osiris make stop short:F of little hour:PL **[in**
Aden]$_{\text{PP}}$
Aden]
'The S.S. Osiris made a stop of a few hours [**in** Aden] [. . .].'
(c) [DeBono 2001: 5]
Matul id-difiża u l-assedju tal-Fortizza l-Ħamra
during DEF-defence and DEF-siege of:DEF-fortress DEF-red:F

23 Cf. Section 4.3.3.2.

[ġewwa Dehli]$_{PP}$ kien ġie ferut gravi ġo saqajh
[**inside** Dehli] be come wound:PART grave in leg:PL:3SG.M
'During the defence and siege of the Red Fort [**in** Dehli] he had been wounded seriously in his legs.'

(d) [Żahra 2008: 62]
Dik il-ħabta [Ø iż-Żejtun]$_{X}$ kellna għadd kbir
DEM:F:DIST DEF-time [**Ø** DEF-Żejtun] have.PAST:1PL number big
ta' parrukiera
of hairdresser:PL
'At that time, we had a huge number of hairdressers [**in** Żejtun] [. . .].'

Since Ø is possible only in combination with place names (and so-called toponouns), the following rule emerges (cf. Figure 5.1).

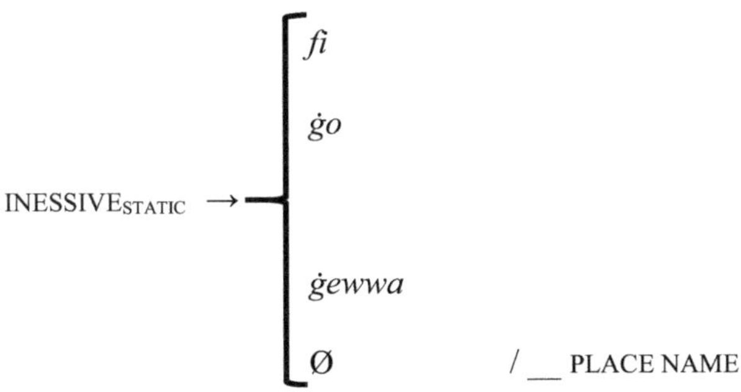

Figure 5.1: Multiple possibilities to express the static INESSIVE$_{SPATIAL}$.

The synonymy does not mean that the three competing ΠS fulfil the INESSIVE function equally often. Anticipating the insights of the section on the ILLATIVE,[24] we look at the competition of the above ΠS independently of the static-dynamic opposition. We have tested the quantitative differences in the Korpus Malti 3.0 with the common nouns *(il-)borża* '(the) bag', *(id-)dar* '(the) house', and the place name *Ruma* 'Rome'. Figure 5.2 leaves no margin for doubt because (Π15) *fi* 'in' is responsible for about 95% of all hits of each of the three Grounds. In contrast, (Π18) *ġewwa* 'inside' and (Π19) *ġo* 'in' divide the remaining 5% differently between them. Zero-marking is excluded from this frequency count.

24 Cf. Section 5.1.2.2.1.

	borża	dar	Ruma
ġewwa	1	230	83
ġo	45	383	10
fi	828	12929	1748

Figure 5.2: Frequency of INESSIVE Πs with selected Grounds (Korpus Malti 3.0).

The topic of a previous section was locative government with pronominal complements.[25] Both (Π18) *ġewwa* 'inside' and (Π19) *ġo* 'in' cannot directly take a pronominal complement but need (Π15) *fi* 'in' as host of the pronominal affix. Figure 5.3 reveals the result of the search we conducted in the Korpus Malti 3.0 to determine how many tokens of locative government can be found for the two competing INESSIVE Πs.

Clearly, (Π19) *ġo* 'in' is attested much more frequently than (Π18) *ġewwa* 'inside' although the differences diminish visibly from left to right. Since (Π15) *fi* 'in' directly takes pronominal complements its frequency is considerably higher than that of its competitors. This means that (Π15) *fi* 'in' is the prototypical marker of the static INESSIVE whereas (Π18) *ġewwa* 'inside' and (Π19) *ġo* 'in' are possible but relatively marginal replacements thereof. The latter Π is stronger than (Π18) *ġewwa* 'inside' in the context of locative government. However, between the two Πs, (Π18) *ġewwa* 'inside' stands out as the number one choice for combinations with place names as shown in Figure 5.4. To avoid skewing, the frequency of *fi*-PPs is not featured in Figure 5.4. The absolute numbers are given in brackets behind the Grounds mentioned in the next sentence. We have chosen the names of three

25 Cf. Section 4.1.2.2.2.1.2.

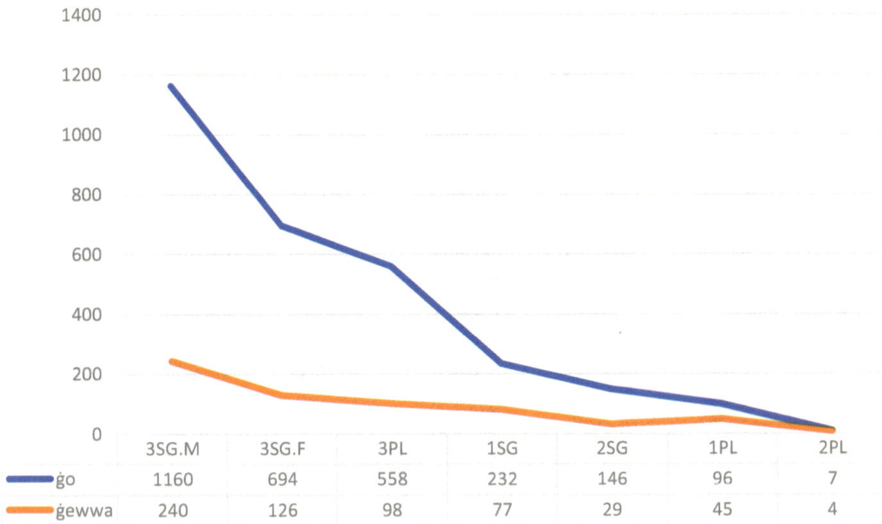

Figure 5.3: Frequencies of LOCATIVE government of (Π18) *ġewwa* 'inside' and (Π19) *ġo* 'in' per person (Korpus Malti 3.0).

international capital cities – *Berlin* 'Berlin' (*f'Berlin* 'in Berlin' = 428 matches), *Londra* 'London' (*f'Londra* 'in London' = 2,598 matches), *Pariġi* 'Paris' (*f'Pariġi* 'in Paris' = 1,122 matches) – a country name – *l-Eġittu* 'Egypt' (*fl-Eġittu* 'in Egypt' = 1,457 matches) –, the name of a continent – *l-Ewropa* 'Europe' (*fl-Ewropa* 'in Europe' = 17,428 matches) –, and an island name – *Sqallija* 'Sicily' (*fi Sqallija* 'in Sicily' = 1,552 matches). Both Πs under closer scrutiny are attested with each of the toponyms. As the absolute numbers given in the brackets show, the *fi*-PPs are overwhelmingly more numerous with each of the toponyms.

However, the share claimed by (Π18) *ġewwa* 'inside' ranges from 78% to 98% whereas that of (Π19) *ġo* 'in' is as small as 22% to 2%. A look back at Figure 4.16 and the accompanying discussion further underlines the fact that (Π18) *ġewwa* 'inside' is particularly prone to combine with proper names. Thus, of this pair of Πs, (Π18) *ġewwa* 'inside' is clearly the preferred option for PPs which involve a place name as complement. The differences observed above notwithstanding, all three Πs – (Π15) *fi* 'in', (Π18) *ġewwa* 'inside', and (Π19) *ġo* 'in' – are markers of the spatial INESSIVE (and ILLATIVE, cf. below).

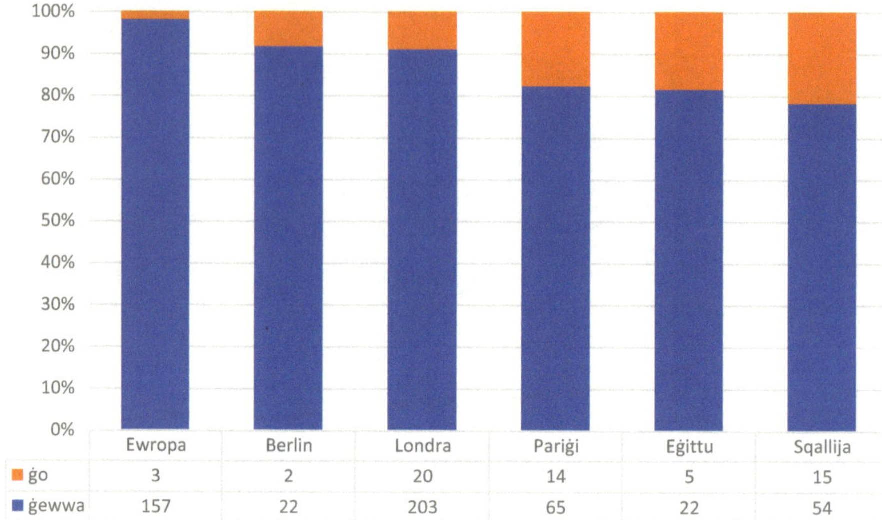

Figure 5.4: Tokens of (Π18) *ġewwa* 'inside' and (Π19) *ġo* 'in' in combination with place names (Korpus Malti 3.0).

5.1.2.1.1.2 Temporal INESSIVE

According to the logic of the metaphor-based approach to grammaticalization the more abstract concepts of TIME are made cognitively accessible by way of employing concepts of SPACE for their expression (Heine/Claudi/Hünnemeyer 1991: 48). On account of this axiom, we expect to find Πs featured in the previous section to surface again here and now when we talk about the temporal INESSIVE. It therefore comes as no surprise that Borg/Azzopardi-Alexander (1997: 169–171) provide many examples of temporal INESSIVEs which involve (Π15) *fi* 'in'. Interestingly, the above synonyms of this Π, i.e. (Π18) *ġewwa* 'inside', and (Π19) *ġo* 'in', are absent from the empirical pieces of evidence. Their exclusion can be taken to mean that the synonymy between (Π15) *fi* 'in' and (Π18) *ġewwa* 'inside', (Π19) *ġo* 'in' is only partial because it is restricted to the spatial domain to the exclusion of the temporal domain.

In (5.8), we present a selection of examples which illustrate the use of (Π15) *fi* 'in' as head of PPs which serve as temporal adverbial adjuncts.

(5.8) Temporal INESSIVE

(a) Year – [Monsarrat 2018]

Iċ-Ċittadella	*kienet*	*assedjata*	*[fl-1551]$_{PP}$*	*mit-Torok*
DEF-citadel	be:3SG.F.PFV	besiege:PART:F	[**in**:DEF-1551]	from:DEF-Turk.PL

'The citadel came under siege from the Turks [**in** 1551] [. . .].'

(b) Date – [Żahra 2008: 15]

L-ewwel tarbija tagħhom twieldet [fid-9 ta' Jannar
DEF-first infant of:3PL be_born:3SG.F.PFV [**in**:DEF-9 of January
1887]$_{PP}$
1887]
'Their first child was born [**on** 9 January 1887] [. . .].'

(c) Time of day (and generic expression) [Ellul Mercer 1986: 24]

Ilbieraħ is-Sibt [f' xi l-ħdax ta' fil-għodu]$_{PP}$
yesterday DEF-Saturday [**in** some DEF-eleven of in:DEF-morning]
mort infittex wieħed li [fiż-żmien]$_{PP}$ kien joqgħod
go:1SG.PFV 1SG.IPFV:seek one that [**in**:DEF-time] be 3SG.M.IPFV:stay
wara s-Suq tal-Belt.
after DEF-wall of:DEF-Valletta
'Yesterday, Saturday, [**at** about eleven o'clock in the morning] I went looking for someone who [**at** the time] lived outside the walls of Valletta.'

(d) [Gauci 1994: 127]

l-Ingliżi jġibu d-draw [fil-81 minuta]$_{PP}$
DEF-English:PL 3.IPFV:bring:PL DEF-draw [**in**:DEF-81 minute]
'[. . .] the English tie the score [**in** the 81st minute] [. . .].'

(e) Season – [Korpus Malti 3.0; parl6206]

għandkom titolbu skuża lil dawn in-nies li
have.IPFV:2PL 2.IPFV:ask:PL excuse to DEM:PL:PROX DEF-people that
llum itertru bil-bard [f' xitwa bħal din]$_{PP}$
today 3.IPFV:shiver:PL with:DEF-cold [**in** winter like DEM:F:PROX]
'[. . .] you must apologise to those people who shiver from the cold [**in** a winter like this] [. . .].'

(f) Month – [Korpus Malti 3.0; culture1442]

Il-kor Għawdxi Fjur Assisi twieled [f' Frar tal-1999]$_{PP}$
DEF-choir Gozitan Fjur Assisi be_born [**in** February of:DEF-1999]
'The Gozitan choir Fjur Assisi was born [**in** February 1999].'

(g) Festival – [Korpus Malti 3.0; literature19]

[Fil-Milied]$_{PP}$ kienu ġew Malta u damu
[**in**:DEF-Christmas] be:3PL.PFV come:3PL.PFV Malta and last:3PL.PFV
tlitt ijiem sħaħ
three day.PL full.PL
'[**At** Christmas] the had come to Malta and stayed for three full days.'

It results from the above examples that (Π15) *fi* 'in' is compatible with time expressions of diverse kinds. It is licit to locate events in a certain year, on a certain day, at a certain hour, at a certain minute, in a given season, in months, or at a certain festival. In (5.8c), *fil-għodu* (currently written <*filgħodu*> without hyphen) 'in the morning' is shaded grey. In contrast to Borg/Azzopardi-Alexander (1997: 169) who analyse this expression as a normal PP, we assume that we are facing a lexicalised univerbation since insertions of further elements are no longer admissible. This observation also holds for *filgħaxija* 'in the evening'. In analogy, we assume that *bil-lejl* 'at/by night' is an unanalysable one-word time expression in contemporary Maltese so that Borg/Azzopardi-Alexander's (1997: 169) idea that (Π15) *fi* 'in' and (Π7) *bi* 'with' are used with periods of the day "as the case may be" has to be taken with a grain of salt. In point of fact, *bil-lejl* 'at/by night' forms a unit which does not allow for the modification of *lejl* 'night' by further attributes whereas this is unproblematic for a PP like *fil-lejl* 'on the night' as shown in (5.9) where the adjective *ħiemed* 'still, (very) silent' modifies the complement of (Π15) *fi* 'in'.

(5.9) *Bil-lejl* vs *fil-lejl*

(a) [Korpus Malti 3.0; culture3265]

Dakinhar	*[bil-lejl]*$_{ADV}$	*fl-istess*	*pjazza*	*se*	*jkunu*
on_that_day	[at_night]	in:DEF-same	square	FUT	3.IPFV:be.FUT:PL

armati	*screens*	*kbar*
set_up:PART:PL	screen:PL	big.PL

'On that day [at night], on the same square, big screens will be set up [. . .].'

(b) [Korpus Malti 3.0; literature20]

Instema'	*grillu*	*hemm*	*barra*	*[fil-lejl*	*ħiemed]*$_{PP}$
REFL:hear	cricket	there	outside	[**in**:DEF-night	silent]

'A field cricket was heard out there [**in** the silent night] [. . .].'

On account of this and similar cases, we count out (Π7) *bi* 'with' as a competitor of (Π15) *fi* 'in' in the domain of the temporal INESSIVE. This domain is however divided in two since to express that a given event takes place on a given day of the week zero-marking is the preferred option. Borg/Azzopardi-Alexander (1997: 169) argue that for days of the week "[e]ither the name of the day is used on its own, without a preposition, or the noun **nhar** 'day' precedes it optionally" [original boldface]. The absence of the Π is illustrated in (5.10).

(5.10) [Ellul Mercer 1986: 25]

[Ø	*Nhar*	*it-Tlieta*	*li*	*għadda]*$_{ADV}$	*fil-ħdax*	*ta'*
[Ø	day	def-third	that	pass]	in:DEF-eleven	of

fil-għodu *kont* *qed* *inkellmu* *qrib* *l-iskola*
in:DEF-morning be:1SG.PFV PROG 1SG.IPFV:speak:3SG.M near DEF-school
'[Last Tuesday] at eleven in the morning I was talking to him near the school.'

The division of labour between *fi*-PPs and naked adverbials is reminiscent of (though by no means identical with) the *fi*-Ø alternation discussed in connection with the spatial INESSIVE in the foregoing section.

5.1.2.1.2 APUDESSIVE

Hagège (2010: 291) describes the APUDESSIVE as "a special type of *inessive* [which] refer[s] to a place in which someone lives or goes as a home" [original italics]. In the same paragraph Hagège (2010: 291) admits to an inconsistency since he has lumped together a number of semantically diverse roles under the cover-term attributive whereas he chooses to keep several types of INESSIVES terminologically distinct. We disagree with the lumping together but accept his decision to treat the APUDESSIVE as a category distinct from the INESSIVE. According to the same author, the APUDESSIVE is cross-linguistically relatively common since it is attested in almost a third of Hagège's sample (Hagège 2010: 292). With reference to Maltese, Borg/ Azzopardi-Alexander (1997: 156) claim that "[f]or general location in association with a person, the preposition **għand** is used" [original boldface]. What can be concluded on the basis of these two short quotes is that the complement of the Maltese APUDESSIVE marker (Π22) *għand* 'at s.o.'s place' is a human noun or a pronominal affix with human reference.

A typical example which fits the above description is given in (5.11a) where the frequent visits of the protagonist at her sister's place are mentioned. Roberto's mother is thus temporarily located at Rose's place. In the (b)- and (c)-sentences however, the complement of (Π22) *għand* 'at s.o.'s place' is not a person in the usual sense of the term.

(5.11) APUDESSIVE
(a) [Bartolo/Vella 2009: 110]
omm *Roberto* *kważi* *dejjem* ***[għand*** *oħtha* *Rose]*$_{PP}$
mother Roberto almost always **[at_s.o.'s_place** sister:3SG.F Rose]
tkun
3SG.F.IPFV:be.FUT
'[. . .] Roberto's mother would almost always be [**at** her sister Rose's **place**].'

(b) [Korpus Malti 3.0; culture1464]
Din *il-mara* *għamlet* *b'* *total* *ta'* *416-il* *siegħa*
DEM:F:PROX DEF-woman make:3SG.F.PFV with total of 416 hour
[għand *il-ħanut* *tat-tattoos* *tal-ħabib* *tagħha]*$_{PP}$
[at_s.o.'s_place DEF-shop of:DEF-tattoo:PL of:DEF-friend of:3SG.F]
'This woman spent altogether 416 hours [**at** the tattoo shop of her friend] [. . .].'

(c) [Korpus Malti 3.0; culture2055]
L-annimali *li* *jinsabu* ***[għand*** *l-SPCA]*$_{PP}$
DEF-animal:PL that 3.IPFV:REFL:find:PL **[at_s.o.'s_place** DEF-SPCA]
ġew *abbandunati* *jew* *ittieħdu* *hemm*
come:3PL.PFV abandon:PART:PL or REFL:take:3PL.PFV there
mis-sidien *tagħhom*
from:DEF-owner:PL of:3PL
'The animals that are [**at** the SPCA] were abandoned or taken there by their owners [. . .].'

There is ample evidence that the restriction of the APUDESSIVE to combinations with Grounds that refer to human individuals does no longer hold for (П22) *għand* 'at s.o.'s place', at least not in the strict interpretation thereof. In (5.11b), the Ground is a tattoo shop owned by a friend of the woman that is talked about. It might be argued that there is a modicum of associative animacy left because the friend is identified by name in the omitted part of the sentence. Yet, syntactically, neither the human noun *ħabib* 'friend' nor the personal name of the friend is directly governed by (П22) *għand* 'at s.o.'s place'. In (5.11c), (П22) *għand* 'at s.o.'s place' forms a PP with the complement SPCA – the abbreviation of *(Malta) Society for the Protection and Care of Animals*. This is the name of an institution on whose premises the abandoned animals temporarily live. The institution as such is of course inanimate. However, it is run by humans and it is a legal entity represented by humans. The notion of place is involved in all of the above cases. The human owner of the place, however, need not be mentioned within the construction itself. This means that the close bond between (П22) *għand* 'at s.o.'s place' and the feature [human] of its complement has been weakened although the diachronic development that led to the present state still needs to be established in a dedicated follow-up study.

It is important to understand that the separation of the APUDESSIVE from the INESSIVE makes sense especially because (П22) *għand* 'at s.o.'s place' and (П15) *fi* 'in' cannot replace each other as can be gathered from the minimal pair in (5.12).

(5.12) Minimal pair
(a) [Korpus Malti 3.0; news79864]
Id-determinazzjoni [f' Borg]$_{\text{PP}}$ *qatt ma naqset tul*
DEF-determination **[in** Borg] ever NEG decrease:3SG.F.PFV during
ħajjitha u fil-karriera
life:POSS:3SG.F and in:DEF-career
'The determination **[in** Borg] never diminished during her life and career.'
(b) [Korpus Malti 3.0; news110904]
wieħed minn dawn l-għasafar li nstabu
one from DEM:PL:PROX DEF-bird.PL that REFL:find:3PL.PFV
[għand *Borg]*$_{\text{PP}}$ *kien għadu kif ġie bbalzmat*
[at_s.o.'s_place Borg] be still:3SG.M as come embalm:PART
'[. . .] one of the birds found [**at** Borg's **place**] had only just been embalmed.'

The PPs in the two examples have their complement in common, viz. the family name *Borg*. If the Figure is (metaphorically or other) located inside of the person identified by this name, only (Π15) *fi* 'in' is licit as head of the PP. (Π15) *fi* 'in' is ruled out, however, when reference is made to someone's place or personal sphere outside of their physical body. As far as we can tell, in cases like (5.12a), the use of the usual replacements of (Π15) *fi* 'in' is blocked.

5.1.2.1.3 ADESSIVE

The ADESSIVE in Hagège's (2010: 261) system poses a number of problems. The problems come to the fore when we look at the English translations suggested in our reference. There, we find the English Πs *at*, *by*, *beside*, and *near*. Once again, we take issue with subsuming all the meanings associated with these Πs under one and the same category label. There are good reasons to separate the Πs from each other. A clear distinction should be made between general location (= ADESSIVE *tout court*) and specific location. In the case of the notion under review, the STATIC PROXIMATE relation represents the specific location. It makes a difference whether there is contact between Figure and Ground or not. Borg/Azzopardi-Alexander (1997: 157) systematically take account of the PROPINQUUS [26] relation to which we will return below.[27] In this section, we address the ADESSIVE as a basic kind of general location.

26 For reasons which will be made clear in Section 5.1.2.2.20.1, we introduce the term *propinquus* (< Latin *propinquus* 'near') at this point to reserve the established term *proximate* (< Latin *proximus* 'the nearest, very near') for a related but nevertheless distinct category.
27 Cf. Section 5.1.2.1.4.

Aquilina (1999: 130) emphasises that in contexts other than that of the APUDESSIVE, English *at* as expression of a spatial relation is not translated by a Π but requires the use of the ersatz-copula *qiegħed*. Similarly, Borg/Azzopardi-Alexander (1997: 155) claim that "[f]or general location in a place which is fairly predictable from the context, no preposition is used." The authors refer to examples like (5.13a) and (5.13c). We contrast the absence of a Π with the use of (Π15) *fi* 'in' in (5.13b).

(5.13) ADESSIVE

(a) (Borg/Azzopardi-Alexander 1997: 49)

Ommu (qiegħda) [Ø l-knisja]$_X$

mother:3SG.M (stay.PART:F) [Ø DEF-church]

'His mother is [**at** church].' [O.T.]

(b) [Korpus Malti 3.0; news121685]

Se nibqgħu niftakru fik kull meta

FUT 1.IPFV:remain:PL 1.IPFV:remember:PL in:2SG each when

nkunu qegħdin [fil-Knisja ta' San Franġisk]$_{PP}$

1.IPFV:be.FUT:PL stay:PART:PL [**in**:DEF-church of St. Francis]

'We will keep remembering you every time we are [**in** St. Francis' Church].'

(c) [Korpus Malti 3.0; parl10116]

hawnhekk m' aħniex qegħdin [Ø il-Knisja]$_X$ *imma*

here NEG 1PL:NEG stay.PART:PL [Ø DEF-church] but

qegħdin [Ø il-Parlament]$_X$

stay.PART:PL [Ø DEF-parliament]

'[. . .] here we are not [at church] but [in parliament].'

Note the (optional) use of the morphologically adequate form of *qiegħed* in all cases. These sentences lead us back once again to the *fi-Ø* alternation as addressed above.[28] In (5.13a), the spatial situation locates the Figure at an unidentified church without determining whether the mother is inside the church building or not. *Il-knisja* 'the church' probably refers to the institution and not to a concrete named church as such. In the case of (5.13b), there can be no doubt that the Figure is envisaged to be inside the church. Zero-marking of the ADESSIVE in (5.13a) is fine because the exact location of the Figure remains unspecified. Example (15.3c) is extracted from a parliamentary speech. The speaker and his audience are definitely inside the parliament as a building but both *il-Knisja* 'the Church' and *il-Parlament* 'the parliament' trigger zero-marking because the NPS are labels for the institutions and not for the

28 Cf. Sections 4.3.3.2 and 5.1.2.1.1.1.

concrete locations. Zero-marking applies if the spatial situation is abstract or highly generic. If these conditions are not met, Πs have to be used. What this also means is that the INESSIVE is already an instance of a specific location. We conclude that the ADESSIVE is usually zero-marked in Maltese.

5.1.2.1.4 PROPINQUUS

Under the section heading "s" in Borg/Azzopardi-Alexander (1997: 157) we read that for the static PROPINQUUS Maltese employs

> **fejn** (or **ħdejn**, but for some speakers the two expressions are not in free variation: the latter tends to be regarded as a dialectal form; and less frequently **qrib** and **viċin**) [original boldface]

In addition, (Π8) *biswit* 'facing' and (Π32) *maġenb* 'close to' must be mentioned. The former is translated as 'next' and 'beside' by Borg/Azzopardi-Alexander (1997: 162–163). The same authors classify both Πs as expressing a lateral position of the Figure in relation to the Ground. Aquilina (1987: 126) gives 'facing' and 'opposite' as primary meanings of (Π8) *biswit* 'facing' adding that "[s]ometimes used in the sense of 'near, close by'." (Π32) *maġenb* 'close to' is registered with the meaning 'close to', and 'beside' in the second volume of the dictionary (Aquilina 1991: 767). In this section, both (Π8) *biswit* 'facing' and (Π32) *maġenb* 'close to' are included exclusively with their PROPINQUUS functions.

There are thus six contestants, namely (Π8) *biswit* 'facing', (Π14) *fejn* 'near', (Π23) *ħdejn* 'beside', (Π32) *maġenb* 'close to', (Π48) *qrib* 'near', and (Π58) *viċin* 'near'. Given that all six of the Πs are (partial) synonyms, we are facing a many-to-one relation with a single function being fulfilled by four different Πs. However, the brackets in the above quote caution us in the sense that (Π23) *ħdejn* 'beside' is marked as dialectal. The frequency count revealed in Table 3.27 does not exactly reflect the assumed differences in the quote. With 20,100 hits in the Korpus Malti 3.0 (Π48) *qrib* 'near' is the most frequent of the four options. (Π23) *ħdejn* 'beside' follows suit with 20,073 matches. The token frequency drops to 8,878 for (Π58) *viċin* 'near' whereas (Π14) *fejn* 'near' is somewhere in the middle with 5,305 tokens. At the bottom, we find (Π8) *biswit* 'facing' with 2,791 matches and (Π32) *maġenb* 'close to' with 2,632 matches. Based on these frequencies, the order in which the Πs are presented in the quote needs to be revised as has the inventory of PROPINQUUS Πs. If we exclusively go by the token frequency (Π48) *qrib* 'near' and (Π23) *ħdejn* 'beside' are the preferred options, (Π58) *viċin* 'near' and (Π14) *fejn* 'near' occupy middle positions but (Π8) *biswit* 'facing' and (Π32) *maġenb* 'close to' are already special cases at the margins. The relatively high frequency of (Π23) *ħdejn* 'beside' does not support the hypothesis that its use is stigmatised as a dialectal feature. The reshuffled order may be an effect of the skewed representation of genres in the Korpus Malti 3.0. Borg/Azzopardi-Alexander (1997) base their statements on their native-speaker intuition and mostly rely on introspection (a licit methodology which however precludes the possibility of exactly quantifying the actual use of a given item across speakers and situations).

If we abstract away from the quantitative differences between the six Πs, what remains to be clarified is the problem of the semantic relation which holds between them. In the quote in the foregoing paragraph, free variation is mentioned although with the proviso that this possibility is probably barred for (Π23) *ħdejn* 'beside' on account of its supposed dialectal character. Is it the case that – dialectal or not – the four Πs can replace each other in every possible context? We try to answer this question in several steps by way of testing whether the six Πs under inspection are compatible to comparable degrees with different kinds of Grounds. Sentences like (5.14) with two coordinated PPS each of which has PROPINQUUS function lend credibility to the free-variation hypothesis.

(5.14) PROPINQUUS I [Korpus Malti 3.0; *parl1238]*

L-uniku	*ħanut*	*tal-kafe*	*kien*	***[qrib***	*il-Knisja]*$_{PP}$	*u*
DEF-unique	shop	of:DEF-coffee	be	**[near**	DEF-church]	and

[viċin *l-iskola]*$_{PP}$
[near DEF-school]
'The only café was [**near** the church] and [(**near**) the school] [. . .].'

Both (Π48) *qrib* 'near' and (Π58) *viċin* 'near' express the same, namely the location of the café in the vicinity of the Ground. The one Π may replace the other and vice versa. However, can it be that certain Grounds are more likely to choose (Π48) *qrib* 'near' while others opt more often for (Π58) *viċin* 'near' (or (Π8) *biswit* 'facing' or (Π14) *fejn* 'near' or (Π23) *ħdejn* 'beside' or (Π32) *maġenb* 'close to')? This possibility calls for being looked into.

For a start, we provide examples of (Π8) *biswit* 'facing', (Π14) *fejn* 'near', (Π23) *ħdejn* 'beside', (Π32) *maġenb* 'close to', (Π48) *qrib* 'near', and (Π58) *viċin* 'near' combining with *is-sodda* 'the bed' as representative of the class of inanimate concrete common nouns (cf. (5.15).

(5.15) PROPINQUUS II (inanimate Ground – common noun)

(a) [Korpus Malti 3.0; news78344]

Għidli x' hemm isbaħ milli ssib ktieb
say:IO:1SG what there beautiful.CMPR from:that 2SG.IPFV:find book
jistenniek fuq il-komodina ***[biswit*** *is-sodda tiegħek]*$_{PP}$
3SG.M.IPFV:wait:2SG on DEF-cupboard [**facing** DEF-bed of:2SG]
'Tell me what is nicer than finding a book waiting for you on the cupboard [**next to** your bed]?'

(b) [Korpus Malti 3.0; literature76]

La Valette kellu ljun ikkatnat ***[fejn*** *is-sodda]*$_{PP}$
La Valette have.PFV:3SG.M lion chain:PART [**near** DEF-bed]
'La Valette had a lion chained [**near** the bed] [. . .].'

(c) [Korpus Malti 3.0; literature18]

Oħti kienet ***[ħdejn*** *is-sodda tar-Re]*$_{PP}$ *tibki*
sister:1SG be:3SG.F.PFV [**beside** DEF-bed of:DEF-king] 3SG.F.IPFV:cry
u lili lanqas biss għarfitni
and to:1SG not_even only recognise:3SG.F.PFV:1SG
'My sister cried [beside the king's bed] and did not even recognise me.'

(d) [Korpus Malti 3.0; literature32]

[Maġenb *is-sodda]*$_{PP}$ *kien hemm komodina ċkejkna*
[**close_to** DEF-bed] be there cupboard small:F
'[**Close to** the bed] there was a small cupboard [. . .].'

(e) [Korpus Malti 3.0; news46235]

Meta l-pulizija marru fuq il-post il-kadavru tal-mara
when DEF-police go:3PL.PFV on DEF-place DEF-corpse of:DEF-woman
nstab fl-art wiċċu 'l fuq ***[qrib*** *is-sodda]*$_{PP}$
REFL:find in:DEF-ground face:3SG.M upwards [**near** DEF-bed]
'When the police went to the place, the woman's body lay on its back on the floor [**near** the bed].'

(f) [Korpus Malti 3.0; *news216150]*

Intqal ukoll li meta nstab il-katavru ta' Winehouse
REFL:say also that when REFL:find DEF-corpse of Winehouse
kien hemm anki żewġ fliexken tal-vodka vojta
be there also two bottle.PL of:DEF-vodka empty:F
fl-art ***[viċin*** *is-sodda tagħha]*$_{PP}$
in:DEF-ground [**near** DEF-bed of:3SG.F]
'It was said also that when Winehouse's body was found there were also two empty bottles of vodka on the floor [**near** her bed].'

Superficially, the data seem to speak in favour of free variation because each of the Πs is licit as head of a PP which involves *is-sodda* 'the bed' as complement. On closer inspection, we notice that the filler of the head slot in the frame [$Π_{PROPINQUUS}$ *is-sodda*]$_{PP}$ is most often (Π23) *ħdejn* 'beside' as shown in Figure 5.5.

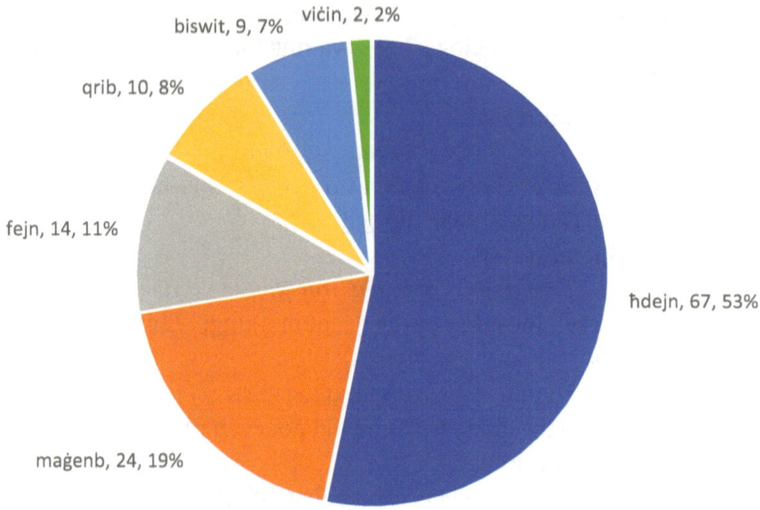

Figure 5.5: Fillers of the head slot in [$Π_{PROPINQUUS}$ *is-sodda*]$_{PP}$ (Korpus Malti 3.0).

With 53% of all tokens of [$Π_{PROPINQUUS}$ *is-sodda*]$_{PP}$ in the Korpus Malti 3.0 (Π23) *ħdejn* 'beside' ousts all other contestants none of which has a share bigger than 19%. To determine whether this discrepancy is incidental we take account of the frequency with which the six Πs are attested in combinations with ten other concrete common nouns in complement function in a static relation of Place, namely in alphabetical order *bieb* 'door', *dar* 'house', *ħajt* 'wall', *knisja* 'church', *mejda* 'table', *siġġu* 'chair', *sptar* 'hospital', *tieqa* 'window', *triq* 'street', and *wied* 'valley' (all of which occur in the singular hosting the definiteness marker). The results of our search in the Korpus Malti 3.0 are revealed in Figure 5.6. The cases are ordered from left to right according to the increasing share of (Π23) *ħdejn* 'beside'. Only in two cases are there other Πs whose shares are bigger than that of (Π23) *ħdejn* 'beside'. A possible extrapolation on the basis of the shares featured in Figure 5.5 assumes that (Π23) *ħdejn* 'beside' is the preferred option with most common nouns. This hypothesis is at least partly borne out by the empirical facts.

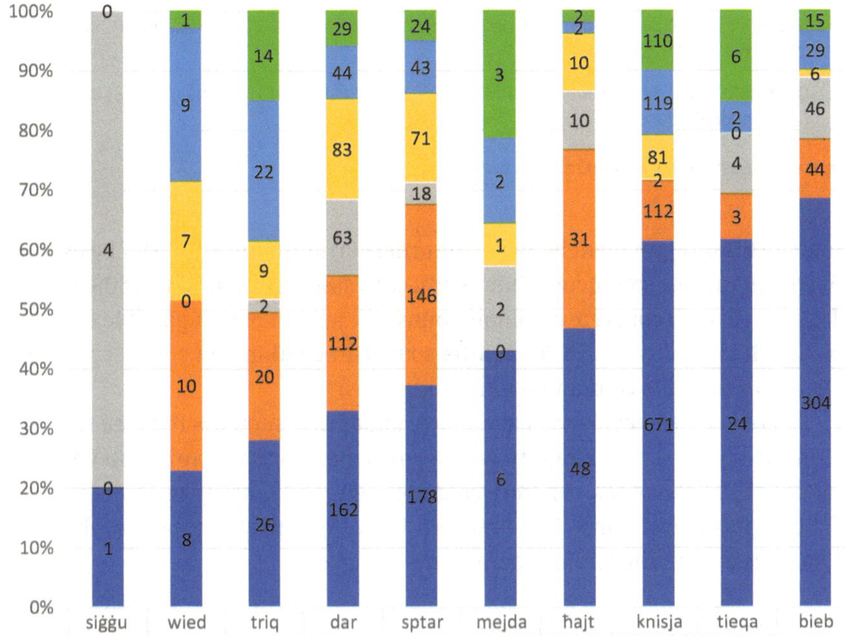

	siġġu	wied	triq	dar	sptar	mejda	ħajt	knisja	tieqa	bieb
maġenb	0	1	14	29	24	3	2	110	6	15
biswit	0	9	22	44	43	2	2	119	2	29
viċin	0	7	9	83	71	1	10	81	0	6
fejn	4	0	2	63	18	2	10	2	4	46
qrib	0	10	20	112	146	0	31	112	3	44
ħdejn	1	8	26	162	178	6	48	671	24	304

Figure 5.6: Fillers of the head slot in [$\Pi_{\text{PROPINQUUS}}$ DEF-N_{COMMON}]$_{\text{PP}}$ (Korpus Malti 3.0).

The columns reflect a continuum which covers a range of variations. The variation notwithstanding, there are only two nouns which pick a Π other than (Π23) *ħdejn* 'beside' as their major partner. With three nouns, (Π23) *ħdejn* 'beside' is responsible for more than 50% of the tokens. Except in one case, no other Π combines with a given noun in more than half of the tokens., (Π23) *ħdejn* 'beside' is the only Π which is attested with each of the ten nouns. The hierarchy which results from the above frequency count (= (a)) and how it diverges from the order according to the overall frequency (= (b)) is disclosed in Figure 5.7.

(a) (Π23) *ħdejn* 'beside' > (Π48) *qrib* 'near' > (Π8) *biswit* 'facing' > (Π58) *viċin* 'near' > (Π32) *maġenb* 'close to' > (Π14) *fejn* 'near'

(b) (Π48) *qrib* 'near' > (Π23) *ħdejn* 'beside' > (Π58) *viċin* 'near' > (Π14) *fejn* 'near' > (Π8) *biswit* 'facing' > (Π32) *maġenb* 'close to'

Figure 5.7: Mismatch between ranking orders.

How can the disagreement between the orders (a) and (b) in Figure 5.7 be explained? Given that the test conducted above suffices to assume that it reflects the behaviour of the six Πs with concrete common nouns as complements, a possible explanation for the change of ranks in (b) can be sought in combinations with complements other than concrete common nouns.

There is indeed evidence from place names that supports the idea that the different Πs tend to associate more with a particular class of nouns than with others. Several of the above ten concrete common nouns frequently form the initial part of a complex place name. This is shown in (5.16) with street names whose initial element is *triq* 'street' without definiteness marker.

(5.16) PROPINQUUS III (place names)

(a) [Korpus Malti 3.0; news21341]

Transport Malta	*tavża*	*wkoll*	*li*	*l-parkeġġ*
Transport Malta	3SG.F.IPFV:give_notice	also	that	DEF-car_park

*[**biswit***	*Triq Vincenzo Dimech*	*fil-Floriana]$_{PP}$*	*se*
[**facing**	Triq Vicenzo Dimech	in:DEF-Floriana]	FUT

jingħalaq	*għal*	*24*	*siegħa*
3SG.M.IPFV:REFL:close	for	24	hour

'Transport Malta also gives notice that the car-park [**near** Triq Vincenzo Dimech in Floriana] will be closed for 24 hours.'

(b) [Korpus Malti 3.0; parl2746]

għalfejn	*ma*	*ttieħdu*	*l-ebda*	*passi*	*mid-Dipartiment*
why	NEG	REFL:take:3PL.PFV	DEF-no	step:PL	from:DEF-department

tas-Saħħa	*biex*	*titneħħa*	*miżbla*	*illegali*	*li*	*hemm*
of:DEF-health	to	3SG.F.IPFV:REFL:remove	dunghill	illegal	that	there

f'	*kantuniera*	*[**fejn***	*Triq San Ġwakkin]$_{PP}$*	*u*	*[Ø*	*Triq is-Sebħ]$_{PP}$*
in	corner	[**near**	Triq San Ġwakkin]	and	[Ø	Triq is-Sebħ]

fil-lokalita'	*ta'*	*Ħal Qormi*
in:DEF-locality	of	Ħal Qormi

'[. . .] why have no steps been taken by the Health Department to remove an illegal dunghill that is there on a corner [**near** Triq San Ġwakkin] and [Triq is-Sebħ] in the village of Ħal Qormi?'

(c) [Korpus Malti 3.0; news89710]
tinsab *fil-Marsa* *eżatt* ***[ħdejn*** *Triq Aldo Moro]*$_{PP}$
3SG.F.IPFV:REFL:find in:DEF-Marsa exact [**beside** Triq Aldo Moro]
'[. . .] it is found exactly [**beside** Triq Aldo Moro].'

(d) [Korpus Malti 3.0; news171499]
Din *tinvolvi* *twessiegħ* *tal-kanal* *eżistenti* *li*
DEM:F:PROX 3SG.F.IPFV:involve widening of:DEF-canal existing that
jinsab ***[maġenb*** *Triq it-Torba* *ġewwa* *Albert Town]*$_{PP}$
3SG.M.IPFV:REFL:find [**close_to** Triq it-Torba inside Albert Town]
'This involves the widening of the existing canal that is located [**close to** Triq it-Torba in Albert Town].'

(e) [Korpus Malti 3.0; news126652]
Anton Agius *kien* *joqgħod* *f'* *dar* ***[qrib*** *Triq Għajn Qajjed]*$_{PP}$
Anton Agizs be 3SG.M.IPFV:stay in house [**near** Triq Għajn Qajjed]
'Anton Agius lived in a house [**near** Triq Għajn Qajjed].'

(f) [Korpus Malti 3.0; news206904]
Il-każ *huwa* *dwar* *xogħol* *li* *sar* ***[viċin*** *Triq San Rokku]*$_{PP}$
DEF-case 3SG.M about work that become [**near** Triq San Rokku]
'The case was about work that was done [**near** Triq San Rokku] [. . .].'

We see that the six Πs are compatible with street names as complements. The same holds for other types of place names. However, it is particularly noteworthy how important place names are for (Π48) *qrib* 'near' and (Π58) *viċin* 'near'. In Figure 5.8, the share of place names in *qrib*-PPS and *viċin*-PPS is calculated for place names with initial *triq* 'street', *wied* 'valley', *knisja* 'church', and *sptar* 'hospital'. Note that in the cases of *triq* and *wied* the absolute numbers exceed those given for the same nouns when used as common nouns in Figure 5.6 whereas those for *knisja* and *sptar* are included already in the count in Figure 5.6. The reason for this difference is the absence of the definiteness marker on street and valley names.

Especially (Π48) *qrib* 'near' is prone to combine with place names of the indicated kind since in each of the four cases more than two-thirds of all *qrib*-PPS contain a place name as complement.

This tendency must be checked against further contexts involving place names before we can generalise the findings. To this end, we look at settlement names like *Londra* 'London' in (5.17). The sentences resemble each other semantically insofar as all of them express that the Figure is located in the vicinity of London. Neither (Π8) *biswit* 'facing' nor (Π32) *maġenb* 'close to' yields any matches for combinations with this settlement name.

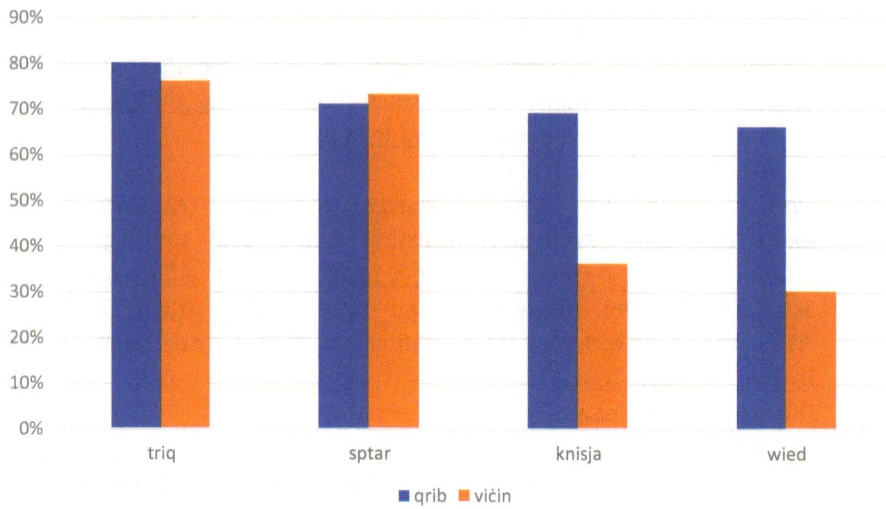

Figure 5.8: Share of place-name Grounds with (Π48) *qrib* 'near' and (Π58) *viċin* 'near'.

(5.17) PROPINQUUS IV (settlement name)

(a) [Korpus Malti 3.0; academic376]

L-Osservatorju Rjali ta' Greenwich [***ħdejn*** <u>*Londra*</u>]$_{PP}$ *hija*
DEF-observatory royal of Greenwich [**beside** <u>London</u>] 3SG.F
l-punt definit bħala l-Meridjan Prim
point define:PART as DEF-meridian prime
'The Royal Observatory of Greenwich [**near** <u>London</u>] is the point that is defined as the prime meridian.'

(b) [Korpus Malti 3.0; news70146]

Huwa żied jgħid li l-aktar li jinħassu
3SG.M add 3SG.M.IPFV:say that DEF-more that 3.IPFV:REFL:feel:PL
l-effetti tan-nixfa huma fl-inħawi [***qrib*** <u>*Londra*</u>]$_{PP}$
DEF-effect:PL of:DEF-draught 3PL in:DEF-side.PL [**near** <u>London</u>]
'He further said that the effects of the draught are felt the most in the areas [**near** <u>London</u>].'

(c) [Korpus Malti 3.0; culture2625]

Dawn il-vetturi li jesgħu żewġ persuni
DEM.PL.PROX DEF-vehicle:PL that 3.IPFV:contain:PL two person:PL
se jkunu qed joperaw fuq toroq
FUT 3.IPFV:be.FUT:PL PROG 3.IPFV:operate:PL on street.PL

iddisinjati *apposta* *fir-raħal* *ta'* *Milton* *Keynes* *li*
design:PART:PL on_purpose in:DEF-village of Milton Keynes that
jinsab ***[vičin*** *Londra]*$_{PP}$
3SG.M.IPFV:REFL:find **[near** London]

'These vehicles which sit two persons will be operating on purposely designed streets in the village of Milton Keynes which is located [**near** London].'

The absence of any examples which involve (Π14) *fejn* 'near' taking *Londra* as complement is no surprise since the three other Πs are attested each only once with this place name, meaning: the above examples in (5.17) are all hapaxes. It is therefore necessary to considerably extend the list of settlement names to determine whether there are any preferences or incompatibilities. For this purpose, we have chosen five settlement names from the Maltese islands and five names of major cities from outside Malta. The results (based on a search of the Korpus Malti 3.0) are shown in Table 5.2. The settlement names referring to geo-objects on Malta or Gozo are highlighted with grey shading.

Table 5.2: Πs with PROPINQUUS meaning in combination with settlement names.

place name	Π						sum
	(Π14) *fejn* 'near'	(Π32) *maġenb* 'close to'	(Π8) *biswit* 'facing'	(Π23) *ħdejn* 'beside'	(Π48) *qrib* 'near'	(Π58) *vičin* 'near'	
Ruma	0	0	0	0	8	6	14
Pariġi	0	0	0	2	7	0	9
Mosta	0	0	1	1	2	4	8
Berlin	0	0	0	1	3	0	4
Nadur	0	0	0	0	3	1	4
New York	0	0	0	0	2	2	4
Sydney	0	0	0	0	4	0	4
Ħal Luqa	0	0	1	1	1	1	4
Naxxar	0	0	0	1	1	1	3
Qrendi	0	0	0	1	1	1	3
total	0	0	2	7	32	16	57

There is no evidence of (Π14) *fejn* 'near' or (Π32) *maġenb* 'close to' governing any of the ten different settlement names. Thus, the first impression that arose in connection with the examples in (5.17) finds corroboration because (Π8) *biswit* 'facing' is attested only twice and its exact reading ('facing' vs 'near to') remains doubtful in these cases. These three Πs are clearly no option for expressing proximity of Figure and Ground provided the latter is a settlement name. In contrast, (Π48) *qrib* 'near' is the only Π that occurs with each of the place names whereas (Π58) *viċin* 'near' and (Π23) *ħdejn* 'beside' combine with seven and six settlement names, respectively. (Π48) *qrib* 'near' always displays a token frequency that exceeds that of (Π23) *ħdejn* 'beside'. Summarily, (Π48) *qrib* 'near' is responsible for 56% of all tokens registered in Table 5.2. (Π48) *qrib* 'near' is attested twice as frequently as (Π58) *viċin* 'near' whose tokens, in turn, outnumber those of (Π23) *ħdejn* 'beside' by a ratio of 2.3-to-one. (Π23) *ħdejn* 'beside' is 3.5 times more frequent than (Π8) *biswit* 'facing'. We read these results as follows. (Π48) *qrib* 'near' and (Π58) *viċin* 'near' are especially fit when it comes to express that the Figure is situated in the vicinity of a major geo-object such as urban agglomerations – a context from which (Π14) *fejn* 'near' and (Π32) *maġenb* 'close to' seem to be excluded and where (Π23) *ħdejn* 'beside' is only a minor option while (Π8) *biswit* 'facing' is attested only marginally. This finding is corroborated by the turnout of another search of the Korpus Malti 3.0. This time we have looked at island names as shown in Table 5.3. (Π8) *biswit* 'facing', (Π14) *fejn* 'near', and (Π32) *maġenb* 'close to' are absent from Table 5.3 because they do not occur in combination with any of the island names. Their frequency is thus nil.

Table 5.3: Πs with PROPINQUUS meaning in combination with island names.

island name	**Π**			**sum**
	(Π23) *ħdejn* 'beside'	**(Π48) *qrib* 'near'**	**(Π58) *viċin* 'near'**	
Sqallija 'Sicily'	0	6	5	11
Ċipru 'Cyprus'	3	0	0	3
Kreta 'Crete'	1	1	0	2
Pantellerija 'Pantelleria'	0	1	0	1
Sardinja 'Sardinia'	0	1	0	1
total	4	9	5	18

The small absolute numbers notwithstanding, Table 5.3 largely replicates the picture painted in Table 5.2. Half of the tokens go to the credit of (Π48) *qrib* 'near'. The second best is (Π58) *viċin* 'near' with about half as many tokens. (Π23) *ħdejn* 'beside' is attested almost equally often but comes only on rank #3. As in the previous test, (Π14) *fejn* 'near' and (Π32) *maġenb* 'close to' fail to produce a single token. The same holds for (Π8) *biswit* 'facing'. On this basis, we postulate a scale of preferences according to which the core of the functional domain of (Π48) *qrib* 'near' and (Π58) *viċin* 'near' is the PROPINQUUS in relation to place names. (Π23) *ħdejn* 'beside' is compatible with place names but to a lesser degree whereas (Π14) *fejn* 'near', (Π32) *maġenb* 'close to', and (Π8) *biswit* 'facing' are practically excluded from PPs in which the names of major geo-objects constitute the complement. The six Πs serve to express the PROPINQUUS but behave differently in terms of their combinatory potential.

5.1.2.1.5 OBESSIVE

Hagège (2010: 286) distinguishes between the "obessive" which applies when the Figure is "in front of a place" and the "preessive" which assumes that the Figure is "before [. . .] a place or person(s)". The English Πs which represent the two spatial meanings are *in front of* and *opposite* in the case of the OBSESSIVE and *before* in the case of the PREESSIVE. In the absence of further explanations as to the conceptual basis of the distinction of OBESSIVE and PREESSIVE we shortcut a potentially lengthy discussion by way of taking Hagège's division of categories for granted.[29] Nevertheless, the term OBESSIVE glosses over a further distinction into two subcategories. It depends on the perspective that is used to verbalise a given spatial situation as either one of the Figure being in front of the Ground or opposite the Ground. In the former case, the Figure metaphorically blocks the view of the observer who looks in the direction of the Ground. In contrast, the situation is construed from the perspective of the Ground when the Figure is said to be situated opposite the Ground. Since Maltese makes use of two different Πs to encode the different viewpoints, we opt for splitting Hagège's OBESSIVE in two, namely the proper OBESSIVE (< Latin *ob* 'opposite') and the FRONTALIS (< Latin *frons* 'front'). The OBESSIVE requires the employment of either (Π12) *faċċata* 'opposite' whereas the FRONTALIS is expressed by (Π49) *quddiem* 'in front of' or (Π8) *biswit* 'facing'. The FRONTALIS is discussed in the subsequent section. Borg/Azzopardi-Alexander (1997: 166) label the OBESSIVE descriptively as "[c]iterior-anterior" whereas the FRONTALIS is termed simply "anterior" (Borg/Azzopardi-Alexander 1997: 159).

Since the morphosyntactic properties of (Π12) *faċċata* 'opposite' have been scrutinised in some detail before on the basis of several sentential examples,[30] it suffices at this point to provide only one additional example from the Korpus Malti 3.0.

29 The PREESSIVE is thus relegated to Section 5.1.2.1.10.

30 E.g. in Section 4.1.2.2.1.2.1.

(5.18) OBESSIVE – [Korpus Malti 3.0; academic29]
Anton Buttigieg ixejjer lill-bastiment minn fuq is-swar
Anton Buttigieg wave to:DEF-ship from on DEF-wall.PL
tal-belt Valletta ***[faċċata** tal-Port il-Kbir]$_{PP}$*
of:DEF-city Valletta [**opposite** of;DEF-harbour DEF-big]
'[. . .] Anton Buttigieg waved to the ship from the city walls of Valletta [**opposite** the Grand Habour].'

The situation is depicted as if seen from the ship that is moving within the Grand Habour. The observer is aboard the ship and watches Anton Buttigieg waving on the city walls. The city walls obstruct the observer's gaze in the sense that s/he cannot see everything that is beyond the obstacle. The Grand Habour is the Ground and Valletta's city walls have the function of the Figure. This is different in the case of the FRONTALIS.

5.1.2.1.6 FRONTALIS

Saari (2002) has dedicated a separate article on aspects of the grammar and the functions of (П49) *quddiem* 'in front of'. In this section, we focus on the differences between FRONTALIS and OBESSIVE. This discussion also involves (П8) *biswit* 'facing'. As in (5.18), the PP of interest for the discussion contains *il-Port il-Kbir* 'the Grand Harbour' as Ground in (5.19), too.

(5.19) FRONTALIS
(a) [Korpus Malti 3.0; parl8367]
Safejn naf jien kellu jsir
as_far_as 1SG.IPFV:know 1SG have.PFV:3SG.M 3SG.M.IPFV:become
impact assessment fuq dawn id-dumping sites li hemm
impact assessment on DEM:PL:PROX DEF-dumping_site:PL that there
fil-baħar li waħda minnhom tinstab eżattament
in:DEF-sea that one:F from:3PL 3SG.F.IPFV:REFL:find exactly
***[quddiem** il-Port il-Kbir]$_{PP}$*
[**in_front_of** of;DEF-harbour DEF-big]
'As far as I know, an impact assessment had to take place on these dumping sites in the sea one of which is located exactly [**in front of** the Grand Harbour].

(b) [Korpus Malti 3.0; news84867]

Iż-żgħażagħ *bdew* *iħallu* *d-djar* *u*
DEF-young_man.PL begin:3PL.PFV 3.IPFV:leave:PL DEF-house.PL and
t-toroq *fejn* *kienu* *jgħixu* *l-ġenituri* *tagħhom*
DEF-street.PL where be.3PL.PFV 3.IPFV:live:PL DEF-parent:PL of:3PL
biex *isibu* *post* *f'* *ċentri* ***[biswit*** *il-Port*
to 3.IPFV:find:PL place in center:PL **[facing** DEF-harbour
il-Kbir]$_{PP}$
DEF-big]

'The young men began to leave the houses and the streets where their parents lived to find a place in centres [**facing** the Grand harbour].'

This time, however, the observer looks towards the Ground from the perspective of the Figure – the dumping sites in the sea – that happens to be located outside the Grand Habour between observer and Ground, in a manner of speaking. If (Π49) *quddiem* 'in front of' and (Π12) *faċċata* 'opposite' switch places in (5.18) and (5.19a) the result might still be grammatically acceptable but the situation would invoke different interpretations as to where the (virtual) observer is located relative to Figure and Ground. The situation is not as clear in the case of (Π8) *biswit* 'facing'. In (5.15a), it is beyond doubt that this Π is used in its PROPINQUUS meaning. In (5.16a), the PROPINQUUS meaning seems to make more sense but it cannot be excluded that the intended meaning is that of the FRONTALIS. Very much the same uncertainty arises with regards to (Π8) *biswit* 'facing' in (5.19b). In this sentence the possibility that the Π expresses the PROPINQUUS cannot be ruled out completely. The residential centres are located in the vicinity of the Grand Harbour. At the same time, they are also facing the Grand Harbour since at least some of them are located on the waterfront. If the latter interpretation is correct, the situation is again construed such that the observer's gaze is directed from the Grand Harbour towards the residential centres.

5.1.2.1.7 SURESSIVE

The SURESSIVE applies in situations where the Figure can be said to be on the Ground (Hagège 2010: 286). This description corresponds to the category of superior-contact for whose expression (Π17) *fuq* 'on' is put to service (Borg/Azzopardi-Alexander 1997: 160). There is no empty space between Figure and Ground. A typical case illustrating the SURESSIVE function of (Π17) *fuq* 'on' is given in (5.20).

(5.20) SURESSIVE I – [Abela 2016: 94]

Teresa	*ħarset*	*lejn*	*it-tifla*	***[fuq***	*id-deckchair]*$_{PP}$
Teresa	look:3SG.F.PFV	towards	DEF-child:F	**[on**	DEF-deckchair]

'Teresa looked towards the girl [**on** the deckchair].'

The Figure (*it-tifla* 'the girl') touches the Ground (*id-deckchair* 'the deckchair') rather than hovering above it. Superficially, the story seems to end right here. However, this conclusion is wrong.

In point of fact, (П17) *fuq* 'on' does not monopolise the expression of the SURESSIVE since (П15) *fi* 'in' is frequently used with the same function. This is visible from (5.21) where a *fi*-PP is contrasted with *fuq*-PPS.

(5.21) SURESSIVE II

(a) [Korpus Malti 3.0; culture3548]

xi	*ħadd*	*sab*	*il-kartiera*	*[fl-art]*$_{PP}$
some	someone	find	DEF-purse	[**in**:DEF-ground]

'[. . .] someone found the purse [**on** the ground] [. . .].'

(b) [Korpus Malti 3.0; academic454]

Huwa	*stimat*	*li*	***[fuq***	*l-Art]*$_{PP}$	*jgħixu*
3SG.M	estimate:PART	that	**[on**	DEF-ground]	3.IPFV:live:PL
6,327,152,352	*persuna*				
6,327,152,352	person				

'It is estimated that 6,327,152,352 human beings live [**on** Earth].'

(c) [Korpus Malti 3.0; academic413]

L-Ammirall	*Horatio Nelson*	*ddeċieda*	*li*	*kien*	*hemm*	*bżonn*
DEF-admiral	Horatio Nelson	decide	that	be	there	need
imblokk	*totali*	*navali*	*u*	***[fuq***	*l-art]*$_{PP}$	*wkoll*
blockade	total	naval	and	**[on**	DEF-ground]	also

'Admiral Horatio Nelson decided that a full blockade, naval as well as [**on** land], was needed.'

The choice of П disambiguates readings of the polysemous noun *art* 'ground'. If it is used in its basic meaning, (П15) *fi* 'in' is appropriate to describe a situation in which the Figure (here: the purse) is located on and has contact with the Ground as in (5.21a). (П15) *fi* 'in' is inadmissible in the contexts involved in the (b)-sentence and the (c)-sentence. Only (П17) *fuq* 'on' can be used if *art* refers to the Earth as in (5.21b). Similarly, (П17) *fuq* 'on' is the only option when reference is made to phenomena located on land (as opposed to seaborne phenomena).

In (5.22), we find another case of disambiguation but of a different kind.

(5.22) SURESSIVE III

(a) [Korpus Malti 3.0; parl6025]

Imma bħal fil-każ tat-Titanic dawk li huma
but like in:DEF-case of:DEF-Titanic DEM:PL:DIST that 3PL
[fil-gverta ta' isfel]$_{PP}$ qed jegħrqu
[**in**:DEF-deck of below] PROG 3.IPFV:sink:PL
'But like in the case of the Titanic, those who are [**on** the lower deck] are drowning.'

(b) [Korpus Malti 3.0; literature32]

***[Fuq** il-gverta tal-Ġinġlija]$_{PP}$ kien hemm diġà ħafna iġsma mejta*
[**on** DEF-deck of:DEF-Ġinġlija] be there already many body.PL dead:PL
'[**On** the deck of the Ġinġlija] there were already many dead bodies [. . .].'

The noun *gverta* 'blanket; deck' triggers the use of (Π15) *fi* 'in' if the ship has multiple decks and the situation is located on one of the lower decks which are conceived of as containers and thus license the use of the INESSIVE Π. If the Figure is on the upper deck (in the open) the Ground is treated like a surface on which the Figure is located. This, in turn, justifies the use of the SURESSIVE (Π17) *fuq* 'on'.

As comes to the fore in (5.23) (like in (5.21)) body parts as Grounds do not necessarily require the use of (Π17) *fuq* 'on'. The body-part term *wiċċ* 'face' is also used in the sense of 'surface'. With this meaning, it may either combine with (Π15) *fi* 'in' as in the (a)-sentence or with (Π17) *fuq* 'on' as in the (b)-sentence.

(5.23) SURESSIVE IV

(a) [Korpus Malti 3.0; literature75]

deher il-katavru ta' Mattew [f' wiċċ il-baħar]$_{PP}$
appear DEF-corpse of Mattew [**in** face DEF-sea]
'[. . .] Mattew's corpse appeared [**on** the surface of the sea] [. . .].'

(b) [Korpus Malti 3.0; *news67228]*

*Ix-xemx tleqq dehbiena [**fuq** wiċċ il- baħar]$_{PP}$*
DEF-sun 3SG.F.IPFV:shine golden:F [**on** face DEF-sea]
'The sun shines golden [**above** the surface of the sea] [. . .].'

The spatial situation is clearly that of a SURESSIVE in (5.23a) but the Π that expresses this notion is (Π15) *fi* 'in' which otherwise is responsible for the INESSIVE. This Π is chosen because there is contact between the Figure and the Ground. Example

(5.23b) does not necessarily invite the SURESSIVE interpretation although it does not exclude the SURESSIVE either (if the reflection of the sunlight on the surface of the water is referred to). (Π17) *fuq* 'on' is also the Π to express the SUPERESSIVE which applies in situations in which the Figure is above the Ground, i.e. Figure and Ground are not touching each other.[31] Borg/Azzopardi-Alexander (1997: 161) have a further distinction involving (Π17) *fuq* 'on', namely that of location on a surface (which is covered by (5.23b) above).

The final aspect to be touched upon in connection with the SURESSIVE is the distinction of horizontal surfaces as Ground vs vertical surfaces as Ground. In the above, we have addressed the former. In (5.24), the common noun *ħajt* 'wall' represents the class of vertically extended Grounds.

(5.24) SURESSIVE V

(a) [Korpus Malti 3.0; literature32]

Ħares	*fil-mera*	*li*	*kien*	*hemm*	*[fil-ħajt]*$_{PP}$
look	in:DEF-mirror	that	be	there	[**in**:DEF-wall]

'He looked in the mirror that was there [**on** the wall].'

(b) [Korpus Malti 3.0; literature59]

Għaddew	*għaxar*	*minuti*	*skont*	*l-arloġġ*	***[fuq***	*il-ħajt]*$_{PP}$
pass:3PL.PFV	ten	minute:PL	according_to	DEF:clock	[**on**	DEF-wall]

'Ten minutes passed according to the clock [**on** the wall].'

The alternation between the two Πs has two possible explanations. On the one hand, in the (a)-sentence the Figure might be integrated into the structure of the Ground and thus understood as being inside the wall. This would explain why (Π15) *fi* 'in' is used in lieu of (Π17) *fuq* 'on'. (Π17) *fuq* 'on' is the correct choice for the (b)-sentence if the clock is hanging on the wall (suspended on a nail, for instance). The second explanation, on the other hand, is that (Π15) *fi* 'in' is optionally eligible in situations which involve Grounds which cover a large extension of space – with large being a matter of variable subjective interpretation.

What the case of the SURESSIVE suggests is that the distribution of Maltese Πs over contexts calls for an in-depth analysis which explores the phenomena in much more detail than can be done in this study. This statement finds corroboration in many of the subsequent sections.

---

31 Cf. Section 5.1.2.1.7.

5.1.2.1.8 Superessive

The SUPERESSIVE relation applies if there is no contact between Figure and Ground with the former occupying a vertically elevated position above the ground (Hagège 2010: 286). Borg/Azzopardi-Alexander (1997: 160) use the label “superior at rest” to characterise this spatial relation. Like in the case of the SURESSIVE discussed in the foregoing section, (Π17) *fuq* ‘on’ functions as marker of the SUPERESSIVE as shown in (5.25).

(5.25) SUPERESSIVE [Korpus Malti 3.0; news179517]

l-Pulizija	*Ingliża*	*xejnet*	*kumplott*	*ta’*	*terroristi*
DEF-police	English:F	nullify:3SG.F.PFV	conspiracy	of	terrorist:PL

biex	*jisplodu*	*ajruplani*	***[fuq***	*il-Baħar*	*Atlantiku]$_{PP}$*
to	3.IPFV:explode:PL	aeroplane:PL	**[on**	DEF-sea	Atlantic]

‘[. . .] the English police destroyed a conspiracy of terrorists to explode aeroplanes [**above** the Atlantic].’

There seems to be no competing Π when it comes to expressing the SUPERESSIVE.

5.1.2.1.9 Subessive

In Hagège’s (2010) system, no distinction is made as to situations in which (a) Figure and Ground are in contact with each other while the Figure is situated under the Ground and (b) the Figure is located somewhere below the Ground without contact. Borg/Azzopardi-Alexander (1997: 161–162) claim that the same Π is used to express inferior without contact and inferior with contact, namely (Π56) *taħt* ‘under’. A typical example is given in (5.26).

(5.26) SUBESSIVE – [Bartolo/Vella 2009: 244]

Estella	*setgħet*	*tisimgħu*	*jqarmeċ*	*u*
Estella	can:3SG.F.PFV	3SG.F.IPFV:hear:3SG.M	3SG.M.IPFV:crunch	and

jitfarfar	***[taħt***	*saqajha]$_{PP}$*
3SG.M.IPFV:shake_off_dust	**[under**	foot:PL:3SG.F]

‘Estella heard [the rusty stairs] crunch and giving off dust [**under** her feet] [. . .].’

As far as we can see, (Π56) *taħt* ‘under’ has the monopoly for expressing the SUBESSIVE.

5.1.2.1.10 PREESSIVE

5.1.2.1.10.1 PREESSIVE – spatial

Borg/Azzopardi-Alexander (1997: 159) assume a category anterior which seems to cover the FRONTALIS and the spatial PREESSIVE. The latter is described by Hagège (2010: 286) as applying to situations in which the Figure is placed before the Ground. This configuration is not easily told apart from the FRONTALIS usage of (П49) *quddiem* 'in front of' as mentioned above.[32] There is however evidence that (П46) *qabel* 'before' can be used to express differences between ranks in a hierarchy which in turn can be interpreted as a succession of different segments of space as in (5.27a). A proper spatial context is illustrated in (5.27b) where the place where the accident happened is situated in space relative to a street that serves as Ground.

(5.27) PREESSIVE spatial

(a) [Korpus Malti 3.0; news106530]
L-erba' timijiet ***[qabel*** *Malta]*$_{PP}$ *huma Latvja Palestina*
DEF-four team:PL **[before** Malta] 3PL Latvia Palestine
Singapore u El Salvador
Singapore and El Salvador
'The four teams [**above** Malta] are Latvia, Palestine, Singapore, and El Salvador [. . .].'

(b) [Korpus Malti 3.0; news116898]
L-aċċident seħħ [ftit ***qabel*** *it-triq li tagħti*
DEF-accident happen [little **before** DEF-street that 3SG.F.IPFV:give
għal Għargħur]$_{PP}$
for Għargħur]
'The accident happened [shortly **before** the street that leads to Għargħur] [. . .].'

The close conceptual ties with the temporal PREESSIVE are obvious: what comes before X in space is also there prior to X in time. Contact does not seem to be a parameter for further distinctions in the domain of the spatial PREESSIVE.

32 Cf. Section 5.1.2.1.6.

5.1.2.1.10.2 PREESSIVE – temporal

Borg/Azzopardi-Alexander (1997: 172) mention that "[t]he preposition **qabel** 'before' precedes the temporal expression with respect to which anteriority is calculated" [original boldface]. This means that (Π46) *qabel* 'before' serves as marker of all kinds of PREESSIVES, be they spatial or temporal. The temporal PREESSIVE is documented in (5.28).

(5.28) PREESSIVE temporal – [Korpus Malti 3.0; literature25]

Kienet	*l-Iżrael*	*[**qabel***	*il-Gwerra*	*tas-Sitt*	*Ijiem]*$_{PP}$
be:3SG.F.PFV	DEF-Israel	[**before**	DEF-war	of:DEF-six	day.PL]

'She was in Israel [**before** the Six Day War].'

In this function, (Π46) *qabel* 'before' is the only option for speakers of Maltese.

5.1.2.1.11 POSTESSIVE

5.1.2.1.11.1 POSTESSIVE – spatial

The domains of the two POSTESSIVES – spatial and temporal – form the logical equivalent of those of the two PREESSIVES since the former focus on the opposite relation which can be expressed in English by *behind* and *after*, respectively (Hagège 2010: 286). Borg/Azzopardi-Alexander (1997: 159) use the label "posterior at rest" for the spatial POSTESSIVE. In Maltese, the appropriate choice for expressing this relation is (Π60) *wara* 'after'. Its typical use is illustrated in (5.29).

(5.29) POSTESSIVE spatial [Bartolo/Vella 2009: 244]

Id-dawl	*ta'*	*barra*	*li*	*kienu*	*ħallew*	*[**warajhom**]*$_{PP}$
DEF-light	of	outside	that	be:3PL.PFV	leave:3PL.PFV	[**after**:3PL]

meta	*qabdu*	*neżlin*	*fil-bir*	*bilkemm*	*kien*	*qed*
when	catch:3PL.PFV	descend.PART:PL	in:DEF-well	hardly	be	PROG

jasal	*sa*	*fejn*	*kienu*	*huma*
3SG.M.IPFV:arrive	till	near	be:3PL.PFV	3PL

'The external light they left [**behind** them] when they began to descend into the well hardly reached as far as where they were [. . .].'

Except for (Π60) *wara* 'after', no other Π is involved in the expression of the spatial POSTESSIVE. As in the case of the spatial PREESSIVE, it is irrelevant whether Figure and Ground are in contact with each other or not.

5.1.2.1.11.2 POSTESSIVE – temporal

Borg/Azzopardi-Alexander (1997: 172) mention (Π60) *wara* 'after' as the only option for the expression of the TEMPORAL POSTESSIVE (= "general posterior" according to their terminology). The sentence in (5.30) shows the typical employment of this Π within a temporal adverbial.

(5.30) POSTESSIVE temporal – [Ellul Mercer 1986: 60]

*[**wara***	*ġurnata*	*oħra*	*li*	*ma*	*ninsewhiex*	*kif*
[**after**	day	other:F	that	NEG	1.IPFV:forget:PL:3SG.F:NEG	like

ġieb	*u*	*laħaq]*$_{PP}$	*issa*	*li*	*waqgħet*	*ftit*	*tal-ħemda*
bring	and	reach]	now	that	fall:3SG.F.PFV	little	of:DEF-calm

ninsab	*waħdi*
1SG.IPFV:REFL:find	one:1SG

'[. . .] [**after** another day that we won't forget easily] now that it became somewhat calmer I find myself to be alone [. . .].'

Once again, there is no other Maltese Π competing with (Π60) *wara* 'after' in this domain. For the next three relations, Hagège (2010) does not distinguish between spatial and temporal categories. Similarly, the possibility of contact and its absence plays no role in these categories.

5.1.2.1.12 CIRCUMESSIVE – spatial and temporal

According to Hagège (2010: 286), the CIRCUMESSIVE covers situations in which the Figure is "around [. . .] a place or person(s)." The term Borg/Azzopardi-Alexander (1997: 166) use for configurations of this kind is "circumferential at rest". The authors of the Maltese reference grammar only mention (Π31) *madwar* 'around' as marker. A genuine spatial context for the use of this Π is featured in (5.31).

(5.31) CIRCUMESSIVE spatial – [Korpus Malti 3.0; academic12]

fil-kampanja	*[**madwar***	*Sao Paulo]*$_{PP}$	*aktar*	*minn*	*80*
in:DEF-countryside	[**around**	Sao Paulo]	more	from	80

fil-mija	*tan-nies*	*kienu*	*immigrati*	*Taljani*
in:DEF-hundred	of:DEF-people	be:3PL.PFV	immigrant:PL	Italian:PL

'[. . .] in the country [**around** Sao Paulo], more than 80% of the people were Italian immigrants [. . .].'

Example (5.32) suggests that the CIRCUMESSIVE is not always necessarily spatial in the strict sense of the term.

(5.32) CIRCUMESSIVE temporal – [Korpus Malti 3.0; news145048]

nfissru	*ruħna*	*dwar*	*l-infiq*	*li*	*jsir*
1.IPFV:explain:PL	self:1PL	about	DEF-spending	that	3SG.M.IPFV:become

f'	*dawn*	*il-jiem*	***[madwar***	*il-Milied]*$_{PP}$
in	DEM:PL:PROX	DEF-day.PL	**[around**	DEF-Christmas]

'[. . .] we explain ourselves as to the expenditure that happens during these days [**around** Christmas] [. . .].'

It is clear that the basically spatial Π is used metonymically to refer to a period of time that starts before Christmas and ends thereafter so that the Ground is sandwiched between the two subperiods in which commercial activities are to be expected. However, the metonymy that is at work here is not so much different from other cases of metonymy like those of (Π46) *qabel* 'before' and (Π60) *wara* 'after' discussed above.[33] We therefore split the CIRCUMESSIVE into a spatial and a temporal category both of which, however, involve the same Π, namely (Π31) *madwar* 'around'. We add that when (Π31) *madwar* 'around' is used temporarily more often than not it conveys a sense of vagueness or uncertainty.[34]

5.1.2.1.13 MEDIOESSIVE I

For the MEDIOESSIVE and the INTERESSIVE, Hagège (2010: 287) assumes that the governed term of the Π may refer to either things or people whereas places are ruled out as complements of the same Π. The MEDIOESSIVE is said to apply if the Figure is situated "among [. . .] people or things located in a place" (Hagège 2010: 286). Interestingly, Borg/Azzopardi-Alexander (1997) make no mention of the appropriate (Π16) *fost* 'amongst' (Saari 2003: 137–138)[35] and (Π47) *qalb* 'amidst'. The peculiarities of the former in the morphological domain and its government properties have been discussed already.[36] The complement of this Π must either be a genuine plural or a collective (which is singular morphologically). The complement of (Π16) *fost* 'amongst' may be animate or inanimate. An example of the latter possibility is given in (5.33a). In (5.33b), an example of the largely synonymous (Π47) *qalb* 'amidst' is presented.

---

33 Cf. Sections 5.1.2.1.10 and 5.1.2.1.11.

34 We will address this issue again in Section 5.2.

35 In contrast, Borg/Azzopardi-Alexander (1997: 165–166) provide examples for *f' nofs* 'in (the) middle' as representative of the "medial at rest". *F' nofs* 'in (the) middle' is employed if the Ground comprises three or more entities. Since *f' nofs* 'in (the) middle' does not meet the definition of the canonical Π we skip discussing it in this study.

36 Cf. Sections 3.1.2.3.1.2 (iii) and 4.1.1.2.2.1 (x).

(5.33) MEDIOESSIVE

(a) [Korpus Malti 3.0; academic12]

Hubert Opperman li ***[fost*** *karigi oħra]*$_{PP}$ *okkupa*
Hubert Opperman that **[amongst** commitment:PL other:F] occupy
dik ta' ministru għall-immigrazzjoni
DEM:F:DIST of minister for:DEF-immigration
'Hubert Opperman who [**among** other commitments] fulfilled that of minister for immigration.'

(b) [Korpus Malti 3.0; literature78]

Farrugia li kien ***[qalb*** *il-folla]*$_{PP}$ *ma ħadha xejn*
Farrugia that be **[amidst** DEF-crowd] NEG take:3SG.F nothing
bi kbira
with big:F
'Farrugia who was [**in** the crowd] took it lightly [. . .].'

Were it not for the isolated temporal expression *fost il-ġimgħa* 'on weekdays' (cf. (5.34)) which is attested 420 times in the Korpus Malti 3.0 and *jum fost l-oħrajn* 'one day, once upon a time' (Aquilina 1987: 358), (Π16) *fost* 'amongst' could be registered as a Π that is semantically restricted to spatial relations (just as (Π47) *qalb* 'amidst').

(5.34) MEDIOESSIVE – fossilised expression [Korpus Malti 3.0; culture2096]

It-tfajliet jaħbtu l-aktar ***[fost*** *il-ġimgħa]*$_{PP}$ *u*
DEF-girl:PL 3.IPFV:turn_up:PL DEF-more **[amongst** DEF-week] and
l-ġuvintur fil-weekends
DEF-young_man:PL in:DEF-weekend:PL
'The girls show up the most [**on** weekdays] and the boys on the weekends.'

Twenty-two hits go to the credit of the idiom *ħalla X bħall-Erbgħa fost il-ġimgħa* 'leave X in the lurch' (Aquilina 1991: 1169) which literally translates as 'leave X like Wednesday in the week'. We assume that these cases are remnants of an erstwhile more widespread use of (Π16) *fost* 'amongst' also in temporal contexts. This possibility is no longer productive. Therefore, we classify (Π16) *fost* 'amongst' as spatial Π. As transpires from the next subsection, this Π is not the only choice for expressing MEDIOESSIVE functions.

5.1.2.1.14 INTERESSIVE – spatial and temporal / MEDIOESSIVE II

Hagège (2010: 286) mentions English *between* as a good representative of a Π that fulfils the function of INTERESSIVE. The Figure is located in the space between two entities which together constitute the discontinuous Ground. Borg/Azzopardi-Alexander (1997: 165) speak of the category "medial at rest with" two Ground elements. (Π4) *bejn* 'between' is the sole candidate for the expression of this configuration in Maltese. Like (Π16) *fost* 'amongst', (Π4) *bejn* 'between' displays morphological and morphosyntactic idiosyncrasies.[37] These special traits are relevant also for the semantic interpretation of the constructions.

In this section, we note that the INTERESSIVE – like the CIRCUMESSIVE – can be spatial as in (5.35a) or temporal as in (5.35b).

(5.35) INTERESSIVE

(a) spatial – [Korpus Malti 3.0; news191851]
Il-fatt li l-ħames bastimenti nstabu f' fond
DEF-fact that DEF-five vessel:PL REFL:find:3PL.PFV in depth
ta' 100 metru qrib ħafna ta' xulxin qrib il-gżira ta'
of 100 meter near very of each_other near DEF-island of
Ventotene nofs triq ***[bejn*** *Ruma u Napli]$_{PP}$*
Ventotene half street **[between** Rome and Naples]
jindika l-valur storiku li kellha
3SG.M.IPFV:indicate DEF-value historical that have.PFV:3SG.F
din il-gżira fi żmien l-imperu Ruman
DEM:F:PROX DEF-island in time DEF-empire Roman
'The fact that the five vessels were found very close to each other at a depth of 100 meters near the island of Ventotene half the way [**between** Rome and Naples] shows the historical value the island had in the time of the Roman Empire.'

(b) temporal – [Korpus Malti 3.0; academic8]
Nina Zita ħadmet fuq bażi volontarja mal-Koperattiva
Nina Zita work:3SG.F.PFV on basis voluntary with:DEF-cooperative
Kummerċ Ġust ***[bejn*** *Settembru 2006 u Ġunju 2007]$_{PP}$*
commerce just **[between** September 2006 and June 2007]
'Nina Zita worked on a voluntary basis with the Cooperative Fair Trade [**between** September 2006 and June 2007].'

On the same grounds as in the previous cases of a division into a spatial and a temporal relation of the same category, we postulate the distinction between the spatial and the TEMPORAL INTERESSIVE. Like in the majority of the other cases of this

37 Cf. Sections 3.1.2.3.1.2 (ii) and 4.1.1.2.2.1 (iii).

kind of split, the same Π is used for both functions. In this case, we are alluding to (Π4) *bejn* 'between'.

On the basis of the distinction of medial location with relation to two Ground elements, on the one hand, and more than two Ground elements, on the other hand, proposed by Borg/Azzopardi-Alexander (1997: 165) and – implicitly – also by Hagège (2010: 286), one might assume that (Π4) *bejn* 'between' is reserved for configurations which involve two Ground elements whereas (Π16) *fost* 'amongst' is required if more than two Ground elements are present in a spatial situation. On closer inspection, we notice that it is true that (Π16) *fost* 'amongst' can only be used if the Figure is located in a more extended group of Ground elements. It is equally true that more often than not the use of (Π4) *bejn* 'between' implies that the Figure is placed in the space between two Ground elements. However, as examples like (5.36) suggest, (Π4) *bejn* 'between' can also be used in spatial situations which must be classified as MEDIOESSIVE because the Ground elements are more numerous than a pair of such elements.

(5.36) MEDIOESSIVE spatial Π – [Korpus Malti 3.0; academic12]

il-Maltin	*ta'*	*Toronto*	*kienu*	*jgħoddu*	*biss*	*xi*
DEF-Maltese:PL	of	Toronto	be:3PL.PFV	3.IPFV:count:PL	only	some

mitejn	*persuna*	*maqsumin*	***[bejn***	*erbatax-il*	*familja]*$_{PP}$
hundred:DU	person	PART:divide:PL	**[between**	fourteen	family]

li	***[bejniet**hom]*$_{PP}$	*kellhom*	*seba' u għoxrin*	*tifel*	*u*	*tifla*
that	[**between**:3PL]	have.PFV:3PL	twenty_seven	boy	and	girl

'[. . .] the Maltese of Toronto numbered only some two hundred persons distributed [**over** fourteen families] which [**between** them] had twenty-seven male and female children [. . .].'

There are two *bejn*-PPs which do not conform to the pattern that is believed to be crucial for the INTERESSIVE. The Ground is the same for both *bejn*-PPs, namely the fourteen Maltese families. In the first situation, the Figure consists of some two hundred persons and in the second situation, the Figure counts twenty-seven children. These absolute numbers do not meet the criteria of a prototypical INTERESSIVE. To the contrary, the *bejn*-PPs in (5.36) resemble the MEDIOESSIVE as sketched in the previous section. The Figure is distributed and its component parts are scattered across the space defined by the different Ground elements. Outside of those cases where (Π4) *bejn* 'between' occurs in a coordinative construction with two complements, this Π conveys the meaning of the MEDIOESSIVE provided its complement refers to a group of entities larger than two.[38]

5.1.2.1.15 EXTRAESSIVE

Hagège (2010: 286) introduces the category ABESSIVE for a spatial situation in which the Figure is "at the starting point of a move." The English Πs illustrating this meaning are *out of* (Hagège 2010: 261) and *outside of* (Hagège 2010: 287). We have difficulties to handle the notion of starting of a move. To our mind, what is more important than the possibility that a motion event starts from where the Figure is located is where the Figure is located, namely (somewhere) outside the Ground. Borg/Azzopardi-Alexander (1997: 158) identify (Π3) *barra* 'outside' for this function. We add (Π44) *oltre* 'beyond' to the short list of EXTRAESSIVE Πs whose usage is illustrated in (5.37).

(5.37) EXTRAESSIVE

(a) [Korpus Malti; culture1336]

Matul	*il-perjodu*	*li*	*kien*	*[**barra***	***minn***	*Malta]$_{PP}$*	*huwa*
during	DEF-period	that	be	[**outside**	**from**	Malta]	3SG.M

ġie	*nnominat*	*ukoll*	*għall-Premju*	*tal-Volontarjat*
come	nominat:PART	also	for:DEF-award	of:DEF-volontary_service

'During the time he was [**outside** Malta], he was also nominated for the Award of the Voluntary Service.'

(b) [Korpus Malti 3.0; european222]

ma	*jikkawżawx*	*danni*	*lill-ambjent*	*ta'*	*Stati*
NEG	3.IPFV:cause:PL:NEG	damage:PL	to:DEF-environment	of	State:PL

oħrajn	*jew*	*ta'*	*żoni*	*[**oltre***	*l-limiti*	*tal-ġurisdizzjoni*
other:PL	or	of	zone:PL	[**beyond**	DEF-limit:PL	of:DEF-jurisdiction

nazzjonali]$_{PP}$
national]

'[. . .] they do not cause damage to the environment of other stats or zones [**beyond** the limits of the national jurisdiction].'

38 Dixon (2021: 345) discusses similar phenomena with reference to the English Π *between* for which he assumes that it "always indicates a relation between two things." It does not seem to be possible to apply his analysis to the Maltese facts since the examples cannot be understood as a succession of binary relations between two Grounds.

In both the (a)-sentence and the (b)-sentence the Figure is situated outside the Ground without any motion event being involved. The relation is static. In (5.37a), the protagonist is nominated for an award in Malta while he is away from the island. In (5.37b) the damage should not happen in territories which are outside the national jurisdiction of a member-state of the E.U. Whether it makes sense to distinguish sub-types of the EXTRAESSIVE (viz. one which implies that the Figure is outside the Ground = (Π3) *barra* 'outside' and one that suggests that the Figure is located beyond a given limit = (Π44) *oltre* 'beyond') needs to be determined in a dedicated follow-up study. For the purpose of this study, it is sufficient to state that the INESSIVE has an adequate counterpart in Maltese, namely the EXTRAESSIVE.

5.1.2.2 Dynamic relations

There is a remarkable asymmetry in Hagège's (2010: 285–287) catalogue of spatial relations expressed by adpositions. We have reviewed and sometimes modified his inventory of static spatial meanings in the foregoing section. As to the dynamic relations, however, it is striking that the only representative of a relation that focuses on the Source in a spatial situation is the ABLATIVE. All other non-static categories are clearly limited to instances of motion (in)to or towards a Ground, meaning: they reflect Goal-oriented processes. In contrast, Borg/Azzopardi-Alexander (1997) consistently distinguish rest at, motion to, motion from, and motion past a given Ground. In light of the data presented by Borg/Azzopardi-Alexander (1997), we consider Hagège's taxonomy to be insufficient. Accordingly, in what follows, we will complement this taxonomy by repeatedly filling the gaps, especially in connection with Source and Path.[39] It is additionally noteworthy that the distinction of spatial and temporal meanings with some of the static categories does not apply to the corresponding dynamic relations whereas the distinction is relevant (according to Hagège 2010) with dynamic relations that have no equivalent in the static domain.

What additionally needs to be kept in mind is the leitmotif of Section 4.1.3, i.e., the binary combination of Πs one of which expresses configuration and the other directionality. Π-Π combinations of this kind abound wherever the dynamic relations of Goal and Source are involved. Moreover, Goal is frequently

---

39 We reserve Section 5.1.2.2.20 for the discussion of motion past whereas Goal and Source relations are treated one by one in the subsequent sections.

not made explicit so that the Π that serves the purpose of expressing the static meaning suffices to also express Goal. We assume that in these cases directionality is subject to zero-marking. These facts render the account of the dynamic meanings more demanding than it was the case with the corresponding static relations.

5.1.2.2.1 ILLATIVE

The ILLATIVE is the Goal-oriented equivalent of the static INESSIVE.[40] According to Hagège's (2010: 286) wording, the ILLATIVE applies if the Figure moves to a place. However, the English Πs illustrating the ILLATIVE are *to* and *into* (Hagège 2010: 261, 287). To our mind, the term ILLATIVE is suggestive of a spatial situation in which the Figure moves into a container-like Ground whereas motion to a Ground without entering into it seems to be conceptually different from the ILLATIVE. The latter situation is perhaps better subsumed under the label allative.[41] In what follows in this section, we focus on evidence for situations in which the Figure (at least potentially) enters the Ground.

In contrast to the static counterpart, the ILLATIVE displays only marginal evidence of temporal usage as shown in (5.38).[42]

(5.38) ILLATIVE – temporal [Korpus Malti 3.0; parl117]
Hawnhekk morna [fi ż<u>mien</u> <u>Byzantium</u>]$_{PP}$
here go:1PL.PFV [**in** <u>time</u> <u>Byzantium</u>]
'At this point we have gone (back) [**to** <u>the Byzantine era</u>].'

We are in doubt about the status of examples of this kind. The issue requires an in-depth study of its own. For the time being, we therefore do not propose the existence of a distinct temporal ILLATIVE.

Except for the absence of hard evidence for the existence of a temporal ILLATIVE, the empirical facts illustrating the ILLATIVE closely resemble those discussed in connection with the spatial INESSIVE.[43] In (5.39), we encounter familiar patterns.

---

40 Cf. Section 5.1.2.1.1.
41 Cf. Section 5.1.2.2.5.
42 But cf. Section 5.1.2.2.7 on the TERMINATIVE.
43 Cf. Section 5.1.2.1.1.1.

(5.39) ILLATIVE

(a) [Casha 1999: 41]

Fl-aħħar jiem ta' ħajjitha kienu ħaduha
in:DEF-last day.PL of life:POSS:3SG.F be:3PL.PFV take:3PL.PFV:3SG.F
[Ø l-isptar]ₓ
[Ø DEF-hospital]
'On the last days of her life, they had taken her [**to** the hospital].'

(b) [Korpus Malti 3.0; literature15]

Mort [f' ristorant li suppost isajjar
go:1SG.PFV [**in** restaurant that suppose:PART 3SG.M.IPFV:cook
ikel Malti]$_{PP}$
food Maltese]
'I went [**(in)to** a restaurant that supposedly cooked Maltese food] [. . .].'

(c) [Camilleri 2013: 63]

Poġġietha bl-akbar galbu [ġo kontenitur
put:3SG.F.PFV:3SG.F with:DEF-big.CMPR good_manners [**in** container
tal-istainless steel]$_{PP}$
of:DEF-stainless steel]
'She put it with the best of manners [**into** a container of stainless steel].'

(d) [Korpus Malti 3.0; parl10108]

Snin ilu kont mort [ġewwa l-Kenya]$_{PP}$
year.PL ago:3SG.M be:1SG.PFV go:1SG.PFV [**inside** DEF-Kenia]
'Years ago, I had gone [**to** Kenia] [. . .].'

Borg/Azzopardi-Alexander (1997: 156) assume that "[c]ertain verbs of motion, followed by the place nominal (without a preposition) express motion to a place, including arrival at the place." The general motion verb *mar* 'go' is one of those verbs which frequently take bare NPs when these NPs refer to the destination of the motion event (Stolz/Lestrade/Stolz 2014: 267). For motion into the Ground, they postulate the functional equivalence of the three Πs (Π15) *fi* 'in', (Π18) *ġewwa* 'inside', and (Π19) *ġo* 'in'. The four options are covered by the sentences (a) to (c) in (5.39). As to (5.39a), we assume that the Figure was not only taken to the hospital but into it. Nevertheless, the construction without Π is not as explicit as the overt PPs about where exactly the Figure winds up in relation to the Ground.

It is interesting to see that the three Πs behave differently in terms of how readily they are used with toponyms as complements in constructions whose motion verb is *mar* 'go'. In Figure 5.9, we compare the results of a search of the Korpus Malti 3.0 for constructions with the perfective forms of *mar* 'go' with the above three Πs. For (Π15) *fi* 'in', only unreduced full forms without cliticization are taken account of.

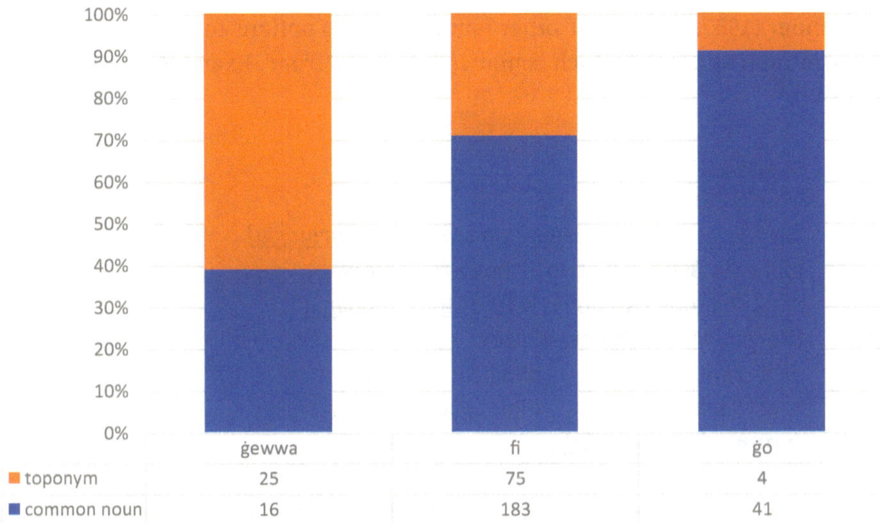

Figure 5.9: Complement types with *mar* + Π to express the ILLATIVE (Korpus Malti 3.0).

The higher absolute number of *fi*-PPs with a toponymic complement notwithstanding, it is evident that taking a toponym as complement is typical for (Π18) *ġewwa* 'inside' whereas combinations of this kind are exceptional in the case of (Π19) *ġo* 'in'. In light of the discussion in connection with the data presented in Figures 5.2 and 5.4 in the section on the spatial INESSIVE,[44] these results are expected. (Π18) *ġewwa* 'inside' displays a strong predilection for combinations with toponyms. In contrast, (Π19) *ġo* 'in' seems to avoid combinations of this kind as far as possible.

5.1.2.2.2 ELATIVE

If it is correct that the notions of INESSIVE and ILLATIVE require the Figure to be inside a container at some point in the spatial situation, it is surprising to see that there is no mention of a category that covers situations in which the Figure exits the Ground. In Hagège's (2010: 261 and 286–287) system of spatio-temporal meanings, there is the ABLATIVE whose English translation *from* is however suggestive of motion from a place (Hagège 2010: 286) in lieu of an event in the course of which the Figure emerges from inside the Ground.[45] Borg/Azzopardi-

44 Cf. Section 5.1.2.1.1.1.

45 It is not at all unlikely that Hagège (2010) had this in mind when he defined the EXTRAESSIVE statically as discussed in Section 5.1.2.1.15.

Alexander (1997: 158) on the other hand, offer two options for the expression of interior motion both of which contain (П41) *minn* 'from'. Examples are provided in (5.40).

(5.40) ELATIVE

(a) [Abela 2016: 11]

ħarġet *żewġ* *skutelli* *[mill-armarju]*$_{PP}$
exit.CAUS:3SG.F.PFV two bowl:PL [**from**:DEF-cupboard]
'[. . .] she took two bowls [**out of** the cupboard].'

(b) [Korpus Malti 3.0; news82068]

nitfarrku *bħal* *tazza* *tal-kristall* *li* *taqa'*
1.IPFV:REFL:crumble:PL like glass of:DEF-crystal that 3SG.F.IPFV:fall
[minn ***ġo*** *armarju]*$_{PP}$
[**from** **in** cupboard]
'[. . .] we will be smashed to pieces like a crystal glass that falls [**out of** a cupboard].'

(c) [Korpus Malti 3.0; literature32]

[minn ***ġewwa*** *armarju* *wiesa'* *mdendel* *mal-ħajt]*$_{PP}$ *ġab*
[**from** **inside** cupboard wide PART:hang with:DEF-wall] bring
dik *li* *skont* *hu* *kienet* *iċ-ċavetta*
DEM:F:DIST that according_to 3SG.M be:3SG.F.PFV DEF-key
'[. . .] [**out of** a large cupboard hanging on the wall] he brought what according to him was the key [. . .].'

In (5.40a), we are facing a case of unexpressed configuration. The directionality marker (П41) *minn* 'from' suffices to express the ELATIVE. The combination of (П41) *minn* 'from' and (П15) *fi* 'in' is generally blocked. There can be no doubt that the Figure (= the two bowls) was inside the Ground (= the cupboard) prior to the protagonist fetching the objects to move them elsewhere. Similarly, the Figure is removed from inside the Ground to a location outside the Ground in (5.40b–c), but in both cases, the configuration is made explicit.

In combination with the directionality marker which remains the same for all instances of the ELATIVE (П18) *ġewwa* 'inside' and (П19) *ġo* 'in' form typical П-П combinations. As we will have ample opportunity to notice throughout the subsequent sections, this pattern holds for numerous combinations in the sense that (П41) *minn* 'from' serves as the directionality marker for motion from a Ground for many kinds of configurations. It is unnecessary to attach distinctive names to each of these combinations. Separate labels are reserved only for those cases in which (П41) *minn* 'from' fulfils a given function without an additional configuration marker. In contrast, all Π-Π combinations which involve (Π41) *minn* 'from' as directionality marker of Source are registered under *motus ex loco* in the synopsis. Examples of the combinations are given at the appropriate place in the course of the discussion of the individual dynamic meanings.

5.1.2.2.3 APUDLATIVE

Another striking fact about Hagège's (2010) proposed system of adpositional meanings is the absence of dynamic meanings in connection with a Ground that refers to somebody's place. This absence is especially remarkable since the author postulates the static category APUDESSIVE.[46] Interestingly, Borg/Azzopardi-Alexander (1997: 156) do not broach this issue either.

Given that Place-Goal syncretism applies generally in Maltese, motion to somebody's place is expected to be expressed by the same Π as the APUDESSIVE, namely (Π22) *għand* 'at s.o.'s place'. This is usually the case as shown in (5.41) where the APUDLATIVE takes the same Π as the APUDESSIVE.

(5.41) APUDLATIVE – [Casha 1999: 24]

Kevin	*u*	*Romina*	*marru*	***[għand***	*wieħed*	*li*
Kevin	and	Romina	go:3PL.PFV	**[at_s.o.'s_place**	one	that

kien	*jinnegozja*	*fi*	*statwetti]$_{PP}$*
be	3SG.M.IPFV:trade	in	statuette:PL]

'Kevin and Romina went [**to** someone who was trading in statuettes] [...].'

We assume that the use of (Π22) *għand* 'at s.o.'s place' without preceding directionality marker implies that the Figure actually reaches the Goal. Spatial situations in which arrival at the Ground is only an option but not a necessity are discussed separately below.[47]

5.1.2.2.4 APUDABLATIVE

If the Figure can be at or go to someone's place, it can also be removed from there. Thus, it makes sense to postulate a meaning that we label APUDABLATIVE which expresses motion away from someone's place. In point of fact, Maltese boasts a distinct Π for the APUDABLATIVE which is (Π38) *mingħand* 'from s.o.'. This category is not subsumed under *motus ex loco* because the univerbation of (Π41) *minn* 'from' and (Π22) *għand* 'at s.o.'s place' has become fixed. As comes to the fore in example

---

46 As discussed in Section 5.1.2.1.2.
47 In Section 5.1.2.2.6.

(5.42), reference to a concrete place owned or inhabited by someone is not a prerequisite for the use of this Π.

(5.42) APUDABLATIVE – [Camilleri 2013: 54]

kienu	*xtrawha*	*bir-ribass*	***[mingħand***	*armla*
be:3PL.PFV	buy:3PL.PFV:3SG.F	with:DEF-discount	**[from_s.o.**	widow
Skoċċiża]$_{PP}$				
Scottish:F]				

'[. . .] they had bought it at a reduced price [**from** a Scottish widow] [. . .].'

Like in the cases of the APUDESSIVE and the APUDLATIVE, (Π38) *mingħand* 'from s.o.' takes a complement that ranges high on the animacy hierarchy. The complement can either be definite or indefinite. Where (Π22) *għand* 'at s.o.'s place' mostly invokes genuine spatial situations, (Π38) *mingħand* 'from s.o.' is frequently used in contexts where the wording *from s.o.'s place* fails to capture the situation appropriately. (Π38) *mingħand* 'from s.o.' presupposes a human (or highly animate) Source which comprises the personal sphere but excludes the body as physiological object. Examples like (5.43) suggest that (Π38) *mingħand* 'from s.o.' that the spatial relation of Source is metonymically applied to describe a situation of information transfer.

(5.43) APUDABLATIVE – [Bartolo/Vella 2009: 175]

Ħaġa	*waħda*	*biss*	*irrid*	*inkun*	*naf*
thing	one:F	only	1SG.IPFV:want	1SG.IPFV:be.FUT	1SG.IPFV:know
[mingħandek]$_{PP}$					
[from_s.o.:2SG]					

'I only want to know one thing [**from** you] [. . .].'

We come back to phenomena of this kind in the section on the ABLATIVE.[48]

5.1.2.2.5 ("Allative" >) PROPINQUELATIVE

With the ALLATIVE, we enter another territory that has been charted by Hagège (2010) in a peculiar way which calls for being commented upon. According to Hagège (2010: 286), the ALLATIVE requires that the Figure "move[s] towards a position in contiguity with [the Ground]." The crucial point in this definition is the notion of contiguity. The Figure is contiguous to the Ground if the distance between the two is minimal. Put differently, the ALLATIVE fits the description of a dynamic spatial situation in which the Figure has reached the Ground (without entering into it).

48 Cf. Section 5.1.2.2.8.

This leaves us with the question of how situations are classified in which the Figure moves in the direction of the Ground without necessarily arriving at the destination. Hagège (2010) does not seem to take account of the latter possibility. We reserve the following section for its discussion. Note that Hagège (2010: 261) translates the ALLATIVE into English as *towards* whereas elsewhere no translation is provided (Hagège 2010: 287).

Going by Hagège's above definition of the ALLATIVE, we realise that what he describes is the Goal-oriented equivalent of the PROPINQUUS.[49] To emphasise this relationship, we have coined the term PROPINQUELATIVE. Accordingly, we expect to find some of the Πs presented in connection with the static meaning to be involved also in the expression of the PROPINQUELATIVE. Borg/Azzopardi-Alexander (1997: 157) argue that motion to within the domain of the PROPINQUUS is expressed by (Π48) *qrib* 'near' and (Π58) *viċin* 'near' with both taking (Π55) *ta'* 'of' as a linker if the Ground is animate. Moreover, animate Goals frequently require the use of (Π32) *maġenb* 'close to'. In contrast, (Π14) *fejn* 'near' and (Π23) *ħdejn* 'beside' which are options for expressing the static PROPINQUUS are not mentioned as markers of the PROPINQUELATIVE. In the next paragraph, we will learn that there are no restrictions which preclude any of the PROPINQUUS Πs from being used in PROPINQUELATIVE function, too.

In (5.44), it is shown that all of the above-mentioned Πs are licit options as markers of configuration in constructions with a PROPINQUELATIVE interpretation. Recall that directionality is usually left unexpressed if the situation is one of motion to the Ground.

(5.44) PROPINQUELATIVE

(a) (Π14) *fejn* 'near' – [Korpus Malti 3.0; literature75]

Imbagħad	*reġa'*	*mar*	***[fejn***	<u>*il-ħbieb*</u>	<u>*tiegħu*</u>*]*$_{PP}$
then	do_again	go	**[near**	<u>DEF-friend.PL</u>	<u>of:3SG.M</u>]

'Then he went again [**close to** <u>his friends</u>] [. . .].'

(b) (Π23) *ħdejn* 'beside' – [Korpus Malti 3.0; literature24]

kien	*telaq*	*minn*	*ħdejn*	*il-kompjuter*	*u*	*mar*	***[ħdejn***
be	leave	from	beside	DEF-computer	and	go	**[beside**

<u>*kuġinuh*</u>*]*$_{PP}$
<u>cousin:M:3SG.M</u>]

'[. . .] he left the computer and went [**close to** <u>his cousin</u>].'

49 Discussed in Section 5.1.2.1.4.

(c) (П32) *maġenb* 'close to' – [Korpus Malti 3.0; literature78]

beda riesaq bil-mod lejha sakemm wasal
begin approach.PART with:DEF-manner towards:3SG.F until arrive
***[maġenb** il-bank tal-ġebel]*$_{PP}$
[**close_to** DEF-bench of:DEF-stone]
'[. . .] he began to slowly approach her until he arrived [**close to** the stone bench] [. . .].'

(d) (П48) *qrib* 'near' – [Korpus Malti 3.0; news139118]

Louis Deguara mar ***[qrib** il-bieb]*$_{PP}$
Louis Deguara go [**near** DEF-door]
'[. . .] Louis Deguara went [**near** the door] [. . .].'

(e) (П58) *viċin* 'near' – [Korpus Malti 3.0; news18894]

il-kaptan dawwar ir-rotta għal Sqallija u mar ***[viċin***
DEF-captain turn:CAUS DEF-course for Sicily and go [**near**
il-Port ta' Pozzallo]$_{PP}$
DEF-port of Pozzallo]
'[. . .] the captain changed the course towards Sicily and went [**near** the port of Pozzallo].'

The motion verbs *mar* 'go' and *wasal* 'arrive' are eligible for combinations with each of the competing Пs. Neither (П14) *fejn* 'near' nor (П23) *ħdejn* 'beside' is barred from PROPINQUELATIVE constructions where we also find evidence for the use of (П32) *maġenb* 'close to' although none of these Пs is registered as marker of (PROPINQUUS) motion to in the Maltese reference grammar.

Since it is necessary to modify the picture painted by the authors of said reference grammar, it might also be the case that the postulated dependence of the use of (П55) *ta'* 'of' in highly animate contexts needs to be revised. In (5.45), we demonstrate that (П48) *qrib* 'near' and (П58) *viċin* 'near' do not have exclusive rights when it comes to combining with animate complements.

(5.45) PROPINQUELATIVE – animate Ground I

(a) [Korpus Malti 3.0; literature77]

Ma kienx għadu bin Lippu wasal ***[fejn** Ġużeppi]*$_{PP}$
NEG be:NEG still:3SG.M son Lippu arrive [**near** Ġużeppi]
meta dan għajjat Sibnieha 'l Ineż għadha ħajja
when DEM:M:PROX shout find:1PL.PFV:3SG.F to Ineż still:3SG.F alive:F
'Lippu's son had just arrived [**near** Ġużeppi] when the latter shouted: 'We found Ineż; she is still alive!'."

(b) [Korpus Malti 3.0; news106999]
beda miexi sakemm wasal ***[ħdejn*** *Ellul li kien qiegħed*
begin go.PART until arrive **[beside** Ellul that be stay.PART
ħdejn il-karozza]$_{PP}$
beside DEF-car]
'[. . .] he began to walk until he arrived [**beside** Ellul who was standing beside the car].'

(c) [Korpus Malti 3.0; literature21]
Riku qam u mar ***[maġenb*** *Marku]*$_{PP}$
Riku rise and go **[close_to** Marku]
'Riku rose up and went [**close to** Marku] [. . .].'

To guarantee a high degree of animacy we have picked examples which involve a personal name as complement. In accordance with what we have learned already,[50] (Π14) *fejn* 'near', (Π23) *ħdejn* 'beside', and (Π32) *maġenb* 'close to' are compatible with all kinds of (in)definite complements including proper names. Animacy is therefore no criterion to separate these three Πs from (Π48) *qrib* 'near' and (Π58) *viċin* 'near'.

In Figure 5.10, we reveal the token frequencies of the five Πs under review for combinations with the perfective forms of the motion verb *mar* 'go' in the Korpus Malti 3.0. There are 339 matches altogether more than half of which go to the credit of (Π23) *ħdejn* 'beside' whereas (Π32) *maġenb* 'close to' is only marginally attested in the context we searched for. The second best is (Π14) *fejn* 'near' which occurs between 1.9 and 2.5 times more often than (Π48) *qrib* 'near' and (Π58) *viċin* 'near'. It is therefore clear that the latter two Πs are not privileged when it comes to expressing the ALLATIVE meaning.

What we also know already is that (Π48) *qrib* 'near' and (Π58) *viċin* 'near', to some extent, are subject to variation in the domain of government.[51] It is no longer necessary to illustrate the behaviour of these Πs when they take pronominal complements. More importantly, the examples in (5.46)–(5.47) show that (Π55) *ta'* 'of' may function as a linker even if the complement is an inanimate common noun.

50 In Section 4.1.1.2.2.1 (ix), (xv), and (xxi).
51 Cf. Sections 4.1.2.2.1.2.1 (ix) and (xi) and 4.1.2.2.2.2.1 (iv).

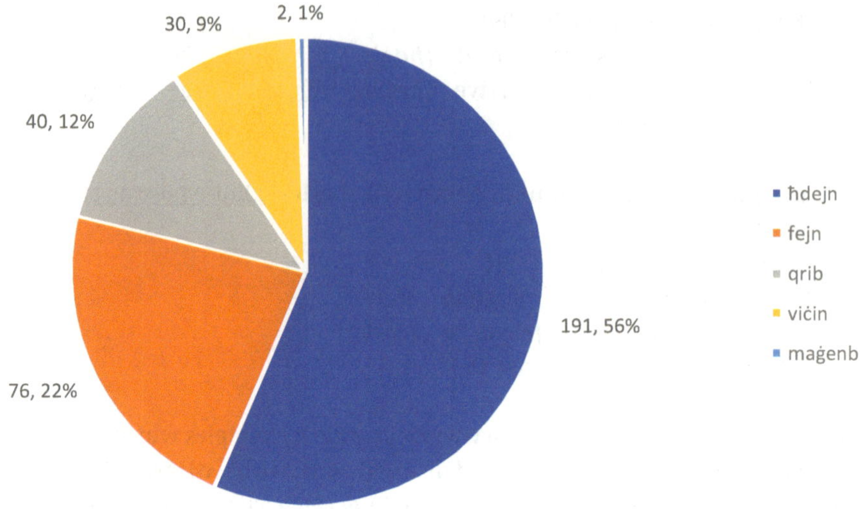

Figure 5.10: Hits for *mar* 'go' (perfective) combining with (П14) *fejn* 'near', (П23) *ħdejn* 'beside', (П32) *maġenb*, (П48) *qrib* 'near', and (П58) *viċin* 'near' in the Korpus Malti 3.0.

(5.46) (П48) *qrib* 'near' + (П55) *ta'* 'of'

(a) [Korpus Malti 3.0; literature11]

Bruno ressaq is-siġġu [aktar ***qrib ta'*** *Simone]*$_{PP}$
Bruno approach:CAUS DEF-chair [more **near of** Simone]
'Bruno moved the chair [**closer to** Simone].'

(b) [Korpus Malti 3.0; literature20]

seta' jirkeb l-ewwel karozza tal-linja li
can 3SG.M.IPFV:get_on DEF-first car of:DEF-line that
tieħdu ***[qrib*** *tad-dar]*$_{PP}$
3SG.F.IPFV:take:3SG.M **[near of**:DEF-house]
'[. . .] he could get on the first bus that takes him [**close to** the house] [. . .].'

(5.47) (П58) *viċin* 'near' + (П55) *ta'* 'of'

(a) [Korpus Malti 3.0; culture1674]

dawn il-kanzunetti nkitbu bil-għan li
DEM:PL:PROX song:DIM:PL REFL:write:3PL.PFV with:DEF-goal that
jqarreb aktar nies ***[viċin ta'*** *Ġesù]*$_{PP}$
3SG.M.IPFV:bring_near.CAUS more people **[near of** Jesus]
'[. . .] these songs were written with the aim of bringing more people [**close to** Jesus].'

(b) [Korpus Malti 3.0; literature24]
Estella nižlet kokka u ressqet
Estella descend:3SG.F.PFV squatting and approach.CAUS:3SG.F.PFV
idejha t-tnejn ***[viċin*** *tal-ħamrija xotta]*$_{PP}$
hand:PL:3SG.F DEF-two **[near** **of**:DEF-red_soil dry:F]
'Estelle squatted down and brought her two hands [**close to** the dry soil].'

It seems that the use of (Π55) *ta'* 'of' is triggered regularly when either (Π48) *qrib* 'near' and (Π58) *viċin* 'near' takes an animate complement. However, the linking Π is also admissible if the complement is inanimate. This latter option is chosen less frequently as opposed to nominative government. In sum, animacy plays a role but fails to absolutely determine the behaviour of the Πs.

5.1.2.2.6 DIRECTIONAL (including: "apuddirectional")

This section deals with a rather intricate situation. The domain of the DIRECTIONAL poses a number of problems not all of which can be solved in this study. It is worthwhile repeating that Hagège (2010) does not mention the category under scrutiny. However, Borg/Azzopardi-Alexander (1997: 156) claim that "[t]he preposition **lejn** is used for the expression of a certain direction" [original boldface]. We argue that (Π28) *lejn* 'towards' is not the only Maltese Π that is adequate for the description of a spatial situation in which the Figure moves in the direction of the Ground but whether the Goal is reached is either irrelevant or the motion event is never carried through, in the first place. In a way, the DIRECTIONAL presupposes that the Figure has started to move without implying that the Figure winds up close to the Ground. The second candidate for the DIRECTIONAL function is (Π21) *għal* 'for'.[52]

52 As to the potential third candidate for the directional function, the 228 hits for (Π57) *versu* 'towards' in the Korpus Malti 3.0 clearly show that this Π is hardly ever employed in properly spatial contexts. The two matches which feature spatial (Π57) *versu* 'towards' are found in one and the same sentence, namely in (5.i):

(5.i) [Korpus Malti 3.0; opinion1568]
minflok erġajt tlajt Triq d'Odessa ***[versu***
instead_of do_again:1SG.PFV ascend:1SG.PFV Rue d'Odessa **[towards**
l-Boulevard Edgar-Quinet]$_{PP}$ *kont qiegħed ninżilha 'l isfel*
DEF-Boulevard_Edgar_Quinet] be:1SG.PFV PROG 1SG.IPFV:descend:3SG.F to down
[versu *l-Boulevard Montparnasse]*$_{PP}$
[towards DEF-Boulevard_Montparnasse]
'[. . .] instead of going again up Rue d'Odessa [**towards** Boulevard Edgar Quinet], I walked down [**towards** Boulevard Montparnasse].'

In (5.48), we present examples of DIRECTIONAL (Π28) *lejn* 'towards'. In the (a)-sentence, the utterance of the direct speech happens while the children are on their way home. Their arrival there is of no significance to the situation described in the sentence. The (b)-sentence is similar insofar as the notary walks to his study and continues to talk on his way there – and the statements he makes while walking are important not his reaching the study. The (c)-sentence shows that (Π28) *lejn* 'towards' may also be used when the Ground is reached by the Figure before the Figure returns to the starting point of the to-and-fro motion event. The three sample sentences are taken from one and the same literary text.

(5.48) DIRECTIONAL I – (Π28) *lejn* 'towards'

(a) [Casha 1999: 29–30]

Jien	*m'*	*għedtlekx*	*li*	*kollu*	*għal*	*xejn*	
1SG	NEG	say:1SG.PFV:IO:2SG:NEG	that	all:3SG.M	for	nothing	
qaltlu	*Romina*	*darba*	*waħda*	*huma*	*u*	*sejrin*	
say:3SG.F.PFV:IO:3SG.M	Romina	time	one:F	3PL	and	go.PART:PL	
lura	***[lejn***	*id-dar]$_{PP}$*					
back	**[towards**	DEF-house]					

'"Haven't I told you that everything is in vain?", Romina said to him while they were on their way back [home] [. . .].'

(b) [Casha 1999: 30]

Tajjeb	*tajjeb*	*kompla*	*n-nutar*	*u*	*beda*	*miexi*	***[lejn***
good	good	continue	DEF-notary	and	begin	go.PART	**[towards**
il-kamra	*fejn*	*kellu*	*l-istudju]$_{PP}$*				
DEF-room	where	have.PFV:3SG.M	DEF-study]				

'"Well, well!", the notary continued to say and began walking [**toward** the room where he had his study].'

(c) [Casha 1999: 40]

Tereża	*kienet*	*ġa*	*qamet*	*u*	*marret*		
Tereża	be:3SG.F.PFV	already	rise:3SG.F.PFV	and	go:3SG.F.PFV		
[lejn	*il-kċina*	*minn*	*fejn*	*ġiet*	*lura*	*ftit*	
[towards	DEF-kitchen	from	where	come:3SG.F.PFV	back	little	

The vast majority of the occurrences of (Π57) *versu* 'towards' belong in different semantic categories outside the sptaio-temporal domain. The functions of this Π will therefore be discussed in Section 5.1.3.13

tal-ħin *wara]$_{PP}$*
of:DEF-time after]
'Tereża had risen already and gone [**towards** the kitchen from where she returned shortly after] [. . .].'

In (5.48c), the character Tereża enters the kitchen to fetch a plate from there. (Π28) *lejn* 'towards' can be understood as being neutral as to the successful completion of the motion event.

In (5.49), it is shown how (Π21) *għal* 'for' fulfils APUDDIRECTIONAL functions with animate complements in Maltese. In the (a)-sentence, the focus is on the early phase of Little Red Riding Hood's walk towards her grandmother's place. What is important is that she starts her journey – not that she gets near her destination. In contrast, the (b)-sentence describes a situation in which the Figure actually reaches the Goal since in the sentence following the example the armed person fires several shots at the victim, i.e., the shooter must have come close to Pietru Borg's home.

(5.49) DIRECTIONAL II – (Π21) *għal* 'for'
(a) [Korpus Malti 3.0; news149411]
Little Red Riding Hood *qabdet* *il-qoffa* *u* *telqet*
Little Red Riding Hood grab:3SG.F.PFV DEF-basket and leave:3SG.F.PFV
[għal ***għand*** *in-nanna]$_{PP}$*
[for **at s.o.'s place** DEF-grandmother]
'Little Red Riding Hood took the basket and left [**for** her grandmother's place].'
(b) [Korpus Malti 3.0; news131868]
Huwa *mar* *id-dar* *tiegħu* *u* *minn* *hemmhekk* *ġab*
3SG.M go DEF-house of:3SG.M and from there bring
senter *u* *mar* ***[għal*** *Pietru Borg* *fir-residenza* *tiegħu]$_{PP}$*
rifle and go **[for** Pietru Borg in:DEF-residence of:3SG.M]
'He went to his home and brought a gun from there and went [**to** Pietru Borg in his residence].'

Superficially, the two examples in (5.49) pose a problem because (Π22) *għand* 'at s.o.'s place' – the expected Π in the domain of the APUDLATIVE – is attested only in (5.49a) whereas it is absent from (5.49b) although, in the latter sentence, the complement is a personal name and thus a typical trigger for (Π22) *għand* 'at s.o.'s place'. On closer inspection, the case turns out to be a non-issue. It is more logical to consider *mar għal X* as a translation of English *go for X* 'attack X'. Accordingly, (5.49b) is no example of the DIRECTIONAL at all.

Section 4.1.3 has familiarised us with the fact that the Π-Π combination *għal għand* is attested frequently enough to require further scrutiny. The question arises whether it makes sense to postulate an APUDDIRECTIONAL, i.e. a meaning category which captures spatial situations in which the Figure moves towards the personal sphere of an animate being that functions as Ground with the Ground being either reached or not. We are cautious not to open Pandora's box by postulating too many distinct categories so that the semantic system of Maltese Πs becomes unrecognisable. However, if the structural facts support the existence of the APUDDIRECTIONAL they should be taken seriously. A first argument in favour of the APUDDIRECTIONAL is the impossibility of combining (Π28) *lejn* 'towards' and (Π22) *għand* 'at s.o.'s place'. There is no evidence of **lejn għand* in either of our corpora. It is however possible for (Π28) *lejn* 'towards' to take an animate complement as shown in (5.50).

(5.50) (Π28) *lejn* 'towards' + animate complement

(a) [Bartolo/Vella 2009: 75]

ix-xiħ	*kien*	*qed*	*iħares*	*fiss*	***[lej**<u>hom</u>]$_{PP}$*	*forsi*
DEF-old	be	PROG	3SG.M.IPFV:look	fixed	[**towards**:<u>3PL</u>]	perhaps
l-iktar	***[lejn***	*<u>Estella</u>]$_{PP}$*				
DEF-more	[**towards**	<u>Estella</u>]				

'The old man was looking steadily [**at** <u>them</u>], perhaps the most [**at** <u>Estella</u>] [. . .].'

(b) [Bartolo/Vella 2009: 31]

Estella	*daret*	***[lejn***	*<u>Tommy</u>]$_{PP}$*
Estella	turn:3SG.F.PFV	[**towards**	<u>Tommy</u>]

'Estella turned [**towards** <u>Tommy</u>] [. . .].'

(c) [Korpus Malti 3.0; *news81407]*

ħareġ	*mill-karozza*	*u*	*mar*	***[lejn***	*<u>l-aggressuri</u>*	*<u>li</u>*
exit	from:DEF-car	and	go	[**towards**	<u>DEF-aggressor:PL</u>	<u>that</u>
<u>kienu</u>	*<u>għadhom</u>*	*<u>fil-karozza</u>]$_{PP}$*				
<u>be:3PL.PFV</u>	<u>still:3PL</u>	<u>in:DEF-car</u>]				

'[. . .] he got out of the car and went [**towards** <u>the attackers who were still in the car</u>].'

Examples (5.50a) and (5.50b), (Π28) *lejn* 'towards' is governed by two verbs – *ħares* 'look' and *dar* 'turn' – whose cooccurrence with the Π is frequently attested in the Korpus Malti 3.0. The situation described in connection with the perception event in (5.50a) cannot be classified as spatial unless directing one's gaze towards an entity is counted as a metonymical motion. In (5.50b), motion is involved as the agent changes her position to face Tommy. The spatial nature of the situation described in the (c)-sentence, however, is uncontroversial since the Figure moves physically in the direction of the aggressors. In none of these cases would (Π22) *għand* 'at s.o.'s place' be appropriate because in (5.50) it is always the physical body of the human participants that is the Goal of the action. If (Π22) *għand* 'at s.o.'s place' had been used additionally, the meaning would change from physical body to a person's personal sphere – a notion that cannot be expressed by way of using (Π28) *lejn* 'towards'. It should not go unmentioned that of the ninety-four instances of *mar lejn* in the Korpus Malti 3.0, only two (= 2%) feature an animate Ground. In the bulk of the cases, the complement of (Π28) *lejn* 'towards' is a place name. On account of the ungrammaticality of the cooccurrence of (Π28) *lejn* 'towards' and (Π22) *għand* 'at s.o.'s place' and given that in cases where directionality is not marked overtly it can be assumed that the Figure reaches the Ground, the only way to leave it open whether the animate Ground is reached is by way of using the Π-Π combination *għal għand.* On this basis, one could be led to assume that Maltese boasts a distinct APUDDIRECTIONAL. However, the combination *għal għand* reflects a more general pattern according to which (Π21) *għal* 'for' serves as directionality marker for a long list of other Maltese Πs only one of which will be talked about in the next paragraph. All other instances involving the directionality marker (Π21) *għal* 'for' will be discussed for each configuration separately.

In (5.51), two further patterns of DIRECTIONAL (Π21) *għal* 'for' are presented. In these cases, the complement is inanimate. In (5.51a–b), we have further instances of (Π21) *għal* 'for' functioning as directionality marker – this time with (Π19) *ġo* 'in' and (Π18) *ġewwa* 'inside' as markers of the configuration. Both (Π19) *ġo* 'in' and (Π18) *ġewwa* 'inside' alone are fine for the expression of either INESSIVE or ILLATIVE. Together with (Π21) *għal* 'for' however, they signal that the intended destination of the motion has not been rewerence.

(5.51) DIRECTIONAL II
(a) [Korpus Malti 3.0; literature20]
Il-vann kien diġà reġa' beda miexi ***[għal***
DEF-van be already do_again begin go.PART **[for**
ġol-main road]$_{PP}$ *meta dan l-anzjan donnu qam*
in:DEF-main_road] when DEM:M:PROX DEF-senior seem:3SG.M rise
mir-raqda
from:DEF-sleep
'The van had already begun again to drive [**into** the main road] when this senior seemed to wake up from sleep [. . .].'

(b) [Korpus Malti 3.0; news111887]
Kien b' dawn is-sentimenti li soqt ***[għal***
be with DEM:PL:PROX DEF-feeling:PL that drive:1SG.PFV **[for**
ġewwa *l-parkeġġ]*$_{PP}$ *nhar it-3 ta' Frar*
inside DEF-car park] day DEF-3 of February
'It was with these feelings that I drove [**towards** the car-park] on the 3rd of February.'

In the (a)-sentence the Figure is about to enter the main road which represents the Ground. However, the motion event is not concluded, meaning the van is still in the process of moving into the main road. This fact motivates the use of (П21) *għal* 'for' which, as mentioned above, signals that the Goal has not been reached yet. Without (П21) *għal* 'for' the interpretation of the spatial situation would suggest that the van has already entered the main road. Similarly, in (5.51b), the focus is on the beginning of the protagonist's drive towards the car-park which he will eventually enter into (as explained in sentences following the example in the original text). Thus, in the early phase of the journey, the event of entering into the Ground is only a possibility not a fact since on the way to the intended destination the Figure could be made not to continue the journey.

In (5.52), (П21) *għal* 'for' takes a place name as its complement.

(5.52) DIRECTIONAL general – [Korpus Malti 3.0; literature77]
Minn hawn immur Katanja għax triq oħra ma
from here 1SG.IPFV:go Catania because street other:F NEG
għandix nitma f' Alla li nkun nista'
have:1SG:NEG 1SG.IPFV:hope in God that 1SG.IPFV:be.FUT 1SG.IPFV:can
minn hemm nibqa' tiela' ***[għal*** *Venezja fejn*
from there 1SG.IPFV:remain ascend.PART **[for** Venice where
irrid immur ngħix]$_{PP}$
want go 1SG.IPFV:live]
'From here I go to Catania because I have no other choice; I hope to God that I will be able to continue to go up from there [**to** Venice where I want to go and live].'

In this sentence, the employment of (П21) *għal* 'for' indicates that on the one hand, the distance from Catania in the extreme south of Italy to Venice in the northeast of the same country is great and the final destination might not be reached in the end. On the other hand, the final relative clause in (5.52) suggests that the speaker intends to enter Venice because she has planned to settle down there.

Both (Π21) *għal* 'for' and (Π28) *lejn* 'towards' can be used for the expression of the DIRECTIONAL. Their competition calls for a separate study at a later point in time. In this section, we make do with checking the occurrences of the above Πs with five place names, namely the island *Sqallija* 'Sicily' and the four capital cities *Berlin* 'Berlin', *Londra* 'London', *Pariġi* 'Paris', and *Ruma* 'Rome' in the Korpus Malti 3.0. The results are disclosed in Figure 5.11.

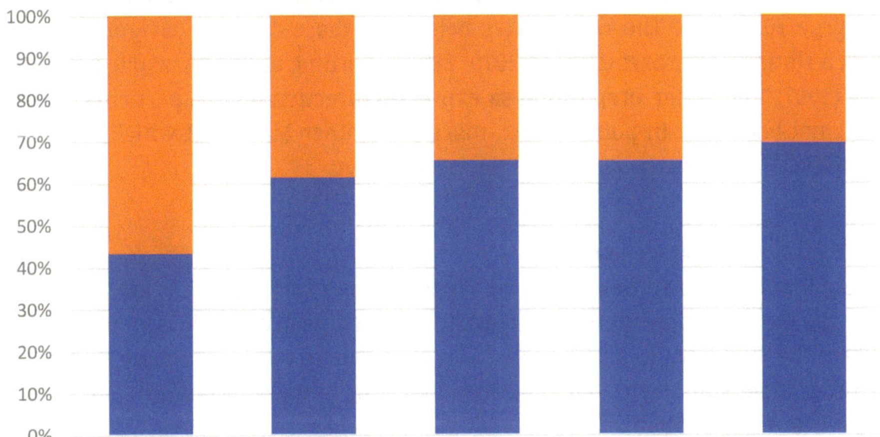

	Berlin	Sqallija	Ruma	Londra	Pariġi
għal	17	201	92	66	28
lejn	13	319	174	124	64

Figure 5.11: Competition of (Π21) *għal* 'for' and (Π28) *lejn* 'towards' with selected place names in the Korpus Malti 3.0.

In four out of five cases, (Π28) *lejn* 'towards' produces more matches than its competitor. Only in the case of *Berlin* 'Berlin', it is the other way around. However, *Berlin* 'Berlin' counts the lowest number of occurrences, namely only thirty, whereas the last but one place namely *Pariġi* 'Paris' is attested ninety-two times, i.e. three times as often as *Berlin* 'Berlin'. We conclude that for the DIRECTIONAL (Π28) *lejn* 'towards' is the preferred option. In contrast, for combinations with (Π22) *għand* 'at s.o.'s place', (Π21) *għal* 'for' is the only option. It is also worth noting that the two Πs behave differently in combination with certain verbs. The Korpus Malti 3.0 provides evidence of a draw between the competitors in the case of *daret* 'she turned' + $\Pi_{\text{DIRECTIONAL}}$. (Π28) *lejn* 'towards' is attested twenty-nine times as opposed to twenty-eight occurrences of (Π21) *għal* 'for'. If we take account of *telaq* 'he left' + $\Pi_{\text{DIRECTIONAL}}$, the picture changes since (Π28) *lejn* 'towards' fills the Π-slot 325 times whereas (Π21) *għal* 'for' yields only 130 hits. The combination *ħares* 'he looked' + $\Pi_{\text{DIRECTIONAL}}$ (which fails the criteria of a genuinely spatial situation) is the exclusive monopoly of (Π28) *lejn* 'towards'. This Π accounts for all of the 678 occurrences of the pattern. How these discrepancies can be interpreted is a matter that we cannot go into further in this study.

5.1.2.2.7 TERMINATIVE

Hagège (2010: 293) only illustrates non-spatial uses of the TERMINATIVE which otherwise is equated with English *down to* and *as far as* (Hagège 2010: 261) and *up to* (Hagège 2010: 287). The TERMINATIVE belongs to those dynamic categories which lack a static counterpart (Hagège 2010: 287). According to Borg/Azzopardi-Alexander (1997: 156), "[t]he preposition **sa** expresses direction up to a particular point" [original boldface]. In point of fact, there is no other Maltese Π which competes with (Π51) *sa* 'till' in the domain of the TERMINATIVE.

5.1.2.2.7.1 TERMINATIVE – spatial

The spatial TERMINATIVE applies if the Figure moves exactly as far as the Ground so that the movement comes to an end at the Goal. A typical example is given in (5.53) where the Figure walks a certain distance to an Indian takeaway where he intends to order food. Thus, he has no intention of walking past the restaurant. The Salaam Namaste is his destination.

(5.53) TERMINATIVE – spatial [Camilleri 2013: 72]

mexa	*l-erba'*	*mitt*	*metru*	*[**sa**s-Salaam Namaste]$_{PP}$*
walk	DEF-four	hundred	metre	[**till**:DEF-Salaam Namaste]

'[. . .] he walked the 400 metres [**up to** the Salaam Namaste].'

The domain of the TERMINATIVE is not limited to cases like (5.53) where the Π directly governs the Ground NP. As a matter of fact, (Π51) *sa* 'till' behaves like the DIRECTIONAL (Π21) *għal* 'for' insofar as it can express directionality in combination with Πs which are otherwise responsible for the INESSIVE or the ILLATIVE as shown in (5.54).

(5.54) (Π51) *sa* 'till' + $\Pi_{\text{ILLATIVE/INESSIVE}}$

(a) [Korpus Malti 3.0; news207117]

L-ibbuljar	*jista'*	*ma*	*jkunx*	*biss*
DEF-bullying	3SG.M.IPFV:can	NEG	3SG.M.IPFV:be.FUT:NEG	only

dak	*fiżiku*	*izda*	*jista'*	*jkun*	*ukoll*
DEM:M:DIST	physical	but	3SG.M.IPFV:can	3SG.M.IPFV:be.FUT	also

verbali *emozzjonali* *u* *anke* *minn* *fuq* *l-internet* *fejn*
verbal emotional and even from on DEF-internet where
l-ibbuljar *jasal* ***[sa*** ***<u>ġo</u>*** *<u>d-dar</u>* *<u>tal-vittma</u>]$_{PP}$*
DEF-bullying 3SG.M.IPFV:arrive **[till** **<u>in</u>** <u>DEF-house</u> <u>of:DEF-victim</u>]
'The bullying cannot only be physical but it can also be verbal, emotional and even via the internet where the bullying arrives [**in** <u>the victim's house</u>].'

(b) [Korpus Malti 3.0; literature78]
deherlu *li* *kliemu* *ma* *kienx* *jista'*
appear:IO:3SG.M that word.COLL:3SG.M NEG be:NEG 3SG.M.IPFV:can
jasal ***[sa*** ***<u>ġewwa</u>*** *<u>s-sala</u>]$_{PP}$*
3SG.M.IPFV:arrive **[till** **<u>inside</u>** <u>hall</u>]
'[. . .] it seemed to him that his words could not carry [**as far as** **<u>inside</u>** <u>the hall</u>].'

The difference between the DIRECTIONAL as characterised in the foregoing section and the TERMINATIVE is that the former is insensitive to the motion event's termination. In contrast, the use of (Π51) *sa* 'till' is crucially associated with the Figure arriving at the Ground no matter that the negative polarity in (5.54b) suggests that the Figure (= *kliemu* 'his words') does not reach the Ground (= *(i)s-sala* 'the hall').

The parallels between the DIRECTIONAL and the TERMINATIVE go further. In Table 5.4, we see that (Π51) *sa* 'till' combines freely with each of those Πs which are otherwise employed to express the PROPINQUUS or the PROPINQUELATIVE.[53] All examples are from the Korpus Malti 3.0.

Table 5.4: (Π51) *sa* 'till' as directionality marker with PROPINQUUS Πs.

source	example / translation	configuration
news105455	*dakinhar li seħħ l-aċċident, mar* ***[sa <u>fejn</u>*** *<u>il-Bajja</u> <u>ta'</u> <u>Spinola</u>]$_{PP}$ biex jixtri l-ħut lill-mara* '[. . .] on the day the accident happened he went [**as far as <u>near</u>** <u>Spinola Bay</u>] to buy fish for his wife.'	(Π14) *fejn* 'near'
news49537	*għen lil martu taħsel il-ħwejjeġ u wara mar* ***[sa <u>ħdejn</u>*** *<u>sħabu</u>]$_{PP}$ l-klabb tal boċċi* '[. . .] he helped his wife with the washing and then went [**to** <u>his friends</u>] at the boccia club.'	(Π23) *ħdejn* 'beside'
literature79	*iwassluhom* ***[sa <u>maġenb</u>*** *<u>il-kaless</u>]$_{PP}$ u telqu flimkien għall-Belt* '[. . .] they brought him [**as far as** <u>the coach</u>] and left together for Valletta [. . .].'	(Π32) *maġenb* 'close to'

53 As discussed in Sections 5.1.2.1.4 and 5.1.2.2.5, respectively.

Table 5.4 (continued)

source	example / translation	configuration
news147538	*huwa rnexxielu jasal [**sa** **qrib** Malta]$_{PP}$* '[. . .] he succeeded to get [**as far as** **close to** Malta].'	(Π48) *qrib* 'near'
parl9611	*ħadt sehem f' mixja li mill-Għammieri wasslet [**sa** **viċin** il-knisja ta' San Bastjan f' Ħal Qormi]$_{PP}$* '[. . .] I took part in a walk that led us from Għammieri [**as far as** **close to** the church of San Bastjan in Ħal Qormi].'	(Π58) *viċin* 'near'

In Table 5.4, the use of (Π51) *sa* 'till' as indicator of directionality implies that the Figure arrives close to the Ground. If (Π51) *sa* 'till' had been replaced with (Π21) *għal* 'for' this reading would not be coerced.

There are also very infrequent examples of (Π51) *sa* 'till' combining with (Π15) *fi* 'in' conveying a TERMINATIVE meaning as is the case in (5.55).

(5.55) (Π51) *sa* 'till' + (Π15) *fi* 'in' – [Korpus Malti 3.0; academic306]

L-Opossum sufi ta' Derby [. . .] huwa mammiferu marsupjal
DEF-opposum woolly of Derby 3SG.M mammifer marsupial
tal-familja Didelphidae [. . .] nativ tal-kontinenti
of:DEF-family Didelphidae native of:DEF-continent:PL
Amerikani b' firxa minn Nofs in-nhar tal-Messiku
American:PL with spread from South of:DEF-Mexico
*[**sa** fit-Tramuntana ta' l-Ekwador]$_{PP}$*
[**till** in:DEF-North of DEF-Ecuador]
'The woolly opossum of Derby [. . .] is a mammifer marsupial of the family Dedelphidae [. . .] native of the Americas with a diffusion from southern Mexico [**into** the north of Ecuador].'

Even if *sa-fi* combinations are relatively uncommon according to the data gathered from the Korpus Malti 3.0, the findings support the hypothesis that the TERMINATIVE marker (Π51) *sa* 'till' is largely compatible with most of the Πs that serve the purpose of expressing configuration.

Accordingly, it cannot surprise us to find that like (Π21) *għal* 'for', (Π51) *sa* 'till' too can combine with (Π22) *għand* 'at s.o.'s place' as illustrated in (5.56).

(5.56) TERMINATIVE – spatial [Korpus Malti 3.0; news66688]
Duminku ħaseb li marret ***[sa*** ***għand*** *ommha]*$_{PP}$
Duminku think that go:3SG.F.PFV **[till** **at s.o.'s place** mother:3SG.F]
'[. . .] Duminku thought that she went [**to** her mother's] [. . .].'

Duminku's wife never arrived at her mother's place. Nevertheless, (Π51) *sa* 'till' is used and not (Π21) *għal* 'for' probably because Duminku thought that she arrived at her destination, meaning: the situation is described from Duminku's vantage point. We assume that (Π51) *sa* 'till' is optional in contexts like that in (5.55) because (Π22) *għand* 'at s.o.'s place' alone is compatible with an APUDLATIVE meaning. What motivates the employment of (Π51) *sa* 'till' is the long duration of the wife's absence in the sense that (Π51) *sa* 'till' suggests that the distance between Duminku's home and the place where his mother-in-law lives is so great that it takes time to travel between the two locations.

5.1.2.2.7.2 TERMINATIVE – temporal

Borg/Azzopardi-Alexander (1997: 171) mention two temporal contexts in which (Π51) *sa* 'till' can be used, namely "duration up to a point which may be in the past" or "in the future". The Korpus Malti 3.0 hosts many examples of (Π51) *sa* 'till' taking a temporal expression as its complement as shown in (5.57). There are examples for different calendaric and other temporal categories which suggest that there are no limits to the compatibility of this Π with all kinds of temporal expressions.

(5.57) TERMINATIVE – temporal
(a) time of day – [Korpus Malti 3.0; news162983]
baqgħet sejra ***[sa*** *nofs il-lejl nieqes ħames*
remain:3SG.F.PFV go.PART:F **[till** half DEF-night lack.PART five
minuti]$_{PP}$
minute:PL]
'[. . .] she went on [**till** five minutes to midnight].'
(b) day of the week – [Korpus Malti 3.0; culture1939]
Il-ġbid tal-film mistenni jibda min-nhar
DEF-pulling of:DEF-film PART:expect 3SG.M.IPFV:begin from:DEF-day
it-Tnejn li ġej u jibqa' għaddej
DEF-Monday that come.PART and 3SG.M.IPFV:remain pass:PART
[sa*t-Tnejn 10 ta' Novembru 2014]*$_{PP}$
[till:DEF-Monday 10 of November 2014]
'The shooting of the expected movie starts next Monday and keeps going [**till** Monday, 10 November 2014].'

(c) month – [Korpus Malti 3.0; news2112]
Dan *it-test* *issa* *jista'* *jiġi*
DEM:M:PROX DEF-test now 3SG.M.IPFV:can 3SG.M.IPFV:come
pospost ***[sa*** *Mejju* *li* *ġej]*$_{PP}$
postpone:PART [**till** May that come.PART]
'This test can now be postponed [**till** next May].'

(d) season – [Kopus Malti 3.0; academic13]
Din *hi* *pjanta* *li* *tagħmel* *il-fjuri* *tagħha*
DEM:F:PROX 3SG.F plant that 3SG.F.IPFV:make DEF-blossom:PL of:3SG.F
fir-rebbiegħa *u* *tibqa'* *sejra* *[****sa****s-sajf]*$_{PP}$
in:DEF-spring and 3SG.F.IPFV:remain go.PART:F [**till**:DEF-summer]
issebbaħ *il-post* *fejn* *tkun* *bl-ilwien*
3SG.F.IPFV:beautify DEF-place where 3SG.F.IPFV:be.FUT with:DEF-colour.PL
tal-fjuri *tagħha*
of:DEF-blossom:PL of:3SG.F
'This is a plant that grows its blossoms in spring and [**till** summer] goes on to beautify the place where it is with the colours of its blossoms.'

(e) year – [Korpus Malti 3.0; news5684]
Cameron *ikun* *jista'* *jmexxi*
Cameron 3SG.M.IPFV:be.FUT 3SG.M.IPFV:can 3SG.M.IPFV:go:CAUS
*[****sa****s-sena* *d-dieħla]*$_{PP}$
[**till**:DEF-year DEF-enter.PART:F]
'[. . .] Cameron will be able to lead [**till** the coming year].'

Section 5.1.2.1.1.2 has revealed that the INESSIVE markers (Π18) *ġewwa* 'inside' and (Π19) *ġo* 'in' are excluded from the domain of the temporal INESSIVE. Their absence from that of the temporal TERMINATIVE is only to be expected. However, there is marginal evidence of Π-Π combinations involving (Π51) *sa* 'till' for directionality and (Π15) *fi* 'in' for configuration. One of these rare cases is (5.58).

(5.58) (Π51) *sa* 'till' + (Π15) *fi* 'in' – [Korpus Malti 3.0; news47036]
Oħrajn *ħarġu* *bit-teorija* *li* *din* *is-sena*
other:PL exit:3PL.PFV with:DEF-theory that DEM:F:PROX DEF-year
għandna ***[sa*** ***f'*** *Frar]*$_{PP}$ *bix-xemx*
have.IPFV:1PL [**till** **in** February] with:DEF-sun
'Others put forward the theory that this year we will have sun [**till** **in** February].'

Instances of temporally employed *sa-fi* combinations are by no means more numerous than their spatial equivalents illustrated in (5.55). What these combinations express is that the Figure moves into the Ground but comes to a stop somewhere inside the Ground. Used without configuration marker, (Π51) *sa* 'till' indicates that the Ground is reached but not necessarily entered into.

5.1.2.2.8 Ablative

Under this rubric, we treat the category that Hagège (2010: 286–287) calls ABLATIVE and which presupposes that the Figure "move[s] from" the Ground. It is not entirely clear to us how this category relates to the ADESSIVE and ALLATIVE in Hagège's system. If a motion event starts near the Ground but not necessarily within the Ground, the category could alternatively be labelled PROXIABLATIVE. We keep Hagège's term but include the PROPINQUUS categories in this section. Our source does not postulate a separate temporal ABLATIVE. On account of other cases of the spatial-temporal distinction, however, we speak also about the spatial ABLATIVE in separate subsections [54]

5.1.2.2.8.1 Ablative – spatial

In general, motion from is expressed by (Π41) *minn* 'from' (Borg/Azzopardi-Alexander 1997: 156). As we know already,[55] this Π is also responsible for the ELATIVE. It is not always easy to tell ELATIVE and ABLATIVE apart. As to (Π41) *minn* 'from' in (5.59), it may be argued that the Figure has not been inside a container-like Ground before he moved away from it.

(5.59) ABLATIVE spatial – [Attard 2019: 294]

Spiċċa	*qam*	*[**minn***	*<u>postu</u>]*$_{PP}$	*u*	*ħaffef*	*telaq*	*'il*	*barra*
finish	rise	[**from**	<u>place:3SG.M</u>]	and	hurry	leave	to	outside

'He finally rose [**from** <u>his place</u>] and hurriedly went outside.'

Yet, it is clear that the Figure was not only in the vicinity of the Ground but Figure and Ground must have been in contact prior to the motion event. This is why examples of this kind do not fit the description of a PROXIABLATIVE.

This description would be fine in cases where (Π41) *minn* 'from' serves as directionality marker in combination with most of those Πs which otherwise express the static PROPINQUUS as shown in (5.60). These cases will be registered under *motus ex loco*.

---

54 In Section 5.1.2.2.8.1 first and then we continue with the temporal ABLATIVE in Section 5.1.2.2.8.2.

55 From Section 5.1.2.2.2.

(5.60) (PROPINQUE)ABLATIVE

(a) [Korpus Malti 3.0; news146799]

Qal li wara li Bone telaq ***[minn*** ***fejn*** *ir-razzett]*$_{PP}$
say that after that Bone leave **[from** **near** DEF-farmhouse]
hu beda jsegwih bil-karozza
3SG.M begin 3SG.M.IPFV:follow:3SG.M with:DEF-car
'He said that after Bone departed [**from** **the vicinity** of the farmhouse] he began to follow him in the car.'

(b) [Korpus Malti 3.0; literature35]

kulħadd telaq ***[minn*** ***ħdejn*** *il-bieb]*$_{PP}$ *u mar*
everybody leave **[from** **beside** DEF-door] and go
jilgħab band' oħra
3SG.M.IPFV:play side other:F
'[. . .] everybody moved away [**from** **the vicinity** of the door] and went to play elsewhere.'

(c) [Korpus Malti 3.0; news162787]

il-vetturi kollha li jitilqu mill-ajruport
DEF-vehicle:PL all:3SG.F that 3.IPFV:leave:PL from:DEF-airport
se jkollhom jgħaddu mill-parkeġġ
FUT 3.IPFV:have.FUT:3PL 3.IPFV:pass:PL from:DEF-parking_space
tal-kowċis u joħorġu ***[minn*** ***qrib*** *il-Mc Donald's]*$_{PP}$
of:DEF-coach:PL and 3.IPFV:exit:PL **[from** **near** DEF-McDonald's]
fid-direzzjoni lejn il-mina ta' Ħal Kirkop
in:DEF-direction towards DEF-tunnel of Ħal Kirkop
'[. . .] all the cars which depart from the airport will have to pass through the parking space of the busses and exit [**close to** the McDonald's] in the direction of the tunnel of Ħal Kirkop.'

(d) [Korpus Malti 3.0; news145398]

Ingħad li dawn it-tiri kienu sparati
REFL:say that DEM:PL:PROX DEF-shot:PL be:3PL.PFV fire:PART:PL
[minn ***viċin*** *l-ajruport]*$_{PP}$
[from **near** DEF-airport]
'It is said that these shots were fired [**from** **near** the airport].'

In the reference grammar of Maltese, only the combinations *minn fejn* and *minn maġenb* are mentioned with regard to motion from a PROPINQUUS Ground (Borg/Azzopardi-Alexander 1997: 157). As far as *minn maġenb* is concerened, we have found evidence of this combination exclusively with Path meanings[56].[57] The distinction between ABLATIVE and PERLATIVE functions of (Π41) *minn* 'from' is frequently blurred. Example (5.60c) is ambiguous and might alternatively be interpreted as a situation in which the Figure passes near the Ground in lieu of exiting from it.

5.1.2.2.8.2 ABLATIVE – temporal

Posterior duration in the past and future requires the presence of (Π41) *minn* 'from' (Borg/Azzopardi-Alexander 1997: 172). This Π freely combines with all kinds of temporal expressions to mark the temporal ABLATIVE. The endpoint of the period need not be made explicit (as in (5.61a)) or can be expressed e.g. by a *sa*-PP (as in (5.61b)).

(5.61) ABLATIVE temporal

(a) [Korpus Malti 3.0; news118257]

tibda	*topera*	*żewġ*	*titjiriet*	*kuljum*	*lejn*
3SG.F.IPFV:begin	3SG.F.IPFV:operate	two	flight:PL	everyday	towards

Shanghai	***[minn***	*Frar*	*2008]*$_{PP}$
Shanghai	**[from**	February	2008]

'[. . .] it [= the airline] will begin to carry out two daily flights to Shanghai [**from** February 2008 (onwards)].'

---

56 Cf. Section 5.1.2.2.20.

57 In the Korpus Malti 3.0, one comes across examples like (5.ii) which superficially seem to invite an ABLATIVE interpretation but can be shown to mean something different.

(5.ii) [Korpus Malti 3.0; literature59]

Bdew	*jispontaw*	*nies*	*minn*	*karozzi*	*warajhom*	***[minn***
begin:3PL.PFV	3.IPFV:sprout:PL	people	from	car:PL	after:3PL	**[from**

[Ø	***maġenbhom]***$_{PP]PP}$	*minn*	*kullimkien*
[Ø	**close_to:3PL]**	from	everywhere

'People began to pop up from cars behind them, [**from** [Ø **on their side**]], from everywhere.'

In (5.ii), the second *minn*-PP provides an example of EQUI-N-deletion, i.e. the noun *karozzi* 'cars' is not repeated after its first occurrence. The zero sign indicates the position it would have had it not been deleted. According to this analysis *maġenbhom* 'close to them' is NP-internal and modifies the omitted noun. There is thus no instance of a Π-Π combination in the first place.

(b) [Korpus Malti 3.0; news128008]

Tliet	*xhur*	*itiru*	*malajr*	*daqs*	*kemm*
three	month.PL	3.IPFV:fly:PL	quickly	equal_to	how_much

it-tliet	*xhur*	*[**mi**s-sajf]$_{PP}$*	*[**sa**l-Milied]$_{PP}$*
DEF-three	month.PL	[**from**:DEF-summer]	[**till**:DEF-Christmas]

għaddew	*b'*	*leħħa*	*ta'*	*berqa*
pass:3PL.PFV	with	flash	of	lightning

'Three months are over quickly just as the three months [**from** the summer] [**till** Christmas] passed like a flash of lightning.'

As these and further examples strongly suggest there is no reason to pass tacitly over the existence of temporal meanings associated with the ABLATIVE.

5.1.2.2.9 Motion (to and from) opposite the Ground

Beginning with this section, the presentation of the spatio-temporal meanings follows a different pattern. Since Hagège (2010) only offers terms for the Goal-oriented motion events, the need arises to create distinctive terms for the Source-oriented motion events. As our prior attempts in the foregoing sections show, the new terms are often unwieldy – and some of them distort the Latin language to such a degree that it seems to be better to revert to descriptive paraphrases in lieu of coining one-word terms. We seise the opportunity to organise the sections to come along the following lines. For each configuration, we first present evidence of motion to the Ground. Motion to the Ground will be divided into three subsections, namely a section looking at the use of the bare single Π followed by a section on the DIRECTIONAL and another on the TERMINATIVE. There will also be a separate section on motion from the Ground to round off the phenomenology of a given configuration.

5.1.2.2.9.1 OBVERSIVE

Hagège (2010: 286) circumscribes the OBVERSIVE as applying to situations in which the Figure "moves[s] to a position in front of [a place]". According to the ideas expressed in connection with the distinction of the OBESSIVE and the FRONTALIS,[58] we differentiate between the OBVERSIVE tout court which involves a Figure which moves to a position facing or opposite the Ground and the dynamic FRONTALIS where the motion event carries the Figure to a position which is in front of the Ground. The distinguishing factor is the perspective: the position of the Figure opposite the Ground applies if the situation is looked at from the point of view of the Ground

58 Cf. Sections 5.1.2.1.5 and 5.1.2.1.6.

(or that of an observer who ideally is placed behind the Ground) whereas being in front of a Ground means that the situation is interpreted from the point of view of the Figure (or an observer who ideally is situated behind the Figure). Motion in front of the Ground is the topic of a later section.[59] In this section, however, we focus on spatial situations involving motion to and from opposite a Ground.

5.1.2.2.9.1.1 Bare OBVERSIVE

Like in the case of the static OBESSIVE, Borg/Azzopardi-Alexander (1997: 162 and 167) do not classify (Π8) *biswit* 'facing' and (Π12) *faċċata* 'opposite' the same. In their view, only the latter expresses the notion of moving to a position opposite the Ground whereas (Π8) *biswit* 'facing' is mentioned under the rubric of motion to a lateral position relative to a Ground. No example for this particular usage of (Π8) *biswit* 'facing' is given in the reference grammar. In our interpretation, the above two Πs are largely synonymous and serve the purpose of identifying the configuration – in this case the opposite side of the Ground. Directionality needs not be made explicit as shown in (5.62).

(5.62) Bare OBVERSIVE

(a) [Korpus Malti 3.0; literature33]

Mar ***[biswit*** *il-knisja]$_{PP}$*

go **[facing** DEF-church]

'He went [**opposite** the church].'

(b) [Korpus Malti 3.0; news131958]

Id-diżel ħareġ minn scammel li ġratlu

DEF-diesel exit from scammel that happen:3SG.F.PFV:IO:3SG.M

l-ħsara kif wasal ***[faċċata*** *tal-pompa tal-petrol*

DEF-damage how arrive **[opposite** of:DEF-pump of:DEF-petrol

fid-direzzjoni ta' Raħal Ġdid]$_{PP}$

in:DEF-direction of Raħal Ġdid]

'The diesel came out of a scammel that was damaged when it arrived [**opposite** the petrol pump in the direction of Raħal Ġdid].'

Not much needs to be said in terms of explanation since the bare OBVERSIVE reflects the general pattern of zero-marking of directionality in Goal-oriented constructions. Constructions in which directionality is made explicit in combination with either (Π8) *biswit* 'facing' or (Π12) *faċċata* 'opposite' are very infrequent in the

59 Namely Section 5.1.2.2.10.

Korpus Malti 3.0. Some of the cases we present in the next two subsections are hapaxes while others belong to a handful of similar cases.

5.1.2.2.9.1.2 Directional obversive

In contrast to the bare OBVERSIVE, overtly marked DIRECTIONAL OBVERSIVE s are attested only in small numbers. It is therefore no surprise that no mention is made of them in Borg/Azzopardi-Alexander (1997). The examples in (5.63) almost exhaust our harvest that results from a search in the Koprus Malti 3.0. Note that we have been unable to identify examples of (Π28) *lejn* 'towards' being used as directionality marker. There is evidence exclusively of (Π21) *għal* 'for' being employed in this function.

(5.63) DIRECTIONAL OBVERSIVE

(a) [Korpus Malti 3.0.; literature77]

mexa	*mbagħad*	*mat-triq*	***[għal***	***<u>biswit</u>***	*<u>id-dar</u>*	*<u>ta'</u>*	*<u>Lippu]</u>*$_{PP}$
walk	then	with:DEF-street	**[for**	**<u>facing</u>**	<u>DEF-house</u>	<u>of</u>	<u>Lippu]</u>

'[. . .] he then walked along the street [**to <u>opposite</u>** <u>Lippu's house</u>].'

(b) [Korpus Malti 3.0; parl177]

Nissuġġerixxi [. . .]	*li*	*tinfetaħ*	*triq*	*ġdida*	*min-naħa*
1SG.IPFV:suggest	that	3SG.F.IPFV:REFL:open	street	new:F	from:DEF-side

tal-Mina	*tas-Sultan*	*jiġifieri*	*min-naħa*	*ta'*	*Bormla*
of:DEF-tunnel	of:DEF-king	that_is	from:DEF-side	of	Bormla

[għal	***<u>faċċata</u>***	*<u>ta'</u>*	*<u>Lourdes]</u>*$_{PP}$	*minn*	*wara*	*Triq Santa Duminka*
[to	**<u>opposite</u>**	<u>of</u>	<u>Lourdes]</u>	from	after	Triq Santa Duminka

'I suggest [. . .] that a new street is opened from the side of the King's Tunnel, that is from the Bormla side [**to <u>opposite</u>** <u>Lourdes</u>] from behind Triq Santa Duminka.'

In both sentences in (5.63) a Figure moves or is moved metaphorically to a position which is situated opposite the Ground. In the (a)-sentence it is the subject referent that walks towards Lippu's house and stops opposite the Goal. In the (b)-sentence a new street is envisaged to be built from the city of Bormla to an endpoint which is located opposite the church named after Lourdes.

5.1.2.2.9.1.3 Terminative obversive

The TERMINATIVE OBVERSIVE is as infrequent as its directional equivalent. The examples in (5.64) cover already a good part of the entire turnout of hits found in the Korpus Malti 3.0.

(5.64) TERMINATIVE OBVERSIVE
(a) [Korpus Malti 3.0; literature77]

il-għarajjes	*telgħu*	*flimkien*	*[sa*	***biswit***	*l-altar*
DEF-bride_and_groom	ascend:3PL.PFV	together	[**till**	**facing**	DEF-altar
il-kbir]$_{PP}$					
DEF-big]					

'[. . .] bride and groom together ascended [**to opposite** the main altar].'

(b) [Korpus Malti 3.0; parl3759]

bejn	*Ottubru*	*2000*	*u*	*Ġunju*	*2001*	*tlestiet*
between	October	2000	and	June	2001	REFL:get_ready:3SG.F.PFV

it-tieni	*fażi*	*minn*	*fejn*	*it-Torri*	*[sa*	***faċċata***
DEF-second	phase	from	near	DEF-tower	[**till**	**opposite**
tal-Kunvent	*tal-Karmnu]*$_{PP}$					
of:DEF-Convent	of:DEF-Karmnu]					

'[. . .] between October 2000 and June 2001, the second phase was finished from near the Tower [**to opposite** the Convent of Karmnu].'

The differences between the DIRECTIONAL OBVERSIVE and the TERMINATIVE OBVERSIVE are subtle. The DIRECTIONAL emphasises the process which is characteristic of the motion event whereas the TERMINATIVE highlights the endpoint of the very same process.

5.1.2.2.9.2 Motion from opposite the Ground

Borg/Azzopardi-Alexander (1997: 162 and 167) mention the Π-Π combinations *minn biswit* and *minn faċċata* for the expression of motion from although in the former case, it is motion from a lateral position with respect to the Ground. Like in the previous cases, motion from opposite a given Ground is rarely encountered in the Korpus Malti 3.0. We provide two examples in (5.65) under the proviso that in (5.65a) it remains unclear to us where exactly the water exits from the farms. It cannot be excluded that motion from a lateral position is the intended reading.

(5.65) Motion from opposite the Ground
(a) [Korpus Malti 3.0; parl3111]

Mhux	*l-ewwel*	*darba*	*meta*	*tagħmel*	*ix-xita*
NEG:3SG.M:NEG	DEF-first	time	when	3SG.F.IPFV:make	DEF-rain

li	*l-ilma*	*tax-xita*	*li*	*jkun*	*ħiereġ*
that	DEF-water	of:DEF-rain	that	3SG.M.IPFV:be.FUT	exit.PART

[minn ***biswit*** *il-farms]$_{PP}$ ikun fih riħa qawwija*
[**from** **facing** DEF-farm:PL] 3SG.M.IPFV:be.FUT in:3SG.M smell strong
ta' awrina u ta' ħmieġ
of urine and of dirt
'It is not the first time that when it rains the rainwater that would come out [(**from**) **opposite** the farms] has a strong odour of urine and dirt.'

(b) [Korpus Malti 3.0; news152940]
qegħdin imorru jistennew biex jaqbdu tal-linja
PROG:PL 3.IPFV:go:PL 3.IPFV:wait:PL to 3.IPFV:grab:PL of:DEF-line
[minn ***faċċata*** *tal-Università]$_{PP}$*
[**from** **opposite** **of**:DEF-university]
'[. . .] they are going to wait and catch a bus [**from** **opposite** the university].'

According to Borg/Azzopardi-Alexander (1997: 167), the linking (Π55) *ta'* 'of' is absent from the PPs which express motion from with the configuration represented by (Π12) *faċċata* 'opposite' – a topic we have addressed already above.[60] There is variation as to the presence and absence of (Π55) *ta'* 'of' in contexts like that illustrated by (5.65b).

5.1.2.2.10 Motion in front of the Ground

We are now in the domain of the FRONTALIS, i.e. the front region of the Ground as viewed from the Figure's perspective (or that of an observer behind the back of the Figure). Borg/Azzopardi-Alexander (1997: 159) mention (Π49) *quddiem* 'in front of' as the appropriate Π for expressing configuration in connection to the front region.

5.1.2.2.10.1 Goal-oriented motion

5.1.2.2.10.1.1 Bare configuration

As can be expected on the basis of the information obtained hitherto, motion to the front region does not require the presence of a Π which marks directionality. Given that a fitting motion verb is involved, (Π49) *quddiem* 'in front of' alone suffices to convey the meaning of the Goal-directed FRONTALIS as shown in (5.66).

60 Cf. Section 4.1.2.2.1.2.1 (ii).

(5.66) Goal-directed FRONTALIS – [Korpus Malti 3.0; literature33]

Għalhekk reġa' niżel isfel bix-xlief
for_this_reason do_again descend down with:DEF-fishing_line
*f' idu u mar [**quddiem** il-mera]$_{PP}$*
in hand:3SG.M and go [**in_front_of** DEF-mirror]
'For this reason, he went down again with the fishing line in his hand and went [**in front of** the mirror].'

The cooccurrence of bare (Π49) *quddiem* 'in front of' after the 3rd person singular masculine perfective *mar* 'go' is attested 325 times in the Korpus Malti 3.0 whereas there is no evidence of this motion verb to form part of constructions of the DIRECTIONAL FRONTALIS or the TERMINATIVE FRONTALIS.

5.1.2.2.10.1.2 DIRECTIONAL FRONTALIS

The possibility of (Π49) *quddiem* 'in front of' combining with the directionality markers (Π21) *għal* 'for' and (Π28) *lejn* 'towards' is ignored in the oft-cited reference grammar. In the Korpus Malti 3.0, the combination of the two Πs (Π21) *għal* 'for' and (Π49) *quddiem* 'in front of' yields 554 hits. However, many of these hits involve adverbial *quddiem* and thus do not count as instances of the Π. In addition, in these combinations, (Π21) *għal* 'for' is frequently used in its PURPOSIVE meaning so that no spatial situation is described, in the first place. This means that properly spatial instances of *għal quddiem* are not as numerous as the above absolute numbers might suggest. The combination of (Π28) *lejn* 'towards' and (Π49) *quddiem* 'in front of' is comparatively infrequent. The adverbial usages are the most numerous also for this combination. Only three instances can count as PPs according to the definition of the canonical Π.

Bona fide examples of genuine spatial situations are given in (5.67) with the (a)-sentence documenting the use of (Π21) *għal* 'for' whereas the (b)-sentence features (Π28) *lejn* 'towards'.

(5.67) DIRECTIONAL FRONTALIS

(a) [Korpus Malti 3.0; literature21]

*Marku qabeż minn fuq il-blata [**għal quddiem** Mikiel]$_{PP}$*
Marku jump from on DEF-rock [**for in front of** Mikiel]
'Marku jumped down from the rock [**in front of** Mikiel].'

(b) [Korpus Malti 3.0; news151711]
Wieħed jissuġġerixxi wkoll li l-pedestrian crossings
one 3SG.M.IPFV:suggest also that DEF-pedestrian_crossing:PL
eżistenti jitressqu ***[lejn*** ***quddiem***
exist:PART:PL 3.IPFV:REFL:approach:CAUS:PL **[towards** **in_front_of**
*il-bieb prinċipali taż-żewġ lukandi kbar]*PP
DEF-door principal of:DEF-two hotel:PL big.PL]
'Someone also suggests that the extant pedestrian crossings are moved closer [**towards** the main entrance of the two hotels] [. . .].'

5.1.2.2.10.1.3 TERMINATIVE FRONTALIS

Like in the previous case, the possibility of (Π49) *quddiem* 'in front of' combining with the directionality marker (Π51) *sa* 'till' to express the TERMINATIVE FRONTALIS is no topic in the reference grammar. There is however evidence of the Π-Π combination. We have found 187 matches for the combination in the Korpus Malti 3.0. In contrast to the DIRECTIONAL FRONTALIS, the bulk of these cases can be classified as spatial.

A typical example is given in (5.68). Note that the complement involved in the PP of interest is the same as in (5.67).

(5.68) TERMINATIVE FRONTALIS – [Korpus Malti 3.0; literature43]
Kull tlett ijiem Jason kien qed iqum mis-sodda
each three day.PL Jason be PROG 3SG.M.IPFV:rise from:DEF-bed
u jimxi ***[sa*** ***quddiem*** *il-mera]*PP *jitlob*
and 3SG.M.IPFV:walk **[till** **in front of** DEF-mirror] 3SG.M.IPFV:pray
'l Alla
to God
'Every third day Jason rose from the bed and walked [**up to** the mirror] to pray to God.'

We assume that the use of (Π51) *sa* 'till' is optional since the bare configuration marker in combination with a motion verb like *mexa* 'walk' could achieve the same.

5.1.2.2.10.2 Motion away from the front region

Borg/Azzopardi-Alexander (1997: 159) state that (Π49) *quddiem* 'in front of' must be preceded by (Π41) *minn* 'from' if motion from the front region is the intended meaning of the PP. With 2,733 matches in the Korpus Malti 3.0, the Π-Π combination *minn quddiem* is by far more frequent than the DIrectional and the TERMINATIVE FRONTALIS taken together. What is more, the vast majority of the hits yielded by *minn quddiem* receive a spatial interpretation (excluding cases of adverbial *quddiem*).

In (5.69) we provide an example in which the Figure (= the file) is removed from its prior position in front of the protagonist.

(5.69) Motion away from the front region – [Attard 2019: 83]

Jieħu	*l-fajl*	*tiegħu*	*f'*	*idu*	***[minn***
3SG.M.IPFV:take:3SG.M	DEF-file	of:3SG.M	in	hand:3SG.M	**[from**
***quddiemu]**$_{PP}$*	*u*	*jiftħu*			
in front of:3SG.M]	and	3SG.M.IPFV:open:3SG.M			

'He takes his file in his hand [**from in front of** him] and opens it.'

No alternative ways of expressing motion away from the front region are reported in the descriptive linguistic literature on Maltese.

5.1.2.2.11 Motion onto and down from the Ground

SURLATIVE is Hagège's (2010: 286) term for the meaning of a Π that is used in a spatial situation in which the Figure moves to a position on the Ground. The English translation provided by the same author varies from *onto, above, over* to *on* (Hagège 2010: 261 and 287). How English *above* fits the description of the motion event is not clear to us.

5.1.2.2.11.1 Motion onto the Ground

5.1.2.2.11.1.1 Bare SURLATIVE

Borg/Azzopardi-Alexander (1997: 160) argue that (Π17) *fuq* 'on' can be used without further directionality marker to indicate motion to a position on the Ground (which involves contact between Figure and Ground). The same Π is used for the SURESSIVE.[61] It reflects the usual pattern of distribution according to which if configuration is made explicit directionality need not be expressed overtly too.

This is exactly the case in (5.70) where the Figure goes onto the bridge with suicidal intentions. There is no overt marking of directionality.

(5.70) bare SURLATIVE – [Korpus Malti 3.0; culture1106]

Kevin Berthia	*spjega*	*li*	*f'*	*dik*	*il-ġurnata*	*fl-2005*
Kevin Berthia	explain	that	in	DEM:F:DIST	DEF-day	in:DEF-2005

61 As shown in Section 5.1.2.1.7.

hu *mar* ***[fuq*** *dak* *il-pont]*$_{PP}$ *għall-ewwel* *darba*
3SG.M go **[on** DEM:M:DIST DEF-bridge] for:DEF-first time
biex *itemm* *ħajtu*
to 3SG.M.IPFV:end life:3SG.M
'Kevin Berthia explained that on this day in 2005 he went [**onto** that bridge] for the first time to end his life.'

Many instances of *mar fuq* in the Korpus Malti 3.0 reflect the idiomatic meaning 'attack' and thus cannot be accepted as instances of spatial (Π17) *fuq* 'on'. However, even after removing cases of this kind from the turnout, there remain several hundreds of properly spatial usages of *mar fuq*.

5.1.2.2.11.1.2 DIRECTIONAL SURLATIVE

Nevertheless, it is possible to make directionality explicit also in combination with (Π17) *fuq* 'on' although this option is not mentioned in the relevant literature on Maltese. For this purpose, (Π21) *għal* 'for' combines with (Π17) *fuq* 'on' as shown in (5.71a). Example (5.71b) is the only piece of evidence of (Π28) *lejn* 'towards' being used as directionality marker for the DIRECTIONAL SURLATIVE.

(5.71) DIRECTIONAL SURLATIVE
(a) [Korpus Malti 3.0; european2419]
jservi *biex* *jidderieġi* *l-fluwidu*
3SG.M.IPFV:serve to 3SG.M.IPFV:direct DEF-liquid
tal-windscreenwasher ***[għal*** ***fuq*** *il-windscreen]*$_{PP}$
of:DEF-windscreenwasher **[for** **on** DEF-windscreen]
'[. . .] it serves to direct the liquid of the windscreenwasher [**onto** the windscreen].'
(b) [Korpus Malti 3.0; european4032]
il-linja [. . .] *tgħaddi* *għan-nofsinhar* *tal-gżira* *ta'* *Öland*
DEF-line 3SG.F.IPFV:pass for:DEF-south of:DEF-island of Öland
[lejn ***fuq*** *il-kosta* *tal-lvant]*$_{PP}$
[towards **on** DEF-coast of: DEF-east]
'[. . .] the line [. . .] passes to the south of the island of Öland [**onto** the eastern coast] [. . .].'

It is very likely that neither the use of (Π21) *għal* 'for' nor especially that of (Π28) *lejn* 'towards' is compulsory in contexts of this kind either.

5.1.2.2.11.1.3 Terminative surlative

The same seems to be true when we look at the combination of (Π51) *sa* 'till' and (Π17) *fuq* 'on' to express the TERMINATIVE SURLATIVE which is also absent from other descriptions of Maltese Πs. The presence of (Π51) *sa* 'till' in (5.72) is a case in point.

(5.72) TERMINATIVE SURLATIVE – [Korpus Malti 3.0; literature32]

telgħu	*[sa*	***fuq***	*il-bejt]*$_{PP}$	*biex*	*jaraw*	*it-temp*
climb:3SG.PFV	**[till**	**on**	DEF-roof]	to	3.IPFV:see:PL	DEF-weather
kif	*inhu*					
how	3SG.M					

'[. . .] they climbed [**up** **onto** the roof] to see how the weather was.'

The TERMINATIVE marker (Π51) *sa* 'till' is clearly only a minority option when the motion verb *tela'* 'climb' is involved. Figure 5.12 reveals that the bare SURLATIVE is almost ten times as frequent as the DIRECTIONAL SURLATIVE in combination with this verb.

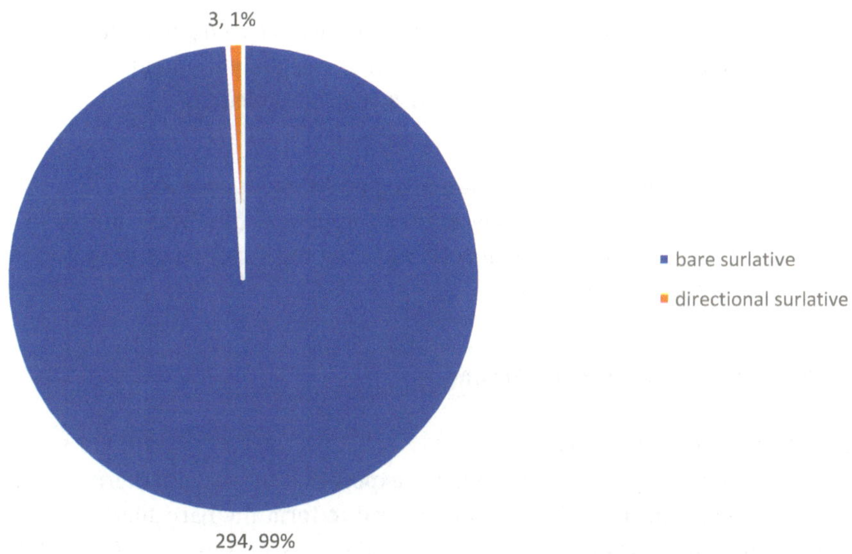

Figure 5.12: Bare VS DIRECTIONAL SURLATIVE with *telgħu* 'they climbed' in the Korpus Malti 3.0.

The quantitative discrepancy is striking although the results need to be refined in the future since we have only taken account of the 3rd person plural perfective of the motion verb.

5.1.2.2.11.2 Motion down from the Ground (DELATIVE)

For Borg/Azzopardi-Alexander (1997: 161), the only way to express motion down from a Ground is by combining the directionality marker (Π41) *minn* 'from' with (Π17) *fuq* 'on'. This combination conforms to the familiar pattern of using Π-Π combinations involving (Π41) *minn* 'from' as Π_1 when it comes to expressing motion from.

In (5.73), the motion event is to be understood metonymically since the signal does not physically move from its source position to a place located below it.

(5.73) DELATIVE – [Korpus Malti 3.0; academic12]
*[**Minn** **fuq** it-turretta tal-Palazz]*$_{PP}$ *ħareġ sinjal*
[**from** **on** DEF-tower:DIM of:DEF-palace] exit signal
'[**From** (**on**) the little tower of the Palace], there came a signal [. . .].'

In the Korpus Malti 3.0, there are 25,744 tokens of the DELATIVE, meaning: we are not facing a marginal combination. The DELATIVE is relatively frequent – more frequent than the DIRECTIONAL and TERMINATIVE SURLATIVE added up.

5.1.2.2.12 Motion to and from the superior region (without contact)

Hagège's (2010: 261 and 268–269) SUPERVERSIVE applies if the Figure moves to a position above the Ground. The English Πs fulfilling this function as proposed by the same author are *above* and *over*.

5.1.2.2.12.1 Motion to above the Ground

5.1.2.2.12.1.1 Bare SUPERVERSIVE

Based on the knowledge acquired so far, we expect that the usual pattern repeats itself in the sense that (Π17) *fuq* 'on' can be used to form the bare SUPERESSIVE in analogy to the bare SURLATIVE.[62] However, Borg/Azzopardi-Alexander (1997: 160) depict the situation differently. The configuration marker suffices to also express the SURLATIVE relation, but for the SUPERVERSIVE the authors assume that motion to requires the use of *'il fuq* 'upwards'. However, their example (792) does not involve

62 Cf. Section 5.1.2.2.11.1.1.

a proper PP; *'il fuq* 'upwards' is used adverbially.[63] What we are looking for is a PP headed by (Π17) *fuq* 'on' involving a (pro)nominal complement.

In contrast to the numerous bare SUPERESSIVES, bare SUPERVERSIVES are hard to find in the Korpus Malti 3.0. One of the infrequent examples is (5.74) where the Figure (= the sun) rises over the Ground (= the island of Gozo) to take in a position above the Ground without touching it.

(5.74) Bare SUPERVERSIVE – [Korpus Malti 3.0; news166658]

x-xemx	*bdiet*	*tiela'*	***[fuq***	*Għawdex]*$_{PP}$	
DEF-sun	begin:3SG.F.PFV	climb.PART	**[on**	Gozo]	

'[. . .] the sun began to rise [**over** Gozo].'

Given that the bare SUPERVERSIVE is a relatively minor option, does this imply that the DIRECTIONAL SUPERVERSIVE is a relative major option?

5.1.2.2.12.1.2 DIRECTIONAL SUPERVERSIVE

The problems multiply as soon as we take account of the DIRECTIONAL SUPERVERSIVE. What we expect to find is evidence for the SUPERVERSIVE function of the Π-Π combination *għal fuq*. This combination is registered by Borg/Azzopardi-Alexander (1997: 161) but in connection with motion (on)to a surface with (Π21) *għal* 'for' being optional. Since the bare SUPERVERSIVE is not a common phenomenon, the question arises as to why motion to a position above the Ground is so rarely expressed. One might assume that flying events which would typically result in motion to a position above the Ground are not as common as events which involve the positioning of a Figure on a surface with contact.

We cannot answer this question in this study. What we can say nevertheless is that there is the occasional example of the DIRECTIONAL SUPERVERSIVE which involves (Π21) *għal* 'for' as shown in (5.75a). The absence of the directionality marker in a similar context is illustrated in (5.75b). The competitor of (Π21) *għal* 'for', i.e. (Π28) *lejn* 'towards', is never attested as directionality marker for the DIRECTIONAL SUPERVERSIVE.

---

63 The original example is reproduced in (5.iii):

(5.iii) (Borg/Azzopardi-Alexander 1997: 160)

Minħabba	*s-sħab*	*l-ajruplan*	*tela'*	*iżjed*	***'il***	***fuq***
on_account_of	DEF-cloud.PL	DEF-plane	climb	more	**to**	**on**

'Because of the clouds, the plaine climbed **up** higher.' [O.T.]

The adverbial modifies the motion verb but takes no complement of its own. Thus, this example does not belong to the same category as other examples from the same reference grammer which represent bona fide PPS.

(5.75) DIRECTIONAL SUPERVERSIVE
(a) [Korpus Malti 3.0; literature80]
jittajru ***[għal*** ***fuq*** *Lascaris]$_{PP}$*
3.IPFV:REFL:fly:PL **[for** **on** Lascaris]
'[. . .] they are flown [**to** **above** Lascaris].'
(b) [Korpus Malti 3.0; news117204]
seba' ajruplani kienu qegħdin jittajru ***[fuq***
seven plane:PL be:3PL.PFV PROG:PL 3.IPFV:REFL:fly:PL **[on**
il-Port ta' Marsamxett]$_{PP}$
DEF-port of Marsamxett]
'[. . .] seven planes were being flown [**over** the Port of Marsamxett] [. . .].'

Sentence (5.75b) is ambiguous. One interpretation is that of the bare SUPERVERSIVE, it is the planes were being flown to a position above the bastions beneath Upper Barracca named *Laskri* or *Lascaris* (Aquilina 1991: 732) or we are facing the SUPERESSIVE in the sense that the planes circle over the Ground with their circling remaining stable in a location.

5.1.2.2.12.1.3 TERMINATIVE SUPERVERSIVE

It is equally difficult to find more than a dozen examples of the TERMINATIVE SUPERVERSIVE scattered over the Korpus Malti 3.0. This is mainly due to the fact that the binary combination *sa fuq* is frequently used like *'il fuq* 'upwards', namely adverbially. As results from (5.76) however, it is by no means impossible to combine (Π51) *sa* 'till' and (Π17) *fuq* 'on' for the purpose of expressing that a Figure moves to a position above the Ground. In (5.76), the Figure is the cloud of volcanic ash that is envisaged to move as far as the sky above Malta.

(5.76) TERMINATIVE SUPERVERSIVE – [Korpus Malti 3.0; parl9351]
Jista' l-Ministru jgħid x' effett jista'
3SG.M.IPFV:can DEF-minister 3SG.M.IPFV:say what effect 3SG.M.IPFV:can
jkollha l-isħaba tal-volcanic ash fuq il-gżejjer
3SG.M.IPFV:have.FUT:3SG.F DEF-cloud of:DEF-volcanic_ash on DEF-island.PL
u l-poplu Malti jekk din tasal ***[sa***
and DEF-people Maltese if DEM:F:PROX 3SG.F.IPFV:arrive **[till**
fuq *Malta]$_{PP}$*
on Malta]
'Can the minister say what effect the cloud of volcanic ash might have on the islands and the Maltese people if it arrives [**above** Malta]?'

5.1.2.2.12.2 Motion from above the Ground (SUPRADELATIVE)

For motion from a position above the Ground, Borg/Azzopardi-Alexander (1997: 160) provide an example which involves the expected directionality marker (Π41) *minn* 'from' which forms the binary Π-Π *minn fuq* together with the configuration marker (Π17) *fuq* 'on'. In the Korpus Malti 3.0, the token frequency of this combination is high. However, the SUPRADELATIVE is responsible only for a small amount of the total turnout.

Example (5.77) is taken from the on-going discussion about Maltese orthography. There is the letter <ċ> which involves a diacritic dot over the basic letter. The dot and the basic letter are not in contact with each other so that an observer could interpret the position of the dot as being above the basic letter.

(5.77) SUPRADELATIVE – [Korpus Malti 3.0; academic340]

Allura	*għaliex*	*ma*	*naħħux*	*it-tikka*	***[minn***	***fuq***	*iċ-Ċe]*$_{PP}$
so	why	NEG	1.IPFV:remove:PL	DEF-dot	**[from**	**on**	DEF-Ċ]

u	*nħaffu*	*l-kitba*	*tal-Malti*
and	1.IPFV:make_light:PL	DEf-writing	of:DEF-Maltese

'So, why don't we remove the dot [**from above** the Ċ] and make the writing of Maltese easier?'

The removal of the dot from the complex grapheme illustrates the metonymical motion of the dot as Figure from the basic letter as Ground.

5.1.2.2.13 Motion to and from under the Ground

As emphasised in the foregoing presentation of motion to and from above the Ground, the search for examples in the Korpus Malti 3.0 has not yielded a rich turnout. This is different with the opposite region. If we look for proof of spatial situations whose Goal or Source is located beneath the Ground, the Korpus Malti 3.0 is noticeably less parsimonious in terms of hits and matches. This difference between two regions or configurations is worthwhile investigating further but not in this purely descriptive study of Maltese Πs.

5.1.2.2.13.1 Motion under the Ground

5.1.2.2.13.1.1 Bare SUBLATIVE

Hagège (2010: 286) speaks of the SUBLATIVE in situations where the Figure moves to a position under the Ground. Borg/Azzopardi-Alexander (1997: 161–162) distinguish motion to an inferior position without contact between Figure and

Ground from a motion event at whose endpoint the Figure winds up under the Ground with which it is in contact. Only in the latter case do they present an example which involves a combination of genuine Πs. When there is no contact between Figure and Ground, however, the adverbial *'l isfel* 'downwards' without complement is employed.[64] The adverbial is outside the scope of this study because it does not represent the canonical Π. This criterion is met however by (Π56) *taħt* 'under' which is the candidate of Borg/Azzopardi-Alexander (1997: 161–162) for expressing configuration with reference to the inferior region of the Ground.

As results from (5.78), it is not too difficult to find evidence of (Π56) *taħt* 'under' being used for the bare SUBLATIVE in situations where there is no contact between Figure and Ground.

(5.78) Bare SUBLATIVE (without contact) – [Korpus Malti 3.0; parl4493]
lagħabha ta' Sherlock Holmes mar ***[taħt*** *xi pont*
play:3SG.F of Sherlock Holmes go **[under** some bridge
f' San Ġiljan]$_{PP}$ *jagħmel l-investigazzjonijiet ma'*
in San Ġiljan] 3SG.M.IPFV:do DEF-investigation:PL with
din il-persuna kriminali magħrufa
DEM:F:PROX DEF-person criminal PART:recognise:F
'[. . .] he played Sherlock Holmes, he went [**under** a bridge in San Ġiljan] to conduct investigations with this known criminal person.'

The person whose actions are described in (5.78) moves to a position under a bridge. The position under the bridge does not mean that Figure and Ground are in contact with each other.

Contact between Figure and Ground applies in (5.79) where the handle of a broom is introduced beneath the sheets so that it can be assumed that both objects touch each other.

---

64 The original example given by Borg/Azzopardi-Alexander (1997: 161) is reproduced in (5.iv).

(5.iv) *Aqbeż* ***'l isfel*** *minn din il-quċċata*
jump.IMP **to down** from DEM:F:PROX DEF-summit
'Throw yourself **down** from this summit.' [O.T.]

If we think back to what the authors assume with reference to the bare SUPERVERSIVE (cf. Section 5.1.2.2.12.1.1), one might get the impression that the absence of an explicit Ground-NP is equated with the absence of contact between Figure and Ground.

(5.79) Bare SUBLATIVE (with contact) – [Korpus Malti 3.0; literature14]

taħt	*is-sodda*	*kont*	*inżomm*	*manku*	*ta'*	*xkupa*
under	DEF-bed	be:1SG.PFV	1SG.IPFV:keep	handle	of	broom

li	*kont*	*indaħħal*	***[taħt***	*il-liżar]*$_{PP}$
that	be:1SG.PFV	1SG.IPFV:enter:CAUS	**[under**	DEF-sheet]

'[. . .] I kept the handle of a broom under the bed which I pushed [**under** the sheet].'

On the basis of examples like (5.78)–(5.79), we hypothesise that there is no formal distinction in the Π-domain between the contactless SUBLATIVE and the SUBLATIVE which involves contact between Figure and Ground.

5.1.2.2.13.1.2 DIRECTIONAL SUBLATIVE

Besides the bare SUBLATIVE there is also the directional SUBLATIVE which presupposes the presence of (Π21) *għal* 'for' as directionality marker. Borg/Azzopardi-Alexander (1997: 162) acknowledge this possibility as an option for motion to an inferior position with contact between Figure and Ground whereas for the contactless situation, they again mention only adverbials, namely *'l isfel* and *għal isfel* 'downwards'. Like in the case of the bare SUBLATIVE, there is evidence for the use of the Π-Π combination *għal taħt* not only for situations without contact but also for situations with contact. The Π-Π combination *lejn taħt* is attested too but in only very small numbers.

In (5.80a), the animals are moved to a position under the branches of the tree without touching the tree. In (5.80b), the person named Erkole gets under the wheels of a car, i.e., he is injured by the impact of the wheels which must have been in physical contact with Erkole's body. In (5.80c), we add one of the infrequent examples of (Π28) *lejn* 'towards' functioning as directionality marker in combination with (Π56) *taħt* 'under'. Like in the case of (5.78), we assume that (5.80c) involves no contact between Figure and Ground.

(5.80) DIRECTIONAL SUBLATIVE

(a) without contact – [Korpus Malti 3.0; literature21]

B'	*idu*	*l-leminija*	*beda*	*jkaxkarhom*	*wieħed*
with	hand:3SG.M	DEF-right:F	begin	3SG.M.IPFV:drag:3PL	one

wieħed	***[għal***	***taħt***	*dik*	*il-ħarruba]*$_{PP}$
one	**[for**	**under**	DEM:F:DIST	DEF-carob tree]

'He began to drag them with one by one [**under** the carob tree].

(b) with contact – [Korpus Malti 3.0; literature7]

U baqa' jidwi f' widnejhom l-għajjat u
and remain 3SG.M.IPFV:echo in ear:PL:3PL DEF-shouting and
l-inbiħ ta' Erkole minuti qabel ma ġera ***[għal***
DEF-barking of Erkole minute:PL before that run **[for**
taħt *ir-roti tal-karozza]*$_{PP}$
under DEF-wheel:PL of:DEF-car]
'And Erkole's shouting and barking minutes before he ran [**under** the wheels of the car] kept echoing in their ears.'

(c) without contact – [Korpus Malti 3.0; news200102]

Williams u żewġt irġiel oħrajn kienu niżlin
Williams and two man.PL other:PL be:3PL.PFV descend.PART:PL
minn Għawdex u fejn il-Mistra kienet bdiet
from Gozo and near DEF-Mistra be:3SG.F.PFV begin:3SG.F.PFV
ġlieda bl-Ingliż isuq ***[lejn taħt*** *il-pont*
fight with:DEF-English 3SG.M.IPFV:drive **[towards under** DEF-bridge
tal-Mistra]$_{PP}$ *u kien hemm li Williams spiċċa aggredit*
of:DEF-Mistra] and be there that Williams finish attack:PART
'Williams and two other men came down from Gouo and near Mistra a fight began with the Englishman driving [**under** the bridge of Mistra] and it was there that Williams was finally attacked.'

The supposed distinction of SUBLATIVES with and without contact does not play any role in the domain under scrutiny.

5.1.2.2.13.1.3 TERMINATIVE SUBLATIVE

We encounter a similar situation when we look at the TERMINATIVE SUBLATIVE. It almost goes without saying that directionality is expressed by (Π51) *sa* 'till'. In combination with the configuration marker (Π56) *taħt* 'under' it is compatible with spatial situations in which there is either a spatial gap between Figure and the Ground or the two touch each other. The former situation is illustrated in (5.81a) whereas the TERMINATIVE SUBLATIVE with contact is exemplified in (5.81b).

(5.81) TERMINATIVE SUBLATIVE

(a) without contact – [Korpus Malti 3.0; literature32]

Pass wara pass waslu ***[sa taħt*** *it-torri]*$_{PP}$
step after step arrive:3PL.PFV **[till under** DEF-tower]
'Step by step they arrived [**under** the tower] [. . .].'

(b) with contact – [Korpus Malti 3.0; literature19]
*Lonzu kellu qalziet twil imxammar [sa **taħt***
Lonzu have.PFV:3SG.M trouser:PL long PART:fold_back [**till under**
irkopptejh]$_{PP}$
knee:PL:3SG.M]
'Lonzu had long trousers rolled up [**to under** his knees].'

We interpret the two spatial situations as follows. In (5.81a), the Figure moves as far as the Ground which extends vertically so that the Figure winds up in a position under it (= at the foot of the tower) without necessarily touching it. This is different for (5.81b) since the trouser legs are certainly in contact with Lonzu's knees since he wears these trousers. Thus, also in the case of the TERMINATIVE SUBLATIVE, contact between Figure and Ground is no criterion for formal distinctions.

5.1.2.2.13.2 Motion from under the Ground (SUBELATIVE)

For motion from under the Ground with contact, Borg/Azzopardi-Alexander (1997: 162) do not provide an example but register the Π-Π combination *minn taħt* which involves the expected directionality marker (Π41) *minn* 'from'. They also mention it in the context of motion from under the Ground without contact (Borg/Azzopardi-Alexander 1997: 161).

The pair of examples in (5.82) confirms that the SUBELATIVE is insensitive to the distinction of situations with and without contact, too.

(5.82) SUBELATIVE
(a) without contact – [Korpus Malti 3.0; literature21]
*Imbagħad tahom l-aħħar tislima u ħareġ [**minn taħt***
then give:3PL DEF-last greeting and exit [**from under**
dik is-siġra]$_{PP}$
DEM:F:DIST DEF-tree]
'Then he gave them a last farewell and went off [**from under** that tree] [. . .].'
(b) with contact – [Korpus Malti 3.0; literature5]
*Qam ħareġ [**minn taħt** il-friex]*$_{PP}$ *u beda*
rise exit [**from under** DEF-coverlet.PL] and begin
jdur mal-kamra
3SG.M.IPFV:turn with:DEF-room
'He rose, came out [**from under** the coverlets] and started to walk about the room [. . .].'

The (a)-sentence describes a spatial situation in which the Figure leaves a position under (the branches of) a tree. At no point is there any indication of the Figure being in contact with the Ground. What we see in the (b)-sentence is a different situation because the one who exits from under the coverlets must have been in physical contact with them before.

5.1.2.2.14 Motion to and from a position before the Ground

5.1.2.2.14.1 Problems in the spatial domain

In Hagège's (2010: 286) system there is the PRELATIVE which is said to apply in a spatial situation in which the Figure moves into a position before the Ground. English *before* is the Π which is used to illustrate this category. The section on the PREESSIVE – the static equivalent of the PRELATIVE – reveals that (Π46) *qabel* 'before' can be used spatially and temporally to indicate that the Figure is located before the Ground either in space or in time.[65] Since we know from many other cases that the static Π can also be used without directionality marker in Goal-oriented contexts, we expect the same to hold also for (Π46) *qabel* 'before'. However, we have been unable to identify uncontroversial cases of the SPATIAL PRELATIVE being expressed by (Π46) *qabel* 'before' alone. Exactly the same can be said as to the Π-Π combinations *għal qabel*, *sa qabel*, and *minn qabel* none of which is attested with a spatial meaning in the Korpus Malti 3.0.[66] There is an abundance of hits but these matches usually invoke a temporal (or PURPOSIVE) reading. We, therefore, assume that (Π46) *qabel* 'before' deviates from the general pattern insofar as it is excluded from the expression of properly spatial meanings with the feature [+dynamic]. In accordance with Borg/Azzopardi-Alexander (1997: 159), we postulate that the expression of motion to or from a position before the Ground is the responsibility of (Π49) *quddiem* 'in front of', i.e. there is neutralisation of the FRONTALIS and the SPATIAL PRELATIVE / PREABLATIVE.[67]

5.1.2.2.14.2 Temporal meanings

In stark contrast to the absence of dynamically employed (Π46) *qabel* 'before' in the spatial domain, there is ample evidence of the same Π being used for temporal meanings.

---

65 Cf. Section 5.1.2.1.10.

66 But cf. Sections 5.1.2.2.20.2 and 5.1.2.2.20.3 on the PERLATIVE and the PROLATIVE.

67 Cf. Section 5.1.2.2.10.1.

5.1.2.2.14.2.1 Goal-oriented meanings

5.1.2.2.14.2.1.1 Bare PRELATIVE

Examples of the bare PRELATIVE with temporal function are difficult to identify. It is doubtful that cases like those in (5.83) are good representatives of a probably only theoretically possible category.

(5.83) Bare PRELATIVE

(a) [Korpus Malti 3.0; news132432]

Il-gowl	*deċiżiv*	*wasal*	***[qabel***	*il-mistrieħ]*$_{PP}$
DEF-goal	decisive	arrive	**[before**	DEF-rest]

'The decisive goal cam [**before** the half time break] [. . .].'

(b) [Korpus Malti 3.0; news74341]

L-Arċisqof	*wasal*	*tard*	*imma*	*xorta*	*wasal*	***[qabel***
DEF-archbishop	arrive	late	but	sort	arrive	**[before**

l-Isqof]$_{PP}$
DEF-bishop]

'The archbishop arrived late but he still arrived [**before** the bishop] [. . .].'

What the two examples show is that the Figure – the goal in (5.83a) and the archbishop in (5.83b) – arrived on the scene before it was time for the Ground – the half-time break in (5.83a) and the (arrival of) the bishop in (5.83b) – to take over.

5.1.2.2.14.2.1.2 DIRECTIONAL PRELATIVE

The combination of the directionality marker (Π21) *għal* 'for' and the configuration marker (Π46) *qabel* 'before' yields relatively common cases of the DIRECTIONAL PRELATIVE. This Π-Π combination occurs most frequently as adverbial adjunct of the motion verb *mar* 'go' in combination with the adverb *lura* 'back' as shown in (5.84a).[68] (Π28) *lejn* 'towards' as directionality marker in combination with (Π46) *qabel* 'before' is attested only four times in the Korpus Malti 3.0. Note that the adverb *lura* 'back' shows up also in (5.84b).

---

68 We present this longish sentence without shortening it in order to show that the entire context of the bracketed part of the example is indicative of a temporal situation.

(5.84) DIRECTIONAL PRELATIVE – temporal
(a) [Korpus Malti 3.0; culture3418]
Dan l-ark li hu l-aktar wieħed antik fost
DEM:M:PROX DEF-arch that 3SG.M DEF-more one old amongst
dawk li għadhom jintramaw fil-Festi
DEM:PL:DIST that still:3PL 3.IPFV:REFL:set_up:PL in:DEF-feast:PL
Maltin imur lura ***[għal*** ***qabel*** *l-1844]*$_{PP}$ *bl-ewwel*
Maltese:PL 3SG.M.IPFV:go back **[for** **before** DEF-1844] with:DEF-first
referenza għalih instabet fl-Osservatore Maltese ta'
reference for:3SG.M REFL:find:3SG.F.PFV in:DEF-Osservatore_Maltese of
dik is-sena
DEM:F:DIST DEF-year
'This arch which is the oldest amongst those which are still used in the Maltese feasts goes back [**to** **before** 1844] with the first reference to it found in the Osservatore Maltese of that year.'
(b) [Korpus Malti 3.0; news166058]
Tonio Fenech jaqaw iridna mmorru lura
Tonio Fenech perhaps 3SG.M.IPFV:want:1PL 1.IPFV:go:PL back
[lejn ***qabel*** *l-1971]*$_{PP}$
[towards **before** DEF-1971]
'[. . .] perhaps Tonio Fenech wants us to go back [**to** **before** 1971] [. . .].'

These are typical examples insofar as they reflect the retrospective orientation. The Figure is taken back in time to a *terminus ante quem* which in (5.84a) is the year 1844 in which the arch was mentioned for the first time in print – but this first mention is taken to mean that the arch existed already prior to the quoted edition of the Osservatore Maltese. It is clear that the motion events described by *għal qabel* are abstract and often only imagined processes because time travelling is still not feasible in real life. In (5.84b), the Figure is the group of Maltese parliamentarians who the politician mentioned by name is suspected to take back to the early days of independent Malta.

5.1.2.2.14.2.1.3 TERMINATIVE PRELATIVE

The combination of the directionality marker (Π51) *sa* 'till' and the configuration marker (Π46) *qabel* 'before' seems to be largely synonymous with that of (Π21) *għal* 'for' and (Π46) *qabel* 'before' presented in the previous section. There may be subtle meaning differences which however are difficult to pinpoint. An example like (5.85) might differ from examples like (5.84) in the sense that the TERMINATIVE PRELATIVE determines the exact period of time which one has to go back to whereas the DIRECTIONAL PRELATIVE is more flexible as to the interpretation of the exact position on the time arrow the Figure has to wind up in.

(5.85) TERMINATIVE PERLATIVE temporal – [Korpus Malti 3.0; news60012]
l-istorja tiegħu tmur lura ***[sa qabel*** *l-Aragoniżi]*$_{PP}$
DEF-story of:3SG.M 3SG.F.IPFV:go back **[till before** DEF-Aragonese:PL]
'[. . .] its story goes back [**as far as before** the Aragonese] [. . .].'

It remains a task for future studies to investigate the functional similarities and dissimilarities of the Π-Π combinations featured in this and the foregoing section.

5.1.2.2.14.2.2 Source-oriented meaning

In (5.86), we provide an example of a temporal situation that involves a Figure (= the construction work) that started its journey through time before May 1987. This date represents the Ground.

(5.86) PREABLATIVE Temporal – [Korpus Malti 3.0; parl28]
Ix-xogħol fuq il-ground tal-Ħamrun ilu sejjer
DEF-work on DEF-ground of:DEF-Ħamrun last:3SG.M go.PART
[minn qabel *Mejju 1987]*$_{PP}$
[from before May 1987]
'The work on the grounds of Ħamrun has been going on [**since before** May 1987] [. . .].'

Like in practically all examples from the temporal domain, it is impossible to speak of a motion event unless we allow for metonymy and metaphor to come into play.

5.1.2.2.15 Motion to and from behind the Ground

In analogy to what was said above,[69] the BEFORE-BEHIND axis displays a remarkable asymmetry. If the focus is on the region before the Ground, genuinely spatial concepts are underrepresented or even completely absent from the Maltese system. Being in or moving to a position before the Ground is predominantly associated

69 Cf. Sections 5.1.2.1.10 and 5.1.2.1.11.

with temporality. In contrast, being in or moving to a position behind the Ground is open to a spatial as well as to a temporal interpretation.

5.1.2.2.15.1 Spatial meanings

5.1.2.2.15.1.1 Motion to behind the Ground

Hagège (2010: 261 and 287) mentions the three English Πs *behind*, *after*, and *beyond* for illustration of the POSTLATIVE. We assume that *after* is more typical of the TEMPORAL POSTLATIVE whereas *beyond* evokes the notion of the EXTRALATIVE.[70] The POSTLATIVE applies if the Figure moves to a position behind the Ground. Borg/Azzopardi-Alexander (1997: 159) refer to (Π60) *wara* 'after' as obligatory marker of configuration to which the optional marker of directionality (Π21) *għal* 'for' may be added.

5.1.2.2.15.1.1.1 Bare POSTLATIVE – spatial

The bare spatial POSTLATIVE is frequently attested in the Korpus Malti 3.0. In (5.87), the protagonist Jason can be equated conceptually with the Figure. Jason moves to a position behind the chapel which, in turn, serves as Ground in the spatial situation. At the end of the motion event, the Figure is located at the backside of the Ground without physical contact between the two participants.[71]

(5.87) Bare PROLATIVE spatial – [Korpus Malti 3.0; literature24]

Jason	*mar*	***[wara***	*l-kappella]*$_{PP}$	*u*	*ntefa'*	*bilqiegħda*
Jason	go	**[after**	DEF-chapel]	and	REFL:throw	sitting
mal-art						
with:DEF-ground						

'Jason went [**behind** the chapel] and sat down on the ground [. . .].'

Directionality remains unexpressed. The bare POSTLATIVE fits into the general picture of Goal-oriented constructions which do not make directionality explicit provided the verbal predicate is a motion verb.

---

70 The EXTRALATIVE will be discussed in Section 5.1.2.2.19.

71 The distinction of WITH CONTACT VS WITHOUT CONTACT does not seem to play any role in the domain of the POSTLATIVE.

5.1.2.2.15.1.1.2 DIRECTIONAL POSTLATIVE – spatial

As mentioned by Borg/Azzopardi-Alexander (1997: 159), motion to a posterior position can be expressed also by the Π-Π combination *għal wara.* The presentation of this option is such that the bare POSTLATIVE and the DIRECTIONAL POSTLATIVE seem to be fully synonymous. This means that the directionality marker could be added to (5.87) just like it could be removed from (5.88a) without altering the meaning of the constructions. In the (b)-sentence, we add one of the rare examples of (Π28) *lejn* 'towards' replacing (Π21) *għal* 'for' as directionality marker for the spatial PROLATIVE.

(5.88) DIRECTIONAL PROLATIVE spatial

(a) [Korpus Malti 3.0; news161055]

PC Vassallo reġa' rikeb il-mutur u mar ***[għal wara*** *l-vann]$_{PP}$*
PC Vassallo do_again mount DEF-bike and go **[for after** DEF-van]
'PC Vassallo got on his motorbike again and went [**behind** the van] [. . .].'

(b) [Korpus Malti 3.0; news111040]

dawn iż-żewġ persuni ħarbu ***[lejn wara*** *l-Knisja]$_{PP}$*
DEM:PL:PROX DEF-two person:PL run_away:3PL.PFV **[towards after** DEF-church]
'[. . .] these two persons fled [**behind** the church].'

It cannot be decided whether the presence/absence of the directionality has a bearing on the interpretation of the spatial situation. As argued in previous sections, it might be the case that (Π21) *għal* 'for' highlights the process of the motion event whereas the bare POSTLATIVE puts special emphasis on the Goal of the motion event. We assume that a similar explanation is also possible for the use of (Π28) *lejn* 'towards'.

5.1.2.2.15.1.1.3 TERMINATIVE POSTLATIVE – spatial

The terminus of the motion event is certainly the focus of constructions which involve (Π51) *sa* 'till'. The TERMINATIVE POSTLATIVE is expressed by way of combining this directionality marker with (Π60) *wara* 'after' as can be gathered from (5.89).

(5.89) TERMINATIVE POSTLATIVE spatial – [Korpus Malti 3.0; literature54]

imma	*ġriet*	*warajh*[72]	*mad-dar*	*kollha*
but	run:3SG.F.PFV	after:3SG.M	with:DEF-house	all:3SG.F

mill-intrata	*[sa*	***wara***	*l-bieb*	*tal-bejt]*$_{PP}$
from:DEF-entrance	**[till**	**after**	DEF-door	of:DEF-roof]

'[. . .] but she ran after him through the entire house from the entrance [**as far as** the door to the roof].'

In Maltese, the use of (П60) *wara* 'after' with Grounds like *bieb* 'door' often receives an interpretation that is different from that expected for a BEHIND relation (but cf. (5.90)). The Figure is located behind a door when the Figure is on the outside of the place to which the door gives access (indirectly claimed by Aquilina 1987: 117). The appropriate English П would thus be *in front of (the door)*. Independent of this inverse organisation on the BEFORE-BEHIND axis, the facts are clear for (5.89) because the Figure (= the anonymous female agent) follows the male participant through the house to stop eventually at the door to the roof. The directionality marker (П51) *sa* 'till' tells us that the motion event comes to an end at exactly this place. If (П21) *għal* 'for' had replaced (П51) *sa* 'till' in (5.89), it would be possible that the Figure never reached the door but moved only in its direction.

5.1.2.2.15.1.2 Source-oriented meanings – spatial

In line with what we have learned sofar about the expression of motion from any kind of configuration, it is no surprise to find that (П41) *minn* 'from' combines with (П60) *wara* 'after' to express the POSTABLATIVE (Borg/Azzopardi-Alexander 1997: 159). This П-П combination is featured in (5.90).

(5.90) POSTABLATIVE spatial – [Korpus Malti 3.0; literature32]

Najàr	*tfaċċa*	***[minn***	***wara***	*l-bieb]*$_{PP}$	*u*	*ħares*	*'l*	*fuq*
Najàr	REFL:emerge	**[from**	**after**	DEF-door]	and	look	to	on

lejn	*is-sema*
towards	DEF-sky

'Najàr emerged [**from behind** the door] and looked upwards to the sky [. . .].'

72 This instance of (П60) *wara* 'after' counts as an example of Hagège's (2010: 261) meaning category with the label SECUTIVE. This category will be looked into in Section 5.1.2.2.20.1.

Note that in this example it makes sense to locate the Figure's point of departure behind the Ground since Najàr seems to go outside to look at the sky. Unfortunately, it is not possible to verify this hypothesis because the text from which (5.90) has been extracted offers no further clues as to where Najàr was when the motion event started.

5.1.2.2.15.2 Temporal meanings

As transpires from the subsequent sections, there is a temporal equivalent for each of the spatial meanings in the domain of the POSTLATIVE and the POSTABLATIVE.

5.1.2.2.15.2.1 Goal-oriented meanings – temporal

5.1.2.2.15.2.1.1 Bare POSTLATIVE – temporal

Borg/Azzopardi-Alexander (1997: 172) do not specifically refer to dynamic relations in the domain of the posterior. The bare TEMPORAL POSTLATIVE is illustrated in (5.91).

(5.91) Bare POSTLATIVE temporal – [Korpus Malti 3.0; sport2919]

Il-ftehim	*ma'*	*Ecclestone*	*wasal*	*[**wara***	*laqgħa*	*ma'*
DEF-agreement	with	Ecclestone	arrive	[**after**	meeting	with

dawk	*imsejħa*	*timijiet*	*żgħar]$_{PP}$*
DEM:PL:DIST	PART:call:PL	team:PL	small.PL]

'The agreement with Ecclestone came [**after** a meeting with the so-called minor teams].'

Once more, we are in the realm of metonymy and metaphor since the Figure is abstract (= the agreement) and speaking of its arrival is of course an example of figurative speech. The Ground is the meeting that took place prior to the establishment of the agreement. This meeting happened at a certain point in time after which the negotiators came to an agreement. The POSTLATIVE is thus temporal.

5.1.2.2.15.2.1.2 DIRECTIONAL POSTLATIVE – temporal

The use of (Π21) *għal* 'for' as directionality marker in (5.92) is interesting for two reasons. On the one hand, without the directionality marker, the primary interpretation of the sentence would be that the extension of the deadline will be made public after the mentioned date. On the other hand, the Π-Π combination *għal wara* is translated by the English Π *beyond* and thus constitutes an instance of the

EXTRALATIVE.[73] In the Korpus Malti 3.0, there is no evidence of (Π28) *lejn* 'towards' as directionality marker for the TEMPORAL POSTLATIVE.

(5.92) DIRECTIONAL POSTLATIVE temporal – [Korpus Malti 3.0; european8167]

L-iskadenza	*b' dankollu*	*m'*	*għandhiex*
DEF-term_of_payment	nevertheless	NEG	have.IPFV:3SG.F:NEG

tkun	*estiża*	***[għal***	***wara***	*il-31*	*ta'*	*Diċembru*
3SG.F.IPFV:be.FUT	extend.PART:F	**[for**	**after**	DEF-31	of	December

2001]$_{PP}$
2001]

'The term of payment however must not be extended [**beyond** the 31st of December, 2001].'

One may say that the Figure (= the term of payment) cannot be postponed until after the date indicated in (5.92), i.e., it is excluded that the Figure is shifted to a phase that follows the Ground.

5.1.2.2.15.2.1.3 TERMINATIVE POSTLATIVE – temporal

The TERMINATIVE POSTLATIVE fixes the endpoint of a process at a time after a given date. To this end, the directionality marker (Π51) *sa* 'till' combines with the configuration marker (Π60) *wara* 'after' according to the by now familiar pattern. In (5.93), *il-Milied* 'Christmas' fulfils the function of Ground. The process of waiting takes so much time that it will end only after Christmas.

(5.93) TERMINATIVE POSTLATIVE temporal – [Korpus Malti 3.0; literature35]

Irid	*jistenna*	*almenu*	***[sa***	***wara***	*l-Milied]*$_{PP}$
3SG.M.IPFV:need	3SG.M.IPFV:wait	at_least	**[till**	**after**	DEF-Christmas]

'He must wait at least [**until after** Christmas].'

In a way, this construction with two Πs resembles the previous one insofar as the process is conceptualised as exceeding a certain limit and thus as going beyond the line represented by the Ground.[74] We will take up this issue again below.[75]

73 Addressed in Section 5.1.2.2.19.
74 Cf. Section 5.1.2.2.15.2.1.2.
75 Cf. Section 5.1.2.2.19.

5.1.2.2.15.2.2 Source-oriented meanings – temporal

Syntactically, (5.94) is an intriguing example of the TEMPORAL POSTABLATIVE. The unsurprising fact about the TEMPORAL POSTABLATIVE is that directionality is expressed in the usual way in the sense that motion from a temporal Source is indicated by (Π41) *minn* 'from' in combination with (Π60) *wara* 'after'. What needs to be explained specifically however is the clausal composition of the example at hand.

(5.94) POSTABLATIVE temporal – [Korpus Malti 3.0; news30941]

llum	*it-tim*	*ta'*	*22*	*ċiklist*	*tal-Actavis LifeCycle*	*wasal*
today	DEF-team	of	22	cyclist	of:DEF-Actavis_LifeCycle	arrive
*[**minn***	***wara***	*għaxart*	*ijiem*	*jaqdfu*	*r-roti*	
[**from**	**after**	ten	day.PL	3.IPFV:row:PL	DEF-wheel:PL	

fil-Ġappun]$_{PP}$
in:DEF-Japan]

'[. . .] today the team of twenty-two cyclists of Actavis LifeCycle arrived [**after** ten days of rotating the wheels in Japan].'

Superficially, the sentence looks to consist of several clauses so that one might suspect some kind of subordination to apply. If *jaqdfu* 'they row' was the predicate nucleus of a temporal clause, *minn wara* would not count as a Π-Π combination but as a complex conjunction introducing a subordinate clause. However, the usual subordinator *li* 'that' is absent from (5.94). The absence of the subordinator precludes the interpretation of *minn wara* as conjunctional. Furthermore, *jaqdfu r-roti fil-Ġappun* 'they turn the wheels in Japan' is a relative clause that modifies *għaxart ijiem* 'ten days', i.e., the relative clause is inside the NP which constitutes the complement of *(minn) wara*. The temporal situation involves a Figure – the cyclists – whose return happens after their sojourn of ten days in Japan is finished. Accordingly, the point of departure of the motion event is the period after the end of the bicycle ride.

5.1.2.2.16 Motion to and from around the Ground

At this point, we enter a heterogeneous domain populated by a multitude of meaning categories à la Hagège (2010) which pose a variety of analytical and conceptual problems. Some of these problems are probably caused by the schematic character of Hagège's (2010: 287) own taxonomy. No TEMPORAL CIRCUMLATIVE is postulated.

5.1.2.2.16.1 Goal-oriented motion

5.1.2.2.16.1.1 There is no bare CIRCUMLATIVE

In contrast to Hagège (2010: 286) who assumes that the CIRCUMLATIVE applies in situations where the Figure moves into a position that surrounds the Ground, Borg/Azzopardi-Alexander (1997: 166) skip motion to in connection with the circumferential configuration. This gap in their inventory of spatial notions is worth noting. Why this is so comes to the fore when we look at (5.95) and try to analyse the example afterwards. The Π of interest is the same as in the case of the CIRCUMESSIVE, namely (Π31) *madwar* 'around'.[76]

(5.95) Bare CIRCUMLATIVE – [Korpus Malti 3.0; literature24]
kellha tqum minn bilqiegħda u timxi
have.PFV:3SG.F 3SG.F.IPFV:rise from sitting and 3SG.F.IPFV:walk
b' pass mgħaġġel ***[madwar*** *il-mejda]$_{PP}$*
with step PART:hurry **[around** DEF-table]
'[. . .] she had to stand up and walk hurriedly [**around** the table].'

The spatial situation described in (5.95) involves a motion event which leads the Figure from her sitting position (presumably at the table) around the Ground. However, at no point in the process does the Figure occupy a position that can be characterised as surrounding the Ground. Rather, the table as Ground serves as an extended path along which the Figure moves to reach the destination. In this sense, (5.95) fails to instantiate the bare CIRCUMLATIVE but should better be treated under the heading of PERLATIVE or SECUTIVE – meaning categories which are discussed below.[77]

In point of fact, our search of the Korpus Malti 3.0 confirms Borg/Azzopardi-Alexander (1997: 166) at least for what concerns the absence of tangible evidence of the existence of the bare CIRCUMLATIVE. The absence of this category does not however mean that there is no way to express motion to a circumferential position.

5.1.2.2.16.1.2 DIRECTIONAL CIRCUMLATIVE

It is generally difficult to find uncontroversial evidence of spatial situations in the Korpus Malti 3.0 that describe a prototypical motion event with the Figure changing place to occupy a position to encircle the Ground. Those cases which come closest to

76 Cf. Section 5.1.2.1.12.
77 Cf. Section 5.1.2.2.20.

this pattern involve the expected directionality marker (Π21) *għal* 'for' in combination with (Π31) *madwar* 'around'.

Of the two examples presented in (5.96) only the Π-Π combination *għal madwar* in the (a)-sentence takes a complement whose referent can be said to be surrounded by the Figure (= the dark stripe). The situation is dynamic only if we allow for a metonymical/metaphorical reading since the observer's gaze moves from the animal's nose via its eyes towards the region beneath the ears. In contrast, the (b)-sentence describes a bona fide motion event because the blood is pumped through the body. However, the blood does not really surround the body but rather reaches all internal parts thereof in the course of a blood cycle. It is therefore doubtful to classify (5.96b) as a good illustration of the DIRECTIONAL CIRCUMLATIVE. Since the blood travels across the body, the same PP could also be interpreted as an instance of the PERLATIVE.[78] Properly spatial cases of (Π28) *lejn* 'towards' combining with (Π31) *madwar* 'around' are absent from our corpus.

(5.96) DIRECTIONAL CIRCUMLATIVE

(a) [Korpus Malti 3.0; academic311]

strixxa skura minn fuq l-imnieħer ***[għal madwar*** *l-għajnejn]*$_{PP}$
stripe dark:F from on DEF-nose **[for around** DEF-eye:PL]
sa taħt il-widnejn tagħmel wiċċ dan il-marsupjal
till under DEF-ear:PL 3SG.F.IPFV:make face DEM:M:PROX DEF-marsupial
jidher kważi iswed
3SG.M.IPFV:appear almost black
'[. . .] a dark stripe from the nose [**around** the eyes] to below the ears makes the face of this marsupial appear almost black.'

(b) [Korpus Malti 3.0; academic248]

id-demm jiġi ippompjat b' saħħa biżżejjed
DEF-blood 3SG.M.IPFV:come pump:PART with strenght enough
minn ġewwa l-qalb ***[għal madwar*** *il-ġisem u l-pulmuni]*$_{PP}$
from inside DEF-heart **[for around** DEF-body and DEF-lung:PL]
'[. . .] the blood is pumped with sufficient force from inside the heart [**around** the body and the lungs].'

What we learn from these examples is that motion to a position around the Ground is not a very common spatial situation. Moreover, (Π31) *madwar* 'around' when used in directional constructions seems to invite other interpretations which are not properly circumferential but rather PERLATIVE.

78 Cf. Section 5.1.2.2.20.2.

5.1.2.2.16.1.3 TERMINATIVE CIRCUMLATIVE

The TERMINATIVE CIRCUMLATIVE involves the usual directional marker (Π51) *sa* 'till' that combines with the configuration marker (Π31) *madwar* 'around'. As in the previous cases of Goal-oriented CIRCUMLATIVES, convincing examples of this Π-Π combination are scarce when they are used to capture the change of position of a Figure such that at the endpoint of the motion event it encircles the Ground.

In (5.97) one of the very few cases from the Korpus Malti 3.0 that realise this pattern is presented.

(5.97) TERMINATIVE CIRCUMLATIVE – [Korpus Malti 3.0; parl256]

Jista'	*l-Ministru*	*jgħid*	*meta*	*se*	*tiġi*
3SG.M.IPFV:can	DEF-minister	3SG.M.IPFV:say	when	FUT	3SG.F.IPFV:come

asfaltata	*Marsa Road*	*Qormi*	*minn*	*taħt*	*il-pont*
asphalt:PART:F	Marsa Road	Qormi	from	under	DEF-bridge

[sa	***madwar***	*ir-roundabout]$_{PP}$*
[**till**	**around**	DEF-roundabout]

'Can the minister say when Marsa Road in Qormi will be asphalted from under the bridge [**to around** the roundabout]?'

The situation is again less than straightforward. In this particular case, the identity of the Figure poses problems. To our mind, the Figure is the road (work) that is conceived as moving along the Marsa Road from the bridge to the roundabout with the latter being surrounded by the asphalted road in the end. Metonymically, the asphalted road moves into a position that encircles the centre of the roundabout. There remain conceptual problems because the road itself forms part of the roundabout and thus cannot properly encircle it. No matter how this and related problems can be solved, we conclude that there is indeed a TERMINATIVE CIRCUMLATIVE in Maltese.

5.1.2.2.16.2 Motion from around the Ground

We are not telling the reader anything new when we say that the CIRCUMABLATIVE is expressed by way of combining the directionality marker (Π41) *minn* 'from' with the configuration marker (Π31) *madwar* 'around'. Another hardly surprising fact about the CIRCUMABLATIVE is the possibility of interpreting the configuration marker in at least two different ways.

This possibility is visible in the pair of examples given in (5.98).

(5.98) CIRCUMABLATIVE

(a) [Korpus Malti 3.0; literature40]

qed tneżża' l-karta tal-fidda pulit pulit
PROG 3SG.F.IPFV:take_off DEF-paper of:DEF-silver neat neat
***[minn madwar** għonq flixkun oħxon]*$_{PP}$
[from around neck bottle thick]
'[. . .] she is very neatly taking off the silver paper [**from around** the neck of a thick bottle] [. . .].'

(b) [Korpus Malti 3.0; academic376]

l-bżonnijiet tal-ekonomija li kienet qegħda
DEF-need:PL of:DEF-economy that be:3SG.F.PFV PROG:F
tirkupra ġabet aktar nies ***[minn madwar***
3SG.F.IPFV:recover bring:3SG.F.PFV more people **[from around**
il-Commonwealth]$_{PP}$
DEF-Commonwealth]
'[. . .] the needs of the economy that was recovering brought more people [**from all over** the Commonwealth] [. . .].'

The (a)-sentence is a bona fide illustration of the CIRCUMABLATIVE. The Figure (= the silver paper) is removed from around the Ground (= the bottleneck). What once was in a position encircling the Ground is now somewhere else. The (b)-sentence is difficult to accommodate with the notion of circumference. The people who migrated to the UK had not surrounded the Commonwealth before they started their journey to Great Britain. They used to live scattered over different places within the Commonwealth. Accordingly, the association with English *across* is stronger than that with English *around.* Once again, we have a piece of evidence that points towards the PERLATIVE as a potential functional domain of *(minn) madwar.*[79]

5.1.2.2.17 Motion to and from amongst the Ground

5.1.2.2.17.1 Goal-oriented motion

According to Hagège (2010: 286) the MEDIOLATIVE is the label for a meaning category of a Π that is used to express that the Figure moves into a position amongst "things or persons" (Hagège 2010: 287) which constitutes the Ground. As mentioned in connection with the MEDIOESSIVE,[80] Borg/Azzopardi-Alexander (1997: 165) do not reg-

79 Cf. Section 5.1.2.2.20.2.
80 In Section 5.1.2.1.13.

ister (Π16) *fost* 'amongst' at all. However, this Π is critical for the expression of the MEDIOLATIVE. As we will see below,[81] there is also an overlap of the MEDIOABLATIVE and the INTERABLATIVE which can be connected to the distinction of MEDIOESSIVE I and MEDIOESSIVE II.[82] Our search for evidence of the TERMINATIVE MEDIOLATIVE yielded not a single hit for the combination **sa fost* and only a very doubtful one for ${}^{?}$*sa qalb*. This is why no subsection is dedicated to this absent category.

5.1.2.2.17.1.1 Bare MEDIOLATIVE

The bare MEDIOLATIVE requires only the presence of the configuration marker (Π16) *fost* 'amongst' whereas directionality is subject to zero-marking. This pattern is in line with the bulk of the previously mentioned cases of bare LATIVES.

In (5.99), the Ground is the ethnic group of the Macuxi in Brazil. The president of Brazil pays a visit to the Macuxi so that the spatial situation involves a Figure (= the visitor) that changes its position by way of moving to the Ground which in turn is formed by a group of people amongst whom the Figure winds up at the end of the motion event.

(5.99) Bare MEDIOLATIVE – [Korpus Malti 3.0; news142309]

Milux	*Lula*	*mar*	***[fost***	*l-Indios*	*Macuxi]*$_{PP}$
NEG:ago:3SG.M:NEG	Lula	go	**[amongst**	DEF-Indio:PL	Macuxi]

biex	*jagħtihom*	*lura*	*l-art*	*ta'*	*Raposa Serra do Sol*
to	3SG.M.IPFV:give:3PL	back	DEF-land	of	Raposa Serra do Sul

'Not long ago, Lula went [**amongst** the Macuxi Indios] to return the territory of Raposa Serra do Sol to them.'

5.1.2.2.17.1.2 DIRECTIONAL MEDIOLATIVE

Besides the bare MEDIOLATIVE, there is also the DIRECTIONAL MEDIOLATIVE which conforms to patterns which are already familiar to us from a plethora of other cases of directional constructions for LATIVES. The directionality marker is as usual (Π21) *għal* 'for' which combines with the configuration marker (Π16) *fost* 'amongst' as shown in (5.100a) and with the configuration marker (Π47) *qalb* 'amidst' in (5.100b).

---

81 In Section 5.1.2.2.17.2.

82 Cf. Sections 5.1.2.1.13–5.1.2.1.14.

(5.100) DIRECTIONAL MEDIOLATIVE

(a) [Korpus Malti 3.0; parl713]

daħlet *il-mentalita'* *minn* *fuq* *għal* *isfel* ***[għal***
enter:3SG.F.PFV DEF-mentality from on for down **[for**
fost *il-ħaddiema* *nfushom]$_{PP}$*
amongst DEF-worker:PL self:3PL]
'[. . .] the mentality descended from above downwards [**amongst** the workers themselves] [. . .].'

(b) [Korpus Malti 3.0; literature21]

flimkien *imxew* *aktar* *'il* *ġewwa* *sa* *tarf* *l-isqaq* *minn*
together walk:3PL.PFV more to inside till end DEF-lane from
fejn *kien* *hemm* *trejqa* *dejqa* *tagħti* ***[għal***
where be there street.DIM narrow:F 3SG.F.IPFV:give **[for**
qalb *l-għelieqi]$_{PP}$*
amidst DEF-field:PL]
'[. . .] together they walked further to the end of the lane where there was a narrow alley that lead [**into** the fields].'

In theses examples, the motion event is to be understood metonymicality. In the (a)-sentence, what is described is the diffusion of a mentality problem among social groups in Malta. The mentality that emerged on the higher social levels descends downwards on the social scale to eventually reach the working class amongst whose members it takes roots, too. The Figure (= the mentality) travels towards the Ground (= the working class) which is composed of all representatives of the working class. In contrast to (5.99), the endpoint of the event does not consist of an individual surrounded by members of a different group of people but by the spread of a mindset across the members of a group of people. In this way, we are again facing a potential PERLATIVE interpretation. The (b)-sentence depicts the alley to move right into the middle of the fields although it functions only as access to the destination for the people who use the alley.

The employment of (Π21) *għal* 'for' in lieu of the bare MEDIOLATIVE has probably to do with the focus being on the process of diffusion. Note also that the adverbial *għal isfel* 'downwards' that immediately precedes the Π-Π combination *għal fost* in (5.100) already involves (Π21) *għal* 'for'. The co-presence of a second (Π21) *għal* 'for' is not blocked by *għal isfel.* It must also be mentioned that the combination **lejn fost* is unattested in the Korpus Malti 3.0.

5.1.2.2.17.2 Motion from amongst the Ground

Situations in which the Figure moves from a position within a Ground that consists of a group of people or objects may be termed MEDIOABLATIVE. As results from the two examples in (5.101), there are three Maltese Πs that compete for the function of configuration markers, namely (Π16) *fost* 'amongst' (= (a)-sentence), (Π47) *qalb* 'amidst' (= (b)-sentence), and (Π4) *bejn* 'between' (= (c)-sentence). This competition is reminiscent of the distinction of the MEDIOESSIVES I and II made in the appropriate subsections on the static spatial meanings above.[83] It almost goes without saying that the directionality marker for motion from amongst the Ground is the ubiquitous (Π41) *minn* 'from'.

(5.101) MEDIOABLATIVE

(a) [Attard 2019: 82]

Is-segretarja tal-Brigadier Camilleri magħżula
DEF-secretary of:DEF-Brigadier_Camilleri PART:choose:F
*[**minn** **fost** ir-rekluti reċenti]$_{PP}$ kienet imżejna*
[**from** **amongst** DEF-recruit:PL recent:PL] be:3SG.F.PFV PART:adorn:F
bi ġmiel u doni naturali
with beauty and gift:PL natural:PL
'Brigadier Camilleri's secretary, chosen [**amongst** the recent recruits], was adorned with beauty and natural gifts.'

(b) [Korpus Malti 3.0; literature3]

*rnexxielu joħroġ [**minn** **qalb** dawk*
succeed:IO:3SG.M 3SG.M.IPFV:exit [**from** **amidst** DEM:PL:DIST
it-tfal irrabjati]$_{PP}$
DEF-child.PL angry:PART:PL]
'[. . .] he succeeded in exiting [**from** **amidst** those angry children] [. . .].'

(c) [Korpus Malti 3.0; news116976]

bħala poplu Malti [. . .] nfittxu dejjem dak li
as people Maltese 1.IPFV:seek:PL always DEM:M:DIST that
*jgħaqqadna u nneħħu [**minn** **bejnietna**]$_{PP}$*
3SG.M.IPFV:unite:1PL and 1.IPFV:take_off:PL [**from** **between:PL**:1PL]
dak kollu li jifridna
DEM:M:DIST all:3SG.M that 3SG.M.IPFV:separate:1PL
'[. . .] as Maltese people [. . .] we always seek what unites us and remove [**from** **amongst** us] all that separates us.'

83 Cf. Sections 5.1.2.1.13–5.1.2.1.14.

Example (5.101a) provides the most typical context for the use of the Π-Π combination *minn fost*. A verb – finite or a passive participle – that belongs to the semantic field of CHOSING, ELECTING, or SELECTING (e.g. *għażel* 'choose', *eleġġa* 'elect', *ħatar* 'elect') preferably combines with *minn fost* if the group is mentioned from which the chosen one is taken. However, in (5.101b–c), we notice that the Π-Π combinations *minn qalb* and *minn bejn* are also licit. In (5.101b), the group from which the Figure emerges consists of more than two children. In (5.101c), the group is formed by the entire Maltese people, so that (Π4) *bejn* 'between' cannot refer to a spatial situation with just two elements forming the Ground. In cases like (5.101c), (Π4) *bejn* 'between' is a synonym of (Π16) *fost* 'amongst'. Interestingly, the situation described in (5.101c) does not involve the removal of a member of the group from the group but that of an abstract concept that causes dissent within the community. Whether this difference between the (a/b)-sentences and the (c)-sentence explains why the one Π is given preference over the other in this but not in that sentence, is a question that we cannot answer in this study but keep in the back of our minds for future investigations.

5.1.2.2.18 Motion to and from between the Ground

The appearance of (Π4) *bejn* 'between' in the domain of the MEDIOABLATIVE forms the bridge to this section which addresses the inter(ab)lative relations for the expression of which (Π4) *bejn* 'between' is the appropriate (and only) candidate.

5.1.2.2.18.1 Goal-oriented motion

Hagège (2010: 286) speaks of the INTERLATIVE when the Figure moves to a position between (presumably two) entities that together constitute the Ground. At the terminus of the motion event, the Figure thus separates the constituents of the Ground from each other. Like in the case of the INTERESSIVE,[84] Borg/Azzopardi-Alexander (1997: 165) postulate the category "medial" with a Ground-membership of two entities in which case (Π4) *bejn* 'between' is made use of (for larger groups of entities which form the Ground, the authors propose *f' nofs* 'in (the) middle' in lieu of mentioning (Π16) *fost* 'amongst').

5.1.2.2.18.1.1 Bare INTERLATIVE

For motion to the medial position, Borg/Azzopardi-Alexander (1997: 165) exclusively refer to the bare INTERLATIVE. In (5.102), directionality remains unexpressed whereas the configuration marker (Π4) *bejn* 'between' is mandatory.

---

84 Cf. Section 5.1.2.1.14.

(5.102) Bare INTERLATIVE – [Korpus Malti 3.0; news109423]
hekk kif l-ajruplan kien wasal [***bejn*** *Għawdex u*
so how DEF-plane be arrive [**between** Gozo and
Sqallija]$_{PP}$ *il-pilota informa lill-passiġġieri li se*
Sicily] DEF-pilot inform to:DEF-passenger:pl that FUT
jagħmel inżul fl-Ajruport ta' Ħal Luqa
3SG.M.IPFV:make descent in:DEF-airport of Ħal Luqa
'[. . .] as soon as the pane had arrived [**between** Gozo and Sicily] the pilot informed the passengers that he was going to land at the airport of Ħal Luqa.'

The Figure (= the plane) reaches a position between the islands of Gozo and Sicily (= the distributed Ground). The motion verb *wasal* 'arrive' suggests that the motion event is "frozen" at this point of the journey so that no PERLATIVE meaning is invoked.

5.1.2.2.18.1.2 DIRECTIONAL INTERLATIVE

Contrary to Borg/Azzopardi-Alexander's (1997: 156) focus on the bare INTERLATIVE alone, we can state that there is evidence not only of the DIRECTIONAL INTERLATIVE but also of the TERMINATIVE INTERLATIVE with the latter being the topic of the next section. In this section, we touch upon (Π21) *għal* 'for' combining with (Π4) *bejn* 'between' as in (5.103).

(5.103) DIRECTIONAL INTERLATIVE – [Korpus Malti 3.0; news160212]
Ngħaddi issa [***għal bejn*** *il-linji ta' din*
1SG.IPFV:pass now [**for between** DEF-line:PL of DEM:F:PROX
it-tweġiba]$_{PP}$
DEF-answer]
'I now move over [**to between** the lines of this answer].'

Like in many previous instances, the motion event is of a metonymical nature. The speaker (in the virtual function of Figure) announces that he is about to turn his attention to another topic, namely what can be read from between the lines of an answer given to a previous question posed in the parliament. The lines serve as the distributed Ground and the Figure metonymically moves to a position between these lines. The use of the Π-Π combination *għal bejn* instead of the bare INTERLATIVE is most probably motivated by the verb *għadda* 'pass' that normally requires a directionality marker to be present. (Π28) *lejn* 'towards' never serves as directionality marker for (Π4) *bejn* 'between'.

5.1.2.2.18.1.3 Terminative interlative

In (5.104),[85] we present an example of the TERMINATIVE INTERLATIVE. The construction involves (П51) *sa* 'till' in combination with (П4) *bejn* 'between' – a combination that meets our expectations when we talk about Maltese TERMINATIVES, in general.

(5.104) TERMINATIVE INTERLATIVE – [Korpus Malti 3.0; literature20]

Ħriġt	*ilsienek*	*u*	*pinġejt*	*linja*	*bżieq*
exit.CAUS:2SG.PFV	tongue:2SG	and	paint:2SG.PFV	line	saliva

dritta	*miż-żokra*	*[sa*	***bejn***	*iż-żewġ*	*ħbejżiet*
straight:F	from:DEF-navel	**[till**	**between**	DEF-two	loaf:DIM:PL

gustużi	*ta'*	*sidirha]*$_{PP}$
nice:PL	of	breast:3SG.F]

'You brought out your tongue and painted a straight line of saliva from the navel [**as far as between** the two lovely loaves of her breast].'

In this situation, the line of saliva represents the Figure. It is not the case that the line changes position. What happens is that the line grows in length until it stops in the cleavage of the anonymous woman. Her two breasts and the gap between them constitute the Ground. The use of (П51) *sa* 'till' is motivated by the author's wish to emphasise that the motion event is discontinued for good at this point.

5.1.2.2.18.2 Motion from between the Ground

Borg/Azzopardi-Alexander (1997: 165) mention the П-П combination *minn bejn* as the usual way of expressing motion from the medial position between the two parts of a distributed Ground. This combination is exactly what one would predict on the basis of the knowledge that (П41) *minn* 'from' is used almost always when motion from a Source is expressed.

The example (5.105) is not as straightforward as it might seem superficially.

(5.105) INTERABLATIVE – [Korpus Malti 3.0; news30456]

Ir-raġel	*ħareġ*	***[minn***	***bejn***	*żewġ*	*karozzi*	*f'*	*din*
DEF-man	exit	**[from**	**between**	two	car:PL	in	DEM:F:PROX

it-triq]$_{PP}$
DEF-street]

'The man appeared [**from between** two cars in this street] [. . .].'

85 The example stems from a text with slightly erotic touches. We were unable to find any better illustration of the TERMINATIVE INTERLATIVE than this.

The example stems from a report on a street accident. A man (= the Figure) steps unexpectedly onto the street and is hit by a car. The Figure emerges from a Ground that consists of two cars between which the man had to pass to reach the street. As we will see in the subsequent section, *minn bejn* is also used for the PERLATIVE. This potential ambiguity notwithstanding, we assume that (5.105) is a good example of the INTERABLATIVE because the motion verb *ħareġ* 'exit' focuses on the whereabouts of the Figure immediately before the man appears on the street. How he came to be between the two cars is irrelevant to the motion event. Otherwise, the motion verb *għadda* 'pass' could have been chosen to convey the idea that the man's movement began before he passed between the two cars.

5.1.2.2.19 Motion to and from outside the Ground

The EXTRAESSIVE is an addition to Hagège's taxonomy.[86] The same holds for the dynamic equivalent of the EXTRAESSIVE – the EXTRALATIVE – which is subsumed under Hagège's (2010: 261) notion of POSTLATIVE.[87]

5.1.2.2.19.1 Motion to outside the Ground

5.1.2.2.19.1.1 Bare EXTRALATIVE

Borg/Azzopardi-Alexander (1997: 158) mention a number of options for expressing motion to the exterior but in their list of three (= (Π28) *lejn* 'towards', (Π51) *sa* 'till', and (Π29) *lil* 'to'), there is no mention of the bare EXTRALATIVE. It is worth noting that the EXTRALATIVE is the only category for which the Πs in the above brackets are mentioned as directionality markers in the reference grammar. We have found ample evidence of the configuration marker (Π3) *barra* 'outside' (either with NOMINATIVE government or ABLATIVE government) in combination with a motion verb but in the absence of a directionality marker as shown in (5.106).

(5.106) Bare EXTRALATIVE I

(a) NOMINATIVE government – [Korpus Malti 3.0; news113089]

Ironikament	*l-uniku*	*Ingliż*	*li*	*mar*	***[barra***	*l-pajjiż]*$_{PP}$
ironically	DEF-sole	English	that	go	**[outside**	DEF-country]

ma	*kienx*	*parti*	*mill-iskwadra*	*ta'*	*Roy Hodgson*
NEG	be:NEG	part	from:DEF-team	of	Roy Hodgson

'Ironically, the only Ennglishman who left [the county] was not part of Roy Hodgson's team.'

---

86 Cf. Section 5.1.2.1.15.

87 Cf. Section 5.1.2.2.15.

(b) ABLATIVE government – [Korpus Malti 3.0; parl3583]
Niftakru *li* *kien* *hemm* *ministru* *li* *kien* *mar*
1.IPFV:remember:PL that be there minister that be go
[barra ***minn*** *Malta]*$_{PP}$
[outside **from** Malta]
'We remember that there was a minister who had left [Malta] [. . .].'

It is also possible to use (П44) *oltre* 'beyond' for the bare EXTRALATIVE – again either with NOMINATIVE or ABLATIVE government.[88] This variation is captured by the (a)- and (b)-sentences under (5.107).

(5.107) Bare EXTRALATIVE II
(a) [Korpus Malti 3.0; news123648]
dan *mar* ***[oltre*** *l-kompetenza* *tagħha]*$_{PP}$
DEM:M:PROX go **[beyond** DEF-competence of:3SG.F]
'[. . .] this went [**beyond** her competence].'
(b) [Korpus Malti 3.0; news220156]
l-Prim Ministru *Joseph Muscat mar* ***[oltre*** ***mill**-mandat*
DEF-prime_minister Joseph Muscat go **[beyond** **from**:DEF-mandate
tiegħu]$_{PP}$
of:3SG.M]
'[. . .] the prime minister Joseph Muscat went [**beyond** his mandate] [. . .].'

As far as we can tell on the basis of the data collected from the Korpus Malti 3.0, (П3) *barra* 'outside' is preferably used in spatial situations in which there is a concrete motion event involving a Figure that moves physically to the outside of the Ground. In the case of (П44) *oltre* 'beyond', however, the hits are predominantly abstract in the sense that an action, opinion, or attitude is depicted as exceeding the legal (or similar) limits imposed on people's rights.

Another difference between (П3) *barra* 'outside' and (П44) *oltre* 'beyond' is the impossibility of the latter to combine with a directionality marker. Thus, there is no evidence of a DIRECTIONAL EXTRALATIVE or a TERMINATIVE EXTRALATIVE construction in which (П44) *oltre* 'beyond' functions as configuration marker.

5.1.2.2.19.1.2 Directional extralative

There are two ways to form the DIRECTIONAL EXTRALATIVE. The first option is the expected combination of (П21) *għal* 'for' with (П3) *barra* 'outside' (+/– (П41) *minn*

88 Cf. Section 4.1.2.2.1.2.2.2 (iii).

'from') not mentioned in the reference grammar. Typical examples of this Π-Π combination are given in (5.108). In (5.108b), we have a ternary combination of Πs because in addition to the presence of the directionality marker and the mandatory configuration marker, there is also ABLATIVE government signalled by (Π41) *minn* 'from'.

(5.108) DIRECTIONAL EXTRALATIVE I

(a) nominative government – [Korpus Malti 3.0; news145985]

kollox *jitmexxa* ***[għal*** ***barra*** *Bieb* *il-Belt]*$_{PP}$
everything 3SG.M.IPFV:REFL:walk.CAUS **[for** **outside** gate DEF-city]
'[. . .] everything is conducted [**outside** the City Gate].'

(b) ABLATIVE government – [Korpus Malti 3.0; news132936]

Il-fattur *prinċipali* *li* *wassal* *għal* *dan*
DEF-factor principal that arrive:CAUS for DEM:M:PROX
it-tnaqqis *kien* *ammont* *inqas* *ta'* *posta* *indirizzata*
DEF-reduction be amount less.CMP of mail direct:PART:F
[għal ***barra*** ***minn*** *Malta]*$_{PP}$
[for **outside** **from** Malta]
'The main factor which led to this reduction is the smaller amount of mail directed [**to** (addresses) **outside** Malta].'

Like in several other cases of directional meanings, there is also evidence of the alternative directionality marker (Π28) *lejn* 'towards' in combination with (Π3) *barra* 'outside' (+/– (Π41) *minn* 'from'). The examples in (5.109) illustrate this possibility.

(5.109) DIRECTIONAL EXTRALATIVE II

(a) [Korpus Malti 3.0; news164981]

Mintoff *ried* *litteralment* *jibni* *bihom* *lil* *Malta*
Mintoff need literally 3SG.M.IPFV:build with:3PL to Malta
biex *tkun* *magna* *ta'* *produzzjoni* *li*
to 3SG.F.IPFV:be.FUT machine of production that
tesporta ***[lejn*** ***barra*** *l-pajjiż]*$_{PP}$
3SG.F.IPFV:export **[towards** **outside** DEF-country]
'Mintoff literally wanted to build up Malta with them so that it would be a machine of production that exports [**to** (recipients) **outside** the country].'

(b) [Korpus Malti 3.0; news205158]

Għolew *ukoll* *il-prezzijiet* *ta'* *pakketti* *mibgħuta*
become_dearer:3PL.PFV also DEF-price:PL of parcel:PL PART:send:PL
[lejn ***barra*** ***minn*** *Malta]*$_{PP}$
[towards **outside** **from** Malta]
'Also the prices for parcels sent [**to** (addresses) **outside** Malta] increased.'

The П-П combination *lejn barra (minn)* is attested only sixteen times in the Korpus Malti 3.0 and thus constitutes a minor option for expressing the DIRECTIONAL EXTRALATIVE. The (b)-sentences in (5.108) and (5.109) contain further examples of П-П-П combinations.

5.1.2.2.19.1.3 TERMINATIVE EXTRALATIVE

In the domain of the TERMINATIVE EXTRALATIVE, we have nothing out of the way to report. Directionality marking is the task of (П51) *sa* 'till' whereas configuration is expressed by (П3) *barra* 'outside' (+/– (П41) *minn* 'from') as shown in (5.110).

(5.110) TERMINATIVE EXTRALATIVE

(a) [Korpus Malti 3.0; news147538]

il-kaptan	*irnexxielu*	*jwassal*	*il-vapur*
DEF-captain	succeed:IO:3SG.M	3SG.M.IPFV:arrive.CAUS	DEF-ship

tal-merkanzija	***[sa***	***barra***	*l-Port*	*il-Kbir]$_{PP}$*
of:DEF-goods	**[till**	**outside**	DEF-harbour	DEF-big]

'[. . .] the captain succeeded in steering the merchantman [**out of** the Grand Habour] [. . .].'

(b) [Korpus Malti 3.0; news205282]

Ma'	*l-impatt*	*is-sewwieq*	*baqa'*	*jsuq*	***[sa***	***barra***
with	DEF-impact	DEF-driver	remain	3SG.M.IPFV:drive	**[till**	**outside**

minn	*Ħ'Attard]$_{PP}$*	*u*	*baqa'*	*sejjer*	*il-Mosta*
from	Ħ'Attard]	and	remain	go.PART	DEF-Mosta

'With the impact the driver kept driving [**out of** Ħ'Attard] and kept going to Mosta.'

In both situations, the Figure – the merchantman in the (a)-sentence and the driver (or the car) in the (b)-sentence – continue their journey after they have reached a temporary position outside the Ground – the Grand Habour in the (a)-sentence and the village of Ħ'Attard in the (b)-sentence. Being outside the port or the village is an important stage of the entire motion event because in both situations the Figure tries to escape a situation of danger. The danger diminishes as soon as the Figure is outside the Ground.

5.1.2.2.19.2 Motion from beyond the Ground

Borg/Azzopardi-Alexander (1997: 158) state that motion from the exterior is expressed by the combination of (П41) *minn* 'from' and (П3) *barra* 'outside'. This is exactly what we expect on the basis of what we have learned sofar about the expression of Source-relations. In (5.111a), we witness a case of NOMINATIVE government whereas (5.111b) gives evidence of ABLATIVE government which results in a construction with a sequence of three ΠS.

(5.111) EXTRAABLATIVE

(a) [Korpus Malti 3.0; news115170]

festa li fiha jipparteċipaw iż-żebbuġin u
feast that in:3SG.F 3.IPFV:participate:PL DEF-żebbuġi:PL and
numru kbir ta' nies li jiġu ***[minn*** ***barra***
number big of people that 3.IPFV:come:PL **[from** **outside**
r-raħal]$_{PP}$
DEF-village]
'[. . .] a feast in which the inhabintants of Żebbuġ participate and a high number of people who come [**from** **outside** the village].'

(b) [Korpus Malti 3.0; academic446]

Persentaġġ kbir tal-plejers tat-tim kienu
percentage big of:DEF-player:PL of:DEF-team be:3PL.PFV
jiġu ***[minn*** ***barra*** *mir-Russja]*$_{PP}$
3.IPFV:come:PL **[from** **outside** from:DEF-Russia]
'A big percentage of the team's players came [**from** **outside** Russia].'

5.1.2.2.20 Path

This section is organised differently from the previous ones because according to Hagège's (2010: 287) dictum "four functions, namely TERMINATIVE, PERLATIVE, PROLATIVE and SECUTIVE, have no static counterpart." The TERMINATIVE was the topic of Section 5.1.2.2.7 already where it was concluded that (Π51) *sa* 'till' more often than not serves the purpose of directionality marker in Π-Π combinations. If this can be said with reference to the TERMINATIVE, one should not lose sight of the possibility that what is labelled PERLATIVE, PROLATIVE, and SECUTIVE might turn out to involve prolific directionality markers, too. Hagège (2010: 291) assumes that

> the perlative/prolative [. . .] semantic functions correspond [. . .] to routes that involve the act of passing and directions that involve any phase of motion in a frame of reference. English examples are *through the kitchen, past the entrance, across/over the roof* for the perlative/prolative. [original italics]

PERLATIVE and PROLATIVE belong to a category within the domain of general location that is potentially compatible with almost every configuration, viz. Path. To explore this possibility, we have to approach the domains of these meaning categories from a different angle than in the bulk of the previous sections. This deviation from the usual pattern is fully justified on account of the many intricate characteristics of the domain to be explored in what follows. We start by way of deconstructing the SECUTIVE. – a category whose exact relationship to the other Path categories is explained nowhere in Hagège's (2010) study.

5.1.2.2.20.1 SECUTIVE (including PROXIMATE/PROXI(AB)LATIVE)

The SECUTIVE is illustrated with the English ΠS *with*, *along* and the gerund *following* (Hagège 2010: 261). In the chart of functions, Hagège (2010: 287) reduces these options to English *along*. Moreover, Hagège (2010: 286) claims that in a spatial situation that involves the SECUTIVE the Figure "follow[s] a place or person." Apart from taking issue with the questionable restriction of the animate Ground to human beings,[89] we argue that the motion event of following is categorically different for Grounds that are human or animate as opposed to Grounds that represent locations. If the Ground that is being followed bears the feature [+human] it can be assumed that the Ground is on the move, too. The Ground thus determines the route that the Figure has to take in order to keep following the Ground. This means that both Figure and Ground are changing their positions in the same direction. If the Ground is a "place" however, it is difficult to conceive of the Ground as being in the process of moving. The location is static and remains stable at a given place. Following a "place" cannot be the same process as following a "person".

What links the different kinds of following events nevertheless, is the virtual trail, i.e. the course of the motion event that is determined either by the animate Ground moving in a given direction ahead of the Figure or the shape and extension of the "place" along which the Figure moves. With reference to the latter scenario, one understands why Borg/Azzopardi-Alexander (1997: 167–168) dedicate eleven subsections of their catalogue of spatial categories to Grounds that are defined as "long object[s]". To clarify the issue of whether the following events can be lumped together, we first look at a typical situation that involves a human Figure and a human Ground. In (5.112), it is shown that the fact that a Figure moves in the direction of the likewise moving Ground is normally expressed by (Π60) *wara* 'after' in combination with a motion verb like *mexa* 'walk'.

89 Not only other animate beings such as animals can be followed (on occasion of a hunting party, for instance) but also inanimate objects such as cars, etc. Moreover, the act of following is not limited to situations which involve concrete Grounds since abstract concepts such as ideas can be followed virtually, too.

(5.112) Following a human Ground – [Korpus Malti 3.0; literature7]

Qam bilwieqfa u mexa ***[wara*** *l-oħrajn]*$_{PP}$
rise standing and walk **[after** DEF-other:PL]
'He stood up and followed [the others].'

The use of this Π associates example (5.112) with the SPATIAL POSTLATIVE as described above.[90] The motion event takes place behind the Ground no matter whether the Ground is changing its position or not. In contrast, the situation described in (5.113) which also involves a human Figure and a human Ground both of which are moving in one and the same direction is different from the previous situation

(5.113) Accompanying a human Ground [Korpus Malti 3.0; literature43]

Is-surġent dendel it-tabella Out on Duty mal-bieb
DEF-sergeant hang DEF-table Out_on_Duty with:DEF-door
tal-għassa u mexa ***[ma'*** *Albert]*$_{PP}$ *sa ħdejn*
of:DEF-police_station and walk **[with** Albert] till beside
il-latrina
DEF-latrine
'The sergeant put the sign-board "Out on Duty" on the door of the police station and walked [**with** Albert] to the public lavatory [. . .].'

It is conceivable that the two characters mentioned in (5.113) are moving alongside each other and that one of them is leading the other to their destination so that the one who is led by the other could be seen as following the other. This interpretation notwithstanding, we interpret the presence of (Π30) *ma'* 'with' as an instance of the COMITATIVE,[91],[92] i.e. a situation of accompaniment that is conceptually very much like genuine spatio-temporal meaning categories but is counted as a member of the set of non-spatio-temporal categories by Hagège (2010: 262). The COMITATIVE interpretation of (5.113) does not mean, however, that (Π30) *ma'* 'with' could generally be filed away like that.

In (5.113), there are two *ma'*-PPS. Besides the PP with COMITATIVE meaning, there is also the properly spatial PP *mal-bieb* 'on the door'. Borg/Azzopardi-Alexander (1997: 163) classify this usage of (Π30) *ma'* 'with' as a case of motion to create lateral contact with the Ground. Before we go into the problems that are posed by the criterion of contact, we present a selection of examples of (Π30) *ma'* 'with' being used

90 Cf. Section 5.1.2.2.15.1.

91 Cf. Section 5.1.3.4.

92 The semantic network which links COMITATIVE and other categories expressed by Maltese (Π30) *ma'* 'with' is described in Stolz/Stroh/Urdze (2006: 294–304).

according to Hagège's (2010) above-mentioned idea of a spatial situation in which the Figure "follows a place". A good example to start the discussion with is (5.114).

(5.114) SECUTIVE (Π30) *ma'* 'with' – [Camilleri 2013: 133]

Ma	*kienx*	*fi*	*ħsiebu*	*jagħmel*	*dawra*	*oħra*
NEG	be:NEG	in	thought:3SG.M	3SG.M.IPFV:make	turning	other:F
*[**ma**l-pjazza*	*ta'*	*Raħal Ġdid]ₚₚ*				
[**with**:DEF-square	of	Raħal Ġdid]				

'He did not intend to make another turn [**around** the square of Raħal Ġdid] [. . .].'

Inspector Gallo has driven to the centre of Raħal Ġdid searching for suspects of a crime. He has seen enough on the square of the village and in lieu of driving around it again he reverses the car to leave the place (as transpires from the further context). The imagined movement of the car around the square is of interest here because the use of (Π30) *ma'* 'with' cannot causally be connected to the fact that there is contact between the Figure and the Ground. What motivates the presence of the Π is the necessity of the car following the path that is determined by the architectural properties of the square. The Figure is required to move along this path to execute the turning.

In our corpus – electronic or other – cases of this kind are abundantly attested. It is necessary therefore to restrict the discussion to a selection of pertinent cases. The seven examples in Table 5.5 are discussed subsequently. Each of the examples is drawn from the Korpus Malti 3.0.

Table 5.5: (Π30) *ma'* 'with' in SECUTIVE and related functions.

#	source	example / translation
1	literature21	*Imbagħad reġgħu bdew mexjin [**ma**l-wied]$_{PP}$ lejn Tal-Qadi [. . .]* 'Then they started again to walk [**along/through** the valley] towards Tal-Qadi [. . .].'
2	literature20	*Imxejtu [**ma**x-xatt]$_{PP}$ lejn Marsalforn [. . .]* 'You walked [**along** the beach] towards Marsalforn [. . .].'
3	literature74	*[. . .] baqa' tiela' [**ma**t-triq il-kbira]$_{PP}$ u lewa għan-naħa ta' l-imtieħen [. . .]* '[. . .] he kept going up [**along** the main street] and turned to the side of the windmills [. . .].'
4	news162350	*L-Istati Uniti tal-Amerika bnew ħitan [**ma**l-fruntiera li għandhom mal-Messiku]$_{PP}$ [. . .]* 'The United States of America built walls [**along** the frontier they have with Mexico] [. . .].'
5	academic2	*[. . .] kien ikolli nitlaq mill-port ta' Lisbona, indur [**ma**l-Cape of Good Hope]$_{PP}$ u naqsam il-Baħar Għarbi [. . .]* '[. . .] I would have to leave from the harbour of Lisbon, turn [**around** the Cape of Good Hope] and cross the Arabian Sea [. . .].'

Table 5.5 (continued)

#	source	example / translation
6	news108426	*Għax biex wieħed imur jixxabbat [**ma**l-Muntanja Everest]$_{PP}$ mhux biss irid iħallas il-passaġġ [. . .]* 'Because for someone to go climbing [the Mount Everest] he not only needs to pay for the passage [. . .].'
7	news89205	*Illum dan l-istrument infirex [**ma**l-Afrika kollha]$_{PP}$, kif ukoll fi Franza u fir-Renju Unit [. . .].* 'Today this instrument is diffused [all **over** Africa] and also in France and the United Kingdom [. . .].'

The complements of (Π30) *ma'* 'with' in Table 5.5 are horizontally or vertically extended geo-objects along which the Figure moves to reach the Goal. In #1, the Figure walks along a valley to arrive at the destination. In #2, the Figure walks along the beach to reach Marsalforn. In #3, the Ground is the main street along which the Figure walks before turning to the Goal. In #4, the frontier between the USA and Mexico forms the Ground for the construction of walls that cover extended sections of the border. In #5, the route of an imagined voyage leads along the Cape of Good Hope. In #6, the slopes of Mount Everest represent the Ground along which the climbing event takes place. In #7, Africa is the Ground for the continent-wide diffusion of a certain musical instrument. In this last case, the SECUTIVE interpretation is perhaps better replaced by a PERLATIVE interpretation (similarly #1).

It results from the above that (Π30) *ma'* 'with' is indeed a good candidate for the role of SECUTIVE marker – to a minor degree this also holds for (Π33) *matul* 'during' whose spatial functions will be addressed very briefly below.[93] For (Π30) *ma'* 'with' this function is not registered in the reference grammar where the Π is mentioned only in connection with lateral contact including the static category at rest (Borg/Azzopardi-Alexander 1997: 163). Remember that a static counterpart of the SECUTIVE is assumed to be non-existent by Hagège (2010: 291). How is it possible, if at all, to accommodate Borg/Azzopardi-Alexander's (1997) analysis with Hagège's (2010) restriction? This question can only be answered by way of assessing the validity of the proposal made by Borg/Azzopardi-Alexander (1997: 163). To achieve this, we have to return to the domain of the static meaning categories. The proper SECUTIVE resurfaces as our topic below.[94]

93 In Section 5.1.2.2.20.1.4 (ii).

94 Cf. Section 5.1.2.2.20.1.4.

5.1.2.2.20.1.1 PROXIMATE

The first point of dissent of ours with regard to what the reference grammar suggests is the lateral position of the Figure in relation to the Ground. In point of fact, it is difficult to interpret the bulk of the occurrences of static (Π30) *ma'* 'with' as referring to a position at the side of the Ground. If we look at cases like (5.115) we notice that the position of the Figure cannot be lateral, in the first place. In the (temporary) absence of a better term, we label this meaning category "lateralis" in inverted commas. A suitable replacement for "lateralis" will be introduced after the discussion of (5.115).

(5.115) "Lateralis" – [Korpus Malti 3.0; literature32]

L-oħrajn	*intasbu*	*bilqiegħda*	***[mal-art]**$_{PP}$*	*taħt*
DEF-other:PL	REFL:find:3PL.PFV	sitting	[**with**:DEF-ground]	under

ħoġor	*it-tieqa*	*daharhom*	***[mal-ħajt]**$_{PP}$*
sill	DEF-window	back:3PL	[**with**:DEF-wall]

'The others were sitting [**on** the ground] under the window sill, their backs [**to** the wall].'

Example (5.115) contains two *ma'*-PPs none of which lends itself to an interpretation according to which the Figure is located in a lateral position to the Ground. One might argue that *mal-ħajt* 'at the wall' fits the description of the lateral positioning of the Figure if the Ground is construed conceptually in such a way that the surface of the wall is identical to a lateral part of the same object that has a vertical and a horizontal dimension. As to *mal-art* 'on the ground', there is no vertical dimension. The surface of the ground would have to be interpreted as a lateral part of the Ground. This interpretation is even more hazardous than the previous one.

Borg/Azzopardi-Alexander (1997: 162–163) distinguish two major categories of laterality whose distinction rests on the criterion of contact. If contact between Figure and Ground does not apply, (Π8) *biswit* 'facing' and (Π32) *maġenb* 'close to' are said to serve as configuration markers. We have argued against laterality and replaced this criterion with that of proximity.[95] For Borg/Azzopardi-Alexander (1997: 163), in a lateral spatial situation with contact between Figure and Ground, (Π30) *ma'* 'with' indicates configuration except if Path has to be expressed. To express lateral Path, they assume that the Π-Π combination *minn maġenb* is employed. The same combination is also one of the two options for expressing PROXIMATE Path (the supposed alternative being *minn biswit*). The necessity of making use of a different configuration marker in the domain of Path – a marker which otherwise serves the

95 Cf. Section 5.1.2.1.4.

same purpose in the domain of the PROPINQUUS circumstantially corroborates our doubts as to the criterial status of laterality.

The category label PROPINQUUS has been introduced above.[96] It covers a multitude of Πs which express that the Figure is situated in the vicinity of the Ground. The distance between Figure and Ground remains undetermined so being near the Ground is a relative notion. In contrast to the PROPINQUUS Πs, (Π30) *ma'* 'with' – at least according to Borg/Azzopardi-Alexander (1997: 163) – implies that the Figure and the Ground physically touch one another. The distance between the two participants ideally is zero. As we will see in the next paragraph, contact in the narrow interpretation of the term is not a necessary condition. What is important is that the distance between Figure and Ground is minimal as opposed to the distance that applies in the case of the PROPINQUUS. This difference motivates the introduction of the term PROXIMATE which invokes a position of the Figure that is particularly close to the Ground.

To defend the PROXIMATE as a spatial category that is based on the (extreme) closeness of Figure and Ground in space, we provide two examples in (5.116) which involve Figure-Ground pairs for which physical contact is excluded or at least irrelevant.

(5.116) PROXIMATE

(a) [Korpus Malti 3.0; european6248]

jinżammu kemm jista' jkun qrib
3.IPFV:REFL:keep:PL how_much 3SG.M.IPFV:can 3SG.M.IPFV:be.FUT near

*[**ma**l-ħajt li fih ikunu mwaħlin*
[**with**:DEF-wall that in:3SG.M 3.IPFV:be.FUT:PL PART:fix.CAUS:PL

il-bibien]$_{PP}$
DEF-door:PL]

'[. . .] they are kept as close as possible [**to** the wall into which the doors are built] [. . .].'

(b) [Korpus Malti 3.0; literature10]

l-irġiel donnhom jinqasmu f' żewġ kategoriji
DEF-man.PL seem:3PL 3.IPFV:REFL:split:PL in two category:PL

dawk miżżewġin u li jkollhom x'
DEM:PL:DIST PART:marry:PL and that 3SG.M.IPFV:have.FUT:3PL what

miżżewġin u li allura jorqdu b' wiċċhom
PART:marry:PL and that then 3.IPFV:sleep:PL with face:3PL

96 In Section 5.1.2.2.20.1.

[mal-ħajt]$_{PP}$
[**with**:DEF-wall]
'[. . .] the men seem to divide into two categories: those who are married and who have an interest in a woman and those who are unmarried and thus sleep with their faces [**to** the wall].'

In (5.116a), the combination of the adverb *qrib* 'near' (cf. (Π48) *qrib* 'near') with the PROXIMATE marker (Π30) *ma'* 'with' is especially telling because their cooccurrence depicts the spatial situation as one in which the location of the Figure (= technical implements to measure the hydro pressure within a ship below deck) is defined as close as possible to the Ground (= the wall in which the doors are integrated) but no mention is made of contact between Figure and Ground. This means that the implements could also be placed somewhere near the wall but not necessarily in or on it. Contact is thus not presupposed. Similarly, in (5.116b), the Figure (= the faces of the bachelors) are turned towards the wall in their sleep but it is not implied that there is physical contact between the face and the wall. However, Figure and Ground are spatially very close to each other. No other object is inserted in the gap between the face and the wall. What is additionally interesting with regards to (5.116a) is the *fi*-PP *fih* 'in it' whose pronominal complement anaphorically refers back to *il-ħajt* 'the wall'. The doors are described as being fixed (passive participle *mwaħħlin* 'fixed' ← *waħħal* 'fix') in the wall, i.e., they are an integrated part of the entire construction. If, however, the object that is "added" to the wall is not properly integrated into the structure the same verb requires a *ma'*-PP as in (5.117).

(5.117) PROXIMATE – [Korpus Malti 3.0; news179564]
il-programm se jsir kif hemm reklamat
DEF-program FUT 3SG.M.IPFV:become how there advertise:PART
ukoll fuq billboard li Silvio Parnis waħħal [mal-ħajt
also on billboard that Silvio Parnis fix [**with**:DEF-wall
tal-Ħabs ta' Kordin fil-Pjazza ta' Raħal Ġdid]$_{PP}$
of:DEF-prison of Kordin in:DEF-square of Raħal Ġdid]
'[. . .] the program will be as advertised also on a billboard that Silvia Parnis attached [**to** the wall of the prison of Kordin in the square of Raħal Ġdid].'

The billboard playing the role of Figure is fixed to the prison wall but covers only part of the wall's surface without being physically integrated into the wall's body. There is contact between Figure and Ground. However, we argue that it is not so much the contact that triggers the employment of (Π30) *ma'* 'with' but the fact that Figure and Ground do not fuse.

Even if the Figure is an object that partially enters the Ground but parts of the Figure still protrude from the Ground, (Π30) *ma'* 'with' is chosen as can be gathered from (5.118).

(5.118) PROXIMATE – [Korpus Malti 3.0; literature37]

Meta	*wasal*	*id-dar*	*waħħal*	*grampun*	*[mal-ħajt]$_{PP}$*	*u*
when	arrive	DEF-house	fix	hook	[**with**:DEF-wall]	and
dendel	*il-mera*	*fejn*	*xtaqet*	*Lugrezja*		
hang	DEF-mirror	where	want:3SG.F.PFV	Lugrezja		

'When he arrived home, he fixed a hook [**to** the wall] and hung the mirror where Lugrezja wanted it to hang.'

If it is important that Figure and Ground are positioned very close to each other but do not become one, it is clear that as such is not crucial for the employment of the Π under inspection. The hook and the wall remain two separate entities although they are physically interconnected.

In Table 5.6, we provide a selection of nine additional examples of the PROXIMATE. In three examples each, the nouns *saqaf* 'ceiling', *swar* 'bastions', and *faċċata* 'façade' represent the Ground. The examples stem from the Korpus Malti 3.0.

Table 5.6: Further examples of PROXIMATE (Π30) *ma'* 'with'.

#	source	example / translation
1	literature21	*L-istalaktiti [**ma**s-saqaf]$_{PP}$ kienu jleqqu* 'The stalactites [**on** the ceiling] were shining [. . .].'
2	literature43	*Għax għalkemm kien hemm tliet fannijiet iduru [**ma**s-saqaf]$_{PP}$, kulħadd kien għarqan xraba*. 'Because, in spite of the existence of three fans ventilating [**on** the ceiling], everybody was bathed in sweat.'
3	literature20	*Il-ħoss tax-xita [**ma**s-saqaf tal-karozza]$_{PP}$ kien qawwi* 'The sound of the rain [**on** the car's hood] was loud [. . .].'
4	news137550	*kull fejn tħares tara massa ta' scaffolding [**ma**s-swar]$_{PP}$* '[. . .] everywhere you look you see a lot of scaffolding [**on** the bastions] [. . .].'
5	news224541	*jinkludi wkoll spettaklu ta' dwal u laser [**ma**s-swar]$_{PP}$* '[. . .] it includes also a light and laser show [**on** the bastions] [. . .].'
6	parl5166	*ma jistax ikun hemm fencing [**ma**s-swar]$_{PP}$* '[. . .] there can be no fencing [**on** the bastions] [. . .].'
7	literature45	*Il-ġibs [**ma**l-faċċata]$_{PP}$ kien ilu li tqaxxar kważi kollu* 'The plaster [**on** the façade] has long since been peeled off almost completely [. . .].'

Table 5.6 (continued)

#	source	example / translation
8	news106502	*hemm dar antika li t-tabella [**ma**l-faċċata tagħha]*$_{PP}$ *tixhed li fiha għal bosta snin għex bniedem kbir* '[. . .] there is an old house, the plaque [**on** its façade] testifies that for many years an important person lived there [. . .].'
9	news128332	*[**Ma**l-faċċata tiegħu]*$_{PP}$ *nsibu żewġ armi tal-Gran Mastru Vilhena u oħra tal-Muniċipalità ta' Malta* '[**On** its façade] there were two coats of arms of the Grand Master Vilhena and another of the municipality of Malta [. . .].'

How is the Figure positioned in relation to the Ground in the nine examples? In #1–#3, the Ground is the ceiling of a room, a cave, or the hood of a car. The stalactites and the fans are hanging from the ceiling to which they are tied either geologically or by screws. These Figures constitute objects that could be removed from the Ground to which they are connected either by (natural) force or by way of using appropriate tools. Example #3 is different in the sense that the sound of the rain falling on the car's hood is not a concrete removable object. It is temporary, nevertheless, because it stops as soon as the rain ceases to fall. We interpret #3 as an instance of the PROXIMATE (i.e. of a static relation) because no action or motion verb is involved. There is a NP that objectifies the sound on the hood. Stalactites and fans are in physical contact with the cave's ceiling and the room's ceiling, respectively. The sound of the falling rain, however, is caused by the raindrops hitting the metal hood of the car. What all three examples #1–#3 have in common is that the Figure consists of several parts which are distributed over the Ground. This also holds for the sound provoked by the falling rain since the raindrops hit the hood at different places. The three Grounds represent extended surfaces.

The six remaining examples also involve Grounds that come in the shape of a sizable surface. There are concrete objects which cover part of this surface, namely the scaffolding in #4, the plaque in #8, and the coats of arms in #9. Similarly, the fencing in #6 would manifest itself in the presence of fences close to the bastions. In #7, it is the material that has been peeling off from the façade that represents the Figure. The last case is #5 where a light and laser show is projected onto the bastions. The Figures – including the plaster on the façade – belong to kinds of removable objects or temporary phenomena. The plaques and coats of arms can be screwed off, the scaffolding can be dismantled, the light show turned off, and the plaster peeled off further, but the Grounds would remain largely unaffected by the removal of the Figures. Except for #6, the examples given in Table 5.6 refer to situations of contact between Figure and Ground – for #3 only metonymically speaking. The fences in #6, however, would be erected close to but certainly not in physical contact with the bastions.

Superficially, this abundance of cases which feature a Figure and a Ground that are in contact attributes a high degree of importance to the criterion [+contact]. The examples in (5.119) counter this claim insofar as there is no physical contact between the two participants of the spatial situation. The Ground in these examples is always *bieb* 'door'.

(5.119) PROXIMATE I

(a) [Korpus Malti 3.0; culture1045]

Id-dħul huwa b' xejn iżda kull min jixtieq
DEF-entrance 3SG.M with nothing but every who 3SG.M.IPFV:wish
jista' jagħti donazzjoni [mal-bieb]ₚₚ
3SG.M.IPFV:can 3SG.M.IPFV:give donation [**with**:DEF-door]
'The admission is for free but everybody who wishes to give a donation can do this [**at** the door].'

(b) [Korpus Mlti 3.0; literature21]

Iż-żewġ suldati għassa [mal-bieb]ₚₚ ma kinux
DEF-two soldier:PL sentinel [**with**:DEF-door] NEG be:3PL.PFV:NEG
iċċaqalqu minn posthom
move:3PL.PFV from place:3PL
'The two soldiers on guard [**at** the door] had not moved from their places.'

(c) [Korpus Malti 3.0; literature21]

Miegħu kienu daħlu wkoll żewġt irġiel goffi
with:3SG.M be:3PL.PFV enter:3PL.PFV also two man.PL rude:PL
li baqgħu lura jistennew [mal-bieb]ₚₚ
that remain:3PL.PFV back 3.IPFV:wait:PL [**with**:DEF-door]
'With him two rude men had entered too who remained in the back waiting [**at** the door].'

(d) [Korpus Malti 3.0; news118443]

B' din is-sistema qegħdin imorru tajjeb il-ħwienet
with DEM:F:PROX DEF-system PROG:PL 3.IPFV:go:PL good DEF-shop.PL
tal-kafè u l-kumpanija privata li tipprovdi
of:DEF-coffee and DEF-company private:F that 3SG.F.IPFV:provide
s-sigurtà [mal-bieb tal-Qorti]ₚₚ
DEF-security [**with**:DEF-door of:DEF-court]
'With this system, the cafés and the private company that provides security [**at** the door of the court of justice] are doing fine.'

In none of the spatial situations featured in (5.119) does the Figure touch the Ground. The Figure is located in close vicinity of the Ground, but there remains a physical gap between the two participants. In (5.119a), the act of donating happens at the door but the donor does not necessarily touch the door. In (5.119b), the two soldiers are on guard at the door but are presumably standing some steps away from the door. The two characters mentioned in (5.119c) are certainly close to the door while they are waiting for things to happen but nothing requires them to be in physical contact with the door. Similarly, in (5.119d), the security service is provided at the door to the court of justice but this does not imply that any of the guards touches the door.

We know from previous examples (cf. Table 5.6) that (Π30) *ma'* 'with' is often employed when Figure and Ground touch each other. This also holds in cases in which *bieb* 'door' is the complement of this Π. If we compare the examples in (5.120) with those in (5.119), we immediately notice that there are potential minimal pairs which result from the differences in the kind of the Figures that are positioned relative to the Ground.

(5.120) PROXIMATE Π – contact

(a) [Kopus Malti 3.0; news129287]
rajt *avviż* *[**ma**l- bieb* *tal-Knisja]*$_{PP}$
see:1SG.PFV notice [**with**:DEF-door of:DEF-church]
'[. . .] I saw a notice [**on** the church door] [. . .].'

(b) [Korpus Malti 3.0; news75658]
Meta *mbagħad* *mar* *jagħmel* *rapport* *l-għassa*
when then go 3SG.M.IPFV:make report DEF-police_station
ta' *Marsalforn* *kien* *hemm* *tabella* *[**ma**l-bieb* *tgħid*
of Marsalforn be there signboard [**with**:DEF-door 3SG.F.IPFV:say
li *fl-għassa* *ma* *kien* *hemm* *ħadd]*$_{PP}$
that in:DEF-police_station NEG be there nobody]
'When he then went to make a report at Marsalforn police station there was a signboard [**on** the door saying that there was nobody in the police station] [. . .].'

(c) [Korpus Malti 3.0; literature12]
Bexxaqt *il-purtiera* *u* *niżżilt* *in-numru* *li* *kien*
open_half DEF-curtain and write_down:1SG.PFV DEF-number that be
hemm *fuq* *l-istiker* *imwaħħla* *[**ma**l-bieb]*$_{PP}$
there on DEF-sticker PART:fix.CAUS:F [**with**:DEF-door]
'I opened the curtain partly and wrote down the number that was there on the sticker fixed [**on** the door].'

The three examples in (5.120) reflect situations in which the Figure is fixed to the Ground. The function of Figure is fulfilled by a piece of paper (or other material) that is fixed to the door to make an announcement addressed to potential callers.

These notices are relatively small inanimate objects whereas those in (5.119) are human beings – if we allow for the donation being carried out by a human donor and the security service being provided by human guards, who are not fixed physically to the door. The interpretation of the spatial situation, as either involving contact or not, depends on the size of the Figure and animacy. We will find similar phenomena also in the domain of the dynamic relation.

5.1.2.2.20.1.2 PROXILATIVE

The next subsections reveal that the domain of the PROXIMATE resembles those of other spatio-temporal domains discussed previously insofar as there is evidence of the bare PROXILATIVE, the DIRECTIONAL PROXILATIVE, and the TERMINATIVE PROXILATIVE as well as the PROXIABLATIVE. What distinguishes the PROXIMATE domain from those we already are familiar with is the category of the SECUTIVE itself to which we will return below.[97]

5.1.2.2.20.1.2.1 Bare PROXILATIVE

Borg/Azzopardi-Alexander (1997: 163) provide an example of the bare PROXILATIVE which involves an action verb (*waħħal* 'fix') whose direct object (*granpun* 'hook') is meant to be fixed to the Ground (*ħajt* 'wall'). Two comparable cases are presented in (5.121).

(5.121) Bare PROXILATIVE I – contact

(a) [Korpus Malti 3.0; literature11]

Mill-bogħod jara 'l Ernest għadu jasal
from:DEF-distance 3SG.M.IPFV:see to Ernest still:3SG.M 3SG.M.IPFV:arrive
iserraħ il-basket [mal-ħajt]$_{PP}$
3SG.M.IPFV:give_rest DEF-basket [**with**:DEF-wall]
'From far he sees Ernest just arriving and resting the basket [**against** the wall].'

(b) [Vella 2014: 38]

lil Francesco ġabu ċatt [mal-ħajt]$_{PP}$
to Francesco bring:3PL.PFV flat [**with**:DEF-wall]
'[. . .] they brought Francesco flat [**to** the wall].'

(c) [Vella 2014: 39]

jissiktawh [mal-bieb]$_{PP}$ u jibdew
3.IPFV:tighten:PL:3SG.M [**with**:DEF-door] and 3.IPFV:begin:PL

97 Cf. Section 5.1.2.2.20.3.

ituh *bil-ponn*
3.IPFV:give:PL:3SG.M with:DEF-fist
'[. . .] they tighten him [**to** the door] and begin to hit him with their fists.'

In the (a)-sentence, the Figure is a basket which is temporarily leaning against the wall. A small object touches the Ground. As to the (b)- and (c)-sentences, the facts are different insofar as the Figure is a human being named Francesco who is bullied by his classmates. The bullies attack Francesco by way of pushing him flat against the wall in (5.121b). In (5.121c), the same victim is immobilised against the door and punched by his enemies. In all three of the examples, the Figures are moved into a position that involves their being in contact with the Ground albeit only temporarily.

The temporary nature of the spatial situation can be overridden by the criterion of contact as shown in (5.122) where the stalactites and stalagmites have developed over time into columns which reach from the ground to the ceiling of the cave and vice versa.

(5.122) PROXILATIVE II – contact – [Korpus Malti 3.0; news81684]
Marku kien sab ruħu quddiem għar kbir mimli ġebliet
Marku be find self:3SG.M in_front_of cave big full rock:PL
weqfin donnhom sħaħar imwaħħlin [mal-art]$_{PP}$
stand.PART:PL like:3PL wizard.PL PART:fix:PL [**with**:DEF-ground]
sas-saqaf u telgħin ħoxnin ħoxnin sa ma
till:DEF-ceiling and ascend.PART:PL thick:PL thick:PL till that
jraqqgħu f' nofs tat-triq u jerġgħu
3.IPFV:grow_thin:PL in half of:DEF-street and 3.IPFV:do_again:PL
jiħxienu meta jaslu [mas-saqaf]$_{PP}$.
3.IPFV:grow_thick:PL when 3.ipfv:arrive:pl [**with**:DEF-ceiling]
'Marku had found himself in front of a big cave full of erect rocks like wizards fixed [**to** the ground] as far as the ceiling and going upwards really thick until they become thinner halfway and get thicker again when they arrive [**at** the ceiling].'

There are two *ma'*-PPs but only the one after the motion verb highlighted in grey is of interest for the discussion. The stalagmites and stalactites move only metonymically in the sense that they slowly grow from above and upwards to eventually fuse to yield a column which connects the cave's ground to the cave's ceiling. In a way, these natural columns reach both the ground and the ceiling and touch them because they are physically connected to them.

In (5.123), events are reported that involve short (in one case also intermittent) contact between the Figure and the Ground.

(5.123) Bare PROXILATIVE III – contact

(a) [Korpus Malti 3.0 literature14]

Kien waħdu fis-sodda u xi ħadd kien qed
be one:3SG.M in:DEF-bed and some someone be PROG
iħabbat ***[ma****l-bieb* *tal-kamra]*$_{PP}$
3SG.M.IPFV:knock [**with**:DEF-door of:DEF-room]
'He was alone in the bed and someone was knocking [**on** the door of the room] [. . .].'

(b) [Korpus Malti 3.0; literature24]

Dan qam bilwieqfa f' daqqa u ħabat rasu
DEM:M:PROX rise standing in stroke and hit head:3SG.M
[ma*s* *saqaf]*$_{PP}$
[**with**:DEF ceiling]
'This one rose up suddenly and hit his head [**on** the ceiling].'

It is clear that the Figure is briefly in contact with the Ground. In (5.123a), the knocking event implies that the agent uses his knuckles to produce sound by way of hitting them against the door. In (5.123b), the presumably hurtful contact between head and door is involuntary and lasts only for some seconds.

From longterm via short-term contacts we move further towards those situations in which Figure and Ground are not in contact although (П30) *ma'* 'with' is used, nevertheless. The examples in (5.124) involve genuine motion verbs (marked in grey) which specify the kind of movement the Figure carries out to approach the Ground.

(5.124) Bare PROXILATIVE IV – no contact

(a) [Korpus Malti 3.0; literature77]

Minn fejn il-knisja ta' San Lawrenz telgħu mbagħad
from near DEF-church of San_Lawrenz ascend:3pl.pfv then
[ma*t-triq]*$_{PP}$ *'il ġewwa u waqfu biswit dar fejn*
[**with**:DEF-street] to inside and stop:3PL.PFV facing house where
il-vistuż wara li ħares 'il fuq u ***[ma****l-ġenb]*$_{PP}$
DEF-mourning after that look to on and [**with**:DEF-side]
resaq ***[ma****l-bieb]*$_{PP}$ *iħabbat*
approach [**with**:DEF-door] 3SG.M.IPFV:knock

'From near the church of San Lawrenz they then ascended [**along** the street] inwards, they stopped opposite a house whose [door] the mourner, after he had looked up and [**to** the side], approached to knock.'

(b) [Korpus Malti 3.0; culture1721]

Meta	*Al-Amari*	*mar*	*[**ma**t-tieqa]*$_{PP}$	*biex*	*jordna*	*burger*
when	Al-Amari	go	[**with**:DEF-window]	to	3SG.M.IPFV:order	burger

il-ħaddiema	*tal-ħanut*	*tal-fast food*	*baqgħu*
DEF-worker:PL	of:DEF-shop	of:DEF-fast_food	remain:3PL.PFV

mistgħaġba
PART:wonder:PL

'When Al-Amari went [**to** the window] to order a burger, the employees of the fast food restaurant were surprised [. . .].'

The (a)-sentence hosts three *ma'*-PPs which depend on motion verbs. While *telgħu mat-triq* 'they walked up the street' instantiates the SECUTIVE and thus belongs to a domain we will address below,[98] *resaq mal-bieb* 'he approached the door' is important for what we have to say in this section. The mourner (= Figure) approaches the door (= Ground), meaning: the situation as captured by (5.124a) is such that no contact between Figure and Ground has taken place yet. The entire scene is pictured before the contact can be made in the first place. The PP *mal-bieb* thus does not require the mourner to touch the Ground. Similarly, the Figure in the (b)-sentence walks up to the window via which the fast food is sold. Nothing requires the Figure to touch the window. In both sentences, the Figure moves to a position that is very close to the Ground but there is still some empty space left between the Figure and the Ground. We conclude that (Π30) *ma'* 'with' is a configuration marker just as many of the Πs featured in the previous sections. In accordance with these other Πs, (Π30) *ma'* 'with' need not be accompanied by a dedicated directionality marker in Goal-oriented constructions. This does not mean, however, that (Π30) *ma'* 'with' is exempt from combining with directionality markers.

5.1.2.2.20.1.2.2 Directional proxilative

It suggests itself that (Π30) *ma'* 'with' is not only involved in constructions of the bare PROXILATIVE but behaves like so many other spatially employed Maltese Πs insofar as it combines with the usual directionality marker (Π21) *għal* 'for' as shown in (5.125).

98 Cf. Section 5.1.2.2.20.1.4.

(5.125) DIRECTIONAL PROXILATIVE

(a) [Vella 2014: 38]

Francesco jxejjer idu kważi ***[għal ma'*** *wiċċ Kevin]*$_{PP}$
Francesco 3SG.M.IPFV:wave hand:3SG.M almost **[for with** face Kevin]
'[. . .] Francesco waved his hand almost [**into** Kevin's face] [. . .].'

(b) [Korpus Malti 3.0; literature23]

Minn hemm Qrempuċu qabeż għal fuq siġġu u mis-siġġu ***[għal ma****l-purtiera tal-qasab]*$_{PP}$
from there Qrempuċu jump for on chair and from:DEF-chair **[for with**:DEF-curtain of:DEF-cane.PL]
'From there, Qrempuċu jumped onto a chair and from the chair [**towards** the reed curtain].'

(c) [Korpus Malti 3.0; news122147]

biex jiddefendi ruħu qabad wieħed minnhom minn għonqu u imbottah ***[għal ma****l-ħajt taċ-ċella]*$_{PP}$
to 3SG.M.IPFV:defend self:3SG.M grab one from:3PL from neck:3SG.M and push:3SG.M **[for with**:DEF-wall of:DEF-cell]
'[. . .] to defend himself he grabbed one of them by his neck and pushed him [**towards** the wall of the cell].'

Except for the (a)-sentence, it remains an open question whether the Figure makes physical contact with the Ground. In (5.125a), Francesco is waving his hands in self-defence in the direction of Kevin's face but does not touch his adversary. In (5.125b), Qrempuċu jumps from the chair in the direction of the reed curtain, but the movement might come to a stop close to the Ground. Similarly, it is clear that the person acts against those who are threatening him by way of attacking one of them violently. However, whether the victim of his attack bumps into the wall or is thrown only in the direction of the wall cannot be determined with certainty. On these grounds, we conclude that the DIRECTIONAL PROXILATIVE applies in situations in which the Figure moves towards a position that is located very close to the Ground (with contact between Figure and Ground being an option).

5.1.2.2.20.1.2.3 TERMINATIVE PROXILATIVE

Tangible evidence for the existence of the TERMINATIVE PROXILATIVE is scarce. The directionality marker (П51) *sa* 'till' combines hardly ever with the configuration marker (П30) *ma'* 'with'. In the Korpus Malti 3.0, we have come across only a single uncontroversial case which we reproduce in (5.126).

(5.126) TERMINATIVE PROXILATIVE – [Korpus Malti 3.0; news108877]
Mhux *talli* *n-NATO* *ma* *żżarmatx*
NEG:3SG.M:NEG because DEF-NATO NEG disarm:3SG.F.PFV:NEG
avolja *m'* *hemmx* *aktar* *bżonnha* *talli*
even_though NEG there:NEG more need:3SG.F because
kompliet *tespandi* ***[sa*** ***ma****<u>l-fruntieri</u>*
continue:3SG.F.PFV 3SG.F.IPFV:expand **[till** **<u>with</u>**<u>:DEF-frontier:PL</u>
<u>Russi</u>]$_{PP}$
<u>Russian:PL</u>]
'Not only didn't NATO disarm, even though it was no longer needed, but it laso continued to expand [**as far as** <u>the Russian borders</u>].'

Since we know from recent history that NATO accepted new members from the European East who formerly belonged to the Soviet Union or were counted as allies of the Soviet Union, there can be no doubt that the process of expansion went as far as touching the boundary of the Russian Federation. The spatial situation in (5.126) thus implies contact between the Figure and the Ground.

5.1.2.2.20.1.3 PROXIABLATIVE

As to the expression of motion from a PROXIMATE position, Borg/Azzopardi-Alexander (1997: 163) consider obligatory the use of the directionality marker (П41) *minn* 'from' whereas the configuration marker (П30) *ma'* 'with' is treated as optional component of the construction. This is remarkable when we think back to the previously discussed cases of motion from a Ground. The way the reference grammar depicts the situation it seems that the general ABLATIVE without specification of configuration is applicable in the case of the PROXIABLATIVE. This, however, is not the monopoly of the PROXIMATE domain as can be understood from what is said above.[99]

In stark contrast to the TERMINATIVE PROXILATIVE, there is ample evidence of the PROXIABLATIVE. It is necessary to divide the phenomena into two classes according to the degree of animacy of the Ground. In (5.127), we have a look at PROXILATIVE constructions which involve an inanimate Ground.

(5.127) PROXIABLATIVE – inanimate Ground
(a) [Vella 2014: 40]
Jaqbad *il-qmis* ***[minn*** ***<u>ma'</u>*** *<u>qaddu</u>]*$_{PP}$ *u*
3SG.M.IPFV:grab DEF-shirt **[from** **<u>with</u>** <u>hip:3SG.M</u>] and

99 Cf. Section 5.1.2.2.8.1.

jipprova *jġebbidha* *lejh*
3SG.M.IPFV:try 3SG.M.IPFV:pull:3SG.F towards:3SG.M
'He grabs the shirt [**from around** his hips] and tries to pull it towards him [. . .].'

(b) [Korpus Malti 3.0; culture1781]
meta *kienu* *qed* *ineħħu* *ż-żebgħa* ***[minn***
when be:3PL.PFV PROG 3.IPFV:take_off:PL DEF-paint **[from**
mal-ħitan]$_{PP}$ *tfaċċaw* *diversi* *graffiti*
with:DEF-wall:PL] emerge:3PL.PFV diverse:PL graffito:PL
'[. . .] when the were taling off the paint [**from** the walls] there suddenly emerged various graffitis [. . .].'

(c) [Korpus Malti 3.0; literature21]
Mikiel *dar* *fuq* *ir-raġel* *li* *kien* *għadu* *ma* *ċċaqlaqx*
Mikiel turn on DEF-man that be still:3SG.M NEG move:NEG
[minn ***mal-bieb]***$_{PP}$
[from **with**:DEF-door]
'Mikiel turned to the man who had not moved [**from** the door], yet.'

The first two examples involve Figures which are moved from a position that was so close to the Ground that there was physical contact. In (5.127a), the shirt that the person following the victim tries to pull towards himself was originally around the victim's hips. Similarly, the paint that was peeled off in (5.127b) originally covered the walls. Both spatial situations describe how an original Figure-Ground contact is discontinued. As to (5.127c), however, contact is not required because the man Mikiel turns to is at the door but whether or not the man touches the door is neither made explicit nor relevant to the context.

In (5.128), we turn our attention to human complements of (Π30) *ma'* 'with' accompanied by the directionality marker (Π41) *minn* 'from'.

(5.128) PROXIABLATIVE – human Ground
(a) [Korpus Malti 3.0; culture2101]
iċ-ċiklisti *ffurmaw* *gruppi* *ta'* *ħames* *persuni* *biex* *ma*
DEF-cyclist:PL form:3PL.PFV group:PL of five person:PL to NEG
jintilfux *u* *ma* *jinqatawx* ***[minn***
3.IPFV:REFL:lose:PL:NEG and NEG 3.IPFV:REFL:cut:PL:NEG **[from**
mal-grupp]$_{PP}$
with:DEF-group]
'[. . .] the cyclists formed groups of five persons in order not to get lost and not to be cut off [**from** the group].'

(b) [Korpus Malti 3.0; law1028]
dak *ir-raġel* *għandu* *jinqata'* ***[minn***
DEM:M:DIST DEF-man have.IPFV:3SG.M 3SG.M.IPFV:REFL:cut **[from**
***ma**l-corps* *tiegħu]*$_{PP}$
with:DEF-corps of:3SG.M]
'[. . .] that man has to be removed [**from** his corps] [. . .].'

(c) [Korpus Malti 3.0; news110899]
Infridt ***[minn*** ***ma**l-familja]*$_{PP}$ *u* *ma* *kellix*
REFL:separate:1SG.PFV **[from** **with**:DEF-family] and NEG have.PFV:1SG
edukazzjoni
education
'I was separated [**from** the family] and had no education [. . .].'

(d) [Korpus Malti 3.0; literature45]
issepara ***[minn*** ***ma**l-mara]*$_{PP}$
separate **[from** **with**:DEF-woman]
'[. . .] he separated [**from** his wife] [. . .].'

(e) [Korpus Malti 3.0; literature20]
Infridtu ***[minn*** ***ma'*** *Roger]*$_{PP}$ *ħdejn* *il-BMW*
REFL:separate:2PL.PFV **[from** **with** Roger] beside DEF-BMW
tiegħu *pparkjata* *fit-telgħa*
of:3SG.M park:PART:F in:DEF-slope
'You separated [**from** Roger] near his BMW which was parked on the slope [. . .].'

In none of the examples is it necessary to translate (Π30) *ma'* 'with' into English. Directionality suffices. It is questionable how many of the above examples involve contact. As to (5.128e), this criterion is certainly inapplicable since we presume that the Figure was in a position close to Roger before the motion away from the Ground started but standing in the vicinity of a person does not automatically mean that there is physical contact to said person. The cases (5.128a), (5.128b), and (5.128c) are different in the sense that the Figure forms part of a larger group from which it separates. Metonymically, belonging to a group might be interpreted as a relation that involves contact. In point of fact, these examples conceptionally invoke the MEDIOABLATIVE.[100] They will surface again when we discuss the COMITATIVE[101] and the PARTITIVE.[102] The COMITATIVE also plays a role when we take account of (5.128d).

100 Cf. Section 5.1.2.2.17.2.
101 Cf. Section 5.1.3.4.
102 Cf. Section 5.1.3.19.

The example refers to a typical scenario of a divorce. The husband separates from his wife. As a married couple husband and wife form a unit which in analogy to the previous three groups might metonymically be taken to involve contact. In our analysis, the presence of (Π30) *ma'* 'with' in contexts like those presented in (5.128) emphasises the especially high degree of closeness between Figure and Ground. It does not, however, require that Figure and Ground touch one another.

5.1.2.2.20.1.4 Back to the SECUTIVE

At this point, it makes sense to return to the SECUTIVE to briefly complement the sketch we have given thereof in the introduction to Section 5.1.2.2.20.1 above. We continue the discussion in connection to (Π30) *ma'* 'with' in (i) and close this section with a sketch of the SECUTIVE properties of (Π33) *matul* 'during' in (ii).

(i) (Π30) *ma'* 'with': There is a conceptual link between the PROXIMATE and the SECUTIVE. In what follows, we try to explain this connection by way of analysing the examples given in (5.129).

(5.129) SECUTIVE

(a) with contact – [Korpus Malti 3.0; literature20]

F'	*daqqa*	*ġiek*	*suspett*	*li*	*jekk*	*tfittex*
in	stroke	come:2SG	suspicion	that	if	2SG.IPFV:search

[mal-ħajt]$_{PP}$	*b'*	*idejk*	*issib*	*swiċċ*
[**with**:DEF-wall]	with	hand:PL:2SG	2SG.IPFV:find	switch

'All of a sudden you suspected that if you searched [**on** the wall] with your hands you would find a switch.'

(b) with contact – [Korpus Malti 3.0; literature24]

Sadanittant	*il-vann*	*inħass*	*jitkaxkar*	*[mal-ħajt*
meanwhile	DEF-van	REFL:feel	3SG.M.IPFV:REFL:drag	[**with**:DEF-wall

fil-ġenb	*tat-triq]$_{PP}$*
in:DEF-side	of:DEF-street]

'Meanwhile the van was felt to be dragged [**along** the wall at the side of the street] [. . .].'

(c) without contact – [Korpus Malti 3.0; literature5]

fetaħ	*il-bieb*	*mingħajr*	*ħsejjes*	*u*	*beda*	*miexi*
open	DEF-door	without	sound.PL	and	begin	walk.PART

[mal-ħajt	*tal-kuritur]$_{PP}$*	*lejn*	*it-taraġ*
[**with**:DEF-wall	of:DEF-corridor]	towards	DEF-stair.PL

'[. . .] he opened the door without noise and began walking [**along** the wall of the corridor] towards the stairs [. . .].'

The three spatial situations in (5.129) are dynamic in the sense that they feature motion events. The motion events differ from those of the (bare / DIRECTIONAL / TERMINATIVE) PROXILATIVE and the PROXIABLATIVE insofar as the Figure neither moves towards nor away from the Ground but along the Ground which is a horizontally extended object. There is contact between Figure and Ground in (5.129a) and (5.129b). In the (a)-sentence, the agent searches manually for a switch on the surface of a wall. This activity involves repeated or even permanent contact of the hands with the surface of the wall. The movement of the hands over the wall's surface is thus not semelfactive-punctual but either continuous or iterative. The hands change position but always keep near the wall. In the (b)-sentence, the van (= Figure) moves while scratching the wall's surface. Similar to the previous spatial situation, Figure and Ground are either constantly or repeatedly in contact with each other but at the same time, the Figure changes its position but without moving away from the Ground or approaching it further. In (5.129c), contact between Figure and Ground is not at issue. The agent moves along the wall towards the stairs. Whether or not he touches the wall during the motion event is not made explicit. What is important is that the Figure is close to the Ground while moving. The closeness is the same as in the PROXIMATE domain, namely such that ideally no other object fills the space between Figure and Ground so that contact could optionally be made. The distance between Figure and Ground is very small.

The SECUTIVE belongs to the wider PROXIMATE domain in the sense that its application implies that the small distance between Figure and Ground remains ideally the same throughout the motion event. In (5.130), the agent walks across the roof terrace. We must not dwell on the fact that the person's feet are in contact with the terrace's surface during the process of crossing the terrace.

(5.130) SECUTIVE with contact – [Korpus Malti 3.0; literature19]

Bil-kamera	*mdendla*	***[ma'***	*spalltu]*$_{PP}$	*qagħad*
with:DEF-camera	PART:hang:F	**[with**	shoulder:POSS:3SG.M]	PROG

jippassiġġa	*[**ma**l-bejt*	*kollu]*	*u*	*jtawwal*
3SG.M.IPFV:stroll	[**with**:DEF-roof	all:3SG.M]	and	3SG.M.IPFV:lengthen

għonqu	*lejn*	*ir-razzett*	*tal-Qerqni*
neck:3SG.M	towards	DEF-farm_house	of:DEF-Qerqni

'With the camera hanging [**on** his shoulder] he is walking [**across** the entire roof terrace] and stretches his neck towards the farm-house of the Qerqni.'

The first of the two *ma'*-PPs in (5.130) is no concern of ours in this context because it is involved in a spatial situation that can be classified as STATIC PROXIMATE (with contact). The second *ma'*-PP however, is associated with a dynamic spatial situation in which the Figure moves over the surface of the Ground. As in the many previously discussed cases, we assume that the employment of (Π30) *ma'* 'with' in (5.130) is appropriate not only because Figure and Ground are maximally close to each other but also because the movement is carried out along the Ground.

The Figure may move along the Ground without coming into contact with it, as results from (5.131).

(5.131) SECUTIVE without contact – [Korpus Malti 3.0; literature19]
l-Professur lewa [mal-knisja]$_{PP}$ u ma deherx aktar
DEF-professor turn [**with**:DEF-church] and NEG appear:NEG more
'[. . .] the professor turned [**around** the church] and was no longer visible [. . .].'

In this example, the Figure moves along the church building to disappear from view behind it. The church represents the Ground along which the motion event is realised. Nowhere is it mentioned that the professor touches the church building's structure while passing it. The distance between Figure and Ground during the motion event is relatively small, but not small enough to allow us to speak of the professor's body scratching along the church's outer walls. In conclusion, it can be assumed that the SECUTIVE is not absolutely without static counterpart since it shares many properties with the PROXIMATE. At the same time, the PROXIMATE-SECUTIVE connection is special because no other spatio-temporal meaning category allows for the use of the Π that expresses the static meaning to be employed also without directionality marker to express movement along a Ground.

(ii) (Π33) *matul* 'during' (+ (Π42) *mintul* 'all along'): The SECUTIVE cannot only be expressed by (Π30) *ma'* 'with' but also by (Π33) *matul* 'during' as shown in (5.132).

(5.132) SECUTIVE – second option [Korpus Malti 3.0; literature26]
Imbagħad inti sibt il-kamra s-sigrieta kompla
then 2SG find:2SG.PFV DEF-room DEF-secret:F continue
t-tfajjel waqt li qam jippassiġġa [matul
DEF-boy.DIM at_the_time_of that rise 3SG.M.IPFV:stroll [**during**
il-bejt]$_{PP}$
DEF-roof]
'"Then you found the secret chamber," the little boy continued while he rose to stroll [**across** the roof terrace].'

The functional home-ground of (Π33) *matul* 'during' is the temporal domain as will come to the fore below.[103] We know already from previous discussions that (Π33) *matul* 'during' and its short allomorph *tul* occasionally fulfil spatial tasks, too.[104] This is the case in (5.132) which contains a parallel to the second *ma'*-PP in (5.130). Spatially employed (Π33) *matul* 'during' occurs infrequently in the Korpus Malti 3.0. Yet, the turnout is sizable enough to require further inspection. In Table 5.7, we present a selection of examples for the use of the long and the short allomorph of (Π33) *matul* 'during' in SECUTIVE function. The examples are taken from the Korpus Malti 3.0.

Table 5.7: (Π33) *matul* 'during' in SECUTIVE function.

#	source	example / translation
1	european10172	*is-sonda tista tkun imdeffsa b' mod lonġitudinali sa l-assi tat-tubu jew [**matul** il-ġemb tiegħu]$_{PP}$* '[. . .] the probe could be pushed longitudinally as far as the tube's axis or [**along** its side].'
2	european10334	*l-isfera tiġi mgħoddija [**matul** il-wiċċ frontali ta' fuq]$_{PP}$, waqt li jinżamm kuntatt mal-windscreen* '[. . .] the sphere is moved [**along** the upper front window] while contact with the windscreen is kept [. . .].'
3	news151726	*hi u binha bdew mexjin [**matul** ix-xatt]$_{PP}$ fuq il-bankina* '[. . .] she and her son started walking [**along** the beach] on the side pavement.'
4	news83665	*nimxi [**matul** it-triq]$_{PP}$* '[. . .] I walk [**along** the street] [. . .].'
5	literature70	*kien jippassiġġa waħdu [**tul** it-triq ta' l-imħażen]$_{PP}$* '[. . .] he strolled alone [**along** the street of wickedness].'
6	news124732	*jħobbu jagħmlu xi passiġġata [**tul** ix-xatt f' Wied il-Għajn]$_{PP}$* '[. . .] they love to take a stroll [along the bank of Wied il-Għajn].'
7	literature7	*Bdiet timxi [**tul** il-ħajt kollu]$_{PP}$* 'She began to walk [**along** the entire wall] [. . .].'

In Table 5.7, #1–4 feature the long allomorph whereas the short allomorph occurs in examples #5–7. Both allomorphs can be employed for the same purpose. The complement nouns come from a variety of semantic classes. Superficially, it appears to be the case that (Π33) *matul* 'during' is a strong competitor of (Π30) *ma'* 'with' when it comes to expressing the SECUTIVE. Figure 5.13 reveals, however, that there are quantitative differences which speak against an egalitarian distribution of the two Πs. We compare the token frequencies in the Korpus Malti 3.0 for combinations of (Π30) *ma'* 'with' and (Π33) *matul* 'during' with five different nouns in Ground function, namely *ħajt* 'wall', *kosta* 'coast', *triq* 'street', *wied* 'valley', and *xatt* 'beach'.

103 Cf. Section 5.1.2.2.20.2.2.
104 Cf. Sections 1.6.4.2 (ix), 3.2.3, 4.1.1.2.2.1 (xxii), and 4.1.2.2.1.2.1 (vi).

The shares calculated for (Π33) *matul* 'during' range from maximally 20.5% to the minimum of 0.1%. In other words, (Π30) *ma'* 'with' is always the winner. In some cases, the employment of its competitor is absolutely exceptional. On the basis of this selection of Ground-nouns, it is of course not possible to generalise over the two Πs's relationship. However, there seems to be a slight preference for (Π33) *matul* 'during' to combine with Grounds whose referent is particularly extended horizontally. However, with Grounds of this kind, (Π33) *matul* 'during' is only the minority option.

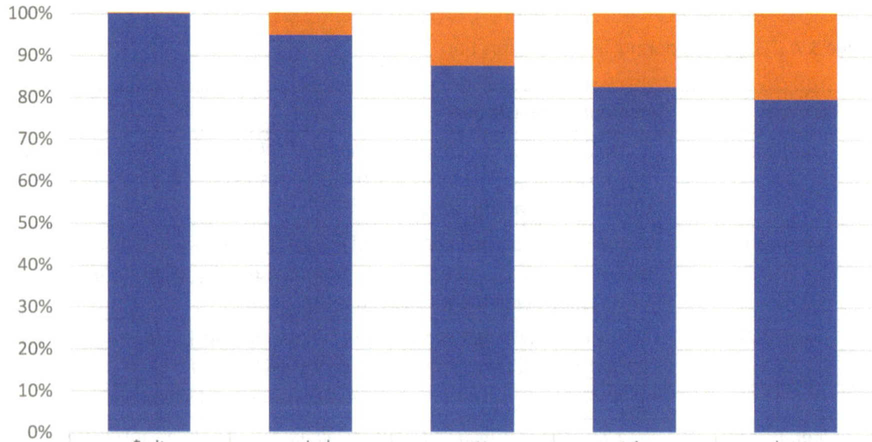

	ħajt	wied	xatt	triq	kosta
matul	2	4	75	225	140
ma'	2116	72	524	1059	545

Figure 5.13: Competition of SECUTIVE markers with five selected Ground-nouns.

In (5.133), a striking fact is waiting for us. The Ground *il-Promenade* refers to a geo-object, namely the seaside promenade in Sliema. What is especially remarkable about this example is the fact that the situation is static. There is no motion event during which the Figure changes its position.

(5.133) (Π33) *matul* 'during' – static [Korpus Malti 3.0; culture3261]

Il-Ħadd *filgħodu* *għall-ħabta* *tal-10.00* ***[matul***
DEF-Sunday in_the_morning for:DEF-collision of:DEF-10.00 **[during**
il-Promenade]$_{PP}$ *se* *jkun* *hemm* *esebizzjoni* *ta* *muturi*
DEF-Promenade] FUT 3SG.M.IPFV:be.FUT there exhibition of motor:PL
'Sunday morning at 10 o'clock sharp there will be an exhibition of motor-bikes [**along** the Promenade].'

The motorcycle exhibition occupies the full length of the seaside promenade without moving from one place to the other. The Figure is located along the Ground. The static use of (Π33) *matul* 'during' disagrees with Hagège's (2010: 287) above quoted assumption that the SECUTIVE has no static counterpart. As results from the data in Table 5.8, the example (5.133) is by no means an isolated case. The use of (Π33) *matul* 'during' in existential and thus static constructions is relatively common. The pattern [Figure (COP) *hemm* [***matul*** Ground]$_{PP}$]$_{EXIST}$ recurs throughout Table 5.8.

Table 5.8: (Π33) *matul* 'during' in existential constructions.

#	source	example / translation
1	news131913	*jista' s-Sur Buttiġieġ jgħidilna kemm minn dawk il-ħwienet li hemm [**matul** ix-xatt inkluż ħdejn il-playing field in-naħa ta' Grabiel]$_{PP}$ ħalli żgur niftehemu, kellhom l-art tagħhom* '[. . .] can Mr Buttiġieġ tell us so that we surely understand how many of those shops that are there [along the beach including near the playing field on the side of Grabiel] own their land [. . .]?'
2	news154914	*Huwa l-ħsieb tal-Armata li 'l quddiem il-kanuni li hemm [**matul** r-runway]$_{PP}$ jiġu rtirati* 'The army intends to later on withdraw the canons that are there [**along** the runway].'
3	news82334	*Kull meta nmur nimxi fil-walking track li hemm [**matul** il-mitjar]$_{PP}$ nara ħafna ġrieden iktar minn għoxrin ċentimetru twal* 'Every time I go walking on the walking track that is there [**along** the airfield] I see many rats of more than twenty centimetres of length [. . .].'
4	parl10668	*l-ħajt li hemm [**matul** Triq id-Dwejra f' San Lawrenz, Għawdex]$_{PP}$ jinsab fi stat ħażin ħafna* '[. . .] the wall that is there [**along** Triq id-Dwejra in San Lawrenz/Gozo] is in a very bad state [. . .].'
5	parl9642	*Jista' l-Ministru jagħti tagħrif jekk il-masġar taż-żebbuġ li hemm [**matul** il-bajja tax-Xemxija]$_{PP}$ fil-fatt hux kollu amministrat mill-BirdLife?* 'Can the minister give us an explanation as to whether the olive groves that are there [**along** Xemxija Bay] are in fact all administered by BridLife?'
6	literature32	*F' xi bnadi [**matul** il-ħajt]$_{PP}$, kien hemm saħansitra rqajja' ta' demm imċarċar mal-ħajt.* 'In some places [**along** the wall], there even were patches of blood spilled onto the wall.'

There is a paradox that calls for being explained. Not only is there no slot for spatial situations like those in Table 5.8. in Hagège's (2010: 261) system of spatio-temporal meaning categories nor do Borg/Azzopardi-Alexander (1997) account for the existence of (Π33) *matul* 'during', in the first place. The nearest one can get to our paradoxical case is the mention of the exterior of a long object serving as Ground for which Borg/Azzopardi-Alexander (1997: 167) assume that (Π32) *maġenb* 'close to' is the appropriate Π.[105] Their example features a dynamic situation. Static location of a Figure along a horizontally or vertically extended Ground is not covered in the reference grammar. A possible ad-hoc term for this missing category is EXTENTUS (< Latin *extentus* 'extended') which covers the static *along*-situations.

The postulation of the STATIC EXTENTUS helps us to understand the subsequent phenomenon. In the Korpus Malti 3.0, there is evidence of the combination of the directionality marker (Π21) *għal* 'for' with (Π33) *matul* 'during'. The two examples in (5.134) illustrate this possibility for both of the allomorphs of (Π33) *matul* 'during'.

(5.134) DIRECTIONAL EXTENTOLATIVE

(a) [Korpus Malti; news203281]

Il-kjoskijiet tas-siment li bħalissa hemm fuq il-ġnub
DEF-kiosk:PL DEF-cement that presently there on DEF-side.PL
tad-daħla se jiċċaqilqu ***[għal tul*** *il-passaġġ*
of:DEF-entrance FUT 3.IPFV:move:PL **[for during** DEF-passage
li jwassal għad-daħla]$_{PP}$
that 3SG.M.IPF:arrive.CAUS for:DEF-entrance]
'The kiosks made of concrete that right now are there on the sides of the entrance will move [**to** (positions) **along** the passage that leads to the entrance].'

(b) [Korpus Malti 3.0; academic70]

Popolazjonijiet żgħar oħra tal-liepru tal-ħaxix ta'
population:PL small.PL other:PL of: DEF-hare of:DEF-grass of
l-Uganda huma preżenti direttament mit-Tramuntana ta'
DEF-Uganda 3PL present:PL directly from:DEF-north of
L-Għadira Victoria ***[għal matul*** *il-baċir tax-xmara Uganda]*$_{PP}$
DEF-lake Victoria **[for during** DEF-shipyard of:DEF-river Uganda]
'Other small populations of the Uganda bush hare are present from directly north of Lake Victoria [**to (along)** the shipyards of the Uganda river].'

In these examples, (Π33) *matul* 'during' is responsible for encoding configuration. The connection to motion is established via the directionality marker (Π21) *għal* 'for'. As a matter of fact, the possibility of expressing motion to a position along

---

105 The example provided by Borg/Azzopardi-Alexander (1997: 167) is reproduced as (5.v) in this footnote.

(5.v) *Il-linji tat-telefown jgħaddu* ***[maġenb*** *il-kanen ta' l-ilma]*$_{PP}$
DEF-line:PL of:DEF-telephone 3.IPFV:pass:PL **[close_to** DEF-pipe.PL of DEF-water]
'The telephone lines are laid [**along** the water pipes].' [O.T.]

an extended Ground in the shape of a bare EXTENTOLATIVE is blocked. We have no evidence for its existence.

There is no TERMINATIVE EXTENTOLATIVE, i.e. the combination of (Π51) *sa* 'till' and (Π33) *matul* 'during' conveying a spatial meaning is not attested in our sources. It is also impossible to prove the existence of an EXTENTOABLATIVE that involves (Π33) *matul* 'during'. Situations in which the Figure that originally occupied a position along an extended Ground is removed from that position are rarely found. In contexts of this kind, (Π41) *minn* 'from' suffices to express the intended meaning. Combinations of (Π41) *minn* 'from' with (Π33) *matul* 'during' seem to invite a PERLATIVE interpretation.[106] The problems posed by (Π42) *mintul* 'all along' have been exposed above.[107] As example (5.135) suggests, (Π42) *mintul* 'all along' is another candidate for the function of the EXTENTUS,

(5.135) EXTENTUS – second option – [Korpus Malti 3.0; law865]
l-immarkar tas-salib għandu jkun
DEF-marking of:DEF-cross have.IPFV:3SG.M 3SG.M.IPFV:be.FUT
***[mintul** it-tarf]$_{PP}$ u għandu jsir b'*
[all_along DEF-extremity] and have.IPFV:3SG.M 3SG.M.IPFV:become with
żebgħa jew b' lapes tal-kulur
paint or with pencil of:DEF-colour
'[. . .] the marking with a cross has to be [**all along** the extreme end] and has to be executed with paint or with a coloured pencil.'

Examples of (Π42) *mintul* 'all along' combining with the directionality markers (Π21) *għal* 'for' and (Π51) *sa* 'till' could not be found just as there is no evidence of a bare EXTENTOLATIVE that involves (Π42) *mintul* 'all along' as configuration marker.

Some of the intricacies of the Path domain and related domains have come to the fore in the previous paragraphs. The picture is much more diversified than one could have predicted on the basis of Borg/Azzopardi-Alexander (1997) and Hagège (2010). As we will see already in the next section, this is not the end of the story of Path in Maltese.

5.1.2.2.20.2 PERLATIVE

According to Hagège (2010: 261), the PERLATIVE comes in two varieties, namely the spatial PERLATIVE on the one hand and on the other hand, the TEMPORAL PERLATIVE. The spatial PERLATIVE is addressed first[108] whereas its temporal counterpart is the

106 Cf. Section 5.1.2.2.20.2.1.
107 Cf. Section 3.1.2.3.1.2.
108 Cf. Section 5.1.2.2.20.2.1.

topic of the next but one section.[109] The discussion reveals that PERLATIVE meanings are not always clearly distinguished inter alia from SECUTIVE meanings.

5.1.2.2.20.2.1 SPATIAL PERLATIVE

The SPATIAL PERLATIVE is associated to the English Πs *across*, *through*, and *via* (Hagège 2010: 261). A spatial situation is termed PERLATIVE if the Figure "moves across a place or person" (Hagège 2010: 286). Moving across the Ground is a situation that we have encountered already in (5.130) and (5.132). These examples illustrate (Π33) *matul* 'during' in SECUTIVE function. PERLATIVE meanings have also been alluded to in the case of the CIRCUMLATIVE and CIRCUMABLATIVE,[110] the MEDIOLATIVE and MEDIOABLATIVE,[111] and the INTERLATIVE and INTERABLATIVE.[112] There is an extended zone of overlap that calls for an in-depth analysis. For the time being, we focus on the main exponent of the PERLATIVE, viz. (Π41) *minn* 'from'.

As far as we can see, Borg/Azzopardi-Alexander (1997) do not distinguish the PERLATIVE from the PROLATIVE, i.e. motion through and motion past the Ground are lumped together under the rubric of motion past. This practice is probably based on the fact that Maltese employs the same Π for both functions. It is, however, necessary to address PERLATIVE functions and PROLATIVE functions separately as is done in this study. In this section, we focus on the SPATIAL PERLATIVE whose characteristics are presented by way of interpreting the combinability of PERLATIVE (Π41) *minn* 'from' with different kinds of Grounds.

In (5.136), we provide a typical example of (Π41) *minn* 'from' in PERLATIVE function. In this example, the Figure (= *ħajta* 'thread') passes through the Ground (= *it-tubu* 'the tube') to reach its destination.[113]

(5.136) PERLATIVE I – [Korpus Malti 3.0; european3720]

Għaddi	*ħajta*	***[mi**<u>t-tubu</u>]$_{PP}$*	*minn*	*kull*	*waħda*	*mill-erba'*
pass.IMP	thread	[**from**:<u>DEF-tube</u>]	from	each	one:F	from:DEF-four
ħolqiet						
ring:PL						

'Pass a thread [**through** <u>the tube</u>] from each of the four rings.'

---

109 Cf. Section 5.1.2.2.20.2.2.

110 Cf. Section 5.1.2.2.16.

111 Cf. Section 5.1.2.2.17.

112 Cf. Section 5.1.2.2.18.

113 In the sections on the PERLATIVE and the PROLATIVE, to avoid semantic ambiguities, we deliberately give preference to examples which involve the motion verb *għadda* 'pass' wherever it is attested in combination with the Πs under discussion.

The process that is captured by the PERLATIVE construction requires that the Figure initially occupies a position outside the object through which it then passes to exit at the other end before reaching a potential Goal. This Goal might also be reached by just passing through the object which is equipped with an entrance and an exit. Situations of this kind are well represented in the Korpus Malti 3.0. The selection of examples (taken from the electronic resource) presented in Table 5.9 gives evidence of this.

Table 5.9: Examples of PERLATIVE uses of (П41) *minn* 'from' I.

#	source	example / translation
1	news156491	*Għaliex taħsbu li aħna ma tistax tinżlilna [**mi**l-gerżuma tagħna]$_{PP}$ din tal-21 ta' Settembru?* 'Why do you think that we cannot believe this event of 21st September (lit. this event of 21st September cannot go down [**through** our throat]?'
2	european6248	*Kull pompa tas-sentina bl-enerġija għandha tkun meħtieġa li tippompja l-ilma [**mi**l-pajp prinċipali]$_{PP}$* 'Each bilge pump with energy must be required to pump the water [**through** the main pipe].'
3	news113553	*It-tibdiliet kienu jinkludu żieda ta' pont dejjaq li jestendi min-naħa għal oħra tal-wisa' tal- bastiment meta jgħaddi [**mi**l-Kanal tal-Panama]$_{PP}$.* 'The changes included the addition of a bridge that extends from one side to the other of the ship's body when it passes [**through** the Panama Canal].'
4	opinion3938	*U meta kien hemm bżonn jiġi mxammar ħalli jgħaddi [**mi**x-xmara]$_{PP}$* 'And when it was necessary he came with his sleeves tucked up to pass [**through** the river] [. . .].'
5	news149831	*Issa kollox irid jgħaddi [**mi**l-mina li saret f' Tignè mill-proġett tal-MIDI]$_{PP}$* 'Now everything must pass [**through** the tunnel that was realised at Tignè by the project of the MIDI].'
6	literature24	*Beda għaddej [**minn** passaġġ tal-fidda]$_{PP}$ u fuq kull naħa kien hemm filliera siġar tal-fidda bil-frott tal-fidda.* 'He began to pass [**through** a passage of silver] and on each side there was a row of silver trees with silver fruit.'
7	culture953	*Jekk tarah għaddej [**mi**t-triq]$_{PP}$, taħsbu li hu anzjan bħall-oħrajn* 'When you see him passing [**through** the street] you think that he is an old man like the others.'
8	news132166	*meta tipparaguna l-ammont ta' ilma li jkun għaddej [**mi**l-kanna]$_{PP}$, ma' dak li ħiereġ barra, dan tal-aħħar hu negliġibbli* '[. . .] when you compare the amount of water that passes [**through** the pipe] with that coming out (of it), (then) the latter is negligible.'

In Table 5.9, the Ground is always a relatively narrow passage through which the Figure passes. There is a body-part like the throat in #1, a pipe in #2 and #8, a channel (#4), a tunnel (#5), a passage (#6), and a street (#7). Example #3 is different insofar as the Figure presumably crosses the river from one bank to the other but does not move downriver or upriver. The Figures are also of various kinds, namely abstract ideas (#1), water (#2 and #8), a ship (#3), persons (#6 and #7), and simply everything (#5). Streets as Grounds show up not only in Table 5.9 #7 but also as complement of (П33) *matul* 'during' in Table 5.7 #4 and Table 5.8 #4 as well as complement of (П30) *ma'* 'with' in Table 5.5 #3. What these parallels entail semantically must be explored in future research.

Examples like (5.137) tell us that Grounds of kinds different from those featured in Table 5.9 are equally eligible for PERLATIVE situations.

(5.137) PERLATIVE II – [Korpus Malti 3.0; news4707]

L-Iżlanda	*ħarġet*	*twissija*	*ġdida*	*wara*	*li*	*reġa'*
DEF-Iceland	exit:3SG.F.PFV	warning	new:F	after	that	do_again

żbroffa	*lava*	*[**minn***	*xaqq*	*qrib*	*il-vulkan*	*Bardarbunga]*$_{PP}$
erupt	lava	[**from**	crack	near	DEF-volcano	Bardabunga]

'Iceland gave out a new warning after lava erupted again [**from** a crack near the volcano Bardabunga].'

The lava emerges through a natural opening in the ground. Before the eruption, the lava was beneath the ground, after the eruption the lava covered part of the ground's surface. Thus, the lava passed through the crack from a position below the crack to wind up in a position above the crack. Spatial situations in which the Figure passes through a natural or artificial opening in an otherwise solid object are commonly found in the Korpus Malti 3.0. Table 5.10 presents only a very small selection of examples which illustrate that many different nouns are candidates for the function of Ground in these situations.

Table 5.10: Examples of PERLATIVE uses of (П41) *minn* 'from' II.

#	source	example / translation
1	news186328	*ma setax jgħaddi [**mi**l-bieb tal-Qorti]*$_{PP}$ *minħabba l-ħxuna tiegħu* '[. . .] he could not pass [**through** the door of the court of justice] on account of his fatness [. . .].'
2	literature20	*Ħareġ rasu [**mi**t-tieqa]*$_{PP}$ 'He put his head [**through** the window] [. . .].'
3	literature24	*Imma malli tgħaddi [**mi**d-daħla]*$_{PP}$ *ssib iktar dawl* 'But as soon as you pass [**through** the entrance] you find more light [. . .].'
4	news89487	*Wara tliet kwarti fuq il-karozza tal-linja, il-passiġġieri kienu tniżżlu [**mi**ll-ħruġ tal-emerġenza]*$_{PP}$ 'After three quarters of an hour on the bus, the passengers were made to descend [**via** the emergency exit].'
5	parl4805	*t-traffiku minn u għal ġol-iskola qed jgħaddi kollu [**mi**ll-exit gate ta' l-istess skola]*$_{PP}$ '[. . .] all the traffic from inside and into the school passes [**through** the exit gate of the said school].'

Examples #1–#5 involve Grounds whose referents are doors, entrances, and exits, i.e. artificial openings which facilitate access to and/or departure from a given object. They are made to allow people, animals, vehicles, or other mobile objects to pass through them. The pair of examples in (5.138) shows, however, that the Ground can also be an entity that is not meant to serve as passage for the Figure to pass through.

(5.138) PERLATIVE III – [Korpus Malti 3.0; news61019]

(a) *Fost il-midrubin f' Ruma kien hemm tifla milquta*
amongst DEF-PART:wound:PL in Rome be there girl PART:hit:F
minn balla li għaddiet [mill-ġisem ta'
from bullet that pass:3SG.F.PFV [**through**:DEF-body of
Fr Wasim Sabieh li nqatel waqt l-attakk]$_{PP}$
Brother Wasim Sabieh that REFL:kill at_the_time_of DEF-attack]
'Amongst the injured in Rome there was a girl hit by a bullet that passed [**through** the body of Brother Wasim Sabieh who died during the attack].'

(b) *l-vittma ntlaqat minn żewġ tiri li wieħed laqatlu*
DEF-victim REFL:hit from two shot:PL that one hit:IO:3SG.M
idu x-xellugija [. . .] waqt li l-ieħor għadda
hand:3SG.M DEF-left [. . .] at_the_time_of that DEF-other pass
[minn dahru]$_{PP}$ *u daħal fil-pulmun u ż-żaqq*
[**through** back:3SG.M] and enter in:DEF-lung and DEF-belly
'[. . .] the victim was hit by two shots one of which hit his left hand [. . .] while the other passed [**through** his back] and penetrated the lung and the belly.'

In both examples, bullets enter the human body. In the (a)-sentence, the bullet does not remain in the first victim's body but passes through it to hit a second victim. The (b)-sentence reflects a different scenario. In this case, the bullet is said to enter the body via the victim's back and then continues to travel inside the body until it reaches the lung and the belly. The Ground through which the Figure passes is Brother Wasim Sabieħ's body in (5.138a) and the anonymous victim's back in (5.138b).

Examples like (5.139) where the Figure passes across the Ground which has the shape of a line (also metonymically) are not particularly frequent in the Korpus Malti 3.0.

(5.139) PERLATIVE IV – [Korpus Malti 3.0; news222212]

ċ-ċiklisti damu madwar sitt sigħat biex
DEF-cyclist:PL take_time:3PL.PFV around six hour:PL to
jidħlu [mill-fruntiera tar-Russja]$_{\text{PP}}$
3.IPFV:enter:PL [**from**:DEF-border of:DEF-Russia]
'[. . .] it took the cyclists about six hours to cross [the frontier of Russia].'

The Russian borderline represents the Ground that the Figure must pass across to reach its destination.

Example (5.140) illustrates the combination of PERLATIVE (Π41) *minn* 'from' with a place name to yield a construction that corresponds to an English *via*-PP.

(5.140) PERLATIVE V – [Korpus Malti 3.0; academic12]

Meta kien sejjer lura lejn Malta huwa reġa' għadda
when be go.PART back towards Malta 3SG.M do_again pass
*[**minn** Perth fejn reġa' mar iżur lit-tfal]*$_{\text{PP}}$
[**from** Perth where do_again go 3SG.M.IPFV:visit to:DEF-child.PL]
'When he was going back to Malta he again passed [**through** Perth where he again went to visit the children].'

The city of Perth is only an intermediate stop on the Figure's journey back to Malta. The traveller thus passes through Perth. He goes to Malta via Perth. The example represents a recurrent pattern in the Korpus Malti 3.0 as the selection of cases in Table 5.11 suggests. The Ground is the name of a city, a village, an island, or a country all of which refer to stops between the point of departure and the destination of the Figure's itinerary.

Table 5.11: Examples of PERLATIVE uses of (Π41) *minn* 'from' III.

#	source	example / translation
1	academic236	*Fl-1796 Beethoven mar fuq dawra jagħti l-kunċerti li ħaditu minn Vjenna sa Berlin u għadda [**minn** Dresden, Leipzig, Nuremberg u Praga]*$_{\text{PP}}$. 'In 1796, Beethoven went on a tour to give concerts that took him from Vienna to Berlin and he passed [**through** Dresden, Leipzig, Nuremberg, and Prague].'
2	news6695	*Il-Korteo, li għadda [**minn** Ħal Lija, Birkirkara, Fleur-de-Lys, Santa Venera, u l-Ħamrun]*$_{\text{PP}}$*, wasal quddiem il-Konkatidral ta' San Ġwann*. 'The cortege, that passed [**through** Ħal Lija, Birkirkara, Fleur-de-Lys, Santa Venera, and Ħamrun], arrived in front of the Co-Cathedral of San Ġwann.'

Table 5.11 (continued)

#	source	example / translation
3	news117296	*xi ħadd bħali li żar jew għadda [**minn** Manchester]$_{PP}$ qabel nofs is-snin tmenin bil-kemm jemmen il-bidla drastika li jara quddiem għajnejh* '[. . .] someone who like me visited or passed [**through** Manchester] before the mid 1980's hardly believes the drastic change that he sees before his eyes [. . .].'
4	news113878	*Il-bastiment segwa vjaġġ twil u għadda [**minn** Madeira, Rio de Janeiro]$_{PP}$, u wara madwar Cape Horn* 'The ship followed a long voyage and passed [**via** Madeira, Rio de Janeiro] and afterwards around Cape Horn [. . .].'
5	news22566	*l-Papa Franġisku huwa l-ewwel wieħed li vvjaġġa lejn ix-Xatt tal-Punent direttament minflok għadda [**minn** Iżrael]$_{PP}$* '[. . .] Pope Franciscus is the first that travelled directly towards the eastern shore instead of passing [**via** Israel].'

The frequent occurrence of (Π41) *minn* 'from' in combination with Grounds of the types featured in Tables 5.9–5.11 contrasts with the relative scarcity of evidence in support of the same Π being used to express motion across a surface. This means that (Π41) *minn* 'from' cannot freely replace (Π33) *matul* 'during' in contexts like that featured in (5.132). However, there are examples like (5.141a) and (5.141b) which clearly describe situations in which the Figure moves across a surface, namely the killer's property in the (a)-sentence and, more specifically, the yard of the same residence in the (b)-sentence. Note that the Figure never had the chance to exit from the Ground – which means that the motion event was aborted before the ideally required endpoint was reached.

(5.141) PERLATIVE VI – [Korpus Malti 3.0; news4195]

(a) *Dan seħħ hekk kif sid il-garaxx Markus Kaarma*
DEM:M:PROX happen so how owner DEF-garage Markus Kaarma
spara fuq student Ġermaniż meta dan għadda
shoot on student German when DEM:M:PROX pass
[mill-propjetà tiegħu]$_{PP}$
[**from**:DEF-property of:3SG.M]
'This happened when Markus Kaarma, the owner of the garage, shot at a German student when he passed [**across** his property].'

(b) *Qal li qatt ma kien jaħseb li ibnu se*
say that ever NEG be 3SG.M.IPFV:think that son:3SG.M FUT
jinqatel sempliċiment għaliex dan għadda
3SG.M.IPFV:REFL:kill simply because DEM:M:PROX pass

[minn bitħa ta' residenza]
[**from** yard of residence]
'He said that he had never thought that his son would be killed simply because he passed [**across** the yard of a residence].'

The two examples refer to the same situation. The victim is shot by the owner of a garage when the victim moves across the yard that belongs to the garage. This means that the victim trespasses by way of physically entering the owner's property. This is important to note because otherwise the victim would just be passing by the property – a situation that would fall under the rubric of PROLATIVE that is hard to tell apart from the PERLATIVE on formal grounds as is shown below.[114]

We also find examples of the PERLATIVE use of (П41) *minn* 'from' with Grounds that refer to geo-objects which cover a huge area. This is the case with the Sahara in (5.142a) as well as the Red Sea and the Mediterranean in (5.142b).

(5.142) PERLATIVE VII
(a) [Korpus Malti 3.0; news66786]
Biex joħorġu mill-pajjiż il-biċċa l-kbira minnhom
to 3.IPFV:exit:PL from:DEF-country DEF-part DEF-big:F from:3PL
jgħaddu [mid-Deżert Sahara]$_{PP}$
3.IPFV:pass:PL [**from**:DEF-desert Sahara]
'To leave the country the majority of them passes [**through** the Sahara Desert].'
(b) [Korpus Malti 3.0; parl455]
huwa prattikament l-itwal fibre optic cable ta' taħt
3SG.M practically DEF-long.COMP fibre_optic_cable of under
il-baħar fid-dinja li jibda mill-Ġappun,
DEF-sea in:DEF-world that 3SG.M.IPFV:begin from:DEF-Japan
jgħaddi [mill-Baħar l-Aħmar]$_{PP}$ jgħaddi
3SG.M.IPFV:pass [**from**:DEF-sea DEF-red] 3SG.M.IPFV:pass
[mill-Mediterran]$_{PP}$ u jibqa' sejjer sa
[**from**:DEF-Mediterranean] and 3SG.M.IPFV:remain go.PART till
l-Ingilterra
DEF-England
'[. . .] it is practically the longest submarine fibre optic cable in the world that starts from Japan, passes [**through** the Red Sea], passes [**through** the Mediterranean] and keeps going as far as England.'

114 Cf. Section 5.1.2.2.20.3.

The (a)-sentence reflects a motion event, namely that of mass migration from an African country via the desert towards the Mediterranean. The Figure (= the migrants) actually move across the Ground. The situation is thus dynamic. Dynamicity does not apply in the (b)-sentence since the underwater cable rests on the bottom of the sea where it is immobile. However, the use of the motion verb *għadda* 'pass' tells us that the situation is metonymically interpreted as a motion event with its starting point in Japan and its endpoint in England. To arrive there the cable is conceived of as a Figure that passes through the Red Sea and the Mediterranean – with the two seas functioning as the Ground.

Moreover, (Π41) *minn* 'from' is attested in combination with different Π_2s in constructions which serve to express the PERLATIVE. In this context, (5.143) is particularly interesting since it involves the Π-Π combination *minn matul.*

(5.143) PERLATIVE VIII – [Korpus Malti 3.0; news6695]

L-Isqof	*il-ġdid*	*jieħu*	*l-mitra*	*u*	*l-baklu*
DEF-bishop	DEF-new	3SG.M.IPFV:take	DEF-mitre	and	DEF-pastoral_staff

u	*jgħaddi*	***[minn***	***<u>matul</u>***	*<u>il-Knisja]</u>*$_{PP}$	*ibierek*
and	3SG.M.IPFV:pass	**[from**	**<u>during</u>**	<u>DEF-church]</u>	3SG.M.IPFV:bless

lil	*dawk*	*preżenti*
to	DEM:PL:DIST	present:PL

'The new bishop takes the mitre and the pastoral staff and passes [**through** <u>the church</u>] blessing those who are present [. . .].'

It seems that neither (Π41) *minn* 'from' nor (Π33) *matul* 'during' alone suffices to express the PERLATIVE in this case. In the Korpus Malti 3.0, all instances of *mil-knisja* 'from the church' identify the church as the Ground in Source function. In contrast, there is no hit for **matul il-knisja*, in the first place.

There are further Π-Π combinations with (Π41) *minn* 'from' as Π_1 that can be classified as instances of the PERLATIVE although some of them appear to be ambiguous semantically. In these cases, the configuration markers are those of the MEDIOESSIVE and INTERESSIVE introduced above.[115] In (5.144), we provide two examples of the PERLATIVE constructions which involve the MEDIOESSIVE markers (Π16) *fost* 'amongst' and (Π47) *qalb* 'amidst'.

---

115 In Sections 5.1.2.1.13 and 5.1.2.1.14, respectively.

(5.144) PERLATIVE IX

(a) [Korpus Malti 3.0; news102164]

biex	*ħarab*	*mill-kustodja*	*tal-Pulizija*	*għadda*
to	escape	from:DEF-custody	of:DEF-police	pass
[minn	***qalb***	*il-ħġieġ*	*imkisser*	*tal-antiporta]*$_{PP}$
[from	**amidst**	DEF-glass	PART:break	of:DEF-inner_door]

'[. . .] to escape from police custody he passed [**through** the broken glass of the inner door] [. . .].'

(b) [Korpus Malti 3.0; culture528]

Il-quddiesa	*din*	*is-sena*	*se*	*ssir*	*fil-Bażilika*
DEF-Mass	DEM:F:PROX	DEF-year	FUT	3SG.F.IPFV:become	in:DEF-Basilica
ta'	*San Pietru*	*sabiex*	*jiġi*	*evitat*	*li*
of	Saint_Peter	in_order_to	3SG.M.IPFV:come	avoid:PART	that
l-Papa	*jgħaddi*	***[minn***	***fost***	*ġemgħat*	*kbar]*$_{PP}$
DEF-Pope	3SG.M.IPFV:pass	**[from**	**amongst**	crowd:PL	big.PL]

'This year the Mass will take place in the Basilica of Saint Peter to avoid that the Pope passes [**through** huge crowds].'

The Ground through which the Figure passes is represented by the pieces of broken glass in the (a)-sentence and the huge crowds of people in the (b)-sentence. The Figure has or would have to pass through in order to reach the destination, namely freedom in (5.1144a) and the church in (5.144b). The configuration markers serve to express that the Ground consists of an agglomeration of many elements.

In (5.145), it is the turn of (Π4) *bejn* 'between' to function as Π_2 in combination with (Π41) *minn* 'from'. The possibility of (Π4) *bejn* 'between' to fulfil the task of the INTERESSIVE marker as well as that of the MEDIOESSIVE marker transpires from the comparison of (5.145a) and (5.145b).[116]

(5.145) PERLATIVE X

(a) [Bartolo/Vella 2009: 100]

għaddiet	***[minn***	***bejn***	*il-purtieri*	*li*	*kienu*
pass:3SG.F.PFV	**[from**	**between**	DEF-curtain:PL	that	be:3PL.PFV
jagħtu	*għall-kċina]*$_{PP}$				
3.IPFV:give:PL	for:DEF-kitchen]				

'[. . .] she passed [**between** the curtains that led to the kitchen].'

116 Cf. Section 5.1.2.1.14.

(b) [Camilleri 2013: 73]
*Għadda [**minn** **bejniethom**]$_{PP}$ u waqaf ġo nofshom*
pass [**from** **between**:3PL] and stop in middle:3PL
'[He passed [**between** them] and stopped in their middle.'

In the (a)-sentence, the situation captures the movement of the elderly lady who departs from a position in front of the closed curtains that separate her from the destination, viz. the kitchen. To get there she has to open the curtains and pass through the opening between the two halves of the curtain. The Ground consists of two parts. This example differs crucially from the INTERABLATIVE in example (5.105) where the prehistory of the motion of the Figure emerging from the space between two cars parked at the curb is irrelevant and accordingly tacitly passed over. In (5.145a) however, the entire itinerary of the Figure from a position outside the kitchen via the parted curtains into the kitchen is described.

In the (b)-sentence of (5.145) however, the context is as follows. The protagonist encounters a group of Maltese tourists at the foot of the Etna so that (Π4) *bejn* 'between' receives a MEDIOLATIVE interpretation rather than an INTERLATIVE one since the group is larger than two persons. What is additionally interesting about this example is the fact that strictly speaking the motion event is not fully in line with the ideal form of a PERLATIVE situation because the Figure ceases to move right in the middle of the group of tourists. Thus, the Figure has not passed through but only moved into. Of course, to reach the middle of the group the Figure must have passed at least some of the members of this group. Generalising, it can be concluded that a situation is PERLATIVE even in those cases in which the motion through or across the Ground is not fully carried through. What suffices is the intention of exiting from the Ground at a place which is not identical to that of entering the Ground.

5.1.2.2.20.2.2 TEMPORAL PERLATIVE

Hagège (2010: 261) offers the English Πs *for*, *during*, and *in* as markers of the TEMPORAL PERLATIVE. In the reference grammar, Borg/Azzopardi-Alexander (1997: 171) mention only (Π21) *għal* 'for' under the heading duration.[117] They argue that "[d]uration is normally expressed by the preposition **għal** 'for' [. . .] but with verbs lexicalizing duration [. . .] the preposition does not occur [original bold-

117 The combination of categories such as anterior + duration + past/future and posterior + duration + past/future (Borg/Azzopardi-Alexander 1997: 171–172) are interpreted by us as belonging to the domains of the temporal TERMINATIVE and the temporal ABLATIVE.

face].” This observation holds for cases in which the length of time is measured for which (and therefore during which) a certain state of affairs is valid as in (5.146).

(5.146) PERLATIVE – temporal I – [Korpus Malti 3.0; parl70]

ikollok	*bżonnhom*	*biss*	***[għal***	*nofs*	*sena]*$_{PP}$
3SG.M.IPFV:have.FUT:2SG	need:3PL	only	[**for**	half	year]

‘[. . .] you will need them only [**for** half a year] [. . .].’

The need for the objects in the future is described as temporary. The state of affairs (= being needy of the objects) lasts exactly half a year.

As transpires from (5.147), (П15) *fi* ‘in’ may be used as a translation equivalent of English *within* defining the length of a period of time during which something takes place or has to take place.

(5.147) PERLATIVE – temporal II

(a) [Korpus Malti 3.0; culture63]

[F'	*sagħtejn]*$_{PP}$	*rajt*	*lil*	*Beyonce*	*Aerosmith*	*Paul McCartney*
[**in**	hour:DU]	see:1SG.PFV	to	Beyonce	Aerosmith	Paul McCartney

Julia Roberts	*Madonna*	*Metallica*	*Daft Punk*	*Pharrell*	*fost*
Julia Roberts	Madonna	Metallica	Daft Punk	Pharrell	amongst

l-oħrajn
DEF-other:PL

‘[**Within** two hours] I saw Beyonce, Aerosmith, Paul McCartney, Julia Roberts, Madonna, Metallica, Daft Punk, Pharrell, among others.’

(b) [Korpus Malti 3.0; news161278]

Naf	*persuna*	*li*	***[f'***	*sagħtejn]*$_{PP}$	*qatlet*	*aktar*
1SG.PFV:know	person	that	[**in**	hour:DU]	kill:3SG.F.PFV	more

minn	*70*	*wirdiena*	*minn*	*quddiem*	*ir-residenza*	*tagħha*
from	70	cockroach	from	in_front_of	DEF-residence	of:3SG.F

‘I know a person who [**within** two hours] killed more than 70 cockroaches in front of her residence.’

In (5.147a), the speaker reports about the transmission of the Grammy Award on TV – a show of two hours during which the person could see many artists appearing on stage one after the other. In (5.147b), the story of a lady killing numerous cockroaches within just two hours is told. The events – the parading of the artists in the (a)-sentence and the hunt for cockroaches in the (b)-sentence – last for two hours each.

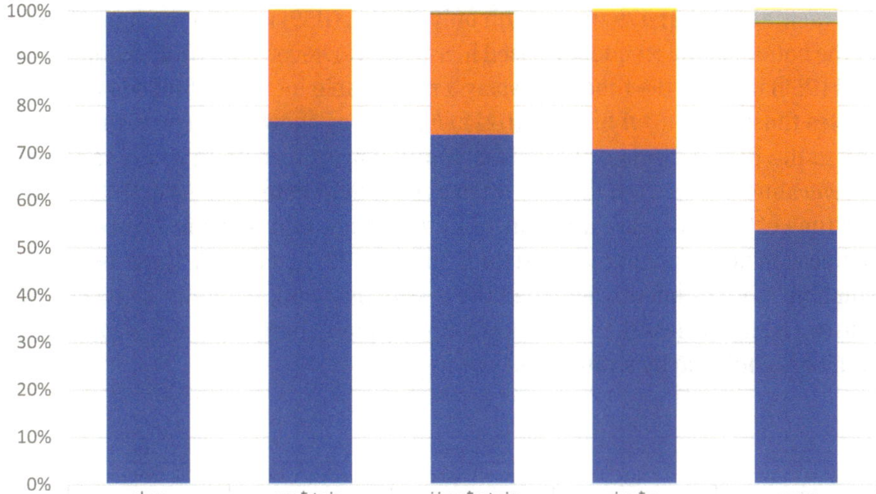

	xhur	sagħtejn	ġimgħatejn	siegħa	sena
waqt	0	0	0	2	6
matul	2	0	1	0	225
fi	1	83	40	217	3915
għal	730	271	115	525	4769

Figure 5.14: (Π15) *fi* 'in', (Π21) *għal* 'for', (Π33) *matul* 'during', and (Π59) *waqt* 'at the time of' competing for the expression of the temporal PERLATIVE.

In point of fact, (Π15) *fi* 'in' and (Π21) *għal* 'for' are attested by far more frequently than (Π33) *matul* 'during' and (Π59) *waqt* 'at the time of' in certain temporal constructions. This discrepancy comes clearly to the fore in Figure 5.14 where we compare the cooccurrences of the above four Πs with the nouns *siegħa* 'hour', *sagħtejn* 'two hours', *ġimgħatejn* 'two weeks', *xhur* 'months', and *sena* 'year'. The quantities refer to the hits in the Korpus Malti 3.0.

The low turnout for (Π59) *waqt* 'at the time of' has the effect that the share is too small to be visible in Figure 5.14. Only in combination with *sena* 'year' does (Π33) *matul* 'during' claim a share that is visible to the naked eye in the very same Figure 5.14. However, with only 2.5% of all TEMPORAL PERLATIVES involving this noun, (Π33) *matul* 'during' is only marginally represented in contrast to its contestants. These quantities seem to speak against attributing a major role to (Π33) *matul* 'during' or (Π59) *waqt* 'at the time of' in the domain of the spatial PERLATIVE. The picture changes drastically if, for instance, we replace the indefinite complement *sena* 'year' with its definite counterpart *is-sena* 'the year'. For *fis-sena* the Korpus Malti 3.0 yields 38,803 hits, for *għas-sena* the turnout is 25,951 hits, whereas *matul is-sena* is attested 21,179 times. With only 97 hits, (Π59) *waqt* 'at the time of' is still at the bottom of the frequency-based hierarchy. However, both (Π33) *matul* 'during' and (Π59) *waqt* 'at the time of' display a remarkable increase of tokens which surpasses those of (Π15) *fi* 'in' and (Π21) *għal* 'for' as shown in Figure 5.15. According to the data presented there (Π33) *matul* 'during' occurs in combination with *is-sena* almost a hundred times as often as with indefinite *sena*. As to (Π59) *waqt* 'at the time of', the increase amounts to sixteen times that noted in Figure 5.14. The increase is much smaller with (Π15) *fi* 'in' and (Π21) *għal* 'for'. These two Πs also combine more frequently with definite *is-sena* than with indefinite *sena*. However, *fis-sena* is slightly less than ten times more frequent than *f'sena* and *għas-sena* outnumbers *għal sena* by a ratio of 5.4-to-one only.

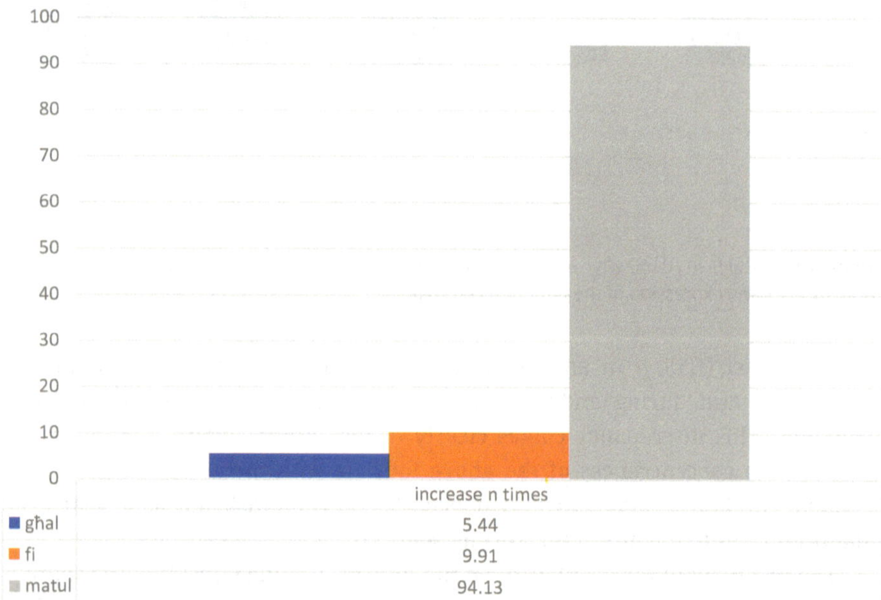

Figure 5.15: Increase of token frequencies for (Π15) *fi* 'in', (Π21) *għal* 'for', (Π33) *matul* 'during', and (Π59) *waqt* 'at the time of' in combination with *is-sena* 'the year'.

What is more, the high absolute numbers calculated for *fis-sena* and *għas-sena* cover a range of meaning categories such as temporal INESSIVE, BENEFACTIVE, PURPOSIVE, etc. so that the properly PERLATIVE usages account only for a part of the total turnout. In the case of *matul is-sena*, however, it is safe to assume that almost all of the tokens represent PERLATIVES. On this basis, we assume that (Π33) *matul* 'during' is a dedicated marker of the TEMPORAL PERLATIVE. In this sense, (5.148) contains typical representatives of the TEMPORAL PERLATIVE.

(5.148) PERLATIVE – temporal III

(a) [Korpus Malti 3.0; literature5]

Fra Mudest	*kellu*	*bosta*	*ħbieb*	*li*	*ltaqa'*
Brother Mudest	have.PFV:3SG.M	many	friend.PL	that	meet
magħhom	***[matul***	*ħajtu*	*fil-kunvent]$_{PP}$*		
with:3PL	**[during**	life:POSS:3SG.M	in:DEF-convent]		

'Brother Mudest had many friends with whom he met [**during** his life in the convent].'

(b) [Korpus Malti 3.0; news138400]

il-Ħadd	*ma*	*tistax*	*issir*	*kaċċa*	*u*
DEF-Sunday	NEG	3SG.F.IPFV:can:NEG	3SG.F.IPFV:become	hunt	and
[matul	*il-ġranet*	*l-oħrajn]$_{PP}$*	*jistgħu*	*joħroġu*	
[during	DEF:day:PL	DEF-other:PL]	3.IPFV:can:PL	3.IPFV:exit:PL	

għall-kaċċa	*nofs*	*ta'*	*nhar*	*biss*	*kuljum*
for:DEF-hunt	half	of	day	only	daily

'[. . .] there can be no hunting on a Sunday and [**during** the other days] they can go out for a hunt only half a day every day [. . .].'

In the (a) sentence, the extended period of time that Brother Mudest spent in the convent provides the frame for the events (= meetings with friends) to happen. In the (b)-sentence, the role of Ground is played by the days of the week during which hunting is subject to restrictions. Going on a hunting party is the Figure.

In point of fact, (П33) *matul* 'during' is compatible with a wide range of complements to express the TEMPORAL PERLATIVE. Table 5.12 covers only a small segment of the combinatory potential of this П. Like many other tables, Table 5.12 exclusively contains data taken from the Korpus Malti 3.0.

Table 5.12: (П33) *matul* 'during' in temporal PERLATIVE constructions.

#	source	example / translation
1	academic109	*Sostituzzjoni tista' issir jekk gowler minnhom ikun weġġa'* ***[matul*** *l-għoti tal-penalties]$_{PP}$* 'A substitution can be made if one of the gowlers is injured [**during** the penalty shoout-out].'
2	academic4	***[Matul*** *is-sena 2007–2008]$_{PP}$* *ser jitniedu Schools Materials, sit interattiv fuq l-internett u programm għat-taħriġ ta' l-għalliema* '[**During** the year 2007–2008] there will be announced School Materials, an interactive site on the internet and a program of teacher training.'
3	academic12	*Dan is-saċerdot kien twieled il-Birgu u* ***[matul*** *l-Ewwel Gwerra Kbira]$_{PP}$* *kien serva bħala kappillan militari* 'This priest was born in Birgu and [**during** the 1st Great War] he had served als military chaplain.'

Table 5.12 (continued)

#	source	example / translation
4	academic12	*il-kaptan ammetta li [**matul** l-vjaġġ]$_{PP}$ mietu żewġ trabi* '[. . .] the captain admitted that [**during** the voyage] two infants died [. . .].'
5	academic13	*Is-sħana żejda wkoll iddejjaqha u minħabba f' hekk [**matul** is-sajf]$_{PP}$, din il-pjanta tmur lura ħafna, jekk ma tmutx* 'The excess heat too annoys it and because of that [**during** the summer] this plant declines a lot, if it does not die.'
6	academic138	*Meta Paganini ra l-ewwel abbozzi ta' dan ix-xogħol biddel il ħsieb u qal li xtaq idoqq il-ħin kollu [**matul** is-sinfonija]$_{PP}$.* 'When Paganini saw the first proof of this work he changed his mind and said that he wanted to play all the time [**during** the symphony].'
7	academic233	*L-unika konsolazzjoni kienet ir-rebħa ta' 4 – 0 kontra ċ-ċittadini rivali Everton [**matul** it-Tazza Ingliża 1954 – 55]$_{PP}$.* 'The only consolation was the 4–0 victory against the local rival Everton [**during** the English Football Cup 1954–55].'
8	academic256	*Problemi oħra qamu barra l-istadju [**matul** il-lejl]$_{PP}$* 'Other problems came up outside the stadium [**during** the night].'

In (5.149), we provide sentential examples involving (П59) *waqt* 'at the time of' which is semantically very close to (П33) *matul* 'during'.

(5.149) PERLATIVE IV

(a) [Korpus Malti 3.0; academic114]
*omm Buddha mietet [**waqt** il-ħlas]$_{PP}$*
mother Buddha die:3SG.F.PFV [**at_the_time_of** DEF-childbirth]
'[. . .] Buddha's mother died [**during** childbirth] [. . .].'

(b) [Korpus Malti 3.0; academic399]
sbatax-il Malti ġew akkużati bi tradiment
seventeen Maltese come:3PL.PFV accuse:PART:PL with treason
*kontra Malta [**waqt** il-gwerra]$_{PP}$*
against Malta [**at_the_time_of** DEF-war]
'[. . .] seventeen Maltese were accused of treason against Malta [**during** the war].'

(c) [Korpus Malti 3.0; culture528]
*[**Waqt** id-diskors tiegħu]$_{PP}$ l-Papa kellu*
[**at_the_time_of** DEF-speech of:3SG.M] DEF-Pope have.PFV:3SG.M
jieqaf u jerġa jibda diversi
3SG.M.IPFV:stop and 3SG.M.IPFV:do_again 3SG.M.IPFV:begin diverse:PL
drabi minħabba ċ-ċapċip tal-pubbliku
time:PL on_account_of DEF-applause of:DEF-public
'[**During** his speech], the Pope had to stop and start again several times because of the applause of the auditorium.'

Frequently, it is possible to give (П59) *waqt* 'at the time of' a punctual reading. The event happened at some point in time during a certain period but did not extend over the entire period. Readings of this kind are also possible with (П33) *matul* 'during'. In (5.149a), death occurred at a certain point during childbirth but the process of dying coincided with an extended section of the event. In (5.149b), the supposed traitors have probably collaborated with the enemy not just once but over a longer period during the war. In (5.149c), the interruptions of the Pope's speech happened repeatedly and thus can be considered to extend the duration of the speech.

In Table 5.13, examples are provided that feature (П59) *waqt* 'at the time of' in combination with complements which are either identical with or similar to those mentioned for (П33) *matul* 'during' in Table 5.12.

Table 5.13: (П59) *waqt* 'at the time of' in TEMPORAL PERLATIVE constructions.

#	source	example / translation
1	academic41	*[**waqt** l-għoti tal-penalties]*$_{PP}$ *huwa salva tlieta wara xulxin* '[. . .] [**during** the penality shoot-out] he saved three in a row [. . .].'
2	european10925	*L-Istati Nordiċi għandhom ikunu integrati malajr kemm jista' jkun [**waqt** is-sena 2000]*$_{PP}$. 'The Nordic states have to be integrated as quickly as possible [**during** the year 2000].'
3	news168749	*l-Imperatur Karl kien inkoronat [**waqt** l-Ewwel Gwerra Dinjija]*$_{PP}$ *u miet fl-eżilju fl-1922* '[. . .] the emperor Karl was crowned [**during** the 1st World War] and died in exile in 1922.'
4	academic12	*[**Waqt** il-vjaġġ lejn Santos]*$_{PP}$ *twieldet tarbija* '[**During** the journey towards Santos] a baby was born [. . .].'
5	parl12758	*ż-żjara tal-ministru li se ssir qabel jew [**waqt** is-sajf]*$_{PP}$ *irridu niżviluppawha iktar* '[. . .] we must develop further the visit of the minister which will take place before or [**during** the summer] [. . .].'
6	culture1069	*[**Waqt** ir-reċita tal-Angelus]*$_{PP}$ *il-Papa awgura l-festa t-tajba ad unur it-twelid tal-Madonna anki lill-Knisja Ortodossa.* '[**During** the recital of the Angelus], the Pope also wishes the Orthodox Church a good feast to the honour of the birth of the Madonna.'
7	academic256	*Hu kien magħżul biex ikun referì [**waqt** it-Tazza tad-Dinja tal-Futbol 2006 fil-Ġermanja]*$_{PP}$. 'He was chosen to be a referee [**during** the Football World Cup 2006 in Germany].'
8	news201836	*[**Waqt** il-lejl]*$_{PP}$ *bejn it-Tnejn u t-Tlieta, il-belt ta' Christchurch, fi New Zealand, kienet mheżża b'terremot* '[**During** the night] between two and three, the town Christchurch in New Zealand was shaken by an earthquake.'

Thus, (Π59) *waqt* 'at the time of' and (Π33) *matul* 'during' display largely the same behaviour. The (potential) omnicompatibility of these two Πs with nouns from diverse semantic domains justifies our decision to give them the status of prototype or default option when it comes to expressing the TEMPORAL PERLATIVE. Whether their prototypicality differs by degrees must be ascertained in a separate study.

5.1.2.2.20.3 PROLATIVE

Hagége (2010: 261 and 287) presents English *past* as illustration of the PROLATIVE. In a spatial situation which is classified as PROLATIVE, the Figure is said to "go past a place or person." As mentioned above, Borg/Azzopardi-Alexander (1997) do not distinguish terminologically between PERLATIVE and PROLATIVE, but subsume both notions under "motion past". They add the important information, however, that for the expression of motion past it is common to use a Π-Π combination with (Π41) *minn* 'from' as Π_1 followed by "another appropriate expression" (Borg/Azzopardi-Alexander 1997: 156). For the individual configurations, however, it does not seem to be mandatory to employ (Π41) *minn* 'from' as the brackets around this Π suggest in the following cases:

- motion past anterior: *(minn) quddiem* (Borg/Azzopardi-Alexander 1997: 159),
- motion past posterior: *(minn) wara* (Borg/Azzopardi-Alexander 1997: 160),
- motion past superior(-contact): *(minn) fuq* (Borg/Azzopardi-Alexander 1997: 160–161 and 167),
- motion past surface: *(minn) fuq* (Borg/Azzopardi-Alexander 1997: 161 and 168),
- motion past medial (2): *(minn) bejn* (Borg/Azzopardi-Alexander 1997: 165),
- inferior (long object): *(minn) taħt* (Borg/Azzopardi-Alexander 1997: 168).

In addition, (Π41) *minn* 'from' is excluded from the expression of motion past with the circumferential; here the configuration marker (Π31) *madwar* 'around' is self-sufficient (Borg/Azzopardi-Alexander 1997: 166). And finally, in the domain of the interior, (Π41) *minn* 'from' expresses motion past without any accompanying Π_2 (Borg/Azzopardi-Alexander 1997: 158). In all other cases, (Π41) *minn* 'from' obligatorily combines with a second Π – at least according to the reference grammar.

The picture resulting from the above observations about the distribution of the Πs is heterogeneous. In this section, we intend to tidy up things where this is necessary. To this end, we first check whether it is indeed possible to use (Π41) *minn* 'from' alone to express the PROLATIVE. (5.150) is one of the very few pieces of evidence that support this possibility.

(5.150) Bare PROLATIVE – [Korpus Malti 3.0; news12755]

Attard	*xehed*	*li*	*meta*	*għadda*	*[**mi**ll-post]$_{PP}$*
Attard	give_evidence	that	when	pass	[**from**:DEF-place]

bil-mutur *kien* *imbeżża'*
with: DEF-motorcycle be PART:fear
'Attard gave evidence that when he passed [the place] on the motorcycle he was in fear [. . .].'

The motorcyclist is in fear of dogs but he has to pass in the vicinity of a dog tied to a chain. The *minn*-PP is ambiguous since it invites two competing interpretations. Either the Figure passes the place where the dog is chained to the wall or the Figure's passing across the property guarded by the dog is described. In the former case, we may speak of a PROLATIVE, but in the latter case, we have another example of the PERLATIVE. As a matter of fact, *minn*-PPs following the verb *għadda* 'pass' are overwhelmingly often PERLATIVE. In stark contrast to the dominance of the PERLATIVE, there is hardly any convincing evidence of the PROLATIVE expressed in this way. The normal shape an indubitable PROLATIVE construction takes is that of combining (Π41) *minn* 'from' with a Π_2 as in (5.151).

(5.151) PROLATIVE I
(a) [Korpus Malti 3.0; literature77]
Qormi baqa' miexi sa tarf is-sur dar lura
Qormi remain walk.PART till extremity DEF-bastion turn back
għadda ***[minn*** ***biswit*** *id-dar]*$_{PP}$ *u raġa' ra 'l Ineż*
pass **[from** **facing** DEF-house] and do_again see to Ineż
tistennieh maġenb it-tieqa
3SG.F.IPFV:wait:3SG.M close_to DEF-window
'Qormi kept walking to the end of the bastion, turned back, passed [**opposite** the house] and saw Ineż again waiting for him near the window.'
(b) [Attard 2019: 64]
L-uffiċjal żagħżugħ kien diġà marbut xħin għaddew
DEF-officer young be already PART:tie when pass:3PL.PFV
iż-żewġ ħallelin ***[minn*** ***ħdejh]***$_{PP}$ *fi triqthom*
DEF-two thief:PL **[from** **beside**:3SG.M] in street:POSS:3PL
lejn il-karozza.
towards DEF-car
'The young officer was already tied up when the two thieves passed [him] on their way to the car.'

The Π_2 and thus the configuration marker is (Π8) *biswit* 'facing' in the (a)-sentence and (Π23) *ħdejn* 'beside' in the (b)-sentence. The combinations *minn biswit* and *minn ħdejn* have occurred under different rubrics before, namely in (5.60b) and (5.65a) when motion away from near and motion away from opposite were under discussion. The Π-Π combinations as such are thus ambiguous. They are disambiguated by the context – most notably by the motion verb they depend on. As far as we can judge, only *għadda* 'pass' straightforwardly triggers a PROLATIVE or PERLATIVE interpretation. Which of these two options is the correct one in a given spatial situation can be recognised on account of the Π_2.

From (5.152), we learn that the presence of *għadda* 'pass' with its semantically incorporated Path meaning renders the employment of the directionality marker (Π41) *minn* 'from' superfluous. We represent the absent directionality marker by Ø.

(5.152) PROLATIVE without directionality marker

(a) [Korpus Malti 3.0; news34256]

dan *l-ilma* *kien* *xi* *ramel* *u* *żrar* *li*
DEM:M:PROX DEF-water be some sand and grit.COLL that
kienu *maħżunin* *f'* *sit* *barra* *mil-lukanda* *li*
be PART:store:PL in site outside from:DEF-hotel that
tkaxkru *bil-maltemp* *u* *tferra* *fil-baħar*
REFL:drag:3PL.PFV with:DEF-stormy_weather and REFL:pour in:DEF-sea
minn *culvert* *għall-ilma* *tax-xita* *li* *jgħaddi*
from culvert for:DEF-water of:DEF-rain that 3SG.M.IPFV:pass
*[Ø **biswit** il-lukanda]$_{PP}$*
[Ø **facing** DEF-hotel]
'[. . .] this water was some sand and grit that had accumulated in a site outside the hotel that was dragged by the stormy weather and poured into the sea through a rain water culvert that passes [**opposite** the hotel].'

(b) [Korpus Malti 3.0; parl9014]

kulħadd *qiegħed* *hemmhekk* *u* *kulħadd* *bix-xewqa* *li*
everbody stay.PART there and everybody with:DEF-wish that
l-Papa *jgħaddi* *[Ø **ħdejh**]$_{PP}$* *jew* *ikellmu*
DEF-Pope 3SG.M.IPFV:pass [Ø **beside**:3SG.M] or 3SG.M.IPFV:speak:3SG.M
'[. . .] everybody is there and everybody has the wish that the Pope passes [**near** him] or speaks to him [. . .].'

We cannot detect any semantic differences between the constructions with and without (Π41) *minn* 'from'. There is unconditioned *minn*-Ø alternation in the case of the PROLATIVE whereas for the identical Π-Π combinations with the function of encoding motion from the Ground the omission of (Π41) *minn* 'from' is ungrammatical. The question arises whether dropping the directionality marker is an option for each of the configurations that are semantically qualified to appear in PROLATIVE situations.

The qualifying criterion is the physical possibility of the Figure passing by the Ground without making any contact with it. On the one hand, this criterion excludes INESSIVE configurations as expressed by (Π15) *fi* 'in'; (Π18) *ġewwa* 'inside', and (Π19) *ġo* 'in' and MEDIOESSIVES like (Π16) *fost* 'amongst' and (Π47) *qalb* 'amidst'. Πs that are responsible for directionality themselves on the other hand, can also be counted out (e.g. (Π21) *għal* 'for', (Π28) *lejn* 'towards', (Π38) *mingħand* 'from s.o.', (Π42) *mintul* 'all along'; (Π43) *mnejn* 'from near', and (Π57) *versu* 'towards'). Moreover, (Π22) *għand* 'at s.o.'s place', (Π31) *madwar* 'around'; (Π33) *matul* 'during'; (Π44) *oltre* 'beyond'; and (Π46) *qabel* 'before' are not attested in PROLATIVE contexts in the Korpus Malti 3.0. We also know already that combinations of (Π41) *minn* 'from' and (Π4) *bejn* 'between' (yielding the INTERABLATIVE e.g. in (5.105))) are difficult to distinguish from PERLATIVES (as in (5.145)). Similar difficulties arise in connection with other configuration markers that combine with the Π_1 (Π41) *minn* 'from'.

In what follows, we ask whether the *minn*-Ø alternation documented for (Π8) *biswit* 'facing' and (Π23) *ħdejn* 'beside' holds also for other configuration markers. For each of the configuration markers, we provide two examples one of which involves a given marker in combination with (Π41) *minn* 'from' and the second example without directionaility marker. Where one of the options is not attested in our sources, its absence is dutifully registered. In contrast to the previous examples in this section, the absence of the directionality marker is not specifically marked. To guarantee comparability of the examples, only those which involve the motion verb *għadda* 'pass' are used. The configuration markers are ticked off in the alphabetical order established for BLOMP 2.0. The examples are accompanied by short interpretations of the spatial situations depicted by them. We assume that where examples come pairwise the optionality of the direction marker has no bearing on the meaning, i.e., PROLATIVE constructions with and without (Π41) *minn* 'from' are synonymous.

The point of departure of our review of the PROLATIVE constructions is marked by (Π3) *barra* 'outside' in (5.153). Note that the reference grammar does not mention the possibility of leaving out the directionality marker (Borg/Azzopardi-Alexander 1997: 159) as is done however in (5.153b).

(5.153) PROLATIVE with (Π3) *barra* 'outside'

(a) [Korpus Malti 3.0; parl6958]

it-traffiku kollu li kien jiġi mit-tramuntana ta'
DEF-traffic all:3SG.M that be 3SG.IPFV:come from:DEF-north of
Malta [. . .] u li s-soltu jgħaddi ***[minn***
Malta and that DEF-usual_practice 3SG.M.IPFV:pass **[from**
barra *Ħaż-Żebbuġ]$_{PP}$ u minn Triq l-Imdina [. . .] bilfors*
outside Ħaż-Żebbuġ] and from Triq l-Imdina perforce

kellu *jiġi* *devjat* *għal* *ġo* *Ħaż-Żebbuġ*
have.PFV:3SG.M 3SG.M.IPFV:come deviate.PART for in Ħaż-Żebbuġ
'[. . .] all the traffic that came from the north of Malta [. . .] and that usually passes [**outside** Ħaż-Żebbuġ] and through Triq l-Imdina [. . .] had forcibly to be diverted into Ħaż-Żebbuġ [. . .].'

(b) [Korpus Malti 3.0; news53362]
Fl-64 *minuta* *xutt* *tal-player* *Taljan* *ta'* *Floriana*
in:DEF-64 minute shot of:DEF-player Italian of Floriana
Matteo Piciollo *għadda* *[**barra*** ***mil-lasti*** *difiżi*
Matteo Piciollo pass [**outside** **from**:DEF-post:PL defend.PART:PL
mill-goalkeeper *Mosti]*$_{PP}$
from:DEF-goalkeeper Mosta:ADJR]
'In the 64th minute, a shot by Floriana's Italian player Matteo Piciollo passed [**outside** the posts defended by Mosta's goalkeeper].'

In the (a)-sentence, the Figure is the traffic from other parts of Malta that usually passes outside the village of Ħaż-Żebbuġ (= Ground). Owing to roadworks, the traffic had to be temporarily diverted into (and through) the village. The Figure in the (b)-sentence is the shot (metonymically representing the ball) that went wide. The posts of the goal represent the Ground. What is interesting about (5.153b) is the absence of (Π41) *minn* 'from' as directionality marker while, at the same time, (Π41) *minn* 'from' follows (Π3) *barra* 'outside' as marker of ABLATIVE government.[118]

In (5.154), we encounter the sole case of a Π that encodes configuration but does not tolerate the absence of the directionality marker (Π41) *minn* 'from' when it comes to expressing the PROLATIVE. What is more, the Korpus Malti 3.0 gives no evidence of (Π12) *faċċata* 'opposite' being used in PROLATIVE situations. This is why we have to adapt an example provided by the reference grammar.

(5.154) PROLATIVE with (Π12) *faċċata* 'opposite' – (Borg/Azzopardi-Alexander 1997: 167)
Għaddiet *[**minn*** ***faċċata*** *tiegħu]*$_{PP}$ *u* *ma* *ndunax*
pass:3SG.F.PFV [**from** **opposite** of:3SG.M] and NEG notice:NEG
biha
with:3SG.F
'She passed [**opposite** him] and he didn't notice her.' [O.T.]

118 Cf. Section 4.1.2.2.1.2.2.2.

The Figure (= the woman = subject of the motion verb) passes somewhere across from the Ground (= the man = subject of the perception verb) but within the Ground's potential field of vision.

The first Π from the domain of the PROPINQUUS is (Π14) *fejn* 'near'. As is shown in (5.155), this Π can be used in PROLATIVE constructions with as well as without (Π41) *minn* 'from'. Borg/Azzopardi-Alexander (1997: 157) do not mention the possibility of dropping the directionality marker in this context.

(5.155) PROLATIVE with (Π14) *fejn* 'near';

(a) [Korpus Malti 3.0; news112466]

Hekk	*kif*	*Bennetti*	*għadda*	*[**minn***	***fejn***	*it-tifla]*$_{PP}$	*din*
so	how	Bennetti	pass	[**from**	**near**	DEF-girl]	DEM:F:PROX

qamet	*u*	*marret*	*ħdejn*	*Joyce*
rise:3SG.F.PFV	and	go:3SG.F.PFV	beside	Joyce

'As soon as Bennetti passed [**near** the girl] she rose and went beside Joyce.'

(b) [Korpus Malti 3.0; european3575]

il-ħaddiem	*għandu*	*jkun*	*provdut*	*b'*
DEF-worker	have.IPFV:3SG.M	3SG.M.IPFV:be.FUT	provide.PART	with

biżżejjed	*spazju*	*ħieles*	*minn*	*fejn*	*ikun*	*jista'*
enough	space	free	from	where	3SG.M.IPFV:be.FUT	3SG.M.IPFV:can

jgħaddi	*[**fejn***	*l-istazzjoni*	*tax-xogħol*	*tiegħu]*$_{PP}$
3SG.M.IPFV:pass	[**near**	DEF-station	of:DEF-work	of:3SG.M]

'[. . .] the worker has to be provided with enough free space which he may pass through [**near** his work station].'

In the (a)-sentence, the Figure (= the person bearing the name Bennetti) is described as moving past close to the Ground (= the girl). In the (b)-sentence, the workers (= Figure) are provided with sufficient space to pass close to their work stations (= Ground) on their way elsewhere.

The reference grammar depicts the use of the directionality marker (Π41) *minn* 'from' as optional in combination with (Π17) *fuq* 'on'. The only example provided by Borg/Azzopardi-Alexander (1997: 161) to illustrate the optional status of (Π41) *minn* 'from' is perhaps better classified as a case of the PERLATIVE.[119] In another example, Borg/Azzopardi-Alexander (1997: 160) do not employ brackets around (Π41) *minn*

---

119 The original example of Borg/Azzopardi-Alexander (1997: 161) is reproduced here as (5.vi).

(5.vi)

Il-barbier	*għadda*	*l-mus*	*ħafif*	*[(**minn**)*	***fuq***	*wiċċu]*$_{PP}$
DEF-barber	pass	DEF-razor	light	[(**from**)	**on**	face:3SG.M]

'The barber passed the razor lightly [**on** his face].' [O.T.]

'from' although the subtitle of the section suggests that the directional marker's use is not mandatory. In (5.156), we present two examples in which the Figure passes (high) above the Ground without making physical contact with it.

(5.156) PROLATIVE with (Π17) *fuq* 'on'

(a) [Korpus Malti 3.0; parl217]

l-ajruplan jgħaddi ***[minn fuq*** *Malta]*$_{PP}$ *mingħajr ma*
DEF-aeroplane 3SG.M.IPFV:pass **[from on** Malta] without that
jinżel l-art
3SG.M.IPFV:descend DEF-ground
'[. . .] the aeroplane passes [**over** Malta] without landing [. . .].'

(b) [Korpus Malti 3.0; news125470]

Mixja oħra li għamilt kienet tibda minn
walk other:F that make:1SG.PFV be:3SG.F.PFV 3SG.F.IPFV:begin from
Triq il-Brolli f' Birżebbuġa u għaddejt minn fuq pont li
Triq_il_Brolli in Birżebbuġa and pass:1SG.PFV from on bridge that
jgħaddi ***[fuq*** *Wied Ħanak]*$_{PP}$
3SG.M.IPFV:pass **[on** Wied Ħanak]
'Another walk I took started from Triq il-Brolli in Birżebbuġa and I passed over a bridge that passes [**over** Wied Ħanak].'

In both cases, the Figure passes across the Ground but at a considerable distance from it so that there is no physical contact between Figure and Ground. The spatial situation therefore is not that of a PERLATIVE, but fits the description of the PROLATIVE.

In (5.157), the configuration marker is (Π30) *ma'* 'with'. As mentioned above in connection with the SECUTIVE, Borg/Azzopardi-Alexander 1997: 163) do not register motion past constructions which involve this Π.[120] To their mind, (Π32) *maġenb* 'close to' must be used as configuration marker (together with (Π41) *minn* 'from') to express this spatial meaning. However, in the Korpus Malti 3.0, there is evidence supporting the idea that (Π30) *ma'* 'with' too is admissible in PROLATIVE constructions.

We consider (5.vi) to represent a PERLATIVE situation since the Figure (= the knife) touches the Ground (= the face) while moving across it. There is thus contact between Figure and Ground. The PROLATIVE however would require that Figure and Ground remain at a (however small) distance from each other.

120 Cf. Section 5.1.2.2.20.1.4 (i).

(5.157) PROLATIVE with (Π30) *ma'* 'with'

(a) [Korpus Malti 3.0; parl508]

tgħaddi *[minn* ***ma'*** *karozza* *li* *kellha* *l-bieba*
3SG.F.IPFV:pass **[from** **with** car that have.PFV:3SG.F DEF-door
miftuħa]PP *ftit* *'il* *bogħod* *minnha*
PART:open:F] little to distance from:3SG.F
'[. . .] she passes [a car that had its door open] at a little distance from her.'

(b) [Korpus Malti 3.0; news166887]

x-xutt *tagħha* *għadda* ***[ma'*** *sieq* *il-lasti]PP*
DEF-shot of:3SG.F pass **[with** foot of:DEF-post:PL]
'[. . .] her shot passed [**by** the lower end of the goalpost].'

In both the (a)-sentence and the (b)-sentence, the Figure moves past the Ground at only a little distance but no contact is made between the two participants of the spatial situation. In (5.157a), the car can be considered to be a horizontally extended Ground whereas in (5.157b) the goalpost is too narrow to fit the same description. We assume that the use of (Π30) *ma'* 'with' as configuration marker is motivated by the temporary proximity of Figure and Ground.[121]

Borg/Azzopardi-Alexander (1997: 163) mention the Π-Π combination *minn maġenb* as one of two options to express motion past in the lateral domain (the other option being *minn biswit* as shown in (5.151a)). The cooccurrence of the configuration marker (Π32) *maġenb* 'close to' and the directionality marker (Π41) *minn* 'from' is documented in (5.158a). The absence of the directionality marker is illustrated in (5.158b) – an option that is not mentioned in the reference grammar.

(5.158) (Π32) *maġenb* 'close to'

(a) [Korpus Malti 3.0; parl2986]

hemm *wied* *u* *hemm* *xmara* *li* *tgħaddi* ***[minn***
there valley and there river that 3SG.F.IPFV:pass **[from**
maġenb *dan* *il-post]PP*
close_to DEM:M:PROX DEF-place]
'[. . .] there is a valley and there is a river that passes [**close to** this place] [. . .].'

(b) [Korpus Malti 3.0; parl5881]

xi *snin* *ilu* *kienu* *saru* *modifikazzjonijet* *meta*
some year.PL ago be:3PL.PFV become:3PL.PFV modification:PL when

121 Cf. Section 5.1.2.2.20.1.1.

kienet *saret* *it-triq* *li* *tgħaddi*
be:3SG.F.PFV become:3SG.F.PFV DEF-street that 3SG.F.IPFV:pass
[maġenb *is-Salini]$_{PP}$*
[**close_to** DEF-Salini]
'[. . .] some years ago, modifications had been realised when the street was built that passes [**close to** Salini].'

(Π48) *qrib* 'near' is another PROPINQUUS Π that is partly synonymous with the previous one. The examples in (5.159) prove that the occurrence of PROPINQUUS Πs in PROLATIVE constructions is not the privilege of (Π14) *fejn* 'near' as suggested by Borg/Azzopardi-Alexander (1997: 157).

(5.159) PROLATIVE with (Π48) *qrib* 'near'
(a) [Korpus Malti 3.0; news150262]
raw *tifel* *tiela'* *wkoll* *it-taraġ* *mekkanizzat* *u*
see:3PL.PFV boy ascend.PART also DEF-stairway mechanise:PART and
kif *għadda* ***[minn*** ***qrib*** *l-anzjana]$_{PP}$* *din*
how pass [**from** **near** DEF-eldery_woman] DEM:F:PROX
tilfet *il-bilanċ*
lose:3SG.F.PFV DEF-balance
'[. . .] they saw a boy who too was ascending the escalator and when he passed [**near** the elderly woman] she lost her balance [. . .].'
(b) [Korpus Malti 3.0; news198144]
Fi *triqtu* *lejn* *Malta* *il-katamaran* *għadda*
in street:POSS:3SG.M towards Malta DEF-catamaran pass
[qrib *is-Somalja]$_{PP}$* *f'* *żona* *li* *hi* *perikoluża* *minħabba*
[**near** DEF-Somalia] in zone that 3SG.F dangerous:F on_account_of
l-pirati *tal-baħar*
DEF-pirate:PL of:DEF-sea
'On its way towards Malta, the catamaran passed [**near** Somalia] in a zone that is dangerous because of the pirates.'

In the (a)-sentence, the Figure (= the boy) passes so close to the Ground (= the elderly lady) that the latter loses her balance (and eventually drops on her back). The distance between the boy and the elderly lady is certainly much smaller than that between the catamaran (= Figure) and Somalia (= Ground) in the (b)-sentence. The coastline of the country extends over hundreds of kilometres whereas the human being put out of balance is like a dot on a small-scale map.

As to motion past in the anterior domain, Borg/Azzopardi-Alexander (1997: 159) treat the directionality marker (Π41) *minn* 'from' as optional since the configuration marker (Π49) *quddiem* 'in front of' alone is sufficient when it comes to expressing the PROLATIVE. This is corroborated by the data found in the Korpus Malti 3.0. Two examples are given in (5.160).

(5.160) PROLATIVE with (Π49) *quddiem* 'in front of';

(a) [Korpus Malti 3.0; literature24]

Jason għadda ***[minn quddiem*** *l-iskola tan-Nadur]$_{PP}$*
Jason pass **[from in_front_of** DEF-school of:DEF-Nadur]
l-iskola li kien jattendi missieru meta kien tifel
DEF-school that be 3SG.M.IPFV:attend father:3SG.M when be boy
'Jason passed **[in front of** the school of Nadur], the school that his father attended when he was a boy.'

(b) [Korpus Malti 3.0; news122175]

Dan kollu għadda ***[quddiem*** *għajnejja]$_{PP}$ din*
DEM:M:PROX all:3SG.M pass **[in_front_of** eye:PL:1SG] DEM:F:PROX
il-ġimgħa
DEF-week
'All this passed [**before** my eyes] this week.'

In the (a)-sentence, the Figure does not enter the Ground but passes in front of it. There is thus no physical contact between Figure and Ground. In the case of the (b)-sentence, the motion event is imaginary and abstract in the sense that the speaker refers to events that he witnessed during the week. We accept this example as a metonymy.

Borg/Azzopardi-Alexander (1997: 162) provide an example of a ternary combination of Πs that is supposed to express motion past a Ground in the inferior domain with contact between Figure and Ground. The combination of three involves *minn* + *ma'* + *taħt*.[122] Presumably, the latter functions as configuration marker while the leftmost Π is responsible for directionality while (Π30) *ma'* 'with' could fulfil the task of encoding contact. Not only does (Π30) *ma'* 'with' not necessarily signal that there is contact between Figure and Ground,[123] but we have not identified any *minn* + *ma'* + *taħt* combination in the Korpus Malti 3.0. Since the supposed contact

---

122 Borg/Azzopardi-Alexander (1997: 162) provide the following example:

(5.vii) *Biex seta' kompla jitla' kellu jixxabbat*
to can continue 3SG.M.IPFV:ascend have.PFV:3SG.M 3SG.M.IPFV:crawl
[minn ma' taħt *il-blata]$_{PP}$*
[from with under DEF-rock]
'In order to be able to continue his climb, he had to creep [**from under** the rock].' [O.T.]

123 Cf. Section 5.1.2.2.20.1.1.

between Figure and Ground violates the definition of the PROLATIVE, examples of this kind would have to be classified as instances of the PERLATIVE, in the first place. What we have found instead are examples of the binary Π-Π combination *minn taħt* and (Π56) *taħt* 'under' without directionality marker to express the PROLATIVE as documented in (5.161). Borg/Azzopardi-Alexander (1997: 168) mention similar cases only in connection with long objects as Grounds.

(5.161) PROLATIVE with (Π56) *taħt* 'under'

(a) [Korpus Malti 3.0; news105365]
Kif Abela għadda ***[minn taħt*** *il-pont]$_{PP}$ u ġie*
how Abela pass **[from under** DEF-bridge] and come
fid-dawra sab ma' wiċċu t-track
in:DEF-turning find with face:3SG.M DEF-truck
'When Abela passed [**under** the bridge] and came to the turning, he found the truck right in front of him.'

(b) [Korpus Malti 3.0; news105819]
Dan il-wied sabiħ jgħaddi ***[taħt*** *ġebel*
DEM:M:PROX DEF-valley beautiful 3SG.M.IPFV:pass **[under** stone
Għomor li hu blat għoli fuq il-lemin li
Għomor that 3SG.M rock high on DEF-right that
għandu l-istess isem tal-Wied]$_{PP}$
have.IPFV:3SG.M DEF-same name of:DEF-valley]
'This valley is beautiful passing [**under** the Għomor stone that is a tall rock on the right that has the same name as the valley].'

In both the (a)-sentence and the (b)-sentence, the motion of the Figure follows a path that leads the Figure past the Ground but Figure and Ground are never in physical contact with each other. In (5.161a), the Figure (= Abela) passes under the Ground (= the bridge) to face a new situation at the exit. In (5.161b), the valley moves metonymically through the landscape passing a tall rock that is placed on a higher level so that it is appropriate to say that the valley passes under or below the rock.

With (Π58) *viċin* 'near' it is the turn of the last of the PROPINQUUS Πs. Like its synonyms, this Π occurs in PROLATIVE constructions which either involve or exclude the directionality marker (Π41) *minn* 'from'. Both possibilities are featured in (5.162).

(5.162) PROLATIVE with (Π58) *viċin* 'near'

(a) [Korpus Malti 3.0; parl3925]
nagħmlu użu mill-gas pipeline li se jgħaddi
1.IPFV:make:PL use from:DEF-gas_pipeline that FUT 3SG.M.IPFV:pass

[minn ***viċin*** *gżiritna]$_{PP}$*
[**from** **near** island:POSS:1PL]
'[. . .] we make use of the gas pipeline that will pass [**near** our island].'

(b) [Korpus Malti 3.0; parl7302]
Dan l-ajruplan jgħaddi minn fuq l-Isla u
DEM:M:PROX DEF-aeroplane 3SG.M.IPFV:pass from on DEF-Isla and
jgħaddi ***[viċin*** *tal-bini]$_{PP}$*
3SG.M.IPFV:pass [**near** DEF-building]
'This aeroplane passes over Isla and passes [**near** the building].'

In (5.162a), it is clear that the underwater pipeline (again metonymically) passes near the island of Malta without touching it. Similarly, in (5.162b), the aeroplane does not come in direct contact with the building it flies over. The important point is that during the motion event, the Figure is near the Ground at least temporarily.

The last configuration marker to discuss in this section is (Π60) *wara* 'after'. Borg/Azzopardi-Alexander (1997: 160) assume that in PROLATIVE constructions expressing motion past in the posterior domain, the directionality marker is optional. This assumption is confirmed by the data we have collected from the Korpus Malti 3.0. Two examples are given in (5.163).

(5.163) PROLATIVE with (Π60) *wara* 'after'
(a) [Korpus Malti 3.0; news14746]
l-vapur tal-Gozo Channel tas-sitta ta' dalgħodu ħa
DEF-ship of:DEF-Gozo_Channel of:DEF-six of this_morning take
madwar 50 minuta biex wasal minn Għawdex għal Malta
around 50 minute to arrive from Gozo for Malta
meta kellu jgħaddi ***[minn wara*** *Kemmuna]$_{PP}$*
when have.PFV:3SG.M 3SG.M.IPFV:pass [**from** **behind** Comino]
'[. . .] this morning's six o'clock Gozo Channel ferry took about 50 minutes to arrive from Gozo in Malta when it had to pass [**behind** Comino].'

(b) [Korpus Malti 3.0; news118304]
parti mit-triq li tgħaddi ***[wara*** *t-Technopark*
part from:DEF-street that 3SG.F.IPFV:pass [**behind** DEF-Technopark
tal-Mosta]$_{PP}$ inbniet fuq barriera antika
of:DEF-Mosta] REFL:build:3SG.F.PFV on quarry old:F
'[. . .] a part of the street that passes [**behind** the Technopark of Mosta] was built on an old quarry.'

In (5.163a), the Gozo Channel ferry is forced to deviate from its usual route towards its destination. It is delayed because it had to take a detour passing behind Comino. No stop was made on the little island between Gozo and Malta. Comino was passed by. In (5.163b), we have another case of metonymical motion in the sense that the Figure is a street that is said to pass behind the Ground (= the Technopark). There is no direct contact between Figure and Ground.

The above examples prove that except (Π12) *faċċata* 'opposite', all configuration markers may occur either together with (Π41) *minn* 'from' or without the directionality marker. In qualitative terms, the different Πs behave the same. The picture changes when we take account of the quantities. In Figure 5.16, we contrast per configuration marker the turnouts and shares of PROLATIVE constructions with and without (Π41) *minn* 'from'. The Πs are ordered top-down according to the increasing share of constructions which involve the directionality marker. Owing to the absence of variation, (Π12) *faċċata* 'opposite' is not included in the survey. We have exclusively counted PROLATIVE PPs that immediately follow an inflected form of *għadda* 'pass'.[124]

The shares of the two options are distributed over the Πs such that a continuum emerges. At the two extremes, (Π58) *viċin* 'near' (the only Italo-Romance Π in Figure 5.16) and (Π14) *fejn* 'near' attested to the almost exclusive use of the configuration marker alone and of the Π-Π combination involving (Π41) *minn* 'from', respectively. Five Πs prefer the *minn*-less construction over the Π-Π combination, for seven Πs it is exactly the other way around. Interestingly, the two Πs at the top and the two Πs at the bottom belong to the PROPINQUUS domain. Equally interesting is the fact that the antonymic Πs (Π60) *wara* 'after' and (Π49) *quddiem* 'in front of' display different preferences. The same holds for the antonyms (Π56) *taħt* 'under' and (Π17) *fuq* 'on'. Thus, the label *minn*-Ø alternation glosses over the strikingly individualistic behaviour of the Πs. The PROLATIVE stands out among the categories subsumed under the heading Path insofar as it is the only category that systematically tolerates the absence of the directionality marker. The occasional omission of (Π41) *minn* 'from' in perlative constructions mentioned in passing by Borg/Azzopardi-Alexander (1997) could not be confirmed on the basis of the Korpus Malti 3.0. The failure to identify instances of *minn*-dropping in perlative constructions in the electronic resource does not mean that *minn*-less perlatives are generally uncom-

124 For practicality, only uninflected forms of the Πs are included in the count. Similarly, no case of fusion with the definite article clitic have been taken account of. The frequencies must be taken with a grain of salt because for several Πs it was necessary to extract the PROLATIVE cases from a considerably larger number of other cases (idioms, loan translations from English affecting the meaning of *għadda* 'pass', etc.).

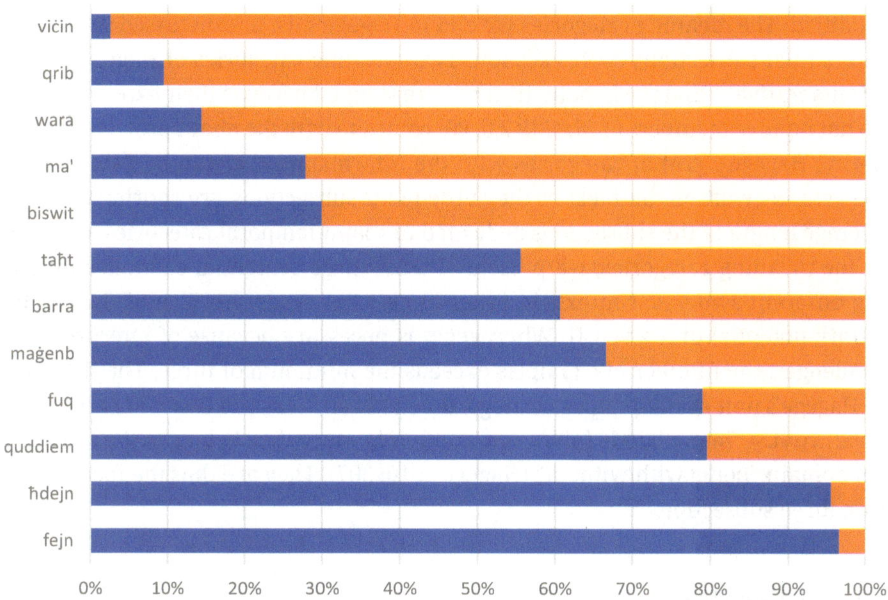

	fejn	ħdejn	quddiem	fuq	maġenb	barra	taħt	biswit	ma'	wara	qrib	vičin
■ minn	58	284	359	252	6	17	421	3	7	52	6	1
■ Ø	2	13	92	67	3	11	336	7	18	306	57	37

Figure 5.16: *minn*-Ø alternation (PROLATIVE) in combination with forms of *għadda* 'pass'.

mon also in the spoken register of Maltese. Whether or not this is the case needs to be investigated in a follow-up study.

The discussion of the PROLATIVE marks the endpoint of the presentation not only of the Path categories but also of the spatio-temporal meanings. We will put them into a new order below.[125] Before we get there, however, it is necessary to look beyond the limits of space and time in the next subsection.

5.1.3 Beyond space and time

At this point, we enter Hagège's (2010: 262) non-spatio-temporal domain. As in the foregoing sections, we take Hagège's taxonomy of meaning categories as our guideline with the proviso that wherever we deem it necessary, the terminology, the order in which the categories are presented, and the number of distinctions will be

125 Cf. Section 5.2.1.

modified. The meaning categories we add to Hagège's list are discussed separately in the final subsections of this section. Two meaning categories of Hagège's – the adnumerative[126] and the hypothetical[127] – are skipped because their expressions do no not involve members of BLOMP 2.0. We continue with the above practice of comparing Hagège's grid of categories with the information that can be gained from the reference grammar. Most of the subsequent subsections are shorter than the average length of the subsections dedicated to spatio-temporal categories. For each of the meaning categories, we present at least three sentential examples involving different syntactic contexts in order to give the reader an idea of the combinatory potential of a given Π. Wherever it is necessary because of interpretative challenges, the number of examples exceeds the minimum of three. The first four of Hagège's non-spatio-temporal categories – PROPRIETIVE, EXCLUSIVE, ECEPTIVE, and COMITATIVE – "immediately follow one another because they belong to the semantic subdomain "being with/without" (Hagège 2010: 307). They are thus the focus of the first three subsections.[128]

5.1.3.1 PROPRIETIVE

Hagège (2010: 262) proposes the English gerund *having* to illustrate the meaning of the PROPRIETIVE. The PROPRIETIVE is termed "ornative" in Stolz/Stroh/Urdze (2006: 288–289) and "circumstance" in Borg/Azzopardi-Alexander (1997: 149). The different terms notwithstanding, we are facing cases of prepositional attribution[129] in which (Π7) *bi* 'with' heads a PP that ascribes a physico-mental property (permanent or other) to the modified noun or indicates that the latter is equipped with or wears or carries a portable object. The attributive PP may also take abstract nouns as its complement. The modification takes place NP-internally. A selection of examples illustrating the use of PROPRIETIVE (Π7) *bi* 'with' is given in (5.164). Some of the uses will be reconsidered in the section on the COMITATIVE.[130]

---

126 The English expression provided by Hagège (2010: 262) for the adnumerative is *in proportion to*. Aquilina (1999: 2440) gives *approporzjon ta'* as Maltese translation equivalent. However, there is no entry for *approporzjon* either in Aquilina (1987, 1991), in E. Serracino-Inglott (1975) or in M. Serracino-Inglott (2003). We have therefore decided not to accept this item as candidate for the status of Π.

127 Hagège (2010: 262) illustrates the hypothetical with English *in case of*. The corresponding Maltese expression is *f' każ ta'* (Aquilina 1999: 387) – a combination of three syntactic words which fails to meet the criteria of the canonical Π and thus falls outside the scope of this study.

128 Cf. Sections 5.1.3.1–5.1.3.4.

129 Cf. Section 4.2.1.

130 Cf. Section 5.1.3.4.

(5.164) PROPRIETIVE I

(a) [Korpus Malti 3.0; news14455]

Keith Schembri	*staqsieh*	*ta'*	*min*	*hi*	*l-karozza*	
Keith Schembri	ask:3SG.M	of	who	3SG.F	DEF-car	
*[**bil**-pjanċi*		*GM 14]*$_{PP}$				
[**with**:DEF-number_plate:PL		GM_14]				

'[. . .] Keith Schembri asked him whom the car [**with** the number plate GM 14] belonged to.'

(b) [Korpus Malti 3.0; news21385]

semmiet	*numru*	*ta'*	*każijiet*	*fosthom*	*ta'*	*mara*
hear.CAUS:3SG.F.PFV	number	of	case:PL	amongst:3PL	of	woman
*[**bil**-kanċer]*$_{PP}$						
[**with**:DEF-cancer]						

'[. . .] she mentioned a number of cases among which that of a woman [**with** cancer] [. . .].'

(c) [Korpus Malti 3.0; religion205]

Kellna	*naqsmu*	*x-xmara*	*[**bil**-kukkudrilli]*$_{PP}$
have.PFV:1PL	1.IPFV:cross:PL	DEF-river	[**with**:DEF-crocodile:PL]

'We had to cross the river [(infested) **with** crocodiles] [. . .].'

Borg/Azzopardi-Alexander (1997: 149) mention (П55) *ta'* 'of' as marker in the domain of quality assignment. We have discussed *ta'*-PPs above.[131] As the examples in (5.165) suggest, (П55) *ta'* 'of' may indeed function as PROPRIETIVE marker.

(5.165) PROPRIETIVE II

(a) [Korpus Malti 3.0; news140441]

Iż-żewġ	*nisa*	*daħlu*	*f'*	*karozza*	*[**ta'***	*kulur*	*blu]*$_{PP}$
DEF-two	woman.PL	enter:3PL.PFV	in	car	[**of**	colour	blue]

'The two women got into a [blue] car [. . .].'

(b) [Korpus Malti 3.0; literature7]

Kienet	*mara*	*[**ta'***	*ftit*	*kliem]*$_{PP}$
be:3SG.F.PFV	woman	[**of**	little	word.COLL]

'She was a woman [**of** few words] [. . .].'

131 Cf. Section 4.2.2.

(c) [Korpus Malti 3.0; news112085]
ż-żoni f' Malta fejn jista' jsir
DEF-zone:PL in Malta where 3SG.M.IPFV:can 3SG.M.IPFV:become
bini ***[ta'*** *għoli medju]*$_{PP}$ *huma San Ġiljan Tas-Sliema Gżira*
building [**of** height medium] 3PL San Ġiljan Tas-Sliema Gżira
'[. . .] the zones in Malta where construction [**of** medium height] can be realised are San Ġiljan, Tas-Sliema, Gżira [. . .].'

What we can state right now as to the competition between the two Πs is that (Π7) *bi* 'with' and (Π55) *ta'* 'of' do not seem to replace each other freely in any given context. However, example (5.166) features (Π7) *bi* 'with' in a context which involves (Π55) *ta'* 'of' in (5.165c).

(5.166) PROPRIETIVE III – [Korpus Malti 3.0; european4806]
is-siġar jagħmlu parti minn klassijiet predominanti jew
DEF-tree.PL 3.IPFV:make:PL part from class:PL predominant or
dominanti [. . .] jew ta' siġar ***[b'*** *għoli medju ta' ±20%]*$_{PP}$
dominant or of tree.PL [**with** height medium of ±20%]
'[. . .] the trees form part of predominant or dominant classes or of tress [**with** a medium height of ±20%] [. . .].'

In terms of their token frequency, attributive PPs of the type *[ta' għoli]* are attested 278 times in the Korpus Malti 3.0 where we also find 103 matches for the competitor *[b' għoli]*. To determine how the two Πs divide the PROPRIETIVE domain between each other, a dedicated in-depth study is required.

5.1.3.2 Exclusive

The EXCLUSIVE is in an antonymic relation to the PROPRIETIVE because it indicates that a given entity is lacking a certain property. Borg/Azzopardi-Alexander (1997: 149) speak of "negative circumstance" or "negative quality". Irrespective of the label, there are two Πs that express the EXCLUSIVE, namely (Π9) *bla* 'without' and (Π36) *mingħajr* 'without' whose formal differences have been addressed already.[132] The existence of two Πs with EXCLUSIVE function does not seem to form a parallel to the distribution of the two Πs discussed in connection with the PROPRIETIVE, i.e., there is no way to link either of the two EXCLUSIVE Πs to any of the two PROPRIETIVE Πs in particular. If there are meaning differences or other they are neutralised in the EXCLUSIVE. As in the case of (Π7) *bi* 'with' and (Π55) *ta'* 'of', (Π9) *bla* 'without' and (Π36) *mingħajr* 'without' are the heads of NP-internal prepositional attributes.

132 Cf. Section 4.1.1.1.

The two EXCLUSIVE Πs are illustrated separately. In (5.167), we provide three examples of (Π9) *bla* 'without' whereas its synonym (Π36) *mingħajr* 'without' is featured in (5.168).

(5.167) EXCLUSIVE I

(a) [Korpus Malti 3.0; culture2686]

hu	*possibbli*	*tara*	*live*	*x'*	*qed*	*jiġri*
3SG.M	possible	2SG.IPFV:see	live	what	PROG	3SG.M:run

fir-raħal	*jew*	*belt*	*donnu*	*bħal*	*qisek*	*qed*
in:DEF-village	or	town	as_if:3SG.M	like	measure:2SG	PROG

tgħaddi	*mit-toroq*	*ġo*	*karozza*	***[bla***	*saqaf]$_{PP}$*
2SG.IPFV:pass	from:DEF-street.PL	in	car	**[without**	ceiling]

'[. . .] it is possible that you see what is going on in the village or town, as if you consider yourself passing through the streets in a convertible.'

(b) [Korpus Malti 3.0; news83657]

Mara	***[bla***	*xagħar]$_{PP}$*	*li*	*qiegħda*	*tbati*
woman	**[without**	hair]	that	PROG:F	3SG.F.IPFV:suffer

bil-kanċer	*iżda*	*ma*	*tgorr*	*xejn*
with:DEF-cancer	but	NEG	3SG.F.IPFV:grumble	nothing

'A woman [**without** hair] who is suffering from cancer but does not complain at all.'

(c) [Korpus Malti 3.0; news50447]

In-nies	*baqgħu*	*telgħin*	*mill-istiva*	*qishom*
DEF-people	remain:3PL.PFV	ascend.PART:PL	from:DEF-hold	measure:3PL

xmara	***[bla***	*tmiem]$_{PP}$*
river	**[without**	end]

'The people kept climbing up the hold like an [end**less**] stream.'

The nouns that are eligible as complements of (Π9) *bla* 'without' stem from a wide range of semantic classes. The only restriction that applies is already known,[133] namely the impossibility of this Π to take definite complements.

In contrast, (Π36) *mingħajr* 'without' may cooccur with definite and indefinite NPs. In (5.168) we nevertheless give preference to illustrating (Π36) *mingħajr* 'without' in combinations with indefinite complements.

133 From Section 4.1.1.1.

(5.168) EXCLUSIVE II

(a) [Korpus Malti 3.0; news201449]

Huwa ammetta wkoll li saq karozza ***[mingħajr*** *numru*
3SG.M admit also that drive car **[without** number
ta' reġistrazzjoni]$_{PP}$
of registration]
'He also admitted to driving a car [**without** number plates].'

(b) [Korpus Malti 3.0; news184557]

Stephanie Butler kienet mara ***[mingħajr*** *ebda xogħol fiss]*$_{PP}$
Stephanie Butler be:3SG.F.PFV woman **[without** no work steady]
'Stephanie Butler was a woman [**without** a steady job].'

(c) [Korpus Malti 3.0; news162859]

U wieħed irid jiftakar ukoll li
and one 3SG.M.IPFV:need 3SG.M.IPFV:REFL:remember also that
għandna sptar ***[mingħajr*** *teatru prinċipali*
have.IPFV:1PL hospital **[without** theatre principal
tal-operazzjoni]$_{PP}$
of:DEF-operation]
'And one has to remember too that we have a hospital [**without** main operating theatre] [. . .].'

Except that based on definiteness, there do not seem to be any clear differences as to the combinability of the two EXCLUSIVE Πs with nouns. Both allow for combinations with concrete and abstract nouns, etc. We thus classify (Π9) *bla* 'without' (Π36) *mingħajr* 'without' as two of a kind. Future in-depth investigations might reveal however that there are subtle meaning nuances which have escaped our notice.

5.1.3.3 EXCEPTIVE

For the EXCEPTIVE, Hagège (2010: 262) provides three English examples, namely *except*, *save*, and *bar(ring)*. When the same author discusses the relation of the EXCEPTIVE and EXCLUSIVE to the PROPRIETIVE and the COMITATIVE, he seems to confuse the notions EXCEPTIVE and EXCLUSIVE (Hagège 2020: 307–308). Borg/Azzopardi-Alexander (1997: 154) only mention (Π34) *minbarra* 'except' as representative of the EXCEPTIVE (which they term "exclusive"). There are however, another three largely synonymous EXCEPTIVE Πs, viz. (Π1) *apparti* 'apart from', (Π20) *għajr* 'except', and (Π24) *ħlief* 'except'. Two Semitic Πs compete with one Italo-Romance Π.

In (5.169)–(5.172), we present two examples for each of the four EXCEPTIVE Πs. At leat two of the Πs may also function as ADDITIVE Πs as we will see below.[134] The first example is connected to (Π1) *apparti* 'apart from'. In (5.169a), we have nominative government whereas in (5.169b), the same place name as complement is governed by the same Π according to the ABLATIVE pattern.

(5.169) EXCEPTIVE I

(a) [Korpus Malti 3.0; parl10770]

Dehret fl-ewwel paġna ta' ħafna mill-gazzetti
appear:3SG.F.PFV in:DEF-first page of many from:DEF-newspaper:PL
ewlenin fid-dinja ***[apparti*** *Malta]*$_{PP}$
first:PL in:DEF-world **[apart_from** Malta]
'It [the article] appeared on the frontpage of many leading newspapers in the world [**except** in Malta].'

(b) [Korpus Malti 3.0; news138672]

Sal-21 ta' April il-linja tal-ajru Emirates mhix
till:DEF-21 of April DEF-line of:DEF-air Emirates NEG:3SG.F.NEG
taċċetta passiġġieri li jridu jivvjaġġaw
3SG.F.IPFV:accept passenger:PL that 3.IPFV:need:PL 3.IPFV:travel:PL
lejn xi destinazzjoni fl-Ewropa ***[apparti minn***
towards some destination in:DEF-Europe **[apart_from from**
Malta Larnaca Moska Ateni u Istanbul]$_{PP}$
Malta Larnaca Moscow Athens and Istanbul]
'Until 21st April, the airline Emirates does not accept passengers who want to travel to a destination in Europe [**other than** Malta, Larnaca, Moscow, Athens, and Istanbul].

What is said in both sentences is that Malta does not belong to those countries in which certain events take place. In the (a)-sentence, Malta is characterised as the only country worldwide whose leading newspapers fail to feature a certain story. In the (b)-sentence, an airline temporarily refuses to accept passengers going to the vast majority of the European airports. There are exceptions to this rule, namely the airports of Malta, Larnaca, Moscow, Athens, and Istanbul. This means that (Π1) *apparti* 'apart from' is used in EXCEPTIVE function and not in its ADDITIVE function.

The next candidate to consider is (Π20) *għajr* 'except'. The examples in (5.170) show that this Π has EXCEPTIVE function like the foregoing Π.

134 Cf. Section 5.1.3.15.

(5.170) EXCEPTIVE II

(a) [Korpus Malti 3.0; news146046]

Xi twieġeb għal min jgħid li ***[għajr***
what 2SG.IPFV:answer for who 3SG.M.IPFV:say that **[except**
Malta u l-Filippini]_{PP} l-pajjiżi l-oħrajn
Malta and DEF-Philippines] DEF-country:PL DEF-other:PL
tad-dinja daħħlu d-divorzju
of:DEF-world enter.CAUS:3PL.PFV DEF-divorce
'What do you answer to someone who says that [**except** Malta and the Philippines] the other countries of the world have introduced divorce?'

(b) [Korpus Malti 3.0; news83431]

Ħadd ma ssogra joħroġ ordni bħal din u
nobody NEG risk 3SG.M.IPFV:exit order like DEM:F:PROX and
ħadd ma azzarda jmur minn jeddu ***[għajr***
nobody NEG dare 3SG.M.IPFV:go from right:3SG.M **[except**
il-mara tal-kaptan li telgħet fuq il-vapur
DEF-woman of:DEF-captain who ascend:3SG.PFV on DEF-ship
biex iġġib qattus li kellha fil-kabina]_{PP}
to 3SG.F.IPFV:bring cat that have.PFV:3SG.F in:DEF-cabin]
'Nobody risked to give out an order like this and nobody dared to go on his own free will [**except** the captain's wife who got on board the ship to fetch a cat that she had in her cabin].'

In (5.170a), it is assumed that all countries have introduced divorce – except Malta and the Philippines. The wider context of (5.170b) pictures a ship being in danger of exploding. The crew does not dare to enter the lower decks. However, the captain's wife behaves contrary to reason because she goes to her cabin to fetch a cat from there. Everybody acts the same except her.

The facts are very much the same also for (Π24) *ħlief* 'except' as shown in (5.171).

(5.171) EXCEPTIVE III

(a) [Korpus Malti 3.0; parl10444]

Ir-rapport ikompli jgħid li organizzazzjonijiet
DEF-report 3SG.M.IPFV:continue 3SG.M.IPFV:say that organisation:PL
sportive minn pajjiżi membri kollha ***[ħlief***
sportive:PL from country:PL member:PL all:3SG.F **[except**

Malta *u* *l-Litwanja]*$_{PP}$ *inġabru* *flimkien* *sabiex*
Malta and DEF-Lithuania] 3.IPFV:REFL:collect:PL together to
flimkien *jagħmlu* *proposti* *għal* *12-il* *proġett* *Madwar*
together 3.IPFV:make:PL proposal:PL for 12 project around
l-Ewropa
DEF-Europe
'The report continues to say that sports organizations from all member countries [**except** Malta and Lithuania] collect their thoughts in order to jointly put forward proposals for twelve projects around Europe.'

(b) [Korpus Malti 3.0; news167322]
ħadd *ma* *għandu* *titolu* *ta'* *din* *l-art*
nobody NEG have.IPFV:3SG.M title of DEM:F:PROX DEF-land
[ħlief *il-mara* *u* *niesha]*$_{PP}$
[**except** DEF-woman and people:3SG.F]
'[. . .] nobody is entitled to this land [**except** the wife and her people].'

In the (a)-sentence, Malta together with Lithuania does not participate in an otherwise pan-E.U. initiative. In the (b)-sentence, it is stated that nobody has the right to own a certain piece of land except the wife (of the deceased) and her family. The EXCEPTIVE meaning of (Π24) *ħlief* 'except' is beyond doubt.

Finally, there is (Π34) *minbarra* 'except'. In (5.172), the same nouns – *Malta* and *il-mara* 'the woman' – as for the previous two EXCEPTIVE Πs are involved as complements.

(5.172) EXCEPTIVE IV
(a) [Korpus Malti 3.0; news107893]
Simon Busuttil *li* *kieku* *kien* *jimmilita* *fi* *kwalunkwe*
Simon Busuttil who if be 3SG.M.IPFV:militate in whatever
pajjiż *tal-Punent* ***[minbarra*** *Malta]*$_{PP}$ *kien* *jisparixxi*
country of:DEF-west [**except** Malta] be 3SG.M.IPFV:disappear
mix-xena *politika* *l-għada* *stess* *li*
from:DEF-scene political:F DEF-tomorrow same that
joħroġ *b'* *dawn* *it-teoriji* *perikolużi*
3SG.M.IPFV:exit with DEM:PL:PROX DEF-theory:PL dangerous:PL
'Simon Busuttil who, if he had militated in any western country [**except** Malta] had disappeared from the political scene on the day after he put forward these dangerous theories [. . .].'

(b) [Korpus Malti 3.0; parl149]
qabel ir-Rivoluzzjoni Franċiża kull min kien aristokratiku
before DEF-revolution French:F all who be aristocrat
*[**minbarra** l-mara]$_{PP}$ kien ikollu l-konkubini*
[**except** DEF-woman] be 3SG.M.IPFV:have.FUT:3SG.M DEF-concubine:PL
'[. . .] before the French Revolution, every aristocrat [**except** the wife] would have concubines [. . .].'

In (5.172a), Malta is depicted as an exception in the western world since it is the only country that allows political activities like those associated with the name of Simon Busuttil. In (5.172b), the pre-revolutionary custom of French aristocrats to have concubines is described as a commonality from which only the noble women were excluded. In both examples, the EXCEPTIVE function of (Π34) *minbarra* 'except' is clear. No other interpretation is possible.

What is striking about Hagège's taxonomy is the absence of the meaning category INCLUSIVE, i.e. the antonym of the EXCEPTIVE does not show up on the list of categories. We dedicate a separate section to the INCLUSIVE.[135]

5.1.3.4 COMITATIVE

The COMITATIVE and its relation to the INSTRUMENTIVE fill a number of pages in Hagège (2010: 309–315). Borg/Azzopardi-Alexander (1997: 148) mention (Π30) *ma'* 'with' as the sole COMITATIVE Π in Maltese. According to Stolz/Stroh/Urdze (2006: 294), this Π can certainly be considered to be the principal exponent of the prototypical COMITATIVE, i.e., it is used to describe situations of accompaniment which involve two human participants – the accompanee and the companion. The two participants are invested with (ideally) the same degree of control over the situation. Situations of this type are given in (5.173).

(5.173) COMITATIVE I
(a) [Korpus Malti 3.0; news216597]
*Bajada ċaħad kompletament li kien mar [**ma'** Camilleri]$_{PP}$*
Bajada deny completely that be go [**with** Camilleri]
Rotterdam
Rotterdam
'[. . .] Bajada denied absolutely that he had gone to Rotterdam [**with** Camilleri].'

135 Cf. Section 5.1.3.22.

(b) [Korpus Malti 3.0; literature20]

Kien hemm raġel ieħor Richard li wasal ***[ma'***
be there man other Richard that arrive **[with**
bintu]
<u>*bintu*</u>*]*$_{PP}$ *ftit warajkom*
<u>daughter:POSS:3SG.M</u>] little after:2PL

'There was another man, Richard, who arrived [**with** <u>his daughter</u>] shortly after you.'

(c) [Korpus Malti 3.0; literature24]

Meta Jason tkellem ***[ma'*** <u>*missieru*</u>*]*$_{PP}$ *dwar dawn*
when Jason speak **[with** <u>father:3SG.M</u>] about DEM:PL:PROX
il-ħsibijiet kollha dan spjegalu li kienet
DEF-thought:PL all:3SG.F DEM:M:PROX explain:IO:3SG.M that be:3SG.F.PFV
ħaġa normali
thing normal

'When Jason talked [**with** <u>his father</u>] about all the thoughts, the father explained to him that it was a normal phenomenon [. . .].'

In (5.173a–b), motion events are described in which two human participants are involved. In the wider context of the (a)-sentence, the person named Bajada is suspected of having accompanied another person with Camilleri to Rotterdam. If the suspicion was correct, both Bajada and Camilleri went to the destination as equal partners. Similarly, in the (b)-sentence, Richard and his daughter arrive together, meaning: the use of (Π30) *ma'* 'with' implies that there is no asymmetrical control of the situation between the two participants. In the (c)-sentence, motion does not play a role. However, Jason is said to have talked about his thoughts with his father. The conversation between father and son involves both participants as active speakers and passive listeners. There is thus equal control of the situation for both Jason and his father.

However, even if both participants in an accompaniment situation are animate, the control of the situation must not always be mutual. In those cases where the control over the situation is the exclusive privilege of only one of the participants, a different Π has to be employed. This Π is (Π7) *bi* 'with'. Compare the examples of bona fide COMITATIVES in (5.174) with those in (5.175).

(5.174) COMITATIVE II

(a) [Korpus Malti 3.0; news8567]

Margaret tirrakkonta li kienet tħobb ħafna
Margaret 3SG.F.IPFV:narrate that be:3SG.F.PFV 3SG.F.IPFV:love very

toħroġ *[mal-ħbieb]_{PP}* *u* *tilgħab* *it-tombla*
3SG.F.IPFV:exit [**with**:DEF-friend.PL] and 3SG.F.IPFV:play DEF-tombola
'Margret reports that she loved very much to go out [**with** the friends] and play tombola.'

(b) [Korpus Malti 3.0; news8442]
Qal li iben l-Imnieħru ħareġ mill-karozza u telaq [ma' Jason]_{PP}
say that son l-Imnieħru exit from:DEF-car and leave [**with** Jason]
'He said that Imnieħru's son got out of the car and left [**with** Jason] [. . .].'

(c) [Korpus Malti 3.0; news33582]
kien qiegħed joqgħod [mat-tfajla tiegħu]_{PP} f' residenza f' Ħaż -Żabbar
be PROG 3SG.M.IPFV:stay [**with**:DEF-girl of:3SG.M] in residence in Ħaż -Żabbar
'[. . .] he lived [**with** his girl] in a flat in Ħaż-Żabbar [. . .].'

(5.175) COMITATIVE III (asymmetrical control)

(a) [Korpus Malti 3.0; literature81]
Kont qed nara 'l dik il-mara li toħroġ [bil-kelb]_{PP}
be:1SG.PFV PROG 1SG.IPFV:see to DEM:F:DIST DEF-woman that 3SG.F.IPFV:exit [**with**:DEF-dog]
'I saw this woman who came out [**with** the dog] [. . .].'

(b) [Korpus Malti 3.0; religion423]
mar ħareġ il-karrozza antika tagħhom mill-maħżen u telaq [b' Michael passiġġier]_{PP}
go exit DEF-car old:F of:3PL from:DEF-warehouse and leave [**with** Michael passenger]
'[. . .] he went and fetched their old car from the warehouse and left [**with** Michael as passenger].'

(c) [Korpus Malti 3.0; news74838]
U dan l-ispazju jerġa' jkompli jonqos jekk is-sieħeb il-ġdid tal-missier jew l-omm se jiġi joqgħod [bit-tfal
and DEM:M:PROX DEF-space 3SG.M.IPFV:do_again 3SG.M.IPFV:continue 3SG.M.IPFV:diminish if DEF-partner DEF-new of:DEF-father or DEF-mother FUT 3SG.M.IPFV:come 3SG.M.IPFV:stay [**with**:DEF-child.PL

tieghu]$_{PP}$	*fl-istess*	*dar*	*taghna*
of:3SG.M]	in:DEF-same	house	of:1PL

'And this space will again shrink further if the new partner of the father or the mother comes to live [**with** his children] in our house.'

The six situations described in (5.175) and (5.176) feature situations which involve two animate participants. In (5.175), the appropriate choice as a linker between accompanee and companion is (Π30) *ma'* 'with'. This Π is chosen because Margret and her friends (= (a)-sentence), Jason and Imnieħru's son (= (b)-sentence) as well as the anonymous delinquent and his girlfriend (= (c)-sentence) exert the same degree of control over the situation since, ideally, they are free to terminate the accompaniment situation at any time. In contrast, the situations in (5.175) are asymmetrical as to who has control over the event. In (5.175a), the woman goes out to walk the dog with the latter being under her control. In (5.175b), the unnamed person drives away with Michael who as a passenger in the car is not in control of where the driver is going. In (5.175c), the children of the parent's new partner are dependent on where their parent is going to live and thus have no control over the situation. The lack of control is signalled by the use of (Π7) *bi* 'with'.

In connection to the choice of either (Π30) *ma'* 'with' or (Π7) *bi* 'with', example (5.176) is instructive.

(5.176) COMITATIVE IV (mixed control) – [Korpus Malti 3.0; literature25]

Imbagħad	*fil-weekend*	*joħorġu*	*[mal-familja]*$_{PP}$
then	in:DEF-weekend	3.IPFV:exit:PL	[**with**:DEF-family]

[bit-tfal]$_{PP}$	*fil-każ*	*ta'*	*Dareen*	*u*	*[mal-ħbieb]*$_{PP}$
[**with**:DEF-child.PL]	in:DEF-case	of	Dareen	and	[**with**:DEF-friend.PL]

fil-każ	*ta'*	*Shireen*
in:DEF-case	of	Shireen

'Then, on the weekend, they go out [**with** the family], [**with** the children] in Dareen's case and [**with** the friends] in Shireen's case.'

This sentence contains three COMITATIVE PPS. In two cases, (Π30) *ma'* 'with' is made use of whereas (Π7) *bi* 'with' occurs once. The motivation for the differential use of the Πs is the control exerted by the companion. (Π30) *ma'* 'with' takes nouns as complements which refer to groups of people that consist at least partly of adults, i.e., participants who are able to exert control over the situation. This is not the case with the children who are taken along on an outing by Dareen. Accordingly, (Π30) *ma'* 'with' does not fit and is replaced by (Π7) *bi* 'with'.

The companions in (5.176) are treated like inanimate objects that are transported or worn in motion events (= CONFECTIVE). Typical CONFECTIVES as a subcategory of the COMITATIVE are presented in (5.177).

(5.177) COMITATIVE V (CONFECTIVE)

(a) [Korpus Malti 3.0; literature43]

*Imbagħad tiġi [bil-ħames liri]*PP

then 2SG.IPFV:come [with:DEF-five pound:PL]

'Then you come [**with** the five pounds].'

(b) [Korpus Malti 3.0; news59534]

biex turini l-ferħ tagħha riedet

to 3SG.F.IPFV:show:1SG DEF-joy of:3SG.F want:3SG.F.PFV

*tiġi [bil-kappell]*PP

3SG.F.IPFV:come [with:DEF-hat]

'[. . .] to show me her joy she wanted to come [**with** the hat **on**] [. . .].'

(c) [Korpus Malti 3.0; parl1211]

qalet li ma rridux il-bucket brigade il-brigata

say:3SG.F.PFV that NEG 1.IPFV:want:PL:NEG DEF-bucket_brigade DEF-brigade

*li tiġi [bil-pala]*PP *u [bil-barmil*

that 3SG.F.IPFV:come [with:DEF-shovel] and [with:DEF-bucket

*tar-ramel]*PP

of:DEF-sand]

'[. . .] she said that we do not want the bucket brigade, the brigade that comes [**with** the shovel] and [(**with**) the sand bucket].'

The referents of the inanimate nouns that function as complements of (Π7) *bi* 'with' in (5.177) have no autonomy. They are fully controlled by human participants. The human participant carries the five pounds to the creditor (= (a)-sentence), the lady plans to go to an event wearing the hat (= (b)-sentence), and the fire brigade is described as bringing along shovel and sand bucket (= (c)-sentence). The money, the hat, the shovel, and the sand bucket are unable to move unless a human participant carries them from a to b.

What we have argued in the foregoing paragraphs boils down to the conclusion that the animacy of the companion is not sufficient to determine which of the two Πs is suitable for a given context. The crucial factor is control. By definition, inanimates do not exert control. Nonhuman animates are probably borderline cases as to the exertion of control. In the case of human companions, the degree of control may vary according to the exact power relations between accompanee and companion.

5.1.3.5 INSTRUMENTIVE

It is only a small step from the CONFECTIVE variety of the COMITATIVE to the instrumental. The latter category goes by the name of INSTRUMENTIVE in Hagège's (2010: 262) system. Borg/Azzopardi-Alexander (1997: 148) mention (Π7) *bi* 'with' as sole Π with INSTRUMENTIVE function in Maltese. As we will see below, there is also (Π45) *permezz* 'by means of' which claims a segment of the INSTRUMENTIVE domain.

In (5.178), we illustrate the INSTRUMENTIVE use of (Π7) *bi* 'with' with scenes in which an inanimate object serves as tool for the successful realisation of an action by human agents.

(5.178) INSTRUMENTIVE I

(a) [Korpus Malti 3.0; news97966]

*Ciantar kien qed iqaċċat iz-zkuk [**b'** sega elettrika]*$_{PP}$
Ciantar be PROG 3SG.M.IPFV:cut DEF-stem.PL [**with** saw electric:F]
'[. . .] Ciantar was cutting the stems [**with** an electric saw] [. . .].'

(b) [Korpus Malti 3.0; news5679]

żewġ suspettati ġew arrestati wara li
two suspect:PL come:3PL.PFV arrest:PART:PL after that
*ħarbu minn fuq il-post [**b'** karozza tal-pulizija]*$_{PP}$
flee:3PL.PFV from on DEF-place [**with** car of:DEF-police]
'[. . .] two suspects were arrested after they had fled from the place [**in** a police car].'

(c) [Korpus Malti 3.0; news773]

*L-attriċi ta' 69 sena waħħlet pinna [**b'** labra*
DEF-actress of 69 year stuck:3SG.F.PFV pen [**with** needle
tas-sarwan]$_{PP}$ *mal-libsa ħamra eleganti tagħha*
of:DEF-sarwan] with:DEF-dress red elegant of:3SG.F
'The 69-years-old actress fastened a pen [**with** a sarwan needle] to her elegant red dress.'

(d) [Korpus Malti 3.0; news141796]

*xi ħadd informa lill-pulizija [**b'** telefonata*
some somebody inform to:DEF-police [**with** phone_call
anonima]$_{PP}$
anonymous]
'[. . .] somebody informed the police [**by means of** an anonymous phone call].'

In the (a)-sentence, the actor uses an electric saw to cut the wood. In the (b)-sentence, a police car is used by two suspects to get away from the scene of the crime. In the (c)-sentence, it is the needle that allows the actress to fasten the pen on her dress. And finally, the (d)-sentence shows that even something that is not tangible can serve as instrument. In this case, an anonymous phone call conveys the information to the police. The saw, the car, the needle, and the phone call are bona fide instruments no matter whether their referents are tangible or not.

As to (Π45) *permezz* 'by means of', the examples in (5.179) are meant to show that this INSTRUMENTIVE Π also allows for combinations with nouns from different semantic domains.

(5.179) INSTRUMENTIVE II

(a) [Korpus Malti 3.0; academic1]
*[**Permezz** **ta**l-mezzi elettroniċi]*$_{PP}$ *l-istudenti*
[**by_means_of** **of**:DEF-mean:PL electronic:PL] DEF-student:PL
ingħataw l-opportunità li jiktbu b' mod
3.IPFV:REFL:give:PL DEF-opportunity that 3.IPFV:write with manner
ġurnalistiku fuq diversi aspetti
journalistic on diverse:PL aspect:PL
[**By means of** the electronic tools] the students are given the opportunity to write in the journalistic style about various aspects [. . .].'

(b) [Korpus Malti 3.0; academic12]
*Kien f' dik is-sena li [**permezz** **ta**d-diplomazija]*$_{PP}$
be in DEM:F:DIST DEF-year that [**by_means_of** **of**:DEF-diplomacy]
l-Ingilterra ħatfet Ċipru
DEF-England grab:3SG.F.PFV Cyprus
'It was in this year that England seised Cyprus [**by** diplomatic means] [. . .].'

(c) [Korpus Malti 3.0; academic232]
Patri Messuccio ipprova jwassal l-ilma lejn
Father Messuccio try 3SG.M.IPFV:arrive.CAUS DEF-water towards
*il-Belt [**permezz** **ta'** kanali minn taħt l-art]*$_{PP}$
DEF-city [**by_means_of** **of** canal:PL from under DEF-ground]
'Father Messuccio tried to bring water to Valletta [**by means of** subterranean canals].'

(d) [Korpus Malti 3.0; academic233]
din il-logħba spiċċat b' telfa ta' 1–0
DEM:F:PROX DEF-match end:3SG.F.PFV with defeat of 1-0
għat-tim Ingliż b' gowl skurjat minn Platini
for:DEF-team English with goal shoot:PART from Platini
*[**permezz** **ta'** penalty]*$_{PP}$
[**by_means_of** **of** penalty]
'[. . .] this match ended with a 1-0 defeat for the English team because of a goal scored by Platini [**by means of** a penalty].'

There are concrete physical objects that serve as instruments for human agents to achieve a goal. This is the case in (5.179c) where the priest is trying to guarantee the water supply for Malta's capital city by way of underground canals. In the other three examples, the instruments are of different kinds. The electronic media (= (a)-sentence) are not as tangible as the underground canals. The same characterisation holds for the penalty in (5.179d) which is an event that can be located in time. Similarly, diplomacy is an abstract political means which in this case serves to secure the possession of Cyprus for England (= (c)-sentence).

Impressionistically, (Π45) *permezz* 'by means of' seems to be used predominantly with abstract concepts that function as instruments whereas (Π7) ***bi*** 'with' is stronger in the domain of concrete concepts. However, these supposed preferences do not mean that there is no overlap in the distribution of the two INSTRUMENTIVE Πs. In (5.180), two examples are contrasted with each other in both of which *nar* 'fire' is the instrument.

(5.180) INSTRUMENTIVE III

(a) [Korpus Malti 3.0; parl356]

qed	*jirreferi*	*għal*	*dak*	*tradizzjonali*	*li*
PROG	3SG.M.IPFV:refer	for	DEM:M:DIST	traditional	that

jaħmi	*l-ħobż*	*[**bi**n-nar]$_{PP}$*
3SG.M.IPFV:bake	DEF-bread	[**with**:DEF-fire]

'[...] he is referring to the traditional (saying) that bread is baked [**with** fire].'

(b) [Korpus Malti 3.0; news49567]

Huwa	*kien*	*infurmah*	*li*	***[permezz***	***ta'***	*telefonata*
3SG.M	be	inform:3SG.M	that	**[by_means_of**	**of**	phone_call

anonima]$_{PP}$	*huwa [...]*	*kien*	*ġie*	*infurmat*	*li*	*f'*	*qasir*
anonymous:F]	3SG.M	be	come	inform:PART	that	in	short

żmien	*id-dgħajsa*	*ta'*	*Baldacchino*	*kienet*	*se*	*tiġi*
time	DEF-boat	of	Baldacchino	be:3SG.F.PFV	FUT	3SG.F.IPFV:come

misruqa	*jew*	*tiġi*	*distrutta*	***[permezz***	***ta**n-nar]$_{PP}$.*
PART:steal:F	or	3SG.F.IPFV:come	destroy:PART:F	**[by_means_of**	**of**:DEF-fire]

'He informed him that he had been informed [**via** an anonymous phone call] that soon Baldacchino's boat would be stolen or destroyed [**by** fire].'

It cannot be ruled out completely that the choice of the appropriate INSTRUMENTIVE Π is determined to some extent by the resultative state emerging from the action. In the (a)-sentence, fire serves the purpose of baking bread, i.e., we have an affected object at the end of the process. In the (b)-sentence, the second *permezz*-PP names the means that help to destroy an object, i.e., the object suffers a change-of-state, it is negatively affected. We doubt however that the affected-effected distinction is crucial for the choice of the INSTRUMENTIVE Πs.

The first *permezz*-PP in (5.180b) contrasts with the *bi*-PP in (5.178d). In both cases, an anonymous phone call is depicted as being instrumental for conveying information. This means that the noun *telefonata* is generally compatible with either of the INSTRUMENTIVE Πs. This assumpation is corroborated by the identical token frequency of *b'telefonata* and *permezz ta' telefonata* in the Korpus Malti 3.0. Both options are attested 105 times. The situation is similar in the case of *ħelikopter* 'helicopter' albeit not exactly the same. Before we reveal the absolute numbers, we have a look at another two examples in (5.181).

(5.181) INSTRUMENTIVE IV

(a) [Korpus Malti 3.0; academic6]

il-pestiċidi	*jitferrxu*	***[permezz***	***ta'***	*ħelikopters]$_{PP}$*
DEF-pesticide:PL	3.IPFV:REFL:disperse:PL	**[by_means_of**	**of**	helicopter:PL]

'[. . .] the pesticides are distributed [**by means of** helicopters] [. . .].'

(b) [Korpus Malti 3.0; news172610]

il-midruba	*ttieħdu*	***[b'***	*helicopters*	*militari]$_{PP}$*
DEF-PART:wound:PL	REFL:take:3PL.PFV	**[with**	helicopter:PL	military]
sa	*l-isptarijiet*	*f'*	*Iquitos*	
till	DEF-hospital:PL	in	Iquitos	

'[. . .] the wounded were taken [**with** military helicopters] to the hospitals in Iquitos.'

We assume that the two INSTRUMENTIVE Πs may replace each other freely in the two examples without invoking any meaning differences between the versions. In point of fact, the 333 hits for INSTRUMENTIVE PPs involving the noun *ħelikopter* 'helicopter' in the Korpus Malti 3.0 reflect an interesting distribution as shown in Figure 5.17. If we skip the three cases of other Πs (twice (Π15) *fi* 'in' and once (Π17) *fuq* 'on') being used for the purpose of expressing the INSTRUMENTIVE relation, we recognise immediately that a solid 62% majority of all tokens goes to the credit of (Π7) *bi* 'with', but (Π45) *permezz* 'by means of' covers more than a third of all hits. (Π45) *permezz* 'by means of' is thus not a negligible option. Since meaning differences do not transpire from the examples, it is possible that style and register are responsible for the distribution of the two INSTRUMENTIVE Πs over the texts.

A domain where (Π45) *permezz* 'by means of' ousts its competitor so that the latter is completely absent is the compatibility of the INSTRUMENTIVE Π with animate (preferably human) nouns. In (5.182), we present three examples of (Π45) *permezz* 'by means of' taking a human noun as complement. The human referent functions as the instrument that has to be used in order to reach a certain goal.

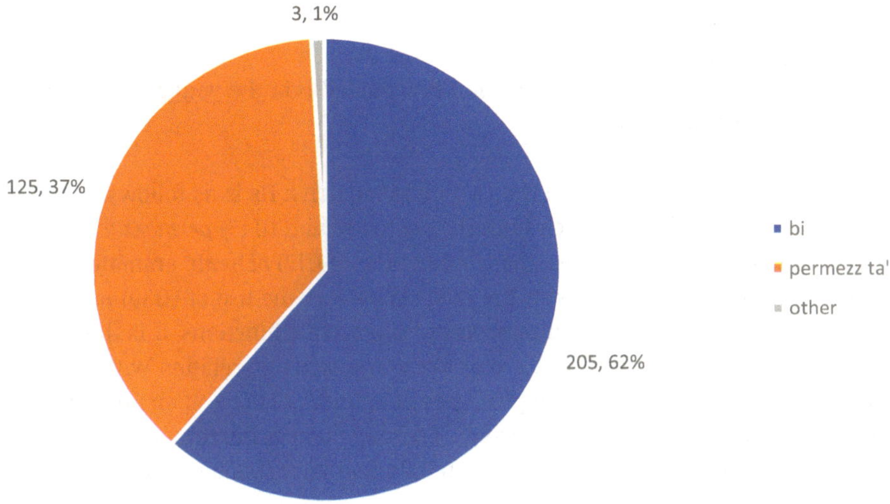

Figure 5.17: Frequency of INSTRUMENTIVE Пs in combination with *ħelikopter* 'helicopter' in the Korpus Malti 3.0.

(5.182) INSTRUMENTIVE V

(a) [Korpus Malti 3.0; parl8748]

L-iskop tagħna hu li nqassmuh b' xejn
DEF-scope of:1PL 3SG.M that 1.IPFV:distribute:PL:3SG.M with nothing
u ***[permezz ta****t-tfal]*$_{PP}$ *nilħqu wkoll lill-kbar*
and **[by_means_of of**:DEF.child.PL] 1.IPFV:reach:PL also to:DEF-big.PL
'It is our goal that we distribute it for free and [**via** the children] we will reach the adults, too.'

(b) [Korpus Malti 3.0; news105971]

nipprovdu għarfien lokali ***[permezz ta'*** *nies li*
1.IPFV:provide:PL knowledge local **[by_means_of of** people that
għandhom dik l-esperjenza lokali tant meħtieġa]$_{PP}$
have.IPFV:3PL DEM:F:DIST DEF-experience local so_much PART:need:F]
'[. . .] we provide local knowledge [**via** people that have this so much needed local experience].'

(c) [Korpus Malti 3.0; news220980]

Fir-raħal ta' Asif l-edukazzjoni tingħata
in:DEF-village of Asif DEF-education 3SG.F.IPFV:REFL:give
[permezz ta*l-ommijiet żgħażagħ li jedukaw*
[by_means_of of:DEF-mother:PL young that 3.IPFV:educate:PL

lil *uliedhom]*$_{PP}$
to children:3PL]
'In the village of Asif, education is provided [**via** the young mothers who educate their children].'

The division of labour between the two INSTRUMENTIVE Πs is as follows. There is a huge area of overlap in which both (Π7) *bi* 'with' and (Π45) *permezz* 'by means of' are equally acceptable. Since the combination of (Π7) *bi* 'with' animate/human nouns is strongly associated with the CONFECTIVE and the use of (Π30) *ma'* 'with' with human nouns is restricted to properly COMITATIVE functions, it is the task of (Π45) *permezz* 'by means of' to express the INSTRUMENTIVE relation with humans functioning as instruments. The high degree of animacy of the human nouns would otherwise qualify them as passive agents. Treating them as instruments is an unexpected alternative. It is interesting that an Italo-Romance Π takes over this task.

5.1.3.6 MOTIVATIVE

The MOTIVATIVE captures situations in which the referent of the complement noun of a given Π is the cause of what is predicated in the clause. A possible alternative label for the meaning category could have been causalis. In point of fact, Borg/Azzopardi-Alexander (1997: 151) use the term "cause". Hagége (2010: 262) lists the English Πs *because of, for*, and *thanks to* as examples of the MOTIVATIVE. In BLOMP 2.0, there are four MOTIVATIVE Πs, namely (Π7) *bi* 'with', (Π21) *għal* 'for', (Π39) *minħabba* 'on account of', and (Π52) *sforz* 'thanks to'.[136] In what follows in this

136 There is evidence of a potential fifth candidate in this domain, namely (Π45) *permezz* 'by means of'. Occasionally, this otherwise INSTRUMENTIVE Π seems to be used with a MOTIVATIVE sense as in (5.viii):

(5.viii) [Korpus Malti 3.0; news161809]
kull mara f' Malta għandha tonora l-emanċipazzjoni
each woman in Malta have.IPFV:3SG.F 3SG.F.IPFV:honour DEF-emancipation
tagħha emanċipazzjoni li qiegħda titgawda [permezz ta'
of:3SG.F emancipation that PROG:F 3SG.F.IPFV:REFL:enjoy **[by_means_of of**
nisa kuraġġużi li jiġġieldu l-mentalità konservattiva ta'
women courageous:PL that 3.IPFV:fight:PL DEF-mentality conservative:F of
żmienhom]$_{PP}$
time:3PL]
'Every woman in Malta must honour her emancipation, an emancipation that is being enjoyed [**because of** courageous women who fought the conservative mentality of their times].'

It is clear that the pironeers of women's liberation are not the means for today's women in Malta to enjoy their emancipation. The women who struggled against the conservative mindset of their section, the use of each of the MOTIVATIVE Πs will be illustrated with three annotated examples. After the discussion of these examples, we try to determine in what way the four Πs divide the functional domain of the MOTIVATIVE between them.

The point of departure for the empirical illustration is marked by (Π7) ***bi*** 'with' (cf. (5.183)). The conceptual proximity of the INSTRUMENTIVE and the MOTIVATIVE transpires from several of the examples not only in the case of (Π7) ***bi*** 'with'. For the purpose of this section and contrary to stylistic aptness, we translate this Π as 'because of' but keep 'with' in the glosses.

(5.183) MOTIVATIVE I

(a) [Korpus Malti 3.0; parl4135]

dan *it-tifel* *miet* ***[bi*** *traskuraġni]*$_{PP}$

DEM:M:PROX DEF-boy die **[with** carelessness]

'[. . .] this boy died [**because of** carelessness] [. . .].'

(b) [Korpus Malti 3.0; news108303]

Jirriżulta *li* *din* *it-tfajla* *mardet*

3SG.M.IPFV:result that DEM:F:PROX DEF-girl fall_ill:3SG.F.PFV

[bi*l-bulimja]*$_{PP}$

[**with**:DEF-bulimia]

'It results that this girl fell ill [**because of** bulimia].'

(c) [Korpus Malti 3.0; news42736]

Il-Kunsill *feraħ* ***[bi****l-ftuħ* *taċ-Ċentru* *tal-Ilsien*

DEF-council rejoice [**with**:DEF-opening of:DEF-Centre of:DEF-language

Malti *li* *sar* *dan* *l-aħħar* *fl-Università* *ta'*

Maltese that become DEM:M:PROX DEF-last in:DEF-university of

Bremen]$_{PP}$

Bremen]

'The Kunsill rejoices [**because of** the opening of the Centre of the Maltese Language that took place recently at the University of Bremen].'

In (5.183a), the lack of care is the reason for the boy's death. Similarly, in (5.183b), bulimia is the reason for the girl's falling ill. In (5.183c), the opening of the Centre of the Maltese Language is the reason for the Kunsill's joy. One might argue that what the *bi*-PP adds to the sentence's meaning is just putting a name to the illness the girl suffered from.

---

times are the reason why there is a certain degree of emancipation in the contemporary society of Malta. Thus, (Π45) *permezz* 'by means of' does not introduce the instrument but the cause for the situation described in (5.viii). Since cases of this kind are rare in the Korpus Malti 3.0, we refrain from counting (Π45) *permezz* 'by means of' as one of the genuine MOTIVATIVE Πs.

According to the alphabetical order, the second MOTIVATIVE marker is (П21) *għal* 'for'. The three examples of its use are given in (5.184). Throughout these examples, the verb *għamel* 'do' is employed because below it will serve as a means to compare the frequencies of the different MOTIVATIVE markers in the Korpus Malti 3.0.

(5.184) MOTIVATIVE II

(a) [Korpus Malti 3.0; news143310]

Ftit kienu jsemmu li Gvern Laburista
little be:3PL.PFV 3.IPFV:mention:PL that government Labour
għamel dan ***[għal*** *żewġ raġunijiet validi]*$_{PP}$
do DEM:M:PROX **[for** two reason:PL valid:PL]
'Few have mentioned that the Labour government did this [**for** two good reasons] [. . .].'

(b) [Korpus Malti 3.0; news149882]

Huwa għamel dan ***[għal*** *sensiela ta' mistoqsijiet*
3SG.M do DEM:M:PROX **[for** series of question:PL
parlamentari li saru waqt is-seduta
parliamentary that become:3PL.PFV at_the_time_of DEF-sitting
parlamentari tal-bieraħ]$_{PP}$
parliamentary of:DEF-yesterday]
'He did this [**because of** a series of parliamentary questions that were posed during yesterday's session of parliament].'

(c) [Korpus Malti 3.0; parl2701]

Jien ngħid li l-ministru għamel dan ***[għal***
1SG 1SG.IPFV:say that DEF-minister do DEM:M:PROX **[for**
konsiderazzjonijiet politiċi]$_{PP}$
consideration:PL political:PL]
'I say that the minister did this [**because of** political considerations] [. . .].'

The complement nouns of the *għal*-PPs represent the reason for what a given participant has done in the (recent) past. In the (a)-sentence, it is the plural form of *raġuni* 'reason' itself that functions as part of the complement. In the (b)-sentence, a number of parliamentary questions impel the anonymous agent to do something. In the (c)-sentence, the motivation for the minister's action are political considerations.

(П39) *minħabba* 'on account of' is the third MOTIVATIVE П. The examples given in (5.185) involve complement nouns which are semantically close to those used in combination with (П7) *bi* 'with' in (5.183).

(5.185) MOTIVATIVE III

(a) [Korpus Malti 3.0; news223694]

t-tabib ġie lliberat minn akkużi li r-raġel miet
DEF-doctor come acquit:PART from charge:PL that DEF-man die
***[minħabba** traskuraġni min-naħa tiegħu]$_{PP}$*
[**on_account_of** carelessness from:DEF-side of:3SG.M]
'[...] the doctor was acquitted of the charges that the man died [**because of** carelessness on his side].'

(b) [Korpus Malti 3.0; news133144]

mijiet ta' tfal sfaw morda mentalment
hundred:PL of child.PL fall:3PL.PFV ill.PL mentally
***[minħabba** t-trawma li għaddew minnha]$_{PP}$*
[**on_account_of** DEF-trauma that pass:3PL.PFV from:3SG.F]
'[. . .] hundreds of children fell mentally ill [**because of** the trauma they went through].'

(c) [Korpus Malti 3.0; news128115]

beda jgħid li hu mhuwiex jifraħ
begin 3SG.M.IPFV:say that 3SG.M NEG:3SG.M:NEG 3SG.M.IPFV:rejoice
***[minħabba** rebħa Laburista]$_{PP}$ iżda **[minħabba** rebħa*
[**on_account_of** victory Labour] but [**on_account_of** victory
fuq il-kaċċa]$_{PP}$
on DEF-hunting]
'[. . .] he began by saying that he does not rejoice [**because** of a Labour victory] but [**because** of a victory over the hunting].'

The three sentences in (5.185) involve four *minħabba*-PPs whose complement nouns identify the reason for the death of a man (= (a)-sentence), the mental illness of children (= (b)-sentence), and the joy of the speaker (= (c)-sentence). The reasons are carelessness on the part of the doctor (= (a)-sentence), the traumatic experiences (= (b)-sentence), and the victory in a vote (= (c)-sentence).

With (Π52) *sforz* 'thanks to' we come to the fourth and last of the MOTIVATIVE Πs. (5.186) contains three examples of the MOTIVATIVE use of (Π52) *sforz* 'thanks to'.

(5.186) MOTIVATIVE IV

(a) [Korpus Malti 3.0; academic139]

***[Sforz** il-biża' li kellu li r-reliġjon għad*
[**thanks_to** DEF-fear that have.PFV:3SG.M that DEF-religion still
tispiċċa minn Malta]$_{PP}$ mbarka ruħu fl-istampa billi
3SG.F.IPFV:end from Malta] embark self:3SG.M in:DEF-press by_way

beda jeditja Leħen is-Sewwa
begin 3SG.M.IPFV:edit Leħen is-Sewwa
'[**Because** of his fear that the religion will sometime come to an end in Malta] he embarked on the press by way of starting to edit Leħen is-Sewwa [. . .].'

(b) [Korpus Malti 3.0; academic15]
il-psikodramm bejn Regnaud minn naħa u Vaubois
DEF-psycho_drama between Regnaud from side and Vaubois
u Ransijat mill-oħra ma setax iħalli lil
and Ransijat from:DEF-other:F NEG can:NEG 3SG.M.IPFV:leave to
Vassalli indifferenti u ***[sforz*** *l-istmellija lejn naħa]*$_{PP}$
Vassalli indifferent and [**thanks_to** DEF-contempt towards side]
seta' jitqarreb lejn l-oħra
can 3SG.M.IPFV:REFL:bring_closer towards DEF-other
'[. . .] the psycho-drama between Regnaud on one side and Vaubois and Ransijat on the other could not leave Vassalli indifferent and [**on account of** the contempt for one side] he could get close to the other [. . .].'

(c) [Korpus Malti 3.0; news103365]
ir-rata ta' mortalità fit-tfal żdiedet b' aktar
DEF-rate of mortality in:DEF-child.PL increase:3SG.F.PFV with more
minn 40 % ***[sforz*** ***ta****n-nuqqas ta' staff u riżorsi*
from 40% [**thanks_to** **of**:DEF-lack of staff and resource:PL
għall-kura adekwata]$_{PP}$
for:DEF-care adequate:F]
'[. . .] the mortality rate of children has increased by more than 40% [**because of** the lack of staff and resources for the adequate care].'

In (5.186a), the motivation for the action is the unnamed character's fear for the future of the Christian faith in Malta. The same role of motivating factor is played by Vassalli's contempt of one of the two French parties in (5.186b) whereas, in (5.186c), the increase in the mortality rate is said to be caused by a lack of personnel and resources. The *sforz*-PPs contain the NPs which identify the reason for the actions and events described in the clause.

Looking back on the examples given in (5.182) through (5.186), we notice that there are no clear indicators that would allow us to determine on what grounds Maltese native speakers make their choice of the four available MOTIVATIVE ΠS. At least superficially, they seem to be fully synonymous with each other. Their supposed synonymy could mean that they are used with comparable frequency in certain contexts. To verify this assumption, we look at the token frequency of the MOTIVATIVE ΠS in the context sketched in Figure 5.18.

$$\left\{\begin{array}{l} \textit{għamel}_{\text{3SG.M.PFV}} \\ \textit{għamlet}_{\text{3SG.F.PFV}} \\ \textit{għamlu}_{\text{3PL.PFV}} \end{array}\right\} + \quad \textit{dan}_{\text{DEM:M:SG}} \quad + \quad \Pi_{\text{MOTIVATIVE}}$$

Figure 5.18: Contexts for which the frequency of MOTIVATIVE Πs is checked.

In the leftmost slot, there is the verb *għamel* 'do' with its perfective forms in the 3rd person of both grammatical genders and numbers. The direct object of the verb is always the demonstrative *dan* 'this' which is followed by a PP in which the reason for an action is given. Figure 5.19 reveals how often a given MOTIVATIVE Π fills the rightmost slot in the schema in the Korpus Malti 3.0. What transpires from Figure 5.19 is that each of the MOTIVATIVE Πs is repeatedly attested in the above context. There are, however, quantitative differences. With a turnout of three hits, (Π52) *sforz* 'thanks to' is used only exceptionally in the context under review. Its share is as small as 4% of the 78 tokens reported for all MOTIVATIVE Πs added up. (Π7) *bi* 'with' and (Π21) *għal* 'for' together take the lion's share with (Π21) *għal* 'for' alone accounting for 40% of all hits. The frequency of (Π39) *minħabba* 'on account of' is much smaller, but with seventeen hits this Π still occurs 5.7 times more frequently than (Π52) *sforz* 'thanks to'. There is thus reason to assume that the four Πs are not absolutely equal partners when it comes to expressing the MOTIVATIVE.

This tentative conclusion finds corroboration when we compare the above results with those disclosed in Figure 5.20. Figure 5.20 accounts for the frequency of cooccurrences of the MOTIVATIVE Πs with the nouns *motivi* 'motives', *raġunijiet* 'reasons', and *skopijiet* 'purposes' as indefinite complements.

The results for the complement nouns are ordered from left to right according to the shrinking share of *għal*-PPs. The decrease of the share claimed by (Π21) *għal* 'for' notwithstanding, there can be hardly any doubt that this Π is the numerically strongest of the four options. In each of the combinations, it accounts for more than half of all matches. If *skopijiet* 'purposes' is the complement, (Π21) *għal* 'for' is responsible for 99% of all PPs. In contrast, (Π52) *sforz* 'thanks to' is attested just once failing to produce any hit for two of the complement nouns. Neither (Π7) *bi* 'with' nor (Π39) *minħabba* 'on account of' is a serious competitor of (Π21) *għal* 'for' although in combination with *raġunijiet* 'reasons', (Π39) *minħabba* 'on account of' yields the sizable turnout of 2,477 matches. We take these quantitative differences to mean that the four MOTIVATIVE Πs are largely synonymous. The preference for (Π21) *għal* 'for' is probably motivated by stylistic factors, the same can be said with regards to the infrequent use of (Π52) *sforz* 'thanks to'. It is worth noting that (Π52) *sforz* 'thanks to' is the only Italo-Romance Π in the domain of the MOTIVATIVE.

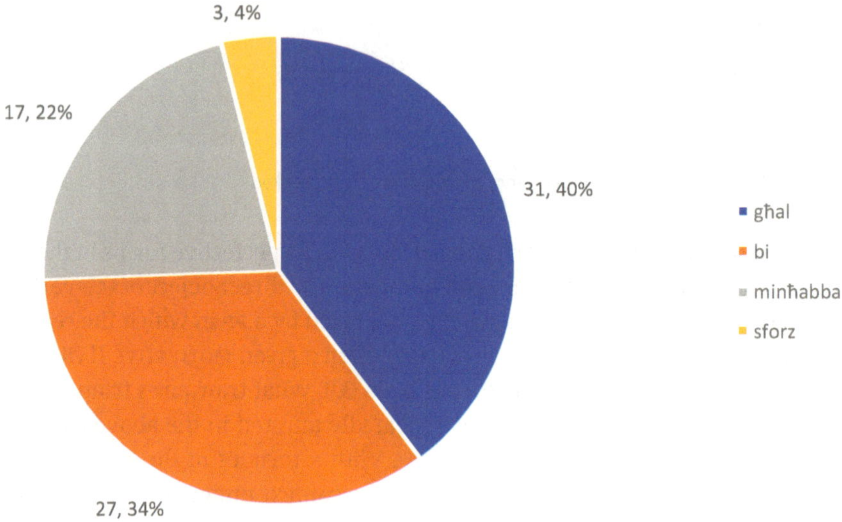

Figure 5.19: Frequency of MOTIVATIVE Πs in the schema of Figure 5.18.

	skopijiet	raġunijiet	motivi
sforz	0	1	0
bi	101	62	9
minħabba	26	2477	7
għal	6042	7342	21

Figure 5.20: Token frequency of MOTIVATIVE Π with three different nouns.

5.1.3.7 Concessive

The CONCESSIVE forms a kind of antonym to the MOTIVATIVE insofar as it identifies a reason that could have impeded something happening but was ignored or overcome by those who carried out the action. Hagège (2010: 262) mentions English *despite* as a CONCESSIVE Π. In BLOMP 2.0, the sole candidate for this function is (Π40) *minkejja* 'in spite of' as illustrated in (5.187). This Π is also given in Borg/Azzopardi-Alexander (1997: 154).

(5.187) CONCESSIVE

(a) [Korpus Malti 3.0; culture1492]

Madankollu din l-anzjana dejjem baqgħet
however DEM:F:PROX DEF-old_woman always remain:3SG.F.PFV
tħares 'il quddiem ***[minkejja*** *l-marda]*$_{PP}$
3SG.F.IPFV:look to in_front_of **[in_spite_of** DEF-illness]
'However, this elderly woman always kept looking forward [**in spite of** the illness] [. . .].'

(b) [Korpus Malti 3.0; news106507]

Sfortuntament il-priġunier miet ***[minkejja*** *t-tentattivi*
unfortunately DEF-prisoner die **[in_spite_of** DEF-attempt:PL
tat-tobba fl-isptar]$_{PP}$
of:DEF-doctor.PL in:DEF-hospital]
'Unfortunately, the prisoner died [**in spite of** the doctors' attempts in the hospital].'

(c) [Korpus Malti 3.0; academic438]

[Minkejja *l-uġiegħ li sofra Battiston]*$_{PP}$ *Schumacher ma*
[in_spite_of DEF-pain that suffer Battiston] Schumacher NEG
kienx penalizzat
be:NEG penalise:PART
'[**In spite of** the pain Battiston suffered], Schumacher was not penalised.'

The three *minkejja*-PPs identify potential reasons for acting or behaving differently. In the (a)-sentence, the elderly woman is optimistic although she suffers from an illness. In the (b)-sentence, the prisoner dies although the doctors tried hard to save his life. In the (c)-sentence, the football player Schumacher escapes being penalised although he injured his opponent seriously. (Π40) *minkejja* 'in spite of' fulfils no other tasks besides that of the CONCESSIVE.

5.1.3.8 MEDIATIVE

The MEDIATIVE is equated with English *à la* and *in the manner of* by Hagège (2010: 262). These expressions are complex with the first option being borrowed from French. In Section 1.4.1.3, we have excluded *la*, the Maltese equivalent of *à la*, from being included in BLOMP 2.0. The direct Maltese translation of English *in the manner of* is *bil-mod ta'*, i.e., an equally complex phrasal construction that does not conform to the definition of the canonical Maltese Π.[137] It is possible though to refer to (Π5) *bħal* 'like' as MEDIATIVE Π. It is by no means easy to distinguish the MEDIATIVE use of (Π5) *bħal* 'like' from its other functions, viz. the EQUATIVE and the ASSIMILATIVE.[138]

In (5.188), three potential instances of the MEDIATIVE are provided. In these examples, we translate the Π under scrutiny with English *in the manner of.*

(5.188) MEDIATIVE

(a) [Korpus Malti 3.0; news90891]

Għamel	***[bħal***	*Pilatu*	*li*	*ħasel*	*idejh]*$_{PP}$	*u*
do	**[like**	Pilate	that	wash	hand:PL:3SG.M]$_{PP}$	and

mingħalih	*salva*	*l-kuxjenza*
in_s.o.'s_opinion:3SG.M	save	DEF-conscience

'He acted [**in the manner of** Pilate who washed his hand] and in his opinion saved his conscience!'

(b) [Korpus Malti 3.0; academic12]

Hemm	*xi*	*Maltin*	*li*	*qed*	*jgħixu*	***[bħal****l-annimali]*$_{PP}$
there	some	Maltese:PL	that	PROG	3.IPFV:live:PL	[**like**:DEF-animal:PL]

'There are some Maltese who are living [**in the manner of** animals].'

(c) [Korpus Malti 3.0; parl6329]

Il-kunsinnatarju	*uffiċjali*	*għandu*	*jaġixxi*	***[bħal***
DEF-consignee	official	have.IPFV:3SG.M	3SG.M.IPFV:act	**[like**

missier	*tajjeb]*$_{PP}$
father	good]

'The consignee has to act [**in the manner of** a good father].'

Whether it makes sense to postulate this distinct meaning category for Maltese, in the first place, will be discussed in the next section that is dedicated to the COMPARATIVE and its subdivisions.

137 Given in Section 1.6.1.

138 Cf. Sections 5.1.3.9.1.1–5.1.3.9.1.2.

5.1.3.9 COMPARATIVE

The COMPARATIVE domain is divided into two subdomains, namely that of equality as opposed to that of inequality (Hagège 2010: 315–320). We start with the discussion of the former and close this section with a description of the COMPARATIVE of inequality.

5.1.3.9.1 Equality

Hagège (2010: 262) divides the domain of equality in two. There is the EQUATIVE whose English translation equivalent is given as *as*. We assume that the intended meaning is that of absolute equality of the comparee and the standard. In the case of the ASSIMILATIVE, the equivalence is only relative.

5.1.3.9.1.1 EQUATIVE

For the usual reasons, we cannot say much about (Π26) *kif* 'as' as EQUATIVE Π since it is generally doubtful that this is indeed a Π, in the first place. No such doubts arise in connection with (Π5) ***bħal*** 'like' and (Π10) *daqs* 'equal to'. The former has been featured already in the section dedicated to the MEDIATIVE.[139] The distinction of MEDIATIVE, EQUATIVE, and ASSIMILATIVE uses of (Π5) ***bħal*** 'like' constitutes a semantic-conceptual challenge. In the English-Maltese dictionary (Aquilina 1999: 117), (Π5) ***bħal*** 'like' is the first translation under the lexicon entry *as*. The cases we mention in (5.189) might easily be registered elsewhere.[140]

(5.189) EQUATIVE I

(a) [Korpus Malti 3.0; news120578]

Kontra	*tim*	*tajjeb*	***[bħal***	*Inter*	*li*	*huma*	*organizzati*
against	team	good	**[like**	Inter	that	3PL	organise:PART:PL

ħafna]$_{PP}$	*trid*	*tkun*	*paċenzjuż*
very]	2SG.IPFV:want	2SG.IPFV:be.FUT	patient

'Against a team as good [**as** Inter who are very organised] you must be patient [. . .].'

(b) [Korpus Malti 3.0; news87380]

Il-melħ	*li*	*jintuża*	*huwa*	*abjad*	***[bħal***	*dak*
DEF-salt	that	3SG.M.IPFV:REFL:use	3SG.M	white	**[like**	DEM:M:DIST

li	*nużaw*	*fuq*	*il-mejda*	*tal-ikel]$_{PP}$*
that	1.IPFV:use:PL	on	DEF-table	of:DEF-food]

'The salt that is used is as white [**as** that we use on the dinner table].'

---

139 Cf. Section 5.1.3.8.

140 Namely in Section 5.1.3.8 or Section 5.1.3.9.1.2.

(c) [Korpus Malti 3.0; parl2536]
Is-sena l-oħra kienet aħjar u s-sena ta' qabel
DEF-year DEF-other:F be:3SG.F.PFV better and DEF-year of before
kienet ħażina ***[bħal*** *din is-sena]*$_{PP}$
be:3SG.F.PFV bad:F **[like** DEM:F:PROX DEF-year]
'The other year was better and the year before was as bad [**as** this year].'

The standard of comparison is identical with the complement in the *bħal*-NPs. What is expressed by the entire construction is that the comparee displays a certain property to exactly the same degree as the standard. In the (a)-sentence, the football team of Inter Milan defines the standard so that a team that is as good as the standard requires a lot of patience from its contestants. The (b)-sentence is about the whiteness of two sorts of salt. The salt used at the dinner-table of the speaker serves as the standard so that the comparee must have the same degree of whiteness as the salt the speaker is used to. In the (c)-sentence, the quality of three years is compared with the current year functioning as standard. Two years ago, there was a year that was as bad as the current one. The quality of the two years must ideally be the same.

In contrast to the often ambiguous (Π5) *bħal* 'like', (Π10) *daqs* 'equal to' is a bona fide case of an EQUATIVE Π. This is illustrated in (5.190).

(5.190) EQUATIVE Π
(a) [Korpus Malti 3.0; literature23]
Kienet iebsa ***[daqs*** *l-azzar]*$_{PP}$
be:3SG.F.PFV hard:F **[equal_to** DEF-steel]
'It was as hard [**as** steel] [. . .].'
(b) [Korpus Malti 3.0; literature20]
kien hemm ħanut tat-toys Londra jismu
be there shop of:DEF-toy:PL London 3SG.M.IPFV:name:3SG.M
Hamleys li kien kbir ***[daqs*** *il-Belt Valletta kollha]*$_{PP}$
Hamleys that be big **[equal_to** DEF-city Valletta all:3SG.F]
'[. . .] there was a toy shop in London named Hamleys that was as big [**as** the entire city of Valletta].'
(c) [Korpus Malti 3.0; literature3]
meta tikbru tmorru tistudjaw barra u
when 2.IPFV:grow:PL 2.IPFV:go:PL 2.IPFV:study:PL outside and
ssiru bravi ***[daqsi]***$_{PP}$
2.IPFV:become:PL good:PL **[equal_to**:1SG]
'[. . .] when you grow up, you will go to study abroad and you will become as good [**as** me].'

The structural and functional facts are comparable to those mentioned in connection with the examples in (5.189). In (5.190a), (5.190b), and (5.190c), (Π10) *daqs* 'equal to' introduces the standard to which the comparee is put in relation. This relation is that of absolute equality as to a certain gradable property. The property is hardness in the (a)-sentence, size in the (b)-sentence, and quality in the (c)-sentence. In each of the cases, it is assumed that comparee and standard yield identical degrees. The identity of the degree also holds in the case of (5.190b) where hyperbole is used as a rhetorical means. The London toy shop is certainly not of the same size as the capital city of Malta.

According to Borg/Azzopardi-Alexander (1997: 261)

> where equality involves size or intensity the expression **daqs** 'size' is used [. . .]. When a quality or manner is involved, the expression **bħal** 'like' is used [original boldface].

Our own examples do not point in the direction of a categorical distinction of this kind. The domains of the two Πs do not seem to be strictly separate from each other. To determine more precisely how the two EQUATIVE Πs are related to each other, an in-depth study is required that also involves native-speaker interviews. This is a task that cannot be fulfilled in the context of our grammar of Maltese Πs. We therefore relegate it to later research projects.

5.1.3.9.1.2 ASSIMILATIVE

The ASSIMILATIVE applies if comparee and standard in a COMPARATIVE construction display similar degrees as to a given property without however reaching full equivalence. For this purpose, Maltese makes use of (Π5) *bħal* 'like' (Borg/Azzopardi-Alexander 1997: 110–111). The examples in (5.191) have to be taken with some caution. With the repeated proviso that MEDIATIVE, EQUATIVE, and ASSIMILATIVE uses of this Π are not always easily distinguished, we can propose three candidates for the ASSIMILATIVE function.

(5.191) ASSIMILATIVE

(a) [Korpus Malti 3.0; european753]

huma	*mibnija*	*minn*	*materjal*	*tajjeb*	*li*	*mhux*
3PL	PART:build:PL	from	material	good	that	NEG:3SG.M:NEG

manjetiku	***[bħal**l-azzar]*$_{PP}$
magnetic	[**like**:DEF-steel]

'[. . .] they are built from good material that is not magnetic [**like** steel].'

(b) [Korpus Malti; news90196]

Taf	*li*	*sirt*	*titkellem*	***[bħal***
2SG.IPFV:know	that	become:2SG.PFV	2SG.IPFV:speak	**[like**

Manwel Cuschieri]$_{PP}$
Manwel Cushieri]
'You know that you have begun to speak [**like** Manwel Cushieri].'

(c) [Korpus Malti; academic340]

Jien	*nawgura*	*li*	*anka*	*intkom*	*tkunu*	*ottimisti*
1SG	1SG.IPFV:wish	that	also	2PL	2.IPFV:be.FUT:PL	optimist:PL

[bħali]$_{PP}$
[**like**:1SG]
'I hope that you too will be optimistic [**like** me].'

In the (a)-sentence, we are facing a difficult problem. Some sorts of steel are magnetic while others are not. It remains uncertain what kind of steel the author is referring to. Either the good materials are not as magnetic as ferrite-free steel or differ from ferrite steel because they are not magnetic, unlike the steel. Apart from this ambiguity, one thing is clear, namely that the steel serves as the standard to which the comparee is compared. The quality is that of magnetism. In the (b)-sentence too, we could speak of a problem because it is possible that the intended meaning is that the addressee speaks exactly like or the same as Manwel Cushieri. In this case, the relation would be EQUATIVE. It is ASSIMILATIVE only if there is just a degree of resemblance between the speech style of the comparee and that of the standard. As to the (c)-sentence, one might argue that there are again two possible readings. Either the addressees are hoped to become as optimistic as the speaker (= EQUATIVE) or at least take an optimistic look at things that is partly in line with the speaker's worldview (= ASSIMILATIVE). In the absence of formal criteria that could help to separate the EQUATIVE from the ASSIMILATIVE, we state that the two meaning categories are neutralised with (Π5) *bħal* 'like'.

5.1.3.9.2 Inequality

In contrast to the categories discussed in the foregoing sections, the COMPARATIVE of inequality does not pose any problem. (Π41) *minn* 'from' is the marker of inequality which precedes the standard in the COMPARATIVE construction (Borg/Azzopardi-Alexander 1997: 108). To the left of the Π, there is the slot for the synthetic or analytic COMPARATIVE of the adjective as shown in (5.192).

(5.192) Inequality

(a) [Chetcuti 1990: 9]

Kellha erbatax-il sena tmien snin iċken ***[minn**-i]$_{PP}$*

have.PFV:3SG.F fourteen year eight year.PL small.CMPR [**from**-1SG]

'She was fourteen years old, eight years younger [**than** me].'

(b) [Korpus Malti 3.0; literature34]

l-uniformijiet bojod kienu saru isbaħ

DEF-uniform:PL white.PL be:3PL.PFV become:3PL.PFV beautiful.CMPR

*[**mill**-kuluri jgħajtu]$_{PP}$*

[**from**:DEF-colour:PL 3.IPFV:call:PL]

'[. . .] the white uniforms had become more beautiful [**than** the bright colours].'

(c) [Korpus Malti 3.0; european10139]

Din l-awtorizzazzjoni proviżorja tista'

DEM:F:PROX DEF-authorization provisional:F 3SG.F.IPFV:can

tingħata għal perjodu li ma jkun

3SG.F.IPFV:REFL:give for period that NEG 3SG.M.IPFV:be.FUT

*itwal [**minn** erba' snin]$_{PP}$*

long.CMPR [**from** four year.PL]

'This provisional authorization can be given for a period that is not longer [**than** four years].'

The COMPARATIVE of inequality is the monopoly of (Π41) *minn* 'from'. It realises the so-called Source Schema according to Heine's (1997: 64–65) typology, i.e., the relator that connects comparee and standard is identical to a relator that is used to express the spatial relation of Source.

5.1.3.10 ESSIVE/TRANSLATIVE/MUTATIVE

Hagège (2010: 262) subsumes three terminologically distinct meaning categories under an umbrella without name. The ESSIVE, the TRANSLATIVE, and the MUTATIVE cover a variety of concepts one of which is static (= ESSIVE) whereas the other two are dynamic (= TRANSLATIVE and MUTATIVE) (Hagège 2010: 320–325). We doubt that it is fully justified to lump these categories together since at least the distinction of static vs dynamic is grammatically relevant in Maltese because different Πs are used.

For a start, we look at the ESSIVE which applies if a participant in a situation is assigned a certain function or role in a given scenario. It is not implied that a change of state takes place. In situations of this kind, Maltese makes use of two different Πs, namely (Π6) *bħala* 'as' and (Π55) *ta'* 'of' (Borg/Azzopardi-Alexander

1997: 152). We start with comments on (П6) *bħala* 'as' whose typical employment is illustrated in (5.193).

(5.193) ESSIVE I

(a) [Korpus Malti; academic10]

l-idea li t-tkattir ta' l-ekonomija hija s-soluzzjoni
DEF-idea that DEF-growth of DEF-economy 3SG.F DEF-solution
għad-diffikultajiet ta' dan il-mudell soċjo-ekonomiku
for:DEF-difficulties of DEM:M:PROX DEF-model socio-economic
għandna nagħrfuha ***[bħala*** *żball kbir strateġiku]*$_{PP}$
have.IPFV:1PL 1.IPFV:recognise:PL **[as** mistake big strategic]
'[. . .] we must recognise the idea that the growth of the economy is the solution for the difficulties of this socio-economic modell [**as** a big strategic mistake].'

(b) [Korpus Malti 3.0; academic101]

r-Rumani oriġinarjmanet ikkunsidraw ix-xitwa
DEF-Roman:PL originally consider:3PL.PFV DEF-winter
[bħala *staġun mingħajr xhur]*$_{PP}$
[as season without month.PL]
'[. . .] originally the Romans considered the winter (to be) [a season without months].'

(c) [Korpus Malti 3.0; academic113]

Gary Neville [. . .] hu plejer tal-futbol Ingliż li
Gary Neville 3SG.M player of:DEF-football English that
jilgħab ***[bħala*** *difensur]*$_{PP}$
3SG.M.IPFV:play **[as** defender]
'Gary Neville is an English football player who plays [**as** defender].'

In (5.193a), at the time of utterance, the socio-economic idea of constant growth is exposed as a strategic mistake. This means that it is a mistake at the reference time. Whether and how this insight has come about is irrelevant to the situation. Similarly, in (5.193b), the interpretation of the winter to be a period without distinct months belonged to the early stages of Roman history. Nothing is said about the development which led to this conception of the winter season. At the reference time, the interpretation was valid. In (5.193c), the nowadays somewhat dated characterisation of a former English football professional as defender was valid in the years 1995–2007. As in the previous case of (П6) *bħala* 'as', the state of affairs at a given point in time is described. The function of the player during his active career was that of a defender.

However, states are one thing, and another are changes of state. To express that a participant is changing from one state to another, Maltese does not make use of (Π6) *bħala* 'as' but resorts to (Π15) *fi* 'in' in combination with change-of-state verbs like those involved in the four examples in (5.194). We do not distinguish the TRANSLATIVE from the MUTATIVE because we lack sufficient native-speaker input to decide whether the process is voluntary or involuntary. Borg/Azzopardi-Alexander (1997: 152) claim that for the TRANSLATIVE no Π is used.[141]

(5.194) TRANSLATIVE/MUTATIVE

(a) [Korpus Malti 3.0; news135600]
Wieħed mill-irħula l-aktar pittoreski tal-gżejjer
one from:DEF-village.PL DEF-more picturesque of:DEF-island.PL
Maltin se jinbidel [f' villaġġ tal-Milied]$_{PP}$
Maltese:PL FUT 3SG.M.IPFV:REFL:change [in village of:DEF-Christmas]
'One of the most picturesque villages of the Maltese islands will change [**into** a Christmas village].'

(b) [Korpus Malti 3.0; news4603]
minn dejjem kien jaffaxxinah id-deheb u
from always be 3SG.M.IPFV:fascinate:3SG.M DEF-gold and
maż-żmien dan l-interess żviluppa [f' passjoni]$_{PP}$
with:DEF-time DEM:M:PROX DEF-interest develop [in passion]
'[. . .] gold always fascinated him and with the time this interest developed [**into** a passion].'

(c) [Korpus Malti 3.0; news47446]
Fil-pajjiżi l-oħra kollha ir-Rebbiegħa Għarbija
in:DEF-country:PL DEF-other:PL all:3SG.F DEF-spring Arabic:F
daret [f' ħarifa krudili]$_{PP}$
turn:3SG.F.PFV [in autumn cruel]
'In all the other countries, the Arabic Spring turned [**into** a cruel autumn].'

(d) [Korpus Malti 3.0; news83433]
Trasforma partit [f' movement]$_{PP}$
transform party [in movement]
'He transformed a party [**into** a movement].'

---

141 The example for the Π-less TRANSLATIVE provided by Borg/Azzopardi-Alexander (1997: 152) is reproduced as (5.ix) in this footnote.

(5.ix) *Innominawh ambaxxatur*
nominate:3PL.PFV:3SG.M ambassador
'They nominated him ambassador.' [O.T.]

In the (a)-sentence, a Maltese village renowned for its beauty turns into a Christmas village – a characterization that formerly did not hold for the village. In the (b)-sentence, the change of state is a drawn-out process which leads from the initial interest of the anonymous person in gold to the emergence of a real passion for the precious metal. Thus, over time, the erstwhile interest turns into a passion. The (c)-sentence tells us what happened to the positively connoted Arabic Spring that is said to have turned into a cruel autumn. In the (d)-sentence, we have a transitive verb *transforma* 'transform'. The agent manipulates a political party in such a way that it turns into a movement.

With (Π55) *ta'* 'of', a second ESSIVE Π is available to Maltese speakers. It resembles (Π6) ***bħala*** 'as' insofar as its domain is restricted to static situations. In (5.195), we provide three examples in which the verb *aġixxi* 'act' is involved (Borg/Azzopardi-Alexander (1997: 152) note that ESSIVE (Π55) *ta'* 'of' is used particularly often with *għamel* 'do'.

(5.195) ESSIVE II

(a) [Korpus Malti 3.0; european8626]

*Il-Viċi-President għandu jaġixxi **[ta'** President*
DEF-vice_president have.IPFV:3SG.M 3SG.M.IPFV:act **[of** president
tal-Kumitat jew tal-Panel ta' Konsulenza]$_{PP}$ fin-nuqqas
of:DEF-committee or of:DEF-panel of consultancy] in:DEF-absence
tal-President elett
of:DEF-president elect:PART
'The vice-president must act [**as** chairman of the committee or of the advisory panel] in the absence of the president-elect.'

(b) [Korpus Malti 3.0; parl10155]

Fl-Amerka għandek elezzjoni separata għall-kungress
in:DEF-America have.IPFV:2SG election separate:F for:DEF-congress
*li mbagħad jaġixxi **[ta'** check and balance]$_{PP}$ fuq*
that then 3SG.M.IPFV:act **[of** check and balance] on
l-eżekuttiv
DEF-executive
'In America you have a separate election for the congress that then acts [**as** check and balance] for the executive.'

(c) [Korpus Malti 3.0; sport1618]

Anki barra l-logħba tal-futbol kien jaġixxi
also outside DEF-play of:DEF-football be 3SG.M.IPFV:act
***[ta'** plejer professjonali]$_{PP}$*
[of player professional]
'Outside the football game too, he acted [**as** a professional player].'

The three examples capture situations in which someone has to temporarily take over the function of someone else (= (a)-sentence)) or function as a control mechanism for a political body (= (b)-sentence) or act like a representative of sports also outside the domain of sports (= (c)-sentence).

Since the two ESSIVE Πs are attested in identical contexts, the question arises as to which of them is the default or whether they are on an equal footing, in a manner of speaking. We have counted the occurrences of (Π6) *bħala* 'as' and (Π55) *ta'* 'of' in the slot to the right of *jaġixxi* 'he acts' and *taġixxi* 'she acts' in the Korpus Malti 3.0. As is evident from Figure 5.21, the 1,095 hits for ESSIVE Πs do not distribute evenly over (Π6) *bħala* 'as' and (Π55) *ta'* 'of'.

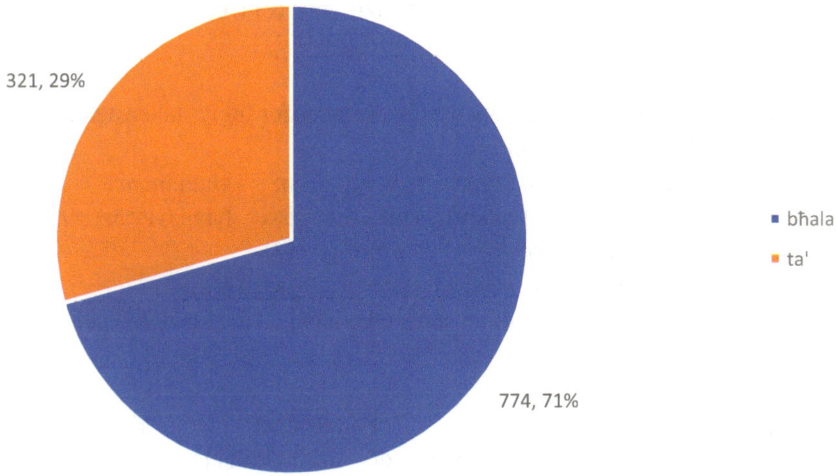

Figure 5.21: Competition between two ESSIVE Πs in the Korpus Malti 3.0.

In combination with the two verb forms, (Π6) *bħala* 'as' is attested 2.4 times as frequently as (Π55) *ta'* 'of'. Apart from this quantitative difference, we notice that (Π55) *ta'* 'of' in ESSIVE function never takes a definite complement whereas the combination with a definite NP is unproblematic (albeit infrequent) in the case of (Π6) *bħala* 'as'.[142]

Maltese not only gives evidence of an ESSIVE expressed by (Π6) *bħala* 'as' whose functional domain is restricted to static situations but also of a TRANSLATIVE/MUTATIVE expressed by (Π15) *fi* 'in' whose distribution is connected to the presence of change-of-state verbs. The two Πs cannot replace each other.

142 As shown in Section 4.1.1.2.2.1 (iv).

5.1.3.11 PURPOSIVE

Hagège (2010: 262) illustrates the PURPOSIVE with English *for*. The PURPOSIVE is represented by (Π21) *għal* 'for' in the Maltese reference grammar (Borg/Azzopardi-Alexander 1997: 151). We discuss a potential second candidate for the PURPOSIVE function below. This second PURPOSIVE Π is (Π57) *versu* 'towards'.

The PURPOSIVE function of (Π21) *għal* 'for' is clearly identifiable in the examples in (5.196).

(5.196) PURPOSIVE I

(a) [Korpus Malti 3.0; academic10]

jerġa' *jittieħed* *il-kontroll* *fuq* *element*
3SG.M.IPFV:do_again 3SG.M.IPFV:REFL:take DEF-control on element
indispensabbli ***[għall-ħajja]***$_{PP}$
indispensable [**for**:DEF-life]
'[. . .] the control over an indispensable element [**for** life] is taken again [. . .].'

(b) [Korpus Malti 3.0; academic10]

Hemm *iktar* *minn* *2.4 biljun* *ruħ* *li* *m'* *għandhomx*
there more from 2.4 billion soul that NEG have.IPFV:3PL:NEG
aċċess *għall-ilma* *meħtieġ* ***[għall-iġene]***$_{PP}$
access for:DEF-water PART:need [**for**:DEF-hygiene]
'There are more than 2.4 billion souls who have no access to the necessary water [**for** hygiene].'

(c) [Korpus Malti 3.0; academic7]

jippromwovi *l-interessi* *ta'* *l-azjendi* *l-kbar* *u*
3SG.M.IPFV:promote DEF-interest:PL of DEF-holding:PL DEF-big.PL and
jiffaċilita *t-tfittxija* *tagħhom* ***[għall-qligħ]***$_{PP}$
3SG.M.IPFV:facilitate DEF-search of:3PL [**for**:DEF-gain]
'[. . .] he promotes the interests of the big holdings and facilitates their search [**for** gain].'

In the three sentences, the complements of (Π21) *għal* 'for' name the purpose for what is predicated in the remainder of the sentence or phrase. In the (a)-sentence, the purpose for which the element is indispensable is life. Hygiene is the purpose for which water is needed in the (b)-sentence whereas the financial gains are what the big holdings are searching for. We gloss over the difficulties that tend to arise when it comes to distinguishing cause from purpose since the latter might also be conceived as a kind of temporally "displaced" cause.

The case of (Π57) *versu* 'towards' is by no means straightforward. We have argued against counting this Π among the directional Πs of Maltese because (Π57) *versu* 'towards' is only exceptionally attested in properly spatial contexts.[143] Its use is much more common elsewhere as can be gathered from the examples in (5.197).

(5.197) PURPOSIVE II

(a) [Korpus Malti 3.0; news208695][144]

droga kienet inbiegħet minnu stess ***[versu***
drug be:3SG.F.PFV REFL:sell:3SG.F.PFV from:3SG.M self **[towards**
l-prezz ta' madwar € 640]$_{PP}$
DEF-price of around € 640]
'[. . .] a drug was sold by himself [**at** the price of approximately € 640].'

(b) [Korpus Malti 3.0; parl11670]

dan hu r-rispett li dan il-Gvern
DEM:M:PROX 3SG.M DEF-respect that DEM:M:PROX DEF-government
għandu ***[versu*** *l-poplu Malti u Għawdxi]*$_{PP}$
have.IPFV:3SG.M **[towards** DEF-people Maltese and Gozitan]
'[. . .] this is the respect that this government has [**for** the Maltese and Gozitan people].'

(c) [Korpus Malti 3.0; parl3885]

Naturalment dan il-gvern iżewweġ
naturally DEM:M:PROX DEF-government 3SG.M.IPFV:marry
dan il-prinċipju mal-prinċipju tad-dmirijiet li
DEM:M:PROX DEF-principle with:DEF-principle of:DEF-duty:PL that
mara jew raġel għandhom ***[versu*** *l-familja tagħhom]*$_{PP}$
woman or man have.IPFV:3PL **[towards** DEF-family of:3PL]
'Naturally, this government links this principle to the principle of the duties that a woman or a man has [**towards** their family].'

In (5.197a), drugs are sold at a certain price. We assume that the sum of € 640 constitutes the purpose on account of which the drug deal takes place. The two other examples belong to one and the same domain. They involve ethical categories such as respect and duties. The people and the family represent the participants towards which the government or women and men are expected to show respect or fulfil duties. In neither case is there a concrete motion event which would justify classifying the *versu*-PPs as indicating the Goal towards which the Figure moves. If the *versu*-PPs are not spatial, what are they then?

---

143 Cf. Section 5.1.2.2.6.

144 In their sections on price and value, Borg/Azzopardi-Alexander (1997: 153–154) mention only construction without IIs.

As the examples in (5.198) suggest, (Π21) *għal* 'for' may replace (Π57) *versu* 'towards' in the contexts in which the latter is attested in (5.197).

(5.198) PURPOSIVE III

(a) [Korpus Malti 3.0; culture1978]

L-ewwel kopja tal-komiks ta' Superman inxtrat fuq
DEF-first copy of:DEF-comics of Superman REFL:buy:3SG.F.PFV on
*eBay [**għal**l-prezz ta' € 2.43 miljun]_{PP}*
eBay [**for**:DEF-price of € 2,430,000]
'The first copy of the Superman comics was bought on eBay [**for** the price of € 2,430,000].'

(b) [Korpus Malti 3.0; parl420]

*jien għandi kull rispett [**għal**l-poplu Żejtuni]_{PP}*
1SG have.IPFV:1SG all respect [**for**:DEF-people of Żejtun]
'I have full respect [**for** the people of Żejtun].'

(c) [Korpus Malti 3.0; news116322]

Għandu jkun li l-bniedem wara li
have.IPFV:3SG.M 3SG.M.IPFV:be.FUT that DEF-man after that
*jmur jaqdi dmiru [**għal**l-familja]_{PP}*
3SG.M.IPFV:go 3SG.M.IPFV:serve duty:3SG.M [**for**:DEF-family]
ikollu daqsxejn ħin biex jagħmel
3SG.M.IPFV:have.FUT:3SG.M a_bit time to 3SG.M.IPFV:do DEM:M:DIST
dak li jħobb
that 3SG.M.IPFV:love
'Should it be that the man after he goes fulfilling his duty [**for** the family] will have a bit of time to do what he loves to do?'

In (5.198a), (Π21) *għal* 'for' introduces the price at which the first edition of a comic is bought whereas in (5.197a), it is (Π57) *versu* 'towards' that heads a functionally equivalent PP. In (5.198b), paying one's respect to the people of Żejtun is expressed via an *għal*-PP. In contrast, a *versu*-PP is used in (5.197b) to identify the Maltese and Gozitan people as the participants the respect is paid to. In both (c)-sentences, the family is the participant to which someone has to serve their duties. In (5.198a), the head of the PP is (Π21) *għal* 'for', but in (5.198b), it is (Π57) *versu* 'towards'. We conclude that the two Πs are functionally identical in the domain of the PURPOSIVE. The price, the people, and the family are the purposes for which something is bought or sold, respect is paid, and duties are served. We do not deny that our interpretation stretches the notion of purpose. This means that future studies are required to settle this issue on a better empirical basis and with native-speaker intuition. This necessity applies also with regards to possible ADVERSATIVE readings of (П57) *versu* 'towards' for which we have not found tangible evidence in the Korpus Malti 3.0.

5.1.3.12 ADVERSATIVE

The ADVERSATIVE is equated with English *against* in Hagège (2010: 262). The Maltese reference grammar does not feature this category. It is nevertheless clear that the only bona fide candidate for this meaning category is the Italo-Romance (П27) *kontra* 'against'. Examples are provided in (5.199).

(5.199) ADVERSATIVE

(a) [Korpus Malti 3.0; academic113]

Hu rritorna mat-tim Ingliż f' Marzu 2006 f' logħba
3SG.M return with:DEF-team English in March 2006 in play
ta' ħbiberija ***[kontra*** *l-Urugwaj]$_{PP}$*
of friendship **[against** DEF-Uruguay]
'He returned to the English team in March 2006 in a friendly match [**against** Uruguay].'

(b) [Korpus Malti 3.0; academic12]

Fis-South Australia saru dimostrazzjonijiet ***[kontra***
in:DEF-South_Australia become:3PL.PFV demonstration:PL **[against**
l-immigrazzjoni tal-Maltin]$_{PP}$
DEF-immigration of:DEF-Maltese:PL]
'In South Australia, demonstrations took place [**against** the immigration of the Maltese].'

(c) [Korpus Malti 3.0; european10792]

l-ebda vaċċinazzjoni ma kienet twettqet
DEF-no vaccination NEG be:3SG.F.PFV REFL:carry_out:3SG.F.PFV
[kontra *l-marda ta' l-ilsien u d-dwiefer]$_{PP}$*
[against DEF-desease of DEF-tongue and DEF-nail.PL]
'[. . .] no vaccination was applied [**against** the foot and mouth disease].'

The *kontra*-PPs host complement NPs which refer to those entities that another participant acts against. In (5.199a), the scenario is that of a friendly football match between England and Uruguay. From the English perspective, England plays against its opponent from South America. In (5.199b), the demonstrations are directed against Maltese immigration to South Australia. The immigration is fought against by the local inhabitants. In (5.199c), the vaccination is meant to have a negative effect on the disease, meaning the disease is the opponent to be fought against.

5.1.3.13 Pertentive

As to the PERTENTIVE, Hagège (2010: 262) proposes English *about* and *with respect to* as illustration. Borg/Azzopardi-Alexander (1997: 152) use the term reference for the same meaning category. They mention only (Π11) *dwar* 'about' as PERTENTIVE marker. There is evidence in the Korpus Malti 3.0 however, that (Π17) *fuq* 'on' and (Π50) *rigward* 'concerning' are also used in the same function. In what follows we address all three of these Πs in the context of the PERTENTIVE. No matter which of the Πs is employed, the PP is the head and introduces the topic that is talked, written, or thought about.

We begin with the presentation of examples illustrating the use of (Π11) *dwar* 'about' in (5.200).

(5.200) PERTENTIVE I

(a) [Korpus Malti 3.0; academic2]

*Hi mexxiet żewġ diskussjonijiet [**dwar** il-kummerċ*
3SG.F go.CAUS:3SG.F.PFV two discussion:PL [**about** DEF-trade
ġust u l-ikel]$_{PP}$
fair and DEF-food]
'She directed two discussions [**about** fair trade and food] [. . .].'

(b) [Korpus Malti 3.0; academic12]

iżda ma ridux jidħlu fi spejjeż u
but NEG want:3PL.PFV:NEG 3.IPFV:enter:PL in expense.PL and
lanqas ma kienu jagħtu informazzjoni soda
neither NEG be:3PL.PFV 3.IPFV:give:PL information firm:F
*[**dwar** il-kundizzjonijiet ta' ħajja u xogħol]*$_{PP}$
[**about** DEF-condition:PL of life and work]
'[. . .] but they did not want to go into the expenses and neither gave sound information [**about** the conditions of life and work].'

(c) [Korpus Malti 3.0; academic254]

*Dan l-artiklu huwa [**dwar** il-pjaneta Marte]*$_{PP}$
DEM:M:PROX DEF-article 3SG.M [**about** DEF-planet Mars]
'This article is [**about** the planet Mars].'

The examples are uncontroversial. (Π11) *dwar* 'about' is the head of the PP that tells us what the discussions are about (= (a)-sentence), what the information is about (= (b)-sentence), and what the article is about (= (c)-sentence). As results from the examples in (5.201), the functions that (Π11) *dwar* 'about' fulfils in (5.200) can easily be expressed by (Π17) *fuq* 'on'.

(5.201) PERTENTIVE II

(a) [Korpus Malti 3.0; academic4]

kienet	*membru*	*minn*	*tim*	*li*	*għamel*	*riċerka*	***[fuq***
be:3SG.F.PFV	member	from	team	that	make	research	**[on**

il- Kummerċ	*Ġust*	*fil-Kambodja]*PP
DEF-trade	fair	in:DEF-Cambodia]

'[. . .] she was a member of a team that conducted research [**about** Fair Trade in Cambodia].'

(b) [Korpus Malti 3.0; academic6]

jagħtu	*informazzjoni*	***[fuq***	*il-prodotti]*PP
3.IPFV:give:PL	information	**[on**	DEF-product:PL]

'[. . .] they give information [**about** the products].'

(c) [Korpus Malti 3.0; parl5532]

Dan	*l-artiklu*	*huwa*	***[fuq***	*il-famużi*	*living wills]*PP
DEM:M:PROX	DEF-article	3SG.M	**[on**	DEF-famous:PL	living wills]

'This article is [**about** the famous living wills].'

What we see is that the *fuq*-PPs occur in similar (and even largely identical) syntactic context as the *dwar*-PPs. In (5.201a), research is conducted about Fair Trade in Cambodia. In (5.201b), information is given about some products. In (5.201c), the topic the article is about is specified in the *fuq*-PP. Thus, (Π11) *dwar* 'about' and (Π17) *fuq* 'on' can replace each other in most of the PERTENTIVE constructions.

The functional equivalence of PERTENTIVE Πs extends to (Π50) *rigward* 'concerning'. In (5.202), we find examples which bear close resemblance to those given in (5.200)–(5.201).

(5.202) PERTENTIVE III

(a) [Korpus Malti 3.0; news196143]

jerġaw	*jibdew*	*id-diskussjonijiet*	***[rigward***
3.IPFV:do_again:PL	3.IPFV:begin:PL	DEF-discussion:PL	**[concerning**
id-diżarm	*nukleari]*PP		
DEF-disarmament	nuclear]		

'[. . .] the discussions [**about** the nuclear disarmament] begin again [. . .].'

(b) [Korpus Malti 3.0; european10449]

L-industrija	*issa*	*ipprovdiet*	*informazzjoni*
DEF-industry	now	provide:3SG.F.PFV	information
[rigward	*ċerti*	*sustanzi]*PP	
[concerning	certain:PL	substance:PL]	

'The industry has now provided information [**about** certain substances] [. . .].'

(c) [Korpus Malti 3.0; parl83]
It-tieni *qasam* *huwa* ***[rigward*** *il-ġlieda* *kontra*
DEF-second field 3SG.M **[concerning** DEF-fight against
l-korruzzjoni]$_{PP}$
DEF-corruption]
'The second part is [**about** the fight against corruption] [. . .].'

As in the two previous cases, (Π50) *rigward* 'concerning' is compatible with *diskussjonijiet* 'discussions', *informazzjoni* 'information', and *qasam* 'field' which can also be used to refer to subdivisions of texts (and thus is comparable to *artiklu* 'article' in the previous (c)-sentences). There can be no doubt that (Π50) *rigward* 'concerning' in (5.202) can be replaced either by (Π11) *dwar* 'about' or (Π17) *fuq* 'on' just as (Π50) *rigward* 'concerning' may fill the slots of the other two PERTENTIVE Πs in (5.200) and (5.201). There is a very high degree of synonymy.

To determine whether the functional equivalence of the three PERTENTIVE Πs means that they are employed with similar token frequencies, we have searched the Korpus Malti 3.0 for attestations of the ternary combinations in Figure 5.22.

{ *diskussjoni* 'discussion'
informazzjoni 'information'
ktieb 'book'
mistoqsija 'question'
taħdita 'talk' } + $\Pi_{\text{PERTENTIVE}}$ + NP

Figure 5.22: Contexts for which the frequency of PERTENTIVE Πs is checked.

The turnouts and shares are revealed in Figure 5.23.

What comes to the fore immediately is the relatively infrequent use of (Π50) *rigward* 'concerning'. This Π fails to show up in combination with two of the five nouns mentioned in Figure 5.22. Except for the cooccurrence with *informazzjoni* 'information', *rigward*-PPs are responsible for only very small shares, namely around 1% or less. Even in the case of *informazzjoni* 'information', (Π50) *rigward* 'concerning' accounts for slightly less than 3% of all PERTENTIVE PPs. The relative infrequency of (Π50) *rigward* 'concerning' is perhaps explicable in terms of stylistic preferences. The choice of this Italo-Romance Π is probably motivated by the speaker's attempt to render the utterance compatible with the exigences of high-

	mistoqsija	diskussjoni	ktieb	informazzjoni	taħdita
rigward	55	40	0	498	0
fuq	3199	5945	119	3391	18
dwar	1470	4470	272	12898	141

Figure 5.23: Frequency of PERTENTIVE Πs in the schema of Figure 5.22.

style written Maltese. The competition between (Π11) *dwar* 'about' and (Π17) *fuq* 'on' is such that (Π11) *dwar* 'about' is the majority option for three of the nouns whereas two of the nouns clearly prefer (Π17) *fuq* 'on' over (Π11) *dwar* 'about'. The shares range from 89% – the top result for (Π11) *dwar* 'about' – to 31% at the opposite end of this Π's frequency count whereas the biggest share for (Π17) *fuq* 'on' is 68% with 11% at the bottom for the same Π. On the basis of these heterogeneous results, it is impossible to pinpoint the decisive factor for the distribution of the two most frequent PERTENTIVE Πs. A dedicated in-depth study is called for to clarify this problem.

5.1.3.14 ROBORATIVE

The ROBORATIVE is illustrated with English *according to*, *depending on*, and *following* (Hagège 2010: 262). This meaning category applies if someone is "speaking to, or depending on, certain considerations" (Hagège 2010: 308). The reference grammar of Maltese keeps silent about the ROBORATIVE.[145] Yet, there is ample evidence of the

145 Borg/Azzopardi-Alexander (1997: 47) make mention only of the use of (Π53) *skont* 'according to' (=<skond>) in combination with (Π26) *kif* 'as' to form the subordinate conjunction *skond kif* 'according to'. They further assume that this binary conjunction is reduced to *skond* if it is followed by existence of two ROBORATIVE ΠS, namely (Π37) *mingħala* 'in s.o.'s opinion' and (Π53) *skont* 'according to'.

The special status of (Π37) *mingħala* 'in s.o.'s opinion' has been referred to already.[146] It is the only member of BLOMP 2.0 that only takes suffixed pronominal complements. Accordingly, the examples in (5.203) reflect this pattern without variation.

(5.203) ROBORATIVE I

(a) [Korpus Malti 3.0; literature4]
Matul eżami minnhom **[*mingħalija*]**$_{PP}$ *dak*
during exam from:3PL **[in_s.o.'s_opinion**:1SG] DEM:M:DIST
tas-Social Aspects bilkemm kont naf x' jiena
of:DEF-Social_Aspects hardly be:1SG.PFV 1SG.IPFV:know what 1SG
nagħmel
1SG.IPFV:do
'During one of the exams, [**according to** me] that of Social Aspects, I hardly knew what I was doing.'

(b) [Korpus Malti 3.0; news110774]
Tixtri laptop ta' ditta tajba **[*mingħalik*]**$_{PP}$ *ġdid*
2SG.IPFV:buy laptop of firm good:F [**in_s.o.'s_opinion**:2SG] new
u wara sitt xhur jiżviluppa ħsara
and after six month.PL 3SG.M.IPFV:develop damage
'You buy a laptop from a good firm, [**to** your **mind**] new, and after six months it develops a damage.'

(c) [Korpus Malti 3.0; literature15]
jekk ministru jirrakkonta dik li **[*mingħalih*]**$_{PP}$
if minister 3SG.M.IPFV:tell DEM:F:DIST that [**in_s.o.'s_opinion**:3SG.M]
hi ċajta kulħadd jidħaklu imma jekk
3SG.F joke everybody 3SG.M.IPFV:laugh:IO:3SG.M but if

---

a non-finite clause. Their example for the latter phenomenon is however an undisputable instance of a PP as shown in (5.x). We keep the original orthography for <skond>.

(5.x) [Borg/Azzopardi-Alexander 1997: 47]
Dejjem imxejna ***[skond*** *it-tagħlim ta' missirijetna]*$_{PP}$
always walk:1PL.PFV [**according_to** DEF-teaching of father:PL:1PL]
'We have always acted [**according to** the teaching of our ancestors].' [O.T.]

We disagree with the grammarians of Maltese that the *skond*-PP has the status of a (reduced) clause although the analysis is certainly a matter of adherence to one syntactic theory and not to the other.

146 Cf. Section 1.6.4.2 (v).

nirrakkonta *ċajta* *bis-sens* *jien* *ma* *jidħakli*
1SG:tell joke with:DEF-sense 1SG NEG 3SG.M.IPFV:laugh:IO:1SG
ħadd
nobody
'[. . .] if a minister tells what [**to** his **mind**] is a joke, everybody laughs about it, but if I tell a joke that makes sense nobody laughs about it [. . .].'

In all three of the examples in (5.203), the complement of (Π37) *mingħala* 'in s.o.'s opinion' identifies the participant who thinks that something is the case irrespective of the truth of the opinion. It is only logical that the restriction to pronominal complements and opinions is tantamount to the highest degree of animacy of the admitted complements. They have to be human.

This restriction does not apply to (Π53) *skont* 'according to' as transpires from (5.204c). (5.204a–b) show that this Π allows for lexical NPs to fill the slot of the complement.

(5.204) ROBORATIVE II
(a) [Korpus Malti 3.0; culture1622]
[Skont *Murasi]PP* *huwa* *ħadem* *l-Indja* *bħala* *skarpan*
[**according_to** Murasi] 3SG.M work DEF-India as cobbler
sal-1957
till:DEF-1957
'[**According to** Murasi], he worked in India as a shoemaker until 1957.'
(b) [Korpus Malti 3.0; culture2054]
[Skont *l-anzjana]PP* *r-Re* *Federiku II* *huwa*
[**according_to** DEF-old_woman] DEF-king Frederic II 3SG.M
l-bużnannu *tagħha*
DEF-great_grandfather of:3SG.F
'[**According to** the elderly lady], Kind Frederic II is her great grandfather.'
(c) [Korpus Malti 3.0; culture192]
[Skont *il-ġurnal* *Canadian Medical Association]PP* *hemm*
[**according_to** DEF-journal Canadian Medical Association] there
numru *kbir* *ta'* *mediċini* *li* *wieħed* *assulutament* *m'*
number big of medicine:PL that one absolutely NEG
għandux *jikkonsma* *jekk* *ikun* *daq*
have.IPFV:3SG.M:NEG 3SG.M.IPFV:consume if 3SG.M.IPFV:be.FUT taste
il-grejpfrut
DEF-grapefruit

'[**According to** the journal Canadian Medical Association], there is a great number of medicines that one should absolutely not consume if there is the taste of grapefruit.'

In the (a)- and (b)-sentences, a human being is identified as the source of an opinion. This is different in the (c)-sentence where an assumption can be traced back to a journal. The latter possibility is ruled out for (Π37) *mingħala* 'in s.o.'s opinion'. In (5.205), it is shown that (Π53) *skont* 'according to' is free to combine with many kinds of inanimate complements.

(5.205) ROBORATIVE III

(a) [Korpus Malti 3.0; culture1197]

l-uffiċjali tal-pulizija kienu sfurzawh
DEF-officer:PL of:DEF-police be:3PL.PFV force:3PL.PFV:3SG.M
jagħti x-xhieda ***[skont*** *il-verżjoni li*
3SG.M.IPFV:give DEF-testimony **[according_to** DEF-version that
tawh huma]$_{PP}$
give:3PL.PFV:3SG.M 3PL]

'[. . .] the police officers had forced him to testify [**according to** the version they gave him].'

(b) [Korpus Malti 3.0; culture1785]

tul il-karriera tagħha hija qatt ma ġiet
during DEF-career of:3SG.F 3SG.F never NEG come:3SG.F.PFV
deskritta ***[skont*** *is-sigħat li taħdem]*$_{PP}$
describe:PART:F **[according_to** DEF-hour:PL that 3SG.F.IPFV:work]
imma ***[skont*** *ix-xogħol li tipproduċi]*$_{PP}$
but **[according_to** DEF-work that 3SG.F.IPFV:produce]

'[. . .] during her career, she was never described [**according to** the hours she worked] but [**according to** the work she produced].

(c) [Korpus Malti 3.0; culture2619]

l-kundizzjonijiet tal-ħaddiema huma
DEF-condition of:DEF-worker:PL 3PL
[skont *il-liġi industrijali fir-Renju Unit]*$_{PP}$
[according_to DEF-law industrial in:DEF-United_Kingdom]

'[. . .] the conditions of the workers are [**in accordance with** the industrial law in the United Kingdom].'

(5.205a) and (5.205c) are comparable to (5.204c) insofar as the *skont*-PP contains a complement that refers to some kind of (written or oral) text. It is to speak of opinions that stem from these texts. The texts serve as a kind of yard-stick against which a state-of-affairs is measured such as the testimony in the (a)-sentence that is supposed to correspond to the version defended by the police. In the (c)-sentence, the conditions under which the workers are doing their jobs are compared to the legal regulations in the UK. Moreover, in the (b)-sentence, there are two *skont*-PPS both of which identify criteria according to which the speaker has been judged by others. It is more than likely that the functional domain of (Π53) *skont* 'according to' is by far bigger than that of (Π37) *mingħala* 'in s.o.'s opinion' not the least because the identification of the origin of a given opinion is only one of the tasks of (Π53) *skont* 'according to' whereas (Π37) *mingħala* 'in s.o.'s opinion' has no other function than exactly this one.

At this point, it makes sense to zoom in on the area of overlap of the two ROBORATIVE Πs. They overlap when the complement is pronominal. We have searched the Korpus Malti 3.0 for ROBORATIVE PPS whose complement is pronominal. In the case of (Π53) *skont* 'according to', this means that PPS with POSSESSIVE and nominative government have to be taken account of whereas (Π37) *mingħala* 'in s.o.'s opinion' only allows for POSSESSIVE government.[147] In Figure 5.24, the number of hits per Π and person category is disclosed. There are striking differences. The quantitative relation between the two Πs is not the same for all person categories, but varies considerably. If the complement is a speech-act participant, (Π37) *mingħala* 'in s.o.'s opinion' is clearly preferred over (Π53) *skont* 'according to'. This preference is most pronounced in the 1st persons of both numbers where (Π37) *mingħala* 'in s.o.'s opinion' covers 93%–95% of all ROBORATIVE PPS. For the 2nd persons, the predominance of (Π37) *mingħala* 'in s.o.'s opinion' is less overwhelming since the shares computed for this Π range from 67% to 69%. The turning point is reached when we look at the 3rd person plural. Both Πs are represented almost evenly since their shares minimally exceed (= (Π37) *mingħala* 'in s.o.'s opinion') the 50%-mark or fail to reach it by a very small margin (= (Π53) *skont* 'according to'). The 3rd person of both genders in the singular clearly favours (Π53) *skont* 'according to' with shares of 69% with masculine complements and 80% with feminine complements.

147 Cf. Section 4.1.2.2.2.2.2.

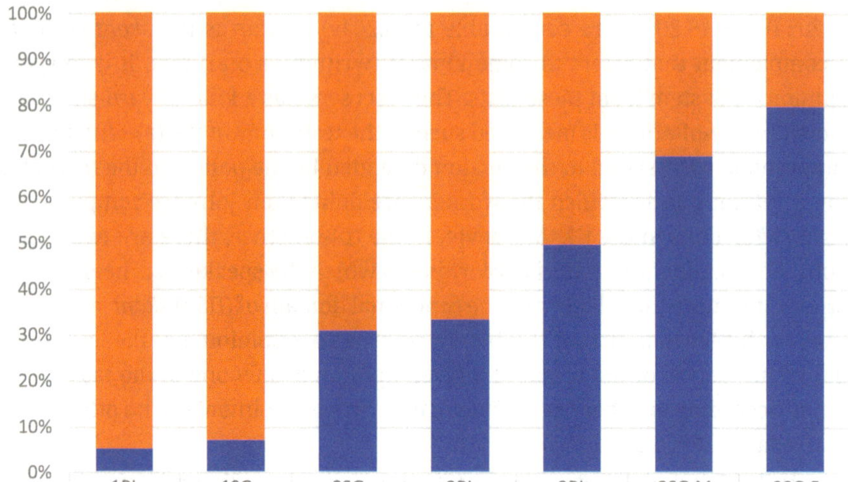

	1PL	1SG	2SG	2PL	3PL	3SG.M	3SG.F
■ mingħala	455	1501	275	50	951	1559	174
■ skont	25	114	123	25	939	3458	679

Figure 5.24: Frequency of (Π37) *mingħala* 'in s.o.'s opinion' vs (Π53) *skont* 'according to' with pronominal complements (per person).

On this basis, we hypothesise that (Π37) *mingħala* 'in s.o.'s opinion' is pragmatically loaded and thus particularly suited to mark an opinion as that of the speaker(s) and to a lesser degree also that of the hearer(s). (Π53) *skont* 'according to' seems to lack this pragmatic property. This is why it is the better solution for neutrally reporting the opinion of a non-speech-act participant. The choice of the appropriate ROBORATIVE Π is thus partly determined by pragmatic factors. Semantics is also crucial since (Π53) *skont* 'according to' is admitted to more contexts than (Π37) *mingħala* 'in s.o.'s opinion' is. How exactly the two Πs divide the ROBORATIVE domain between them calls for a separate study in the future.

5.1.3.15 ADDITIVE

For Hagège (2010: 262), the best English representatives of the ADDITIVE are *in addition to* and *besides*. As to this meaning category in Maltese, Borg/Azzopardi-Alexander (1997: 155) provide an example that involves (Π34) *minbarra* 'except'. Our work with the Korpus Malti 3.0 has revealed, however, that (Π1) *apparti* 'apart from', too can be used as ADDITIVE Π. This ADDITIVE use of (Π1) *apparti* 'apart from' is documented in (5.206).

(5.206) ADDITIVE I

(a) [Korpus Malti 3.0; culture3538]
*Ix-xuffier raġel ta' 24 sena issa [**apparti** akkuża*
DEF-driver man of 24 year now [**apart_from** accusation
ta' sewqan perikoluż]$_{PP}$ *se jkun akkużat*
of driving dangerous] FUT 3SG.M.IPFV:be.FUT accuse:PART
ukoll b' attentat ta' qtil
also with attempt of murder
'The driver, a man of 24 years, [**apart from** an accuse of dangerous driving], will now also be accused of attempted murder.'

(b) [Korpus Malti 3.0; academic29]
*[**Apparti** l-Partit Nazzjonalista]*$_{PP}$ *il-Christian Workers Party*
[**apart_from** DEF-National_Party] DEF-Christian Workers Party
in-Nationalist Democratic Party il-Progressive Constitutional Party
DEF-Nationalist Democratic Party DEF-Progressive Constitutional Party
u l-Christian Democratic Party kienu kjarament kontra
and DEF-Christian Democratic Party be:3PL.PFV clearly against
l-Partit Laburista
DEF-Labour_Party
'[**Apart from** the National Party], the Christian Workers Party, the Nationalist Democratic Party, the Progressive Constitutional Party, and the Christian Democratic Party clearly were against the Labour Party.'

(c) [Korpus Malti 3.0; culture1163]
*[**Apparti** l-ikel]*$_{PP}$ *se jkun hemm ukoll żfin*
[**apart_from** DEF-food] FUT 3SG.M.IPFV:be.FUT there also dancing
minn 12-il skola kanzunetti country ma' Marisa D'Amato u
from 12 school song:DIM:PL country with Marisa D'Amato and
anke ftit daħk ma' Gable ta' Deċeduti
also little laughing with Gable ta' Deċeduti
'[**Apart from** the food], there will also be dancing from twelve schools, country songs with Marisa D'Amato and also some laughing with Gable ta' Deċeduti.'

In these examples, the *apparti*-PP introduces a participant that forms part of a more or less extended list of participants for which the same predication holds. This participant is added to the list – in some cases, it is the other way around insofar as the *apparti*-PP marks the basic entity to which the others are added. In the (a)-sentence, there is already the accusation of dangerous driving to which that of attempted murder is added. In the (b)-sentence, the political position of the National Party is the reference point for the enumeration of four other parties which share the same views as the Labour Party. In the (c)-sentence, it is said that, on the occasion of a festival, there is not only the usual food on the program of the event but also a number of other highlights. (П1) *apparti* 'apart from' is used in the ADDITIVE function when it occurs in lists or enumerations which comprise at least two units.

The same can be said about (П34) *minbarra* 'except' in (5.207).

(5.207) ADDITIVE II

(a) [Korpus Malti 3.0; news22008]

*[**Minbarra** Malta]*$_{PP}$ *Ċipru u l-Lussemburgu wkoll*
[**except** Malta] Cyprus and DEF-Luxembourg also
għandhom kwalità eċċellenti ta' ilma
have.IPFV:3PL quality excellent of water
'[**Apart from** Malta], Cyprus and Luxembourg too have an excellent water quality.'

(b) [Korpus Malti 3.0; academic17]

*[**Minbarra** l-lingwi u l-letteratura]*$_{PP}$ *studja l-liġi u*
[**except** DEF-language:PL and DEF-literature] study DEF-law and
kien għaref ukoll fit-tmexxija tan-negozju
be wise also in:DEF-leadership of:DEF-business
'[**Apart from** languages and literature], he studied law and was also knowledgeable in business management.'

(c) [Korpus Malti 3.0; culture1015]

*[**minbarra** spejjeż oħra fosthom it-trobbija ta'*
[**except** expense.PL other:PL amongst:3PL DEF-upbringing of
żewġt itfal]$_{PP}$ *għandu € 200 f' kera kull tliet*
two child.PL] have.IPFV:3SG.M € 200 in rent each three
xhur
month.PL
'[. . .] [**apart from** other expenses with among them the upbringing of two children], he has € 200 in rent every three months.'

In (5.207a), Malta is shown not to be the only country whose water quality is high. With Cyprus and Luxembourg, Malta forms a list of three. The *minbarra*-PP in (5.207b) introduces two subject matters the anonymous person has studied to which law studies and managerial knowledge are added so that a list of four emerges. The list is shorter in the case of (5.207c) where it is reported that someone complaints about the high costs of living enumerating those occasioned by raising two children and the rent. We hypothesise that one may substitute either of the two ADDITIVE IIs for the other in (5.206) and (5.207) since they seem to be full equivalents of each other in this domain.

Both (Π1) *apparti* 'apart from' and (Π34) *minbarra* 'except' have been presented as EXCEPTIVE Πs.[148] However, the EXCEPTIVE function is only a minor option in comparison to the ADDITIVE function of these Πs. Our search in the Korpus Malti 3.0 reveals that of the 1st 100 hits of (Π1) *apparti* 'apart from' and (Π34) *minbarra* 'except', the overwhelming majority reflects the ADDITIVE function. According to Figure 5.25, the ADDITIVE function predominates to the extreme in both cases. Only two *apparti*-PPS cannot be classified as ADDITIVE. As to the *minbarra*-PPS, those which fulfil functions other than that of the ADDITIVE yield a turnout of ten hits. Accordingly, ADDITIVE is the preferred or default reading of both (Π1) *apparti* 'apart from' and (Π34) *minbarra* 'except'. In contrast to the ADDITIVE function, the EXCEPTIVE function of these Πs is only of marginal relevance. It is also worth noting that ADDITIVE and EXCEPTIVE are antonyms in the sense that the ADDITIVE adds a unit to a list whereas the EXCEPTIVE excludes a unit from a list. These meanings are in conflict with each other. We still ignore how potential ambiguities are resolved by Maltese native speakers. It is clear that the communicative context provides clues for the correct interpretation of (Π1) *apparti* 'apart from' and (Π34) *minbarra* 'except'. Yet, this phenomenon needs to be looked into more deeply. This is why only a dedicated follow-up study can answer the questions that arise from the double-faced semantics of the two Πs under scrutiny.

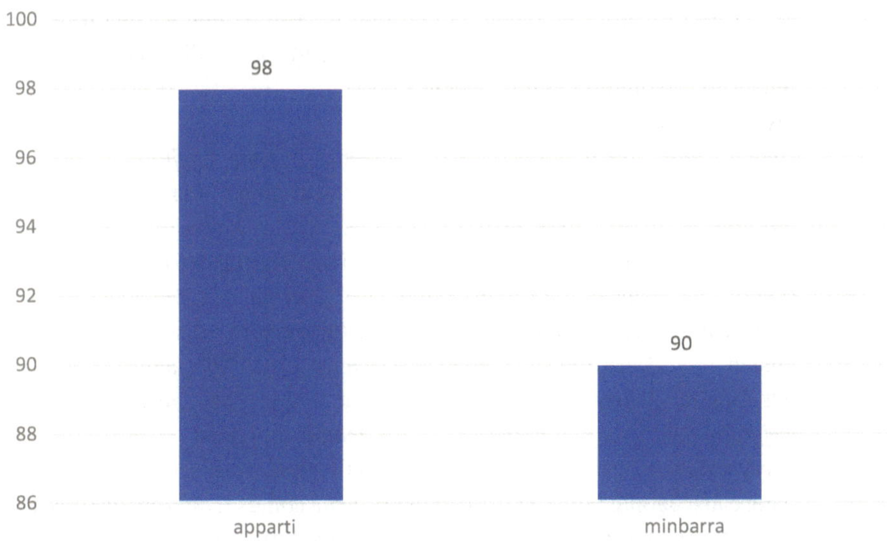

Figure 5.25: ADDITIVE (Π1) *apparti* 'apart from' and (Π34) *minbarra* 'except' in the 1st 100 tokens.

148 Cf. Section 5.1.3.3.

5.1.3.16 SUBSTITUTIVE

The SUBSTITUTIVE is the last meaning category in Hagége's (2010) inventory to be discussed. All subsequent sections feature meaning categories that are absent from this inventory. English *instead of* is Hagège's (2010: 262) representative of the SUBSTITUTIVE. The SUBSTITUTIVE applies in situations where a participant (or event or property) replaces another one. The Maltese equivalent of the SUBSTITUTIVE is expressed by (Π35) *minflok* 'instead of'. The category as such is absent from the reference grammar. This absence does not mean that there is no empirical evidence of its existence. Three pertinent cases are presented in (5.208).

(5.208) SUBSTITUTIVE

(a) [Korpus Malti 3.0; academic6]

il-konsumaturi	*li*	*meta*	*jagħżlu*	*prodott*	*tal-Kummerċ*
DEF-consumer:PL	that	when	3.IPFV:choose:PL	product	of:DEF-Trade

Ġust	*[**minflok***	*prodott*	*simili*	*maħdum*	*f'*	*suq*
Fair	[**instead_of**	product	similar	PART:work	in	market

tradizzjonali]$_{PP}$	*ikunu*	*qed*	*jagħrfu*	*l-poter*
traditional]	3.IPFV:be.FUT:PL	PROG	3.IPFV:recognise:PL	DEF-power

tagħhom	*fil-qasam*	*tas-suq*
of:3PL	in:DEF-field	of:DEF-market

'[. . .] the consumers who when they choose a Fair-Trade product [**instead of** a similar product made on a traditional market] will recognise their power in the domain of the market [. . .].'

(b) [Korpus Malti 3.0; academic2009

Hemm	*min*	*jgħid*	*li*	*l-kelma*	*pepè*	*ġejja*
there	who	3SG.M.IPFV:say	that	DEF-word	pepè	come.PART:F

mill-fatt	*li*	*s-Slimiżi*	*kienu*
from:DEF-fact	that	DEF-inhabitant_of_Sliema:PL	be:3PL.PFV

jsejjħu	*lil*	*missierhom*	*papà*	*b'*	*ċertu*	*tgħawwiġ*	*ta'*
3.IPFV:call:PL	to	father:3PL	papa	with	certain	distortion	of

l-ilsien	*[**minflok***	*missier*	*pà*	*jew*	*tà]$_{PP}$*	*kif*	*kienu*
DEF-tongue	[**instead_of**	missier	pà	or	tà]	as	be:3PL.PFV

jsejjħulu	*n-nies*	*t'*	*irħula*	*oħra*
3.IPFV:call:PL:IO:3SG.M	DEF-people	of	village.PL	other:PL

'There are people who say that the word *pepè* comes from the fact that the inhabitants of Sliema called their father *papa* with a certain distortion of the language [**instead of** *missier, pà* or *tà*] as the people of other villages called him.'

(c) [Korpus Malti 3.0; academic283]

Infatti fl-istess sena Boieldieu ġie elett
in_fact in:DEF-same year Boieldieu come elect.PART
***[minflok** Nicolò Isouard]$_{PP}$ biex jimla l-post*
[instead_of Nicolò Isouard] to 3SG.M.IPFV:fill DEF-post
fl-Institut de France
in:DEF-Institut_de_France
'In fact, in the same year, Boieldieu was elected **[instead of** Nicolò Isouard] to fill the position in the Institut de France.'

In the (a)-sentence, the consumer chooses a Fair-Trade product instead of a traditional product. The latter is replaced with the former. In the (b)-sentence, the etymology of the Maltese expression *pepè* 'dandyism' (Aquilina 1991: 1047) is explained as originating from the local pronunciation of *papà* 'dad' in Sliema. The speakers of Sliema Maltese use *pepè* instead of the elsewhere widespread *missier*, *pà*, or *tà*. The local (Sliema) expression replaces those of other varieties of Maltese. In the (c)-sentence, Boieldieu is elected instead of his competitor. In this sense, the elected candidate replaces the runner-up.

5.1.3.17 Benefactive

Hagège (2010) does not consider the BENEFACTIVE to be a meaning category in its own right. We do not subscribe to this opinion. The BENEFACTIVE is the functional monopoly of (Π21) *għal* 'for' according to Borg/Azzopardi-Alexander (1997: 147–148). However, there is the Italo-Romance (Π13) *favur* 'in favour of' that also expresses that something is done or happens to the benefit of someone. The use of the latter BENEFACTIVE Π is illustrated in (5.209).

(5.209) BENEFACTIVE I
(a) [Korpus Malti 3.0; academic12]

L-amministrazzjoni kienet ***[favur** il-pjan]$_{PP}$ u*
DEF-administration be:3SG.F.PFV **[in_favour_of** DEF-plan] and
fil-fatt applikaw madwar 440 ruħ għalkemm dawn
in:DEF-fact apply:3PL.PFV around 440 soul although DEM:PL:PROX
qatt ma telqu
ever NEG leave:3PL.PFV
'The administration was **[in favour of** the plan] and in fact about 440 persons applied although they never left.'

(b) [Korpus Malti 3.0; academic267]
Michael Mifsud skorja għal ħames darbiet f' logħba ta'
Michael Mifsud score for five time:PL in play of
ħbiberija li Malta lagħbet kontra l-Liechtenstein u
friendship that Malta play:3SG.F.PFV against DEF-Liechtenstein and
li spiċċat 7–1 ***[favur*** *il-Maltin]*$_{PP}$ *l-ikbar*
that end:3SG.F.PFV 7–1 [**in_favour_of** DEF-Maltese:PL] DEF-big.CMPR
rebħa li Malta qatt għamlet
victory that Malta ever make:3SG.F.PFV
'Michael Mifsud scored five times in a friendly match that Malta played against Liechtenstein and that ended 7–1 [**for** the Maltese] – the biggest victory that Malta ever made.'

(c) [Korpus Malti 3.0; news10319]
Il-Partit Nazzjonalista fl-Oppożizzjoni mistenni jivvota
DEF-National_Party in:DEF-opposition PART:wait 3SG.M.IPFV:vote
[favur *il-Baġit]*$_{PP}$
[**in_favour_of** DEF-budget]
'The National Party in the opposition is expected to vote [**for** the budget].'

In (5.209a), the position the administration takes in connection to a given plan is characterised as favourable. The behaviour of the administration is thus to the benefit of the plan. The example in (5.209b) tells us that the result of a football match against Liechtenstein was to the benefit of the Maltese national team. And in (5.209c), the vote the National Party is expected to make is to the benefit of the budget. The plan, Malta, and the budget are benefitting from the other participants' behaviour.

In (5.210), the second BENEFACTIVE Π is the focus of attention. (Π21) *għal* 'for' is the head of PPs whose complement refers to the beneficiary (as is the case with the complements of (Π13) *favur* 'in favour of' in the foregoing examples).

(5.210) BENEFACTIVE II
(a) [Korpus Malti 3.0; literature24]
Xbajtu tibagħtuni taħt l-art naqta'
fed_up:2PL.PFV 2.IPFV:send:PL:1SG under DEF-ground 1SG.IPFV:cut
l-għeruq ***[għal*** *ommi]*$_{PP}$
DEF-root.PL [**for** mother:1SG]
'Were you fed up with sending me under the ground to cut the roots [**for** my mother]?'

(b) [Korpus Malti 3.0; news122355]

Dak	*li*	*huwa*	*tajjeb*	***[għal***	*<u>Malta</u>]*$_{PP}$	*mhux*	*dejjem*
DEM:M:DIST	that	3SG.M	good	**[for**	<u>Malta</u>]	NEG:3SG.M:NEG	always

jgħodd	*għal*	*gżiritna*
3SG.M.IPFV:count	for	island:POSS:1PL

'What is good [**for** <u>Malta</u>] does not always count for our island.'

(c) [Korpus Malti 3.0; news173768]

l-effetti	*tal-kafè*	*huma*	*pożittivi*	*mmens*
DEF-effect:PL	of:DEF-coffee	3PL	positive:PL	immensely

[għal	*<u>bniedem</u>]*$_{PP}$
[for	<u>human being</u>]

'[. . .] the effects of coffee are immensely positive [**for** <u>a human being</u>] [. . .].'

In the (a)-sentence, the speaker's mother is the beneficiary of the action of her child. The roots are cut for her. In the (b)-sentence, a representative of the island of Gozo claims that not everything that is to the benefit of Malta is also to the benefit of the smaller sister island. In the (c)-sentence, it is claimed that coffee has effects which are to the benefit of human beings.

Functionally, the two BENEFACTIVE Πs are not absolutely identical. Their differences result mostly from the fact that (Π13) *favur* 'in favour of' is limited to the BENEFACTIVE meaning category whereas (Π21) *għal* 'for' is multifunctional so that it occurs in many more contexts than (Π13) *favur* 'in favour of'. Furthermore, (Π13) *favur* 'in favour of' is semantically more explicit or loaded than (Π21) *għal* 'for'. The former emphasises the BENEFACTIVE relation. In contrast, (Π21) *għal* 'for' does not seem to highlight the BENEFACTIVE relation to any noticeable degree.[149] We conclude that there are also pragmatic differences between the two Πs.

5.1.3.18 MALEFACTIVE

Given that Hagège (2010) does not deem the BENEFACTIVE worth reserving a distinct meaning category, it does not come as a surprise that the author does not mention the MALEFACTIVE either. This category is also absent from the Maltese reference grammar. (Π2) *a skapitu* 'at the expense of' is the antonym of BENEFACTIVE (Π13) *favur* 'in favour of' and (Π21) *għal* 'for'. The use of the MALEFACTIVE Π is illustrated in (5.211).

149 Cf. Section 5.2.2.3.

(5.211) MALEFACTIVE

(a) [Korpus Malti 3.0; academic344]

Il-problema ġejja mill-għalliema u mill-medja
DEF-problem come.PART:F from:DEF-teacher:PL and from:DEF-media
li qed jabbużaw mill-Ingliż ***[a skapitu***
that PROG 3.IPFV:abuse:PL from:DEF-English **[at_the_expense_of**
tal-Malti]$_{PP}$
of:DEF-Maltese]
'The problem comes from the teachers and the media that are abusing English **[at the expense of** Maltese].'

(b) [Korpus Malti 3.0; news112416]

Mhux aċċettabbli li jkun hawn min
NEG:3SG.M:NEG acceptable that 3SG.M.IPFV:be.FUT here who
jagħmel il-flus ***[a skapitu*** *tat-tfal*
3SG.M.IPFV:make DEF-money **[at_the_expense_of** of:DEF-child.PL
tagħna]$_{PP}$
of:1PL]
'It is not acceptable that there will be someone here who makes money **[at the expense of** our children].'

(c) [Korpus Malti 3.0; news136899]

Ried jeleva lil Chris Said fil-kariga ta' ministru
want 3SG.M.IPFV:elevate to Chris Said in:DEF-post of minister
[a skapitu *ta' Debono]*$_{PP}$
[at_the_expense_of of Debono]
'He wanted to elevate Chris Said to the post of minister **[at the expense of** Debono].'

What we see in these examples is that the complement noun of the *a skapitu*-PP identifies the maleficient, i.e., the participant to whose detriment some action is carried out. In the (a)-sentence, teachers and the media are blamed for giving preference to English to the detriment of Maltese. In the (b)-sentence, the children are depicted as the participants who are in danger of being exploited by other persons. The (c)-sentence mentions Debono as the candidate at whose expense someone else is promoted to the rank of minister.

5.1.3.19 PARTITIVE

Borg/Azzopardi-Alexander (1997: 152–153) present (Π41) *minn* 'from' as the PARTITIVE Π of Maltese. It is used in situations where a subset from a numerically larger group of entities is singled out. This is what happens throughout the examples in (5.212).

(5.212) PARTITIVE

(a) [Korpus Malti 3.0; academic4]

jitolbuhom	*jiktbu*	*esej*	*qasir*	*dwar*	*xi*	*wieħed*
3.IPFV:ask:PL:3PL	3.IPFV:write:PL	essay	short	about	some	one
***[minnhom]**$_{PP}$*						
[**from**:3PL]						

'[. . .] they ask them to write a short essay about one [**of** them].'

(b) [Korpus Malti 3.0; academic12]

qatel	*tnejn*	***[minn***	*nies*	*tiegħu]$_{PP}$*
kill	two	[**from**	people	of:3SG.M]

'[. . .] he killed two [**of** his people].'

(c) [Korpus Malti 3.0; culture1572]

l-akbar	*dulur*	*li*	*jista'*	*jkollha*	
DEF-big.CMPR	pain	that	3SG.M.IPFV:can	3SG.M.IPFV:have.FUT:3SG.F	
Marija	*hu*	*meta*	*xi*	*ħadd*	***[minn** uliedha]$_{PP}$*
Marija	3SG.M	when	some	someone	[**from** children:3SG.F]
jdawwar	*spallejh*	*għall-offerta*	*tagħha*		
3SG.M.IPFV:turn.CAUS	shoulder:PL:3SG.M	for:DEF-offer	of:3SG.F		

'[. . .] the biggest pain that Maija could have is that one [**of** her children] shrugs his shoulders at her offer.'

In (5.212a), the students are asked to write an essay about someone who belongs to their group. The *minn*-PP in (5.212b) tells us that two members of an unnamed criminal's gang were killed. It can be inferred from the choice of the PARTITIVE that the gang counted more than just two members. The situation in (5.212c) involves a mother who has several children and who lives in fear of one of her children not accepting an offer she is making.

5.1.3.20 Inclusive

In Maltese, the INCLUSIVE is expressed by (Π25) *inkluż* 'including' (Borg/Azzopardi-Alexander 1997: 154). We have argued that (Π25) *inkluż* 'including' is not a straightforward example of a Π and that only the morphologically invariable form of this item can be included in BLOMP 2.0.[150] Accordingly, the examples in (5.213) lack number and gender agreement on the Π. The INCLUSIVE applies in situations where we have a group or collective as a participant whose membership is described as also involving further specific participants. The INCLUSIVE is the antonym of the EXCEPTIVE.

150 Cf. Section 1.6.6 (vi).

(5.213) INCLUSIVE

(a) [Korpus Malti 3.0; culture3791]

l-istaff tal-Isptar ***[inkluż*** *l-infermieri]*$_{PP}$ *ta' kuljum*
DEF-staff of:DEF-hospital **[including** DEF-nurse:PL] of every_day
qed jiffaċċjaw pressjoni dejjem tikber
PROG 3.IPFV:face:PL pression always 3SG.F.IPFV:grow
'[. . .] the hospital's staff [**including** the nurses] is daily facing an always growing pressure.'

(b) [Korpus Malti 3.0; culture3089]

Fiha qed jieħdu sehem nies ta' kull età
in:3SG.F PROG 3.IPFV:take:PL part people of each age
[inkluż *tfal]*$_{PP}$ *bil-biċċa kbira minn Birkirkara*
[including child.PL] with:DEF-piece big:F from Birkirkara
'In it, people of all ages [**including** children] mostly from Birkirkara take part.'

(c) [Korpus Malti 3.0; sport1648]

Kulħadd jgħid li jrid jirbaħ
everyone 3SG.M.IPFV:say that 3SG.M.IPFV:want 3.SG.M.IPFV:win
[inkluż *jien]*$_{PP}$
[including 1SG]
'Everybody says that he wants to win, [**including** me].'

In (5.213a), it is made explicit that the staff of the hospital also includes the nurses. In (5.213b), the group of participants at a festival is described as comprising people of all age groups including children. Finally, (5.213c) is a supposedly universal statement about everybody wanting to win. The speaker reveals that the group referred to as everybody also includes him or her.

5.1.3.21 MATERIAL

The final two sections of this part of the chapter on semantics address meaning categories which are probably and tacitly subsumed under the umbrella of the INSTRUMENTIVE in Hagège's (2010) system. In the light of Borg/Azzopardi-Alexander's (1997: 150–151) presentation of three different Πs as markers in the domains of MATERIAL and MANNER, we prefer to treat these categories as distinct from the INSTRUMENTIVE although one of the MATERIAL/MANNER Πs is also employed in the latter function. This applies to (Π7) *bi* 'with'. This Π competes with (Π41) *minn* 'from' and (Π55) *ta'* 'of' when it comes to expressing of which MATERIAL a given entity is made.

The case of (Π7) *bi* 'with' is illustrated in (5.214). The *bi*-PPs depend on the passive participles of the verbs *ħadem* 'work' and *għamel* 'do'.

(5.214) MATERIAL I

(a) [Korpus Malti 3.0; news167812]

L-artista Stephanie Borg qiegħda ttella' t-tieni
DEF-artist:F Stephanie Borg PROG.F 3SG.F.IPFV:put_up DEF-second
wirja personali tagħha li tinkludi 22 tpinġija
exhibition personal of:3SG.F that 3SG.F.IPFV:include 22 drawing
*maħdumin [**bi**l-linka]*$_{PP}$
PART:work:PL [**with**:DEF-ink]

'The artist Stephanie Borg is putting up her second personal exhibition that includes 22 drawings made [**in** ink].'

(b) [Korpus Malti 3.0; news86715]

Installati fuq il-bjut tagħhom dawn l-iskejjel
install.PART:PL on DEF-roof.PL of:3PL DEM:PL:PROX DEF-school.PL
*bejniethom għandhom ħmistax-il tank maħduma [**b'***
between:3PL have.IPFV:3PL fifteen tank PART:work [**with**
dan il-materjal]$_{PP}$
DEM:M:PROX DEF-material]

'Installed on their roofs these schools together have fifteen tanks made [**of** this material].'

(c) [Korpus Malti 3.0; *news67237*]

*iduq ix-xarba oriġinali magħmula [**bi**l-cocoa]*$_{PP}$
3SG.M.IPFV:taste DEF-drink original PART:make:F [**with**:DEF-cocoa]

'[. . .] he tastes the original drink made [**with** cocoa] [. . .].'

Arguably, in these examples, the material can also be interpreted as a kind of instrument that is used to produce something. In the (a)-sentence, the drawings are made in ink. The ink is a kind of second-order tool which needs a first-order tool (the pen) to be put on the drawing canvas. Other second-order tools in this sense are the material used to build the tanks in the (b)-sentence and the cocoa that serves as an ingredient for a drink in the (c)-sentence. What is important here however is the fact that we are not dealing with first-order tools.

The same situation is met in (5.215) where (Π41) *minn* 'from' is the head of the PP in which the material is identified. As in the previous case, the *minn*-PPs depend on passive participles of the verbs *ħadem* 'work' and *għamel* 'do'.

(5.215) MATERIAL II

(a) [Korpus Malti 3.0; academic6]

jinsabu dejjem iżjed prodotti fil-ħwienet
3.IPFV:REFL:find:PL always more product:PL in:DEF-shop.PL

magħmula *[minn* *qoton* *organiku]*$_{PP}$
PART:made:PL **[from** cotton organic]
'[. . .] ever more products made [**of** organic cotton] are found in the shops.'

(b) [Korpus Malti 3.0; culture2350]
L-iskulturi *kollha* *huma* *magħmulin* *[mill-injam*
DEF-sculpture:PL all:3SG.F 3PL PART:made:PL [**from**:DEF-wood
ġebel *u* *rħam]*$_{PP}$
stone and marble]
'All the sculptures are made [**of** wood, stone, and marble].'

(c) [Korpus malti 3.0; literature39]
Għalkemm *kienet* *kbira* *daqs* *il-pala* *ta'* *idi*
although be:3SG.F.PFV big:F equal_to DEF-palm of hand:1SG
kienet *madankollu* *ħafifa* *rixa* *għax* *kienet*
be:3SG.F.PFV however light:F feather because be:3SG.F.PFV
maħduma *[mill-aluminju]*$_{PP}$
PART:make:F [**from**:DEF-aluminium]
'Although it was as big as the palm of my hand it was nevertheless light as a feather because it was made [**of** aluminium].'

In (5.215a), products made of cotton are talked about. In (5.215b), the sculptures are said to be made of wood whereas in (5.215c), the unidentified object is surprisingly light because it is made of aluminium. We assume that (Π41) *minn* 'from' can replace (Π7) *bi* 'with' in (5.214) just as the latter Π may replace (Π41) *minn* 'from' in (5.215).

Replacement is not possible with the third Π in the domain of MATERIAL. In line with the logic exposed above,[151] (Π55) *ta'* 'of' heads an attributive PP that requires a noun it modifies. As explained earlier in this study,[152] this noun may be absent under certain conditions. (Π55) *ta'* 'of' cannot take in the position of either (Π7) *bi* 'with' or (Π41) *minn* 'from' in constructions with passive participles, for instance. Similarly, neither (Π7) *bi* 'with' nor (Π41) *minn* 'from' are suitable substitutes for (Π55) *ta'* 'of' in cases like those presented in (5.216).

(5.216) MATERIAL III
(a) [Korpus Malti 3.0; academic12]
Ħaseb *biex* *jiġi* *lura* *lejn* *Malta* *fejn* *ħadem*
think to 3SG.M.IPFV:come back towards Malta where work

151 Cf. Section 4.2.2.
152 Cf. Section 4.2.2.

dar	***[ta'***	*l-injam]$_{PP}$*	*u*	*bagħatha*	*fuq*	*vapur*	*lejn*
house	**[of**	DEF-wood]	and	send:3SG.F	on	ship	towards

l-Awstralja
DEF-Australia
'He thought of coming back to Malta where he built a [**wooden**] house and sent it on a ship to Australia.'

(b) [Korpus Malti 3.0; academic420]

Illum	*il-ġurnata*	*tas-Sliema*	*tilfet*	*ftit*
today	DEF-day	of:DEF-Sliema	lose:3SG.F.PFV	little

mill-karateristika	*tagħha*	*minħabba*	*li*	*ħafna*
from:DEF-characteristics	of:3SG.F	on_account_of	that	many

mid-djar	*ġew*	*imwaqqa*	*biex*
from:DEF-house.PL	come:3PL.PFV	PART:knock_down	to

inbnew	*appartamenti*	*moderni*	***[ta****l-konkrit]$_{PP}$*
REFL:build:3PL.PFV	apartment:PL	modern:PL	[**of**:DEF-concrete]

'Nowadays, Sliema has lost a little of its characteristics because many of the houses were knocked down so that modern appartments [**of** concrete] were built.'

(c) [Korpus Malti 3.0; parl3026]

Ġeneralment	*meta*	*jsir*	*dan*	*it-tip*	*ta'*
generally	when	3SG.M.IPFV:become	DEM:M:PROX	DEF-type	of

sajd	*ikun*	*hemm*	*cable*	***[ta'***	*l-azzar]$_{PP}$*	*imqabbad*
fishing	3SG.M.IPFV:be.FUT	there	cable	**[of**	DEF-steel]	PART:connect

max-xibka
with:DEF-net
'Generally, when this kind of fishing takes place there will be a [**steel**] cable connected to the net.'

As can be expected on the basis of what was exposed in Section 4.2.2., the *ta'*-PPs are always NP-internal attributes. We have a wooden house in the (a)-sentence, apartments made of concrete in the (b)-sentence, and a steel cable in the (c)-sentence. In contrast, the *bi*-PPs and *minn*-PPs may also occur VP-internally. They are also possible as adverbial modifiers of participles. For the passive participle of *ħadem* 'work' with the forms SG.M *maħdum* ≠ SG.F *maħdum* ≠ PL *maħdumin*, there are eight combinations with (Π7) *bi* 'with' as opposed to five hits for (Π41) *minn* 'from' and the complement noun *injam* 'wood' (yielding translation equivalents of English *wooden* or *made of wood*). There are no hits for (Π55) *ta'* 'of' in this context. On account of the multifunctionality of both (Π7) *bi* 'with' and (Π41) *minn* 'from' it is a demanding task to determine which of the two Πs is the first choice when it comes to expressing MATERIAL. Therefore, we restrict the search in the Korpus Malti 3.0 to four nouns referring to the MATERIAL a given object is made of, namely *aluminju* 'aluminium', *azzar* 'steel', *injam* 'wood', and *konkrit* 'concrete'. Figure 5.26 shows that the two Πs do not yield identical results since with each of the above four nouns, (Π41) *minn* 'from' ousts (Π7) *bi* 'with'. The shares of the minority option range from minimally 3% to maximally 34%, meaning: (Π41) *minn* 'from' is the major option and the default Π in the domain of MATERIAL.

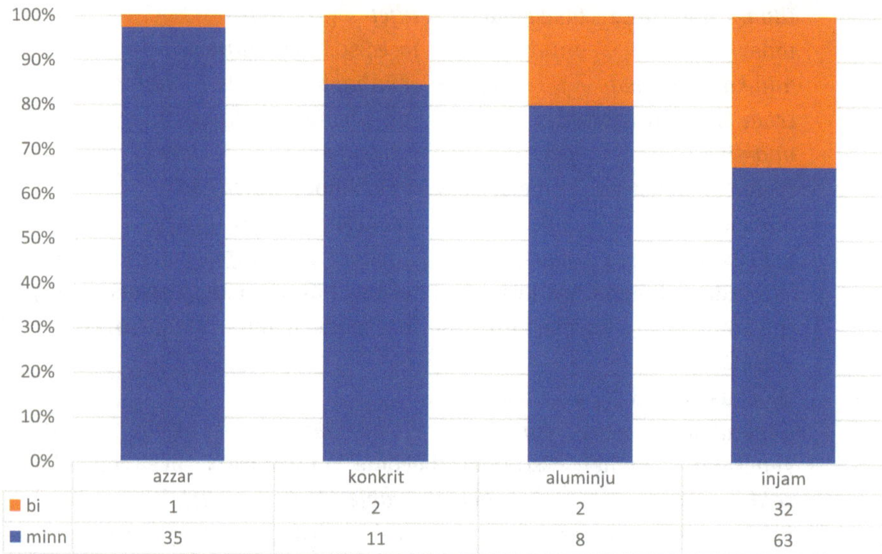

Figure 5.26: Competition of (Π7) *bi* 'with' and (Π41) *minn* 'from' in the domain of MATERIAL.

5.1.3.22 MANNER

As to MANNER, Borg/Azzopardi-Alexander (1997: 151) mention only (Π7) *bi* 'with'. In point of fact, MANNER adverbials very often come in the shape of a *bi*-PP.[153] The complement in these PPs refers to the manner in which a given action is carried out. We illustrate this pattern with the three examples in (5.217).

(5.217) MANNER

(a) [Korpus Malti 3.0; news12875]

dejjem għamel mill-aħjar li seta' biex jaqdi
always make from:DEF-better that can to 3SG.M.IPFV:serve
d-dmirijiet tiegħu għalkemm xi kultant għamel

153 As argued in Section 4.3.3.1.

DEF-duty:PL of:3SG.M although some sometimes make
dan *[bi* *tbatija]*$_{PP}$
DEM:M:PROX **[with** suffering]
'[. . .] he always did his best to serve his duties although sometimes he did this [**painfully**].'

(b) [Korpus Malti 3.0; culture405]
Clark għamel dan ***[b'*** *ħeffa liema bħalha]*$_{PP}$ *kif*
Clark make DEM:M:PROX **[with** speed which like:3SG.F] as
jidher fil-vidjow
3SG.M.IPFV:appear in:DEF-video
'Clark made this [**with** uncomparable speed] as is shown in the video.'

(c) [Korpus Malti 3.0; news149379]
Għamel dan ***[b'*** *għaqal kbir]*$_{PP}$ *ħalli jħares*
make DEM:M:PROX **[with** wisdom big] so_that 3SG.M.IPFV:protect
l-interessi ta' dawk kollha li ħassew li
DEF-interest:PL of DEM:PL:DIST all:3SG.F that feel:3PL.PFV that
setgħu jibbenefikaw minn dan il-pass
can:3PL.PFV 3.IPFV:benefit:PL from DEM:M:PROX DEF-step
'He did this [very **wisely**] so that he protects the interests of all those who feel that they could benefit from this step.'

As we see, the nouns *tbatija* 'suffering', *ħeffa* 'speed', and *għaqal* 'wisdom' are involved in *bi*-PPs to indicate that something is done painfully (= (a)-sentence), speedily (= (b)-sentence), or wisely (= (c)-sentence). The above examples are by no means isolated cases. In the Korpus Malti 3.0, they yield the following number of hits as MANNER adverbials:

- *bi tbatija* 'painfully' = 240 hits
- *b' ħeffa* 'speedily' = 417 hits
- *b' għaqal* 'wisely' = 946 hits

Given that there ideally is no limit as to the fillers of the complement slot in these *bi*-PPs, one must expect to find MANNER adverbials of this kind all over the texts covered by the Korpus Malti 3.0. MANNER adverbials are one of the strongest factors that make (Π7) *bi* 'with' a highly frequent Π in Maltese.

We have now reached the final paragraph of the catalogue of meaning categories and the Πs used to express them. It is necessary at this point to systematise the many details mentioned in the previous sections. This systematization is the task of Section 5.2.

5.2 The semantic network

The subsections of Section 5.1 served the purpose of crosschecking the list of adpositional functions proposed by Hagège (2010) and the services to which the members of BLOMP 2.0 are put in Maltese. We have therefore looked at the semantics of the Πs from the onomasiological perspective, i.e., functions came first and then expressions were allotted to them. However, it is not sufficient to exclusively approach the Maltese Πs from the viewpoint of preestablished categories whose existence is cast doubt upon by Haspelmath (2007b). To avoid misunderstanding we emphasise that Hagège's (2010) catalogue of functions that are cross-linguistically attested for adpositions does not require of a given language that each of the functions is expressed by adpositions. Neither does the catalogue assume that there are (always) one-to-one mappings of Π and function. We have seen that for several of Hagège's meaning categories, one and the same Maltese Π is employed and at the same time, a given meaning is frequently expressed by more than just one Π. The existence of one-to-many as well as many-to-one relations between Πs and functions calls for a second round of relating the content side and the expression side of Maltese Πs – this time from the semasiological perspective. This means that we take the concrete Πs as our point of departure (although with certain deviations from the semasiological ideal as explained below). The goal is to determine the functional domain of each Π and to show where the domains border on each other or overlap so that we are eventually able to draw up a semantic map of BLOMP 2.0.

As to the semasiological approach, we apply a compromise in the sense that the Πs are addressed in two steps with each step being semantically based so that an onomasiological element comes into play. In analogy to the practice employed in the previous section, we once more follow Hagège (2010) by way of treating spatio-temporal meaning categories separately from other categories. Accordingly, the first subsection is dedicated to those Πs which are known to (also) fulfil spatial and/or temporal functions.[154] Several of these Πs are not restricted to the spatio-temporal domain. This is why they re-enter the scene again in the second part where we address the non-spatio-temporal meaning categories and those termed core meanings by Hagège (2010).[155] Section 5.2.3 takes up the issue of polysemy which directly takes us to Section 5.2.4 where we present the semantic map alluded to above.

154 Cf. Section 5.2.1.
155 Cf. Section 5.2.2.

5.2.1 The semasiology of Πs in the spatio-temporal domain

5.2.1.1 The composition of the spatio-temporal subset of BLOMP 2.0

Maltese Πs with spatio-temporal functions constitute the biggest group within BLOMP 2.0. As shown in Figure 5.27, far more than half of the inventory is covered by spatio-temporal Πs whereas Maltese Πs that do not fulfil any spatio-temporal function claim a smaller share of BLOMP 2.0.

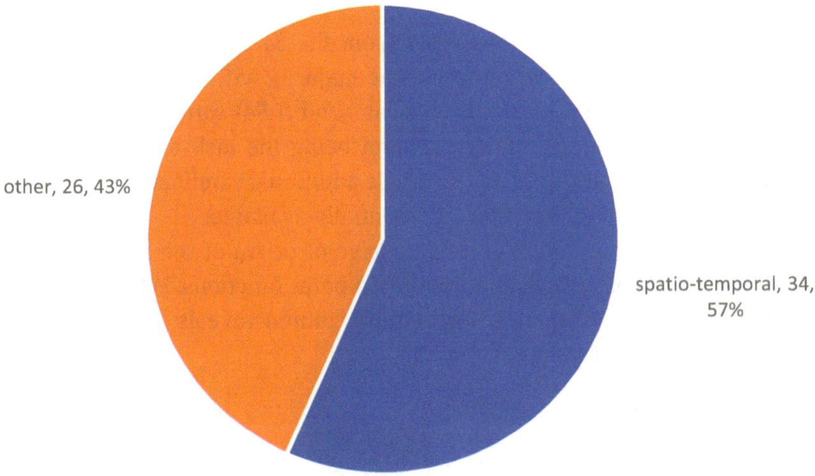

Figure 5.27: Shares of Πs with spatio-temporal meanings vs Πs without spatio-temporal meaning.

The absolute numbers are as follows. Of the sixty members of BLOMP 2.0, 34 can be classified as Πs with spatio-temporal functions. In contrast, only 26 Πs have no spatio-temporal function at all. Thirty Πs are bona fide instances of spatio-temporal Πs. They are listed alphabetically in (5.218).

(5.218) Bona fide spatio-temporal Πs included in BLOMP 2.0

(Π3) *barra* 'outside'; (Π4) *bejn* 'between'; (Π8) *biswit* 'facing'; (Π12) *faċċata* 'opposite'; (Π14) *fejn* 'near'; (Π15) *fi* 'in'; (Π16) *fost* 'amongst'; (Π17) *fuq* 'on'; (Π18) *ġewwa* 'inside'; (Π19) *ġo* 'in'; (Π21) *għal* 'for'; (Π22) *għand* 'at s.o.'s place'; (Π23) *ħdejn* 'beside'; (Π28) *lejn* 'towards'; (Π30) *ma'* 'with'; (Π31) *madwar* 'around'; (Π32) *maġenb* 'close to'; (Π33) *matul* 'during'; (Π38) *mingħand* 'from s.o.'; (Π41) *minn* 'from'; (Π44) *oltre* 'beyond'; (Π46) *qabel* 'before'; (Π47) *qalb* 'amidst'; (Π48) *qrib* 'near'; (Π49) *quddiem* 'in front of'; (Π51) *sa* 'till'; (Π56) *taħt* 'under'; (Π58) *viċin* 'near'; (Π59) *waqt* 'at the time of'; (Π60) *wara* 'after'

In addition to the above thirty uncontroversial cases, there are a further four Пs whose meaning is associated with the spatial domain but which for one reason or another pose problems so that, in the ensuing discussion, they are referred to only with reservations. The four problematic cases are given in (5.219).

(5.219) Problematic additions to (5.218)
(П42) *mintul* 'all along'; (П43) *mnejn* 'from near'; (П54) *sotta* 'under'; (П57) *versu* 'towards'

Even if we subtract the controversial cases from the 34 spatio-temporal Пs the remaining thirty types would still represent the majority within BLOMP 2.0 since (П42) *mintul* 'all along', (П43) *mnejn* 'from near', and (П54) *sotta* 'under' display spatial semantics nevertheless – their problem being the lack of attestations – whereas with (П57) *versu* 'towards' only a single additional candidate for the class of Пs without spatio-temporal functions is potentially available.

It is worthwhile taking a look at the language of origin of the Пs in (5.218)–(5.219) as opposed to those which lack spatio-temporal functions. What we see in Figure 5.28 clearly is a discrepancy. The middle column reveals the shares that result from including the cases mentioned in (5.129).

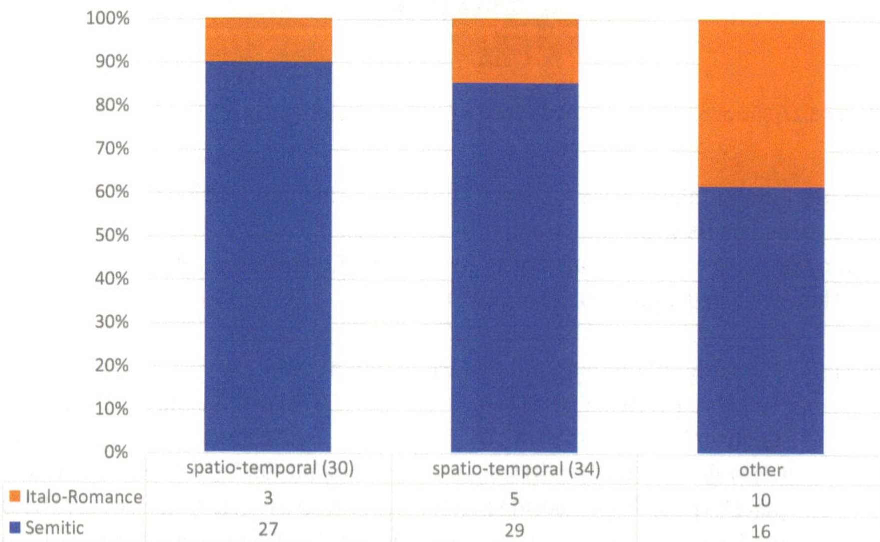

Figure 5.28: Shares of Semitic and Italo-Romance Пs within and outside the spatio-temporal domain.

With shares of 10%–15%, Italo-Romance Πs are underrepresented in the spatio-temporal domain. In contrast, they contribute ten types to the 26 non-spatio-temporal Πs which equals a share of 38%. We remind the reader that exactly a quarter of BLOMP 2.0 consists of Italo-Romance Πs.

In Figure 5.29, we look at the same situation from a different vantage point. This time we determine the shares of spatio-temporal and other functions per language of origin.

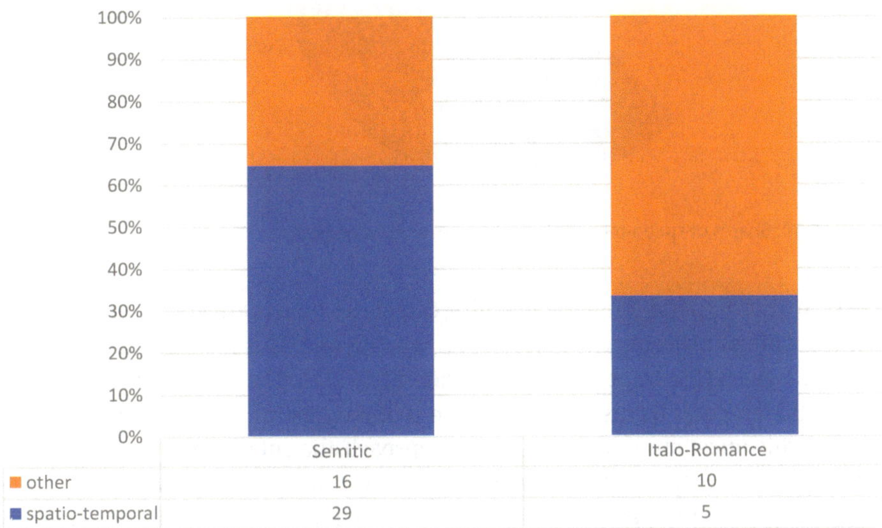

Figure 5.29: Shares of spatio-temporal and other functions per language of origin.

Only a minority of the Semitic Πs is barred from spatio-temporal functions. The seventeen cases equal 36% of all Semitic Πs. As to the Italo-Romance Πs, the opposite applies in the sense that 67% of the Italo-Romance Πs fulfil functions other than spatio-temporal ones. Thus, the probability of an Italo-Romance Π expressing a non-spatio-temporal meaning is higher than that of a Semitic Π doing the same.

As we have learned from the discussion of Hagège's (2010) categories,[156] it is possible to identify 77 meaning categories for Maltese. Of these 77 meaning categories, 48 belong to the spatio-temporal domain. Spatio-temporal meaning categories are thus responsible for the majority of the semantic distinctions in the realm of Maltese Πs as shown in Figure 5.30.

---

156 In Section 5.1.

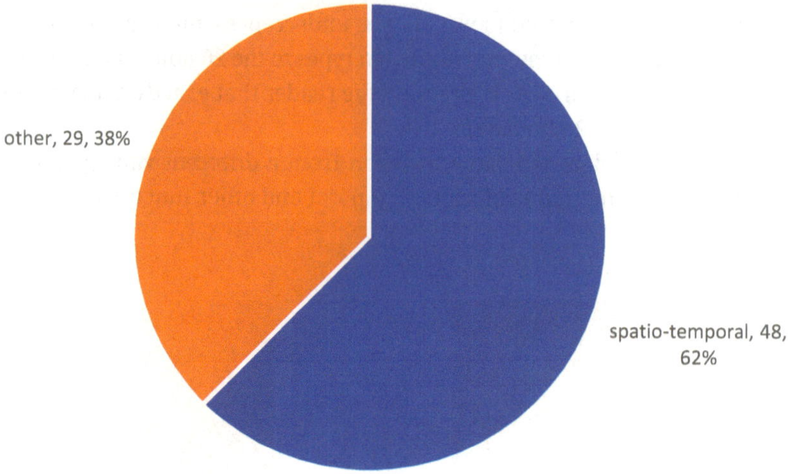

Figure 5.30: Shares of spatio-temporal vs other meaning categories.

There are more spatio-temporal meaning categories than there are spatio-temporal Πs. The ratio of meaning categories vs Πs is 1.5-to-one. This means that the probability of a given Π to be polysemous is moderately high. The question arises of how the spatio-temporal Πs distribute over the spatio-temporal meaning categories. To answer this question, it is necessary to approach the subject from an angle that differs from that chosen for the presentation in the onomasiological part.

It is clear that Hagège's labelled categories do not provide the best foundation for a semasiological account of the Maltese Πs not the least because certain systematic phenomena cannot properly be captured. The main reason for this failure is Hagège's focus on singleton Πs. The previous discussion has revealed however that many meaning categories require the combination of two Πs in Maltese.[157] These binary combinations generally reflect the pattern DIRECTIONALITY + CONFIGURATION that we introduced above.[158] The differentiation between directionality marking Π and configuration marking Π is crucial for the internal organisation of the spatio-temporal domain. Moreover, to adequately account for the systematicity of this domain, it is essential to take into consideration also the strategy of zero-marking. We distinguish Πs which are exclusively used as directionality markers from those which are only employed for the expression of configuration. Zero-marking is possible both with directionality and configuration. In Table 5.14, we oppose the minority of four directionality markers and the majority of 25 configuration markers. The

157 Cf. Section 5.1.2.
158 Cf. Section 4.1.3.

problematic cases of (5.219), (Π38) ***mingħand*** 'from s.o.', and zero-marking are not included in Table 5.14 which therefore hosts only 25 instead of 26 configuration markers.

Table 5.14: Directionality markers vs configuration markers.

directionality	configuration
(Π21) *għal* 'for' (Π28) *lejn* 'towards' (Π41) *minn* 'from' (Π51) *sa* 'till'	(Π3) *barra* 'outside'; (Π4) *bejn* 'between'; (Π8) *biswit* 'facing'; (Π12) *faċċata* 'opposite'; (Π14) *fejn* 'near'; (Π15) *fi* 'in'; (Π16) *fost* 'amongst'; (Π17) *fuq* 'on'; (Π18) *ġewwa* 'inside'; (Π19) *ġo* 'in'; (Π22) *għand* 'at s.o.'s place'; (Π23) *ħdejn* 'beside'; (Π30) *ma'* 'with'; (Π31) *madwar* 'around'; (Π32) *maġenb* 'close to'; (Π33) *matul* 'during'; (Π44) *oltre* 'beyond'; (Π46) *qabel* 'before'; (Π47) *qalb* 'amidst'; (Π48) *qrib* 'near'; (Π49) *quddiem* 'in front of'; (Π56) *taħt* 'under'; (Π58) *viċin* 'near'; (Π59) *waqt* 'at the time of'; (Π60) *wara* 'after'

The overwhelming majority of the Maltese Πs is dedicated to encoding configuration whereas only a small share of the Πs mentioned in (5.218) marks directionality as shown in Figure 5.31.

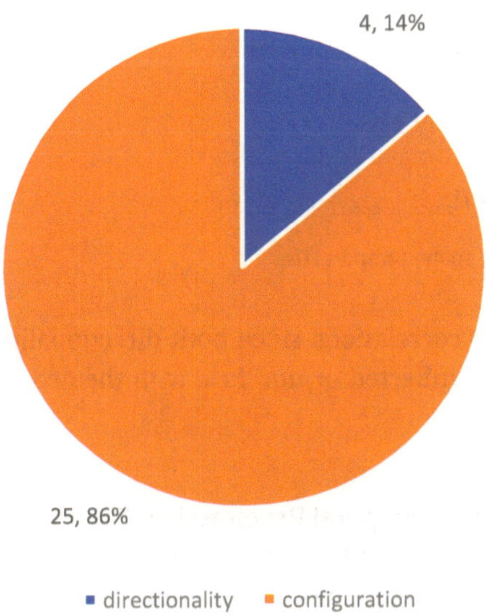

Figure 5.31: Shares of configuration markers and directionality markers.

If we continue to ignore zero-marking, all directionality markers in Table 5.14 bear the feature [dynamic] whereas this applies to no Π under the rubric of configuration since these Πs are insensitive to the distinction of Place and Goal. A special case is (Π38) *mingħand* 'from s.o.' because it combines directionality and configuration in one form. In contrast to the configuration markers, it is excluded from the expression of the spatial relations of Place and Goal but serves to express the Source relation (normally) with human grounds.[159] Historically, we are facing a univerbation of the directionality marker (Π41) *minn* 'from' and the configuration marker (Π22) *għand* 'at s.o.'s place'. Similarly, (Π43) *mnejn* 'from near' can be understood as integrating both directionality (= (Π41) *minn* 'from') and configuration (= (Π14) *fejn* 'near') in one no longer analysable form.

The Πs that are registered under directionality in Table 5.14 are Semitic monosyllables. All twelve polysyllables belong to the domain of configuration where we also find the three Italo-Romance Πs mentioned in (5.218). If (Π57) *versu* 'towards' is accepted as forming part of the class of spatio-temporal Πs it would be the only example of an Italo-Romance Πs expressing directionality because the feature [dynamic] applies. However, (Π57) *versu* 'towards' does not allow a configuration marker to occupy the slot to its right and thus behaves differently from genuine directionality markers. Figure 5.32 can be read as implication if only (5.218) is taken account of or as preference if (5.219) is taken account of, too. The same holds for Figure 5.33.

DIRECTIONALITY ⊃ Semitic + monosyllabic

Figure 5.32: Implication/preference with directionality markers.

polysyllabic ⊃ CONFIGURATION / $\Pi_{\text{spatio-temporal}}$

Figure 5.33: Implication/preference with polysyllabic spatio-temporal Πs.

There is no evidence of further structural correlations since both directionality markers and configuration markers can be inflected or not, fuse with the article proclitic or not, cliticise or not, etc.

5.2.1.2 The division of labour between spatio-temporal Πs: directionality

With a view to systematising further the richness of information provided above,[160] we need to separate the directionality markers from the configuration markers in a first step before we eventually reunite the two classes of spatio-temporal Πs to finalise the semasiological account of the spatio-temporal domain.

159 Cf. Section 5.1.2.2.4 – APUDABLATIVE.

160 Cf. Section 5.1.2.

For a start, we extract the directionality markers from Table 5.14 and add the zero-marker to which we pay more attention below.[161] At this point, it suffices to note that $\varnothing_{\text{dynamic}}$ has to be distinguished from $\varnothing_{\text{static}}$. In Figure 5.34, reference is made to both $\varnothing_{\text{dynamic}}$ and $\varnothing_{\text{static}}$ with the former being addressed in the immediately ensuing discussion whereas the discussion of $\varnothing_{\text{static}}$ is relegated to a later section.[162] The third variety of zero – $\varnothing_{\text{PROLATIVE}}$ – will be discussed in connection with Table 5.16 below. The functions fulfilled by the five directionality markers are revealed in Figure 5.34. The terminology employed in Figure 5.34 differs slightly from Hagège's (2010).[163] The boxes contain identical forms with different functions.

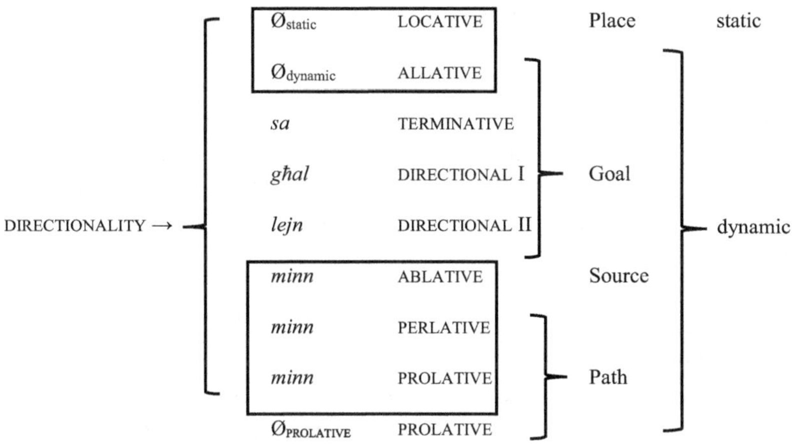

Figure 5.34: Differentiation of functions in the domain of directionality.

There are four ways of expressing Goal and three to express Path as opposed to a single option for the expression of Source. Motion towards a Ground allows for fine-grained distinctions whereas motion from a Ground is not susceptible to further formal differentiations on the side of the directionality marker. This asymmetry is striking but expected because of ample cross-linguistic evidence thereof (Papafragou 2010).

As explained above,[164] we assume that, for the expression of spatio-temporal meanings, Maltese generally makes use of a constructional schema [[DIRECTIONAL-

161 Cf. Section 5.2.1.2.
162 Cf. Section 5.2.1.3.
163 Henceforth, we use ALLATIVE and ABLATIVE as terms that refer to motion towards and motion from a Ground, respectively. The term LOCATIVE is used for general location at rest.
164 Cf. Section 4.1.3.

ITY CONFIGURATION] COMPLEMENT]$_{PP}$ with two separate slots for directionality and configuration (always in this linear order) either of which can be empty (or: filled by zero, cf. below). This means several things, namely that

(a) Ø is used both for directionality and configuration,
(b) Ø is used both for static (Place) and dynamic (Goal, Path) spatial relations,
(c) a directionality marker on its own is sufficient to realise the above schema,
(d) a configuration marker on its own is sufficient to realise the above schema,
(e) combinations of Ø$_{DIRECT}$ and Ø$_{CONFIG}$ are licit so that spatial adverbials are grammatical also without overt Π,
(f) zero as well as (Π41) *minn* 'from' are multifunctional with the latter serving to encode spatial relations in motion events which are not Goal-directed.

In accordance with (a), (c), and (e), the directionality markers presented in Figure 5.34 combine with Ø$_{CONFIG}$, i.e., spatial constructions may involve only a directional marker as shown in (5.130). The markers of directionality and configuration are indexed. The Ø-symbol re-emerges for practical reasons to be contextualised in the subsequent section. Note that Ø$_{PROLATIVE}$ requires the presence of a Π$_{CONFIG}$. It is therefore not featured in (5.220).

(5.220) Spatial constructions with only a directionality marker

(a) [Korpus Malti 3.0; news211105]

t-tabib *Paul Chetcuti Caruana* *kien* *mar* ***[Ø**$_{DIRECT}$* ***Ø**$_{CONFIG}$*
DEF-doctor Paul Chetcuti Caruana be go **[Ø** **Ø**
l-uffiċju *tiegħu]*$_X$ *l-bierah* *filgħaxija*
DEF-office of:3SG.M] DEF-yesterday in_the_evening
'[. . .] the doctor Paul Chetcuti Caruana had gone [**to** his office] they previous evening.'

(b) [Korpus Malti 3.0; news7645]

L-inċident *ġara* *wara* *li* *t-tifel* *ried* *imur*
DEF-incident happen after that DEF-boy want 3SG.M.IPFV:go
***[sa**$_{DIRECT}$* ***Ø**$_{CONFIG}$* *l-kamra* *l-baxxa]*$_{PP}$ *imma* *l-għalliema* *ma*
[till **Ø** DEF-room DEF-low:F] but DEF-teacher:F NEG
ħallietux.
let:3SG.F.PFV:NEG
'The incident happened after the boy wanted to go [**to** the toilets] but the teacher did not let him go.'

(c) [Korpus Malti 3.0; news504]

dakinhar *qalli* *li* *ma* *kellux* *aptit*
on_the_day say:IO:1SG that NEG have.PFV:3SG.M:NEG appetite

imur *[għal*$_{\text{DIRECT}}$ *Ø*$_{\text{CONFIG}}$ *l-għassa]*$_{\text{PP}}$
3SG.M.IPFV:go [**for** **Ø** DEF-police_station]
'[. . .] on that day he told me that he did not feel like going [**to** the police station].'

(d) [Korpus Malti 3.0; literature2]
Is-Surġent Mangion *kien* *qed* *isuq* *[lejn*$_{\text{DIRECT}}$ *Ø*$_{\text{CONFIG}}$
DEF-sergeant Mangion be PROG 3SG.M.IPFV:drive [**towards** **Ø**
id-dar]$_{\text{PP}}$
DEF-house]
'Sergeant Mangion was driving [home**wards**].'

(e) [Korpus Malti 3.0; news14387]
x-xufier [. . .] *kien* *għadu* *kif* *telaq* *[minn*$_{\text{DIRECT}}$ *Ø*$_{\text{CONFIG}}$
DEF-driver be still:3SG.M as leave [**from** **Ø**
bus stop]$_{\text{PP}}$ *meta* *raġel* *beda* *jipprova* *jwaqqaf*
bus_stop] when man begin 3SG.M.IPFV:try 3SG.M.IPFV:CAUS.stop
il-karozza *tal-linja* *biex* *ikun* *jista'*
DEF-car of:DEF-line to 3SG.M.IPFV:be.FUT 3SG.M.IPFV:can
jirkeb
3SG.M.IPFV:ride
'[. . .] the driver had just left [the bus stop] when a man started trying to stop the bus to ride on it.'

(f) [Korpus Malti 3.0; literature20]
Għajnejh *flew* *il-karozzi* *għaddejjin*
eye:PL:3SG.M inspect_closely:3PL.PFV DEF-car:PL pass.PART:PL
[mit-triq *quddiem* *il-lukanda* *Preluna]*$_{\text{PP}}$ *fid-dlam*
[**from**:DEF-street in_front_of DEF-hotel Preluna] in:DEF-darkness
'His eyes flashed at the cars passing [**through** the road in front of the Hotel Preluna] in the dark.'

(g) [Korpus Malti 3.0; news119842]
Kull *min* *inzerta* *għaddej* *[mix-Xatt]*$_{\text{PP}}$ *kellu*
each who happen pass.PART [**from**:DEF-shore] have.PFV:3SG.M
joqgħod *attent* *li* *ma* *jitkaxkarx* *mir-riħ*
3SG.M.IPFV:stay careful that NEG 3SG.M.IPFV:REFL:drag from:DEF-wind
qawwi
strong
'Anyone who happened to stroll [**along** the shore] had to be careful not to be dragged by the strong wind.'

In cases like those in (5.220), i.e., in constructions which involve no phonologically realised configuration marker, Ø$_{\text{CONFIG}}$ does not specify any particular part of the

Ground. The Grounds in (a)–(d) are certainly meant to be entered into by the Figure whereas in (5.220e), the Figure was temporarily located at the Ground before the motion event started. In (5.220b–d), the Figure never reaches the Ground because no permission was granted or the agent had no intention to go there or something happened to the Figure on its way to the Ground. In the Path constructions (= (f)-sentence and (g)-sentence), the Figure passes through the Ground and is therefore physically inside it (= PERLATIVE) or moves along the Ground and thus remains outside the Ground (= PROLATIVE).

What is important is not so much the fact that the directionality markers need no configuration marker to combine with but the general availability of the directionality marker to combine with configuration markers other than $Ø_{\text{CONFIG}}$. In point of fact, there are only three Πs with configurational meaning (components) which are never attested in combination with phonologically realised directionality markers, namely the special case (Π38) *mingħand* 'from s.o.' as well as the Italo-Romance (Π44) *oltre* 'beyond' and the temporal (Π59) *waqt* 'at the time of'. The majority of 23 configurational Πs however combine with directionality markers in the Korpus Malti 3.0. In Table 5.15, the combinations of these configuration markers with directionality markers are identified. Grey shading marks unattested combinations. The colour green is used for configurational Πs that combine with all four material directionality markers. The exclamation mark together with yellow shading highlights the position where the special case (Π38) *mingħand* 'from s.o.' could be placed provided it is interpreted as compositional. The cells in the bottom line tell us how many configuration markers there are and how many of them combine with a given directionality marker. Sentential examples of the attested combinations will be presented and evaluated below.[165]

No phonologically realised directionality marker is attested with all 23 configuration markers. (Π41) *minn* 'from' ranks at the top with 21 combinations – or 22 if (Π38) *mingħand* 'from s.o.' is added to the list. (Π21) *għal* 'for', too displays a similarly high degree of combinability whereas (Π51) *sa* 'till' co-occurs only with a smaller number of configuration markers and (Π28) *lejn* 'towards' seems to be restricted to a very limited set of combinations. There are only two combinations which are unique to a given directionality marker, namely *minn faċċata*, and *sa fi*. Five configuration markers are combinable with all four directionality markers. These omni-compatible configuration markers are: (Π3) *barra* 'outside', (Π17) *fuq* 'on', (Π49) *quddiem* 'in front of', (Π56) *taħt* 'under', and (Π60) *wara* 'after'.

165 Cf. Section 5.2.1.4.3.

Table 5.15: Combinations of directionality markers and configuration markers in the Korpus Malti 3.0.

configuration	directionality			
	(Π41) *minn* 'from'	**(Π21) *għal* 'for'**	**(Π51) *sa* 'till'**	**(Π28) *lejn* 'towards'**
(Π3) *barra* 'outside'	yes	yes	yes	yes
(Π4) *bejn* 'between'	yes	yes	yes	no
(Π8) *biswit* 'facing'	yes	yes	no	no
(Π12) *faċċata* 'opposite'	yes	no	no	no
(Π14) *fejn* 'near'	yes	yes	yes	no
(Π15) *fi* 'in'	no	no	yes	no
(Π16) *fost* 'amongst'	yes	yes	no	no
(Π17) *fuq* 'on'	yes	yes	yes	yes
(Π18) *ġewwa* 'inside'	yes	yes	no	no
(Π19) *ġo* 'in'	yes	yes	no	no
(Π22) *għand* 'at s.o.'s place'	!	yes	yes	no
(Π23) *ħdejn* 'beside'	yes	yes	yes	no
(Π30) *ma'* 'with'	yes	yes	yes	no
(Π31) *madwar* 'around'	yes	yes	yes	no
(Π32) *maġenb* 'close to'	yes	yes	yes	no
(Π33) *matul* 'during'	yes	yes	no	no
(Π46) *qabel* 'before'	yes	yes	no	yes
(Π47) *qalb* 'amidst'	yes	yes	no	no
(Π48) *qrib* 'near'	yes	yes	yes	no
(Π49) *quddiem* 'in front of'	yes	yes	yes	yes
(Π56) *taħt* 'under'	yes	yes	yes	yes
(Π58) *viċin* 'near'	yes	yes	yes	no
(Π60) *wara* 'after'	yes	yes	yes	yes
23	21 (22)	21	15	6

In Table 5.15, it is not difficult to recognise the implicational pattern presented in Figure 5.35.

The relatively small number of configurational Πs which are attested in combination with the DIRECTIONAL (Π28) *lejn* 'towards' also allow for combinations with (Π21) *għal* 'for' and (Π41) *minn* 'from'. Since (Π41) *minn* 'from' is multifunctional, it is interesting to see which of its three functions – ABLATIVE, PERLATIVE, PROLATIVE – are attested with which Π_{CONFIG}. We extract the 21 combinations involving (Π41) *minn* 'from' from Table 5.15 to disentangle the three possible functions of this Π in Table 5.16. Except for the orange shading to be explained below the same conventions as in the case of Table 5.15 also apply to Table 5.16.

$$\textit{lejn}_{\text{DIRECTIONAL2}} \supset \left\{ \begin{array}{c} \textit{għal}_{\text{DIRECTIONAL1}} \\ \\ \textit{minn} \end{array} \right\} / \_\_\_\ \Pi_{\text{CONFIG}}$$

Figure 5.35: Implicational pattern for directionality markers.

Table 5.16: Combinations of $minn_{ABLATIVE}$, $minn_{PROLATIVE}$, and $minn_{PERLATIVE}$ and configuration markers in the Korpus Malti 3.0.

configuration	Function of (Π41) *minn* 'from'		
	ABLATIVE	PROLATIVE	PERLATIVE
(Π3) *barra* 'outside'	yes	yes	no
(Π4) *bejn* 'between'	yes	no	yes
(Π8) *biswit* 'facing'	yes	yes	no
(Π12) *faċċata* 'opposite'	yes	yes	no
(Π14) *fejn* 'near'	yes	yes	no
(Π16) *fost* 'amongst'	yes	no	yes
(Π17) *fuq* 'on'	yes	no	no
(Π18) *ġewwa* 'inside'	yes	no	yes
(Π19) *ġo* 'in'	yes	no	yes
(Π22) *għand* 'at s.o.'s place'	!	no	no
(Π23) *ħdejn* 'beside'	yes	yes	no
(Π30) *ma'* 'with'	yes	yes	no
(Π31) *madwar* 'around'	yes	yes	no
(Π32) *maġenb* 'close to'	yes	yes	no
(Π33) *matul* 'during'	no	no	yes
(Π46) *qabel* 'before'	yes	no	no
(Π47) *qalb* 'amidst'	yes	no	yes
(Π48) *qrib* 'near'	yes	yes	no
(Π49) *quddiem* 'in front of'	yes	yes	no
(Π56) *taħt* 'under'	yes	yes	no
(Π58) *viċin* 'near'	yes	yes	no
(Π60) *wara* 'after'	yes	yes	no
22	20 (21)	13	6

The three functions of (Π41) *minn* 'from' are combinable with configurational Πs to different degrees. The highest number of patterns goes to the credit of $minn_{ABLATIVE}$ which is involved in nineteen out of twenty-one patterns of combination of (Π41) *minn* 'from'. There is a gap for *minn*$_{ABLATIVE}$ filled by *minn*$_{PERLATIVE}$. Thirteen combinations are registered for *minn*$_{PROLATIVE}$ with none being unique to this function of (Π41) *minn* 'from'. Only six patterns have been identified for *minn*$_{PERLATIVE}$ with one pattern being exclusively attested with this function of (Π41) *minn* 'from'. Interestingly, a given configurational Π can combine either with *minn*$_{PROLATIVE}$ or *minn*$_{PERLATIVE}$ but never with both. The two functions are thus mutually exclusive whereas no restriction of this kind applies in the case of *minn*$_{ABLATIVE}$. A configurational Π is compatible with only one of two possible Path categories. This incompatibility of the two Path categories is put into a simple formula in Figure 5.36.

$$Path_x \quad \rightarrow \quad \neg Path_y$$

Figure 5.36: Mutual exclusion of Path categories.

In Table 5.16, the middle column hosts eleven of twelve filled cells that are highlighted in orange. This colour indicates that the use of (Π41) *minn* 'from' is optional. The bare configuration marker alone suffices when the PROLATIVE is to be expressed. In contrast, the use of *minn*$_{ABLATIVE}$ and *minn*$_{PERLATIVE}$ is mandatory or so it seems. This is where the third zero (= $Ø_{PROLATIVE}$) enters the stage. As mentioned above, $Ø_{PROLATIVE}$ differs from the other zeroes insofar as it always calls for the presence of a configurational Π. The phenomenon has been illustrated sufficiently already.[166] Further examples are therefore unnecessary in this section. Nevertheless, we will come back to this issue below.[167]

We return to Table 5.15. In fourteen out of fifteen cases, a configurational Π that combines with the TERMINATIVE (Π51) *sa* 'till' also combines with the DIRECTIONAL$_1$ (Π21) *għal* 'for'. This is a very strong preference (cf. Figure 5.37 where the symbol ⊃ has to be read as indicator of a preference).

sa$_{TERMINATIVE}$ ⊃ *għal*$_{DIRECTIONAL1}$

Figure 5.37: Preferred partner of the TERMINATIVE.

Another pattern that emerges from Table 5.15 requires that if either (Π21) *għal* 'for' or (Π41) *minn* 'from' is excluded from combining with a given $Π_{CONFIG}$ at least one other directionality marker is also barred from combining with this $Π_{CONFIG}$.

It is most probably still premature to generalise over the above findings because the gaps in 5.15 may be caused by the insufficient coverage of phenomena in the Korpus Malti 3.0. We speculate that at least some of the supposedly illicit combinations of directionality and configuration markers might turn out to be perhaps

166 Cf. Section 5.1.2.2.20.3.
167 Cf. Section 5.2.1.4.4.1.2.

infrequent but nevertheless grammatically acceptable, nevertheless. We register this possibility as one of the many tasks to be addressed in the future.

5.2.1.3 Zeroes and what they mean

Hitherto, we have looked at phonologically realised directionality markers. It is now time to face the questions that arise in connection to $Ø_{\text{dynamic}}$ and $Ø_{\text{static}}$ or $Ø_{\text{DIRECT}}$ and $Ø_{\text{CONFIG}}$ and further distinctions that can be made in connection with zero-marking. These questions are intricate and thus require considerable space to be adequately answered.

Before we go into the details, it is indispensable to clarify what is meant when we refer to zero(es) and employ the symbol Ø with or without index. The *fi*-Ø alternation was already in the focus of attention.[168] This phenomenon involves an excellent example of what Stolz/Levkovych (2019) call AOME, i.e., the absence of material exponence. AOME applies if a slot in a constructional frame that could host a phonologically realised marker remains empty. With reference to the *fi*-Ø alternation, this means that the position occupied by (Π15) *fi* 'in' in a given PP can alternatively be left unfilled. Put differently, no material is there to fill the slot. In our parlance, zero is a terminological shortcut making it possible for us to refer to AOME without having to resort to lengthy explanations. Similarly, the Ø-symbol only functions as a handy tool to mark cases of AOME in the examples. Neither the term zero nor the symbol Ø presupposes the existence of an invisible structural entity. The indexes added to Ø seem to suggest that different zeroes can be distinguished and that these zeroes are equipped with different meanings and functions. However, this interpretation is not ours. What we intend the indexes to make explicit is the function/meaning of phonologically realised Πs that are attested in the same slot. We assume that Ø is not a "silent" Π. Thus, Ø cannot head a PP which, in turn means, that in the absence of a Π there is no PP, in the first place. Accordingly, Ø is no member of BLOMP 2.0. Yet, AOME requires further scrutiny in order to describe the grammar of Maltese Πs properly. In what way Ø is in a paradigmatic relation with Πs is the topic for a dedicated follow-up study that has to situate the phenomena in the wider framework of the research on the linguistic concept of zero(es).

On the basis of what the previous paragraphs have revealed, it suggests itself to ask how many and which configurational Πs allow for dynamic directionality being zero-coded. Is $Ø_{\text{dynamic}}$ an option for more or less combinations when compared to the phonologically realised directionality markers? Table 5.17 contains examples of

168 Cf. Section 4.3.3.2.

combinations of $Ø_{dynamic}$ with twenty different configuration markers. All examples stem from the Korpus Malti 3.0.[169]

Table 5.17: $Ø_{dynamic}$ in combination with $Π_{CONFIG}$.

#	source	example / translation	Π
1	parl98	*Meta delegazzjoni ministerjali marret [**Ø**$_{dynamic}$ **barra** minn Malta]$_{PP}$, inżammew minuti dwar ma' min iltaqgħet?* 'When a ministerial delegation went [abroad], were minutes kept about who they met?'	(Π3) *barra* 'outside'
2	news99441	*Kif wasal [Ø$_{dynamic}$ **bejn** il-fabbrika Baxter u l- istamperija Union Print fi Triq Valletta]$_{PP}$, Cassar waqaf mal-ġenb u talab għall-għajnuna medika.* 'As he arrived [**between** the Baxter factory and the Union Print printing house in Triq Valletta], Cassar stopped at the side and asked for medical help.'	(Π4) *bejn* 'between'
3	parl8030	*morna [Ø$_{dynamic}$ **biswit** il-lokalità ta' Ħal Għaxaq]$_{PP}$* '[. . .] we went [**next to** the locality of Għaxaq].'	(Π8) *biswit* 'facing'
4	news167434	*Kif wasal [**Ø**$_{dynamic}$ **faċċata ta**d-daħla tal-Kumpless Sportiv tal-Marsa]$_{PP}$, Borg tilef il-kontroll tal-karozza* 'When he arrived [**in front of** the entrance of the Marsa Sports Complex], Borg lost control of the car [. . .].'	(Π12) *faċċata* 'opposite'
5	news82618	*Ħriġt 'il barra bħal leħħa ta' berqa, mort [**Ø**$_{dynamic}$ **fejn** sħabi l-ġurnalisti]$_{PP}$* 'I went out like a flash of lightning, I went [**to** my fellow journalists]'	(Π14) *fejn* 'near'
6	parl441	*l-ballun minflok baqa' fil-ground mar [**Ø**$_{dynamic}$ **fost** in-nies]$_{PP}$* '[. . .] the ball instead of staying in the ground and went [**among** the people].'	(Π16) *fost* 'amongst'
7	news159012	*saħansitra marret [**Ø**$_{dynamic}$ **fuq** għolja fejn ħasbet biex ittemm ħajjitha]$_{PP}$* '[. . .] she even went [**to** a hill where she thought of ending her life].'	(Π17) *fuq* 'on'
8	parl408	*Jien ma kelli l-ebda skruplu ta' kuxjenza li mmur [**Ø**$_{dynamic}$ **ġewwa** Spanja]$_{PP}$* 'I had no qualms about going [**to** Spain]'	(Π18) *ġewwa* 'inside'
9	news75089	*Se jkolli mmur [**Ø**$_{dynamic}$ **ġo** dar tal-anzjani]$_{PP}$?* 'Will I have to go [**into** a nursing home]?'	(Π19) *ġo* 'in'
10	news8027	*qassis Ġiżwita mar jiġri [**Ø**$_{dynamic}$ **għand** il-Pulizija]$_{PP}$ biex isejjaħ għall-għajnuna* '[. . .] a Jesuit priest ran [**to** the police] to call for help.'	(Π22) *għand* 'at s.o.'s place'

169 For the combinations involving $Ø_{PROLATIVE}$, we refer the reader to Section 5.1.2.2.20.3.

Table 5.17 (continued)

#	source	example / translation	Π
11	news15859	*X'ħin naslu [Ø$_{dynamic}$* ***ħdejn*** *Pjazza Regina]$_{PP}$ nilmaħ stakkament ta' Pulizija.* 'When we arrive [**near** Pjazza Regina] I see a detachment of Police.'	(Π23) *ħdejn* 'beside'
12	news148006	*Hu telaq minn Tripli fit-22 ta' Marzu u wasal [Ø$_{dynamic}$* ***mal-*** *fruntiera tat-Tuneżija]$_{PP}$, wara li għamel aktar minn ġurnata mixi.* 'He left Tripoli on March 22 and arrived [**at** the Tunisian border], after walking for more than a day.'	(Π30) *ma'* 'with'
13	*news123309*	*minnufih niżel minn fuq il-palk u mexa minn qalb id-delegati sakemm wasal [Ø$_{dynamic}$* ***maġenb*** *Abela]$_{PP}$* '[. . .] he immediately got off the stage and walked through the delegates until he got [**to** Abela's **side**] [. . .].'	(Π32) *maġenb* 'close to'
14	parl839	*Malta hija gżira u allura ma tistax taqbad karozza u tmur [Ø$_{dynamic}$* ***oltre*** *l-fruntiera]$_{PP}$ u tasal f'pajjiż ieħor.* 'Malta is an island and therefore you cannot take a car and go [**over** the border] and arrive in another country.'	(Π44) *oltre* 'beyond'
15	parl11983	*L-aħħar darba li kont hemm jien inżilt fil-pjazza u mort [Ø$_{dynamic}$* ***qalb*** *il-protestanti]$_{PP}$* 'The last time I was there I went down to the square and went [**amidst** the protesters].'	(Π47) *qalb* 'amidst'
16	news124234	*Il-patrol boat dawret ir-rotta u marret [Ø$_{dynamic}$* ***qrib*** *id-dgħajsa tal-klandestini]$_{PP}$* 'The patrol boat turned the course and went [**close to** the boat of the clandestines]'	(Π48) *qrib* 'near'
17	news223656	*l-akkużata marret [Ø$_{dynamic}$* ***quddiem*** *l-Ambaxxata Russa]$_{PP}$ u qalbet il-mutur tiegħu mal-art* '[. . .] the accused went [**in front of** the Russian Embassy] and overturned his motorcycle on the ground.'	(Π49) *quddiem* 'in front of'
18	news79496	*Mort [Ø$_{dynamic}$* ***taħt*** *it-tieqa tiegħu]$_{PP}$ u qalli xi ħaġa imma ma fhimtx x'kien.* 'I went [**under** his window] and he said something to me but I didn't understand what it was'	(Π56) *taħt* 'under'
19	news142073	*Sussegwentament mar [Ø$_{dynamic}$* ***viċin*** *it-tfajla]$_{PP}$ biex jipprova jikkonvinċiha tixtri mingħandu d-droga kokaina.* 'Next, he went [**near** the girl] to try to convince her to buy the drug cocaine from him'	(Π58) *viċin* 'near'
20	parl122	*l-Ministru Galea mhux se joqgħod imur [Ø$_{dynamic}$* ***wara*** *l-bibien tad-discos]$_{PP}$ jara min ikun dieħel* '[. . .] Minister Galea will not be going [**to** the doors of the discos] to see who is entering.'	(Π60) *wara* 'after'

Two observations are necessary with regards to restrictions on the combinatory potential of $Ø_{dynamic}$ with $Π_{CONFIG}$. Two kinds of combinations are ruled out. First of all, if the Π_{CONFIG} is used to express or introduce a temporal concept, the use of $\varnothing_{\text{dynamic}}$ is blocked. For Πs which like (Π60) *wara* 'after' in Table 5.17 #20 can be used either with spatial or temporal meaning, combinations with $\varnothing_{\text{dynamic}}$ are attested only with a spatial reading. Accordingly, (Π46) *qabel* 'before' and (Π59) *waqt* 'at the time of' are missing from Table 5.17. Their absence from the list can be explained as follows. If a *qabel*-PP or a *waqt*-PP co-occur with a motion verb they usually do not indicate the Goal of the motion event but situate it in time. The Goal can be a different participant in the spatial situation.

As to (Π33) *matul* 'during' – a Π with both temporal and spatial meanings, there is no compatibility with $\varnothing_{\text{dynamic}}$ because this Π_{CONFIG} characterises its complement as being extended on one dimension so movement in relation to the Ground is associated with the SECUTIVE as argued above.[170] The dimensional extension in space disqualifies the Ground as Goal for a motion event of the general-location type. A specification is needed, meaning: a phonologically realised directionality marker is called for. The same holds for (Π31) *madwar* 'around' as Π_{CONFIG}. Probably, (Π30) *ma'* 'with' steps in where neither (Π33) *matul* 'during' nor (Π31) *madwar* 'around' combine with $\varnothing_{\text{dynamic}}$. A schematic rendering of these observations is given in Figure 5.38.

$$\Pi_{\text{CONFIG}} \quad \rightarrow \left\{ \begin{array}{c} [\text{-temporal}] \\ \\ [\text{-extended}] \end{array} \right\} \; / \; \varnothing_{\text{dynamic}} \_\_\_$$

Figure 5.38: Exclusion of configurational Πs from combinations with $\varnothing_{\text{dynamic}}$.

The second observation to be made with regards to the combinations of $\varnothing_{\text{dynamic}}$ and $\Pi\text{s}_{\text{CONFIG}}$ has to do with the INESSIVE and the ILLATIVE discussed above.[171] What we are going to say about (Π15) *fi* 'in' also applies to (Π18) *ġewwa* 'inside' and (Π19) *ġo* 'in' (cf. Table 5.17 #8–9). We argue that (Π15) *fi* 'in' should be added to the list of configurational Πs that combine with the dynamic zero. Accordingly, Table 5.17 virtually hosts 21 types. In the domain of general location, Maltese displays Place-Goal syncretism. Accordingly, (Π15) *fi* 'in' may be translated as 'into' in the appropriate context. However, it is not entirely compelling to assume a polysemous Π. To make our point, we present (5.221) and subsequently provide an analysis to support our hypothesis.

170 Cf. Section 5.1.2.2.20.1.4 (ii).

171 In Sections 5.1.2.1.1.1 and 5.1.2.2.1, respectively.

(5.221) [Korpus Malti 3.0; culture1740]

l-aktar	*faċli*	*hu*	*li*	*tmur*	*[Ø$_{dynamic}$*	***fi***	*klinika*
DEF-more	easy	3SG.M	that	2SG.IPFV:go	[Ø	**in**	clinic
barra	*minn*	*Malta]$_{PP}$*					
outside	from	Malta]					

'[. . .] the easiest for you is to go [**to** a clinic outside Malta].'

The motion verb *mar* 'go' defines the spatial situation as dynamic. Accordingly, the ILLATIVE reading of (Π15) *fi* 'in' is activated. This is a possible account of what is the case in (5.221). An alternative interpretation of the same spatial situation, in lieu of postulating the polysemy of (Π15) *fi* 'in', assumes that (Π15) *fi* 'in' exclusively functions as configuration marker and is thus devoid of meaning components of the directionality kind. Put differently, the contribution of (Π15) *fi* 'in' to the constructional meaning is limited to indicating the part of the Ground that is involved in the motion event. In this case, it is the Ground's interior. Directionality as such need not be made explicit because of the main verb's semantics. The symbol Ø$_{dynamic}$ is but a placeholder whose position could be taken over by phonologically realised directionality markers to replace general location with specified location. In analogy to (Π15) *fi* 'in', (Π18) *ġewwa* 'inside' and (Π19) *ġo* 'in', too can be classified as purely configurational.

After elaborating on the dynamic variant of the directional zeroes, we now turn our attention to Ø$_{static}$ and its properties. Table 5.18 is organised according to the same principles as Table 5.17. This time, however, the number of licit combinations with configurational Πs rises to 24 patterns. The bulk of the examples involve forms of the ersatz-copula *qiegħed* 'staying' which is commonly used when being at rest at a given Ground is to be expressed in the imperfective. Other examples involve a personal pronoun in copula function or the perfective form *kien* 'was' so that it is guaranteed that the spatial situation is indeed static.

Table 5.18: Ø$_{static}$ in combination with Π$_{CONFIG}$.

#	source	example / translation	Π
1	news161470	*l-imputat reġa' rċieva telefonata mingħand dan Freddie li kien qiegħed [**Ø$_{static}$** **barra** r-restorant]$_{PP}$* '[. . .] the defendant received a call again from this Freddie who was [**outside** the restaurant] [. . .].'	(Π3) *barra* 'outside'
2	news103613	*Ġeografikament, iż-Żejtun qiegħed [**Ø$_{static}$** **bejn** tlett portijiet]$_{PP}$: dawk tal-Marsa, Marsaxlokk u Wied il-Għajn* 'Geographically, Żejtun is [**between** three ports]: those of Marsa, Marsaxlokk and Wied il-Għajn.'	(Π4) *bejn* 'between'

Table 5.18 (continued)

#	source	example / translation	П
3	parl2483	*rajt stampat ritratt ta' pulizija b'revolver f'idejh li qiegħed [$Ø_{static}$* ***biswit*** *villa ġewwa Ħ'Attard]$_{PP}$* '[. . .] I saw a printed photo of a policeman with a revolver in his hand [**opposite** a villa in Attard].'	(П8) *biswit* 'facing'
4	parl294	*F'dan ir-ritratt hemm ukoll Nerik Mizzi li qiegħed [$Ø_{static}$* ***faċċata ta'*** *John Cole]$_{PP}$ jisma' bil-paċenzja kollha l-introduzzjoni ta' dan il-Bill* 'In this photo there is also Nerik Mizzi who is (standing) [**in front of** John Cole] patiently listening to the introduction of this Bill.'	(П12) *faċċata* 'opposite'
5	literature40	*Ir-raġel qiegħed [$Ø_{static}$* ***fejn*** *it-tifel]$_{PP}$ u qed jgħidlu li dalwaqt se mmorru għall-ġelat.* 'The man is [**near** the boy] and is telling him that we will go for ice cream soon.'	(П14) *fejn* 'near'
6	news61328	*Missier it-tarbija qiegħed [$Ø_{static}$* ***fost*** *il-folla]$_{PP}$* 'The father of the girl is [**amongst** the crowd].'	(П16) *fost* 'amongst'
7	literature29	*Sthajlu li kien qiegħed [$Ø_{static}$* ***fuq*** *il-bejt tal-knisja]$_{PP}$* 'He thought he was [**on** the roof of the church].'	(П17) *fuq* 'on'
8	parl11516	*Riċentement kont qiegħed [$Ø_{static}$* ***ġewwa*** *Brussell]$_{PP}$ fil-bini tal-Kummissjoni Ewropea* 'I was recently [**in** Brussels] at the European Commission building.'	(П18) *ġewwa* 'inside'
9	news80884	*Kont qiegħed [$Ø_{static}$* ***ġo*** *spiżerija tal-għażla tiegħek]$_{PP}$* 'You were [**in** a pharmacy of your choice].'	(П19) *ġo* 'in'
10	academic30	*Bħalissa però qiegħed [$Ø_{static}$* ***għand*** *ommi]$_{PP}$, fejn m'għandix ktieb wieħed fuq dan is-suġġett* 'At the moment, however, I am [**at** my mother'**s**], where I do not have a single book on this subject.'	(П22) *għand* 'at s.o.'s place'
11	literature19	*Albert xtaq li dak il-ħin kien qiegħed [$Ø_{static}$* ***ħdejn*** *l-għelieqi ta' Kola]$_{PP}$* 'Albert wished he was [near the fields of Kola] at that time.'	(П23) *ħdejn* 'beside'
12	news132692	*Il-vittma rrakkonta lill-Pulizija kif waqt li kien qiegħed [$Ø_{static}$* ***ma****l-bar]$_{PP}$, ħdejh kien hemm erbat irġiel oħrajn* 'The victim told the Police how while he was [**at** the bar], there were four other men next to him.'	(П30) *ma'* 'with'
13	parl10228	*Anzi jekk se mmur f'restaurant biex niekol ma rridx li jkun hemm l-istorbju għax inkun irrid nikkomunika b'mod tajjeb ma' min qiegħed [$Ø_{static}$* ***madwar*** *il-mejda tiegħi]$_{PP}$!* 'On the contrary, if I'm going to go to a restaurant to eat I don't want there to be noise because I want to communicate in a good way with those [**around** my table]!'	(П31) *madwar* 'around'

Table 5.18 (continued)

#	source	example / translation	П
14	parl444	*t-triq qiegħda [Ø$_{static}$ **matul** il-kosta]$_{PP}$* '[. . .] the road is [**along** the coast].'	(П33) *matul* 'during'
15	parl10928	*Mela ejjew ngħidu li Edwin jirrifjutaha, u jien, li qiegħed [Ø$_{static}$ **maġenb**u]$_{PP}$, niġi second best u jien naċċettaha.* 'So, let's say that Edwin rejects it, and I, who am [**next to** him], come second best and I accept it.'	(П32) *maġenb* 'close to'
16	parl235	*il-miżuri li se jittieħdu huma [Ø$_{static}$ **oltre** l-iskop tal-liġi li qed nitkellmu dwarha hawn]$_{PP}$* '[. . .] the measures that will be taken are [**beyond** the scope of the law we are talking about here].'	(П44) *oltre* 'beyond'
17	parl8723	*Huwa veru li pajjiżna qiegħed [Ø$_{static}$ **qabel** l-Italja]$_{PP}$* 'It is true that our country is [**ahead of** Italy].'	(П46) *qabel* 'before'
18	news81684	*Kien qiegħed [Ø$_{static}$ **qalb** siġar]$_{PP}$ u [Ø$_{static}$ **qalb** ħitan għoljin]$_{PP}$* 'He was (standing) [**among** trees] and [(**among**) high walls].'	(П47) *qalb* 'amidst'
19	news133327	*Il-container qiegħed [Ø$_{static}$ **qrib** il-Knisja]$_{PP}$ fejn hemm ukoll il-palk* 'The container is [**close to** the Church] where there is also the stage.'	(П48) *qrib* 'near'
20	literature35	*Kien qiegħed [Ø$_{static}$ **quddiem** it-tieqa l-kbira]$_{PP}$ b'ħafna xemx dieħla ġo fiha* 'He was standing [**in front of** the big window] with lots of sun coming in.'	(П49) *quddiem* 'in front of'
21	parl1374	*l-area fejn hemm dak il-bini fi stat dilapidat – li qiegħed [Ø$_{static}$ **taħt** il-Forti Sant' Anġlu]$_{PP}$ u li kellu jkun parti mill-iżvilupp – f'daqqa waħda inqatgħet barra* '[. . .] the area where that dilapidated building – which is [**under** Fort Sant' Anġlu] and which was supposed to be part of the development – was suddenly cut off.'	(П56) *taħt* 'under'
22	news178166	*Il-villaġġ ta' Namache qiegħed [Ø$_{static}$ **viċin** il-kamp bażi tal-muntanja Everest]$_{PP}$* 'The village of Namache is [close to the base camp of Mount Everest].'	(П58) *viċin* 'near'
23	news21034	*it-tielet darba kien [Ø$_{static}$ **waqt** tieġ]$_{PP}$* '[. . .] the third time was [**during** a wedding] [. . .].'	(П59) *waqt* 'at the time of'
24	parl8825	*Dan għallimni li meta inti tkun qiegħed [Ø$_{static}$ **wara** mikrofonu]$_{PP}$ u [Ø$_{static}$ **wara** sett ta' cameras]$_{PP}$ tkun qiegħed hemmhekk biex twassal messaġġ lin-nies.* 'This taught me that when you are [**behind** a microphone] and [(**behind**) a set of cameras] you are there to deliver a message to people.'	(П60) *wara* 'after'

The bigger turnout for combinations of $Ø_{static}$ and Π_{CONFIG} in contrast to those of $Ø_{dynamic}$ and Π_{CONFIG} finds its explanation in the possibility of the static zero co-occurring with temporal Πs like (Π59) *waqt* 'at the time of' in Table 5.18 #23. Combinations of this kind are licit because the period of time covered by the complement noun serves as metaphorical container for the Figure. Moreover, $Ø_{static}$ is also compatible with Πs which introduce a Ground that is extended on a given dimension (be it spatially or in terms of time) as shown e.g. by (Π31) *madwar* 'around' in Table 5.18 #13. The restriction presented in Figure 5.38 thus does not hold for $Ø_{static}$. Sentence #13 in Table 5.18 is of further interest. The combination of $Ø_{static}$ and (Π31) *madwar* 'around' is licit because the Figure is distributed in the sense that it consists of a group of people whose members are placed at (= around) the table. With reference to the absence of (Π31) *madwar* 'around' from Table 5.17, we emphasise that we have not found any example of a distributed Figure moving to a Ground so that the different members of the group wind up in a position encircling the Ground. To settle this issue, native speaker judgements are required. Collecting these judgements is a task for a follow-up study.

For the sake of comprehensiveness, we address once more the problems raised by (Π15) *fi* 'in'. As argued in connection with (5.221), we assume that (Π15) *fi* 'in' is purely configurational also in static spatial situations like that depicted in (5.222). The conclusions we draw with regards to (Π15) *fi* 'in' also hold for (Π18) *ġewwa* 'inside' and (Π19) *ġo* 'in' – two Πs that are featured already in Table 5.18 #8 and #9, respectively.

(5.222) [Korpus Malti 3.0; parl2387]

Dr Mario Grixti	*li*	*jaħdem*	*fil-qasam*	*tal-Onkoloġija*	
Dr Mario Grixti	that	3SG.M.IPFV:work	in:DEF-field	of:DEF-oncology	
ġie	*mneħħi*	*mill-Isptar Boffa*	*biex*	*illum*	*qiegħed*
come	PART:take_off	from:DEF-Hospital_Boffa	to	today	stay.PART
[$Ø_{static}$	***fi**l-klinika*	*tal-Gżira]$_{PP}$*			
[Ø	**in**:DEF-clinic	of:DEF- Gżira]			

'Dr. Mario Grixti who works in the field of Oncology was removed from Boffa Hospital so that today he is [**in** the Gżira clinic].'

It is assumed that (Π15) *fi* 'in' is insensitive to the distinction dynamic vs static. The predicate kernel is responsible for providing the necessary information for the adequate interpretation of the utterance. Since *qiegħed* 'staying' bears the feature [+static] the entire predication is static and so is the spatial situation. The semantic contribution of the Π boils down to identifying the part of the Ground that is affected. This is the interior of the clinic. The presence of $Ø_{static}$ in the bracketed part of (5.222) is motivated in the same way as the presence of $Ø_{dynamic}$ in the bracketed part of (5.221), i.e., $Ø_{static}$ marks the slot, where semantically more specific phonologically realised directionality markers are placed. We may now add (Π15) *fi* 'in' to the list of combinations in Table 5.18. The number of admissible combinations of $Ø_{static}$ with $Π_{CONFIG}$ amounts to 25. This means that the static zero is not subject to any restrictions when it comes to combining with configurational Πs. In contrast, the dynamic zero is a licit partner only for 21 configurational Πs.

To conclude this excursus on zeroes, we state that $Ø_{static}$ occurs in constructions that express general location at rest – a notion for which we employ the unspecific term locative in what follows. The zeroes as a grammarian's construct are helpful when it comes to determining the functional load of Maltese Πs in the spatio-temporal domain. The zeroes do not only make it easier for us to separate directionality markers from configuration markers but they also call for a revision of the hypothesis that those spatial Πs of Maltese which do not encode Source are generally instances of Place-Goal syncretism. If, however, Πs like (Π15) *fi* 'in' are basically configurational, they cannot display Place-Goal syncretism because these notions are relevant only for Πs that function as directionality markers. Where AOME applies, directionality is not made explicit. This is where the zeroes enter the scene as purely technical means to help visualise the absence of dedicated markers. On the basis of this conclusion, we can now proceed to the class of configurational Πs.

5.2.1.4 The division of labour between spatio-temporal Πs: configurationality

The right column of Table 5.14 hosts the 25 configurational Πs we have identified for the spatio-temporal domain in Maltese. Figure 5.31 tells us that configurational Πs outnumber directionality-marking Πs by a ratio of 4-to-1. It is thus normal for a Maltese spatio-temporal Π to serve configurational purposes whereas it is far less normal that a spatio-temporal Π expresses directionality. Accordingly, the configurational function can be considered to be a feature of the canonical Π in our object language.[172]

We divide the class of configurational Πs into two subclasses, namely those with only a single spatio-temporal function and those with multiple spatio-temporal functions. To this end, we first tidy up the previous results.[173] This action is taken because the classification (not only) of the spatio-temporal Πs according to Hagège's (2010) meaning categories is by no means the best way to reconstruct the systematicity of the Maltese Π-system. The number of distinctions can be considerably reduced if we separate directionality from configuration. An important factor that contributes to this reduction is treating $Ø_{dynamic}$, $Ø_{static}$, and $Ø_{PROLATIVE}$ as directionality markers. In this way, the configurational Πs are stripped of their putative function

172 Cf. Section 5.3.
173 Cf. Section 5.1.2.

of co-encoding Place and Goal. Accordingly, a given Π_{CONFIG} is no longer the marker of a X-ESSIVE, X-LATIVE, X-VERSIVE, but is exclusively responsible for identifying the part of the Ground that is involved in the spatial situation. The labels X-ESSIVE, X-LATIVE, X-VERSIVE are appropriate only for the binary package of directionality and configuration markers, no matter whether slots are left empty. This is a radical reorganisation of our approach to the matter at hand. It has serious repercussions in the terminological domain. Moreover, spatio-temporal Πs that[174] might give the impression of being multifunctional become monofunctional under the configuration analysis because there is no need any more to differentiate between static and dynamic readings of the Π. In point of fact, polysemy in the spatio-temporal domain can be shown to be the minor option as opposed to the predominant monofunctionality of configurational Πs. To ensure that the previous results can be easily related to those of this section we will constantly cross-reference the phenomena we discuss subsequently. Those parameters (e.g. the distinction of with vs without contact) which are not integrated into the general matrix of the spatio-temporal categories will be touched upon separately.[175]

5.2.1.4.1 Monofunctional configurational Πs

As sketched in the previous paragraph, for configurational Πs the dynamic-static distinction is irrelevant. The Π can be used both in constructions which express rest at a Ground (Place) or motion towards a Ground (Goal). By way of example, we mention (Π12) *faċċata* 'opposite' which is mentioned in connection with the meaning categories OBESSIVE, DIRECTIONAL_1, TERMINATIVE (spatial), and OBVERSIVE.[176] None of these meanings is, however, conveyed by (Π12) *faċċata* 'opposite' alone. The directionality associated with these meaning categories is the exclusive responsibility of the directionality marker. The function of (Π12) *faċċata* 'opposite' is the identification of the part of the Ground or its surroundings that is relevant to the spatial situation. In this case, it is the opposite side of the Ground, i.e., a region that does not form a proper part of the Ground but is oriented towards it. On these grounds, it makes more sense to use the Latin-derived label OPPOSITUM 'opposite' for the function fulfilled by (Π12) *faċċata* 'opposite'.

With a view to make our point as clear as possible, we illustrate the configurational function of the Πs presented in this and the subsequent section with (Π56) *taħt* 'under' – a Π that is mentioned in the sections on the SUBESSIVE and the SUBLA-

174 Going by Section 5.1.2.

175 Cf. Section 5.2.1.4.4.

176 In Sections 5.1.2.1.5, 5.1.2.2.6, 5.1.2.2.7.1, and 5.1.2.2.9.1.

TIVE.[177] In analogy to what we have argued in connection with (Π12) *faċċata* 'opposite' in the foregoing paragraph, we assume that (Π56) *taħt* 'under' is the marker neither of the SUBESSIVE nor the SUBLATIVE but is responsible for identifying the bottom part of the Ground. To prove that this is indeed the case, we refer to the examples in (5.223).

(5.223) (Π56) *taħt* 'under' as Π_{CONFIG}

(a) [Korpus Malti 3.0; academic13] – LOCATIVE

[$Ø_{static}$ ***taħt*** *il-weraq]$_{PP}$ u maz-zkuk ikun*
[Ø **under** DEF-leaf.COLL] and with:DEF-stem.PL 3SG.M.IPFV:be.FUT
hemm għadd ta' insetti
there number of insect:PL
'[**Under** the leaves] and on the stems there will be a number of insects [. . .].'

(b) [Korpus Malti 3.0; culture2948] – ALLATIVE

Ċippa li tiddaħħal [$Ø_{dynamic}$ ***taħt*** *il-ġilda]$_{PP}$*
chip that 3SG.F.IPFV:enter:CAUS [Ø **under** DEF-skin]
bdiet tieħu post il-cards li jintużaw
begin:3SG.F.PFV 3SG.F.IPFV:take post DEF-card:PL that 3.IPFV:REFL:use:PL
biex impjegati jidħlu fl-uffiċju
to employee:PL 3.IPFV:enter:PL in:DEF-office
'A chip that is inserted [**under** the skin] has started to replace the cards that are used for employees to enter the office.'

(c) [Korpus Malti 3.0; literature21] – DIRECTIONAL_1

resqu ***[għal taħt*** *il-ħarrub]$_{PP}$*
approach:3PL.PFV [**for under** DEF-locust.COLL]
'[. . .] they moved [**under** the locusts] [. . .].'

(d) [Korpus Malti 3.0; news152022] – DIRECTIONAL_2

Kemm l-Avukat [. . .] u kemm l-Ispettur [. . .]
how_much DEF-lawyer and how-much DEF-inspector
resqu ***[lejn taħt*** *il-bank tal-Maġistrat]$_{PP}$*
approach:3PL.PFV [**towards under** DEF-bench of:DEF-magistrate]
'Both the Lawyer [. . .] and the Inspector [. . .] approached [the Magistrate's bench] [. . .].'

(e) [Korpus Malti 3.0; literature71] – TERMINATIVE

jibqa' sejjer bil-għawm ***[sa taħt*** *il-għolja*
3SG.M.IPFV:remain go.PART with:DEF-swimming [**till under** DEF-hill

177 Cf. Sections 5.1.2.1.9 and 5.1.2.2.13.1.

ta' Sciberras]$_{PP}$
of Sciberras]
'[. . .] he keeps swimming [**as far as** Mount Sciberras] [. . .].'

(f) [Korpus Malti 3.0; academic13] – ABLATIVE
Infakkar li għalkemm il-kompost li nġibu
1SG.IPFV:remind that although DEF-compost that 1.IPFV:get:PL
***[minn taħt** il-ħarrub]*$_{PP}$ *huwa tajjeb għall-pjanti*
[**from under** DEF-locust.COLL] 3SG.M good for:DEF-plant:PL
mhux tajjeb għall-azalea
NEG:3SG.M:NEG good for:DEF-azalea
'I remind you that although the compost we get [**from under** the locusts] is good for the plants, it is not good for the azalea [. . .].'

(g) [Korpus Malti 3.0; literature7] – PERLATIVE
kien żamm in-nifs biex għadda ***[minn taħt** il-ħarruba]*$_{PP}$
be hold DEF-breath to pass [**from under** DEF-locust]
'[. . .] he had held his breath to pass [**under** the locust] [. . .].'

(h) [Korpus Malti 3.0; literature39] – PROLATIVE
kien jimxi bil-mod ***[minn taħt** il-ħajt]*$_{PP}$
be 3SG.M.IPFV:walk with:DEF-manner [**from under** DEF-wall]
bla ma jsellem lil ħadd
without that 3SG.M.IPFV:greet to someone
'[. . .] he walked slowly [**along** the wall] without greeting anyone [. . .].'

As results from (5.223), (П56) *taħt* 'under' is attested in combination with each directionality marker, be it overt or covert. In (5.223a), the combination with $Ø_{static}$ serves to locate the insects under the leaves. In (5.223b), the combination with $Ø_{dynamic}$ serves to identify the endpoint of the motion event to be under the skin of the employee. (П56) *taħt* 'under' is preceded by (П21) *għal* 'for' in (5.223c) to mark the Goal of the motion event which is located under the locust trees. Similarly, (П28) *lejn* 'towards' functions as directionality marker in (5.223d) whereas the configuration marker (П56) *taħt* 'under' indicates that the magistrate's bench is placed on an elevated level so that the Figure moving there winds up in front of the bench on a lower level. The Figure is thus metonymically under the Ground. In (5.223e), Toni Bajjada – a national hero from Malta's distant past – swims as far as Mount Sciberras, i.e. the Figure (= the swimmer) moves through the water until he arrives under the highest elevation of the Sciberras Peninsula. Directionality is encoded by (П51) *sa* 'till', the position below the hill is indicated by (П56) *taħt* 'under'. In (5.223f), the compost was originally placed under the trees from where it was removed. (П41) *minn* 'from' encodes the ABLATIVE whereas (П56) *taħt* 'under' identifies the part of the Ground relative to which the Figure was found before the motion event started. In (5.223g), the Figure had to pass under the trees. Directionality is encoded by *minn*$_{PERLATIVE}$. In (5.223h), directionality is encoded by *minn*$_{PROLATIVE}$. The Figure moves along the Ground (wall) which is described as towering over the Figure as (Π56) *taħt* 'under' is used. In all of these examples, the Ground occupies a position such that its vertical extension exceeds that of the Figure. The Figure is thus closer to the bottom of the Ground than it is to the top of the Ground. The bottom part of the Ground is identified by (Π56) *taħt* 'under'. This is the sole function of this Π$_{CONFIG}$. It is in no way involved in the expression of directionality. At the same time, no directionality marker can take over the function of (Π56) *taħt* 'under'.

What we have argued in connection with the uses to which (Π56) *taħt* 'under' is put in (5.223) can be visualised schematically in Figure 5.39.

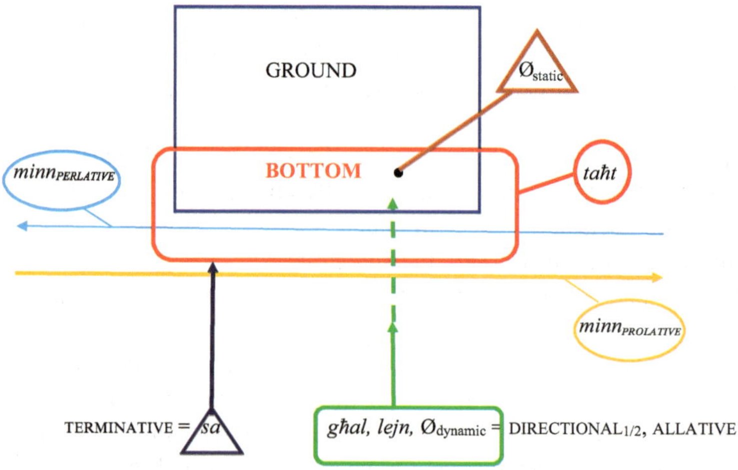

Figure 5.39: Schema of the semantics of combinations of (Π56) *taħt* 'under' with directionality markers.

The black dot locates the Figure somewhere in the bottom region when (Π56) *taħt* 'under' is preceded by Ø$_{static}$, i.e., the locative is realised. The bottom region extends over the physical bottom of the Ground and its immediate neighbourhood beyond the boundaries of the Ground. The bottom is referred to by the Π$_{CONFIG}$ which remains constant whereas the directionality markers vary in correlation to the spatial situation. No other task is fulfilled by (Π56) *taħt* 'under'. It is therefore fully justified to introduce the label INFERIOR 'low(er)' for this Π.

What we have shown for (Π56) *taħt* 'under' also holds for the other configurational Πs, namely that they are solely responsible for identifying a given part of the Ground and the region associated with this part. Given that this assumption is water-tight, the following catalogue of monofunctional spatio-temporal Πs can be drawn up. In Table 5.19, Πs with identical functions are placed next to each other.

Table 5.19: Labels for monofunctional configurational Πs.

Π_{CONFIG}	label	previous mention(s) under
(Π14) *fejn* 'near'	PROPINQUUS	PROPINQUUS, POXILATIVE
(Π23) *ħdejn* 'beside'	PROPINQUUS	PROPINQUUS, POXILATIVE
(Π32) *maġenb* 'close to'	PROPINQUUS	PROPINQUUS, POXILATIVE
(Π48) *qrib* 'near'	PROPINQUUS	PROPINQUUS, POXILATIVE
(Π58) *viċin* 'near'	PROPINQUUS	PROPINQUUS, POXILATIVE
(Π18) *ġewwa* 'inside'	INTERIOR	INESSIVE (spatial), ILLATIVE
(Π19) *ġo* 'in'	INTERIOR	INESSIVE (spatial), ILLATIVE
(Π16) *fost* 'amongst'	MEDIAL$_1$	MEDIOESSIVE, MEDIOLATIVE
(Π47) *qalb* 'amidst'	MEDIAL$_1$	MEDIOESSIVE
(Π3) *barra* 'outside'	EXTERIOR	EXTRAESSIVE
(Π44) *oltre* 'beyond'	ULTERIOR	EXTRAESSIVE
(Π12) *faċċata* 'opposite'	OPPOSITUM	OBESSIVE, DIRECTIONAL$_1$, TERMINATIVE (spatial), OBVERSIVE
(Π49) *quddiem* 'in front of'	FRONS	FRONTALIS, FRONTIVERSIVE, PRELATIVE (spatial)
(Π46) *qabel* 'before'	ANTERIOR	PREESSIVE (spatial/temporal), PRELATIVE
(Π60) *wara* 'after'	POSTERIOR	POSTESSIVE (spatial/temporal), POSTLATIVE (spatial/temporal)
(Π59) *waqt* 'at the time of'	LONGITUDO$_1$	PERLATIVE (temporal)
(Π17) *fuq* 'on'	SUPERIOR	SURESSIVE, SUPERESSIVE
(Π56) *taħt* 'under'	INFERIOR	SUBESSIVE, SUBLATIVE
(Π22) *għand* 'at s.o.'s place'	DOMI/DOMUM	APUDESSIVE, APUDLATIVE

Most of the labels in the middle column bear resemblance to those used by Borg/Azzopardi-Alexander (1997). In the column on the right-hand side, the terms of the spatial relations are mentioned in association with which a given Π_{CONFIG} has previously been mentioned.[178] With FRONS 'front' we introduce another Latinate label. The same holds for LONGITUDO 'length (spatial/temporal)'. The label DOMI/DOMUM is based on the residual LOCATIVE and the directional ACCUSATIVE of Latin *domus* 'house' meaning 'at home' and 'homewards', respectively. For monofunc-

178 Cf. Section 5.1.2.

tional configurational Πs, it is a relatively straightforward thing to create a label for their functions. In the subsequent subsection, we will see whether multifunctional configurational Πs are as easily labelled as their monofunctional partners in BLOMP 2.0.

5.2.1.4.2 Multifunctional configurational Πs

As Figure 5.40 shows multifunctional Πs constitute a minority in the domain of the configurational Πs.

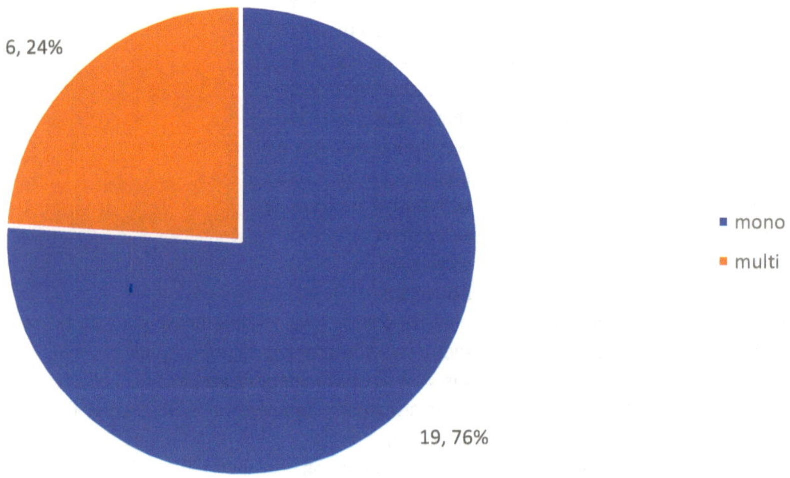

Figure 5.40: Shares of monofunctional and multifunctional configurational Πs

No Italo-Romance Π counts among the multifunctional configurational Πs so it is possible to put forward the implicational patterns in Figures 5.41–5.42.

Italo-Romance → monofunctional / configuration

Figure 5.41: Italo-Romance origin implies monofunctionality.

multifunctional → Semitic / configuration

Figure 5.42: Multifunctionality implies Semitic origin.

A glimpse at Table 5.15 additionally tells us that none of the multifunctional configuration Πs gives evidence of the DIRECTIONAL$_2$, i.e., no cases of (Π28) *lejn* 'towards' combining with any of the Πs listed in Table 5.20 are attested. This is the basis for the implicational patterns in Figures 5.43–5.44. The motivation for the restriction imposed on the combinability of (Π28) *lejn* 'towards' is not clear to us at the moment.

multifunctional → ¬directional$_2$

Figure 5.43: Multifunctionality excludes directional$_2$.

directional$_2$ → monofunctional

Figure 5.44: Directional$_2$ implies monofunctionality.

As we will see shortly, multifunctionality is a kind of terminological hyperbole because all attested patterns of configurational Πs fulfilling more than just one function are binary, i.e., there are always two functions involved. The two functions are not always closely related to each other conceptually. No multifunctional configurational Π is attested with each of the directionality markers as comes to the fore in Table 5.15. Some of the multifunctional Πs fulfil functions which are absent from those registered for the monofunctional Πs and thus call for coining extra-labels for them, cf. Table 5.20.

Table 5.20: Labels for multifunctional configurational Πs.

Π_{CONFIG}	label$_1$	label$_2$	previous mention(s) under
(Π8) *biswit* 'facing'	PROPINQUUS	FRONS	PROPINQUUS, FRONTALIS, DIRECTIONAL$_1$, TERMINATIVE (spatial), OBVERSIVE
(Π31) *madwar* 'around'	PROPINQUUS	CIRCUM	PROPINQUUS, CIRCUMESSIVE (spatial/temporal)
(Π30) *ma'* 'with'	PROPINQUUS	SEQUENS	PROXIMATE, SECUTIVE
(Π15) *fi* 'in'	INTERIOR	SUPERIOR	INESSIVE (spatial/temporal), ILLATIVE, SURESSIVE
(Π4) *bejn* 'between'	MEDIAL$_1$	MEDIAL$_2$	MEDIOESSIVE, INTERESSIVE (spatial/temporal), INTERLATIVE
(Π33) *matul* 'during'	LONGITUDO$_1$	LONGITUDO$_2$	EXTENTUS, PERLATIVE (temporal)

There are two middle columns. The cells host the labels for each of the categories which are expressed by the Π. The order from left to right is not hierarchical. The two Πs at the bottom of Table 5.20 display identical labels with distinctive indexes. MEDIAL$_1$ applies in situations in which the Figure is situated in a distributed Ground whereas MEDIAL$_2$ refers to a situation in which the Figure fills the otherwise empty space between two entities. We speak of LONGITUDO$_1$ if the Ground is a temporal concept. The label LONGITUDO$_2$ is employed if the Ground is spatial. It is worth

Table 5.21: Synopsis of configurational Πs and their functions.

Π	configurations																
	PROPINQUUS	INTERIRO	MEDIAL$_1$	FRONS	LONGITUDO$_1$	SUPERIOR	EXTERIOR	ULTERIOR	OPPOSITUM	ANTERIOR	POSTERIOR	INFERIOR	DOMI	CIRCUM	SEQUENS	MEDIAL$_2$	LONGITUDO$_2$
(Π14) *fejn* 'near'	X																
(Π23) *ħdejn* 'beside'	X																
(Π32) *maġenb* 'close to'	X																
(Π48) *qrib* 'near'	X																
(Π58) *viċin* 'near'	X																
(Π8) *biswit* 'facing'	X			X													
(Π31) *madwar* 'around'	X													X			
(Π30) *ma'* 'with'	X														X		
(Π18) *ġewwa* 'inside'		X															
(Π19) *ġo* 'in'		X															
(Π15) *fi* 'in'		X				X											
(Π16) *fost* 'amongst'			X														
(Π47) *qalb* 'amidst'			X														

(П4) *bejn* 'between'			X													X	
(П49) *quddiem* 'in front of'				X													
(П59) *waqt* 'at the time of'					X												
(П33) *matul* 'during'					X												X
(П17) *fuq* 'on'						X											
(П3) *barra* 'outside'							X										
(П44) *oltre* 'beyond'								X									
(П12) *faċċata* 'opposite'									X								
(П46) *qabel* 'before'										X							
(П60) *wara* 'after'											X						
(П56) *taħt* 'under'												X					
(П22) *għand* 'at s.o.'s place'													X				
Xs	8	3	3	2	2	2	1	1	1	1	1	1	1	1	1	1	1

noting that the shared feature PROPINQUUS yields a class of eight Πs composed of five monofunctional and three multifunctional Πs. The share of PROPINQUUS-Πs is as big as 32% of all configurational Πs. No other function comes close to this share.

The synopsis of the observations made in this and the previous section is given in Table 5.21. An X and yellow shading are indicative of the association of a given function with a given configurational Π. It strikes the eye that the vast majority of the cells are empty. No row hosts more than two Xs. Only six out of seventeen functions are fulfilled by two or more configurational Πs. CIRCUM, SEQUENS, MEDIAL_2, and LONGITUDO_2 are only attested with multifunctional Πs. The latter two only occur with Πs that also attest to MEDIAL_1 and LONGITUDO_1, respectively.

The knowledge we have gained separately in connection with directional and configurational Πs provides us with a picture of the individual components of the Π-Π combinations. What is needed at this point is determining how the two components interact. This is why, in the next subsection, we elaborate on the surveys of combinations of directionality and configuration markers given in Tables 5.15 and 5.16.

5.2.1.4.3 $\Pi_{\text{DIRECT}} + \Pi_{\text{CONFIG}}$

The presentation of the empirical facts is divided in two. We start with the phonologically realised Goal-oriented (Π21) *għal* 'for', (Π28) *lejn* 'towards', and (Π51) *sa* 'till' and their combinations with configurational Πs because no multifunctional aspects have to be taken account of.[179] This is different in the case of (Π41) *minn* 'from' because of the three possible readings of this Π whose discussion will therefore conclude the presentation of Π-Π combinations.[180]

5.2.1.4.3.1 Goal

The three Goal-oriented directional markers are presented in the order of the decreasing number of combinations with configuration markers. Thus, the first candidate to be looked at is (Π21) *għal* 'for'. Sentential examples for the 21 attested combinations of this Π with configurational Πs are given in Table 5.22.

---

179 In Section 5.2.1.4.3.1.

180 In the separate Section 5.2.1.4.3.2.

Table 5.22: Combinations of (Π21) *għal* 'for' with configurational Πs.

#	source	example / translation	$Π_{CONFIG}$
1	news184006	*kienu jgħaddu l-flus mid-droga [**għal barra** l-Kolombja]$_{PP}$* '[. . .] they transferred drug money [**outside of** Colombia] [. . .]'	(Π3) *barra* 'outside'
2	news191851	*Dan l-aħħar il-kumpanija sabet ħames bastimenti anti-ki mgħarrqa f' qiegħ il-baħar li jmorru lura [**għal bejn** l-ewwel seklu Qabel Kristu u l-ħames seklu Wara Kris-tu]$_{PP}$.* 'Recently the company found five ancient sunk-en ships at the bottom of the sea dating back [**to between** the first century BC and the fifth century AD].'	(Π4) *bejn* 'between'
3	literature77	*Mal-biki u l-għajat tat-tifla n-nies tad-dar kollha ġrew [**għal fejn** il-marid]$_{PP}$.* 'With the girl's cries and screams, all the people of the house rushed [**to** the sick man].'	(Π14) *fejn* 'near'
4	news41571	*se tpoġġi lil Malta mill-inqas [**għal fost** l-aqwa fl-Unjoni Ewropea]$_{PP}$* '[. . .] it will put Malta at least [**among** the best in the European Union].'	(Π16) *fost* 'amongst'
5	culture3662	*L-iskorpjun l-ewwel niżel fuq spallet persuna li min-naħa tagħha xejritha direttament [**għal fuq** il-mara]$_{PP}$.* 'The scorpion first landed on the shoulders of a person who in turn threw it directly [**onto** the woman]'.	(Π17) *fuq* 'on'
6	news115665	*Kif resqu lejna dawwarnihom u sforzajnihom [**għal ġewwa** l-isqaq]$_{PP}$.* 'As they approached us we made them turn around and forced them [**into** the alley].'	(Π18) *ġewwa* 'inside'
7	european10051	*jibqa sejjer [**għal ġo**l-Brażil Amażonjan]$_{PP}$* '[. . .] it keeps going [**into** Amazonian Brazil] [. . .].'	(Π19) *ġo* 'in'
8	*news75221*	*xi lanċ ta' salvataġġ telqu minn Ragusa [**għal biswit** il-vapur]$_{PP}$* '[. . .] some rescue boats left from Ragusa [**for near** the ship].'	(Π8) *biswit* 'facing'
9	literature10	*Għamlitlu xenata nobis, ippakkjat sorra ħwejjeġ, u reġgħet telqet [**għal għand** ommha]$_{PP}$.* 'She made a real scene for him, packed some clothes, and left again [**for** her mother].'	(Π22) *għand* 'at s.o.'s place'
10	literature32	*Sadanittant Xon kien mexa [**għal ħdejn** bużnannuh]$_{PP}$* 'Meanwhile Xon had gone [**to** his great-grandfather's **side**] [. . .].'	(Π23) *ħdejn* 'beside'
11	literature35	*Imma tant beda jitħatteb biex jilqa' d-daqqiet li s-siġġu nqaleb bih b'kollox [**għal ma**l-art]$_{PP}$.* 'But he began to struggle so much to receive the blows that the chair was completely overturned [**on** the floor].'	(Π30) *ma'* 'with'

Table 5.22 (continued)

#	source	example / translation	Π_{CONFIG}
12	news117404	*Servizz ċirkolari b'xejn se jitħaddem mir-Razzett tal-Ħbiberija għal dawk b'diżabbiltà u għax-xjuħ minn Triq l-Ordinanza [**għal madwar** il-Belt]$_{PP}$.* 'A free circular service will be operated from the Friendship Farm for the disabled and the elderly from Triq l-Ordinanza [**to all around** Valletta].'	(Π31) *madwar* 'around'
13	news81231	*Il-borża fl-aħħar iċċaqalqet sewwa u ġġenbet imgerrba [**għal maġenb** il-vann]$_{PP}$.* 'The bag finally just moved and was rolled [**to the side of** the van].'	(Π32) *maġenb* 'close to'
14	news192898	*Il-unions minn naħa tagħhom ħarġu ordni li l-protesti jiġu estiżi [**għal matul** il-jumejn li ġejjin]$_{PP}$.* 'The unions for their part issued an order that the protests be extended [**over** the next two days].'	(Π33) *matul* 'during'
15	news107340	*Il-każ imur lura [**għal qabel** l-Elezzjoni Ġenerali li għaddiet]$_{PP}$* 'The case goes back [**to before** the last General Election] [. . .].'	(Π46) *qabel* 'before'
16	news150357	*Minn hemm hu ħarab [**għal qalb** il-fabbriki]$_{PP}$ bil-Pulizija jiġru warajh fejn daħal f'għalqa biex jipprova jistaħba qalb il-ħaxix.* 'From there he fled [**among** the factories] with the Police chasing him where he entered a field to try to hide among the grass.'	(Π47) *qalb* 'amidst'
17	news75280	*kellha d-dmir biss tikkoordina s-salvataġġ f'wisa' ta' baħar li twassal minn ħdejn Lampedusa [**għal qrib** il-gżira ta' Kreta]$_{PP}$.* '[. . .] it only had the duty to coordinate the rescue in a width of sea that reaches from near Lampedusa [**to near** the island of Crete].'	(Π48) *qrib* 'near'
18	news22138	*Il-gwardja tal-unur u l-banda tal-Korp tal-Pulizija jitilqu minn Pjazza San Ġorġ [**għal quddiem** il-binja l-ġdida tal-Parlament]$_{PP}$* 'The guard of honor and the band of the Police Force leave from St. George Square [**for** the new Parliament building] [. . .].'	(Π49) *quddiem* 'in front of'
19	literature74	*telaq ġirja waħda [**għal taħt** il-kastell]$_{PP}$* '[. . .] he left running [**to under** the castle] [. . .].'	(Π56) *taħt* 'under'
20	parl469	*Mela kollox huwa miexi [**għal viċin** l-elezzjoni]$_{PP}$* 'So everything is moving [**closer to** the election] [. . .].'	(Π58) *viċin* 'near'
21	literature77	*tnejn jibqgħu sejrin [**għal wara** l-knisja]$_{PP}$* '[. . .] two keep going [**to the back of** the church] [. . .].'	(Π60) *wara* 'after'

Inspected in isolation, the data in Table 5.22 do not tell us much beyond the fact that (Π21) *għal* 'for' is a prolific directionality marker that allows for a plethora of combinations with configurational Πs. (Π21) *għal* 'for' is combinable with 87% of all configuration markers. This observation becomes more interesting when we compare the behaviour of (Π21) *għal* 'for' with that of the other two Goal-oriented directional Πs. In terms of the number of types of combinations with configurational Πs, (Π51) *sa* 'till' is second best. The fifteen attested combinatory patterns are illustrated in Table 5.23.

Table 5.23: Combinations of (Π51) *sa* 'till' with configurational Πs.

#	source	example / translation	Π_{CONFIG}
1	literature20	*filgħodu l-vann wassalna bħas-soltu [**sa barra** l-grada tal-iskola]$_{PP}$* '[. . .] in the morning the van brought us as usual [**up to** the school gate].'	(Π3) *barra* 'outside'
2	news86594	*Fuħħar li nstab fihom jagħtu x'jifhem li l-oqbra jmorru lura [**sa bejn** is-sena 3,000 QK u l-2,500 QK]$_{PP}$.* 'Pottery found in them suggests that the tombs date back [**to between** 3,000 BC and 2,500 BC].	(Π4) *bejn* 'between'
3	literature21	*Jiġru kemm jifilħu, baqgħu sejrin bla nifs [**sa fejn** il-ġnien]$_{PP}$* 'Running as hard as they could, they breathlessly kept going [**as far as** the garden] [. . .].'	(Π14) *fejn* 'near'
4	news57416	*Hija fidi li tidħol [**sa fi**l-qiegħnett tar-realtà umana tagħna]$_{PP}$* 'It is a faith that penetrates [**to** the bottom of our human reality] [. . .].'	(Π15) *fi* 'in'
5	literature32	*telgħu [**sa fuq** il-bejt]$_{PP}$ biex jaraw it- temp kif inhu* 'they went [**up to** the roof] to see how the weather was'	(Π17) *fuq* 'on'
6	european6816	*Il-bejjiegħ għandu jwassal [**sa għand** il-konsumatur]$_{PP}$ oġġetti li huma f'konformità mal-kuntratt ta' bejgħ.* 'The seller must deliver [**to** the consumer] goods that comply with the sales contract.'	(Π22) *għand* 'at s.o.'s place'
7	literature24	*Kaxkar mejda tqila tal-fidda sewda [**sa ħdejn** il-ħajt]$_{PP}$ biex jitla' fuqha.* 'He dragged a heavy black silver table [**up to** the wall] for him to climb onto.	(Π23) *ħdejn* 'beside'
8	news108877	*Mhux talli n-NATO ma żżarmatx, avolja m'hemmx aktar bżonnha, talli kompliet tespandi [**sa ma**l-fruntieri Russi]$_{PP}$* 'Not because NATO has not been dismantled, even though it is no longer needed, because it has continued to expand [**up to** the Russian borders].'	(Π30) *ma'* 'with'

Table 5.23 (continued)

#	source	example / translation	Π_{CONFIG}
9	parl256	*Jista' l-ministru jgħid meta se tiġi asfaltata Marsa Road, Qormi minn taħt il- pont [**sa madwar** ir-roundabout magħrufa ta' ħdejn l-erba' qaddisin]$_{PP}$* 'Can the minister tell when Marsa Road, Qormi will be asphalted from under the bridge [**to around** the known roundabout near the four saints] [. . .]?'	(Π31) *madwar* 'around'
10	parl853	*Jista' l-Viċi Prim Ministru jpoġġi fuq il-mejda tal- Kamra kopja tal-pjanta ta' kif inhuwa maħsub illi jsir żvilupp fl-area [**sa maġenb** id-dar ta' l-anzjani fl-Imsida]$_{PP}$* 'Can the Deputy Prime Minister put on the table of the House a copy of the plan of how it is intended that a development is realised in the area [**up to** the old people's home in Msida] [. . .]?'	(Π32) *maġenb* 'close to'
11	news116434	*Sophie se tibda l-ġirja minn quddiem il-Lukanda Suncrest tul ix-xatt kollu [**sa qrib** il-Kappella ta' San Pawl]$_{PP}$ u lura minn fejn bdiet.* 'Sophie will start the run from in front of the Suncrest Hotel along the entire waterfront [**to near** the Chapel of St. Paul] and back from where she started.'	(Π48) *qrib* 'near'
12	literature73	*ħareġ mill-Imdina u baqa' sejjer [**sa quddiem** il-Kunvent]$_{PP}$* '[. . .] he came out of Mdina and kept going [**as far as in front of** the Convent] [. . .].'	(Π49) *quddiem* 'in front of'
13	news138629	*Spjega li sa mijiet ta' snin ilu l-baħar kien jidħol [**sa taħt** il-Victoria Lines]$_{PP}$, u sal-1560 il-baħar kien għadu jidħol sa ħdejn il-Kappella ta' San Pawl Milqi f'Burmarrad.* 'He explained that until hundreds of years ago the sea reached [**as far as** the Victoria Lines], and until 1560 the sea was still reached as far as the Chapel of San Paul Milqi in Burmarrad.'	(Π56) *taħt* 'under'
14	parl375	*madwar 6,000 ħamiema tal-ġiri ittieħdu fuq barkun [**sa viċin** il-Port ta' Reggio Calabria]$_{PP}$ minn fejn kellu jintelaq* '[. . .] around 6,000 racing pigeons were taken on a barge [**to near** the Port of Reggio Calabria] from where they were to be released.'	(Π58) *viċin* 'near'
15	news114805	*tibqa' sejjer, maltemp jew mhux, sħana jew bard, [**sa wara** l-iskrivanija tiegħek f'Tas-Sliema]$_{PP}$* '[. . .] you keep going, stormy weather or not, hot or cold, [**behind** your desk in Tas-Sliema][. . .].'	(Π60) *wara* 'after'

(Π51) *sa* 'till' is attested in combination with 65% of the configurational Πs. This means that its combinatory potential is considerably smaller than that of (Π21) *għal* 'for'. The latter is excluded from combining with (Π15) *fi* 'in' – a combination that is unique to (Π51) *sa* 'till'. In contrast, there are six configurational Πs which are incompatible with the TERMINATIVE marker but combine with (Π21) *għal* 'for', namely (Π16) *fost* 'amongst', (Π18) *ġewwa* 'inside', (Π19) *ġo* 'in', (Π33) *matul* 'during', (Π46) *qabel* 'before'; and (Π47) *qalb* 'amidst' (cf. the preference rule in Figure 5.37). This half a dozen impossible partners of (Π51) *sa* 'till' comprises four Πs that are linked to each other conceptually. This is self-explanatory in the paired cases of, on the one hand, (Π16) *fost* 'amongst' and (Π47) *qalb* 'amidst' and, on the other hand, (Π18) *ġewwa* 'inside' and (Π19) *ġo* 'in'. The first mentioned pair of Πs introduces a Ground that is distributed in the sense that the Figure has to enter the space that is occupied by a group of entities that constitute the Ground. Since the marker of the DIRECTIONAL$_1$, (Π21) *għal* 'for', marks motion towards the Goal with the possibility of the Figure winding up at or in the Ground, it is a good choice for combinations with (Π16) *fost* 'amongst' and (Π47) *qalb* 'amidst' whereas (Π51) *sa* 'till' invokes the idea that the Figure never enters the Ground because the motion stops at the outer boundary of the Ground. Similarly, (Π18) *ġewwa* 'inside' and (Π19) *ġo* 'in' as configuration markers are employed for spatial situations in which the Figure at the end of the motion event occupies a position inside the Ground. The position inside the Ground is not normally associated with the TERMINATIVE marker. In connection to this argument, example #4 in Table 5.23 poses a problem because it involves (Π51) *sa* 'till' and (Π15) *fi* 'in' and thus violates the restriction against using the TERMINATIVE in situations that feature the Figure inside the Ground. This *sa-fi* combination is a hapax in the Korpus Malti 3.0. On a basis this small, it is impossible to put forward any argument that makes sense. Further research is needed to determine whether *sa-fi* combinations are more common in Maltese and whether this requires a modification of the above restriction.

Why (Π51) *sa* 'till' fails to combine with either (Π33) *matul* 'during' or (Π46) *qabel* 'before' cannot be explained with absolute certainty. First of all, neither of the two configurational Πs is a frequent partner of (Π21) *għal* 'for'. Second, both configuration markers introduce temporal concepts as complements. The TERMINATIVE is not generally incompatible with temporal concepts. However, temporality is the trait shared by (Π33) *matul* 'during' and (Π46) *qabel* 'before'. As in the previous case, further research on this matter is called for before any robust conclusions can be drawn.

The smallest number of combinations with configuration Πs goes to the credit of the directional$_2$, (Π28) *lejn* 'towards', for which only six patterns are registered. These six patterns are illustrated in Table 5.24.

Table 5.24: Combinations of (Π28) *lejn* 'towards' with configurational Πs.

#	source	example / translation	Π_{CONFIG}
1	news10451	*Filfatt, 3,881, minn total ta' kważi 30,000 [. . .] kienu turisti li marru fuq destinazzjoni [**lejn barra mil**l-Ewropa]$_{PP}$* 'In fact, 3,881, out of a total of almost 30,000 thousand were tourists who went to a destination [**outside of** Europe] [. . .].'	(Π3) *barra* 'outside'
2	european4032	*wara li tgħaddi għan-nofsinhar tal-gżira ta' Öland [**lejn fuq** il-kosta tal-lvant]$_{PP}$ fuq 56° 30' latitudni tat-tramuntana, imbagħad f'direzzjoni tal-punent matul il-kosta tal- Polonja* '[. . .] after passing south of the island of Öland [**to** the east coast] at 56° 30' north latitude, then in a westerly direction along the coast of Poland [. . .].'	(Π17) *fuq* 'on'
3	news93216	*noqgħod inħares lura [**lejn qabel** l-aħħar elezzjoni ġenerali]$_{PP}$* 'I look back [**to before** the last general election] [. . .].'	(Π46) *qabel* 'before'
4	news151952	*imxew tul Triq ir-Repubblika, daru [**lejn quddiem** il-Kon-Kattidral ta' San Ġwann]$_{PP}$ u komplew lejn Kastilja* '[. . .] they walked along Triq ir-Repubbblica, turned [**to the front of** the Co-Cathedral of San Ġwann] and continued towards Castile [. . .].'	(Π49) *quddiem* 'in front of'
5	news159372	*Jew ma waddbuh xejn iżda ġarrewh [**lejn taħt** is-sur wara d-depot]$_{PP}$ biex jgħidu li sabuh hemmhekk wara li qabeż?* 'Or they didn't throw him at all but they carried him [**to** the wall behind the depot] to say they found him there after he jumped?'	(Π56) *taħt* 'under'
6	news111040	*X'ħin ġara hekk, dawn iż-żewġ persuni ħarbu [**lejn wara** l-Knisja]$_{PP}$* 'When that happened, these two people fled [**behind** the Church] [. . .].'	(Π60) *wara* 'after'

Combinations with (Π28) *lejn* 'towards' are possible only for 26% of the configurational Πs. From Figure 5.35, we know already that if a given Π_{CONFIG} combines with (Π28) *lejn* 'towards' it also always allows for combining with (Π21) *għal* 'for', i.e., DIRECTIONAL$_2$ presupposes DIRECTIONAL$_1$. This implication raises the question of whether the coexistence of DIRECTIONAL$_1$ and DIRECTIONAL$_2$ is a case of synonymy or of subtle meaning differences between the two directionals. The configurational Πs that go together with (Π28) *lejn* 'towards' can be ordered as follows: (Π17) *fuq* 'on' and (Π56) *taħt* 'under' are antonyms representing opposing spatial concepts on the vertical dimension, (Π49) *quddiem* 'in front of' and (Π60) *wara* 'after' form a comparable pair of concepts on the horizontal dimension to which (Π46) *qabel* 'before' might be added as antonym for the temporal readings of (Π60) *wara* 'after'. (Π3) *barra* 'outside' cannot be categorised with any of these antonymic pairs. It is worth noting that (Π28) *lejn* 'towards' is excluded from five of the combinations attested for (Π21) *għal* 'for' but not for (Π51) *sa* 'till', viz. (Π16) *fost* 'amongst', (Π18) *ġewwa* 'inside', (Π19) *ġo* 'in', (Π33) *matul* 'during', and (Π47) *qalb* 'amidst'. This parallel between the TERMINATIVE and the DIRECTIONAL_2 seems to speak in favour of a meaning difference between the directionals. Where DIRECTIONAL_1 does not exclude the possibility of the Figure entering the Ground, this scenario is less likely in the case of DIRECTIONAL_2, meaning: (Π28) *lejn* 'towards' most probably is a pure directional in the sense that it only indicates that a motion is Goal-oriented without implying that the Goal is reached by the Figure. We will address this issue again within the wider context of the spatio-temporal meaning categories.[181]

5.2.1.4.3.2 Source and Path

In this section, we try to separate the three meaning categories covered by (Π41) *minn* 'from' from each other. In analogy to the previous section, we first give a look at the meaning of this Π which is that of the ABLATIVE. Table 5.25 contains examples of the twenty attested combinations with configurational Πs. For the usual reasons, no example of (Π38) *mingħand* 'from s.o.' is included in Table 5.25.

Table 5.25: Combinations of $minn_{\text{ABLATIVE}}$ with configurational Πs.

#	source	example / translation	Π_{CONFIG}
1	academic281	*wara li nfiltra min-nofs tal-grawnd hu għamel xutt [**minn barra** <u>l-kaxxa</u>]$_{PP}$ li ma ta l-ebda ċans lil Mark Schwarzer biex isalvah* '[. . .] after infiltrating from the middle of the field he made a shot [**from outside** <u>the box</u>] that didn't give Mark Schwarzer any chance to save it [. . .].'	(Π3) *barra* 'outside'
2	literature45	*Waqgħet [**minn bejn** <u>il-paġni tal- ktieb</u>]$_{PP}$ biċċa karta li niżlet mal-art b'toqol.* 'A piece of paper fell [**from between** <u>the pages of the book</u>] and landed heavily on the floor.'	(Π4) *bejn* 'between'
3	parl2485	*tneħħew il-bandli [**minn biswit** <u>iċ-ċimiterju ta' Ħal Qormi</u>]$_{PP}$* '[. . .] the swings have been removed [**from next to** <u>the Qormi cemetery</u>] [. . .].'	(Π8) *biswit* 'facing'

181 Cf. Section 5.2.1.4.4.

Table 5.25 (continued)

#	source	example / translation	Π_{CONFIG}
4	parl1504	*dan l- uffiċjal tant għandu arja li issa wasal biex anke qala' l-bus-stop sign [**minn faċċata ta**t-triq tiegħu]$_{PP}$ biex ikun jista' jipparkja l-van personali tiegħu* '[. . .] this officer so full of himself that now he even removed the bus-stop sign [**from across** his street] so he could park his personal van [. . .].'	(Π12) *faċċata* 'opposite'
5	literature77	*Qam [**minn fejn** il-mejda]$_{PP}$ malli qara li Ineż ma kenitx maqtula* 'He got up [**from** the table] as soon as he read that Inez had not been killed [. . .].'	(Π14) *fejn* 'near'
6	news109009	*Schulz hu persuna li telgħet [**minn fost** il-ħaddiema]$_{PP}$* 'Schulz is a person who rose [**from among** the workers] [. . .].'	(Π16) *fost* 'amongst'
7	academic12	*F'Lulju 1913 sitt Maltin niżlu Seattle fl-Istati Uniti [**minn fuq** il-vapur Dunblane]$_{PP}$* 'In July 1913 six Maltese got off [**from** the ship Dunblane] in Seattle in the United States [. . .].'	(Π17) *fuq* 'on'
8	literature77	*ħareġ għalhekk [**minn ġewwa** l-ġnien]$_{PP}$ u mar lejn ir-raħal biex jara* '[. . .] so he came [out of the garden] and went to the village to see [. . .].'	(Π18) *ġewwa* 'inside'
9	european4335	*jkunu mibgħuta bil-vapur [**minn ġo**l-portijiet Kanadiżi]$_{PP}$* '[. . .] they will be sent via ship [**from** the Canadian ports].'	(Π19) *ġo* 'in'
10	culture1469	*Il-maratona se tibda fl-10 am [**minn ħdejn** il-Knisja ta' San Girgor iż-Żejtun]$_{PP}$ u se tgħaddi minn tlett lokalitajiet* 'The marathon will start at 10 am [**from** the Church of St. Gregory in Żejtun] and will pass through three localities [. . .].'	(Π23) *ħdejn* 'beside'
11	culture1781	*meta kienu qed ineħħu ż-żebgħa [**minn ma**l-ħitan]$_{PP}$ tfaċċaw diversi graffiti* '[. . .] when they were removing the paint [**from** the walls] several graffiti appeared [. . .].'	(Π30) *ma'* 'with'
12	news51585	*ovvjament nitbiegħdu [**minn madwar** il-mejda]$_{PP}$* '[. . .] obviously we move away [**from** the table] [. . .].'	(Π31) *madwar* 'around'
13	literature7	*Klaus ġabar mazz karti [**minn maġenb** il-munzell trab]$_{PP}$* 'Klaus gathered a pack of cards [**from beside** the pile of dust] [. . .].'	(Π32) *maġenb* 'close to'
14	news109642	*Gvern Laburista [**minn qabel** il-kampanja elettorali]$_{PP}$ saħaq li gvern ġdid se jkun wieħed meritokratiku* 'A Labor government [**from before** the election campaign] stressed that the new government will be a meritocratic one [. . .].'	(Π46) *qabel* 'before'

Table 5.25 (continued)

#	source	example / translation	Π_{CONFIG}
15	literature21	*Ftit tal-ħin wara tfaċċa Xandru [**minn qalb** is-siġar]$_{PP}$* 'A little later, Xandru appeared [**from among** the trees] [. . .].'	(Π47) *qalb* 'amidst'
16	news124234	*hija ġejja lura [**minn qrib** Lampedusa]$_{PP}$* '[. . .] she is coming back [**from near** Lampedusa] [. . .].'	(Π48) *qrib* 'near'
17	culture2338	*l-ħbieb l-oħra wkoll ħarbu [**minn quddiem** is-salib]$_{PP}$ u ħallew lil Ġesù waħdu.* '[. . .] the other friends also fled [**from** the cross] and left Jesus alone.'	(Π49) *quddiem* 'in front of'
18	literature29	*Kellha troffa xagħar ħamrani ħierġa [**minn taħt** il-maktur]$_{PP}$* 'She had a tuft of red hair coming out [**from under** the headscarf] [. . .].'	(Π56) *taħt* 'under'
19	news72525	*Nisimgħu tir ieħor ġej [**minn viċin** il-Kon-Kattidral ta' San Ġwann]$_{PP}$.* 'We hear another shot coming [**from near** the Co-Cathedral of San Ġwann].'	(Π58) *viċin* 'near'
20	literature16	*Meta ttawwalt [**minn wara** l-purtieri tat-tieqa]$_{PP}$ lmaħt kuntistabbli fuq il-bankina faċċata tad-dar.* 'When I peeked out [**from behind** the window curtains] I saw a constable on the sidewalk in front of the house.'	(Π60) *wara* 'after'

There are altogether 22 configurational Πs that combine with (Π41) *minn* 'from'. *Minn*$_{ABLATIVE}$ is attested with twenty (= 91%) or, if (Π38) *mingħand* 'from s.o.' is counted in, 21 configurational Πs (= 95%). Like in the case of (Π21) *għal* 'for', these shares and absolute numbers become meaningful only when they are compared to the shares and absolute numbers calculated for other categories.

The second most frequent meaning of (Π41) *minn* 'from' is the PROLATIVE. *Minn*$_{PROLATIVE}$ is attested with thirteen different configurational Πs (= 59%) as shown in Table 5.26.

Table 5.26: Combinations of *minn*$_{PROLATIVE}$ with configurational Πs.

#	source	example / translation	Π_{CONFIG}
1	parl1765	*ninterċettaw lil kulmin ikun għaddej [**minn barra** l-ibħra territorjali tagħna]$_{PP}$ għal skopijiet ta' traffikar ta' droga* '[. . .] we intercept anyone passing [**outside** our territorial waters] for drug trafficking purposes [. . .].'	(Π3) *barra* 'outside'

Table 5.26 (continued)

#	source	example / translation	Π_{CONFIG}
2	news152058	*Il-vetturi kollha meta jgħaddu [**minn biswit** il-Kappella tal-Ħġejjeġ]_{PP} u jiġbdu lejn Dawriet il-Gżejjer ma jistgħux iduru lejn Plajjiet Bognor* 'All vehicles when they pass [**next to** the Kappella tal-Ħġejjeġ] and pull towards Dawriet il-Gżejjer cannot turn towards Bognor Beaches.'	(Π8) *biswit* 'facing'
3	(Borg/Azzopardi-Alexander 1997: 167)	= (5.154)	(Π12) *faċċata* 'opposite'
4	literature40	*Ipparkja fit-tarf tat-triq u kien għoddu wasal quddiemu meta għadda [**minn fejn** il-vann tal-pastizzi]_{PP}.* 'He parked at the end of the road and was about to get in front of him when he passed [the pastry van].'	(Π14) *fejn* 'near'
5	literature80	*Għadda [**minn ħdejn** il-bieb tal-kamra tiegħi]_{PP}* 'He passed [**by** the door of my room] [. . .].'	(Π23) *ħdejn* 'beside'
6	literature34	*ma jidhrux fid-dlam tal-platea mixgħula biss bid-dawl jiżżerżaq [**minn ma**l-palk]_{PP}.* '[. . .] they do not appear in the darkness of the stage lit only by the light sliding [**across** the stage].'	(Π30) *ma'* 'with'
7	news152258	*jonqos it-traffiku li jgħaddi [**minn madwar** il-pjazza]_{PP} fi kwalunkwe ħin tal-ġurnata* '[. . .] he reduces the traffic that passes [**around** the square] at any time of the day [. . .].'	(Π31) *madwar* 'around'
8	news15453	*Din it-triq tgħaddi [**minn maġenb** il-park nazzjonali]_{PP}* 'This road passes [**next to** the national park] [. . .].'	(Π32) *maġenb* 'close to'
9	news150647	*n-nies li jkunu għaddejjin [**minn qrib** it-Tokk]_{PP} jieqfu jaraw x'qed jiġri.* '[. . .] the people who are passing [**near** the Tokk] stop to see what is happening.'	(Π48) *qrib* 'near'
10	culture3141	*Bniedem adult jiġrilu l-istess meta jgħaddi [**minn quddiem** post fejn jinżammu l-klieb bla dar]_{PP}* 'The same thing happens to a grown man when he passes [**in front of** a place where homeless dogs are kept].'	(Π49) *quddiem* 'in front of'
11	literature59	*qabad iressaq siequ bil-mod [**minn taħt** il-mejda]_{PP} sakemm messha ma' tagħha* '[. . .] he began to move his foot slowly [**under** the table] until it touched hers.'	(Π56) *taħt* 'under'

Table 5.26 (continued)

#	source	example / translation	Π_{CONFIG}
12	news55851	*jispjega l-avventura mhux tas-soltu li s-sajjieda esperjenzaw, bix- xark jgħaddi għal numru ta' drabi [**minn viċin** ħafna tad-dgħajsa]$_{PP}$.* '[. . .] it explains the unusual adventure the fishermen experienced, with the shark passing a number of times [very **close to** the boat].'	(Π58) *viċin* 'near'
13	literature20	*Minn bejnhom dehret it-triq li tgħaddi [**minn wara** l-iskola]$_{PP}$* 'From between them the road was visible that passes [**behind** the school].'	(Π60) *wara* 'after'

Each Π_{CONFIG} that is attested to combine with *minn*$_{PROLATIVE}$ is also attested in combination with *minn*$_{ABLATIVE}$. There is thus often a need for disambiguation. This task is the responsibility of the verb and/or the wider context. What is interesting when we compare Tables 5.25 and 5.26 are not so much the thirteen configurational Πs that are compatible both with *minn*$_{ABLATIVE}$ and *minn*$_{PROLATIVE}$, but the set of seven configurational Πs that allow for combining with *minn*$_{ABLATIVE}$ but not with *minn*$_{PROLATIVE}$. They are interesting because with (Π16) *fost* 'amongst', (Π18) *ġewwa* 'inside', (Π19) *ġo* 'in', (Π46) *qabel* 'before', and (Π47) *qalb* 'amidst', five correspond to configurational Πs that are excluded from combinations with (Π51) *sa* 'till' and (Π28) *lejn* 'towards' as shown in the foregoing section. Conceptually, the PROLATIVE does not allow for the Figure to be inside the Ground at any point of the motion event. Therefore, configurations such as $\text{MEDIAL}_{1/2}$ and INTERIOR do not lend themselves to combinations with *minn*$_{PROLATIVE}$. This explanation can be extended also to the failure of (Π4) *bejn* 'between' to combine with *minn*$_{PROLATIVE}$. The absence of examples of (Π17) *fuq* 'on' in combination with *minn*$_{PROLATIVE}$ can probably be attributed to the inadequate size of our corpus. Nothing conceptual seems to stand in the way of combining (Π17) *fuq* 'on' with *minn*$_{PROLATIVE}$, not the least because #11 in Table 5.26 proves that (Π56) *taħt* 'under', the antonym of (Π17) *fuq* 'on', has no problems as to combining with *minn*$_{PROLATIVE}$. Whether the temporality of (Π46) *qabel* 'before' is an obstacle to a combination with *minn*$_{PROLATIVE}$ needs to be investigated in the future.

The third and last meaning of (Π41) *minn* 'from' is the PERLATIVE. We have found evidence in the Korpus Malti 3.0 of only six (= 27%) different patterns of combinations with configurational Πs. These combinations are illustrated in Table 5.27.

Table 5.27: Combinations of *minn*$_{PERLATIVE}$ with configurational Πs.

#	source	example / translation	Π$_{CONFIG}$
1	literature21	*Id-dawl tax-xemx dieħel [**minn bejn** il-frieghi tas-siġar]$_{PP}$* 'The sunlight comes in [**through** the branches of the trees].'	(Π4) *bejn* 'between'
2	news119355	*Il-kontinġent mhux se jgħaddi [**minn fost** il-pubbliku]$_{PP}$ iżda se joħroġ mill-gates tas-sigurtà ta' l-MIA.* 'The contingent will not pass [**through** the public] but will exit through the security gates of the MIA.'	(Π16) *fost* 'amongst'
3	european197	*kienu użati għall-operazzjoni ta' ġarr in transitu fit-triq [**minn ġewwa** l-Awstrija]$_{PP}$* '[. . .] they were used for the transit operation on the way [**through** Austria] [. . .].'	(Π18) *ġewwa* 'inside'
4	european197	*għamlu operazzjonijiet ta' transitu [**minn ġo**l-Awstrija]$_{PP}$* '[. . .] they carried out transit operations [**through** Austria] [. . .].'	(Π19) *ġo* 'in'
5	news6695	= (5.143)	(Π33) *matul* 'during'
6	culture559	*Qabel ma tibda l-udjenza, il-Papa mistenni jgħaddi bil-Popemobil [**minn qalb** il-pellegrini]$_{PP}$* 'Before the audience begins, the Pope is expected to pass in the Popemobil [**through** the (crowd of) pilgrims] [. . .].'	(Π47) *qalb* 'amidst'

It is hardly surprising that *minn*$_{PERLATIVE}$ is possible with configurational Πs that do not combine with *minn*$_{PROLATIVE}$. On the conceptual level, the PERLATIVE requires the Figure to move through or across the Ground so that it is temporarily contained in the Ground. This possibility is categorically ruled out for the prolative. This also holds for (Π33) *matul* 'during' provided the configuration is connected to a temporal concept. The distribution of *minn*$_{PERLATIVE}$ and *minn*$_{PROLATIVE}$ over the combinations with configurational Πs is thus complementary. In a given context, only one of the meanings is the correct choice. On account of the mutual exclusion of PROLATIVE and PERLATIVE, one might doubt that it makes sense to distinguish the meanings terminologically, in the first place. Would it not be more adequate to lump prolative and PERLATIVE together under the umbrella of Path? The subsequent section will give an answer to this and further questions that arise from synonymy and polysemy of the Πs under scrutiny.

5.2.1.4.4 Polysemy vs synonymy in the spatio-temporal domain

In this section, we scrutinise in some detail form-function relationships of the one-to-many kind (= polysemy) and the many-to-one kind (= synonymy) attested with the spatio-temporal segment of BLOMP 2.0. The first subsection is dedicated to polysemy[182] while synonymy is treated afterwards.[183] The discussion of these semantic phenomena provides an opportunity for us to also look into the possibility of certain semantic parameters determining the choice of Π. We provide schemas that are meant to capture the internal organisation of the two main components of the spatio-temporal domain.[184]

5.2.1.4.4.1 Polysemy

Thanks to the strict differentiation of directionality and configuration markers and also thanks to the introduction of the distinction $Ø_{static}$ vs $Ø_{dynamic}$, it has been shown above that polysemy is not as common across the spatio-temporal domain as it might have appeared to be the case on superficial inspection. In connection with this problem, we have addressed zeroes and their meanings.[185] Multifunctionality among configurational Πs was in the focus of attention[186] whereas the potential polysemy of (Π41) *minn* 'from' was the topic of another section.[187] It is the latter two issues we will raise again in the subsequent paragraphs.

5.2.1.4.4.1.1 Configurational Πs

Table 5.20 features the six configurational Πs that are reported to fulfil two functions. Three of these Πs are used for the expression of PROPINQUUS, namely (Π8) *biswit* 'facing', (Π31) *madwar* 'around', and (Π30) *ma'* 'with'. They differ only as to the second function for which they are recruited. It is this second function that renders it difficult for the analyst to speak of full synonymy in the case of these three Πs. Their potential synonymy will be explored subsequently.[188] In this section however, we aim at clarifying whether the pairings of the functions PROPINQUUS-FRONS, PROPINQUUS-CIRCUM, and PROPINQUUS-SEQUENS instantiate polysemy or can be explained differently.

What can be said in reference to (Π8) *biswit* 'facing' is that FRONS is a possible (though perhaps frequent) reading of PROPINQUUS no matter what the English translation equivalent of the lexicon form of this Π suggests. In spatial situations whose description involves (Π8) *biswit* 'facing', the position the Figure can occupy is not restricted to that in front of the Ground. In the Korpus Malti 3.0, the bulk of

182 Cf. Section 5.2.1.4.4.1.
183 Cf. Section 5.2.1.4.4.2.
184 Cf. Section 5.2.1.4.4.3.
185 Cf. Section 5.2.1.3.
186 In Section 5.2.1.4.2.
187 Cf. Section 5.2.1.4.3.2.
188 Cf. Section 5.2.1.4.4.2.

the hits for (Π8) *biswit* 'facing' is semantically ambiguous insofar as the Π serves to locate the Figure near the Ground with the Ground's front being an option for the placement of the Figure. This situation comes clearly to the fore when we compare the (a)- sentence and the (b)-sentence in (5.224).

(5.224) PROPINQUUS-FRONS

(a) [Korpus Malti 3.0; news118966]

Huwa	*nstab*	*mejjet*	***[biswit***	*il-vettura*	*tiegħu]*$_{PP}$
3SG.M	REFL:find	dead	**[facing**	DEF-vehicle	of:3SG.M]

'He was found dead [**next to** his vehicle].'

(b) [Korpus Malti 3.0; news128991]

Is-sewwieq	*waqaf*	*preċiż*	***[biswit***	*il-karozza*	*tal-bank]*$_{PP}$
DEF-driver	stop	exact	**[facing**	DEF-car	of:DEF-bank]

'The driver stopped exactly [**in front of** the bank car].'

Two different translations of (Π8) *biswit* 'facing' are used in (5.224). As to the (a)-sentence, only English *next to* makes sense since in the immediately preceding context in the news report the victim is located on the sidewalk, i.e., the dead body is not at the front end of the car. For the (b)-sentence, the most obvious choice for translating (Π8) *biswit* 'facing' is English *in front of* since the driver's car comes to a stop such that the car of the bank cannot move from its position. It is therefore logical to assume that the driver's car stops at the bonnet of the car of the bank. Alternatively, even in the (b)-sentence, (Π8) *biswit* 'facing' could mean that the driver's car stops next to the bank's car – close enough to stop the latter from driving off.

As shown in (5.225), the dedicated FRONS marker (Π49) *quddiem* 'in front of' is semantically much more explicit than (Π8) *biswit* 'facing' when the intended meaning is that the Figure is in front of the Ground.

(5.225) [Korpus Malti 3.0; news184510]

huwa	*kellu*	*jieqaf*	*minnufih*	*meta*	*ftit*
3SG.M	have.PFV:3SG.M	3SG.M.IPFV:stop	immediately	when	little

metri	***[quddiem***	*il-karozza*	*tiegħu]*$_{PP}$	*ra*	*xi*	*ħaġa*	*kbira*	*taqa'*
meter:PL	**[in_front_of**	DEF-car	of:3SG.M]	see	some	thing	big:F	3SG.F.IPFV:fall

'[. . .] he had to stop abruptly when he saw something big falling a few meters [**in front of** his car] [. . .].'

The scenario is probably like this. The driver looked through the windscreen of his car when something fell down on the road in front of the car. This must have happened at the car's bonnet. There is no doubt that this is a case of FRONS. In contrast, (5.224b) may also be interpreted such that the Figure is not at the front of the Ground. On account of this ambiguity, we conclude that the main function of (Π8) *biswit* 'facing' is the coding of PROPINQUUS with FRONS being a context-induced reading of PROPINQUUS. Whether other readings of PROPINQUUS are possible with (Π8) *biswit* 'facing' (such as POSTERIOR) can best be determined with the help of native speakers and is thus a question that needs to be addressed in the future.

(Π8) *biswit* 'facing' has been shown not to be multifunctional, in the first place, because the two supposedly distinct meanings of this Π turn out to be of different kinds. The higher-level meaning is that of PROPINQUUS whereas FRONS is a secondary interpretation of PROPINQUUS according to the context. Mutatis mutandis, this explanation also holds for PROPINQUUS-CIRCUM as associated with (Π31) *madwar* 'around'. In analogy to the examples given for ambiguous (Π8) *biswit* 'facing' in (5.224), we provide two examples of (Π31) *madwar* 'around' in (5.226).

(5.226) PROPINQUUS-CIRCUM

(a) [Korpus Malti 3.0; literature34]
l-art ***[madwar*** *il-karozza]*$_{PP}$ *kienet* *tinkesa*
DEF-ground **[around** DEF-car] be:3SG.F.PFV 3SG.F.IPFV:REFL:cover
biċċiet *tas-sigaretti*
piece:PL of:DEF-cigarette
'[. . .] the ground [**around** the car] was covered with cigarette butts.'

(b) [Korpus Malti 3.0; culture2450]
l-bniedem *jista'* *jsalva* *wkoll* *billi*
DEF-man 3SG.M.IPFV:can 3SG.M.IPFV:save also by_way_of
joħloq *kolonji* *fl-ispazju* ***[madwar*** *il-pjaneta*
3SG.M.IPFV:create colony:PL in:DEF-space **[around** DEF-planet
tagħna]$_{PP}$
of:1PL]
'[. . .] man can also save (humankind) by way of founding colonies in the space [**around** our planet].'

Proximity is a relative notion. The interpretation depends to some extent on the size of the Ground. Compared to the Earth mentioned in the (b)-sentence, the car mentioned in the (a)-sentence is very small. Accordingly, the distance between the car and the ground surrounding it is small, too. The space around the Earth, however, is theoretically unlimited and the human colonies in space might be located very far from the planet. How near the Figure is to the Ground is largely irrelevant for the use of (Π31) *madwar* 'around'. This Π is chosen not because of Figure and Ground being close to each other but because we have a distributed Figure whose constituent parts occupy different positions which ideally encircle the Ground – whether nearby it or far away from it. The position of a distributed Figure around a Ground frequently means that the Figure is located in the vicinity of the Ground. PROPINQUUS is thus a secondary reading of CIRCUM. This means that like (Π8) *biswit* 'facing', (Π31) *madwar* 'around' does not attest to multifunctionality.

How is the pair PROPINQUUS-SEQUENS to be analysed? The functions to which (Π30) *ma'* 'with' is put in Maltese have extensively been discussed.[189] The examples in (5.227) help us to find out whether we are facing a genuine case of polysemy.

(5.227) PROPINQUUS-SEQUENS

(a) [Korpus Malti 3.0; literature43]

bdew	*jibnu*	*ruxxmata*	*flettijiet*	*kull*	*fejn*
begin:3PL.PFV	3.IPFV:build:PL	large_quantity	flat:PL	each	where

akkwistaw	*biċċa*	*art*	*tmiss*	***[mat-triq]**$_{PP}$*
acquire:3PL.PFV	piece	ground	3SG.F.IPFV:touch	[**with**:DEF-street]

'[. . .] they began building flats wherever they bought patches of land bordering [**on** the street].'

(b) [Korpus Malti 3.0; literature77]

mexa	*mbagħad*	***[mat-triq]**$_{PP}$*	*għal*	*biswit*	*id-dar*	*ta'*	*Lippu*
walk	then	[**with**:DEF-street]	for	facing	DEF-house	of	Lippu

'[. . .] he then walked [**along** the street] to opposite Lippu's house [. . .].'

On the basis of examples like those in (5.227), we assume that (Π30) *ma'* 'with' is basically a marker of the PROPINQUUS provided the Ground is a spatio-temporal entity that is extended on at least one dimension. This is a requirement that is met by *triq* 'street' in both (5.227a) and (5.227b). SEQUENS is not a basic meaning of (Π30) *ma'* 'with' but a reading of PROPINQUUS in accordance with the semantics of the predicate kernel. As argued above already,[190] a Figure that moves along an extended Ground is metonymically close to it. A Figure that is near an extended Ground covers part of the Ground's extension. Thus, even in the case of (Π30) *ma'* 'with', there is no reason to defend multifunctionality because one meaning is derived from the other according to the context in which the PP occurs.

189 In Section 5.1.2.2.20.1 and its subdivisions.
190 Cf. Section 5.1.2.2.20.1.4 (i).

The situation is not much different in the case of (Π15) *fi* 'in'. This Π is registered in Table 5.20 as exhibiting polysemy with INTERIOR-SUPERIOR. In (5.228), we contrast the (a)-sentence as bona fide example of INTERIOR with the (b)-sentence for which a SUPERIOR-reading seems to be the best solution.

(5.228) INTERIOR- SUPERIOR

(a) [Korpus Malti 3.0; literature1]

Poġġiet *iċ-ċrieket* *u* *l-brazzuletti* *[fil-kaxxa* *żgħira]$_{PP}$*
put:3SG.F.PFV DEF-ring:PL and DEF-bracelet:PL [**in**:DEF-box small:F]
'She put the rings and bracelets [**into** the little box] [. . .].'

(b) [Korpus Malti 3.0; news54925]

Ferrex *ftit* *icing sugar* *minn* *ġo* *passatur* *fin* *[fil-wiċċ*
spread.IMP little icing_sugar from in sieve fine [**in**:DEF-face
tal-kejk]$_{PP}$
of:DEF-cake]
'Spread a little icing sugar through a fine sieve [**on** the surface of the cake].'

Since there is a distinct Π_{CONFIG} that is dedicated to the expression of SUPERIOR this category cannot be downgraded to the rank of a reading. The Π we are alluding to is (Π17) *fuq* 'on'. A context similar to that in (5.228b) is given in (5.229) where (Π17) *fuq* 'on' is used instead of (Π15) *fi* 'in'.

(5.229) [Korpus Malti 3.0; news186628]

Itfa' *naqra* *ġobon* *Parmeġjan* *[fuq* *il-wiċċ]$_{PP}$*
throw.IMP a_bit cheese Parmesan [**on** DEF-face]
'Sprinkle a bit of Parmesan cheese [**on** top].'

As is clearly visible from Figure 5.45, (Π15) *fi* 'in' is the primary option not only when it comes to expressing SUPERIOR in combination with *il-wiċċ* 'the face' as complement but also in combination with *il-qiegħ* 'the bottom' and *il-gżira* 'the island'. In the Korpus Malti 3.0, (Π17) *fuq* 'on' takes these complements between five and thirty-four times less frequently than its competitor (Π15) *fi* 'in'. However, other complement nouns such as *is-siġġu* 'the chair', *il-pjaneta* 'the planet', and *il-mejda* 'the table' give evidence of a strong preference for (Π17) *fuq* 'on'. With these complements, (Π17) *fuq* 'on' is attested from two to 3,510 times more often than (Π15) *fi* 'in'.

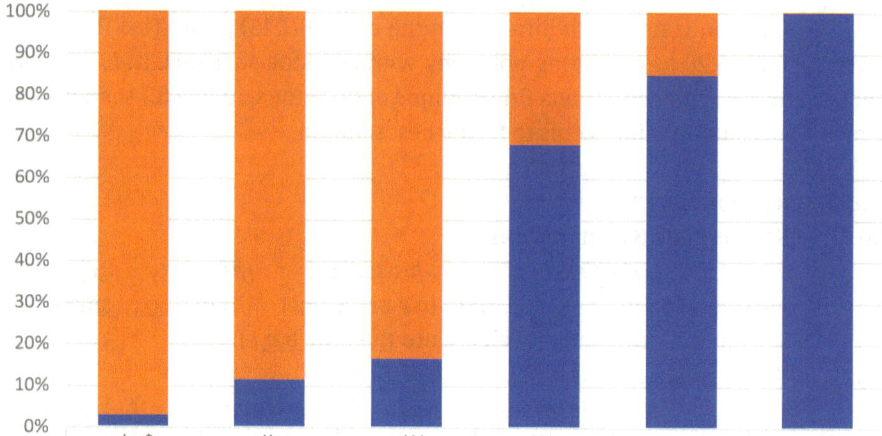

	qiegħ	gżira	wiċċ	siġġu	pjaneta	mejda
fi	1093	3841	2096	135	13	17
fuq	32	493	414	288	73	59680

Figure 5.45: Competition between (Π17) *fuq* 'on' and (Π15) *fi* 'in' in combination with six different complements (Korpus Malti 3.0).

The referents of the complement nouns mentioned in Figure 5.45 can be conceived of as objects with a relatively huge surface that functions as Ground in the spatial situation. The Figure is placed on(to) the surface. No clear picture emerges from Figure 5.45. Some surfaces display a preference for (Π17) *fuq* 'on' and others for (Π15) *fi* 'in'. The shares of the two contestants are always separated by a huge gap. This differential distribution over complement nouns calls for a dedicated in-depth study that must be conducted separately in the future. For the time being, we tentatively conclude that (Π15) *fi* 'in' attests to polysemy and that this Π can express SUPERIOR provided the Ground is a sizable horizontal surface. The criterion SURFACE will return in the section on synonymy below.

A subsection was dedicated to a sketch of the functional overlap between (Π4) *bejn* 'between' (otherwise associated with the INTERESSIVE) and the MEDIOESSIVE markers (Π16) *fost* 'amongst' and (Π47) *qalb* 'amidst'.[191] Since both the INTERESSIVE and the MEDIOESSIVE situate the Figure in a position in the middle of a distributed Ground that may consist of two or more entities, the configurations MEDIAL_1 – MEDIAL_2 are distinguished. The examples in (5.230) reflect this distinction.

191 Cf. Section 5.1.2.1.14.

(5.230) MEDIAL$_1$ – MEDIAL$_2$

(a) [Korpus Malti 3.0; academic376]

L-Art *prinċipali* *tinsab* ***[bejn*** *il-latitudni*
DEF-ground principal 3SG.F.IPFV:REFL:find **[between** DEF-latitude
ta' *49°* *u* *ta'* *59°* *N]$_{PP}$*
of 49° and of 59° N]
'The mainland is located [**between** latitudes 49° and 59° N].'

(b) [Korpus Malti 3.0; literature59]

lanqas *kien* *laħaq* *wasal* *nofs* *triq* *jimbotta*
not_even be reach arrive half street 3SG.M.IPFV:push
[bejn *in-nies]$_{PP}$* *u* *jipprova* *jikkonċentra* *fuq*
[**between** DEF-people] and 3SG.M.IPFV:try 3SG.M.IPFV:concentrate on
x' *qed* *jagħmel* *imbeżża'* *li* *jinsa* *li* *ma*
what PROG 3SG.M.IPFV:do PART:fear that 3SG.M.IPFV:forget that NEG
bediex *jaqta'* *qalbu*
begin:NEG 3SG.M.IPFV:cut heart:3SG.M
'[. . .] he hadn't even reached halfway, pushing [**between** the people] and trying to concentrate on what he was doing, afraid that he would forget, that he didn't start to give up.'

Conceptually, being located between two entities or amongst a larger group of entities are two situations that are related to each other. However, neither the INTERESSIVE nor the MEDIOESSIVE can be claimed to represent but a reading of the one or the other. This results from the impossibility of replacing (Π4) *bejn* 'between' with either (Π16) *fost* 'amongst' or (Π47) *qalb* 'amidst' when the intended meaning is that of the INTERESSIVE, i.e., when MEDIAL$_2$ applies. In contrast, (Π4) *bejn* 'between' is a licit replacement of (Π16) *fost* 'amongst' and (Π47) *qalb* 'amidst' in situations that can be labelled MEDIOESSIVE. The MEDIAL$_1$ constitutes the zone of competition of the three Πs. This competition will concern us again.[192] What is important in the context of polysemy is the fact that in the case of (Π4) *bejn* 'between', the distinction of MEDIAL$_1$ vs MEDIAL$_2$ has repercussions on the expression side because MEDIAL$_1$ triggers a construction that differs structurally from that triggered by MEDIAL$_2$. All Πs expressing MEDIAL$_1$ – be it (Π4) *bejn* 'between', (Π16) *fost* 'amongst', or (Π47) *qalb* 'amidst' – take a complement that is plural or collective. All entities that make up the distributed Ground are representatives of the same ontological category. The constructions used for the MEDIAL$_2$ involve two distinct identifiable entities that together form the distributed Ground but do not have to belong to the same onto-

192 In Section 5.2.1.4.4.2 on synonymy.

logical class. This is a characteristic of (Π4) *bejn* 'between' in INTERESSIVE function. The complements can be singular. More often than not, the MEDIAL$_2$ goes along with the overt coordination of the two complements. There are thus functional as well as formal reasons for considering (Π4) *bejn* 'between' a genuine case of polysemy.

The last case to be inspected in this section is (Π33) *matul* 'during' and its employment for LONGITUDO$_1$ and LONGITUDO$_2$. In (5.231), the two contexts in which this Π is used are illustrated again.

(5.231) LONGITUDO$_1$ – LONGITUDO$_2$

(a) [Korpus Malti 3.0; news86289]

Dawn it-tabelli jinsabu fil-passaġġ riservat
DEM:PL:PROX DEF-sign:PL 3.IPFV:REFL:find:PL in:DEF-passage reserve:PART
għaċ-ċiklisti ***[matul*** *il-kosta ta' Wied il-Għajn]*$_{PP}$
for:DEF-cyclist:PL [**during** DEF-coast of Wied il-Għajn]
'These signs are found in the passage reserved for cyclists [**along** the coast of Wied il- Għajn].'

(b) [Korpus Malti 3.0; literature3]

Kien jidher ċar li xi ħadd kien daħal
be 3SG.M.IPFV:be_visible clear that some someone be enter
fl-iskola ***[matul*** *il-weekend]*$_{PP}$ *imma ħadd ma kien*
in:DEF-school [**during** DEF-weekend] but someone NEG be
jaf min
know who
'It was clear that someone had entered the school [**during** the weekend], but no one knew who.'

As is known from the previous discussion,[193] LONGITUDO$_1$ and LONGITUDO$_2$ reflect the distinction of spatial (= (a)-sentence) and temporal (= (b)-sentence) uses of (Π33) *matul* 'during'. In both cases, the referent of the complement noun is conceived of as an entity that is extended on a dimension – an extension that can be interpreted as length. The Ground in (5.231a) stretches as far as the topographic object reaches. In (5.231b), the break-in happened in a period of time that spans over a weekend. There is no compelling reason to assume that *matul*$_{SPATIAL}$ and *matul*$_{TEMPORAL}$ are categorically distinct to classify this as a case of polysemy. What can be done instead is interpreting the one as a special reading of the other. Given the conceptual primacy of space over time (Heine/Claudi/Hünnemeyer 1991), we assume that *matul*$_{TEMPORAL}$ is a metonymical reading of *matul*$_{SPATIAL}$. Thus, (Π33) *matul* 'during' is not involved in polysemy.

193 Cf. Section 5.2.1.4.2.

5.2.1.4.4.1.2 Treble *minn*

Three earlier subsections contain information about the ABLATIVE, the PERLATIVE, and the PROLATIVE all of which are expressed by (Π41) *minn* 'from'.[194] The question was asked whether it makes sense to distinguish the PERLATIVE from the PROLATIVE since the distribution of the two meanings over combinations with configurational Πs is complementary.[195] An economic alternative account would subsume both meanings under the general category Path. The complementary distribution is a strong argument in favour of the alternative analysis. However, there is a formal property of the PROLATIVE that suggests that treating PERLATIVE and PROLATIVE as two distinct functions is at least as compelling a solution as lumping all Path categories together. The formal difference between *minn*$_{PERLATIVE}$ and *minn*$_{PROLATIVE}$ has been mentioned already in the discussion of the data presented in Table 5.16. As with *minn*$_{ABLATIVE}$, (Π41) *minn* 'from' can never be omitted from a PERLATIVE construction (but cf. below). In contrast to this mandatory use of the Π, (Π41) *minn* 'from' is optional in PROLATIVE constructions.

In what follows we skip the ABLATIVE use of (Π41) *minn* 'from' because it poses no analytical problems. In contrast, the PERLATIVE and PROLATIVE uses of the same Π do. In (5.232), we provide two further examples of the verb *għadda* 'pass' combined with a *minn*-PP whose interpretation is context-dependent.

(5.232) PERLATIVE-PROLATIVE I

(a) PERLATIVE – [Korpus Malti 3.0; news6697]

Minn hemmhekk beda l-korteo lejn
from there begin DEF-funeral_procession towards
*il-Belt Valletta li għadda [**minn** diversi rħula]*$_{PP}$
DEF-city Valletta that pass [**from** diverse:PL village.PL]
'From there the funeral procession started towards Valletta that passed [**through** several villages] [. . .].'

(b) PROLATIVE – [Korpus Malti 3.0; parl217]

Nixtieq nigbed l-attenzjoni tal-Ministru li
1SG.IPFV:want 1SG.IPFV:draw DEF-attention of:DEF-minister that
quddiem fejn kien il-Comet il-Furjana fil-lane ta'
in_front_of where be DEF-Comet DEF-Floriana in:DEF-lane of
barra hemm ħofra kbira li kuljum ikolli
outside there hole big:F that every_day 3SG.M.IPFV:have.FUT:1SG

194 Cf. Sections 5.1.2.2.8, 5.1.2.2.20.2 and 5.1.2.2.20.3.
195 In the final paragraph of Section 5.2.1.4.3.2.

ngħaddi *[mill-ġenb]*$_{PP}$ *biex* *nevitaha*
1SG.IPFV:pass [**from**:DEF-side] to 1SG.IPFV:avoid:3SG.F
'I'd like to draw the minister's attention [to the fact] that in front of the Comet, in Floriana, there is a big hole in the outer lane that I will have to [**pass by** sideways] to avoid it.'

As to the (a)-sentence, it is unlikely that the funeral cortege passes the villages by. On the contrary, there is independent evidence of the march passing through the village streets since this is what happens traditionally in Malta. This is different in the case of the (b)-sentence. The speaker does not want his car to be damaged by way of passing through the big hole in the street. This is why he drives along the side of the hole. Admittedly, examples of the latter kind are definitely rare in the Korpus Malti 3.0. PERLATIVES abound whereas PROLATIVES without additional configuration marker are attested only infrequently.

The different behaviour of PERLATIVE and PROLATIVE comes to the fore when combinations with configurational Πs are looked at. In (5.233), we see that *minn*$_{PROLATIVE}$ can either be absent or present if the motion verb *għadda* 'pass' is used together with the Π_{config} (Π49) *quddiem* 'in front of'.

(5.233) PROLATIVE without and with (Π41) *minn* 'from'
(a) [Korpus Malti 3.0; news130490]
qabel *ma* *bdew* *it-tiġrija* *ħadu* *sehem* *fi* *sfilata*
before that begin:3PL.PFV DEF-race take:3PL.PFV part in parade
fejn *għaddew* *[Ø* ***quddiem*** *il-pubbliku* *preżenti]*$_{PP}$
where PASS:3PL.PFV [**Ø** **in_front_of** DEF-public present]
rekbin *fuq* *żwiemel*
ride.PART:PL on horse.PL
'[. . .] before they started the race they took part in a parade where they passed [**in front of** the present public] riding on horses [. . .].'
(b) [Korpus Malti 3.0; news169292]
It-tiri *fuq* *il-Każin* *Nazzjonalista [. . .]* *kienu* *sparati*
DEF-shot:PL on DEF-club nationalist be:3PL.PFV shot:PART:PL
minn *karozza* *li* *għaddiet* ***[minn*** ***quddiem*** *il-Każin]*$_{PP}$.
from car that pass:3SG.F.PFV [**from** **in_front_of** DEF-club]
'The shots on the Nationalist Club [. . .] were fired from a car that passed [**in front of** the Club].

It does not seem to matter whether (Π41) *minn* 'from' is there or not since the combination of motion verb and configuration marker only allows for a PROLATIVE interpretation. The configuration expressed by (П49) *quddiem* 'in front of' is incompatible with a PERLATIVE reading. As claimed repeatedly above, *minn*$_{PERLATIVE}$ is not normally omitted from combinations with configuration markers so that we assume it to be mandatory. However, there is marginal (and perhaps also doubtful) evidence of dropped *minn*$_{PERLATIVE}$ too in the Korpus Malti 3.0 as the examples in (5.234) show.

(5.234) PERLATIVE without and with (П41) *minn* 'from'

(a) [Korpus Malti 3.0; parl12197]

Jista' l-Ministru jgħin biex tinbidel
3SG.M.IPFV:can DEF-minister 3SG.M.IPFV:help to 3SG.F.IPFV:REFL:change
ir-rotta 136 minn kif inhi bħalissa u tibda
DEF-route_136 from how 3SG.F presently and 3SG.F.IPFV:start
tgħaddi ***[minn ġewwa*** *il-Mosta]*$_{PP}$ *wara li titlaq*
3SG.F.IPFV:pass **[from inside** DEF-Mosta] after that 3SG.F.IPFV:leave
mill-Għargħur qabel ma tinżel Birkirkara
from:DEF-Għargħur before that 3SG.F.IPFV:descend Birkirkara

'Can the minister help to change [the] route [of bus line] 136 from what it is now so that it starts to go [**through** Mosta] after it leaves Għargħur before it gows down to Birkirkara?'

(b) [Korpus Malti 3.0; parl10052]

Ninsab infurmat li r-residenti tal-Kappara
1SG.IPFV:REFL:find inform:PART that DEF-resident:PL of:DEF-Kappara
se jkollhom rotta ċirkulari bin-numru 123
FUT 3SG.M.IPFV:have.FUT:3PL route circular with:DEF-number_123
li titlaq mill-Park & Ride ta' Pembroke
that 3SG.F.IPFV:leave from:DEF-Park_&_Ride of Pembroke
tgħaddi ***[Ø ġewwa*** *l-Kappara]*$_{PP}$ *tibqa' sejra*
3SG.F.IPFV:pass **[Ø inside** DEF-Kappara] 3SG.F.IPFV:remain go.PART:F
lejn l-Isptar Mater Dei u terġa' lura
towards DEF-Hospital Mater Dei and 3SG.F.IPFV:do_again back
lejn il-Park & Ride ta' Pembroke
towards DEF-Park_&_Ride of Pembroke

'I am informed that the residents of Kappara will have a circular bus service with the number 123 that starts from the Park & Ride in Pembroke, passes [**through** Kappara], keeps going towards the Mater Dei Hospital and returns back to the Park & Ride in Pembroke.'

Both examples are situated in the context of parliamentary debates about the new system of public transportation in Malta. In most cases, the situation that involves a route passing through a village is described by the combination of the directionality marker (Π41) *minn* 'from' and the configuration marker (Π18) *ġewwa* 'inside'. The former is absent which is clearly exceptional. Its absence is certainly by far less frequent than in the case of *minn*$_{\textit{PROLATIVE}}$. If the different behaviour of *minn*$_{\textit{PERLATIVE}}$ and *minn*$_{\textit{PROLATIVE}}$ turns out not to be as categorical as originally suggested by us, there is still the possibility that the two directionality markers are prone to being omitted to noticeably different degrees.

In sum, the situation is not entirely clear and thus calls for further in-depth investigations. For the time being, we assume that it is possible to distinguish between *minn*$_{\textit{PERLATIVE}}$ and *minn*$_{\textit{PROLATIVE}}$ with the proviso that future research might reveal that there is only *minn*$_{path}$ with two context-dependent readings.

5.2.1.4.4.2 Synonymy

This section is dedicated to the discussion of potential cases of synonymy as featured in Table 5.21 for configuration markers and separately for directionality markers.[196] The latter are addressed first[197] whereas the more numerous cases of supposed or real synonymy in the domain of configuration are discussed afterwards.[198]

5.2.1.4.4.2.1 Synonymy of directionality markers

According to the schema presented in Figure 5.34, (Π21) *għal* 'for' and (Π28) *lejn* 'towards' are functionally very close to each other as markers of DIRECTIONAL$_1$ and DIRECTIONAL$_2$. Superficially, the examples in (5.235) seem to support the idea of full synonymity between the two Πs.

(5.235) DIRECTIONAL$_1$ = DIRECTIONAL$_2$ – #1

(a) (Π21) *għal* 'for' – [Korpus Malti 3.0; news126071][199]

Dak	*it-traffiku*	*li*	*jkun*	*ġej*	*minn*
DEM:M:DIST	DEF-traffic	that	3SG.M.IPFV:be.FUT	come.PART	from

Triq Rudolfu	*biex*	*imur*	***[għal***	*<u>Triq il-Katidral</u>]*$_{PP}$
Triq Rudolfu	to	3SG.M.IPFV:go	**[for**	<u>Triq il-Katidral</u>]

jgħaddi	*Triq l-Annunzjata*	*Triq San Publiju*	*Triq San Pietru*
3SG.M.IPFV:pass	Triq l-Annunzjata	Triq San Publiju	Triq San Pietru

196 Cf. Section 5.2.1.2.

197 Cf. Section 5.2.1.4.4.2.1.

198 In the six subsections of Section 5.2.1.4.4.2.2.

199 Note that we interpret the use of *jgħaddi* 's/he passes' in combination with the bare street names as PROLATIVE constructions as the motion event involves a Figure (the traffic) that moves along a horizontally extended Ground (the street(s)).

u *jidħol* *fi* *Triq il- Katidral*
and 3SG.M.IPFV:enter in Triq il-Katidral
'This traffic that will be coming from Triq Rudolfu to go [**towards** Triq il-Katidral] passes through Triq l-Annunzjata, Triq San Publiju, Triq San Pietru and enters Triq il-Katidral.'

(b) (Π28) *lejn* 'towards' – [Korpus Malti 3.0; news192722]
il-grupp *ta'* *partitarji* *Laburisti* *mar* ***[lejn*** *Triq il-Kbira*
DEF-group of partisan:PL Labour go [**towards** Triq il-Kbira
f' *Birkirkara]$_{PP}$*
in Birkirkara]
'[. . .] the group of Labour partisans went [**towards** Triq il-Kbira in Birkirkara] [. . .].'

In both sentences, the predicate nucleus is the motion verb *mar* 'go' which is indicative of motion Goal-directed motion. The Ground is represented by a street name in (a) as well as in (b). The motion is telic in the sense that the Figure eventually reaches the Ground. In (5.235a), (Π21) *għal* 'for' is employed to link the verb to the street name whereas it is (Π28) *lejn* 'towards' that is chosen for the same task in (5.235b). We assume that in these sentences the two Πs may replace each other without causing semantic changes.

However, the interchangeability of the two directionality markers does not apply blindly. It seems to be licit only if the Ground is a concrete physical entity such as the toponymically identified settlements in (5.236).

(5.236) DIRECTIONAL$_1$ = DIRECTIONAL$_2$ – #2

(a) (Π21) *għal* 'for' – [Korpus Malti 3.0; news83254]
Minn *Port Said* *l-HMS Cricket* *telqet* ***[għal*** *Lixandra]$_{PP}$*
from Port Said DEF-HMS Cricket leave:3SG.F.PFV [**for** Alexandria]
minn *fejn* *imbagħad* *kellha* *titlaq* ***[għal***
from where then have.PFV:3SG.F 3SG.F.IPFV:leave [**for**
Derna *fil-Libja]$_{PP}$*
Derna in:DEF-Libya]
'From Port Said HMS Cricket left [**for** Alexandria] from where she then had to leave [**for** Derna in Libya] [. . .].'

(b) (Π28) *lejn* 'towards' – [Korpus Malti 3.0; news119253]
mar *għand* *il-parrukkier* *biex* *jaqta'* *xagħru*
go at_s.o.'s_place DEF-hairdresser to 3SG.M.IPFV:cut hair:3SG.M
qabel *jitlaq* ***[lejn*** *Ruma]$_{PP}$*
before 3SG.M.IPFV:leave [**towards** Rome]
'[. . .] he went to the hairdresser to cut his hair before leaving [**for** Rome].'

If, however, the Ground belongs to a different ontological class, the possibility of replacing one Π with the other diminishes drastically. In (5.237) for instance, we find (Π21) *għal* 'for' being used twice with indefinite complements which represent the Ground of a motion event. The Ground is not a proper place but an event, namely an interview and a debate.

(5.237) DIRECTIONAL$_1$ with abstract Ground – [Korpus Malti 3.0; news33986]
Jiena mort ***[għal*** *intervista]$_{PP}$* *ma mortx* ***[għal***
1SG go:1SG.PFV **[for** interview] NEG go:1SG.PFV:NEG **[for**
dibattitu]$_{PP}$
debate]
'I went [**to** an interview], I did not go [**to** a debate].'

In cases of this kind, the replacement of (Π21) *għal* 'for' with (Π28) *lejn* 'towards' is subject to restrictions. In the Korpus Malti 3.0, there is no evidence of the latter taking *intervista* 'interview' as a definite or indefinite complement when the predicate nucleus is *mar* 'go'. We only find it occasionally in combination with *ħares* 'look' as in (5.238a). The complement *dibattitu* 'debate' on the other hand, is attested in *lejn*-PPs which depend on *mar* 'go' but only if the metonymy applies, i.e., the motion event is abstract as in (5.238b).

(5.238) DIRECTIONAL$_2$ with different Grounds
(a) [Korpus Malti 3.0; parl4709]
naħseb li tkun xi ħaġa loġika li ma
1SG.IPFV:think that 3SG.F.IPFV:be.FUT some thing logical:F that NEG
tħarisx biss ***[lejn*** *l-intervista]$_{PP}$* *imma*
2SG.IPFV:look:NEG only **[towards** DEF-interview] but
tħares ukoll ***[lejn*** *it-track record]$_{PP}$*
2SG.IPFV:look also **[towards** DEF-track record]
'[. . .] I think that it will be logical that you do not only look [**at** the interview] but that you also look [**at** the track record] [. . .].'
(b) [Korpus Malti 3.0; news83872]
Meta smajt bil-ħatra ta' Dr Tonio Portughese bħala
when hear:1SG.PFV with:DEF-bet of Dr Tonio Portughese as
Ċermen tal-Bord tad-Diretturi tal-PBS ħsiebi
chairman of:DEF-Board of:DEF-director:PL of:DEF-PBS thought:1SG
dlonk mar ***[lejn*** *id-dibattitu li kien sar qabel*
suddenly go **[towards** DEF-debate that be become before
l-elezzjoni tal-1987]$_{PP}$
DEF-election:PL of:DEF-1987]

'When I heard of the bet of Dr Tonio Portughese as chairman of the Board of Directors of the PBS my thoughts suddenly went [**to** the debate that had taken place before the elections of 1987] [. . .].'

According to Aquilina (1987: 507), *ħares* 'look' normally requires (Π28) *lejn* 'towards' to introduce the participant that is looked at. In this context, the use of (Π21) *għal* 'for' is not permitted.[200] There are thus no examples of this Π in constructions as that given in the (a)-sentence. As to the (b)-sentence, we have only one piece of evidence of (Π28) *lejn* 'towards' being used in a construction that expresses a metonymical motion event involving the verb *mar* 'go'. In this case, the speaker's thoughts are described as going back in his memory to an earlier event that resembles the one that the speaker is currently witnessing. (Π21) *għal* 'for' occurs twelve times as often as linker between *mar* 'go' and *dibattitu* 'debate' – and in each of these cases the debate is an event that takes place at a venue to which the Figure has to go in order to participate in the debate. Neither metonymy nor metaphor can be made responsible for the use of this Π.

To determine whether there is a pattern according to which DIRECTIONAL$_1$ and DIRECTIONAL$_2$ differ in terms of their distribution, we look at eight different nouns that serve as complement in DIRECTIONAL PPS and which refer to events and/or places where certain events take place, namely *audition* 'audition', *intervista* 'interview', *konferenza* 'conference', *lezzjoni* 'lecture', *meeting* 'meeting', *party* 'party', *tournament* 'tournament', and *training camp* 'training camp' (five of which are recent loans from English). We have searched the Korpus Malti 3.0 for cooccurrences of these nouns (with and without definiteness proclitic) and forms of the verb *mar* 'go'. The results are telling as becomes immediately clear when we look at Figure 5.46. In seven out of eight cases, the results show that (Π21) *għal* 'for' is preferred over (Π28) *lejn* 'towards' with the latter being excluded from combinations with five of the complement nouns. With each of the three remaining complement nouns, (Π28) *lejn* 'towards' is attested only once. *Tournament* 'tournament' is the only noun that combines equally infrequently with both of the Πs. In all other cases, (Π21) *għal* 'for' is attested several times peaking at thirty-six attested combinations

200 The rare occasions in which (Π21) *għal* 'for' is attested in combination with *ħares* 'look' trigger a different reading of the verb, namely as translation equivalent of English *keep, save* as in (5.xi).

(5.xi) [Korpus Malti 3.0; parl6802]

Il-Polonja	*per eżempju*	*wkoll*	*ipposponew*	*id-data*	*tagħhom*	*u*
DEF-Poland	for example	also	postpone:3PL.PFV	DEF-date	of:3PL	and

ħarsu	***[għal***	*data*	*aktar*	*'il*	*quddiem]$_{PP}$*
look:3PL.PFV	**[for**	date	more	to	in_front_of]

'Poland for instance, postponed its date too and saved it [**for** a later date].'

with *konferenza* 'conference'. On the basis of these frequencies, we conclude that (Π21) *għal* 'for' is the preferred option when concrete movement to an event has to be expressed whereas (Π28) *lejn* 'towards' is given preference when the motion event is abstract or metonymic.

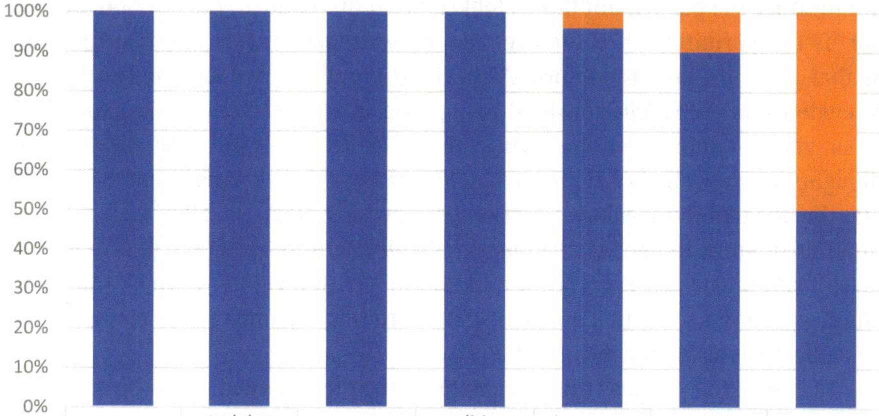

	konferenza	training camp	party	audition, lezzjoni	intervista	meeting	tournament
lejn	0	0	0	0	1	1	1
għal	36	26	11	3	24	9	1

Figure 5.46: Competition of DIRECTIONAL I + II with eight complement nouns (verb: *mar* 'go') in the Korpus Malti 3.0.

Accordingly, we further assume that DIRECTIONAL$_1$ and DIRECTIONAL$_2$ are only partially synonymous. The domains of (Π21) *għal* 'for' and (Π28) *lejn* 'towards' overlap only to a certain extent. The issue is not straightforward as results from the brief discussion of two further areas of competition of (Π21) *għal* 'for' and (Π28) *lejn* 'towards'. In (5.239), examples of these Πs are presented where they depend on the higher PP *fi triqtu* 'on his way'.

(5.239) DIRECTIONAL$_1$ = DIRECTIONAL$_2$ – #3

(a) (Π21) *għal* 'for' – [Korpus Malti 3.0; news77696]

Il-pilota ta' ajruplan tar-Royal Air Maroc mexxa
DEF-pilot of aeroplane of:DEF-Royal_Air_Maroc go.CAUS
l-ajruplan inkwistjoni direttament għal ġol-Muntanji Atlas
DEF-aeroplane in_question directly for in:DEF-Mountain:PL_Atlas
ftit wara li kien tela' minn Agadir fi triqtu
little after that be ascend from Agadir in street:POSS:3SG.M

[għal	*Casablanca]$_{PP}$*								
[**for**	Casablanca]								

'The pilot of the aeroplane of Royal Air Maroc directed the said aeroplane directly into the Atlas Mountains shortly after he had left from Agadir on his way [**to** Casablanca].'

(b) (Π28) *lejn* 'towards' – [Korpus Malti 3.0; news13404]

L-ajruplan	*kien*	*inħataf*	*fit-23*	*ta'*	*Novembru*	*1985*	*ftit*		
DEF-aeroplane	be	REFL:snatch	in:DEF-23	of	November	1985	little		
wara	*li*	*telaq*	*minn*	*Ateni*	*fi*	*triqtu*	***[lejn***		
after	that	leave	from	Athens	in	street:POSS:3SG.M	[**towards**		
il-Kajr]$_{PP}$									
DEF-Cairo]									

'The aeroplane was captured on 23 November 1985 shortly after it left Athens on its way [**to** Cairo].'

Both competitors are admitted to the same construction with no apparent meaning difference. However, there is a pronounced preference for (Π28) *lejn* 'towards' that comes to the fore in Figure 5.47. The token frequencies reported there are those of (Π21) *għal* 'for' and (Π28) *lejn* 'towards' following *fi triqtu* 'on his way'. We have not taken account of the construction involving a possessor other than the 3rd person singular masculine.

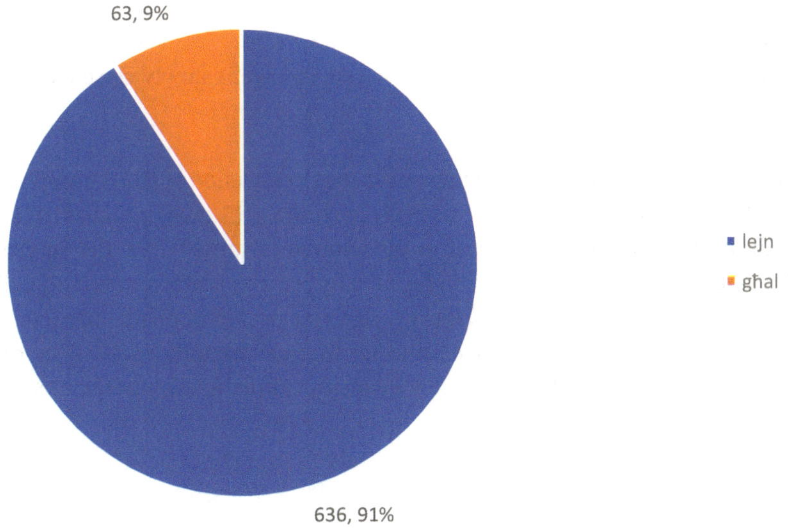

Figure 5.47: Token frequencies of DIRECTIONAL I + II in combination with *fi triqtu* 'on his way' in the Korpus Malti 3.0.

(П28) *lejn* 'towards' is attested ten times as often as (П21) *għal* 'for' in this context. It can be considered the default option when it comes to expressing directionality in association with a planned route.

The second case to comment upon shortly is connected to the verb *resaq* 'approach'.[201] Aquilina (1991: 1207) clearly distinguishes *resaq* + *lejn*-PP from *resaq* + *għal*-PP. In the former case, (tentative) motion towards a concrete or abstract Goal whereas (П21) *għal* 'for' is exclusively used in constructions that refer to examinations or similar events. Examples are given in (5.240).

(5.240) DIRECTIONAL$_1$ ≠ DIRECTIONAL$_2$

(a) (П21) *għal* 'for' – [Korpus Malti 3.0; parl7924]

Jista' *l-Ministru* *jgħid* *kemm-il* *persuna*
3SG.M.IPFV:can DEF-minister 3SG.M.IPFV:say how_many person
resqet ***[għal*** *eżami* *tas-sewqan]*$_{PP}$ *kategorija* *b'*
approach:3SG.F.PFV [**for** exam of:DEF-driving] category with
kategorija *xahar* *b'* *xahar* *din* *is-sena*
category month with month DEM:F:PROX DEF-year
'Can the minister say how many persons sat [**for** the driving exam] category by category, month by month this year?'

(b) (П28) *lejn* 'towards' – [Korpus Malti 3.0; literature29]

Niżlet *it-taraġ* *bla* *ħoss* *u* *resqet*
descend:3SG.F.PFV DEF-stairs without sound and approach:3SG.F.PFV
[lejn *il-presepju]*$_{PP}$ *bil-lajma*
[**towards** DEF-Christmas_crib] with:DEF-slowness
'She went down the stairs without a sound and went slowly close [**to** the Christmas crib].'

We take (5.240a) as a further piece of evidence for the predilection of (П21) *għal* 'for' to take complement nouns which refer to events. In contrast, (5.240b) involves (П28) *lejn* 'towards' as head of a PP that identifies the Ground towards which the Figure moves. The data that we have looked at so far are suggestive of different functional preferences of the two directionals. In Figure 5.48, these preferences are characterised as physical places and events. We emphasise that we are talking about more or less strong preferences and not about strictly disjunct functional subdomains.

201 Cf. also Section 5.2.1.4.4.2.2.1.

DIRECTIONAL	*għal*	/	Ground = event
	lejn	/	Ground = concrete place

Figure 5.48: Different preferences of the two DIRECTIONALS.

The supposed synonymy of (Π21) *għal* 'for' and (Π28) *lejn* 'towards' is only partial. The exact nature of the relation between the two DIRECTIONALS can be determined only in the course of a dedicated in-depth study which is beyond the goals of our own project.

5.2.1.4.4.2.2 Synonymy of configuration markers

In the previous section, we have shown that the extent of the synonymy of DIRECTIONAL$_1$ and DIRECTIONAL$_2$ is limited. It is time now to put synonymy to the test also with configuration markers. In previous sections, we have learned already that there indeed are contexts in which different Πs compete with each other to express PROPINQUUS, INTERIOR, MEDIAL$_1$, FRONS, LONGITUDO$_1$, or SUPERIOR. The guiding question for the subsequent section is however, whether this functional equivalence is punctual or covers the entire range of contexts in which the individual Πs are attested.

5.2.1.4.4.2.2.1 PROPINQUUS

There are eight Πs for which physical closeness to the Ground is registered as main or secondary feature, namely (Π8) *biswit* 'facing', (Π14) *fejn* 'near', (Π23) *ħdejn* 'beside', (Π30) *ma'* 'with', (Π31) *madwar* 'around', (Π32) *maġenb* 'close to', (Π48) *qrib* 'near', and (Π58) *viċin* 'near'. If we go beyond the basic fact that the above Πs convey the idea that the distance between Figure and Ground is relatively small, the behaviour of the Πs is too heterogeneous to define them as a set of full synonyms. That homogeneity does not apply can be shown inter alia by way of determining which member of the group of eight is compatible with what kind of complement (in terms of the complement's semantics).

For a start, we search the Korpus Malti 3.0 for instances of $\Pi_{\text{PROPINQUUS}}$ + *(il-) ġenb* '(the) side' in order to see whether the lateral position of the Figure relative to the Ground plays any role in the choice of Π. The examples in (5.241) prove that only five out of eight PROPINQUUS Πs are compatible with the complement noun *ġenb* 'side'.

(5.241) PROPINQUUS Πs in combination with *(il-)ġenb* '(the) side'

(a) (Π8) *biswit* 'facing' – [Korpus Malti 3.0; european2998]
id-dispożizzjonijiet li jeżistu kienu inadekwati
DEF-disposition:PL that 3.IPFV:exist:PL be:3PL.PFV inadequate:PL
fir-rigward tal-kamp ta' viżjoni fiż-żona
in:DEF-regard of:DEF-field of vision in:DEF-zone
***[biswit** il-ġenb ta' l-istruttura tal-kabina faċċata tas-sewwieq]$_{PP}$*
[**facing** DEF-side of DEF-structure of:DEF-cabin opposite of:DEF-driver]
'[. . .] the extant dispositions were inadequate as to the field of vision in the zone [**near** the side of the cabin's structure opposite the diver].'

(b) (Π14) *fejn* 'near' – [Korpus Malti 3.0; news165677]
*il-vann ukoll sofra daqqiet kbar **[fejn** il-ġenb tan-naħa*
DEF-van also suffer stroke:PL big.PL [**near** DEF-side of:DEF-side
tas-sewwieq]$_{PP}$
of:DEF-driver]
'[. . .] the van too suffered heavy blows [**near** the driver's side] [. . .].'

(c) (Π23) *ħdejn* 'beside' – [Korpus Malti 3.0; news78120]
spikka fost oħrajn is-Sindku ta' Ħal Safi Pietru Pawl Busuttil
stand_out among other.PL DEF-mayor of Ħal Safi Pietru Pawl Busuttil
*li mar **[ħdejn** il-ġenb]$_{PP}$ bil-ħġieġ tal-karozza u bi*
that go [**beside** DEF-side] with:DEF-glass of:DEF-car and with
fjura f' idu beda jibki
flower in hand:3SG.M begin 3SG.M.IPFV:weep
'[. . .] among others the mayor of Ħal Safi, Pietru Pawl Busuttil, stood out who went [**near** the side] with the glass of the car and a flower in his hands began to weep [. . .].'

(d) (Π32) *maġenb* 'close to' – [Korpus Malti 3.0; european8062]
Għall-vetturi li jkollhom kapaċità li ma
for:DEF-vehicle:PL that 3SG.M.IPFV:have.FUT:3PL capacity that NEG
teċċedix 22 passiġġier fil-każ ta' sedili
3SG.F.IPFV:exceed:NEG 22 passenger in:DEF-case of seat:PL
***[maġenb** il-ġenb tal-vettura]$_{PP}$ l-ispazju disponibbli*
[**close_to** DEF-side of:DEF-vehicle] DEF-space available
m' għandux jinkludi fil-parti ta' fuq
NEG have:3SG.M:NEG 3SG.M.IPFV:include in:DEF-part of above
tiegħu żona triangolari 20 mm wiesa bi 100 mm għolja
of:3SG.M zone triangular 20 mm wide:F with 100 mm high:F
'For the vehicles whose capacity does not exceed 22 passengers, in the case of seats [**close to** the side of the vehicle], the available space must not include in its upper part a triangular zone 22 mm wide and 100 mm high.'

(e) (Π30) *ma'* 'with' – [Korpus Malti 3.0; literature21]

*[**Ma**l-ġenb*	*tas-sodda]*$_{PP}$	*kien*	*hemm*	*bħal*	*komodina*	*maħduma*
[**with**:DEF-side	of:DEF-bed]	be	there	like	cupboard	PART:make:F

raffa
rough:F
'[**At** the side of the bed] there was something like a bedside cupboard with a rough surface [. . .].'

It is especially remarkable that in the (d)-sentence, (Π32) *maġenb* 'close to' takes *ġenb* 'side' as complement although the Π itself comprises this noun as one of its two components. This means that (Π32) *maġenb* 'close to' is no longer transparent as to its internal makeup to the Maltese speakers. As to the (e)-sentence, it remains doubtful whether the use of (Π30) *ma'* 'with' implies that Figure and Ground are two pieces of furniture that are placed in such a way that they touch each other. Discounting this uncertainty for the time being, what can be said about the different Πs in (5.241) is that they are probably interchangeable in constructions that involve the above complement noun. Lateral position of the Figure can be expressed by each of them. It must be strongly emphasised, however, that cases of this kind are generally infrequent in the Korpus Malti 3.0.

The absentees from (5.241) are also interesting. The absence of (Π31) *madwar* 'around' is largely self-explanatory since this Π typically requires a complement noun whose referent is an object that can be surrounded or circled by a Figure. This is not the case with *ġenb* 'side' because a side always forms part of a larger entity and thus can neither be surrounded nor circled in isolation from the larger unit. More striking is the absence of (Π48) *qrib* 'near' and (Π58) *viċin* 'near' because, superficially, their meaning seems to be rather neutral so that one would expect them to be compatible also with *ġenb* 'side' in analogy to the other Πs in (5.241). This noun is also a body-part term which may be translated as *flank* or *hip* in English (Aquilina 1987: 386). Its partial synonym *naħa* 'side' however, does not directly associate with a given part of the body (Aquilina 1991: 882). The former lexeme is clearly referring to the lateral part of an entity whereas the latter is not specialised for indicating the lateral part of a given object. Accordingly, the three Πs which never show up in combination with *ġenb* 'side' are attested (albeit infrequently) with *naħa* 'side' in the Korpus Malti 3.0 as the examples in (5.242) confirm.

(5.242) PROPINQUUS Πs combining with *naħa* 'side' I

(a) (Π31) *madwar* 'around' – [Korpus Malti 3.0; culture1782]

Bit-tluq	*mix-xatt*	*ta'*	*tas-Sliema*	*id-dgħajsa*
with:DEF-departure	from:DEF-beach	of	Sliema	DEF-boat

imbagħad *tbaħħar* ***[madwar*** *in-naħa* *t'* *isfel* *ta'* *Malta]*$_{PP}$
then 3SG.F.IPFV:navigate **[around** DEF-side of down of Malta]
'With the departure from the beach of Sliema the boat then navigates [**around** the southern part of Malta] [. . .].'

(b) (Π48) *qrib* 'near' – [Korpus Malti 3.0; parl224]
għadd *kbir* *ta'* *residenti* *li* *jabitaw* *u/jew* *qegħdin* *jibnu* ***[qrib*** *in-naħa* *taċ-Ċimiterju* *ta'* *Ħaż Żabbar]*$_{PP}$ *qegħdin* *joġġezzjonaw* *għal* *estensjoni* *ta'* *l-istess* *Ċimiterju*
number big of resident:PL that 3.IPFV:inhabit:PL and / or PROG:PL 3.IPFV:build:PL **[near** DEF-side of:DEF-cemetery of Ħaż Żabbar] PROG:PL 3.IPFV:object:PL for extension of DEF-same cemetery
'[. . .] numerous residents who live and / or are constructing houses [**near** the cemetery of Ħaż Żabbar] are objecting against the extension of said cemetery.'

(c) (Π58) *viċin* 'near' – [Korpus Malti 3.0; parl325][202]
B' *referenza* *għad-domanda* *tiegħi* *numru* *21699* *biex* *fix-Xatt* *tal-Kalkara* *l-parti* ***[viċin*** *in-naħa* *tal-bandli* *fejn* *kien* *beda* *xogħol]*$_{PP}$ *biex* *parti* *mill-baħar* *tiġi* *reklamata* *għal* *xi* *raġuni* *ftit* *ġimgħat* *wara* *li* *nbeda* *ix-xogħol* *twaqqaf* *kollox* *jista'* *l-Ministru* *jgħidli* *meta* *ħa* *jerġa'* *jitkompla* *dan* *ix-xogħol*
with reference for:DEF-question of:1SG number 21699 to in:DEF-shore of:DEF-Kalkara DEF-part **[near** DEF-side of:DEF-swing:PL where be begin work] to part from:DEF-sea 3SG.F.IPFV:come claim:PART:F for some reason little week:PL after that REFL:begin DEF-work REFL.CAUS.stop everything 3SG.M.IPFV DEF-minister 3SG.M.IPFV:say:IO:1SG when FUT 3SG.M.IPFV:do_again 3SG.M.IPFV:REFL:complete DEM:M:PROX DEF-work
'Referring to my question number 21699 because on the shore of Kalkara in the part [**near** the swings where work had started] so that part of the Sea is claimed for some reason, a few weeks after the work began it was stopped altogether, can the minister tell me when this work will be completed?'

202 On account of the first subordinate clause introduced by the PURPOSIVE conjunction *biex* '(in order) to', this sentence poses problems making it difficult to parse it correctly. Independent of the correctness of our analysis, the *viċin*-PP remains unaffected by any modifications that might be necessary.

The choice of (Π31) *madwar* 'around' for describing the situation in the (a)-sentence makes sense because the Figure executes a circular motion to navigate along the southern coastline of the island. Since the purpose of the journey is touristic, the boat always stays close to the shore so that it is correct to associate the Figure-Ground relation with the PROPINQUUS although expressing the PROPINQUUS is not the main task of the Π. Since the PROPINQUUS is only a secondary feature of (Π31) *madwar* 'around', it cannot replace (Π48) *qrib* 'near' nor (Π58) *viċin* 'near' in the (b)-sentence and (c)-sentence, respectively. In contrast, the latter two Πs are freely interchangeable in (5.242a–b). Note that in these examples, the translation of *in-naħa* 'the side' seems to be superfluous.

In the Korpus Malti 3.0; (Π23) *ħdejn* 'beside' and (Π32) *maġenb* 'close to' are not attested in combination with *in-naħa* 'the side', whereas the remaining three PROPINQUUS Πs are as results from (5.243).

(5.243) PROPINQUUS Πs combining with *naħa* 'side' II

(a) (Π8) *biswit* 'facing' – [Korpus Malti 3.0; parl4760]

*L-istess f' Ħal-Luqa [**biswit** in-naħa ta' Ħal Farruġ*
DEF-same in Ħal-Luqa [**facing** DEF-side of Ħal Farruġ
fejn hemm industrial estate]$_{PP}$ *nsibu kull kwalità ta'*
where there industrial estate] 1.IPFV:find:PL all quality of
tniġġiż inkluż noise pollution
pollution including noise pollution
'The same in Ħal-Luqa [**near** Ħal Farruġ where there is an industrial estate], we find every kind of pollution including noise pollution.'

(b) (Π14) *fejn* 'near' – [Korpus Malti 3.0; news181625]

huwa qabad taxi għall-ajruport ta' Katanja minn fejn mar
3SG.M seize taxi for:DEF-airport of Catania from where go
*[**fejn** in-naħa tal-parkeġġ tal-karozzi]*$_{PP}$
[**near** DEF-side of:DEF-parking lot of:DEF-car:PL]
'[. . .] he took a taxi to the airport of Catania from where he went [**near** the car park] [. . .].'

(c) (Π30) *ma'* 'with' – [Korpus Malti 3.0; news12033]

hemm ordni tal-MEPA stess li ħadd ma jista'
there order of:DEF-MEPA self that nobody NEG 3SG.M.IPFV:can
joqgħod bil-karavan jew bit-tined
3SG.M.IPFV:stay with:DEF-caravan or with:DEF-tent.PL
*[**ma**n-naħa tal-kosta]*$_{PP}$ *b' mod permanenti*
[**with**:DEF-side of:DEF-coast] with fashion permanent
'[. . .] there is an order by the MEPA itself that nobody can stay permanently with a caravan or tents [**near** on coast].'

Except (5.243b), the examples pose again analytical problems. What is clear nevertheless is that, like in the previous cases, there is no necessity for translating *in-naħa* 'the side'. As to (5.243a), the question arises whether (Π8) *biswit* 'facing' must be interpreted as PROPINQUUS in the first place. A likely alternative is to interpret the spatial situation such that the Figure (= the effects of pollution) is looking at the Ground (= the village of Ħal Farruġ). The distance between Figure and Ground may be small, but more importantly, the Figure is located in a certain position relative to the Ground. In (5.243b), (Π14) *fejn* 'near' conveys the meaning that the Figure is located in the vicinity of the Ground – but not necessarily in a lateral position to the latter. In (5.243c), the employment of (Π30) *ma'* 'with' is probably motivated by the fact that the Figure (= people with caravans and/or tents) live very close to the shore so that the distance between Figure and Ground can be understood as minimal with the two participants almost touching each other. At the same time, the coastline is itself a horizontally extended Ground. This is why (Π30) *ma'* 'with' is chosen over any of the other Πs which would not invite the reading that the situation extends over a longer stretch of the coast.[203]

What we learn from the above is that the Πs under review in this section display overlapping distributions without, however, yielding absolutely identical patterns. The synonymy between them is restricted to certain contexts outside of which the Πs tend to behave individually (though to different degrees). Not every Π that expresses the PROPINQUUS either primarily or secondarily can replace every other PROPINQUUS Π in every context although (Π48) *qrib* 'near' and (Π58) *viċin* 'near' come very close to full interchangeability. To determine where replacements are possible and where they are blocked a fine-grained in-depth study of the phenomena is called for. This desirable research project has to be postponed until after the completion of this monograph.

5.2.1.4.4.2.2.2 INTERIOR

The situation in the domain of the INTERIOR resembles the one we mentioned for (Π48) *qrib* 'near' and (Π58) *viċin* 'near' in the foregoing paragraph. (Π15) *fi* 'in', (Π18) *ġewwa* 'inside', and (Π19) *ġo* 'in' may replace each other in many contexts but not everywhere. The degree of functional equivalence is high nevertheless. To determine the extent of the correspondence between the three INTERIOR Πs, we proceed as follows. We identify the first ten common nouns with which the Πs are attested in the Korpus Malti 3.0. These common nouns are disclosed in Table 5.28. The cells that host common nouns that recur in several columns are marked out in distinctive

203 Cf. Section 5.1.2.2.20.1.1.

colours. Then, the frequencies of the co-occurrences of the Πs with these common nouns are determined to see whether there are preferences and/or restrictions.

Table 5.28: INTERIOR Πs and the first ten common nouns attested as complements in the Korpus Malti 3.0.

#	**(Π15) *fi* 'in'**	**(Π18) *ġewwa* 'inside'**	**(Π19) *ġo* 'in'**
1	*żmien* 'time'	*klassi* 'class'	*forum* 'forum'
2	*ħdan* 'bosoms'	*pajjiż* 'country'	*sptar* 'hospital'
3	*sħab* 'partners; clouds'	*post* 'place'	*dar* 'house'
4	*tmiem* 'end'	*dar* 'house'	*pajjiż* 'country'
5	*kliem* 'words'	*call centre* 'call centre'	*morsa* 'screw-vice'
6	*kriżi* 'crisis'	*ħanut* 'shop'	*belt* 'city'
7	*banda* 'side'	*ġisem* 'body'	*skola* 'school'
8	*flett* 'flat'	*marsupju* 'marsupial'	*ġnien* 'garden'
9	*skola* 'school'	*arloġġ* 'watch'	*istitut* 'institute'
10	*proċess* 'process'	*ċentru* 'centre'	*familja* 'family'

There are twenty-seven different common nouns (types). Three nouns are attested among the ten first complements of two different INTERIOR Πs, all other types occur only once in Table 5.28. The twenty-seven types are taken as input for the next step. We search the Korpus Malti 3.0 for attestations of each of the three INTERIOR Πs with each of the complement nouns. For each of the nouns, the definite and indefinite singular and plural are taken account of.[204] Only those constructions are counted

204 We have excluded pronominally possessed complement nouns from the frequency count. We assume that the inclusion of complement nouns hosting the possessor suffix would not have changed the general picture much. Examples (5.xii)–(5.xiv) involve the possessed noun *daru* 'his house' which is attested 749 times in *fi*-PPS, sixty-six times in combination with (Π19) *ġo* 'in' whereas it is the complement of (Π18) *ġewwa* 'inside' forty-nine times. This quantitative distribution over INTERIOR Πs resembles that of *dar* 'house' without possessor suffix.

(5.xii) [Korpus Malti 3.0; culture1054]

Joe Bonello	*laqgħana*	*[f'*	*daru]*$_{PP}$	*u*	*kellimna*	*dwar*	*ix-xorti*
Joe Bonello	receive:1PL	**[in**	house:3SG.M]	and	speak:1PL.PFV	about	DEF-luck

kbira	*li*	*kellu*
big:F	that	have.PFV:3SG.M

'Joe Bonello welcomed us [**in** his home] and we talked about the good fortune he had [. . .].'

(5.xiii) [Korpus Malti 3.0; news181102]

Il-ġisem	*mejjet*	*tal-kantant*	*tal-band*	*mettalika*	*Quiet Riot*	*Kenvin DuBrow*
DEF-body	dead	of:DEF-singer	of:DEF-band	metallic:F	Quiet Riot	Kenvin DuBrow

in which the Π_{INTERIOR} and its complement noun are direct syntagmatic neighbours. The token frequencies are revealed in Table 5.29. Cells that host a zero are additionally highlighted in grey. Hapaxes are identified by yellow shading of the cell. The complement nouns are ordered top-down according to the decreasing overall frequency.

Table 5.29: Co-occurrences of INTERIOR Πs with complement nouns as of Table 5.28 in the Korpus Malti 3.0 (tokens).

noun	(Π15) *fi* 'in'	(Π18) *ġewwa* 'inside'	(Π19) *ġo* 'in'
pajjiż 'country'	51347	228	190
żmien 'time'	48457	0	0
ħdan 'bosoms'	29723	1	16
sptar 'hospital'	27449	1745	150
skola 'school'	27374	172	121
post 'place'	26091	100	192
dar 'house'	20509	356	643
belt 'city'	13874	638	149
tmiem 'end'	10922	0	0
ċentru 'centre'	10338	124	68
proċess 'process'	8716	0	1
kliem 'words'	6666	0	1
familja 'family'	5042	32	142
klassi 'class'	4996	20	39
ħanut 'shop'	3873	37	136
istitut 'institute'	2523	26	21
ġnien 'garden'	1987	40	15
kriżi 'crisis'	1927	0	0
ġisem 'body'	634	11	23

---

instab *[ġewwa* *daru* *f'* *Las Vegas]*$_{PP}$ *nhar* *il-Ħadd* *li* *għadda*
REFL:find [**inside** house:3SG.M in Las Vegas] day DEF-Sunday that pass
'The dead body of the singer of the metal band Quiet Riot, Kenvin DuBrow, was found [**in** his house in Las Vegas] last Sunday.'

(5.xiv) [Korpus Malti 3.0; academic112]
Il-Knisja *xorta* *ġiegħelitu* *jibqa'* *maqful*
DEF-Church no_matter compel:3SG.F.PFV:3SG.M 3SG.M.IPFV:remain PART:lock_up
[ġo *daru]*pp *sakemm* *miet*
[**in** house:3SG.M] until die
'The Church nevertheless compelled him to stay locked up [**in** his house] until he died.'

Table 5.29 (continued)

noun	(Π15) *fi* 'in'	(Π18) *ġewwa* 'inside'	(Π19) *ġo* 'in'
forum 'forum'	409	2	4
sħab 'partners; clouds'	370	2	9
banda 'side'	298	1	0
morsa 'screw-vice'	275	1	6
call centre 'call centre'	91	1	0
arloġġ 'watch'	65	1	0
flett 'flat'	24	1	1
marsupju 'marsupial'	1	1	0

On account of the well-known high frequency of (Π15) *fi* 'in',[205] it is no surprise to see that this Π outnumbers the INTERIOR Πs in all but one of the cases. The complement *marsupju* 'marsupial' is exceptional since (Π15) *fi* 'in' and (Π18) *ġewwa* 'inside' yield identical minimal turnouts. In 26 out of 27 cases, however, (Π15) *fi* 'in' boasts token frequencies which neither (Π18) *ġewwa* 'inside' nor (Π19) *ġo* 'in' can compete with. Moreover, (Π15) *fi* 'in' is attested with each of the complement nouns whereas (Π18) *ġewwa* 'inside' fails to show up in combination with five complement nouns while the number of gaps in the *ġo*-column amounts to seven. Furthermore, for seven other complement nouns (Π18) *ġewwa* 'inside' is attested only once as head of the PP. (Π19) *ġo* 'in' produces three hapaxes. We observe also that if a given combination is either unattested or attested only once with (Π19) *ġo* 'in', it also has zero frequency or is a hapax in the case of (Π18) *ġewwa* 'inside'. What additionally strikes the eye is the fact that if either (Π18) *ġewwa* 'inside' or (Π19) *ġo* 'in' boasts a frequency of n ≥ 20 the other counts at least ten attestations (but usually more). These two INTERIOR Πs do not behave quantitatively as one. However, there are striking parallels which are indicative of the close semantic and functional relation between (Π18) *ġewwa* 'inside' and (Π19) *ġo* 'in'.

The two INTERIOR Πs (Π18) *ġewwa* 'inside' and (Π19) *ġo* 'in' display relatively high token frequencies when they combine with complement nouns whose referents are prototypical Grounds in the sense that their referents are three-dimensional container-like physical entities such as *sptar* 'hospital', *skola* 'school', *post* 'place', *dar* 'house', and *belt* 'city' (although it still needs to be ascertained whether the first two types refer to concrete buildings or to abstract institutions). At the same time, (Π18) *ġewwa* 'inside' and (Π19) *ġo* 'in' are excluded from or only exceptionally attested in contexts where the complement noun represents a temporal or

205 Cf. Table 3.27.

abstract concept, an event, or a process. We conclude from this that (Π18) *ġewwa* 'inside' and (Π19) *ġo* 'in' are semantically confined to the expression of properly spatial configurations whereas the domain of (Π15) *fi* 'in' reaches far beyond the purely spatial situations. This difference is captured by the diagram in Figure 5.49.

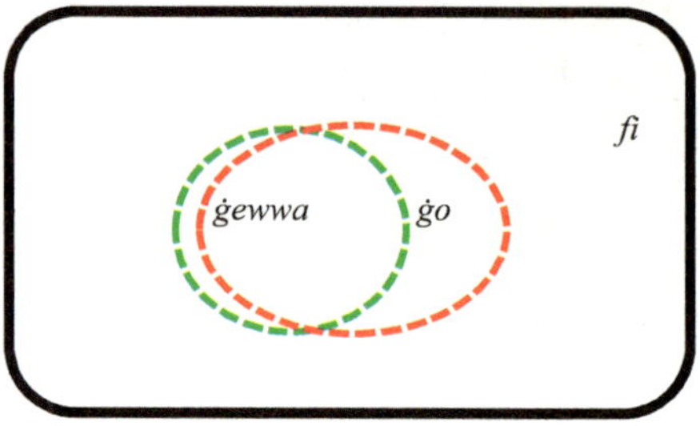

Figure 5.49: Inclusion of functional domains (INTERIOR Πs).

(Π15) *fi* 'in' has many non-spatial functions to fulfil whereas (Π18) *ġewwa* 'inside' and (Π19) *ġo* 'in' only fulfil spatial functions. INTERIOR is thus only one among a plethora of meanings that can be expressed by (Π15) *fi* 'in'. In contrast, INTERIOR is the primary and possibly only meaning that can be expressed by either (Π18) *ġewwa* 'inside' or (Π19) *ġo* 'in'.

5.2.1.4.4.2.2.3 MEDIAL$_1$

The topic of this section has been addressed three times already.[206] It is worthwhile looking at the facts again but from a different angle. The Πs that are under discussion in what follows are (Π4) *bejn* 'between', (Π16) *fost* 'amongst', and (Π47) *qalb* 'amidst'. The examples in (5.244) show once again that there is a zone where the functional domains of these Πs overlap.

(5.244) MEDIAL$_1$ expressed in three different ways

(a) (Π4) *bejn* 'between' – [Korpus Malti 3.0; news66133]

Dan	*qed*	*joħloq*	*ħafna*	*tensjoni*	***[bejn***
DEM.M.PROX	PROG	3SG.M.IPFV:create	much	tension	**[between**
in-nies	*tal-post]$_{PP}$*				
DEF-people	of:DEF-place]				

'This is creating a lot of tension [**amongst** the local people].'

206 In Sections 5.1.2.1.13, 5.1.2.1.14, and 5.1.2.2.17.

(b) (Π16) *fost* 'amongst' – [Korpus Malti 3.0; culture1352]

Dan	*jista'*	*jkun*	*riżultat*	*tal-eċitament*
DEM.M.PROX	3SG.M.IPFV:can	3SG.M.IPFV:be.FUT	result	of:DEF-excitement

li	*qam*	***[fost***	*in-nies]*$_{PP}$	*meta*	*skoprew*
that	rise	**[amongst**	DEF-people]	when	discover:3PL.PFV

dan	*il-fatt*
DEM:M:PROX	DEF-fact

'This can be the result of the excitement that emerged [**amongst** the people] when they discovered this fact.'

(c) (Π47) *qalb* 'amidst' – [Korpus Malti 3.0; news128235]

Hawn	*min*	*jgħid*	*li*	*l-mexxejja*	*tal-Hamas*
here	who	3SG.M.IPFV:say	that	DEF-leader.PL	of:DEF-Hamas

joqogħdu	***[qalb***	*in-nies]*$_{PP}$	*biex*	*jekk*	*Iżrael*
3.IPFV:stay:PL	**[amidst**	DEF-people]	to	if	Israel

jattakkahom	*imutu*	*n-nies*	*innoċenti*	*magħhom*
3SG.M.IPFV:attack:3PL	3.IPFV:die:PL	DEF-people	innocent	with:3PL

'Here are those who say that the leaders of Hamas live [**amidst** the people] so that in case Israel attacks them innocent people die together with them.'

It is clear that in all three sentences in (5.244), the complement *in-nies* 'the people' refers to a group of human beings of undetermined size for which it is nevertheless clear that it is made up of more than two individuals. The various Figures – the tension in the (a)-sentence, the excitement in the (b)-sentence, and the leaders of Hamas in the (c)-sentence – are located metaphorically or physically on the inside of a distributed Ground. In cases of this kind, all three Πs express the MEDIAL$_1$. We assume that it is possible to indiscriminately use (Π4) *bejn* 'between', (Π16) *fost* 'amongst', and (Π47) *qalb* 'amidst' in each of the three sentences.

Yet, (5.245a–b) suggest that the possibility of replacing the one Π with any of the others and vice versa is limited.

(5.245) Coordinated Πs

(a) [Korpus Malti 3.0; news41169]

Neħtieġu	*solidarjetà*	***[fost***	*Ø]*$_{PP}$	*u*	***[bejn***
1.IPFV:need:PL	solidarity	**[amongst**	Ø]	and	**[between**

il-ġnus	*tagħna]*$_{PP}$
DEF-people.PL	of:1PL]

'We need solidarity [**amongst**] and [**between** our peoples].'

(b) [Korpus Malti 3.0; european6768]
jistabbilixxi *relazzjonijiet* ***[bejn*** *Ø]$_{PP}$* *u* ***[fost***
3SG.M.IPFV:establish relation:PL **[between** Ø] and **[amongst**
il-ġirien]$_{PP}$ *li* *huma* *tajbin* *u* *durabbli* *u*
DEF-neighbour:PL] that 3PL good:PL and durable and
mal-Komunitàinternazzjonali
with:DEF-community international
'[. . .] it establishes relations which are good and durable [**between**] and [**amongst** the neighbours] and with the international community.'

In both the (a)-sentence and the (b)-sentence (Π4) *bejn* 'between' and (Π16) *fost* 'amongst' are coordinated (in two different orders) sharing the same complement. We have no evidence of (Π47) *qalb* 'amidst' being involved in coordinative constructions of this kind. We take the coordination to mean that (Π4) *bejn* 'between' and (Π16) *fost* 'amongst' do not exactly mean the same. In point of fact, (Π4) *bejn* 'between' is used if the relation between two participants (individuals or groups of entities) is focused upon. (Π16) *fost* 'amongst' however, is the better choice when the intended meaning is that of group-internal relationships. This semantic difference comes nicely to the fore under coordination because if both Π were full synonyms one of them would be sufficient to convey the message.

Even in cases like those in (5.246) where superficially no semantic distinction is apparent, (Π4) *bejn* 'between' refers to a configuration of participants that is different from those referred to by (Π16) *fost* 'amongst' and (Π47) *qalb* 'amidst'.

(5.246) Hidden nuances
(a) (Π4) *bejn* 'between' – [Korpus Malti 3.0; academic404]
Il-perspettivi *ta'* *konflitt* *jaqblu* *li* *hemm*
DEF-perspective:PL of conflict 3.IPFV:agree:PL that there
differenza *ta'* *interessi* ***[bejn*** *il-gruppi* *f'* *soċjetà]$_{PP}$*
difference of interest:PL **[between** DEF-group:PL in society]
'The conflict-based perspectives agree that there is a difference of interests [**between** the groups in society].'
(b) (Π16) *fost* 'amongst' – [Korpus Malti 3.0; academic3]
jħeġġu *l-iskambju* *ta'* *informazzjoni* ***[fost***
3.IPFV:kindle:PL DEF-exchange of information **[amongst**
il-gruppi *ta'* *artiġġjani* *differenti]$_{PP}$*
DEF-group:PL of artisan:PL different]
'[. . .] they encourage the exchange of information [**amongst** the different groups of artisans].'

(c) (Π47) *qalb* 'amidst' – [Korpus Malti 3.0; news177524]

l-preżenza	*ta'*	*l-Iran*	*fit-Tazza*	*tad-Dinja*	*qed*
DEF-presence	of	DEF-Iran	in:DEF-Cup	of:DEF-World	PROG
toħloq		*kontroversja*	***[qalb***	*il-gruppi*	*Lhud*
3SG.F.IPFV:create		controversy	**[amidst**	DEF-group:PL	Jewish
fil-Ġermanja]$_{PP}$					
in:DEF-Germany]					

'[. . .] Iran's presence in the World Cup is creating a controversy [**amongst** Jewish groups in Germany].'

In each of the relevant PPs in (5.246), the complement noun is *il-gruppi* 'the groups'. Superficially, the parallels between the three Πs seem to be strong enough to assume that replacing the one with the other in any of the sentences would not trigger a meaning change. However, to our mind, this assumption can be disproved easily. If (Π4) *bejn* 'between' is replaced by either (Π16) *fost* 'amongst' or (Π47) *qalb* 'amidst' in the (a)-sentence, the situation described would no longer regard the differences between societal groups but shift the focus to the differences within these groups. Likewise, the use of (Π4) *bejn* 'between' in the (b)-sentence and the (c)-sentence would result in a different interpretation. This is so because, by default, (Π4) *bejn* 'between' invokes a scenario with exactly two participants whose interrelation is strictly binary. This also applies to (5.246a) because the example is taken from a paragraph in which the classical Marxist dichotomy of a society divided into two – capitalists and proletarians – is summarised. Thus, in this sociological conception, there are only two groups. In contrast, both in the (b)-sentence and the (c)-sentence, it can safely be assumed that the number of groups is bigger than just two. If this is the case, the use of (Π4) *bejn* 'between' in these sentences is blocked. Furthermore, there are two possible interpretations in connection with the *fost*-PP and the *qalb*-PP in (5.246a-b), namely either the exchange of information and the controversy take place within each of the mentioned groups or in the interaction between the various groups. In the latter case, we repeat, (Π4) *bejn* 'between' is no option since there are many groups which are involved in the situations.

We now turn our attention back to coordination. What is the case semantically when a MEDIAL$_1$ Π has scope over several coordinated complement nouns? A first idea in relation to this question takes shape when we look at the examples in (5.247) where we find three instances of [$\Pi_{medial1}$ NP *u* Ø NP]$_{PP}$, i.e., there is EQUI-Π-deletion. For purely expository reasons, we use only one pair of square brackets for the entire coordinated structure which otherwise could also be represented as [$\Pi_{medial1}$ NP]$_{PP}$ *u* [Ø NP]$_{PP}$.

(5.247) EQUI-Π-deletion with MEDIAL$_1$

(a) (Π4) *bejn* 'between' – [Korpus Malti 3.0; academic12]
is-saċerdot kien il-ħolqa ewlenija ***[bejn*** *il-grupp etniku*
DEF-priest be DEF-link principal:F **[between** DEF-group ethnic
Malti u ***Ø*** *l-awtoritajiet Awstraljani]*$_{PP}$
Maltese and **Ø** DEF-authority:PL Australian:PL]
'[. . .] the priest was the principal link [**between** the Maltese ethnic group and the Australian authorities].'

(b) (Π16) *fost* 'amongst' – [Korpus Malti 3.0; academic40]
[Fost *il-lingwi u* ***Ø*** *kulturi oħra li studja*
[**amongst** DEF-language:PL and **Ø** culture:PL other that study
Sapir]$_{PP}$ *ngħoddu il-Wishram Chinook Navajo Nootka*
Sapir] 1.IPFV:count:PL DEF-Wishram Chinook Navajo Nootka
Paiute Takelma u Yana
Paiute Takelma and Yana
'[**Amongst** the other languages and cultures that Sapir studied] we count Wishram Chinook, Navajo, Nootka, Paiute, Takelma and Yana.'

(c) (Π47) *qalb* 'amidst' – [Korpus Malti 3.0; literature20]
Iżda għall-aħħar ftit metri kellna nidħlu
but for:DEF-last little meter:PL have.PFV:1PL 1.IPFV:enter:PL
[qalb *il-ħaxix u* ***Ø*** *l-ħurrieq li tela' sa qaddna*
[**amidst** DEF-grass and **Ø** DEF-nettle that rise till waist:1PL
imxarrab għasra bin-nida]$_{PP}$
PART:wet drench with:DEF-dew]
'But for the last few meters we had to walk [**amidst** the grass and the nettles which rose up to our waist, drenched with dew].'

The situations described in (5.247a–c) differ in the following way. In the (a)-sentence, the coordinated complements of (Π4) *bejn* 'between' identify the two parties that are involved in the relationship. It is this binary relation that is expressed by the Π. Neither in the (b)-sentence nor in the (c)-sentence is there a binary relation. The scientific interests of Sapir are the topic of (5.247b). (Π16) *fost* 'amongst' does not determine the relation between languages and cultures but its use implies that there is an undetermined sizable number of languages and an equally undetermined number of cultures that were of interest to Sapir. The list of glossonyms which also stand for the cultures associated with the ethnic groups comprises half a dozen items so that also with regards to the languages and cultures no binary relation applies. In (5.247c), (Π47) *qalb* 'amidst' does not express that there is a relation between the grass and the nettles but that there was so much vegetation that the Figure had to move through a lot of grass and through a lot of nettles. Meaning:

(Π47) *qalb* 'amidst' serves to separately characterise the density of grass as well as that of nettles as considerable.

The differences remain the same even if EQUI-Π-deletion does not apply as in (5.248) where the Π is repeated with each conjunct so that the structure $[\Pi_{\text{MEDIALX}}\ \text{NP}]_{\text{PP}}$ *u* $[\Pi_{\text{MEDIALX}}\ \text{NP}]_{\text{PP}}$ emerges.

(5.248) Coordinated PPs with overt second MEDIAL_1 Π

(a) (Π4) *bejn* 'between' – [Korpus Malti 3.0; literature72]

*[**Bejn** Ras il-Wardija]*$_{PP}$ *u* *[**bejn** il-Madliena]*$_{PP}$ *hemm*
[**between** Ras il-Wardija] and [**between** DEF-Madliena] there
irdum ħurrieqi fejn qalb dak il-blat kbir
cliff nettle_infested where amidst DEM:M:DIST DEF-rock.COLL big
li jagħti għall-baħar jinsabu ħafna għerien
that 3SG.M.IPFV:give for:DEF-sea 3.IPFV:REFL:find:PL many cave:PL
'[**Between** Ras il-Wardija] and [Madliena], there is a nettle-infested cliff where amidst those big rocks that look out on the sea, many caves are found.'

(b) (Π16) *fost* 'amongst' – [Korpus Malti 3.0; european7369]

*dawn jistgħu jvarjaw [**fost** l-ispeċi]*$_{PP}$
DEM:PL:PROX 3.IPFV:can:PL 3.IPFV:vary:PL [**amongst** DEF-species:PL]
*[**fost** it-tessuti]*$_{PP}$ *u [**fost** il-punti ġenetiċi*
[**amongst** DEF-fabric:PL] and [**amongst** DEF-point:PL genetic:PL
ta' konklużjoni]$_{PP}$
of conclusion]
'[. . .] these can vary [**amongst** the species], [**amongst** the fabrics], and [**amongst** the conclusive genetic points].'

(c) (Π47) *qalb* 'amidst' – [Korpus Malti 3.0; news81684]

Imbagħad in<t>ebaħ li minflok miexi lejn
then <REFL>become_aware that instead_of walk.PART towards
fejn il-blat jinferaq u
where DEF-rock.COLL 3SG.M.IPFV:REFL:separate and
*jitbaxxa biex isib ruħu [**qalb***
3SG.M.IPFV:REFL:bend_down to 3SG.M.IPFV:find self:3SG.M [**amidst**
in-nies]$_{PP}$ *u [**qalb** il-karozzi]*$_{PP}$ *kien sab ruħu*
DEF-people] and [**amidst** DEF-car:PL] be find self:3SG.M
dieħel u dieħel aktar 'il ġewwa
enter.PART and enter.PART more to inside
'Then he realised that in lieu of walking to the rocks separating and lowering himself down to be [**amidst** the people] and [**amidst** the cars] he had been going ever more towards the inside.'

In (5.248a), (Π4) *bejn* 'between' occurs twice.[207] The double occurrence of the Π is correlative in the sense that it overtly marks the two participants between which a given relation is established. In the case of the (a)-sentence, the two *bejn*-PPs identify the two extreme points of the stretch of the coast that hosts the nettle-infested cliff. This is a peculiarity of (Π4) *bejn* 'between'. The multiple occurrences of (Π16) *fost* 'amongst' and (Π47) *qalb* 'amidst' in the (b)-sentence and the (c)-sentence, respectively, do not have the same effect as in the case of (Π4) *bejn* 'between'. Each instance of (Π16) *fost* 'amongst' and each instance of (Π47) *qalb* 'amidst' has but local scope, meaning: neither of the Πs establishes a relation between the different complements with which it combines. What the Π does is convey the meaning that the Figure is located (perhaps only metaphorically) within a distributed Ground identified by the complement noun. Put differently, in (5.247c), the Figure is depicted as being amidst people on the one hand, and amidst cars on the other hand. The coordination of the two *qalb*-PPs does not express that the Figure might find itself amidst a distributed Ground that is constituted by people and cars alike. We concede that whoever reads a sentence like (5.247c) might picture the situation in their minds such that the Figure is surrounded by people and cars at the same time. We argue however that this is a secondary interpretation of the entire situation whereas the two *qalb*-PPs are semantically as well as syntactically independent of each other. Note that this is completely different with (Π4) *bejn* 'between' because in cases like that in the (a)-sentence both *bejn*-PPs are interdependent.

In Figure 5.50, we try to schematise the relationship that holds between the three Πs we have discussed in this section.

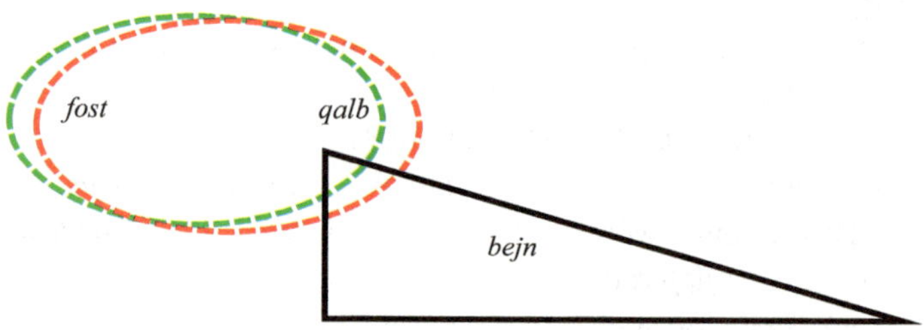

Figure 5.50: Inclusion and overlap of functional domains (MEDIAL$_1$ Πs).

The diagram shows that synonymy is relatively strong for (Π16) *fost* 'amongst' and (Π47) *qalb* 'amidst' so that it is hard to find any context where one could not replace the other freely. (Π16) *fost* 'amongst' and (Π47) *qalb* 'amidst' represent MEDIAL$_1$ par excellence. In contrast, (Π4) *bejn* 'between' only exceptionally expresses the same

207 The phenomenon was mentioned already in Section 4.1.4.

category. The functional overlap with (Π16) *fost* 'amongst' and (Π47) *qalb* 'amidst' is thus comparatively small. The proper domain of (Π4) *bejn* 'between' lies outside that of the MEDIAL$_1$. It is the MEDIAL$_2$ whose expression is the uncontested monopoly of (Π4) *bejn* 'between'.

5.2.1.4.4.2.2.4 FRONS

The PROPINQUUS properties of (Π8) *biswit* 'facing' were discussed above.[208] In this section, it is the possible synonymy with (Π49) *quddiem* 'in front of' that is of interest to us. Both Πs are associated with FRONS as comes to the fore in the examples presented under (5.249).

(5.249) Competition within the domain of FRONS

(a) (Π8) *biswit* 'facing' – [Korpus Malti 3.0; european3027]

Kumitat	*ta'*	*Kuntatt [. . .]*	*għandu*	*jiġi*
committee	of	contact	have.IPFV:3SG.M	3SG.M.IPFV:come

stabbilit	*[**biswit***	*<u>il-Kummissjoni</u>]$_{PP}$*
establish:PART	[**facing**	<u>DEF-Commission</u>]

'A contact committee [. . .] has to be established [**before** <u>the Commission</u>].'

(b) (Π49) *quddiem* 'in front of' – [Korpus Malti 3.0; news180434]

Ir-riżoluzzjoni	*issa*	*tkun*	*ippreżentata*
DEF-resolution	now	3SG.F.IPFV:be.FUT	present:PART:F

*[**quddiem***	*<u>il-Kummisjoni</u>*	*<u>u</u>*	***Ø***	*<u>l-Kunsill</u>*	*<u>Ewropew</u>]$_{PP}$*
[**in_front_of**	<u>DEF-Commission</u>	<u>and</u>	**Ø**	<u>DEF-Council</u>	<u>European</u>]

biex	*tkun*	*ikkunsidrata*	*waqt*	*it-teħid*
to	3SG.F.IPFV:be.FUT	consider:PART:F	at_the_time_of	DEF-taking

tad-deċiżjonijiet
of:DEF-decision:PL

'The resolution will now be presented [**before** <u>the Commission and the European Council</u>] so that it will be considered during the decision-making.'

Above we mentioned the possibility that the choice of appropriate Π might depend on the perspective.[209] Is the Figure looking in the direction of the Ground or is it the other way around? It seems that (Π8) *biswit* 'facing' is chosen when the observer's gaze is directed from the location of the Figure towards the Ground whereas (Π49) *quddiem* 'in front of' allows for construing the spatial situation from the point of view of the Ground.

---

208 The phenomenon was mentioned already in Section 4.1.4.
209 Cf. Sections 5.1.2.15 and 5.1.2.16.

The Figure-Ground perspective is less important if (Π8) *biswit* 'facing' is given a PROPINQUUS interpretation implying a lateral position of the Figure relative to the Ground. In (5.250a), the *biswit*-PP refers to a spatial situation that is not in line with the one expressed by the *quddiem*-PP in the (b)-sentence.

(5.250) (Π8) *biswit* 'facing' ≠ (Π49) *quddiem* 'in front of' I

(a) (Π8) *biswit* 'facing' – [Korpus Malti 3.0; news144184]

Biex wettqu din is-serqa il-ħallelin
to implement:3PL.PFV DEM.F.PROX DEF-robbery DEF-thief:PL
daħlu minn residenza ***[biswit*** *il-ħanut ta'*
enter:3PL.PFV from residence **[facing** DEF-shop of
Sterling Jewellers]$_{PP}$
Sterling Jewellers]
'To carry out this robbery, the thieves entered from a house [**next to** the shop of Sterling Jewellers].'

(b) (Π49) *quddiem* 'in front of' – [Korpus Malti 3.0; news128964]

Joseph Zammit kellu jistenna lil sħabu
Joseph Zammit have.PFV:3SG.M 3SG.M.IPFV:wait to comrade:3SG.M
[quddiem *il-ħanut Downtown]*$_{PP}$
[in_front_of DEF-shop Downtown]
'[. . .] Joseph Zammit had to wait for his accomplice [**in front of** the shop Downtown] [. . .].'

As to the spatial situation in (5.250a), it is clear that the thieves must have entered the jewellers' shop not from the opposite side of the street but through an empty building next door to the shop because they could access the latter by breaking down a wall separating the empty house from the shop. Thus, the Figure is located at the side of the Ground and not in front of it. This is different in the (b)-sentence since the named delinquent waited for his partner-in-crime not somewhere near the shop but exactly in front of it. The two Πs cannot replace each other in (5.250a–b) unless a change of meaning is intended.

Similarly, in (5.251), (Π8) *biswit* 'facing' and (Π49) *quddiem* 'in front of' locate the Figure differently in relation to the Ground.

(5.251) (Π8) *biswit* 'facing' ≠ (Π49) *quddiem* 'in front of' II

(a) (Π8) *biswit* 'facing' – [Korpus Malti 3.0; news169952]

Il-każ seħħ għall-ħabta tad-0930 am meta l-mara
DEF-case happen for:DEF-hit of:DEF-0930 a.m. when DEF-woman
kienet għadha kemm daħħlet il-karozza
be:3SG.F.PFV still:3SG.F how_much enter.CAUS:3SG.F.PFV DEF-car

tagħha *fil-garaxx* [***biswit*** *id-dar* *tagħha]*$_{PP}$
of:3SG.F in:DEF-garage [**facing** DEF-house of:3SG.F]
'The case happened at exactly 9.30 a.m. when the woman was still driving her car in the garage [**next to** her house].'

(b) (Π49) *quddiem* 'in front of' – [Korpus Malti 3.0; news97745]
qatel *lill-Avukat Grech* *f'* *garaxx* [***quddiem*** *il-blokka* *ta'* *appartamenti* *f'* *Marsalforn]*$_{PP}$ *fejn* *kien* *joqgħod* *il-vittma* *u* *familtu*
kill to:DEF-lawyer Grech in garage [**in_front_of** DEF-block of apartment:PL in Marsalforn] where be 3SG.M.IPFV:stay DEF-victim and family:POSS:3SG.M
'[. . .] he killed the lawyer Grech in a garage [**in front of** an apartment block in Marsalforn] where the victim lived with his family.'

In (5.251a), the garage is placed at the side of the woman's house. It is not detached from the house standing on the other side of the street. This however is exactly the position of the garage in (5.251b) where it is the site of a crime. The garage is placed in front of the Ground and not on its side. The choice of Π is thus motivated by the different positions of the Figure in relation to the Ground. We have informally searched the Korpus Malti 3.0 for *biswit*-PPs and *quddiem*-PPs which contain *id-dar* 'the house' as complement. There were 548 hits for *quddiem id-dar* 'in front of the house' all of which are bona fide instances of FRONS. *Biswit id-dar* yielded forty-four matches forty of which can be understood as reflecting a situation in which the Figure is placed in front of the Ground ('facing the house'). In four cases, however, the Figure is located laterally on one of the sides of the house ('next to the house'). Interestingly, in these cases, the Figure is always a garage that is an extension of an individual's house.

It is therefore not possible to classify (Π8) *biswit* 'facing' and (Π49) *quddiem* 'in front of' as full synonyms. Figure 5.51 is meant to schematise the relationship of the two Πs.

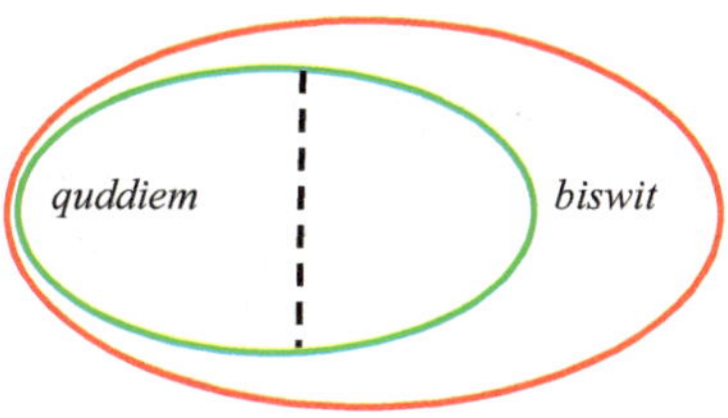

Figure 5.51: Inclusion of functional domains (FRONS Πs).

The red circle circumscribes the functional domain of (Π8) *biswit* 'facing' which happens to be larger than that of (Π49) *quddiem* 'in front of' that is totally included in the former.

The interrupted vertical line within the space surrounded by the green circle indicates that the choice of Π reflects the different perspectives of Figure and Ground.

5.2.1.4.4.2.2.5 LONGITUDO$_1$

With LONGITUDO$_1$ we enter the temporal domain where (Π33) *matul* 'during' and (Π59) *waqt* 'at the time of' are in competition with each other. How they are interrelated semantically is difficult to determine. In principle, both Πs can be used in identical or at least similar contexts. In (5.252), we show this for situations in which the entire extension over time of a given event is expressed.

(5.252) (Π33) *matul* 'during' = (Π59) *waqt* 'at the time of' I

(a) (Π33) *matul* 'during' – [Korpus Malti 3.0; academic426]

l-LTU Arena	*ta'*	*Düsseldorf [. . .]*	*Weserstadion*	*ta'*	*Bremen [. . .]*	
DEF-LTU Arena	of	Düsseldorf	Weserstadion	of	Bremen	
u	*Borussia-Park*	*ta'*	*Mönchengladbach [. . .]*	*ma*		
and	Borussia-Park	of	Mönchengladbach	NEG		
ntużawx	*ukoll*	***[matul***	*il-kampjonat]$_{PP}$*			
REFL:use:3PL.PFV:NEG	also	**[during**	DEF-championship]			

'[. . .] the LTU Arena in Düsseldorf, the Weserstadion in Bremen, and the Borussia-Park in Mönchengladbach were not used too [**during** the Championship].'

(b) (Π59) *waqt* 'at the time of' – [Korpus Malti 3.0; academic171]

Fl-2003	*Kaká*	*kien*	*il-captain*	*tat-tim*	*Brażiljan*
in:DEF-2003	Kaká	be	DEF-captain	of:DEF-team	Brazilian
[waqt	*il-kampjonat*	*tal-Gold Cup]$_{PP}$*			
[at_the_time_of	DEF-championship	of:DEF-Gold Cup]			

'In 2003, Kaká was the captain of the Brazilia team [**during** the Gold Cup Championship].'

The two sentences describe situations in which the Figure played a given role during the entire period covered by a sports event. In (5.252a), the football stadiums of several German cities are not used for the matches that took place during the championship. This means that their exclusion from the list of venues was valid for the entire tournament. Similarly, the footballer Kaká served as captain of the Brazilian team for the entire duration of the Gold Cup Championship in 2003 (= (b)-sentence)). Thus, the temporal Figure-Ground relation to which the two Πs refer spans over an extended period of time.

In (5.253), the meaning of both (Π33) *matul* 'during' and (Π59) *waqt* 'at the time of' is different because the Figure is present in the situation only for part of the time during the event that serves as Ground.

(5.253) (Π33) *matul* 'during' = (Π59) *waqt* 'at the time of' II

(a) (Π33) *matul* 'during' – [Korpus Malti 3.0; news139056]

*[**Matul** il-kunċert]_{PP} indaqqet ukoll kompożizzjoni*
[**during** DEF-concert] REFL:play:3SG.F.PFV also composition
oriġinali mill-kompożitur / pjanist Malti Norman Cristina
original from:DEF-composer / pianist Maltese Norman Cristina
'[**During** the concert], also an original composition was performed by the Maltese composer / pianist Norman Cristina.'

(b) (Π59) *waqt* 'at the time of' – [Korpus Malti 3.0; news176894]

Il-mużiċisti l-oħra li se jakkumpanjaw lil
DEF-musician:PL DEF-other:PL that FUT 3.IPFV:accompany:PL to
Julie u Ludwig huma Joe Vella (kuntrabaxx) Jes Baldacchino
Julie and Ludwig 3PL Joe Vella (bass) Jes Baldacchino
*(drums) u Ivan Spiteri Lucas (saxophone) li [**waqt***
(drums) and Ivan Spiteri Lucas (saxophone) that [**at_the_time_of**
il-kunċert]_{PP} ħa jkanta wkoll xi kanzunetti
DEF-concert] FUT 3SG.M.IPFV:sing also some song:PL
mir-repertorju tiegħu
from:DEF-repertoire of:3SG.M
'The other musicians who will accompany Julie and Ludwig are Joe Vella (bass), Jes Baldacchino (drums) and Ivan Spiteri Lucas (saxophone) who will also sing some songs from his repertoire [**during** the concert].'

In (5.253a-b), the concert defines the time frame for the activities of the musicians. The activities that are explicitly mentioned in the examples form part of the concert's program but do not exhaust the program. This means that the performance of an original composition (= (a)-sentence) and of songs from Ivan Spieri Lee's own repertoire (= (b)-sentence) cover only a small part of the entire concert. In this case and in stark contrast to the previous one, (Π33) *matul* 'during' and (Π59) *waqt* 'at the time of' display a punctual reading.

In the temporal domain, (Π33) *matul* 'during' is a possible replacement of (Π59) *waqt* 'at the time of' and vice versa. These Πs differ, however, when we look beyond the boundaries of the temporal domain because outside the time-related concepts only (Π33) *matul* 'during' is admitted (to express LONGITUDO$_2$). Where (Π33) *matul* 'during' allows for being employed in temporal as well as in spatial contexts,[210] (Π59) *waqt* 'at the time of' is banned from the latter. To our mind, we are facing a situation of total inclusion of the functional domain of (Π59) *waqt* 'at the time of' within the more extended functional domain of (Π33) *matul* 'during' as shown in Figure 5.52.

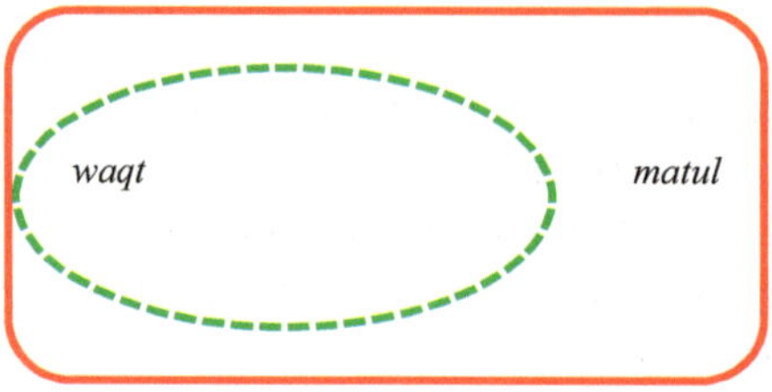

Figure 5.52: Inclusion of functional domains (LONGITUDO$_1$ Πs).

Accordingly, the semantic relation between the two Πs does not fit the description of full synonymy. Figure 5.52 is suggestive of partial synonymy instead.

5.2.1.4.4.2.2.6 SUPERIOR

The final pair of potentially synonymous Πs comprises (Π15) *fi* 'in' and (Π17) *fuq* 'on' as markers of SUPERIOR. The choice of the appropriate Π seems to depend (albeit not strictly) upon the physical properties of the referent of the Ground noun. Both Πs are fine with *muntanja* 'mountain' as shown in (5.254).

(5.254) (Π15) *fi* 'in' = (Π17) *fuq* 'on' I

(a) (Π15) *fi* 'in' – [Korpus Malti 3.0; news3811]

Ajruplan partikolari Boeing 727 li kien intilef lura
aeroplane particular Boeing 727 that be REFL:lose back
fl-1985 kien instab fl-2006 [f' muntanja fil-Bolivja]$_{PP}$
in:DEF-1985 be REFL:find in:DEF-2006 [**in** mountain in:DEF-Bolivia]
'A particular Boeing 727 that got lost back in 1985 was found in 2006 [**on** a mountain in Bolivia].'

(b) (Π17) *fuq* 'on' – [Korpus Malti 3.0; news3508]

sajjetta partikolari laqtet uħud mis-swaba'
lightning particular hit:3SG.F.PFV some.PL from:DEF-finger.PL
tal-id il-leminija tal-Istatwa Ġesu Kristu li
of:DEF-hand DEF-right:F of:DEF-statue Jesus Christ that

210 Cf. Section 5.1.2.2.20.1.4 (ii).

tinsab *[**fuq*** *il-muntanja* *Corcovado* *f'* *Rio de Janiero*
3SG.F.IPFV:REFL:find [**on** DEF-mountain Corcovado in Rio de Janeiro
fil-Brażil]$_{PP}$
in:DEF-Brazil]
'[. . .] an individual lightning hit some of the fingers of the right hand of the statue of Jesus Christ that stands [**on** the Corcovado Hill in Rio de Janeiro in Brazil].'

The complement noun was specifically chosen for the comparison of the behaviour of the two ΠS because it refers to a vertically extended geo-object whose uppermost surface is prototypically seen as a pointed peak. Accordingly, it might be hypothesised that (Π17) *fuq* 'on' is preferred over (Π15) *fi* 'in'. This hypothesis is corroborated in the singular (= *muntanja* 'mountain'). In the plural (= *muntanji* 'mountains'), however, (Π15) *fi* 'in' outnumbers (Π17) *fuq* 'on' as results from Figure 5.53. The token count is based on a search of the Korpus Malti 3.0.

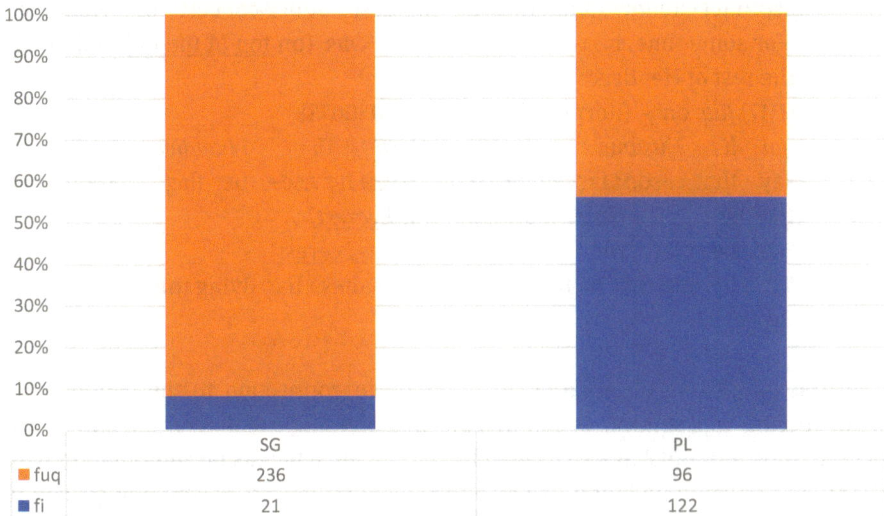

Figure 5.53: Hits for (Π15) *fi* 'in' and (Π17) *fuq* 'on' combining with *muntanja* 'mountain' in SG and PL (Korpus Malti 3.0).

We interpret this number-based difference between (Π15) *fi* 'in' and (Π17) *fuq* 'on' as follows. In the singular, the degree of individualisation of the geo-object is high and thus allows for precisely locating the Figure relative to the Ground. That is why the semantically more specialised (Π17) *fuq* 'on' is made use of much more frequently than (Π15) *fi* 'in'. In the plural, the complement noun does no longer refer to a single geo-object but to a group of similar objects (sometimes forming a mountain range). The location of the Figure can no longer be determined with reference to an individual mountain but with reference to several mountains which taken together cover an extended stretch of land so that the vertical position of the Figure is less important than its horizontal position within the area constituted by the group of mountains.

To put this interpretation to the test, we have run another search in the Korpus Malti 3.0, this time with the complement noun *quċċata* 'top; summit'. Our expectation was to find a distribution of the Πs that resembles the one we showed in Figure 5.53. First of all, (5.255) proves that both (Π15) *fi* 'in' and (Π17) *fuq* 'on' are licit with *quċċata* 'top; summit'.

(5.255) (Π15) *fi* 'in' = (Π17) *fuq* 'on' II

(a) (Π15) *fi* 'in' – [Korpus Malti 3.0; literature59]

Għal	*xi*	*ħin*	*baqa'*	*fejn*	*kien*	*[fil-quċċata*	*tal-għolja]*$_{PP}$
for	some	time	remain	where	be	[**in**:DEF-top	of:DEF-hill]

isegwi	*taħtu*	*l-bqija*	*tal-kunċert*	*ta'*	*Hendrix*
3SG.M.IPFV:follow	under:3SG.M	DEF-rest	of:DEF-concert	of	Hendrix

'For some time, he remained where he was, [**on** top of the hill], following the rest of Hendrix's concert below him.'

(b) (Π17) *fuq* 'on' – [Korpus Malti 3.0; news130073]

qal	*li*	*jittama*	*li*	*jara*	*l-bandiera*	*Maltija*
say	that	3SG.M.IPFV:hope	that	3SG.M.IPFV:see	DEF-flag	Maltese:F

tperper	***[fuq***	*il-quċċata*	*tal-Everest]*$_{PP}$
3SG.F.IPFV:flap	[**on**	DEF-top	of:DEF-Everest]

'[. . .] he said that he hopes to see the Maltese flag flying [**on** the top of the Everest].'

The first interesting observation to be made in connection to the complement noun *quċċata* 'top; summit' is that there is only one single match for its plural form *quċċati* 'tops; summits' in a PP headed by (Π15) *fi* 'in' whereas no combination with (Π17) *fuq* 'on' could be found. We give the single hit for a *fi*-PP containg *quċċati* in (5.256) because it is a telling example whose relevance will become clear in the next paragraph.

(5.256) On top of the list I – [Korpus Malti 3.0; news204055]

dawn	*iż-żewġ*	*gruppi*	*għandhom*	*ukoll*	*numri*	*ta'*
DEM:PL:PROX	DEF-two	group:PL	have.IPFV:3PL	also	number:PL	of

suċċessi	*li*	*telgħu*	*[fil-quċċati*	*tal-klassifiċi*
success:PL	that	ascend:3PL.PFV	[**in**:DEF-top:PL	of:DEF-classification:PL

<u>*tal-mużika]*</u>$_{PP}$
<u>of:DEF-music]</u>
'[. . .] these two groups also have a number of successes that made it [**to** <u>the top of the music charts</u>].'

Example (5.256) features *quċċati* 'tops; summits' which is metonymically used in the sense that it does not refer to the highest peaks of hills, mountains, or other geo-objects but to the top rank in a hierarchy (here: the music charts). Metonymy and metaphor are most probably responsible for the unexpected discrepancy of the token frequencies given in Figure 5.54.

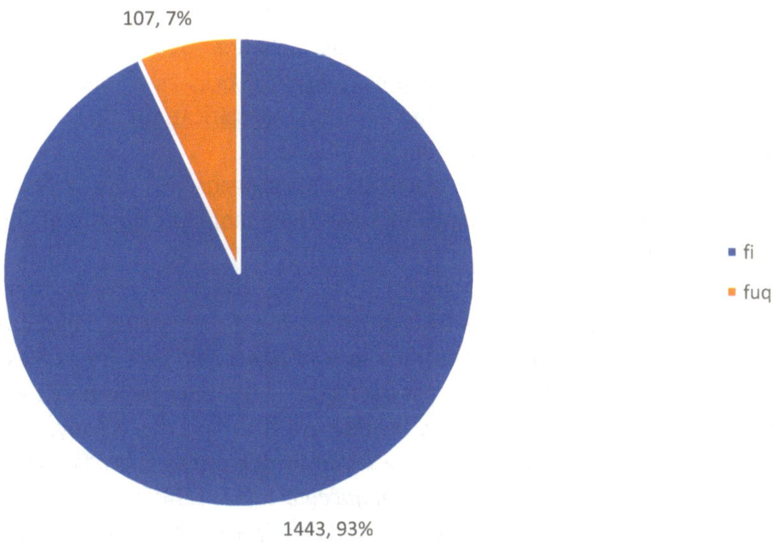

Figure 5.54: Token frequencies of (Π15) *fi* 'in' and (Π17) *fuq* 'on' cooccurring with *quċċata* 'top; summit' in the Korpus Malti 3.0.

In Figure 5.54, there is only a small niche for (Π17) *fuq* 'on' whereas the lion's share of the tokens goes to (Π15) *fi* 'in'. There is a reason for this enormous difference between the two shares. To understand this better, we first have to filter out a host of cases for different reasons. For what we are going to argue all those constructions are irrelevant in which *quċċata* is not followed by a genitival attribute or an apposition. In the Korpus Malti 3.0, this is the case for 476 hits reported for the PP *fil-quċċata* whereas thirty-two matches of *fuq il-quċċata* are counted out on the basis of the same criterion. This means that between 30% (= *fuq*) and 33% (= *fi*) of the hits given in Figure 5.54 are invalid for the next step we are going to take. We are left with 967 hits for *fil-quċċata ta(X)* as opposed to seventy-five cases of *fuq il-quċċata ta(X)* with X functioning as variable for all morphonological changes that (Π55) *ta'* 'of' can be subject to.

We seize the opportunity offered by example (5.256) to determine whether or not the complement *klassifika* 'classification' yields any significant results when we compare *fuq*-PPS with *fi*-PPS. In point of fact, there are 477 hits for *fil-quċċata tal-klassifika* 'at the top of the class(ification)', i.e., 49% of the 967 relevant hits of *fil-quċċata ta(X)* involve this complement noun. In contrast, there is only one example (= slightly more than 1%) of *fuq il-quċċata ta(X)* where *klassifika* 'classification' takes in the position of X as shown in (5.257).

(5.257) On top of the list II – [Korpus Malti 3.0; news63780]

Kull	*darba*	*li*	*jkollhom*	*ċans*	*jagħmlu*
each	time	that	3SG.M.IPFV:have.FUT:3PL	chance	3.IPFV:make:PL
l-marka	*tagħhom*	***[fuq***	*il-quċċata*	*tal-klassifika]*$_{PP}$	*huma*
DEF-mark	of:3PL	**[on**	DEF-top	of:DEF-classification]	3PL
jonqsu	*milli*	*jagħmlu*	*dan*		
3.IPFV:miss:PL	from:that	3.IPFV:make:PL	DEM:M:PROX		

'Every time they have the chance to make their mark [**at** the top of the league] they fail to do that.'

The single attestation of *klassifika* 'classification' in combination with *fuq il-quċċata ta(X)* notwithstanding, it is predominantly the task of *fil-quċċata ta(X)* to express that the Figure is placed at or moves towards the top of a hierarchy. In point of fact, there is a plethora of other complement nouns that are either partial synonyms of *klassifika* 'classification' or belong to the same semantic domain as *klassifika* 'classification' that appear in combination with *fil-quċċata ta(X).* Table 5.30 contains only a small selection of typical examples embracing a variety of list-like organised hierarchies which include inter alia several sports leagues (mostly football) from different countries. The examples are ordered alphabetically according to the initial letter of the complement noun.

Table 5.30: Selection of recurrent complement nouns combining with *fil-quċċata ta(X).*

source	example / translation	complement
news178350	*r-Russja poġġiet is-sigurta' tal-enerġija [**fi**l-quċċata tal-aġenda tal-laqgħa tal-G8]*$_{PP}$ 'Russia put the security of the energy [**on** top of the agenda of the G8 meeting] [. . .].'	*aġenda* 'agenda'
news52350	*Dortmund huma biss punt wara Bayern Munich [**fi**l-quċċata tal-Bundesliga]*$_{PP}$ 'Dortmund is only one point behind Bayern München [**at** the top of the Bundesliga].'	*Bundesliga* 'German Federal League'

Table 5.30 (continued)

source	example / translation	complement
culture659	*it-tim Malti rebaħ lil-Lussemburgu 6–0 u b' hekk it-tim Malti spiċċa [**fi**l-quċċata tal-grupp tiegħu]$_{PP}$* 'The Maltese team beat Luxemburg 6–0 and thus the Maltese team wound up [**at** the top of its group].'	*grupp* 'group'
news135707	*Qormi jkunu jridu jżommu posthom [**fi**l-quċċata tal-kampjonat]$_{PP}$* '[. . .] Qormi will want to keep their psoition [**at** the top of the championship].'	*kampjonat* 'championship'
culture1654	*L-isem ta' Noah qabeż lill-ismijiet Michael u Jacob li kienu dejjem [**fi**l-quċċata tal-lista tal-iktar ismijiet popolari]$_{PP}$ mill-1960 sal-2013.* 'The name Noah overtook the names Michael and Jacob that always were [**at** the top of the most popular names] from 1960 to 2013.'	*lista* 'list'
news149574	*issa ilhom tmien partiti ma jitilfu biex jinsabu wkoll [**fi**l-quċċata tan-North American Soccer League]$_{PP}$* '[. . .] they now haven't lost the last eight matches so that they are also [**at** the top of the NSL].'	*North American Soccer League* 'NSL'
news162488	*issa Federer reġa' tela' [**fi**l-quċċata tar-rankings mondjali]$_{PP}$* '[. . .] Federer now ascended again [**to** the top of the world rankings].'	*ranking* 'ranking'
news86042	*B' din ir-rebħa [. . .] Tarxien qabżu [**fi**l-quċċata tar-Relegation Pool]$_{PP}$* 'With this victory [. . .] Tarxien jumped [**to** the top of the Relegation Pool].'	*Relegation Pool* 'relegation pool'
news120738	*b' dawn ir-riżultati t-tim ta' Ruma jinsab [**fi**l-quċċata tas-Serie B]$_{PP}$* '[. . .] with these results the team of Rome finds itself [**at** the top of Serie B].'	*Serie B* 'Italian 'Second Division'
news103860	*Tarxien Rainbows telgħu waħedhom [**fi**l-quċċata tat-Tieni Diviżjoni]$_{PP}$ b' rebħa ta' 5–0 fuq Vittoriosa Stars* 'With 5–0 victory over Vittoriosa Stars, Tarxien Rainbows climbed alone [**to** the top of the Second Division].'	*Tieni Diviżjoni* 'Maltese Second Division'

Cases of these kinds are responsible for the vast majority of hits for *fil-quċċata ta(X)* whereas they are infrequent in the case of *fuq il-quċċata ta(X)*. The differences between the two options come to the fore in Figure 5.55.

On the one hand, it is true that *fil-quċċata ta(X)* and *fuq il-quċċata ta(X)* are compatible with complement nouns of both classes, i.e., geo-objects (= all kinds of physical objects in space) are fine and hierarchies as well. However, the two options are not randomly made use of. In the case of *fil-quċċata ta(X)*, there is a solid 94% majority of the tokens involving a complement noun that belongs to the class of hierarchically ordered lists. In contrast, *fuq il-quċċata ta(X)* prefers combinations with nouns referring to geo-objects – a preference that accounts for 83% of all instances. This means that (Π17) *fuq* 'on' displays a stronger inclination towards proper spatial situations whereas (Π15) *fi* 'in' is leaning towards metonymical interpretations of space.

In (5.258), two examples of the above Πs are given in which the relevant spatial situation involves a Ground that can be conceived of as an extended surface. Note

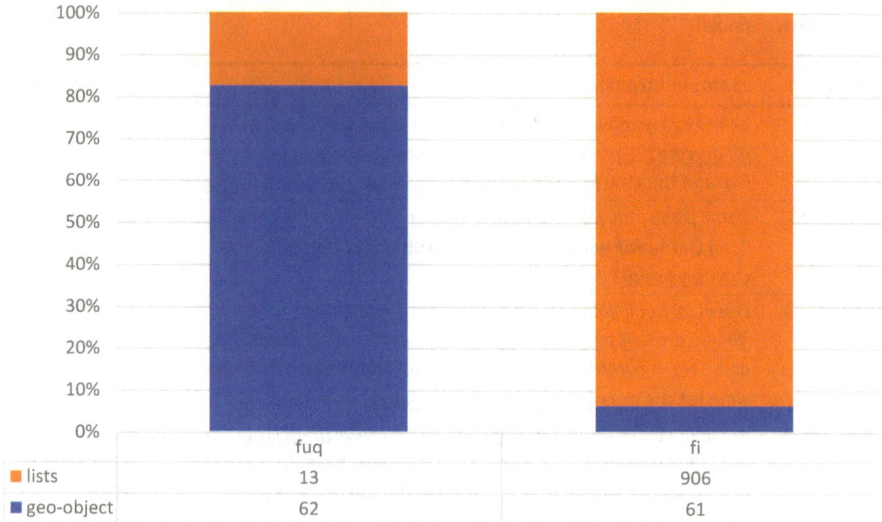

Figure 5.55: Token frequencies of *fil-quċċata ta(X)* / *fuq il-quċċata ta(X)* with complements of two different classes (Korpus Malti 3.0).

that the sentences are almost identical. They are taken from E.U. regulations with regard to certain aspects of cheese-making. The complement noun is *wiċċ* '(sur)face'.

(5.258) (Π15) *fi* 'in' = (Π17) *fuq* 'on' III

(a) (Π15) *fi* 'in' – [Korpus Malti 3.0; european2555]

Ġobon artab u matur huwa kkurat jew immaturat b'
cheese soft and ripe 3SG.M cure:PART or ripen:PART with
aġenti kuranti bijoloġiċi bħal moffa ħmira u organiżmi
agent:PL curing biological:PL like mould yeast and organism:PL
oħra li jkunu jiffurmaw qoxra prominenti
other:PL that 3.IPFV:be.FUT:PL 3.IPFV:form:PL peel prominent
*[**fi**l-wiċċ tal-ġobon]*$_{PP}$
[**in**:DEF-face of:DEF-cheese]
'Soft and ripe cheese is taken care of or ripened with biological curative agents like mould, yeast, and other organisms that will form a prominent layer [**on** the surface of the cheese].'

(b) (Π17) *fuq* 'on' – [Korpus Malti 3.0; european2615]

Ġobon artab u matur huwa kkurat jew immaturat
cheese soft and ripe 3SG.M care:PART or ripen:PART
permezz ta' aġenti ta' kura bijoloġika bħall-moffa
by_means_of of agent:PL of cure biological:F like:DEF-mould

ħmira *u* *organiżmi* *oħra* *li* *jiffurmaw*
yeast and organism:PL other:PL that 3.IPFV:be.FUT:PL
qoxra ***[fuq*** *il-wiċċ* *tal-ġobon]*$_{PP}$
3.IPFV:form:PLpeel **[in** DEF-face of:DEF-cheese]
'Soft and ripe cheese is taken care of or ripened with agents of biological curing like the mould, yeast, and other organisms that will form a layer [**on** the surface of the cheese].'

These examples suggest again that the two Πs are functionally equivalent. What the examples do not tell us is whether they are preferred for expressing certain concepts or not. To clarify this issue, we proceed in analogy to what we have done above in connection to *quċċata* 'top; summit'. First, we determine how often we find evidence of *fil-wiċċ ta(X)* 'on the surface of X' and *fuq il-wiċċ ta(X)* 'on the surface of X' in the Korpus Malti 3.0. Afterwards, we will see which nouns are attested as X. Figure 5.56 reveals that *fil-wiċċ ta(X)* is the majority solution.

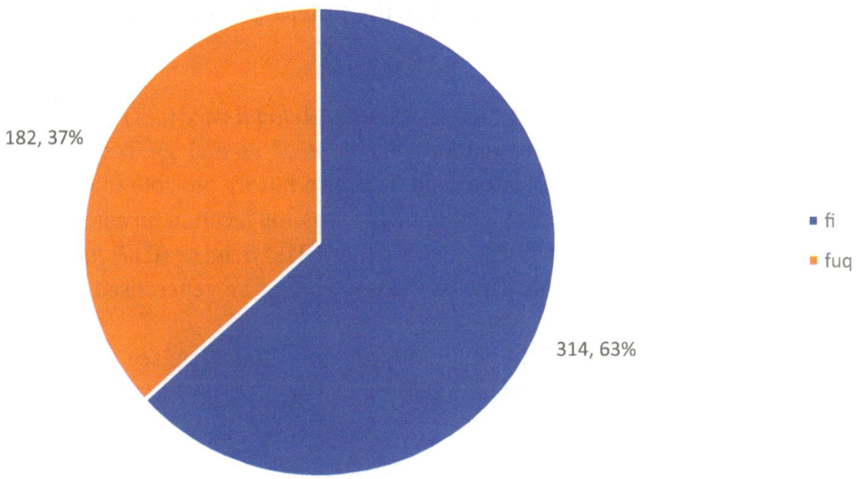

Figure 5.56: Frequencies and shares of *fil-wiċċ ta(X)* vs *fuq il-wiċċ ta(X)* in the Korpus Malti 3.0.

In contrast to the previous case, the search that involves *wiċċ* '(sur)face' does not yield results that point towards principled differences between the two options if we discount the expected quantitative predominance of the construction with (Π15) *fi* 'in'. It is also hardly of relevance whether the Ground is a human face or a surface of some other kind. In (5.259), two examples are provided that show that both options are possible if *wiċċ* '(sur)face' refers to the body part of a living being.

(5.259) (П15) *fi* 'in' = (П17) *fuq* 'on' III

(a) (П15) *fi* 'in' – [Korpus Malti 3.0; news83437]

Hemm hu l-White Rocks monument ħaj tad-daħq
there 3SG.M DEF-White Rocks monument alive of:DEF:laughter
[fil-wiċċ ta' Gonzi PN]$_{PP}$
[**in**:DEF-face of Gonzi PN]
'There is the White Rocks – a living monument of the smile [**on** the face of PN Gonzi].'

(b) (П17) *fuq* 'on' – [Korpus Malti 3.0; news196202]

Sal-iċken dettall imnaqqax f' din il-vara
till:DEF-small.CMPR detail PART:sculpt in DEM:F:PROX DEF-statue
b' paċenzja kbira ir-riżultat finali huwa wieħed impressjonanti
with patience big:F DEF-result final 3SG.M one impressive
*bit-tbatija evidenti [**fuq** il-wiċċ ta' Kristu]*$_{PP}$
with-DEF-suffering evident [**on** DEF-face of Christ]
'Sculpted in this statue with great patience till the smallest detail, the final result is an impressive one, with the suffering evident [**on** the face of Christ].'

The body part *wiċċ* '(sur)face' is attested six times with *fuq il-wiċċ ta(X)* and seventeen times with *fil-wiċċ ta(X)* corresponding to shares of 3% and 5%, respectively. None of the options is particularly frequent. It is also hardly possible to consider one of them to be the default when it comes to expressing location on somebody's face. It does not seem to make a difference whether (П15) *fi* 'in' or (П17) *fuq* 'on' is used with this complement noun. This conclusion cannot be generalised over all possible contexts.

As a matter of fact, (П15) *fi* 'in' and (П17) *fuq* 'on' cannot replace each other everywhere. Consider the examples in (5.260).

(5.260) (П15) *fi* 'in' ≠ (П17) *fuq* 'on'

(a) (П15) *fi* 'in' – [Korpus Malti 3.0; literature19]

*Qalu li raw qisu qamar [**fis**-sema]*$_{PP}$
say:3PL.PFV that see:3PL.PFV measure:3SG.M moon [**in**:DEF-sky]
tond tond ileqq b' dawl ħadrani
round round 3SG.M.IPFV:shine with light greenish
'They said that they saw something like a moon [**in** the sky] really round shining with greenish light.'

(b) (Π17) *fuq* 'on' – [Korpus Malti 3.0; literature59]

anke	*l-qamar*	***[fuq***	*<u>rashom</u>]*$_{PP}$	*kellu*	*jneħħi*
also	DEF-moon	**[on**	<u>head:3PL</u>]	have.PFV:3SG.M	3SG.M.IPFV:remove

l-kappell	*u*	*japplawdi*
DEF-hat	and	3SG.M.IPFV:applaud

'[. . .] also the moon [**above** <u>their head</u>] had to take off the hat and applaud [. . .].'

(Π15) *fi* 'in' and (Π17) *fuq* 'on' are often interchangeable if Figure and Ground are in contact with each other. However, this possibility is blocked when the Figure is described as being detached from the Ground. In this case, (Π17) *fuq* 'on' is interpreted as the translation equivalent of English *above* as in (5.260b). It is impossible to put (Π15) *fi* 'in' in the slot of (Π17) *fuq* 'on'. Similarly, (Π15) *fi* 'in' is not replaceable in (5.260a) although the Ground (= the sky) could be understood as an extended surface on which the Figure is located.

In conclusion, the relation between the two Πs is again one of partial overlap of their domains. Both (Π15) *fi* 'in' and (Π17) *fuq* 'on' claim larger territories great parts of which are outside the reach of the competing Π. This is what Figure 5.57 is meant to tell us.

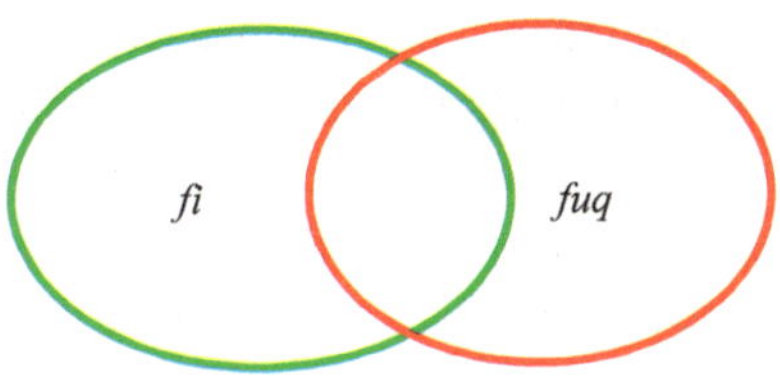

Figure 5.57: Overlap of functional domains (SUPERIOR Πs).

We emphasise once again that a lot still needs to be done before we can declare this and the previous cases closed. We have only scratched the surface, in a manner of speaking. The exact spell out of the interrelationship between the Πs requires much more research that we cannot afford to conduct within the framework of this study.

5.2.1.4.4.3 The internal organisation of the spatio-temporal domain

On the basis of the previous discussion, we conclude that polysemy and synonymy are only relatively weakly represented in the spatio-temporal domain. As to the latter phenomenon, we argue that the best candidates for the status of (full) synonyms are listed in (5.261).

(5.261) Synonymous spatial Πs

a. (Π14) *fejn* 'near' = (Π23) *ħdejn* 'beside' = (Π32) *maġenb* 'close to' (PROPINQUUS),

b. (Π16) *fost* 'amongst' = (Π47) *qalb* 'amidst' ($MEDIAL_1$),

c. (Π18) *ġewwa* 'inside' = (Π19) *ġo* 'in' (INTERIOR),

d. (Π48) *qrib* 'near' = (Π58) *viċin* 'near' (PROPINQUUS).

For the binary relations (b)–(d), we are confident that the Πs that form a given pair are indeed largely synonymous. In the case of the triple chain in (a) however, further investigations are called for to determine whether the functional equivalence is as assumed by us. The necessity of further research holds for practically all our sketches of the semantics of Maltese Πs. It would be too repetitive to keep emphasising this fact time and again in what follows. We therefore refrain from drawing the reader's attention again to the hypothetical nature of our conclusions (not only in this section).

5.2.1.4.4.3.1 Spatial vs temporal

We briefly review the different parameters which are mentioned in Borg/Azzopardi-Alexander (1997) as to their relevance in the system of spatio-temporal Πs in Maltese. For a start, we address the distinction of spatial vs temporal. The evidence we have proves the obvious in the sense that the vast majority of the Πs evaluated throughout the semantics section fall under the rubric of exclusively or predominantly spatial Πs the only case of a purely temporal Π being (Π59) *waqt* 'at the time of'. This temporal Π is excluded from the expression of spatial relations. It was shown above that (Π46) *qabel* 'before' may be used to indicate a position of the Figure in a hierarchy on a rank that is higher than that occupied by the Ground.[211] This is a metonymical and thus secondary use of (Π46) *qabel* 'before'. One might therefore claim that (Π46) *qabel* 'before' basically is a temporal Π that tolerates spatial readings, too. The case of (Π31) *madwar* 'around' is similar although this time it is a basically spatial Π that allows for being used with a temporal meaning.[212] Further Maltese spatial Πs that allow for being also employed in temporal contexts are (Π4) *bejn* 'between', (Π15) *fi* 'in', (Π21) *għal* 'for', (Π33) *matul* 'during', (Π41) *minn* 'from', (Π51) *sa* 'till', and (Π60) *wara* 'after'. Temporal meanings are never expressed by the Πs in (5.262).

211 Cf. Section 5.1.2.1.10.1.

212 Cf. Section 5.1.2.1.12.

(5.262) Spatial Πs without temporal usages
(Π3) *barra* 'outside'; (Π8) *biswit* 'facing'; (Π12) *faċċata* 'opposite'; (Π14) *fejn* 'near'; (Π17) *fuq* 'on'; (Π18) *ġewwa* 'inside'; (Π19) *ġo* 'in'; (Π22) *għand* 'at s.o.'s place'; (Π23) *ħdejn* 'beside'; (Π28) *lejn* 'towards'; (Π30) *ma'* 'with'; (Π32) *maġenb* 'close to'; (Π38) *mingħand* 'from s.o.'; (Π44) *oltre* 'beyond'; (Π47) *qalb* 'amidst'; (Π48) *qrib* 'near'; (Π49) *quddiem* 'in front of'; (Π56) *taħt* 'under'; (Π57) *versu* 'towards'; (Π58) *viċin* 'near'.

An additional case needs to be mentioned, namely (Π16) *fost* 'amongst'. Except for one frequent idiomatic expression,[213] (Π16) *fost* 'amongst' is never used in temporal contexts. We, therefore, lean towards including this Π in the class of those Πs which only serve spatial purposes.

In Figure 5.58, we reveal the distribution over three categories of the 31 Πs that have been taken into account for the purposes of Section 5.2. The categories are (a) exclusively temporal, (b) mixed spatio-temporal, and (c) exclusively spatial Πs.

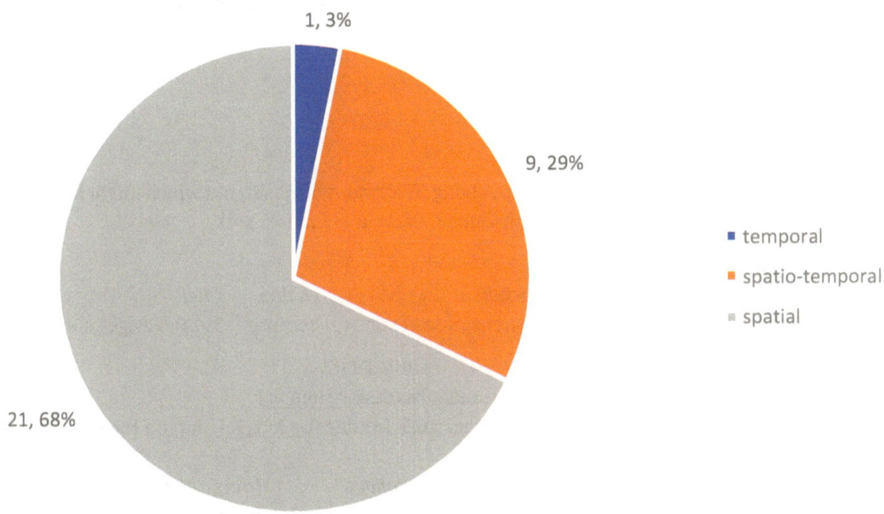

Figure 5.58: Shares of temporal, spatio-temporal, and spatial Πs (types).

The prominence of exclusively spatial Πs is clearly visible. Interestingly, without exception, the four Italo-Romance Πs belong to the class of exclusively spatial Πs. Those Πs which (also) have temporal functions are Semitic.

213 Discussed in Section 5.1.2.1.13.

5.2.1.4.4.3.2 With(out) contact

The second criterion to be looked at is that of contact between Figure and Ground. This criterion is irrelevant with regard to the temporal dimension. It is exclusively relevant for spatial relations. Borg/Azzopardi-Alexander (1997: 160–164) distinguish between SUPERIOR VS SUPERIOR-CONTACT, INFERIOR VS INFERIOR-CONTACT, LATERAL VS LATERAL-CONTACT, CITERIOR VS CITERIOR-CONTACT, and ULTERIOR VS ULTERIOR-CONTACT. The latter two pairs of distinctions are skipped in what follows because the means used for their expression violate the definition of the canonical Π.[214]

The only Π within BLOMP 2.0 that can express contact on its own is (Π30) *ma'* 'with'. It is mentioned by Borg/Azzopardi-Alexander (1997: 163) as marker of LATERAL-CONTACT (with the notable exception of Path). That (Π30) *ma'* 'with' serves as Π_2 in certain Π-Π combinations to indicate that Figure and Ground touch each other was remarked upon already.[215] This possibility is acknowledged by Borg/Azzopardi-Alexander (1997: 162–163) only for INFERIOR-CONTACT MOTION PAST (= *minn ma' taħt*) and LATERAL-CONTACT MOTION FROM (= *minn (ma')*). In our corpus, there is evidence of further combinations[216].[217] More importantly, the presence of a dedicated

214 As exposed in Section 1.6.1.

215 Cf. Section 4.1.3.1.

216 As exposed in Section 4.1.3.1.

217 There are doubtful cases supposedly involving (Π30) *ma'* 'with' as Π_1 but which might be better filed under *ma'*-PPS with a multiword complement. Compare (5.xv)–(5.xvi).

(5.xv) [Korpus Malti 3.0; academic13]

Dan	*l-insett*	*nagħrfuh*	*għaliex*	*jeħel*
DEM.M.PROX	DEF-insect	1.IPFV:recognise:PL:3SG.M	because	3SG.M.IPFV:get_stuck

[ma'	***taħt***	*il-weraq]*$_{PP}$	*u*	*[**ma**z-zkuk]*$_{PP}$
[**with**	**under**	DEF-leaves.COLL]	and	[**with**:DEF-stem.PL]

'We recognise this insect because it gets stuck [**under** the foliage] and [**on** the stems].

(5.xvi) [Korpus Malti 3.0; academic13]

naraw	*bħal*	*imħar*	*kannella*	*mqabbad*	*[**ma**z-zokk]*$_{PP}$	*jew*
1.IPFV:see:PL	like	limpet	brown	PART:CAUS.catch	[**with**:DEF-stem]	or

imwaħħal	***[ma'***	*taħt*	*tal-weraq]*$_{PP}$
PART:CAUS.stick	[**with**	under	of:DEF-leaves.COLL]

'[. . .] we see something like brown limpets fixed [**onto** the stem] or stuck [**on** the underside of the leaves].'

Examples (5.xv)–(5.xvi) are taken from one and the same academic text. There are two *ma'*-PPS per sentence. Those PPS whose complement is a form of *zokk* 'stem' make do with one Π to express contact with a vertically extended Ground. In these cases, (Π30) *ma'* 'with' is self-sufficient, in a manner of speaking. This is different in those cases where the complement is *weraq* 'leaves; foliage'. The Figure is not directly visible to the eye since the insects live on the underside of the leaves. The status of (Π56) *taħt* 'under' seems to differ between (5.xv) and (5.xvi). In example (5.xv), marker of contact does not seem to be obligatory. For the vast majority of the spatial Πs, no special construction is needed to express contact. Therefore, it is logical to conclude that this parameter is of limited importance for the system under scrutiny. Whether Figure and Ground are in physical contact or not is relevant only in connection with Path.[218]

5.2.1.4.4.3.3 Surface

Surface is a separate spatial category in the reference grammar (Borg/Azzopardi-Alexander 1997: 161). According to this source, the surface marker is always the same, namely (Π17) *fuq* 'on'. As we have seen above,[219] the situation is not as straightforward because there is also (Π15) *fi* 'in' that can be used to express location of a Figure on a surface. The competition between (Π17) *fuq* 'on' and (Π15) *fi* 'in' comes to the fore especially when the spatial situation is static as is the case in (5.263).

(5.263) (Π15) *fi* 'in' = (Π17) *fuq* 'on' IV

(a) (Π15) *fi* 'in' – [Korpus Malti 3.0; literature24]

il-flixkun *kien* *[fl-ixkaffa* *f'* *postu]*$_{PP}$
DEF-bottle be [**in**:DEF-shelf in place:3SG.M]
'[. . .] the bottle was [**on** the shelf at its place].'

(b) (Π17) *fuq* 'on' – [Korpus Malti 3.0; literature45]

Il- ktieb *kien* *għadu* *jistenna* *għalxejn*
DEF-book be still:3SG.M 3SG.M.IPFV:wait uselessly
jiġbor *it-trab* *[**fuq*** *l-ixkaffa]*$_{PP}$
3SG.M.IPFV:collect DEF-dust [**on** DEF-shelf]
'The book was still uselessly collecting dust [**on** the shelf] [. . .].'

Xkaffa 'shelf' is also the complement in Borg/Azzopardi-Alexander's (1997: 161) example for SURFACE AT REST. Depending on the actual build of the shelf, it can be conceived of as either an object with a surface or as a container-like object. In the vast majority of the hits for *xkaffa* 'shelf' as complement in a PP that locates a Figure at rest on the shelf, the Korpus Malti 3.0 gives evidence of the use of (Π17) *fuq* 'on'

---

it is possible to speak of a Π-Π combination whose second Π governs its complement according to the nominative pattern. In (5.xvi), the presence of (Π55) *ta'* 'of' defines the government pattern as genitival. In light of idioms like *minn taħt it-taħt* 'stealthily', one might regard the second PP in (5.xvi) as residually related to a (hypothetical) noun **taħt* 'underside'. Problems of this kind can only be solved within the framework of a study that analyses Maltese Πs from the vantage point of grammaticalization theory.

218 As shown in Section 5.1.2.2.20.

219 Cf. Section 5.2.1.4.4.2.2.6.

with (Π15) *fi* 'in' occurring only infrequently. The replacement of (Π17) *fuq* 'on' with (Π15) *fi* 'in' is blocked if dynamic spatial relations are overtly expressed by a directionality marker. Since both (Π21) *għal* 'for' (DIRECTIONAL$_1$) and (Π41) *minn* 'from' (PROLATIVE) are optional according to Borg/Azzopardi-Alexander (1997: 160–161), the configuration marker (Π17) *fuq* 'on' can express motion to and motion past on its own. Or more precisely: no formal distinction needs to be made between Place, Goal, and Path – and, what is more, the distinctions surface vs other Ground and contact vs no contact are (partially) neutralised, too. Three examples capturing part of this situation are provided in (5.264).

(5.264) Neutralisation

(a) DIRECTIONAL$_1$ (+ ABLATIVE) – [Korpus Malti 3.0; news131022]

kienu	*qegħdin*	*itellgħu*	*lill-anzjan*				
be	PROG:PL	3.IPFV:ascend.CAUS:PL	to:DEF-senior				
[minn	*<u>fuq</u>*	*<u>is-siġġu</u>*	*<u>tar-roti</u>]$_{PP}$*	*[għal*	*<u>fuq</u>*	*<u>is-sodda</u>]$_{PP}$*	
[from	**<u>on</u>**	<u>DEF-seat</u>	<u>of:DEF-wheel:PL</u>]	**[for**	**<u>on</u>**	<u>DEF-bed</u>]	

'[. . .] they were lifting the elderly man [**from** <u>the wheelchair</u>] [**<u>on</u>to** <u>the bed</u>].'

(b) Motion to without directional marker – [Korpus Malti 3.0; literature32]

Wara	*ftit*	*reġa'*	*daħal*	*ġewwa*	*u*	*poġġa*	*[fuq*	*<u>is-sodda</u>]$_{PP}$*
after	little	do_again	enter	inside	and	sit	**[on**	<u>DEF-bed</u>]

'After a short while he came back in and sat down [**on** <u>the bed</u>].'

(c) At rest / no contact marker – [Korpus malti 3.0; literature43]

Kienet	*qiegħda*	*[fuq*	*<u>is-sodda</u>]$_{PP}$*
be:3SG.F.PFV	stay:PART:F	**[on**	<u>DEF-bed</u>]
b'	*għajnejha*	*miftuħa*	
with	eye:PL:3SG.F	PART:open:PL	

'She was staying [**on** <u>the bed</u>] with her eyes open.'

Example (5.264a) contains a *minn*-PP and a *għal*-PP where (Π41) *minn* 'from' and (Π21) *għal* 'for' precede (Π17) *fuq* 'on'. The first PP represents the spatial relation of Source which implies that the directionality marker can never be dropped whereas the use of the configuration marker is by no means compulsory. The second PP is an instance of DIRECTIONAL$_1$ whose directionality marker is optional. Proof of the optionality can be found in (5.264b). Since the Place relation in (5.264c) is zero-marked, too, the disambiguation of the *fuq*-PP rests entirely on the verb and contextual information. Accordingly, contact only plays a very minor role in giving structure to the system of Maltese Πs. The criterion can be activated if necessary but its overt expression is not compulsory. Contact will enter the scene again in the subsequent section.

5.2.1.4.4.3.4 Laterality

For Borg/Azzopardi-Alexander (1997: 162–163), coding the lateral position of a Figure relative to the Ground is the task of (Π8) *biswit* 'facing' and (Π32) *maġenb* 'close to' if contact does not apply. If LATERAL-CONTACT is the case, (Π30) *ma'* 'with' comes into play. It is said to be optional in constructions which refer to a spatial relation of Source (whereas the directionality marker is mandatory). For Path, (Π41) *minn* 'from' is said to combine with (Π32) *maġenb* 'close to'. Above we discussed the interrelation of (Π8) *biswit* 'facing' and (Π32) *maġenb* 'close to' with other PROPINQUUS Πs.[220] We also know that (Π8) *biswit* 'facing' (but not (Π32) *maġenb* 'close to') may fulfil the function of FRONTALIS marker.[221] It is thus clear that extensionally, (Π8) *biswit* 'facing' and (Π32) *maġenb* 'close to' have different meanings although their zone of overlap is huge.

In (5.265), there is a *minn-maġenb* combination which, taken in isolation, is ambiguous.

(5.265) Lateral with or without contact? – [Korpus Malti 3.0; news3651]

L-asterojde 2000 EM 26	*għaddiet*	***[minn***	***<u>maġenb</u>***	*<u>id-dinja</u>]*$_{PP}$	
DEF-asteroid 2000 EM 26	pass:3SG.F.PFV	**[from**	**<u>close to</u>**	<u>DEF-world</u>]	
bla	*problemi*	*hekk*	*kif*	*kien*	*skedat*
without	problem:PL	so	how	be	schedule:PART

'The asteroid 2000 EM 26 passed [**close by** <u>the Earth</u>] without problems as scheduled.'

Were it not for the *bla*-PP to the right of the closing square bracket, one would not know whether the asteroid scratches the Earth on its voyage in space. Of course, world knowledge tells us that no such contact has taken place in reality. However, the fact remains, that the combination of (Π41) *minn* 'from' and (Π32) *maġenb* 'close to' leaves it to the recipient of the message to find the correct interpretation. The information conveyed by (Π32) *maġenb* 'close to' is that Figure and Ground were not too distant from each other when the former passed by the latter. Since the Earth is conceived of as a globe, there is no way to tell whether the asteroid passed the Earth laterally, frontally, above it, beneath it, or behind it. Laterality can be determined only in those cases where the Ground has a referent whose structure allows us to identify and distinguish different sides thereof.

We argue that laterality – like contact – is not as important a criterion in Maltese as the category grid of the LDS-format used by Borg/Azzopardi-Alexander

220 Cf. Section 5.1.2.1.4.
221 Cf. Section 5.1.2.1.6.

(1997) suggests. As mentioned already,[222] the reference grammar concedes that both Π8) *biswit* 'facing' and (Π32) *maġenb* 'close to' can be used as synonyms of (Π14) *fejn* 'near', (Π23) *ħdejn* 'beside', (Π48) *qrib* 'near', and (Π58) *viċin* 'near'. At this point, it is worthwhile to reopen the case of the semantic equivalence of these Πs once again. To this end, we divide the group of Πs of interest into two subsets. Examples of the members of the first subset are given in (5.266).

(5.266) Three options – one meaning

(a) (Π14) *fejn* 'near' – [Korpus Malti 3.0; literature35]
Georgie taħrab tiġri u tasal
Georgie 3SG.F.IPFV:escape 3SG.F.IPFV:run and 3SG.F.IPFV:arrive
***[fejn** ix-xmara]$_{PP}$*
[**near** DEF-river]
'Georgie runs away and arrives [**near** the river].'

(b) (Π23) *ħdejn* 'beside' – [Korpus Malti 3.0; literature81]
Għalqet il-pitazz u ħaditni magħha
close:3SG.F.PFV DEF-copybook and take:3SG.F.PFV:1SG with:3SG.F
***[ħdejn** ix-xmara]$_{PP}$*
[**beside** DEF-river]
'She closed the copybook and took me with her [**to (the vicinity of)** the river].'

(c) (Π48) *qrib* 'near' – [Korpus Malti 3.0; news5663]
kien qed jikkampja mal-familja tiegħu
be PROG 3SG.M.IPFV:camp with:DEF-family of:3SG.M
***[qrib** ix-xmara Eildon f' Victoria]$_{PP}$*
[**near** DEF-river Eildon in Victoria]
'[. . .] he was camping with his family [**near** the river Eildon in Victoria].'

(d) (Π58) *viċin* 'near' – [Korpus Malti 3.0; news57756]
se jsir festival bil-logħob tan-nar
FUT 3SG.M.IPFV:become festival with:def-playing of:DEF-fire
***[viċin** ix-xmara ta' Hudson]$_{PP}$*
[**near** DEF-river of Hudson]
'[. . .] a festival with fireworks will take place [**near** the river Hudson].'

The Ground is represented by the common noun *xmara* 'river'. In the four sentences, the Figure is located in or moved to the vicinity of the river. Since the geo-ob-

222 Cf. Section 5.1.2.1.4.

ject extends horizontally with a left bank and a right bank defining the sides of the object, the above Figures are always put in a lateral position relative to the Ground. This laterality also applies in (5.267) where a further three Πs are involved.

(5.267) Three options – one meaning
(a) (Π8) *biswit* 'facing' – [Korpus Malti 3.0; news58652]
Il-lukanda tiegħi kienet ***[biswit*** *ix-xmara Nil]*$_{PP}$
DEF-hotel of:1SG be:3SG.F.PFV **[facing** DEF-river Nile]
'My hotel was [**looking on** the river Nile].'
(b) (Π30) *ma'* 'with' – [Korpus Malti 3.0; news181991]
Minn hawn noħorġu nagħmlu mixja
from here 1.IPFV:exit:PL 1.IPFV:make:PL stroll
*[****ma****x-xmara Tiber]*$_{PP}$
[**with**:DEF-river Tiber]
'From here we go out for a stroll [**along** the river Tiber].'
(c) (Π32) *maġenb* 'close to' – [Korpus Malti 3.0; news183872]
din il-koppja kienu qegħdin passiġata
DEM.F.PROX DEF-couple be:3PL.PFV PROG:PL stroll
[maġenb *ix-xmara Jinshi]*$_{PP}$
[**close_to** DEF-river Jinshi]
'[. . .] this couple was on a stroll [**beside** the river Jinshi] [. . .].'

First of all, the spatial situations in (5.266) and (5.267) have in common that the Figure is close to one side of the Ground. This means that laterality applies no matter whether Georgie arrives at the river (= (5.266a)), the speaker is taken down to the river (= (5.266b)), the family camps near the river (= 5.266c)), the fireworks take place near the river (= (5.266d)), the hotel is facing the Nile (= (5.267a)), the protagonists take a walk along the Tiber (= (5.267b)), or along the Jinshi (= (5.267c)). There is, however, a difference insofar as the examples (5.266c–d) and (5.267a) reflect a static spatial relation. The Figure is (at least temporarily) immobile and so small that its relation to the Ground can be termed punctual. The punctual relation holds also for (5.266a–b) because the motion event has its endpoint at the river, i.e., the river is the Goal. In contrast, the examples in (5.267b–c) capture motion events which involve a Figure that moves along one side of the Ground so that the Figure-Ground relation holds for a longer stretch of the total length of the river. These situations invoke the notion of PROLATIVE.

We argue that laterality is especially relevant for spatial situations that involve a Ground that prominently extends on one dimension. This observation guides us directly to the next section where long objects are in the focus of interest.

5.2.1.4.4.3.5 Long object

Borg/Azzopardi-Alexander (1997: 167–168) have separate sections for Grounds that refer to long objects for the following categories: INTERIOR, EXTERIOR, SUPERIOR, SUPERIOR-CONTACT, SURFACE, INFERIOR, INFERIOR-CONTACT, INTERIOR, SUPERIOR, SUPERIOR-CONTACT, and SURFACE-CONTACT.[223] Two things are remarkable about this long list. We start from the back because in this way lengthy discussions of irrelevant cases can be avoided. The shared marker of INTERIOR (LONG OBJECT), SUPERIOR (LONG OBJECT), SUPERIOR-CONTACT (LONG OBJECT), and SURFACE-CONTACT (LONG OBJECT) is not a Π, but the fully-fledged transitive verb *qasam* 'break' that is also used as translation equivalent of English *cross (a road, etc.)*. Since this verb belongs to a different word class from Πs displaying a completely different syntactic behaviour, *qasam* 'break' does not meet the criteria defined for the canonical Π. This is why we stroke INTERIOR (LONG OBJECT), SUPERIOR (LONG OBJECT), SUPERIOR-CONTACT (LONG OBJECT), and SURFACE-CONTACT (LONG OBJECT) from the list of items we need to discuss further.

The second fact of interest with the long-object criterion comes to surface when we inspect the markers that are used to express certain configurations with Grounds whose referents fall into the class of long objects and compare them to the markers that are used with objects of different sizes. Table 5.31 contains the necessary information. Identically filled cells are marked out in identical colours.

Table 5.31: Markers for configurations with unspecified and long-object Grounds.

configuration	Ground	
	unspecified	long
INTERIOR	*minn*	*minn*
EXTERIOR	*minn barra*	*maġenb*
SUPERIOR	*minn fuq*	*minn fuq*
SUPERIOR-CONTACT	*minn fuq*	*(minn) fuq*
SURFACE	*minn fuq*	*(minn) fuq*
INFERIOR	*minn taħt*	*(minn) taħt*
INFERIOR-CONTACT	*minn taħt(?)*	*minn taħt*

223 In the source, the label SURFACE (LONG OBJECT) is used twice (Sections 2.1.1.5.24a and 2.1.1.5.29a in the reference grammar). We have corrected the second mention of the label to SURFACE-CONTACT (LONG OBJECT) because the exponent used in this case is identical to the one used for the expression of SUPERIOR-CONTACT (LONG OBJECT).

There is an abundance of cells filled either by *minn*$_{\text{PERLATIVE}}$ or *minn*$_{\text{PROLATIVE}}$. These and similar cases have been extensively discussed repeatedly already – along with the possibility of omitting (П41) *minn* 'from' in certain contexts.[224] We do not intend to take up these issues again in this section. It should not go unmentioned, however, that the colour code reveals that in Table 5.31, several potential distinctions are neutralised (and we count the optional absence of (П41) *minn* 'from' as distinctive). For INTERIOR, SUPERIOR, and INFERIOR-CONTACT, the distinction between unspecified objects and long objects is irrelevant because the same marker is used for both object types. At the same time, in the column of unspecified Grounds, SUPERIOR, SUPERIOR-CONTACT, and SURFACE are expressed identically as are INFERIOR and INFERIOR-CONTACT. In the column of long objects, identical expression holds for SUPERIOR-CONTACT and SURFACE. This leaves us with only three distinct expressions, i.e., those which are uniquc to a given cell, namely (according to the specifications given in the reference grammar) *minn barra* (EXTERIOR MOTION FROM / unspecified object),[225] *maġenb* (EXTERIOR long object), and *(minn) taħt* (INFERIOR long object). In what follows, we focus on the role which (П32) *maġenb* 'close to' is assigned to in Table 5.31.

To make our point, we deviate from the usual practice applied in the previous sections where relatively short examples sufficed to support the arguments we wanted to make. In (5.268), we present a longish passage from a parliamentary speech which contains a number of interesting PPS (marked out in boldface) to which we will refer back subsequently. The verbs on which some of these PPS depend are underlined. We refrain from glossing the lengthy example since the marked parts thereof will be discussed separately in the next paragraph where their internal structure is explained, too.

(5.268) Waterways – [Korpus Malti 3.0; parl3213]
Sal-lum il-pożizzjoni hi li l-ilma jasal ***ħdejn il-ground ta' Burmarrad*** *– iżda issa minflok* ***mill-wiċċ tat-triq*** *se jkun għaddej* ***minn ġol-pajpijiet*** *– imma sfortunatament imbagħad* ***minn ħdejn il-ground sa ħdejn Kennedy Grove, matul il-wied*** *għandna problema kbira li akkumulat* ***matul is-snin,*** *fejn dawk li kienu l-passaġġi ta' l-ilma antiki spiċċaw – hemmhekk il-bużnanniet kellhom il-kanen ta' l- ilma jgħaddu* ***minn ġor-raba'*** *– u allura hemm il-probabbilta' li bħalma ġara kull darba li għamlet ix-xita, meta l-ilma jasal* ***ħdejn il-ground tal-football ta' Burmarrad*** *jerġa' jiżbokka fl-inħawi tar-raba' għax l-ilma m' għandux* ***minn fejn*** *jgħaddi u allura jerġa' jkollna għarar fir-raba' ta' dawk l- inħawi.*

224 Cf. Sections 5.1.2.2.20.2.1, 5.1.2.2.20.3, and 5.2.1.4.4.1.2.
225 Cf. Section 3.2.1.

'Until today the sitation is such that the water arrives **near the Burmarrad football pitch** – but now, in lieu of **across the surface of the street**, it will pass **through the pipes** – but unfortunately then **from near the football ground to near Kennedy Grove** we have a big problem **along the valley** that has accumulated **over the years** where what used to be the old passages of the water came to an end – where the great-grandfathers had the water-pipes passiing **through the fields** – and thus there is the probability that like each time it rained, when the water arrives **at the Burmarrad football ground** it again flows into the fields because the water has **nowhere** to pass **through** and thus will we again have floods in the fields of these areas.'

In (5.268), ten PPs are highlighted. Which connection do they have to the long-object criterion? The following list shows that the long-object criterion needs to be revised.

- *jasal ħdejn il-ground ta' Burmarrad* 'it arrives near the Burmarrad (football) ground' – the Figure (= the water) runs as far as close to the Ground that is a rectangular object with two long and two shorter sides. Neither the width nor the length is of any importance for the choice of the PROPINQUUS Π;
- *jasal ħdejn il-ground tal-football ta' Burmarrad* 'it arrives near the Burmarrad football ground' – the previous description is valid also for this example;
- *minn ħdejn il-ground* 'from near the (football) ground' – this is the starting point whence the problematic terrain stretches out till the endpoint to be mentioned next; the spatial situation involves an ABLATIVE use of (Π41) *minn* 'from'; the size of the Ground on any of the three dimensions is of no avail for the choice of the configuration marker because what counts is the proximity of the Figure to the Ground;
- *sa ħdejn Kennedy Grove* 'until near Kennedy Grove' – this is the endpoint of the problematic terrain; the combination of (Π51) *sa* 'till' and (Π23) *ħdejn* 'beside' forms a parallel to that of (Π41) *minn* 'from' and (Π23) *ħdejn* 'beside' in the previous PP; the properties of the Ground are irrelevant for the choice of the configuration marker;
- *mill-wiċċ tat-triq* 'across the street's surface' – originally the water flowed above the ground across the street which can be understood as a long object, meaning: we are facing a PERLATIVE or PROLATIVE use of (Π41) *minn* 'from'; according to Table 5.31, this could either be an instance of motion within an object of unspecified size or within a long object;
- *għaddej minn ġol-pajpijiet* 'passing through the pipes' – the pipes are bona fide examples of long objects; as in the foregoing case, we are witnessing the use of *minn*$_{PERLATIVE}$ with the water moving on the inside of the long object; according to the reference grammar, "**minn** followed by **ġo/ġewwa** [original boldface]" (Borg/Azzopardi-Alexander 1997: 158) is used (only?) for motion from the

interior of an object (=ELATIVE); our example however, describes a situation in which the water runs through the pipes, i.e., on their inside; note that the combination of (Π41) *minn* 'from' and (Π19) *ġo* 'in' is absent from Table 5.31;

- *jgħaddu minn ġor-raba'* 'they pass through the fields' – since the fields constitute a referent whose flanks are long; this case resembles the previous one except that (Π41) *minn* 'from' and (Π19) *ġo* 'in' combine to depict the spatial situation as one that affects the interior of the Ground although in real life the water passes through the fields in the open air;
- *minn fejn jgħaddi* 'where it passes through' – the object through which the water can pass is unidentifiable (because it is absent, in the first place), so that one cannot be sure about the applicability of the long-object criterion; it is clear however, that (Π41) *minn* 'from' functions as PERLATIVE;
- *matul il-wied* 'along the valley' – the valley is a good candidate for a long object; the problems (= Figure) arise at different points along the valley (whether inside or outside of it cannot be determined); one might ask why (Π33) *matul* 'during' does not appear in Table 5.31;
- *akkumulat matul is-snin* 'it accumulated over the years' – the PP forms a temporal adverbial and is thus not included in the ensuing discussion, although metonymically speaking, a succession of many years might be taken to represent a long object, too.

The gist of the data in (5.268) is as follows. The putative paradigm in Table 5.31 fails to cover the entire range of phenomena that are connected to the criterion of long objects. On the one hand, there are gaps in the inventory. On the other hand, the fact that an object that represents the Ground in a spatial situation happens to be long does not automatically trigger the use of a given Π or combination of Πs. This is a fact that calls for being elaborated upon further.

In partial repetition of what was said already in connection with laterality,[226] we first present two examples that demonstrate that the length of the object serving as Ground need not dictate what Π the speaker has to use. In (5.269) and (5.270), the Ground is represented by the common noun *wied* 'valley'. In the majority of the examples in (5.269)–(5.270), the Figure is represented by expressions that refer to types of streets/roads. The reason for this choice of Figure will be disclosed below.

226 Cf. Section 5.2.1.4.4.3.4.

(5.269) Long Figures combining with long Grounds I

(a) (Π48) *qrib* 'near' – [Korpus Malti 3.0; parl1659]

Din it-triq għadha qatt ma ġiet
DEM:F:PROX DEF-street still:3SG.F never NEG come:3SG.F.PFV
livellata tinsab ***[qrib*** *il-wied]*$_{PP}$ *u tista'*
level:PART:F 3SG.F.IPFV:REFL:find **[near** DEF-valley] and 2SG.IPFV:can
taħseb x' jgħaddi ta' l-ilma minn$_{HA}$
2SG.IPFV:think what 3SG.M.IPFV:pass of DEF-water from:3SG.F
'This street has never been levelled, it is situated [**near** the valley] and you can imagine how much water passes over it [. . .].'

(b) (Π58) *viċin* 'near' – [Korpus Malti 3.0; parl5444]

din it-triq hija ***[viċin*** *il-wied]*$_{PP}$ *u kemm-il darba*
DEM:F:PROX DEF-street 3SG.F **[near** DEF-valley] and how_many time
sofriet minn problemi ta' l-għargħar
suffer:SG.F.PFV from problem:PL of DEF-flood
'[. . .] this street is [**near** the valley] and how many times has it suffered from problems of flooding.'

We assume that the Ground is indeed long. The valley seasonally may turn into a torrent because of the rain that pours down over an extended period of time on the island. The masses of water that are too much for the valley to transport to the sea create problems for everything that is built on the slopes of the valleys or near them because the slopes temporarily turn into river banks. The rising water floods nearby streets causing damages, etc. Interestingly, the streets mentioned in (5.269) are on the exterior of a long object but the expected (Π32) *maġenb* 'close to' is not used to express the relation between Figure and Ground although this is what Table 5.31 requires to happen. Accordingly, the length of the object in Ground function cannot be decisive alone for the choice of the appropriate Π.

The LDS-format does not make explicit which participant in a spatial situation is the long object, but it can be inferred from the context in the reference grammar that it is the Ground that counts. We argue that it is not enough for the Ground to have a longish referent. We assume that both Figure and Ground need to be taken account of. In (5.269), the streets are long objects just like the valley is a long object. In (5.270), the combination of a Figure and a Ground both of which fulfil the criterion of being long objects affects the choice of Π without however privileging any Π in particular.

(5.270) Long Figures combining with long Grounds II

(a) (Π8) *biswit* 'facing' – [Korpus Malti 3.0; news123901]

t-trejqa li hemm ***[biswit*** *il-wied]*$_{PP}$ *hija taħt*
DEF-street.DIM that there **[facing** DEF-valley] 3SG.F under

ir-responsabilità *tal-Gvern* *Ċentrali*
DEF-responsibility of:DEF-government central
'[. . .] the small road that is there [**near** the valley] is the responsibility of the central government [. . .].'

(b) (Π23) *ħdejn* 'beside' – [Korpus Malti 3.0; news79702]
Meta ngħaddi ***[minn ħdejn*** *il-wied]$_{PP}$* *ir-riħa tinten*
when 1SG.IPFV:pass [**from beside** DEF-valley] DEF-odour 3SG.F.IPFV:stink
tal-ħmieġ tal-klieb tispikka
of:DEF-dirt of:DEF-dog.PL 3SG.F.IPFV:stand_out
'When I pass [**near** the valley] the malodour of the dogs' dirt stands out [. . .].'

(c) (Π30) *ma'* 'with' – [Korpus Malti 3.0; literature21]
Imbagħad reġgħu bdew mexjin
then do_again:3PL.PFV begin:3PL.PFV walk:PART:PL
[ma*l-wied]$_{PP}$* *lejn Tal-Qadi*
[**with:** DEF-valley] towards Tal-Qadi
'Then the began again to walk [**along** the valley] towards Tal-Qadi.'

(d) (Π32) *maġenb* 'close to' – [Korpus Malti 3.0; news138247]
mit-triq ta' ***[maġenb*** *il-wied]$_{PP}$* *kien għaddej ħafna traffiku*
from:DEF-street of [**close_to** DEF-valley] be pass:PART many traffic
'[. . .] through the street [**close to** the valley] a lot of traffic was passing [. . .].'

(e) (Π33) *matul* 'during' – [Korpus Malti 3.0; literature23]
Iżda l-Pitirross kien jieħu gost jilgħab
but DEF-robin be 3SG.M.IPFV:take pleasure 3SG.M.IPFV:play
magħhom u f' tebqa t' għajn rikkeb tnejn
with:3PL and in one_of_a_pair of eye ride.CAUS two
minn$_{HOM}$ fuq dahru u rħielha jtir bihom
from:3PL on back:3SG.M and let_go:IO.3SG.F 3SG.M.IPFV:fly with:3PL
[matul *il-wied]$_{PP}$*
[**during** DEF-valley]
'But the robin had fun playing with them and in a twinkling of an eye he lifted them onto his back and went off flying with them [**along** the valley].'

(f) (Π41) *minn* 'from' – [Korpus Malti 3.0; literature20]
Quddiemkom it-triq li tgħaddi sserrep
in_front_of:2PL DEF-street that 3SG.F.IPFV:pass 3SG.F.IPFV:meander
[mil *il-wied]$_{PP}$*
[**from**:DEF-valley]
'In front of you is the road that meanders [**through** the valley] [. . .].'

There are six different ways of expressing that the Figure and the Ground form a parallel at least for an extended stretch of space. In (5.270a), the road is said to be small but it runs parallel to the valley for part of the latter's length. In (5.270b–c), the protagonists walk along or through the valley, i.e., their stroll covers part of the length of the valley. Example (5.270d) can be compared to (5.270a) provided the street is seen as the Figure whose length and that of the valley coincide at least partly. In (5.270d), the bird takes two other characters of a fairy story on a flight along the valley, i.e., their itinerary is identical with part of the length of the valley. And finally, in (5.270f), the road is depicted as meandering through the valley. Once again, the length of the Figure is partly identical to that of the Ground. To our mind, it is this parallel between the Figure's extension and the Ground's extension that allows the Maltese speakers to select from a set of Πs that trigger associations with length. The examples in (5.269) clearly show that this is not a must. Πs that are associated with proximity rather than length may also be chosen. It is also possible (though less likely) that the parallel between the Figure and the Ground in (5.269) is too insignificant to invoke the notion of length. In this case, the streets would not run alongside the valley but punctually come close to one slope of the valley. This reasoning remains speculative. Whether or not arguments of this kind can be upheld can only be determined interactively, i.e., in collaboration with Maltese native speakers.

There is also an urgent need to investigate the possibility that it makes a difference whether the long object is prominently extended horizontally (as is the case in all examples hitherto discussed) or vertically. As far as we can tell at the moment, the wealth of possible markers reported above diminishes considerably when the longest side of the Ground can be defined as its height. A typical example is *arblu* 'mast'. In (5.271), we provide an example of this noun in Ground function.

(5.271) Length replaced by height – [Korpus Malti 3.0; culture971]

kulħadd	*jipprova*	*xortih*	*jitla'*
everybody	3SG.M.IPFV:try	luck:3SG.M	3SG.M.IPFV:ascend
[m'	*arblu*	*mimli*	*xaħam]$_{PP}$*
[with	mast	full	grease]

'[...] everybody tries his luck by way of climbing up [a mast full of grease].'

Whenever the Ground noun refers to an object that extends vertically like *lasta* 'pole', *kolonna* 'column', *pilastru* 'pillar', etc. the Π of choice is always (Π30) *ma'* 'with'. If the mast happens to lie on the ground, however, (Π30) *ma'* 'with' is replaced with (Π33) *matul* 'during' as in (5.182).

(5.172) Back to length – [Korpus Malti 3.0; literature39]

Fit-tarf	*kien*	*ikollu*	*bandiera*	*u*
in:DEF-end	be	3SG.M.IPFV:have.FUT:3SG.M	flag	and

l-kompetituri *kienu* *jridu* *jiġru* *wieħed*
DEF-competitor:PL be:3PL.PFV 3.IPFV:want:PL 3.IPFV:run:PL one
wieħed *minn* *fuq* *il-barkun* ***[matul*** *l-arblu]*$_{PP}$ *sal-bandiera*
one from on DEF-pontoon **[during** DEF-mast] till:DEF-flag
'At the end, there would be a flag and the competitors needed to run one by one from the pontoon's deck [**over** the mast] to the flag.'

(5.271)–(5.272) are taken from descriptions of funny games. In both games, a mast is an important prop. In (5.271), the mast is erected vertically so that people have to climb it in order to fulfil their task. The situation in (5.272) is different insofar as the mast lies on the deck of the pontoon from which it protrudes out onto the sea. The participants in the game have to walk to the end of the mast to reach a flag. The choice of Π is most probably motivated by the position of the mast and its natural orientation. (Π30) *ma'* 'with' seems to make more sense with the latter. Thus, (Π30) *ma'* 'with' would cause misunderstanding in a context like the one in (5.272). In both cases, Figure and Ground are physically in touch with each other because the competitors' hands and/or feet automatically touch the mast during the game. We do not know yet how exactly the parallel position or movement of Figure and Ground without contact is expressed in Maltese when there is an extended overlap in length.

The remaining uncertainties notwithstanding, it can be concluded that Table 5.31 tells only part of the story. There are many more options for expressing certain relations than the table has on offer. The fragmentary character of Table 5.31 and the open questions raised in this section underline the necessity of further in-depth work in this domain.

5.2.1.4.4.3.6 Animacy

The last parameter that we comment upon is animacy. We keep the discussion as short as possible. There is no need to go further into the uncontroversial cases of (Π22) *għand* 'at s.o.'s place' and (Π38) *mingħand* 'from s.o.' since their complement must be [+human] (under personification also [+animate] with further possibilities for metonymy as e.g. the literary work of an author functioning as indirectly animate Ground).[227] It remains to be determined to what extent the degree of animacy of the Ground (or perhaps also of the Figure) influences the choice of Π in spatial situations.

It is also clear that the purely temporal (Π59) *waqt* 'at the time of' does not tolerate an animate Ground. In contrast, the temporal (Π46) *qabel* 'before' is compat-

227 Examples of this restriction were given already in Sections 5.1.2.1.2, 5.1.2.2.3 and 5.1.2.2.4.

ible with animate Grounds. Independent of its being read as spatial or temporal Π, (Π33) *matul* 'during' does not seem to take animate Grounds. Animacy is no obstacle for either (Π18) *ġewwa* 'inside' or (Π19) *ġo* 'in' to combine with a complement noun although most of the time (Π15) *fi* 'in' is intercalated because there is a strong tendency towards locative government. In point of fact, the bulk of the spatial Πs may take inanimate complements as well as animate complements. There are only gradual differences affecting the frequency of the combinations. These differences can be drastic at times. However, we leave this topic for consideration in follow-up studies that apply a much more sophisticated quantitative methodology. Animacy is relevant for the choice of Π only in a small minority of cases.

5.2.1.4.4.3.7 Excursus: a Π that sometimes isn't a Π

At this point, it is in order to come back shortly to an issue that was raised above already.[228] With reference to (Π31) *madwar* 'around', we are uncertain as to its Π-status when used in combination with countable concepts like in (5.273) no matter to which ontological class they belong.

(5.273) Adverb or Π? – [Korpus Malti 3.0; academic1]

Dawn *l-attivitajiet* *jiffukaw* *fuq* *suġġett* *speċifiku*
DEM:PL:PROX DEF-activity:PL 3.IPFV:focus:PL on subject specific
huma *żviluppati* *b'* *mod* *adattat* *għal* *età*
3PL develop:PART:PL with manner suitable for age
partikolari *u* *jistgħu* *isiru* *fi* *żmien* *xahar*
particular and 3.IPFV:can:PL 3.IPFV:become:PL in time month
b' ***madwar*** ***sagħtejn*** *xogħol* *fil-ġimgħa*
with **around** **hour:DU** work in:DEF-week
'These activities focus on a specific subject, they are developed in a suitable manner for a particular age and they can be realised within a month with **about two hours** work per week.'

In examples of this kind, (Π31) *madwar* 'around' serves the purpose of characterizing a proposed quantity as being approximate. The Π adds vagueness to the statement. In this function, (Π31) *madwar* 'around' is in a paradigmatic relationship with adverbials like *eżatt* 'exact(ly)' in (5.274).

228 In Section 4.1.1.2.2.1 (xx) (and Section 4.1.2.2.1.2.1 (v) in passing).

(5.274) Adverb – [Korpus Malti 3.0; culture3139]
Eżatt 600 sena *wara s-Serb Slobodan Miloseviç ħabat*
exactly 600 year after DEF-Serbian Slobodan Miloseviç strike
għall-Kosovo u wettaq massakru
for:DEF-Kosovo and implement massacre
'**Exactly 600 years** later, the Serbian Slobodan Miloseviç attacked the Kosovo and committed a massacre.'

When used in contexts of this kind, (Π31) *madwar* 'around' resembles *(in)ċirka* 'approximately' – a candidate we have cancelled from the list of Maltese Πs[229] because it does not fit the description of the canonical Π as given in Figure 1.5. Where (Π31) *madwar* 'around' precedes a quantitative expression, it does not link the latter to an external head nor does it function as the head of a PP. What it does instead is modify the quantity adverbially. According to this criterion, (Π31) *madwar* 'around' is no longer a Π in (5.273) and similar cases. It keeps its status as Π in those contexts where a CIRCUMESSIVE spatial or temporal relation is expressed. This status also holds in many cases in which (Π31) *madwar* 'around' is followed by a numerically quantified complement as in (5.275).

(5.275) (Π31) *madwar* 'around' as Π – [Korpus Malti 3.0; news174314]
Il-Mexxej Laburista ddawwar ***[madwar*** *it-tmien*
DEF-leader labour REFL.CAUS.turn_around **[around** DEF-eight
sulari]$_{PP}$ *minn fejn illum qed topera l-kumpanija*
floor:PL] from where today PROG 3SG.F.IPFV:operate DEF-company
tal-media tal-PN
of:DEF-media of:DEF-PN
'The leader of the Labour Party was shown [**around** the eight floors] from where today the media company of the PN operates.'

This is not an estimation of the approximate number of storeys the party leader is made to inspect. He is shown around on all eight storeys of the building. Thus, (Π31) *madwar* 'around' does not modify the numeral but connects the complement to the preceding verb. This time, (Π31) *madwar* 'around' is a bona fide Π and not an adverb. The presence of the definiteness marker is an additional sign for this status because definite quantified nouns are incompatible with approximation. Since the definiteness marker is not always involved in the contexts in which (Π31) *madwar* 'around' occurs, it is by no means easy to decide whether it is used prepositionally or adverbi-

229 As early as Section 1.4.1.5.

ally. What can be said nevertheless is that there is no general restriction that excludes the possibility of (Π31) *madwar* 'around' governing a quantified complement.

5.2.1.4.4.3.8 Wrap-up

This section contains the recapitulation of the many details we have provided above in order to systematise the findings. Except in the case of LONGITUDO, we have not found evidence for potential differences between spatial and temporal functions of configurational Πs. Beyond this category, this distinction does not seem to be exploited to any noticeable degree in Maltese. Similarly, we have seen that the parameter of contact between Figure and Ground is relevant only in a niche within the system. In most of the cases, [±contact] seems to be irrelevant for the choice of Π_{CONFIG}. The distinction between PROPINQUUS and PROXIMATE has been introduced.[230] For the purpose of simplification, we have chosen to ignore this distinction at this point to facilitate the systematisation of the Πs' semantics. The variable distance between Figure and Ground is only marginally relevant on the same grounds as the distinction between with and without contact. Furthermore, we have only limited evidence for laterality being relevant for the formal expression of the relation between Figure and Ground. We do not claim that on closer inspection, these criteria might turn out to be by far more important than assumed in this study. However, our database does not suggest that this is the case. We therefore propose to look into this matter under the linguistic microscope in future research.

On account of the above survey of configuration markers and directionality markers, it can be concluded that the Maltese system is relatively clearly structured in the spatio-temporal domain since there is a transparent division of labour between Πs which encode directionality and those which encode configuration. Multifunctionality is attested in both sub-classes of Πs but this phenomenon is restricted to a relatively small number of Πs and functions. In addition to the idea that the canonical Maltese Π is configurational, it can be assumed that in the domain of spatio-temporal meaning categories, the canonical Maltese Π is monofunctional.[231]

Figure 5.59 sums up what we have learned about the systematicity of the directionality markers. It hardly needs to be mentioned that the schema is an idealisation that glosses over a number of details whose relevance for the overall picture of the grammar of Maltese Πs has yet to be determined.directionality

There is no equivalent tree structure for the configurational Πs. Therefore, we have to outline this part of the system without relying on a diagram. For each possible configuration, there is at least a dedicated Π. The Figure can be inside or at the

230 Cf. Section 5.1.2.2.20.1.1.

231 We come back to this issue in Section 5.3.

Ground ((Π15) *fi* 'in', (Π18) *ġewwa* 'inside', (Π19) *ġo* 'in'; (Π22) *għand* 'at s.o.'s place'). The Ground may consist of a multitude (n > 2) of parts of a kind so that the Figure can be placed in the midst of these components ((Π16) *fost* 'amongst', (Π47) *qalb* 'amidst'). If there are only two components which constitute a distributed Ground (that does not have to consist of parts which belong to the same ontological class) the Figure may occupy a position in between the two parts ((Π4) *bejn* 'between'). The bulk of the configurational Πs is fit to encode spatial situations in which the Figure is outside the Ground either with or without ((Π3) *barra* 'outside', (Π44) *oltre* 'beyond') contact between the participants. The Figure may be placed at the front ((Π49) *quddiem* 'in front of') or back ((Π60) *wara* 'after') of the Ground. It may be on top or above ((Π17) *fuq* 'on') the Ground, at the bottom or below ((Π56) *taħt* 'under') the Ground. What we notice in any case is the high number of PROPINQUUS Πs for which it is difficult to pinpoint clearcut semantic differences. The Figure may be placed in the vicinity ((Π14) *fejn* 'near', (Π23) *ħdejn* 'beside', (Π32) *maġenb* 'close to', (Π48) *qrib* 'near', (Π58) *viċin* 'near') of the Ground either being detached from it or in contact with it ((Π30) *ma'* 'with'). The Figure can be described as being located opposite the Ground ((Π8) *biswit* 'facing', (Π12) *faċċata* 'opposite'). A Figure that is a long object or moving along a Path can be pictured as extending over a stretch of space parallel to the Ground ((Π33) *matul* 'during'). The Figure may also surround the Ground ((Π31) *madwar* 'around').

To close the discussion of the spatio-temporal component of the system of Maltese Πs, we state that apart from the usual uncertainties on the micro-level, the systematicity of the functional network of the Πs is clearly visible. Directional and configurational Πs interact systematically to cover all possible spatial situations. Polysemy and synonymy are relatively restricted phenomena. How important certain parameters like laterality are for the internal organisation of the Π-system cannot be determined in this study. As of now, it seems that their importance is severely limited. However, this impression is probably only an effect of our reliance on corpus data and the written register. This proviso holds of course also for our subsequent account of the semasiology of Maltese Πs in the non-spatio-temporal domain.

5.2.2 The semasiology of Πs outside the spatio-temporal domain

The structure of this section is not an exact replica of the organisation of the foregoing section because of the multitude of different semantic relations that characterise the non-spatio-temporal domain. Besides the absence of the typical division of labour found with spatio-temporal Πs, we also notice that the prominent role played by zero-marking in the expression of spatial relations has no direct equivalent in connection with the categories this section is meant to address from the

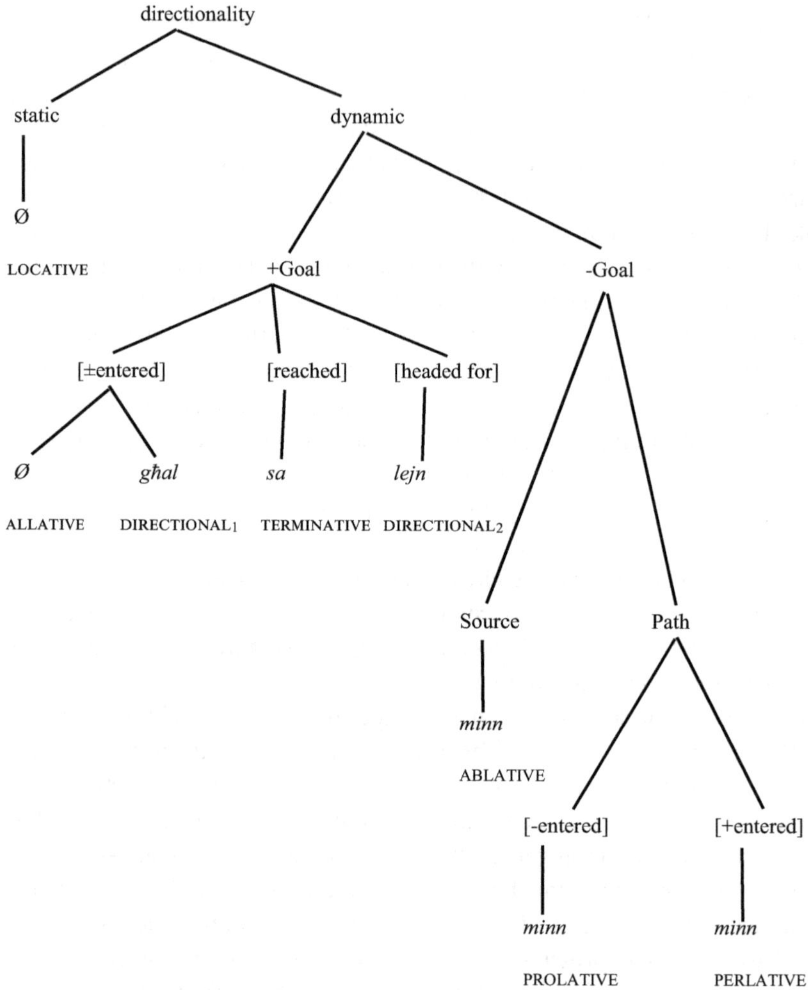

Figure 5.59: The system of directionality markers.

semasiological point of view. There obviously is no need to address phenomena which are not attested in the non-spatio-temporal domain. It therefore makes sense to only discuss polysemy, synonymy, and antonymy.[232] However, before we take this step it is in order to call in mind a number of important insights about non-spatio-temporal ΠIs we have gained already at the beginning of this section. Given that

232 In Sections 5.2.2.1, 5.2.2.2, and 5.2.2.3, respectively.

the non-spatio-temporal domain is intriguingly heterogeneous internally, we can offer only preliminary results.[233]

In point of fact, we already know many facts about the non-spatio-temporal Πs. Figure 5.27 shows that they constitute the minority within BLOMP 2.0 where they account for 43% of the inventory. The twenty-six non-spatio-temporal Πs are listed alphabetically in (5.276). In accordance with our general practice, (Π26) *kif* 'as' will not be discussed further. We emphasise that if we use the term non-spatio-temporal Π, we only refer to those Πs that are not also featured in the spatio-temporal domain.[234]

(5.276) Non-spatio-temporal Πs in BLOMP 2.0
(Π1) *apparti* 'apart from'; (Π2) *a skapitu* 'at the expense of'; (Π5) *bħal* 'like'; (Π6) *bħala* 'as'; (Π7) *bi* 'with'; (Π9) *bla* 'without'; (Π10) *daqs* 'equal to'; (Π11) *dwar* 'about'; (Π13) *favur* 'in favour of'; (Π20) *għajr* 'except'; (Π24) *ħlief* 'except'; (Π25) *inkluż* 'including'; (Π26) *kif* 'as'; (Π27) *kontra* 'against'; (Π29) *lil* 'to'; (Π34) *minbarra* 'except'; (Π35) *minflok* 'instead of'; (Π36) *mingħajr* 'without'; (Π37) *mingħala* 'in s.o.'s opinion'; (Π39) *minħabba* 'on account of'; (Π40) *minkejja* 'in spite of'; (Π45) *permezz* 'by means of'; (Π50) *rigward* 'concerning'; (Π52) *sforz* 'thanks to'; (Π53) *skont* 'according to'; (Π55) *ta'* 'of'

Thanks to the information conveyed by Figures 5.28–5.29 we further know that Italo-Romance Πs are underrepresented in the spatio-temporal domain where they account for maximally 15% of all cases whereas their share rises to 38% if we look at the non-spatio-temporal cases. The relation between Italo-Romance origin and non-spatio-temporal meaning comes to the fore even more strongly when we consider that 67% of all Italo-Romance Πs fulfil non-spatio-temporal functions. In the case of the Semitic Πs, the percentage is considerably lower, namely only 36%.

The special position of the Italo-Romance Πs must be remembered when we evaluate the findings.[235] Furthermore, we will face several cases of Πs that were treated as members of the spatio-temporal subset of BLOMP 2.0 above but that are also employed for non-spatio-temporal functions.[236] In this way, they compete with genuinely non-spatio-temporal Πs. The reverse case is also attested though less frequently. Maltese Πs that feel at home on both sides of the dividing line between spatio-temporal and non-spatio-temporal domains are of particular interest for the discussion of polysemy and synonymy. The same holds for those Πs that are

233 Cf. Section 5.2.2.4.
234 The latter will be investigated separately in Section 5.2.3.
235 Cf. Section 5.2.2.4.
236 Cf. Section 5.2.3.

employed in the domain of the core meanings but have a second (and perhaps a third) stronghold in the non-spatio-temporal domain. In contrast to the multifunctional spatio-temporal Πs that also express non-spatio-temporal meanings, the Πs from the core domain are always counted in when we speak about multifunctionality of non-spatio-temporal Πs. In this and the subsequent subsections we exclusively take account of Πs that are excluded from the spatio-temporal domain.

5.2.2.1 Polysemy in the non-spatio-temporal domain

As the foregoing sections amply show, the semantics of Πs is a slippery terrain. We are fully aware of the necessity to gloss over many meaning-related aspects of the Maltese Πs since accounting for all of the semantic nuances would not only require native-speaker knowledge and intuition but also demand ample time to write a separate dissertation on this subject that cannot be exhaustively treated in a summary section. As an incentive for others to take up this issue in the format of a dedicated in-depth study, we make the daring move of putting forward several hypotheses based on a rather coarse-grained semantic analysis that can be proved wrong in the future.

This section is divided into two subsections. In the first one monofunctionality is addressed[237] and the second is dedicated to multifunctionality.[238] It cannot be counted out completely that future research will reveal that, on the micro level, polysemy is much more common than the following paragraphs suggest.

5.2.2.1.1 Monofunctional non-spatio-temporal Πs

First of all, in analogy to what we have stated above with reference to configurational markers,[239] we assume that there are monofunctional non-spatio-temporal Πs. If we look close enough at BLOMP 2.0 we recognise that the vast majority of the non-spatio-temporal Πs forms part of the group of monofunctional Πs. As results from the list in (5.277), twenty of the non-spatio-temporal Πs are characterised by monofunctionality. This leaves not more than half a dozen non-spatio-temporal Πs that fall under the rubric of polysemy. Note that for the time being, we conceive of the latter group as being comprised of only those Πs that have not been mentioned in connection with spatio-temporal meanings already. In Section 5.2.3, for reasons to be explained there, it will be necessary to deviate from this practice.

---

237 Cf. Section 5.2.2.1.1.
238 Cf. Section 5.2.2.1.2.
239 Cf. Section 5.2.1.4.1.

(5.277) Monofunctional non-spatio-temporal Πs in BLOMP 2.0
(Π2) *a skapitu* 'at the expense of'; (Π6) *bħala* 'as'; (Π9) *bla* 'without'; (Π10) *daqs* 'equal to'; (Π11) *dwar* 'about'; (Π13) *favur* 'in favour of'; (Π20) *għajr* 'except'; (Π24) *ħlief* 'except'; (Π25) *inkluż* 'including'; (Π26) *kif* 'as'; (Π27) *kontra* 'against'; (Π35) *minflok* 'instead of'; (Π36) *mingħajr* 'without'; (Π37) *mingħala* 'in s.o.'s opinion'; (Π39) *minħabba* 'on account of'; (Π40) *minkejja* 'in spite of'; (Π45) *permezz* 'by means of'; (Π50) *rigward* 'concerning'; (Π52) *sforz* 'thanks to'; (Π53) *skont* 'according to'

Eight of the twenty Πs are monosyllabic, i.e., monosyllabicity is characteristic for 40% of the monofunctional non-spatio-temporal Πs. Five of the Πs in (5.277) involve the initial element *min(n)-*. On the formal side, there is thus, no property which is shared by all of the monofunctional non-spatio-temporal Πs.

Figure 5.60 captures the disproportion between the shares of the minority of multifunctional non-spatio-temporal Πs and the majority of monofunctional non-spatio-temporal Πs.

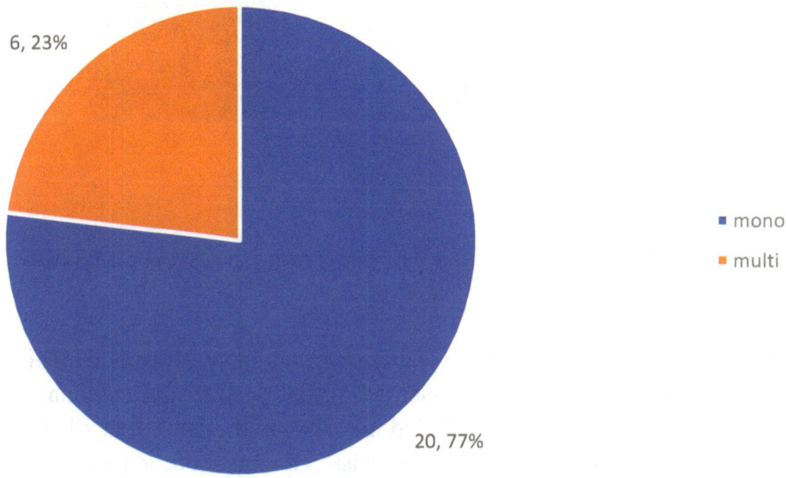

Figure 5.60: Shares of multifunctional vs monofunctional Πs in the non-spatio-temporal domain.

Even more striking is the particularly big share of Italo-Romance Πs within the monofunctional subset of non-spatio-temporal Πs. With nine out of twenty such Πs, the Italo-Romance component is as large as 45% as opposed to the small share of 17% that Italo-Romance Πs claim in the group of multifunctional non-spatio-temporal Πs as shown in Figure 5.61. The situation is reminiscent of the implicational pattern presented in Figure 5.41 where the Italo-Romance origin of a configuration marker correlates monolaterally with monofunctionality. In the non-spatio-temporal domain, too, the Italo-Romance preference for monofunctionality is pronounced as can be gathered from Figure 5.62. The shares differ considerably between Semitic and Italo-Romance Πs. Only 10% of the latter are polysemous whereas the share of polysemous non-spatio-temporal Πs equals 25% in the Semitic component. As to monofunctionality, 90% of all Italo-Romance non-spatio-temporal Πs fulfil this criterion – but only 69% of the Semitic non-spatio-temporal Πs do the same.

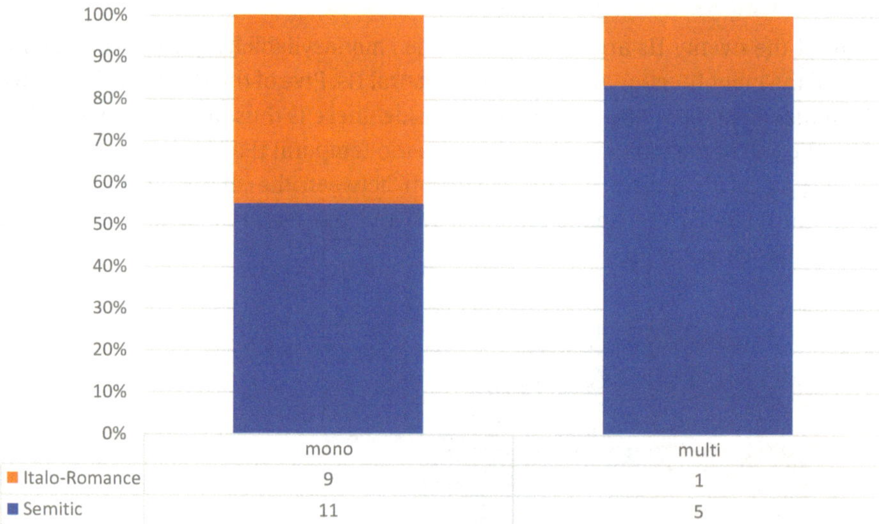

Figure 5.61: Shares of Italo-Romance / Semitic Πs in the monofunctional and multifunctional subset, respectively.

Semantically, the Πs in (5.277), as expected, do not form a homogeneous set. There are all kinds of meaning categories represented in the list of monofunctional Πs. Several meaning categories are fully included in (5.277) in the sense that all of the Πs that serve to express these meanings are registered as monofunctional. Those cases which involve more than one Π are two, viz. EXCLUSIVE (= (Π9) *bla* 'without', (Π36) *mingħajr* 'without'), and PERTENTIVE which is represented by (Π11) *dwar* 'about' and (Π50) *rigward* 'concerning' in (5.277). Other meaning categories either have only a single representative or are divided between monofunctional and multifunctional Πs. The discussion of the latter cases is reserved for later.

In sum, monofunctionality is generally preferred in the non-spatio-temporal domain. Yet, this preference is much stronger with Italo-Romance Πs than it is with Semitic Πs. Accordingly, we postulate the two-layered preference rule in Figure 5.63 where the double arrow >> indicates that the tendency is particularly strong.

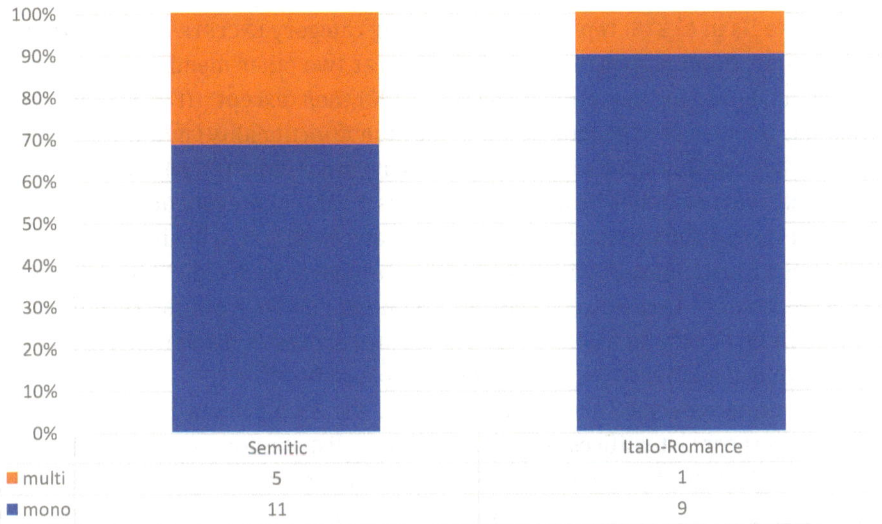

Figure 5.62: Monofunctional / multifunctional Πs across language of origin.

monofunctional	>	multifunctional	/ non-spatio-temporal
monofunctional	>>	multifunctional	/ non-spatio-temporal$_{\text{Italo-Romance}}$

Figure 5.63: Unmarked monofunctionality.

At this point, we can close the discussion of monofunctionality and turn our attention to multifunctionality or polysemy *tout court*.

5.2.2.1.2 Multifunctional non-spatio-temporal Πs

The multifunctional non-spatio-temporal meaning categories pose a number of problems because there is no straightforward way to separate them from the spatio-temporal ones and those of the core. In (5.278), we enumerate the six Πs that remain after subtracting the above monofunctional non-spatio-temporal Πs. The inventory as such does not tell us the whole story of the non-spatio-temporal meaning categories.

(5.278) Multifunctional non-spatio-temporal Πs in BLOMP 2.0
(Π1) *apparti* 'apart from'; (Π5) *bħal* 'like'; (Π7) *bi* 'with'; (Π29) *lil* 'to'; (Π34) *minbarra* 'except'; (Π55) *ta'* 'of'

Two of the Πs in (5.278) represent the meaning category EXCEPTIVE (= (Π1) *apparti* 'apart from', (Π34) *minbarra* 'except')[240] whereas two other members of this category count among the monofunctional Πs (= (Π20) *għajr* 'except'; (Π24) *ħlief* 'except'). Thus, it is impossible to generalise about the monofunctionality / multifunctionality of EXCEPTIVE Πs. This is the same for COMPARATIVE since only (Π5) *bħal* 'like' is multifunctional whereas other Πs involved with this meaning category are registered as monofunctional. Furthermore, we have encountered (Π7) *bi* 'with' in many diverse functions already most of which might be subsumable under the umbrella of the INSTRUMENTIVE.[241] Last but not least, (5.278) hosts two Πs from the core domain, namely (Π29) *lil* 'to' and (Π55) *ta'* 'of'. The former expresses the PATIENTIVE and the DATIVE while the latter is responsible for the POSSESSIVE.[242]

It is unnecessary to provide examples of the employment of the above Πs for said purposes. It suffices to determine their distribution across the different meaning categories by way of referring back to the examples given elsewhere in this study. The polysemic patterns we were able to determine in our database are exposed in (5.279). They are ordered top-down according to the decreasing number of meaning categories in which a given Π is involved. Identical meaning categories that are associated with several Πs are highlighted in colour.

(5.279) Polysemic patterns

(Π7) *bi* 'with' – PATIENTIVE (5.3), PROPRIETIVE (5.164), COMITATIVE (5.175), INSTRUMENTIVE (5.178), MOTIVATIVE (5.183), MATERIAL (5.214), and MANNER (5.217)

(Π55) *ta'* 'of' – POSSESSIVE (5.5), ESSIVE (5.195), and MATERIAL (5.216)

(Π5) *bħal* 'like' – MEDIATIVE (5.188), COMPARATIVE (equality) (5.189), ASSIMILATIVE (5.191)

(Π1) *apparti* 'apart from' – EXCEPTIVE (5.169) and ADDITIVE (5.205)

(Π34) *minbarra* 'except' – EXCEPTIVE (5.172) and ADDITIVE (5.207)

(Π29) *lil* 'to' – PATIENTIVE (5.2) and DATIVE (5.4)

For a Π to be multifunctional, the required minimum is its use in two different meaning categories. Three of six Πs exceed this minimum – with (Π7) *bi* 'with' and (Π55) *ta'* 'of' rank high on the frequency hierarchy in Table 3.27. Their high token frequency correlates with their particularly heavy functional load. It is worth noting that (Π29) *lil* 'to' and (Π5) *bħal* 'like', too, are close to the top of the same

240 Cf. Section 5.1.3.3.

241 Cf. Section 5.1.3.5.

242 The basic information on these core Πs are provided in Sections 5.1.1–5.1.1.4.

hierarchy. High frequency and monosyllabicity often associate with each other in BLOMP 2.0 as Figure 2.30 and its evaluation have shown.[243] Monosyllabicity holds for 67% of the multifunctional non-spatio-temporal Πs, i.e., this criterion is much stronger than it is in the case of the monofunctional Πs.

In contrast, the polysyllables (Π34) *minbarra* 'except' and (Π1) *apparti* 'apart from' occupy ranks 37 and 42, respectively. This means that polysemy in the non-spatio-temporal cannot be explained sweepingly by way of invoking a frequency effect. It is additionally interesting that the three highly frequent Πs count at least one core meaning among the meanings they express.

No meaning is expressed by more than two of the above Πs. (Π34) *minbarra* 'except' and (Π1) *apparti* 'apart from' display identical patterns of polysemy. Except for this pair of Πs, there is no proof of implications which bind categories of a given polysemy pattern to each other. This identity is important for the semantic map we will draw.[244] Only (Π5) *bħal* 'like' is a multifunctional non-spatio-temporal Π that displays a polysemy pattern all of whose members are unique to this Π. The conceptual closeness of the meanings expressed by (Π5) *bħal* 'like' will be of interest again below.[245] Eleven meaning categories are monopolised in the sense that they occur in the polysemic pattern of only one of the Πs. With a chain of seven meaning categories, the polysemy pattern of (Π7) *bi* 'with' ousts every other pattern in (5.189). Superficially, (Π7) *bi* 'with' might look like an all-purpose Π. This impression is wrong, of course, since there are many meaning categories also in the non-spatiotemporal domain where (Π7) *bi* 'with' has no responsibility for any function.

We conclude that multifunctionality is relatively weakly developed among non-spatio-temporal Πs. One Π stands out because of the number of meanings it is used for, namely (Π7) *bi* 'with'. For the bulk of the non-spatio-temporal Πs, monofunctionality is the rule. Multifunctionality is a minority option that privileges a small subset of the Πs under review. The next subsection will tell us whether similar conclusions can be drawn with regard to synonymy.

5.2.2.2 Synonymy in the non-spatio-temporal domain

In analogy to the previous subsection, this subsection is based on the inventory of non-spatio-temporal Πs given in (5.186). Synonymic patterns that involve Πs from outside the non-spatio-temporal will be inspected more closely below.[246] In this section, however, only their non-spatio-temporal members will be paid attention.

243 Cf. Section 2.3.
244 Cf. Section 5.2.4.
245 Cf. Section 5.2.4.
246 Cf. Section 5.2.3.

There are five non-spatio-temporal Πs that are never involved in synonymy. They are listed in (5.280) together with the meaning category they express and the reference to the section in which the meaning category has been scrutinised.

(5.280) Non-spatio-temporal Πs without synonymy
MALEFACTIVE (Π2) *a skapitu* 'at the expense of' (Section 5.1.3.18); INCLUSIVE (Π25) *inkluż* 'including' (Section 5.1.3.20); ADVERSATIVE (Π27) *kontra* 'against' (Section 5.1.3.12); SUBSTITUTIVE (Π35) *minflok* 'instead of' (Section 5.1.3.16); CONCESSIVE (Π40) *minkejja* 'in spite of' (Section 5.1.3.7)

From Figure 5.64, we learn that the absence of synonymy is relatively uncommon in the domain of non-spatio-temporal meaning categories. Not even a fifth of these Πs is exempt from synonymy.

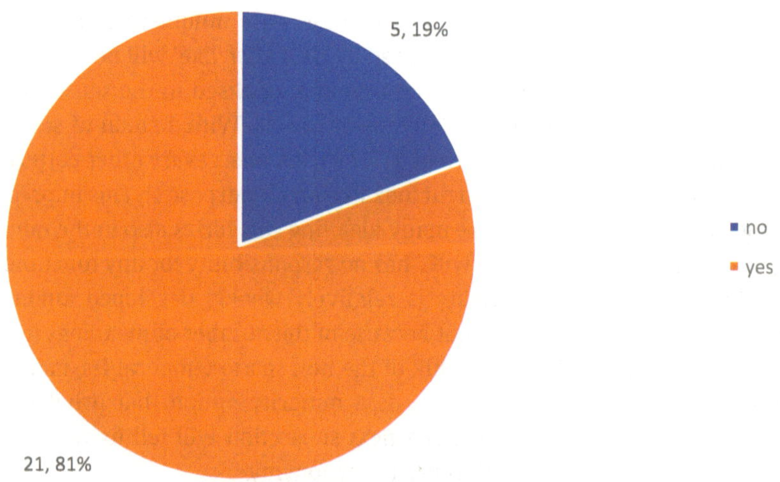

Figure 5.64: Shares of non-spatio-temporal Π with and without synonymy.

It further strikes the eye that four out of five Πs in (5.280) are Italo-Romance Πs, i.e., 80% of the non-spatio-temporal Πs that escape synonymy are borrowings from Italo-Romance. In contrast, with six types, Italo-Romance Πs that attest to synonymy cover only 29% of the Π mentioned in (5.281) – all of which are involved in synonymic relations with one or several other Πs on the same list.

(5.281) Non-spatio-temporal Πs with synonymy
(Π1) *apparti* 'apart from'; (Π5) *bħal* 'like'; (Π6) *bħala* 'as'; (Π7) *bi* 'with'; (Π9) *bla* 'without'; (Π10) *daqs* 'equal to'; (Π11) *dwar* 'about'; (Π13) *favur*

'in favour of'; (Π20) *għajr* 'except'; (Π24) *ħlief* 'except'; (Π26) *kif* 'as'; (Π29) *lil* 'to'; (Π34) *minbarra* 'except'; (Π36) *mingħajr* 'without'; (Π37) *mingħala* 'in s.o.'s opinion'; (Π39) *minħabba* 'on account of'; (Π45) *permezz* 'by means of'; (Π50) *rigward* 'concerning'; (Π52) *sforz* 'thanks to'; (Π53) *skont* 'according to'; (Π55) *ta'* 'of'

Figures 5.65–5.66 help to better visualise the disproportion between the two languages of origin.

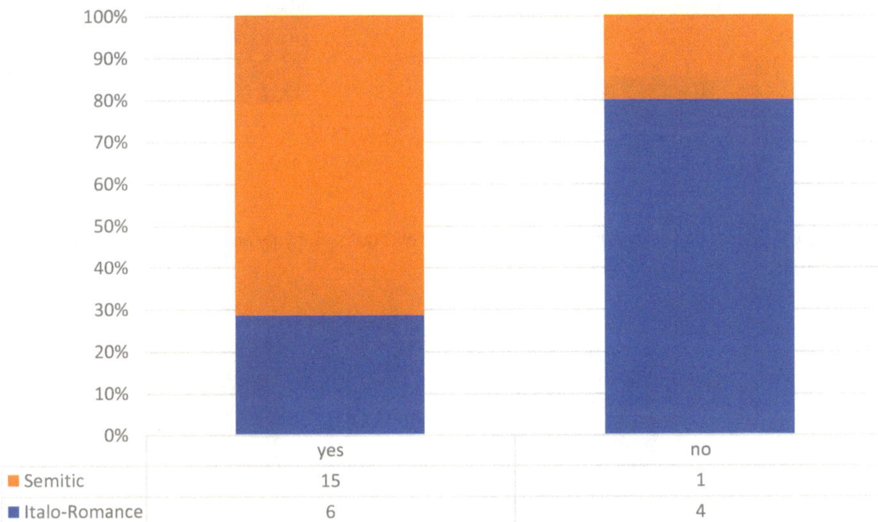

Figure 5.65: Shares of Semitic / Italo-Romance Πs in the domains of non-spatio-temporal Πs with / without synonymy.

According to Figure 5.65, the absence of synonymy is preferably associated with the Italo-Romance origin of a given Π. This is but a recapitulation of what we have stated earlier in this section. Figure 5.66, however, clearly shows that Semitic non-spatio-temporal Πs almost without exception are involved in synonymy. Even for Italo-Romance non-spatio-temporal Πs, the absence of synonymy is the minority option. Yet, this minority covers 40% of the Italo-Romance Πs in this domain. Being exempt from synonymy is thus more normal for Italo-Romance Πs than it is for Semitic Πs.

On the parameter of syllabicity, (5.280) and (5.281) also yield different results. All those Πs that are excluded from synonymy are polysyllabic. In the case of the Πs that are involved in synonymy, monosyllabic Πs outnumber polysyllabic Πs by twelve to nine. This means that 57% of the Πs mentioned in (5.281) consist only of one syllable. For convenience, Figure 5.67 presents these results in the format of a diagram.

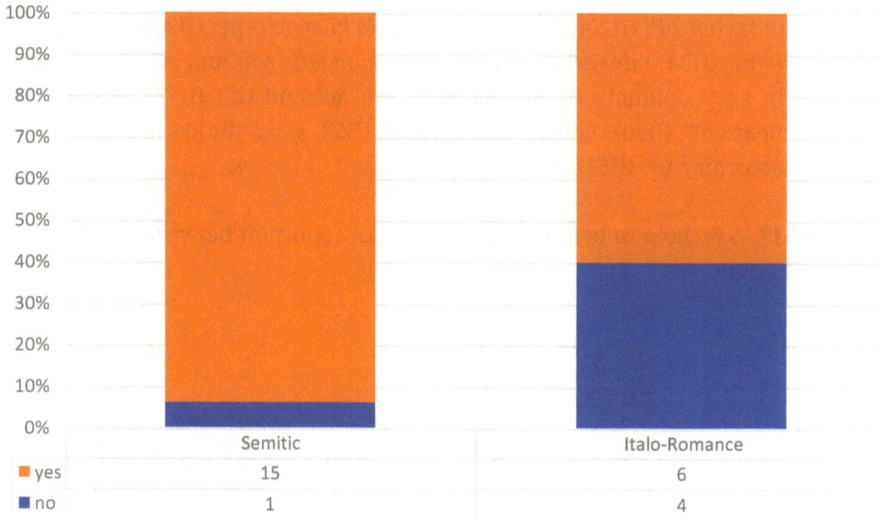

Figure 5.66: Shares of non-spatio-temporal Πs with / without synonymy across Semitic / Italo-Romance Πs.

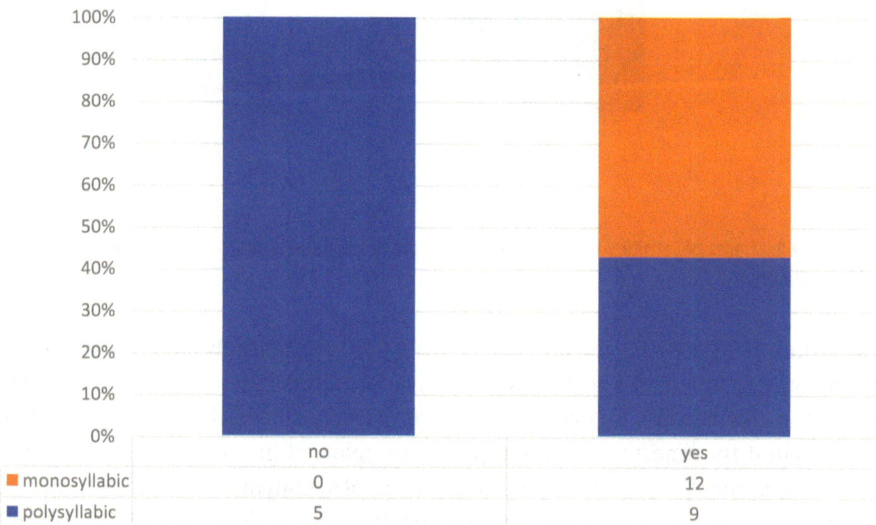

Figure 5.67: Syllabicity and synonymy I.

In Figure 5.68, we change the vantage point in order to ask whether there is a bilateral correlation between syllabicity and synonymy in the non-spatio-temporal domain. One may guess from the information conveyed by Figure 5.67 that the answer is yes.

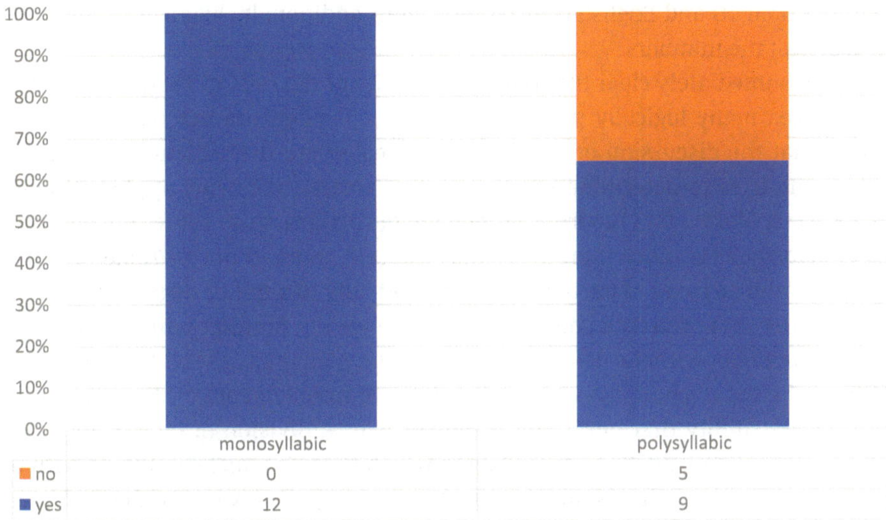

Figure 5.68: Syllabicity and synonymy II.

Monosyllabicity is strictly limited to non-spatio-temporal Πs with synonymy. At the same time, the lack of synonymy requires that monosyllabicity applies. These observations form the basis for two more implicational patterns given in Figures 5.69–5.70.

Σ $\supset$ synonymy / $\Pi_{\text{non-spatio-temporal}}$

Figure 5.69: Monosyllabicity implies synonymy.

~~synonymy~~ $\supset$ $\Sigma\Sigma(\Sigma)$ / $\Pi_{\text{non-spatio-temporal}}$

Figure 5.70: Absence of synonymy implies polysyllabicity.

From (5.280), we know already that certain meaning categories are excluded from synonymy. The majority of the non-spatio-temporal meaning categories allow for synonymy to apply. To address this issue in more detail, it is necessary to provide a survey of the combinations of non-spatio-temporal Πs in synonymic patterns. This is the task of Table 5.32. Only those cases of synonymy are taken account of which involve two or more non-spatio-temporal Πs. The contribution of spatio-temporal

Πs to polysemy and synonymy will be looked at more closely below.[247] In Table 5.32, the non-spatio-temporal meaning categories are identified by bracketed numbers. The number code is spelled out below. For reasons of space, the Πs are given without their ID and English translation. Colour additionally highlights those cells which host the numbers.

It is immediately clear that the vast majority of the cells in Table 5.32 remain empty, i.e., many logically possible synonymic pairings are not realised. Moreover, what the discussion of the data has revealed are the difficulties that arise when one tries to determine whether full or partial synonymy applies.[248] Even the Korpus Malti 3.0 is far from large enough to allow us to settle this question satisfactorily. It is more than likely that the cells marked out in Table 5.32 frequently refer to cases of partial synonymy and only to a minor degree to full synonyms. The EXCLUSIVE is a good example since we are familiar with the fact that (Π36) *mingħajr* 'without' takes all kinds of complements whereas its synonym (Π9) *bla* 'without' combines only with indefinite complements.[249] We cannot rule out the possibility that our analysis glosses over a division of labour and very subtle semantic distinctions according to which there is a niche in the domain of a meaning category that is the uncontested territory of only one of the supposed synonyms (e.g. the TRANSLATIVE expressed by (Π15) *fi* 'in' in (5.194)) so that in this case doubts arise as to the suitability of the term synonymy, in the first place. It is this uncertainty that should trigger follow-up studies conducted by semanticists who are familiar with Maltese.

To better interpret Table 5.32, we extract all cases of synonymy from the table and list them in (5.282). The meaning categories are ordered according to the succession of subsections in Section 5.1.3. The bracketed numbers are identical to the final number used in the title of the subsection. Certain numbers do not appear in (5.282) either because the meaning category does not attest to synonymy (e.g. [16] SUBSTITUTIVE) or one of two Πs is a spatio-temporal Π (e.g. [4] COMITATIVE), or because the meaning category is exclusively expressed by otherwise spatio-temporal Πs (e.g. [11] PURPOSIVE). Different colours are used to highlight those Πs which are involved in several synonymic patterns. Spatio-temporal Πs that are involved in spatio-temporal categories are added in square brackets with added boldface.

247 Cf. Section 5.2.3.
248 Throughout Section 5.1.3.
249 Cf. Section 4.1.1.1.

Table 5.32: Combinations of non-spatio-temporal Πs in synonymic patterns.

	apparti	*bħal*	*bħala*	*bi*	*bla*	*daqs*	*dwar*	*favur*	*għajr*	*ħlief*	*lil*	*minbarra*	*mingħajr*	*mingħala*	*minħabba*	*permezz*	*rigward*	*sforz*	*skont*	*ta'*
apparti									[3]	[3]		[3], [15]								
bħal						[9]														
bħala																				[10]
bi															[6]	[5]		[6]		[21]
bla													[2]							
daqs		[9]																		
dwar																	[13]			
favur																				
għajr	[3]									[3]		[3]								
ħlief																				
lil																				
minbarra	[3], [15]																			
mingħajr					[2]															
mingħala																			[14]	
minħabba				[6]														[6]		
permezz				[5]																
rigward							[13]													
sforz				[6]											[6]					
skont														[14]						
ta'			[10]	[21]																

(5.282) Synonymic patterns

[2] EXCLUSIVE: (Π9) *bla* 'without' + (Π36) *mingħajr* 'without'

[3] EXCEPTIVE: (Π1) *apparti* 'apart from' + (Π20) *għajr* 'except' + (Π24) *ħlief* 'except' + (Π34) *minbarra* 'except'

[5] INSTRUMENTIVE: (Π7) *bi* 'with' + (Π45) *permezz* 'by means of'

[6] MOTIVATIVE: (Π7) *bi* 'with' + (Π39) *minħabba* 'on account of' + (Π52) *sforz* 'thanks to' **[+ (Π21) *għal* 'for']**

[9] EQUATIVE: (Π5) *bħal* 'like' + (Π10) *daqs* 'equal to'

[10] ESSIVE: (Π6) *bħala* 'as' + (Π55) *ta'* 'of'

[13] PERTENTIVE: (Π11) *dwar* 'about' + (Π50) *rigward* 'concerning' **[+ (Π17) *fuq* 'on']**

[14] ROBORATIVE: (Π37) *mingħala* 'in s.o.'s opinion' + (Π53) *skont* 'according to'

[15] ADDITIVE: (Π1) *apparti* 'apart from' + (Π34) *minbarra* 'except'

[21] MATERIAL: (Π7) *bi* 'with' + (Π55) *ta'* 'of' **[+ (Π41) *minn* 'from']**

Except (Π29) *lil* 'to', all multifunctional non-spatio-temporal Πs (as of (5.279) are also subject to synonymy. This fact is hardly surprising since polysemy of a Π increases the chances that it competes with another Π in one of the meaning categories it expresses. On the same account, it can be expected that multifunctional Πs like (Π7) *bi* 'with', (Π55) *ta'* 'of', (Π1) *apparti* 'apart from', and (Π34) *minbarra* 'except' occur more than just once in (5.282). (Π7) *bi* 'with' – the record holder in the domain of polysemy – also counts the highest number of synonymic relations in (5.282). We conclude that being polysemous is a favourable condition for being synonymous, too. It should not go unmentioned that thirteen of the non-spatio-temporal Πs in (5.279) are monofunctional and form a 76% majority of the altogether eighteen synonymic Πs featured in (5.279).

Six of the ten synonymic patterns involve pairs of non-spatio-temporal Πs, viz. [2], [5], [9], [10], [14], and [15]. Pattern [3] is the only example of an extended synonymic chain involving only non-spatio-temporal Π. In the case of the ternary synonymic chains [6], [13], and [21], one of the Πs belongs to the spatio-temporal domain.[250]

5.2.2.3 Antonymy in the non-spatio-temporal domain

The third phenomenon that needs to be addressed in the non-spatio-temporal domain is antonymy. Antonymy is irrelevant for the core meanings. As to its role in the spatio-temporal domain, one might argue that the distinctions of [static]

250 We return to these cases in Section 5.2.3.

vs [dynamic] and [ALLATIVE] vs [ABLATIVE] can be understood as constituting antonymic pairs. However, in our opinion, the home ground of antonymy is with non-spatio-temporal categories. Since antonymy as such has not been discussed before in this study, we add sentential examples for each of the Πs involved. The examples are taken from the Korpus Malti 3.0. Our comments on these examples are meant to show that the situation in the realm of antonymy is much more intricate than the simplifying view of two opposing members of an antonymic relation suggests.

This conviction of ours notwithstanding, there are only three antonymic patterns in the non-spatio-temporal domain involving twelve different non-spatio-temporal Πs (plus a spatio-temporal one). We face one-to-many and many-to-many relationships. The latter kind supposedly applies in the first configuration presented in Figure 5.71. As will come to the fore shortly, this impression does not stand the test. In Figure 5.71, the red box singles out the spatio-temporal Π (Π21) *għal* 'for' whose intrusion into the non-spatio-temporal domain will be discussed again below.[251] In this section, we anticipate part of this discussion by way of giving an example of the BENEFACTIVE use of this Π in (5.283b) while the putative synonym (Π13) *favur* 'in favour of' is featured in (5.283a).

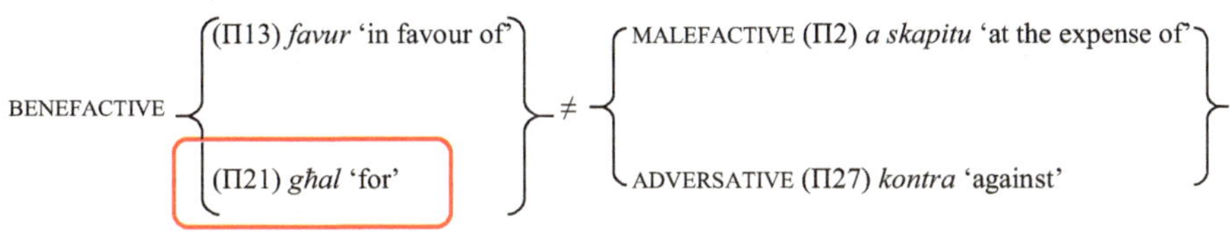

Figure 5.71: Antonymy BENEFACTIVE VS MALEFACTIVE / ADVERSATIVE (preliminary version).

The opposition in Figure 5.71 is termed preliminary for a reason. In the paragraphs below, it is shown why the diagram needs to be modified.

There are two Πs on each side of the antonymic relation. Several questions arise from this confrontation of two BENEFACTIVE Πs and two MALEFACTIVE Πs one of which is labelled ADVERSATIVE. One might want to know whether the two Πs with identical labels are indeed full synonyms of each other. Are the equations in Figure 5.72 tenable?

(a)	(Π13) *favur* 'in favour of'	=	(Π21) *għal* 'for'?
(b)	(Π2) *a skapitu* 'at the expense of'	=	(Π27) *kontra* 'against'?

Figure 5.72: Two potential equations.

251 Cf. Section 5.2.3.

We claim that the equation (a) in Figure 5.72 is incorrect insofar as the domains of the two Πs overlap only partly. (Π21) *għal* 'for' claims the much bigger domain which ranges over a number of meaning categories that cannot be expressed by (Π13) *favur* 'in favour of'.[252] As to the BENEFACTIVE, (Π21) *għal* 'for' is by no means as dedicated to this category as is (Π13) *favur* 'in favour of' as can be seen from a comparison of the examples in (5.283).

(5.283) BENEFACTIVE (+ NEGATIVE BENEFACTIVE?)

(a) (Π13) *favur* 'in favour of' [Korpus Malti 3.0; news214152]

din hija wegħda oħra li qed iwettaq
DEM.F.PROX 3SG.F promise other:F that PROG 3SG.M.IPFV:realise
Gvern Nazzjonalista ***[favur*** *il-Maltin li*
government nationalist **[in_favour_of** DEF-Maltese:PL that
jgħixu barra minn Malta]$_{PP}$
3.IPFV:live:PL outside from Malta]

'[. . .] this is another promise that a Nationalist government is realizing [**in favour of** the Maltese who are living outside Malta].'

(b) (Π21) *għal* 'for' [Korpus Malti 3.0; culture3018]

Il-forum se jkun ukoll opportunità tajba
DEF-forum FUT 3SG.M.IPFV:be.FUT also opportunity good:F
*[**għal**l-Maltin]$_{PP}$*
[**for**:DEF-Maltese:PL]

'The forum will also be a good opportunity [**for** the Maltese].'

(c) (Π21) *għal* 'for' [Korpus Malti 3.0; academic12]

*Iżda aħbar kerha [**għal**l-Maltin]$_{PP}$ kienet it-telfa*
but news ugly:F [**for**:DEF-Maltese:PL] be:3SG.F.PFV DEF-loss
mhix mistennija ta' Dun Guliermu Bonett
NEG:3SG.F:NEG PART:expect:F of Dun Guliermu Bonett

'But bad news [**for** the Maltese] was the unexpected loss of Dun Guliermu Bonett.'

The examples given in (5.283) refer back to what we have said in connection to the BENEFACTIVE above.[253] In the subsequent paragraph, we argue along the same lines but emphasise the fact that (Π21) *għal* 'for' is not a fully-blown BENEFACTIVE Π.

Ideally, a BENEFACTIVE situation involves a BENEFACTOR who does something (= BENEFACTUM) to the benefit of another participant (= BENEFICIARY). This configura-

252 Cf. Section 5.2.3.

253 In Sections 5.1.1.3 and 5.1.3.17.

tion applies in both the (a)-sentence and the (b)-sentence. In (5.283a), the Nationalist government plays the BENEFACTOR's part. It acts in such a way that the BENEFICIARY (= the ex-pat Maltese) takes advantage of the government's activities. The BENEFACTUM is the promise (or more precisely: its realisation). Similarly, the Maltese function as BENEFICIARY in (5.283b) where the good opportunity can be understood as the BENEFACTUM and the forum as the (metonymical) BENEFACTOR. What happens is to the benefit of the Maltese. In contrast, it is difficult to speak of a benefit for the Maltese in the situation described by (5.283c). In point of fact, neither the demise of Dun Guliermu Bonett nor the bad news thereof are good candidates for the roles of BENEFACTOR and BENEFACTUM, respectively, because the death of the cleric has detrimental effects on the Maltese community. Thus, the Maltese cannot be BENEFICIARIES in the strict sense of the term. What we learn from these examples is that the use of (Π13) *favur* 'in favour of' implies that something is done to the benefit of somebody. This Π is genuinely BENEFACTIVE. In the case of (Π21) *għal* 'for', it is appropriate to assume neutrality as to the opposition of BENEFACTIVE VS MALEFACTIVE because the use of this Π is always possible no matter whether the event is to the benefit or to the detriment of a participant. (Π21) *għal* 'for' does not distinguish positive from negative effects. It simply introduces the participant who undergoes the effect of an action that is carried out by a different participant. It is therefore necessary to reformulate the (a)-equation of Figure 5.72. The reformulation is given in Figure 5.73.

(Π13) *favur* 'in favour of' = (Π21) *għal* 'for' / BENEFACTIVE

Figure 5.73: Conditioned overlap I.

The two Πs are semantically synonymous solely under the condition that the intended meaning is BENEFACTIVE. If this criterion is not fulfilled, the equation of (Π13) *favur* 'in favour of' and (Π21) *għal* 'for' does not hold any more. The true BENEFACTIVE Π is (Π13) *favur* 'in favour of'.

Do we need to tidy up the situation also on the other side of the above antonymic opposition? Discounting syntactic constraints, which regulate the distribution of the two representatives of the MALEFACTIVE when conceived of as an umbrella, we are inclined to accept the equation (b) in Figure 5.72 on purely semantic grounds since both Πs are used in situations where a MALEFACTOR commits a MALEFACTUM to the detriment of a MALEFICIARY as shown in (5.284).

(5.284) MALEFACTIVE

(a) (Π2) *a skapitu* 'at the expense of' [Korpus Malti 3.0; news105086]

Għax hemm min jibża' li ċertu employers qed
because there who 3SG.M.IPFV:fear that certain employers PROG
jabbużaw minnhom ***[à skapitu*** *tal-Maltin u*
3.IPFV:abuse:PL from:3PL **[at_the_expense_of** of:DEF-Maltese:PL and
l-Għawdxin]$_{PP}$
DEF-Gozitan:PL]
'Because there are people who fear that certain employers are abusing them [**at the expense of** the Maltese and the Gozitans].'

(b) (Π27) *kontra* 'against' [Korpus Malti 3.0; academic12]

Anzi r-restrizzjonijiet ***[kontra*** *l-Maltin]*$_{PP}$ *kienu agħar*
even DEF-restricion:PL **[against** DEF-Maltese:PL] be:3PL.PFV bad.CMPR
minn hekk
from so
'The restrictions [**against** the Maltese] were even worse than that.'

The situations described in (5.284a) and (5.284b) are both to the detriment of the Maltese (and in one case the Gozitans, too). The MALEFICIARY is thus identified. The MALEFACTOR is made explicit only in the (a)-sentence where the employers are suspected of illegally employing foreign workers at low wages. This practice is the MALEFACTUM. The MALEFACTOR is absent from the (b)-sentence but can be retrieved from the context. The US Congress is made responsible for imposing restrictions on the immigration of Maltese migrants in the 1920's. The restrictions are the MALEFACTUM. Formally, ADVERSATIVE[254] and MALEFACTIVE[255] are distinct but conceptually they belong to the same domain. In the absence of a better term, we use the label MALEFACTIVE also as cover term for the two subcategories.

Since (Π21) *għal* 'for' is no longer considered a genuine BENEFACTIVE Π, it is no longer required to ask whether there are two strictly separate binary antonymic relations involving only one of the BENEFACTIVE Πs and one of the MALEFACTIVE Πs to the exclusion of the other Πs mentioned in Figure 5.71. Figure 5.74 presents the updated version of Figure 5.72.

The antonymy is asymmetric in the sense that there are no equal numbers on the different sides of the opposition. We have a one-to-many relation in Figure 5.74.

Next in line is the antonymic opposition of INSTRUMENTIVE and EXCLUSIVE featured in Figure 5.75. As in the previous case, things are not as simple as that since

254 Cf. Section 5.1.3.12.
255 Cf. Section 5.1.3.18.

BENEFACTIVE (Π13) *favur* 'in favour of' ≠ { MALEFACTIVE (Π2) *a skapitu* 'at the expense of' / ADVERSATIVE (Π27) *kontra* 'against' }

Figure 5.74: Antonymy BENEFACTIVE VS MALEFACTIVE / ADVERSATIVE (final version).

only two of the four Πs included in the diagram are bona fide cases of synonymy. The schematic representation of the antonymic relations in Figure 5.75 needs to undergo modification as will result from the ensuing discussion. It is therefore characterised as preliminary.

INSTRUMENTIVE { (Π7) *bi* 'with' / (Π45) *permezz* 'by means of' } ≠ EXCLUSIVE { (Π9) *bla* 'without' / (Π36) *mingħajr* 'without' }

Figure 5.75: Antonymy INSTRUMENTIVE VS EXCLUSIVE (preliminary version).

Like in the case of (Π21) *għal* 'for' and (Π13) *favur* 'in favour of', the pair of INSTRUMENTIVE Πs constituted by (Π7) *bi* 'with' and (Π45) *permezz* 'by means of' fails to fit the description of full synonymy. On the one hand, there is the multi-purpose (Π7) *bi* 'with' for which INSTRUMENTIVE is only a segment within its extended functional domain. On the other hand, there is (Π45) *permezz* 'by means of' for which INSTRUMENTIVE is coextensive with its functional domain. This means that the domain of (Π45) *permezz* 'by means of' is fully included in the much wider domain of (Π7) *bi* 'with'. The situation is reminiscent of that captured by Figure 5.73 for the BENEFACTIVE as can be gathered from a comparison of Figure 5.73 with 5.76.

(Π45) *permezz* 'by means of' = (Π7) *bi* 'with' / INSTRUMENTIVE

Figure 5.76: Conditioned overlap II.

Figure 5.76 tells us that (Π45) *permezz* 'by means of' and (Π7) *bi* 'with' can replace each other only in genuinely INSTRUMENTIVE contexts and nowhere else. The conditioned overlap of the two Πs is illustrated by the examples in (5.285). Further details on the interrelationship of (Π45) *permezz* 'by means of' and (Π7) *bi* 'with' have been provided already above.[256]

256 Cf. Section 5.1.3.5.

(5.285) INSTRUMENTIVE

(a) (Π7) *bi* 'with' [Korpus Malti 3.0; european10025]

L-istatuti jistgħu jippermettu li voti mibgħuta
DEF-statute:PL 3.IPFV:can:PL 3.IPFV:permit:PL that vote:PL PART:send:PL
*[**bi**l-posta]*$_{PP}$ *u* *[**b'** mezzi elettroniċi]*$_{PP}$
[**with**:DEF-mail] and [**with** means:PL electronic:PL]
'The statues may allow for votes being submitted [**by** mail] and [**via** electronic channels] [. . .].'

(b) (Π45) *permezz* 'by means of' [Korpus Malti 3.0; european10177]

Dik l-informazzjoni għandha ukoll issir
DEM:F:DIST DEF-information have:3SG.F also 3SG.F.IPFV:become
*pubblika [**permezz** ta' mezzi elettroniċi]*$_{PP}$
public:F [**by_means_of** of means:PL electronic:PL]
'That information must also become public [**by means of** electronic channels].'

The semantic equivalence of the two EXCLUSIVE Πs is not at issue since it has already been established. We only add the examples in (5.286)–(5.287) to pose the question of whether any of the EXCLUSIVE Πs may pick out only one of the INSTRUMENTIVE Πs to form an antonymic pair, and vice versa.

(5.286) EXCLUSIVE – (Π9) *bla* 'without'

(a) [Korpus Malti 3.0; opinion877]

Il-kunsill hu limitat kemm-il triq jista'
DEF-council 3SG.M limit:PART how_many street 3SG.M.IPFV:can
*jagħmel [**bla** għajnuna mill-Gvern Ċentrali]*$_{PP}$
3SG.M.IPFV:make [**without** help from:DEF-government central]
'The council is limited as to what it can do [**without** help from the central government].'

(b) [Korpus Malti 3.0; parl5748]

Però jidher li issa dan il-kumitat
but 3SG.M.IPFV:appear that now DEM:M:PROX DEF-committee
*spiċċa [**bla** mezzi finanzjarji]*$_{PP}$
wind_up [**without** means:PL financial:PL]
'However, it seems now that this committee has ended [**without** financial means].'

(5.287) EXCLUSIVE – (Π36) *mingħajr* 'without'

(a) [Korpus Malti 3.0; news118684]

Kien proġett personali li bdejt xi snin ilu
be project personal that begin:1SG.PFV some year.PL ago
u li ma kien qatt jirnexxieli
and that NEG be ever 3SG.M.IPFV:succeed:IO:1SG
nagħmlu ***[mingħajr*** *l-għajnuna ta' Michel]$_{PP}$*
1SG.IPFV:make:3SG.M **[without** DEF-help of Michel]

'It was a personal project that I began some years ago and that I would never have succeeded in carrying out [**without** Michel's help].'

(b) [Korpus Malti 3.0; news63110]

huwa kważi impossibbli li timmaġina l-ħajja
3SG.M almost impossible that 2SG.IPFV:imagine DEF-life
tal-familja umana ***[mingħajr*** *il-mezzi*
of:DEF-family human:F **[without** DEF-means:PL
tal-komunikazzjoni soċjali]$_{PP}$
of:DEF-communication social]

'[. . .] it is almost impossible to imagine the life of a human family [**without** the social media].'

The two (a)-sentences illustrate how the EXCLUSIVE Πs serve to signal that an abstract instrument that is necessary for the successful completion of an activity is (imagined to be) absent in a given situation. In this function, both (Π9) *bla* 'without' and (Π36) *mingħajr* 'without' form a conceptual contrast to the INSTRUMENTIVE meanings expressed by (Π45) *permezz* 'by means of' and (Π7) *bi* 'with'. In the (b)-sentences however, neither of the EXCLUSIVE Πs is the antonym of (Π45) *permezz* 'by means of' because the situation described does not involve a participant (present or absent) that functions as instrument. The *bla*-PP in (5.286b), as well as the *mingħajr*-PP in (5.287b), introduce a temporary property of one of the participants in the situation. In the case of (5.286b) we are facing a resultative: being without financial means is the state the committee finds itself in at the end of a process. In the case of (5.287b), we are facing an attributive PP that modifies *familja umana* 'human family'. This imagined family is characterised as not being equipped with social media. In both these cases, the EXCLUSIVE Πs are in an antonymic relation with (Π7) *bi* 'with' outside of the INSTRUMENTIVE part of its functional domain. Like (Π7) *bi* 'with', (Π9) *bla* 'without' and (Π36) *mingħajr* 'without' distribute over a wide range of contexts among which stating the absence of an instrument is only one.

On account of the limited parallels of the distribution profiles of the above Πs, a revision of Figure 5.75 is called for. The revision results in the division of the original antonymic pattern into two separate patterns featured in Figures 5.77–5.78. (Π9) *bla* 'without'

(Π9) *bla* 'without'
(Π7) *bi* 'with' ≠ EXCLUSIVE { } / X
(Π36) *mingħajr* 'without'

Figure 5.77: Antonymy (Π7) *bi* 'with' vs EXCLUSIVE.

(Π9) *bla* 'without'
(Π45) *permezz* 'by means of' ≠ EXCLUSIVE { } / INSTRUMENTIVE
(Π36) *mingħajr* 'without'

Figure 5.78: Antonymy (Π45) *permezz* 'by means of' vs EXCLUSIVE.

According to Figure 5.77, (Π7) *bi* 'with' is in opposition to either the one or the other EXCLUSIVE Π in a variety of contexts. This multitude of contexts is represented by X in the formula. The INSTRUMENTIVE is but one of the possible realisations of X. Figure 5.78 suggests that a relation of the antonymic kind exists for (Π45) *permezz* 'by means of' and the EXCLUSIVE Πs only where the context can be characterised as INSTRUMENTIVE.

The story is about to repeat itself, in a manner of speaking. What on superficial inspection looks more or less like a straightforward relation between meaning categories or between their representatives turns out to be an intricate crisscrossing of several relations that cannot easily be put in a watertight order. This characterisation also holds for the final case to be looked into in this section. In Figure 5.79, the preliminary version of the antonymic relation between the INCLUSIVE on one side and the EXCEPTIVE on the other side is presented. It is again a one-to-many relation, or so it seems. As will turn out in the course of the critical evaluation, the formula cannot remain as is but needs to be thoroughly worked over.

To base our argumentation on firm ground, we start with the uncontroversial component of the antonymic pattern, namely with (Π25) *inkluż* 'including' in (5.288).

INCLUSIVE (Π25) *inkluż* 'including' ≠ EXCEPTIVE {
(Π1) *apparti* 'apart from'
(Π20) *għajr* 'except'
(Π24) *ħlief* 'except'
(Π34) *minbarra* 'except'
}

Figure 5.79: Antonymy INCLUSIVE VS EXCEPTIVE (preliminary version).

(5.288) INCLUSIVE (Π25) *inkluż* 'including' [Korpus Malti 3.0; news42530]

Fl-imgħoddi	*l-iskart*	*kollu*	***[inkluż***	*l-ikel]$_{PP}$*	*kien*
in:DEF-past	DEF-waste	all:3SG.M	**[including**	DEF-food]	be

jintrema	*f'*	*miżbla*
3SG.M.IPFV:REFL:throw_away	in	dump

'In the past, all the waste, [**including** food], was thrown into a dump.'

The complement of (Π25) *inkluż* 'including', the definite noun *l-ikel* 'the food', is described as forming part of the list of objects that together constitute the referents of the mass-noun *l-iskart* 'the waste'. The other items that belong to the category of waste are not specified. (5.288) describes a situation of inclusion whereas the examples in (5.289) do the opposite by way of identifying a participant who is not part of the group of other participants that are talked about.

(5.289) EXCEPTIVE

(a) (Π1) *apparti* 'apart from' [Korpus Malti 3.0; academic6]

Bħala	*parti*	*mill-proġett*	*tal-Gospel House*	*se*
as	part	from:DEF-project	of:DEF-Gospel House	FUT

jinfetaħ	*ċentru*	*ta'*	*taħriġ*	***[apparti***	*l-kamra*
3SG.M.IPFV:REFL:open	center	of	training	**[apart_from**	DEF-room

tal-ħjata	*u*	*dik*	*tal-qtugħ*	*tad-drapp]$_{PP}$*
of:DEF-sewing	and	DEM:F:DIST	of:DEF-cutting	of:DEF-cloth]

'As part of the Gospel House project a training centre will be opened [**except** the room for sewing and that for cutting of the cloth].'

(b) (Π34) *minbarra* 'except' [Korpus Malti 3.0; academic113]

Fl-1996	*kien*	*il-plejer*	*l-iktar*	*żagħżugħ*	*li*	*pparteċipa*
in:DEF-1996	be	DEF-player	DEF-more	young	that	participate

fil-Kampjonati	*Ewropej*	*1996*	*fejn*	*lagħab*
in:DEF-Championship:PL	European:PL	1996	where	play

fil-partiti *kollha* ***[minbarra*** *s-semi-finali]*$_{PP}$
in:DEF-match:PL all:3SG.F **[except** DEF-semi-final:PL]
minħabba *skwalifikazzjoni*
on_account_of disqualification
'In 1996 he was the youngest player that participated in the European Championships 1996 where he played in all matches [**except** the semi-finals] because of a disqualification.'

(c) (П20) *għajr* 'except' [Korpus Malti 3.0; european10386]
Dawn *jikkonċernaw* *dħul* *mill-pajjiżi* *terzi*
DEM:PL:PROX 3.IPFV:concern:PL entrance from:DEF-country:PL third:PL
kollha *fi* *Stat* *Membru* ***[għajr*** *l-Irlanda]*$_{PP}$
all:3SG.F in state member **[except** DEF-Ireland]
'Those concern income from all third countries in a member state [**except** Ireland].'

(d) (П24) *ħlief* 'except' [Korpus Malti 3.0; academic413]
ħabtu *għal* *Għawdex* *u* *kaxkru* *l-Għawdxin* *kollha*
collide:3PL.PFV for Gozo and drag:3PL.PFV DEF-Goitan:PL all:3SG.F
[ħlief *ix-xjuħ* *u* *l-morda]*$_{PP}$
[except DEF-old.PL and DEF-sick.PL]
'[. . .] they attacked Gozo and dragged all the Gozitans with them [**except** the old and the sick].'

In (5.289a), certain rooms are excluded from those parts of the training centre that are said to be going to be opened. In (5.289b), the semi-finals of the championship are identified as those matches that do not belong to the matches in which the young player made an appearance during the championships. In (5.289c), Ireland is mentioned as the only E.U. member state in which the regulations do not apply. And finally, in (5.289d), the entire population is taken captive by the pirates except the elderly and sick. These are clear instances of exclusion that form a contrast to the inclusion expressed by (П25) *inkluż* 'including' in (5.288). Accordingly, one might conclude that INCLUSIVE (П25) *inkluż* 'including' boasts four synonymous EXCEPTIVE antonyms.

However, the situation is once more not what you see. In connection to polysemy and synonymy of non-spatio-temporal Пs,[257] we have repeatedly mentioned that (П1) *apparti* 'apart from' and (П34) *minbarra* 'except' do not only serve the purpose of expressing the EXCEPTIVE but may also be used to express the ADDITIVE. Compare the examples in (5.290) with (5.289a-b).

257 Cf. Sections 5.2.2.1.2 and 5.2.2.2.

(5.290) ADDITIVE

(a) (Π1) *apparti* 'apart from' [Korpus Malti 3.0; news140570]

[Apparti	*l-ikel]*$_{PP}$	*li*	*kien*	*jinkludi*	*wkoll*
[apart_from	DEF-food]	that	be	3SG.M.IPFV:include	also

dak	*mhux*	*tradizzjonali*	*Malti*	*kien*	*hemm*
DEM:M:DIST	NEG:3SG.M:NEG	traditional	Maltese	be	there

għadd	*ta'*	*wirjiet*	*agrikoli*
number	of	show:PL	agricultural

'[**Besides** the food] that included also non-traditional Maltese items there was a number of agricultural exhibits.'

(b) (Π34) *minbarra* 'except' [Korpus Malti 3.0; news134258]

[Minbarra	*l-ikel]*$_{PP}$	*il-bakterja*	*tista'*	*tinstab*
[except	DEF-food]	DEF-bacterium	3SG.F.IPFV:can	3SG.F.IPFV:REFL:find

ukoll	*fl-ilma*	*il-ħamrija*	*u*	*ħmieġ*	*tal-annimali*
also	in:DEF-water	DEF-soil	and	excrement	of:DEF-animal:PL

[**Beside** food] the bacterium can be found also in water, soil, and animal excrements.'

In (5.290a), (Π1) *apparti* 'apart from' does not exclude the food from the list of attractions of the event but instead, the use of this Π implies that the food forms part of a larger group of items that count as assets of the event. Similarly, in (5.290b), (Π34) *minbarra* 'except' is not used in its EXCEPTIVE function. It serves to convey the ADDITIVE meaning according to which food is only one item on a list of habitats in which the bacterium can be found. In other words, (Π1) *apparti* 'apart from' and (Π34) *minbarra* 'except' are ambiguous since they may receive an EXCEPTIVE or an ADDITIVE interpretation. In the latter case, their meaning is not too far from that of INCLUSIVE (Π25) *inkluż* 'including'. At the same time, when (Π1) *apparti* 'apart from' and (Π34) *minbarra* 'except' express the ADDITIVE they can be considered to function as antonyms of EXCEPTIVE (Π20) *għajr* 'except' and (Π24) *ħlief* 'except'.[258]

On the basis of the ambiguity of (Π1) *apparti* 'apart from' and (Π34) *minbarra* 'except', Figure 5.79 must be refined. To this end, it is necessary to divide it in two. For each of the functions of the ambiguous Πs, a different scenario emerges. In Figure 5.80, it is shown that (Π1) *apparti* 'apart from' and (Π34) *minbarra* 'except' are opposed to (Π25) *inkluż* 'including' only in those contexts where (Π1) *apparti* 'apart from' and (Π34) *minbarra* 'except' unambiguously indicate that a given participant is excluded from a given set of items.

258 Cf. Sections 5.1.3.3 and 5.1.3.15.

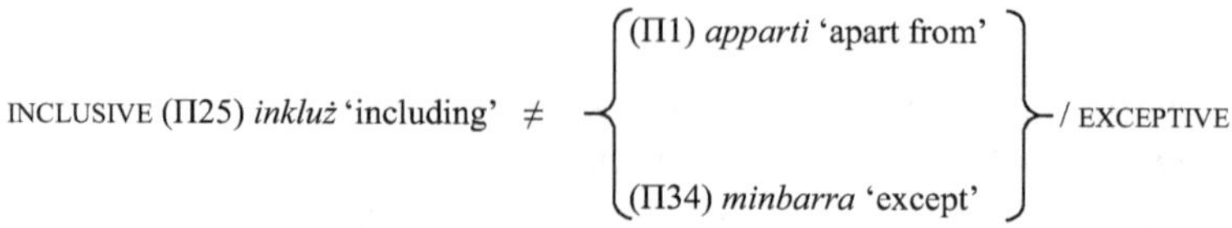

Figure 5.80: Antonymy INCLUSIVE VS EXCEPTIVE (for ambiguous Πs).

If, however, the ambiguous Πs are analysed as ADDITIVE Πs, their antonymic counterparts are (Π20) *għajr* 'except' and (Π24) *ħlief* 'except' as transpires from Figure 5.81.

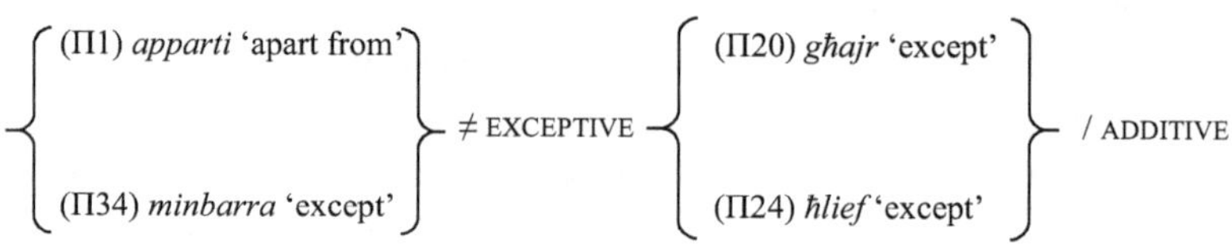

Figure 5.81: Antonymy ADDITIVE VS EXCEPTIVE (for ambiguous Πs).

We do not claim that the ADDITIVE Πs are full synonyms of the INCLUSIVE Π. ADDITIVE and INCLUSIVE look at situations from different perspectives. In the case of the ADDITIVE, the list of items that form a bigger unit is opened with the participant mentioned in the ADDITIVE PP. This participant is the basic element from which the extension of the set is developed. The INCLUSIVE functions differently. Normally, the participant involved in the INCLUSIVE PP is mentioned only after a set of participants is introduced into the discourse. The INCLUSIVE PP often comes as a kind of afterthought. There is a conceptual link between ADDITIVE and INCLUSIVE since both of these meaning categories single out one member of a set without cancelling it from the set. The cancelling function is proper to the EXCEPTIVE, though. The ambiguity of (Π1) *apparti* 'apart from' and (Π34) *minbarra* 'except' is suggestive of another conceptual link – this time between diametrically opposed meanings, namely those of the ADDITIVE and the EXCEPTIVE to which we turn our attention again below.[259]

5.2.2.4 Preliminary conclusions on the non-spatio-temporal domain

In this section, we can be very brief since our analysis of the semasiology of Πs within the non-spatio-temporal domain of necessity remains superficial and more often than not also inconclusive. What we have argued so far does not give answers to questions, but mostly raises questions that call for being answered in future research. The principal shortcoming of the previous sections is caused by

259 Cf. Section 5.2.4.

our limited ability to make decisions as to the semantics of the non-spatio-temporal Πs on the micro-level of their distribution over contexts. This inability can only be overcome if there is access to native-speaker intuition. We have made this point repeatedly already in this study, but it is worthwhile repeating it again, especially in connection to semantic matters.

In compensation for the lack of detail, we have sketched some of the most important macro-level aspects of this component of BLOMP 2.0. It has turned out that neither polysemy nor synonymy is particularly common across the non-spatio-temporal Πs, although this conclusion has to be taken with a grain of salt, too, since in-depth studies on the micro-level might reveal that both multifunctionality and (partial) semantic equivalence is more common than our results seem to suggest. On the basis of our own findings, we hypothesise that

(a) when polysemy applies, the Π in question normally expresses primarily only one of the meanings for which it can be employed whereas all other meanings are but secondary options for the same Π;
(b) when synonymy applies, it is normally partial synonymy that is realised while fully convincing examples of full synonymy are infrequent and at times also doubtful;
(c) it is typical of Maltese non-spatio-temporal Πs to be monofunctional and without a competitor in the same meaning category;
(d) Italo-Romance Πs and Semitic Πs display different preferences, especially with regard to polysemy.

The third phenomenon that was scrutinised in the foregoing sections is antonymy. Only a selection of antonymic relations was reviewed.[260] The examples given served to show that the semantic intricacies are such that many supposedly clear cases become problematic if investigated intensively. The complexity of the situation in the domain of antonymy crucially depends on the interaction with the two aforementioned phenomena, namely polysemy and synonymy. It seems that the usual configuration is that of a one-to-many relation. It is a relatively common constellation that the domains of the Πs overlap only partially in the sense that two supposed synonyms on one side of the antonymic relation turn out to share only part of their range of meanings or the meaning range of the one is fully included in the much more extended meaning range of the other. Accordingly, the Πs on the opposite side of the relation may be antonymic to different degrees with their partners in the opposition.

260 Cf. Section 5.2.2.3.

The little we can say about polysemy, synonymy, and antonymy of non-spatio-temporal Πs leaves no place for doubt. There is an urgent need for a dedicated in-depth study of the semantics of Maltese Πs. Many issues that must be addressed in this study-to-be have been raised here for the first time. We consider this to be the one achievement of this section.

5.2.3 Polysemy across domains

Throughout the foregoing treatment of the semantics of Maltese Πs, we had occasion time and again to allude to Πs whose semantic profile covers meaning categories which belong to different domains, i.e., they cross the borders between core, spatio-temporal, and non-spatio-temporal meaning categories. This section has the task of determining in what way these Πs can be characterised. Do they have certain properties that render them especially fit for being employed for the expression of meanings that, superficially, are remote from each other? Or is there something about the co-expressed categories that binds them together? In this section, we will look at some general formal aspects of the Πs under scrutiny.[261] As is known from the previous discussion, polysemy often goes along with synonymy. This is also the case when polysemy stretches over several semantic domains.

We have encountered some pertinent cases already. An example of a Π that is associated with core meaning categories and takes over also functions from the non-spatio-temporal domain is (Π55) *ta'* 'of' that is not only used for the expression of the POSSESSIVE[262] but also for those of ESSIVE and MATERIAL (cf. (5.282) [10] + [21]). In (5.282), further polysemic patterns are mentioned in which spatio-temporal Πs are involved, namely (Π21) *għal* 'for' employed for the MOTIVATIVE (= (5.282) [6]) and for the BENEFACTIVE (= Figure 5.71), (Π17) *fuq* 'on' for the PERTENTIVE (= (5.282) [13]), and (Π41) *minn* 'from' for MATERIAL (= (5.282) [21]). The previous discussion has revealed that the BENEFACTIVE use of (Π21) *għal* 'for' is not primary.[263] This Π can be interpreted as BENEFACTIVE only because its relatively neutral and unspecific relational meaning allows (Π21) *għal* 'for' to occur in many different contexts one of which the BENEFACTIVE happens to be. It is likely that this incidental fit applies also in other situations of polysemy and synonymy. Under this proviso, we take the daring step to identify those members of BLOMP 2.0 that are made use of for the

261 The question that relates to the content side of the Πs is in the focus of interest of Section 5.2.4.
262 Cf. Section 5.1.1.4.
263 Cf. Section 5.2.2.3.

expression of meaning categories in at least two different domains according to Hagège's (2010) model.

The seven Πs that fulfil the criterion of occurring in several semantic domains are presented in (5.291). They are ordered top-down according to the decreasing number of meaning categories that form part of the polysemic chain. Meaning categories whose label is given in red belong to the core domain. Those in green identify meaning categories from the spatio-temporal domain. No special colour is used for the non-spatio-temporal categories.

(5.291) Polysemic Πs across domains
- (a) (Π41) *minn* 'from' – AGENTIVE, ABLATIVE, PERLATIVE, PROLATIVE, COMPARATIVE (INEQUALITY), PARTITIVE, MATERIAL
- (b) (Π21) *għal* 'for' – DIRECTIONAL, MOTIVATIVE, PURPOSIVE, BENEFACTIVE
- (c) (Π15) *fi* 'in' – INTERIOR, SUPERIOR, TRANSLATIVE
- (d) (Π30) *ma'* 'with' – PROPINQUUS, SEQUENS, COMITATIVE
- (e) (Π55) *ta'* 'of' – POSSESSIVE, ESSIVE, MATERIAL
- (f) (Π17) *fuq* 'on' – SUPERIOR, PERTENTIVE

This group of six represents 10% of BLOMP 2.0. This means that the vast majority of the Maltese Πs is semantically specified for only one domain. The unequal relation of the two subsets of BLOMP 2.0 comes nicely to the fore in Figure 5.82.

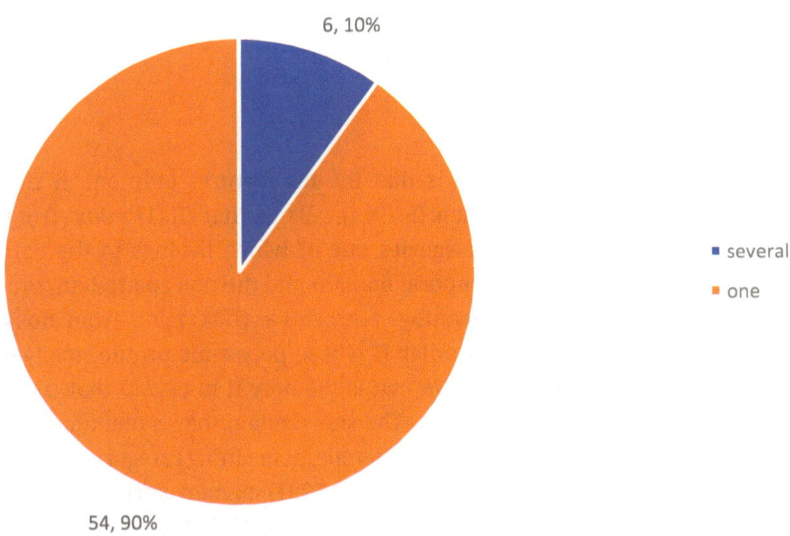

Figure 5.82: Shares of Πs used in several dimensions vs Πs used in one dimension.

Several further characteristics hold for the entire half dozen of Πs in (5.291). First of all, we notice that there is not a single Italo-Romance Π among them. Crossing the border between domains is thus a privilege of Semitic Πs. Accordingly, we can postulate the implicational pattern in Figure 5.83.

Π → Semitic / dimension_x + dimension_y

Figure 5.83: Multidimensional Πs are Semitic.

Moreover, each of the six multidimensional Πs consists of only one syllable with either CV or CVC structure. The use of a given Π in several domains correlates with this Π's monosyllabicity. This observation can be used as the basis for the next implicational pattern in Figure 5.84.

Π → Σ / dimension_x + dimension_y

Figure 5.84: Multidimensional Πs are monosyllabic.

Monosyllabicity invokes high frequency. If we go by the data exposed in Table 3.27, we recognise immediately that the six Πs in (5.291) occupy positions between the 1st and the 8th rank of the frequency-based hierarchy, i.e., each of them forms part of the set of high-frequency Πs. We may therefore add a third implicational pattern in Figure 5.85.

Π → highly frequent / dimension_x + dimension_y

Figure 5.85: Multidimensional Πs are highly frequent.

Another noteworthy fact about (5.291) is that the list features only one Π that is attested in all three domains. Position (5.291a) tells us that (Π41) *minn* 'from' expresses seven different meaning categories one of which belongs to the core dimension, three each to the spatio-temporal domain and the non-spatio-temporal domain. No other Π counts as many meaning categories as (Π41) *minn* 'from' does. Furthermore, (Π55) *ta'* 'of' is the only other Π whose polysemic profile involves a core category. At the same time, (Π55) *ta'* 'of' is the only Π in (5.291) that is not represented in the spatio-temporal domain. The sole domain that is mentioned in each polysemic profile is that of the non-spatio-temporal meaning categories. There is only one meaning category that occurs twice in (5.291), namely SUPERIOR which forms part of the polysemic patterns of (Π15) *fi* 'in' and (Π17) *fuq* 'on'. All other meaning categories are attested only once.

Figures 5.83–5.85 give us insights into properties that are shared by all of the Πs in (5.291). What the figures as such do not make clear is the causality that links the different properties to each other. Monosyllabicity can be ruled out as the trigger for high frequency and/or polysemy across several domains. What about the possibility that high frequency is a necessary condition for polysemy to arise? Or is it the other way around in the sense that being employed for a variety of different meaning categories epiphenomenally increases the frequency of a Π? If frequency fails to explain polysemy, one might want to claim that the meaning categories themselves are responsible for their co-expression by one and the same Π. The next section is dedicated to determining whether co-expression and synonymy are semantically motivated in the domain of the Maltese Πs.

5.2.4 Towards the semantic map of Maltese Πs

Given the preliminary character of our above semantic analyses, it does not make sense to attempt to provide a fully-blown semantic map for BLOMP 2.0 in its entirety. This is a task that can be envisaged for the future. Here and now, we focus on those Πs that are involved in polysemy and/or synonymy. This means that monofunctional Πs that lack a competitor in the domain of the meaning category they express are counted out from the ensuing discussion. To be admitted to the discussion, a Π has to be multifunctional or it must be involved in a synonymic pattern. Polysemy and synonymy are equally favoured or avoided by spatio-temporal Πs and Πs from the other two domains. The differences between the domains are relatively small. This is visible in Figure 5.86.

The shares exclusively refer to cases of polysemy and synonymy that apply within a given semantic domain. We still have to establish what polysemy and synonymy across domains look like.

To this end, we proceed as follows. Each Π's distribution over meaning categories is individually assessed. To this end, the presentation is divided in two. First,[264] we pay attention to those Πs whose distribution is limited to one of the three domains (core meanings, spatio-temporal meanings, non-spatio-temporal meanings). Then,[265] it is the turn of the Πs which are attested in more than one of these domains. The presentation of the individual Πs within these two sections starts from the binary patterns and continues stepwise along the increasing number

264 In Section 5.2.4.1.

265 In Section 5.2.4.2.

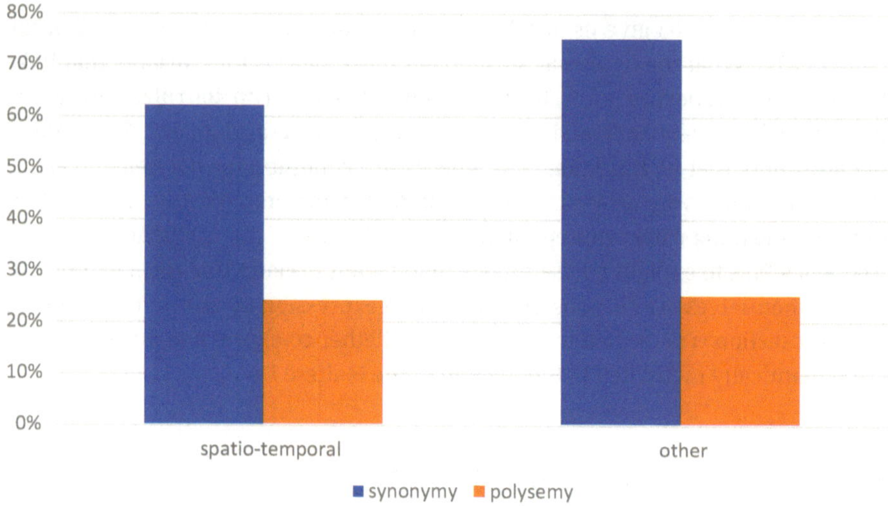

Figure 5.86: Polysemy vs synonymy within domains (compared).

of meaning categories towards the maximum size of a synonymic/polysemic pattern.[266]

5.2.4.1 Πs limited to one domain

In the core domain, there is only (Π29) *lil* 'to' that is attested as exponent of two different meaning categories within the same domain. As shown above,[267] this Π marks the PATIENTIVE as well as the DATIVE. In Hagège's (2010) taxonomy, these two meaning categories are immediate neighbours as schematised in Figure 5.87. Similar schemas will be used for all of the Πs discussed in this and the following section.

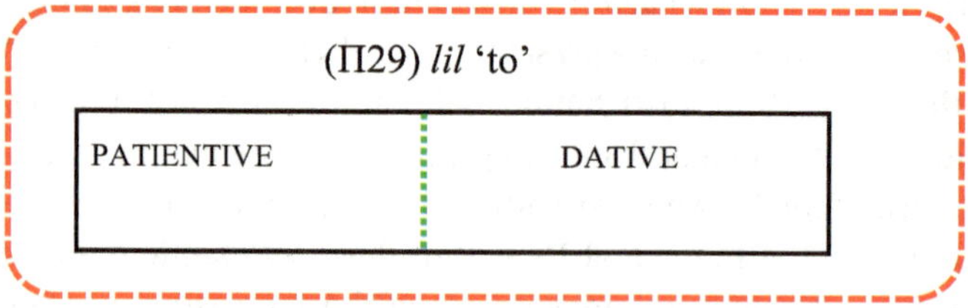

Figure 5.87: The distribution of (Π29) *lil* 'to' over meaning categories.

266 A short summary is given in Section 5.2.4.3.
267 Cf. Sections 5.1.1.2–5.1.1.3.

The red dashed line surrounds the functional domain of the Π under discussion. The labels of the meaning categories that form part of the Π's functional domain are given in rectangular boxes. If these categories are directly adjacent in the semantic space as conceived by Hagège (2010), their labels share a box that is further divided by a green dotted line. If adjacency does not apply, the labels occupy separate boxes. Under adjacency, the neighbourhood relation reflects conceptual similarity so that the employment of one and the same Π for two interrelated categories is unsurprising.

In the case of PATIENTIVE and DATIVE, it may be argued that these meaning categories are close to each other since they both relate to the semantic macro-role of undergoer. The participants that are characterised as PATIENTIVE or DATIVE are affected by the action undertaken by the actor. Their degree of affectedness is not identical, but, independent of the differences between PATIENTIVE and DATIVE, it can be assumed that these labels refer to categories that presuppose a degree of affectedness that surpasses that of all other semantic roles subsumed under the macro-role undergoer.

We enter the non-spatio-temporal domain. The ROBORATIVE is expressed by either (Π37) *mingħala* 'in s.o.'s opinion' or (Π53) *skont* 'according to'. We argued above that the domain of the former is fully included in the domain of the latter.[268] For the present purpose, it suffices that the two Πs are in some kind of synonymic relation. In contrast to polysemous (Π29) *lil* 'to', we observe a combination of two monofunctional Πs that share the same meaning category as shown in Figure 5.88.

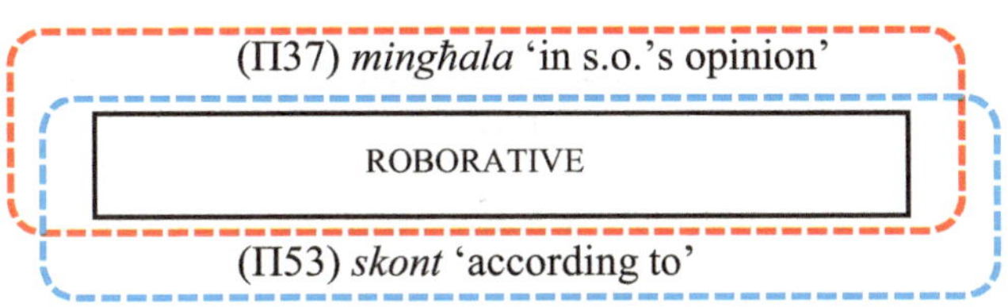

Figure 5.88: The distribution of (Π37) *mingħala* 'in s.o.'s opinion' and (Π53) *skont* 'according to' over meaning categories.

The picture does not change if we take account of the EXCLUSIVE and its two exponents (Π9) *bla* 'without' and (Π36) *mingħajr* 'without' whose properties have been referred to repeatedly in previous sections. The synonymy of (Π9) *bla* 'without' and (Π36) *mingħajr* 'without' is beyond doubt. Accordingly, Figure 5.89 is a look-alike of Figure 5.88 with the sole difference being the new labels.

268 In Section 5.1.3.14.

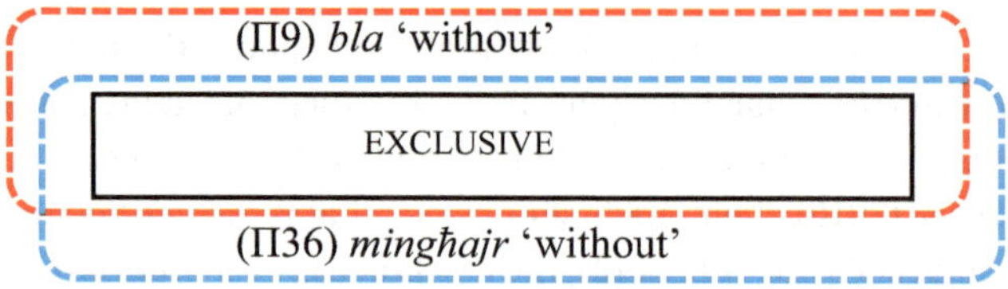

Figure 5.89: The distribution of (Π9) *bla* 'without' and (Π36) *mingħajr* 'without' over meaning categories.

As will transpire from the presentation of further Πs, the complexity of the diagrams increases parallel to the number of synonyms and members of a polysemic chain.

A case in point is the network of the meaning categories ASSIMILATIVE, COMPARATIVE (EQUATIVE), and MEDIATIVE. (Π5) *bħal* 'like' covers all three of these meanings whereas (Π10) *daqs* 'equal to' expresses only the COMPARATIVE (EQUATIVE). Meaning: a multifunctional Π combines with a monofunctional Π to shape a segment of the semantic map. The three meaning categories are direct neighbours in Hagège's (2010) taxonomy.[269] We can thus safely assume that the meaning categories are conceptually linked to each other. The question remains how exactly they should be ordered in conceptual space. Figure 5.90 reflects our interpretation for which the partial synonymy of (Π5) *bħal* 'like' and (Π10) *daqs* 'equal to' is the most important factor.

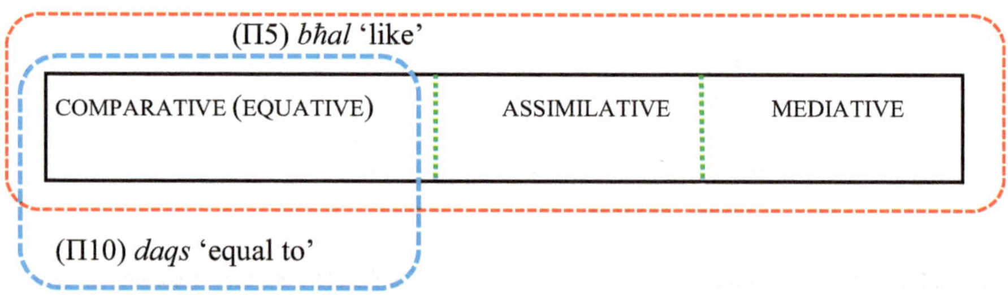

Figure 5.90: The distribution of (Π5) *bħal* 'like' and (Π10) *daqs* 'equal to' over meaning categories.

The by now familiar connection of ADDITIVE and EXCEPTIVE through the polysemy of (Π1) *apparti* 'apart from' and (Π34) *minbarra* 'except' forms the background of the next scenario where also monofunctional (Π20) *għajr* 'except' and (Π24) *ħlief* 'except' play their part. In Figure 5.91, two synonymous Πs yield identical polysemic patterns.

269 They are the topic of Sections 5.1.3.8–5.1.3.9.1.2.

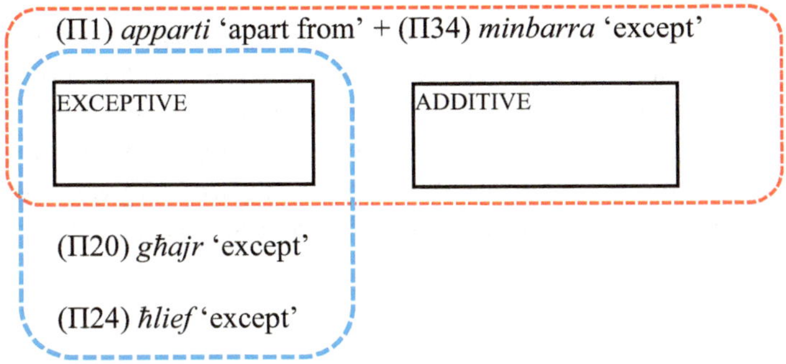

Figure 5.91: The distribution of ((Π1) *apparti* 'apart from', (Π20) *għajr* 'except', (Π24) *ħlief* 'except', and (Π34) *minbarra* 'except' over meaning categories.

The two meaning categories are found at different ends of Hagège's (2010) taxonomy, i.e., a conceptual link between them is not immediately evident. We have highlighted this problem above.[270] The discussion accompanying Figures 5.80–5.81 makes it clear that this issue cannot be clarified without further in-depth research dedicated to it.

The next cases only remotely resemble the previous ones insofar as they involve relatively limited polysemic patterns. What renders them different from (Π29) *lil* 'to', (Π37) *mingħala* 'in s.o.'s opinion', (Π53) *skont* 'according to', (Π9) *bla* 'without', (Π36) *mingħajr* 'without', (Π5) *bħal* 'like', and (Π10) *daqs* 'equal to' is not only their location within the spatio-temporal domain but also the fact that they display a combination of polysemy and synonymy. On account of this combination, the schemas cannot be as simple as the ones in Figures 5.87–5.89. Several graphic renderings of the polysemic and synonymic relations that are characteristic for certain Πs and meaning categories have already been presented above.[271] The relevant figures depict the different situations in accordance with the detailed analysis of the distribution of the Πs that are involved in expressing a given meaning category. In what follows, new diagrams are proposed that cover the same situations but gloss over details to better capture the general picture in a way that allows comparison between the profiles of several Πs.

The first Πs to be jointly looked at are (Π33) *matul* 'during' and (Π59) *waqt* 'at the time of'.[272] Both (Π33) *matul* 'during' and (Π59) *waqt* 'at the time of' express LONGITUDO$_1$ but only (Π33) *matul* 'during' is also responsible for the expression

270 In Sections 5.1.3.2 and 5.1.3.15.

271 Cf. Sections 5.2.1.4.4.1–5.2.1.4.4.2.

272 The synonymic relation between these Πs was addressed in Section 5.2.1.4.4.2.2.5 and schematically presented in Figure 5.52.

of LONGITUDO$_2$. As in the previous case, a multifunctional Π goes together with a monofunctional Π. The conceptual bond between LONGITUDO$_1$ and LONGITUDO$_2$ is evident since we have extension along a stretch of time in one case and along a stretch of space in the other. The metonymical/metaphorical bridge between time and space is well-known from grammaticalization research. Figure 5.92 gives us an idea of how the partially synonymous Πs are ordered in conceptual space relative to the meaning categories.

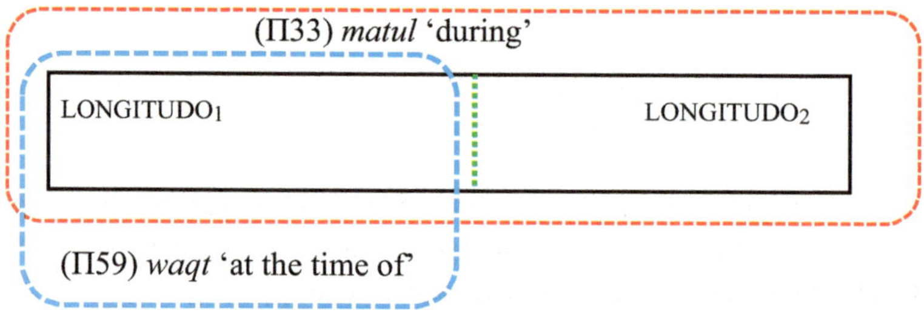

Figure 5.92: The distribution of (Π33) *matul* 'during' and (Π59) *waqt* 'at the time of' over meaning categories.

The next segment of the semantic network comprises (Π4) *bejn* 'between', (Π16) *fost* 'amongst', and (Π47) *qalb* 'amidst'. We have learned above that the three Πs display profiles that are not fully identical.[273] From the synopsis in Table 5.20, it is known that (Π4) *bejn* 'between' covers the meaning categories of MEDIAL$_1$ and MEDIAL$_2$. This means that a multifunctional Π interacts with two monofunctional ones. As argued at length above,[274] the two meaning categories are conceptually closely related – their closeness being reflected by their adjacency in Hagège's (2010) taxonomy. (Π16) *fost* 'amongst' and (Π47) *qalb* 'amidst', however, are restricted to the expression of MEDIAL$_1$. This situation is captured in Figure 5.93.

To continue with the topic at hand we have to look beyond the boundaries of the individual semantic domains in the next section.

5.2.4.2 Πs attested in several domains

Crossing the border between domains causes the complexity of the networks to rise considerably. For a start, we look at the least complex case first in which the meaning categories INTERIOR and SUPERIOR from the spatio-temporal domain and TRANSLATIVE and PERTENTIVE from the non-spatio-temporal domain are involved.

273 Cf. Section 5.2.1.4.4.2.2.3 and Figure 5.50.

274 In Sections 5.1.2.1.13–5.1.2.1.14 and in connection to (5.230).

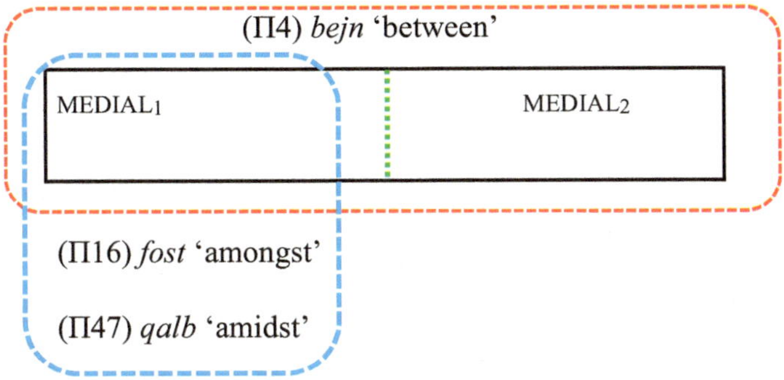

Figure 5.93: The distribution of (Π4) *bejn* 'between', (Π16) *fost* 'amongst', and (Π47) *qalb* 'amidst' over meaning categories.

Their involvement means that the profiles of no less than six Πs have to be situated relative to each other in conceptual space. These Πs are: (Π11) *dwar* 'about', (Π15) *fi* 'in', (Π17) *fuq* 'on', (Π18) *ġewwa* 'inside', (Π19) *ġo* 'in', and (Π50) *rigward* 'concerning'. The meaning categories they represent are not all direct neighbours. There are thus mediated relationships as results from Figure 5.94.

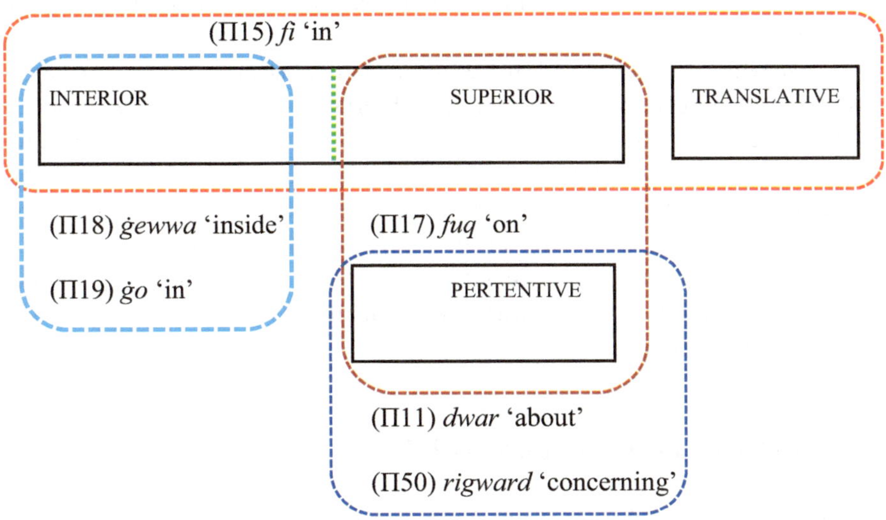

Figure 5.94: The distribution of (Π11) *dwar* 'about', (Π15) *fi* 'in', (Π17) *fuq* 'on', (Π18) *ġewwa* 'inside', (Π19) *ġo* 'in', and (Π50) *rigward* 'concerning' over meaning categories.

We see that two polysemic patterns and three synonymic patterns render the network complex. Neither the INTERIOR nor the TRANSLATIVE is directly connected to the PERTENTIVE. However, they are all connected to the SUPERIOR and thus indirectly connected between them. The conceptual side of this link is an aspect that needs to be looked into in-depth in a separate study.

The final network is the most extended since it embraces fifteen meaning categories which are represented by eighteen different Πs, viz. (Π6) *bħala* 'as', (Π7) *bi* 'with', (Π8) *biswit* 'facing', (Π14) *fejn* 'near', (Π21) *għal* 'for', (Π23) *ħdejn* 'beside', (Π28) *lejn* 'towards', (Π30) *ma'* 'with', (Π31) *madwar* 'around', (Π32) *maġenb* 'close to', (Π39) *minħabba* 'on account of', (Π41) *minn* 'from', (Π45) *permezz* 'by means of', (Π48) *qrib* 'near', (Π49) *quddiem* 'in front of', (Π52) *sforz* 'thanks to', (Π55) *ta'* 'of', and (Π58) *viċin* 'near'. The network covers exactly 30% of BLOMP 2.0. The previous case has shown that the meaning categories involved in a complex network may not necessarily form a conceptual neighbourhood. This observation holds to an even higher degree for the network we are about to sketch in this paragraph.

Figure 5.95 is replete with indirect relations between meaning categories whose link is always a polysemous Π mostly with a synonymous Π that is involved in yet another polysemous pattern. Given the intricacy of the network, we must stipulate a point of departure from where the further relations are developed step by step. The starting point of our choice is the PROPINQUUS on account of the particularly high number of synonyms this meaning category comprises. For expository reasons, Figure 5.95 deviates slightly from the principles applied for designing the previous figures in this section.

The network in Figure 5.95 is not self-explanatory because DIRECTIONAL, PARTITIVE and FRONS for instance, do not invite the observer to assume that these meaning categories are naturally associated with each other. For lack of time and space, we relegate the detailed interpretation of this jigsaw puzzle to future research. Nevertheless, Figure 5.95 might serve as a preliminary basis for the future semantic map of Maltese Πs. To create this map, it is necessary not only to find a way to connect the different smaller and bigger semantic networks presented in this section but also to account for the Πs that neither participate in polysemy nor synonymy and therefore are not featured here.

5.3 Semantics and the canonical Π

To close the semantics chapter of this study, the semantic component of the canonical Π is characterised as follows. According to the usual practice that was applied to define the properties of the canonical Π in phonology, morphology, and syntax, we largely follow the quantitative principle that accepts the most frequently attested phenomenon as canonical. There are three canonical properties in the semantic domain as disclosed in (5.292).

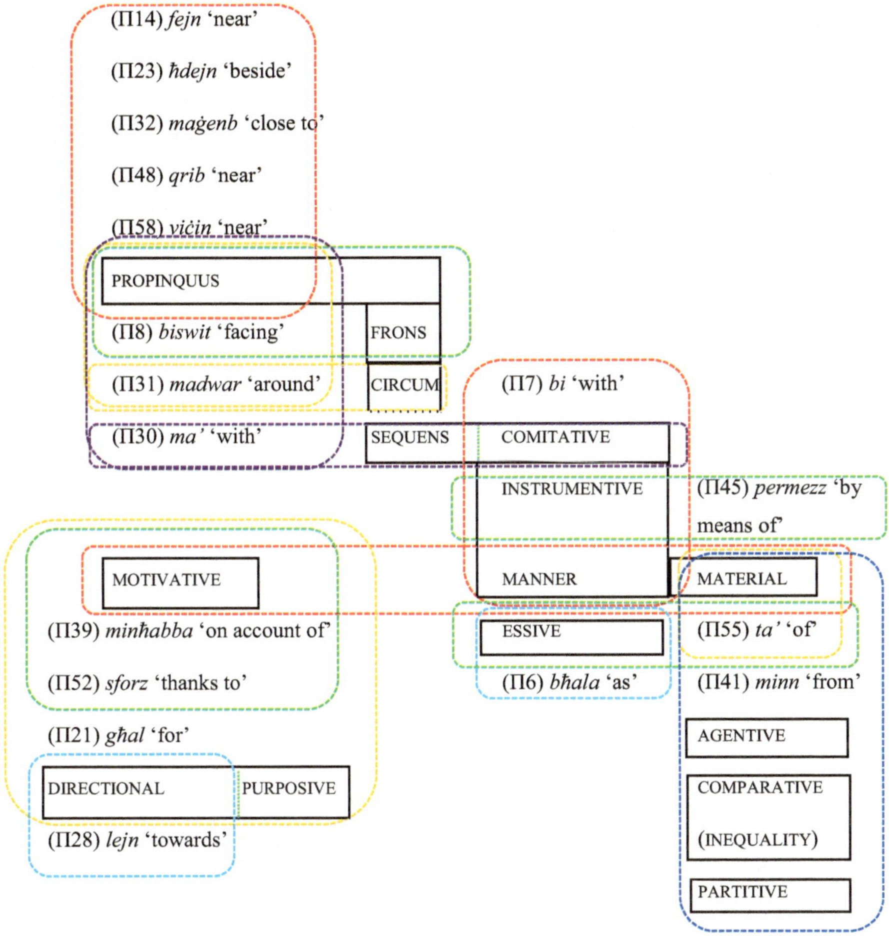

Figure 5.95: The most extended network.

(5.292) Semantic properties of the canonical Π
- (a) spatio-temporal
- (b) configurational
- (c) monofunctional
- (d) synonymous

A relatively good (but not the only) exemplar meeting these criteria is (Π56) *taħt* 'under'. It expresses the spatio-temporal meaning category of INFERIOR.[275] It has

275 Cf. e.g. Section 5.1.2.1.9.

a configurational function (cf. Table 5.15) and combines with all directionality markers (cf. (5.233)). It does not express any other meaning (cf. Table 5.19). It fails, however, to fulfil criterion (5.292d) since it is the sole Π to express the meaning category for which it is employed. Thus, this Π conforms only to some of the requirements of the canon on several other levels of grammar. Where exactly (Π56) *taħt* 'under' stands as to its degree of canonicity is a question that will be answered together with others in the conclusions in Section 6.

PART C: **Terminus**

6 Conclusions and outlook

At this point, we have reached the outer limit of what we are able to say about the grammar of the Maltese Πs. This section is therefore reserved for complementing what we have said in Sections 1–5 with a small number of further important observations without which this study would not be properly completed. Since it is time to draw conclusions, we try to be as brief as possible in what follows. In Section 6.1, we take up the issue of the canonical Maltese Π by way of synoptically summarising how the members of BLOMP 2.0 fare on the relevant parameters. Section 6.2 gives an overview of the basic insights we have gained while studying the Maltese facts. The many phenomena we have addressed notwithstanding, there remain numerous topics that could either be looked at more closely or even for the first time in the future. This is what Section 6.3 is about that also marks the endpoint of this project.

6.1 Finalising the canon

Throughout this study, we have utilised the concept of the canon as a frame of reference for calibrating the behaviour of the Maltese Πs on the different levels of grammar. This has been done separately for phonology, morphology, syntax, and semantics. We seize the opportunity offered by the conclusions to add up the individual results to yield a comprehensive picture of the canonical Π. To this end, we first recapitulate the criteria of canonicity postulated for each of the above levels of grammar. In a second step, we then determine the degree of canonicity of each member of BLOMP 2.0 by way of synoptically reviewing their scores in Table 6.1.

In the course of the previous sections, a plethora of canonical properties have been put forward. They are recapitulated in (6.1) with (6.1c) unifying two originally separate criteria used in connection with syllable structure. The additional factor of frequency will be added later as a control criterion.

(6.1) Canonical properties

(a)	Semitic	cf. Section 2.4
(b)	monosyllabic	
(c)	covered-closed	
(d)	[n < 5 segments]	
(e)	inflected	cf. Tables 3.1–3.2 and 3.26–3.27
(f)	no short-long alternation	
(g)	no segmental reduction	
(h)	no procliticisation	
(i)	no fusion	

https://doi.org/10.1515/9783111331591-006

(j)	takes complements irrespective of definiteness	cf. Section 4.4
(k)	nominative government with lexical NPs as complement	
(l)	possessive government with pronominal complements	
(m)	Π2	
(n)	EQUI-P-deletion	
(o)	attributive	
(p)	resumptive within relative clauses	
(q)	spatio-temporal	cf. Section 5.3
(r)	configurational	
(s)	monofunctional	
(t)	synonymous	

Of these twenty criteria, the four marked in yellow are not applicable to the entire set of Πs because (6.1b–d) and (6.1r) further narrow down the size of a set of canonical cases established on the basis of previous criteria. They are thus not taken account of in the resumé. Accordingly, the highest possible score of canonicity is 16. For a criterion to count as canonical we go by the majority principle which means that a given phenomenon must apply to at least twenty-nine Πs because, as to BLOMP 2.0, the usual problematic candidates are excluded from the evaluation, namely (Π26) *kif* 'as', (Π42) *mintul* 'all along'; (Π43) *mnejn* 'from near'; and (Π54) *sotta* 'under', so that only fifty-six Πs are relevant for what follows. The scores for individual levels of grammar have already been calculated in Tables 2.9, 3.1–3.2, and 3.26–3.27. Table 6.1 integrates these pieces of information with evidence of canonicity from other parts of the study to provide a comprehensive overview.

Table 6.1 is organised as follows. In the leftmost column, the fifty-six Πs to which the evaluation is applicable come in the order of decreasing canonicity. The degree of canonicity is calculated on the basis of the number of Xs that occupy cells in the respective row from column (n) to column (q). The letters at the head of these columns refer back to the criteria of canonicity mentioned in (6.1). If a Π fulfils a given criterion, this is marked by an X. In the opposite case, the cell hosts a 0 and is additionally shaded grey. Italo-Romance Πs are identified by yellow shading of the leftmost cell in a row. In the column headed by X, the sum of Xs per row is given. The rightmost column identifies the frequency class to which the Π belongs with A (additionally shaded blue in the leftmost cell) covering the ranks 1–14 of Table 3.27, B refers to the ranks 15–18, C to 19–42, and D to 43–56. At the bottom of Table 6.1, the total of Xs per column is given.

Table 6.1: Synopsis of canonical properties.

Π	n	j	g	h	f	p	o	i	a	s	t	e	k	l	m	q	X	
(Π16) *fost* 'amongst'	X	X	X	X	X	X	X	X	X	X	X	X	X	X	X	X	16	B
(Π49) *quddiem* 'in front of'	X	X	X	X	X	X	X	X	X	X	X	X	X	X	X	X	16	B
(Π59) *waqt* 'at the time of'	X	X	X	X	X	X	X	X	X	X	X	X	X	X	X	X	16	C
(Π47) *qalb* 'amidst'	X	X	X	X	X	X	X	X	X	X	X	X	X	X	X	X	16	D
(Π32) *maġenb* 'close to'	X	X	X	X	X	X	X	X	X	X	X	X	X	X	X	X	16	D
(Π17) *fuq* 'on'	X	X	X	X	X	X	X	X	X	0	X	X	X	X	X	X	15	A
(Π60) *wara* 'after'	X	X	X	X	X	X	X	X	X	X	0	X	X	X	X	X	15	A
(Π28) *lejn* 'towards'	X	X	0	X	X	X	X	X	X	X	X	X	X	X	X	X	15	B
(Π46) *qabel* 'before'	X	X	X	X	X	X	X	X	X	X	0	X	X	X	X	X	15	B
(Π56) *taħt* 'under'	X	X	X	X	X	X	X	X	X	X	0	X	X	X	X	X	15	B
(Π22) *għand* 'at s.o.'s place'	X	X	X	X	X	X	X	X	X	X	0	X	X	X	X	X	15	C
(Π23) *ħdejn* 'beside'	X	X	0	X	X	X	X	X	X	X	X	X	X	X	X	X	15	C
(Π48) *qrib* 'near'	X	X	X	X	X	X	X	X	X	X	X	X	0	X	X	X	15	C
(Π14) *fejn* 'near'	X	X	X	X	X	X	0	X	X	X	X	X	X	X	X	X	15	D
(Π4) *bejn* 'between'	X	X	0	X	X	X	X	X	X	0	X	X	X	X	X	X	14	A
(Π15) *fi* 'in'	X	X	X	0	X	X	X	0	X	X	X	X	X	X	X	X	14	A
(Π41) *minn* 'from'	X	X	X	X	X	X	X	0	X	0	X	X	X	X	X	X	14	A
(Π30) *ma'* 'with'	X	X	X	X	X	X	X	0	X	0	X	X	X	X	X	X	14	A
(Π31) *madwar* 'around'	X	X	X	X	X	X	X	X	X	0	X	X	0	X	X	X	14	C
(Π38) *mingħand* 'from s.o.'	X	X	X	X	X	X	X	X	X	X	0	X	X	X	0	X	14	C
(Π10) *daqs* 'equal to'	X	X	X	X	X	X	X	X	X	X	X	X	X	X	0	0	14	D

(continued)

Table 6.1 (continued)

П	n	j	g	h	f	p	o	i	a	s	t	e	k	l	m	q	X	
(П20) *għajr* 'except'	X	X	X	X	X	X	X	X	X	X	X	X	X	X	0	0	14	D
(П24) *ħlief* 'except'	X	X	X	X	X	X	X	X	X	X	X	X	X	X	0	0	14	D
(П21) *għal* 'for'	X	X	X	X	X	X	X	0	X	0	X	X	X	X	0	X	13	A
(П5) *bħal* 'like'	X	X	X	X	X	X	X	0	X	0	X	X	X	X	X	0	13	A
(П6) *bħala* 'as'	X	X	X	X	X	0	X	X	X	X	X	0	X	X	X	0	13	A
(П11) *dwar* 'about'	X	X	X	X	X	X	X	X	X	X	X	X	0	X	0	0	13	A
(П33) *matul* 'during'	X	X	X	X	0	X	X	X	X	0	X	X	0	X	X	X	13	B
(П36) *mingħajr* 'without'	X	X	X	X	X	X	X	X	X	X	X	X	0	X	0	0	13	B
(П53) *skont* 'according to'	X	X	X	X	X	X	X	X	X	X	X	X	X	0	0	0	13	B
(П18) *ġewwa* 'inside'	X	X	X	X	X	X	X	X	X	X	X	0	0	0	X	X	13	C
(П19) *ġo* 'in'	X	X	X	X	X	X	X	0	X	X	X	0	X	0	X	X	13	C
(П8) *biswit* 'facing'	X	X	X	X	X	X	X	X	X	0	X	X	0	0	X	X	13	D
(П3) *barra* 'outside'	X	X	X	X	X	0	X	X	X	X	0	0	X	0	X	X	12	B
(П58) *viċin* 'near'	X	X	X	X	X	X	X	X	0	X	X	0	0	0	X	X	12	D
(П7) *bi* 'with'	X	X	X	0	X	X	X	0	X	0	X	X	X	X	0	0	11	A
(П51) *sa* 'till'	X	X	X	0	X	X	X	0	X	X	0	0	X	0	X	X	11	A
(П55) *ta'* 'of'	X	X	X	0	X	X	X	0	X	0	X	X	X	X	0	0	11	A
(П9) *bla* 'without'	X	0	X	X	X	0	X	X	X	X	X	0	X	0	0	0	10	B
(П27) *kontra* 'against'	X	X	X	X	X	X	X	X	0	X	0	X	0	0	X	0	11	B
(П13) *favur* 'in favour of'	X	X	X	X	0	X	X	X	0	X	X	X	0	0	X	0	11	C
(П12) *faċċata* 'opposite'	X	X	X	X	X	X	X	X	0	X	0	0	0	0	X	X	11	D

П	n	j	g	h	f	p	o	i	a	s	t	e	k	l	m	q	X	
(П37) *mingħala* 'in s.o.'s opinion'	X	0	X	X	X	X	0	X	X	X	X	X	0	X	0	0	11	D
(П39) *minħabba* 'on account of'	X	X	X	X	0	X	X	X	X	X	X	0	0	0	0	0	10	B
(П45) *permezz* 'by means of'	10	X	X	X	X	X	X	X	0	X	X	0	0	0	0	0	10	B
(П50) *rigward* 'concerning'	X	X	X	X	X	X	X	X	0	X	X	0	0	0	0	0	10	B
(П40) *minkejja* 'in spite of'	X	X	X	X	X	X	0	X	X	X	0	0	X	0	0	0	10	C
(П2) *a skapitu* 'at the expense of'	X	X	X	X	X	X	0	X	0	X	0	X	X	0	0	0	10	D
(П44) *oltre* 'beyond'	X	X	X	X	X	X	X	X	0	X	0	0	0	0	0	X	10	D
(П52) *sforz* 'thanks to'	X	X	X	X	X	X	0	X	0	X	X	0	X	0	0	0	10	D
(П57) *versu* 'towards'	X	X	X	X	X	0	X	X	0	X	X	0	X	0	0	0	10	D
(П29) *lil* 'to'	X	X	0	X	X	X	0	0	X	0	0	X	X	X	0	0	9	A
(П1) *apparti* 'apart from'	X	X	X	X	X	X	X	X	0	0	X	0	0	0	0	0	9	C
(П25) *inkluż* 'including'	X	X	X	X	X	0	X	X	0	X	0	0	X	0	0	0	9	C
(П34) *minbarra* 'except'	X	X	X	X	0	X	X	X	X	0	X	0	0	0	0	0	9	C
(П35) *minflok* 'instead of'	X	X	X	X	0	X	0	X	0	X	0	X	0	X	0	0	9	C
total	56	54	52	52	51	51	49	46	43	42	41	38	37	34	31	30	707	

The canonicity values range from the realised maximum of sixteen to the relatively low result of nine. The average is 12.6 rounded to 13. Thirty-three Πs (= 59% of the Πs in Table 6.1) reach this threshold, whereas twenty-three (= 41%) remain below it. As to the criteria, there is (6.1n) at the top because it is met by each of the Πs whereas at the opposite end, we find criterion (6.1q) that is fulfilled by 54% of the Πs under investigation. The average number of Πs that receive an X for a criterion is 44 (eight criteria each exceed or fail to reach this value, respectively). The averages are marked with extra thick lines in Table 6.1. The upper left quarter of Table 6.1 is where the Πs with the highest canonicity scores and the most frequently fulfilled criteria meet. In contrast, the lower right quarter forms that part of Table 6.1 where zeroes abound.

The most obvious result that can be directly read from the colour code used for Table 6.1 is that fourteen out of fifteen Italo-Romance Πs are in the lower part which means that their canonicity scores are below the average. In contrast, the only Italo-Romance representative in the upper part is (Π53) *skont* 'according to' that has to compete with thirty-two Semitic Πs. In Figure 6.1, this disproportion based on the language of origin is better visualised. In Figure 6.2, we look at the same givens but this time from the point of view of the positions relative to the average score of canonicity.

	above	below
Italo-Romance	1	14
Semitic	32	9

Figure 6.1: Shares of Semitic vs Italo-Romance Πs above and below the average.

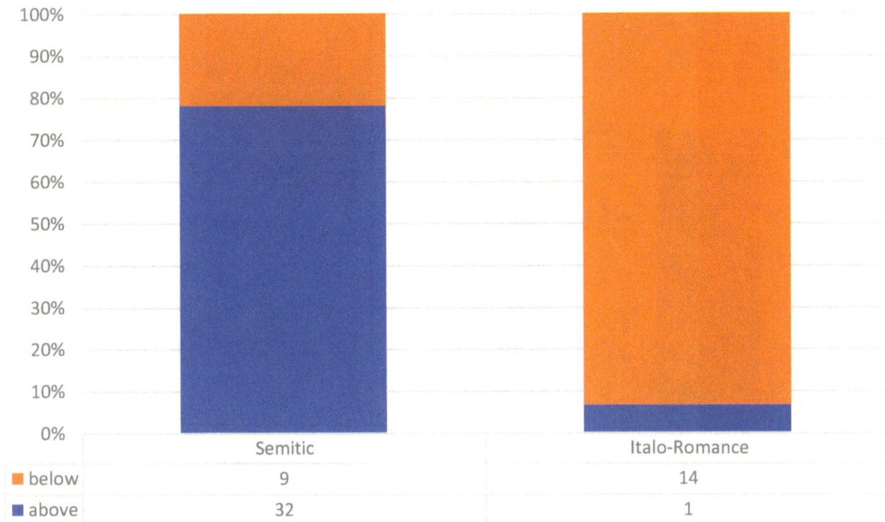

Figure 6.2: Shares of positions above / below the average over language of origin.

The overall result meets our expectations since the different behaviour of Semitic ΠS and Italo-Romance ΠS is an issue that has come up in each of the major sections of this study. To sum up, we state the obvious: Semitic ΠS are overwhelmingly close to the canon whereas Italo-Romance ΠS are generally more remote from the canon.

In Figure 6.3, we take account of the distribution of the positions of ΠS above and below the average in connection to their frequency class. Figure 6.4. determines the shares of the individual frequency classes per positions above/below the average.

The frequency classes in the middle range – B and C – do not yield interesting results. For frequency classes A and D however, it comes to the fore that ΠS with relatively high token frequency associate particularly strongly with a relatively high score of canonicity whereas low frequency ΠS form the biggest group below the average canonicity score. Interestingly, neither do members of frequency class A yield the highest canonicity scores nor do members of frequency class B wind up at the absolute bottom of Table 6.1. Note also that very frequent ΠS like (Π55) *ta'* 'of' and (Π7) *bi* 'with' are associated with scores that remain below the average, but low frequent ΠS like (Π32) *maġenb* 'close to' and (Π47) *qalb* 'amidst' fulfil all sixteen canonicity criteria. Considering all arguments, we conclude that frequency is a factor that enhances canonicity. However, in terms of reliability, the frequency factor produces far less reliable results than the language of origin.

What additionally transpires from Table 6.1 is the dominant position of spatio-temporal ΠS at the top of the canonicity scale. The ΠS with the two highest scores

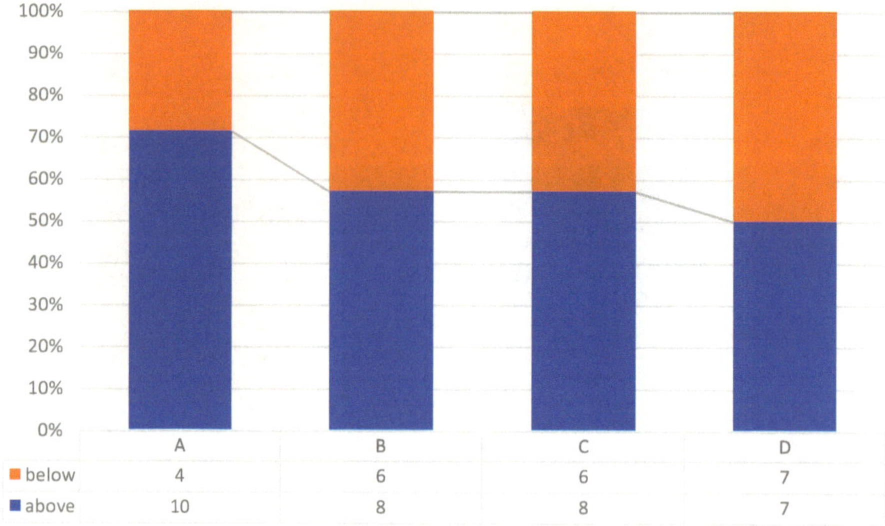

Figure 6.3: Shares of positions above/below the average over frequency class.

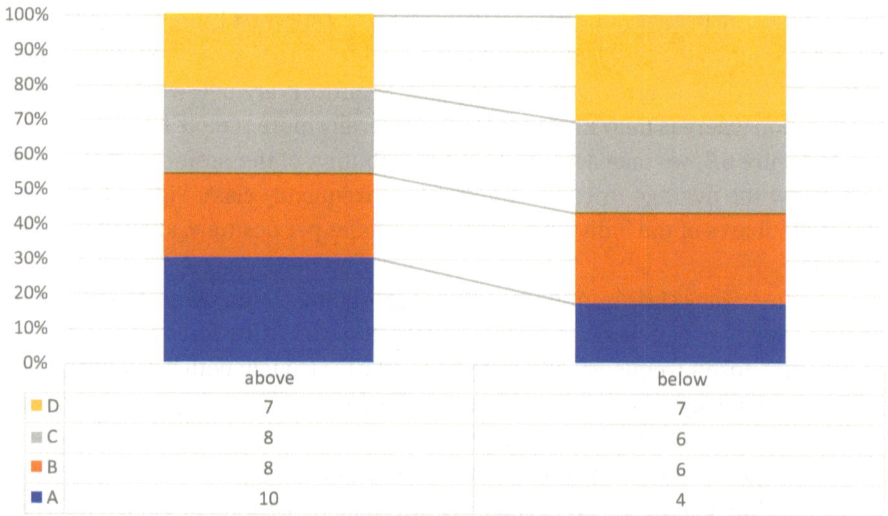

Figure 6.4: Shares frequency classes across Πs above and below the average.

of sixteen and fifteen fulfilled criteria are all representatives of the spatio-temporal domain which is largely unsurprising since with (6.1q) one of the criteria explicitly refers to spatio-temporal meaning categories. The privileged status of spatio-temporal Πs is pictured in Figures 6.5–6.6.

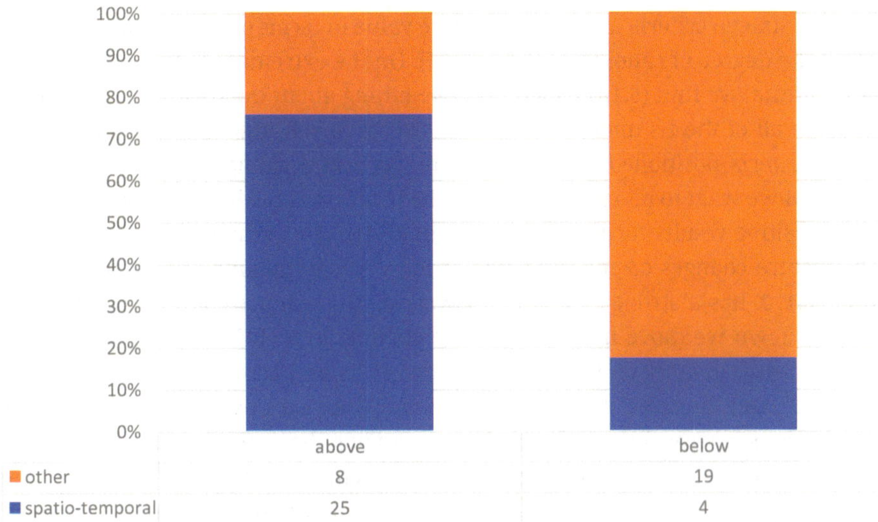

Figure 6.5: Shares of meaning domains above and below the average.

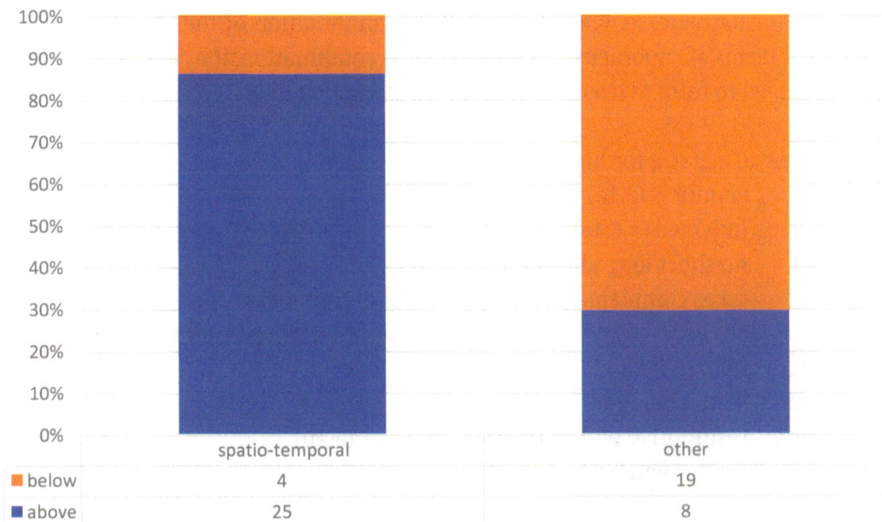

Figure 6.6: Shares of ПS above/below the average across meaning domains.

It cannot be denied that there are strong bilateral attractors. Spatio-temporal meanings and a high degree of canonicity go together as wells as the absence of spatio-temporal meanings tends to combine with lower degrees of canonicity.

The sixteen criteria are not of the same value in terms of their informativeness about the degree of canonicity of a given Π. On the extreme side of the less important criteria, we find (6.1n) which tells us nothing about the relative canonicity of Πs since all of the Πs under review fulfil this criterion. One might add that none of the criteria positioned on the left of the vertical dividing line are particularly helpful if we want to hierarchise the Πs. The problem is the high number of Πs that yield positive results for the criteria (6.1n)–(6.1i). Too many Πs behave similarly. The picture changes on the right side of the vertical dividing line. In this part of Table 6.1, X has a strong competitor, viz. zero. The competition gets stronger the further down we move in Table 6.1. It is possible to identify many patterns that arise from a comparison of the distribution of Xs and zeroes in several columns to the left and the right of the vertical dividing line. Various (positive or negative) implications emerge. Identifying and systematizing these implications is a task for future work on Maltese Πs. In these concluding remarks, it suffices to mention that the first eighteen Πs in Table 6.1 – from (Π16) *fost* 'amongst' down to (Π31) *madwar* 'around' – display uninterrupted parallel Xs for four criteria on the left side (= criteria (6.1n, j, p, f)) and five criteria on the right side (= criteria (6.1a, e, l, m, q). The presence of (6.1n) is trivial as this is the criterion that is met by each Π. Apart from this uninformative case, we assume that the combination of the above criteria is crucial in terms of canonicity. According to this combination, the canonical Maltese Π is required to fulfil at least the criteria made explicit in (6.2).

(6.2) Crucial criteria for high degrees of canonicity
- i. Semitic = (6.1a)
- ii. inflected (= 6.1e)
- iii. no short-long alternation = (6.1f)
- iv. takes complements irrespective of definiteness = (6.1j)
- v. possessive government with pronominal complements = (6.1l)
- vi. Π2 = (6.1m)
- vii. EQUI-P-deletion = (6.1n)
- viii. resumptive within relative clauses = (6.1p)
- ix. spatio-temporal = (6.1q)

In what ways these criteria are interrelated, if at all, remains a question that needs to be answered in a separate study. (6.3) provides the full list of those members of BLOMP 2.0 that are fully in line with what the criteria in (6.2) require.

(6.3) Πs meeting the entire set of criteria in (6.2)
(Π4) *bejn* 'between'; (Π16) *fost* 'amongst'; (Π14) *fejn* 'near'; (Π15) *fi* 'in'; (Π17) *fuq* 'on'; (Π22) *għand* 'at s.o.'s place'; (Π23) *ħdejn* 'beside'; (Π28) *lejn* 'towards'; (Π30) *ma'* 'with'; (Π31) *madwar* 'around'; (Π32) *maġenb* 'close to'; (Π41) *minn* 'from'; (Π46) *qabel* 'before'; (Π47) *qalb* 'amidst'; (Π48) *qrib* 'near'; (Π49) *quddiem* 'in front of'; (Π56) *taħt* 'under'; (Π59) *waqt* 'at the time of'; (Π60) *wara* 'after'

This is where we find (Π56) *taħt* 'under' – the Π whose canonicity was sketched as unclear at the end of Section 5.3. The bulk of the Πs in Table 6.1 (i.e. thirty-eight further members of BLOMP 2.0) fails to yield Xs for the entire set of criteria in (6.2). However, they meet (ever-diminishing) subsets of the combined criteria so that it can be safely assumed that they all display at least a minimal degree of canonicity. This is also true of the four problem-children to some extent at least. For convenience, we show them again in (6.4).

(6.4) Possible candidates for exclusion from BLOMP 2.0
(Π26) *kif* 'as'; (Π42) *mintul* 'all along'; (Π43) *mnejn* 'from near';(Π54) *sotta* 'under'

Given the scarcity of data as to the Πs in (6.4), it is impossible to check which of the criteria for canonicity they fulfil or fail to fulfil – except (6.2) (i) since the first three of them are of Semitic origin. In a revised version of BLOMP 2.0, there will probably be no place for them any more unless new evidence comes up that proves that they are bona fide Πs in contemporary Maltese.

6.2 Achievements and findings

The overview of the canonicity scores and their evaluation in Section 6.1 marks the end of the argumentative part of this study. At this point, we can look back on the contents of the previous sections in order to recapitulate very briefly what the main steps forward are that we have made to add to the general knowledge about Maltese Πs. To avoid unnecessary repetitions of less important details, we only mention those aspects of the grammar of Maltese Πs that are new, one way or another, to the interested scholar. With this goal in mind, we proceed chronologically according to the table of contents of this study.

In Section 1, we have shown on the basis of a critical assessment of the previous literature on our subject matter that it is not only necessary but also feasible to define what exactly counts as a Π in Maltese. The concept of the canonical

Maltese Π has enabled us to set up BLOMP 2.0 as the inventory of those sixty Πs that served as research object in Sections 2–5. We also disclosed our methodology and the mixed data-base from which the data have been taken. Except for Saari (2003), this is the first study of Maltese Πs that approaches the topic comprehensively so that the systematicity of the behaviour of Maltese Πs emerges. At the same time, Section 1 contributes to the history of research dedicated to a special segment of Maltese grammar.

Section 2 accounted for the phonological properties of the members of BLOMP 2.0. In this section, the division of BLOMP 2.0 in a Semitic majority and an Italo-Romance minority was made explicit. The data lend themselves to a systematic evaluation with a special focus on syllabicity and segmental phonology of the Πs. It showed that the Semitic or Italo-Romance background of the Πs very frequently correlates with markedly different preferences on the phonological level including the length of the segmental chains that constitute the expression side of a given Π. Monosyllabicity and moderate size of the segmental chains, more often than not, characterise Semitic Πs whereas Italo-Romance Πs are more tolerant as to polysyllabicity and longer segmental chains. Similar divergent preferences between the languages of origin recur throughout the subsequent sections dedicated to other levels of grammar. The majority options together with further phonological properties tied to them form the basic ingredients of the catalogue of canonical phonological features. Furthermore, Section 2 strongly suggests that token frequency is an important (but not necessarily decisive) factor for a Π to display canonical properties.

The topic of Section 3 was morphology. We could empirically prove that the possibility to inflect for person forms part of the catalogue of canonical properties of Maltese Πs. Person inflection on Πs is common to the majority of the members of BLOMP 2.0. It is mostly but not exclusively found on Semitic Πs. Under inflection, the morphological behaviour is such that several inflection classes could be identified one of which covers the vast majority of the inflected Πs whereas the other classes are only thinly populated. It could further be shown that there is a certain degree of variation of stem classes. A small number of polysyllabic and originally polymorphic Πs proved to be subject to variation between short and (sometimes several) long forms which differ from each other as to the presence or absence of an erstwhile functional initial morpheme that does no longer contribute to the meaning of the Π. It resulted from the corpus-based research that short and long forms do not yield identical turnouts in terms of tokens. Contrary to expectation, it is not always the short alloform that wins out quantitatively. External sandhi with segmental reduction, fusion, and pro-cliticisation were discussed as well. Italo-Romance Πs are not involved in these processes the latter two of which could be shown to correlate very strongly with the high token frequency of the Πs involved.

At the end of Section 3, the modified ranking order of the Πs in BLOMP 2.0 according to their token frequency in the Korpus Malti 3.0 was presented. The canonical properties in the morphological domain were identified.

Morphosyntax was addressed in Section 4. The three phrase types PP, NP, and VP were scrutinised as to the syntactic behaviour of Πs within these phrases. It turned out that the overwhelming majority of the Πs is compatible with definite and indefinite complements including proper names although there are at times huge differences as to how often a given Π is attested in combination with a given complement type. In Section 4.1.2 and its subdivisions, half a dozen different patterns of government were identified with nominative government being typical with lexical nouns as complements and possessive government being typical with pronominal complements. It was shown that variation between patterns of government occurs frequently enough to be explored further. So-called compound Πs and the problems they pose were discussed at length in order to prove that many of the supposed compounds can better be analysed as combinations of two syntactically autonomous Πs in a hierarchically organised construction. On the level of the VP, several phenomena were paid attention, namely differential object marking where (Π29) *lil* 'to' plays the most important role, valency enhancement where (Π7) *bi* 'with' is in the focus of interest (the same holds for the multitude of *bi*-adverbials discussed in the same section), and the *fi-Ø* alternation in Place and Goal constructions. This is where the importance of the concept of zero surfaced for the first time in our study. The morphosyntactic properties of the canonical Π were proposed in accordance with the majority solutions discussed throughout Section 4.

Section 5 looked at the semantics of the Maltese Π from two different perspectives. We applied an onomasiological approach to determine how the meaning categories established by Hagège (2010) are expressed by Maltese Πs. Hagège's original taxonomy had to be adjusted on several occasions. Most of the meaning categories could be proved to be prepositionally expressed in the object language. It came clearly to the fore that an adequate description of the onomasiology of the Πs cannot be achieved without taking account of the important role that zero-marking has in the grammar of Maltese Πs. In the second part of Section 5, the Maltese givens were scrutinised again but this time from the point of view of semasiology, i.e., we determined for what purposes the different Πs are employed in Maltese. This was done with special focus on the phenomena polysemy, synonymy, and antonymy. The differential behaviour of Semitic and Italo-Romance Πs is also discernible in these domains. More generally, polysemy is largely avoided without being absolutely marginal whereas synonymy is widespread. This difference was integrated into the semantic part of the canon.

Finally, the canonicity scores were unified, ranked, and evaluated in Section 6.1. We do not exaggerate when we claim that Maltese Πs have never before been studied as meticulously as has been done in this study. The many previously ignored phenomena and regularities we have unearthed notwithstanding, we dare not pretend that the subject has been exhausted for good. To the contrary, every new finding generates new questions – and there are equally many problems we have been unable to solve.

6.3 Remaining issues

At many points throughout the text, finding the answers to questions that came up is relegated to future research. The reasons for this practice are manifold. First and foremost, there are our own limitations. On account of the lack of native-speaker intuition and knowledge, we ignore many important facts about Maltese language structure and use so that we often could not make head or tail of a given phenomenon. This is especially true of our treatment of the semantics of the ΠIs. Moreover, we repeatedly got stuck because the Korpus Malti 3.0 – despite its size – is too skewed in terms of its coverage of genres that certain phenomena simply could not be investigated properly. A third reason for leaving certain things undone is that going into these issues would have meant to further inflate the size of this volume with the danger of making it too difficult to read. We sincerely hope that our study triggers other linguists to fill the numerous gaps that remain.

Similarly, we put our hope on future scholars to set things right where we have solved problems by way of simply stipulating a solution. This applies for instance to our decision to consider the *għajn* a consonantal phoneme on a par with uncontroversial consonants. One might perhaps also cast doubt on the inclusion in or exclusion from BLOMP 2.0 of certain candidates such as (Π2) *a skapitu* 'at the expense of'. We admit that taking these and further decisions was by no means easy. This is why we feel that a second opinion on these matters could be very helpful if one envisages drawing up a new list of ΠIs for BLOMP 2.1 or BLOMP 3.0.

Contrary to our original plan, this study is strictly descriptive, synchronic, monolingual, and purely empirical. Taking account of topics that go far beyond the horizon defined by phonology, morphology, syntax, and semantics would have had the effect of keeping us busy for the rest of our academic lives. However, the economic solution we have grudgingly accepted in the end does not imply that other topics are of no interest. To our mind, this study has the merit of providing solid foundations for research on those linguistically relevant issues that dropped from our agenda. To give an example: Semitic and Italo-Romance ΠIs display different preferences for each of the levels of grammar. This differential behaviour makes language contact a prime candidate for follow-up studies dedicated to the Maltese ΠIs – not the least because we have not addressed the possibility that the semantics and perhaps also some morphosyntactic properties of Maltese Πs might be influenced by English and/or Italian. Another topic worth looking into is diachrony. On the one hand, grammaticalization seems to provide fertile ground for research since several members of BLOMP 2.0 are the result of the univerbation of erstwhile syntagmata (e.g. (Π34) *minbarra* 'except') whereas other Πs have been grammaticalized from nouns retaining some of the original semantics (e.g. (Π10) *daqs* 'equal to'). On the other hand, there are now texts available that date back to the early days of written Maltese in the 18th century. The properties of Πs used at this stage of the language could be compared longitudinally with those of later stages. The reconstruction of the internal development of the system of Maltese Πs could be complemented with a pan-Semitic comparison of the Maltese results with those from other Neo-Arabic varieties and others (including diachronic data) along the lines of Procházka (1993). In the next step, the comparison can be extended beyond the Afro-Asiatic domain in the sense that the Maltese facts are integrated into what typologists know about Πs and their cross-linguistic distribution. This is the stage at which enough information has been accumulated to move from the predominantly empirical orientation to theory. What is the contribution of Maltese to the general linguistic theory of Πs? We are content if our study constitutes the first stepping stone towards this still-distant goal.

PART D: **Appendix**

Lexicon of Maltese Πs

The Lexicon of Maltese Πs provides the reader with a convenient way to look up the characteristics of each Π described and discussed in this book. The information is delivered for each member of BLOMP 2.0 in a concise format to provide easy access to data scattered throughout Sections 1–5.

A phonetic realisation in IPA (cf. Section 2) and the syllable boundaries for polysyllables (cf. Section 3) are given after the Π's citation form. We then provide the language of origin – Semitic vs Italo-Romance, inflectional class, realisations (cf. Section 3), government class[1] (cf. Section 4) and semantics (cf. Section 5).[2]

The Πs are listed in this section in alphabetical order starting with (Π1) *apparti* 'apart from' and ending with (Π60) *wara* 'after'.

(Π1) *apparti* [ɐppɐrtɪ] /ap.par.ti/ 'apart from'

Origin:	Italo-Romance
Realisations:	*apparti*
Government class:	13
Semantic relation:	non-spatio-temporal; ADDITIVE, EXCEPTIVE

(Π2) *a skapitu* [ɐskɐpɪtʊ] /as.ka.pi.tu/ 'at the expense of'

Origin:	Italo-Romance
Realisations:	*a skapitu (ta'), bi skapitu, għal skapitu, għas-skapitu, skapitu*
Government class:	8
Semantic relation:	non-spatio-temporal; MALEFACTIVE

(Π3) *barra* [bɐrrɐ] /bar.ra/ 'outside'

Origin:	Semitic
Realisations:	*barra*
Government class:	9
Semantic relation:	non-spatio-temporal; EXTRAESSIVE

(Π4) *bejn* [bɛɪn] /bejn/ 'between'

Origin:	Semitic
Inflectional class:	Id

1 For those Πs that are attested both in the nominal and the pronominal domain.
2 Π-Π combinations are not included. A question mark indicates unclear or problematic cases.

https://doi.org/10.1515/9783111331591-007

Realisations: *bej, bejn, bejni, bejnek, bejnu, bejnha, bejnna, bejnietna, bejnkom, bejnietkom, bejnhom, bejniethom*
Government class: 1
Semantic relation: spatio-temporal; INTERESSIVE (spatial/temporal), INTERLATIVE, MEDIOESSIVE

(Π5) *bħal* [phɐːl] /bħal/ 'like'
Origin: Semitic
Inflectional class: Ia
Realisations: *bħal, bħal-, bħall-, bħaċ-, bħad-, bħan-, bħar-, bħas-, bħat-, bħax-, bħaż-, bħaz-, bħali, bħalek, bħalu, bħalha, bħalna, bħalkom, bħalhom*
Government class: 1
Semantic relation: non-spatio-temporal; ASSIMILATIVE, COMPARATIVE (EQUATIVE), MEDIATIVE

(Π6) *bħala* [phɐːlɐ] /bħa.la/ 'as'
Origin: Semitic
Realisations: *bħala*
Government class: 5
Semantic relation: non-spatio-temporal; ESSIVE

(Π7) *bi* [bɪ] /bi/ 'with'
Origin: Semitic
Inflectional class: II
Realisations: *b', bi, bl-, bil-, biċ-, bid-, bin-, bir-, bis-, bit-, bix-, biż-, biz-, bija, bini, bik, biha, bih, bina, bikom, bihom*
Government class: 1
Semantic relation: non-spatio-temporal; COMITATIVE, INSTRUMENTIVE, MANNER, MATERIAL, MOTIVATIVE, PATIENTIVE, PROPRIETIVE

(Π8) *biswit* [bɪswiːt] /bis.wit/ 'facing'
Origin: Semitic
Inflectional class: Ia
Realisations: *biswit, biswiti, biswitek, biswitu, biswitha, biswitna, biswitkom, biswithom*
Government class: 3
Semantic relation: spatio-temporal; DIRECTIONAL_1, FRONTALIS, OBVERSIVE, PROPINQUUS, TERMINATIVE (spatial)

(Π9) *bla* [blɐ] /bla/ 'without'

Origin:	Semitic
Realisations:	*bla*
Semantic relation:	non-spatio-temporal; EXCLUSIVE

(Π10) *daqs* [dɐʔs] /daqs/ 'equal to'

Origin:	Semitic
Inflectional class:	Ia
Realisations:	*daqs, daqsi, daqsek, daqsu, daqsha, daqsna, daqskom, daqshom*
Government class:	1
Semantic relation:	non-spatio-temporal; COMPARATIVE (EQUATIVE)

(Π11) *dwar* [dwɐr] /dwar/ 'about'

Origin:	Semitic
Inflectional class:	Ia
Realisations:	*dwar, dwari, dwarek, dwaru, dwarha, dwarna, dwarkom, dwarhom*
Government class:	2
Semantic relation:	non-spatio-temporal; PERTENTIVE

(Π12) *faċċata* [fɐʧʧɐtɐ] /faċ.ċa.ta/ 'opposite'

Origin:	Italo-Romance
Realisations:	*faċċata*
Government class:	4
Semantic relation:	spatio-temporal; DIRECTIONAL_1, OBSESSIVE, OBVERSIVE, TERMINATIVE (spatial)

(Π13) *favur* [fɐvʊr] /fa.vur/ 'in favour of'

Origin:	Italo-Romance
Inflectional class:	IIIb
Realisations:	*affavur, a favur, għaffavur, favur, favurija, favurik, favurih, favuriha, favurha, favurina, favurna, favurikom, favurkom, favurihom, favurhom*
Government class:	3
Semantic relation:	non-spatio-temporal; BENEFACTIVE

(Π14) *fejn* [fɛɪn] /fejn/ 'near'

Origin:	Semitic
Inflectional class:	Ia

Realisations: *fejn, fejni, fejnek, fejnu, fejnha, fejnna, fejnkom, fejnhom*
Government class: 1
Semantic relation: spatio-temporal, static; PROPINQUUS, PROXILATIVE

(Π15) *fi* [fɪ] /fi/ 'in'
Origin: Semitic
Inflectional class: II
Realisations: *f', fi, fil-, fiċ-, fid-, fin-, fir-, fis-, fit-, fix-, fiż-, fiz-, fl-, fija, fini, fik, fih, fiha, fina, fikom, fihom*
Government class: 1
Semantic relation: spatio-temporal; ILLATIVE, INESSIVE (STATIC/temporal), MUTATIVE, SURESSIVE

(Π16) *fost* [fɔst] /fost/ 'amongst'
Origin: Semitic
Inflectional class: Ie
Realisations: *fost, fostna, fostkom, fosthom*
Government class: 1
Semantic relation: spatio-temporal; MEDIOESSIVE, MEDIOLATIVE

(Π17) *fuq* [fʊʔ] /fuq/ 'on'
Origin: Semitic
Inflectional class: Ia
Realisations: *fuq, fuqi, fuqek, fuqu, fuqha, fuqna, fuqkom, fuqhom*
Government class: 1
Semantic relation: spatio-temporal; PERTENTIVE, SUPERESSIVE, SUPERVERSIVE, SURESSIVE, SURLATIVE

(Π18) *ġewwa* [ʤɛʊwɐ] /ġew.wa/ 'inside'
Origin: Semitic
Realisations: *ġew, ġewwa*
Government class: 7
Semantic relation: spatio-temporal; ILLATIVE, INESSIVE (STATIC)

(Π19) *ġo* [ʤɔː] /*ġo*/ 'in'
Origin: Semitic
Realisations: *ġo, ġol-, goċ-, ġod-, ġon-, ġor-, ġos-, ġot-, ġox-, ġoż-, ġoz-*
Government class: 6
Semantic relation: spatio-temporal; ILLATIVE, INESSIVE (STATIC)

(Π20) *għajr* [ɐːɪr] /għajr/ 'except'

Origin:	Semitic
Inflectional class:	Ia
Realisations:	*għajr, għajri, għajrek, għajru, għajrha, għajrna, għajrkom, għajrhom*
Government class:	1
Semantic relation:	non-spatio-temporal; EXCEPTIVE

(Π21) *għal* [ɐːl] /għal/ 'for'

Origin:	Semitic
Inflectional class:	IIIb
Realisations:	*għal, għalija, għalik, għalih, għaliha, għalina, għalikom, għalihom, għal-, għall-, għaċ-, għad-, għan-, għar-, għas-, għat-, għax-, għaz-, għaż-*
Government class:	1
Semantic relation:	non-spatio-temporal; BENEFACTIVE, DIRECTIONAL_1, MOTIVATIVE, PURPOSIVE

(Π22) *għand* [ɐːnt] /għand/ 'at s.o.'s place'

Origin:	Semitic
Inflectional class:	Ia
Realisations:	*għand, għandi, għandek, għandu, għandha, għandna, għandkom, għandhom*
Government class:	1
Semantic relation:	spatio-temporal; APUDESSIVE, APUDLATIVE

(Π23) *ħdejn* [hdɛɪn] /*ħdejn*/ 'beside'

Origin:	Semitic
Inflectional class:	IIIa
Realisations:	*ħdej, ħdejn, ħdeja, ħdejk, ħdejh, ħdejha, ħdejna, ħdejkom, ħdejhom*
Semantic relation:	spatio-temporal; PROPINQUUS, PROXILATIVE

(Π24) *ħlief* [hlɪːf] /ħlief/ 'except'

Origin:	Semitic
Inflectional class:	Ia
Realisations:	*ħlief, ħliefi, ħliefek, ħliefha, ħliefu, ħliefna, ħliefkom, ħliefhom*
Government class:	1
Semantic relation:	non-spatio-temporal; EXCEPTIVE

(Π25) *inkluż* [ɪnklʊːs] /in.kluż/ 'including'
Origin: Italo-Romance
Realisations: *inkluż*
Government class: 5
Semantic relation: non-spatio-temporal; INCLUSIVE

(Π26) *kif* [kɪf] /kif/ 'as'
Origin: Semitic
Realisations: *kif*
Semantic relation: non-spatio-temporal; EQUATIVE?

(Π27) *kontra* [kɔntrɐ] /kon.tra/ 'against'
Origin: Italo-Romance
Inflectional class: IIIb
Realisations: *kontra, kontrija, kontrik, kontrih, kontriha, kontrina, kontrikom, kontrihom*
Government class: 3
Semantic relation: non-spatio-temporal; ADVERSATIVE

(Π28) *lejn* [lɛɪn] /lejn/ 'towards'
Origin: Semitic
Inflectional class: IIIa
Realisations: *lej, lejn, lejja, lejk, lejh, lejha, lejna, lejkom, lejhom*
Government class: 1
Semantic relation: spatio-temporal; DIRECTIONAL$_2$

(Π29) *lil* [lɪl] /lil/ 'to'
Origin: Semitic
Inflectional class: Ia
Realisations: *'il, 'l, lill, lil-, lill-, liċ-, lid-, lin-, lir-, lis-, lit-, lix-, liż-, liz-, lili, lilek, lilu, lilha, lilna, lilkom, lilhom*
Government class: 1
Semantic relation: core; DATIVE, PATIENTIVE

(Π30) *ma'* [mɐ] /ma/ 'with'
Origin: Semitic
Inflectional class: IIIc
Realisations: *m', ma', mal-, maċ-, mad-, man-, mar-, mas-, mat-, max-, maż-, maz-, miegħi, miegħek, miegħu, magħha, magħna, magħkom, magħhom*

Government class: 1
Semantic relation: non-spatio-temporal; COMITATIVE, PROXIMATE, SECUTIVE

(Π31) *madwar* [mɐdwɐr] /ma.dwar/ 'around'
Origin: Semitic
Inflectional class: Ia
Realisations: *madwar, madwari, madwarek, madwaru, madwarha, madwarna, madwarkom, madwarhom*
Government class: 2
Semantic relation: spatio-temporal; CIRCUMESSIVE (spatial/temporal), PROPINQUUS

(Π32) *maġenb* [mɐʤɛmp] /ma.ġenb/ 'close to'
Origin: Semitic
Inflectional class: Ia
Realisations: *maġenb, maġenbi, maġenbek, maġenbu, maġenbha, maġenbna, maġenbkom, maġenbhom*
Government class: 1
Semantic relation: spatio-temporal; PROXILATIVE

(Π33) *matul* [mɐtʊːl] /ma.tul/ 'during'
Origin: Semitic
Realisations: *matul, matuli, matulek, matulu, matulha, matulna, matulkom, matulhom, tul, tuli, tulek, tulu, tulha, tulna, tulna, tulkom, tulhom*
Government class: 2
Semantic relation: spatio-temporal; EXTENTUS, PERLATIVE (temporal)

(Π34) *minbarra* [mɪnbɐrrɐ] /min.bar.ra/ 'except'
Origin: Semitic
Inflectional class: If
Realisations: *mbarra, minbarra*
Government class: 14
Semantic relation: non-spatio-temporal; ADDITIVE, EXCEPTIVE

(Π35) *minflok* [mɪnflɔk] /min.flok/ 'instead of'
Origin: Italo-Romance
Inflectional class: Ib
Realisations: *flok, floki, flokok, floku, flokha, flokna, flokkom, flokhom, minflok, minfloki, minflokok, minfloku, minflokha,*

	minflokna, minflokkom, minflokhom, minnflok, minnfloki, minnflokok, minnfloku, minnflokha, minnflokna, minnflokkom, minnflokhom
Government class:	2
Semantic relation:	non-spatio-temporal; SUBSTITUTIVE

(Π36) *mingħajr* [mɪnɐːɪr] /min.għajr/ 'without'

Origin:	Semitic
Inflectional class:	Ia
Realisations:	*mingħajr, mingħajri, mingħajrek, mingħajru, mingħajrha, mingħajrna, mingħajrkom, mingħajrhom*
Government class:	2
Semantic relation:	non-spatio-temporal; EXCLUSIVE

(Π37) *mingħala* [mɪnɐːlɐ] /min.għa.la/ 'in s.o.'s opinion'

Origin:	Semitic
Inflectional class:	IIIb
Realisations:	*mingħala, mingħalija, mingħalik, mingħalih, mingħaliha, mingħalina, mingħalikom, mingħalihom*
Semantic relation:	non-spatio-temporal; ROBORATIVE

(Π38) *mingħand* [mɪnɐːnt] /min.għand/ 'from s.o.'

Origin:	Semitic
Inflectional class:	Ia
Realisations:	*mingħand, mingħandi, mingħandek, mingħandu, mingħandha, mingħandna, mingħandkom, mingħandhom*
Government class:	1
Semantic relation:	spatio-temporal; APUDABLATIVE

(Π39) *minħabba* [mɪnhɐbbɐ] /min.ħab.ba/ 'on account of'

Origin:	Semitic
Realisations:	*ħabba, mħabba, minħabba*
Government class:	7
Semantic relation:	non-spatio-temporal; MOTIVATIVE

(Π40) *minkejja* [mɪnkɛjjɐ] /min.kej.ja/ 'in spite of'

Origin:	Semitic
Realisations:	*minkejja*
Government class:	6
Semantic relation:	non-spatio-temporal; CONCESSIVE

(П41) *minn* [mɪnn] /minn/ 'from'

Origin: Semitic
Inflectional class: Ia
Realisations: *minn, minni, minnek, minnu, minnha, minna, minnkom, minnhom, mil-, mill-, miċ-, mid-, min-, mir-, mis-, mit-, mix-, miż-, miz-*
Government class: 1
Semantic relation: spatio-temporal; ABLATIVE (spatial/temporal), AGENTIVE, ELATIVE, INEQUALITY, MATERIAL, PARTITIVE, PERLATIVE (spatial), PROLATIV

(П42) *mintul* [mɪntʊ:l] /min.tul/ 'all along'

Origin: Semitic
Realisations: *mintul, min tul, minn tul*
Government class: 1
Semantic relation: spatio-temporal; EXTENTUS?

(П43) *mnejn* [mnɛɪn] /mnejn/ 'from near'

Origin: Semitic
Realisations: *mnejn*
Government class: 1
Semantic relation: spatio-temporal; ?

(П44) *oltre* [oltrɛ] /ol.tre/ 'beyond'

Origin: Italo-Romance
Realisations: *oltre, oltri*
Government class: 15
Semantic relation: spatio-temporal; ILLATIVE

(П45) *permezz* [pɛrmɛtsts] /per.mezz/ 'by means of'

Origin: Italo-Romance
Realisations: *permezz*
Government class: 8
Semantic relation: non-spatio-temporal; INSTRUMENTIVE

(П46) *qabel* [ʔɐbɛl] /qa.bel/ 'before'

Origin: Semitic
Inflectional class: Ic
Realisations: *qabel, qabli, qablek, qablu, qabilha, qabilna, qabilkom, qabilhom*

Government class: 1
Semantic relation: spatio-temporal; PREESSIVE (spatial/temporal), PRELATIVE (temporal)

(Π47) *qalb* [ʔɐlp] /qalb/ ‘amidst’
Origin: Semitic
Inflectional class: Ie
Realisations: *qalb, qalbna, qalbkom, qalbhom*
Government class: 1
Semantic relation: spatio-temporal; MEDIOESSIVE

(Π48) *qrib* [ʔrɪp] /qrib/ ‘near’
Origin: Semitic
Inflectional class: Ia
Realisations: *qrib, qribi, qribek, qribu, qribha, qribna, qribkom, qribhom*
Government class: 3
Semantic relation: spatio-temporal; PROPINQUUS, PROXILATIVE

(Π49) *quddiem* [ʔʊddɪ:m] /qud.diem/ ‘in front of’
Origin: Semitic
Inflectional class: Ia
Realisations: *quddiem, quddiemi, quddiemek, quddiemu, quddiemha, quddiemna, quddiemkom, quddiemhom*
Government class: 1
Semantic relation: spatio-temporal; FRONTALIS, FRONTIVERSIVE, PERLATIVE (spatial)

(Π50) *rigward* [rɪgwɐrt] /ri.gward/ ‘concerning’
Origin: Italo-Romance
Realisations: *rigward*
Government class: 4
Semantic relation: non-spatio-temporal; PERTENTIVE

(Π51) *sa* [sɐ] /sa/ ‘till’
Origin: Semitic
Realisations: *s’, sa, sal-, saċ-, sad-, san-, sar-, sas-, sat-, sax-, saż-, saz-*
Government class: 11
Semantic relation: spatio-temporal; TERMINATIVE (spatial/temporal)

(Π52) *sforz* [sfɔrts] /sforz/ 'thanks to'

Origin: Italo-Romance
Realisations: *sforz*
Semantic relation: non-spatio-temporal; MOTIVATIVE

(Π53) *skont* [skɔnt] /skont/ 'according to'

Origin: Italo-Romance
Inflectional class: Ib
Realisations: *skond/skont, skondi/skonti, skondok/skontok, skondu/skontu, skondha/skontha, skondna/skontna, skondkom/skontkom, skondhom/skonthom*
Government class: 12
Semantic relation: non-spatio-temporal; ROBORATIVE

(Π54) *sotta* [sɔttɐ] /sot.ta/ 'under'

Origin: Italo-Romance
Realisations: *sotta*
Semantic relation: spatio-temporal; SUBESSIVE?

(Π55) *ta'* [tɐ] /ta/ 'of'

Origin: Semitic
Inflectional class: IIIc
Realisations: *t', ta', tiegħi, tiegħek, tiegħu, tagħha, tagħna, tagħkom, tagħhom, tal-, taċ-, tad-, tan-, tar-, tas-, tat-, tax-, taż-, taz-*
Government class: 1
Semantic relation: core; ESSIVE, MATERIAL, POSSESSIVE, PROPRIETIVE

(Π56) *taħt* [tɐht] /taħt/ 'under'

Origin: Semitic
Inflectional class: Ia
Realisations: *taħt, taħti, taħtek, taħtu, taħtha, taħtna, taħtkom, taħthom*
Government class: 1
Semantic relation: spatio-temporal; SUBLATIVE

(Π57) versu [vɛrsu] /ver.su/ 'towards'

Origin: Italo-Romance
Realisations: *versu*
Government class: 10
Semantic relation: spatio-temporal; PURPOSIVE

(Π58) *viċin* [vɪʧɪn] /vi.ċin/ 'near'

Origin:	Italo-Romance
Realisations:	*viċin*
Government class:	4
Semantic relation:	spatio-temporal; PROPINQUUS, PROXILATIVE

(Π59) *waqt* [wɐʔt] /waqt/ 'at the time of'

Origin:	Semitic
Realisations:	*waqt, waqti, waqtek, waqtu, waqtha, waqtna, waqtkom, waqthom*
Government class:	1
Semantic relation:	spatio-temporal; PERLATIVE (temporal)

(Π60) *wara* [wɐrɐ] /wa.ra/ 'after'

Origin:	Semitic
Inflectional class:	IIIa
Realisations:	*wara, warajja/urajja, warajk/urajk, warajh/urajh, warajha/urajha, warajna/urajna, warajkom/urajkom, warajhom/urajhom*
Government class:	1
Semantic relation:	spatio-temporal; POSTESSIVE (spatial/temporal), POSTLATIVE (spatial/temporal)

Primary sources

Abela, Deborah [adapted by Leanne Ellul & Clare Azzopardi]. 2016. *Teresa*. Blata l-Bajda: Merlin.
Agius, Ben. 2002. *Ulied il-kwiekeb*. Valletta: Klabb Kotba Maltin.
Arana, Marie [translated by Mateo Cardona & Marta Cecilia Mesa]. 2020. *Libertador de América: Bolívar.* Barcelona: Debate.
Attard, Lawrence E. 1999. *L-emigrazzjoni Maltija. Is-seklu dsatax u għoxrin.* Il-Pjetà: PIN.
Attard, Richard. 2019. *Difrejn il-falkun*. S.l. Horizons.
Baldacchino, Carmel. 1999. *Il-futbol Malti*. Il-Pjetà: PIN.
Bartolo, Simon & Loranne Vella. 2009. *Il-ġnin tad-dmugħ. Il-fiddien: it-tielet ktieb*. Blata l-Bajda: Merlin.
Briffa, Charles. 1992. *Teżor fl-imħabba*. Valletta: Bugelli.
Buhagiar, Victor. 2001. *Louise*. Valletta: Klabb Kotba Maltin.
Camilleri, John Joseph. 2014. *Malta li għext fiha*. Il-Ħamrun: SKS.
Camilleri, Mark. 2013. *Volens*. Blata l-Bajda: Merlin.
Camilleri, Mark. 2021. *Alias*. Blata l-Bajda: Merlin.
Camilleri, Mark. 2022. *Ġaħan fl-aqwa żmien*. Malta: Dar Camilleri.
Casha, Charles. 1999. Avventura fl-"Għar il-Kbir". Blata l-Bajda: Merlin.
Casha, Charles. 2008. *Il-prinċep u t-trogloditi*. Santa Venera: Merlin.
Cauchi, Carmel G. 1997. *F'Ħal Busbies ma' Mastru Gerfex*. Blata l-Bajda: Merlin.
Chetcuti, Ġużè. 1990. *L-isqaq*. Blata l-Bajda: Merlin.
Cooper, James Fennimore [translated by Micheal Grech]. 1985. *L-aħħar Moħoki.* Valletta: Klabb Kotba Maltin.
DeBono, Paul J. 2001. *Il-pendent tas-seba' rubini.* St. Julians: Bieb Bieb Enterprises.
De Saint-Exupéry, Antoine [translated by Katherine Woods]. 1991. *The Little Prince*. London: Mammoth.
De Saint-Exupéry, Antoine [translated by Ieva Lase]. 1995a. *Mazais Princis*. Rīga: Sprīdītis.
De Saint-Exupéry, Antoine [translated by Bonifacio del Carril]. 1995b. *El Principito*. Buenos Aires, Madrid: Alianza-Emece.
De Saint-Exupéry, Antoine [translated by Toni Aquilina]. 2000. *Iċ-Ċkejken Prinċep*. Msida: Mireva.
Ellul Mercer, Ġużè. 1986. *Taħt in-nar*. Valletta: Klabb Kotba Maltin.
Farrugia, Aleks. 2022. *Ir-re Borg*. Malta: SKS.
Friggieri, Oliver. 2000. *It-tfal jiġu bil-vapuri*. Malta: Mireva.
Galea, Ġużè. 2001[1938]. *Żmien l-ispanjoli.* Valletta: Klabb Kotba Maltin.
Gauci, Manwel. 1994. *It-Tazza tad-Dinja. Ġrajja ta' 64 sena*. Pietà: Indipendenza.
Korpus Malti 2.0, beta: https://mlrs.research.um.edu.mt/CQPweb/malti02/
Korpus Malti 3.0, 2016: https://mlrs.research.um.edu.mt/CQPweb/malti03/
Korpus Malti 4.2, 2023: https://mlrs.research.um.edu.mt/CQPweb/malti04/
Monsarrat, Nicholas [translated by Wistin Born]. 2018. *Il-kappillan ta' Malta*. Tas-Sliema: SKS.
Q – Sandra Abela, personal communication and questionnaire.
Schembri, Karl. 2002. *Taħt il-Kappa tax-Xemx*. S.l. Minima.
Teuma, Edmund [translator]. 2015. *L-Iljieli Għarbin*. San Ġwann: BDL.
Vella, Mark. 2014. *X'seta' ġralu lil Kevin Cacciatolo?* Il-Blata l-Bajda: Merlin.
Żahra, Trevor. 2008. *Il-ġenn li jżommni f'sikti*. Blata l-Bajda: Merlin.
Zammit, George. 1976. *Id-dinja ta' Wenzu u Rozi.* Valletta: Klabb Kotba Maltin.

https://doi.org/10.1515/9783111331591-008

References

Akkademja tal-Malti. 1998. *Regoli tal-kitba tal-Malti*. Valletta: Klabb Kotba Maltin.

Ambros, Arne A. 1998. *Bonġornu, kif int? Einführung in die maltesische Sprache*. Wiesbaden: Reichert.

Andersen, Henning. 1986. Introduction: sandhi. In Henning Andersen (ed.), *Sandhi Phenomena in the Languages of Europe*, 1–10. Berlin/New York: Mouton de Gruyter.

Aquilina, Joseph. 1959. *The Structure of Maltese*. Valletta: Progress Press.

Aquilina, Joseph. 1976. Prepositional verbs in Maltese. In Joseph Aquilina (ed.), *Maltese Linguistic Survey*, 67–80. Malta: The University of Malta.

Aquilina, Joseph. 1987. *Maltese-English Dictionary*. Volume One A-L. Malta: Midsea.

Aquilina, Joseph. 1991. *Maltese-English Dictionary*. Volume Two M–Z and Addenda. Malta: Midsea.

Aquilina, Joseph. 1999. *English-Maltese dictionary. Four Volumes*. Malta: Midsea.

Altmann, Gabriel & Werner Lehfeldt. 1973. *Allgemeine Sprachtypologie*. München: Fink.

Azzopardi, Kirsty. 2023. The argument structure of motion verbs in Maltese. Talk on occasion of the *8th International Conference of Maltese Linguistics "The next century of Maltese Linguistics"*, Bremen/Germany, 18–20 September, 2023.

Baerman, Matthew. 2007. Morphological typology of deponency. In Matthew Baerman, Greville G. Corbett, Dunstan Brown & Andrew Hippisley (eds.), *Deponency and Morphological Mismatches*, 1–20. Oxford: Oxford University Press.

Baerman, Matthew, Dunstan Brown & Greville G. Corbett. 2005. *The Syntax-Morphology Interface. A Study of Syncretism.* Cambridge: Cambridge University Press.

Baerman, Matthew & Greville G. Corbett. 2010. Introduction: Defectiveness: typology and diachrony. In Matthew Baerman, Greville G. Corbett & Dunstan Brown (eds.), *Defective Paradigms. Missing Forms and What They Tell Us*, 1–18. Oxford: Oxford University Press.

Bakker, Dik. 2005. Person marking on adpositions. In Matthew Dryer & Martin Haspelmath (eds.), *The World Atlas of Language Structures*, 198–201. Oxford: Oxford University Press.

Biber, Douglas. 2010. Corpus-based and corpus-driven analyses of language variation and use. In Bernd Heine & Heiko Narrog (eds.), *The Oxford Handbook of Linguistic Analysis*, 193–224. Oxford: Oxford University Press.

Borg, Albert. 1988. *Ilsienna. Studju grammatikali.* Ħas Sajjied: self-published.

Borg, Albert. 1996. The structure of the noun phrase in Maltese. *Rivista di Linguistica* 8 (1). 5–28.

Borg, Albert & Marie Azzopardi-Alexander. 1997. *Maltese*. London/New York: Routledge.

Borg, Albert & Marie Azzopardi-Alexander. 2009. Topicalisation in Maltese. In Bernard Comrie, Ray Fabri, Elizabeth Hume, Manwel Mifsud, Thomas Stolz & Martine Vanhove (eds.), *Introducing Maltese Linguistics*, 71–82. Amsterdam/Philadelphia: Benjamins.

Borg, Albert & Bernard Comrie. 1984. Object diffuseness in Maltese. In Frans Plank (ed.), *Objects*, 109–126. London: Academic Press.

Borg, Albert & Manwel Mifsud. 2002. Maltese object marking in a Mediterranean context. In Paolo Ramat & Thomas Stolz (eds.), *Mediterranean languages. Papers from the MEDTYP workshop, Tirrenia, June 2000*, 33–46. Bochum: Brockmeyer.

Borg, Alexander. 1978. *A Historical and Comparative Phonology and Morphology of Maltese.* Jerusalem: University of Jerusalem PhD-Thesis.

Borg, Alexander. 1997. Maltese phonology. In Alan S. Kaye (ed.), *Phonologies of Asia and Africa*, vol. 1, 245–286. Winona Lake, Indiana: Eisenbrauns.

Bovingdon, Roderick. 2001. *The Maltese Language of Australia: Maltraljan. A lexical compilation with linguistic notations and a social, political and historical background*. München: Lincom Europa.

https://doi.org/10.1515/9783111331591-009

Brincat, Joseph M. 2000. *Il-Malti. Elf sena ta' storja*. Il-Pjetà: PIN.

Brother Henry. 1980. *Grammatika Maltija*. Gżira: Stella Maris College.

Cachia, Lawrenz. 1994. *Grammatika ġdida tal-Malti*. Żabbar: Veritas.

Cachia, Lawrenz. 2000. *Ħabbew l-ilsien Malti*. Malta: Veritas.

Camilleri, Maris. 2014. The Maltese restrictive relative clause. In Albert Borg, Sandro Caruana & Alexandra Vella (eds.), *Perspectives on Maltese Linguistics*, 161–200. Berlin: Akademie.

Caruana, Antonio Emanuele. 1903. *Vocabolario della lingua maltese preceduto da un compendio di grammatica della stessa favella*. Valletta: Muscat.

Cassar, Mario. 2000. *Ir-reqqa tal-kitba*. Il-Blata l-Bajda: Merlin.

Cassola, Arnold. 2014. A note on the dating of *ħ*, *għ* and *x* in Maltese. In Albert Borg, Sandro Caruana & Alexandra Vella (eds.), *Perspectives on Maltese Linguistics*, 13–22. Berlin: Akademie.

Čéplö, Slavomír. 2018. *Constituent Order in Maltese: A Quantitative Analysis*. Prague: Charles University PhD-thesis.

Čéplö, Slavomír & Erika Just. 2023. A corpus-driven account of indirect object indexing in Maltese. Talk on occasion of the *8th International Conference of Maltese Linguistics "The next century of Maltese Linguistics"*, Bremen/Germany, 18–20 September, 2023.

Cohen, David. 1970. Le système phonologique du maltais: aspects synchroniques et diachroniques. In David Cohen (ed.), *Études de linguistique arabe et sémitique*, 126–149. La Haye/Paris: Mouton.

Cohen, David. 1988. Arabe. In Jean Perrot (ed.), *Les langues dans le monde ancient et modern. Langues chamito-sémitiques*, 105–124. Paris: CNRS.

Comrie, Bernard. 1982. Syntactic-semantic discrepancies in Maltese sentence structure. *Communication and Cognition* 15(3/4). 719–743.

Comrie, Bernard & Michael Spagnol. 2016. Maltese loanword typology. In Gilbert Puech & Benjamin Saade (eds.), *Shifts and Patterns in Maltese*, 315–330. Berlin/Boston: De Gruyter Mouton.

Corbett, Greville G. 2005. The canonical approach in typology. In Zygmunt Frajzyngier, Adam Hodges & David S. Rood (eds.), *Linguistic Diversity and Language Theories*, 25–50. Amsterdam/Philadelphia: Benjamins.

Corbett, Greville G. 2006. *Agreement*. Cambridge: Cambridge University Press.

Corbett, Greville G. 2007a. Deponency, syncretism, and what lies in between. In Matthew Baerman, Greville G. Corbett, Dunstan Brown & Andrew Hippisley (eds.), *Deponency and Morphological Mismatches*, 21–44. Oxford: Oxford University Press.

Corbett, Greville G. 2007b. Canonical Typology, Suppletion, and Possible Words. *Language* 83(1). 8–42.

Corbett, Greville G. & Matthew Baerman. 2006. Prolegomena to a typology of morphological features. *Morphology* 16. 231–246.

Cremona, Antonio. 1970. *Tagħlim fuq il-kitba Maltija. L-ewwel ktieb*. Malta: Aquilina.

Croft, William. 1991. *Syntactic Categories and Grammatical Relations. The Cognitive Organization of Information*. Chicago/London: The University of Chicago Press.

Croft, William. 2013. Radical Construction Grammar. In Thomas Hoffmann & Graeme Trousdale (eds.), *The Oxford Handbook of Construction Grammar*, 211–232. Oxford University Press.

Croft, William. 2022. *Morphosyntax. Constructions of the world's languages*. Cambridge: Cambridge University Press.

DeLancey, Scott. 2005. Adpositions as a non-universal category. In Zygmunt Frajzyngier, Adam Hodges & David S. Rood (eds.), *Linguistic Diversity and Language Theories*, 185–202. Amsterdam/Philadelphia: Benjamins.

De Soldanis, Giovanni Pietro Francesco Agius. 1750. *Della lingua punica presentemente usata da maltesi*. Roma: Generoso Salomoni.

Dixon, R.M.W. 2010a. *Basic Linguistic Theory. Volume 1: Methodology*. Oxford: Oxford University Press.

Dixon, R.M.W. 2010b. *Basic Linguistic Theory. Volume 2: Grammatical Topics.* Oxford: Oxford University Press.

Dixon, R.M.W. 2013. *Basic Linguistic Theory. Volume 3: Further Grammatical Topics.* Oxford: Oxford University Press.

Dixon, R.M.W. 2021. *English Prepositions. Their meanings and uses*. Oxford: Oxford University Press.

Döhla, Hans-Jörg. 2016. The origin of differential object marking in Maltese. In Gilbert Puech & Benjamin Saade (eds.), *Shifts and Patterns in Maltese*, 149–174. Berlin/Boston: De Gruyter Mouton.

Doleschal, Ursula. 1997. Indeklinabilität, Semideklinabilität und Flexionsklassen: Russisch und Slowakisch. *Wiener Slavistisches Jahrbuch* 43. 41–52.

Fabri, Ray. 1993. *Kongruenz und die Grammatik des Maltesischen.* Tübingen: Niemeyer.

Fabri, Ray. 1996. The construct state and the pseudo-construct state in Maltese. *Rivista di Linguistica* 8(1). 229–244.

Fabri, Ray. 2022. The lexical *ta'* construction. In Przemysław Turek & Julia Nintemann (eds.), *Maltese: Contemporary Changes and Historical Innovations*, 23–46. Berlin/Boston: De Gruyter Mouton.

Fabri, Ray. 2023. Secondary predication in Maltese. Talk on occasion of the *8th International Conference of Maltese Linguistics "The next century of Maltese Linguistics"*, Bremen/Germany, 18–20 September, 2023.

Fabri, Ray and Albert Borg. 2002. Topic, focus and word order in Maltese. In Abderrahim Youssi, Fouzia Benjelloun, Mohamed Dahbi & Zakia Iraqui-Sinaceur (eds.), *Aspects of dialects of Arabic today: Proceedings of the 4th Conference of the International Arabic Dialectology Association (AIDA), Marrakesh, April 1–4, 2000*, 354–363. Rabat: Amapatril.

Fagard, Benjamin, José Pinto de Lima & Dejan Stosic (eds.). 2019. *Les prépositions complexes dans les langues romanes*. Amsterdam/Philadelphia: Benjamins.

Falzon, Giovanni Battista. 1882. *Dizionario italiano-inglese-maltese / Dizionario maltese-italiani-inglese Preceduto da una breve esposizione grammaticale della lingua maltese.* Malta: s.l.

Fenech, Edward. 1978. *Contemporary Journalistic Maltese. An Analytical and Comparative Study.* Leiden: Brill.

Fenech, Karm. 1984. *Idjomi Maltin*. Malta: Ġiżwiti Maltin.

Fillmore, Charles J., Paul Kay & Mary Catherine O'Connor. 1988. Regularity and idiomaticity in grammatical constructions: the case of *let alone*. *Language* 64(4). 501–538.

Gaeta, Livio & Davide Ricca. 2006. Productivity in Italian word formation: a variable corpus approach. *Linguistics* 44(1). 57–89.

Galea, Luke & Adam Ussishkin. 2018. Onset clusters, syllable structure and syllabification in Maltese. In Patrizia Paggio & Albert Gatt (eds.), *The Languages of Malta*, 55–80. Berlin: Language Science Press.

Garzonio, Jacopo & Silvia Rossi (eds.). 2020. *Variation in P. Comparative approaches to adpositional phrases*. Oxford: Oxford University Press.

Gatt, Albert & Slavomír Čéplö. 2013. Digital corpora and other electronic resources for Maltese. In Andrew Hardie & Robbie Love (eds.), *Corpus Linguistics 2013, Abstract Book*, 96–97. Lancaster: UCREL.

Gatt, Albert & Ray Fabri. 2018. Borrowed affixes and morphological productivity: a case study of two Maltese nominalisations. In Patrizia Paggio & Albert Gatt (eds.), *The Languages of Malta*, 143–170. Berlin: Language Science Press.

Givón, Talmy. 2001. *Syntax. An Introduction*. 2 vols. Amsterdam/Philadelphia: Benjamins.

Givón, Talmy. 2021. *The Life Cycle of Adpositions*. Amsterdam/Philadelphia: Benjamins.

Gordon, Matthew K. 2016. *Phonological Typology*. Oxford: Oxford University Press.

Guimier, Claude. 1981. *Prepositions. An Analytical Bibliography*. Amsterdam/Philadelphia: Benjamins.
Hagège, Claude. 1975. *Le problème linguistique des prepositions et la solution chinoise (avec un essai de typologie à travers plusieurs groups de langues)*. Paris/Louvain: Peeters.
Hagège, Claude. 2010. *Adpositions*. Oxford: Oxford University Press.
Hall, T. Allan. 2011. *Phonologie. Eine Einführung*. Berlin/New York: De Gruyter.
Haspelmath, Martin. 2007a. Coordination. In Timothy Shopen (ed.), *Language Typology and Language Description, vol. 2: Complex Constructions*, 1–51. Cambridge: Cambridge: University Press.
Haspelmath, Martin. 2007b. Preestablished categories don't exist: Consequences for language description and typology. *Linguistic Typology* 11(1). 119–132
Haspelmath, Martin. 2010. Framework-free grammatical theory. In Bernd Heine & Heiko Narrog (eds.), *The Oxford Handbook of Linguistic Analysis*, 287–310. Oxford: Oxford University Press.
Haspelmath, Martin. 2013. Argument indexing: a conceptual framework for the syntactic status of bound person forms. In Dik Bakker & Martin Haspelmath (eds.), *Languages across bounadries. Studies in the memory of Anna Siewierska*, 197–226. Berlin, Boston: De Gruyter Mouton.
Haspelmath, Martin. 2019. Differential place marking and differential object marking. *STUF – Language Typology and Universals* 72(3). 313–334.
Heine, Bernd. 1997. *Possession. Cognitive sources, forces, and grammaticalization*. Cambridge: Cambridge University Press.
Heine, Bernd, Ulrike Claudi & Friederike Hünnemeyer. 1991. *Grammaticalization. A conceptual framework*. Chicago/London: The University of Chicago Press.
Heine, Bernd & Heiko Narrog. 2010. Introduction. In Bernd Heine & Heiko Narrog (eds.), *The Oxford Handbook of Linguistic Analysis*, 1–25. Oxford: Oxford University Press.
Hume, Elizabeth, Jennifer Venditti, Alexandra Vella & Samantha Gett. 2009. Vowel duration and Maltese 'għ'. In Bernard Comrie, Ray Fabri, Elizabeth Hume, Manwel Mifsud, Thomas Stolz & Martine Vanhove (eds.), *Introducing Maltese Linguistics*, 15–46. Amsterdam/Philadelphia: Benjamins.
Il-Kunsill Nazzjonali tal-Ilsien Malti. 2008. *Deċiżjonijiet 1 tal-Kunsill Nazzjonali tal-Ilsien Malti dwar il-Varjanti Ortografiċi*. [Decisions 1 of the National Council for the Maltese Language about orthographic variants]. Malta: Il-Furjana.
Johnson, Keith. 2008. *Quantitative methods in linguistics*. Oxford: Blackwell.
Just, Erika & Slavomír Čéplö. 2022. Differential object indexing in Maltese. In Przemysław Turek & Julia Nintemann (eds.), *Maltese: Contemporary Changes and Historical Innovations*, 105–132. Berlin/Boston: De Gruyter Mouton
Kontzi, Reinhold. 1999. *Wort und Schrift. Des Kanonikus Fortunato Panzavecchia Bibelübersetzung ins Maltesische, nach den Handschriften des Kathedralarchivs in Mdina*. Tübingen: Narr.
Koptjevskaja-Tamm, Maria. 1996. Possessive noun phrases in Maltese: alienability, iconicity, and grammaticalization. *Rivista di Linguistica* 8(1). 245–274.
Krier, Fernande. 1975. Analyse syntaxique de la phrase nominale en maltais. *La Linguistique* 11. 93–116.
Krug, Manfred, Lisa Bonnici & Michaela Hilbert. 2011. Maltese English. In Bernd Kortmann & Kerstin Lunkenheimer (eds.), *The electronic World Atlas of Varieties of English* [eWAVE]. Leipzig: Max Planck Institute for Evolutionary Anthropology.
Kuteva, Tania, Bernd Heine, Bo Hong, Heiping Long, Heiko Narrog & Seongha Rhee. 2019. *World lexicon of grammaticalization*. Cambridge: Cambridge University Press.
Landmann, Fred. 1989. Group I & Group II. *Linguistics and Philosophy* 12. 559–605.
Lasersohn, Peter. 1990. Group action and spatio-temporal proximity. *Linguistics and Philosophy* 13. 179–206.
Lehmann, Christian. 1995. *Thoughts on grammaticalization*. München/Newcastle: Lincom Europa.

Lehmann, Christian & Christel Stolz. 1992. *Bildung von Adpositionen im Deutschen*. ASSidUE 6. Erfurt: Institut für Sprachwissenschaft.

Lehmann, Christian. 2004. Funktionale Grammatikographie. In Waldfried Premper (ed.), *Dimensionen und Kontinua. Beiträge zu Hansjakob Seilers Universalienforschung*, 147–166. Bochum: Brockmeyer.

Lestrade, Sander. 2010. *The space of case*. Nijmegen: Radboud Universiteit Dissertation.

Lucas, Christopher & Slavomír Čéplö. 2020. Maltese. In Christopher Lucas & Stefano Manfredi (eds.), *Arabic and contact-induced change*, 265–302. Berlin: Language Science Press.

Maddieson, Ian. 2005. Syllable structure. In Matthew Dryer & Martin Haspelmath (eds.), *The World Atlas of Language Structures*, 54–57. Oxford: Oxford University Press.

Matras, Yaron. 2011. Universals of structural borrowing. In Peter Siemund (ed.), *Linguistic universals and language variation*, 204–234. (Trends in Linguistics. Studies and Monographs 231). Berlin: Mouton de Gruyter.

Mel'čuk, Igor. 2006. *Aspects of the Theory of Morphology*. Berlin/New York: Mouton de Gruyter.

Mel'čuk, Igor. 2018. Le prépositions zero en français. *Lingvisticae investigationes* 41(2). 269–283.

Müller, Stefan. 2016. *Grammatical Theory. From transformational grammar to constraint-based approaches*. 2 vols. Berlin: Language Science Press.

Munaro, Nicola & Cecilia Poletto. 2020. Prepositional "where" in Southern Italian dialects. In Jacopo Garzonio & Silvia Rossi (eds.), *Variation in P. Comparative approaches to adpositional phrases*, 84–113. Oxford: Oxford University Press

Nichols, Johanna. 1993. Heads in discourse: structural versus functional centricity. In Greville G. Corbett, Norman M. Fraser & Scott McGlashan (eds.), *Heads in Grammatical Theory*. 164–185. Cambridge: Cambridge University Press.

Nübling, Damaris, Fabian Fahlbusch & Rita Heuser. 2015. *Namen. Eine Einführung in die Onomastik*. Tübingen: Narr.

Panzavecchia, Fortunato. 1843. *Grammatica della lingua maltese spiegata secondo i principj delle lingue orientali e della lingua italiana*. Malta: Weiss.

Papafragou, Anna. 2010. Source-goal asymmetries in motion representation: Implications for language production and comprehension. *Cognitive Science* 34. 1064–1092.

Peterson, John. 2009. "Pseudo-verbs": an analysis of non-verbal (co-)predication in Maltese. In Bernard Comrie, Ray Fabri, Elizabeth Hume, Manwel Mifsud, Thomas Stolz & Martine Vanhove (eds.), *Introducing Maltese Linguistics*, 181–204. Amsterdam/Philadelphia: Benjamins.

Plank, Frans & Edith Moravcsik. 1996. The Maltese article: language-particulars and universals. *Rivista di Linguistica* 8(1), 183–212.

Portelli, Sergio & Sandro Caruana. 2018. Observing Eurolects: the case of Maltese. In Laura Mori (ed.), *Observing Eurolects: corpus analysis of linguistic variation in EU law*, 267–293. Amsterdam: Benjamins

Procházka, Stephan. 1993. *Die Präpositionen in den neuarabischen Dialekten*. Wien: Verband der wissenschaftlichen Gesellschaften Österreichs.

Puech, Gilbert. 1994. *Ethnotextes maltais*. Wiesbaden: Harrassowitz.

Puech, Gilbert. 2018. Loss of emphatic and guttural consonants: from medieval to contemporary Maltese. In Patrizia Paggio & Albert Gatt (eds.), *The Languages of Malta*, 7–54. Berlin: Language Science Press.

Rauh, Gisa (ed.). 1991. *Approaches to Prepositions*. Tübingen: Narr.

Rizzi, Luigi. 1988. Il sintagma preposizionale. In Lorenzo Renzi, Giampaolo Salvi & Anna Cardinaletti (eds.), *Grande grammatica italiana di consultazione. I. La frase. I sintagmi nominale e preposizionale*, 521–545. Bologna: Il Mulino.

Rosner, Michael. 2009. Electronic resources for Maltese. In Bernard Comrie, Ray Fabri, Elizabeth Hume, Manwel Mifsud, Thomas Stolz & Martine Vanhove (eds.). *Introducing Maltese Linguistics*, 252–276. Amsterdam/Philadelphia: Benjamins.

Round, Erich R. 2015. Rhizomorphomes, meromorphomes, and metamorphomes. In Matthew Baerman, Dunstan Brown & Greville G. Corbett (eds.), *Understanding and measuring morphological complexity*, 29–52, Oxford: Oxford University Press.

Saade, Benjamin. 2012. Pro-drop systems in Maltese and Italian: a quantitative approach. *Ilsienna – Our Language* 2. 67–78.

Saade, Benjamin. 2018. The distribution of short and long pronouns in Maltese. *STUF –Language Typology and Universals* 71(2). 199–220.

Saade, Benjamin. 2019. Assessing productivity in contact: Italian derivation in Maltese. *Linguistics* 57(1). 59–86.

Saade, Benjamin. 2020. Quantitative approaches to productivity and borrowing in Maltese derivation. *Morphology* 30. 447–467.

Saari, Rami. 2002. *Quddiem* and some remarks on grammatical aspects of Maltese prepositions. In Susanne Feigenbaum & Dennis Kurzon (eds.), *Prepositions in their Syntactic, Semantic and Pragmatic Context*, 269–282. Amsterdam/Philadelphia: Benjamins.

Saari, Rami. 2003. *Milot Hayaxas Hamalteziyot* [Maltese prepositions]. Jerusalem: Carmel.

Saari, Rami. 2007. Maltese *ilu* – the only postposition in a language of prepositions? In Tali Bar & Eran Cohen (eds.), *Studies in Semitic and General Linguistics in Honor of Gideon Goldenberg*, 75–88. Münster: Ugarit-Verlag.

Saari, Rami. 2009. Some remarks on Maltese prepositions of Italian origin. In Gideon Goldenberg & Ariel Shisha-Halevi (eds.), *Egyptian, Semitic and General Grammar: Studies in the Memory of H.J. Polotsky*, 270–276. Jerusalem: The Israel Academy of Sciences and Humanities.

Saydon, Pietru Paul 1931. *Tifsir il-Kliem fil-Kotba Ġeneżi, Eżodu, Levitku, Numri, Dewtoronomju*. Malta: The Empire Press.

Schabert, Peter. 1976. *Laut- und Formenlehre des Maltesischen anhand zweier Mundarten*. Erlangen: Palm & Enke.

Schmidt, Emeli, Maike Vorholt & Nele Witt. 2020. Form and behavior of Maltese prepositions – a usage-based approach. In Slavomír Čéplö & Jaroslav Drobný (eds.), *Maltese Linguistics on the Danube*, 241–270. Berlin/Boston: De Gruyter Mouton.

Schwarze, Christoph (ed.). 1981. *Analyse des prépositions. IIIme Colloque franco-allemand de linguistique théorique*. Tübingen: Niemeyer.

Serracino-Inglott, Erin. 1975a. *Il-miklem Malti. L-Ewwel volum A–Ċ.* Malta: Klabb Kotba Maltin.

Serracino-Inglott, Erin. 1975b. *Il-miklem Malti. Il-Tieni volum D–F.* Malta: Klabb Kotba Maltin.

Serracino-Inglott, Erin. 1976. *Il-miklem Malti. It-Tielet volum Ġ–H.* Malta: Klabb Kotba Maltin.

Serracino-Inglott, Erin. 1977. *Il-miklem Malti. Ir-Raba' volum Ħ–I.* Malta: Klabb Kotba Maltin.

Serracino-Inglott, Erin. 1978. *Il-miklem Malti. Il-Ħames volum J–L.* Malta: Klabb Kotba Maltin.

Serracino-Inglott, Erin. 1979. *Il-miklem Malti. Is-Sitt volum M–O.* Malta: Klabb Kotba Maltin.

Serracino-Inglott, Erin. 1981. *Il-miklem Malti. Is-Seba' volum P–R.* Malta: Klabb Kotba Maltin.

Serracino-Inglott, Erin. 1984. *Il-miklem Malti. It-Tmien volum S–U.* Malta: Klabb Kotba Maltin.

Serracino-Inglott, Erin. 1989. *Il-miklem Malti. Id-Disa' volum V–Ż.* Malta: Klabb Kotba Maltin.

Serracino-Inglott, Mario. 2000. *Dizzjunarju Malti u Teżawru ta' Malti Mħaddem*. Blata l-Bajda: Merlin.

Serracino-Inglott, Mario. 2003. *Il-miklem Malti. L-Għaxar volum: Addenda.* Malta: Klabb Kotba Maltin.

Siewierska, Anna. 2004. *Person*. Cambridge: Cambridge University Press.

Stoebke, Renate. 1968. *Die Verhältniswörter in den ostseefinnischen Sprachen*. Bloomington: Indiana University.

Stolz, Thomas. 1984. Para-Präpositionen, Präpositionen, Adverbien und (Pseudo)Postpositionen im Lettischen. *Indogermanische Forschungen* 89. 251–269.

Stolz, Thomas. 1991. Il passivo, la negazione e una tipologia nuova. In Dieter Kremer (ed.), *Actes du XVIIIe Congrès International de Linguistique et de Philologie Romanes. Université de Trèves 1986. Tome II*, 542–550. Tübingen: Max Niemeyer.

Stolz, Thomas. 2009. Splitting the verb chain in modern literary Maltese. In Bernard Comrie, Ray Fabri, Elizabeth Hume, Manwel Mifsud, Thomas Stolz & Martine Vanhove (eds.), *Introducing Maltese Linguistics*, 133–180. Amsterdam/Philadelphia: John Benjamins.

Stolz, Thomas. 2011a. Maltese. In Bernd Kortmann & Johan van der Auwera (eds.), *The Languages and Linguistics of Europe. A Comprehensive Guide*, 241–256. Berlin: De Gruyter Mouton.

Stolz, Thomas. 2011. The possessive relative clause in Maltese. In Sandro Caruana, Ray Fabri & Thomas Stolz (eds.), *Variation and Change. The Dynamics of Maltese in Space, Time and Society*, 183–232. (Studia Typologica 9). Berlin: Akademie.

Stolz, Thomas. 2020. A chapter on compound prepositions in Maltese: PREP-PREP combinations and related issues. In Benjamin Fagard, José Pinto de Lima, Dejan Stosic & Elena Smirnova (eds.), *Complex Adpositions in European Languages. A Micro-Typological Approach to Complex Nominal Relators*, 439–470. Berlin/Boston: De Gruyter Mouton.

Stolz, Thomas & Carolin Ahrens. 2017. On prepositional ellipsis and the factors which block its application in Maltese. In Benjamin Saade & Mauro Tosco (eds.), *Advances in Maltese Linguistics*, 121–147. (Studia Typologica 21). Berlin/ Boston: De Gruyter Mouton.

Stolz, Thomas, Sander Lestrade & Christel Stolz. 2014. *The Crosslinguistics of Zero-Marking of Spatial Relations.* (Studia Typologica 15). Berlin/ Boston: De Gruyter Mouton.

Stolz, Thomas & Nataliya Levkovych. 2019. Absence of material exponence. *STUF – Language Typology and Universals* 72(3). 373–400.

Stolz, Thomas & Nataliya Levkovych. 2020. From variation towards the grammar of Maltese prepositions – first steps. In Slavomír Čéplö & Jaroslav Drobný (eds.), *Maltese Linguistics on the Danube*, 199–240. Berlin/Boston: De Gruyter Mouton.

Stolz, Thomas, Nataliya Levkovych & Aina Urdze. 2017. When zero is just enough… In support of a Special Toponymic Grammar in Maltese. *Folia Linguistica* 51(2). 453–482.

Stolz, Thomas & Nataliya Levkovych [in cooperation with Beke Seefried]. 2021. *Areal linguistics within the Phonological Atlas of Europe: Loan phonemes and their distribution*. Berlin/ Boston: De Gruyter Mouton.

Stolz, Thomas, Nataliya Levkovych & Maike Vorholt. 2023. Variable overt marking of Place/Goal with place names as complements: On the competition between *fi*, *ġewwa*, and *ġo*. Journal of Maltese Studies 30. 177–215.

Stolz, Thomas & Benjamin Saade. 2016. On short and long forms of personal pronouns in Maltese. In Gilbert Puech & Benjamin Saade (eds.), *Shifts and Patterns in Maltese*, 199–268. Berlin/Boston: De Gruyter Mouton.

Stolz, Thomas, Cornelia Stroh & Aina Urdze. 2006. *On Comitatives and related categories. A typological study with special focus on the languages of Europe: Empirical Approaches to Language Typology 33.* Berlin/ New York: Mouton de Gruyter.

Stolz, Thomas, Cornelia Stroh & Aina Urdze. 2011. *Total Reduplication. The Areal Linguistics of a Potential Universal*. Berlin: Akademie.

Stolz, Thomas & Maike Vorholt. 2022. On WH-PREPS. A contrastive grammatical sketch of Maltese *fejn* and Spanish *donde/dónde*. In Przemysław Turek & Julia Nintemann (eds.), *Maltese: Contemporary Changes and Historical Innovations*, 163–196. Berlin/Boston: De Gruyter Mouton.

Stolz, Thomas, Sonja Kettler, Cornelia Stroh & Aina Urdze. 2008. *Split possession. An areal-linguistic study of the alienability correlation and related phenomena in the languages of Europe*. Amsterdam/ Philadelphia: John Benjamins.

Sutcliffe, Edmund F. 1936. *A Grammar of the Maltese Language*. Oxford: Oxford University Press.

Thornton, Anna M. 2019. Overabundance: a canonical typology. In Franz Rainer, Francesco Gardani, Wolfgang U. Dressler & Hans Christian Luschützky (eds.), *Competition in inflection and word-formation*, 223–258. Berlin: Springer.

Vandeloise, Claude. 1991. *Spatial Prepositions. A Case Study from French*. Chicago/London: The University of Chicago Press.

Vanhove, Martine. 1993. *La langue maltaise. Ètudes syntaxiques d'un dialecte arabe »périphérique«*. Wiesbaden: Harrassowitz.

Vanhove, Martine. 2009. The nominal quantifier *xi* in Maltese. In Ray Fabri (ed.), *Maltese Linguistics: A Snapshot. In Memory of Joseph A. Cremona (1922–2003)*, 17–34. Bochum: Brockmeyer.

Vassalli, Mikiel Anton. 1796. *Ktyb yl Klym Mâlti mfysser byl-Latin u byt-Taljan*. Rome: Antonio Fulgonio.

Vassalli, Mikiel Anton. 1827. *Grammatica della lingua maltese*. Malta: self published.

Vella, Olvin (ed.). 2010. Essays on de Soldanis. *Journal of Maltese Studies* 27.

Vennemann, Theo. 1988. *Preference Laws for Syllable Structure and the Explanation of Sound Change*. Berlin, New York/Amsterdam: Mouton de Gruyter.

Vorholt, Maike. 2022a. Can frequency predict length? A crosslinguistic investigation of Zipf's law for European adpositions. In Przemysław Turek & Julia Nintemann (eds.), *Maltese: Contemporary Changes and Historical Innovations*, 197–228. Berlin/Boston: De Gruyter Mouton.

Vorholt, Maike. 2022b. Textsorte und Präpositionen im Maltesischen. Eine quantitative Analyse des Korpus Malti 3.0. Talk on occasion of the workshop *Text matters: Looking (far) beyond German*, Bremen/Germany, 13 May, 2022.

Vorholt, Maike. 2023a. The relevance of origin: Exploring the connection between the origin and frequency of Maltese prepositions. In Nataliya Levkovych (ed.), *Diversity in contact)*, 165–194. (Colonial and Postcolonial Linguistics 21). Berlin/Boston: De Gruyter.

Vorholt, Maike. 2023b. Bound pronouns in coordinating constructions – The case of Maltese prepositions. Talk on occasion of the *56th Annual meeting of the Societas Linguistica Europaea*, Athens/Greece, 29 August–1 September, 2023.

Vorholt, Maike. 2023c. Coordinated compound prepositions in Maltese. Talk on occasion of the *8th International Conference of Maltese Linguistics "The next century of Maltese Linguistics"*, Bremen/ Germany, 18–20 September, 2023.

Vorholt, Maike. 2023d. Maltese prepositions in coordinating constructions – On the influence of definiteness. Talk on the occasion of *Vielfaltslinguistik V*, Hamburg/Germany, 19–20 May, 2023.

Wälchli, Bernhard. 2005. *Co-compounds and natural coordination*. Oxford: Oxford University Press.

Weber, Heinz J. 1992. *Dependenzgrammatik. Ein Arbeitsbuch*. Tübingen: Narr.

Zipf, George Kingsley. 1935. *The psycho-biology of language. An introduction to dynamic philology*. Boston: Houghton Mifflin Company.

Index of authors

https://doi.org/10.1515/9783111331591-010

Index of subjects

https://doi.org/10.1515/9783111331591-011

Detailed contents

PART A: **Introduction**

Towards defining and identifying Maltese prepositions

PART B: **Grammar**

From phonology to semantics

PART C: **Terminus**

PART D: Appendix